Selected Titles in This Series

66 **Robert S. Doran, Ze-Li Dou, and George T. Gilbert, Editors,** Automorphic forms, automorphic representations, and arithmetic (Texas Christian University, Fort Worth, 1996)

65 **M. Giaquinta, J. Shatah, and S. R. S. Varadhan, Editors,** Differential equations: La Pietra 1996 (Villa La Pietra, Florence, Italy, 1996)

64 **G. Ferreyra, R. Gardner, H. Hermes, and H. Sussmann, Editors,** Differential geometry and control (University of Colorado, Boulder, 1997)

63 **Alejandro Adem, Jon Carlson, Stewart Priddy, and Peter Webb, Editors,** Group representations: Cohomology, group actions and topology (University of Washington, Seattle, 1996)

62 **János Kollár, Robert Lazarsfeld, and David R. Morrison, Editors,** Algebraic geometry—Santa Cruz 1995 (University of California, Santa Cruz, July 1995)

61 **T. N. Bailey and A. W. Knapp, Editors,** Representation theory and automorphic forms (International Centre for Mathematical Sciences, Edinburgh, Scotland, March 1996)

60 **David Jerison, I. M. Singer, and Daniel W. Stroock, Editors,** The legacy of Norbert Wiener: A centennial symposium (Massachusetts Institute of Technology, Cambridge, October 1994)

59 **William Arveson, Thomas Branson, and Irving Segal, Editors,** Quantization, nonlinear partial differential equations, and operator algebra (Massachusetts Institute of Technology, Cambridge, June 1994)

58 **Bill Jacob and Alex Rosenberg, Editors,** K-theory and algebraic geometry: Connections with quadratic forms and division algebras (University of California, Santa Barbara, July 1992)

57 **Michael C. Cranston and Mark A. Pinsky, Editors,** Stochastic analysis (Cornell University, Ithaca, July 1993)

56 **William J. Haboush and Brian J. Parshall, Editors,** Algebraic groups and their generalizations (Pennsylvania State University, University Park, July 1991)

55 **Uwe Jannsen, Steven L. Kleiman, and Jean-Pierre Serre, Editors,** Motives (University of Washington, Seattle, July/August 1991)

54 **Robert Greene and S. T. Yau, Editors,** Differential geometry (University of California, Los Angeles, July 1990)

53 **James A. Carlson, C. Herbert Clemens, and David R. Morrison, Editors,** Complex geometry and Lie theory (Sundance, Utah, May 1989)

52 **Eric Bedford, John P. D'Angelo, Robert E. Greene, and Steven G. Krantz, Editors,** Several complex variables and complex geometry (University of California, Santa Cruz, July 1989)

51 **William B. Arveson and Ronald G. Douglas, Editors,** Operator theory/operator algebras and applications (University of New Hampshire, July 1988)

50 **James Glimm, John Impagliazzo, and Isadore Singer, Editors,** The legacy of John von Neumann (Hofstra University, Hempstead, New York, May/June 1988)

49 **Robert C. Gunning and Leon Ehrenpreis, Editors,** Theta functions – Bowdoin 1987 (Bowdoin College, Brunswick, Maine, July 1987)

48 **R. O. Wells, Jr., Editor,** The mathematical heritage of Hermann Weyl (Duke University, Durham, May 1987)

47 **Paul Fong, Editor,** The Arcata conference on representations of finite groups (Humboldt State University, Arcata, California, July 1986)

46 **Spencer J. Bloch, Editor,** Algebraic geometry – Bowdoin 1985 (Bowdoin College, Brunswick, Maine, July 1985)

45 **Felix E. Browder, Editor,** Nonlinear functional analysis and its applications (University of California, Berkeley, July 1983)

(Continued in the back of this publication)

Automorphic Forms, Automorphic Representations, and Arithmetic

Professor Goro Shimura at the CBMS Conference
"Euler Products and Eisenstein Series,"
held at Texas Christian University, May 1996

Proceedings of Symposia in PURE MATHEMATICS

Volume 66, Part 1

Automorphic Forms, Automorphic Representations, and Arithmetic

NSF-CBMS Regional Conference in Mathematics
on Euler Products and Eisenstein Series
May 20–24, 1996
Texas Christian University

Robert S. Doran
Ze-Li Dou
George T. Gilbert
Editors

American Mathematical Society
Providence, Rhode Island

PROCEEDINGS OF AN NSF-CBMS REGIONAL CONFERENCE ON EULER PRODUCTS AND EISENSTEIN SERIES HELD AT TEXAS CHRISTIAN UNIVERSITY, FORT WORTH, TEXAS, MAY 20–24, 1996

Any opinions, findings, and conclusions or recommendations expressed in this material are those of the authors and do not necessarily reflect the view of the National Science Foundation.

1991 *Mathematics Subject Classification.* Primary 11Fxx; Secondary 14Gxx, 22Exx.

Library of Congress Cataloging-in-Publication Data

NFS-CBMS Regional Conference in Mathematics on Euler Products and Eisenstein Series (1996 : Texas Christian University)

Automorphic forms, automorphic representations, and arithmetic : NFS-CBMS Regional Conference in Mathematics on Euler Products and Eisenstein Series, May 20–24, 1996, Texas Christian University / Robert S. Doran, Ze-Li Dou, George T. Gilbert, editors.

p. cm. — (Proceedings of symposia in pure mathematics, 0082-0717 ; v. 66, pts. 1–2)

Includes bibliographical references.

ISBN 0-8218-0659-9 (set)

ISBN 0-8218-1050-2 (pt. 1)

ISBN 0-8218-1051-0 (pt. 2)

1. Automorphic functions Congresses. 2. Automorphic forms Congresses. 3. Representations of groups Congresses. I. Doran, Robert S., 1937– . II. Dou, Ze Li, 1962– . III. Gilbert, George T., 1957– . IV. Title. V. Series.

QA353.A9 N74 1996

511.3′3—dc21

99-28916

CIP

∞ The paper used in this book is acid-free and falls within the guidelines
established to ensure permanence and durability.
Visit the AMS home page at URL: http://www.ams.org/

10 9 8 7 6 5 4 3 2 1 04 03 02 01 00 99

8/11/99 JB

Dedicated to Goro Shimura

Table of Contents–Part 1

Table of Contents–Part 2

Foreword

Professor Goro Shimura was the principal speaker at the CBMS regional conference held in May, 1996 at Texas Christian University, entitled "Euler Products and Eisenstein Series." His ten lectures resulted in a monograph bearing the same title, which appeared as volume 93 of the CBMS Regional Conference Series in Mathematics. In these volumes we have collected papers based on most of the other talks presented at the conference, as well as many additional contributions.

Professor Shimura's stature as a mathematical giant, based on the plethora of seminal ideas he has introduced into number theory, on the numerous and varied theories that have matured in his hands, and on the high quality of the mathematicians he has cultivated and continues to influence, is certainly secure and will long endure. It is neither our place nor our intention to comment at length on Professor Shimura's achievements, but we believe that many aspects of Professor Shimura's work as well as that of his students are reflected in these volumes. It is therefore highly fitting that they be dedicated to Professor Shimura.

We have selected a rather general title, but an informed reader will find after a glance at the tables of contents that even this title does not adequately encompass the spectrum of the topics discussed in these volumes. We have also endeavored to place the papers which consider related topics and use similar techniques in the same volume. However, the division of the papers into two volumes cannot be based on a scientific principle. All in all, we must simply allow the mathematics to speak for itself.

We thank Professor Shimura for making the conference a success, and for making these volumes possible. We also acknowledge the hard work and patience of our authors and referees. We are especially grateful to Professor Hiroyuki Yoshida for his advice and for his lending a helping hand on several occasions. We are indebted to Shirley Doran, our technical secretary, for a thoroughly professional job. Finally, our heartfelt gratitude is reserved for our families, for their cheerful long-suffering.

Robert S. Doran
Ze-Li Dou
George T. Gilbert
February 1999

Proceedings of Symposia in Pure Mathematics
Volume **66.1**, 1999

On Period Torsors

Don Blasius and Mikhail Borovoi

For Goro Shimura, with great admiration

ABSTRACT. We study the period torsor of the category of motives for absolute Hodge cycles generated by abelian varieties defined over the rational numbers. Using a Hasse principle, this torsor is characterized, for a natural subcategory, by its real completion and its projection to the Taniyama group. The subcategory may be defined as that generated by abelian varieties with simply connected Mumford-Tate group. If instead one considers ALL motives for absolute cycles, then our method gives the same characterization of the period torsor, provided Serre's conjecture that the Mumford-Tate group of this category is simply-connected, is true.

1. Introduction

1.1. Let $\mathcal{A}$ be the category of abelian varieties defined over $\mathbf{Q}$ and let $A \in \mathcal{A}$. Deligne's theory of absolute Hodge cycles enables the construction of a Tannakian category of motives $\otimes_{\mathrm{AH}}A$ from A [**DM**, Sect. 6]. The functors $\omega_{\mathrm{B}}\colon M \mapsto M_{\mathrm{B}}$ (topological cohomology) and $\omega_{\mathrm{DR}}\colon M \mapsto M_{\mathrm{DR}}$ (de Rham cohomology) are *fiber functors* on $\otimes_{\mathrm{AH}}A$. The Tannakian formalism then provides us with reductive algebraic groups $\mathcal{G}_{\mathrm{B}}^A$ and $\mathcal{G}_{\mathrm{DR}}^A$ over $\mathbf{Q}$, the tensor automorphism groups of the respective fiber functors. Further, there is a natural principal homogeneous space (torsor) $\mathcal{P}^A$ for $\mathcal{G}_{\mathrm{B}}^A$, namely $\mathcal{P}^A = \mathrm{Iso}^{\otimes}(\omega_{\mathrm{B}}, \omega_{\mathrm{DR}})$, the tensor compatible isomorphisms of ω_{B} with ω_{DR} . We call $\mathcal{G}_{\mathrm{B}}^A$ the *motivic Galois group* of A and $\mathcal{P}^A$ the *period torsor* of A. Similar definitions and constructions can be made for categories $\mathcal{C}$ of abelian varieties, instead of single varieties A; for notation, we simply replace A by $\mathcal{C}$.

1.2. In this paper we characterize $\mathcal{P}^{\mathcal{C}}$ for the subcategory $\mathcal{C} = \mathcal{H}$ of abelian varieties which satisfy the following non-trivial hypothesis.

Let $((\mathcal{G}_{\mathrm{B}}^A)^\circ)^{\mathrm{ad}}$ be the adjoint group of the connected component, in the sense of algebraic groups over $\mathbf{Q}$, of $\mathcal{G}_{\mathrm{B}}^A$. We say that A is of type (H) if $((\mathcal{G}_{\mathrm{B}}^A)^\circ)^{\mathrm{ad}}_{\mathbf{R}}$ has no factor of type D_n^H, with notation of [**D79**, 1.3.9].

1991 *Mathematics Subject Classification.* Primary 11F75, 11G10, 12G05, 14K15.
Key words and phrases. Hasse principle, torsor, abelian variety, motives.
NSF Grant DMS 9304289 (Blasius), BSF Grant 92-00278 (both authors).

Let $\mathcal{H} \subset \mathcal{A}$ be the subcategory of varieties of type (H). Of course, $((\mathcal{G}_{\mathrm{B}}^{A})^{\circ})$ is the Mumford-Tate group of A [**D82**, Sect. 3]. Hence the condition that A be of type (H) depends only on the Hodge structure of $H_B^1(A_{\mathbf{C}})$, i.e., upon the complex isogeny class of A.

1.3. Relation with the CM case. Let $CM \subset \mathcal{H}$ be the subcategory of all A such that $((\mathcal{G}_{\mathrm{B}}^{A})^{\circ})^{\mathrm{der}} = 1$; these are abelian varieties of CM-type, and $\mathcal{G}_{\mathrm{B}}^{\mathrm{CM}}$ is normally called the Taniyama group [**D82b**]. We have a natural morphism $\pi_{\mathcal{P}} \colon \mathcal{P}^{\mathcal{H}} \to \mathcal{P}^{\mathrm{CM}}$ which identifies $\mathcal{P}^{\mathrm{CM}}$ with the quotient of $\mathcal{P}^{\mathcal{H}}$ by $K_B = ((G_{\mathrm{B}}^{\mathcal{H}})^{\circ})^{\mathrm{der}} = \ker[\pi_{\mathrm{B}} \colon \mathcal{G}_{\mathrm{B}}^{\mathcal{H}} \to \mathcal{G}_{\mathrm{B}}^{\mathrm{CM}}]$. The torsor $\mathcal{P}^{\mathrm{CM}}$ is described in [**W**]; see below (4.2).

1.4. Description of $\mathcal{P}^{A}$ and $\mathcal{P}^{\mathcal{H}}$ at infinity. The torsor $P_{\mathbf{R}}^{A}$ is defined by a cohomology class $c_\infty \in H^1(\mathrm{Gal}(\mathbf{C}/\mathbf{R}), \mathcal{G}_{\mathrm{B}}^{A}(\mathbf{C}))$ as follows. Let F_∞ denote the complex conjugation. Then F_∞ defines a smooth map $A(\mathbf{C}) \to A(\mathbf{C})$, whose induced action on $H^1(A(\mathbf{C}), \mathbf{R})$ has image in $\mathcal{G}_{\mathrm{B}}^{A}(\mathbf{R})$. Let c_∞ be the associated cohomology class. The class of $P_{\mathbf{R}}^{\mathcal{H}}$ is similarly determined. We can now state our main result.

THEOREM 1.5. *Let $\pi \colon \mathcal{P} \to \mathcal{P}^{\mathrm{CM}}$ be any $\mathcal{G}_{\mathrm{B}}^{\mathcal{H}}$-torsor for which $\mathcal{P}_{\mathbf{R}}$ belongs to the class of c_∞. Then there exists an isomorphism $\lambda \colon \mathcal{P}^{\mathcal{H}} \xrightarrow{\sim} \mathcal{P}$ of $\mathcal{G}_{\mathrm{B}}^{\mathcal{H}}$-torsors such that $\pi \circ \lambda = \pi_P$.*

The proof of this result is given in Sections 2–5.

1.6. We ask whether a variant of Theorem 1.5 holds for $\mathcal{P}^{\mathcal{A}}$. Our argument relies upon the fact that, for the category $\mathcal{H}$, the group $((\mathcal{G}_{\mathrm{DR}}^{\mathcal{H}})^{\circ})^{\mathrm{der}}$ is a projective limit of simply connected groups. However, for $\mathcal{A}$ itself $((\mathcal{G}_{\mathrm{DR}}^{\mathcal{A}})^{\circ})^{\mathrm{der}}$ is *not* a limit of simply connected groups, and an analogue of Theorem 1.5, if true, will depend upon a closer analysis of the Galois cohomology of the non-simply-connected groups involved. Even so, one would need additionally a local characterization of $\mathcal{P}_{\mathbf{Q}_p}$ for each prime p. See [**RZ**, Chapter 1], for a conjecture. Finally, it is natural to ask what happens if, instead of $\mathcal{A}$ or $\mathcal{H}$, we take the category $\mathcal{M}$ of *all* motives for absolute Hodge cycles. Here it seems natural to ask (as has Serre [**Se**, 8.1]) whether $((\mathcal{G}_{\mathrm{B}}^{\mathcal{M}})^{\circ})^{\mathrm{der}}$ is a projective limit of simply connected groups. If so, then an analog of Theorem 1.5 holds for $\mathcal{M}$ and $\mathcal{P}^{\mathcal{M}}$.

2. A Hasse principle

Let k be a number field. We write k_v for the completion of k at a place v. We write $\mathcal{V}_\infty(k)$ or just ∞ for the set of infinite places of k.

Let G be a reductive k-group. We denote by G^{der} the derived group of G, and set $G^{\mathrm{ab}} = G/G^{\mathrm{der}}$. We write G^{ad} for the adjoint group of G.

We will need the following result.

THEOREM 2.1 *Let G be a connected reductive group over a number field k. Assume that G^{der} is simply connected and that $H^1(k, G^{\mathrm{ab}}) = 1$. Then the following Hasse principle for G holds: if $\xi \in H^1(k, G)$ and $\mathrm{loc}_v(\xi) = 1$ for all $v \in \mathcal{V}_\infty(k)$, then $\xi = 1$, where $\mathrm{loc}_v \colon H^1(k, G) \to H^1(k_v, G)$ is the localization map. In particular $\text{Ш}(k, G) = 0$.*

PROOF. By the result of Kneser–Harder–Chernousov, we have

$$H^1(k, G^{\mathrm{der}}) = \prod_{v \in \infty} H^1(k_v, G^{\mathrm{der}})$$

because G^{der} is simply connected. Consider the commutative diagram with exact rows

$$\begin{array}{ccccccccc} G(k) & \xrightarrow{\beta} & G^{\mathrm{ab}}(k) & \xrightarrow{\delta} & H^1(k,G^{\mathrm{der}}) & \xrightarrow{\alpha_*} & H^1(k,G) & \longrightarrow & 1 \\ \downarrow & & \downarrow{\scriptstyle \mathrm{loc}_\infty} & & {\scriptstyle \mathrm{loc}_\infty}\downarrow{\scriptstyle \sim} & & \downarrow{\scriptstyle \mathrm{loc}_\infty} & & \\ \prod_\infty G(k_v) & \xrightarrow{\beta} & \prod_\infty G^{\mathrm{ab}}(k_v) & \xrightarrow{\delta} & \prod_\infty H^1(k_v,G^{\mathrm{der}}) & \xrightarrow{\alpha_*} & \prod_\infty H^1(k_v,G) & & \end{array}$$

Let $\xi \in H^1(k,G)$ and assume that $\mathrm{loc}_\infty(\xi)=1$. Since $H^1(k,G^{\mathrm{ab}})=1$, we see that $\xi=\alpha_*\zeta$ for some $\zeta\in H^1(k,G^{\mathrm{der}})$. Since $\mathrm{loc}_\infty(\xi)=1$, we have $\alpha_*(\mathrm{loc}_\infty\zeta)=1$, hence $\mathrm{loc}_\infty\zeta$ is the image of some $t_\infty=(t_v)\in\prod_\infty G^{\mathrm{ab}}(k_v)$.

The map $G\to G^{\mathrm{ab}}$ is smooth, hence the image of $\prod_\infty G(k_v)$ in $\prod_\infty G^{\mathrm{ab}}(k_v)$ is open. On the other hand, by the real approximation theorem (Serre, cf. [**Sa**, Cor. 3.5(iii)]), $G^{\mathrm{ab}}(k)$ is dense in $\prod_\infty G^{\mathrm{ab}}(k_v)$. It follows that $t_\infty = t\cdot\beta(g_\infty)$ for some $g_\infty\in\prod_\infty G(k_v)$ and $t\in G^{\mathrm{ab}}(k)$. Hence $\mathrm{loc}_\infty(\zeta)=\mathrm{loc}_\infty(\delta(t))$. But the map

$$\mathrm{loc}_\infty\colon H^1(k,G^{\mathrm{der}})\to\prod_\infty H^1(k_v,G^{\mathrm{der}})$$

is bijective, hence $\zeta=\delta(t)$, and $\xi=\alpha_*(\delta(t))=1$. □

We remark that it is easy to give a variant of this result which assumes only the Hasse principle for G^{ab}, and that this result would suffice for the proof of Theorem 1.5.

3. Abelian varieties

3.1. Let A be an abelian variety over $\mathbf{C}$, and let $\mathrm{MT}(A)$ be the Mumford-Tate group of A, cf. [**D82**, Sect. 3]. It is a connected reductive group over $\mathbf{Q}$ and comes with a homomorphism $h_A\colon \mathbb{S}\to\mathrm{MT}(A)_{\mathbf{R}}$, where $\mathbb{S}=R_{\mathbf{C}/\mathbf{R}}\mathbb{G}_{m\mathbf{C}}$, and which defines the Hodge structure of $H^1_B(A)$ in a well-known way.

Let K be a CM-field, i.e., a totally imaginary quadratic extension of a totally real field F, and let c denote the image of complex conjugation in $\mathrm{Gal}(\mathbf{C}/\mathbf{R})$ in $\mathrm{Aut}(K)$. Define (cf. [**D82b**, p. 268]) S^K to be a torus over $\mathbf{Q}$,the quotient of $R_{K/\mathbf{Q}}(\mathbb{G}_m)$, whose character group is the subgroup of $I_K=\mathbf{Z}[\mathrm{Hom}(K,\mathbf{C})]$ consisting of all $\Psi=\sum\Psi(\sigma)\sigma$ such that $\Psi(\sigma)+\Psi(\sigma c)$ is independent of σ. The torus S^K is called the Serre group of K, and the limit $S=\varprojlim S^K$ is called the Serre group.

THEOREM 3.2. *Let A be an abelian variety of type (H), defined over $\mathbf{Q}$. Then there exists an abelian variety A' of type (H) over $\mathbf{Q}$ such that $\mathrm{MT}(A\times A')^{\mathrm{der}}$ is simply connected and $\mathrm{MT}(A\times A')^{\mathrm{ab}}\simeq S^K$ for some CM-field K.*

We need three lemmas. For any A we define h_A^{ad} to be the composition

$$\mathbb{S}\to\mathrm{MT}(A)_{\mathbf{R}}\to\mathrm{MT}(A)^{\mathrm{ad}}_{\mathbf{R}}.$$

LEMMA 3.3. *Let A_1, A_2 be abelian varieties over $\mathbf{C}$. Assume that the pairs $(\mathrm{MT}(A_1)^{\mathrm{ad}},h^{\mathrm{ad}}_{A_1})$ and $(\mathrm{MT}(A_2)^{\mathrm{ad}},h^{\mathrm{ad}}_{A_2})$ are isomorphic, i.e., there exists an isomorphism $\lambda\colon\mathrm{MT}(A_1)^{\mathrm{ad}}\to\mathrm{MT}(A_2)^{\mathrm{ad}}$ (defined over $\mathbf{Q}$) such that $\lambda\circ h^{\mathrm{ad}}_{A_1}=h^{\mathrm{ad}}_{A_2}$. Suppose that A_1 can be defined over a number field. Then A_2 can be defined over a number field as well.*

PROOF. We set $G_i=\mathrm{MT}(A_i)$, $h_i=h_{A_i}$ for $i=1,2$. Let X_i denote the $G_i(\mathbf{R})$-conjugacy class of h_i. The point $h_1\in X_1$ defines a $\mathbf{C}$-point $x_1=[h_1,1]\in\mathrm{Sh}(G_1,X_1)$ where $\mathrm{Sh}(G_1,X_1)$ is the corresponding Shimura variety, see [**D79**, 2.1.2, 2.1.4]. Since A_1 is defined over $\bar{\mathbf{Q}}$, the point x_1 is defined over $\bar{\mathbf{Q}}$.

The map

$$\mathrm{ad}_*\colon \mathrm{Sh}(G_1, X_1) \to \mathrm{Sh}(G_1^{\mathrm{ad}}, X_1^{\mathrm{ad}})$$

is defined over a number field, hence the image $x_1^{\mathrm{ad}} \in \mathrm{Sh}(G_1^{\mathrm{ad}}, X_1^{\mathrm{ad}})$ of x_1 is defined over $\bar{\mathbf{Q}}$.

We define $x_2^{\mathrm{ad}} \in \mathrm{Sh}(G_2^{\mathrm{ad}}, X_2^{\mathrm{ad}})$ similarly, starting from A_2. By assumption there is an isomorphism $\mathrm{Sh}(G_1^{\mathrm{ad}}, X_1^{\mathrm{ad}}) \to \mathrm{Sh}(G_2^{\mathrm{ad}}, X_2^{\mathrm{ad}})$ taking x_1^{ad} to x_2^{ad}. Hence x_2^{ad} is defined over $\bar{\mathbf{Q}}$. Since the map $\mathrm{ad}_*\colon \mathrm{Sh}(G_2, X_2) \to \mathrm{Sh}(G_2^{\mathrm{ad}}, X_2^{\mathrm{ad}})$ is etale, we conclude that x_2 is defined over $\bar{\mathbf{Q}}$. It follows that A_2 is defined over $\bar{\mathbf{Q}}$. □

LEMMA 3.4. *Let $G = \prod_{s\in S} G_s$ where S is a finite set and G_s are simply connected semisimple groups over a field k of characteristic 0. Let H be a semisimple subgroup of G and suppose that for each $s \in S$ the map $p_s|_H\colon H \to G_s$ is surjective, where $p_s\colon H \to G_s$ is the projection. Then H is simply connected.*

PROOF. We may and will assume that k is algebraically closed and all the groups G_s are simple. Let L denote the Lie algebra of H. Then $L \simeq \bigoplus_{i=1}^r L_i$ with simple Lie algebras L_i. For each i, let S_i be the set of $s \in S$ such that $p_s(L_i) \neq 0$. Then $S_i \neq \emptyset$. Let $j : L \to \mathrm{Lie}\, G$ be the inclusion. Then $j(L_i) \subset \prod_{s\in S_i} L_s$.

Pick $s_i \in S_i$ for each i. Then $(p_{s_i} \circ j)(L_i) = \mathrm{Lie}\, G_{s_i}$ for each i. Hence $\mathrm{Lie}\, H$ is isomorphic to $\oplus_{i=1}^r \mathrm{Lie}\, G_{s_i}$ via $\pi = \prod_{i=1}^r p_i \circ j$. Hence the corresponding map $H \to \prod_{i=1}^r G_{s_i}$ is an isogeny. Since $\prod_{i=1}^r G_{s_i}$ is simply connected, the map $H \to \prod_{i=1}^r G_{s_i}$ is an isomorphism. Thus H is simply connected. □

LEMMA 3.5. *Let K be a Galois CM-field, and S^K the corresponding Serre group. There exists an abelian variety A over $\mathbf{Q}$ such that $MT(A) \simeq S^K$.*

PROOF. Recall that I_K is spanned by the CM-types, i.e., the Ψ such that $\Psi(\sigma) \in \{0, 1\}$ and $\Psi(\sigma) + \Psi(\sigma c) = 1$ for all σ. If Φ is a CM-type, define $\Phi^* = \sum_\sigma \Phi(\sigma^{-1})\sigma$, and define a complex torus $A_\Phi = \mathbf{C}^{|\Phi^*|}/\Phi^*(L)$ where

a) $|\Phi^*| = \{\sigma|\ \Phi(\sigma^{-1}) \neq 0\}$,

b) $L \subset K$ is a $\mathbf{Z}$-lattice,

c) $\Phi^*\colon K \to \mathbf{C}^{|\Phi^*|}$ is defined by $(\Phi^*(\alpha))(\tau) = \tau(\alpha)$ for $\tau \in |\Phi^*|$, $\alpha \in K$.

Define a homomorphism $[\Phi]\colon S^K \to \mathrm{GL}(H^1_{\mathrm{B}}(A_\Phi))$ by the rule

$$[\Phi](s)_\tau = (\tau\Phi)(s)$$

for $s \in S^K$ and where, for $\tau \in \mathrm{Hom}(K, \mathbf{C})$, the subscript τ denotes the τ eigenspace for the action of K on $\mathrm{GL}(H^1_{\mathrm{B}}(A_\Phi) \otimes \mathbf{C})$. Clearly $[\Phi]$ is defined over $\mathbf{Q}$. Each A_Φ is of CM-type (K, Φ^*) in Shimura's terminology. In particular, each A_Φ is isomorphic to an abelian variety B_Φ defined over a number field F_Φ. Hence $B = \prod B_\Phi$ is defined over F, the compositum of the F_Φ. If τ is an automorphism of $\mathbf{C}$, τB is isogenous to B. Hence $A = R_{F/\mathbf{Q}}(B)$ is isogenous to $B^{[F:\mathbf{Q}]}$. Clearly, for any B, $\mathrm{MT}(B^n) = \mathrm{MT}(B)$. Hence it is sufficient to show that $\mathrm{MT}(B) = S^K$.

Clearly, the representation $\prod[\Phi]$ of S^K is faithful. For each Φ, let h_Φ : $\mathbb{S} \to \mathrm{MT}(A_\Phi)_{\mathbf{R}}$ be the morphism defining the Hodge structure on $H^1(A_\Phi)$. Let $h_{\mathrm{can}}\colon \mathbb{S} \to S^K_{\mathbf{R}}$ be the morphism defined by

$$\Phi \circ h_{\mathrm{can}}(z, w) = z^{-\Phi(1)} w^{-\Phi(c)}$$

for all Φ. (Here we identify $S(\mathbf{C})$ with $\mathbf{C}^* \times \mathbf{C}^*$ in the evident way.) Then it is easy to see that $[\Phi] \circ h_{\mathrm{can}} = h_\Phi$ for every Φ. Since $\mathrm{MT}(B)$ is the $\mathbf{Q}$-Zariski closure of $\mathrm{Im}(\prod_\Phi h_\Phi)$ inside $\mathrm{GL}(H^1_B(B)\otimes \mathbf{R})$, it is enough to check that the smallest $\mathbf{Q}$-subtorus T of S^K such that $T_{\mathbf{R}}$ contains $\mathrm{Im}(h_{\mathrm{can}})$ is S^K itself. But $\mathrm{Im}(h_{\mathrm{can}})$ is the subtorus of S^K corresponding to the quotient of $X^*(S^K)$ by the submodule $W = \{\Psi | \Psi(1) = \Psi(c) = 0\}$. The submodule associated to its $\mathbf{Q}$-closure is $\bigcap_\tau \tau W$ where $\tau \in \mathrm{Hom}(K, \mathbf{C})$, and this module is zero. This proves the lemma. □

3.6. Proof of Theorem 3.2. Let A be an abelian variety of type (H) over $\mathbf{Q}$. Let G_1 be a $\mathbf{Q}$-simple factor of $G = \mathrm{MT}(A)^{\mathrm{der}}$ and let $h_1 \colon \mathbb{S} \to G^{\mathrm{ad}}_{1\mathbf{R}}$ be the projection of $h_A \colon \mathbb{S} \to \mathrm{MT}(A)_{\mathbf{R}}$. Let G_1^{sc} denote the universal covering of G_1^{der}. Because A is of type (H), we can construct an abelian variety A_1 over $\mathbf{C}$ such that the pair $(MT(A_1)^{\mathrm{der}}, h^{\mathrm{ad}}_{A_1} \colon \mathbb{S} \to \mathrm{MT}(A_1)^{\mathrm{ad}}_{\mathbf{R}})$ is isomorphic to (G_1^{sc}, h_1) (see [**D79**, Prop. 2.3.10]). Do this for each $\mathbf{Q}$-factor G_i of $G = \mathrm{MT}(A)^{\mathrm{der}}$, $i = 1, \dots, n$. Define $A' = A_1 \times \cdots \times A_n$; it is an abelian variety over $\mathbf{C}$. By construction $MT(A')^{\mathrm{der}}$ is simply connected, and $MT(A')^{\mathrm{ad}} \simeq \mathrm{MT}(A)^{\mathrm{ad}}$. Moreover the pairs $(\mathrm{MT}(A)^{\mathrm{ad}}, h_A^{\mathrm{ad}})$ and $(\mathrm{MT}(A')^{\mathrm{ad}}, h_{A'}^{\mathrm{ad}})$ are isomorphic. Since A is defined over $\mathbf{Q}$, Lemma 3.3 shows that A' can be defined over some number field $k \subset \mathbf{C}$.

Set $A'' = R_{k/\mathbf{Q}} A'$. Then $\mathrm{MT}(A'')$ is a subgroup of $\prod \mathrm{MT}(\sigma A')$ where σ runs over all the embeddings $\sigma \colon k \to \mathbf{C}$. The projection map $p_\sigma \colon \mathrm{MT}(A'') \to \mathrm{MT}(\sigma A')$ is surjective for any σ. Set $H = \mathrm{MT}(A'')^{\mathrm{der}}$ and $G_\sigma = \mathrm{MT}(\sigma A')^{\mathrm{der}}$. Then H is a subgroup of $\prod G_\sigma$, and the map $p_\sigma \colon H \to G_\sigma$ is surjective for any σ. The groups G_σ are simply connected. By Lemma 3.4 H is simply connected. Thus $\mathrm{MT}(A'')^{\mathrm{der}}$ is simply connected.

Since the pairs $(\mathrm{MT}(A)^{\mathrm{ad}}, h_A^{\mathrm{ad}})$ and $(\mathrm{MT}(A')^{\mathrm{ad}}, h_{A'}^{\mathrm{ad}})$ are isomorphic, the pair $(\mathrm{MT}(A \times A')^{\mathrm{ad}}, h^{\mathrm{ad}}_{A\times A'})$ is isomorphic to $(\mathrm{MT}(A')^{\mathrm{ad}}, h^{\mathrm{ad}}_{A'})$, where the isomorphism is induced by the projection $\mathrm{MT}(A \times A') \to \mathrm{MT}(A')$. This projection is surjective and it induces a surjective map $\mathrm{MT}(A\times A')^{\mathrm{der}} \to \mathrm{MT}(A')^{\mathrm{der}}$. By construction, the induced map $\mathrm{MT}(A \times A')^{\mathrm{ad}} \to \mathrm{MT}(A')^{\mathrm{ad}}$ is bijective. Therefore $\mathrm{MT}(A \times A')^{\mathrm{der}}$ is a covering of $\mathrm{MT}(A')^{\mathrm{der}}$. But $\mathrm{MT}(A')^{\mathrm{der}}$ is simply connected (by construction), hence $\mathrm{MT}(A \times A')^{\mathrm{der}}$ is simply connected.

Let $\sigma_0 \colon k \to \mathbf{C}$ denote the inclusion map. Then $A \times A'' = (A \times A') \times \prod_{\sigma \neq \sigma_0} \sigma A$. The groups $\mathrm{MT}(A \times A')^{\mathrm{der}}$ and $MT(\sigma A)^{\mathrm{der}}$ are simply connected. By Lemma 3.4 $\mathrm{MT}(A \times A'')^{\mathrm{der}}$ is simply connected.

Set $T = \mathrm{MT}(A \times A'')^{\mathrm{ab}}$ and let $h_T \colon \mathbb{S} \to T_{\mathbf{R}}$ be the projection of $h_{A\times A''}$ to $T_{\mathbf{R}}$. There exists a Galois CM-field $K \subset \mathbf{C}$ such that (T, h_T) is a quotient of (S^K, h_{S^K}). By Lemma 3.5 there exists an abelian variety A''' of CM-type defined over $\mathbf{Q}$ such that $MT(A''', h''')$ is isomorphic to (S^K, h_{S^K}). We set $A^\sharp = A \times A'' \times A'''$. Then $A^\sharp$ is defined over $\mathbf{Q}$, $MT(A^\sharp)^{\mathrm{der}}$ is simply connected and $MT(A^\sharp)^{\mathrm{ab}}$ is isomorphic to S^K. □

4. Real points

Let c be the complex conjugation in $\Gamma = \mathrm{Gal}(\bar{\mathbf{Q}}/\mathbf{Q})$. Let $Z_\Gamma(c)$ denote the centralizer of c in Γ.

LEMMA 4.1. $Z_\Gamma(c) = \{c, 1\}$.

PROOF. Let $\sigma \in \Gamma$, and assume that σ commutes with c. We have $\sigma(\bar{\mathbf{Q}} \cap \mathbf{R}) = \bar{\mathbf{Q}} \cap \mathbf{R}$. If $\sigma(\sqrt{-1}) = -\sqrt{-1}$, we take $c\sigma$ for σ. We wish to prove that $\sigma = 1$. It suffices to show that $\sigma|_{\bar{\mathbf{Q}}\cap\mathbf{R}} = 1$.

Let $x \in \bar{\mathbf{Q}} \cap \mathbf{R}$, $x > 0$. Then σ takes the real numbers $\pm\sqrt{x}$ to real numbers. It follows that $\sigma x > 0$. We prove that $\sigma x = x$ for any $x \in \bar{\mathbf{Q}} \cap \mathbf{R}$. Indeed, assume that $\sigma x = y$, $y \neq x$. We may assume, replacing σ by σ^{-1} if necessary, that $y < x$. Choose $a \in \mathbf{Q}$ such that $x > a > y$. Then σ takes $x - a$ to $y - a$. But $x - a > 0$ and $y - a < 0$ which is impossible. □

4.2. We write $\mathcal{G}_{\mathrm{B}}$ for $\mathcal{G}_{\mathrm{B}}^{\mathcal{H}}$, $\mathcal{G}_{\mathrm{DR}}$ for $\mathcal{G}_{\mathrm{DR}}^{\mathcal{H}}$, $\mathcal{P}$ for $\mathcal{P}^{\mathcal{H}}$. The torsor $\mathcal{P}$ defines a cohomology class $\xi_{\mathcal{P}} \in H^1(\mathbf{Q}, \mathcal{G}_{\mathrm{B}})$. Let z be a cocycle representing $\xi_{\mathcal{P}}$. There is an exact sequence

$$1 \to \mathcal{G}_{\mathrm{B}}^{\circ} \to \mathcal{G}_{\mathrm{B}} \to \mathrm{Gal}(\bar{\mathbf{Q}}/\mathbf{Q}) \to 1 ,$$

so z defines a cocycle $z' \in Z^1(\mathbf{Q}, \mathrm{Gal}(\bar{\mathbf{Q}}/\mathbf{Q}))$.

The corresponding de Rham sequence is

$$1 \to \mathcal{G}_{\mathrm{DR}}^{\circ} \to \mathcal{G}_{\mathrm{DR}} \to \mathrm{Gal}(\bar{\mathbf{Q}}/\mathbf{Q})_{\mathrm{DR}} \to 1 .$$

Here $\mathrm{Gal}(\bar{\mathbf{Q}}/\mathbf{Q})_{\mathrm{DR}}$ is the profinite group scheme attached to the class in

$$H^1(\mathbf{Q}, \mathrm{Gal}(\bar{\mathbf{Q}}/\mathbf{Q}))$$

defined by the identity map of $\mathrm{Gal}(\bar{\mathbf{Q}}/\mathbf{Q})$ (regarding the latter as a constant group scheme). Hence, if $x \in \mathrm{Gal}(\bar{\mathbf{Q}}/\mathbf{Q})_{\mathrm{DR}}(\bar{\mathbf{Q}}) = \mathrm{Gal}(\bar{\mathbf{Q}}/\mathbf{Q})$ and $\sigma \in \mathrm{Gal}(\bar{\mathbf{Q}}/\mathbf{Q})$ with action $x \mapsto \sigma_{\mathrm{DR}} x$ on $\mathrm{Gal}(\bar{\mathbf{Q}}/\mathbf{Q})_{\mathrm{DR}}(\bar{\mathbf{Q}})$, we have $\sigma_{\mathrm{DR}} x = \sigma x \sigma^{-1}$, regarding x as an element of $\mathrm{Gal}(\bar{\mathbf{Q}}/\mathbf{Q})$. The $\mathrm{Gal}(\bar{\mathbf{Q}}/\mathbf{Q})$-torsor $\mathcal{P}^{\mathrm{Art}} = \mathcal{P}/\mathcal{G}_{\mathrm{DR}}^{\circ}$ is canonically identified with $\mathrm{Iso}_{\mathbf{Q}}(\bar{\mathbf{Q}}, \mathbf{C})$; $\mathcal{P}^{\mathrm{Art}}$ is called the *Artin torsor*. By [**W**], $\mathcal{P}^{\mathrm{CM}}$ is the unique torsor for $\mathcal{G}^{\mathrm{CM}}$ which projects to $\mathcal{P}^{Art}$.

LEMMA 4.3. *${}_{z'}\mathrm{Gal}(\bar{\mathbf{Q}}/\mathbf{Q})(\mathbf{R}) = \{1, c\}$, where c is the nontrivial element of* $\mathrm{Gal}(\mathbf{C}/\mathbf{R})$.

PROOF. The cocycle z' is the Artin cocycle, i.e., $z'(\sigma) = \sigma$ for every $\sigma \in \mathrm{Gal}(\bar{\mathbf{Q}}/\mathbf{Q})$, cf. [**W**]. The action of the Galois group on ${}_{z'}\mathrm{Gal}(\bar{\mathbf{Q}}/\mathbf{Q})(\bar{\mathbf{Q}})$ is given by ${}^{\sigma}\tau = \sigma\tau\sigma^{-1}$, $\sigma, \tau \in \mathrm{Gal}(\bar{\mathbf{Q}}/\mathbf{Q})$. Hence $\mathrm{Gal}(\mathbf{C}/\mathbf{R})$ acts on ${}_{z'}\mathrm{Gal}(\bar{\mathbf{Q}}/\mathbf{Q})(\mathbf{C})$ by ${}^{c}\tau = c\tau c^{-1}$. It follows that ${}_{z'}\mathrm{Gal}(\bar{\mathbf{Q}}/\mathbf{Q})(\mathbf{R}) = \{\tau | c\tau c^{-1} = \tau\}$. By Lemma 4.1 $\{\tau | c\tau c^{-1} = \tau\} = \{1, c\}$. Thus ${}_{z'}\mathrm{Gal}(\bar{\mathbf{Q}}/\mathbf{Q})(\mathbf{R}) = \{1, c\}$. □

PROPOSITION 4.4. *The map $\mathcal{G}_{\mathrm{DR}}(\mathbf{R}) \to {}_{z'}\mathrm{Gal}(\bar{\mathbf{Q}}/\mathbf{Q})(\mathbf{R})$ is surjective.*

PROOF. Let $Hod_{\mathbf{R}}$ be the Tannakian category of real Hodge structures with F_∞ action, i.e., of pairs $(H_{\mathbf{R}}, F_\infty)$ such that $H_{\mathbf{R}}$ is a real vector space with Hodge structure and $F_\infty \in \mathrm{End}_{\mathbf{R}}(H_{\mathbf{R}})$ satisfies $F_\infty^2 = 1$ and $F_\infty(H^{p,q}) = H^{q,p}$ on $H_{\mathbf{C}}$. This category has two fiber functors into real vector spaces: 1. ω_B, the functor which maps a pair $(H_{\mathbf{R}}, F_\infty)$ to the underlying vector space of $H_{\mathbf{R}}$, and ω_{DR}, defined by $\omega_{\mathrm{DR}}((H_{\mathbf{R}}, F_\infty)) = (H_{\mathbf{R}} \otimes \mathbf{C})^{(F_\infty \otimes c)}$. Let $\mathcal{F}_{\mathrm{B}}$ and $\mathcal{F}_{\mathrm{DR}}$ be the groups of tensor automorphisms of these functors. These are groups defined over $\mathbf{R}$. We have short exact sequences

$$1 \to \mathbb{S} \to \mathcal{F}_{\mathrm{B}} \to \mathrm{Gal}(\mathbf{C}/\mathbf{R}) \to 1$$

and

$$1 \to \mathbb{S}_{\mathrm{DR}} \to \mathcal{F}_{\mathrm{DR}} \xrightarrow{\pi} \mathrm{Gal}(\mathbf{C}/\mathbf{R}) \to 1.$$

Here $\mathbb{S} \simeq \mathbf{C}^* = R_{\mathbf{C}/\mathbf{R}}\mathbb{G}_m$, because this group is the one attached to real Hodge structures without F_∞-action, and its arrow to $\mathcal{F}_{\mathrm{B}}$ is that gotten from the functor to the category of real Hodge structures which forgets the F_∞ action. On the other hand, the second arrow (to $\mathrm{Gal}(\mathbf{C}/\mathbf{R})$) is obtained by Tannakian duality from the functor to $\mathrm{Gal}(\mathbf{C}/\mathbf{R})$ modules which forgets the Hodge structure and identifies the F_∞ action with one of $\mathrm{Gal}(\mathbf{C}/\mathbf{R})$. With this identification the finite dimensional real $\mathrm{Gal}(\mathbf{C}/\mathbf{R})$ modules form a full subcategory of $Hod_{\mathbf{R}}$ and the short exact sequence is correspondingly split; the action of c on S is via the algebraic complex conjugation:

$$\mathcal{F}_{\mathrm{B}} = \mathbb{S} \rtimes \mathrm{Gal}(\mathbf{C}/\mathbf{R}).$$

Finally, to understand $\mathcal{F}_{\mathrm{DR}}$, note that the cocycle in $Z^1(\mathbf{R}, \mathcal{H}_{\mathrm{B}})$ defining it is given by the map $F_\infty \mapsto (1, F_\infty) \in S \rtimes \mathrm{Gal}(\mathbf{C}/\mathbf{R}) = \mathcal{F}_{\mathrm{B}}$. Hence $\mathbb{S}_{\mathrm{DR}} = (\mathbf{R}^*)^2$. Since $H^1(\mathbf{R}, (\mathbf{C}^*)^2) = 1$, the torsor $\pi^{-1}(c) \subset \mathcal{H}_{\mathrm{DR}}$ is trivial, i.e., there exists $\tilde{c} \in \mathcal{F}_{\mathrm{DR}}(\mathbf{R})$ such that $\pi(\tilde{c}) = c$. Let $\mathcal{M}_{\mathbf{Q}}^{\mathrm{av}} \otimes \mathbf{R}$ be the category of motives for absolute Hodge cycles with coefficients in $\mathbf{R}$ generated by abelian varieties defined over $\mathbf{Q}$; this is the Tannakian envelope of the category gotten by tensoring each Hom space in $\mathcal{M}_{\mathbf{Q}}^{\mathrm{av}}$ with $\mathbf{R}$. The automorphism group of its De Rham functor is just $\mathcal{F}_{\mathrm{DR}}(\mathbf{R})$. The evident functor from $\mathcal{M}_{\mathbf{Q}}^{\mathrm{av}} \otimes \mathbf{R}$ to $Hod_{\mathbf{R}}$ defines a commutative diagram

$$\begin{array}{ccc} \mathcal{F}_{\mathrm{DR}}(\mathbf{R}) & \longrightarrow & \mathcal{G}_{\mathrm{DR}}(\mathbf{R}) \\ \downarrow & & \downarrow \\ \mathrm{Gal}(\mathbf{C}/\mathbf{R}) & = & \mathrm{Gal}(\mathbf{C}/\mathbf{R}) \end{array}$$

and the claim follows. □

5. Proof of the main theorem

In this section we prove Theorem 1.5.

5.1. We write $\mathcal{G}_{\mathrm{B}}$ for $\mathcal{G}_{\mathrm{B}}^{\mathcal{H}}$, $\mathcal{G}_{\mathrm{DR}}$ for $\mathcal{G}_{\mathrm{DR}}^{\mathcal{H}}$, $\mathcal{P}$ for $\mathcal{P}^{\mathcal{H}}$. The torsor $\mathcal{P}$ clearly has the required properties, so we must prove only the uniqueness.

The torsor $\mathcal{P}$ defines a Galois cohomology class $\xi_{\mathcal{P}} \in H^1(\mathbf{Q}, \mathcal{G}\mathrm{B})$. Let $\mathcal{P}'$ be another torsor with the same properties, and $\xi'_{\mathcal{P}} \in H^1(\mathbf{Q}, \mathcal{G}_{\mathrm{B}})$ the corresponding cohomology class. Then $\xi'_{\mathcal{P}}$ differs from $\xi_{\mathcal{P}}$ by some cohomology class $\xi \in H^1(\mathbf{Q}, {}_z\mathcal{G}_{\mathrm{B}})$ where z is a cocycle representing ξ_P. By the definitions of $\mathcal{P}$ and $\mathcal{G}^{\mathrm{DR}}$ we have ${}_z\mathcal{G}_{\mathrm{B}} = \mathcal{G}_{\mathrm{DR}}$. Since $\mathcal{P}_{\mathbf{R}} \simeq \mathcal{P}'_{\mathbf{R}}$, the localization $\mathrm{loc}_\infty \xi \in H^1(\mathbf{R}, \mathcal{G}_{\mathrm{DR}})$ equals 1. Since $\mathcal{P}$ and $\mathcal{P}'$ project to the same torsor $\mathcal{P}^{\mathrm{CM}}$ of $\mathcal{G}_{\mathrm{B}}^{\mathrm{CM}}$, the image of ξ in $H^1(\mathbf{Q}, {}_z\mathcal{G}^{\mathrm{CM}})$ is 1. We must prove that $\xi = 1$.

There are exact sequences

$$1 \to \mathcal{G}_{\mathrm{DR}}^\circ \to \mathcal{G}_{\mathrm{DR}} \to {}_z\mathrm{Gal}(\bar{\mathbf{Q}}/\mathbf{Q}) \to 1$$
$$1 \to (\mathcal{G}_{\mathrm{DR}}^\circ)^{\mathrm{der}} \to \mathcal{G}_{\mathrm{DR}} \to \mathcal{G}_{\mathrm{DR}}^{\mathrm{CM}} \to 1$$

Since the image of ξ in $H^1(\mathbf{Q}, \mathcal{G}_{\mathrm{DR}}^{\mathrm{CM}})$ is zero, ξ is the image of some

$$\zeta \in H^1(\mathbf{Q}, (\mathcal{G}_{\mathrm{DR}}^\circ)^{\mathrm{der}}).$$

By Proposition 4.4 the homomorphism $\mathcal{G}_{\mathrm{DR}}(\mathbf{R}) \to {}_z\mathrm{Gal}(\bar{\mathbf{Q}}/\mathbf{Q})(\mathbf{R})$ is surjective, hence the kernel of the map $H^1(\mathbf{R}, \mathcal{G}_{\mathrm{DR}}^\circ) \to H^1(\mathbf{R}, \mathcal{G}_{\mathrm{DR}})$ is trivial. It follows that the image of $\mathrm{loc}_\infty \zeta$ in $H^1(\mathbf{R}, \mathcal{G}_{\mathrm{DR}}^\circ)$ is trivial.

5.2. We prove that $\xi = 1$. Let $\eta \in H^1(\mathbf{Q}, \mathcal{G}_{\mathrm{DR}}^\circ)$ be the image of ζ. Then ξ is the image of η in $H^1(\mathbf{Q}, \mathcal{G}_{\mathrm{DR}})$. It suffices to prove that $\eta = 1$.

By Theorem 3.2 $\mathcal{G}_{\mathrm{B}}^\circ = \varprojlim (\mathcal{G}_{\mathrm{B}}^A)^\circ$ where A runs over some set of abelian varieties, $((\mathcal{G}_{\mathrm{B}}^A)^\circ)^{\mathrm{der}}$ is simply connected and $((\mathcal{G}_{\mathrm{B}}^A)^\circ)^{\mathrm{ab}} = S^K$ for some CM-field K containing an imaginary quadratic field. By [**W**, Proof of Theorem 1], $H^1(\mathbf{Q}, ((\mathcal{G}_{\mathrm{DR}}^A)^\circ)^{\mathrm{ab}}) = 1$. Set $G^A = (\mathcal{G}_{\mathrm{DR}}^A)^\circ$. Then $\mathcal{G}_{\mathrm{DR}}^\circ = \varprojlim G^A$ where $(G^A)^{\mathrm{der}}$ is simply connected and $H^1(\mathbf{Q}, (G^A)^{\mathrm{ab}}) = 1$.

Now we prove that $\eta = 1$. By construction $\eta \in H^1(\mathbf{Q}, \mathcal{G}^\circ_{\mathrm{DR}})$, hence $\eta = \varprojlim \eta_A$ where $\eta_A \in H^1(\mathbf{Q}, G^A)$. We have seen that $\mathrm{loc}_\infty \eta = 1$, hence $\mathrm{loc}_\infty \eta_A = 1$. Since $(G^A)^{\mathrm{der}}$ is simply connected and $H^1(\mathbf{Q}, (G^A)^{\mathrm{ab}}) = 1$, by Theorem 3.2 $\eta_A = 1$ for every A. Hence $\eta = 1$ and $\xi = 1$. This proves the theorem. $\square$

References

[D79] P. Deligne, *Variétés de Shimura: interprétation modulaire, et techniques de construction de modéles canoniques*, Automorphic Forms, Representations and L-Functions, Proc. Symp. Pure Math. **33**, part 2, Providence, RI: Amer. Math. Soc. (1979), 247–290.

[D82] P. Deligne (notes by J. S. Milne), *Hodge cycles on abelian varieties*, Lect. Notes in Math. **900**, Berlin: Springer (1982), 9–100.

[D82b] P. Deligne, *Motifs et group de Taniyama*, Hodge Cycles, Motives, and Shimura Varieties, Lect. Notes in Math. **900**, Berlin: Springer (1982), 261–279.

[DM] P. Deligne, J. S. Milne, *Tannakian categories*, Hodge Cycles, Motives, and Shimura Varieties, Lect. Notes in Math. **900**, Berlin: Springer (1982), 101–228.

[RZ] M. Rapoport, T. Zink, Period Spaces for p-Divisible Groups, Annals of Mathematics Studies **141**, Princeton: Princeton University Press (1996).

[Sa] J.-J. Sansuc, *Groupe de Brauer et arithmétique des groupes algébriques linéaires sur un corps de nombres*, J. für die Reine und Angew. Math. **327** (1981), 12–80.

[Se] J.-P. Serre, *Propriétés conjecturales des groups de Galois motiviques et des représentations ℓ-adiques*, Motives, Proc. Symp. Pure Math. **55**, part 1, Providence, RI: Amer. Math. Soc. (1994), 777–800.

[W] J.-P. Wintenberger, *Torseurs pour les motifs et pour les représentations p-adiques potentiellement de type CM*, Math. Ann. **288** (1990), 1–8.

Department of Mathematics, UCLA, Los Angeles, CA 90024–1555, USA
E-mail address: blasius@math.ucla.edu

Raymond and Beverly Sackler School of Mathematical Sciences, Tel Aviv University, 69978 Tel Aviv, Israel
E-mail address: borovoi@math.tau.ac.il

Proceedings of Symposia in Pure Mathematics
Volume **66.1**, 1999

Near Holomorphy, Arithmeticity, and the Theta Correspondence

Antonia W. Bluher

Dedicated to Goro Shimura

Abstract. Let G_1, G_2 be two subgroups of Sp_N which centralize each other. The theta correspondence from G_1 to G_2 has long been known to number theorists as an effective way to produce automorphic forms on G_2 by integrating known automorphic forms on G_1 against a kernel function which is defined on $G_1 \times G_2$. Roughly speaking, this kernel function is the restriction to $G_1 \times G_2$ of a theta function defined on Sp_N. We will prove that when $G_1 = \mathrm{Sp}_1$, $G_2 = \mathrm{O}(n, 2)$, the theta correspondence often preserves analytic properties (holomorphy, near holomorphy) and arithmetic properties (special values at CM-points) of automorphic forms. Actually, the analytic and arithmetic properties are closely related to one another as shown by work of Shimura. The statement about holomorphy is not obvious since the kernel function is nonholomorphic in both variables. Our methods rely on some remarkable algebraic properties of the summands of the theta function, which we expect will generalize to other groups.

The main result of this paper is that to a certain extent the theta correspondence from Sp_1 to $\mathrm{O}(n, 2)$ preserves near holomorphy and arithmeticity. Our emphasis is on the technique of proving near holomorphy, as the methods are likely to generalize to theta correspondences between other groups. The problem of near holomorphy has important applications to arithmeticity; see Section 10.

First we give two definitions of nearly holomorphic automorphic forms. We will work over a totally real number field E, but for now we assume $E = \mathbf{Q}$. Let G be the identity component of the real orthogonal group of signature $(n, 2)$, K a maximal compact subgroup of G, and $\mathcal{Z}$ the bounded symmetric domain associated to G; thus $\mathcal{Z}$ is an open region in $\mathbf{C}^n$, G acts transitively on $\mathcal{Z}$, and K is the isotropy group of a point $\mathbf{0} \in \mathcal{Z}$. If f is a smooth function from $\mathcal{Z}$ to $\mathbf{C}$, we define $f_k : G \to V$ by $f_k(g) = \mu(g, \mathbf{0})^{-k} f(g\mathbf{0})$ for $g \in G$, where $\mu : G \times \mathcal{Z} \to \mathrm{GL}_1(\mathbf{C})$ is a factor of automorphy which will be defined in Section 9. (In Section 9 we consider a more general factor of automorphy.) Denote by $\mathcal{C}_{[k,0]}$ the space of all smooth functions $f : \mathcal{Z} \to \mathbf{C}$ such that $f_k(\gamma g) = f_k(g)$ for all γ in a congruence

1991 *Mathematics Subject Classification.* Primary 11F37; Secondary 11G18, 11F55.

Key words and phrases. Near holomorphy, theta correspondence, arithmeticity, automorphic form, orthogonal group.

subgroup Γ of G. Let $\mathfrak{g}$ denote the complexification of the Lie algebra of G and $\mathfrak{p}_-$ the abelian subalgebra of $\mathfrak{g}$ corresponding to the space of antiholomorphic tangent vectors on $\mathcal{Z}$ at $\mathbf{0}$ (see [**He**, Ch. VIII, §7]). A smooth function φ on G is called nearly holomorphic of degree $\le t$ if $E_1E_2\dots E_{t+1}\varphi = 0$ for all $E_1, \dots, E_{t+1} \in \mathfrak{p}_-$. A smooth function f on $\mathcal{Z}$ is called nearly holomorphic of degree $\le t$ (written $f \in \mathcal{N}^t(\mathcal{Z})$) if f_k is nearly holomorphic for any (hence every) k. In particular, f is nearly holomorphic of degree 0 if and only if it is holomorphic. In the text we give a more convenient definition for near holomorphy on $\mathcal{Z}$ in terms of differential operators $\partial/\partial r_i$ which will be defined in Section 2, namely a smooth function f on $\mathcal{Z}$ is nearly holomorphic of degree $\le t$ if and only if $p(\partial/\partial r_1, \dots, \partial/\partial r_n)f = 0$ for all homogeneous polynomials p of degree $t+1$. The equivalence of the two definitions is proven in [**S90**, Prop. 7.3] and [**S86**, Prop. 2.4(*ii*)]. We will be interested in the space $\mathcal{N}^t_{[k,0]}(\mathcal{Z})$ of nearly holomorphic automorphic forms of weight k and degree $\le t$; this is by definition the space of all $f \in \mathcal{C}_{[k,0]} \cap \mathcal{N}^t(\mathcal{Z})$ which are finite at cusps of $\Gamma\backslash\mathcal{Z}$.

We make similar definitions for the group $\mathrm{SL}_2(\mathbf{R})$. An excellent exposition in this case can be found in [**H**]. Here the symmetric domain is the complex upper half plane $\mathfrak{H}$, and a function f on $\mathfrak{H}$ is nearly holomorphic of degree $\le t$, written $f \in \mathcal{N}^t(\mathfrak{H})$, if and only if $\varepsilon^{t+1}f = 0$, where $\varepsilon = -y^2\partial/\partial\bar{z}$, $y = \Im(z)$. We also utilize the differential operator $\delta_k = (2\pi i)^{-1}(k/(2iy) + \partial/\partial z)$. It is well-known that if f and g are automorphic forms on $\mathfrak{H}$ of weights k and $k-2$, resp., then εf and $\delta_{k-2}g$ have weights $k-2$ and k, resp., and if one of f or g decreases rapidly at cusps, then $\langle \varepsilon f, g\rangle = 2\pi i\,\langle f, \delta_{k-2}g\rangle$, where $\langle\ ,\ \rangle$ denotes the Petersson inner product. If $k' \in (1/2)\mathbf{Z}$, we denote by $\mathcal{N}^t_{k'}(\mathfrak{H})$ the space of all automorphic forms on $\mathfrak{H}$ of weight k' which belong to $\mathcal{N}^t(\mathfrak{H})$ (see Section 3).

Let k be a positive integer and H a harmonic polynomial on $\mathbf{C}^n$ which is homogeneous of degree h. Put $k' = k + 1 - n/2$. Note that k' is half-integral if and only if n is odd. We will define a certain theta function $\Theta_{k,H}(z, w)$ on $\mathfrak{H} \times \mathcal{Z}$ for $h \le k$ which satisfies a transformation formula in each variable. In fact, the function $G_{k,H}(z, w) = y^{n/2+h}\overline{\Theta_{k,H}(z, w)}$ has weight $k' - 2h$ as a function of z, and Θ_k is automorphic in the w variable of weight k, where $\Theta_k = \Theta_{k,1}$. We do not give the formula for $\Theta_{k,H}$ here because it would introduce too much notation, but we remark that it can be obtained by first restricting a theta function defined on the metaplectic cover of a large symplectic group into which $\mathrm{Sp}_1 \times G$ can be embedded as a dual reductive pair and then taking derivatives to introduce harmonic factors into the summands; see [S87a, §11]. We define a theta correspondence from the space of automorphic forms on $\mathfrak{H}$ of weight k' to the space of automorphic forms on $\mathcal{Z}$ of weight k by the formula $\Theta_k(F)(w) = \langle F, G_k(\cdot, w)\rangle$, where $G_k = G_{k,1}$. Here the Petersson inner product converges because G_k is rapidly decreasing at cusps. Now $\Theta_k(z, w)$ is nonholomorphic in both variables. Nonetheless, because of some remarkable algebraic properties of the summands of Θ_k and $\Theta_{k,H}$ we are able to prove the following theorems.

THEOREM 1. *Let k be a positive integer, $k' = k + 1 - n/2$, and*

$$\mu = \begin{cases} k/2 - 1 & \textit{if } k > 2 \\ k - 1 & \textit{if } k \le 2. \end{cases}$$

Let $F \in \mathcal{N}^t_{k'}(\mathfrak{H})$, where $0 \leq t \leq \mu$. If k is even and $t = \mu$, we assume that F decreases rapidly at cusps. Then $\Theta_k(F) \in \mathcal{N}^q_{[k,0]}(\mathcal{Z})$, where $q = 2t$ or t according as $k > 2$ or $k \leq 2$.

If F is holomorphic, Theorem 1 implies that $\Theta_k(F)$ is holomorphic, a result which has been proven by Friedberg [**F**]. See also [**KM**] for a related problem when F is holomorphic.

THEOREM 2. *Under the hypotheses of Theorem 1, if $F \in \mathcal{N}^t_{k'}(\mathfrak{H})$ and F is arithmetic then $\pi\Theta_k(F)$ is arithmetic.*

We actually prove these theorems for a more general type of theta correspondence, and over an arbitrary totally real number field. We also prove that if the hypotheses on t are weakened in Theorem 2, $\Theta_k(F)$ is still arithmetic on a dense set of CM-points. Arithmeticity and CM-points on $\mathcal{Z}$ will be defined in Chapter 2, and Theorem 2 will be proven using Theorem 1 and a technique of Shimura of reducing the problem to the study of special values at a single CM-point in $\mathcal{Z}$ for which $\Theta_k(F)$ decomposes into a certain Rankin product.

Let us now discuss the proof of Theorem 1. For illustration, we take $t = 1$; thus let $F \in \mathcal{N}^1_{k'}(\mathfrak{H})$, where $k \geq 2$. We must prove that $p(\partial/\partial r_1, \ldots, \partial/\partial r_n)\Theta_k(F) = 0$ for all homogeneous polynomials p of degree $q+1$, where $q = 1$ or 2 according as $k = 2$ or $k > 2$. By a well-known theorem, we can write $p = H_1 + H_2 L$ with H_1, H_2 homogeneous harmonic polynomials of degree $q+1, q-1$, respectively and $L(X) = \sum_{i=1}^n X_i^2$. Let $\partial/\partial r = (\partial/\partial r_1, \ldots, \partial/\partial r_n)$ and let $\overline{p}(\partial/\partial r)$ be the operator: $\overline{p}(\partial/\partial r)g = \overline{p(\partial/\partial r)\overline{g}}$, where g is a smooth function on $\mathcal{Z}$. Using algebraic properties of the summands of $\Theta_{k,H}(z,w)$ we can prove (Lemma 2) that for any homogeneous harmonic polynomial H on $\mathbf{C}^n$ of degree h,

$$\overline{H}(\partial/\partial r)G_k(z,w) = (4\pi)^h \delta_{k'-2}\delta_{k'-4}\ldots\delta_{k'-2h}G_{k,H}(z,w) \text{ if } h \leq k,$$

$$(*)\quad \overline{L}(\partial/\partial r)G_k(z,w) = -8\pi i\delta_{k'-2}\delta_{k'-4}\varepsilon G_{k-2} - (n/2)(k-2)\delta_{k'-2}G_{k-2} \text{ if } k \geq 2.$$

Thus,

$$\begin{aligned} p(\partial/\partial r)\Theta_k(F) &= \langle F, \overline{H}_1(\partial/\partial r)G_k\rangle + H_2(\partial/\partial r)\langle F, \overline{L}(\partial/\partial r)G_k\rangle \\ &= (4\pi)^{q+1}\langle F, \delta_{k'-2}\delta_{k'-4}\ldots\delta_{k'-2q-2}G_{k,H_1}\rangle \\ &\quad -8\pi i H_2(\partial/\partial r)\langle F, \delta_{k'-2}\delta_{k'-4}\varepsilon G_{k-2}\rangle \\ &\quad -(n/2)(k-2)H_2(\partial/\partial r)\langle F, \delta_{k'-2}G_{k-2}\rangle \\ &= (-2i)^{q+1}\langle \varepsilon^{q+1}F, G_{k,H_1}\rangle + (2i/\pi)H_2(\partial/\partial r)\langle \varepsilon^2 F, \varepsilon G_{k-2}\rangle \\ &\quad -(n/4\pi i)(k-2)H_2(\partial/\partial r)\langle \varepsilon F, G_{k-2}\rangle. \end{aligned}$$

The first two terms vanish because $\varepsilon^2 F = 0$. The third term clearly vanishes when $k = 2$. If $k \neq 2$, then the degree of H_2 is one and hence $H_2(\partial/\partial r)\langle \varepsilon F, G_{k-2}\rangle = 4\pi\langle \varepsilon F, \delta_{k'-2}G_{k-2,H_2}\rangle = -2i\langle \varepsilon^2 F, G_{k-2,H_2}\rangle = 0$. The reason for the extra hypothesis on F when $k = 2$ is that G_0 might not decrease rapidly at cusps, so that $\langle \varepsilon F, G_{k-2}\rangle$ could diverge if F does not decrease rapidly at cusps. Notice that our results are stronger for $k \leq 2$ than for $k > 2$. For $k = 2$ this is explained by the vanishing of the second term in (*), and for $k = 1$ the explanation is that all polynomials of degree ≤ 1 are harmonic, so that the equation (*) does not enter into the calculation.

The author wishes to express her appreciation to Goro Shimura for his dedication as a teacher and example as a scholar, and for his continued interest in the author's well-being, even a decade after she completed her studies at Princeton.

Table of Contents

I. Near holomorphy

1. Automorphic forms on $O(n,2)$

Let E be a totally real number field, V a vector space over E of dimension $n+2$ with $0 < n \in \mathbf{Z}$, and S a symmetric bilinear form on V of signature $(n,2)$ at every archimedean place of E. We sometimes write S for the matrix of S with respect to a fixed basis of V, and we write $S[v] = S(v,v)$. We denote by J the set of embeddings of E into $\mathbf{R}$ and by $\mathfrak{o}$ the integer ring of E. For any set (respectively group or ring) A, A^J denotes the Cartesian product (respectively direct sum) of $[E:\mathbf{Q}]$ copies of A. If A is a ring, A^n_m denotes the $n \times m$ matrices with entries in A. If $a, b \in \mathbf{R}^J$, $a \leq b$ shall mean $a_\tau \leq b_\tau$ for all $\tau \in J$. Let $V_\tau = V \otimes_{\tau(E)} \mathbf{R}$, S_τ be the extension of S to V_τ, $R = \mathrm{diag}\{1_n, \left(\begin{smallmatrix} 0 & -1 \\ -1 & 0 \end{smallmatrix}\right)\}$, and $Q = \mathrm{diag}\{1_n, -1_2\} \in \mathrm{GL}_{n+2}(\mathbf{Z})$, where 1_n is the identity matrix in GL_n and $\mathrm{diag}\{A, B\}$ is the matrix $\left(\begin{smallmatrix} A & 0 \\ O & B \end{smallmatrix}\right)$ when $A \in \mathrm{GL}_n$, $B \in \mathrm{GL}_m$. We fix once and for all a matrix $A_\tau \in \mathrm{GL}_{n+2}(\overline{\mathbf{Q}})$ for each $\tau \in J$ such that $S_\tau[v] = R(A_\tau v, A_\tau v)$ and $S_\tau(\overline{v}, v) = Q(\overline{A_\tau v}, A_\tau v)$ for all $v \in V_\tau \otimes \mathbf{C}$.

The orthogonal group of S_τ acts naturally on the set

$$X_\tau = \{\, x \in \mathbf{C}^{n+2} : S_\tau(x,x) = 0,\ S_\tau(\overline{x}, x) < 0 \,\}.$$

As shown in [**S80**, §2], this set has two connected components, given by

$$X^0_\tau = \{\, \lambda w_- : \lambda \in \mathbf{C}^\times,\ w \in \mathcal{Z}_\tau \,\}, \quad X^1_\tau = A_\tau^{-1} R A_\tau X^0_\tau, \tag{1.1}$$

where

$$\mathcal{Z}_\tau = \{\, w \in \mathbf{C}^n : {}^t\overline{w}w < (1/4)|{}^tww|^2 + 1 < 2 \,\},$$
$$w_- = A_\tau^{-1}\,{}^t({}^tw, {}^tww/2, 1) \in \mathbf{C}^{n+2} \quad \text{for } w \in \mathcal{Z}_\tau.$$

Note that for each $w \in \mathcal{Z}_\tau$, $\mathrm{Re}(w_-)$ and $\mathrm{Im}(w_-)$ span a two-dimensional totally negative subspace of V_τ. Put

$$G_+(S_\tau) = \{\gamma \in \mathrm{GL}(V_\tau, \mathbf{R}) : {}^t\gamma S_\tau \gamma = S_\tau \text{ and } \gamma X_\tau^0 = X_\tau^0\},$$
$$G_+(S) = \mathrm{GL}(V, E) \cap \prod_{\tau \in J} G_+(S_\tau).$$

Since $\gamma w_- \in X_\tau^0$ for $\gamma \in G_+(S_\tau)$ and $w \in \mathcal{Z}_\tau$, we can define an action $w \mapsto \gamma w$ of $G_+(S_\tau)$ on $\mathcal{Z}_\tau$ and a factor of automorphy $\mu_\tau(\gamma, w) \in \mathrm{GL}_1(\mathbf{C})$ by the formula

$$\gamma w_- = \mu_\tau(\gamma, w)(\gamma w)_-.$$

If $w \in \mathcal{Z}_\tau$, let

$$p(w) = \begin{pmatrix} w \\ {}^t w w/2 \\ 1 \end{pmatrix} = A_\tau w_-, \qquad p'(w) = \begin{pmatrix} w \\ 1 \\ {}^t\overline{w}\overline{w}/2 \end{pmatrix},$$
$$\eta(w) = -S_\tau(\overline{w_-}, w_-) = 1 + (1/4)|{}^t w w|^2 - {}^t\overline{w} w > 0,$$
$$r(w) = \partial/\partial w\{\log \eta(w)\} = \eta(w)^{-1t}(({}^t\overline{w}\overline{w}/2)w - \overline{w}) \in \mathbf{C}_n^1,$$
$$q(w) = \partial p(w)/\partial w - p(w) r(w) \in \mathbf{C}_n^{n+2}, \quad w_+ = A_\tau^{-1} q(w),$$
$$\mathbf{B}(w) = (q(w) \;\; p'(w) \;\; p(w)), \quad B_\tau(w) = A_\tau^{-1}\mathbf{B}(w) = (w_+ \;\; A_\tau^{-1}p'(w) \;\; w_-),$$

where $\partial/\partial w = (\partial/\partial w_1, \dots, \partial/\partial w_n)$. As shown in [**S80**, §2], there is a factor of automorphy $\psi_\tau(\gamma, w)$ with values in $\mathrm{SO}(n, \mathbf{C})$ defined by $\gamma w_+ = (\gamma w)_+ \psi_\tau(\gamma, w)$; thus

$$\gamma B_\tau(w) = B_\tau(\gamma w) \mathrm{diag}\{\psi_\tau(\gamma, w), \overline{\mu_\tau(\gamma, w)}, \mu_\tau(\gamma, w)\}. \tag{1.2}$$

If $w = (w_\tau) \in \mathcal{Z} = \prod_\tau \mathcal{Z}_\tau$ and $\gamma \in \prod_\tau G_+(S_\tau)$, we write $w_- = (w_{\tau-}) \in (\mathbf{C}^{n+2})^J$, $\gamma w = (\gamma_\tau w_\tau) \in \mathcal{Z}$, $\psi(\gamma, w) = (\psi_\tau(\gamma_\tau, w_\tau)) \in \mathrm{SO}(n, \mathbf{C})^J$, *etc.*

Let $\mathfrak{K} = (\mathrm{GL}_1(\mathbf{C}) \times \mathrm{SO}(n, \mathbf{C}))^J$ and let $\rho : \mathfrak{K} \to \mathrm{GL}(W)$ be a polynomial representation into a finite-dimensional complex vector space W. In Chapter 2 we will assume that W has a $\overline{\mathbf{Q}}$-rational structure and that ρ is $\overline{\mathbf{Q}}$-rational. We denote by $\mathcal{C}_\rho(\mathcal{Z})$ the space of smooth functions f from $\mathcal{Z}$ into W such that $f(\gamma w) = \rho(\mu(\gamma, w), \psi(\gamma, w)) f(w)$ for all γ in a congruence subgroup of $G_+(S)$. Now we consider a special case of a representation ρ. For $k, h \in \mathbf{Z}^J$ with $0 \le h \le k$, let $\mathcal{H}_h$ denote the complex vector space of homogeneous harmonic polynomials on $(\mathbf{C}^n)^J$ of degree h, and let $[k, h] : \mathfrak{K} \to \mathrm{GL}(\mathcal{H}_h^*)$ be the representation $[k, h](a, b)(\varphi)(H) = a^{k-h}\varphi(H \circ b)$ for $(a, b) \in \mathfrak{K}$, $\varphi \in \mathcal{H}_h^*$, $H \in \mathcal{H}_h$. Here we write $a^x = \prod_{\tau \in J} a_\tau^{x_\tau} \in \mathbf{C}$ for $a, x \in \mathbf{C}^J$ when this is unambiguous. We also put $|a| = (|a_\tau|) \in (\mathbf{R}_{\ge 0})^J$, $u = (1, \dots, 1) \in \mathbf{Z}^J$, and $\{a\} = \sum_\tau a_\tau \in \mathbf{C}$.

2. Near holomorphy on $\mathcal{Z}$

A function $f : \mathcal{Z}_\tau \to W$ is called nearly holomorphic of degree $\le t$ on $\mathcal{Z}_\tau$ (where $0 \le t \in \mathbf{Z}$) if its coordinate functions are polynomials of degree $\le t$ in the components of $r(w_\tau)$ with $\mathcal{Z}_\tau$-holomorphic coefficients. This is equivalent to the condition (see [**S86**, §3]):

$$\partial^{t+1}/\partial r_{\tau i_1} \dots \partial r_{\tau i_{t+1}} f(w) \equiv 0 \quad \text{for all } (i_1, \dots, i_{t+1}) \in \{1, \dots, n\}^{t+1},$$

where $\partial/\partial r_{\tau i}$ is defined by:

$$\partial/\partial\overline{w}_j = \sum_{i=1}^{n} \partial r_{\tau i}/\partial\overline{w}_j\, \partial/\partial r_{\tau i}$$

with $w_1, \dots, w_n$ the standard coordinates on $\mathcal{Z}_\tau$. For $0 \le t \in \mathbf{Z}^J$, we denote by $\mathcal{N}^t(\mathcal{Z}, W)$ the space of functions from $\mathcal{Z}$ into W which are nearly holomorphic of degree $\le t_\tau$ on $\mathcal{Z}_\tau$ for each $\tau \in \delta$; thus $\mathcal{N}^0(\mathcal{Z}, W)$ is the space of holomorphic functions from $\mathcal{Z}$ into W. If $\rho : \mathfrak{K} \to \mathrm{GL}(W)$, we define $\mathcal{N}^t_\rho(\mathcal{Z})$ to be the space of all $f \in \mathcal{C}_\rho(\mathcal{Z}) \cap \mathcal{N}^t(\mathcal{Z}, W)$ which are finite at cusps. The cusp condition is necessary only when $n = 1$ and $E = \mathbf{Q}$. We put also $\mathrm{M}_\rho(\mathcal{Z}) = \mathcal{N}^0_\rho(\mathcal{Z})$.

3. Nearly holomorphic automorphic forms on $\mathfrak{H}^J$

Let $0 \le t \in \mathbf{Z}^J$ and let k be an integral or half-integral weight of E; that is, $k \in (1/2)\mathbf{Z}^J$ and $k_\tau \bmod \mathbf{Z}$ is independent of $\tau \in J$. If $f : \mathfrak{H}^J \to W$ and $\gamma = \left(\begin{smallmatrix} a & b \\ c & d \end{smallmatrix}\right) \in \mathrm{SL}_2(E)$, let

$$f|_k\gamma(z) = (cz+d)^{-k} f(\gamma z),$$

where $(\gamma z)_\tau = (a^\tau z_\tau + b^\tau)/(c^\tau z_\tau + d^\tau)$. We must assume that γ belongs to the set C'' of [**S85**, p. 294] if k is half-integral, in which case we define $(cz+d)^k = (cz+d)^{k-u/2} h(\gamma, z)$ with $h(\gamma, z)$ as in [**S85**, Proposition 3.2]. We write $j(\gamma, z) = cz + d$. The spaces $\mathcal{C}_k(\mathfrak{H}^J)$, $\mathcal{N}^t(\mathfrak{H}^J, W)$, $\mathcal{N}^t_k(\mathfrak{H}^J)$, and $\mathrm{M}_k(\mathfrak{H}^J)$ are defined in the same manner as for $\mathcal{Z}$, except that $\mathfrak{K}$ should be replaced by $\mathrm{GL}_1(\mathbf{C})^J$, ρ by the representation $a \in \mathrm{GL}_1(\mathbf{C})^J \mapsto a^k$, $r(w_\tau)$ by y_τ^{-1}, and $\partial/\partial r(w_\tau)$ by $\varepsilon_\tau = -y_\tau^2 \partial/\partial\overline{z_\tau}$, where $y_\tau = \mathrm{Im}(z_\tau)$. The precise definitions are given in Section 7. When $W = \mathbf{C}$ we sometimes omit W from the notation.

4. The theta correspondence and statement of Theorem 1

We define a theta correspondence from $\mathcal{C}_{k'-2h}(\mathfrak{H}^J)$ (with a cusp condition) to $\mathcal{C}_{[k,h]}(\mathcal{Z})$, where $h, k \in \mathbf{Z}^J$, $0 \le h \le k$, and $k' = k + (1 - n/2)u$. Let $\Theta_{k,h} : \mathfrak{H}^J \times \mathcal{Z} \to \mathcal{H}_h^*$ be given by

$$\text{(1.3)}\quad \Theta_{k,h}(z, w)(H) = \sum_{v \in V} C(v)\, X^{k-h}\, H(W^+)\, e_E(zS[v] + 4iyXW^-), \qquad H \in \mathcal{H}_h,$$

where $z \in \mathfrak{H}^J$, $y = \mathrm{Im}(z) \in \mathbf{R}^J$, $w \in \mathcal{Z}$, C is a complex-valued locally constant function on V, $e_E(x) = e^{2\pi i \{x\}}$ for $x \in \mathbf{C}^J$, and

$$\text{(1.4)}\quad W^- = S(v, w_-) \in \mathbf{C}^J, \quad X = \eta(w)^{-1}\overline{W^-} \in \mathbf{C}^J, \quad W^+ = {}^t w_+ S v \in (\mathbf{C}^n)^J.$$

This satisfies a transformation formula ([**S87a**, §11]; see also [**B90**, (6.9)]):

$$\text{(1.5)}\qquad \Theta_{k,h}(\gamma z, w) = h(\gamma, z)^{n+2}\, j(\gamma, z)^{h-u}\, \overline{j(\gamma, z)}^{u+k-h}\, \Theta_{k,h}(z, w)$$

for γ in a sufficiently small congruence subgroup Γ of $\mathrm{SL}_2(E)$. If $k + h \ne 0$, then $\Theta_{k,h}$ decreases rapidly at cusps of $\Gamma\backslash\mathfrak{H}^J$; this can be seen from Lemma 6.2 of [**B90**]. Let F be an element of $\mathcal{C}_{k'-2h}(\mathfrak{H}^J)$ which is finite at cusps. We define the theta transform $\Theta_{k,h}(F)$ of F to be the function from $\mathcal{Z}$ to $\mathcal{H}_h^*$ given by:

$$\Theta_{k,h}(F)(w) = \mathrm{vol}(\Phi)^{-1} \int_\Phi F(z)\, \Theta_{k,h}(z, w)\, y^{k-h+u}\, \{dz\},$$

where $\{dz\} = \prod_\tau y_\tau^{-2}\, dx_\tau\, dy_\tau$ is a Haar measure on $\mathfrak{H}^J$, $\Phi = \Gamma\backslash\mathfrak{H}^J$, and Γ is a sufficiently small congruence subgroup of $\mathrm{SL}_2(\mathfrak{o})$ for the integral to be defined. We see easily that $\Theta_{k,h}(F) \in \mathcal{C}_{[k,h]}(\mathcal{Z})$. When $h = 0$ we identify $\mathcal{H}_h^*$ with $\mathbf{C}$ and write Θ_k in place of $\Theta_{k,0}$. Also we write $\Theta_{k,H}(z,w) = \Theta_{k,h}(z,w)(H)$ and $\Theta_{k,H}(F) = \Theta_{k,h}(F)(H)$ when $H \in \mathcal{H}_h$.

THEOREM 1. *Let $h, k \in \mathbf{Z}^J$, $0 \le h \le k - u$, $k' = k + (1 - n/2)u$, and*

$$\mu_\tau = \begin{cases} k_\tau/2 - h_\tau - 1 & \text{if } k_\tau > 2 \\ k_\tau - h_\tau - 1 & \text{if } k_\tau \le 2. \end{cases}$$

Let $F \in \mathcal{N}_{k'-2h}^t(\mathfrak{H}^J)$, where $0 \le t \le \mu$. If $E = \mathbf{Q}$, k is even, and $t = \mu$, we assume that F decreases rapidly at cusps. Then $\Theta_{k,h}(F) \in \mathcal{N}_{[k,h]}^q(\mathcal{Z})$, where

$$q_\tau = \begin{cases} 2t_\tau + h_\tau & \text{if } k_\tau > 2 \\ k_\tau & \text{if } k_\tau \le 2. \end{cases}$$

In particular, if F is holomorphic and $h = 0$ then $\Theta_k(F)$ is holomorphic.

Fix $\tau \in J$, and put $\partial/\partial r(w_\tau) = {}^t(\partial/\partial r_{\tau 1}, \dots, \partial/\partial r_{\tau n})$. We must show that if F satisfies the hypothesis, then $p(\partial/\partial r(w_\tau))\Theta_{k,h}(F) = 0$ for all homogeneous polynomials p on $\mathbf{C}^n$ of degree $q_\tau + 1$. For $0 \le \ell \in \mathbf{Z}$, let $\mathcal{S}_\ell$ denote the space of homogeneous polynomials on $\mathbf{C}^n$ of degree ℓ, and $L \in \mathcal{S}_2$ the polynomial $L(u) = {}^t uu$, $u \in \mathbf{C}^n$. We identify $\mathbf{C}^n$ with the subspace $\{x \in (\mathbf{C}^n)^J : x_\sigma = 0 \quad \text{for all } \sigma \ne \tau\} \subset (\mathbf{C}^n)^J$. In particular, the space of harmonic polynomials in $\mathcal{S}_\ell$ is identified with $\mathcal{H}_{\ell\tau}$. Then $\mathcal{S}_\ell = \oplus_{d,e\ge 0, 2d+e=\ell} L^d \mathcal{H}_{e\tau}$, and $\mathcal{H}_{e\tau}$ is spanned by polynomials of the form $\zeta(u) = ({}^t cu)^e$ with $c \in \mathbf{C}^n$, where ${}^t cc = 0$ if $e \ge 2$. The first fact is well-known, the second is due to Shimura [**S84**, Theorem 2.B]; see also [**B90**, Remark 4.5]). Thus we need formulas for $\zeta(\partial/\partial r(w_\tau))\Theta_{k,h}(z,w)$ and $L(\partial/\partial r(w_\tau))\Theta_{k,h}(z,w)$. These will be derived in Lemmas 1 and 2. We complete the proof of the theorem after Lemma 2.

5. Duality formulas for differential operators on the theta kernel

By [**S86**, p. 380] we have for $w \in \mathcal{Z}$, $v \in V$:

$$X = -(\partial W^-/\partial w){}^t r + W^-({}^t rr/2) - c, \tag{1.6}$$

$$W^+ = {}^t(\partial W^-/\partial w) - W^-\,{}^t r, \quad W^- = {}^t aw - c{}^t ww/2 - b, \tag{1.7}$$

where $a \in (\mathbf{C}^n)^J$ and $b, c \in \mathbf{C}^J$ are defined by $A_\tau v = {}^t({}^t a_\tau, b_\tau, c_\tau)$. (The symbols W^-, W^+, r of this paper are denoted respectively by $\mu[v,w]$, ${}^t\lambda[v,w]$, and ${}^t r$ in [**S86**].) From these formulas we easily derive:

$$\begin{aligned} \partial X_\tau/\partial r(w_\tau) &= -W_\tau^+ \\ \partial({}^t W_\tau^+)/\partial r(w_\tau) &= -W_\tau^- 1_n \\ {}^t W^+ W^+ &= S[v] + 2W^- X. \end{aligned} \tag{1.8}$$

Let $Y = (Y_\sigma) \in \mathbf{C}^J$, where Y_σ are variables and $\partial Y_\sigma/\partial r(w_\tau) = 0$. Later we will put $Y = -8\pi W^- y$ (assuming $W_\sigma^- \ne 0$ for all $\sigma \in J$), where $y = \mathrm{Im}(z)$, $z \in \mathfrak{H}^J$.

For $0 \le p \in \mathbf{Z}^J$ and k a weight of E, let

$$(1.9)\quad d_k^{Y,p} = \prod_{\tau\in J} d^{Y,\tau}_{k_\tau+2p_\tau-2}\, d^{Y,\tau}_{k_\tau+2p_\tau-4}\cdots d^{Y,\tau}_{k_\tau}, \quad d^{Y,\tau}_{k_\tau} = \frac{k_\tau}{Y_\tau} + \partial/\partial Y_\tau,$$

$$(1.10)\quad \delta_k^p = \prod_{\tau\in J} \delta^\tau_{k_\tau+2p_\tau-2}\, \delta^\tau_{k_\tau+2p_\tau-4}\cdots \delta^\tau_{k_\tau}, \quad \delta^\tau_{k_\tau} = (1/2\pi i)\left(\frac{k_\tau}{2iy_\tau} + \partial/\partial z_\tau\right).$$

LEMMA 1. *If* $\zeta \in \mathcal{H}_{e\tau}$ *and* $L(u) = {}^t uu \in \mathcal{S}_2$, *then*

a) $\zeta(\partial/\partial r(w_\tau))(X^k \exp\{XY\}) = \zeta(-W^+_\tau)\, X^{k-e\tau}\, Y^{e\tau}\, d^{Y,e\tau}_{k+(1-e)\tau}(\exp\{XY\})$
$= \zeta(-W^+_\tau)\, X^{k-e\tau}\, d^{Y,e\tau}_{k+(1-2e)\tau}(Y^{e\tau}\exp\{XY\})$

b) $L(\partial/\partial r(w_\tau))(X^k \exp\{XY\}) = nW^-_\tau\, X^{k-\tau}\, Y^\tau\, d^{Y,\tau}_k(\exp\{XY\})$
$+ (2W^-_\tau X_\tau + S_\tau[v])\, X^{k-2\tau}\, Y^{2\tau}\, d^{Y,2\tau}_{k-\tau}(\exp\{XY\}).$

PROOF. As we remarked earlier, we may assume $\zeta(u) = ({}^t cu)^e$ with $c \in \mathbf{C}^n$, where ${}^t cc = 0$ if $e \ge 2$. Also, since

$$(1.11)\qquad Y^{s\tau}\, d^{Y,p}_k = d^{Y,p}_{k-s\tau}\, Y^{s\tau} \quad \text{if } s \text{ is a weight of } E,$$

the two expressions on the right side in (a) are equal. If $e = 1$, (a) can be verified by a simple computation, using (1.8). If $e \ge 2$ then, by induction on e,

$$\begin{aligned}
&\zeta(\partial/\partial r(w_\tau))(X^k \exp\{XY\}) = \\
&\sum_i c_i\partial/\partial r_{\tau i}\{(-{}^t cW^+_\tau)^{e-1}\, X^{k-(e-1)\tau}\, Y^{(e-1)\tau}\, d^{Y,(e-1)\tau}_{k+(2-e)\tau}(\exp\{XY\})\} \\
&= (e-1)\sum c_i(-{}^t cW^+_\tau)^{e-2}\, c_i\,(-W^-_\tau)\cdot(\text{terms independent of i}) \\
&\quad + (-{}^t cW^+_\tau)^{e-1}\, Y^{(e-1)\tau} \sum c_i\partial/\partial r_{\tau i}\{X^{k-(e-1)\tau}\, d^{Y,(e-1)\tau}_{k+(2-e)\tau}(\exp\{XY\})\} \\
&= 0 + (-{}^t cW^+_\tau)^{e-1}\, Y^{(e-1)\tau}\, d^{Y,(e-1)\tau}_{k+(2-e)\tau}\{-{}^t cW^+_\tau\, X^{k-e\tau}\, Y_\tau\, d^{Y,\tau}_{k+(1-e)\tau}(\exp\{XY\})\}.
\end{aligned}$$

Using (1.11), we obtain the formula (a). For (b), we have

$$\begin{aligned}
&L(\partial/\partial r(w_\tau))(X^k \exp\{XY\}) = \sum_{i=1}^n -\partial/\partial r_{\tau i}\{W^+_{\tau i}\, X^{k-\tau}\, Y_\tau\, d^{Y,\tau}_k \exp\{XY\}\} \\
&= Y_\tau\, d^{Y,\tau}_k \sum_{i=1}^n \{\partial/\partial r_{\tau i}(-W^+_{\tau i})\, X^{k-\tau}\exp\{XY\} - W^+_{\tau i}\,\partial/\partial r_i(X^{k-\tau}\exp\{XY\})\} \\
&= nY_\tau\, W^-_\tau\, X^{k-\tau}\, d^{Y,\tau}_k \exp\{XY\} + Y^2_\tau L(W^+_\tau)\, X^{k-2\tau}\, d^{Y,2\tau}_{k-\tau}\exp\{XY\}.
\end{aligned}$$

Since $L(W^+_\tau) = 2W^-_\tau X_\tau + S_\tau[v]$ by (1.8), this completes the proof. □

For $h, k \in \mathbf{Z}^J$ with $0 \le h \le k$ and $H \in \mathcal{H}_h$, let $G_{k,h} : \mathfrak{H}^J \times \mathcal{Z} \to \mathcal{H}^*_h$, $G_{k,H} : \mathfrak{H}^J \times \mathcal{Z} \to \mathbf{C}$, and $G_k : \mathfrak{H}^J \times \mathcal{Z} \to \mathcal{H}^*_0 = \mathbf{C}$ be given by

$$(1.12)\qquad \begin{aligned}
G_{k,h}(z,w) &= y^{(n/2)u+h}\overline{\Theta_{k,h}(z,w)}, \\
G_{k,H}(z,w) &= G_{k,h}(z,w)(H), \\
G_k(z,w) &= G_{k,0}(z,w).
\end{aligned}$$

Put $k' = k+(1-n/2)u$. Then $G_{k,h}(\cdot, w) \in \mathcal{C}_{k'-2h}(\mathfrak{H}^J, \mathcal{H}_h^*)$, $G_{k,H}(\cdot, w) \in \mathcal{C}_{k'-2h}(\mathfrak{H}^J)$, and $G_k(\cdot, w) \in \mathcal{C}_{k'}(\mathfrak{H}^J)$ by (1.5). For $p \in \mathcal{S}_\ell$, let $\overline{p}(\partial/\partial r(w_\tau))$ be the operator:

$$\overline{p}(\partial/\partial r(w_\tau))G = \overline{p(\partial/\partial r(w_\tau))\overline{G}}, \tag{1.13}$$

where G is a smooth function on $\mathcal{Z}$ or $\mathcal{Z}_\tau$. Observe that if $F \in \mathcal{C}_{k'-2h}(\mathfrak{H}^J)$, then

$$p(\partial/\partial r(w_\tau))\Theta_{k,h}(F) = p(\partial/\partial r(w_\tau))\,\langle F, G_{k,h}\rangle = \langle F, \overline{p}(\partial/\partial r(w_\tau))G_{k,h}\rangle\,, \tag{1.14}$$

where $\langle\ ,\ \rangle$ denotes the Petersson inner product. The following gives a "duality" formula relating the effect on G_k of differential operators in the z variable with the effect of differential operators in the w variable.

LEMMA 2. *Let $0 \le k \in \mathbf{Z}^J$, $k' = k + (1 - n/2)u \in (1/2)\mathbf{Z}^J$.*

a) If $H \in \mathcal{H}_h$ and $0 \le h \le k$, then

$$\overline{H}(\partial/\partial r)G_k(z, w) = (4\pi)^{\{h\}}\,\delta^h_{k'-2h}\,G_{k,H}(z, w).$$

b) If $k_\tau \ge 2$, then

$$\overline{L}(\partial/\partial r(w_\tau))G_k(z, w) = -8\pi i\,\delta^{2\tau}_{k'-4\tau}\,\varepsilon_\tau\,G_{k-2\tau} - (n/2)(k_\tau - 2)\delta^\tau_{k'-2\tau}G_{k-2\tau},$$

where $L(u) = {}^t u u$ and $\varepsilon_\tau = -y_\tau^2\,\partial/\partial\overline{z}_\tau$.

PROOF. Since $\mathcal{H}_h$ is the tensor product of $\mathcal{H}_{h_\tau\tau}$ for $\tau \in J$, it suffices to prove (a) when $H \in \mathcal{H}_{e\tau}$, $0 \le e \in \mathbf{Z}$. Let $Y = -8\pi W^- y$. Since $Y^p d_k^{Y,p} = y^p d_k^{y,p}$ and $d_k^{y,\tau} = -4\pi\delta_k^\tau - i\partial/\partial x_\tau$, we see from Lemma 1(a) and (1.11) that if s is a weight of E,

$$y^{s\tau}\,H(\partial/\partial r)(X^k \exp\{XY\}) = (4\pi)^e\,H(W^+)\,X^{k-e\tau}\,\delta^{e\tau}_{k+(1-s-2e)\tau}(y^{(s+e)\tau}\exp\{XY\}).$$

Hence,

$$\begin{aligned} H(\partial/\partial r)\overline{G_k(z, w)} &= \sum_{v\in V} C(v)\,e_E(zS[v])\,y^{(n/2)u}\,H(\partial/\partial r)(X^k \exp\{XY\}) \\ &= (4\pi)^e \sum_{v\in V} C(v)\,H(W^+)\,X^{k-e\tau}\,e_E(zS[v])\,\delta^{e\tau}_{k'-2e\tau}(y^{(n/2)u+e\tau}\exp\{XY\}). \end{aligned}$$

By taking the conjugate, we obtain (a), since XY is real. For (b), let

$$A = \overline{L}(\partial/\partial r(w_\tau))G_k(z, w), \quad B = \delta^{2\tau}_{k'-4\tau}\,\varepsilon_\tau G_{k-2\tau}, \quad C = \delta^\tau_{k'-2\tau}G_{k-2\tau}.$$

By Lemma 1(b),

$$\begin{aligned} \overline{A} &= \sum_{v\in V} C(v)\,e_E(S[v]z)\,y^{(n/2)u}\,L(\partial/\partial r(w_\tau))(X^k \exp\{XY\}) \\ &= \sum_{v\in V} C(v)\,e_E(S[v]z)\,y^{(n/2)u}\,X^{k-2\tau}\{nW_\tau^-\,X_\tau Y_\tau d_k^{Y,\tau}(\exp\{XY\}) \\ &\qquad + U_\tau\,Y^{2\tau}\,d^{Y,2\tau}_{k+u-2\tau}\exp\{XY\}\}, \end{aligned}$$

where $U = 2W^-X + S[v] \in \mathbf{R}^J$. Since XY, W^-X, and $S[v]$ are real, we obtain:

$$\begin{aligned} A = \sum_{v\in V} \overline{C(v)}\,e_E(-\overline{z}S[v])\,y^{(n/2)u}\overline{X}^{k-2\tau}\{&-4\pi nW_\tau^-\,X_\tau\,y_\tau\,\delta_k^\tau\exp\{XY\} \\ &+ 16\pi^2 U_\tau y^{2\tau}\,\delta^{2\tau}_{k+u-2\tau}\exp\{XY\}\}. \end{aligned}$$

Now we compute B. First,

$$\begin{aligned}\varepsilon_\tau G_{k-2\tau}(z,w) &= -y^{2\tau}\partial/\partial\overline{z}_\tau\{y^{(n/2)u}\sum_{v\in V}\overline{C(v)}\,\overline{X}^{k-2\tau}\,e_E(-\overline{z}S[v]+4iyW^-X)\}\\ &= y^{(n/2)u}\sum\overline{C(v)}\overline{X}^{k-2\tau}\left\{\frac{n}{4i}y_\tau+2\pi iy^{2\tau}U_\tau\right\}e_E(-\overline{z}S[v]+4iyW^-X).\end{aligned}$$

Since $y_\tau\delta_k^\tau=\delta_{k-\tau}^\tau y_\tau$, this implies

$$\begin{aligned}B = y^{(n/2)u}&\sum_{v\in V}\overline{C(v)}\overline{X}^{k-2\tau}\,e_E(-\overline{z}S[v])\\ &\cdot\left\{\frac{ny_\tau}{4i}\delta^{2\tau}_{k+u-3\tau}\mathrm{exp}\{XY\}+2\pi iy_\tau^2U_\tau\delta^{2\tau}_{k+u-2\tau}\mathrm{exp}\{XY\}\right\}.\end{aligned}$$

Thus,

$$\begin{aligned}(A+8\pi iB)/n = 2\pi\,y^{(n/2)u+\tau}&\sum_{v\in V}\overline{C(v)}\overline{X}^{k-2\tau}\,e_E(-\overline{z}S[v])\\ &\cdot\{-2W_\tau^-\,X_\tau\,\delta_k^\tau\mathrm{exp}\{XY\}+\delta^{2\tau}_{k+u-3\tau}\mathrm{exp}\{XY\}\}.\end{aligned}$$

Now

$$\begin{aligned}\delta^{2\tau}_{k+u-3\tau}\mathrm{exp}\{XY\} &= \delta_k^\tau\left\{\left(\frac{-(k_\tau-2)}{4\pi y_\tau}+2W_\tau^-\,X_\tau\right)\mathrm{exp}\{XY\}\right\}\\ &= -\frac{k_\tau-2}{4\pi y_\tau}\delta^\tau_{k-\tau}\mathrm{exp}\{XY\}+2W_\tau^-\,X_\tau\,\delta_k^\tau\mathrm{exp}\{XY\},\end{aligned}$$

so that

$$\begin{aligned}2(A+8\pi iB)/n &= y^{(n/2)u}\sum_{v\in V}\overline{C(v)}\,\overline{X}^{k-2\tau}\,e_E(-\overline{z}S[v])\,(2-k_\tau)\,\delta^\tau_{k-\tau}\mathrm{exp}\{XY\}\\ &= -(k_\tau-2)C.\end{aligned}$$

This completes the proof. □

6. Proof of Theorem 1

Let $F\in\mathcal{N}^t_{k'-2h}(\mathfrak{H}^J)$, where $k'=k+(1-n/2)u$. Since $\varepsilon_\tau(f(z)y^{-k})=(ik_\tau/2)f(z)y^{\tau-k}$ for $\tau\in J$ when f is holomorphic, we see that $\varepsilon_\tau:\mathcal{N}^t_k\to\mathcal{N}^{t-\tau}_{k-2\tau}$ is surjective, therefore we can find $F'\in\mathcal{N}^{t+h}_{k'}$ such that $(-2i)^{\{h\}}\,\varepsilon^hF'=F$, and F' is rapidly decreasing at cusps if and only F is rapidly decreasing at cusps. Let $H\in\mathcal{H}_h$. By (1.14) and Lemma 2,

$$\begin{aligned}\Theta_{k,H}(F) &= (-2i)^{\{h\}}\left\langle\varepsilon^hF',G_{k,H}\right\rangle=(4\pi)^{\{h\}}\left\langle F',\delta^h_{k'-2h}G_{k,H}\right\rangle\\ &= \left\langle F',\overline{H}(\partial/\partial r)G_k\right\rangle=H(\partial/\partial r)\left\langle F',G_k\right\rangle\\ &= H(\partial/\partial r)\Theta_k(F').\end{aligned}$$

We have used the well-known duality formula:

$$2\pi i\left\langle F,\delta_k^\tau G\right\rangle=\left\langle\varepsilon_\tau F,G\right\rangle\tag{1.15}$$

whenever $F\in\mathcal{C}_{k+2\tau}(\mathfrak{H}^J)$, $G\in\mathcal{C}_k(\mathfrak{H}^J)$, F and G are finite at cusps, and either F or G decreases rapidly at cusps. Our task now is to show that $\Theta_k(F')\in\mathcal{N}^{q+h}(\mathcal{Z}')$; thus we have reduced the problem to the case $h=0$, which we now assume.

Let $\tau \in J$, $2d+e = q_\tau + 1$, $p = L^d\zeta \in \mathcal{S}_{2d+e}$ with $\zeta \in \mathcal{H}_{e\tau}$. As pointed out before the proof of Lemma 1, we want to show that $p(\partial/\partial r(w_\tau))\Theta_k(F) \equiv 0$. First suppose $d = 0$. Then by (1.14) and Lemma 2(a), since $e = q_\tau + 1 \leq k_\tau$,

$$(1.16) \qquad \zeta(\partial/\partial r(w_\tau))\Theta_k(F) = (4\pi)^e \left\langle F, \delta_{k'-2e\tau}^{e\tau} G_{k,\zeta}\right\rangle = (-2i)^e \left\langle \varepsilon_\tau^e F, G_{k,\zeta}\right\rangle = 0.$$

Now suppose $d \neq 0$. Then $k_\tau \geq 2$, and by (1.14) and Lemma 2(b),

$$(1.17) \qquad \begin{aligned} &L(\partial/\partial r(w_\tau))\Theta_k(F) \\ &= 8\pi i \left\langle F, \delta_{k'-4\tau}^{2\tau}\, \varepsilon_\tau G_{k-2\tau}\right\rangle - (n/2)(k_\tau - 2)\left\langle F, \delta_{k'-2\tau}^{\tau} G_{k-2\tau}\right\rangle. \end{aligned}$$

Note that all terms converge: if $k - 2\tau \neq 0$, then $G_{k-2\tau}$ decreases rapidly at cusps, and the operators ε_τ and δ_*^* preserve this property. If $k - 2\tau = 0$, then (since $k \geq u$) we conclude that $E = \mathbf{Q}$, $k = 2$, $q = 1$, $t = 1$, so that F decreases rapidly at cusps by hypothesis. By (1.15),

$$L(\partial/\partial r(w_\tau))\Theta_k(F) = -4\left\langle \delta_{k'-4\tau}^{\tau}\varepsilon_\tau^2 F, G_{k-2\tau}\right\rangle - (n/4\pi i)(k_\tau - 2)\left\langle \varepsilon_\tau F, G_{k-2\tau}\right\rangle.$$

If $k_\tau = 2$ then both terms vanish, since $\varepsilon_\tau^2 F = 0$ and $k_\tau - 2 = 0$. This settles the case when $k_\tau \leq 2$. Now assume $k_\tau > 2$. In general, if $G \in \mathcal{N}_k^t$, then $\varepsilon_\tau G \in \mathcal{N}_{k-2\tau}^{t-\tau}$ and $\delta_k^\tau G \in \mathcal{N}_{k+2\tau}^{t+\tau}$. For this reason, $\delta_{k'-4\tau}^\tau \varepsilon_\tau^2 F$ and $\varepsilon_\tau F$ both belong to $\mathcal{N}_{k'-2\tau}^{t-\tau}$. After d steps, we see that $L(\partial/\partial r(w_\tau))^d\, \Theta_k(F)$ is a linear combination of $\langle F', G_{k-2d\tau}\rangle$ with $F' \in \mathcal{N}_{k'-2d\tau}^{t-d\tau}$, hence $p(\partial/\partial r(w_\tau))\Theta_k(F)$ is a linear combination of $\zeta(\partial/\partial r(w_\tau))\Theta_{k-2d\tau}(F')$ with $\zeta \in \mathcal{H}_{e\tau}$. We check for convergence: if $k \neq 2d\tau$ then the $G_{k-2j\tau}(z,w)$, $0 \leq j \leq d$, decrease rapidly at cusps, and if $k = 2d\tau$ then $E = \mathbf{Q}$, k is even, and $q + 1 \geq k$. This forces $t = \mu$, so that F (and hence also F') decreases rapidly at cusps by hypothesis. Thus in both cases all the Petersson inner products as in (1.17) converge. Finally, since $e \leq k_\tau - 2d$, $\zeta(\partial/\partial r(w_\tau))\Theta_{k-2d\tau}(F')$ is equal to $(-2i)^e \langle \varepsilon_\tau^e F', G_{k-2d\tau,\zeta}\rangle$ by (1.16); and $\varepsilon_\tau^e F' = 0$ because $t_\tau - d = q_\tau/2 - d = (e-1)/2 \leq e - 1$. (Note that $e \geq 1$ because q_τ is even.) Thus $p(\partial/\partial r(w_\tau))\Theta_k(F) \equiv 0$, and the proof is complete. □

II. Arithmeticity

7. Arithmetic automorphic forms on $\mathfrak{H}^J$

Let δ_k^p be as in (1.10), where $0 \leq p \in \mathbf{Z}^J$ and k is a weight of E. We interpret that δ_k^0 is the identity operator. Then $\delta_k^p(f|_k\gamma) = (\delta_k^p f)|_{k+2p}\gamma$ for all smooth functions f on $\mathfrak{H}^J$ and all $\gamma \in \Gamma[2\mathfrak{d}^{-1}, 2\mathfrak{d}]$, hence $\delta_k^p : \mathcal{C}_k \to \mathcal{C}_{k+2p}$, where

$$\mathcal{C}_k = \{f \in C^\infty(\mathfrak{H}^J, \mathbf{C}) : f|_k\gamma = f \text{ for all } \gamma \text{ in some congruence subgroup of } \mathrm{SL}_2(E)\}.$$

Let $0 \leq t \in \mathbf{Z}^J$, k a weight of E. We define subspaces of $C^\infty(\mathfrak{H}^J, \mathbf{C})$:

$$\mathcal{N}^t(\mathfrak{H}^J) = \{f \in C^\infty(\mathfrak{H}^J, \mathbf{C}) : f = \sum_{0 \leq i \leq t} (\pi y)^{-i} f_i \ \text{ with } f_i \text{ holomorphic}\}$$

$$\mathcal{N}_k^t = \{\, f \in \mathcal{C}_k \cap \mathcal{N}^t(\mathfrak{H}^J) : f_i \ \text{ is holomorphic at cusps for each } 0 \leq i \leq t\}.$$

These are the spaces of nearly holomorphic functions and nearly holomorphic automorphic forms which were defined in Section 3. We put also $\mathrm{M}_k(E) = \mathcal{N}_k^0$, the space of Hilbert modular forms of weight k, and $\mathrm{M}_k(E, \overline{\mathbf{Q}})$ the subspace of arithmetic Hilbert modular forms. Let CM$= \{\, z \in \mathfrak{H}^J : (\mathbf{Q}(z_\sigma) : \mathbf{Q}) = 2 \text{ for all } \sigma \in J\,\}$.

We can define arithmetic subspaces of $\mathcal{C}_k$ and $\mathcal{N}_k^t$ by

$$\mathcal{C}_k(\overline{\mathbf{Q}}) = \{\, f \in \mathcal{C}_k : f(z)/g(z) \in \overline{\mathbf{Q}} \text{ whenever } z \in \mathrm{CM},\ g \in \mathrm{M}_k(E, \overline{\mathbf{Q}}), \text{ and } g(z) \neq 0\},$$
$$\mathcal{N}_k^t(\overline{\mathbf{Q}}) = \mathcal{C}_k(\overline{\mathbf{Q}}) \cap \mathcal{N}_k^t.$$

Observe that if $f = \sum_{0 \le i \le t} (\pi y)^{-i} f_i \in \mathcal{N}_k^t$ with f_i holomorphic, then $f_i\left(\left(\begin{smallmatrix} 1 & N \\ 0 & 1 \end{smallmatrix}\right) z\right) = f_i(z+N) = f_i$ for some $0 < N \in \mathbf{Z}$, therefore f_i has a Fourier series

$$f_i(z) = \sum_{\xi \in E} a_i(\xi; f)\, e_E(\xi z).$$

It was shown in [**S87b**, Proposition 3.9] that

$$f \in \mathcal{N}_k^t(\overline{\mathbf{Q}}) \iff a_i(\xi; f) \in \overline{\mathbf{Q}} \quad \text{for all } i \text{ and } \xi \tag{2.1}$$

when k is an integral weight. If k is half-integral, let $\theta \in \mathrm{M}_{u/2}(E, \overline{\mathbf{Q}})$. Then $f \in \mathcal{N}_k^t(\overline{\mathbf{Q}}) \iff \theta f \in \mathcal{N}_{k+u/2}^t(\overline{\mathbf{Q}}) \iff a_i(\xi; \theta f) \in \overline{\mathbf{Q}}$ for all i and $\xi \iff a_i(\xi; f) \in \overline{\mathbf{Q}}$ for all i and ξ; therefore (2.1) is true for half-integral weights as well. Now (2.1) implies $\delta_k^p : \mathcal{N}_k^t(\overline{\mathbf{Q}}) \to \mathcal{N}_{k+2p}^{t+p}(\overline{\mathbf{Q}})$. Put

$$\tilde{\mathcal{N}}_k^t = \{f \in \mathcal{N}_k^t : f = \sum_{\substack{r \in \mathbf{Z}^J \\ 0 \le 2r \le k}} \delta_{k-2r}^r g_r \ \text{ with } g_r \in \mathrm{M}_{k-2r}(E)\}, \tag{2.2}$$
$$\tilde{\mathcal{N}}_k^t(\overline{\mathbf{Q}}) = \tilde{\mathcal{N}}_k^t \cap \mathcal{C}_k(\overline{\mathbf{Q}}).$$

LEMMA 3. *1. If $f \in \tilde{\mathcal{N}}_k^t$, then $f \in \tilde{\mathcal{N}}_k^t(\overline{\mathbf{Q}})$ if and only if f has an expression as in (2.2) with all the g_r arithmetic.*
2. If $E \neq \mathbf{Q}$, $k \notin 2\mathbf{Z}^J$, or $k \ge 2t + u$, then $\tilde{\mathcal{N}}_k^t = \mathcal{N}_k^t$.
3. If $f \in \tilde{\mathcal{N}}_k^t$ and $g \in \tilde{\mathcal{N}}_\ell^p$, then $fg \in \tilde{\mathcal{N}}_{k+\ell}^{t+p}$. A similar statement is true for $\tilde{\mathcal{N}}_k^t(\overline{\mathbf{Q}})$, $\mathcal{N}_k^t$, and $\mathcal{N}_k^t(\overline{\mathbf{Q}})$.

PROOF. For (1), we know that $\delta_\ell^r g_r$ is arithmetic if g_r is arithmetic by the comment preceding this proposition. The converse uses the same argument as in the proof of [**S87b**, Theorem 3.7], which we repeat here for the reader's convenience. Suppose $f = \sum \delta_{k-2r}^r g_r$ is arithmetic, where $g_r \in \mathrm{M}_{k-2r}(E)$. Since $\mathrm{M}_{k-2r}(E) = \mathrm{M}_{k-2r}(\overline{\mathbf{Q}}) \otimes \mathbf{C}$, we can write f as a sum: $f = \sum_c c f_c$, where c runs through a set of representatives for $\mathbf{C}^\times$ mod $\overline{\mathbf{Q}}^\times$ with the identity coset represented by 1, and each f_c is of the form $\sum \delta_{k-2r}^r g_{c,r}$ with $g_{c,r} \in \mathrm{M}_{k-2r}(\overline{\mathbf{Q}})$. If $f_{c_0} \neq 0$, choose a CM-point $w \in \mathfrak{H}^J$ such that $f_{c_0}(w) \neq 0$. Then, since f and f_c are arithmetic, $c_0 + \sum_{c \neq c_0} c f_c(w)/f_{c_0}(w) = f(w)/f_{c_0}(w) \in \overline{\mathbf{Q}}$. This is a linear relation between the c's with coefficients in $\overline{\mathbf{Q}}$, therefore $c_0 = 1$ and $f_c(w) = 0$ for $c \neq 1$. So $f = f_1$ has the required form. Now we prove (2). That $E \neq \mathbf{Q}$ or $k \notin 2\mathbf{Z}^J$ implies $\tilde{\mathcal{N}}_k^t = \mathcal{N}_k^t$ is contained in the result [**S87b**, Theorems 5.2 and 5.5]. Now assume that $E = \mathbf{Q}$ and $k > 2t$. Then $\delta_{k-2r}^r = A_r y^{-r} +$ terms of lower degree in y^{-1}, where $A_r \in \mathbf{C}$, and the hypothesis on k implies that $A_r \neq 0$ when $r \le t$. Now (2) can be proven exactly as in [**S76**, Lemma 7]. For (3), it suffices to prove that $\delta_{k-2t}^t f \cdot \delta_{\ell-2p}^p g \in \tilde{\mathcal{N}}_{k+\ell}^{t+p}$ if $f \in \mathrm{M}_{k-t}(E)$ and $g \in \mathrm{M}_{\ell-2p}(E)$. (The remaining statements follow immediately from definitions.) If $k_\sigma \le 0$ for some $\sigma \in J$, then either $\mathrm{M}_k(E) = 0$ or $k = 0$, $\mathrm{M}_k(E) = \mathbf{C}$. Thus we may assume that $k_\sigma > 2t_\sigma$ and $\ell_\sigma > 2p_\sigma$ for all $\sigma \in J$. But then $\mathcal{N}_{k+\ell}^{t+p} = \tilde{\mathcal{N}}_{k+\ell}^{t+p}$ by (2). This implies the result, since obviously $\tilde{\mathcal{N}}_k^t \tilde{\mathcal{N}}_\ell^p \subset \mathcal{N}_k^t \mathcal{N}_\ell^p \subset \mathcal{N}_{k+\ell}^{t+p}$. □

8. CM-points on $\mathcal{Z}$

Suppose we are given a representation $\vartheta : Y \to \mathrm{End}(V)$, where Y is an E-algebra. Suppose further that:

(i) Y is a direct sum of L and some copies of E, where L is a direct sum of CM-fields containing E, and $[Y : E] = n+2$;

(ii) $S(\vartheta(a)v_1, v_2) = S(v_1, \vartheta(a^c)v_2)$ for all $v_1, v_2 \in V$ and $a \in Y$, where c denotes complex conjugation on Y; and

(iii) $\vartheta(a) \in G_+(S)$ for all $a \in Y^u$, where $Y^u = \{a \in Y : aa^c = (1, 1, \dots, 1)\}$

By virtue of (ii) it is automatic that $\vartheta(a) \in G(S)$ for all $a \in Y^u$; thus (iii) essentially states that for each $\tau \in J$ and $a \in Y^u$, $\vartheta(a)^\tau$ preserves the connected components of X_τ (see (1)). Since the closure of $\vartheta(Y^u)$ is compact, $\vartheta(Y^u)$ has a fixed point on $\mathcal{Z}$. As shown in [**S80**, Proposition 5.1], the conditions (i)-(iii) ensure that the fixed point is unique, hence it has algebraic coordinates. A CM-point in $\mathcal{Z}$ is defined as the fixed point of $\vartheta(Y^u)$ for some ϑ and Y as above. The set of all CM-points in $\mathcal{Z}$ is denoted by CM.

REMARK. If $w^0 \in \mathrm{CM}$ then so is γw^0, for all $\gamma \in G_+(S)$. For if $\vartheta(Y^u)$ fixes w^0, then $\vartheta^\gamma(Y^u)$ fixes γw^0, where $\vartheta^\gamma(a) = \gamma \circ \vartheta(a) \circ \gamma^{-1}$ for $a \in Y$. Hence one example of a pair (ϑ, Y) will show that the CM-points are dense in $\mathcal{Z}$. Such an example is given below.

Given ϑ and $Y = E^m \oplus L$ satisfying (i)-(iii) above and $\tau \in J$, we let $\vartheta_\tau : Y \to \mathrm{End}(V_\tau)$ be defined by $\vartheta_\tau(y) = \vartheta(y)^\tau$. By (1.2) we have:

$$B_\tau(w^0)^{-1}\vartheta_\tau(a)B_\tau(w^0) = \mathrm{diag}\{\phi_\tau(a), \overline{\lambda_\tau(a)}, \lambda_\tau(a)\} \tag{2.3}$$

for all $a \in Y^u$ and $\tau \in J$, where w^0 is the fixed point of

$$\vartheta(Y^u), \phi_\tau(a) = \psi_\tau(\vartheta_\tau(a), w^0_\tau),$$

and $\lambda_\tau(a) = \mu_\tau(\vartheta_\tau(a), w^0_\tau)$. Since the E-linear span of Y^u is Y, we see that this matrix has the same block form for all $a \in Y$, therefore we can define representations $\phi_\tau : Y \to M_n(\overline{\mathbf{Q}})$ and $\lambda_\tau : Y \to \overline{\mathbf{Q}}$ by the same formula. Now ϑ is essentially a regular representation because $[Y : E] = [V : E]$, hence $\lambda_\tau \neq c\lambda_\tau$. Thus $L \neq 0$ and $\lambda_\tau|_L$ is a nontrivial element of $J(L)$, which we continue to denote by λ_τ, $\mathrm{Res}_E(\lambda_\tau) = \tau$, and $\lambda_\tau(b) = \mu_\tau(\vartheta_\tau(1, \dots, 1, b), w^0_\tau)$ if $b \in L$, $bb^c = 1$. Let $\lambda = \sum_{\tau \in J} \lambda_\tau \subset J(L)$ and $\phi = (\phi_\tau) : Y \to \prod_{\tau \in J} M_n(\overline{\mathbf{Q}})$. Then $\mathrm{Res}_E(\lambda) = J$.

The following example of a pair (ϑ, Y) is constructed in [**S80**, Lemma 10.1]. Fix a coordinate system for V such that S is diagonal, say $S = \mathrm{diag}\{s_1, \dots, s_{n+2}\}$, with s_j totally positive for $j \le n$. Let $L = E(\zeta)$, where $\zeta^2 = -s_{n+1}/s_{n+2}$, and let $Y = E^n \oplus L$. Define $\vartheta : Y \to \mathrm{End}(V)$ by the formula

$$\vartheta(u_1, u_2, \dots, u_n, a + b\zeta) = \mathrm{diag}\left\{u_1, \dots, u_n, \begin{pmatrix} a & b \\ \zeta^2 b & a \end{pmatrix}\right\} \tag{2.4}$$

for $a + b\zeta \in L$ and $u_1, \dots, u_n \in E$. Properties (i) and (ii) above are obvious, and property (iii) is verified in [**S80**, Lemma 10.1].

9. Arithmeticity on $\mathcal{Z}$

Let $L = \oplus L_i$ be a finite direct sum of CM-fields. We will have need of the bilinear map $p_L : \mathbf{Z}^{J(L)} \times \mathbf{Z}^{J(L)} \to \mathbf{C}^\times/\overline{\mathbf{Q}}^\times$ of [**S80**, §1], which is related to periods of abelian varieties with L_i as the field of complex multiplication. This map satisfies for $\xi, \eta \in \mathbf{Z}^{J(L)}$ the following properties (see [**S80**, Theorem 1.11]):

(2.5) If $\gamma : L' \to L$ is an isomorphism, then $p_{L'}(\gamma\xi, \gamma\eta) = p_L(\xi, \eta)$.

$$p_L(c\xi, \eta) = p_L(\xi, c\eta) = p_L(\xi, \eta)^{-1}. \tag{2.6}$$

Let $L^u = \{\, a \in L : aa^c = 1 \,\}$, considered as a $\mathbf{Q}$-rational algebraic group. The complex points of L^u can be identified with the group

$$L^u_{\mathbf{C}} = \{a \in \mathbf{C}^{J(L)} : a_\tau a_{c\tau} = 1 \quad \text{for all } \tau \in J(L)\,\}.$$

We define a homomorphism $\mathcal{P}_L : \mathbf{Z}^{J(L)} \to L^u_{\mathbf{C}} \bmod (\overline{\mathbf{Q}}^\times)^{J(L)}$ by the formula $\mathcal{P}_L(\beta) = \big(p_L(\tau, \beta)\big)_{\tau \in J(L)}$ for $\beta \in \mathbf{Z}^{J(L)}$, which is well-defined because of (2.6). By fixing coset representatives, we will consider $\mathcal{P}_L$ to take values in $L^u_{\mathbf{C}}$.

Let w^0, Y, L, ϑ, and $\lambda \subset J(L)$ be as in Section 8 and $\mathfrak{K}$, W as in Section 1. Define a map $\Lambda : \prod_{\tau \in J} G_+(S_\tau) \times \mathcal{Z} \to \mathfrak{K}$ by

$$\Lambda(\gamma, w) = \prod_{\tau \in J} \big(\mu_\tau(\gamma_\tau, w_\tau), \psi_\tau(\gamma_\tau, w_\tau)\big)$$

for $\gamma = (\gamma_\tau) \in \prod G_+(S_\tau)$ and $w = (w_\tau) \in \mathcal{Z}$. We define a map $\vartheta^u : L^u \to G_{\mathbf{Q}}$ by

$$\vartheta^u(a) = \vartheta(1, 1, \ldots, 1, a) \quad \text{for } a \in L^u.$$

Given a $\overline{\mathbf{Q}}$-rational representation $\rho : \mathfrak{K} \to \mathrm{GL}(W)$, let $\xi_\rho : L^u_{\mathbf{C}} \to \mathrm{GL}(W)$ be the holomorphic representation such that

$$\xi_\rho(a) = \rho\big(\Lambda(\vartheta^u(a), w^0)\big) \quad \text{for all } a \in L^u,$$

and let $P_\rho(w^0) = \xi_\rho\big(\mathcal{P}_L(2\lambda)\big)$.

LEMMA 4. *1. $P_\rho(w^0)$ depends only on ρ and w^0, not on the choice of ϑ, Y, and L.*
2. If $\gamma \in G_{\mathbf{Q}}$, then $P_\rho(\gamma w^0) = \rho\big(\Lambda(\gamma, w^0)\big) P_\rho(w^0) \rho\big(\Lambda(\gamma, w^0)\big)^{-1}$.
3. If $\rho'(a, b, c) = U\rho(a, b, c)U^{-1}$ with $U \in \mathrm{GL}(W)$, then $P_{\rho'}(w^0) = UP_\rho(w^0)U^{-1}$.

PROOF. To prove (1), suppose that $\vartheta'(Y'^u)$ also has w^0 as its fixed point and $Y' = E^{m'} \oplus L'$. By [**S80**, Proposition 5.3], there is an isomorphism $\gamma : L \to L'$ such that $\vartheta(0, b) = \vartheta'(0, b^\gamma)$ for all $b \in L$. Extend γ to an isomorphism $\gamma : L \otimes \mathbf{C} \to L' \otimes \mathbf{C}$; thus $(a^\gamma)_\tau = a_{\gamma\tau}$ for $a \in L \otimes \mathbf{C} \cong \mathbf{C}^{J(L)}$ and $\tau \in J(L')$. Let λ', ξ'_ρ, and P'_ρ denote the objects λ, ξ_ρ, and P_ρ, respectively when ϑ and Y are replaced by ϑ' and Y'. Then $\lambda = \gamma\lambda'$ by (2.3). For $a \in L^u$ we have

$$\xi'_\rho(a^\gamma) = \rho\big(\Lambda(\vartheta'^u(a^\gamma), w^0)\big) = \rho\big(\Lambda(\vartheta^u(a), w^0)\big) = \xi_\rho(a),$$

therefore $\xi'_\rho(a^\gamma) = \xi_\rho(a)$ for all $a \in L^u_{\mathbf{C}}$. Also $\big(\mathcal{P}_L(2\lambda)\big)_{\gamma\tau} = p_L(\gamma\tau, 2\gamma\lambda') = p_{L'}(\tau, 2\lambda') = \big(\mathcal{P}_{L'}(2\lambda')\big)_\tau$ for $\tau \in J(L')$ by (2.5), so $P'_\rho(w^0) = \xi'_\rho\big(\mathcal{P}_{L'}(2\lambda')\big) = \xi'_\rho\big(\mathcal{P}_L(2\lambda)^\gamma\big) = \xi_\rho\big(\mathcal{P}_L(2\lambda)\big) = P_\rho(w^0)$, as was to be shown. The second statement is a simple consequence of the remark in Section 8, and the third follows immediately from definitions. □

We say an automorphic form $f \in \mathcal{C}_\rho(\mathcal{Z})$ is *arithmetic at*

$$w^0 \in \mathrm{CM} \text{ if } P_\rho(w^0)^{-1} f(w^0)$$

is algebraic, and we say f is *arithmetic* if f is arithmetic at every CM-point $w^0 \in \mathcal{Z}$. The set of arithmetic elements of $\mathcal{C}_\rho(\mathcal{Z})$, $\mathcal{M}_\rho(\mathcal{Z})$, $\mathcal{N}_\rho^e(\mathcal{Z})$, $\mathcal{N}_\rho(\mathcal{Z})$ is denoted by $\mathcal{C}_\rho(\mathcal{Z}, \overline{\mathbf{Q}})$, $\mathcal{M}_\rho(\mathcal{Z}, \overline{\mathbf{Q}})$, $\mathcal{N}_\rho^e(\mathcal{Z}, \overline{\mathbf{Q}})$, and $\mathcal{N}_\rho(\mathcal{Z}, \overline{\mathbf{Q}})$, respectively.

10. Significance of near holomorphy to the study of arithmeticity

Now that we have defined all the terms in the title of this paragraph for the spaces $\mathfrak{H}^J$ and $\mathcal{Z}$, we find it appropriate to insert some remarks on the importance of near holomorphy to the study of arithmeticity. It has long been known that some classical Eisenstein series of weight 2 have Fourier series of the form $\pi^{-2}E_2(z,s) = a_0(\pi y)^{-1} + \sum_{n\geq 1} a_n\, e^{2\pi i n z}$ with all a_n belonging to the maximal abelian extension of $\mathbf{Q}$; thus $\pi^{-2}E_2(z,s)$ is nearly holomorphic but not holomorphic, and by (2.1) it is arithmetic. This example shows that nearly holomorphic arithmetic automorphic forms arise naturally. Moreover, this example is not sporadic; see the introduction to [**S86**]. In addition, nearly holomorphic forms can be constructed naturally from holomorphic ones by applying iterates of weight-raising operators, and in fact this is the only way to obtain nearly holomorphic automorphic forms when the weight is large compared to the degree; see [**S87b**, Proposition 3.3]. For $\mathfrak{H}^J$ these are the operators δ_k^e given by (1.10), and for $\mathcal{Z}$ they are the operators D_ρ^e defined in [**S80**, §11]. In both cases these operators correspond to elements of $\mathfrak{p}_+$, where $\mathfrak{p}_+$ corresponds to the holomorphic tangent vectors at a point of $\mathfrak{H}$ or $\mathcal{Z}$; see [**S90**]. The remarkable fact is that these operators preserve arithmeticity. A simple proof of this for $\mathfrak{H}^J$ is given in [**S78**] and the more general case is worked out in [**S87b**]. This fact leads to the following theorem of Shimura [**S87b**, Theorem 3.7], which is actually stated there for automorphic forms on any hermitian symmetric space.

THEOREM (Shimura) *If a nearly holomorphic automorphic form f on $\mathfrak{H}^J$ (resp. $\mathcal{Z}$) is arithmetic on a dense set of* CM*-points then $\delta_k^e f$ (resp. $\pi^{-\sum e_\tau} D_\rho^e f$), as well as f itself, is arithmetic at every* CM*-point.*

11. Arithmeticity and the theta correspondence

Our next theorem states that to a certain extent the theta correspondence preserves arithmeticity. The proof is based on ideas of Shimura. The idea is to reduce the problem to proving arithmeticity at a single CM-point w^0 by letting the locally constant function C in the definition of $\Theta_{k,h}$ vary and then choosing the CM-point w^0 in such a way that $\Theta_{k,h}(F)(w^0)$ can be expressed as a known Rankin product.

THEOREM 2. *Let $t,k,h \in \mathbf{Z}_{\geq 0}^J$, $k-h \geq u$, $k' = k+(1-n/2)u$, $F \in \mathcal{N}_{k'-2h}^t(\overline{\mathbf{Q}})$. If $E = \mathbf{Q}$ and k is even, assume $k-h \geq 2t$ or $F \in \tilde{\mathcal{N}}_{k'-2h}^t(\overline{\mathbf{Q}})$. Then $\pi^{[E:\mathbf{Q}]}\Theta_{k,h}(F)$ is arithmetic on a dense set of* CM*-points.*

COROLLARY. *If t,k,h and F satisfy the hypotheses of Theorem 1 and if F is arithmetic then $\pi^{[E:\mathbf{Q}]}\Theta_{k,h}(F)$ is nearly holomorphic and arithmetic.*

To deduce the corollary from the theorem, we observe that $2\mu \leq k-h$, therefore the hypothesis $0 \leq t \leq \mu$ in Theorem 1 implies that $2t \leq k-h$. Now $\pi^{[E:\mathbf{Q}]}\Theta_{k,h}(F)$ is nearly holomorphic by Theorem 1 and arithmetic on a dense set of CM-points by Theorem 2, therefore it is arithmetic by the theorem of Shimura in Section 10.

PROOF OF THEOREM 2. The first step will be to replace the matrices A_τ of Section 1 by matrices A'_τ which we will give explicitly. Note that a change in the matrices A_τ affects the definitions of w_-, w_+, $G_+(S_\tau)$, μ_τ, ψ_τ, and the action of $G_+(S_\tau)$ on $\mathcal{Z}$; see Section 1. Take a basis for V such that $S = \operatorname{diag}\{s_1, \dots, s_{n+2}\}$ and s_{n+1}, s_{n+2} are totally negative elements of E, and let

$$A'_\tau = C_\tau \operatorname{diag}\left\{(s_1^\tau)^{1/2}, \dots, (s_n^\tau)^{1/2}, \left(\begin{smallmatrix} \alpha_1 & i\alpha_2 \\ a_1 & -i\alpha_2 \end{smallmatrix}\right)\right\},$$

where $\alpha_i = (-s_{n+i}^\tau/2)^{1/2}$ for $i = 1, 2$ and $C_\tau = R$ or 1_{n+2}. C_τ should be chosen in such a way that $A_\tau^{-1} A'_\tau X_\tau^0 = X_\tau^0$, which is possible by (1.1). Then $D_\tau = {A'_\tau}^{-1} A_\tau \in G_+(S_\tau)$ and hence A'_τ determines the same set $G_+(S_\tau)$ as does A_τ. Let $\Theta'_{k,h}$ be defined like $\Theta_{k,h}$, but with the matrices A_τ replaced by A'_τ. We will show that the theorem for $\Theta'_{k,h}$ and A'_τ implies the theorem for $\Theta_{k,h}$ and A_τ. Let ϑ, Y be as in Section 8. We claim that if ϑ, Y determine w^0, λ, ϕ with respect to the matrices A_τ in the sense described in Section 8 then they determine $D^{-1}w^0, \lambda, \phi'$ with respect to the matrices A'_τ, where $D = (D_\tau) \in \prod_{\tau \in J} G_+(S_\tau)$ and $\phi'(a) = \psi(D, D^{-1}w^0)^{-1}\phi(a)\psi(D, D^{-1}w^0)$ for $a \in Y$. Here $D^{-1}w^0$ and $\psi(D, D^{-1}w^0)$ are defined with respect to the matrices A_τ. To see that the claim holds, let $a \in Y^u$ and $\alpha_\tau = \vartheta_\tau(a) \in G_+(S_\tau)$. If $B_\tau(w)$ is determined as in Section 1 by the matrices A_τ, then the matrix "$B_\tau(w)$" determined by A'_τ is ${A'_\tau}^{-1}\mathbf{B}(w_\tau) = D_\tau B_\tau(w)$. Put

$$\psi^0 = \psi(D, D^{-1}w^0), \qquad \mu^0 = \mu(D, D^{-1}w^0).$$

By (1.2) and (2.3),

$$\begin{aligned} \alpha_\tau D_\tau B_\tau(D^{-1}w^0) &= \alpha_\tau B_\tau(w^0) \operatorname{diag}\{\psi_\tau^0, \overline{\mu_\tau^0}, \mu_\tau^0\} \\ &= B_\tau(w^0) \operatorname{diag}\{\phi_\tau(a)\psi_\tau^0, \overline{\lambda_\tau(a)\mu_\tau^0}, \lambda_\tau(a)\mu_\tau^0\} \\ &= DB_\tau(D^{-1}w^0) \operatorname{diag}\{\phi'_\tau(a), \overline{\lambda_\tau(a)}, \lambda_\tau(a)\}, \end{aligned} \tag{2.7}$$

which proves the claim. For $\rho = [k, h]$, let P_ρ, ξ_ρ and P'_ρ, ξ'_ρ be as defined in Section 9 with respect to the matrices A_τ and A'_τ, respectively. These take values in $\mathrm{GL}(\mathcal{H}_h^*)$. Define $U = \rho(\mu^0, \psi^0)^{-1} \in \mathrm{GL}(\mathcal{H}_h^*)$. Note that $\psi_\tau(D, D^{-1}w)$ and $\mu_\tau(D, D^{-1}w)$ are algebraic in the coordinates of D and w by (1.2) and D, w^0 have algebraic entries, therefore U is algebraic. We will show that $P'_\rho(D^{-1}w^0) = U \circ P_\rho(w^0) \circ U^{-1}$. Since λ is the same with respect to A_τ and A'_τ, it suffices to verify that $\xi'_\rho(a) = U \circ \xi_\rho(a) \circ U^{-1}$ for all $a \in L^u$. Indeed, $\xi'_\rho(a) \circ U = \rho(\lambda(a), \phi'(a)) \circ \rho(\mu^0, \psi^0)^{-1} = \rho(\mu^0, \psi^0)^{-1} \circ \rho(\lambda(a), \phi(a)) = U \circ \xi_\rho(a)$. Also, we see from the definition of $\Theta_{k,h}$ that $\Theta'_{k,h}(z, D^{-1}w^0) = U\big(\Theta_{k,h}(z, w^0)\big)$. Thus $\pi^{[E:\mathbf{Q}]}\Theta_{k,h}(F)$ is arithmetic at w^0 with respect to the matrices A_τ $\iff$ $\pi^{[E:\mathbf{Q}]}P_\rho(w^0)^{-1}\Theta_{k,h}(F)(w^0)$ is $\overline{\mathbf{Q}}$-rational $\iff$ $\pi^{[E:\mathbf{Q}]}U^{-1} \circ P'_\rho(D^{-1}w^0)^{-1} \circ U \circ U^{-1}\big(\Theta'_{k,h}(F)(D^{-1}w^0)\big)$ is $\overline{\mathbf{Q}}$-rational $\iff$ $\pi^{[E:\mathbf{Q}]}P'_\rho(D^{-1}w^0)^{-1}\big(\Theta'_{k,h}(F)(D^{-1}w^0)\big)$ is $\overline{\mathbf{Q}}$-rational $\iff$ $\pi^{[E:\mathbf{Q}]}\Theta'_{k,h}(F)$ is arithmetic at $D^{-1}w^0$ with respect to the matrices A'_τ.

Thus we may replace $A_\tau, \Theta_{k,h}$ by $A'_\tau, \Theta'_{k,h}$. Let $\zeta^2 = -s_{n+1}/s_{n+2}$, $L = E(\zeta)$, $Y = E^n \oplus L$, and $\vartheta_0 : Y \to E_{n+2}^{n+2}$ be given by (2.4). Then the fixed point w^0 of $\vartheta_0(Y^u)$ is of the form $w_{\tau-}^0 = {}^t(0, \dots, 0, c, c\zeta^{\lambda_\tau})$ with $c \in \overline{\mathbf{Q}}$, from which it easily follows from the explicit expression for A_τ that $w^0 = 0$. Furthermore, if $a \in L^u$ and $\gamma_\tau = \vartheta_0(1, \dots, 1, a_\tau)$ then by a direct calculation $\mu_\tau(\gamma, 0) = a^{\lambda_\tau}$, where $\lambda_\tau \in J(L)$ and $\lambda_\tau|_E = \tau$. Thus, for ξ_ρ as in §9 we have $\xi_{[k,h]}(a) = [k, h](a^\lambda, I_n) = a^q$, where $Res(\lambda) = u$ and $q = \sum_{\tau \in J}(k_\tau - h_\tau)\lambda_\tau \in \mathbf{Z}^{J(L)}$.

Suppose we know that $\pi^{[E:\mathbf{Q}]}\Theta_{k,h}(F)$ is arithmetic at $w = 0$ for all locally constant functions $C \in \mathcal{L}(V)$. We will show that this implies that $\pi^{[E:\mathbf{Q}]}\Theta_{k,h}(F)$

is arithmetic on the dense set of CM-points $\{\gamma 0 : \gamma \in G_+(S)\}$. By Lemma 4 we have $P_\rho(\gamma 0) = \rho(\Lambda(\gamma, 0))\, P_\rho(0)\, \rho(\Lambda(\gamma, 0))^{-1}$. Also, if $\tilde{\Theta}_{k,h}$ is defined with $C \circ \gamma$ in place of C, we easily see that $\Theta_{k,h}(z, \gamma w) = \rho(\Lambda(\gamma, w))\big(\tilde{\Theta}_{k,h}(z, w)\big)$, where $\rho = [k, h]$. Thus, $\pi^{[E:\mathbf{Q}]} P_\rho(\gamma 0)^{-1} \Theta_{k,h}(F)(\gamma 0) = \pi^{[E:\mathbf{Q}]} \rho(\Lambda(\gamma, 0)) P_\rho(0)^{-1} \tilde{\Theta}_{k,h}(z, 0)$. Now $\pi^{[E:\mathbf{Q}]} P_\rho(0)^{-1} \tilde{\Theta}_{k,h}(F)(0)$ is $\overline{\mathbf{Q}}$-rational since we are assuming $\pi^{[E:\mathbf{Q}]} \tilde{\Theta}_{k,h}(F)$ is arithmetic at 0, and $\rho\big(\Lambda(\gamma, 0)\big)$ is algebraic, therefore $\pi^{[E:\mathbf{Q}]} \Theta_{k,h}(F)$ is arithmetic at $\gamma 0$ as claimed.

Now we show that $\pi^{[E:\mathbf{Q}]} \Theta_{k,h}(F)$ is arithmetic at 0. We already observed that $\xi_\rho(a) = a^q$ for $a \in L^u$, where $q = \sum_{\tau \in J} (k_\tau - h_\tau) \lambda_\tau$. It follows directly from the definitions in Section 9 that $P_\rho(0) = p_L(q, 2\lambda)$. Let $\{ H_i \}$ be a $\overline{\mathbf{Q}}$-rational basis of $\mathcal{H}_h$ and $\{\varphi_i\} \in \mathcal{H}_h^*$ its dual basis. If $\pi^{[E:\mathbf{Q}]} \Theta_{k,h}(F)(0) = \sum c_i \varphi_i$ with $c_i \in \mathbf{C}$, then $\pi^{[E:\mathbf{Q}]} P_\rho(0)^{-1} \Theta_{k,h}(F)(0)(H_i) = c_i p_L(q, 2\lambda)^{-1}$, thus our task is to show that $c_i = \pi^{[E:\mathbf{Q}]} \Theta_{k,H_i}(F)(0) \in p_L(q, 2\lambda) \overline{\mathbf{Q}}$.

Let $v_{0\tau} = {}^t(1, \ldots, 1, 0, 1) \in V_\tau$, $V' = \vartheta_0(E^n) v_0$, and $S' : V' \times V' \to E$ be the restriction of S to V'. If $v = \vartheta_0(b) v_0 + v' \in V$ with $b \in L$ and $v' \in V'$, then by a direct calculation using the explicit formula for A_τ and that $0_+ = A_\tau^{-1}{}^t(1_n\ 0\ 0)$, we have $W^+(v, 0) = W^+(v', 0) = \sqrt{S'} v'$, where $\sqrt{S'}_\tau = \operatorname{diag}\{(s_1^\tau)^{1/2}, \ldots, (s_n^\tau)^{1/2}\}$ and V' is identified with $(\mathbf{C}^n)^J$ by the projection $(v_1, \ldots, v_n, 0, 0) \mapsto (v_1, \ldots, v_n)$. We also compute that $2 X_\tau(v, 0)\, W_\tau^-(v, 0) = 2 |S_\tau(v, 0_-)|^2 = -b^{(1+c)\tau} s_{n+2} = -S_\tau[\vartheta_0(b) v_0]$, hence

$$zS[v] + 4iyX(v, 0)\, W^-(v, 0) = zS'[v'] + \overline{z} s_{n+2} \mathrm{N}_{L/E}(b). \tag{2.8}$$

Also, $X_\tau(v, 0) = d'_\tau b^{\lambda_\tau}$, where $d'_\tau \in \overline{\mathbf{Q}}$ and d'_τ is independent of b and v'. Now $C(\vartheta_0(b) v_0 + v')$ is a $\overline{\mathbf{Q}}$-linear combination of $B(b)\, D(v')$ with $B \in \mathcal{L}(L)$ and $D \in \mathcal{L}(V')$, so by (2.8), $\Theta_{k,H}(z, 0)$ is a $\overline{\mathbf{Q}}$-linear combination of $g_1(z)\, g_2(-\overline{z})$, where

$$g_1(z) = \sum_{v' \in V'} D(v')\, H(\sqrt{S'} v')\, e_E(z\, S'[v'])$$

$$g_2(z) = \sum_{b \in L} B(b)\, b^q\, e_E(-s_{n+2} \mathrm{N}_{L/E}(b) z).$$

It is well-known that $g_1 \in \mathrm{M}_{(n/2)u+h}(E, \overline{\mathbf{Q}})$ and $g_2 \in \mathrm{M}_{k-h+u}(E, \overline{\mathbf{Q}})$; these are special cases of [**S87a**, Proposition 11.8] and see also [**B90**, §6]. In fact, g_2 is a cusp form since $q \neq 0$. Therefore $\Theta_{k,H}(F)$ is a $\overline{\mathbf{Q}}$-linear combination of $\langle F g_1, g_2^c \rangle$, where $g_2^c(z) = \overline{g_2(-\overline{z})}$. Since $k - h \geq u$, we have

$$\pi^{[E:\mathbf{Q}]}\, \langle \varphi, g_2^c \rangle \in p_L(q, 2\lambda) \overline{\mathbf{Q}} \tag{2.9}$$

whenever $\varphi \in \tilde{\mathcal{N}}_{k-h+u}(\overline{\mathbf{Q}}) = \cup_e \tilde{\mathcal{N}}_{k-h+u}^e(\overline{\mathbf{Q}})$, as explained in [**S88**, p. 256]. (The Petersson inner product of that paper is the complex conjugate of ours. To reconcile this difference, note that $\overline{\langle F, G \rangle} = \langle F^c, G^c \rangle$.) Thus we'll be done once we establish that $F g_1 \in \tilde{\mathcal{N}}_{k-h+u}^t(\overline{\mathbf{Q}})$. Now $F g_1 \in \mathcal{N}_{k'-2h}^t(\overline{\mathbf{Q}}) \cdot \mathcal{N}_{(n/2)u+h}^0(\overline{\mathbf{Q}}) \subset \mathcal{N}_{k-h+u}^t(\overline{\mathbf{Q}})$. If $F \in \tilde{\mathcal{N}}_{k'-2h}^t(\overline{\mathbf{Q}})$ then $F g_1 \in \tilde{\mathcal{N}}_{k-h+u}^t(\overline{\mathbf{Q}})$ by Lemma 3(3). Otherwise, $\mathcal{N}_{k-h+u}^t(\overline{\mathbf{Q}}) = \tilde{\mathcal{N}}_{k-h+u}^t(\overline{\mathbf{Q}})$ by the hypothesis and Lemma 3(2). This completes the proof that $\pi^{[E:\mathbf{Q}]} \Theta_{k,h}(F)$ arithmetic on a dense set of CM-points. □

12. Concluding remarks

Theorems 1 and 2 have applications to the study of the automorphic forms $f(w, s)$ of [**S80**] and [**S86**] and to related automorphic forms $\tilde{f}(w, s)$ and $\Phi(w, s)$

which were defined in [**B90**], since these come from a theta correspondence. Near holomorphy of these functions has been proven at $s = 0$ in [**S86**] and at certain negative integer values of s in [**B90**] by applying $\partial/\partial r(w_\tau)$ to a series expansion for f, $\tilde{f}$, or Φ. This method works well at integers, but has no hope of providing results at some half-integers where one might expect to have near holomorphy. By the methods of this paper, on the other hand, we are able to prove in [**B94**] that $\tilde{f}$ and Φ are nearly holomorphic (and arithmetic) at some half-integers in the left half of a critical strip when a certain invariant associated to $\tilde{f}$ or Φ is half-integral.

References

[B90] A. Bluher, *Near holomorphy of some automorphic forms at critical points*, Invent. Math. **102** (1990), 335-376.

[B94] A. Bluher, *Arithmeticity of some automorphic forms at critical points*, American J. of Math. **116** (1994), 1283–1335.

[F] S. Friedberg, *Differential operators and theta series*, Trans. Amer. Math. Soc. **287** no. 2 (1985), 569–589.

[H] M. Harris, *A note on three lemmas of Shimura*, Duke Math. J. **46** no. 4 (1979), 871-879.

[He] S. Helgason, Differential Geometry, Lie Groups, and Symmetric Spaces, New York-London: Academic Press (1978).

[KM] S. S. Kudla and J. J. Millson, *Intersection numbers of cycles on locally symmetric spaces and Fourier coefficients of holomorphic modular forms in several complex variables*, Inst. Hautes Etudes Sci. Publ. Math. No. 71 (1990), 121–172.

[S76] G. Shimura, *The special values of the zeta functions associated with cusp forms*, Comm. Pure and Appl. Math. **29** (1976), 783–805.

[S78] G. Shimura, *The special values of the zeta functions associated with Hilbert modular forms*, Duke Math. J. **45** (1978), 637–679.

[S80] G. Shimura, *The arithmetic of certain zeta functions and automorphic forms on orthogonal groups*, Ann. of Math. **111** (1980), 313–375.

[S84] G. Shimura, *On differential operators attached to certain representations of classical groups*, Invent. Math. **77** (1984), 463–488.

[S85] G. Shimura, *On Eisenstein series of half-integral weight*, Duke Math. J. **52** (1985), 281–324.

[S86] G. Shimura, *On a class of nearly holomorphic automorphic forms*, Ann. of Math. **123** (1986), 347–406.

[S87a] G. Shimura, *On Hilbert modular forms of half-integral weight*, Duke Math. J. **55** (1987), 765–838.

[S87b] G. Shimura, *Nearly holomorphic functions on Hermitian symmetric spaces*, Math. Ann. **278** (1987), 1–28.

[S88] G. Shimura, *On the critical values of certain Dirichlet series and periods of automorphic forms*, Invent. Math. **94** (1988), 245–305.

[S90] G. Shimura, *Invariant differential operators on Hermitian symmetric spaces*, Ann. of Math. **132** (1990), 237–272.

Department of Defense, 9800 Savage Road, Fort George G. Meade, MD 20755

E-mail address: `bluher@afterlife.ncsc.mil`

Proceedings of Symposia in Pure Mathematics
Volume **66.1**, 1999

Torsion Points on $X_0(N)$

Robert Coleman, Bruce Kaskel, and Kenneth A. Ribet

To professor Goro Shimura

Abstract. Let N be a prime number, and let X be the modular curve $X_0(N)$, considered over the field of complex numbers. Suppose that the genus of X is at least 2, i.e., that N is at least 23. Using the cusp at infinity on X, we identify X with a subvariety of positive codimension of the Jacobian of X. The set of points on X which map to torsion points on the Jacobian is a finite set (Manin-Mumford conjecture) which always contains the two cusps on X and which contains the hyperelliptic branch points on X in the special case where X is hyperelliptic and N is different from 37. We conjecture that this set contains no other points, and verify our conjecture when $N = 37$. For general N, we accumulate information about this set in the direction of the conjecture. With this information in hand, M. Baker has been able to prove the conjecture for certain other values of N.

1. Introduction

Let X be a complex algebraic curve of genus > 1. Choosing a base point $x_0 \in X$, we view X as a subvariety of its Jacobian $J = J(X)$ via the Albanese map which sends each $x \in X$ to the class of the divisor $(x) - (x_0)$. Let J_{tor} be the group of torsion points on J. The Manin-Mumford conjecture states that the intersection

$$\mathcal{S} := X \cap J_{\text{tor}}$$

is finite.

This conjecture was studied first by S. Lang [**L65**], who showed that it is implied by a statement about the action of $\text{Gal}(\overline{K}/K)$ on J_{tor}, where K is a finitely generated subfield of $\mathbf{C}$ over which X and x_0 are defined. Lang's hypothesis states that the image of $\text{Gal}(\overline{K}/K)$ in Aut J_{tor} contains an open subgroup of the homothety group $\hat{\mathbf{Z}}^* \subseteq \text{Aut} J_{\text{tor}}$. The statement remains unproved, despite partial results by Bogomolov[**B**] and Serre [**Se**]. (See also [**H**], which may be regarded as a sequel to [**L65**].)

1991 *Mathematics Subject Classification.* Primary 11F11, 11G18, 14G25.

Key words and phrases. Manin-Mumford conjecture, torsion points on curves, modular curves, Galois action on torsion points of abelian varieties.

Coleman's work was partially supported by grant DMS 93 07195. Ribet's work was partially supported by grant DMS 96 22801.

Meanwhile, the Manin-Mumford conjecture was proved by M. Raynaud in a 1983 article [**R**]. Subsequently, a second proof of the conjecture was given by R. Coleman in [**C87**]. Generalizations and strengthenings of the Manin-Mumford conjecture have been proposed in various quarters. For example, a recent article by E. Ullmo [**U**] studies points on X which map to points of J with small canonical height. (See also [**MiU**].)

This article concerns the intersection $X \cap J_{\text{tor}}$ in the special case where X is a modular curve. More specifically, we limit our attention to the curve $X = X_0(N)$ where N is a prime number. The requirement that X have genus > 1 becomes the condition $N \geq 23$, which we now impose. We take x_0 to be the standard "cusp at ∞," which we call C_∞.

A general theorem of Manin-Drinfeld implies that $\mathcal{S}$ contains the set of cusps on X. There are two cusps, the standard cusp C_∞, which maps to 0 in $J_0(N)$, and a second cusp, C_0, which maps to a point on $J_0(N)$ of order $n := \text{num}(N-1)/12$. (See [**M77**, Ch. II, §11] for a discussion of the group C generated by C_0 on $J_0(N)$; this group is called the cuspidal subgroup of $J_0(N)$. Theorem 1.2 of [**M77**, Ch. III] states that C is the torsion subgroup of the Mordell-Weil group of J.) The inclusion $\{C_0, C_\infty\} \subseteq \mathcal{S}$ is the first piece of information about $\mathcal{S}$ at our disposal. Because of it, we describe $\mathcal{S}$ as the "cuspidal torsion packet" on X. (As the reader may recall Coleman calls the intersection of a curve embedded in its Jacobian with the set of torsion points a "torsion packet.")

A second constraint on $\mathcal{S}$ arises from the fixed points of the hyperelliptic involution of X if $X = X_0(N)$ is a hyperelliptic curve:

PROPOSITION 1.1. *Suppose that X is hyperelliptic, and let P be a fixed point of the hyperelliptic involution of X. Then we have $P \in \mathcal{S}$ if and only if N is different from* 37. *In this latter case,* $2P$ *belongs to the cuspidal subgroup of J.*

PROOF. Let h be the hyperelliptic involution. Since h operates on $J = J_0(N)$ as -1, the images in J of the divisors $(P) - (C_\infty)$ and $\big(h(C_\infty)\big) - (P)$ are equal. Adding $(P) - (C_\infty)$, we find that $2\big((P) - (C_\infty)\big)$ and $\big(h(C_\infty)\big) - (C_\infty)$ are equal on J. Hence we have $P \in \mathcal{S}$ if and only if the point $h(C_\infty)$ has finite order in J. A theorem of Ogg states that X is hyperelliptic if and only if N is either 37 or one of the seven primes 23, 29, 31, 41, 47, 59 and 71. (See [**O74a**, Th. 2] and the article [**O74b**].) For the last seven primes on the list, h is the Atkin-Lehner involution $w = w_N$ of X and hence $h(C_\infty)$ is the cusp C_0. In particular, $h(C_\infty)$ has finite order on J, so that P belongs to $\mathcal{S}$. Further, the point $2P$ on J is the class of the divisor $(C_0) - (C_\infty)$, which is a generator of the cuspidal group C. For the exceptional case $N = 37$, Mazur and Swinnerton-Dyer [**MaS**, §5] have observed that $h(C_\infty)$ has infinite order on J. Hence P is not an element of $\mathcal{S}$. □

Let $\mathcal{S}_0$ be the set which is defined as follows: If X is not hyperelliptic or if $N = 37$, $\mathcal{S}_0 = \{C_0, C_\infty\}$; if X is hyperelliptic and $N \neq 37$, $\mathcal{S}_0$ is the set consisting of the two cusps of X and the hyperelliptic fixed points on X. We then have in all cases

$$\mathcal{S}_0 \subseteq \mathcal{S}.$$

It seems plausible to us that this inclusion is an equality, i.e., that one has

$$\mathcal{S} \stackrel{?}{=} \mathcal{S}_0$$

for all $N \geq 23$[1]. Despite the somewhat ad hoc definition of $\mathcal{S}_0$, the equality $\mathcal{S} = \mathcal{S}_0$ has a simple restatement:

PROPOSITION 1.2. *We have $\mathcal{S} = \mathcal{S}_0$ if and only if $2P$ is a rational point of J for all $P \in \mathcal{S}$.*

PROOF. Assume that $\mathcal{S} = \mathcal{S}_0$, and let P be an element of $\mathcal{S}$. If P is one of the two cusps of $X_0(N)$, then P is rational. If P is not a cusp, then P is a hyperelliptic branch point of $X_0(N)$ and N is different from 37. We have seen in this case that $2P$ lies in the cuspidal group C; in particular, $2P$ is rational. Assume now that $2P$ is rational for each $P \in \mathcal{S}$. Take $P \in \mathcal{S}$ and suppose first that P is not itself a rational point. Then there is a $g \in \mathrm{Gal}(\overline{\mathbf{Q}}/\mathbf{Q})$ such that $gP \neq P$. The difference $gP - P$, computed on J, is then a 2-division point because $2P$ is rational. This means that the non-zero divisor $2(gP) - 2(P)$ is the divisor of a rational function on $X_0(N)$. This function defines a degree-two map $X_0(N) \to \mathbf{P}^1$ for which P is a ramified point. We find then that P is a hyperelliptic branch point of $X_0(N)$. Since P lies in $\mathcal{S}$, we must have $N \neq 37$; thus P is a point in $\mathcal{S}_0$. The second possibility is that P is a rational point of $X_0(N)$. In that case, P lies in the intersection $C \cap X_0(N)$, in view of Mazur's theorem [**M77**, p. 33] that C is the torsion subgroup of the Mordell-Weil group of $J_0(N)$. We need to know that this intersection consists of the two cuspidal points C_0 and C_∞ of J. One way to proceed is to cite the well-known theorem of Mazur [**M78**, Th. 7.1] which states for all but four prime numbers $N \geq 23$ that $X_0(N)(\mathbf{Q}) = \{C_0, C_\infty\}$. The four exceptional primes are $N = 37$, 43, 67, and 163. For the first of these primes, the equality $X_0(N) \cap C = \{C_0, C_\infty\}$ is proved by Mazur and Swinnerton-Dyer [**MaS**, §5, Prop. 2]. For the primes 43, 67, and 163, there is a unique non-cuspidal point of $X_0(N)$, which arises from elliptic curves with complex multiplication by $\mathbf{Q}(\sqrt{N})$. By uniqueness, this point is fixed by the Atkin-Lehner involution w of $X_0(N)$. Since w operates on C by multiplication by -1 [**M77**, p. 99], and since $n = \mathrm{num}(N-1)/12$ is odd when $N = 43$, 67 and 163, the desired statement follows. □

In this article, we present an assortment of results about $\mathcal{S}$. These include theorems proved by the p-adic techniques of the first author [**C85, C87**] and an inequality (proved in §4) which is presented "in the spirit of Lang [**L65**]."

To illustrate the scope of our results, we will state a theorem which is valid for all $N \geq 23$ (see Corollary 2.3, Proposition 3.12 and Theorem 3.15) and a complement which concerns the special case $N = 37$ in §6. Before presenting these results, we introduce some notation: First, we let $\mathbf{T}$ be the ring of endomorphisms of $J_0(N)$, i.e., the ring of Hecke operators acting on the space of weight-two cusp forms on $\Gamma_0(N)$. (See [**M77**, Ch. II, Prop. 9.5].) Further, we write P_p for the "p-primary part" of P, whenever P is a torsion point on J and p is a prime number. Thus P is the sum of its p-primary parts (as p ranges over the set of primes) and P_p has p-power order. Using [**M77**, Ch. III, Th. 1.2], we may restate Proposition 1.2 as follows: the equality $\mathcal{S} = \mathcal{S}_0$ holds if and only if $2P_p$ lies in the cuspidal group C for all primes p and all $P \in \mathcal{S}$.

THEOREM 1.3. *Let P be an element of $\mathcal{S}$, and let $p \neq 2, 3$ be a prime for which P_p does not belong to the cuspidal group C. Then at least one of the following holds:*

[1]Through the efforts of Matt Baker we now know this equality is true for all but seven primes up to 600. Note that $X_0(N)$ is hyperelliptic in each of these cases.

(i) $p = N$; (ii) p satisfies $5 \leq p < 2g$, $X_0(N)$ does not have ordinary reduction at p, and p is ramified in $\mathbf{T}$ in the sense that $\mathbf{T}/p\mathbf{T}$ is not a product of fields. Further, suppose that $p = N$ and that N is not ramified in $\mathbf{T}$. Then there is some $\ell \neq N$ such that $P_\ell \notin C$.

THEOREM 1.4. *The intersection $X_0(37) \cap J_0(37)_{\text{tor}}$ consists of the two cusps C_0 and C_∞.*

We prove the latter theorem by combining results such as Theorem 1.3 with some facts which are specific to the case and $N = 37$. In particular, we lean heavily on results which were proved by B. Kaskel in [**K**]; these involve a detailed description of the image of $\text{Gal}(\overline{\mathbf{Q}}/\mathbf{Q})$ in $\text{Aut} J_0(37)_{\text{tor}}$. To obtain this description, Kaskel employs many of the techniques pioneered by Serre [**S**] and Lang-Trotter [**LT**] in the case where $X_0(N)$ has genus one. To the best of our knowledge, no analogous description is available in general for $J_0(N)$; it would certainly be of considerable interest to find such a description.

It strikes us that the general results at our command, when augmented by the description of [**K**], seem to represent overkill: they allow us to prove Theorem 1.4 in several, closely related, ways. Because of our interest in the general case, we will present three proofs of the theorem in §6.

The authors thank Matt Baker for his helpful feedback during the preparation of this paper.

2. Ramification

Let P be a torsion point on $J = J_0(N)$. We wish to study the set of primes which ramify in the extension $\mathbf{Q}(P)/\mathbf{Q}$. First, we record the following statement, which follows directly from the main results of [**C87**] and [**C85**, Th. 5.5].

THEOREM 2.1. *Suppose that P lies in $\mathcal{S}$. Then the extension $\mathbf{Q}(P)/\mathbf{Q}$ can be ramified only at the bad prime N, the two very small primes 2 and 3, and the primes p satisfying $5 \leq p < 2g$. Further, if $\mathbf{Q}(P)/\mathbf{Q}$ is ramified at $p = 3$, then the ramification at 3 is tame. Finally, if $\mathbf{Q}(P)/\mathbf{Q}$ is ramified at p and p satisfies $5 \leq p < 2g$, then X does not have ordinary reduction at p.*

For each prime number p, we again write P_p for the "p-primary component of P." Further, we again use the symbol $\mathbf{T}$ to denote the ring of endomorphisms of $J_0(N)$, i.e., the ring of Hecke operators acting on the space of weight-two cusp forms on $\Gamma_0(N)$. Finally, we recall that a prime number p is said to be *Eisenstein* if it divides the order of the cuspidal subgroup C of J.

THEOREM 2.2. *Suppose that the field extension $\mathbf{Q}(P)/\mathbf{Q}$ is unramified at p. Then either $p = 2$, or P_p belongs to the cuspidal group C.*

PROOF. Assume, to the contrary, that p is an odd prime and that P_p is not contained in the cuspidal group C. Let M be the $\mathbf{T}[\text{Gal}(\overline{\mathbf{Q}}/\mathbf{Q})]$-submodule of $J_0(N)(\overline{\mathbf{Q}})_{\text{tor}}$ which is generated by P_p. Then M is a finite abelian group of p-power order which is not contained in C. As a module for the Galois group $\text{Gal}(\overline{\mathbf{Q}}/\mathbf{Q})$, M is unramified at p.

As explained in [**M77**, Ch. II, §14], the Jordan-Hölder constituents of M can only be of the following three forms: (i) the constant group $\mathbf{Z}/p\mathbf{Z}$; (ii) the group μ_p; (iii) an irreducible module of the form $J_0(N)[\mathfrak{m}]$, where $\mathfrak{m}$ is a non-Eisenstein maximal ideal of $\mathbf{T}$ with residue characteristic p. Types (i) and (ii) occur only if p

is an Eisenstein prime, i.e., if p divides the integer n introduced above. However, since p is odd, the modules of types (ii) and (iii) are visibly ramified at p: note that the determinant of $J_0(N)[\mathfrak{m}]$ is the mod p cyclotomic character. In view of our supposition that M is unramified at p, M is a successive extension of copies of $\mathbf{Z}/p\mathbf{Z}$, so that p is a divisor of n. In particular p and N are distinct.

We view M as an extension of $Q := M/M_C$ by M_C, where M_C is the intersection $M \cap C$. Since M is not contained in C, Q is non-zero. Let $Q' \cong \mathbf{Z}/p\mathbf{Z}$ be a minimal sub-module of Q, and let M' be the inverse image of Q' in M. Thus we have a tautological exact sequence

$$0 \to M_C \to M' \to Q' \to 0$$

in which the groups M_C and Q' are trivial $\mathrm{Gal}(\overline{\mathbf{Q}}/\mathbf{Q})$-modules.

We claim that M' is unramified at N. This claim follows from an argument due to Mazur [**M77**, Ch. II, Lemma 16.5], which we shall now recapitulate. Let σ be an element of an inertia group for the prime N in $\mathrm{Gal}(\overline{\mathbf{Q}}/\mathbf{Q})$, and let m be an element of M'. Let p^r be the order of m. Grothendieck's theory of semistable reduction (together with the Deligne-Rapoport theorem [**DR**] that $J_0(N)$ has multiplicative reduction at N) shows that $\sigma m - m$ belongs to the "toric part" of $J_0(N)[p^r](\overline{\mathbf{Q}}_N)$. The displayed exact sequence forces $\sigma m - m$ to be an element of C. However, reduction mod N induces an isomorphism between C and the group of connected components of the Néron model for $J_0(N)$ in characteristic N; this yields in particular that the intersection of C and the toric part of $J_0(N)[p^r](\overline{\mathbf{Q}}_N)$ is zero. Hence $\sigma m - m = 0$, which proves what was claimed.

Now M' can be ramified a priori only at p and at N. We have just seen that M' is unramified at N, and the hypothesis to the theorem implies that it is unramified at p. It is therefore everywhere unramified, so that the action of $\mathrm{Gal}(\overline{\mathbf{Q}}/\mathbf{Q})$ on M' must be trivial. In other words, we have $M' \subseteq J_0(N)(\mathbf{Q})_{\mathrm{tor}}$. On the other hand, one of the main results of [**M77**] asserts that $C = J_0(N)(\mathbf{Q})_{\mathrm{tor}}$. We thus have $M' = M_C$ and $Q' = 0$, which is a contradiction. □

COROLLARY 2.3. *Let P be an element of $\mathcal{S}$, and assume that P_p does not belong to the cuspidal group C. Then at least one of the following holds: (i) $p = 2$ or $p = 3$; (i) $p = N$; (ii) p satisfies $5 \le p < 2g$ and X does not have ordinary reduction at p.*

This corollary results directly from the two results above. We have stated it in order to record our progress toward proving Theorem 1.3.

THEOREM 2.4. *Suppose that $P \in \mathcal{S}$ and that the order of P is even but not divisible by N. Assume that N is not congruent to 1 mod 8. Then P belongs to $\mathcal{S}_0$.*

PROOF. It will be convenient to write the order of P as $2d$. Let R be the 2-division point dP. We claim that the extension $\mathbf{Q}(R)/\mathbf{Q}$ is ramified at the prime N.

For this, let M be the $\mathbf{T}[\mathrm{Gal}(\overline{\mathbf{Q}}/\mathbf{Q})]$-submodule of $J_0(N)(\overline{\mathbf{Q}})_{\mathrm{tor}}$ which is generated by R; further, let V be a minimal submodule of M. Thus V is a constituent of $J_0(N)[2]$ which belongs to a maximal ideal $\mathfrak{m}$ of $\mathbf{T}$ of residue characteristic 2. (See [**M77**, Ch. II, §14] for the terminology.) In view of the hypothesis $N \not\equiv 1$ mod 8, $\mathfrak{m}$ cannot be an Eisenstein ideal; this means that V is an irreducible 2-dimensional representation of $\mathrm{Gal}(\overline{\mathbf{Q}}/\mathbf{Q})$ over $\mathbf{T}/\mathfrak{m}$. By a theorem of Tate [**T**], the representation V is ramified at N. This proves the claim.

Equivalently: there is an inertia group I for N in $\mathrm{Gal}(\overline{\mathbf{Q}}/\mathbf{Q})$ and a $\sigma \in I$ such that $\sigma R \neq R$. It follows that $\sigma R - R$ has order 2.

Consider the action of σ on $J_0(N)[2d]$. By Grothendieck's semistable reduction theorem (and the results of Deligne-Rapoport [**DR**]), σ acts as the sum $1+A$, where 1 represents the identity operator and A an endomorphism satisfying $A^2 = 0$. Using the binomial theorem, we find

$$\sigma^d(P) - P = (1 + dA)P - P = AdP = \sigma(R) - R.$$

Hence if $g = \sigma^d$, then g is an element of $\mathrm{Gal}(\overline{\mathbf{Q}}/\mathbf{Q})$ such that $gP - P$ has order 2. As we have seen in the proof of Proposition 1.2, it follows that P lies in $\mathcal{S}_0$. □

3. Ramified torsion points revisited

As we indicated above, this article concludes with a list of errata to [**C87**]. In particular, we point out that the proofs of cases (ii) and (iii) of [**C87**, Theorem 20] are incorrect (however, the statements and proofs are correct when the Jacobian is a canonical lifting). These cases concern the primes $p = 2$ and $p = 3$. Although those statements are not needed for the main theorem of [**C87**], we hope that appropriately modified versions of them may be quite useful. Accordingly, we provide in this section some results for the primes $p = 2$ and $p = 3$. In addition, we will complete the proof of Theorem 1.3.

Fix a prime p, and let $\mathbf{C}_p$ be the completion of an algebraic closure of $\mathbf{Q}_p$. Let v be the valuation of $\mathbf{C}_p$ such that $v(p) = 1$. Let K^{ur} be the completion of the maximal unramified extension of $\mathbf{Q}_p$ and let K be a complete subfield of K^{ur}. Let R be the ring of integers of K and let k be the residue field of R. Also let R^{ur} be the ring of integers of K^{ur}, $\mathbf{F}$ the residue field of R^{ur} and σ the Frobenius automorphism of K^{ur}.

Let A be an Abelian scheme over R. Define a pairing

$$(\ ,\ ) : \Omega^1_{A/R^{\mathrm{ur}}}(A) \times A(\mathbf{F}) \to K^{\mathrm{ur}}/pR^{\mathrm{ur}}$$

as follows: Given $\omega \in \Omega^1_{A/R^{\mathrm{ur}}}(A)$ and $P \in A(\mathbf{F})$, let $\tilde{P}$ be a lifting of P to $A(K^{\mathrm{ur}})$ and set

$$(\omega, P) = \int_0^{\tilde{P}} \omega \mod pR^{\mathrm{ur}}.$$

It is easy to check that (ω, P) is independent of the choice of $\tilde{P}$: when $P = 0$, $\int_0^{\tilde{P}} \omega \in pR^{\mathrm{ur}}$. We note that $(\ ,\ )$ is a Galois-invariant bilinear form and that (ω, P) is trivial when the order of P is prime to p.

REMARK 3.1. We can relate $(\ ,\ )$ to the Serre-Tate parameters when A is ordinary. Recall that there is a natural pairing

$$q\colon \mathrm{Ta}_p A(\mathbf{F}) \times \mathrm{Ta}_p \hat{A}(\mathbf{F}) \to 1 + pR.$$

(See [**Katz**].) There is also a natural map $\mathrm{Ta}_p\hat{A}(\mathbf{F}) \to \Omega^1_{A/R^{\mathrm{ur}}}(A)$, $\alpha \mapsto \omega_\alpha$. Then one can show:

PROPOSITION 3.2. *Suppose $\alpha = (\alpha_n) \in \mathrm{Ta}_p\hat{A}(\mathbf{F})$ and $\beta = (\beta_n) \in \mathrm{Ta}_pA(\mathbf{F})$. Then*

$$(\omega_\alpha, \beta_n) \equiv \frac{1}{p^n} \log q(\alpha, \beta) \mod pR^{\mathrm{ur}}.$$

We say that a quadruple (H, W, F, V) is an F-T-crystal over R if H is free R module of finite rank, W is a direct summand of H and F and V are σ and σ^{-1}-linear endomorphisms of H, respectively, such that

$$FV = VF = p, \qquad VH = W + pH.$$

(This was called an F-crystal in [**C87**].) If S is a smooth complete scheme of finite type over R, the first de Rham cohomology of S over R has a natural structure of an F-T-crystal where $H = H^1_{\mathrm{DR}}(S/R)$ and $W = H^0(S, \Omega^1)$ and we call a sub-F-T-crystal of this a *de Rham F-T-crystal* on S.

Suppose (H, W) is a de Rham F-T-crystal on A. Let P be a point of $A(\mathbf{F})$. For a coset B of pR^{ur} in $\mathbf{C}_p$ we set $v(B) = \min\{v(x) \colon x \in B\}$. Let

$$M_P(H) := 1 - \min\{v(\omega, P) \colon \omega \in W\}.$$

Let H be the first de Rham cohomology group of A over R thought of as an F-T crystal. Suppose $P \in A(\mathbf{F})$. We have in general,

$$M_{pP}(H) + 1 \geq M_P(H) \geq M_{pP}(H) \tag{1}$$

and if $M_P(H) \geq 1$

$$M_{pP}(H) + 1 = M_P(H). \tag{2}$$

It follows, from (1) and (2), that if $M_{pP}(H) \geq 1$

$$M_{pP}(H) + 1 = M_P(H). \tag{3}$$

LEMMA 3.3. *Suppose $H = H^1_{\mathrm{DR}}(A, R)$. Let $P \in A(\mathbf{F})$. Then $M_P(H) = 0$ if and only if there is a torsion point of $A(K^{\mathrm{ur}})$ lying over P.*

PROOF. If there is such a torsion point, P, then

$$\int_O^P \omega = 0,$$

for all $\omega \in W$. It follows that $v(\omega, P) = 1$ for all $\omega \in W$ and so $M_P(H) = 0$.

Now suppose $M_P(H) = 0$. Let $U(P)$ be the residue class above P. Let $\omega_1, \ldots, \omega_g$ be a basis for W over R. Let $\omega = (\omega_1, \ldots, \omega_g)$. Define a function on $U(P)$ into $\mathbf{C}_p^g$ by

$$F(X) = \int_0^X \omega,$$

for $X \in U(P)$. Now let $z_1, \ldots, z_g$ be a set of local parameters on $U(P)$ defined over R and let $Z = (z_1, \ldots, z_g)$ be the corresponding multivariable. We may use Z to identify $U(P)$ with $B(0,1)^g$. Once this is done the R^{ur}-valued points of $U(P)$ become identified with $p(R^{\mathrm{ur}})^g$. Also we may now write F as a series in Z,

$$F(Z) = \left(\sum_I A_{1,I} Z^I, \ldots, \sum_I A_{g,I} Z^I \right),$$

where I runs over $\mathbf{N}^g$. Since $dF = \omega$, it follows that

$$G(Z) := (F(pZ) - F(0))/p \in R[[Z]].$$

Also since $\omega, \ldots, \omega_g$ is a basis for W, $G'(0)$ is an invertible matrix. Since $\mathbf{F}$ is algebraically closed, it follows that as a function from $(R^{\mathrm{ur}})^g$ to itself, G is surjective. Now $M_P(H) = 1 - \max\{1, v(F(0))\}$, so since $M_P(H) = 0$, $F(0) \in pR^g$. Thus, there exists a $z \in (R^{\mathrm{ur}})^g$ such that $G(z) = -F(0)/p$ and so there exists an $x \in U(P)(R^{\mathrm{ur}})$

such that $F(x) = 0$. It follows from [**C85**, Theorem 2.11] that x is a torsion point. □

COROLLARY 3.4. *Suppose $P \in A(\mathbf{F})$ and $k \in \mathbf{Z}_{>0}$. Then $M_P(H) = k$ if and only if there exists a torsion point in $U(p^kP)(R^{\mathrm{ur}})$ but not in $U(p^{k-1}P)(R^{\mathrm{ur}})$.*

PROOF. Suppose $M_P(H) = k$. Then it follows from (2) that $M_{p^{k-1}P}(H) = 1$ and $M_{p^kP}(H) = 0$ and so from the lemma there exists a torsion point in $U(p^kP)$ but not in $U(p^{k-1}P)$. Now assume there exists a torsion point in $U(p^kP)(R^{\mathrm{ur}})$ but not in $U(p^{k-1}P)(R^{\mathrm{ur}})$. Then it follows from the lemma that $M_{p^kP}(H) = 0$ and $M_{p^{k-1}P}(H) \neq 0$. It follows from (1) that $M_{p^{k-1}P}(H) = 1$. Now apply (3) repeatedly to deduce the corollary. □

Now suppose Y is a smooth complete curve over R and that (H, W) is a de Rham F-T-crystal on Y, which we may equally well think of as a de Rham F-T crystal on its Jacobian J. We will use a great deal of the notation of [**C87**, §3].

We will say that H is *special* if it is either ordinary or superspecial in the language of [**C87**]. In the ordinary case, as explained in [**C87**] we have bijective σ^{-1} and σ-linear operators operators $\mathcal{C}$ and $\mathcal{C}^*$ on $\overline{W} := W/pW$ and $\overline{W}^* := \mathrm{Hom}(\overline{W}, k)$ satisfying

$$(\mathcal{C}w, v)^\sigma = (w, \mathcal{C}^*v)$$

for $w \in \overline{W}$ and $v \in \overline{W^*}$, where (,) is the natural pairing. (Caution: in [**C87**, §3], the superscript "*" has a different meaning.) In the superspecial case, there are analogous operators, again called $\mathcal{C}$ and $\mathcal{C}^*$, with similar properties. In both cases, we obtain an operator on $\mathbf{P}(\overline{W})$ which we call $\mathbf{P}(\mathcal{C})$.

Let $O \in Y(K)$ and suppose (H, W) is a special de Rham F-T-crystal on Y. Let $q = p$ if H is ordinary and p^2 if H is superspecial. Let

$$\lambda_\nu(Q) = \int_O^Q \nu.$$

That H is special implies that for $P \in Y(\mathbf{F})$ there exists $g_\nu \in T\mathcal{O}_Y(U(P))$ such that

$$dg_\nu \equiv \nu \mod T^{q-1}\Omega_Y(U(P)).$$

Here, T is a uniformizing parameter for $U(P)$. Also, $\overline{g}_\nu$ is the reduction of g_ν modulo p. Set

$$k_P(H) = \min\{q, \mathrm{ord}_P\overline{g}_\nu \ : \nu \in W\}.$$

We note that $k_P(H) \leq 1 + \min\{\ \mathrm{ord}_P\bar{g}_\nu \colon \nu \in W\}$ if the minimum is strictly less than q, with equality if it is strictly less than p. The following two propositions refine Proposition 15 of [**C87**].

PROPOSITION 3.5. *Suppose H is special. Let $P \in Y(\mathbf{F})$, $U = U(P)$ and T be a uniformizing parameter on U over K^{ur}. Assume $k_P(H) < q$. Suppose $b \in U(\mathbf{C}_p)$ is a common zero of $\Lambda = \{\lambda_\nu \mid \nu \in W\}$. Let $k = k_P(H)$ and $M = M_P(H)$ and $v = v(T(b))$. Then one of the following holds:*

(i) *$M = 0$, $v = \infty$ or r/s where r and s are positive integers and $s \leq k$ or*
(ii) *$M > 0$, $v = 1/kq^M$ or*
(iii) *$v = 1/(kq^n(q-1))$ for some integer n such that $q^n(q-1) > q^M$.*

For S a uniformizing parameter on U let h_S be the element of $\overline{W}^*$

$$\omega \to \frac{\omega}{dS}(P)$$

and let $j(P)$ be the image of h_S in $\mathbf{P}(\overline{W})$. (This is independent of the choice of S.) If $M > 0$, we also define $e(P)$ to be the image in $\mathbf{P}(W)$ of the element $d(P)$ in W^* which is given by $\omega \mapsto p^{M-1}(\omega, P)$.

PROPOSITION 3.6. *In addition to the hypotheses of the last proposition suppose* $k_P(H) = 1$ *(i.e., that P is not a base point of W). Then we have:*

In case (i), $b \in U(K^{\mathrm{ur}})$.

In case (ii), every differential in W which vanishes at P vanishes at least $q-2$ times.[2] *Moreover, if* $q \geq 3$,

$$\mathbf{P}(\mathcal{C})(j(P)) = j(P) = e(P). \tag{$*$}$$

If $q = 2$, choose a 2^Mth root of 2, let r be the image of $T(b)/2^{2^{-M}}$ in the residue field of $K(b)$ and let $h_b = rh_T \in \overline{W^*}$. Then

$$\mathcal{C}^{*M+1}(h_b) + \mathcal{C}^{*M}(h_b) = d(P).$$

In case (iii), every differential in W which vanishes at P vanishes at least $q-2$ times if $n = M$ and $q-1$ times if $n > M$; further,

$$\mathbf{P}(\mathcal{C})(j(P)) = j(P).$$

We note that the element h_b defined above is independent of choices.

PROOF OF PROPOSITIONS 3.5 and 3.6. The main ingredient of the proof of these propositions, as discussed in the proof of Proposition 15 of [**C87**], is the fact that for each $\omega \in W$ there exists $f_{\omega,m}(T) \in TR^{\mathrm{ur}}[[T]]$ such that

$$df_{\omega,m} \equiv \mathcal{C}^m\omega \mod (p, T^q dT/T), \tag{4}$$

and for $x \in U$,

$$\lambda_\omega(x) = c_\omega + f_{\omega,0}(T(x)) + \frac{f^\sigma_{\omega,1}(T(x)^q)}{p} + \cdots + \frac{f^{\sigma^m}_{\omega,m}(T(x)^{q^m})}{p^m} + \cdots, \tag{5}$$

where $c_\omega = \lambda_\omega(T^{-1}(0))$. Call the series on the right $L_\omega(T(x))$. What we do is consider the lower convex hull of the Newton polygons of the series $L_\omega(T)$ for $\omega \in W$. This is the lower convex hull of the set of points

$$S \cup \{(kq^i, -i)\}_{i \geq 0},$$

where S is $\{(0, 1-M)\}$ if $M > 0$ and if $M = 0$ is some set of points of the form (x, y) with x-coordinate a non-negative integer at most $k-1$ and y-coordinate an integer at least 1. From this and Lemma 14 of [**C87**, Proposition 3.5], follows.

Parts (i) and (iii) of Proposition 3.6 follow from Proposition 15 of [**C87**]. Part (ii) is new (although the first sentence with $q-2$ replaced by $q-1$ was claimed in [**C87**] and we prove it now.

Assume $M > 0$, $k = 1$.

[2]In fact, using the results proven below, Matt Baker has shown that when $q \geq 3$ this case can only occur if $\dim W = 1$.

First suppose $q \geq 3$. Let $a = T(b)$. Then the assertion $\lambda_\omega(b) = 0$ implies

$$(**) \qquad c_\omega + \frac{f^{\sigma^{M-1}}_{\omega,M-1}(a^{q^{M-1}})}{p^{M-1}} + \frac{\left(f'_{\omega,M}(0)\right)^{\sigma^M} a^{q^M}}{p^M} \equiv 0 \mod p^{2-M}.$$

As

$$f'_{\omega,M}(0) \equiv \left(\mathcal{C}^M \bar{\omega}/dT\right)|_P \mod p,$$

we obtain $\mathbf{P}(\mathcal{C})^M(j(P)) = e(P)$ immediately. Hence, to prove $(*)$, we only have to establish the first equality. It also follows from $(**)$ that $\mathcal{C}^M\bar{\omega}$ vanishes at P implies c_ω is congruent to 0 modulo $a^{q^{M-1}}p^{1-M}$ and hence modulo p^{2-M} as $c_\omega \in K^{\mathrm{ur}}$. Hence

$$f_{\omega,M-1}(a^{q^{M-1}}) \equiv 0 \mod p,$$

which together with (4) implies $\mathcal{C}^{M-1}\bar{\omega}$ vanishes $q-2$ times at P as $v(a^{q^{M-1}}) = 1/q$. In particular, using the facts that $\mathcal{C}$ is bijective on $\bar{W}$ and is σ^{-1} linear we see that the hyperplane of differentials which vanish at P is stable under $\mathcal{C}$, and this is equivalent to the first equation of $(*)$. The fact that ω vanishes $q-2$ times at P follows from the above with ω replaced by an element $\nu \in W$ such that $\mathcal{C}^{1-M}\bar{\omega} = \bar{\nu}$ as we now know such a ν vanishes at P.

Suppose now $q = 2$. Then we see from (5) that, in the residue field of $K(b)$,

$$2^{M-1}c_\omega + \left(r\left(\frac{d}{dT}f_{\omega,M}\right)(0)\right)^{\sigma^M} + \left(r\left(\frac{d}{dT}f_{\omega,M+1}\right)(0)\right)^{\sigma^{M+1}} = 0.$$

The assertion now follows from (4) and the definitions. □

If $Q \in Y(\mathbf{C}_p)$ and T is a uniformizing parameter defined over K^{ur} on the residue disk containing Q, let $v(Q) = v(T(Q) + pR^{\mathrm{ur}}) = \min\{1, v(T(Q))\}$. This is independent of the choice of T.

COROLLARY 3.7 *Suppose that we are in the situation of Proposition 3.6. In cases (i) or (ii), Λ has exactly q^M common zeroes c such that $v(c) = q^{-M}$ unless $q = 2$, in which case it has either 2^M or 2^{M+1} zeroes; in the latter case, $\mathbf{P}(\mathcal{C})(j(P)) = j(P)$. In case (iii), Λ has exactly $q^n(q-1)$ common zeroes c such that $v(c) = (q^n(q-1))^{-1}$.*

PROOF. From Lemma 10 of [**C87**] together with the proofs of Proposition 15 of [**C87**] when $q \geq 3$ and 3.6 when $q = 2$, we see, that Λ has at most the numbers of common zeroes claimed. That it has at least this number follows from Galois theory. The statement about $j(P)$ when $q = 2$ follows by an examination of the proof of Proposition 3.6. □

The following proposition summarizes most of what we know about $p = 2$ and 3 in the ordinary case.

PROPOSITION 3.8. *Suppose Y is embedded in its Jacobian J so that O is the origin. Suppose Q is a ramified torsion point on Y and $\bar{Y}$ is ordinary. Then p equals 2 or 3. Moreover:*

(i) *Suppose $p = 2$. Then either $\bar{Q}$ does not lift to an unramified torsion point of J or $\bar{Y}$ is hyperelliptic and $\bar{Q}$ is a hyperelliptic branch point.*
(ii) *Suppose $p = 3$. Then: (a) $\mathbf{P}(\mathcal{C})(j(\bar{Q})) = j(\bar{Q})$. (b) If $3|(1/v(Q))$, $\bar{Q}$ does not lift to an unramified torsion point on J. (c) If $\bar{Y}$ is hyperelliptic, then $g(Y) = 2$. Moreover, in this case, if $y^2 = f(x)$ is an equation for Y, where*

$f(x)$ is a polynomial of degree 6 in $R[x]$ such that $(\overline{f}(x), \overline{f'}(x)) = 1$ and $\overline{f'}(x)$ has degree 4, then

$$\overline{f'(Q)} = 0.$$

(d) If $g(Y) = 2$ and $v(Q) = 1/2$, then $\bar{Q}$ lifts to an unramified torsion point on J. Moreover, there is an even number, at most 16, of such torsion points on Y.

Note that, by Corollary 3.7, in situation (ii) of the proposition, $v(Q)$ is either of the form $1/3^n$, $n > 0$ or of the form $1/(2 \cdot 3^n)$, $n \geq 0$.

PROOF. Suppose $p = 2$. Then, if $M_{\bar{Q}}(H) > 0$, $\bar{Q}$ does not lift to an unramified torsion point by Lemma 3.3. Suppose $M_{\bar{Q}}(H) = 0$. Then we must be in case (iii), so $\mathbf{P}(\mathcal{C})(j(\bar{Q})) = j(\bar{Q})$ and we can apply the argument in the proof of Theorem 5.5 of [**C85**] for $p = 2$.

Now suppose $p = 3$. Then (a) is contained in Proposition 3.6. If $v(Q) = 1/3^n$, then we are in case (ii) of this proposition and (b) follows from Lemma 3.6. Now suppose $v(Q) = 1/2 \cdot 3^n$ and $n > 0$. Then it follows from this proposition that every differential which vanishes at $\bar{Q}$ must vanish at least 2 times. Thus, $\bar{Y}$ is hyperelliptic and $\bar{Q}$ is a hyperelliptic branch point. By the argument in the proof of Theorem 5.5 of [**C85**], this is impossible, so also in this case we may conclude that $\bar{Q}$ does not lift to an unramified torsion point. Now (c) follows from (a) and the proof of Theorem 5.5 of [**C85**]. Finally, if $v(Q) = 1/2$, then $M_{\bar{Q}}(H) = 0$, so $\bar{Q}$ does lift to an unramified torsion point on J. By (c), $\bar{Q}$ is a zero of a function on $\bar{Y}$ of degree 8 and by Corollary 3.7 there are 0 or 2 such Q's in any residue class. □

NOTE. Theorem 19 of [**C87**] has the appealing corollary:

PROPOSITION 3.9. *Let C be a smooth curve of genus $g > 1$ with good reduction, embedded in its Jacobian J over R. If $p > 2g$ and P is a non-trivial torsion point in the kernel of reduction of $J(\mathbf{C}_p)$, then*

$$C(\mathbf{C}_p) \cap (P + J(K)) = \emptyset.$$

This is, in fact, equivalent to the theorem when the crystalline cohomology of C has no unit root part.

Now fix p and let

$$H_1 = \bigcap_{n \geq 0} T_p^n H^1_{\mathrm{DR}}(X_0(N), \mathbf{Z}_p), \quad H_2 = \{ h \in H^1_{\mathrm{DR}}(X_0(N), \mathbf{Z}_p) \mid \lim_{n \to \infty} T_p^n h = 0 \}.$$

Set

$$W_1 = H_1 \cap H^0(X_0(N), \Omega^1_{X_0(N)/\mathbf{Z}_p}), \quad W_2 = H_2 \cap H^0(X_0(N), \Omega^1_{X_0(N)/\mathbf{Z}_p}).$$

If I is an ideal of $\mathbf{T}$, we let $\mathbf{T}_I$ denote the completion of $\mathbf{T}$ at I. We write $\mathbf{T}_p$ for $\mathbf{T}_{(p)}$.

PROPOSITION 3.10. *Assume that $p \neq N$. Then the pairs $\mathcal{H}_1 := (H_1, W_1)$ and $\mathcal{H}_2 := (H_2, W_2)$ are de Rham F-T-crystals, $\mathcal{H}_1$ is ordinary and*

$$H_1 + H_2 = H^1_{\mathrm{DR}}(X_0(N), \mathbf{Z}_p)$$
$$W_1 + W_2 = H^0(X_0(N), \Omega^1_{X_0(N)/\mathbf{Z}_p}).$$

Further, suppose that $\mathbf{T}$ is unramified at p, or more generally that T_p is divisible by p in $\mathbf{T}_\mathfrak{m}$ for each maximal ideal $\mathfrak{m}$ of $\mathbf{T}$ such that $\mathfrak{m}|p$ and $T_p \in \mathfrak{m}$. Then $\mathcal{H}_2$ is superspecial.

PROOF. Let V and F be the Verschiebung and Frobenius endomorphisms acting on $H^1_{\mathrm{DR}}(X_0(N), \mathbf{Z}_p)$. Since these endomorphisms commute with T_p, it follows immediately that (H_i, W_i) is a sub-F-T-crystal. Since $T_p = V + F$ we see that H_1 is ordinary. Because $\mathbf{T}_p$ is Gorenstein, we have an isomorphism

$$W_2 \approx \oplus \mathbf{T}_{\mathfrak{m}}$$

of $\mathbf{T}_p$-modules in which the sum runs over those maximal ideals $\mathfrak{m}$ of $\mathbf{T}$ which lie over p and contain T_p. Since $FW_2 \equiv 0 \mod p$, it follows from the hypotheses of the proposition that $VW_2 \equiv 0 \mod p$. The fact that T_p is self-adjoint and W_2 is self-orthogonal with respect to the cup product implies that $h := \mathrm{rk}(H_2) = 2\mathrm{rk}(W_2)$. We also know $VH_2 = W_2 + pH_2$. Let $\bar{V}$ and $\bar{F}$ be the reductions of V and F mod p, $\mathrm{rk}\bar{V} = h/2$. Thus the sequence

$$0 \to W_2/pW_2 \to H_2/pH_2 \xrightarrow{\bar{V}} W_2/pW_2 \to 0$$

is exact. Together with the equation $VF = p$, this implies that $FH_2 \subseteq W_2 + pH_2$ and $\mathrm{rk}\bar{F} = h - \mathrm{rk}\bar{V} = h/2$. We conclude that $FH_2 = W_2 + pH_2$, i.e., that H_2 is superspecial. □

REMARKS. (i) It is easy to see that H_2 is not superspecial if $\mathbf{T}$ does not satisfy the hypothesis of the proposition which concerns local divisibilities of T_p by p. (ii) This hypothesis fails to be satisfied in the particular case $N = 37$, $p = 2$. In this case T_2 lies in the unique maximal ideal $\mathfrak{m}$ of $\mathbf{T}$ which divides 2, but T_2 does not lie in $2\mathbf{T}_{\mathfrak{m}}$.

LEMMA 3.11. *Suppose that ω is an element of W_2 and that P lies in $J_0(N)(\mathbf{F})$. Then we have $(\omega, P) = 0$. In particular, $M_P(H_2) = 0$.*

PROOF. This follows from the identity

$$(\omega, T_p Q) = (T_p\omega, Q),$$

for $\omega \in \Omega^1_{J_0(N)/R^{ur}}(J_0(N))$ and $Q \in J_0(N)(\mathbf{F})$, which is evident from the definitions and the fact that the T_p acts bijectively on the p-power torsion points of $J_0(N)(\mathbf{F})$. □

PROPOSITION 3.12. *Suppose $p \geq 5$, $p \neq N$, H_2 is superspecial and $Q \in \mathcal{S}$. Then the extension $\mathbf{Q}(Q)$ of $\mathbf{Q}$ is unramified above p.*

PROOF. We may work locally at p. Let P be an element of $X_0(N)(\mathbf{F})$ and let T be a uniformizing parameter on $U(P)$ defined over R^{ur}. Let $H_i^{\mathrm{ur}} = H_i \otimes R^{\mathrm{ur}}$. Suppose $Q \in U(P)$ and Q is ramified. For $k \in \{1, 2\}$, let $M_k = M_P(H_k^{\mathrm{ur}})$ and $q_k = p^k$. Note that $M_k = 0$ for $k = 2$ by the Lemma above. Then since P is not a base point for $H^0(X_0(N), \Omega^1_{X_0(N)/\mathbf{F}})$, it is not a base point for $W_j \otimes \mathbf{F}$ for some $j \in \{1, 2\}$. Suppose $\{1, 2\} = \{i, j\}$. Then $k_P(H_j^{\mathrm{ur}}) = 1$, and Proposition 3.6, Proposition 3.5 and the Lemma above imply that either (a) $j = 1$, $M_1 > 0$, $v(Q) = 1/p^{M_1}$ and every differential in $W_1 \otimes \mathbf{F}$ which vanishes at P vanishes $p - 2$ times or (b) $v(Q) = 1/q_j^n(q_j - 1)$ where $n \geq M_j$ and all differentials in $W_j \otimes \mathbf{F}$ which vanish at P vanish at least $q_j - 2$ times (or $q_j - 1$ times if $n > M_j$). If $k_P(H_i^{\mathrm{ur}}) \geq p$, every differential in $W_i \otimes \mathbf{F}$ vanishes at least $p - 1$ times at P. If $1 \leq k_P(H_i^{\mathrm{ur}}) < p$ then Proposition 3.5 applies, but none of its conclusions are consistent with what we have concluded above. Thus every differential in $W_i \otimes \mathbf{F}$ vanishes at least $p - 1$ times at P.

Thus every differential in $H^0(X_0(N), \Omega^1_{X_0(N)/\mathbf{F}})$ which vanishes at P vanishes at least $p-2$ times. This is impossible for $p \geq 5$ since $g > 1$. $\square$

In view of Theorem 2.2, Proposition 3.12 completes the proof of Theorem 1.3 for primes p which are different from N. To prove Theorem 1.3 we need only prove that $P_N \notin C$ implies $P - P_N \notin C$, when $\mathbf{T}$ is unramified at N. This question will be addressed below.

The proof of Proposition 3.12 yields:

COROLLARY 3.13. *Suppose $p = 2$ or 3 and $\mathbf{Q}(Q)$ is ramified at some prime $\mathfrak{p}$ above p. Suppose that $\{1,2\} = \{i,j\}$ and the reduction P of Q modulo $\mathfrak{p}$ is not a base point for W_j. Then P is a base point for W_i. Moreover, if $p = 3$ every non-zero differential in W_i must vanish at least two times at P and if at a prime $\mathfrak{p}$ above 3 $j = 2$ or $v(Q) < \left(p^{M_1}(p-1)\right)^{-1}$ and $j = 1$ then $X_0(N) \mod 3$ is hyperelliptic and P is a hyperelliptic branch point.*

We next prove:

PROPOSITION 3.14. *Suppose $Q \in \mathcal{S}$ and N divides the order of Q. Then for all embeddings $\overline{\mathbf{Q}}$ into $\overline{\mathbf{Q}}_N$, the image of Q in $X_0(N)(\overline{\mathbf{Q}}_N)$ reduces to a singular point of the reduction of the Deligne-Rapoport model* [**DR**] *of $X_0(N)$ over $\mathbf{Q}_N$.*

PROOF. Let $\mathcal{N}$ be the Néron model of $J_0(N)$ and let $\mathcal{X}$ be the complement of the singular locus in the minimal regular model of $X_0(N)$. The Néron property of $\mathcal{N}$ implies that the map from $X_0(N)$ to $J_0(N)$ extends to a map from $\mathcal{X}$ to $\mathcal{N}$. The reduction of $\mathcal{X}$ differs from the non-singular locus of the reduction of Deligne-Rapoport model only by at most three components, which correspond to supersingular elliptic curves with complex multiplication by the fourth or third roots of unity. (See [**M77**, Appendix, §1].)

Let M be the $\mathbf{Z}_N$-submodule of $H^0(X_0(N), \Omega^1_{X_0(N)/\mathbf{Q}_N})$ consisting of differentials which correspond to weight-two cusp forms with integral q-expansions at the two cusps of $X_0(N)$. Let D be the special fiber of the Deligne-Rapoport model of $X_0(N)$ over $\mathbf{Z}_N$, and let $X^\times$ represent the irreducible component of D which contains the cusp C_∞, endowed with log-structure at the supersingular points. (See [**Kato**].) Let W_∞ and W_0 be the wide opens in $X_0(N)$ whose points reduce to the irreducible components of D containing C_∞ and C_0. (See [**C89**].) Then the natural map from M into $H^1_{\mathrm{DR}}(W_\infty, \mathbf{Q}_N)$ is an injection. Its image is $H := H^1_{\mathrm{cris}}(X^\times, \mathbf{Z}_N)$. (See [**C84**, §3].) Indeed, let $\mathcal{Y}_\infty$ and $\mathcal{Y}_0$ be the connected components of C_∞ and C_0 in the formal completion of $\mathcal{X}$ along its reduction. Then $\mathcal{Y}_\infty \otimes \mathbf{Q}_N$ may be identified with the underlying affinoid (ibid.) of W_∞, $W_\infty \setminus W_0$, and the natural map of $\mathcal{Y}_\infty$ into the formal completion of $\mathcal{N}$ along its reduction takes $\mathcal{Y}_\infty$ into the connected component of the origin, $\mathcal{N}^0$. Further, the map $\bar{\mathcal{Y}}_\infty \to \bar{\mathcal{N}}^0$ is an Albanese morphism into the generalized Jacobian of X relative to the modulus SS of supersingular points ($\bar{\mathcal{Y}}_\infty$ may naturally be identified with $X - \mathrm{SS}$). (See also [**Gr**, §8] and Theorem 3.1 and Corollary A2.3 of [**C84**].) Now H has natural endomorphisms F and V so that if $W = H$, the quadruple (H, W, F, V) is an ordinary F-T-crystal. It follows by a direct translation of the proof of Theorem 20(i) of [**C87**] as $g = \mathrm{rk}_{\mathbf{Z}_N}(W)$ that as soon as $g > 1$, $\sigma(\mathcal{S}) \cap \mathcal{Y}_\infty(\mathbf{C}_N) \subset \mathcal{Y}_\infty(\mathbf{Q}_N^{\mathrm{ur}})$ for any embedding σ of $\bar{\mathbf{Q}}$ into $\bar{\mathbf{Q}}_N$. Exploiting the Atkin-Lehner involution, we see similarly that if $g > 1$, $\sigma(\mathcal{S}) \cap \mathcal{Y}_0(\mathbf{C}_N) \subset \mathcal{Y}_0(\mathbf{Q}_N^{\mathrm{ur}})$ for any embedding σ of $\bar{\mathbf{Q}}$ into $\bar{\mathbf{Q}}_N$.

The proposition follows since every torsion point of order N is ramified at all primes above N in view of Theorem 2.2 and the fact that the order of cuspidal group is prime to N. □

Our aim now is to prove the following result, which will complete the proof of Theorem 1.3:

THEOREM 3.15. *Suppose N does not divide the discriminant of the Hecke algebra $\mathbf{T}$. Let P be an element of $\mathcal{S}$, written as usual as the sum $\sum P_p$. Assume that $P_p \in C$ for all $p \neq N$. Then $P_N = 0$.*

Our proof of Theorem 3.15 exploits a recent theorem of B. Edixhoven [**E**, Thm. 8.2] which concerns the special fibre $\overline{\mathcal{N}}$ of $\mathcal{N}$. In its statement, the horizontal bar denotes the reduction map to characteristic N and the points 0 and ∞ are the cusps C_0 and C_∞:

Edixhoven proves:

THEOREM 3.16. *Let $\overline{C}$ be the image of C in $\overline{\mathcal{N}}$.* Then $\overline{C} \cap \overline{\mathcal{X}} = \{\overline{0}, \overline{\infty}\}$.

We recall also a simple result from [**Ri**] which concerns the action of $\mathrm{Gal}(\overline{\mathbf{Q}}/\mathbf{Q})$ on the ℓ-power torsion points of $J = J_0(N)$, i.e., the ℓ-adic representation of $\mathrm{Gal}(\overline{\mathbf{Q}}/\mathbf{Q})$ attached to J. This result has been proved only under the hypothesis that ℓ is different from 2 and 3 and is prime to the discriminant of $\mathbf{T}$. We apply it in the case $\ell = N$, which explains the presence of the discriminant hypothesis in the statement of Theorem 3.15. When the hypothesis is satisfied, $\mathbf{T}_N = \mathbf{T} \otimes \mathbf{Z}_N$ is the product of discrete valuation rings; thus, Mazur's theorem [**M77**, Ch. II, §14–§15] to the effect that the N-adic Tate module $\mathrm{Ta}_N(J)$ of J is free of rank 2 over $\mathbf{T}_N$ is especially easy to prove. Choosing a basis for $\mathrm{Ta}_N(J)$ over $\mathbf{T}_N$, we view the N-adic representation of $\mathrm{Gal}(\overline{\mathbf{Q}}/\mathbf{Q})$ defined by J as a continuous homomorphism

$$\rho_N \colon \mathrm{Gal}(\overline{\mathbf{Q}}/\mathbf{Q}) \to \mathbf{GL}(2, \mathbf{T}_N).$$

We obtain from [**Ri**]:

PROPOSITION 3.17. *Suppose that N is prime to the discriminant of $\mathbf{T}$. Then the image of ρ_N contains the group $\mathbf{SL}(2, \mathbf{T}_N)$.*

Here is our proof of Theorem 3.15:

PROOF. Using Proposition 3.17, one shows that there is a $\sigma \in \mathrm{Gal}(\overline{\mathbf{Q}}/\mathbf{Q})$ such that $\sigma(P_N)$ lies in the kernel of the reduction map to characteristic N. Replacement of P by $\sigma(P)$ preserves the hypothesis that $P_p \in C$ for all $p \neq N$ and does not affect the question of whether P_N is non-zero. We may thus assume that P and $P - P_N$ have the same image in characteristic N. It follows that P is a point on $\mathcal{N}^{\mathrm{rig}}$, the generic fiber of the formal completion of $\mathcal{N}$ along its reduction. The connected components of this rigid space are in one-to-one correspondence with the connected components of the reduction of the Néron model, which in turn are in one-to-one correspondence, via reduction, with the elements of C. Thus the reduction of P is a point in $\overline{C} \cap \overline{\mathcal{X}}$. Now it will be shown in [**C**] that the intersection of $\mathcal{N}^{\mathrm{rig}}$ with $X_0(N)$ is the rigid space associated to the formal completion of the minimal model of this curve along its reduction This is also a consequence of general results proven by Edixhoven [**E**, §9] which assert, in particular, that the morphism from the non-singular locus of the minimal model of a curve into the Néron model of its Jacobian is a closed immersion when the curve has genus at least 1 and the reduction of the minimal model is semi-stable and contains no disconnecting double points. Thus it follows from Edixhoven's theorem that $P - P_N$ is either 0 or ∞. But now Proposition 3.14 implies that $P_N = 0$. $\square$

4. Lang's approach

We proceed in the spirit of Lang [**L65**]:

LEMMA 4.1. *Suppose X is a curve of genus $g > 1$ embedded in its Jacobian and n an integer such that $|n| > 1$. Then $\#(nX \cap X) \leq gn^2$.*

PROOF. Let $W_n = X + X + \cdots + X$, where the sum is over $|n|$ copies of X. If $a \in W_{g-2}$, then $X \subseteq W_{g-1} - a$. It follows from Lemma 5.4 of [**C85**] that there exists an $a \in W_{g-2}$ such that $(W_{g-1} - a) \cap nX$ is finite. (Note that condition (ii) in the proof of this lemma should be replaced by the condition: D is not special.) Hence

$$\#(X \cap nX) \leq \deg(W_{g-1} \cdot nX),$$

where $(\ \cdot\ )$ is the intersection pairing. By Theorem 5 of [**L83**, Ch. 4, §3],

$$\deg(W_{g-1} \cdot n(X)) = \deg\left(n^{-1}(W_{g-1}) \cdot X\right).$$

We see from the discussion following Proposition 4 of §3 that $n^{-1}(W_{g-1})$ is numerically equivalent to $n^2 \cdot W_{g-1}$. The result now follows from the fact that $\deg(W_{g-1} \cdot X) = g$. (Proof: We actually compute $W_{g-1} \cdot -X$. Suppose the divisor $Q_1 + \cdots + Q_g$ is not special. Let A be the point on J equal to $\sum Q_i$. Then if $P \in X$ and $-P$ lies in $W_{g-1} - A$, we see that $P = Q_i$ for some i.) $\square$

PROPOSITION 4.2. *Suppose X and the embedding into its Jacobian are defined over $\mathbf{Q}$ and there exists a $\sigma \in \mathrm{Gal}$ which acts on $J[M](\overline{\mathbf{Q}})$ via the homothety n, where n is an integer satisfying $|n| > 1$. Then*

$$\#\left(X(\overline{\mathbf{Q}}) \cap J[M](\overline{\mathbf{Q}})\right) \leq gn^2.$$

PROOF. The proposition follows immediately from the lemma, since the indicated intersection lies both in X and nX. $\square$

Proposition 4.2 and the results of [**C85**] can be combined to deduce consequences in other contexts. For example, let p be a prime, and for each triple of integers (a, b, c) with abc prime to p, let

$$F^p_{a,b,c} : y^p = (-1)^c x^a (1-x)^b$$

denote the indicated quotient of the degree-p Fermat curve $F^p : X^p + Y^p + Z^p = 0$. (See, e.g., [**CM**, §3] for background on Fermat quotients.) The cusps on F^p are the images of points on the Fermat curve for which one of the projective coordinates X, Y, Z is 0; the cusps on $F^p_{a,b,c}$ are the points where the coordinate z is 0, 1 or ∞. According to well-known results of Rohrlich [**Ro**], the cusps of F^p lie in a common torsion packet; i.e., their differences form degree-zero divisors whose images on the Jacobian of F_p are torsion. The corresponding statement then holds for the quotients $F^p_{a,b,c}$.

To prove statements of this sort, it has proved helpful to show first that $\mathcal{S}$ has small cardinality. As an application of Proposition 4.2, we obtain:

PROPOSITION 4.3. *Assume that $F^p_{a,b,c}$ is not hyperelliptic and that $p \geq 5$. Then the cuspidal torsion packet of $F^p_{a,b,c}$ has at most $(p-1)/2 \cdot (1+p)^2$ elements.*

Suppose now that p is a prime ≥ 5. Then Coleman, Tamagawa and Tzermias [**CTT**] have shown, using the proposition and results of Coleman [**C89b**] and Greenberg [**G**], that the cusps on F^p form a complete torsion packet.

5. Galois action on torsion points of $J_0(37)$

We next present some results of [**K**] about the action of $\mathrm{Gal}(\overline{\mathbf{Q}}/\mathbf{Q})$ on the group of torsion points of $J_0(37)$.

The space of weight-two cusp forms on $\Gamma_0(37)$ is two dimensional; it is spanned by the eigenforms with integral coefficients which arise from the two isogeny classes of rational elliptic curves with conductor 37. The action of the Hecke ring $\mathbf{T}$ on these forms induces an inclusion $\mathbf{T} \subseteq \mathbf{Z} \times \mathbf{Z}$. It is well known that the image of $\mathbf{T}$ in $\mathbf{Z} \times \mathbf{Z}$ is the order consisting of pairs of integers (a, b) which satisfy $a \equiv b \bmod 2$. Using the results of [**M77**, Ch. II], one sees that the adelic Tate module

$$\mathrm{Ta_f} J := \varprojlim J[n]$$

of $J = J_0(37)$ is free of rank 2 over $\mathbf{T} \otimes \hat{\mathbf{Z}}$. Equivalently: for each prime p, the p-adic Tate module $\mathrm{Ta}_p J = \mathrm{Ta}_p\left(J(\overline{\mathbf{Q}})\right)$ is free of rank two over the ring $\mathbf{T}_p := \mathbf{T} \otimes \mathbf{Z}_p$.

The image G of $\mathrm{Gal}(\overline{\mathbf{Q}}/\mathbf{Q})$ in $\mathrm{Aut} J(\overline{\mathbf{Q}})_{\mathrm{tors}}$ is the image of $\mathrm{Gal}(\overline{\mathbf{Q}}/\mathbf{Q})$ in

$$\mathrm{Aut}_{\mathbf{T} \otimes \hat{\mathbf{Z}}} \mathrm{Ta_f} \approx \mathbf{GL}(2, \mathbf{T} \otimes \hat{\mathbf{Z}}).$$

Since the determinant of the representation $\mathrm{Gal}(\overline{\mathbf{Q}}/\mathbf{Q}) \to \mathbf{GL}(2, \mathbf{T} \otimes \hat{\mathbf{Z}})$ is the adelic cyclotomic character $\mathrm{Gal}(\overline{\mathbf{Q}}/\mathbf{Q}) \to \hat{\mathbf{Z}}^*$, the image of this representation lies in the group

$$A := \{\, M \in \mathbf{GL}(2, \mathbf{T} \otimes \hat{\mathbf{Z}}) \mid \det M \in \hat{\mathbf{Z}}^* \,\}.$$

In other words, we have

$$G \subseteq A \subset \mathbf{GL}(2, \mathbf{T} \otimes \hat{\mathbf{Z}}).$$

For each prime p, let G_p and A_p be the images of G and A in the group $\mathbf{GL}(2, \mathbf{T}_p)$. One has $G \subseteq \prod_p G_p$. More generally, if S is a set of primes (possibly infinite), let G_S be the image of G in $\mathbf{GL}(2, \prod_{p \in S} \mathbf{T}_p)$, i.e., in the product $\prod_{p \in S} G_p$. Further, let G^S be the image of G in $\prod_{p \notin S} G_p$, and make similar definitions with G replaced by A. For each S, one has trivially $A = A_S \times A^S$, but the analogous decomposition for G is by no means automatic. If $p_1, \ldots, p_n$ are distinct primes, we sometimes write $G(p_1, \ldots, p_n)$ in place of $G_{\{p_1, \ldots, p_n\}}$ and make similar use of $A(p_1, \ldots, p_n)$.

The main result of [**K**] is a description of G as a subgroup of A.

PROPOSITION 5.1. *For all $p \geq 5$, one has*

$$G_p = A_p = \{\, M \in \mathbf{GL}(2, \mathbf{Z}_p \times \mathbf{Z}_p) \mid \det M \in \mathbf{Z}_p^* \,\}.$$

Further, let S be the set $\{2, 3, 37\}$. The natural inclusion $G \hookrightarrow G_S \times G^S$ is an isomorphism.

In the displayed equation, the second equality, which asserts that A_p is a $\mathbf{GL}(2)$, results from the circumstance that $2 = (\mathbf{Z} \times \mathbf{Z} \colon \mathbf{T})$ is prime to p. The first equality is not too difficult to establish, even in the more general case where 37 is replaced by an arbitrary prime number [**Ri**]. The equation $G = G_S \times G^S$ may be summarized by the statement that each group G_p with $p \neq 2, 3, 37$ is a subgroup of G. One can prove it by the methods of Serre [**S**, §6.2].

Proposition 5.1 reduces the determination of G to a determination of its quotient G_S, where $S = \{2, 3, 37\}$. On the other hand, it is the quotient G_S, as opposed to the full group G, which intervenes in our proof of Theorem 1.4. The group G_S is a subgroup of $G_2 \times G_3 \times G_{37}$ which turns out to have index 3 in the product $G_2 \times G_3 \times G_{37}$. In order to describe G_S, it is convenient to discuss first the individual factors G_2, G_3, and G_{37}. The last of the three factors has already been determined: it coincides with its overgroup A_{37}. (One can say that G_{37} is "as large as possible.") The following consequence of this fact will be useful below:

COROLLARY 5.2. *Let P be a torsion point on $J_0(37)$ whose order is divisible by 37. Then P has at least $37^2 - 1 = 1368$ conjugates.*

We next describe G_2. To do this, we shall introduce a certain homomorphism

$$\theta\colon A_2 \to \mathbf{F}_2.$$

View A_2 as the group

$$\{\, (\alpha, \beta) \in \mathbf{GL}(2, \mathbf{Z}_2) \times \mathbf{GL}(2, \mathbf{Z}_2) \mid \det(\alpha) = \det(\beta) \text{ and } \alpha \equiv \beta \mod 2 \,\}.$$

Let $\mathfrak{sl}_2(\mathbf{F}_2)$ be the lie algebra of 2×2 matrices over $\mathbf{F}_2$ with trace 0. The adjoint action $(\alpha, \beta) : M \mapsto aMa^{-1}$, where $a = \alpha \bmod 2 = \beta \bmod 2$, makes $\mathfrak{sl}_2(\mathbf{F}_2)$ into an A_2-module. We regard $\mathbf{F}_2$ as an A_2-module with trivial action. One may verify explicitly that the map

$$\mathfrak{sl}_2(\mathbf{F}_2) \to \mathbf{F}_2, \qquad \begin{pmatrix} a & b \\ c & a \end{pmatrix} \longmapsto a + b + c$$

is a non-zero map of A_2-modules. Meanwhile, it is well known, and not hard to check by direct calculation, that the association

$$A_2 \to \mathfrak{sl}_2(\mathbf{F}_2), \qquad (\alpha, \beta) \mapsto \frac{\alpha\beta^{-1} - I}{2} \mod 2$$

is a 1-cocycle on A_2 with values in $\mathfrak{sl}_2(\mathbf{F}_2)$. On composing this cocycle with the map $\mathfrak{sl}_2(\mathbf{F}_2) \to \mathbf{F}_2$ we have exhibited, we obtain the desired homomorphism θ.

For later use, we define a second non-trivial homomorphism

$$\epsilon\colon A_2 \to \mathbf{F}_2$$

as the composite of the surjection $A_2 \to \mathbf{GL}(2, \mathbf{F}_2)$ defined by $(\alpha, \beta) \mapsto \alpha \bmod 2$ and the unique non-zero homomorphism $\mathbf{GL}(2, \mathbf{F}_2) \to \mathbf{F}_2$. This latter map is the "sign" function $S_3 \to \{\pm 1\}$, viewed as taking values in the additive group $\mathbf{F}_2$. It is clear that ϵ is non-trivial on G_2, since G_2 maps onto $\mathbf{GL}(2, \mathbf{F}_2)$.

PROPOSITION 5.3. *The group G_2 is the kernel of $\theta\colon A_2 \to \mathbf{F}_2$.*

For the proof of this Proposition, see [**K**]. In what follows, we we will present a concrete explanation for the inclusion $G_2 \subseteq \ker \theta$.

View G_2 as the Galois group of the extension K of $\mathbf{Q}$ gotten by adjoining all 2-power division points on the two strong modular elliptic curves of conductor 37. Call these curves E and E'; order them so that the discriminants of E and E' are $\Delta = 37$ and $\Delta' = 37^3$, respectively. The field K contains $\mathbf{Q}(\sqrt{\Delta}) = \mathbf{Q}(\sqrt{\Delta'})$: this field is the quadratic subfield of the S_3-extension of $\mathbf{Q}$ gotten by adjoining the 2-division points on E or E'. On the other hand, K contains $\mathbf{Q}(\sqrt{-1}, \sqrt[4]{\Delta})$ and $\mathbf{Q}(\sqrt{-1}, \sqrt[4]{\Delta'})$: these are subfields of the field of 4-division points of $E \times E'$. In particular, K contains the biquadratic field $\mathbf{Q}(\sqrt{-1}, \sqrt[4]{\Delta'/\Delta})$. Here we note that Δ'/Δ is a perfect square because of the isomorphism $E[2] \approx E'[2]$, so that $\sqrt[4]{\Delta'/\Delta}$ is either rational or a quadratic irrationality. Thus the subfield $\mathbf{Q}(\sqrt{-1}, \sqrt[4]{\Delta'/\Delta}, \sqrt{\Delta})$ of K is a $(2, \ldots, 2)$-extension of $\mathbf{Q}$ whose degree over $\mathbf{Q}$ could a priori be as large as 8. In fact, however, the quantities $\sqrt[4]{\Delta'/\Delta}$ and $\sqrt{\Delta}$ are equal, so that the degree of this field over $\mathbf{Q}$ is at most 4.

COROLLARY 5.4. *Let P be a non-zero point on $J_0(37)$ of 2-power order. Then there is a $g \in G_2$ such that $\epsilon(g) = 0$ and such that $gP - P$ has order 2.*

This result may be deduced from the equality $G_2 = \ker\theta$ by straightforward calculation. Alternatively, we can proceed somewhat more intrinsically as in the proof of Theorem 2.4, considering the action of an inertia subgroup I of $\mathrm{Gal}(\overline{\mathbf{Q}}/\mathbf{Q})$ for the prime 37. Composing ϵ with the projection $\mathrm{Gal}(\overline{\mathbf{Q}}/\mathbf{Q}) \to G_2$, we will regard ϵ as a map $\mathrm{Gal}(\overline{\mathbf{Q}}/\mathbf{Q}) \to \mathbf{F}_2$. We must find a $g \in \mathrm{Gal}(\overline{\mathbf{Q}}/\mathbf{Q})$ in the kernel of ϵ with the property that $gP - P$ has order two.

Suppose first that P itself has order 2. Then it suffices to find a g in the kernel of ϵ which does not fix P. For this, it is helpful to recall the structure of $J_0(37)[2]$ as a $\mathrm{Gal}(\overline{\mathbf{Q}}/\mathbf{Q})$-module. Let $J = J_0(37)$ and let $\mathfrak{m}$ be the unique maximal ideal of $\mathbf{T}$ with residue characteristic 2. The kernel $J[\mathfrak{m}]$ is isomorphic to $E[2]$ (and also to $E'[2]$), and the quotient $J[2]/J[\mathfrak{m}]$ is again isomorphic to $J[\mathfrak{m}]$ (cf. [**M77**, p. 112]). The group $\mathrm{Gal}(\overline{\mathbf{Q}}/\mathbf{Q})$ acts on $E[2]$ via its quotient $\mathbf{GL}(2, \mathbf{F}_2)$, which is the symmetric group on three letters. The image in $\mathrm{Aut}E[2]$ of the kernel of ϵ is the alternating group on three letters, which permutes transitively the three non-zero elements of $E[2]$. An element g in the kernel of ϵ which maps non-trivially to this alternating group can fix no non-zero element of $J[2]$. Indeed, such elements g fix no non-zero elements of the quotient $E[2]$ and no non-zero elements of the submodule $E[2]$.

Suppose now that P has order 2^n, with $n \geq 2$. Then $Q = 2^{n-1}P$ is a point of $J_0(37)$ of order 2. As we have noted, the action of $\mathrm{Gal}(\overline{\mathbf{Q}}/\mathbf{Q})$ on $E[2]$ cuts out an S_3-extension of $\mathbf{Q}$. This extension is ramified at 37; indeed, it contains the quadratic field $\mathbf{Q}(\sqrt{37})$. For each non-zero R in $E[2]$, we may find an inertia subgroup $I \subset \mathrm{Gal}(\overline{\mathbf{Q}}/\mathbf{Q})$ for the prime 37 and a $\sigma \in I$ such that $\sigma R \neq R$. (Indeed, the image of each I in S_3 has order 2; some conjugate of this element will move R.) Using again that $J[2]$ is an extension of $E[2]$ by $E[2]$, we may find an inertia group I for 37 and a $\sigma \in I$ such that $\sigma Q \neq Q$. Now consider the action of σ on $J_0(37)[2^n]$. As in the proof of Theorem 2.4, σ acts as the sum $1 + A$, where 1 represents the identity operator and A an endomorphism satisfying $A^2 = 0$. Since

$$\sigma Q = Q + A\, 2^{n-1}P,$$

the quantity $2^{n-1}AP$ has order 2. Also

$$\sigma^{2^{n-1}}P = (1 + A)^{2^{n-1}}P = P + 2^{n-1}AP$$

by the Binomial Theorem. It thus suffices to take $g = \sigma^{2^{n-1}}$: this element is automatically in the kernel of ϵ because it is the square of an element of $\mathrm{Gal}(\overline{\mathbf{Q}}/\mathbf{Q})$.

We now describe

$$G_3 = \mathrm{Gal}\left(\mathbf{Q}(E[3^\infty], E'[3^\infty])/\mathbf{Q}\right) \hookrightarrow \mathrm{Aut}(\mathrm{Ta}_3 E) \times \mathrm{Aut}(\mathrm{Ta}_3 E').$$

As noted by Serre [**S**, 5.5.8], G_3 maps onto the first factor $\mathrm{Aut}(\mathrm{Ta}_3 E)$. Let X be the image of G_3 in the second factor $\mathrm{Aut}(\mathrm{Ta}_3 E') \approx \mathbf{GL}(2, \mathbf{Z}_3)$. Since $E'[3]$ may be regarded as $J_0(37)[\mathfrak{m}]$, where $\mathfrak{m}$ is the unique Eisenstein prime of $\mathbf{T}$, the module $E'[3]$ is isomorphic to the direct sum $\mu_3 \oplus \mathbf{Z}/3\mathbf{Z}$. In particular, the image of X

mod 3 has order 2. Let $\overline{X}$ be the image of X in $\mathbf{GL}(2, \mathbf{Z}/9\mathbf{Z})$, i.e., in $\mathrm{Aut}E[3^2]$. Then the order of $\overline{X}$ is a divisor of $2 \cdot 81$ which is divisible by 2. According to [**K**], $\overline{X}$ has order $54 = 2 \cdot 27$; thus, $\overline{X}$ is of index 3 in the inverse image in $\mathbf{GL}(2, \mathbf{Z}/9\mathbf{Z})$ of the image of $\overline{X}$ mod 3. On the other hand, X is the full inverse image of $\overline{X}$ in $\mathrm{Aut}(\mathrm{Ta}_3 E') \approx \mathbf{GL}(2, \mathbf{Z}_3)$. Concretely, with respect to a suitable basis of $\mathrm{Ta}_3 E'$ one has:

$$X = \left\{ \begin{pmatrix} 1+3e & 3f \\ 3g & h \end{pmatrix} \in \mathbf{GL}(2, \mathbf{Z}_3) \;\middle|\; g \equiv -fh \mod 3 \right\}.$$

For later use, we introduce the surjective homomorphism

$$X \to \mathbf{Z}/9\mathbf{Z}, \qquad \begin{pmatrix} 1+3e & 3f \\ 3g & h \end{pmatrix} \mapsto e + 3(e^2 - g^2) \mod 9.$$

Composing this map with the projection $G_3 \to X$, we obtain a surjection

$$F\colon G_3 \to \mathbf{Z}/9\mathbf{Z}.$$

The group G_3 is contained in the group of pairs $(g, x) \in \mathbf{GL}(2, \mathbf{Z}_3) \times X$ which satisfy the condition $\det g = \det x$. As Kaskel shows in [**K**], G_3 has index 3 in this latter group; this additional constraint on the elements of G_3 may be traced to the fact that both $\mathbf{Q}(E[3^\infty])$ and $\mathbf{Q}(E'[3^\infty])$ contain the field $\mathbf{Q}(\sqrt[3]{37})$.

PROPOSITION 5.5. *With respect to suitable bases, G_3 is the group of pairs*

$$\left(\begin{pmatrix} a & b \\ c & d \end{pmatrix}, \begin{pmatrix} 1+3e & 3f \\ 3g & h \end{pmatrix} \right) \in \mathbf{GL}(2, \mathbf{Z}_3) \times X$$

which satisfy the determinant condition $ad - bc = (1+3e)h - 9fg$ and the supplementary condition

$$(ac + bd)(c^2 + d^2) \equiv g \equiv -fh \mod 3.$$

COROLLARY 5.6. *The homothety -2 belongs to G_3.*

We now state without proof a description of the full Galois group G. This description involves the homomorphisms $\epsilon\colon G_2 \to \mathbf{Z}/2\mathbf{Z}$ and $F\colon G_3 \to \mathbf{Z}/9\mathbf{Z}$ which were introduced above.

THEOREM 5.7. *The group G is the subgroup of $\prod_\ell G_\ell$ consisting of all elements $(\dots, g_\ell, \dots)$ which are related by the identity*

$$\det(g_{37})^2 \equiv 2^{20F(g_3)+18\epsilon(g_2)} \mod 37$$

in $\mathbf{F}_{37}^*$.

The reader will remark that the groups $G_{\{2,3,37\}}$ and $G^{\{2,3,37\}}$ are "independent" in the sense that G is the product of these two factors of G; we have already noted this point above. In the crucial quotient $G_{\{2,3,37\}}$, the two factors G_2 and G_3 are independent of each other: the group $G_{\{2,3\}}$ is the product of its quotients G_2 and G_3. On the other hand, G_{37} is linked to each of G_2 and G_3. See [**S**, Prop. 22] for a related observation.

The following consequence of Theorem 5.7 can be seen by inspection.

COROLLARY 5.8. *The homothety -5 belongs to $G_{\{2,3,37\}}$.*

6. Torsion points on $X_0(37)$

We now prove Theorem 1.4, which states that $\mathcal{S} = X_0(37) \cap J_0(37)_{\rm tor}$ contains only the two cusps C_∞ and C_0.

Let P be an element of $\mathcal{S}$. Recall that we may decompose P as the sum of its primary components P_p. According to Corollary 2.3, all P_p are zero except perhaps for $p = 2$, $p = 3$, and $p = 37$:

$$P = P_2 + P_3 + P_{37}.$$

We may thus view $\mathcal{S}$ as a subset of $J[M]$, where M is the product of suitable powers of 2, 3, and 37. (With the appropriate definition, one can take $M = 2^\infty 3^\infty 37^\infty$.) As we have just seen, there is an element of $\mathrm{Gal}(\overline{\mathbf{Q}}/\mathbf{Q})$ which acts on $J[M]$ as the homothety -5. By Proposition 4.2 we get

$$\#\mathcal{S} \le 2(-5)^2 = 50,$$

since the genus of $X_0(37)$ is two. On the other hand, any element of $\mathcal{S}$ for which P_{37} is non-zero must have at least 1368 conjugates by Corollary 5.2. Thus each $P \in \mathcal{S}$ may be decomposed as the sum $P_2 + P_3$.

We claim next that $P_2 = 0$, i.e., that $P = P_3$. If not, there is a $g \in \mathrm{Gal}(\overline{\mathbf{Q}}/\mathbf{Q})$ such that $gP_2 - P_2$ has order 2 by Corollary 5.4. Because of the independence of G_2 and G_3, we may assume that g fixes P_3. Therefore, $gP - P$ has order two on J. As we saw in §1, it follows that P is a hyperelliptic branch point of $X_0(37)$. Indeed, the divisor $2\big((gP) - (P)\big)$ is the divisor of a non-constant function, and the point P is forced to ramify in the corresponding degree-two covering of $\mathbf{P}^1$. However, we have seen in Proposition 1.1 that the hyperelliptic branch points of $X_0(37)$ do not belong to $\mathcal{S}$.

Knowing that $\mathcal{S} \subseteq J_0(37)[3^\infty]$, we apply Corollary 5.6 and Proposition 4.2 and deduce the bound $\#\mathcal{S} \le 8$. To conclude, we consider the degree-two covering $\pi\colon X_0(37) \to E$ which is obtained by dividing $X_0(37)$ by its Atkin-Lehner involution. This covering may be viewed as the composite of the Albanese map $X_0(37) \to J_0(37)$ sending C_∞ to 0 with a homomorphism of abelian varieties $J_0(37) \to E$ whose kernel is E' (regarded as an abelian subvariety of $J_0(37)$). As we have remarked, the image of G_3 in $\mathrm{Aut}(E[3^\infty])$ is all of $\mathbf{GL}(2, \mathbf{Z}_3)$. Hence any non-zero element of $E[3^\infty]$ has at least eight conjugates. It follows that the image of $\mathcal{S}$ in $E[3^\infty]$ is the single element 0. Indeed, $\mathcal{S}$ contains the two rational points C_∞ and C_0 and consists of at most eight elements. Therefore, it can contain no point with more than six conjugates. Thus $\mathcal{S}$ is contained in the set $\pi^{-1}(\{0\})$; since π has degree two, $\mathcal{S}$ can have no more than two elements.

With an eye to possible generalizations, we present two variants of the proof of Theorem 1.4. In the first variant, we begin as before, writing $P \in \mathcal{S}$ as $P_2+P_3+P_{37}$, but noting that $P_{37} = 0$ because of Proposition 4.2. Thus $P = P_2 + P_3$.

PROPOSITION 6.1. *The point P_3 belongs to the cuspidal subgroup of $J_0(37)$.*

PROOF. We consider the two normalized newforms

$$F_1(q) = q + 0q^2 + \ q^3 + \cdots$$
$$F_2(q) = q - 2q^2 - 3q^3 + \cdots$$

of weight two on $\Gamma_0(37)$ and let ω_1 and ω_2 be the corresponding regular differentials on $X_0(37)$.

LEMMA 6.2. *If p is an odd prime, neither ω_1 nor ω_2 has a double zero modulo p.*

PROOF. Let w be the Atkin-Lehner involution of $X_0(37)$. Then

$$w^*\omega_1 = -\omega_1$$
$$w^*\omega_2 = \omega_2.$$

It follows, from this and the fact that $X_0(37)$ is hyperelliptic of genus 2, that both of these differentials are pullbacks of invariant differentials on elliptic curves by degree two morphisms. From this and the Riemann-Hurwitz theorem, the lemma follows. □

Continuing the proof of Proposition 6.1, we work in the context of §3, taking $p = 3$. After glancing at the displayed q-expansions, we see that $W_1 = \mathbf{Z}_p\omega_1$ and $W_2 = \mathbf{Z}_p\omega_2$ in that setting. In view of the Lemma we have just proved and Corollary 3.13, we find that the extension $\mathbf{Q}(P)/\mathbf{Q}$ is unramified at 3. The desired conclusion now follows from Theorem 2.2. □

Our second proof of Theorem 1.4 now concludes as follows. We claim as before that $P_2 = 0$. Indeed, if P_2 is non-zero, then we may find as before a $g \in \mathrm{Gal}(\overline{\mathbf{Q}}/\mathbf{Q})$ such that $gP_2 - P_2$ has order 2. Because P_3 lies in $J_0(37)(\mathbf{Q})$, g fixes P_3. Therefore, $gP - P$ has order two on J, and we have seen that this is not the case. Thus we have $P = P_3 \in C$, where C is the cuspidal subgroup of $J_0(37)$. However, one knows that $X_0(37) \cap C$ consists precisely of the two cusps on $X_0(37)$ [**MaS**, §5, Prop. 2].

To make a third proof, we again write $P \in \mathcal{S}$ as the sum $P_2 + P_3 + P_{37}$. As we saw in the second proof, P_3 belongs to C. Once we know that P_{37} is zero, we can conclude as before. To see that this is the case, we argue by contradiction, supposing that P_{37} is not zero. Then P_{37} cannot belong to C, a group of order three. Thus, by Theorem 1.3, P_2 must be non-cuspidal, and hence non-zero. By Theorem 5.4, there is a $g_2 \in G_2$ with $\epsilon(g_2) = 0$ such that $g_2P_2 - P_2$ has order two. In view of Theorem 5.7, there is a $g = (\dots, g_\ell, \dots) \in G$ whose component at 2 is the given g_2 and whose component at 37 is 1. We then have

$$gP - P = (g_2P_2 - P_2) + (g_3P_3 - P_3) + (P_{37} - P_{37}) = (g_2P_2 - P_2).$$

In computing $gP - P$, we have used that P_3 is rational, so that $g_3P_3 = P_3$ for any choice of g_3. (Of course, we could have taken $g_3 = 1$.) Thus $gP - P$ has order two, and we have seen that this is impossible.

References

[B] F. A. Bogomolov, *Sur l'algébricité des représentations l-adiques*, CRAS Paris, série A **290** (1980), 701–703.

[C84] R. F. Coleman, *On a p-adic inner product on elliptic modular forms*, Barsotti Symposium in Algebraic Geometry, New York: Academic Press (1984), 125–151.

[C85] R. F. Coleman, *Torsion points on curves and p-adic abelian integrals*, Annals of Math. **121** (1985), 111–168.

[C86] R. F. Coleman, *Torsion points on Fermat curves*, Compositio Math. **58** (1986), 191–208.

[C87] R. F. Coleman, *Ramified torsion points on curves*, Duke Math. J. **54** (1987), 615–640.

[C89] R. F. Coleman, *Reciprocity laws on curves*, Compositio Math. **72** (1989), 205–235.

[C89b] R. F. Coleman, *Torsion points on abelian étale coverings of* $\mathbf{P}^1 \setminus \{0, 1, \infty\}$, Trans. AMS **311** (1989), 185–208.

[C] R. Coleman, *The monodromy pairing*, to appear.

[CM] R. F. Coleman and W. McCallum, *Stable reduction of Fermat curves and Jacobi sum Hecke characters*, J. Reine Angew. Math. **385** (1988), 41–101.

[CTT] R. F. Coleman, A. Tamagawa and P. Tzermias, *The cuspidal torsion packet on Fermat curves*, Crelle **496** (1998), 73–81.

[DR] P. Deligne and M. Rapoport, *Les schémas de modules de courbes elliptiques*, Lecture Notes in Math. **349** Berlin and New York: Springer-Verlag (1973), 143–316.

[E] B. Edixhoven, *On Néron models, divisors and modular curves*, to appear.

[G] R. Greenberg, *On the Jacobian variety of some algebraic curves*, Compositio Math. **l42** (1991), 345–359.

[Gr] B. H. Gross, *A tameness criterion for Galois representations associated to modular forms* mod p, Duke Math. J. **61** (1990), 445–517.

[H] M. Hindry, *Autour d'une conjecture de Serge Lang*, Invent. Math. **94** (1988), 575–603.

[K] B. Kaskel, *The adelic representation associated to* $X_0(37)$, Ph.D. thesis, UC Berkeley (May, 1996).

[Kato] K. Kato, *Logarithmic structures of Fontaine-Illusie*, Algebra, Analysis, Geometry and Number Theory (J.-I. Igusa, ed.), Baltimore: Johns Hopkins University Press (1989), 191–224.

[Katz] N. M. Katz, *Serre-Tate local moduli*, Lecture Notes in Math. **868**, Berlin and New York: Springer-Verlag (1981), 138–202.

[L65] S. Lang, *Division points on curves*, Ann. Mat. Pura Appl. **70** (1965), 229–234.

[L83] S. Lang, Abelian Varieties, Berlin and New York: Springer-Verlag (1983).

[LT] S. Lang and H. Trotter, Frobenius Distributions in $\mathbf{GL}_2$-Extensions, Lecture Notes in Math. **504**, Berlin and New York: Springer-Verlag (1976).

[M77] B. Mazur, *Modular curves and the Eisenstein ideal*, Publ. Math. IHES **47** (1977), 33–186.

[M78] B. Mazur, *Rational isogenies of prime degree*, Invent. Math. **44** (1978), 129–162.

[MaS] B. Mazur and P. Swinnerton-Dyer, *Arithmetic of Weil curves*, Invent. Math. **25** (1974), 1–61.

[MiU] P. Michel and E. Ullmo, *Points de petite hauteur sur les courbes modulaires* $X_0(N)$, Invent. Math. **131** (1998), 645–674.

[Miy] T. Miyake, Modular Forms, Berlin and New York: Springer-Verlag (1989).

[O74a] A. Ogg, *Hyperelliptic modular curves*, Bull. Soc. Math. France **102** (1974), 449–462.

[O74b] A. Ogg, *Automorphismes de courbes modulaires*, Sém. Delange-Pisot-Poitou (1974/1975, exposé 7).

[R] M. Raynaud, *Courbes sur une variété abélienne et points de torsion*, Invent. Math. **71** (1983), 207–233.

[Ri] K. Ribet, *Images of semistable Galois representations*, Pacific J. of Math. **181** (1997), 277–297.

[Ro] D. Rohrlich, *Points at infinity on Fermat curves*, Invent. Math. **39** (1977), 95–127.

[S] J-P. Serre, *Propriétés galoisiennes des points d'ordre fini des courbes elliptiques*, Invent. Math. **15** (1972), 259–331.

[Se] J-P. Serre, Course at the Collège de France, 1985–1986.

[T] J. Tate, *The non-existence of certain Galois extensions of* **Q** *unramified outside 2*, Contemporary Mathematics **174** (1994), 153–156.

[U] E. Ullmo, *Positivité et discretion des points algébriques des courbes*, Annals of Math. **147** (1998), 167–179.

Errata for [**C87**]

This list of errata for Coleman's article "Ramified torsion points on curves" was compiled by Matt Baker:

p. 615 Conjecture B, part (ii): should read "$K/\mathbb{Q}$ is unramified at $\mathfrak{p}$."

p. 620 line 3: should read " ... we get $p^k w = F^n h_n$ for all $n \geq k$."

p. 623 fifth line from bottom: should read "$\omega_i = \frac{\phi^* \omega^\sigma_{i+1}}{p} + df_i$."

p. 623 fourth line from bottom: "Let $0 < n_0 < n_1 < \cdots$" should read "Let $0 \leq n_0 < n_1 < \cdots$."

p. 623 third line from bottom: "$\tilde{\omega_i} \in \tilde{\Omega}$" should be "$[\omega_i]^* \in \tilde{\Omega}$."

p. 624 line 2: the p^{a_i} in the denominator should be a p.

p. 624 line 3: should read "Let $h_i = g_i - g_i(Q) \in A^0(Y) \otimes R^u$."

p. 624 line 5: should read "$dL(T) = \nu$ on U."

p. 624 line 8 : should read "Let $k_i = \text{ord}_U \tilde{\nu}_i + 1$."
p. 624 Theorem 10: When $p = 2$ in part (iii), one can conclude from the proof that $i = M = 0, k_i = 1, Q_i = (1,0), Q_{i+1} = (2,-1)$.
p. 625 line 6: should read "$s(k_1 p - h) \leq (1+s)(k_0 - h)$."
p. 625 third line from bottom: p^h should be p^n.
p. 627 fifth line from bottom: replace sentence beginning "To finish ..." with "To finish the proof of the proposition, we will eliminate all but one of the above possibilities for D."
p. 628 line 7: "$n_{i+1} - n_i > 1$" should be "$n_{i+1} - n_i \geq 1$."
p. 628 line 11: v^{n_i} should be V^{n_i}.
p. 628 line 12: "for all $\eta \in \tilde{W}^*$" should be "for all $\eta \in W$."
p. 629 line 4: "$k_0\omega$)" should be "$k_0(\omega)$."
p. 630 seventh line from bottom: should read "$\omega \in \tilde{W} \mapsto (\frac{\omega}{d\tilde{T}})(U)$."
p. 630 fourth line from bottom: "let $e = e(a)$" should be "let $e = e(A)$."
p. 631 line 2: should read "$b \in U(\mathbb{C}_p)$."
p. 631 line 13: Proposition 10 should be Theorem 10.
p. 631 Prop. 15, part (ii): $q-1$ should be replaced by $q-2$.
p. 631 line 15: "when $p = 2, A = 1$" should be "when $q = 2, A \leq 1$."
p. 631 middle: "for some $g_{\omega,i}T$)" should read "for some $g_{\omega,i}(T)$."
p. 631 fifth line from bottom: replace "$g_{\omega,i}(0)$" by "$\frac{d}{dT} g_{\omega,i}(0)$."
p. 632 line 3: replace "$a_n = a^{q^n}$" by "$a_n = a^{q^{n-1}}$."
p. 632 line 4: replace "Theorem 4.2.1 of [C-1]" by "Theorem 4.1 of [C-1]."
p. 632 middle: replace "$\frac{1}{p^{n'_{i+1}} - p^{n'_j}}$" by "$\frac{1}{p^{n'_{j+1}} - p^{n'_j}}$."
p. 633 line 10: should say "$\omega \in W$."
p. 633 proof of Theorem 18: reference to Cor. 13.1 should be to Cor. 13.3.
p. 633 proof of Theorem 19: reference to Prop. 10 should be to Prop. 13.
p. 634 Theorem 20: when $p = 2$ or 3, the theorem is incorrect. The present article contains corrected version.
p. 635 proof of Theorem 20: reference to Prop. 19 should be to Prop. 15.

University of California, Berkeley, UC Mathematics Department, Berkeley, CA 94720-3840 USA
E-mail address: coleman@math.berkeley.edu

Adobe Software, Inc., San Jose, CA
E-mail address: bkaskel@adobe.com

University of California, Berkeley, UC Mathematics Department, Berkeley, CA 94720-3840 USA
E-mail address: ribet@math.berkeley.edu

Proceedings of Symposia in Pure Mathematics
Volume **66.1**, 1999

A Fundamental Group for Symbolic Dynamics I: Definitions

Paul Feit

Dedicated to Goro Shimura

ABSTRACT. In previous work, author extended Grothendieck's formulation for a fundamental group to a universal construction. The present paper applies it to a family of categories, which includes dynamical systems with discrete time variable, as well as tilings of the plane or of space. Of especial interest are algebraic situations where natural topologies are totally disconnected. The result is a new invariant, which, we hope, can add to existing structure theory. In this paper, we show that group associated to the full 2 shift (over $\mathbb{Z}$) has a homomorphic image of a semi-direct product of profinite completions.

Our primary role here is translation. Axioms for the universal construction are verified. Concrete analogues are found for general terms. Computations are elementary, but suffice to demonstrate that the new invariant is non-trivial. We shall develop computations in future papers.

Introduction

The present paper is an outgrowth of a universal formulation for construction by descent. Previous papers introduce a homotopy group which applies to a broad class of categories. The model, represented here by the functor π_G, is based on Grothendieck's formulation of a fundamental group for schemes. For that reason, we refer to it as a *fundamental group functor.* In the sequence of papers which begins here, we study the groups in specific cases.

Our intentions are two-fold.

First, analysis (to appear in later papers) suggests that the theory can unify existing observations in Combinatorics and Symbolic Dynamics. Classically, homotopy groups serve as "holding pens" for information. Although defined by general nonsense, these algebras have practical value. Any deep question requires specific analysis, with tools suitable to the particular context. Typically, an interesting issue about an object translates into a structural comment on homotopy. A categorical perspective helps to get the most out of a theorem. Often, it is easy to see, or to make rigorous, new implications by thinking in terms of mapping properties.

1991 *Mathematics Subject Classification.* Primary 54H20; Secondary 18, 18F25.

Key words and phrases. Homotopy, dynamical systems, symbolic dynamics, category theory.

Objects under consideration are (Hausdorff) topological spaces with a continuous action by a(n infinite) group Ω. We denote the category by $\mathbf{HT}[\Omega]$. In later papers, we shall prove that if $X \in \mathbf{HT}[\Omega]$ is a locally simply-connected pathwise connected, then $\pi_G(X)$ is an extension of a quotient of the profinite completion of the classical $\pi_1(X)$ by the profinite completion of Ω. Of greater interest are objects whose underlying topology is totally disconnected, such as shift spaces. Their fundamental group can be complicated. In spirit, the construction is closer to the Galois group of a field than to the topological π_1.

It is difficult to discuss the potential and shape of the theory in this introduction, before language is in place. The ground rules are thrashed out in Section 2. The next section interrupts the formalism with a second "introduction". It discusses elementary examples, provides sample heuristical arguments, and tosses out a conjecture which appears within reach.

Another goal for the study of $\mathbf{HT}[\Omega]$ is to find gaps in the general theory. Explicit calculation points outs useful structure, some of which may be expressible in universal terms. What follows is an illustration of the interplay between very abstract language and specific cases.

Suppose $\mathcal{C}$ is a category in which there exist *Galois morphisms,* in the sense of [**F94**, Definition 36]. These morphisms are analogues to Galois extensions in the category of fields, to covering spaces with enough deck automorphisms in point-set topology, and to certain étale morphisms between schemes. At the universal level, one may ask under what conditions a morphism behaves like a Galois map with *infinite* degree. Based on the techniques of [**F93**] and on the sequence [**F94**] to [**F96b**], it is natural to ask whether $\mathcal{C}$ may be embedded within a formal category $\mathcal{C}^+$ in which such "extensions" exist. The latter would include profinite versions of simply connected covering spaces. Presented this way, the issue seems abstract.

Let us add some structure. Suppose that $\mathcal{C}$ admits a final object, which we denote by E. (For the categories of immediate concern, E is a set with one point.) Each $X \in \mathcal{C}$ admits a unique morphism to E, which we denote by E_X. In examples, the *kernel* of $\pi_G(E_X)$ is the true analogue to the classical $\pi_1(X)$. This similarity is well-known in algebraic geometry. For schemes defined over $\mathbb{Q}$, the spectrum of the field $\mathbb{Q}$ serves as E, and $\pi_G(E)$ is the Galois group Γ of the algebraic closure of $\mathbb{Q}$. For V a scheme, the natural action by Γ on V and on its conjugates is encoded in the general nonsense statement that there is a group homomorphism $\pi_G(V) \longrightarrow \Gamma$ whose kernel is (essentially) $\pi_1(V)$. That relationship is realized for the categories here.

It is difficult to compute a kernel. For a scheme V, this kernel is isomorphic to the fundamental group of $V \times \mathbb{C}$. We conjecture that, in general,

1. in many cases there is a (possibly infinite) Galois morphism $\hat{E} \longrightarrow E$ such that, for X an object, the kernel of $\pi_G(E_X)$ is isomorphic to $\pi_G(X \times \hat{E})$, and
2. if $\hat{E}$ does not exist, the category can be enlarged to include it (even a purely formal $\hat{E}$ provides structure).

Thus, kernels may have an interpretation in terms of π_G on another, rather formal, category. If so, every proposition about π_G has the potential to imply a new proposition on kernels. Other computational concerns relate to this question.

The universal framework may conflict with the reader's intuition. The historical fundamental group has an *intrinsic* definition. Given a pathwise connected space

X and a point $x_1 \in X$, the group $\pi_1(X, x_1)$ consists of equivalence classes of loops at x_1 paired with a binary operation. In theory, one can compute the group without reference to any space other than X.

In contrast, definitions from [**F94**] are *extrinsic*. Given an object A in a category $\mathcal{C}$, the group $\pi_G(A)$ requires classification of all $\mathcal{C}/A$-objects $b : B \longrightarrow A$ of a certain kind. Conceivably, that class could be too "big" to be a set! If $\mathcal{C}$ is contained in a larger category $\mathcal{D}$, it is theoretically possible that

(1.a) $\mathcal{D}$ has so many more morphisms of the appropriate type that $\pi_G(A)$, as defined with respect to $\mathcal{D}$, is much bigger than $\pi_G(A)$ as defined with respect to $\mathcal{C}$, or (even worse) that

(1.b) $\mathcal{C}/A$-objects which are Galois in the context of $\mathcal{C}$ are no longer Galois in $\mathcal{D}$, in which case the two groups bear no relationship at all.

In practice, the program does not spawn classifications beyond set theory. Possibilities (1) are more interesting. They do not occur in existing models, which suggests that common categories are "complete" in some sense yet to be articulated.

Definitions based on classification of all morphisms of a specified type may seem exotic. Its history is recent, but respectable. The Galois group of a field F is defined by classifying all finite field extensions K/F. This process turns out to be equivalent to an intrinsic definition, based on polynomials irreducible over F. Grothendieck [**G**] expanded this approach in formulating the pro-finite fundamental group of algebraic geometry.

Problems that bedevil Galois Field theory originate in the *nature* of the definition. Generation of Galois extension fields is easy in the abstract, but complicated in practice. Given two Galois extensions created by different methods, it can be difficult to determine whether the two are isomorphic or not. Presented with a Galois extension field, determination of the Galois group may be messy. The group is finite, but each member is a function on an (often) infinite set.

These obstructions appear in our program. Galois Field Theory developed by exploiting observations which apply in special cases. Likewise, the power of π_G will be its ability to encode theorems proved by specific (i.e., non-universal) techniques.

Despite these caveats, the approach has unique strengths. Let A be a finite set, and let X be the set of (two-sided) infinite sequences of members of A. Assign to A the discrete topology and to X the resultant product topology. The group $\mathbb{Z}$ acts on X by "shifting" indices. A space X of this form, or a closed subspaces preserved by $\mathbb{Z}$-action, is known as a *shift*. From a classical perspective, a shift is totally disconnected, which means that a group construction based on loops does not apply. From the abstract perspective, certain of these spaces, including entire shift space, are connected. The extrinsic construction creates a non-trivial invariant.

The immediate purpose is to lay groundwork. The primary task is to translate the universal terminology into concrete terms. Section 1 reviews elementary concepts. Section 2 reviews the language of [**F94**], and provides concrete equivalents. The trek through the language is divided into stages, as subsections. Section 3 interrupts the rigor with an informal review. Examples are considered, and intuitions are suggested. Arguments beyond the present effort are sketched. Section 4 relates abstractions to natural manipulations with points. In particular, an Ω-topset E with one member serves as final object. This means that each object X comes with a canonical **Conj**-morphism $\pi_G(X) \longrightarrow \pi_G(E) = \hat{\Omega}$. Interesting invariants should manifest themselves in the kernel of this map. That is, the kernel contains the significant information. Separating it from elementary confusions related to canonical projection is the dominant theme for the remainder of the paper.

A topological space X with an action by a group Ω automatically inherits actions by every subgroup $\Gamma \subseteq \Omega$. Certain aspects of $\pi_G(X)$ can be understood by viewing X as a Γ-set.

Two types of functors, Induction and Restriction exploit this structure. Restriction is introduced in Section 4. Section 5 brings in Induction, and establishes foundations.

Spaces of interest are *absolutely connected.* The concept includes totally disconnected topologies. For X absolutely connected, the canonical $\pi_G(X) \longrightarrow \hat{\Omega}$ is surjective. Section 6 provides gritty analysis. The result is extremely important but terribly messy. We follow with treatment of a special case. For X the full 2-shift, Section 7 classifies the contribution to $\pi_G(X)$ of all additive homomorphisms $X \longrightarrow X$.

For an absolutely connected space, the kernel of the canonical projection is a natural analogue to the classical $\pi_1(X)$. At present, we lack a functorial description of the kernel. The closing section offers a provisional approach.

This paper is dedicated to my advisor, Goro Shimura. He showed how mathematics grows by pursuit and triumph over the grand questions, rather than isolated acts of cleverness.

1. The relevant categories

Let $\mathbb{N}$ and $\mathbb{Z}$ denote the sets of natural numbers and of integers, respectively. For $N \in \mathbb{N}$, let $\mathbb{N}(N)$ denote the natural numbers from 1 to N.

Suppose that $\mathcal{C}$ is a category. The set of morphisms from an object B to another A is denoted by $\mathrm{Mor}_{\mathcal{C}}(B, A)$. We often encounter *slice* categories. For $A \in \mathcal{C}$, the slice category $\mathcal{C}/A$ consists of all morphisms into A (with a standard definition for $\mathcal{C}/A$-morphism). We denote such an object as either a morphism $b : B \longrightarrow A$ or as a pair (B, b). For $(B, b), (C, c) \in \mathcal{C}/A$, indicate the set of $\mathcal{C}/A$-morphisms from $(B, b) \longrightarrow (C, c)$ by $\mathrm{Mor}_{\mathcal{C}/A}(B, b; C, c)$.

Automorphism groups, and projective limits of such, are crucial to our program. Unless stated otherwise, we adopt multiplicative notation for all groups. The identity of a group G is denoted by e or e_G.

Let **Top** denote the category of topological spaces. Categories of immediate relevance rely on topological bases. However, experiences from algebra geometry lie behind many choices made in our formulation.

Occasionally, we allude to Galois theory of fields, or fundamental groups of schemes, to illustrate concepts. These topics are not part of the present formal theory. When we mention fields or rings, we, either, interpret them as algebras, or identify them with their spectra. Morphisms depend on interpretation, since the spectral embedding of commutative rings into the category of schemes is contravariant. Precise meaning should be clear from context.

Let us review **Conj**, the technical category introduced in [**F94**]. Its class of objects consists of all compact Hausdorff topological groups. Suppose G and H are two such groups. Let M be the set of continuous group homomorphisms $G \longrightarrow H$. Define an equivalence relation on M by

$$f \sim g \text{ if and only if } \exists z \in H \ : \ \forall x \in G, \ f(x) = z \cdot g(x) \cdot z^{-1}. \tag{2}$$

We let $[f]$ denote the equivalence class of a homomorphism f. The notation is adopted only when the codomain is clear from context. Define the set of **Conj**-morphisms $G \longrightarrow H$ to be the equivalence classes of M with respect to the relation (2). There is a composition operation defined by the rule

$$[h] \circ [g] = [h \circ g]$$

for $g : G \longrightarrow H$ and $h : H \longrightarrow K$ two continuous group homomorphisms. Similarly, for G an object, the class $[1_G]$ meets the requirements of an identity morphism.

Let $\alpha : G \longrightarrow H$ be a **Conj**-morphism. If one representative of α is surjective, then every representative is surjective. In this case, we simply say that α is *surjective.*

Next, consider a family of categories on which a fundamental functor is to be defined. Let Ω be a group. Let $\mathbf{HT}[\Omega]$ denote the category of topological Ω-sets which are Hausdorff. Formally, a member of $\mathbf{HT}[\Omega]$ is a pair (X, σ) where X is a Hausdorff topological space and σ is a group homomorphism from Ω to the group, under composition written on the left, of topological automorphisms of X. An $\mathbf{HT}[\Omega]$-morphism $(X, \sigma) \longrightarrow (Y, \tau)$ is a continuous map $f : X \longrightarrow Y$ such that $f \circ \sigma[\omega] = \tau[\omega] \circ f$ for each $\omega \in \Omega$.

Sometimes, we simply denote an $\mathbf{HT}[\Omega]$-object by a single symbol X. In this case, we let X represent the underlying topological space and let σ, or σ_X, denote the group homomorphism on Ω. In addition, we use dot notation, writing "$\omega \cdot x$" to represent $\sigma_X[\omega](x)$, for $\omega \in \Omega$ and $x \in X$.

EXAMPLE 1. Reversible discrete time dynamical systems fit this model. Let $\Omega = \mathbb{Z}$. Suppose X represents all possible states of some experiment, and action by $1 \in \mathbb{Z}$ is transition from one state to the result after a unit of time.

Evolution over continuous or differentiable time variable can be modeled in a similar manner. Although our immediate discussion does not include these cases, our method extends in an obvious fashion. Our primary objective in this paper is to convince the reader that the universal framework of [**F94**] has worthwhile application. The heart of this work is translation.

REMARK 1. Fundamental group construction applies to dynamical systems with irreversible time development. That is, one can work with Ω-sets for Ω the monoid $\mathbb{N}$. However, details become more complex. We hope to pursue the irreversible case after the situation $\Omega = \mathbb{Z}$ is better understood.

EXAMPLE 2. A novel aspect of our general nonsense is its ready application to topologies which, in the classical sense, are totally disconnected. Let S be a finite set, assigned the discrete topology. Let S^Ω be the set of all functions $\Omega \longrightarrow S$. This space can be interpreted as a product of copies of S, indexed by Ω. Assign to it the product topology. The result is a compact Hausdorff space. Let Ω act on the left by (in function form)

$$(\omega \cdot f)(\tau) = f(\tau \cdot \omega)$$

If S is a finite abelian group, the space is an abelian $\mathbf{HT}[\Omega]$-group object under componentwise addition.

In traditional terms, the space S^Ω is totally disconnected. Consequently, the language of fundamental groups does not apply. We focus on the case when S is the field of two elements and $\Omega = \mathbb{Z}$. The resultant S^Ω is commonly referred to as the "2 shift".

Let $X \in \mathbf{HT}[\Omega]$. We refer to members of Ω, or to actions by members on X, as *shifts*. Let $U \subseteq X$ be a subset. A *shift* of U is any set of the form

$$s \cdot U = \{s \cdot x \ : \ x \in U\}, \tag{3}$$

where $s \in \Omega$. A subset U is *shift-invariant* if it contains all of its shifts. Equivalently, U is shift-invariant if the action by Ω restricts to an action on U; with this action and the subset topology, U becomes an $\mathbf{HT}[\Omega]$-object. We freely interpret a shift-invariant subset as a member of the category.

Let $X \in \mathbf{HT}[\Omega]$, let $x \in X$ and let Γ be a subgroup of Ω. Define the Γ-*orbit* to be

$$\Gamma \cdot x = \{\gamma \cdot x \ : \ \gamma \in \Gamma\}.$$

When $\Gamma = \Omega$, we call this set the *shift-orbit* of x.

For convenience, we refer to an $\mathbf{HT}[\Omega]$-object as an Ω-*topset*. Specifically, for $X \in \mathbf{HT}[\Omega]$ and Γ a subgroup of Ω, we refer to the $\mathbf{HT}[\Gamma]$-object corresponding to X paired with the restriction of the action map to $\Gamma \times X \longrightarrow X$ as the *underlying* Γ-*topset of* X.

A standard fact is that

PROPOSITION 3. *Let Ω be a group. Define a covariant functor $\Lambda : \mathbf{HT}[\Omega] \longrightarrow$ **Top** by mapping each $\mathbf{HT}[\Omega]$-object to its underlying topological space (and mapping each morphism to itself). Refer to Λ as the* canonical forgetful functor to **Top**. *Then Λ preserves arbitrary inverse limits.*

Let us review the proof, and emphasize certain cases.

Let G be a diagram of $\mathbf{HT}[\Omega]$-objects, indexed by a set J. For each $j \in J$,let (X_j, σ_j) be the j-th object. Let $G^\star$ be the diagram in which each object is interpreted as a set. Formally, $G^\star$ is the image of G under composition of Λ with the forgetful functor from **Top** to sets. Let $\beta : B \longrightarrow G^\star$ be an inverse limit in the sense of sets. Assignment to B of the smallest topology with respect to which each map $\beta[j]$ is continuous makes β an inverse limit in **Top**. Either by inspection or by general nonsense, one finds a unique action by Ω on B with respect to which every $\beta[j]$ is an $\mathbf{HT}[\Omega]$-morphism. It follows that, with this auxiliary structure, β becomes an inverse limit in $\mathbf{HT}[\Omega]$.

The paper [**F94**] developed machinery to handle situations in which "points" are problematic. Consider schemes over a number field or ring R. Formally, a scheme is a sheaf over a topological base space called the "spectrum". The spectrum of a fibered product of schemes is rarely the product of respective spectra. Often, the spectrum of a product is not determined solely by the spectra of its components. To further confuse matters, it is natural to look at the "T-rational" points of a scheme, where T is a finite extension ring of R. The latter model is close to the motivations for algebraic geometry. In terms of sets of points, the model is compatible with fibered product. However, as topological spaces, inverse limits are incompatible.

To include these important cases, our universal theory is cast in the language of morphisms rather than points. In essence, Proposition 3 promises that the topological model accurately represents $\mathbf{HT}[\Omega]$. So, for our immediate applications, it is natural to translate our generalizations back into point-set topology.

Let $A \in \mathbf{HT}[\Omega]$, $n \in \mathbb{N}$ and let $(C_1, c_1), \cdots, (C_n, c_n) \in \mathbf{HT}[\Omega]/A$. The standard set-theoretical model for $(C_1, c_1) \times_A \cdots \times_A (C_n, c_n)$ is

$$P = \{(x_1, \cdots, x_n) \in C_1 \times \cdots \times C_n \ : \ c_i(x_i) = c_j(x_j) \ \forall i, j \in \mathbb{N}(n)\}. \tag{4}$$

To be exact, the fibered product consists of P paired with the n projection maps $(x_1, \cdots, x_n) \mapsto x_j$, $j \in \mathbb{N}(n)$. The space P becomes a topological fibered product if it is assigned the restriction of the product topology on $\prod_{j \in \mathbf{N}(n)} C_j = C_1 \times \cdots \times C_n$. Finally, P becomes a fibered product in $\mathbf{HT}[\Omega]$ with the additional structure that

$$s \cdot (x_1, \cdots, x_n) = (s \cdot x_1, \cdots, s \cdot x_n).$$

Refer to this as the *standard model* for the product.

There is one case in which another model is appropriate. Let $A \in \mathbf{HT}[\Omega]$, let $(B, b) \in \mathbf{HT}[\Omega]/A$, let C be a shift-invariant subset of A, and let $c : C \longrightarrow A$ be subset injection. Let $P = b^{-1}C$ and define p and q on P to be, respectively, injection into B and restriction of b as a function into C. Then $(P; p, q)$ is a fibered product $(B, b) \times_A (C, c)$. (This illustrates why fiber product is commonly denoted $b^{-1}c$.) Again, P becomes fibered product in $\mathbf{HT}[\Omega]$. For convenience, we refer to this as the *subset pullback model.*

The following propositions are elementary. Proofs are left to the reader.

LEMMA 4. *Let Ω be a group. Let $A \in \mathbf{HT}[\Omega]$ and let $S \subseteq A$ be a shift-invariant subset. Then the complement of S and the closure of S are shift-invariant subsets.*

LEMMA 5. *Let Ω be a group. Let $A \in \mathbf{HT}[\Omega]$ and let $S \subseteq A$ be a shift-invariant subset. Let $U \subseteq S$ be a shift-invariant subset.*

(A) If U is open in S, then there is an open shift-invariant subset V of A such that $U = V \cap S$.

(B) If U is closed in S, then there is a closed shift-invariant subset V of A such that $U = V \cap S$.

2. Definition of the fundamental group

Throughout this section, we fix

(5) Let Ω be a group.

This section has two purposes. First, it proves that the framework of [**F94**] applies to $\mathbf{HT}[\Omega]$. Consequently, that category supports a fundamental group functor from its "connected" objects to **Conj**. Next, it translates the generalities of that construction into the language of point-set topology.

Actual verification of the axioms is elementary. Essentially, the universal work applies as long as disjoint unions behave. However, mere existence leaves the user with a practical problem: how does one calculate the monstrosity? The definition is troublesome. The bulk of this section brings abstractions down to earth.

There is a conflict in terminology. In [**F94**], words like "connected", "finite" and "discrete" were appropriated to indicate generalizations of point-set properties involved in covering space theory. Our present context has an obvious topological basis. However, the universal term does not specialize to the standard, purely topological, meaning. The author's fundamental group exists for totally disconnected spaces because our categorical version of "connected" disagrees with the classical. To emphasize that a word is being meant in the sense of [**F94**], we prefix it with "cat" or, if in a specific category $\mathcal{C}$ (such as $\mathbf{HT}[\Omega]$), with "$\mathcal{C}$-".

2.1. $\mathbf{HT}[\Omega]$ is componentwise. The empty set $\emptyset$ admits a unique topology and a unique action by Ω. In this way $\emptyset$ becomes an $\mathbf{HT}[\Omega]$-object. For each $B \in \mathbf{HT}[\Omega]$, the unique map $\emptyset \longrightarrow B$ is an $\mathbf{HT}[\Omega]$-morphism. If $b : B \longrightarrow \emptyset$ is an $\mathbf{HT}[\Omega]$-morphism, then b must be an $\mathbf{HT}[\Omega]$-isomorphism. Consequently, $\emptyset$ is an *empty* or *null* object in the sense of [**F94**, (14.a,b)]. We refer to a morphism as *empty* (respectively, *non-empty*) if its domain is empty (respectively, non-empty).

Let $A, B \in \mathbf{HT}[\Omega]$. Let C be the topological disjoint union of A with B, and let $\iota_A : A \longrightarrow C$ and $\iota_B : B \longrightarrow C$ be canonical injections. By inspection, there is a unique action by Ω on C with respect to which ι_A and ι_B become $\mathbf{HT}[\Omega]$-morphisms. Suppose that $D \in \mathbf{HT}[\Omega]$ and $f : A \longrightarrow D$ and $g : B \longrightarrow D$ are two $\mathbf{HT}[\Omega]$-morphisms. There is a unique function $h : C \longrightarrow D$ such that

$$f = h \circ \iota_A \text{ and } g = h \circ \iota_B.$$

Moreover, this h is an $\mathbf{HT}[\Omega]$-morphism. To paraphrase, the triple $(C; \iota_A, \iota_B)$ is a disjoint union of A with B in the categorical sense.

Let us take a second look at this construction. By inspection, the underlying object of $\iota_A \times_C \iota_B$ is empty. Now suppose $w : W \longrightarrow C$ is another $\mathbf{HT}[\Omega]$-morphism. From the subset pullback model mentioned in Section 1, it follows that $(W; w^{-1}\iota_A, w^{-1}\iota_B)$ is itself a disjoint union. In other words, disjoint unions in $\mathbf{HT}[\Omega]$ satisfy [**F94**, (16.a,b)].

We have proved that, in the language of [**F94**],

PROPOSITION 6. *For any group Ω, the category $\mathbf{HT}[\Omega]$ is componentwise.*

Consequently, the theory of connectedness from [**F94**] applies. Recall that an object is connected if it is not a disjoint union of two non-empty objects. In this context, the property translates as

DEFINITION 7. Assume (5). An $\mathbf{HT}[\Omega]$-object A is said to be *cat-disconnected* (or $\mathbf{HT}[\Omega]$*-disconnected*) if there are two subsets $U, V \subseteq A$ such that

(6.a) U and V are each open, closed, non-empty and shift-invariant,
(6.b) $U \cup V = A$ and $U \cap V = \emptyset$.

An object which is not cat-disconnected is said to be *cat-connected* (or $\mathbf{HT}[\Omega]$*-connected*).

Let us recall other terminology. A *complement* to an $\mathbf{HT}[\Omega]$-morphism $b : B \longrightarrow A$ is another morphism $c : C \longrightarrow A$ such that $(A; b, c)$ is a disjoint union; if such a morphism c exists, then we say that b is complemented. A morphism $b : B \longrightarrow A$ is said to be *cat-connected* (or $\mathbf{HT}[\Omega]$*-connected*) if B is both cat-connected *and* non-empty. (Many properties and definitions apply to a morphism with cat-connected domain except in the pathological case when that domain is empty.) A complemented, cat-connected morphism $b : B \longrightarrow A$ is called a *cat-component* (or $\mathbf{HT}[\Omega]$*-component*).

Let $A \in \mathbf{HT}[\Omega]$ and let $B \subseteq A$ be a shift-invariant subset. Let $b : B \longrightarrow A$ be subset injection. Then

(7.a) b is cat-complemented if and only if B is an open and closed subset, and
(7.b) a complement c to b is an open injection whose image is the set-theoretic complement to B in A.

We apply universal terminology to B, with the understanding that it is interpreted with respect to b.

A shift-invariant subset always has a complement in the set-theoretic sense. However, here we say it is complemented only if it is both open and closed. This will not be an issue in practice.

Point-set topology has a general definition of "component". For X a space and $x \in X$, one can define the component of x to be the largest connected subset of X that contains it. This sort of thing is not available to us.

At the categorical level, it is unknown how, and probably impossible, to discuss components for arbitrary objects. Happily, we concern ourselves only with finite disjoint unions.

REMARK 2. Foundational theory of disjoint unions, in an arbitrary category, is covered in [**F94**]. It is quite easy, without surprises, and published.

The following remarks are elementary.

PROPOSITION 8. *Assume (5). Let* $A \in \mathbf{HT}[\Omega]$ *and let* $S \subseteq A$ *be a shift-invariant subset. Let* $\overline{S}$ *be the closure of* S *in* A, *and let* T *be a shift-invariant subset of* A *such that* $S \subseteq T \subseteq \overline{S}$. *If* S *is cat-connected, then so is* T.

COROLLARY 9. *Assume (5). Let* $A \in \mathbf{HT}[\Omega]$ *and let* $x \in A$. *Let* S *be the shift-orbit of* x, *and let* T *be the closure of* S. *Then* S *and* T *are cat-connected. We call* T *the* closed shift-orbit of x.

PROOF. Trivial. □

PROPOSITION 10. *Assume (5). Let* $A \in \mathbf{HT}[\Omega]$, *and let* $S \subseteq A$ *be a subset which is connected in the topological sense. Suppose* A *is the union of all shifts of* S. *Then* A *is* $\mathbf{HT}[\Omega]$*-connected.*

PROPOSITION 11. *Assume (5). Let* $b : B \longrightarrow A$ *be an* $\mathbf{HT}[\Omega]$*-morphism, and let* $S \subseteq B$ *be a cat-connected shift-invariant subset. Then* $b(S)$ *is a cat-connected, shift-invariant subset of* A.

PROOF. Trivial. □

PROPOSITION 12. *Let* Ω *be an infinite group. Let* S *be a topological space. Let* S^Ω *be the space of functions from* $\Omega \longrightarrow S$, *assigned the product topology. Let* Λ *be a subgroup of finite index in* Ω. *Define an action by* Λ *on* S^Ω *by*

$$(\lambda \cdot f)(\tau) = f(\tau \cdot \lambda),$$

for $f \in S^\Omega$ *and* $\lambda \in \Lambda$. *This action makes* S^Ω *a cat-connected* $\mathbf{HT}[\Lambda]$*-object.*

In addition, every non-empty Λ*-invariant open subset of* S^Ω *is dense.*

PROOF. Suppose S^Ω is a disjoint union of two open and closed shift-invariant subsets U and V. Suppose $u \in U$ and $v \in V$. Then there exist finite subsets $I, J \subseteq \Omega$ such that, for $f \in S^\Omega$,

(8.a) if $u(i) = f(i)$ for each $i \in I$, then $f \in U$, and
(8.b) if $v(j) = f(j)$ for each $j \in J$, then $f \in V$.

The set

$$K = \{j^{-1}i \ : \ i \in I, \ j \in J\}$$

is finite. Fix $s \in \Lambda - K$. Then $(J \cdot s) \cap I = \emptyset$. Clearly, there is some $g \in S^\Omega$ such that

$$g(i) = u(i) \text{ for each } i \in I \text{ and } g(j \cdot s) = v(j) \text{ for each } j \in J.$$

Then $g \in U$, but $s \cdot g \in V$. This contradicts shift-invariance.

Consider the above logic, but in a modified context. Suppose that U is a non-empty Λ-invariant open subset, $u \in U$, $v \in S^{\Omega}$ is arbitrary and V is any neighborhood of v. The previous reasoning implies that $U \cap V \neq \emptyset$. The last claim of the proposition follows. □

REMARK 3. We have shown that (up to a set-theoretic nicety) the program of [**F94**] defines a fundamental group on $\mathbf{HT}[\Omega]$! However, its definition exists only in universal terms. The following subsections flesh out the terminology.

2.2. Discreteness and finiteness. Let $b : B \longrightarrow A$ be an $\mathbf{HT}[\Omega]$-morphism. Two properties, which are defined separately at the categorical level, have elementary point-set interpretation in combination. The properties are that of being *finite* and of being *discrete*.

Let $b : B \longrightarrow A$ and $c : C \longrightarrow A$ be $\mathbf{HT}[\Omega]$-morphisms into a common codomain. A *lifting*, or *factoring* of c through b is a morphism $f : C \longrightarrow B$ such that $c = b \circ f$. (The latter is equivalent to $f \in \mathrm{Mor}_{\mathrm{HT}[\Omega]/A}(C, c; B, b)$.) Now let $b : B \longrightarrow A$ be a fixed morphism, and let $N \in \mathbb{N}$. Then b is *finite of order N*, in the sense of [**F94**, Definition 13], if each connected morphism $c : C \longrightarrow A$ has at most N liftings through b. As usual, we refer to the property here as either *cat-finiteness* or $\mathbf{HT}[\Omega]$-*finiteness*. If, in addition, there is some connected morphism which admits exactly N liftings then we say that N is the *index* of b. We denote the index by $[b : A]$ or $[B : b : A]$.

PROPOSITION 13. *Assume (5). Let $b : B \longrightarrow A$ be an $\mathbf{HT}[\Omega]$-morphism, and let $N \in \mathbb{N}$. If b is of order N, then for each $y \in A$, the fiber $b^{-1}\{y\}$ is finite of size $\leq N$.*

PROOF. Let $y \in A$ and let $x_1, \cdots, x_M$ be a finite and irredundant list of some (but not necessarily all) members of $b^{-1}\{y\}$. Let $(P; \pi_1, \cdots, \pi_M)$ be the standard model for the self-product of M copies of (B, b). Note that the product comes with a canonical projection $p : P \longrightarrow A$ such that $b \circ \pi_j = p$ for each $j \in \mathbb{N}(M)$. Let D be the closed shift-orbit of $(x_1, \cdots, x_M)$ in P. Then the restriction of p to D is a cat-connected morphism, and the the restrictions of the projections are M distinct liftings of it through b. It follows that $M \leq N$. □

Let $b : B \longrightarrow A$ be an $\mathbf{HT}[\Omega]$-morphism, and let $c : C \longrightarrow A$ be a cat-connected morphism. Suppose that f and g are two liftings of c through b. Let

$$D = \{x \in C \ : \ f(x) = g(x)\}.$$

Clearly D is a closed shift-invariant subset of C. The reader can verify that the condition that b is (cat-)discrete given in [**F94**, Definition 13] is equivalent to the requirement that, for all choices of C, c, f and g in the present hypothesis, the set D must be either empty or all of C.

An existing variation of discreteness is useful. Again, let $b : B \longrightarrow A$ be a morphism. In $\mathbf{HT}[\Omega]$, the self-product $b \times_A b$ always exists. Let P be the standard model for it. Define the *diagonal* to be the subset

$$\Delta = \{(x, x) \ : \ x \in B\}.$$

The diagonal morphism $\delta : B \longrightarrow P$ used in [**F94**, Definition 16] is just the map $x \mapsto (x, x)$. The condition that b be *testably discrete* is equivalent to the requirement that Δ is a cat-complemented subset of P. Since Δ is always closed and shift-invariant, the latter is equivalent to the condition that the diagonal be open.

It is known, by [**F94**, Proposition 18], that a testably discrete morphism is discrete. Furthermore, separable and Galois morphisms, which will be discussed later, are both.

LEMMA 14. *Assume (5). Let $b : B \longrightarrow A$ be a $\mathbf{HT}[\Omega]$-morphism. Then b is testably discrete if and only if it is locally injective.*

PROOF. Consider

$$P = \{(x, y) \in B^2 \ : \ b(x) = b(y)\}$$
$$\Delta = \{(x, x) \ : \ x \in B\}$$

as subsets of the product topological space B^2. Now P becomes a self-product in $\mathbf{HT}[\Omega]$ under the subset topology, and testable discreteness is equivalent to the statement that Δ is open in P. Let $x \in B$ and let U be a neighborhood of x in B. Tautologically, the condition that b is injective on U is equivalent to the requirement that

$$P \cap (U \times U) \subseteq \Delta.$$

The lemma is an easy consequence. □

COROLLARY 15. *Assume (5). Let $b : B \longrightarrow A$ be an $\mathbf{HT}[\Omega]$-morphism. Assume that B is compact, and that ρ is a metric on B which induces its topology. Then b is testably discrete if and only if there exists a positive real number r such that*

$$\forall x, y \in B, \ (b(x) = b(y) \ \text{and} \ \rho(x, y) < r \ \Rightarrow \ x = y)\,. \tag{9}$$

PROOF. Clearly, if such an r exists, then b is locally injective.

Suppose that b is testably discrete. Let P be the standard model for $b \times_A b$. On P, define a metric by

$$\rho'\left((x_1, y_1), (x_2, y_2)\right) = \max\{\rho(x_1, x_2), \rho(y_1, y_2)\}.$$

The topology of P is induced from ρ'. Let Δ be the diagonal, and let C be its complement. Since the two subsets are closed, they are compact. Consequently,

$$r = \min\{\rho'(u, v) \ : \ u \in \Delta, \ v \in C\},$$

is positive.

Suppose $x, y \in B$ such that $b(x) = b(y)$ and $\rho(x, y) < r$. Then $(x, y) \in P$ and $\rho'((x, x), (x, y)) = \rho(x, y) < r$. Since $(x, x) \in \Delta$, it follows that

$$(x, y) \in \Delta,$$

and, therefore, $x = y$. □

PROPOSITION 16. *Assume (5). Let $b : B \longrightarrow A$ be an $\mathbf{HT}[\Omega]$-discrete morphism. Let $N \in \mathbb{N}$. Then b is finite of order N if and only if $b^{-1}\{y\}$ is a finite set of $\leq N$ members for each $y \in A$. Moreover, b has index N if and only if it has order N and there is some $y \in A$ for which $|b^{-1}\{y\}| = N$.*

PROOF. Assume that b is discrete. Proposition 13 states one direction.

Suppose that $|b^{-1}\{y\}| \leq N$ for each $y \in A$. Let $c : C \longrightarrow A$ be a connected morphism, let $x \in C$ and let $y = c(x)$. For each $z \in b^{-1}\{y\}$, discreteness implies that c has at most one lifting through b which takes x to z. Consequently, c has fewer than $|b^{-1}\{y\}|$ liftings, for this particular y. If c has exactly N liftings, then the fiber must have exactly N members. □

2.3. Overlaying and openness. This subsection discusses the various forms of "overlaying". We observe that openness is stronger than overlaying absolutely. In the final subsection, we show that openness is required for separable and Galois morphisms.

Let $b : B \longrightarrow A$ be an $\mathbf{HT}[\Omega]$-morphism. Let $(P; \pi_1, \pi_2)$ be the standard model for $b \times_A b$. We say that b *overlays intrinsically* if, for $f : B \longrightarrow X$ an arbitrary morphism, there is a unique $F : A \longrightarrow X$ for which $f = F \circ b$ if and only if $f \circ \pi_1 = f \circ \pi_2$.

Define an equivalence relation $\sim$ on B by

$$x \sim y \;\Leftrightarrow\; b(x) = b(y). \tag{10}$$

By inspection of the standard model, a morphism $f : B \longrightarrow X$ has the property that $f \circ \pi_1 = f \circ \pi_2$ if and only if $f(x) = f(y)$ for all $x, y \in B$ such that $x \sim y$. Put $Q = B/\sim$, and let $q : B \longrightarrow Q$ be the canonical map. Assign to Q the topology induced from q. Clearly, Q is Hausdorff. Routine arguments verify that there is a unique action by Ω on Q with respect to which Q becomes an $\mathbf{HT}[\Omega]$-object and q becomes an $\mathbf{HT}[\Omega]$-morphism. The universal property of a quotient translates as follows: for $f : B \longrightarrow X$ any continuous function, there is a unique continuous function $F : Q \longrightarrow X$ for which $f = F \circ q$ if and only if $f \circ \pi_1 = f \circ \pi_2$. The analogous factoring property for $\mathbf{HT}[\Omega]$-morphisms is easily verified.

The above argument shows that an $\mathbf{HT}[\Omega]$-morphism $b : B \longrightarrow A$ overlays intrinsically if and only if b is surjective and the topology on A is induced from that of B.

A morphism is said to *overlay absolutely* if

(11.a) it can be pulled back along any morphism, and
(11.b) every pullback of it overlays intrinsically.

These requirements are equivalent to [**F94**, Definition 36]. In general, the pullback of a quotient map for some relation need not be (isomorphic to) a quotient map. Although each pullback is a quotient for the relation in a set-theoretic sense, the induced topology need not agree with the topology induced from the domain.

One particular class of quotient maps is closed under pullback. Suppose that $b : B \longrightarrow A$ is an open surjection. In this case, A has the quotient property with respect to the relation (10). Moreover, every pullback is open and surjective.

LEMMA 17. *Assume (5). An open and surjective* $\mathbf{HT}[\Omega]$*-morphism overlays absolutely.*

Now we can complete the translation of [**F94**, Definition 36]. Let $b : B \longrightarrow A$ be an $\mathbf{HT}[\Omega]$-morphism, and let $c : C \longrightarrow A$ be a cat-connected morphism. Let $p : P \longrightarrow C$ be a pullback $c^{-1}b$. We say that b *overlays absolutely and uniformly* if and only if, for all possible c, the pullback satisfies

(12.a) P is a non-empty finite disjoint union of cat-connected objects, and
(12.b) each restriction of p to a cat-component overlays absolutely.

The definition is thoroughly abstract. It is not clear how to verify whether a morphism overlays absolutely and uniformly.

2.4. Separable and Galois. We begin with the main formulation arising from [**F94**, Theorem 28 and Definition 36].

DEFINITION 18. Assume (5). Let $b : B \longrightarrow A$ be an $\mathbf{HT}[\Omega]$-morphism. Suppose that

(13.a) A and B are $\mathbf{HT}[\Omega]$-connected and non-empty,
(13.b) b is $\mathbf{HT}[\Omega]$-finite and $\mathbf{HT}[\Omega]$-discrete,
(13.c) b overlays absolutely and uniformly.

We say that b is *separable.* If, in addition, $|\mathrm{Mor}_{\mathrm{HT}[\Omega]/A}(B, b; B, b)| = [B : b : A]$, (where $[B : b : A]$ is the number defined in [**F94**, Definition 13]) then we say that b is Galois.

The earlier work lays out the foundational theory for separable morphisms. Certain points to emphasize are

(14.a) If b is Galois, $\mathrm{Mor}_{\mathrm{HT}[\Omega]/A}(B, b; B, b)$ is a finite group under composition. We refer to this as the *Galois group* of b, and denote it by $\mathrm{Gal}(B, b)$.
(14.b) Suppose that b is separable. Then there is a Galois morphism $c : C \longrightarrow A$ such that c factors through b *and* such that every Galois morphism $d : D \longrightarrow A$ which factors through b factors through c. We refer to such a c as a *Galois closure* for b.

Note that factoring maps are *not* unique. Consequently, the Galois closure is determined up to *non-canonical* isomorphism.
(14.c) Suppose that b is Galois and $G = \mathrm{Gal}(B, b)$. Then b is a quotient map for the action by G on B.

There are other, more technical observations which we cite when needed.

Property (14.c) deserves expansion. Let $B \in \mathbf{HT}[\Omega]$, and let G be a group of $\mathbf{HT}[\Omega]$-automorphisms of B. The action is *discrete* in the sense of [**F96**, Section 8] if for any non-empty morphism $f : C \longrightarrow B$ and for any two $\alpha, \beta \in G$,

$$(\alpha \circ f = \beta \circ f) \;\Rightarrow\; \alpha = \beta.$$

The reader can easily translate this nonsense as the requirement that $\alpha(x) = x$ for some $x \in B$ and $\alpha \in G$ if and only if α is the identity element.

The main results of this Section are

THEOREM 19. *Assume (5). Let $b : B \longrightarrow A$ be an $\mathbf{HT}[\Omega]$-morphism, and let $N \in \mathbb{N}$. Then b is separable of index N if and only if*

(15.a) *A and B are non-empty and $\mathbf{HT}[\Omega]$-connected,*
(15.b) *for every $y \in A$, $|b^{-1}\{y\}| = N$,*
(15.c) *b is an open local homeomorphism.*

COROLLARY 20. *Assume (5). Let $B \in \mathbf{HT}[\Omega]$ be a non-empty, cat-connected object, and let G be a group of $\mathbf{HT}[\Omega]$-automorphisms which acts discretely on B. Let $A = G\backslash B$, and let $b : B \longrightarrow A$ be the canonical quotient map. Assign to A the quotient topology, and assign to it the factored action by Ω to make it an Ω-topset and to make b an $\mathbf{HT}[\Omega]$-morphism. Then b is Galois, and G is its Galois group.*

COROLLARY 21. *Assume (5). A separable $\mathbf{HT}[\Omega]$-morphism is a proper, closed continuous mapping.*

PROOF. Proofs of theorem and corollaries are intertwined.

First, assume that $b : B \longrightarrow A$ satisfies (15.b,c). Inspection of the standard pullback model easily yields that every pullback of b satisfies the same two conditions. (Condition (15.a) is not preserved.) Consequently, any remarks verified for b automatically apply to its pullbacks.

Immediately, Lemma 14, Proposition 16 and [**F94**, Proposition 18] imply that b is testably discrete, discrete, and finite of index N.

Let $y \in A$ and let $x_1, \cdots, x_N$ be a complete and irredundant listing of $b^{-1}\{y\}$. It is an exercise in point-set topology that there exists an open neighborhood V of y and, for each $j \in \mathbb{N}(N)$, an open neighborhood U_j of x_j such that

(16.a) for $j, k \in \mathbb{N}(N)$ such that $j \neq k$, $U_j \cap U_k = \emptyset$,

(16.b) for each $j \in \mathbb{N}(N)$, b restricts to a homeomorphism $U_j \longrightarrow V$.

From here, it is elementary to further conclude that b is closed and proper.

Next, suppose that A is cat-connected. Fix $y \in A$ and let $x_1, \cdots, x_N$ be a complete and irredundant listing of $b^{-1}\{y\}$.

Suppose B can be expressed as a disjoint union of non-empty, shift-invariant subsets $C_1, \cdots, C_M$, for some $M \in \mathbb{N}$. First, fix $p \in \mathbb{N}(M)$. The image $b(C_p)$ must be a non-empty, open and closed, shift-invariant subset of A. As A is cat-connected, the image must be all of A! Consequently, there is some $j \in \mathbb{N}(N)$ for which $x_j \in C_p$.

Now let p vary. It follows that $M \leq N$. Since decomposition was taken arbitrarily, it follows that B must be the disjoint union of a finite number of cat-connected subobjects. Moreover, the restriction of b to each is known to overlay absolutely.

At this point, we have verified that (15.a,b,c) implies separability, and that Corollary 21 is true.

Next, suppose that B and G satisfy the hypothesis of Corollary 20, and let $b : B \longrightarrow A$ be as constructed. It is known that b has the universal mapping property of a mod(G) quotient in the sense of point-set topology. Clearly, it has the universal property with respect to $\mathbf{HT}[\Omega]$ as well. Consequently, if we knew that G were the Galois group of some Galois map $c : B \longrightarrow C$, general nonsense promises that c must be isomorphic to b (in the sense of morphisms on a fixed domain). This remark is used shortly.

By Proposition 11, the space A is cat-connected. It follows easily that b satisfies (15.a,b,c), with $N = |G|$. Consequently, b is separable. Tautologically, $G \subseteq \mathrm{Mor}_{\mathbf{HT}[\Omega]/A}(B, b; B, b)$, and the latter has no more than N members. It follows that b is Galois, and that $G = \mathrm{Mor}_{\mathbf{HT}[\Omega]/A}(B, b; B, b)$ is the Galois group.

Finally, suppose that $b : B \longrightarrow A$ is separable. Let $c : C \longrightarrow B$ be a Galois closure. Let $G = \mathrm{Gal}(C, b \circ c)$ and let $H = \mathrm{Gal}(C, c)$ be a subgroup. It has been observed that c must be equivalent to the factoring construction provided in Corollary 20; so must $b \circ c$. Composition of b with appropriate isomorphisms yields the canonical factoring

$$H \backslash C \longrightarrow G \backslash C.$$

Properties (15.a,b,c) for b follow directly from this characterization. □

All that remains is to build the functor π_G from Galois morphisms. The details are in [**F94**]; we do not reprise them here. Below are comments which are to interpreted in the context of the earlier paper.

COROLLARY 22. *Assume (5). Let GCov be the class of Galois morphisms in* $\mathbf{HT}[\Omega]$. *Then GCov satisfies* [**F94**, (65)].

Verification of the axiom set [**F94**, (65)] follows easily from remarks developed here. These postulates, in turn, justify construction, which follows in [**F94**], of a covariant functor from connected objects to **Conj**.

DEFINITION 23. Assume (5). We refer to the covariant functor described in [**F94**, Theorem 46], based on the class of Galois morphisms in $\mathbf{HT}[\Omega]$, as the "fundamental group" functor for $\mathbf{HT}[\Omega]$. Throughout the remainder of this paper, π_G denotes this functor.

Errata. The behavior of π_G on morphisms relies on a construction in [**F94**, Theorem 38]. Its logic is valid, but the result is slightly misstated.

As drafted, Part (B) of the theorem assumes that $c : C \longrightarrow A$ is a quasi-Galois morphism. The conclusion has two parts. First, to each $f \in \mathrm{Mor}_{\mathcal{C}/A}(C, c; B, b)$, a homomorphism $f^* : \mathrm{Gal}(C, c) \longrightarrow \mathrm{Gal}(B, b)$ is assigned. Second, it is proved that homomorphisms derived from different choices are conjugate under $\mathrm{Gal}(B, b)$. In practice, the conclusion is needed under a slightly weaker hypothesis.

In the context of the [**F94**, Theorem 38], suppose that H is a subgroup of the automorphism group (defined with respect to the slice category $\mathcal{C}/A$) of (C, c). The case of interest is when c decomposes as $d \circ c_1$, where

(17.a) $c_1 : C \longrightarrow D$ is quasi-Galois, and
(17.b) $H = \mathrm{Gal}(C, c_1)$.

Then relation [**F94**, (54)], which characterizes f^*, still characterizes a unique group homomorphism, but from $H \longrightarrow \mathrm{Gal}(B, b)$. The stated argument is easily modified to show that the new version f^* is, up to conjugation, independent of choice for f.

The action by π_G on morphisms can be characterized as follows: Let $\mathcal{C}$ be a componentwise category, and let $GCov$ be a suitable choice of Galois covers. Let $d : D \longrightarrow A$ be a morphism between non-empty, connected objects. Let $b : B \longrightarrow A$ be a member of $GCov$. Let $c_1 : C \longrightarrow D$ be any member of $GCov$ for which there is another morphism $f : C \longrightarrow B$ such that $b \circ f = d \circ c_1$. (In particular, any choice of $d^{-1}b$ will serve for c_1.) Associated to (B, b) and (C, c_1) are canonical projections $p_B : \pi_G(A) \longrightarrow \mathrm{Gal}(B, b)$ and $p_C : \pi_G(D) \longrightarrow \mathrm{Gal}(C, c_1)$. The group homomorphism f^* determines a **Conj**-morphism $[f^*]$ which, by [**F94**, Theorem 38], is independent of choice of factoring f for $c = d \circ c_1$ through b. Then, in this context,

$$p_B \circ \pi_G(d) = [f^*] \circ p_C,$$

where $\pi_G(d)$ is the image of morphism d under the fundamental group functor.

3. Examples and partial results

With vocabulary in place, we reflect on examples.

Let $\mathbb{T}$ be the complex unit circle, and let $\Omega = \mathbb{Z}$. Fix a real number θ, and let $1 \in \mathbb{Z}$ act by $z \mapsto e^{\iota\theta} \cdot z$.

Let $b : B \longrightarrow \mathbb{T}$ be an $\mathbf{HT}[\mathbb{Z}\,]$-separable map. If B is connected, Theorem 19 implies that it is a covering space, in the usual sense. Specifically, up to topological isomorphism, b can be represented by $w \mapsto w^N$ on $\mathbb{T} \longrightarrow \mathbb{T}$, for some N.

Conversely, let us begin with the continuous function $b : w \mapsto w^N$ from $B = \mathbb{T}$ to $\mathbb{T}$. Make the domain a $\mathbb{Z}$ -topset by letting $1 \cdot w = e^{\iota\theta/N} \cdot w$; in this way, b becomes an $\mathbf{HT}[\mathbb{Z}]$-morphism. For convenience, let B_1 be $\mathbb{T}$ with this action, and let b_1 denote raising to power N. Let G be the group of complex N-th roots of unity. Each $\omega \in G$ acts on B_1 by multiplication. Classically, G corresponds to the group of deck transformations of b_1. Clearly, these actions preserve the $\mathbb{Z}$ -action. Thus, b_1 is Galois and G is its Galois group.

There are other actions on B with respect to which b remains a morphism. For a fixed $\omega \in G$, let 1 act on $\mathbb{T}$ by

$$w \mapsto e^{\iota\theta/N} \cdot \omega \cdot w,$$

Denote the resulting topset by B_ω, and denote $w \mapsto w^N$ by $b_\omega : B_\omega \longrightarrow \mathbb{T}$. For $\alpha, \beta \in G$, (B_α, b_α) and (B_β, b_β) are isomorphic if and only if $\alpha = \beta$! To see this, note that any such isomorphism must be a deck transformation of N-th power exponentiation. The latter comprise G. By inspection, G preserves every one of these actions! They are all Galois, and all with the same group.

To sort through this confusion, consider yet another action. This time, we consider $c : C \longrightarrow \mathbb{T}$ in which C is connected as a $\mathbb{Z}$ -topset, but not as a topological space. Let

$$C = \mathbb{T} \times (\mathbb{Z}/N\mathbb{Z}),\ c(w, \bar{j}) = w^N,$$

where $\mathbb{Z}$ acts on C by

$$1 \cdot (w, \bar{j}) = (e^{\iota\theta/N} \cdot w, \overline{j+1}).$$

The group $G \times (\mathbb{Z}/N\mathbb{Z})$ (which is, of course, isomorphic to $(\mathbb{Z}/N\mathbb{Z})^2$) acts on (C, c) by

$$(\omega, \bar{r}) \cdot (w, \bar{j}) = (\omega \cdot w, \overline{r+j}).$$

The reader should see that, for $\omega \in G$, (B_ω, b_ω) is a factoring of c through a quotient by a subgroup $< (\omega, \overline{-1}) >$.

This example emphasizes a principle which dominates subsequent sections. Let X be a pathwise connected $\mathbb{Z}$ -topset, and let $N \in \mathbb{N}$. We can create a new topset on the base space $X \times (\mathbb{Z}/N\mathbb{Z})$ by letting

$$n \cdot (x, \bar{j}) = (n \cdot x, \overline{n+j}).$$

As we shall see, $\mathbb{Z}$ and N may be replaced by any group Ω and any subgroup Γ of finite index. Effectively, we replace X, as a $\mathbb{Z}$ topset, by X as an $N\mathbb{Z}$-topset.

This trick feels superficial. Replacing objects by clumps of copies does not address fundamental invariants. It is an atavism of definition, a bookkeeper's pathology. Analysis, in later sections, disposes of the ambiguity by splitting $\pi_G(X)$ into two pieces. The following is a conjecture, but one which will probably be proved soon:

CONJECTURE 1. *Let Ω be a group and let X be a pathwise connected, locally pathwise simply connected Ω-topset. Then there is a short exact sequence of groups*

$$1 \longrightarrow K \longrightarrow \pi_G(X) \longrightarrow \hat{\Omega} \longrightarrow 1, \tag{18}$$

where K is a quotient of the profinite completion of the (classical) fundamental group $\pi_1(X)$. If $\Omega = \mathbb{Z}$ and $\pi_1(X)$ is finitely generated, then K is the profinite completion of $\pi_1(X)$.

This result is both good and bad news. The up side is that π_G recaptures $\pi_1(X)$, which is known to contain vital information. The disappointment is that construction, at least on pathwise connected spaces, adds little to the information within $\pi_1(X)$. The action, by conjugation, of $s \in \Omega \subseteq \hat{\Omega}$ on K, seems to be the automorphism (determined up to conjugation) from $\pi_1(X) \longrightarrow \pi_1(X)$ induced by s^{-1}, interpreted as a continuous function.

It is not surprising that π_1 appears. The category **HT**$[\{e\}]$ is exactly **Top**. In this situation, adjectives "connected", etc., have their standard meaning. By Theorem 19, a map $b : B \longrightarrow A$ between connected spaces is separable if and only if it is a covering space morphism. Such b is Galois if and only if its group of deck transformations, which serves as Galois group, acts transitively on each fiber.

Standard topology identifies covering spaces of finite fiber with subgroups of $\pi_1(A)$ of finite index. In addition, groups of deck transformations represent quotients.

We offer the following heuristic discussion of the term K.

Let Ω be a group and let $A \in \mathbf{HT}[\Omega]$. Assume that the underlying topology of A is connected and locally pathwise-and-simply connected. Classical topology provides a universal, simply connected covering space $\alpha_1 : A_1 \longrightarrow A$ whose group of deck transformations has a canonical identification with $G = \pi_1(A)$. (We omit reference to base point. All constructions are canonical, but only up to inner automorphism.) Let $b : B \longrightarrow A$ be a covering space map of finite fiber, and let

$$\alpha \in \mathrm{Mor}_{\mathbf{Top}/A}(A_1, \alpha_1; B, b).$$

Put

$$H = \{h \in G \ : \ \alpha \circ h = \alpha\}.$$

If $H \lhd G$, then G/H has a canonical identification with the group of $\mathbf{Top}/A$-automorphisms of (B, b).

To contribute to $\pi_G(A)$, B must support an action compatible with b. Let us begin with a simpler issue: fix $s \in \Omega$, and consider whether action on A by s lifts to B. There is an automorphism s_1 of (A_1, α_1) such that

$$\forall x \in A_1, \ \alpha_1(s_1(x)) = s \cdot \alpha_1(x). \tag{19}$$

Moreover, s_1 induces a group automorphism σ of G by

$$\forall g \in G, \ g \circ s_1 = s_1 \circ \sigma(g).$$

Function s_1 factors to an automorphism of (B, b) if and only if $\sigma(H) = H$.

There is some choice here, but it is largely irrelevant. Any automorphism which satisfies (19) must equal $s_1 \circ k$ for some $k \in G$. Change of s_1 replaces σ by its composition with a conjugation. Effectively, σ is defined up to inner automorphism. If we restrict attention to the case where H is normal, validity of $\sigma(H) = H$ is independent of choice.

If there is an Ω-action on B compatible with b, then every $s \in \Omega$ must lift. However, calculation of π_G is complicated by a gimmick previously mentioned: If $\Gamma \subseteq \Omega$ is a subgroup of finite index which supports an action on B compatible with b, then $b \times \mathbf{1}_{\Omega/\Gamma}$ is a $\mathbf{HT}[\Omega]$-Galois.

Consider the choice $\Omega = \mathbb{Z}$. Whether b can be modeled as $\mathbf{HT}[\mathbb{Z}\ ]$-Galois depends on whether action by $s = 1$ lifts. It is easy to construct cases where $\sigma(H) \neq H$. However, b contributes to π_G as long as s^N lifts for *some* $N \in \mathbb{N}$. If we fix $H \lhd G$ of finite index, and if G is finitely generated, there will be a power N such that $\sigma^N(H) = H$.

The next case to study is $\Omega = \mathbb{Z} \times \mathbb{Z}$. There are two canonical generators, which we label s and t, and one key relationship, namely $s \cdot t = t \cdot s$. Commutation complicates lifting. In addition to whether s and t lift individually, the issue of lifting to commuting actions must be considered. At present, the author lacks significant results.

Shift spaces hold tantalizing prospects. Consider the *full 2 shift* (over $\mathbb{Z}$). This consist of all two-sided infinite strings of 0s and 1s. We regard 0 and 1 as members in $\mathbb{F}$, the field $\mathbb{Z}/2\mathbb{Z}$. Let X denote the space. Each member $x \in X$ is regarded as a sequence of "bits" $\{x_i\}_{-\infty}^{\infty}$. We signify a choice by

$$\cdots x_{-2}x_{-1}.x_0x_1\cdots$$

where a decimal point is placed between positions -1 and 0.

What follows are comments without proof. Some propositions are established in later sections of this paper.

Let us begin with a well-known map from $X \longrightarrow X$:

$$\forall i \in \mathbb{Z}, \ b(x)_i = x_i + x_{i+1}$$

This is Galois, of index two. The non-trivial involution is characterized by

$$\forall i \in \mathbb{Z}, \ (\overline{x})_i = 1 + x_i.$$

This operation is sometimes called *conjugation.*

Let us move to a less familiar variation. Consider $c : X \longrightarrow X$ characterized by

$$c(x)_i = x_i + x_{i+2}.$$

In fact, $c = b \circ b$! However, it ceases to be Galois. Instead, it is separable.

Proof of separability follows from Theorem 19. The reader should see that for any $a, b \in \mathbb{F}$ and any $y \in X$, there is a unique $x \in X$ for which

$$x_0 = a, \ x_1 = b, \text{ and } y = c(x).$$

Divide X into four sets $\{U(a,b)\}_{a,b\in\mathbf{F}}$, where

$$U(a,b) = \{x \in X \ : \ x_0 = a, \ x_1 = b\}.$$

Restriction of c to each $U(a,b)$ is a homeomorphism. Consequently, c is separable of index 4.

A similar picture derives from another perspective. Regard X as a group, under componentwise addition. Let $\hat{0}$ be the zero. Map c is a group endomorphism. Let K be the kernel. By inspection, K has four members.

$$\begin{array}{rlrl} \hat{0} = & \cdots 00.00 \cdots, & \hat{1} = & \cdots 11.11 \cdots, \\ \epsilon_1 = & \cdots 101.010 \cdots, & \epsilon_2 = & \cdots 010.101 \cdots. \end{array}$$

In what follows, read the subscript of $\epsilon_\star$ mod(2). As a subgroup, $K \approx \mathbb{F}^2$.

Topologically, c is isomorphic to $X \times \mathbb{F}^2 \longrightarrow X$. As a group homomorphism, it is surjective of kernel $\mathbb{F}^2$. Despite this, c is not Galois. Two members of K induce $\mathbf{HT}[\mathbb{Z}]/X$-automorphisms: the identity is $x \mapsto x + \hat{0}$, and conjugation is $x \mapsto x + \hat{1}$. However, for $i \in \{1, 2\}$, the function $\alpha_i(x) = x + \epsilon_i$ does not commute with action by 1:

$$\begin{array}{l} \alpha_i(1 \cdot x) = (1 \cdot x) + \epsilon_i, \\ 1 \cdot \alpha_i(x) = 1 \cdot (x + \epsilon_i) = (1 \cdot x) + \epsilon_{i+1} \end{array}$$

We claim that a Galois closure of c appears as a component in $c \times c$.

Fibered product $c \times c$ is represented by

$$P = \{(x, z) \in X^2 \ : \ c(x) = c(z)\}.$$

There are two obvious components of P, namely

$$M = \{(x, x) \ : \ x \in X\} \text{ and } N = \{(x, x + \hat{1}) \ : \ x \in X\}.$$

Let D be the $\mathbf{HT}[\mathbb{Z}]$-component of P which contains $(\hat{0}, \epsilon_1)$. Then D also contains

$$1 \cdot (\hat{0}, \epsilon_1) = (\hat{0}, \epsilon_2).$$

It follows that projection $\pi_1(x,y) = x$ restricts to a separable morphism $D \longrightarrow X$ of index 2. Moreover, one can identify D with $X \times \{\epsilon_1, \epsilon_2\}$ by

$$(x,y) \in D \longleftrightarrow (x, y+x).$$

This isomorphism identifies $c \circ \pi_1$ with $(x,\epsilon) \mapsto c(x)$. The latter has Galois group which contains

$$\begin{aligned} \alpha &: (x,\epsilon) \mapsto (x+\hat{1}, \epsilon) \\ \beta &: (x,\epsilon) \mapsto (x+\epsilon, \epsilon) \\ \gamma &: (x,\epsilon) \mapsto (x, 1\cdot\epsilon) \end{aligned} \tag{20}$$

The reader should be able to check that these automorphisms generate a non-commutative group $\mathbb{F}^2 \rtimes \mathbb{F}$.

This analysis extends to all endomorphisms. Recall some standard conventions. Let t be an indeterminate. There is an obvious action by the polynomial rings $\mathbb{F}[t]$ on X on which t acts as 1—that is, by a shift. Augment the ring by adding the inverse to t in the field of quotients.

Every surjective endomorphism admits a unique expression as $x \mapsto p(t)\cdot x$, where $p(t)$ belongs to

$$\mathbb{F}[t,t^{-1}]^* = \mathbb{F}[t,t^{-1}] \setminus \{0\}.$$

Composition corresponds to ring multiplication, and each non-zero member corresponds to a separable morphism.

In Section 7, we prove that a semi-direct product of the additive group of a profinite completion of $\mathbb{F}[t;t^{-1}]$ with $\widehat{\mathbb{Z}}$ is a homomorphic image for $\pi_G(X)$. The image is the contribution to $\pi_G(X)$ from additive endomorphisms.

Other separable morphisms into X exist. Let $m \in \mathbb{N}$ and let $g : \mathbb{F}^m \longrightarrow \mathbb{F}$ be an arbitrary function. Consider $G : X \longrightarrow X$ characterized by

$$\forall i \in \mathbb{Z},\ G(x)_i = x_i + g(x_{i+1}, \cdots, x_{i+m}) + x_{i+m+1}.$$

Again, Theorem 19 easily implies that G is separable. Only preliminary results on the corresponding Galois groups are known to the author. It is likely that calculation is equivalent to analysis already in the literature. In the case $g(a,b) = ab$, the group appears to be S_8.

There is no *a priori* reason why $\pi_G(X)$ should be derived solely from maps with domain X. Formulation offers no hint on how Galois maps may be found or classified. Philosophy reflects the definition of Galois groups over a field. The Galois group of $\mathbb{Q}$ is an object of great interest, but is not expected to be classified (effectively) soon.

4. Points and components

Issues of connectedness complicate our new π_G. For example, consider an $A \in \mathbf{HT}[\Omega]$ whose underlying space is pathwise connected and locally simply connected. Suppose $b : B \longrightarrow A$ is a Galois $\mathbf{HT}[\Omega]$-morphism. Using the fact that b is a local homeomorphism, it is easy to check that B is, topologically, a finite disjoint union of subsets which are connected in the classical sense. The analysis of b divides into two linked questions: what can the restrictions of b to these components be, and how does the multiplicity of components manifest in our fundamental group?

An Ω-topset may be $\mathbf{HT}[\Omega]$-connected and, simultaneously, be $\mathbf{HT}[\Gamma]$-disconnected for a subgroup $\Gamma \subseteq \Omega$. Indeed, it is possible that, for each subgroup Γ, there is an $\mathbf{HT}[\Gamma]$-connected open and closed subset U which is disconnected with respect to a smaller subgroup $\Gamma' \subseteq \Gamma$. As the reader might guess, multiple versions of "connectedness" muddle calculation. Our group π_G depends heavily on how fibered products decompose into components. Luckily, one can isolate this phenomenon.

This section develops an elementary structure theory. It consists of several simple observations. Restriction functors are introduced.

4.1. The homotopy of a point. In what follows,

(21.a) Let Ω be a group and let E be a set consisting of one element, interpreted as an $\mathbf{HT}[\Omega]$-object. For each $X \in \mathbf{HT}[\Omega]$, let E_X denote the unique function $X \longrightarrow E$.

(21.b) Let Sub be the set of all subgroups of Ω of finite index, and let $Norm \subseteq Sub$ be the subset of members which are normal. For each $\Gamma \in Sub$, assign to the left Ω-set Ω/Γ the discrete topology, to make in an $\mathbf{HT}[\Omega]$-object.

From the characterization of Theorem 19,

PROPOSITION 24. *Assume (21).*

(A) For $X \in \mathbf{HT}[\Omega]$, the morphism E_X is separable if and only if it is $\mathbf{HT}[\Omega]/E_X$-isomorphic to $E_{\Omega/\Gamma}$ for some $\Gamma \in Sub$.

(B) Let $\Gamma_1, \Gamma_2 \in Sub$. For $g \in \Omega$, there is an $\mathbf{HT}[\Omega]$-morphism $\alpha : \Omega/\Gamma_1 \longrightarrow \Omega/\Gamma_2$ such that $\alpha(\Gamma_1) = g\cdot\Gamma_2$ if and only if $\Gamma_1 \subseteq g\cdot\Gamma_2\cdot g^{-1}$. Moreover, if such a morphism α exists, then it is unique.

(C) Let $\Gamma \in Sub$. The morphism $E_{\Omega/\Gamma}$ is Galois if and only if $\Gamma \in Norm$. For $\Gamma \in Norm$, there is a group isomorphism $\omega : \Omega/\Gamma \longrightarrow \mathrm{Gal}(E_{\Omega/\Gamma})$ characterized by

$$\omega(g)[t \cdot \Gamma] = tg^{-1} \cdot \Gamma \text{ for all } g, t \in \Omega. \tag{22}$$

PROOF. Trivial. □

COROLLARY 25. *Assume (21). Let $\hat{\Omega}$ denote a profinite completion of Ω, regarded as a topological group. For each $\Gamma \in Norm$,*

(23.a) *let $\rho_\Gamma : \hat{\Omega} \longrightarrow \Omega/\Gamma$ denote the canonical group homomorphism,*

(23.b) *let $r_\Gamma : \pi_G(E) \longrightarrow \mathrm{Gal}(E_{\Omega/\Gamma})$ denote the canonical* **Conj**-*morphism, and*

(23.c) *define $\omega_\Gamma : \Omega/\Gamma \longrightarrow \mathrm{Gal}(E_{\Omega/\Gamma})$ as in Proposition 24.C.*

Then there is a unique **Conj**-*isomorphism $\omega : \hat{\Omega} \longrightarrow \pi_G(E)$ such that, for each $\Gamma \in Norm$, $r_\Gamma \circ \omega$ is represented by $\omega_\Gamma \circ \rho_\Gamma$.*

PROOF. Trivial. □

Let $\Gamma \in Sub$ and let X be a connected and non-empty $\mathbf{HT}[\Omega]$-object. The pullback of $E_{\Omega/\Gamma}$ along E_X can be modeled as projection $\rho : (\Omega/\Gamma) \times X \longrightarrow X$. It is trivial to check that there is a bijection between open and closed Ω-invariant subsets of $(\Omega/\Gamma) \times X$ and open and closed Γ-invariant subsets of X which sends each U to the set U' characterized by

$$U \cap (\{\Gamma\} \times X) = \{\Gamma\} \times U'.$$

As the domain for a pullback of a separable morphism, $(\Omega/\Gamma) \times X$ is a finite disjoint union of connected objects.

DEFINITION 26. There is a forgetful functor from $\mathbf{HT}[\Omega] \longrightarrow \mathbf{HT}[\Gamma]$ which sends each $Y \in \mathbf{HT}[\Omega]$ to the underlying topological space paired with the restriction of the action to Γ. (The functor sends each morphism to itself.) Refer to this as a restriction (from Ω to Γ) functor. We denote it by $\mathrm{Res}, \mathrm{Res}_\Gamma$, or $\mathrm{Res}_\Gamma^\Omega$.

Several foundational results are deduced from the above remark. The second relies on the definition of the π_G functor on morphisms.

PROPOSITION 27. *Assume (21). Let X be a connected, non-empty* $\mathbf{HT}[\Omega]$*-object. Then, for any $\Gamma \in Sub$, $\mathrm{Res}^{\Omega}_{\Gamma}[X]$ is a finite disjoint union of connected* $\mathbf{HT}[\Gamma]$*-objects. In particular, $\mathrm{Res}^{\Omega}_{\Gamma}[X]$ is connected if and only if $(\Omega/\Gamma) \times X$ is* $\mathbf{HT}[\Omega]$*-connected.*

PROPOSITION 28. *Assume (21). Let X be a connected, non-empty* $\mathbf{HT}[\Omega]$*-object. Then $\pi_G(E_X)$ is surjective if and only if $\mathrm{Res}_{\Gamma}(X)$ is* $\mathbf{HT}[\Gamma]$*-connected for every $\Gamma \in Sub$. In this case, we say that X is (*absolutely*)* $\mathbf{HT}[\Omega]$-connected.

REMARK 4. This proof of Proposition 27 is overly sophisticated. It is easy to argue the implication without the machinery of Galois morphisms. The present language sets up Proposition 28.

The latter proposition is similar to a classical remark on the fundamental group of algebraic geometry. Let F be a field, and consider the category of varieties over F. The field is itself a variety, or scheme, often visualized as a space with one point. Let F^{sep} be the separable closure of F.

If F is algebraically closed, such as $F = \mathbb{C}$, then Grothendieck's profinite fundamental group resembles the topological fundamental group. In this case, $\pi_G(F)$ is trivial. But, in general, $\pi_G(F) \approx \mathrm{Gal}(F^{sep}/F)$. For X a scheme over F, and for E_X the unique morphism $X \longrightarrow F$, the *kernel* of $\pi_G(E_X)$ resembles the topological fundamental group. The full $\pi_G(X)$ extends it by $\mathrm{Gal}(F^{sep}/F)$.

In algebraic geometry, it is natural to restrict attention to *absolutely connected* varieties. In our language, these are objects X for which $\pi_G(E_X)$ is surjective.

Suppose that $A \in \mathbf{HT}[\Omega]$ is pathwise connected and locally simply connected. Let $b : B \longrightarrow A$ be Galois. Since b is a local homeomorphism, each connected component (in the classical sense of component) is pathwise connected and locally simply connected. However, the overall object might not be connected topologically. Ultimately, decomposition into classical components can be characterized in the language of morphisms in terms like "absolutely connected". The decomposition leads to a link between the classical fundamental group and the new π_G.

A last remark is useful later.

DEFINITION 29. Assume (21). Let $X \in \mathbf{HT}[\Omega]$ and let $\Gamma \in Sub$. A non-empty, open and closed Γ-invariant subset $U \subseteq X$ is called a Γ-*factor* of X. A Γ-factor which is $\mathbf{HT}[\Gamma]$-connected is called a Γ-*component.* If X is a finite union of its Γ-components, then we say that X *decomposes* (or *has a decomposition*) into Γ-components.

COROLLARY 30. *Assume (21). Suppose $X \in \mathbf{HT}[\Omega]$ has a decomposition into Ω-components. Then, for any $\Gamma \in Sub$, X decomposes into Γ-components.*

4.2. Pointed spaces. The π_G construction is intentionally set in a language independent of points. Topsets have a base space, which is to say they have a good geometric model. We reintroduce points. However, we can not bring in loops.

DEFINITION 31. Define the category of *pointed Ω-topsets* to be the following data. An object is a pair (X, x) where $X \in \mathbf{HT}[\Omega]$ and $x \in X$. A morphism $(X, x) \longrightarrow (Y, y)$ is any $\mathbf{HT}[\Omega]$-morphism $f : X \longrightarrow Y$ such that $y = f(x)$. Composition and identity are defined in the obvious manner.

Actually, the category is of little interest. Elementary notions, such as connectedness and empty object, do not apply. Instead of using the category structure, the objects and morphisms are useful as auxiliary constructs. We use the adjective *pointed* to emphasize when an object or morphism is interpreted in the present sense.

The relevance of pointed objects stems from a proposition of [**F94**]. Let A be a connected $\mathbf{HT}[\Omega]$-object, and let S be a family of Galois morphisms (B, b), interpreted as members of $\mathbf{HT}[\Omega]/A$. Order S by the relation that $(B, b) \geq (C, c)$ if b factors through c. Assume that, with respect to this relation, S is well-ordered.

Consider the situation of [**F94**, Theorem 38]. Suppose $(B, b), (C, c) \in S$. For each $f \in \mathrm{Mor}_{\mathrm{HT}[\Omega]/A}(B, b; C, c)$, define $f^\star : \mathrm{Gal}(B, b) \longrightarrow \mathrm{Gal}(C, c)$ by the condition that

$$\forall g \in \mathrm{Gal}(B, b),\ f \circ g = f^\star(g) \circ f.$$

The map $f^\star$ depends on the specific selection for f. However, the **Conj**-morphism $[f^\star]$ is independent of choice. Consider the inversely directed system of **Conj**-objects, indexed by S, which assigns $\mathrm{Gal}(B, b)$ to each $(B, b) \in S$ and assigns to each pair in the relation $(B, b) \geq (C, c)$ the unique class discussed above. If every Galois morphism into A is isomorphic to a member of S, then $\pi_G(A)$ is the inverse limit of this system, taken with respect to **Conj**. If S is smaller, than the inverse limit represents a quotient of $\pi_G(A)$.

Choose any $x_0 \in A$. For each $(B, b) \in S$, let $x_{B,b}$ be any choice of point from $b^{-1}(x_0)$. If $(B, b) \geq (C, c)$, then there is a unique pointed morphism $\rho : (B, x_{B,b}) \longrightarrow (C, x_{C,c})$ such that $b = c \circ \rho$. Consequently, an inversely directed system can be defined in the category of groups, where the (B, b)-th object is still $\mathrm{Gal}(B, b)$, and each projection is the group homomorphism derived from canonical factoring.

The new system has an inverse limit in the category of profinite groups. This kind of limit is well-understood. The foundational [**F94**, Lemma 44] promises that, in the obvious sense, it is also the inverse limit with respect to **Conj**. This method is the only practical technique for computing $\pi_G(A)$.

To summarize: in abstract discussion, we use the artificial category **Conj** to emphasize that the invariant group is independent of base point. However, in actual computation, and within proofs related to computation, we must select base points in order to choose between morphisms.

4.3. Galois action. Let $b : B \longrightarrow A$ be a separable morphism. Let $x \in A$, and let $y_1, \cdots, y_N$ be a complete and irredundant indexing of $b^{-1}\{x\}$. Let $c : C \longrightarrow A$ be a Galois closure for b, let G be its Galois group, and let $F = \mathrm{Mor}_{\mathrm{HT}[\Omega]/A}(C, c; B, b)$. There is a right action defined by $f \cdot g = f \circ g$ for $f \in F$ and $g \in G$. Let $z \in c^{-1}\{x\}$. Then $f \mapsto f(z)$ is a non-canonical bijection $F \longrightarrow b^{-1}\{x\}$. Any two right actions on $b^{-1}\{x\}$ induced by choices of z are conjugate.

The above process determines an equivalence class of finite G-sets. The procedure is clumsy. It relies on choices. The difficulty is that the action on a fiber merely represents the true, canonical action, which is on the set F. The action on F is analogous, but adjoint, to the classical action by the Galois group of a polynomial on the polynomial's roots.

There is another perspective. Let $b : B \longrightarrow A$, x and $y_1, \cdots y_N$ be as before. Let $M \in \mathbb{N}(N)$, and let $(P; \pi_1, \cdots, \pi_M)$ be the standard model for the product of

M copies of b (in $\mathbf{HT}[\Omega]/A$). The group Σ_M of permutations for $\mathbb{N}(M)$ acts on P on the left by the rule

$$\forall \sigma \in \Sigma_M, \ \sigma \cdot (a_1, \cdots, a_M) = (a_{\sigma^{-1}(1)}, \cdots, a_{\sigma^{-1}(M)}).$$

Each permutation acts as an $\mathbf{HT}[\Omega]$-automorphism. Consequently, each member permutes cat-components.

Let D be the $\mathbf{HT}[\Omega]$-connected component of $(x_1, \cdots, x_M)$, and let $d : D \longrightarrow A$ denote the canonical injection. Let $H \subseteq \Sigma_M$ be the stabilizer of D. If $|H|$ equals the index of d, then (D, d) is a Galois closure for (B, b) and its Galois group is H.

This interpretation is easy validated with the lemmas following [**F94**, Definition 19]. It is truly geometric. The classical version is that for a finite, separable extension field K/F, then for all $M \in \mathbb{N}$, the M-fold *tensor* product $K \otimes_F \cdots \otimes_F K$ is a sum of subfields of the Galois closure field; moreover, for M sufficiently large, the Galois field occurs.

5. Induction functors

Our standing hypothesis for this section is

(24) Let Ω be a group, and let Γ be a subgroup of Ω.

Induced topsets are characterized by a universal property dual to restriction.

DEFINITION 32. Assume (24). Let $X \in \mathbf{HT}[\Gamma]$. For convenience, let $\mathcal{C}$ be the class of pairs (A, α) where

(25.a) $A \in \mathbf{HT}[\Omega]$, and

(25.b) $\alpha : X \longrightarrow \mathrm{Res}^{\Omega}_{\Gamma}[A]$ is an $\mathbf{HT}[\Gamma]$-morphism.

An *induced Ω-topset* for X (or *Ω-topset induced from* X) is a pair $(B, \beta) \in \mathcal{C}$ such that, for each $(A, \alpha) \in \mathcal{C}$, there is a unique $\mathbf{HT}[\Omega]$-morphism $f : B \longrightarrow A$ for which $f \circ \beta = \alpha$.

We represent an induced object either as a pair (B, β) or in "arrow" form $\beta : X \longrightarrow B$. Alternatively, the pair is indicated by a single symbol B, which also represents the topset alone. The morphism is referred to as *canonical.* Also, we let $\mathrm{Ind}^{\Omega}_{\Gamma}[X]$ signify a(n arbitrary) choice of induced topset for X.

Let $X, Y \in \mathbf{HT}[\Gamma]$. Suppose that $\iota_X : X \longrightarrow \mathrm{Ind}^{\Omega}_{\Gamma}[X]$ and $\iota_Y : Y \longrightarrow \mathrm{Ind}^{\Omega}_{\Gamma}[Y]$ are induced topsets. For each $f \in \mathrm{Mor}_{\mathrm{HT}[\Gamma]}(X, Y)$, there is a unique $f_0 \in \mathrm{Mor}_{\mathrm{HT}[\Omega]}(\mathrm{Ind}^{\Omega}_{\Gamma}[X], \mathrm{Ind}^{\Omega}_{\Gamma}[Y])$ such that

$$f_0 \circ \iota_X = \iota_Y \circ f. \tag{26}$$

We refer to f_0 as the *morphism induced from* f, or $\mathrm{Ind}^{\Omega}_{\Gamma}[f]$.

Induced topsets are unique up to an obvious notion of isomorphism.

Let $X \in \mathbf{HT}[\Gamma]$. Assign to Ω the discrete topology. The topological space $\Omega \times X$ becomes an Ω-topset under the

$$\forall r, s \in \Omega, \ \forall x \in X \ r \cdot (s, x) = (rs, x). \tag{27}$$

Let Γ act continuously on $\Omega \times X$ on the right by

$$\forall s \in \Omega, x \in X, t \in \Gamma, \ (s, x) \cdot t = (st, t^{-1} \cdot x). \tag{28}$$

Clearly, the right action commutes with action by Ω. Let $q : \Omega \times X \longrightarrow Q$ be the canonical map to (any choice of) the quotient for equivalence modulo the action by Γ. Assign to Q the topology and Ω-action induced from q. General theory implies that $Q \in \mathbf{HT}[\Omega]$. Adopt the conventions that

(29.a) the image of a pair $(s, x) \in \Omega \times X$ under q is $[s, x]$,
(29.b) the function $\iota : X \longrightarrow Q$ is given by $x \mapsto [e, x]$.

The following comments are elementary:

(30.a) ι is an open and closed injection,
(30.b) ι is an $\mathbf{HT}[\Gamma]$-morphism $X \longrightarrow \mathrm{Res}_\Gamma^\Omega[Q]$,
(30.c) (Q, ι) is an Ω-topset induced from X.

This remark provides existence of induced topsets. That is, there exists a functor $\mathbf{HT}[\Gamma] \longrightarrow \mathbf{HT}[\Omega]$ which sends each object X to its induced topset and each morphism f to the corresponding $\mathrm{Ind}_\Gamma^\Omega[f]$. Any two such constructions are functorially equivalent, so we indicate any choice by Ind[*].

There is an easy characterization of the underlying topological space of an induced topset. Fix a complete and irredundant set of representatives M for Ω/Γ which contains the identity e, and assign to it the discrete topology. For any $X \in \mathbf{HT}[\Gamma]$, there is a function

$$\omega_X : \Omega \times X \longrightarrow M \times X$$

which is uniquely characterized by the property that

$$\omega_X(st, x) = (s, t \cdot x) \text{ for all } s \in M,\ t \in \Gamma,\ x \in X.$$

The map ω_X is a quotient for the right action by Γ on $\Omega \times X$. We may interpret to $M \times X$ as $\mathrm{Ind}[X]$. The canonical injection becomes $x \mapsto (e, x)$.

Next, consider an $\mathbf{HT}[\Gamma]$-morphism $f : X \longrightarrow Y$. The corresponding $\mathrm{Ind}_\Gamma^\Omega[f]$: $M \times X \longrightarrow M \times Y$ is simply $1_M \times f$.

The two remarks can be paraphrased: composition of Ind[*] with the forgetful functor $\mathbf{HT}[\Omega] \longrightarrow \mathbf{Top}$ is functorially equivalent to the composition of the forgetful functor $\mathbf{HT}[\Gamma] \longrightarrow \mathbf{Top}$ with $M \times *$. The latter functor is known to preserve arbitrary inverse limits.

In what follows

(31) Let Ω be a group, and let Γ be a subgroup of Ω. Let Ind[*] denote a choice of induced topset functor.

PROPOSITION 33. *Assume (31).*

(A) $\mathrm{Ind}[*]$ *preserves arbitrary colimits. Specifically, it preserves disjoint unions.*

(B) $\mathrm{Ind}[*]$ *preserves arbitrary inverse limits.*

(C) For $B \in \mathbf{HT}[\Gamma]$, *B is* $\mathbf{HT}[\Gamma]$*-empty if and only if* $\mathrm{Ind}[B]$ *is* $\mathbf{HT}[\Omega]$*-empty.*

(D) For $B \in \mathbf{HT}[\Gamma]$, *B is* $\mathbf{HT}[\Gamma]$*-connected if and only if* $\mathrm{Ind}[B]$ *is* $\mathbf{HT}[\Omega]$*-connected.*

(E) For $A, B \in \mathbf{HT}[\Gamma]$, $\mathrm{Ind}[*]$ *induces an injection*

$$\mathrm{Mor}_{\mathrm{HT}[\Gamma]}(B, A) \longrightarrow \mathrm{Mor}_{\mathrm{HT}[\Omega]}(\mathrm{Ind}[B], \mathrm{Ind}[A]).$$

PROOF. Trivial. □

Our next objective is the proof that $\mathrm{Ind}[*]$ preserves the fundamental group π_G. The latter's construction is based on connectedness, fibered products, and morphisms. In effect, Proposition 33 promises that induction preserves certain of these basic concepts. The next lemmas cover the remaining details.

LEMMA 34. *Assume (24). Let* $X \in \mathbf{HT}[\Omega]$. *Let Y be an open* Γ*-invariant subset. In what follows, regard Y as a* Γ*-topset. Let* $\iota : Y \longrightarrow X$ *be subset injection. In this situation,* (X, ι) *is induced from Y if and only if the following two conditions hold:*

(32.a) $\{s \in \Omega \ : \ s \cdot Y \cap Y \neq \emptyset\} = \Gamma$,
(32.b) $X = \bigcup_{s\in\Omega} s \cdot Y$.

COROLLARY 35. *Assume (31). Let $A \in \mathbf{HT}[\Gamma]$, and let $b{:}\,B \longrightarrow \mathrm{Ind}[A]$ be an arbitrary $\mathbf{HT}[\Omega]$-morphism into its induced topset. Then there is an $\mathbf{HT}[\Gamma]$-morphism $c : C \longrightarrow A$ such that (B, b) is $\mathbf{HT}[\Omega]/\mathrm{Ind}[A]$-isomorphic to* $(\mathrm{Ind}[C], \mathrm{Ind}[c])$.

PROOF. Trivial. □

LEMMA 36. *Assume (31). Let $b : B \longrightarrow A$ be an $\mathbf{HT}[\Gamma]$-morphism. Let ι_A : $A \longrightarrow \mathrm{Ind}[A]$ and $\iota_B : B \longrightarrow \mathrm{Ind}[B]$ be the canonical injections. With respect to $\mathbf{HT}[\Gamma]$, $(B; b, \iota_B)$ is a fibered product for $(A, \iota_A) \times_{\mathrm{Ind}[A]} (\mathrm{Ind}[B], \mathrm{Ind}[b])$.*

COROLLARY 37. *Assume (31). Let $A \in \mathbf{HT}[\Gamma]$ and suppose $(B,b), (C,c) \in \mathbf{HT}[\Gamma]/A$. Then* Ind *induces a bijection*

$$\mathrm{Mor}_{\mathrm{HT}[\Gamma]/A}(B, b; C, c) \longrightarrow \mathrm{Mor}_{\mathrm{HT}[\Omega]/\mathrm{Ind}[A]}(\mathrm{Ind}[B], \mathrm{Ind}[b]; \mathrm{Ind}[C], \mathrm{Ind}[c]).$$

PROOF. Proofs can be completed by analyzing the explicit constructions. However, once Lemma 36 is accepted, the key step of Corollary 37 can be justified in one line: If $f \in \mathrm{Mor}_{\mathrm{HT}[\Omega]/\mathrm{Ind}[A]}(\mathrm{Ind}[B], \mathrm{Ind}[b]; \mathrm{Ind}[C], \mathrm{Ind}[c])$, then the lemma immediately implies existence of an $\mathbf{HT}[\Omega]$-morphism $g : B \longrightarrow C$ such that $c \circ g = b$ and $\iota_C \circ g = f \circ \iota_B$. □

PROPOSITION 38. *Assume (31). Let $b : B \longrightarrow A$ be an $\mathbf{HT}[\Gamma]$-morphism. Then*

(33.a) *b is $\mathbf{HT}[\Gamma]$-finite if and only if* $\mathrm{Ind}[b]$ *is $\mathbf{HT}[\Omega]$-finite. Moreover, if b is finite, then the indices of b and of* $\mathrm{Ind}[b]$ *agree.*
(33.b) *b is $\mathbf{HT}[\Gamma]$-discrete if and only if* $\mathrm{Ind}[b]$ *is $\mathbf{HT}[\Omega]$-discrete.*
(33.c) *b is $\mathbf{HT}[\Gamma]$-separable if and only if* $\mathrm{Ind}[b]$ *is $\mathbf{HT}[\Omega]$-separable.*
(33.d) *b is $\mathbf{HT}[\Gamma]$-Galois if and only if* $\mathrm{Ind}[b]$ *is $\mathbf{HT}[\Omega]$-Galois.*

COROLLARY 39. *Assume (31). Let A be a connected $\mathbf{HT}[\Gamma]$-object. Then $\pi_G(A)$ is $\mathbf{Conj}$-isomorphic to $\pi_G(\mathrm{Ind}[A])$*

PROOF. Trivial. □

6. Absolute components

Let $b : B \longrightarrow A$ be a separable $\mathbf{HT}[\Omega]$-morphism. Let $\Gamma \in \mathit{Sub}$. It is possible for A to be $\mathbf{HT}[\Gamma]$-connected although B is not. That possibility is scrutinized next.

First, take another look at restriction, using facts about the induced topset functor. The universal property states that, for $A \in \mathbf{HT}[\Omega]$ and $B \in \mathbf{HT}[\Gamma]$, each member $b \in \mathrm{Mor}_{\mathrm{HT}[\Gamma]}(B, \mathrm{Res}_\Gamma[A])$ has a unique extension to

$$b_1 \in \mathrm{Mor}_{\mathrm{HT}[\Omega]}(\mathrm{Ind}_\Gamma^\Omega[B], A).$$

Induction sends connected objects to connected objects. Consequently, properties based on morphisms and connectedness can be analyzed categorically.

PROPOSITION 40. *Assume (21). Let $\Gamma \in \mathit{Sub}$ and let $b : B \longrightarrow A$ be an $\mathbf{HT}[\Omega]$-morphism.*

(A) The morphism b is $\mathbf{HT}[\Omega]$-discrete if and only if $\mathrm{Res}_\Gamma[b]$ *is $\mathbf{HT}[\Gamma]$-discrete.*

(B) The morphism b is $\mathbf{HT}[\Omega]$-finite if and only if $\mathrm{Res}_\Gamma[b]$ *is $\mathbf{HT}[\Gamma]$-finite. Moreover, if b is $\mathbf{HT}[\Omega]$-finite, then it and* $\mathrm{Res}_\Gamma[b]$ *have the same index.*

PROOF. Only one non-categorical observation is necessary. Suppose that $A, B \in \mathbf{HT}[\Omega]$ and B is $\mathbf{HT}[\Omega]$-connected and non-empty. Let U be a Γ-component of B. Restriction of function $B \longrightarrow A$ to U induces a map

$$\mathrm{Mor}_{\mathbf{HT}[\Omega]}(B, A) \longrightarrow \mathrm{Mor}_{\mathbf{HT}[\Gamma]}(U, \mathrm{Res}_\Gamma(A)),$$

which is injective. □

PROPOSITION 41. *Assume (31). Let $b : B \longrightarrow A$ be a separable $\mathbf{HT}[\Omega]$-morphism. Let X be a Γ-component of B and let Y be a Γ-component of A. Suppose that $b(X) \cap Y \neq \emptyset$.*

(A) $b(X) = Y$, and, interpreted as an $\mathbf{HT}[\Gamma]$-morphism, restriction b_1 of b to $X \longrightarrow Y$ is $\mathbf{HT}[\Gamma]$-separable.

(B) Suppose, in addition, that

(34.a) *b is Galois with Galois group Π, and*
(34.b) *Γ is normal in Ω.*

Put

$$\begin{aligned} \Pi_X &= \{\sigma \in \Pi \ : \ \sigma(X) = X\}, \\ \Omega_Y &= \{s \in \Omega \ : \ s \cdot Y = Y\}, \\ \Omega_X &= \{s \in \Omega \ : \ s \cdot X = X\}. \end{aligned} \tag{35}$$

Then

(36.a) *$\Pi_X \lhd \Pi$ and $\Omega_X \lhd \Omega_Y$,*
(36.b) *there is a group isomorphism $\theta : \Pi/\Pi_X \longrightarrow \Omega_Y/\Omega_X$ which is determined by the property that, for $\sigma \in \Pi$ and for $s \in \Omega$ representative of $\theta(\sigma\Pi_X)$,*

$$\sigma(X) = s^{-1} \cdot X.$$

COROLLARY 42. *Assume (31). Suppose that $b : B \longrightarrow A$ is a separable $\mathbf{HT}[\Omega]$-morphism. Let $N = [B : b : A]$. Then B has at most N distinct Γ-components.*

PROOF. By Proposition 27, B must decompose into Γ-components. Let

$$X_1, \cdots, X_m$$

be a complete and irredundant listing of Γ-components. Arrange the list so that $X = X_1$, and $X_1, \cdots, X_k$ are all components whose image under b intersects Y. Let $N = [B : b : A]$.

Fix an arbitrary $y_0 \in Y$. For each $j \in \mathbb{N}(m)$, let

$$N_j = |b^{-1}\{y_0\} \cap X_j|.$$

Then

$$N = \sum_{j=1}^{m} N_j.$$

Put

$$W = \{y \in A \ : \ \forall j \in \mathbb{N}(m), |b^{-1}\{y\} \cap X_j| = N_j\}.$$

By Theorem 19, b is a local homeomorphism, and each $y \in A$ has *exactly* N inverse images. It is a simple exercise in point-set topology to show that W is an open and closed, non-empty Γ-invariant subset of A. From here, it follows that

(37.a) $Y \subseteq W$,
(37.b) for each $j \in \mathbb{N}(m) \setminus \mathbb{N}(k)$, $b(X_j) \cap Y = \emptyset$,
(37.c) for each $j \in \mathbb{N}(k)$, $b(X_j) = Y$.

For each $j \in \mathbb{N}(k)$, denote restriction of b to a function $X_j \longrightarrow Y$ by b_j. For the moment, fix $j \in \mathbb{N}(k)$. Clearly, b_j satisfies the criterion of Theorem 19, with Γ in place of Ω. Thus, each b_j is $\mathbf{HT}[\Gamma]$-separable.

For the rest of the argument, assume that (34.a,b) hold.

Tautologically, Π permutes the Γ-components of B. Moreover, the action preserves the subset $S = \{X_1, \cdots, X_k\}$. It is known that Π acts transitively on $b^{-1}\{y_0\}$. Therefore, the action on S is transitive as well.

Let Z be either A or B, and let U be a Γ-invariant subset of Z. Using the assumption that Γ is *normal* in Ω, it is easily proved that, for each $s \in \Omega$, $s \cdot U$ is also Γ-invariant. It follows that the action by Ω permutes the Γ-components of Z.

Consider the groups defined in (35).

Obviously, the union

$$B' = \bigcup_{s \in \Omega} s \cdot X$$

is an open and closed Ω-invariant subset of B. Therefore, $B' = B$. By (37), for $s \in \Omega$, $s \cdot X \in S$ if and only if $s \in \Omega_Y$. Consequently, the action by Ω_Y on S is transitive as well.

We may identify S, as a Π-set, with Π/Π_X. Suppose α is an automorphism of S, as a Π-set. Choose $c \in \alpha(\Pi_X)$. Then α must have the form $p\Pi_X \mapsto pc\Pi_X$. Identification of α with c induces an isomorphism from the automorphism group to the *opposite* group of Ξ/Π_X, where Ξ is the normalizer of Π_X in Π. Each member of Ω_Y induces such a Π-automorphism. Moreover, the image of Ω_Y acts transitively. It follows that $\Xi = \Pi$, or, paraphrasing, that $\Pi_X \lhd \Pi$. By inspection, the automorphism group is isomorphic to Π/Π_X by sending the class of $d \in \Pi$ to $p\Pi_X \mapsto pd^{-1}\Pi_X$. The only subgroup of this which acts transitively is the entire group. Moreover, the action is primitive.

Properties (36) follow easily. □

Let us focus on the fundamental group of an absolutely connected space. Suppose $b : B \longrightarrow A$ is $\mathbf{HT}[\Omega]$-separable, and A is absolutely connected. The domain B might not be absolutely connected. However, one can deduce that it is a finite union of pieces which are, essentially, absolutely connected.

DEFINITION 43. Assume (21). Let $X \in \mathbf{HT}[\Omega]$. Let $U \subseteq X$ be an open and closed subset. We call U a *topset factor* of X if there exists $\Gamma \in Sub$ such that U is Γ-invariant. If $U \neq \emptyset$ and there exists $\Gamma \in Sub$ such that

(38.a) U is Γ-invariant, and

(38.b) U is absolutely $\mathbf{HT}[\Gamma]$-connected.

then we call U an *absolute component* of X. We say that X *finitely decomposes into absolute components* if it is a finite topological disjoint union of absolute components.

PROPOSITION 44. *Assume (21). Let $A \in \mathbf{HT}[\Omega]$.*

(A) If $U, V \subseteq A$ are topset factors of A, then $U \cap V$ is a topset factor of A.

(B) Let U, V be absolute components of A. Then either $U = V$ or $U \cap V = \emptyset$.

(C) Let U be a topset factor of A, and let $s \in \Omega$. Then $s \cdot U$ is a topset factor of A. If U is an absolute component, the so is $s \cdot U$.

(D) Let $N \in \mathbb{N}$. Suppose that, for each $\Gamma \in Sub$, A has fewer than N Γ-components. Then A decomposes into absolute components.

(E) Let U be an absolute component of A. If A is $\mathbf{HT}[\Omega]$*-connected, then* $A = \bigcup_{s\in\Omega} s \cdot U$.

(F) Suppose that A finitely decomposes into absolute components. Then there exists $\Gamma \in Norm$ such that the set of absolute components of A is the set of Γ-components of A.

PROOF. Trivial. □

COROLLARY 45. *Assume (21). Suppose that $b : B \longrightarrow A$ is a separable* $\mathbf{HT}[\Omega]$*-morphism. If A decomposes finitely into absolute components, then B decomposes finitely into absolute components.*

DEFINITION 46. Assume (21). We introduce an auxiliary category, denoted by **Aux**. An object is a pair (Γ, X) where $\Gamma \in Norm$ and X is an absolutely connected $\mathbf{HT}[\Gamma]$-object. For $(\Gamma_1, X_1), (\Gamma_2, X_2) \in \mathbf{Aux}$, a morphism $(\Gamma_1, X_1) \longrightarrow (\Gamma_2, X_2)$ is a continuous function $f : X_1 \longrightarrow X_2$ for which there exists $\Gamma \in Sub$ such that $\Gamma \subseteq \Gamma_1 \cap \Gamma_2$ and

$$\forall x \in X_1, \forall t \in \Gamma,\ f(t \cdot x) = t \cdot f(x). \tag{39}$$

Composition and identity are defined in the usual manner.

Suppose that $(\Gamma, X) \in \mathbf{Aux}$ and $s \in \Omega$. Define a new action by Γ on X, indicated by $\star$, characterized by

$$\forall t \in \Gamma, \forall x \in X,\ t \star x = (s^{-1}ts) \cdot x. \tag{40}$$

We refer to X, paired with this action, as the *s-twist of* (Γ, X), and denote it by $(\Gamma, X)^s$, or simply X^s.

Let A be an absolutely connected Ω-topset. Define another auxiliary category, which is *almost* the slice category $\mathbf{Aux}/(\Omega, A)$. An $\mathbf{Aux}//A$-object is a triple (Γ, B, b) where $(\Gamma, B) \in \mathbf{Aux}$ and $b : B \longrightarrow \mathrm{Res}_\Gamma(A)$ is an $\mathbf{HT}[\Gamma]$-morphism. We freely regard b or (Γ, B, b) as a member of $\mathbf{Aux}/(\Omega, A)$. Define an $\mathbf{Aux}//A$-morphism to be an $\mathbf{Aux}/(\Omega, A)$-morphism.

Let A be as before, let $(\Gamma, B, b) \in \mathbf{Aux}//A$ and let $s \in \Omega$. Then the function $c(x) = s \cdot b(x)$ is an $\mathbf{HT}[\Gamma]$-morphism $B^s \longrightarrow A$. Refer to (Γ, B^s, c) as the *s-twist of* (Γ, B, b), or $(\Gamma, B, b)^s$.

PROPOSITION 47. *Assume (21). Let A be an absolutely connected Ω-topset. Let $(\Gamma, B, b) \in \mathbf{Aux}//A$. Let $b^\star : \mathrm{Ind}_\Gamma^\Omega[B] \longrightarrow A$ be the induced* $\mathbf{HT}[\Omega]$*-morphism. Then $b^\star$ is* $\mathbf{HT}[\Omega]$*-Galois if and only if*

(41.a) *b is* $\mathbf{HT}[\Gamma]$*-Galois, and*

(41.b) *for each $s \in \Omega$, there is an* $\mathbf{HT}[\Gamma]$*-isomorphism $c : B \longrightarrow B^s$ such that*

$$\forall x \in B,\ s \cdot (b[c(x)]) = b(x).$$

In this case, we call (Γ, B, b) a Galois germ *for A.*

PROOF. Let us start with some general observations. Let $\iota : B \longrightarrow \mathrm{Ind}[B]$ be the canonical injection, and identify B with its image under ι. Then an open subset $C \subseteq \mathrm{Ind}[B]$ is a Γ-component if and only if $C = s \cdot B$ for some $s \in \Omega$.

For $N \in \mathbb{N}$, it is clear that b is a local homeomorphism for which the fiber over each point has N members if and only if $b^\star$ is a local homeomorphism for which the fiber over each point has $N \cdot [\Omega : \Gamma]$ members. Extension identifies the group of $\mathbf{HT}[\Gamma]/A$-automorphisms for (B, b) with a subgroup of $\mathbf{HT}[\Omega]/A$-automorphisms of Ind$[B, b]$. The index of the subgroup equals the size of the orbit of B under the action by Aut$(b^\star)$ on Γ-components.

Let $f : B \longrightarrow \text{Ind}[B]$ be a function. We know that f extends to a $\mathbf{HT}[\Omega]/A$-automorphism of Ind$[B]$ if and only if

(42.a) f is an injective $\mathbf{HT}[\Gamma]$-morphism whose image is a Γ-component, and

(42.b) $b^\star \circ f = b$.

Suppose that f can be expressed as $x \mapsto s \cdot c(x)$, where $s \in \Omega$ is fixed and c is a bijection $B \longrightarrow B$. For each $x \in B$ and $t \in \Gamma$,

$$\begin{aligned} & f(t \cdot x) = t \cdot f(x) \\ \Leftrightarrow \quad & s \cdot c(t \cdot x) = t \cdot (s \cdot c(x)) \\ \Leftrightarrow \quad & c(t \cdot x) = (s^{-1} t s) \cdot c(x). \end{aligned} \tag{43}$$

For each $x \in B$,

$$\begin{aligned} & \{b^\star \circ f\}(x) = b(x) \\ \Leftrightarrow \quad & b^\star(s \cdot c(x)) = b(x) \\ \Leftrightarrow \quad & s \cdot b(c(x)) = b(x) \end{aligned} \tag{44}$$

The desirable conditions for f have equivalent forms for c.

In the context of Theorem 19, proof of the equivalence stated in the current proposition is a trivial consequence of the above remarks. □

DEFINITION 48. Assume (21). Let A be an absolutely connected Ω-topset. Let $\mathbf{GalC}//A$ be the subcategory of $\mathbf{Aux}//A$ consisting of all objects (Γ, B, b) for which the induced $\mathbf{HT}[\Omega]$-morphism $\text{Ind}_\Gamma^\Omega[B] \longrightarrow A$ is $\mathbf{HT}[\Omega]$-Galois.

Actual calculation demands a complicated context.

(45.a) Assume (31). Let A be an absolutely connected Ω-topset.

(45.b) Fix a profinite completion $\hat{\Omega}$ for Ω. For $\Gamma \in Norm$, let $\omega[\Gamma] : \hat{\Omega} \longrightarrow \Omega/\Gamma$ denote the canonical projection.

(45.c) To each $(\Gamma, B, b) \in \mathbf{GalC}//A$, assign a choice of induced topset $\iota[\Gamma, B, b] : B \longrightarrow \nu[\Gamma, B, b]$. We also write $\nu[B]$ for $\nu[\Gamma, B, b]$, and let $\nu[b]$ represent the induced morphism $\nu[B] \longrightarrow A$. Denote the Galois group of $\nu[b]$ by Gal(Γ, B, b).

The morphism $\nu[b]$ satisfies the hypothesis of Proposition 41.(B). In its context, $\Pi = \text{Gal}(\Gamma, B, b)$ and (36.b) characterizes an isomorphism $\text{Gal}(\Gamma, B, b)/\Pi_X \longrightarrow \Omega/\Gamma$. Let $\theta[\Gamma, B, b]$ denote composition of the latter with projection $\text{Gal}(\Gamma, B, b) \longrightarrow \text{Gal}(\Gamma, B, b)/\Pi_X$. That is, for $\alpha \in \text{Gal}(\Gamma, B, b)$,

$$\alpha(\iota(B)) = \theta(\alpha)^{-1} \cdot \iota(B),$$

where $\iota = \iota[\Gamma, B, b]$ and $\theta = \theta[\Gamma, B, b]$.

(45.d) For $\Gamma \in Norm$, let $Norm/\Gamma$ be the subset of $\Gamma' \in Norm$ for which $\Gamma' \subseteq \Gamma$. Represent projection $\Omega/\Gamma' \longrightarrow \Omega/\Gamma$ by $\omega_{\Gamma\Gamma'}$, or simply ω.

(45.e) Suppose $\alpha = (\Gamma, B, b) \in \mathbf{GalC}//A$ and $\Gamma' \in Norm/\Gamma$. Define $\text{Res}_{\Gamma'}(\alpha)$ to be $(\Gamma', \text{Res}_{\Gamma'}(B), \text{Res}_{\Gamma'}(b))$. For now, denote the latter by β. Let $f : \nu[\beta] \longrightarrow \nu[\alpha]$ be the unique $\mathbf{HT}[\Omega]$-morphism for which $\iota[\beta] \circ f = \iota[\alpha]$. Let $f^\star$ be the induced homomorphism $\text{Gal}(\beta) \longrightarrow \text{Gal}(\alpha)$, as defined in [**F94**, Theorem 38]. We refer to f as the $\mathbf{HT}[\Omega]$-*morphism induced from* *restriction of* α *to* Γ', and to $f^\star$ as the *group homomorphism induced from restriction of* α *to* Γ'.

(45.f) Suppose $\Gamma \in Norm$ and, for each index $i \in \{1,2\}$, suppose that $\alpha_i = (\Gamma, B_i, b_i) \in \mathbf{GalC}//A$ has subgroup term equal to Γ. We refer to an $\mathbf{HT}[\Gamma]/A$-morphism $(B_1, b_1) \longrightarrow (B_2, b_2)$, interpreted as a $\mathbf{GalC}//A$-morphism, as a *straight equivariant morphism.*

Now suppose $f_0 : \alpha_1 \longrightarrow \alpha_2$ is straight equivariant. Let $f : \nu[\alpha_1] \longrightarrow \nu[\alpha_2]$ be the unique $\mathbf{HT}[\Omega]$-morphism for which $f \circ \iota[\alpha_1] = \iota[\alpha_2] \circ f_0$, and let $f^\star$ be the corresponding group homomorphism. We refer to f as the $\mathbf{HT}[\Omega]$*-morphism induced from* f_0, and to $f^\star$ as the *group homomorphism induced from* f_0.

Begin with an elementary relationship.

PROPOSITION 49. *Assume (45). Let* (Γ, B, b) *be a member of* $\mathbf{GalC}//A$. *Define*

$$\kappa : \mathrm{Gal}(b) \longrightarrow \mathrm{Gal}(\Gamma, B, b)$$

by mapping each automorphism g *to the* $\mathbf{HT}[\Omega]$*-morphism induced from it. Then*

$$1 \to \mathrm{Gal}(b) \xrightarrow{\kappa} \mathrm{Gal}(\Gamma, B, b) \xrightarrow{\theta} \Omega/\Gamma \to 1 \tag{46}$$

is a short exact sequence of groups.

PROOF. Trivial. □

LEMMA 50. *Assume (45). Let* $\alpha = (\Gamma, B, b) \in \mathbf{GalC}//A$, *and suppose* $\Gamma' \in Norm/\Gamma$. *Let* $\beta = \mathrm{Res}_{\Gamma'}(\alpha)$, *and let* $f^\star$ *be the group homomorphism induced from restriction of* α *to* Γ'. *Then* $(\mathrm{Gal}(\beta); f^\star, \theta[\beta])$ *is a fibered product in the category groups for*

$$(\mathrm{Gal}(\alpha), \theta[\alpha]) \times_{\Omega/\Gamma} (\Omega/\Gamma', \omega_{\Gamma\Gamma'}). \tag{47}$$

PROOF. Let f be the $\mathbf{HT}[\Omega]$-morphism induced from restriction of α to Γ. Let $p \in \mathrm{Gal}(\beta)$, and let $s \in \Omega$ represent $\theta[\beta](p)$. Then

$$f^\star(p) \circ f = f \circ p \tag{48}$$

sends $\iota[\beta](B)$ to $s \cdot \iota[\alpha](B)$. Therefore, s represents $\theta[\alpha](f^\star(p))$ as well. Consequently, $\theta[\alpha] \circ f^\star = \omega \circ \theta[\beta]$.

The property of fibered product determines a canonical group homomorphism from $\mathrm{Gal}(\beta)$ into the object (47) which commutes with projections $f^\star$ and $\theta[\beta]$. The two groups are finite. Simple counts show that they have the same size. To finish, it suffices to show that

$$\ker(f^\star) \cap \ker(\theta[\beta]) = \{\mathbf{1}_{\nu[\beta]}\}$$

Suppose

$$p \in \ker(f^\star) \cap \ker(\theta[\beta])$$

Then

(49.a) $f = f \circ p$, and
(49.b) $p(\iota[\beta](B)) = \iota[\beta](B)$.

It follows that $p \circ \iota[\beta] = \iota[\beta]$. The uniqueness property of induction now implies that p is the identity morphism. □

The Lemma reveals how to remove dependence on Γ in each (Γ, B, b). It remains to formalize that independence.

THEOREM 51. *Assume (45). Let* **TGp** *be the category of topological groups. There exists*

(50.a) *a covariant functor* $\Lambda_0 : \mathbf{GalC}//A \longrightarrow \mathbf{TGp}/\hat{\Omega}$, *whose action on objects is expressed as* $\Lambda_0(\alpha) = (\Lambda(\alpha), \lambda(\alpha))$, *and*

(50.b) *a functorial transformation* $\zeta : \Lambda \longrightarrow \mathrm{Gal}(*)$,

such that

(51.a) *for* $\alpha = (\Gamma, B, b) \in \mathbf{GalC}//A$, $(\Lambda(\alpha); \zeta(\alpha), \lambda(\alpha))$ *is a fibered product* $(\mathrm{Gal}(\alpha), \theta[\alpha]) \times_{\Omega/\Gamma} (\hat{\Omega}, \omega[\Gamma])$,

(51.b) *for* $\alpha = (\Gamma, B, b) \in \mathbf{GalC}//A$ *and* $\Gamma' \in Norm/\Gamma$, *let* z *be* $\mathbf{1}_B$ *interpreted as a morphism* $\alpha \longrightarrow \beta = \mathrm{Res}_{\Gamma'}(\alpha)$ *and let* $f^\star$ *be the group homomorphism induced from restriction of* α *to* Γ', *and then*

$$\Lambda_0(z) = \mathbf{1}_{\Lambda(\alpha)} \text{ and } f^\star \circ \zeta(\beta) = \zeta(\alpha),$$

(51.c) *for* $f : \alpha \longrightarrow \beta$ *a straight equivariant morphism,* $\zeta(\beta) \circ \Lambda_0(f) = f^\star \circ \zeta(\alpha)$, *where* $f^\star$ *is induced from* f.

PROOF. Define an equivalence relation on the object class of $\mathbf{GalC}//A$ by $(\Gamma_1, B_1, b_1) \sim (\Gamma_2, B_2, b_2)$ if and only if

(52.a) the underlying topological spaces of B_1 and B_2 are the same,

(52.b) $b_1 = b_2$, and

(52.c) the identity $\mathbf{1}_{B_1} = \mathbf{1}_{B_2}$ is a $\mathbf{GalC}//A$-morphism between the triples.

Let S be a class which contains exactly one representative of each equivalency class for $\sim$. To each $s = (\Gamma, B, b) \in \mathbf{GalC}//A$, assign a choice of fibered product $(F(s); p(s), q(s))$ for $(\mathrm{Gal}(s), \theta[s]) \times_{\Omega/\Gamma} (\hat{\Omega}, \omega[\Gamma])$.

For $\alpha \in \mathbf{GalC}//A$, put

$$\Lambda(s) = F(s) \text{ and } \lambda(\alpha) = q(s)$$

for the unique $s \in S$ which is equivalent to α. We define ζ and the morphism assignment of Λ_0 in stages.

First, consider only $\alpha = \mathrm{Res}_{\Gamma'}(s)$ for some $s \in S$. Clearly, for such an α, s and Γ' are unique. We may define $p = \zeta_1(\alpha)$ to be the unique morphism such that

$$f^\star \circ p = p(s) \text{ and } \theta[\alpha] \circ p = \omega[\Gamma'] \circ q(s).$$

Lemma 50 justifies this formulation.

Now consider an arbitrary $\beta = (\Gamma, B, b) \in \mathbf{GalC}//A$. There is $\Gamma' \in Norm/\Gamma$ such that $\alpha = \mathrm{Res}_{\Gamma'}(\beta)$ belongs to the class discussed in the previous paragraph. Consider the construction $f^\star \circ \zeta_1(\alpha)$, where f^* is induced from restriction. Note that both α and $f^\star$ depend on Γ'. An easy consequence of the construction of ζ is that this composition is unchanged if Γ' is replaced by any member of $Norm/\Gamma'$. It follows that $f^\star \circ \zeta_1(\alpha)$ is independent of Γ'. If β happens to equal $\mathrm{Res}_{\Gamma'}(s)$ for some s and Γ', the latter expression equals $\zeta_1(\beta)$. We define $\zeta(\beta)$ to be this common element.

All properties listed in the theorem, which do not invoke Λ_0 on morphisms, are easy consequences of construction, up to this point. The morphism assignment occupies the remainder of this argument.

Let $\alpha_i = (\Gamma_i, B_i, b_i)$ for each $i \in \{1, 2\}$, and let $f : \alpha_1 \longrightarrow \alpha_2$ be a morphism. Let N be the set of $\Gamma \in Norm$ such that $\Gamma \subseteq \Gamma_1 \cap \Gamma_2$ and f is Γ-equivariant.

For each $\Gamma \in N$, let $f[\Gamma]$ and $f[\Gamma]^\star$ be, respectively, the $\mathbf{HT}[\Omega]$-morphism and group homomorphism induced from f interpreted as a straight equivariant map $\mathrm{Res}_\Gamma(\alpha_1) \longrightarrow \mathrm{Res}_\Gamma(\alpha_2)$. In addition, there is a unique group homomorphism $\Lambda[\Gamma, f] : \Lambda(\alpha_1) \longrightarrow \Lambda(\alpha_2)$ such that

(53.a) $\lambda(\alpha_2) \circ \Lambda[\Gamma, f] = \lambda(\alpha_1)$,

(53.b) $\zeta(\mathrm{Res}_\Gamma(\alpha_2)) \circ \Lambda[\Gamma, f] = f[\Gamma]^\star \circ \zeta(\mathrm{Res}_\Gamma(\alpha_1))$.

Now suppose

(54) $\Lambda[\Gamma, f]$ is independent of Γ for every choice of f.

Under (54), we may define a class-theoretic function by $\Lambda_0(f) = \Lambda[\Gamma, f]$ for any appropriate Γ. We leave it to the reader to verify that the rest of the theorem's claims are satisfied by this assignment. The remaining discussion is dedicated to proof of (54). We gradually reduce the proposition to a tautology.

Clearly, it suffices to show that $\Lambda[\Gamma, f] = \Lambda[\Gamma', f]$ for $\Gamma, \Gamma' \in N$ such that $\Gamma' \subseteq \Gamma$. For the rest of this argument, assume this hypothesis.

For each $i \in \{1, 2\}$, let q_i and $q_i^\star$ be, respectively, the morphism and group homomorphism induced from restriction (Γ, B_i, b_i) to Γ'. From known properties of ζ, (53) follows once we show that

(55) $(\mathrm{Gal}(\Gamma', B_1, b_1); q_1^\star, f[\Gamma']^\star)$ is a fibered product for

$$(\mathrm{Gal}(\Gamma, B_1, b_1), f[\Gamma]^\star) \times_{\mathrm{Gal}(\Gamma, B_2, b_2)} (\mathrm{Gal}(\Gamma', B_2, b_2), q_2^\star).$$

Now, Lemma 50 already characterizes each $\mathrm{Gal}(\Gamma', B_i, b_i)$ and $q_i^\star$ in terms of a product involving $\mathrm{Gal}(\Gamma, B_i, b_i)$. Thus, (55) follows if we demonstrate that

(56.a) $\theta[\Gamma', B_2, b_2] \circ f[\Gamma']^\star = \theta[\Gamma', B_1, b_1]$, and

(56.b) $q_2^\star \circ f[\Gamma']^\star = f[\Gamma]^\star \circ q_1^\star$.

For $i \in \{1, 2\}$, let $t_i : \nu[\Gamma', B_i, b_i] \longrightarrow \Omega/\Gamma'$ be determined by $t_i(B_i) = \{e\}$. Recall that the θ constructions derive from application of [**F94**, Theorem 38] to morphisms of the form t_i. Therefore, (56) reduces to

(57.a) $t_2 \circ f[\Gamma'] = t_1$, and

(57.b) $q_2 \circ f[\Gamma'] = f[\Gamma] \circ q_1$.

The last two equalities follow directly from the commutation relations that determine the morphisms involved. □

COROLLARY 52. *Assume the context of Theorem 51. For each $\alpha \in \mathbf{GalC}//A$, let $r_\alpha : \pi_G(A) \longrightarrow \mathrm{Gal}(\alpha)$ be the canonical* **Conj***-morphism.*

Let $\alpha = (\Gamma, B, b) \in \mathbf{GalC}//A$. Then there is a unique **Conj***-morphism $\hat{r}_\alpha : \pi_G(A) \longrightarrow \Lambda(\alpha)$ such that, for all $\Gamma' \in Norm/\Gamma$, put $\beta = \mathrm{Res}_{\Gamma'}(\alpha)$, and then*

$$\zeta(\beta) \circ \hat{r}_\alpha = r_\beta.$$

In addition, $[\lambda(\alpha)] \circ \hat{r}_\alpha = \pi_G(E_A)$.

COROLLARY 53. *We continue in the context of Corollary 52.*

For $\alpha, \beta \in \mathbf{GalC}//A$, write "$\alpha \geq \beta$" if $\mathrm{Mor}_{\mathrm{GalC}//A}(\alpha, \beta) \neq \emptyset$. If $\alpha \geq \beta$, then there is a unique **Conj***-morphism $\rho_{\alpha,\beta} : \Lambda(\alpha) \longrightarrow \Lambda(\beta)$ which contains $\Lambda_0(f)$ for every $f \in \mathrm{Mor}_{\mathrm{GalC}//A}(\alpha, \beta)$.*

Let S be a set of **GalC**$//A$*-objects such that each* **GalC**$//A$*-object is isomorphic to some member of S. Consider the diagram Ξ in* **Conj** *in which,*

(58.a) *to each $s \in S$ assign the object $\Lambda(s)$,*

(58.b) *to each pair $(s, t) \in S^2$ such that $s \geq t$, assign the morphism $\rho_{s,t}$.*

Then assignment $s \mapsto \hat{r}_s$ *is an inverse limit of* Ξ.

PROOF. Begin with an arbitrary $\alpha \in \mathbf{GalC}//A$. We may define an inversely directed system $\Xi[\alpha]$ in **TGp**, and cone to it $\zeta^\alpha : \Lambda(\alpha) \longrightarrow \Xi[\alpha]$ as follows:

(59.a) The indexing set is $Norm/\Gamma$, ordered by containment.

(59.b) For $\Gamma' \in Norm/\Gamma$, the corresponding object in $\Xi[\alpha]$ is $\mathrm{Gal}(\mathrm{Res}_{\Gamma'}(\alpha))$, and $\zeta^\alpha(\Gamma')$ is $\zeta(\mathrm{Res}_{\Gamma'}(\alpha))$.

(59.c) For $\Gamma_1 \subseteq \Gamma_2$, the corresponding projection in $\Xi[\alpha]$ is the group homomorphism induced from restriction.

General nonsense of fibered products states that ζ^α is an inverse limit. Once this is accepted, [**F94**, Lemma 44] implies that the image of ζ^α under the canonical $\mathbf{TGp} \longrightarrow \mathbf{Conj}$ is also an inverse limit.

This observation immediately justifies existence and uniqueness of $\hat{r}_\alpha$. The conclusion of Corollary 53 is reduced to a diagram chase.

In the same manner, the cone $\Gamma \mapsto \omega[\Gamma]$ is an inverse limit in **Conj**. Equivalence $[\lambda(\alpha)] \circ \hat{r}_\alpha = \pi_G(E_A)$ follows. □

7. Endomorphisms of 2-shift

We continue with a sample application. Proof of certain comments from Section 3 is sketched. Let X be the full 2-shift. Consider action by the ring $\mathbb{F}[t, t^{-1}]$ as indicated earlier. We assume elementary facts concerning shifts (see, for example, [**LM**]).

Each non-zero $p(t) \in \mathbb{F}[t, t^{-1}]$ generates a separable map $X \longrightarrow X$. In fact, each may be considered a member of $\mathbf{GalC}//X$. In the first step of analysis, fix a choice $p(t) \in \mathbb{F}[t, t^{-1}]^*$, and let $b : X \longrightarrow X$ be the corresponding map.

Let K =ker(b). The set K is invariant under $\mathbb{Z}$. Let $N \in \mathbb{N}$ be the order of action by t on K. Consider X as an $\mathbf{HT}[\Gamma]$-object for $\Gamma = N\mathbb{Z}$. For each $k \in K$, the function $x \mapsto x + k$ is an $\mathbf{HT}[\Gamma]/X$-automorphism for $p(t)$! With respect to Γ, b is Galois with Galois group K.

To avoid ambiguity, let $X_b = \mathrm{Res}_\Gamma(X)$. That b is separable with respect to $\mathbb{Z}$ yields internal structure concerning the Galois group.

PROPOSITION 54. *Let* Ω *be a group and let* $\Gamma \lhd \Omega$ *be a normal subgroup of finite index. Let* $b_1 : B_1 \longrightarrow A$ *be an* $\mathbf{HT}[\Omega]$*-morphism such that*

(60.a) A *and* B_1 *are absolutely connected, and*

(60.b) $\mathrm{Res}_\Gamma(b_1)$ *is* $\mathbf{HT}[\Gamma]$*-Galois.*

Let $B = \mathrm{Res}_\Gamma(B_1)$ *and let* $b = \mathrm{Res}_\Gamma(b_1)$. *(That is,* B *and* b *are* B_1 *and* b_1 *interpreted as members of* $\mathbf{HT}[\Gamma]$.*) Note that* $(\Gamma, B, b) \in \mathbf{GalC}//A$. *We use the conventions of (45) and Proposition 49.*

There are several actions. We use the standard dot notation for action by Γ *on* B *or* Ω *on* $\nu[B]$. *The underlying set of* $B = B_1$ *has an action by* Ω, *which we indicate by "*$s \star x$*" for* $s \in \Omega$ *and* $x \in B$.

The short exact sequence (46) has a section. Specifically, for each $\alpha \in \Omega/\Gamma$, *there is a unique* $\rho[\alpha] \in \mathrm{Gal}(\Gamma, B, b)$ *characterized by the property that*

$$\forall s \in \alpha,\ \forall x \in B,\ \rho[\alpha](\iota(x)) = s^{-1} \cdot \iota(s \star x). \tag{61}$$

Then $\rho : \alpha \mapsto \rho[\alpha]$ *is a section. Moreover, for* $s \in \Omega$ *and* $g \in \mathrm{Gal}(b)$,

$$\rho[s\Gamma] \circ \kappa(g) \circ \rho[s\Gamma]^{-1} = \kappa(g^s), \tag{62}$$

where g^s is characterized by

$$\forall x \in B,\ g^s(x) = s \star g(s^{-1} \star x) \tag{63}$$

PROOF. Trivial. □

We return to specific analysis.

Our calculation identifies $\mathbf{HT}[\mathbb{Z}]$-isomorphic endomorphism. By general theory, two elements $p, q \in \mathbb{F}[t, t^{-1}]^*$ represent isomorphic morphisms if and only if there is $M \in \mathbb{Z}$ for which $p = t^M \cdot g$. Let **Mon** denote the multiplicative monoid of *polynomials* in $\mathbb{F}[t, t^{-1}]$ with constant term 1. Each equivalence class of $\mathbb{F}[t, t^{-1}]^*$ contains exactly one member of **Mon**.

For the moment, fix $p \in$**Mon**. Let b be the corresponding endomorphism, let K be its kernel, and let n be the degree of p as a polynomial. For each

$$\alpha = (x_0, \cdots, x_{n-1}) \in \mathbb{F}^n,$$

there is a unique element in K, denoted x^α, such that

$$\text{for } 0 \leq i \leq n-1,\ (x^\alpha)_i = x_i.$$

Consequently, K has 2^{n-1} members.

Let $\epsilon^{(p)} = x^{(1,0,\cdots,0)}$. By inspection, if $q \in$**Mon** has degree $m < n$, then

$$\begin{array}{ll} (q \cdot \epsilon^{(p)})_0 = 1 & \text{and} \\ (q \cdot \epsilon^{(p)})_i = 0 & \text{for } 1 \leq i \leq n-m-1. \end{array} \tag{64}$$

Consequently, for $h \in \mathbb{F}[t]$, $h \cdot \epsilon^{(p)} = 0$ implies that h has degree $\geq n$. We conclude that $h \mapsto h \cdot \epsilon^{(p)}$ induces a bijection

$$f^{(p)} : \mathbb{F}[t]/p\mathbb{F}[t] \longrightarrow K.$$

Moreover, for $\beta \in \mathbb{F}[t]/p\mathbb{F}[t]$,

$$f^{(p)}(t \cdot \beta) = 1 \cdot f^{(p)}(\beta).$$

Identify K with $\mathrm{Gal}(b)$. The conjugation described in (63) becomes $k^1 = 1 \cdot k$ for each $k \in K$.

Next, let $p, m \in$**Mon**, and let b, d be the corresponding endomorphisms. Let $c = b \circ d$ and $q = p \cdot m$. It is a simple conclusion from (64) that d maps $\epsilon^{(q)}$ to $\epsilon^{(p)}$. Untangling identifications, the reader should find that the induced map on kernels becomes the standard

$$w(t) + (p \cdot m) \cdot \mathbb{F}[t] \mapsto w(t) + p \cdot \mathbb{F}[t].$$

Consider the inversely directed system of such congruences. The projective limit is (the additive group of) $\widehat{\mathbb{F}[t, t^{-1}]}$, profinite completion of the ring $\mathbb{F}[t, t^{-1}]$.

Taking projective limits, one obtains a homomorphic image of π_G that is a semi-direct product

$$\widehat{\mathbb{F}[t, t^{-1}]} \rtimes \widehat{\mathbb{Z}}.$$

Conjugation by the canonical generator of $\widehat{\mathbb{Z}}$ is multiplication by t on the normal factor.

References

[F93] P. Feit, *Axiomization of Passage from 'Local' Structure to 'Global' Object*, Memoirs of Amer. Math. Soc. **485** (1993).

[F94] P. Feit, *Existence of Orbifolds I: Universal Galois Theory*, Communications in Algebra, **22**(7) (1994), 2367–2404.

[F94b] P. Feit, *Existence of Orbifolds II: Orbifold Structures*, Communications in Algebra, **22**(7) (1994), 2405–2453.

[F96] P. Feit, *Existence of Orbifolds III: Pseudoétale Topologies,* Communications in Algebra, **24**(4) (1996), 1281–1326.

[F96b] P. Feit, *Existence of Orbifolds IV: Examples*, Communications in Algebra, **24**(4) (1996), 1327–1369.

[G] A. Grothendieck, Revêtements Étales et Groupe Fondamentale, Lect. Notes in Math **224**, New York: Springer-Verlag (1971).

[LM] D. Lind & B. Marcus, An Introduction to Symbolic Dynamics and Coding, New York: Cambridge Univ. Press (1995).

University of Texas of the Permian Basin, 4901 E. University Blvd., Odessa, TX 79762

E-mail address: `feit_ p@utpb.edu`

Proceedings of Symposia in Pure Mathematics
Volume **66.1**, 1999

Critical Values of Twisted Tensor L-Functions over CM-Fields

Eknath Ghate

Dedicated to Professor Shimura with respect

Abstract. The twisted tensor L-function is a certain automorphic L-function attached to a cusp form over a quadratic extension of the underlying number field. We treat here the case the case of CM-extensions, and give a rationality result for this L-function. The result matches well with the predictions of Deligne on the critical values of motivic L-functions, and extends previous results of the author which dealt with imaginary quadratic extensions of $\mathbf{Q}$.

1. Introduction

The purpose of this article is to generalize the main result of the author's thesis: a rationality result for the twisted tensor L-function of a cuspidal automorphic form f over an imaginary quadratic field (see [**G**]). In this note we extend that result to forms over any CM field.

To state the result, let us fix some notation. Say that K is a totally imaginary quadratic extension of the totally real field F, i.e., K is a CM-field. Let f be a cusp form over K, with central character χ. Let E be the number field generated by the Fourier coefficients of f and the values of χ. Let $\mathbb{G}(s,f)$ denote the $E \otimes \mathbf{C}$-valued twisted tensor L-function of f. Then, subject to a rationality assumption on certain Eisenstein series over F at $s=0$, we show that

$$\frac{\mathbb{G}(l,f)}{\mathbb{c}_l(f)} \in E \otimes 1 \subset E \otimes \mathbf{C},$$

for each *even* 'critical' integer l. Here $\mathbb{c}_l(f) \in E \otimes \mathbf{C}$ is a period that depends only on a fixed period $\mathbb{c}(f)$, a certain Gauss sum, and a power of $2\pi i$, depending only on l and the weight of f. We refer the reader to Theorem 2 for the exact details.

The word 'critical' above is used in the sense of Deligne [**D**], i.e., the 'critical integers' are those integers at which all the Gamma factors in 'both sides' of

---

1991 *Mathematics Subject Classification.* Primary 11F67.

Key words and phrases. L-functions, critical values, twisted tensor.

This article was written while visiting the University of California at Los Angeles. I would like to sincerely thank Professor Haruzo Hida for the mathematical guidance and financial support I received from him during my stay there.

the (conjectural) functional equation are simultaneously finite. Indeed, if we interpret $\mathbb{G}(s,f)$ motivically, we see that our rationality result fits nicely into Deligne's framework on the critical values of motivic L-functions. Our result, however, does not yield any information about $\mathbb{G}(s,f)$ at the odd critical integers. To investigate these values, one could first establish the functional equation (see [**G**, Remark 4, Section 8] for the imaginary quadratic case) or, instead, establish the rationality of the Eisenstein series at other suitable integral values of s. In any case, we do not explore this matter in this paper.

In view of the purpose of this volume, it should be pointed out that the conjectures in [**D**] were made only after Shimura had established rationality results a few years earlier for L-functions attached to algebraic Hecke-characters over CM fields (see [**S75**]), and elliptic cusp forms (see [**S76**] and [**S77**]). Moreover, the twisted tensor L-function was also first introduced by Shimura, though not by this name. In fact, Shimura and Asai (see [**A**]) establish its basic analytic properties in the real quadratic case. Later, Shimura even went on to investigate its special values (in the totally real quadratic case) for cusp forms coming from the Jacquet-Langlands correspondence (see [**S81**]). The reader is referred to the introduction of [**G**] for other work (of an analytic nature) on the twisted tensor L-function.

The proof of the rationality result is based on the basic principle that underlies [**H94**]: one may establish rationality results by comparing two different rational structures on the space of cusp forms—the first coming from Fourier expansions, and the other from cohomology. Indeed, in [**H94**], Hida uses this principle to establish rationality results for the standard L-function and the Rankin product L-function over general number fields. For a summary of the basic ingredients used in this method, the reader is referred to [**H96**, Sections 0 and 1] which appears elsewhere in this volume, as well as, to the original computations in [**H94**].

Our proof in the present case is also based on this principle. In particular, the methods used here directly generalize the methods used to establish the rationality result for the twisted tensor L-function in the imaginary quadratic case. It might be useful, therefore, for the reader to first browse through [**G**], for therein he or she will find many of the essential details, with the added simplicity that the base field $F = \mathbf{Q}$ has strict class number 1 and only one archimedean place.

NOTATION: Before we begin, we take this opportunity to set up some notation. Fix a totally positive element $\Delta \in F$. We assume that $K = F(\sqrt{-\Delta})$. For a number field L, let I_L denote the set of imbeddings of L into $\mathbf{C}$. Let c denote complex conjugation. Say $[F:\mathbf{Q}] = r$, so that $[K:\mathbf{Q}] = 2r$. Let $\Sigma_K \subset I_K$ be the subset of those $\sigma \in I_K$, determined by $\sqrt{-\Delta}^{\,\sigma} = i\sqrt{\Delta^\sigma}$ (as opposed to $-i\sqrt{\Delta^\sigma}$), and by the fact that

$$I_K = \{\sigma \,|\, \sigma \in \Sigma_K\} \cup \{\sigma c \,|\, \sigma \in \Sigma_K\}.$$

We identify, Σ_K with the set of archimedean places of K, and sometimes (by restriction) with those of F as well.

Fix formal sums

$$\boldsymbol{n} = \sum n_\sigma \sigma \quad \text{and} \quad \boldsymbol{v} = \sum v_\sigma \sigma \;\in\; \mathbf{Z}[I_K],$$

with $n_\sigma \geq 0$, for all $\sigma \in I_K$. These will play the role of the 'weights' of the cusp form f. For each $\sigma \in \Sigma_K$, let

$$n^*_\sigma = n_\sigma + n_{\sigma c} + 2,$$

and set $\boldsymbol{n}^* = \sum_{\sigma\in\Sigma_K} n^*_\sigma \sigma$. Let $W(\boldsymbol{n}^*, \mathbf{C})$ be the space of polynomials, with coefficients in $\mathbf{C}$, in the $2r$ variables $\boldsymbol{s} = \left\{\boldsymbol{s_\sigma} = \left(\begin{smallmatrix} S_\sigma \\ T_\sigma \end{smallmatrix}\right)\right\}_{\sigma\in\Sigma_K}$, homogeneous of degree n^*_σ in $\boldsymbol{s_\sigma}$. In general, we will consider functions $f : X \to W(\boldsymbol{n}^*, \mathbf{C})$, for various spaces X, in which case we will sometimes write $f(x,\, \boldsymbol{s})$ to emphasize the dependence of f on the variables $\boldsymbol{s}$.

Finally, let $\mathcal{O}_K$ (resp. $\mathcal{O}_F$) be the ring of integers of K (resp. F). Let $\mathfrak{N} \subset \mathcal{O}_K$ a fixed ideal, and let $N = \mathfrak{N} \cap \mathcal{O}_F$ denote the contraction of $\mathfrak{N}$ to $\mathcal{O}_F$.

2. Preliminaries

2.1. Cusp forms associated to K. Let us define what we mean by a cusp form over K. Let $G = \mathrm{Res}_{K/\mathbf{Q}}\mathrm{GL}_{2/K}$, and let Z be its center. Then $G(\boldsymbol{A}) = \mathrm{GL}_2(K_{\boldsymbol{A}})$, where $K_{\boldsymbol{A}}$ is the adèle ring of K, and we identify the center $Z(\boldsymbol{A})$ of $G(\boldsymbol{A})$ with the idele group $K^\times_{\boldsymbol{A}}$.

Let $\chi : K^\times\backslash K^\times_{\boldsymbol{A}} \to \mathbf{C}^\times$ be a Hecke character, whose conductor divides $\mathfrak{N}$, and whose infinity type is $-\boldsymbol{n} - 2\boldsymbol{v}$. Let

$$U_0(\mathfrak{N}) = \left\{\left(\begin{smallmatrix} a & b \\ c & d \end{smallmatrix}\right) \in \mathrm{GL}_2(\widehat{\mathcal{O}_K}) \mid c \equiv 0 \bmod \mathfrak{N}\widehat{\mathcal{O}_K}\right\},$$

where $\widehat{\mathcal{O}_K} = \prod_{\wp \text{ finite}} \mathcal{O}_{K\,\wp}$. Note $\chi|_{\widehat{\mathcal{O}_K}^\times}$ may be considered as a character on $(\mathcal{O}_K/\mathfrak{N}\mathcal{O}_K)^\times$, and we will denote this character by $\chi_\mathfrak{N}$. For $u_f = \left(\begin{smallmatrix} a & b \\ c & d \end{smallmatrix}\right) \in U_0(\mathfrak{N})$ we set

$$\chi_\mathfrak{N}(u_f) = \chi_\mathfrak{N}(d) = \prod_{\wp|\mathfrak{N}} \chi_\wp(d_\wp).$$

Finally, let $C_\infty = \prod_{\sigma\in\Sigma_K} SU_2(\mathbf{C})$ be the standard maximal compact subgroup of $G(\mathbf{R}) = \prod_{\sigma\in\Sigma_K} \mathrm{GL}_2(\mathbf{C})$.

DEFINITION 1. A cusp form (of weight $(\boldsymbol{n}, \boldsymbol{v})$, level $\mathfrak{N}$, and central action χ) is a function $f : G(\boldsymbol{A}) \to W(\boldsymbol{n}^*, \mathbf{C})$ satisfying the following properties

1. For all $\gamma \in \mathrm{GL}_2(K)$, $z \in K^\times_{\boldsymbol{A}}$ and $g \in G(\boldsymbol{A})$,
$$f(\gamma z g,\, \boldsymbol{s}) = \chi(z)\, f(g,\, \boldsymbol{s}).$$
2. For all $u = u_f \cdot \prod_{\sigma\in\Sigma_K} u_\sigma \in U_0(\mathfrak{N}) \cdot C_\infty$,
$$f(gu,\, \boldsymbol{s}) = \chi_\mathfrak{N}(u_f)\, f(g, (u_\sigma \boldsymbol{s_\sigma})_\sigma).$$
3. f is an eigenfunction of the Casimir operators D_σ: For all $\sigma \in I_K$,
$$D_\sigma f = (n_\sigma^2/2 + n_\sigma) f,$$
where we consider $f(g_f g_\infty,\, \boldsymbol{s})$ as a function of $g_\infty \in G(\mathbf{R})$. (see [**H93**, Section 1.3])
4. f satisfies the cuspidal condition
$$\int_{U(\mathbf{Q})\backslash U(\boldsymbol{A})} f(vg,\, \boldsymbol{s})\, du = 0$$
for all $g \in G(\boldsymbol{A})$, where
$U(\mathbf{Q}) = \left\{v = \left(\begin{smallmatrix} 1 & u \\ 0 & 1 \end{smallmatrix}\right) \mid u \in K\right\}$, and $U(\boldsymbol{A}) = \left\{v = \left(\begin{smallmatrix} 1 & u \\ 0 & 1 \end{smallmatrix}\right) \mid u \in K_{\boldsymbol{A}}\right\}$,
and du is the Lebesgue measure on $K_{\boldsymbol{A}}$.

We denote the space of such forms by $\boldsymbol{S}_{(\boldsymbol{n},\boldsymbol{v})}(\mathfrak{N},\chi)$. Note that since we are dealing with cusp forms we may assume (see [**H94**, Section 2.5, Corollary 2.2]) that $n_\sigma = n_{\sigma c}$, for all $\sigma \in \Sigma_K$. In particular $\boldsymbol{n}^* = \sum_{\sigma\in\Sigma_K}(2n_\sigma+2)\sigma$. Also, $\boldsymbol{n}$ and $\boldsymbol{v}$ are not completely independent. Indeed, for each unit $\epsilon \in \mathcal{O}_K{}^\times$, $\chi(\epsilon) = \chi_f(\epsilon)\chi_\infty(\epsilon) = 1$. Since $\chi_f(\epsilon) = 1$, for all ϵ in some finite index subgroup of $\mathcal{O}_F{}^\times$, we conclude that, for all such ϵ

$$\prod_{\sigma\in\Sigma_K} (\epsilon^\sigma)^{2n_\sigma+2v_\sigma+2v_{\sigma c}} = 1.$$

This implies that

(1) $$n_\sigma + v_\sigma + v_{\sigma c} \quad \text{is independent of } \sigma \in \Sigma_K.$$

2.2. Fourier expansions. If $f : G(\boldsymbol{A}) \to W(\boldsymbol{n}^*, \mathbf{C})$ is a cusp form as above, then f has Fourier expansion (cf. [**H94**, Theorem 6.1]):

$$f\left(\begin{smallmatrix} y & x \\ 0 & 1\end{smallmatrix}\right) = |y|_{K_{\boldsymbol{A}}} \sum_{\xi\in K^\times} c\,(\xi y d, f)\, W(\xi y_\infty)\, \mathbf{e}_K(\xi x).$$

Here

- $|\ |_{K_{\boldsymbol{A}}}$ is the usual idele character of $K_{\boldsymbol{A}}^\times$ trivial on $K^\times$.
- d is an idele such that $d\mathcal{O}_K = \vartheta_{K/\mathbf{Q}}$, the different of K.
- The Fourier coefficient $c\,(\ \ , f)$ may be considered as a function on the fractional ideals of K that vanishes outside the integral ideals.
- $W : \mathbf{C}^\times \to W(\boldsymbol{n}^*, \mathbf{C})$ is the Whittaker function

  $$W(s) = \prod_{\sigma\in\Sigma_K} W_\sigma(s)$$

  where

  $$W_\sigma(s) = \sum_{\alpha_\sigma=0}^{n_\sigma^*} \binom{n_\sigma^*}{\alpha_\sigma} \frac{1}{s^{v_\sigma}\bar{s}^{v_{\sigma c}}} \left(\frac{s}{i|s|}\right)^{n_\sigma+1-\alpha_\sigma} K_{\alpha_\sigma-n_\sigma-1}(4\pi|s|)\, S_\sigma^{n_\sigma^*-\alpha_\sigma} T_\sigma^{\alpha_\sigma},$$

  where $K_\alpha(x)$ is the modified Bessel function as in [**G**, Section 2.1].
- $\mathbf{e}_K$ is an additive character of $K_{\boldsymbol{A}}$, trivial on K, defined via

  $$\mathbf{e}_K = \prod_{\wp} (\mathbf{e}_p \circ \mathrm{tr}_{K_\wp/\mathbf{Q}_p}) \cdot \prod_{\sigma\in\Sigma_K} (\mathbf{e}_\infty \circ \mathrm{tr}_{\mathbf{C}/\mathbf{R}}),$$

  where $\mathbf{e}_p(\sum_j c_j p^j) = e^{-2\pi i \sum_{j<0} c_j p^j}$ and $\mathbf{e}_\infty(r) = e^{2\pi i r}$.

2.3. Hyperbolic $3r$-dimensional space, $\mathcal{Z}$. We will need to work with certain functions and differential forms associated to f defined on the product of r copies of hyperbolic upper half three space

$$\mathcal{H} = \left\{ \left(\begin{smallmatrix} x & -y \\ y & \bar{x}\end{smallmatrix}\right) \,\middle|\, x \in \mathbf{C},\ y \in \mathbf{R},\ y > 0 \right\}.$$

Recall that $\mathrm{SL}_2(\mathbf{C})$ acts transitively on $\mathcal{H}$, via

$$\gamma \cdot z = (\rho(a)\, z + \rho(b))\, (\rho(c)\, z + \rho(d))^{-1}$$

where

$$\gamma = \left(\begin{smallmatrix} a & b \\ c & d\end{smallmatrix}\right) \in \mathrm{SL}_2(\mathbf{C}), \quad z = \left(\begin{smallmatrix} x & -y \\ y & \bar{x}\end{smallmatrix}\right) \in \mathcal{H} \quad \text{and} \quad \rho(t) = \left(\begin{smallmatrix} t & 0 \\ 0 & \bar{t}\end{smallmatrix}\right).$$

Fix a point $\epsilon = \left(\begin{smallmatrix} 0 & -1 \\ 1 & 0 \end{smallmatrix}\right) \in \mathcal{H}$. Then the stabilizer of ϵ is $SU_2(\mathbf{C})$ and so we may identify the symmetric spaces

$$\mathrm{SL}_2(\mathbf{C})/SU_2(\mathbf{C}) \simeq \mathcal{H}.$$

Note that the the Poincaré upper half plane H naturally sits in $\mathcal{H}$ via

$$x + iy \hookrightarrow \left(\begin{smallmatrix} x & -y \\ y & x \end{smallmatrix}\right).$$

The above action of $\mathrm{SL}_2(\mathbf{C})$ on $\mathcal{H}$ is a natural extension of the standard action of $\mathrm{SL}_2(\mathbf{R})$ on H via fractional linear transformations.

For $\gamma = \left(\begin{smallmatrix} a & b \\ c & d \end{smallmatrix}\right) \in \mathrm{SL}_2(\mathbf{C})$ and $z \in \mathcal{H}$, we have the automorphy factor

$$j_1(\gamma, z) = \rho(c)\, z + \rho(d).$$

For each $\sigma \in \Sigma_K$, let $\mathcal{H}_\sigma = \mathcal{H}$ and set $\mathcal{Z} := \prod_\sigma \mathcal{H}_\sigma$. For $\gamma = \left(\begin{smallmatrix} a & b \\ c & d \end{smallmatrix}\right) \in \mathrm{SL}_2(K)$ and $z = (z_\sigma) \in \mathcal{Z}$, let

$$\gamma \cdot z := (\gamma^\sigma \cdot z_\sigma) \quad \text{and} \quad j(\gamma, z) := (j_1(\gamma^\sigma, z_\sigma))_\sigma,$$

where $\gamma^\sigma = \left(\begin{smallmatrix} a^\sigma & b^\sigma \\ c^\sigma & c^\sigma \end{smallmatrix}\right)$. We have

$$j(\gamma_1\gamma_2, z) = j(\gamma_1, \gamma_2 z)\, j(\gamma_2, z),$$

for all $\gamma_1, \gamma_2 \in \mathrm{SL}_2(K)$ and $z \in \mathcal{Z}$.

Finally, note that every C^∞ 1-form on $\mathcal{H}_\sigma$ can be written as a $C^\infty(\mathcal{H}_\sigma)$-linear combination of $dx_\sigma, -dy_\sigma, -d\bar{x}_\sigma$, where $C^\infty(\mathcal{H}_\sigma)$ is the space of C^∞ function on $\mathcal{H}_\sigma$. Let ρ^σ_m denote the mth symmetric tensor representation of $\mathrm{GL}_2(\mathbf{C})$. Thus

$$\rho^\sigma_m\left(\left(\begin{smallmatrix} a & b \\ c & d \end{smallmatrix}\right)\right)\left(\begin{smallmatrix} S_\sigma \\ T_\sigma \end{smallmatrix}\right)^m = \left(\left(\begin{smallmatrix} a & b \\ c & d \end{smallmatrix}\right)\left(\begin{smallmatrix} S_\sigma \\ T_\sigma \end{smallmatrix}\right)\right)^m$$

where $\left(\begin{smallmatrix} S_\sigma \\ T_\sigma \end{smallmatrix}\right)^m = ({S_\sigma}^m, {S_\sigma}^{m-1}T_\sigma, \ldots, S_\sigma {T_\sigma}^{m-1}, {T_\sigma}^m)^t$. Note that ρ^σ_1 is the standard representation of $\mathrm{GL}_2(\mathbf{C})$ on $\mathbf{C}^2$ and that $W(\boldsymbol{n}^*, \mathbf{C})$ with $G(\mathbf{R})$-action $(\gamma_\sigma) \cdot P\left(\left(\begin{smallmatrix} X_\sigma \\ Y_\sigma \end{smallmatrix}\right)\right) = P\left(\gamma^\iota_\sigma \left(\begin{smallmatrix} X_\sigma \\ Y_\sigma \end{smallmatrix}\right)\right)$ is a model for the representation dual to $\rho_{n^*} = \otimes_\sigma \rho^\sigma_{n^*}$. Here $\gamma^\iota_\sigma = \det(\gamma_\sigma)\gamma_\sigma^{-1}$ is the adjoint of γ_σ. Also recall that (cf. [**H94**, Equation 2.4], for each $\gamma_\sigma \in \mathrm{SL}_2(\mathbf{C})$, the behavior of of 1-forms on $\mathcal{H}_\sigma$ under 'pull back by γ' is given by

$$\begin{pmatrix} d\,(x_\sigma \circ \gamma) \\ -d\,(y_\sigma \circ \gamma) \\ -d\,(\bar{x}_\sigma \circ \gamma) \end{pmatrix} = \rho^\sigma_2(j(\gamma, z)^t)^{-1} \begin{pmatrix} dx_\sigma \\ -dy_\sigma \\ -d\bar{x}_\sigma \end{pmatrix}. \tag{2}$$

2.4. Relation between forms on $\mathrm{GL}_2(K_A)$ and $\mathcal{Z}$. Let h denote the class number of K. The strong approximation theorem gives us the decomposition

$$\mathrm{GL}_2(K_{\boldsymbol{A}}) = \coprod_{i=1}^{h} \mathrm{GL}_2(K)\, t_i\, U_0(\mathfrak{N})\, G(\mathbf{R})$$

where $t_i = \left(\begin{smallmatrix} a_i & 0 \\ 0 & 1 \end{smallmatrix}\right)$, and the ideles a_i satisfy $a_{i\infty} = 1$.

For any fractional ideal $\mathfrak{a}$ of $\mathcal{O}_K$, we define the discrete subgroup $\Gamma_{\mathfrak{a}}$ of $\mathrm{SL}_2(K)$ via

$$\Gamma_{\mathfrak{a}} = \left\{\left(\begin{smallmatrix} a & b \\ c & d \end{smallmatrix}\right) \,\middle|\, a, d \in \mathcal{O}_K,\ b \in \mathfrak{a},\ c \in \mathfrak{a}^{-1}\mathfrak{N},\ ad - bc = 1\right\}.$$

We see that $\mathrm{SL}_2(K) \cap t_i\, U_0(\mathfrak{N})\, t_i^{-1}\, G(\mathbf{R}) = \Gamma_{\mathfrak{a}_i}$, where $\mathfrak{a}_i = a_i \mathcal{O}_K$.

Set

$$Y_0(\mathfrak{N}) = G(\mathbf{Q})\backslash G(\boldsymbol{A})/U_0(\mathfrak{N}) Z(\mathbf{R}) C_\infty.$$

This is a manifold of dimension $3r$. Indeed, we have the decomposition

$$Y_0(\mathfrak{N}) = \coprod_{i=1}^{h} \Gamma_{\mathfrak{a}_i} \backslash \mathcal{Z}.$$

Now starting with a cusp form $f \in \boldsymbol{S}_{(\boldsymbol{n},\boldsymbol{v})}(\mathfrak{N}, \chi)$, and *any* element $t_i = \left(\begin{smallmatrix} a_i & 0 \\ 0 & 1 \end{smallmatrix}\right) \in G(\boldsymbol{A})$, with $a_{i\infty} = 1$ (we will specify how to choose the $a_i \in K_{\boldsymbol{A}}^{\times}$ later), we show how to associate to it a cusp form f_i on $\mathcal{Z}$. Define

$$F_i : G(\mathbf{R}) \longrightarrow W(\boldsymbol{n}^*, \mathbf{C})$$

by $F_i(g) = f(t_i g)$ and let $f_i : \mathcal{Z} \to W(\boldsymbol{n}^*, \mathbf{C})$ be given by

$$f_i(z, \boldsymbol{s}) = F_i\left(g, j(g_\sigma, \epsilon)^t \left(\begin{smallmatrix} S_\sigma \\ T_\sigma \end{smallmatrix}\right)\right),$$

where $g = (g_\sigma)$, $z = (z_\sigma)$ and $g_\sigma \in \mathrm{SL}_2(\mathbf{C})$ is chosen such that $g_\sigma \cdot \epsilon = z_\sigma$. It is a routine matter to check that f_i satisfies

$$f_i(\gamma z, \boldsymbol{s}) = \chi_{\mathfrak{N}}^{-1}(d)\, f_i(z, j(\gamma, z)^t\, \boldsymbol{s}),$$

for all $\gamma = \left(\begin{smallmatrix} a & b \\ c & d \end{smallmatrix}\right) \in \Gamma_{\mathfrak{a}_i}$. Thus f_i is an element of $\mathcal{S}_{\boldsymbol{n}}(\Gamma_{\mathfrak{a}_i}, \chi_{\mathfrak{N}}^{-1})$, the space of cusp forms on $\mathcal{Z}$ satisfying the automorphy condition above.

The Fourier expansion of f also descends to $\mathcal{Z}$. In fact one may check that each f_i has Fourier expansion

$$(3)\quad f_i(z, \boldsymbol{s}) = |a_i|_{K_{\boldsymbol{A}}} \sum_{\alpha=0}^{n^*} \binom{n^*}{\alpha} \sum_{\xi \in K^\times} c(\xi a_i d) \prod_\sigma \left\{ y_\sigma \left(\frac{\xi^\sigma}{i|\xi^\sigma|}\right)^{n_\sigma+1-\alpha_\sigma} \right.$$
$$\left. \cdot \frac{1}{\xi^{\sigma\, v_\sigma} \overline{\xi^{\sigma}}^{\,v_{\sigma c}}} K_{\alpha_\sigma - n_\sigma - 1}(4\pi y_\sigma |\xi^\sigma|)\, e^{2\pi i \mathrm{Tr}_{\mathbf{C}/\mathbf{R}}(\xi^\sigma x_\sigma)}\, S_\sigma^{n_\sigma^* - \alpha_\sigma} T^{\alpha_\sigma} \right\}.$$

Here we are using 'multi-index' notation: thus the sum over α is really an r-fold sum over $0 \leq \alpha_\sigma \leq n_\sigma^*$, for each $\sigma \in \Sigma_K$, and $\binom{n^*}{\alpha} = \prod_\sigma \binom{n_\sigma^*}{\alpha_\sigma}$.

2.5. The twisted tensor L-function of f. Let $f \in \boldsymbol{S}_{(\boldsymbol{n},\boldsymbol{v})}(\mathfrak{N}, \chi)$ be as in Definition 1. In addition, we will hereafter assume that f is a normalized primitive form. In this section we recall the definition of the twisted tensor L-function of f. Let

$$\chi|_{F_{\boldsymbol{A}}} : F^\times \backslash F_{\boldsymbol{A}}^\times \longrightarrow \mathbf{C}^\times$$

denote the restriction of χ to the ideles of F. We also think of χ and $\chi|_{F_{\boldsymbol{A}}}$ as ideal characters.

Recall that the 'standard' L-function of f is given by

$$D(s, f) = \sum_{\mathfrak{m} \subset \mathcal{O}_K} \frac{c(\mathfrak{m}, f)}{N_{K/\boldsymbol{Q}}(\mathfrak{m})^s}.$$

Set

$$G'(s, f) = \sum_{M \subset \mathcal{O}_F} \frac{c(M\mathcal{O}_K, f)}{N_{F/\boldsymbol{Q}}(M)^s}$$

and, following Shimura and Asai (see [**A**]), define the twisted tensor L-function of f by

$$G(s, f) = L(2s - 2, \chi|_{F_{\boldsymbol{A}}}) \cdot G'(s, f).$$

Thus the twisted tensor L-function is essentially a 'sub' L-function of $D(s,f)$, obtained by restricting the summation to (integral) ideals coming from F, modified by a Hecke L-function to guarantee a good Euler product expansion. The choice of nomenclature 'twisted tensor' becomes more transparent when one interprets $G(s,f)$ in the context of automorphic representations (see for instance [**F**]). In the literature, $G(s,f)$ is sometimes referred to as an L-function of 'Asai type' (no doubt because it first makes its appearance in the literature in [**A**]).

The Euler expansion of $G(s,f)$ is well-known (see [**A**], [**Im**], [**S81**], [**Z**], and its derivation in [**G96**]). We state it now so that we may eventually conjecturally realize $G(s,f)$ as a motivic L-function. For simplicity, write $c(\mathfrak{m})$ for $c(\mathfrak{m},f)$. Recall that

$$D(s,f)=\prod_{\wp\mid\mathfrak{N}}\frac{1}{1-c(\wp)N(\wp)^{-s}}\cdot\prod_{\wp\nmid\mathfrak{N}}\frac{1}{1-c(\wp)N(\wp)^{-s}+\chi(\wp)N(\wp)^{1-2s}}.$$

Let $\alpha_\wp$ and $\beta_\wp$ denote the reciprocal roots of the polynomial

$$1-c(\wp)X+\chi(\wp)\,N(\wp)X^2.$$

Then $G(s,f)=\prod_P G_P(s,f)$ where

$$\frac{1}{G_P(s,f)}=\begin{cases}(1-\alpha_\wp\alpha_{\bar\wp}P^{-s})(1-\alpha_\wp\beta_{\bar\wp}P^{-s})(1-\beta_\wp\alpha_{\bar\wp}P^{-s})(1-\beta_\wp\beta_{\bar\wp}P^{-s}) & \text{if } P\mathcal{O}_K=\wp\bar\wp,\\ (1-\alpha_\wp P^{-s})(1-\chi|_{F_{\mathbf{A}}}(P)P^{-2s+2})(1-\beta_\wp P^{-s}) & \text{if } P\mathcal{O}_K=\wp,\\ (1-\alpha_\wp^2P^{-s})(1-\chi(\wp)P^{-s+1})(1-\beta_\wp^2P^{-s}) & \text{if } P\mathcal{O}_K=\wp^2.\end{cases}$$

REMARK 1. It is possible to write down a 'cousin' $G(s,f,\chi_{K/F})$ of $G(s,f)$, which is defined by twisting $G(s,f)$ by the quadratic character $\chi_{K/F}$ associated to the extension K/F. This L-function also has a degree four Euler product over F and one has the well known relation (see [**A**, Theorem 3]):

$$H(s,f)=G(s,f)\cdot G(s,f,\chi_{K/F}),$$

where $H(s,f)$ is the Rankin product L-function, over K, of f and f^c. Here f^c is the form conjugate to f (with Fourier coefficients $c(\wp,f^c)=c(\wp^c,f)$, where we think of c as the non-trivial element of $\mathrm{Gal}(K/F)$). In this note we concentrate only on $G(s,f)$, although a rationality result could possibly be proven for $G(s,f,\chi_{K/F})$ as well.

3. Motivic interpretation of $G(s,f)$

Let f be as in Section 2.5. Following [**H94**], we hypothesize the existence of a rank 2 motive M over K, associated to f, with coefficients in

$$E=\mathbf{Q}(f,\chi),$$

the number field generated by the Fourier coefficients of f and the values of χ. Moreover, we expect M to be pure of weight $w=n_\sigma+1+v_\sigma+v_{\sigma c}$. By (1), w is indeed independent of σ. We have the usual realizations of M (see [**D**]):

- Betti realization: $H_\sigma(M)$, for all $\sigma \in I_K$. Each is a two dimensional E-vector space.
- de Rham realization: $H_{DR}(M)$, a two dimensional $E \otimes K$-module.
- ℓ-adic realization: $H_\ell(M)$, an $E \otimes \mathbf{Q}_\ell$-module. For each $\lambda|\ell$, $H_\ell(M)$ may be regarded as a two dimensional E_λ-module, equipped with a continuous action of $\mathrm{Gal}(\bar{\mathbf{Q}}/K)$, say ρ. We expect ρ to be unramified outside $\mathfrak{N}\ell$. Let
$$L_\wp(X) = \det(1-\rho(\mathrm{Frob}_\wp)X, H_\ell(M)^{I_\wp})^{-1}.$$
Then $L_\wp(X)$ is conjecturally a polynomial in $E(X) \subset E_\lambda(X)$ and we define (see Remark 2 below)

$$L(s,M) = \prod_\wp L_\wp(N_{F/\mathbf{Q}}(\wp))^{-s}. \tag{4}$$

Conjecturally we expect that
$$\begin{aligned} D_\wp(s,F) &:= (1-\alpha_\wp N(\wp)^{-s})^{-1}(1-\beta_\wp N(\wp)^{-s})^{-1} \\ &= L_\wp(N_{F/\mathbf{Q}}(\wp))^{-s} \end{aligned}$$
holds for all $\wp$. In particular we expect that $D(s,f) = L(s,M)$.

We have Hodge decompositions (see [**H94**, Section 8]); for each $\sigma \in I_K$
$$\begin{aligned} H_\sigma(M) \otimes \mathbf{C} &= H_\sigma^{(n_\sigma+1+v_\sigma, v_{\sigma c})}(M) \oplus H_\sigma^{(v_\sigma, n_\sigma+1+v_{\sigma c})}(M) \\ &= (E\otimes \mathbf{C})\, x_\sigma \oplus (E \otimes \mathbf{C})\, y_\sigma. \end{aligned}$$
If $F_\infty : H_\sigma(M) \to H_{\sigma c}(M)$ is the isomorphism induced functorially from the action of complex conjugation, then $F_\infty \otimes 1$ takes x_σ to $y_{\sigma c}$ and y_σ to $x_{\sigma c}$.

Let M^c be the motive conjugate to M, i.e., the motive associated to the conjugate form f^c. This has realizations conjugate to those of M. In particular, if $\rho^c : \mathrm{Gal}(\bar{\mathbf{Q}}/K) \longrightarrow \mathrm{Aut}(H_\ell(M^c))$ denotes the corresponding ℓ-adic representation of M^c, then we may identify $H_\ell(M^c)$ with $H_\ell(M)$ and ρ^c with the conjugate action of ρ, namely

$$\rho^c(h) = \rho(chc^{-1}). \tag{5}$$

Consider the motive $M \otimes M^c$, a rank 4 motive over K, of weight $2w = 2n_\sigma + 2 + 2v_\sigma + 2v_{\sigma c}$, with coefficients in E, obtained by tensoring M and M^c. This has realizations

- Betti: $H_\sigma(M \otimes M^c) = H_\sigma(M) \otimes_E H_\sigma(M^c)$, for $\sigma \in I_K$,
- de Rham : $H_{DR}(M\otimes M^c) = H_{DR}(M) \otimes_{E\otimes K} H_{DR}(M^c)$,
- ℓ-adic : $H_\ell(M \otimes M^c) = H_\ell(M) \otimes_{E_\lambda} H_\ell(M^c)$, with Galois action $\rho \otimes \rho^c$: $\mathrm{Gal}(\bar{\mathbf{Q}}/K) \longrightarrow \mathrm{Aut}(V \otimes V^c)$, where $V = V^c = H_\ell(M)$, but the action on V^c is as in (5).

By comparing Euler products, one easily sees that
$$H(s,f) = L(s, M \otimes M^c).$$
We may also realize $G(s,f)$ and $G(s,f,\chi_{K/F})$ as motivic L-functions. This is a simple generalization of the imaginary quadratic case, so we only briefly sketch the details here. We first extend the representation $\rho \otimes \rho^c$ of $H = \mathrm{Gal}(\bar{\mathbf{Q}}/K)$ to

$G = \mathrm{Gal}(\bar{\mathbf{Q}}/F)$ in two ways. Define

$$\begin{aligned} R_\pm : G &\longrightarrow \mathrm{Aut}(V \otimes V^c) && \text{via} \\ R_\pm(h)(v_1 \otimes v_2) &= hv_1 \otimes chc^{-1}v_2 && \text{if } h \in H, \\ R_\pm(c)(v_1 \otimes v_2) &= \pm v_2 \otimes h_0 v_1, \end{aligned}$$

where $c^2 = h_0$, and we are thinking of c more generally as any element of G whose image under restriction $G \to \mathrm{Gal}(K/F)$ is a generator. Here we are omitting ρ from the notation. One can easily check that $R_\pm : G \longrightarrow \mathrm{Aut}(V \otimes V^c)$ are well defined homomorphisms, such that $R_\pm|_H = \rho \otimes \rho^c$. Furthermore, R_+ (respectively R_-) is independent of the choice of $c \in G$ up to isomorphism. Regarding $\chi_{K/F}$ as a character of G (trivial on H) we see that $R_- = R_+ \otimes \chi_{K/F}$.

The motive $M \otimes M^c$ descends to F in two corresponding ways, which we denote by $(M \otimes M^c)_+$ and $(M \otimes M^c)_-$. These are rank 4 motives over F, of weight $2w$, with coefficients in E, defined with respect to the descent data

$$\begin{aligned} i_+ : M \otimes M^c &\longrightarrow M^c \otimes M && u \otimes v \mapsto v \otimes u \\ i_- : M \otimes M^c &\longrightarrow M^c \otimes M && u \otimes v \mapsto -v \otimes u. \end{aligned}$$

$(M \otimes M^c)_+$ has realizations

- Betti: $H_\tau((M \otimes M^c)_+) = H_\sigma(M) \otimes H_{\sigma c}(M)$ for all $\tau \in I_F$, where $\sigma\big|_F = \sigma c\big|_F = \tau$,
- de Rham: $H_{DR}((M \otimes M^c)_+) = (H_{DR}(M) \otimes H_{DR}(M^c))^{\mathrm{Gal}(K/F)}$ for a suitable action of $\mathrm{Gal}(K/F)$,
- ℓ-adic: $H_\ell((M \otimes M^c)_+) = H_\ell(M) \otimes_{E_\lambda} H_\ell(M^c)$, with $\mathrm{Gal}(\bar{\mathbf{Q}}/F)$-action given by R_+.

We have analogous realizations for $(M \otimes M^c)_-$. A computation comparing Euler factors, exactly analogous to the one in [**G**, Section 4.1, Proposition 1], yields

$$G(s, f) = L(s, (M \otimes M^c)_+) \quad \text{and} \quad G(s, f, \chi_{K/F}) = L(s, (M \otimes M^c)_-).$$

REMARK 2. Throughout the discussion above, we have fixed an embedding $\iota : E \hookrightarrow \mathbf{C} \in I_E$. Thus for instance in (4) above, we should really have written $L(\iota, s, M)$ instead of $L(s, M)$. Deligne, in fact, defines motivic L-functions that take values in the algebra $E \otimes \mathbf{C} = \mathbf{C}^{I_E}$, and so take into account all the imbeddings of E into $\mathbf{C}$:

$$\mathbb{L}(s, M) := \prod_{\wp} L_\wp(1 \otimes N_{F/\mathbf{Q}}(\wp)^{-s}) = \prod_{\tau} L(\iota, s, M).$$

We may interpret the other L-functions above in a similar manner. In particular, we have the conjectural identity of $E \otimes \mathbf{C}$-valued L-functions

$$\mathbb{G}(s, f) = \mathbb{L}(s, (M \otimes M^c)_+),$$

where

$$\mathbb{G}(s, f) = \mathbb{L}(2s-2, \chi|_{F_{\mathbf{A}}}) \cdot \sum_{M \subset \mathcal{O}_F} c\,(M\mathcal{O}_K, f) \otimes N_{F/\mathbf{Q}}(M)^{-s},$$

with

$$\mathbb{L}(s, \chi|_{F_{\mathbf{A}}}) = \prod_{P \subset \mathcal{O}_F} (1 - \chi|_{F_{\mathbf{A}}}(P) \otimes N_{F/\mathbf{Q}}(P)^{-s})^{-1}.$$

3.1. Critical integers. Let us compute the critical integers of $\mathbb{L}(s,(M\otimes M^c)_+)$. For each $\tau\in I_F$, the Hodge decomposition of $(M\otimes M^c)_\pm$ is given by

$$\begin{aligned}H_\tau((M\otimes M^c)_\pm)\otimes\mathbf{C} &= H_\tau^{(2n_\sigma+2+v_\sigma+v_{\sigma c},v_\sigma+v_{\sigma c})}\oplus H_\tau^{(n_\sigma+1+v_\sigma+v_{\sigma c},n_\sigma+1+v_\sigma+v_{\sigma c})}\\ &\quad\oplus H_\tau^{(n_\sigma+1+v_\sigma+v_{\sigma c},n_\sigma+1+v_\sigma+v_{\sigma c})}\oplus H_\tau^{(v_\sigma+v_{\sigma c},2n_\sigma+2+v_\sigma+v_{\sigma c})}\\ &= (E\otimes\mathbf{C})\,(x_\sigma\otimes x_{\sigma c})\,\oplus\,(E\otimes\mathbf{C})\,(x_\sigma\otimes y_{\sigma c})\\ &\quad\oplus(E\otimes\mathbf{C})\,(y_\sigma\otimes x_{\sigma c})\,\oplus\,(E\otimes\mathbf{C})\,(y_\sigma\otimes y_{\sigma c}).\end{aligned}$$

Note that the (p_τ,p_τ) space is two dimensional. $F_\infty\otimes 1$ acts as follows:

$$\begin{aligned}x_\sigma\otimes x_{\sigma c}&\mapsto y_{\sigma c}\otimes y_\sigma=\pm y_\sigma\otimes y_{\sigma c}\\ x_\sigma\otimes y_{\sigma c}&\mapsto y_{\sigma c}\otimes x_\sigma=\pm x_\sigma\otimes y_{\sigma c}\\ y_\sigma\otimes x_{\sigma c}&\mapsto x_{\sigma c}\otimes y_\sigma=\pm y_\sigma\otimes x_{\sigma c}\\ y_\sigma\otimes y_{\sigma c}&\mapsto x_{\sigma c}\otimes x_\sigma=\pm x_\sigma\otimes x_{\sigma c},\end{aligned}$$

where the last map is $i_\pm$. Thus, for $(M\otimes M^c)_+$, $F_\infty\otimes 1$ acts by $\left(\begin{smallmatrix}1&0\\0&1\end{smallmatrix}\right)$ on the (p_τ,p_τ) part for each $\tau\in I_F$, whereas for $(M\otimes M^c)_-$, it acts by $\left(\begin{smallmatrix}-1&0\\0&-1\end{smallmatrix}\right)$. In view of [**D**, Table 5.3] we see that for each $\tau\in I_F$, the following Gamma factors are involved in the (conjectural) functional equation of $G(s,f)$:

$$\Gamma_{\mathbf{C}}(s-v_\sigma-v_{\sigma c}),\quad \Gamma_{\mathbf{C}}(2n_\sigma+3+v_\sigma+v_{\sigma c}-s),$$
$$\Gamma_{\mathbf{R}}(s+\epsilon-n_\sigma-1-v_\sigma-v_{\sigma c})^2,\quad \Gamma_{\mathbf{R}}(n_\sigma+2+v_\sigma+v_{\sigma c}+(1-\epsilon)-s)^2,$$

where

$$\Gamma_{\mathbf{C}}(s)=2(2\pi)^{-2s}\Gamma(s)\quad\text{and}\quad\Gamma_{\mathbf{R}}(s)=\pi^{-s/2}\Gamma(s/2),$$

and $\epsilon=0$ or 1, is determined by $(-1)^{n_\sigma+1+v_\sigma+v_{\sigma c}+\epsilon}=\pm1$. We conclude that the critical integers for $(M\otimes M^c)_+$ consists of the intersection over all $\tau\in I_F$ of the sets

$$\begin{aligned}\{j\in[v_\sigma+v_{\sigma c}+1,n_\sigma+1+v_\sigma+v_{\sigma c}]\mid j\text{ is odd}\}\ &\cup\\ \{j\in[n_\sigma+2+v_\sigma+v_{\sigma c},2n_\sigma+2+v_\sigma+v_{\sigma c}]\mid j\text{ is even}\}&\end{aligned}\tag{6}$$

where $\sigma\big|_F=\sigma c\big|_F=\tau$. For $(M\otimes M^c)_-$ we get a similar description, except that the parities of the integers j are reversed.

3.2. Deligne's conjecture. We recall Deligne's conjecture on the critical values of $\mathbb{L}(s,(M\otimes M^c)_+)$ which we have conjecturally identified with $\mathbb{G}(s,f)$. For each $\tau\in I_F$, we have the comparison isomorphism

$$I_\tau: H_\tau((M\otimes M^c)_+)\otimes\mathbf{C}\xrightarrow{\sim}H_{DR}((M\otimes M^c)_+)\otimes_\tau\mathbf{C}.$$

Let $H_\tau^+((M\otimes M^c)_+)=\ker\left[F_\infty-1: H_B((M\otimes M^c)_+)\to H_B((M\otimes M^c)_+)\right]$, and let $H_{DR}^+((M\otimes M^c)_+)=H_{DR}((M\otimes M^c)_+)/F^-$, where F^- is an appropriate (1-dimensional) part of the Hodge filtration on $H_{DR}((M\otimes M^c)_+)$ (see [**D**]). Then, for each $\tau\in I_F$, we have the induced map

$$I_\tau^+: H_\tau^+\otimes\mathbf{C}\hookrightarrow H_\tau\otimes\mathbf{C}\xrightarrow{I_\tau}H_{DR}\otimes_\tau\mathbf{C}\to H_{DR}^+\otimes_\tau\mathbf{C}.$$

Let

$$c_\tau^+((M\otimes M^c)_+):=\det(I_\tau^+)\in(E\otimes\mathbf{C})^*,$$

where the determinant is calculated in E-rational bases of H_τ^+ and H_{DR}^+, and so is only defined up to multiplication by an element of $E\otimes 1$. Some elementary linear algebra shows that when $(M \otimes M^c)_+$ is critical (i.e., 0 is a critical value of $(s, (M \otimes M^c)_+)$), then, up to sign

$$c^+(\mathrm{Res}_{F/\mathbf{Q}}(M \otimes M^c)_+) = \left(1 \otimes \sqrt{D_{F/\mathbf{Q}}}^{\,3}\right) \cdot \prod_\tau c_\tau^+((M \otimes M^c)_+).$$

We have the following conjecture (see [**D**, Conjecture 2.8]):

CONJECTURE 1. *(Deligne) If $(M \otimes M^c)_+$ is critical, then*

$$\frac{\mathbb{L}(0, (M \otimes M^c)_+)}{c^+(\mathrm{Res}_{F/Q}(M \otimes M^c)_+)} \in E \otimes 1 \subset E \otimes \mathbf{C}.$$

In view of the fact that l is a critical integer of $(M \otimes M^c)_+ \iff$ the j-fold Tate twist $(M \otimes M^c)_+(j)$ is critical, this conjecture includes all the critical integers within its scope.

4. Integral expression for $G(s, f)$

In this section we realize f as a tuple of harmonic differential r-form $\delta(f_i)$ on $\mathcal{Z}$ with values in a suitable sheaf. After restricting these forms to $H^r \subset \mathcal{Z}$, we integrate each of them against a suitable Eisenstein r-form over F to get an integral expression for $G(s, f)$.

4.1. The Eichler-Shimura-Harder isomorphism. Let $L(\boldsymbol{n}, A)$ denote the space of homogeneous polynomials of degree n_σ in $\boldsymbol{x_\sigma} = \binom{X_\sigma}{Y_\sigma}$ and degree $n_{\sigma c} = n_\sigma$ in $\boldsymbol{x_{\sigma c}} = \binom{X_{\sigma c}}{Y_{\sigma c}}$, as σ varies through Σ_K, with coefficients in a $K(\chi_\mathfrak{N})$-algebra A. We regard $L(\boldsymbol{n}, A)$ as a $\Gamma_{\mathfrak{a}_i}$-module via

$$\gamma \cdot P(\boldsymbol{x_\sigma}, \boldsymbol{x_{\sigma c}}) = \chi_\mathfrak{N}^{-1}(d_\gamma)\, P(\gamma^{\sigma\iota}\, \boldsymbol{x_\sigma}, \gamma^{\sigma c \iota}\, \boldsymbol{x_{\sigma c}}) \tag{7}$$

where $v^\iota = \det(v)\, v^{-1}$ is the adjoint matrix of v. We give $L(\boldsymbol{n}, A)$ the discrete topology, and denote $\widetilde{L(\boldsymbol{n}, A)}$ to be the sheaf of locally constant sections of the projection

$$\Gamma_{\mathfrak{a}_i} \backslash \mathcal{Z} \times L(\boldsymbol{n}, A) \to \Gamma_{\mathfrak{a}_i} \backslash \mathcal{Z}.$$

Recall that there is a Hecke-equivariant isomorphism (outlined by Harder in [**H**]) which generalizes the Eichler-Shimura isomorphism for elliptic cusp forms

$$\delta : \bigoplus_\chi \boldsymbol{S}_{(\boldsymbol{n},\boldsymbol{v})}(\mathfrak{N}, \chi) \xrightarrow{\sim} \mathrm{H}^r_{\mathrm{cusp}}(Y_0(\mathfrak{N}), \widetilde{L(\boldsymbol{n}, \mathbf{C})}). \tag{8}$$

Here the sum is over all Hecke characters χ of $K_{\boldsymbol{A}}^\times$, with infinity type $-\boldsymbol{n} - 2\boldsymbol{v}$, and whose restriction to $\widehat{\mathcal{O}_K}$ is $\chi_\mathfrak{N}$, and the cohomology group is computed using the de Rham resolution. This is the isomorphism of 'minimal degree', in the sense that there is a similar isomorphism into each degree q cuspidal cohomology group, for each $r \leq q \leq 2r$ (see [**H94**, Proposition 3.1]).

For computational purposes it is more convenient to consider, for each i, the isomorphism,

$$\delta : \mathcal{S}_{\boldsymbol{n}}(\Gamma_{\mathfrak{a}_i}, \chi_\mathfrak{N}^{-1}) \xrightarrow{\sim} \mathrm{H}^r_{\mathrm{cusp}}(\Gamma_{\mathfrak{a}_i} \backslash \mathcal{Z}, \widetilde{L(\boldsymbol{n}, \mathbf{C})}),$$

which we write down following Hida [**H94**, Section 2.5]. Introduce, for each $\sigma \in \Sigma_K$, the auxiliary variables $\boldsymbol{u_\sigma} = \binom{U_\sigma}{V_\sigma}$ and $\boldsymbol{a_\sigma} = \binom{A_\sigma}{B_\sigma}$. Set

$$\boldsymbol{Q_\sigma} = \left(\binom{n_\sigma^*}{\alpha_\sigma}(-1)^{n_\sigma^* - \alpha_\sigma}\ U_\sigma^{\alpha_\sigma} V_\sigma^{n_\sigma^* - \alpha_\sigma}\right)_{\alpha_\sigma = 0, 1, \dots, n_\sigma^*}.$$

Let

$$\boldsymbol{\psi_\sigma} = \boldsymbol{\psi_\sigma}(\boldsymbol{x_\sigma}, \boldsymbol{x_{\sigma c}}, \boldsymbol{a_\sigma}) = (\psi_\sigma^0, \psi_\sigma^1, \dots, \psi_\sigma^{n_\sigma^*})^t$$

where each $\psi_\sigma^{\alpha_\sigma} = \psi_\sigma^{\alpha_\sigma}(\boldsymbol{x_\sigma}, \boldsymbol{x_{\sigma c}}, \boldsymbol{a_\sigma})$ is a homogeneous polynomial of degree n_σ in $\boldsymbol{x_\sigma} = \binom{X_\sigma}{Y_\sigma}$, degree n_σ in $\boldsymbol{x_{\sigma c}} = \binom{X_{\sigma c}}{Y_{\sigma c}}$ and degree 2 in $\boldsymbol{a_\sigma} = \binom{A_\sigma}{B_\sigma}$ defined by

$$(9) \qquad (X_\sigma V_\sigma - Y_\sigma U_\sigma)^{n_\sigma} (X_{\sigma c} U_\sigma + Y_{\sigma c} V_\sigma)^{n_\sigma} (A_\sigma V_\sigma - B_\sigma U_\sigma)^2 = \boldsymbol{Q_\sigma} \cdot \boldsymbol{\psi_\sigma}.$$

Let $\boldsymbol{x} = \{\boldsymbol{x_\sigma}\}_{\sigma\in\Sigma_K}$, $\boldsymbol{x_c} = \{\boldsymbol{x_{\sigma c}}\}_{\sigma\in\Sigma_K}$, $\boldsymbol{a} = \{\boldsymbol{a_\sigma}\}_{\sigma\in\Sigma_K}$, and for each multi-index α with $0 \le \alpha_\sigma \le n_\sigma^*$, let

$$\boldsymbol{\psi}^\alpha(\boldsymbol{x}, \boldsymbol{x_c}, \boldsymbol{a}) = \bigotimes_{\sigma\in\Sigma_K} \psi_\sigma^{\alpha_\sigma}(\boldsymbol{x_\sigma}, \boldsymbol{x_{\sigma c}}, \boldsymbol{a_\sigma}).$$

Finally set $\boldsymbol{\psi}(\boldsymbol{x}, \boldsymbol{x_c}, \boldsymbol{a}) = (\boldsymbol{\psi}^\alpha(\boldsymbol{x}, \boldsymbol{x_c}, \boldsymbol{a}))_\alpha$

Let f be as in Definition 1, and let F_i and $f_i \in \mathcal{S}_{\boldsymbol{n}}(\Gamma_{\mathfrak{a}_i}, \chi_{\mathfrak{N}}^{-1})$ be as in Section 2.4. Let $G^1(\mathbf{R}) = \prod_{\sigma\in\Sigma_K} \mathrm{SL}_2(\mathbf{C}) \subset G(\mathbf{R}) = \prod_{\sigma\in\Sigma_K} \mathrm{GL}_2(\mathbf{C})$, and let $F = F_i|_{G^1(\mathbf{R})}$. Note that F takes values in $W(\boldsymbol{n}^*, \mathbf{C})$, and so, for each multi-index α with $0 \le \alpha_\sigma \le n_\sigma^*$, we let F^α denote the 'components' of F:

$$F(g, \boldsymbol{s}) = \sum_{\alpha=0}^{n^*} F^\alpha(g)\, S^{n^*-\alpha} T^\alpha,$$

where we write $S^{n^*-\alpha}T^\alpha$ for $\prod_{\sigma\in\Sigma_K} S^{n_\sigma^*-\alpha_\sigma} T^{\alpha_\sigma}$. Let $\boldsymbol{F} = (F^\alpha)_\alpha$.

Let $\boldsymbol{t} = \sum_{\sigma\in\Sigma_K} 1\cdot\sigma$, and let $L(2\boldsymbol{t}, \mathbf{C})$ denote the space of polynomials over $\mathbf{C}$, in the variables $\boldsymbol{a}$, of degree two in each $\binom{A_\sigma}{B_\sigma}$ Now define $F' : G^1(\mathbf{R}) \to L(\boldsymbol{n}, \mathbf{C}) \otimes L(2\boldsymbol{t}, \mathbf{C})$ by

$$F'(g, \boldsymbol{x}, \boldsymbol{x_c}, \boldsymbol{a}) = \boldsymbol{F} \cdot \boldsymbol{\psi}(\boldsymbol{x}, \boldsymbol{x_c}, \boldsymbol{a})$$

and $F'' : G^1(\mathbf{R}) \to L(\boldsymbol{n}, \mathbf{C}) \otimes L(2\boldsymbol{t}, \mathbf{C})$ by

$$(10) \qquad F''(g, \boldsymbol{x}, \boldsymbol{x_c}, \boldsymbol{a}) = F'(g, g_\sigma^\iota\, \boldsymbol{x_\sigma}, \overline{g_\sigma}^\iota\, \boldsymbol{x_{\sigma c}}, j(g_\sigma, \epsilon)^t\, \boldsymbol{a_\sigma}).$$

Then F'' is invariant on the right by $\prod_\sigma SU_2(\mathbf{C})$, and so can be considered as a function on $\mathcal{Z}$.

Finally, let $\delta(f_i)$ be the $L(\boldsymbol{n}, \mathbf{C})$-valued differential form on $\mathcal{Z}$ obtained from F'' by replacing $(A_\sigma^2, A_\sigma B_\sigma, B_\sigma^2)$ by $(dx_\sigma, -dy_\sigma, -d\bar{x}_\sigma)$. More specifically, if $g\epsilon = z$, then

$$\delta(f_i)(z) = \boldsymbol{F} \cdot \boldsymbol{\psi}(g_\sigma^\iota\, \boldsymbol{x_\sigma}, \overline{g_\sigma}^\iota\, \boldsymbol{x_{\sigma c}}, j(g_\sigma, \epsilon)^t\, \boldsymbol{a_\sigma})$$

where each $(A_\sigma^2, A_\sigma B_\sigma, B_\sigma^2)$ is replaced by $(dx_\sigma, -dy_\sigma, -d\bar{x}_\sigma)$.

Note that here ${A_{\sigma_1}}^2 {A_{\sigma_2}}^2$ is replaced by $dx_{\sigma_1} \wedge dx_{\sigma_2}$ (in the case when $r = 2$) etc. Thus there is an ambiguity of sign, which is removed if we fix once and for all an ordering on the set Σ_K. Note that one may indeed check that

$$\gamma^*\delta(f_i) = \gamma \cdot \delta(f_i) \quad \text{for all } \gamma \in \Gamma_{\mathfrak{a}_i}.$$

Let us now write down $\delta(f_i)$ explicitly. As in [**G**], we simplify the computation in two ways. Firstly, since we are really interested in computing $\delta(f_i)\big|_{H^r}$, we will assume from the outset that $dx_\sigma = d\bar{x}_\sigma$ for all $\sigma \in \Sigma_K$.

Secondly, we would like to compute $\begin{pmatrix} 1 & -x \\ 0 & 1 \end{pmatrix} \cdot \delta(f_i)\big|_{H^r}$, where $x = \sum x_\sigma \in K_{\boldsymbol{A}}$, and x has finite part $= 0$. Since this amounts to setting $x_\sigma = 0$, for all σ, in

$\delta(f_i)\big|_{H^r}$, we do this from the start, and thus actually compute this 'modified' differential form, which we denote by $\widetilde{\delta(f_i)}$. Set

$$A_i = \mathfrak{a}_i \cap \mathcal{O}_F, \qquad \Gamma_{A_i} = \Gamma_{\mathfrak{a}_i} \cap \mathrm{SL}_2(F).$$

Then $\widetilde{\delta(f_i)} \in \mathrm{H}^r_{\mathrm{cusp}}(\Gamma_{A_i}\backslash H^r, \widetilde{L(\boldsymbol{n}, \mathbf{C})})$, and agrees with $\delta(f_i)\big|_{H^r}$ on the locus $x = 0$.

Let us begin the computation. Using the definition of $\boldsymbol{\psi_\sigma}$ in (9), we see that for $\alpha_\sigma = 0, 1, \dots, n^*_\sigma = 2n_\sigma + 2$,

$$\psi_\sigma^{\alpha_\sigma}(\boldsymbol{x_\sigma}, \boldsymbol{x_{\sigma c}}, \boldsymbol{a_\sigma}) = (-1)^{\alpha_\sigma} \frac{{A_\sigma}^2 c_{\alpha_\sigma} - 2A_\sigma B_\sigma c_{\alpha_\sigma - 1} + {B_\sigma}^2 c_{\alpha_\sigma - 2}}{\binom{2n_\sigma+2}{\alpha_\sigma}} \tag{11}$$

where $c_{\alpha_\sigma} = c_{\alpha_\sigma}(X_\sigma, Y_\sigma, X_{\sigma c}, Y_{\sigma c})$ are polynomials given by

$$c_{\alpha_\sigma} = \sum_{\substack{j_\sigma, k_\sigma = 0 \\ n_\sigma - (j_\sigma - k_\sigma) = \alpha_\sigma}} (-1)^{k_\sigma} \binom{n_\sigma}{k_\sigma} \binom{n_\sigma}{j_\sigma} X_\sigma^{n_\sigma - k_\sigma} X_{\sigma c}^{n_\sigma - j_\sigma} Y_\sigma^{k_\sigma} Y_{\sigma c}^{j_\sigma}. \tag{12}$$

Now let $g_\sigma = (1/\sqrt{y_\sigma}) \left(\begin{smallmatrix} y_\sigma & x_\sigma \\ 0 & 1 \end{smallmatrix}\right) \in \mathrm{SL}_2(\mathbf{C})$, with $x_\sigma, y_\sigma \in \mathbf{R}$. Then

$$g_\sigma \epsilon = z_\sigma, \quad g_\sigma^\iota = \overline{g_\sigma}^\iota = \frac{1}{\sqrt{y_\sigma}} \left(\begin{smallmatrix} 1 & -x_\sigma \\ 0 & y_\sigma \end{smallmatrix}\right) \quad \text{and} \quad j(g_\sigma, \epsilon) = \frac{1}{\sqrt{y_\sigma}} \left(\begin{smallmatrix} 1 & 0 \\ 0 & 1 \end{smallmatrix}\right).$$

Let f^α be the components of f_i as a $W(\boldsymbol{n}^*, \mathbf{C})$-valued function on $\mathcal{Z}$. Since $f\left(z, \left(\begin{smallmatrix} S_\sigma \\ T_\sigma \end{smallmatrix}\right)\right) = F\left(g, j(g_\sigma, \epsilon)^t \left(\begin{smallmatrix} S_\sigma \\ T_\sigma \end{smallmatrix}\right)\right)$ we have the following relation between the components of f_i and F:

$$f^\alpha(z) = \left(\prod_\sigma \frac{1}{\sqrt{y_\sigma}}\right)^{2n_\sigma+2} F^\alpha(g). \tag{13}$$

By (10) we see

$$\delta(f_i)\big|_H = \sum_{\alpha=0}^{\boldsymbol{n}^*} F^\alpha(g)\, \boldsymbol{\psi}^\alpha \left(\frac{1}{\sqrt{y_\sigma}} \left(\begin{smallmatrix} 1 & -x_\sigma \\ 0 & y_\sigma \end{smallmatrix}\right) \left(\begin{smallmatrix} X_\sigma \\ Y_\sigma \end{smallmatrix}\right), \frac{1}{\sqrt{y_\sigma}} \left(\begin{smallmatrix} 1 & -x_\sigma \\ 0 & y_\sigma \end{smallmatrix}\right) \left(\begin{smallmatrix} X_{\sigma c} \\ Y_{\sigma c} \end{smallmatrix}\right), \frac{1}{\sqrt{y_\sigma}} \left(\begin{smallmatrix} 1 & 0 \\ 0 & 1 \end{smallmatrix}\right) \left(\begin{smallmatrix} A_\sigma \\ B_\sigma \end{smallmatrix}\right) \right).$$

Now setting $x_\sigma = 0$, and using (13) above, we get

$$\begin{aligned} \widetilde{\delta(f_i)} &= \sum_{\alpha=0}^{n^*} \left(\prod_\sigma \sqrt{y_\sigma}^{2n_\sigma+2} \right) f^\alpha(z) \\ &\quad \cdot \boldsymbol{\psi}^\alpha \left(\frac{1}{\sqrt{y_\sigma}} X_\sigma, \sqrt{y_\sigma}\, Y_\sigma, \frac{1}{\sqrt{y_\sigma}} X_{\sigma c}, \sqrt{y_\sigma}\, Y_{\sigma c}, \frac{1}{\sqrt{y_\sigma}} A_\sigma, \frac{1}{\sqrt{y}} B_\sigma \right) \\ &= \sum_{\alpha=0}^{n^*} f^\alpha(z)\ \boldsymbol{\psi}^\alpha(X_\sigma, y_\sigma Y_\sigma, X_{\sigma c}, y_\sigma Y_{\sigma c}, A_\sigma, B_\sigma) \end{aligned}$$

with $(A_\sigma^2, A_\sigma B_\sigma, B_\sigma^2)$ replaced by $(dx_\sigma, -dy_\sigma, -dx_\sigma)$. Thus by (11),

$$\widetilde{\delta(f_i)} = \sum_{\alpha=0}^{n^*} f^\alpha(z) \prod_\sigma (-1)^{\alpha_\sigma} \frac{(c_{\alpha_\sigma} - c_{\alpha_\sigma - 2}) dx_\sigma + 2c_{\alpha_\sigma - 1} dy_\sigma}{\binom{2n_\sigma+2}{\alpha_\sigma}} \tag{14}$$

where the polynomials c_{α_σ} are evaluated at $(X_\sigma, y_\sigma Y_\sigma, X_{\sigma c}, y_\sigma Y_{\sigma c})$. This gives us an explicit expression for $\widetilde{\delta(f_i)} \in \mathrm{H}^r_{\mathrm{cusp}}(\Gamma_{A_i}\backslash H^r, \widetilde{L(\boldsymbol{n}, \mathbf{C})})$.

Note that $L(\boldsymbol{n},\mathbf{C})$ is an irreducible $\Gamma_{\mathfrak{a}_i}$ module, but not an irreducible Γ_{A_i} module. In fact, for any $\mathbf{Q}$-algebra A, we have the decomposition (see [**H94**, Section 11])

$$L(\boldsymbol{n}, A) \cong \bigoplus_{\boldsymbol{m}=0}^{\boldsymbol{n}} L(2\boldsymbol{n}-2\boldsymbol{m}, A) \tag{15}$$

$$P(\boldsymbol{x},\, \boldsymbol{x_c}) \mapsto \bigoplus_{\boldsymbol{m}=0}^{\boldsymbol{n}} \prod_{\sigma\in\Sigma_K} \frac{1}{m_\sigma!^2}\left(\frac{\partial^2}{\partial X_\sigma \partial Y_{\sigma c}} - \frac{\partial^2}{\partial X_{\sigma c}\partial Y_\sigma}\right)^{m_\sigma} P(\boldsymbol{x},\, \boldsymbol{x_c})\Big|_{\substack{X_{\sigma c}=X_\sigma \\ Y_{\sigma c}=Y_\sigma}}.$$

Here, $L(2\boldsymbol{n}-2\boldsymbol{m}, A)$ is the space of polynomials with coefficients in A, that are homogeneous of degree $2n_\sigma - 2m_\sigma$ in the variables $\left(\begin{smallmatrix} X_\sigma \\ Y_\sigma\end{smallmatrix}\right)$, for each $\sigma\in\Sigma_K$ This, in turn, induces a decomposition on cohomology

$$\mathrm{H}^r_{\mathrm{cusp}}(\Gamma_{A_i}\backslash H^r, \widetilde{L(\boldsymbol{n},\mathbf{C})}) \cong \bigoplus_{\boldsymbol{m}=0}^{\boldsymbol{n}} \mathrm{H}^r_{\mathrm{cusp}}(\Gamma_{A_i}\backslash H^r, \widetilde{L(2\boldsymbol{n}-2\boldsymbol{m},\mathbf{C})})$$

$$\widetilde{\delta(f_i)} = \bigoplus_{\boldsymbol{m}=0}^{\boldsymbol{n}} \widetilde{\delta_{2\boldsymbol{n}-2\boldsymbol{m}}(f_i)}.$$

Let us compute $\widetilde{\delta_{2\boldsymbol{n}-2\boldsymbol{m}}(f_i)}$. Set

$$\nabla_\sigma = \left(\frac{\partial^2}{\partial X_\sigma\partial Y_{\sigma c}} - \frac{\partial^2}{\partial X_{\sigma c}\partial Y_\sigma}\right).$$

Recall that (see [**G**, Lemma 3])

$$\frac{1}{m_\sigma!^2}\nabla_\sigma^{m_\sigma} c_{\alpha_\sigma}(X_\sigma, Y_\sigma, X_{\sigma c}, Y_{\sigma c})\,\Big|_{\substack{X_{\sigma c}=X_\sigma\\ Y_{\sigma c}=Y_\sigma}} = \binom{n_\sigma}{m_\sigma}^2 \sum_{p_\sigma=0}^{m_\sigma} (-1)^{p_\sigma}\binom{m_\sigma}{p_\sigma}$$

$$\cdot \sum_{\substack{j_\sigma,k_\sigma=0\\ n_\sigma-(j_\sigma-k_\sigma)=\alpha_\sigma}}^{n_\sigma} (-1)^{k_\sigma}\binom{n_\sigma-m_\sigma}{k_\sigma-p_\sigma}\binom{n_\sigma-m_\sigma}{j_\sigma-(m_\sigma-p_\sigma)}$$

$$\cdot X_\sigma^{2n_\sigma-(j_\sigma+k_\sigma)-m_\sigma} Y_\sigma^{j_\sigma+k_\sigma-m_\sigma}.$$

Applying this to (14), we obtain, after some algebraic manipulations, that

$$\widetilde{\delta_{2\boldsymbol{n}-2\boldsymbol{m}}(f_i)} = \sum_{\alpha}^{\boldsymbol{n}^*} f^\alpha(z)\cdot \prod_{\sigma\in\sigma_K}(-1)^{\alpha_\sigma}$$

$$\cdot \sum_{l_\sigma=0}^{2n_\sigma-2m_\sigma} \frac{\left(a\left(m_\sigma,l_\sigma,\alpha_\sigma\right)dx_\sigma + 2b\left(m_\sigma,l_\sigma,\alpha_\sigma\right)dy_\sigma\right)}{\binom{2n_\sigma+2}{\alpha_\sigma}} y_\sigma^{2n_\sigma-m_\sigma-l_\sigma} X_\sigma^{l_\sigma} Y_\sigma^{2n_\sigma-2m_\sigma-l_\sigma}$$

where $a\left(m_\sigma,l_\sigma,\alpha_\sigma\right)$ and $b\left(m_\sigma,l_\sigma,\alpha_\sigma\right)$ are constants given by (see the end of [**G**, Section 6])

$$a(m,l,\alpha) = \binom{n}{m}^2 (-1)^{\frac{n+\alpha-l-m}{2}} \sum_{p=0}^{m}(-1)^p\binom{m}{p}$$

$$\left[\binom{n-m}{\frac{n-m-l+\alpha}{2}-p}\binom{n-m}{\frac{3n-3m-l-\alpha}{2}+p} + \binom{n-m}{\frac{n-m-l+\alpha-2}{2}-p}\binom{n-m}{\frac{3n-3m-l-\alpha+2}{2}+p}\right],$$

$$b(m,l,\alpha) = \binom{n}{m}^2 (-1)^{\frac{n+\alpha-1-l-m}{2}} \sum_{p=0}^{m} (-1)^p \binom{m}{p} \cdot \left[\binom{n-m}{\frac{n-m-l+\alpha-1}{2}-p}\binom{n-m}{\frac{3n-3m-l-\alpha+1}{2}+p}\right].$$

Writing this in multi-index notation we have finally that

$$\widetilde{\delta_{2n-2m}}(f_i) \tag{16}$$
$$= \sum_{l=0}^{2n-2m} \left\{ \sum_{\alpha=0}^{n^*} \frac{f^\alpha(z)(-1)^\alpha \prod_\sigma \left(a(m_\sigma,l_\sigma,\alpha_\sigma)\, dx_\sigma + 2b(m_\sigma,l_\sigma,\alpha_\sigma)\, dy_\sigma\right)}{\binom{2n+2}{\alpha}} \right\} \cdot y^{2n-m-l} X^l Y^{2n-2m-l},$$

where

$$(-1)^\alpha = \prod (-1)^{\alpha_\sigma}, \quad y^{2n-m-l} = \prod y_\sigma^{2n_\sigma-m_\sigma-l_\sigma},$$

and

$$X^l Y^{2n-2m-l} = \prod X_\sigma^{l_\sigma} Y_\sigma^{2n_\sigma-2m_\sigma-l_\sigma} \quad \text{etc.}$$

4.2. A pairing. The proof of the rationality result relies heavily on studying a certain pairing, which we introduce now, between certain cohomology groups of the modular manifold $\Gamma_{A_i}\backslash H^r$. Let A be a $\mathbf{Q}$-algebra. Then the pairing (see [**H94**, Equation 3.1b]).

$$L(2n-2m,A) \otimes L(2n-2m,A) \xrightarrow{<,>} A$$

$$\left\langle \sum_{l=0}^{2n-2m} a_l X^{2n-2m-l} Y^l, \sum_{l=0}^{2n-2m} b_l X^{2n-2m-l} Y^l \right\rangle = \sum_{l=0}^{2n-2m} (-1)^l \frac{a_l b_{2n-2m-l}}{\binom{2n-2m}{l}}$$

induces a perfect pairing on cohomology (see [**H94**, Equation 5.3])

$$\mathrm{H}_c^r(\Gamma_{A_i}\backslash H^r, L(\widetilde{2n-2m},A)) \otimes \mathrm{H}^r(\Gamma_{A_i}\backslash H^r, L(\widetilde{2n-2m},A)) \xrightarrow{\cup} \tag{17}$$
$$\mathrm{H}_c^{2r}(\Gamma_{A_i}\backslash H^r, L(\widetilde{2n-2m},A) \otimes L(\widetilde{2n-2m},A)) \xrightarrow{<,>} \mathrm{H}_c^{2r}(\Gamma_{A_i}\backslash H^r, \widetilde{A})$$

where the first map is cup product (wedge product). We continue to denote the pairing on cohomology by $\langle\, ,\, \rangle$.

4.3. Eisenstein series. Let

$$\Gamma_{\mathfrak{a}_i}^\infty = \left\{ \left(\begin{smallmatrix} \varepsilon & b \\ 0 & \varepsilon^{-1}\end{smallmatrix}\right) \,\middle|\, \varepsilon \in \mathcal{O}_K{}^\times,\ b \in \mathfrak{a}_i \right\}, \quad \Gamma_{A_i}^\infty = \Gamma_{\mathfrak{a}_i}^\infty \cap \mathrm{SL}_2(F)$$

and let

$$\omega = \omega_{2n-2m} = (X - xY)^{2n-2m} dx = \prod_\sigma (X_\sigma - x_\sigma Y_\sigma)^{2n_\sigma-2m_\sigma} dx_\sigma$$

be a closed form whose class is in $\mathrm{H}^r(\Gamma_{A_i}^\infty\backslash H^r, L(\widetilde{2n-2m},\mathbf{Q}))$. Let, for any $\sigma \in \Sigma_K$ (see (1)),

$$\psi = \chi|_{F_{\mathbf{A}}} \cdot |\ |_{F_{\mathbf{A}}}^{2n_\sigma+2v_\sigma+2v_{\sigma c}}.$$

Then ψ is a finite order character. We define the Eisenstein differential form $E_{2n-2m}(\omega,s)$ by

$$E_{2n-2m}(\omega,s) = \sum_{\Gamma_{A_i}^\infty\backslash\Gamma_{A_i}} \psi(\gamma)\,\gamma^{-1} \cdot \gamma^*(\omega\, y^s).$$

We assume from now on that

$$n_\sigma - m_\sigma \quad \text{is independent of } \sigma \in \Sigma_K. \tag{18}$$

ASSUMPTION: We make the following assumption about the rationality of $E_{2\boldsymbol{n}-2\boldsymbol{m}}(\omega, s)$ at $s = 0$ (however, see the Remark below). Suppose that

$$\text{class of } E_{2\boldsymbol{n}-2\boldsymbol{m}}(\omega, 0) \in \mathrm{H}^r(\Gamma_{A_i}\backslash H^r, L(\widetilde{2\boldsymbol{n}-2\boldsymbol{m}}, \mathbf{Q}(\psi)). \tag{19}$$

REMARK 3. There are two ingredients that go into verifying this assumption. The first involves checking that $E_{2\boldsymbol{n}-2\boldsymbol{m}}(\omega, 0)$ is a closed form. The second involves showing that the class of $E_{2\boldsymbol{n}-2\boldsymbol{m}}(\omega, 0)$ is $\mathbf{Q}(\psi)$-rational. The second fact follows from the first by work of Harder (see [**H**]), and we do not dwell on this anymore. As for the first fact, note that if $2n_\sigma - 2m_\sigma > 2$, then 0 is in the range of absolute convergence of the Eisenstein series, and so the the closedness of $E_{2\boldsymbol{n}-2\boldsymbol{m}}(\omega, 0)$ follows from the closedness of ω, since the differential d 'commutes' with the sum. So it remains to investigate the case when $2n_\sigma - 2m_\sigma = 0$, for in this case $s = 0$ is outside the range of absolute convergence. Note

$$\begin{aligned} d(\omega y^s) &= d(\omega y_{\sigma_1}^s y_{\sigma_2}^s \cdots y_{\sigma_r}^s) \\ &= d\omega \cdot y^s + s y_{\sigma_1}^{s-1} y_{\sigma_2}^s \cdots y_{\sigma_r}^s dy_{\sigma_1} \wedge \omega + \cdots + s y_{\sigma_1}^s y_{\sigma_2}^s \cdots y_{\sigma_r}^{s-1} dy_{\sigma_r} \wedge \omega. \end{aligned}$$

Since $d\omega = 0$, we deduce that for $\mathrm{Re}(s)$ sufficiently large,

$$\begin{aligned} dE_{2\boldsymbol{n}-2\boldsymbol{m}}(\omega, s) &= \sum_{\Gamma_{A_i}^\infty \backslash \Gamma_{A_i}} \psi(\gamma)\,\gamma^{-1} \cdot \gamma^* d(\omega\, y^s) \\ &= sE\left(\frac{dy_{\sigma_1}}{y_{\sigma_1}} \wedge \omega, s\right) + \cdots + sE\left(\frac{dy_{\sigma_r}}{y_{\sigma_r}} \wedge \omega, s\right). \end{aligned}$$

Since the above equality must continue to hold for the meromorphic continuations of the Eisenstein series above, it follows that if the last r Eisenstein series do not have poles at $s = 0$, then $dE(\omega, 0) = 0$, that is $E(\omega, 0)$ is closed. To check the finiteness of these Eisenstein series, one needs to write down their Fourier expansions, which we do not do. Suffice it to say that if $F \neq \mathbf{Q}$, we expect all the Eisenstein series to be finite at $s = 0$, and in the $F = \mathbf{Q}$ case, we expect the same if $\psi \neq 1$. For the latter case (i.e., $F = \mathbf{Q}$), we refer the reader to [**G**, Section 7.4], where we have investigated these matters in more detail. See also [**H93**, Section 4], Thus we expect the Assumption to hold in all cases when $F \neq \mathbf{Q}$, and if $F = \mathbf{Q}$, when $\psi \neq 1$.

It is also convenient to introduce the modified Eisenstein series

$$E^*_{2\boldsymbol{n}-2\boldsymbol{m}}(\omega, s) = L(2s + 2n_\sigma - 2m_\sigma + 2, \psi)\, E_{2\boldsymbol{n}-2\boldsymbol{m}}(\omega, s). \tag{20}$$

4.4. Integral expression. With these preliminaries, we are ready now to derive an integral expression for $G(s, f)$. First note

$$\begin{aligned} &\int\!\!\int_{\Gamma_{A_i}\backslash H^r} \langle \delta_{2n-2m}(f_i), E_{2\boldsymbol{n}-2\boldsymbol{m}}(\omega, s)\rangle \\ &= \int\!\!\int_{\Gamma_{A_i}\backslash H^r} \langle \left(\begin{smallmatrix} 1 & -x \\ 0 & 1 \end{smallmatrix}\right) \cdot \delta_{2n-2m}(f_i),\ \left(\begin{smallmatrix} 1 & -x \\ 0 & 1 \end{smallmatrix}\right) \cdot E_{2\boldsymbol{n}-2\boldsymbol{m}}(\omega, s)\rangle \\ &= \int\!\!\int_{\Gamma_{A_i}\backslash H^r} \langle \widetilde{\delta_{2\boldsymbol{n}-2\boldsymbol{r}\imath}(f_i)},\ \widetilde{E_{2\boldsymbol{n}-2\boldsymbol{m}}(\omega, s)}\rangle \end{aligned}$$

and a standard unwinding argument shows this last integral

$$= \iint_{\Gamma^{\infty}_{A_i}\backslash H^r} \langle \widetilde{\delta_{2n-2m}(f_i)},\, \widetilde{\omega}\, y^s \rangle.$$

Here $\widetilde{\omega} = X^{2n-2m}\,dx$ since the tilde amounts to setting $x = 0$. Using the expression (16) for $\widetilde{\delta_{2n-2m}(f_i)}$ and the pairing (17), we obtain that

$$\begin{aligned}&\langle \widetilde{\delta_{2n-2m}(f_i)}, \widetilde{\omega}\, y^s \rangle \\ &\qquad = \left\{ \sum_{\alpha=0}^{n^*} \frac{f^{\alpha}(z)(-1)^{\alpha} 2^r \prod_{\sigma} b\,(m_\sigma, 0, \alpha_\sigma)}{\binom{2n+2}{\alpha}} \right\} y^{2n-m+s} \prod_{\sigma} dy_\sigma \prod_{\sigma} dx_\sigma.\end{aligned}$$

Further, if $\left(\begin{smallmatrix} \varepsilon & b \\ 0 & \varepsilon^{-1}\end{smallmatrix}\right) \in \Gamma^{\infty}_{A_i}$, and $z = \left(z_\sigma = \left(\begin{smallmatrix} x_\sigma & -y_\sigma \\ y_\sigma & x_\sigma \end{smallmatrix}\right)\right)_{\sigma\in\Sigma_K} \in H^r$, then

$$\left(\begin{smallmatrix} \varepsilon & b \\ 0 & \varepsilon^{-1}\end{smallmatrix}\right) \cdot \left(\begin{smallmatrix} x_\sigma & -y_\sigma \\ y_\sigma & x_\sigma \end{smallmatrix}\right) = \left(\begin{smallmatrix} \varepsilon^{\sigma}(\varepsilon^{\sigma} x_\sigma + b^{\sigma}) & -(\varepsilon^{\sigma})^2 y_\sigma \\ (\varepsilon^{\sigma})^2 y_\sigma & \varepsilon^{\sigma}(\varepsilon^{\sigma} x_\sigma + b^{\sigma}) \end{smallmatrix}\right).$$

Hence, we may separate variables, using the identification

$$\Gamma^{\infty}_{A_i}\backslash H^r = (\mathcal{O}_F{}^{\times})^2 \backslash \mathbf{R}^r_+ \times A_i\backslash \mathbf{R}^r.$$

Thus, using (3), we get

$$\begin{aligned}&\iint_{\Gamma^{\infty}_{A_i}\backslash H^r} \langle \widetilde{\delta_{2n-2m}(f_i)},\, \widetilde{\omega}\, y^s \rangle \\ &= \sum_{\alpha=0}^{n^*} (-1)^{\alpha}\, 2^r \prod_{\sigma} b\,(m_\sigma, 0, \alpha_\sigma)\, |a_i|_{K_{\mathbf{A}}} \sum_{\xi\in K^{\times}} c\,(\xi a_i d) \prod_{\sigma} \frac{1}{\xi^{\sigma\, v_\sigma}\, \overline{\xi^{\sigma}}^{\,v_{\sigma c}}} \\ &\quad \cdot \int_{(\mathcal{O}_F{}^{\times})^2\backslash \mathbf{R}^r_+} \prod_{\sigma} y^{2n_\sigma - m_\sigma + 1 + s} \left(\frac{\xi^{\sigma}}{i|\xi^{\sigma}|}\right)^{n_\sigma + 1 - \alpha_\sigma} K_{\alpha_\sigma - n_\sigma - 1}(4\pi y_\sigma |\xi^{\sigma}|)\, dy_\sigma \\ &\quad \cdot \int_{A_i\backslash \mathbf{R}^r} \prod_{\sigma} e^{2\pi i \operatorname{Tr}_{\mathbf{C}/\mathbf{R}}(\xi^{\sigma} x_\sigma)}\, dx_\sigma.\end{aligned}$$

The integral in x can be rewritten as

$$\int_{A_i\backslash \mathbf{R}^r} e^{2\pi i \operatorname{Tr}_{F/\mathbf{Q}}(x \operatorname{Tr}_{K/F}\xi)} dx$$

and so is given by

$$\begin{cases} 0, & \text{if } \operatorname{Tr}_{K/F}(\xi) \neq 0, \\ \operatorname{vol}(A_i\backslash \mathbf{R}^r) = N_{F/\mathbf{Q}}(A_i)\sqrt{D_{F/\mathbf{Q}}}, & \text{otherwise} \iff \xi = \eta\sqrt{-\Delta},\ \eta \in F. \end{cases}$$

So far the ideles a_i, which give rise to the f_i, have been arbitrary. We specify them now: assume these are representatives of Cl_F^+, the strict class group of F, thought of as elements of $K_{\mathbf{A}}^{\times}$. Thus $a_i\mathcal{O}_K = \mathfrak{a}_i = A_i\mathcal{O}_K$ is an extended ideal, where $A_i = \mathfrak{a}_i \cap \mathcal{O}_F = a_i\mathcal{O}_F$. Let

$$\mathfrak{b}_i = \sqrt{-\Delta}\,\vartheta_{K/\mathbf{Q}}\,\mathfrak{a}_i \subset \mathcal{O}_K.$$

Since $\sqrt{-\Delta}\vartheta_{K/\mathbf{Q}}$ is also an extended ideal, so is $\mathfrak{b}_i$: say $\mathfrak{b}_i = B_i\mathcal{O}_K$, where $B_i \subset \mathcal{O}_F$.

In view of the fact that $c(\ ,f)$ vanishes outside integral ideals, we see that the above integral vanishes unless $\eta \in \mathfrak{b}_i^{-1} \cap F$. Thus we may rewrite

$$\iint_{\Gamma_{A_i}^{\infty}\backslash H^r} \langle \widetilde{\delta_{2\boldsymbol{n}-2\boldsymbol{m}}(f_i),\, \widetilde{\omega}\, y^s} \rangle = 2^r\, N_{F/\mathbf{Q}}(A_i)^{-1} \sqrt{D_{F/\mathbf{Q}}}$$

$$\cdot \sum_{\alpha=0}^{\boldsymbol{n}^*} (-1)^{\alpha} \prod_{\sigma} b\,(m_\sigma, 0, \alpha_\sigma)\, \frac{1}{i^{v_\sigma - v_{\sigma c}} \sqrt{\Delta}^{\sigma}} \sum_{\eta \in F^{\times}} c\,(\eta \mathfrak{b}_i) \prod_{\sigma} \frac{1}{(\eta^\sigma)^{v_\sigma + v_{\sigma c}}}$$

$$\cdot \int_{(\mathcal{O}_F{}^{\times})^2 \backslash \mathbf{R}_+^r} \prod_{\sigma} y^{2n_\sigma - m_\sigma + 1 + s} \left(\frac{\eta^\sigma}{|\eta^\sigma|}\right)^{n_\sigma + 1 - \alpha_\sigma} K_{\alpha_\sigma - n_\sigma - 1}(4\pi y_\sigma |(\eta\sqrt{-\Delta})^\sigma|)\, dy_\sigma.$$

We need to evaluate the integral in y. Let, for each $\sigma \in \Sigma_K$,

$$l_\sigma = 2n_\sigma + 2 - m_\sigma + v_\sigma + v_{\sigma c}.$$

Note that by (1) and (18), l_σ is independent of σ. We denote this common value simply by l. We now break the sum over $\eta \in F^{\times}$, into a sum over $\eta \in F^{\times}/(\mathcal{O}_F{}^{\times})^2$ and a sum over $\theta^2 \in (\mathcal{O}_F{}^{\times})^2$, and use the relation

$$\prod_{\sigma} (\theta^2)^{2n_\sigma + 2 - m_\sigma + v_\sigma + v_{\sigma c}} = N_{F/\mathbf{Q}}(\theta^2)^l = 1$$

to get

$$\iint_{\Gamma_{A_i}^{\infty}\backslash H^r} \langle \widetilde{\delta_{2\boldsymbol{n}-2\boldsymbol{m}}(f_i),\, \widetilde{\omega}\, y^s} \rangle = 2^r\, N_{F/\mathbf{Q}}(A_i)^{-1} \sqrt{D_{F/\mathbf{Q}}}$$

$$\cdot \sum_{\alpha=0}^{\boldsymbol{n}^*} (-1)^{\alpha} \prod_{\sigma} b\,(m_\sigma, 0, \alpha_\sigma)\, \frac{1}{i^{v_\sigma - v_{\sigma c}} \sqrt{\Delta}^{\sigma}} \sum_{\eta \in F^{\times}/(\mathcal{O}_F{}^{\times})^2} c\,(\eta \mathfrak{b}_i) \prod_{\sigma} \frac{1}{(\eta^\sigma)^{v_\sigma + v_{\sigma c}}}$$

$$\cdot \left(\frac{\eta^\sigma}{|\eta^\sigma|}\right)^{n_\sigma + 1 - \alpha_\sigma} \int_{\mathbf{R}_+^r} \prod_{\sigma} y^{2n_\sigma - m_\sigma + 1 + s}\, K_{\alpha_\sigma - n_\sigma - 1}(4\pi y_\sigma |(\eta\theta^2\sqrt{-\Delta})^\sigma|)\, dy_\sigma.$$

Now the integral in y is easily evaluated by repeated use of the formula (see [**W**, Chapter 13.21, Equation 8], for instance)

$$\int_0^{\infty} K_\nu(ay)\, y^{\mu-1}\, dy = 2^{\mu-2}\, a^{-\mu}\, \Gamma\left(\frac{\mu+\nu}{2}\right) \Gamma\left(\frac{\mu-\nu}{2}\right).$$

Setting $s' = s + l$, we get

$$\iint_{\Gamma_{A_i}^{\infty}\backslash H^r} \langle \widetilde{\delta_{2\boldsymbol{n}-2\boldsymbol{m}}(f_i),\, \widetilde{\omega}\, y^s} \rangle = \frac{N_{F/\mathbf{Q}}(A_i)\, N_{F/\mathbf{Q}}(B_i)^s \sqrt{D_{F/\mathbf{Q}}}}{2^r \sqrt{N_{F/\mathbf{Q}}(\Delta)}^{s'} \prod_\sigma i^{v_\sigma + v_{\sigma c}}}$$

$$\cdot \sum_{\eta \in F^{\times}/(\mathcal{O}_F{}^{\times})^2} \frac{c\,(\eta \mathfrak{b}_i)}{N_{F/\mathbf{Q}}(\eta B_i)^s} \prod_{\sigma} \left(\frac{\eta^\sigma}{|\eta^\sigma|}\right)^{n_\sigma + 1 - \alpha_\sigma} \frac{1}{(\eta^\sigma)^{v_\sigma + v_{\sigma c}} |\eta^\sigma|^{2n_\sigma + 2 - m_\sigma}}\, G_\infty(s, f)$$

where $G_\infty(s,f)$ is a sum of products of Gamma factors

$$=\sum_{\alpha=0}^{n^*}(-1)^\alpha\prod_\sigma \frac{b\left(m_\sigma,0,\alpha_\sigma\right)}{(2\pi)^{2n_\sigma+2-m_\sigma+s}} \cdot\Gamma\left(\frac{n_\sigma+1-m_\sigma+\alpha_\sigma+s}{2}\right)\Gamma\left(\frac{3n_\sigma+3-m_\sigma-\alpha_\sigma+s}{2}\right).$$

For future use we note that by manipulations very similar to those towards the end of [**G**, Section 7.4], we may rewrite

$$G_\infty(s,f)=(-1)^r\prod_\sigma\frac{\binom{n_\sigma}{m_\sigma}^2}{(2\pi)^{2n_\sigma+2-m_\sigma+s}}\frac{\Gamma\left(\frac{s}{2}+n_\sigma-m_\sigma+1\right)^2\Gamma\left(s+2n_\sigma-m_\sigma+2\right)}{\Gamma\left(s+2n_\sigma-2m_\sigma+2\right)}. \tag{21}$$

We may even simplify

$$\sum_{\eta\in F^\times/(\mathcal{O}_F{}^\times)^2}\frac{c\left(\eta\mathfrak{b}_i\right)}{N_{F/\mathbf{Q}}(\eta B_i)^s}\prod_\sigma\left(\frac{\eta^\sigma}{|\eta^\sigma|}\right)^{n_\sigma+1-\alpha_\sigma}\frac{1}{(\eta^\sigma)^{v_\sigma+v_{\sigma c}}}|\eta^\sigma|^{2n_\sigma+2-m_\sigma}$$
$$=[\mathcal{O}_F{}_+^\times:(\mathcal{O}_F{}^\times)^2]N_{F/\mathbf{Q}}(B_i)^l\sum_{e\in(\pm1)^{I_F}}\prod_\sigma(e_\sigma)^{l_\sigma}\,G'_{(B_i,e)}(s',f)$$

where

$$G'_{(B_i,e)}(s,f)=\sum_{\substack{\mathrm{sgn}(\eta)=e\\ \eta\in F^\times/\mathcal{O}_F{}_+^\times}}\frac{c\left(\eta B_i\mathcal{O}_K\right)}{N_{F/\mathbf{Q}}(\eta B_i)^s}$$

is only a 'partial' L-function (note that the summation is only over the ideals in the strict ideal class (B_i,e)).

Taking an aggregate over all the representatives of the strict ideal classes, $B_i\subset\mathcal{O}_F$, we get

$$\sum_{[B_i]\in Cl_F^+}\frac{N_{F/\mathbf{Q}}(A_i)}{N_{F/\mathbf{Q}}(B_i)^{s'}}\iint_{\Gamma_{A_i}^\infty\backslash H^r}\langle\widetilde{\delta_{2\boldsymbol{n}-2\boldsymbol{m}}(f_i)},\,\widetilde{\omega}\,y^s\rangle \tag{22}$$
$$=\frac{\sqrt{D_{F/\mathbf{Q}}}\,[\mathcal{O}_F{}_+^\times:(\mathcal{O}_F{}^\times)^2]}{2^r\sqrt{N_{F/\mathbf{Q}}(\Delta)}^{s'}\prod_\sigma i^{v_\sigma+v_{\sigma c}}}\sum_{[B_i]\in Cl_F^+}\sum_{e\in(\pm1)^{I_F}}\prod_\sigma(e_\sigma)^{l_\sigma}\,G'_{(B_i,e)}(s',f)\,G_\infty(s,f)$$
$$=\frac{\sqrt{D_{F/\mathbf{Q}}}\,[\mathcal{O}_F{}_+^\times:(\mathcal{O}_F{}^\times)^2]}{2^r\sqrt{N_{F/\mathbf{Q}}(\Delta)}^{s'}\prod_\sigma i^{v_\sigma+v_{\sigma c}}}\prod_\sigma(1+(-1)^{l_\sigma})\,G'(s',f)\,G_\infty(s,f)$$
$$=\frac{\sqrt{D_{F/\mathbf{Q}}}\,[\mathcal{O}_F{}_+^\times:(\mathcal{O}_F{}^\times)^2]}{\sqrt{N_{F/\mathbf{Q}}(\Delta)}^{s'}\prod_\sigma i^{v_\sigma+v_{\sigma c}}}\,G'(s',f)\,G_\infty(s,f).$$

Note that the equality (22) above is only valid if each $l_\sigma=l$ is even, (otherwise the sum vanishes!) and hence we get a parity condition on the m_σ, namely

$$m_\sigma\equiv v_\sigma+v_{\sigma c}\ (\mathrm{mod}\ 2),\quad\text{for all }\sigma\in\Sigma_K. \tag{23}$$

Now multiplying the integral expression above by

$$L(2s'-2,\,\chi|_{F_{\boldsymbol{A}}})=L(2s+2n_\sigma-2m_\sigma+2,\,\psi),$$

and using (20), we finally get an integral expression for the twisted tensor L-function of f, namely

$$\sum_{[B_i]\in Cl_F^+} \frac{N_{F/\mathbf{Q}}(A_i)}{N_{F/\mathbf{Q}}(B_i)^{s'}} \int\!\!\int_{\Gamma_{A_i}\backslash H^r} \langle \delta_{2n-2m}(f_i), E^*_{2\boldsymbol{n}-2\boldsymbol{m}}(\omega,s)\rangle \tag{24}$$

$$= \frac{\sqrt{D_{F/\mathbf{Q}}}\,[\mathcal{O}_{F}{}_+^{\times} : (\mathcal{O}_F{}^{\times})^2]}{\sqrt{N_{F/\mathbf{Q}}(\Delta)}^{s'}\prod_\sigma i^{v_\sigma+v_{\sigma c}}}\, G(s',f)\, G_\infty(s,f),$$

where $s' = s + l = s + 2n_\sigma + 2 - m_\sigma + v_\sigma + v_{\sigma c}$. We emphasize again that (24) holds only for those $\boldsymbol{m}$ satisfying the conditions (18) and (23).

5. Rationality result for $G(s,f)$

Let λ be the Hecke algebra character corresponding to f. Thus $T(\mathfrak{m})f = \lambda(T(\mathfrak{m}))f = c(\mathfrak{m},f)f$, for all the Hecke operators $T(\mathfrak{m})$. Then, since the following modules are free of rank 1 over $E = \mathbf{Q}(f,\chi)$ (see [**H94**, Section 8]), we may write

$$\boldsymbol{S}_{(\boldsymbol{n},\boldsymbol{v})}(\mathfrak{N},\chi)[\lambda] = E\cdot f$$

$$\mathrm{H}^r_{\mathrm{cusp}}(Y_0(\mathfrak{N}),\widetilde{L(\boldsymbol{n},E)})[\lambda] = E\cdot \eta(f)$$

where $\eta(f)$ is an E-rational Hecke eigen-differential form. Following Hida, we define the 'period' $\Omega(f)\in\mathbf{C}^\times$ as follows:

$$\delta(f) = \Omega(f)\cdot\eta(f). \tag{25}$$

Note that the value of $\Omega(f)$ depends on the choice of basis $\eta(f)$, but up to multiplication by an element of E, is independent of this choice. Thus, working component wise, we see that

$$\mathcal{S}_{\boldsymbol{n}}(\Gamma_{\mathfrak{a}_i},\chi_{\mathfrak{N}}^{-1})[\lambda] = E\cdot f_i$$

$$\mathrm{H}^r_{\mathrm{cusp}}(\Gamma_{A_i}\backslash H^r,\widetilde{L(\boldsymbol{n},E)})[\lambda] = E\cdot\eta(f_i)$$

and hence that $\delta(f_i) = \Omega(f)\cdot\eta(f_i)$.

We now restrict to H^r. By (15), we may write

$$\eta(f_i) = \oplus_{\boldsymbol{m}=0}^{\boldsymbol{n}}\,\eta_{2\boldsymbol{n}-2\boldsymbol{m}}(f_i),$$

where $\eta_{2\boldsymbol{n}-2\boldsymbol{m}}(f_i)$ is an E-rational form $\in \mathrm{H}^r_{\mathrm{cusp}}(\Gamma_{A_i}\backslash H^r,\widetilde{L(2\boldsymbol{n}-2\boldsymbol{m},E)})$. Thus

$$\delta_{2\boldsymbol{n}-2\boldsymbol{m}}(f_i) = \Omega(f)\cdot\eta_{2\boldsymbol{n}-2\boldsymbol{m}}(f_i).$$

Set

$$\Omega_l(f) \overset{\mathrm{def}}{=} \frac{\Omega(f)\cdot G(\psi)\cdot\prod_{\sigma\in\Sigma_K}(2\pi i)^{3l-2w}}{\prod_{\sigma\in\Sigma_K}(2\pi i)^{v_\sigma+v_{\sigma c}}}.$$

THEOREM 1. *Say ψ is primitive of conductor N, and non-trivial, if $F=\mathbf{Q}$. Also assume that (19) holds. Then, for all even critical integers l (c.f. (6))*

$$\frac{G(l,f)}{\Omega_l(f)} \in E \subset \mathbf{C}.$$

PROOF. The proof follows by analyzing the integral expressions (24) for $G(s,f)$ at $s=0$, as $\boldsymbol{m}=(m_\sigma)$ varies through all possibilities, subject to the constraints (18) and (23). Indeed, as $\boldsymbol{m}$ varies as above, then $l = l_\sigma = 2n_\sigma + 2 - m_\sigma + v_\sigma + v_{\sigma c}$

varies through exactly the set of *even* critical integers. Now note that the L.H.S. of (24) at $s = 0$ is

$$(26) \qquad = \sum_{[B_i]\in Cl_F^+} \frac{N_{F/\mathbf{Q}}(A_i)}{N_{F/\mathbf{Q}}(B_i)^l} \int\!\!\int_{\Gamma_{A_i}\backslash H^r} \langle \delta_{2n-2m}(f_i), E^*_{\mathbf{2n}-\mathbf{2m}}(\omega, 0)\rangle$$

$$= L(2n_\sigma - 2m_\sigma + 2,\, \psi)\, \Omega(f)$$

$$\cdot \sum_{[B_i]\in Cl_F^+} c_i \int\!\!\int_{\Gamma_{A_i}\backslash H^r} \langle \eta_{2n-2m}(f_i),\, E_{\mathbf{2n}-\mathbf{2m}}(\omega, 0)\rangle$$

for some constants $c_i \in \mathbf{Q}$. The values of $L(k,\, \psi)$ for k even and positive, are just the critical values, and are well known. In fact, we have

$$L(2n_\sigma - 2m_\sigma + 2,\, \psi) = \frac{G(\psi)\prod_\sigma (2\pi)^{2n_\sigma - 2m_\sigma + 2}}{\sqrt{D_{F/\mathbf{Q}}}}$$

up to multiplication by an element of $\mathbf{Q}(\psi) \subset E$. Here

$$(27) \qquad G(\psi) = \psi(d')^{-1} \sum_u \psi(u)\, \mathbf{e}_F(d'^{-1}u)$$

is the Gauss sum associated to ψ, where $d'\mathcal{O}_F = \vartheta_{F/\mathbf{Q}}$ is the different of $F/\mathbf{Q}$, and the sum is over all $u \in N^{-1}\mathcal{O}_F/\mathcal{O}_F$, with $\mathrm{Ann}(u) = N$.

By assumption (see (19)), $E_{\mathbf{2n}-\mathbf{2m}}(0)$ is E-rational. Using the pairing (17), we deduce that all the integrals in (26) $\in E$. Here, note that $\eta_{2n-2m}(f_i) \in \widetilde{\mathrm{H}^r_{\mathrm{cusp}}}(\Gamma_{A_i}\backslash H^r, L(\mathbf{2n} - \mathbf{2m}, E))$, so is *not* compactly supported. However, there is a canonical section from cuspidal cohomology to compactly supported cohomology, with respect to which the pairing is invariant. See, for instance, [**H96**, Section 1.1], or the proof of Theorem 1 in [**G**].

Thus, we see that the L.H.S. of (24) at $s = 0$ is

$$= \frac{\Omega(f)\, G(\psi)\prod_\sigma (2\pi)^{2n_\sigma - 2m_\sigma + 2}}{\sqrt{D_{F/\mathbf{Q}}}},$$

up to multiplication by an element in E.

On the other hand, we see from (21) that

$$G_\infty(0, f) = \frac{c'}{\prod_\sigma (2\pi)^{2n_\sigma - m_\sigma + 2}},$$

for some non-zero constant $c' \in \mathbf{Q}$. Hence, the R.H.S. of (24) is

$$= \frac{\sqrt{D_{F/\mathbf{Q}}}\, G(l, f)}{\prod_\sigma i^{v_\sigma + v_{\sigma c}} \prod_\sigma (2\pi)^{2n_\sigma - m_\sigma + 2}},$$

up to multiplication by an element of $\mathbf{Q}$.

Comparing these two expression we see finally that, up to multiplication by an element in E,

$$G(l, f) = \frac{\Omega(f)\, G(\psi)\prod_\sigma (2\pi i)^{3l - 2w}}{\prod_\sigma (2\pi i)^{v_\sigma + v_{\sigma c}}} = \Omega_l(f).$$

□

REMARK 4. In the above discussion, we have implicitly fixed an imbedding of E into $\mathbf{C}$. However, it is possible to refine our analysis, to obtain a rationality result for the $E \otimes \mathbf{C}$-valued function, $\mathbb{G}(s, f)$, which we do now.

For each $\iota : E \hookrightarrow \mathbf{C}$, define $\Omega(f^\iota)$ as in (25), and set

$$\mathbb{c}(f) = (\Omega(f^\iota))_\iota.$$

This is an element of $E \otimes \mathbf{C}$, well defined up to an multiplication by an element of $E \otimes 1$. Similarly, define $G(\psi^\iota)$ as in (27), and set

$$\mathbb{G}(\psi) = (G(\psi^\iota))_\iota.$$

Then it is well known that

$$\frac{\mathbb{L}(2n_\sigma - 2m_\sigma + 2,\, \psi)}{\mathbb{G}(\psi)\cdot(1 \otimes \sqrt{D_{F/\mathbf{Q}}}\prod_\sigma (2\pi)^{2n_\sigma - 2m_\sigma + 2})} \in E \otimes 1 \subset E \otimes \mathbf{C}.$$

Set

$$\mathbb{c}_l(f) \stackrel{\text{def}}{=} \frac{\mathbb{c}(f)\cdot \mathbb{G}(\psi)\,(1 \otimes \prod_{\sigma\in\Sigma_K}(2\pi i)^{3l-2w})}{(1 \otimes \prod_{\sigma\in\Sigma_K}(2\pi i)^{v_\sigma + v_{\sigma c}})}.$$

Then we have the following

THEOREM 2. *With the same hypothesis as in Theorem 1, we have that*

$$\frac{\mathbb{G}(l,f)}{\mathbb{c}_l(f)} \in E \otimes 1 \subset E \otimes \mathbf{C},$$

for all even critical integers l.

PROOF. This follows from the remarks above, and from the fact that

$$\iint_{\Gamma_{A_i}\backslash H^r} \langle \eta_{2n-2m}(f_i),\, E_{2n-2m}(\omega, 0)\rangle \otimes 1 \,\in\, E \otimes \mathbf{C} \subset E \otimes \mathbf{C}.$$

□

References

[A] T. Asai, *On certain Dirichlet series associated with Hilbert modular forms and Rankin's method*, Math Ann. **226** (1977), 81–94.

[D] P. Deligne, *Valeurs de Fonctions L et périodes d'intégrales*, Proc. Sympos. Pure Math. **33** part 2, Providence, RI: Amer. Math. Soc. (1979), 313–346.

[F] Y. Flicker, *Twisted tensors and Euler products*, Bull. Soc. Math. **116** (1988), 295–313.

[G96] E. Ghate, *Critical values of the Asai L-function in the imaginary quadratic case*, Dissertation, University of California at Los Angeles (1996).

[G] E. Ghate, *Critical values of the twisted tensor L-function in the imaginary quadratic case*, to appear in Duke Math. J.

[H] G. Harder, *Eisenstein cohomology of arithmetic groups, the case* GL_2, Invent. Math. **89** (1987), 37–118.

[H93] H. Hida, *p-Ordinary cohomology groups for* SL(2) *over number fields*, Duke Math. J. **69** (1993), 259–314.

[H94] H. Hida, *On the critical values of L-functions of* GL(2) *and* GL(2) × GL(2), Duke Math. J. **74** (1994), 431–528.

[H96] H. Hida, *Non-critical values of adjoint L-functions for* SL(2), this volume.

[Im] J. Im, *Twisted tensor L-functions attached to Hilbert modular forms*, CRM Proceedings & Lecture Notes **4**, Providence, RI: Amer. Math. Soc. (1994), 111–119.

[S75] G. Shimura, *On some arithmetic properties of modular forms of one and several variables*, Ann. of Math. **102** (1975), 491-515.

[S76] G. Shimura, *The special values of the zeta functions associated with cusp forms*, Comm. Pure. Appl. Math. **29** (1976), 783–804.

[S77] G. Shimura, *On the periods of modular forms*, Math. Ann. **229** (1977), 211–221.

[S81] G. Shimura, *On certain zeta functions attached to two Hilbert modular forms: II. The case of automorphic forms on a quaternion algebra*, Ann. of Math. **114** (1981), 569–607.

[W] G. N. Watson, A Treatise on the Theory of Bessel Functions, Cambridge: Cambridge University Press (1948).

[Z] Y. Zhao, *Certain Dirichlet series attached to automorphic forms over imaginary quadratic fields*, Duke Math. J. **72** (1993), 695–724.

University of California, Los Angeles, CA 90095-1555

Current address: School of Mathematics, Tata Institute of Fundamental Research, Homi Bhabha Road, Mumbai 400 005, India.

E-mail address: `eghate@math.tifr.res.in`

Proceedings of Symposia in Pure Mathematics
Volume **66.1**, 1999

Zeta Functions Formed With Modular Symbols

Dorian Goldfeld

Dedicated to Goro Shimura

ABSTRACT. A new class of automorphic functions is introduced where the automorphic relation involves a shift by a modular symbol. In the special case of Eisenstein series, this leads to new zeta functions whose Dirichlet coefficients are themselves modular symbols. On the basis of this new theory a number of conjectures are made concerning the distribution of moments of modular symbols.

1. Introduction and statement of results

For $N = 1, 2, 3\ldots$, let $\Gamma_0(N)$ denote the group of matrices $\left(\begin{smallmatrix} a & b \\ c & d \end{smallmatrix}\right)$ with $a, b, c, d \in \mathbb{Z}$, $ad - bc = 1$, and $c \equiv 0 \pmod{N}$. Fix an even Dirichlet character χ to the modulus N and for a matrix $\gamma = \left(\begin{smallmatrix} a & b \\ c & d \end{smallmatrix}\right) \in \Gamma_0(N)$ define $\chi(\gamma) = \chi(d)$ and $j(\gamma, z) = cz + d$. Let

$$\mathfrak{h} = \{z = x + iy \in \mathbb{C} \mid y > 0\}$$

denote the upper half plane and define $\mathfrak{h}^* = \mathfrak{h} \cup \mathbb{Q} \cup \{\infty\}$ to be the extended upper half–plane including the cusps. Consider the set of automorphic forms $\mathrm{M}_k(N, \chi)$ of weight k and character χ. Thus, if $f \in \mathrm{M}_k(N, \chi)$, then f is a function (not necessarily holomorphic in z) from h^* to $\mathbb{C}$ satisfying

$$f(\gamma z) = \chi(\gamma)\, j(\gamma, z)^k\, f(z), \qquad \gamma \in \Gamma_0(N),\ z \in h^*,$$

which has polynomial growth at the cusps. This last condition simply means that if $s \in \mathbb{Q}$ is a cusp of $\Gamma_0(N)$ and $\rho \in \mathrm{SL}_2(\mathbb{R})$ satisfies $\rho s = \infty$, then $f(\rho^{-1}z)\cdot j(\rho^{-1}, z)^{-k}$ has a polynomial growth in $y = \mathrm{Im}(z)$ as $y \to \infty$. Similarly, let $\mathrm{S}_k(N)$ denote the space of holomorphic cusp forms in $\mathrm{M}_k(N, \chi_0)$ where χ_0 is the trivial character. There is a bilinear pairing

$$\mathrm{S}_2(N) \times \mathrm{H}_1(X_0(N), \mathbb{C}) \longrightarrow \mathbb{C}$$

1991 *Mathematics Subject Classification.* 11F67.

Key words and phrases. Modular symbols, automorphic forms, Eisenstein series, Fourier expansions.

Supported in part by NSF grant DMS9003907.

given by

$$(f,\gamma) \mapsto \langle \gamma, f\rangle = -2\pi i \int_\gamma f(z)\,dz.$$

The modular symbols $\langle \gamma, f\rangle$ have played a crucial role in the study of cusp forms and elliptic curves (see, [**C**], [**M**], [**MSD**]). Very little is known at present on the distribution properties of the values $\langle \gamma, f\rangle$ as γ ranges over $\Gamma_0(N)$. Using results of Shimura [**Sh**], one may write

$$\langle \gamma, f\rangle \;=\; \sum_{i=1}^{2n} m_i \Omega_i$$

where Ω_i are fixed complex numbers (periods of $J_0(N)$), $m_i \in \mathbb{Z}$, and n is the degree (over $\mathbb{Q}$) of the smallest field containing the Fourier coefficients of f. We have conjectured [**G89**] that if $\gamma = \left(\begin{smallmatrix} a & b \\ c & d\end{smallmatrix}\right) \in \Gamma_0(N)$ with $|c| \le N^2$, then

$$m_i \ll N^\kappa \quad (\text{for } i = 1,2,3,\ldots 2n,\ N \to \infty)$$

for some fixed $\kappa > 0$. In the special case that f is associated to an elliptic curve over $\mathbb{Q}$, the above conjecture is equivalent (see [**G89**]) to the well known conjecture of Szpiro [**Sz**],

$$D \ll N^C \quad (\text{for some fixed constant } C > 6)$$

relating the conductor N and the discriminant D of the elliptic curve.

In relation to the problem of obtaining distribution theorems on the values $\langle \gamma, f\rangle$, it is natural to introduce the Epstein type zeta function

$$\sum_{\substack{c \equiv 0 \pmod N \\ (c,d)=1}} \frac{\langle \gamma, f\rangle}{(c^2+d^2)^s}$$

where $\gamma = \left(\begin{smallmatrix} a & b \\ c & d\end{smallmatrix}\right)$. This is a special case (set $z = i$, $\chi = \chi_0 =$ trivial character) of the generalized Eisenstein series

$$E^*(z,s,\chi) = \sum_{\gamma \in \Gamma_\infty \backslash \Gamma_0(N)} \chi(\gamma)\,\langle \gamma, f\rangle\, \mathrm{Im}(\gamma z)^s,$$

where Γ_∞ denotes the group $\left\{\left(\begin{smallmatrix} 1 & n \\ 0 & 1\end{smallmatrix}\right) \,\middle|\, n \in \mathbb{Z}\right\}$ which stabilizes infinity. It is easy to see that $E^*(z,s,\chi)$ satisfies the automorphic relation

$$E^*(\gamma z, s, \chi) = \bar\chi(\gamma) E^*(z,s,\chi) - \bar\chi(\gamma)\langle \gamma, f\rangle E(z,s,\chi)$$

where $E(z,s,\chi) = \sum_{\gamma \in \Gamma_\infty \backslash \Gamma_0(N)} \chi(\gamma)\mathrm{Im}(\gamma z)^s$ is the classical non–holomorphic weight zero Eisenstein series for $\Gamma_0(N)$.

Set $E^*(z,s) = E(z,s,\chi_0)$, $E(z,s) = E(z,s,\chi_0)$. Since $E(z,s)$ has a simple pole at $s = 1$, the automorphic relation for E^* implies that $E^*(z,s)$ must have a pole at $s = 1$. Let $R(z)$ denote the residue at $s = 1$. Since $E^*(z,s)$ is an eigenfunction of the Laplacian it immediately follows that $R(z)$ must be harmonic. Further, $R(z)$ must satisfy the automorphic relation

$$R(\gamma z) = R(z) - \langle \gamma, f\rangle r_N$$

where

$$r_N = \frac{3}{\pi} \prod_{p|N} \left(1 + \frac{1}{p}\right)^{-1}$$

is the residue at $s = 1$ of $E(z, s)$. In this paper, we compute the Fourier expansion of E^* at infinity and show that the constant term vanishes. It follows from this that

$$R(z) = r_N \cdot F(z)$$

where $F(z) = 2\pi i \int_z^{i\infty} f(w)\, dw$ is the anti-derivative of f.

The above calculations suggest that for arbitrary $M > 0$,

$$S_N(X) = S_N(X; f) = \sum_{\substack{c^2M^2+d^2 \le X \\ c \equiv 0 \pmod{N}}} \langle \gamma, f \rangle \sim R(iM) \cdot X$$

for $X \to \infty$. We have made some preliminary computations with the software package Mathematica in the special case $M = 1$. For $N = 11$ there is a Hecke cusp form of weight two associated to the elliptic curve $y^2 - y = x^3 - x^2 - 10x - 20$ of conductor 11 with periods $\Omega_1 = 1.269209304...$ and $\Omega_2 = (1.458816617...) \cdot i$. In this case, we have computed:

$$\begin{aligned} S_{11}(5000) &= 8\Omega_1, \\ S_{11}(20,000) &= 6\Omega_1, \\ S_{11}(100,000) &= 14\Omega_1. \end{aligned}$$

Similarly, for $N = 67$, the elliptic curve $y^2 + y = x^3 + x^2 - 12x - 21$ has periods given by $\Omega_1 = 1.273770037...$ and $\Omega_2 = (3.029968401...) \cdot i$. As before, we have computed:

$$\begin{aligned} S_{67}(5000) &= 10\Omega_1, \\ S_{67}(20,000) &= -25\Omega_1, \\ S_{67}(100,000) &= 39\Omega_1, \\ S_{67}(500,000) &= 35\Omega_1. \end{aligned}$$

In this paper, we compute the Fourier expansion of E^*. Further properties of E^*, such as the analytic continuation and functional equation, should be obtainable by the method of Selberg (see [**S56**]). It would also be of great interest to study even more general Eisenstein series

$$E_p^*(z, s, \chi) = \sum_{\gamma \in \Gamma_\infty \backslash \Gamma_0(N)} \chi(\gamma) \cdot p\Big(\langle \gamma, f \rangle, \overline{\langle \gamma, f \rangle} \Big) \operatorname{Im}(\gamma z)^s,$$

where $p(x, y)$ is an arbitrary polynomial in two variables x, y with complex coefficients. For example, the constant term of the Fourier expansion of the Eisenstein series

$$E^{**}(z, s) = \sum_{\gamma \in \Gamma_\infty \backslash \Gamma_0(N)} |\langle \gamma, f \rangle|^2 \operatorname{Im}(\gamma z)^s$$

has a simple pole at $s = 1$. Let $R^*(z)$ denote the residue of E^{**} at $s = 1$. Then R^* satisfies the automorphic relation

$$R^*(\gamma z) = R^*(z) - \overline{\langle \gamma, f \rangle}\, R(z) - \langle \gamma, f \rangle\, \overline{R(z)} + |\langle \gamma, f \rangle|^2\, r_N.$$

This suggests the asymptotic formula

$$\sum_{c^2+d^2 \le X} \left| \left\langle \begin{pmatrix} * & * \\ c & d \end{pmatrix}, f \right\rangle \right|^2 \sim R^*(i) \cdot X$$

for $X \to \infty$. The precise determination of $R^*(z)$ is a problem of considerable interest.

The meromorphic continuation and functional equations of other such generalized Eisenstein series will provide important information on the higher moment distributions of modular symbols. Since special values of L-functions associated to cusp forms of weight two can be expressed in terms of modular symbols, the analytic study of the proposed generalized Eisenstein series should ultimately give new information on the growth properties and value distribution of special values of L-functions.

The generalized Eisenstein series E^* is a special case of a new class of automorphic functions where the automorphic relation involves a shift by a modular symbol multiplied by a classical automorphic form. We give a rapid introduction to this new theory by specializing to the case of the congruence subgroups $\Gamma_0(N)$. The further study of this class of automorphic functions opens a new horizon in the analytic theory of modular symbols.

Acknowledgment: The author would like to thank Cormac O'Sullivan for several helpful conversations.

2. A new class of automorphic functions

Let M, N be positive integers with $M|N$, and let χ be Dirichlet character modulo N. Fix $f \in \mathrm{S}_2(M)$ and $G \in \mathrm{M}_k(N,\chi)$. We do not assume that G is holomorphic. We shall consider the $\mathbb{C}$–vector space of automorphic forms $G^* : \mathfrak{h}^* \to C$ which have polynomial growth at the cusps and satisfy the relation

$$G^*(\gamma z) = j(\gamma, z)^k \Big(\chi(\gamma)\, G^*(z) + c \cdot \chi(\gamma) \langle \gamma, f\rangle\, G(z) \Big)$$

for some fixed $c \in \mathbb{C}$, all $z \in \mathfrak{h}^*$ and all $\gamma \in \Gamma_0(N)$. Let $\mathrm{M}_k^*(f, G)$ denote the set of automorphic forms as defined above. It is clear that if $G_1^*, G_2^* \in \mathrm{M}_k^*(f,G)$, then there exist constants $c_1, c_2 \in \mathbb{C}$ (not both zero) such that

$$c_1 G_1^*(z) - c_2 G_2^*(z) \in \mathrm{M}_k(N, \chi).$$

The existence of automorphic forms in $\mathrm{M}_k^*(f,G)$ can be shown by explicitly constructing them as Poincaré series. We first consider the case where G is a holomorphic modular form of integer weight $k > 2$. Define the Poincaré series

$$P_n(z, k, \chi, f) = \sum_{\gamma \in \Gamma_\infty \backslash \Gamma_0(N)} \bar{\chi}(\gamma) \langle \gamma, f\rangle j(\gamma, z)^{-k} e^{2\pi i n \gamma(z)}.$$

The absolute convergence of the above series is an immediate consequence of the following lemma:

Lemma 1. *Let* $\gamma = \left(\begin{smallmatrix} a & b \\ c & d \end{smallmatrix}\right) \in \Gamma_0(N)$. *Then* $\langle \gamma, f\rangle \ll |c|^{1/2+\epsilon}$ *where the implied constant depends only on* ϵ *and* f.

Proof. Let

$$f(z) = \sum_{n=1}^{\infty} a_n\, e^{2\pi i n z}$$

be the Fourier expansion of f where $a_n \ll n^{1/2+\epsilon}$. It follows by integration that

$$\begin{aligned}\langle \gamma, f\rangle &= -2\pi i \int_\tau^{\gamma\tau} f(z)\,dz\\ &= \sum_{n=1}^{\infty} \frac{a_n e^{2\pi i n\tau}}{n} - \sum_{n=1}^{\infty} \frac{a_n e^{2\pi i n\gamma\tau}}{n}\end{aligned}$$

is independent of τ. If $c = 0$ then $\langle \gamma, f\rangle = 0$. Otherwise, choose $\tau = -d/c + i/|c|$ so that $\mathrm{Im}(\gamma\tau) = \mathrm{Im}(\tau) = 1/|c|$. Then

$$\langle \gamma, f\rangle \ll \sum_{n \ll |c|^{1+\epsilon}} \frac{1}{\sqrt{n}} \ll |c|^{1/2+\epsilon}.$$

This completes the proof of the lemma. □

By construction, we have

$$P_n(\gamma z, k, \chi, f) = j(\gamma, z)^k \Big(\chi(\gamma)\, P_n(z, k, \chi, f) - \chi(\gamma)\langle\gamma, f\rangle P_n(z, k, \chi) \Big)$$

where

$$P_n(z, k, \chi) = \sum_{\gamma \in \Gamma_\infty \backslash \Gamma_0(N)} \bar{\chi}(\gamma) j(\gamma, z)^{-k} e^{2\pi i n \gamma(z)}.$$

Since every holomorphic modular form $G \in \mathrm{M}_k(N, \chi)$ can be constructed as a linear combination $\sum_n c_n P_n(z, \chi)$ with $c_n \in \mathbb{C}$, the existence of automorphic forms in $\mathrm{M}_k^*(f, G)$ is established. The same argument extends to nonholomorphic forms and forms of weight $k < 2$ if one uses the nonholomorphic Poincaré series

$$P_n(z, k, s, \chi, f) = \sum_{\gamma \in \Gamma_\infty \backslash \Gamma_0(N)} \bar{\chi}(\gamma) \langle \gamma, f\rangle j(\gamma, z)^{-k}\, \mathrm{Im}(\gamma z)^s\, e^{2\pi i n \gamma(z)}$$

instead of $P_n(z, k, \chi, f)$ as above.

An alternative method to construct $G^* \in \mathrm{M}_k^*(f, G)$ is to introduce the antiderivative

$$F(z) = 2\pi i \int_z^{i\infty} f(w)\,dw = \sum_{n=1}^{\infty} \frac{a_n}{n} e^{2\pi i n z}$$

which satisfies the automorphic relation

$$F(\gamma z) = F(z) - \langle \gamma, f\rangle$$

for all $\gamma \in \Gamma_0(N)$. Then it is easily seen that if $G \in \mathrm{M}_k(N, \chi)$, then

$$G^*(z) = G(z) \cdot F(z) \in \mathrm{M}_k^*(f, G).$$

We shall say $G^* \in \mathrm{M}_k^*(f, G)$ is a holomorphic cusp form if for each cusp $s \in \mathbb{Q}$ and $\rho \in \mathrm{SL}(2, \mathbb{R})$ satisfying $\rho s = \infty$ we have $G^*(\rho^{-1} z) \cdot j(\rho^{-1}, z)^{-k}$ vanishes at $z = i\infty$. Let $\mathrm{S}^*(f, G)$ denote the $\mathbb{C}$–vector space of holomorphic cusp forms in $\mathrm{M}^*(f, G)$.

An immediate consequence of the above is the following proposition.

PROPOSITION 2. *Let M, N be positive integers with $M|N$, and let χ be a Dirichlet character modulo N. Fix $f \in S_2(M)$ and $G \in M_k(N, \chi)$. Then we have*

$$\mathrm{Dim}\,(\mathrm{S}^*(f, G)) = 1 + \mathrm{Dim}\,(S_k(N, \chi)).$$

It is also possible to define the space of holomorphic Eisenstein automorphic forms in $\mathrm{M}^*(f,G)$. Again, the dimensions exceed the dimensions of the classical holomorphic Eisenstein modular forms in $\mathrm{M}_k(N,\chi)$ by one.

3. Fourier expansions

Let $G^* \in \mathrm{M}^*(f,G)$. Fix $T = \left(\begin{smallmatrix} 1 & 1 \\ 0 & 1 \end{smallmatrix}\right) \in \Gamma_0(N)$. Then, since

$$\langle T, f\rangle = 0, \quad \chi(T) = 1, \quad j(T,z) = 1,$$

we see that

$$G^*(Tz) = G^*(z+1) = G^*(z)$$

and, therefore, for any fixed $y > 0$, G^* has a Fourier expansion of the form

$$G^*(z) = \sum_{n=-\infty}^{\infty} c(n,y)e^{2\pi i n x}$$

for $c(n,y) \in \mathbb{C}$. If G^* is holomorphic then, the Fourier expansion takes the simpler form

$$G^*(z) = \sum_{n=0}^{\infty} c(n)e^{2\pi i n z}$$

with $c(n) \in \mathbb{C}$.

The Poincaré series

$$P_m(z,k,s,\chi,f) = \sum_{\gamma\in\Gamma_\infty\backslash\Gamma_0(N)} \bar{\chi}(\gamma)\langle\gamma,f\rangle j(\gamma,z)^{-k}\,\mathrm{Im}(\gamma z)^s\, e^{2\pi i m\gamma(z)}$$

has a Fourier expansion

$$P_m(z,k,s,\chi,f) = \sum_{n=-\infty}^{\infty} A_n(k,s,\chi,f,m)e^{2\pi i n x}$$

where

$$A_n(k,s,\chi,f,m)$$
$$= \sum_{\substack{c\equiv 0 \pmod N \\ c\neq 0}} \frac{S(m,n,\chi,f;c)}{|c|^{2s}c^k}\cdot\int_{-\infty}^{\infty} e^{\frac{-2\pi i m}{c^2 y(x+i)}}\cdot\frac{y^{1-s-k}}{(x^2+1)^s(x+i)^k}e^{-2\pi i n x y}\,dx$$

and

$$S(m,n,\chi,f;c) = \sum_{\substack{r=1 \\ (r,c)=1}} \bar{\chi}(r)e^{\frac{2\pi i(mr+n\bar{r})}{c}}\left\langle\begin{pmatrix} * & * \\ c & r\end{pmatrix}, f\right\rangle.$$

The sum $S(m,n,\chi,f;c)$ is a type of generalized Kloosterman sum which has an associated zeta function

$$Z(m,n,\chi,f;s) = \sum_{\substack{c\equiv 0 \pmod N \\ c\neq 0}} \frac{S(m,n,\chi,f;c)}{|c|^{2s}}.$$

These Kloosterman type zeta functions are generalizations of those first studied by Selberg [**S65**], and it would be of great interest to obtain their analytic continuation and polar structure. The zeta function $Z(0,0,\chi,f;s)$ occurs in the constant term in the Fourier expansion of $E^*(z,s,\chi)$. If χ is a nonprincipal character to the modulus N, we have

$$S(0,0,\chi,f;c) = \sum_{\substack{r=1\\(r,c)=1}} \bar{\chi}(r) \left\langle \begin{pmatrix} * & * \\ c & r \end{pmatrix}, f \right\rangle$$
$$= \sum_{\substack{r=1\\(r,c)=1}} \bar{\chi}_c(r) \left\langle \begin{pmatrix} * & * \\ c & r \end{pmatrix}, f \right\rangle$$

where χ_c is a primitive character induced from χ, i.e., $\chi_c(n) = \chi(n)$ if $(n,c) = 1$ and otherwise $\chi_c(n) = 0$. Note that the conductor of χ_c must divide c. It follows that

$$S(0,0,\chi,f;c) = \tau(\chi_c)\, L_f(1,\bar{\chi}_c)$$

where

$$\tau(\chi_c) = \sum_{a=1}^{c} \chi_c(z) e^{\frac{2\pi i a}{c}}$$

is the Gauss sum and

$$L_f(s,\chi_c) = \sum_{n=1}^{\infty} \frac{a(n)\chi_c(n)}{n^s}$$

is the twisted L-series associated to

$$f(z) = \sum_{n=1}^{\infty} a_n e^{2\pi i n z}.$$

Thus, the zeta function $Z(0,0,\chi,f;s)$ can be written in the form

$$Z(0,0,\chi,f;s) = \sum_{\substack{c\equiv 0 \pmod N\\ c\neq 0}} \frac{\tau(\chi_c)\, L_f(1,\bar{\chi}_c)}{|c|^{2s}}.$$

On the other hand, if $\chi = \chi_0$ is the principal character (mod N), then

$$S(0,0,\chi,f;c) = \sum_{\substack{r=1\\(r,c)=1}}^{c} \left\langle \begin{pmatrix} * & * \\ c & r \end{pmatrix}, f \right\rangle = 0$$

because of the identity (choose $\tau = -r/c + i/|c|$ as in the proof of Lemma 1)

$$\left\langle \begin{pmatrix} * & * \\ c & r \end{pmatrix}, f \right\rangle = \sum_{n=1}^{\infty} \frac{A(n)}{n} \exp\left(\frac{-2\pi n}{|cN|}\right) \left[\exp\left(\frac{2\pi \in r}{cN}\right) - \exp\left(\frac{-2\pi \in \bar{r}}{cN}\right)\right]$$

where

$$r\bar{r} \equiv 1 \pmod c.$$

When we sum over r in the identity above, the sum with r and the sum with $\bar{r}$ cancel to give zero.

In a similar manner, we may obtain the Fourier expansion of the more general Eisenstein series:

$$E^{**}(z,s,\chi_0) = \sum_{\gamma\in\Gamma_\infty\backslash\Gamma_0(N)} \chi_0(\gamma) \mid \langle \gamma, f\rangle \mid^2 \operatorname{Im}(\gamma z)^s.$$

For example, the constant term of the Fourier expansion takes the form

$$\int_0^1 E^{**}(z,s,\chi_0)\,dx = \sqrt{\pi}y^{1-s}\frac{\Gamma(s-1/2)}{\Gamma(s)}\,Z(s)$$

where

$$Z(s) = \lim_{w\to 1} \sum_{\substack{c\equiv 0\,(N)\\ c\neq 0}} \frac{1}{c^{2s}} \sum_{d|c} \mu\left(\frac{c}{d}\right)\cdot d \sum_{m\equiv n\,(d)} \frac{a_m\bar{a}_n}{(mn)^w}.$$

The meromorphic continuation and polar structure of $Z(s)$ is not easy to obtain. One approach would be to use the method of Selberg [**S56**]. Preliminary calculations suggest that $Z(s)$ has a simple pole at $s=1$ with residue proportional to the Petersson inner product $\langle f, f\rangle$.

4. L-functions

Let

$$G^*(z) = \sum_{n=0}^{\infty} c(n)e^{2\pi i n z}$$

be in $\mathrm{M}^*(f,G)$. The L-function associated to G^* is defined to be

$$\mathrm{L}_{G^*}(s) = \sum_{n=1}^{\infty} \frac{c(n)}{n^s}.$$

We will show that the Dirichlet series for $\mathrm{L}_{G^*}(s)$ converges absolutely in a half plane, has an analytic continuation to the entire complex s–plane, and in certain cases has a functional equation of simple type.

LEMMA 3. *Let* $G\in M_k(N,\chi)$, *and let* $G^*(z) = \sum\limits_{n=0}^{\infty} c(n)e^{2\pi i n z} \in M^*(f,G)$. *Then if* G *is cuspidal we have*

$$c(n) \ll n^{k/2+\epsilon}.$$

Otherwise

$$c(n) \ll n^{k-1/2+\epsilon}.$$

PROOF. By the results of §2, every holomorphic $G^*\in \mathrm{M}^*(f,G)$ may be expressed in the form

$$G^*(z) = cF(z)G(z) + G_1(z) \qquad c\in\mathbb{C},$$

with $G_1(z)\in \mathrm{M}_k(N,\chi)$ also holomorphic. Recall that $F(z) = \sum\limits_{n=1}^{\infty} a_n e^{2\pi i n z}/n$ where the coefficients $a(n)$ satisfy the Ramanujan bound $a(n)\ll n^{1/2+\epsilon}$. If we write

$$G(z) = \sum_{n=0}^{\infty} b_n e^{2\pi i n z}, \quad G_1(z) = \sum_{n=0}^{\infty} {b'}_n e^{2\pi i n z},$$

then one easily sees that

$$c(n) = c\sum_{m=0}^{n-1} b_m \frac{a_{n-m}}{n-m} + {b'}_n.$$

If G, G_1 are holomorphic cusp forms of weight k, we have the estimate $b_n, {b'}_n \ll n^{(k-1)/2+\epsilon}$. Otherwise, $b_n, {b'}_n \ll n^{k-1+\epsilon}$. Plugging these bounds in the formula for $c(n)$ given above concludes the proof of the lemma. □

It immediately follows from lemma 3 that the Dirichlet series $\sum_{n=1}^{\infty} c(n)n^{-s}$ converges absolutely in the half–plane $\mathrm{Re}(s) > k + 1/2$. Similar estimates can also be obtained in the case that $G^*(z)$ is not holomorphic.

In certain cases, the L-function $\mathrm{L}_{G^*}(s)$ will have a simple functional equation. Let

$$W_N = \begin{pmatrix} 0 & 1 \\ -N & 0 \end{pmatrix}$$

denote the involution on $\Gamma_0(N)$. Let $G \in \mathrm{M}_k(N, \chi)$, with k an even positive integer, satisfy

$$G(W_N z) = \epsilon \cdot N^{k/2} z^k G(z)$$

with $\epsilon = \pm 1$. If we assume that $\langle W_N, f\rangle = 0$ and G^* is of the form

$$G^*(z) = cG(z)F(z) + c'G(z)$$

with $c, c' \in \mathbb{C}$, then we see that G^* satisfies the automorphic relation

$$G^*(W_N z) = \epsilon \cdot N^{k/2} z^k G^*(z).$$

Riemann's method of analytic continuation yields the meromorphic continuation of $\mathrm{L}_{G^*}(s)$ to the entire complex s–plane and the functional equation

$$\psi(s) = (2\pi)^{-s}\Gamma(s)N^{s/2}\mathrm{L}_{G^*}(s) = \epsilon \cdot (-1)^{k/2}\psi(k - s).$$

It can be shown that $\psi(s)$ has at most simple poles at $s = 0$ and $s = k$ with residue a multiple of $c(0)$, the constant term in the Fourier expansion of $G^*(z)$.

If $< W_N, f\rangle \neq 0$, then the automorphic relation takes the form

$$G^*(W_N z) = \epsilon \cdot N^{k/2} z^k G^*(z) + \epsilon'\langle W_N, f\rangle G(z)$$

where ϵ' is a root of unity. In this case, Riemann's method again gives the meromorphic continuation of $\mathrm{L}_{G^*}(s)$. In the most general case, we may express an arbitrary element $G^* \in \mathrm{M}^*(f, G)$ in the form

$$G^*(z) = cG(z)F(z) + g(z)$$

with $g \in \mathrm{M}_k(N, \chi)$ and $c \in \mathbb{C}$. Since we know that $\mathrm{L}_g(s)$ has a meromorphic continuation and satisfies a functional equation, it follows from the above that $\mathrm{L}_{G^*}(s)$ also has a meromorphic continuation. We have thus proved.

PROPOSITION 4. *Fix $f \in S_2(N)$, $G \in M_2(N, \chi)$. Then for every*

$$G^*(z) = \sum_{n=0}^{\infty} c(n) e^{2\pi i n z} \ \in \ M^*(f, G)$$

the L-function associated to G^*,

$$L_{G^*}(s) = \sum_{n=1}^{\infty} \frac{c(n)}{n^s}$$

converges absolutely for $Re(s) > k + 1/2$ and has a meromorphic continuation to the entire complex plane with at most a simple pole at $s = k$ whose residue is proportional to $c(0)$. A similar result holds if G^* is not holomorphic.

5. One–cocycles

Let $G^* \in \mathrm{M}_k^*(f,G)$. For $z \in \mathfrak{h}^*$ and $\gamma \in \Gamma_0(N)$, we define the one–cocycle (see [**G95**])

$$\langle \gamma, z\rangle_{G^*} = 2\pi i \int_z^{\gamma z} G^*(w)\,dw$$

which satisfies the cocycle relation

$$\langle \gamma\gamma', z\rangle_{G^*} = \langle \gamma, \gamma' z\rangle_{G^*} + \langle \gamma', z\rangle_{G^*}$$

for all $\gamma, \gamma' \in \Gamma_0(N)$, $z \in \mathfrak{h}^*$. In the special case that $k=2$ and $G=f$, the above cocycles can be explicitly computed. This is illustrated in the following proposition.

PROPOSITION 5. *Let N be a positive integer. Fix $f \in S_2(N)$. Then for all $G^* \in S^*(f,f)$ and $\gamma \in \Gamma_0(N)$, we have*

$$\langle \gamma, z\rangle_{G^*} = -\langle \gamma, f\rangle F(z) + \frac{\langle \gamma, f\rangle^2}{2} - \langle \gamma, g\rangle,$$

for some $g \in S_2(N)$. Further, for every $G_1 \in S_2(N)$, $\gamma \in \Gamma_0(N)$,

$$G_1(z)\cdot \langle \gamma, z\rangle_{G^*} \in S^*(f, G_1).$$

For fixed $G_1 \in S_2(N)$, the elements $G_1(z)\cdot\langle \gamma, z\rangle_{G^*}$ generate a one–dimensional subspace of $S^*(f,G_1)$ as γ runs over $\Gamma_0(N)$.

PROOF. Recall that every $G^* \in S^*(f,f)$ can be expressed in the form

$$G^*(z) = f(z)F(z) + g(z)$$

with $g \in S_2(N)$ and

$$F(z) = 2\pi i \int_z^{i\infty} f(w)\,dw.$$

It follows that

$$\begin{aligned}\langle \gamma, z\rangle_{G^*} &= 2\pi i \int_z^{\gamma z} \Big(f(w)F(w) + g(w)\Big)\,dw \\ &= 2\pi i \int_z^{\gamma z} f(w)F(w)\,dw - \langle \gamma, g\rangle.\end{aligned}$$

By integration by parts,

$$2\pi i \int_z^{\gamma z} f(w)F(w)\,dw = F(\gamma z)^2 - F(z)^2 - 2\pi i \int_z^{\gamma z} f(w)F(w)\,dw.$$

Consequently,

$$2\pi i \int_z^{\gamma z} f(w)F(w)\,dw = -\langle \gamma, f\rangle F(z) + \frac{\langle \gamma, f\rangle^2}{2}.$$

Hence,

$$\langle \gamma, z\rangle_{G^*} = -\langle \gamma, f\rangle F(z) + \frac{\langle \gamma, f\rangle^2}{2} - \langle \gamma, g\rangle.$$

The transformation Formula

$$F(\gamma z) = F(z) - \langle \gamma, f\rangle$$

immediately implies that $G_1(z)\cdot\langle \gamma, z\rangle_{G^*} \in \mathrm{S}^*(f, G_1)$. The rest of the proposition easily follows.

References

[C] J. E. Cremona, Algorithms for Modular Elliptic Curves, Cambridge: Cambridge University Press (1992).

[G89] D. Goldfeld, *Modular elliptic curves and diophantine problems*, Proc. of the First Canadian Number Theory Assoc., Banff, Canada 1988, New York: de Gruyter (1989).

[G95] D. Goldfeld, *Special values of derivatives of L-series*, Number Theory, Fourth Conference of the Canadaian Number Theory Association, 1994, Dalhousie Univ., Halifax, Nova Scotia, Karl Dilcher Editor, Canadian Math. Soc. Conference Proc. **15**, Providence, RI: Amer. Math. Soc. (1995), 159–173.

[M] Ju. I. Manin, *Parabolic points and zeta-functions of modular curves,* Math. USSR Izvestija **6** (1972), 19–64.

[MSD] B. Mazur and P. Swinnerton–Dyer, *Arithmetic of Weil curves*, Invent. Math. **25** (1974), 1–61.

[S56] A. Selberg, *Harmonic analysis and discontinuous groups in weakly symmetric Riemannian spaces with applications to Dirichlet series*, J. Indian Math. Soc. B **20** (1956), 47–87.

[S65] A. Selberg, *On the estimation of Fourier coefficients of modular forms*, Proc. Symp. Pure Math. (Cal. Tech. Pasadena, CA 1963), **VIII**, Providence, RI: Amer. Math. Soc. (1965), 1–15.

[Sh] G. Shimura, *On the factors of the jacobian variety of a modular function field,* J. Math. Soc. Japan **25** (1973), 523–544.

[Sz] L. Szpiro, *Seminaire sur les pinceaux de courbes de genre au moins deux,* Astérisque, exposé n° 3, **86** (1981), 44–78.

Department of Mathematics, Columbia University, New York, NY 10027
E-mail address: goldfeld@columbia.edu

Proceedings of Symposia in Pure Mathematics
Volume **66.1**, 1999

Non-critical Values of Adjoint L-Functions for SL(2)

Haruzo Hida

To Goro Shimura with admiration

ABSTRACT. For a given system $\lambda(T(\mathfrak{p}))$ of eigenvalues of Hecke operators acting on cohomological cusp forms on GL(2) over a number field F, we look into the adjoint square L-function $L(s, \mathrm{Ad}(\lambda) \otimes \alpha)$ twisted by a Hecke character α. If λ is associated to a 2-dimensional Galois representation φ, the adjoint square $\mathrm{Ad}(\varphi)$ is the three dimensional factor of $\varphi \otimes {}^t\varphi^{-1}$, whose L-function is given by $L(s, \mathrm{Ad}(\lambda))$. The L-value $L(1, \mathrm{Ad}(\lambda) \otimes \alpha)$ is critical if and only if F is totally real and the Hecke character α is totally even. We are interested in both critical and non-critical cases. When α is quadratic, a rationality result: $L(1, \mathrm{Ad}(\lambda)\otimes\alpha)/\Omega \in \mathbb{Q}(\lambda)$ is shown, where Ω is a canonical (topological) period of the base change lift of λ to the quadratic extension K/F associated to α, and $\mathbb{Q}(\lambda)$ is the number field generated by $\lambda(T(\mathfrak{p}))$ for all primes $\mathfrak{p}$. Alongside, we shall give an evidence for the divisibility of the L-value by the order of the Selmer group of $\mathrm{Ad}(\varphi) \otimes \alpha$. Towards the end, a period relation is given, as an application of our main result, when F is totally imaginary.

1. Introduction

If one has two canonical rational structures on a given complex vector space, one can define a period which is the determinant of the linear transformation bringing one rational structure to the other. This principle applied to cohomology groups on a projective variety V (defined over a number field F) yields the classical periods of F-rational differential forms on V. In this case, one rational structure is given by Betti cohomology, and another comes from algebraic de Rham cohomology. The two cohomology groups are put together into one vector space by the comparison isomorphism. This definition extends to motives, and the periods are conjectured to give canonical transcendental factors of the critical values of the motivic L-functions (a conjecture of Deligne; cf. [**Hi94**]).

Even if the manifold V is not algebraic, it is feasible to define periods in a similar way if V is modular, that is,

$$V = H(\mathbb{Q})\backslash H(\mathbb{A})/UZ_H(\mathbb{R})C_\infty,$$

1991 *Mathematics Subject Classification.* Primary 11F03, 11F30, 11F33, 11F37, 11F41, 11F67.

Key words and phrases. L-function, L-value, modular form, cohomology.

The research is partially supported by NSF grant DMS 9701017.

where $H_{/\mathbb{Q}}$ is a classical linear group with center Z_H, C_∞ is the standard maximal compact subgroup of $H(\mathbb{R})$, and U is an open compact subgroup of the finite part $H(\mathbb{A}^{(\infty)})$. For a locally constant sheaf L on V coming from a polynomial representation of H, we have a canonical rational structure on the cuspidal cohomology group $H^q_{\text{cusp}}(V, L)$ coming from the rational structure of L. This cohomology group is often isomorphic to a product of several copies of a space S of cohomological cusp forms on $H(\mathbb{A})$. If one can specify this isomorphism in a canonical way as Eichler and Shimura did for elliptic modular forms [**Sh71**], we have another rational structure on the cohomology groups provided that the Fourier expansion of cusp forms gives a good rational structure on S. If $H = \text{Res}_{F/\mathbb{Q}}\text{GL}(2)_{/F}$, there is an optimal function W, in the Whittaker model of a cuspidal automorphic representation π, giving the standard L-function under the Mellin transform. The Fourier expansion with respect to W gives a rational structure on the space of cohomological cusp forms. Applying the above principle to H and $H \times H$, I proved rationality and integrality theorems for critical values of standard L and Rankin products in [**Hi94**]. This was extended to the so-called twisted tensor L-functions of π in [**Gh**] for imaginary quadratic F. The results for GL(2) described above could be generalized to $\text{GL}(n)$. Anyway hereafter we assume that $H = \text{Res}_{F/\mathbb{Q}}\text{GL}(2)_{/F}$.

In the investigation in [**Hi94**] and [**Gh**], only the minimal degree cohomology group is used, and the minimal degree seems to yield rationality only for critical values. If H has a holomorphic structure yielding a Shimura variety V ($\iff$ F is totally real), the degree of non-trivial cuspidal cohomology is unique (that is $q = [F : \mathbb{Q}]$), which is the minimal degree I meant, and it is natural from the conjecture of Deligne that we can get results only for critical values. If F is not totally real, there are several values of q with non-trivial cohomology. However the space S of cohomological cusp forms is independent of q.

In this paper, we study the rational structures for the maximal degree and some middle degree cohomology groups, and we shall prove a rationality result of the adjoint L-value $L(1, \text{Ad}(\pi) \otimes \alpha)$ for α with $\alpha^2 = 1$ relative to the period $\Omega(\widehat{\lambda})$ of a cohomology class of degree depending on K/F, where K/F is the quadratic extension of F associated to α (Corollaries 3.2 and 4.2 for $F = \mathbb{Q}$, Theorem 6.1 for totally real F and Theorems 7.1 and 8.1 for F with complex places). Here we write λ for the system of Hecke eigenvalues associated to π, that is, $L(s, \pi) = \sum_{\mathfrak{n}} \lambda(T(\mathfrak{n}))N(\mathfrak{n})^{-s}$, and we hereafter write $L(s, \text{Ad}(\lambda))$ for $L(s, \text{Ad}(\pi))$. This value $L(1, \text{Ad}(\lambda) \otimes \alpha)$ is non-critical if either the character α is odd at some real places of F or F is not totally real. Thus in the non-critical case, the automorphic period is close to the Beilinson period [**RSS**], assuming his conjecture and the existence of a motive yielding the adjoint L-function. While in the critical case, our automorphic period should be equal to the Deligne period (see [**Hi94**, Section 1]). Moreover, when $F = \mathbb{Q}$ and α is a quadratic Dirichlet character, we shall prove, under some assumptions, that the L-value gives congruences between non-base-change forms on $\text{GL}(2)_{/K}$ and the base change $\widehat{\lambda}$ of λ (cf. [**J**] and [**L**]) to the quadratic extension K as conjectured in [**DHI**] (see Theorem 5.2). This shows that the p-primary part of the value is close to the order of the Selmer group of $\text{Ad}(\rho) \otimes \alpha$, ρ being the p-adic Galois representation of λ. As is shown by Wiles [**W**, Chapter 4] and [**TW**], the p-primary part of $L(1, \text{Ad}(\lambda))/\Omega$ for many p gives the exact order of the Selmer group of $\text{Ad}(\rho)$. Our result (Theorem 5.2) is a partial generalization of this non-abelian class number formula. Thus we expect the conjectures in [**DHI**] made originally for real cyclic extensions to hold even for imaginary cyclic extensions of $\mathbb{Q}$ (see Conjecture 5.1).

Actually the period $\Omega(\widehat{\lambda})$ giving the transcendental factor of the L-value is defined using the base change lift $\widehat{\lambda}$ to $G = \mathrm{Res}_{K/\mathbb{Q}}(\mathrm{GL}(2))$ for the quadratic extension K/F associated to α. Because of this, we need to assume

(cusp) $\widehat{\lambda}$ *remains cuspidal.*

This condition is equivalent to the condition that the λ-eigenspace in the space of cusp forms on $H(\mathbb{A})$ is orthogonal to any theta series associated to the norm form of the quadratic extension K (see [**L**, Lemma 11.3]). We assume this condition throughout the paper. When F is totally real, the definition of the automorphic period is a little more transparent than the other cases, because we are either in the minimal or maximal degree case where the multiplicity of $\widehat{\lambda}$ in the modular cohomology group for G is basically 1 up to a group action at archimedean places. If F has complex place, things are more complicated, and we need to use (H, λ) and $(G, \widehat{\lambda})$ at the same time (see Sections 7 and 8). In addition to this, the explicit description of the Eichler-Shimura map obtained in [**Hi94**] from a result of Harder [**Ha**] is different depending on the shape of K/F, although the general principle is the same as explained in Section 2.4. This is why we treat the imaginary quadratic case in Section 3, the real quadratic case in Section 4, the case of totally real F in Section 6 and general cases in Sections 7 and 8. It is an interesting problem to study relations among periods of λ of different degrees. We list some of them in Section 9, which follow easily from our main result. It is also interesting to know what type of L-values can be dealt with by looking into middle degrees.

It is not an isolated phenomenon that topological rational structure yields a canonical transcendental factor of an L-value. Starting from a number field K, we induce the trivial Galois character from K to $\mathbb{Q}$. Then $\mathrm{Ind}_{\mathbb{Q}}^{K}\,\mathrm{id} = \mathrm{id} \oplus \chi$ for an Artin Galois representation χ. The classical class number formula is written in terms of $L(1, \chi)$ whose main transcendental factor is the regulator of K. As is obvious from the definition, the regulator is the period of the maximal degree cohomology group of $F_{\mathbb{A}}^{1}/F^{\times}$ for the norm 1 ideles $F_{\mathbb{A}}^{1}$ normalized with respect to the L-function (see [**Hi89**, p. 90]). The fact that $\mathrm{Ind}_{\mathbb{Q}}^{K}\,\mathrm{id}$ contains the identity representation once is essentially used to identify the residue of the Dedekind zeta function $L(s, \mathrm{Ind}_{\mathbb{Q}}^{K}\,\mathrm{id})$ of K with the Artin L-value $L(1, \chi)$ as a product of the regulator and the class number. Computation of the residue tends to be easier than the computation of values. In our case, a similar phenomenon occurs. For the contragredient $\rho^{\vee}$, $\rho \otimes \rho^{\vee} \cong \mathrm{id} \oplus \mathrm{Ad}(\rho)$, and the residue formula of the Rankin product $L(s, \rho \otimes \rho^{\vee})$is essentially used to obtain the non-abelian class number formula for $L(1, \mathrm{Ad}(\rho))/\Omega$ [**Hi81**], [**Hi88b**] and [**Hi89**] (see also [**U**] for a generalization to imaginary quadratic K). The transcendental factor Ω is the period of the maximal degree cohomology group for $\mathrm{GL}(2) \times \mathrm{GL}(2)$ (see [**Hi81**] and Section 8 in the text). Exactly the same phenomenon happens also for $L(1, \mathrm{Ad}(\rho) \otimes \alpha)/\Omega(\widehat{\lambda})$, although the period may not be of maximal degree. The idea of proof is simple, which is summarized in Section 2.4, although the computation, in order to get an effective integral expression of the L-value, is a bit demanding. In the process of obtaining the integral expression, it is necessary to find a Hecke character of K with a prescribed restriction to F. The argument to find such a character is substantially shortened by a suggestion made by the referee of this paper, in particular, Lemma 2.1 is supplied by him along with a concise proof. Here I wish to thank the referee for the suggestion and his careful reading of the manuscript.

Holomorphy of the adjoint L-function was first dealt with by Shimura [**Sh75**] and then generalized to arbitrary π and F by Gelbart-Jacquet [**GeJ**] (see also [**Sh94**] for another integral expression). The rationality result was dealt with for critical values by Sturm for $F = \mathbb{Q}$ and Im for general totally real F [**St**] and [**I**]. Our proof is a cohomological interpretation of the Rankin method studied by Shimura and Asai [**As**], which is generalized to $\mathrm{GL}(n)$ by Flicker [**Fl**] and [**FlZ**]. In the course of the proof of the congruence theorem (Theorem 5.2), we need to use a non-vanishing result of twisted tensor L-functions, which follows from a more general result of Shahidi [**S81**], [**S88**].

Here is general notation. We write $F_{\mathbb{A}}$ for the adele ring of F. When $F = \mathbb{Q}$, we write $\mathbb{A}$ for that. The finite part of $\mathbb{A}$ is written as $\mathbb{A}^{(\infty)}$, and the infinite part of $F_{\mathbb{A}}$ is written as F_∞. As a subring of $\mathbb{A}^{(\infty)}$, we write $\widehat{\mathbb{Z}}$ for the product of the p-adic integer ring $\mathbb{Z}_p$ over all primes p. For the integer ring $\mathfrak{r}$ of F, we put $\widehat{\mathfrak{r}} = \mathfrak{r} \otimes_{\mathbb{Z}} \widehat{\mathbb{Z}}$ as a subring of $F_{\mathbb{A}}^{(\infty)}$. For a number field X, we write I_X for the set of embeddings of X into $\mathbb{C}$. We write Σ_X for the set of archimedean places of X and decompose $\Sigma_X = \Sigma_X(\mathbb{R}) \sqcup \Sigma_X(\mathbb{C})$ for the set of all real places $\Sigma_X(\mathbb{R})$. For a number field denoted by F in the text, which is the base field, we drop the subscript "F" like I for I_F.

CONTENTS

2. Idea of the proof, and preliminaries

In this section, we first describe how we can interpret analytic integration of cuspidal automorphic forms in terms of group and sheaf cohomology theory in an algebraic way (Sections 2.1–2.2). Then we define various modular L-functions, and we study multiplicative relations among the L-functions we defined (Section 2.3). This relation combined with an integral expression gives a key to our proof of rationality theorem (Section 2.4). At the end of this section, we describe Γ-factors of L-functions and criticality of L-values in terms of motives (Section 2.5).

2.1. Integration of cuspidal cohomology classes. Let G be a classical linear algebraic group defined over $\mathbb{Z}$. We consider an open compact subgroup S of $G(\widehat{\mathbb{Z}})$ of $G(\mathbb{A}^{(\infty)})$. We write $G(\mathbb{R})_+$ for the identity component of the Lie group

$G(\mathbb{R})$. We put $G(\mathbb{A})_+ = G(\mathbb{A}^{(\infty)}) \times G(\mathbb{R})_+$ and $G(\mathbb{Q})_+ = G(\mathbb{A})_+ \cap G(\mathbb{Q})$, where $\mathbb{A}^{(\infty)}$ is the finite part of the adele ring. Then we study the modular manifold associated to S

$$Y(S) = G(\mathbb{Q})_+ \backslash G(\mathbb{A})_+ / C_{\infty+} Z(\mathbb{R}) S$$

for the center $Z = Z_G$ of G and the maximal compact subgroup $C_{\infty+}$ of $G(\mathbb{R})_+$. Decompose $Y(S) = \sqcup_a Y_a$ into a finite disjoint union of connected components Y_a. Then we fix a and write $Y = Y_a$. Then $Y_a \cong \Gamma\backslash\mathfrak{Z}$ for a discrete arithmetic subgroup Γ of $G(\mathbb{Q})_+$ and the symmetric space $\mathfrak{Z} = G(\mathbb{R})_+/Z_G(\mathbb{R})C_{\infty+}$. When it is necessary to indicate the dependence on a, we write $\Gamma^{(a)}$ for Γ. Then we suppose that we have a coordinate system $(t, x_1, ..., x_{d-1})$ $(d = \dim(Y))$ of a coordinate neighborhood U_s around the cusp s of Y such that

$$U_s \cong (t_0, \infty) \times (\Gamma_s \backslash \mathbb{R}^{d-1})$$

(for an open interval (t_0, ∞), $t_0 > 0$) with compact quotient $\Gamma_s\backslash\mathbb{R}^{d-1}$ for a discrete subgroup Γ_s of Γ acting on $\mathbb{R}^{d-1}$. This is the case where G has a maximal $\mathbb{Q}$-split torus of rank 1, and the variable t is given by the variable of the unique $\mathbb{Q}$-split torus (of a Levi subgroup) in the minimal parabolic subgroup fixing the cusp s. Let L be a finite dimensional $\mathbb{R}$-vector space with an action of Γ. Let ω be a C^∞-closed p-form on $(t_0, \infty) \times (\mathbb{R}^{d-1})$ with values in L decreasing exponentially as $t \to \infty$. We suppose that $\gamma^*\omega = \gamma\omega$ for $\gamma \in \Gamma_s$. Here $\gamma\omega(x)$ is the image under the action $\gamma : L \to L$ applied to the value $\omega(x)$. We write

$$\omega = \sum_{i_1 < \ldots < i_p} \alpha_{i_1 \ldots i_p} dx_{i_1} \wedge \ldots \wedge dx_{i_p} + \sum_{j_1 < \ldots < j_{p-1}} \beta_{j_1 \ldots j_{p-1}} dt \wedge dx_{j_1} \wedge \ldots \wedge dx_{j_{p-1}}.$$

Since $d\omega = 0$,

$$\frac{\partial \alpha_{i_1 \ldots i_p}(t,x)}{\partial t} = \sum_{k=1}^{p} (-1)^{k-1} \frac{\partial \beta_{i_1 \ldots i_{k-1} i_{k+1} \ldots i_p}(t,x)}{\partial x_{i_k}}.$$

Then we have

$$\alpha_{i_1 \ldots i_p}(t,x) = \int_\infty^t \frac{\partial \alpha_{i_1 \ldots i_p}(t,x)}{\partial t} dt = \sum_{k=1}^{p} (-1)^{k-1} \int_\infty^t \frac{\partial \beta_{i_1 \ldots i_{k-1} i_{k+1} \ldots i_p}}{\partial x_{i_k}} dt.$$

Let $\theta = \sum_{j_1 < \ldots < j_{p-1}} (\int_\infty^t \beta_{j_1 \ldots j_{p-1}}(t,x) dt) dx_{j_1} \wedge \ldots \wedge dx_{j_{p-1}}$. Then $d\theta = \omega$. Note that θ is invariant under Γ_s, that is, $\gamma^*\theta = \gamma\theta$ for $\gamma \in \Gamma_s$ because t is invariant under the action of Γ_s. Thus for a given ω, we can find a canonical lift θ. Let C be the set of all cusps of $Y(S)$. We take an open neighborhood U_s for each $s \in C$ as above. Let us consider the quotient $\underline{L} = \Gamma\backslash(\mathfrak{Z} \times L)$ given by the action $\gamma(z, \lambda) = (\gamma x, \gamma\lambda)$. We write $\pi : \underline{L} \to Y$ for the projection. Then we consider the sheaf $\mathcal{L}$ made of locally constant sections of π. For each cuspidal closed differential p-form ω on Y which is a C^∞-section of $\mathcal{L} \otimes_{\mathbb{R}} \Omega^p$, we take $\theta(\omega|_{U_s})$ as above so that $d\theta(\omega|_{U_s}) = \omega|_{U_s}$. Then we take a C^∞-function $\phi : Y \to \mathbb{R}$ such that ϕ is identically 1 if $t > t_1$ with a $t_1 > t_0$ for every s, and outside $\cup_{s\in C} U_s$, ϕ is identically 0. Then we define $\theta_\phi(\omega) = \sum_{s \in C} \phi\theta(\omega|_{U_s})$. The form $\omega - d\theta_\phi(\omega)$ is compactly supported. The cohomology class $[\omega - d\theta_\phi(\omega)]$ in $H_c^p(Y, \mathcal{L})$ is independent of the choice of U_s and ϕ, because $\theta_\phi(\omega) - \theta_{\phi'}(\omega)$ is compactly supported for any other choice of ϕ'. For $\mathbb{Q}$-rank 1 case, we have a canonical choice of t given by the variable of the

$\mathbb{Q}$-split torus of G and $\{x_i\}$ coming from $\mathbb{Q}$-non-split torus and unipotent radical of the minimal parabolic subgroup fixing s. Thus we have a canonical section

$$i : H^p_{\mathrm{cusp}}(Y, \mathcal{L}) \to H^p_c(Y, \mathcal{L}),$$

which is compatible with Hecke operator action. The compatibility follows from the expression of the Hecke operators on boundary cohomology groups, for example for GL(2), the expression is given in [**Hi93b**, Section 3], and the uniqueness of $i([\omega]) = [\omega - d\theta_\phi(\omega)]$.

We now compactify Y adding the boundary $\infty \times (\Gamma_s\backslash\mathbb{R}^{d-1})$ to U_s for all cusps s. We write the compactification as $\overline{Y}$. Then $\overline{Y}$ is a manifold with boundary $\partial\overline{Y} = \sqcup_s \infty \times (\Gamma_s\backslash\mathbb{R}^{d-1})$. Let C be a C^∞-class p-cycle modulo $\partial\overline{Y}$. By our construction, $\omega - i(\omega) = d\theta_\phi(\omega)$ is rapidly decreasing towards cusps s. We assume that Y has finite volume with respect to the Haar measure on $G(\mathbb{R})$. Then if ω is rapidly decreasing towards cusps s (that is, exponentially decreasing with respect to t), $\int_C \omega$ converges. We see easily from the Stokes theorem

$$\text{(it1)} \qquad \int_C \omega = \int_C i(\omega).$$

If $p = \dim Y = d$, we see that $\mathrm{Tr} : H^d_c(Y, A) \cong A$ by the evaluation at d-relative cycle $\overline{Y}$ modulo $\partial\overline{Y}$. In particular, if $A = \mathbb{C}$,

$$\text{(it2)} \qquad \mathrm{Tr}([\omega]) = \int_Y \omega = \int_Y i(\omega).$$

This is usually stated for Y smooth (for example, if S is sufficiently small), but is valid always, because of the following reason. We take sufficiently small normal subgroup Γ' of Γ of finite index such that $Y' = \Gamma'\backslash\mathfrak{Z}$ is smooth. Then $H^d(Y, \mathbb{C}) = H^d(Y', \mathbb{C})^\Delta$ for $\Delta = \Gamma/\Gamma'$, and Tr with respect to Γ' induces that of Y. This shows the assertion $(it2)$ for Y' implies that for Y.

2.2. Modular cohomology groups. We summarize here the definition of modular cohomology groups and Hecke operator action on them. A detailed exposition can be found in [**Hi94**] and [**Hi88a**]. Let F be a number field with the integer ring $\mathfrak{r}$. We consider the torus $T = \mathrm{Res}_{\mathfrak{r}/\mathbb{Z}}\mathbf{G}_m$. We fix an algebraic closure $\overline{\mathbb{Q}}$ of $\mathbb{Q}$ inside $\mathbb{C}$. The group of characters $X(T) = \mathrm{Hom}_{alg-gr}(T_{/\overline{\mathbb{Q}}}, \mathbf{G}_{m/\overline{\mathbb{Q}}})$ can be identified with the formal free module $\mathbb{Z}[I]$ generated by the set I of all field embeddings of F into $\overline{\mathbb{Q}}$. Since any $\sigma \in I$ induces an algebra homomorphism $\sigma : F \otimes_{\mathbb{Q}} A \to A$ for any $\overline{\mathbb{Q}}$-algebra A by $k \otimes a \mapsto \sigma(k)a$, for each $n = \sum_{\sigma\in I} n_\sigma\sigma$, n as an element of $X(T)$ takes $a \in T(A)$ to $a^n = \prod_\sigma \sigma(a)^{n_\sigma}$. Note that $T(\mathbb{A})$ for the adele ring $\mathbb{A}$ is the idele group $F^\times_{\mathbb{A}}$. Thus $T(\mathbb{A})/T(\mathbb{Q})$ is the idele class group. A Hecke character $\psi : T(\mathbb{A})/T(\mathbb{Q}) \to \mathbb{C}^\times$ is called arithmetic if it induces an element $\infty(\psi)$ in $X(T)$ on the identity connected component $T(\mathbb{R})_+$ of the archimedean part $T(\mathbb{R})$ of $T(\mathbb{A})$. This element $\infty(\psi) \in \mathbb{Z}[I]$ is called the infinity type of ψ. We write Ξ_F for the submodule of $\mathbb{Z}[I] = X(T)$ made of infinity types of arithmetic Hecke characters. For each arithmetic Hecke character ψ, the field generated by $\psi(x)$ for all $x \in T(\mathbb{A}^{(\infty)})$ is a finite extension $\mathbb{Q}(\psi)$ of $\mathbb{Q}$, which is either a totally real or a CM field. A Hecke character ω is called algebraic if $\omega(x) \in \overline{\mathbb{Q}}$ for all $x \in T(\mathbb{A}^{(\infty)})$. There are many algebraic characters which are not arithmetic (cf. [**Hi94**, p. 467]).

We put

$$H = \mathrm{Res}_{\mathfrak{r}/\mathbb{Z}}\mathrm{GL}(2)_{/\mathfrak{r}} \quad\text{and}\quad \mathfrak{H} = H(\mathbb{R})_+/C_{\infty+}Z_H(\mathbb{R}).$$

As S, we take

$$U_0(N) = U_{0,F}(N) = \left\{ \left(\begin{smallmatrix} a & b \\ c & d \end{smallmatrix}\right) \in H(\widehat{\mathbb{Z}}) \middle| c \in \widehat{N} \right\}$$

for an ideal N of $\mathfrak{r}$, where $\widehat{N} = N \otimes_{\mathbb{Z}} \widehat{\mathbb{Z}}$. We write $Y_0(N) = Y_{0,F}(N)$ for $Y(S)$. For L, we take polynomial representations of H. Let F^{cl} be the Galois closure of $F/\mathbb{Q}$, and write $\mathfrak{r}^{cl}$ for the integer ring of F^{cl}. Writing I for the set of all embeddings of F into $\overline{\mathbb{Q}}$, for each $\mathfrak{r}^{cl}$-algebra A, each σ induces a projection $\sigma : H(A) \to \mathrm{GL}_2(A)$ which coincides with σ on $\mathfrak{r}$. In particular, $H(F^{cl}) \cong \prod_{\sigma \in I} \mathrm{GL}_2(F^{cl})$ via $a \mapsto (\sigma(a))_\sigma$. Then over F^{cl}, each irreducible polynomial representation of H is isomorphic to

$$\bigotimes_{\sigma \in I} \det(\sigma(x))^{v_\sigma} \mathrm{Sym}(\sigma(x))^{\otimes n_\sigma}$$

for integer tuples (v_σ) and (n_σ) with $n_\sigma \geq 0$. Here $\mathrm{Sym}(\sigma(\mathrm{x}))^{\otimes \mathrm{n}_\sigma}$ is the symmetric n_σ-th tensor matrix of $\sigma(x)$. Thus irreducible polynomial representations of H are classified by tuples (n, v) of $\mathbb{Z}[I]$, where $n = \sum_{\sigma \in I} n_\sigma \sigma$ and $v = \sum_\sigma v_\sigma \sigma$. We write κ for the pair (n, v) sometimes. We can concretely realize the above polynomial representation on an A-free module $L(\kappa; A)$ made of polynomials of $2[F : \mathbb{Q}]$ variables $(X_\sigma, Y_\sigma)_{\sigma \in I}$ with coefficients in A homogeneous of degree n_σ for each pair (X_σ, Y_σ). We let $\gamma \in H(A)$ act on $P \in L(\kappa; A)$ by

$$\gamma P(X_\sigma, Y_\sigma) = \det(\gamma)^v P((X_\sigma, Y_\sigma)^t \sigma(\gamma)^\iota),$$

where $\gamma^\iota = \det(\gamma)\gamma^{-1}$ and $\det(\gamma)^v = \prod_\sigma \det(\sigma(\gamma))^{v_\sigma}$.

By the approximation theorem, choosing a complete set R of representatives for the class group $Cl(S) = T(\mathbb{Q})_+ \backslash T(\mathbb{A}) / \det(S) T(\mathbb{R})$, we have

$$H(\mathbb{A})_+ = \bigsqcup_{a \in R} H(\mathbb{Q})_+ \left(\begin{smallmatrix} a & 0 \\ 0 & 1 \end{smallmatrix}\right) S H(\mathbb{R})_+.$$

Thus $Y(S) = \sqcup_a Y_a$ and $Y_a = \Gamma^{(a)} \backslash \mathfrak{Z}$, where $\Gamma^{(a)} = tSH(\mathbb{R})t^{-1} \cap H(\mathbb{Q})_+$ for $t = \left(\begin{smallmatrix} a & 0 \\ 0 & 1 \end{smallmatrix}\right)$. When $S = U_0(N)$, for a Hecke character ψ with $\psi_\infty(x) = x^{-n-2v}$ for all $x \in F_\infty^\times$ whose conductor is a factor of N, we twist a little the action of $\Gamma^{(a)}$ on $L(\kappa; A)$ as follows:

$$P(X_\sigma, Y_\sigma) \mapsto \psi_N(d) \det(\gamma)^v P((X_\sigma, Y_\sigma)^t \sigma(\gamma)^\iota)$$

if $\gamma = \left(\begin{smallmatrix} * & * \\ c & d \end{smallmatrix}\right) \in \Gamma^{(a)}$, where ψ_N is the restriction of ψ to $\prod_{\mathfrak{p}|N} F_\mathfrak{p}^\times \subset T(\mathbb{A})$. We write this twisted module as $L(\kappa, \psi; A)$. To have a non-trivial sheaf, the action of $\Gamma^{(a)}$ has to factor through the fundamental group $\pi_1(Y)$ which is $\Gamma^{(a)}/\mathfrak{r}^\times$. Therefore, the center $\mathfrak{r}^\times$ of $\Gamma^{(a)}$ has to act trivially on $L(\kappa, \psi; A)$. The condition: $\psi_\infty(x) = x^{-n-2v}$ for all $x \in F_\infty^\times$ assures this. When ψ is trivial on $T(\widehat{\mathbb{Z}})$, $L(\kappa, \psi; A) = L(\kappa; A)$. Thus $L(\kappa, \psi; A)$ only depends on the restriction of ψ to units $T(\widehat{\mathbb{Z}}) = \widehat{\mathfrak{r}}^\times$.

We describe the cohomology groups

$$H^q(Y_0(N), \mathcal{L}(\kappa, \psi; A)), \quad H_c^q(Y_0(N), \mathcal{L}(\kappa, \psi; A))$$

and

$$H_{\mathrm{cusp}}^q(Y_0(N), \mathcal{L}(\kappa, \psi; A))$$

defined in [**Hi94**, Sections 3 and 5]. Let $\mathbb{Q}(\kappa)$ be the subfield of $\overline{\mathbb{Q}}$ fixed by the subgroup

$$\mathcal{G}(\kappa) = \left\{ \sigma \in \mathrm{Gal}(\overline{\mathbb{Q}}/\mathbb{Q}) \middle| \kappa\sigma = (n\sigma, v\sigma) = \kappa \right\}.$$

Let $\mathbb{Q}(\psi)$ be a subfield of $\overline{\mathbb{Q}}$ generated by $\psi(x)$ for all $x \in T(\mathbb{A}^{(\infty)})$, which is a CM field finite over $\mathbb{Q}$. We write $\mathbb{Q}(\kappa,\psi)$ for the composite of $\mathbb{Q}(\kappa)$ and $\mathbb{Q}(\psi)$. Write

$$P(X_\sigma, Y_\sigma) = \sum_{0\le j\le n} a_j X^{n-j}Y^j \in L(\kappa,\psi,A),$$

where $X^{n-j}Y^j = \prod_\sigma X_\sigma^{n_\sigma - j_\sigma} Y_\sigma^{j_\sigma}$, $a_j \in A$, and $0 \le j \le n$ implies $0 \le j_\sigma \le n_\sigma$ for all $\sigma \in I$. Then we let $\sigma \in \mathrm{Gal}(\overline{\mathbb{Q}}/\mathbb{Q})$ act on $L(\kappa,\psi;\overline{\mathbb{Q}})$ by

$$\sum_{0\le j\le n} a_j X^{n-j}Y^j \mapsto \sum_{0\le j\le n} a_j^\sigma X^{n\sigma - j\sigma}Y^{j\sigma}.$$

Then σ takes $L(\kappa,\psi;\overline{\mathbb{Q}})$ onto $L(\kappa\sigma,\psi^\sigma;\overline{\mathbb{Q}})$, where ψ^σ is the unique Hecke character such that $\psi^\sigma(x) = \psi(x)^\sigma$ for $T(\mathbb{A}^{(\infty)})$ and $\infty(\psi^\sigma) = \infty(\psi)\sigma$. Thus the $\Gamma^{(a)}$-module $L(\kappa,\psi;\mathbb{Q}(\kappa,\psi))$ is well defined, and for the integer ring $\mathbb{Z}(\kappa,\psi)$ of $\mathbb{Q}(\kappa,\psi)$,

$$tL(\kappa,\psi;\mathbb{Z}(\kappa,\psi)) = tL(\kappa,\psi;\mathbb{Z}(\kappa,\psi)\otimes_{\mathbb{Z}}\widehat{\mathbb{Z}}) \bigcap L(\kappa,\psi;\mathbb{Q}(\kappa,\psi))$$

is an $\mathbb{Z}(\kappa,\psi)$-lattice in $L(\kappa,\psi;\mathbb{Q}(\kappa,\psi))$ stable under $\Gamma^{(a)}$ (see the paragraph below (3.5) of [**Hi94**]), where $t = \left(\begin{smallmatrix} a & 0 \\ 0 & 1\end{smallmatrix}\right)$, and the intersection is taken in

$$L(\kappa,\psi;\mathbb{A}^{(\infty)}(\kappa,\psi)) \quad \text{for} \quad \mathbb{A}^{(\infty)}(\kappa,\psi) = \mathbb{Q}(\kappa,\psi)\otimes_{\mathbb{Q}}\mathbb{A}^{(\infty)}.$$

Thus writing $tL(A) = tL(\kappa,\psi;A)$ for $tL(\kappa,\psi;\mathbb{Z}(\kappa,\psi))\otimes_{\mathbb{Z}(\kappa,\psi)} A$, we have the covering $\bigsqcup_a \underline{tL}(A) \to Y_0(N)$. We write $\mathcal{L}(\kappa,\psi;A)$ for the sheaf of locally constant sections of this covering. Then, if $Y_0(N)$ is smooth, the cohomology groups H^q and H^q_c are defined in the usual manner, and the cuspidal cohomology group $H^q_{\mathrm{cusp}}(Y_0(N),\mathcal{L}(\kappa,\psi;\mathbb{C}))$ is defined to be the subspace of $H^q(Y_0(N),\mathcal{L}(\kappa,\psi;\mathbb{C}))$ spanned by cuspidal harmonic forms [**Hi94**, Section 2]. When $Y_0(N)$ is not smooth, we take a normal subgroup $S \subset U_0(N)$ and define $H^q_{\mathrm{cusp}}(Y_0(N),\mathcal{L}(\kappa,\psi;\mathbb{C}))$ by the subspace of $H^q_{\mathrm{cusp}}(Y(S),\mathcal{L}(\kappa,\psi;\mathbb{C}))$ fixed by $U_0(N)/S$. Similarly, we define $H^q_c(Y_0(N),\mathcal{L}(\kappa,\psi;\mathbb{C}))$ for non-smooth $Y_0(N)$. For any $\mathbb{Z}(\kappa,\psi)$-subalgebra A of $\mathbb{C}$, we define $H^q_{\mathrm{cusp}}(Y_0(N),\mathcal{L}(\kappa,\psi;A))$ by the intersection of $H^q_{\mathrm{cusp}}(Y_0(N),\mathcal{L}(\kappa,\psi;\mathbb{C}))$ with the natural image of $H^q_c(Y_0(N),\mathcal{L}(\kappa,\psi;A))$. Anyway, as seen in 2.1, we have a canonical section

$$i : H^q_{\mathrm{cusp}}(Y_0(N),\mathcal{L}(\kappa,\psi;\mathbb{C})) \hookrightarrow H^q_c(Y_0(N),\mathcal{L}(\kappa,\psi;\mathbb{C})),$$

and

$$H^q_{\mathrm{cusp}}(Y_0(N),\mathcal{L}(\kappa,\psi;\mathbb{C})) \cong H^q_{\mathrm{cusp}}(Y_0(N),\mathcal{L}(\kappa,\psi;\mathbb{Z}(\kappa,\psi)))\otimes_{\mathbb{Z}(\kappa,\psi)}\mathbb{C}.$$

The cohomology groups $H^q_{\mathrm{cusp}}(Y_0(N),\mathcal{L}(\kappa,\psi;A))$ have a natural action of Hecke operators $T(\mathfrak{n})$ for integral ideals $\mathfrak{n}$ of $\mathfrak{r}$ and the action of the center $Z(\mathbb{A})$ [**Hi94**, Section 4] as long as either A is a $\mathbb{Q}(\kappa,\psi)$-algebra or $v \ge 0$ ($\iff v_\sigma \ge 0$ for all σ), and i is equivariant under Hecke operators. When A is not a $\mathbb{Q}$-algebra and $v \not\ge 0$, we need to modify $T(\mathfrak{n})$ as in [**Hi94**, Section 4] to preserve integrality. By a result of Harder, the cuspidal cohomology group is trivial if one of the following conditions is satisfied (i) $q < r_1(F)+r_2(F)$; (ii) $q > r_1(F)+2r_2(F)$, and (iii) $n \ne nc$ for complex conjugation c (cf. [**Ha**], [**Hi94**, Section 2]), where $r_1(F)$ (resp. $r_2(F)$) is the number of real (resp. complex) places of F.

We assume that $H^q_{\mathrm{cusp}}(Y_0(N),\mathcal{L}(\kappa,\psi;\mathbb{C})) \ne 0$. Since we have an action of the center $Z_H(\mathbb{A}^{(\infty)}) = T(\mathbb{A}^{(\infty)})$, we can decompose $H^q_{\mathrm{cusp}}(Y_0(N),\mathcal{L}(\kappa,\psi;\mathbb{C}))$ into a product of eigenspaces under this action:

$$H^q_{\text{cusp}}(Y_0(N),\mathcal{L}(\kappa,\psi;\mathbb{C})) = \bigoplus_{\psi'} H^q_{\text{cusp}}(Y_0(N),\mathcal{L}(\kappa,\psi;\mathbb{C}))[\psi'],$$

where ψ' runs over arithmetic Hecke characters such that $\psi' = \psi$ on $T(\widehat{\mathbb{Z}})$ and $\infty(\psi') = \infty(\psi)$. We write $h_\kappa(N,\psi;A)_{/F}$ for the A-subalgebra of

$$\text{End}_{\mathbb{C}}(H^q_{\text{cusp}}(Y_0(N),\mathcal{L}(\kappa,\psi;\mathbb{C}))[\psi])$$

generated by $T(\mathfrak{n})$ for all integral ideals $\mathfrak{n}$. Again by Harder [**Ha**], the algebra $h_\kappa(N,\psi;A)$ is independent of q. Let $\lambda : h_\kappa(N,\psi;\mathbb{C}) \to \mathbb{C}$ be an algebra homomorphism. Then λ-eigenspace is non-trivial in the cohomology group. Its dimension depends on q and F. To describe this, write $\Sigma = \Sigma_F$ (resp. $\Sigma(\mathbb{R}) = \Sigma_F(\mathbb{R}), \Sigma(\mathbb{C}) = \Sigma_F(\mathbb{C})$) for the set of archimedean (resp. real, complex) places of F. We identify $\Sigma(\mathbb{C})$ with a subset of complex embeddings in I so that each place is induced by the corresponding embedding. Note that

$$\mathfrak{H} \cong \prod_{\sigma\in\Sigma(\mathbb{R})} \mathcal{H}_\sigma \times \prod_{\tau\in\Sigma(\mathbb{C})} \mathcal{H}_\tau,$$

where $\mathcal{H}_\sigma$ is a upper half complex plane on which $\sigma(\gamma) \in \text{GL}_2(\mathbb{R})$ ($\sigma \in \Sigma(\mathbb{R})$) acts through a linear fractional transformation, and

$$\mathcal{H}_\tau = \left\{ \begin{pmatrix} x & -y \\ y & \overline{x} \end{pmatrix} \middle| x \in \mathbb{C},\ 0 < y \in \mathbb{R} \right\},$$

on which $\tau(\gamma) \in \text{GL}_2(\mathbb{C})$ ($\tau \in \Sigma(\mathbb{C})$) acts as in [**Hi94**, 2.2]. In [**Hi94**, Section 3 (M1-4)], we defined a space of cohomological cusp forms $S_{\kappa,J}(N;\psi)_{/F}$ for each subset J of $\Sigma(\mathbb{R})$. Actually, we need to assume that functions in $S_{\kappa,J}(N;\psi)_{/F}$ are rapidly decreasing towards cusps, which follows from the cuspidal condition [**Hi94**, M4] if F is different from $\mathbb{Q}$ or imaginary quadratic fields. In [**Hi94**], this condition is implicitly assumed when F is $\mathbb{Q}$ or imaginary quadratic field (see $(m'3)$ of [**Hi94**, p. 460]). An element $f \in S_{\kappa,J}(N;\psi)$ corresponds to a real analytic modular form on $\mathfrak{H}$ holomorphic on the copy $\mathcal{H}_\sigma$ at $\sigma \in J$ and anti-holomorphic at $\sigma \in \Sigma(\mathbb{R}) - J$ (see the remark in [**Hi94**, p. 60] after $(m'3)$). This space, when $F = \mathbb{Q}$, is isomorphic to the classical space of elliptic cusp forms. More precisely, $S_{\kappa,I}(N,\psi)$ is isomorphic to the space $S_k(\Gamma_0(N),\psi_N)$ of holomorphic cusp forms of weight $k = n+2$ with Neben character ψ_N, which is the restriction of ψ to $\widehat{\mathbb{Z}}^\times$ regarded as a Dirichlet character. The isomorphism is given by $f \mapsto \phi(x+iy) = y^{v-1} f\left(\left(\begin{smallmatrix} y & x \\ 0 & 1\end{smallmatrix}\right)\right)$. Thus the space itself does not depends on v, but the Hecke operator $T(n) = T_v(n)$ depends on v in the following way: When $v = 0$, $T_0(n)$ is the classical Hecke operator acting on $S_k(\Gamma_0(N),\psi_N)$ defined by Hecke. Then $T_v(n) = n^v T_0(n)$. In other words, by pulling back classical cusp forms in $S_k(\Gamma_0(N),\psi_N)$ to $H(\mathbb{A})$, we get the space $S_{(n,0),I}(N,\psi)$ and

$$S_{(n,v),I}(N,\psi) = \left\{ f(x)|\det(x)|_{\mathbb{A}}^{-v} \middle| f \in S_{(n,0),I}(N,\psi) \right\}.$$

When $F \neq \mathbb{Q}$, $S_{(n,v),J}(N,\psi)$ does not necessarily have such a simple relation to $S_{(n,0),J}(N,\psi)$.

In [**Hi94**, Sections 2-3], we described a very explicit map

$$\delta_{J,J'} : S_{\kappa,J}(N;\psi) \hookrightarrow H^q_{\text{cusp}}(Y_0(N),\mathcal{L}(\kappa,\psi;\mathbb{C}))[\psi]$$

indexed by $J \subset \Sigma(\mathbb{R})$ and $J' \subset \Sigma(\mathbb{C})$. The linear map $\delta_{J,J'}$ is Hecke equivariant and takes a cohomological cusp form f to a differential form holomorphic of degree

1 in the variables of $\mathcal{H}_\sigma$ for $\sigma \in J$, anti-holomorphic of degree 1 for $\sigma \in \Sigma(\mathbb{R}) - J$, harmonic of degree 1 for $\sigma \in J'$ and harmonic of degree 2 for $\Sigma(\mathbb{C}) - J'$. Thus the total degree of the differential form is $r_1 - |J'| + 2r_2$. We will recall the explicit form of $\delta_{J,J'}$ later in our computation in specific cases. Then, for $d = [F : \mathbb{Q}]$,

$$\delta = \oplus_{J,J'}\delta_{J,J'} : \bigoplus_{J,J':\#(J')=d-q} S_{\kappa,J}(N,\psi) \cong H^q_{\mathrm{cusp}}(Y_0(N), \mathcal{L}(\kappa,\psi;\mathbb{C}))[\psi].$$

Since the λ-eigenspace of $S_{\kappa,J}(N,\psi)$ is 1-dimensional, this completely determines the dimension of the cohomological eigenspace. Since $T(\mathfrak{n})$ leaves stable the cohomology group, $\lambda(T(\mathfrak{n}))$ is an algebraic number in a fixed finite extension. We write $\mathbb{Q}(\lambda)$ for the subfield of $\overline{\mathbb{Q}}$ generated by $T(\mathfrak{n})$ for all $\mathfrak{n}$, which is a CM field or a totally real field containing $\mathbb{Q}(\kappa,\psi)$.

For a standard Whittaker function $W = W_\kappa : T(\mathbb{R})_+ \to L((n^*,0);\mathbb{C})$ with $n^* = \sum_{\sigma\in\Sigma(\mathbb{C})}(n_\sigma + n_{\sigma c} + 2)\sigma$, $f \in S_{\kappa,J}(N,\psi)$ has a Fourier expansion of the following form [**Hi94**, Section 6]:

$$f\left(\begin{pmatrix} y & x \\ 0 & 1\end{pmatrix}\right) = |y|_{F_\mathbb{A}} \sum_{\xi\in F^\times} \mathbf{a}(\xi y\mathfrak{d}; f) W(\xi y_\infty)\mathbf{e}_F(\xi x), \tag{F}$$

where ξ runs over all elements in F with $\xi^\sigma > 0$ for $\sigma \in J$ and $\xi^\sigma < 0$ for $\sigma \in \Sigma(\mathbb{R}) - J$, $y \in T(\mathbb{A})$ with $y_\infty \in T(\mathbb{R})_+$, $\mathfrak{n} \mapsto \mathbf{a}(\mathfrak{n}; f)$ is a function with values in $\mathbb{C}$ supported by the set of integral ideals, $\mathfrak{d} = \mathfrak{d}_F$ is the different of $F/\mathbb{Q}$, and $\mathbf{e}_F : F_\mathbb{A}/F \to \mathbb{C}$ is the standard additive character with $\mathbf{e}_F(x_\infty) = \exp(2\pi\sqrt{-1}\sum_{\sigma\in I} x^\sigma)$. The function W is the optimal element in the Whittaker model at archimedean places, whose Mellin transform gives the exact Γ-factor of the standard L-function. Its explicit form will be recalled later. This Fourier expansion determines the cusp form uniquely. If $f|T(\mathfrak{n}) = \lambda(T(\mathfrak{n}))f$ with $\mathbf{a}(\mathfrak{r}; f) = 1$, then $\mathbf{a}(\mathfrak{n}; f) = \lambda(T(\mathfrak{n}))$. Thus the eigenspace of λ is one dimensional (Multiplicity 1) (cf. [**Hi94**, Theorem 6.4]). For any automorphism $\sigma \in \mathrm{Aut}(\mathbb{C})$, we always have f^σ in $S_{\kappa\sigma,J}(N,\psi^\sigma)$ such that $\mathbf{a}(\mathfrak{n}; f^\sigma) = \mathbf{a}(\mathfrak{n}; f)^\sigma$ by the Hecke equivariance of δ and the rational structure of the cuspidal cohomology groups. In particular, $\lambda^\sigma(T(\mathfrak{n})) = \lambda(T(\mathfrak{n}))^\sigma$ gives an algebra homomorphism of $h_{\kappa\sigma}(N,\psi^\sigma;\mathbb{C})$ into $\mathbb{C}$. Sometimes, we call λ a system of Hecke eigenvalues.

2.3. Modular L-functions. We fix a system of Hecke eigenvalues

$$\lambda : h_\kappa(N,\psi;\mathbb{C}) \to \mathbb{C}$$

and define several L-functions of λ we will study. The standard L-function of λ twisted by a Hecke character $\eta : T(\mathbb{A})/T(\mathbb{Q}) \to \mathbb{C}^\times$ is given by

$$\begin{aligned} L(s,\lambda\otimes\eta) &= \sum_{\mathfrak{n}\subset\mathfrak{r}} \eta(\mathfrak{n})\lambda(T(\mathfrak{n}))N_{F/\mathbb{Q}}(\mathfrak{n})^{-s} \\ &= \prod_{\mathfrak{p}} \left\{(1-\alpha_\mathfrak{p}\eta(\mathfrak{p})N_{F/\mathbb{Q}}(\mathfrak{p})^{-s})(1-\beta_\mathfrak{p}\eta(\mathfrak{p})N_{F/\mathbb{Q}}(\mathfrak{p})^{-s})\right\}^{-1}, \end{aligned}$$

which is continued to an entire function on the whole complex s-plane and has a functional equation if λ is primitive (that is, λ gives eigenvalues of a primitive form of conductor N; cf. [**Mi**]). When η is arithmetic with infinity type $-w \in \mathbb{Z}[I]$, $\lambda\otimes\eta : T(\mathfrak{n}) \mapsto \eta(\mathfrak{n})\lambda(T(\mathfrak{n}))$ for $\mathfrak{n}$ prime to the conductor $C = C(\eta)$ of η gives an algebra homomorphism of $h_{\kappa'}(N\cap C^2,\psi\eta^2;\mathbb{C})$ into $\mathbb{C}$ for $\kappa' = \kappa + (0,w)$. Thus we can view $L(s,\lambda\otimes\eta)$ as the standard L-function of $\lambda\otimes\eta$. The adjoint L-function is given by

$$L(s,\mathrm{Ad}(\lambda)\otimes\eta) = \prod_{\mathfrak{p}}\left\{\left(1-\frac{\alpha_{\mathfrak{p}}\beta_{\mathfrak{p}}^{-1}\eta(\mathfrak{p})}{N_{F/\mathbb{Q}}(\mathfrak{p})^s}\right)\left(1-\frac{\eta(\mathfrak{p})}{N_{F/\mathbb{Q}}(\mathfrak{p})^s}\right)\left(1-\frac{\alpha_{\mathfrak{p}}^{-1}\beta_{\mathfrak{p}}\eta(\mathfrak{p})}{N_{F/\mathbb{Q}}(\mathfrak{p})^s}\right)\right\}^{-1},$$

which again has a meromorphic continuation to the whole complex s-plane [**Sh75**], [**Sh94**] and [**GeJ**]. It has a functional equation after adding finitely many Euler factors if necessary. For another system μ of Hecke eigenvalues, we define the Rankin product of λ and μ. Writing the Euler $\mathfrak{p}$ factor of $L(s,\mu)$ as $(1-\alpha'_{\mathfrak{p}}/N(\mathfrak{p})^s)^{-1}(1-\beta'_{\mathfrak{p}}/N(\mathfrak{p})^s)^{-1}$, we put

$$L(s,\lambda\otimes\mu) = \prod_{\mathfrak{p}}\left\{\left(1-\frac{\alpha_{\mathfrak{p}}\alpha'_{\mathfrak{p}}}{N_{F/\mathbb{Q}}(\mathfrak{p})^s}\right)\left(1-\frac{\alpha_{\mathfrak{p}}\beta'_{\mathfrak{p}}}{N_{F/\mathbb{Q}}(\mathfrak{p})^s}\right)\cdot\left(1-\frac{\beta_{\mathfrak{p}}\alpha'_{\mathfrak{p}}}{N_{F/\mathbb{Q}}(\mathfrak{p})^s}\right)\left(1-\frac{\beta_{\mathfrak{p}}\beta'_{\mathfrak{p}}}{N_{F/\mathbb{Q}}(\mathfrak{p})^s}\right)\right\}^{-1},$$

For each primitive Hecke eigensystem $\lambda : h_\kappa(N,\psi;\mathbb{C})\to\mathbb{C}$, it is a well known conjecture that there exists a compatible system of $\mathfrak{l}$-adic representations $\rho=\rho(\lambda)$ of $\mathrm{Gal}(\overline{F}/F)$ with coefficients in $\mathbb{Q}(\lambda)$ such that $L(s,\rho\otimes\eta) = L(s,\lambda\otimes\eta)$ (cf. [**Hi94**, Section 1]), where $\overline{F}$ is the algebraic closure of F. This conjecture is known for totally real F (see [**BR**]). If such ρ exists, $L(s,\lambda\otimes\mu) = L(s,\rho(\lambda)\otimes\rho(\mu))$ and $L(s,\mathrm{Ad}(\lambda)) = L(s,\mathrm{Ad}(\rho))$, where $\mathrm{Ad}(\rho)$ is a three dimensional representation fitting into the following exact sequence:

$$0\to\mathrm{Ad}(\rho)\to\rho\otimes\rho^\vee\to\mathrm{id}\to 0$$

for the contragredient $\rho^\vee$ of ρ and the identity Galois character id.

We take a semi-simple quadratic extension K/F. We allow $K=F\oplus F$. We write α for the quadratic character of $T(\mathbb{A})/T(\mathbb{Q})$ associated to K/F if K is a field. If $K=F\oplus F$, we simply put $\alpha=\mathrm{id}$ for the identity character id. We write $\mathcal{R}$ for the integral closure of $\mathfrak{r}$ in K. We put $G=\mathrm{Res}_{\mathcal{R}/\mathbb{Z}}\mathrm{GL}(2)_{/\mathcal{R}}$. If K is not a field, we simply agree to put $G=H\times H$. We consider the cohomological modular forms on G. Thus if $G=H\times H$ and $N=C\oplus C\subset\mathcal{R}$, then

$$S_{(\kappa,\mu)}(N,(\psi,\psi')) = S_\kappa(C,\psi)\otimes_{\mathbb{C}} S_\mu(C,\psi'),$$

where $f\otimes g(x,x') = f(x)g(x')$ on $G(\mathbb{A})$. We define $Y_{0,K}(N)$ as above for G if K is a field, and otherwise, $Y_{0,K}(N) = Y_{0,F}(C)\times Y_{0,F}(C)$. We write I_K for the set of all non-trivial algebra homomorphisms of K into $\overline{\mathbb{Q}}$. Thus if $K=F\oplus F$, $I_K\cong I\sqcup I$ canonically. Let $\mu : h_\kappa(N,\chi;\mathbb{C})_{/K}\to\mathbb{C}$ be a system of Hecke eigenvalues for $\kappa\in\mathbb{Z}[I_K]$. Suppose that K is a field and the compatible system $\rho=\rho(\mu)$ exists. Extending a non-trivial automorphism of K/F to $\sigma\in\mathrm{Gal}(\overline{F}/F)$, we put $\rho^\sigma(g)=\rho(\sigma g\sigma^{-1})$. Then $\Psi=\rho\otimes\rho^\sigma$ is equivalent to Ψ^σ, and it extends to $\mathrm{Gal}(\overline{F}/F)$ in two ways. Realizing Ψ on $V\otimes V$ for a two dimensional vector space V on which ρ acts, one of the two extensions, writing Ψ_+, satisfies $\Psi_+(\sigma)(x\otimes y) = y\otimes\rho(\sigma^2)x$ and the other is given by $\Psi_-(\sigma)(x\otimes y) = -y\otimes\rho(\sigma^2)x$ (cf. [**Gh**, 5.1]). We write the L-function of Ψ_+ twisted by a Hecke character η of F as $L(s,(\mu\otimes\mu_\sigma)_+\otimes\eta)$. Then $L(s,\Psi_-\otimes\eta) = L(s,\Psi_+\otimes\alpha\eta) = L(s,(\mu\otimes\mu_\sigma)_+\otimes\alpha\eta)$. See [**Gh**, 5.1] for an explicit description of Euler factors of $L(s,(\mu\otimes\mu^\sigma)_+)$. Although it is assumed in

[**Gh**] that K is an imaginary quadratic field, the computation (and the description) is the same for general K/F. We have

$$L(s, \mu \otimes \mu_\sigma) = L(s, (\mu \otimes \mu_\sigma)_+) L(s, (\mu \otimes \mu_\sigma)_+ \otimes \alpha),$$

where μ_σ is a system of Hecke eigenvalues given by $\mu_\sigma(T(\mathfrak{n})) = \mu(T(\mathfrak{n}^\sigma))$. When $K = F \oplus F$, we just put

$$L(s, (\mu \otimes \mu_\sigma)_+) = L(s, \lambda \otimes \lambda)$$

if μ is given by $\lambda \otimes \lambda'$ taking $(T(\mathfrak{n}), T(\mathfrak{m}))$ to $\lambda(T(\mathfrak{n}))\lambda'(T(\mathfrak{m}))$ for two systems of Hecke eigenvalues λ and λ' of H. Then we have

$$\text{(R1)} \qquad L(s+1, (\mu \otimes \mu_\sigma)_+ \otimes \eta) = L(2s, \chi_F \eta^2) L_{K/F}(s, \mu, \eta),$$

where χ_F is the restriction of χ to $F_{\mathbb{A}}^\times$ and

$$L_{K/F}(s, \mu, \eta) = \sum_{\mathfrak{n} \subset \mathfrak{r}} \eta(\mathfrak{n}) \mu(T(\mathfrak{n})) N_{F/\mathbb{Q}}(\mathfrak{n})^{-s-1}$$

for a Hecke character η of F. Here $\mathfrak{n}$ runs over all ideals of $\mathfrak{r}$ (not $\mathcal{R}$) extended to $\mathcal{R}$.

2.4. Idea of the proof. There is a way to get an integral expression of

$$L(1, \mathrm{Ad}(\lambda) \otimes \alpha)$$

for a Hecke character α with $\alpha^2 = 1$. Here we summarize the idea, and in the following sections, we shall give details of computation. This integral expression gives a key to prove the rationality theorem. Let $\mu = \widehat{\lambda}$ for the base change lift of λ of H to G. Here we need to invoke the assumption that $\widehat{\lambda}$ remains cuspidal. When K is a field, $\widehat{\lambda}$ is given by Jacquet [**J**] so that $L(s, \rho(\lambda)|_K \otimes \eta) = L(s, \widehat{\lambda} \otimes \eta)$ for all arithmetic Hecke character η of K, where $\rho|_K$ is the restriction of ρ to $\mathrm{Gal}(\overline{F}/K)$. If $G = H \times H$, we simply put $\widehat{\lambda} = \lambda \otimes \lambda$. Then $\chi = \psi \circ N_{K/F}$ and $\chi_F = \psi^2$. We write $\widehat{\kappa} = (\widehat{n}, \widehat{v}) \in \mathbb{Z}[I_K]^2$ for the weight of $\widehat{\lambda}$. We have the restriction map $\mathrm{Res}_F^K : \mathbb{Z}[I_K] \to \mathbb{Z}[I]$ which takes σ to its restriction to F, where F is embedded into $F \oplus F$ diagonally if K is not a field. Writing $\mathrm{Inf}_F^K : \mathbb{Z}[I] \to \mathbb{Z}[I_K]$ for the inflation map: $\mathrm{Inf}(\sigma) = \sum_{\tau \in I_K, \tau|_F = \sigma} \tau$, we see that $\widehat{n} = \mathrm{Inf}_F^K(n)$. Then we get, again looking into Euler factorization

$$\begin{aligned} L_{K/F}(s, \widehat{\lambda}, \eta) L(2s, (\psi\eta)^2) &= L(s+1, (\mu \otimes \mu_\sigma)_+ \otimes \eta) \\ &= L(s, \alpha\psi\eta) L(s, \mathrm{Ad}(\lambda) \otimes \psi\eta) \end{aligned}$$

up to finite Euler factors. Let N_0 be the least common multiple of the conductor of η and $N \cap \mathfrak{r}$. Since discrepancy of Euler factors can only occur for primes dividing N_0, we get the following exact identity:

$$\text{(R2)} \qquad L_{N_0}(2s, (\psi\eta)^2) L_{K/F, N_0}(s, \widehat{\lambda}, \eta) = L_{N_0}(s, \alpha\psi\eta) L_{N_0}(s, \mathrm{Ad}(\lambda) \otimes \psi\eta),$$

where we write $L_{N_0}(s)$ for the L-function obtained from the original $L(s)$ by removing Euler factors for primes $\mathfrak{p}$ dividing N_0. Thus assuming $\alpha\psi\eta = \mathrm{id}$, $L_{N_0}(s, \alpha\psi\eta)$ has a simple pole at $s = 1$. Thus we get the following residue formula:

$$\begin{aligned} \text{(Res1)} \qquad &\{\mathrm{Res}_{s=1} \zeta_{F, N_0}(s)\} L_{N_0}(1, \mathrm{Ad}(\lambda) \otimes \alpha) \\ &= \mathrm{Res}_{s=1} \left\{ \zeta_{F, N_0}(2s) L_{K/F, N_0}(s, \widehat{\lambda}, \alpha\psi^{-1}) \right\}. \end{aligned}$$

Thus we need to compute the residue of the right-hand side of $(Res1)$. Keeping the assumption that $\alpha\psi\eta = \mathrm{id}$, we use for that purpose a pull back integration of $\delta(g)$ over $Y_{0,F}(N')$ (for a suitable N') to get an integral expression of $L(s+1,(\mu\otimes\mu_\sigma)_+\otimes\eta)$, where g is a suitable cohomological modular form having Hecke eigenvalues related to $\widehat{\lambda}$. To describe g, we need to express $\eta = \alpha\psi^{-1} = \varphi_F\omega$ for the restriction φ_F of an arithmetic Hecke character φ of K to F and a Hecke character ω of conductor 1. We will show later that this is possible in most cases. We write $w = -\infty(\varphi)$ and C for the conductor of φ. We put $N' = N \cap C^2 \cap \mathfrak{r}$. Then up to finitely many Euler factors

$$\begin{aligned}\zeta_F(s)L(s,\mathrm{Ad}(\lambda)\otimes\alpha) &= L(s+1,(\widehat{\lambda}\otimes\widehat{\lambda}_\sigma)_+\otimes\eta)\\ &= L(s+1,((\widehat{\lambda}\otimes\varphi)\otimes(\widehat{\lambda}\otimes\varphi)_\sigma)_+\otimes\omega).\end{aligned}$$

We choose suitable $J\subset\Sigma_K(\mathbb{R})$ and $J'\subset\Sigma_K(\mathbb{C})$ so that

$$g\in S_{\widehat{\kappa}+(0,w),J}(N\cap C^2,\chi\varphi^2)[\widehat{\lambda}\otimes\varphi]$$

and $\delta_{J,J'}(g)$ gives a cohomology class in $H^q_{\mathrm{cusp}}(Y_{0,K}(N\cap C^2),\mathcal{L}(\widehat{\kappa}+(0,w),\chi\varphi^2;\mathbb{C}))$ for $q=\dim Y_{0,F}(N')$. The cusp form g is the image of λ-eigenvector f under the twisting operator $R(\varphi)$ and the (cohomological) rationality of $\delta(f)$ and $\delta(f|R(\varphi))$ are equal (see [**Hi94**, 6.8] and the proof of Theorem 8.1). Under the assumption: $\alpha\psi\eta=\mathrm{id}$, we have a non-trivial sheaf morphism

$$\pi:\mathcal{L}(\widehat{\kappa}+(0,w),\chi\varphi^2;\mathbb{C})|_{Y_{0,F}(N')}\to\mathcal{L}(0,\omega^{-2};\mathbb{C}).$$

Thus for a suitable Eisenstein series $E(s)$ giving a global section of $\mathcal{L}(0,\omega^2;\mathbb{C})$, we can prove by a Rankin convolution method that the integral

$$\int_{Y_{0,F}(N')}\pi(\delta_{J,J'}(g))E(s)$$

gives

$$L(s+1,((\widehat{\lambda}\otimes\varphi)\otimes(\widehat{\lambda}\otimes\varphi)_\sigma)_+\otimes\omega)$$

up to the canonical Γ-factor and a constant. Actually,

$$E(s)=E(x,s)=\omega(\det(x))E_0(x,s)$$

as a function of $x\in H(\mathbb{A})$, where $E_0(x,s)$ is the weight 0 Eisenstein series attached to the trivial character of the Borel subgroup of H. As is well known, $E_0(x,s)$ has a simple pole at $s=1$ with constant residue equal to $\mathrm{Res}_{s=1}\zeta_F(s)$ up to rational numbers (see Appendix). Then comparing the residue of the two sides of $(Res1)$, we finally for $C'=C(\varphi)\cap\mathfrak{r}$,

$$\text{(Res2)}\qquad L_{C'}(1,\mathrm{Ad}(\lambda)\otimes\alpha)=C_0\int_{Y_{0,F}(N')}\pi(\delta_{J,J'}(g))\det(\omega(x)),$$

where the constant C_0 is the product of the Gauss sum $G(\psi\alpha)$ and a power of π up to rational numbers. Thus if the $\widehat{\lambda}$-eigenspace $H^q_{\mathrm{cusp}}(Y_{0,K}(N),\mathcal{L}(\widehat{\kappa},\chi;\mathbb{C}))[\widehat{\lambda}]$ is one dimensional, we see that $\delta(f)=\Omega(\widehat{\lambda})\xi$ for a $\mathbb{Q}(\widehat{\lambda})$-rational cohomology class ξ, and we know the algebraicity of $L_{N'}(1,\mathrm{Ad}(\lambda)\otimes\alpha)$ up to $\Omega(\widehat{\lambda})$ and a power of π. This happens when K is a CM quadratic extension of totally real F. Let $\Sigma_K(\mathbb{R})$ be the set of real places of K. When F is totally real, we can further decompose the cohomology group through the action of $C_\infty/C_{\infty+}=\{\pm1\}^{\Sigma_K(\mathbb{R})}$ into a direct sum of one dimensional pieces, and hence we can still prove the algebraicity of the L-value. For general K/F, in place of $H^q_{\mathrm{cusp}}(Y_{0,K}(N),\mathcal{L}(\widehat{\kappa},\chi;\mathbb{C}))$, we can use

$H^q_c(Y_{0,F}(N'), \mathbb{C}) \cong \mathbb{C}$, which is one dimensional. Therefore similarly we can define the period $\Omega(\widehat{\lambda})$ and get the algebraicity, although the definition of the transcendental factor Ω is not so transparent as the case where F is totally real. This is one of the reasons why we have divided our argument according to the shape of K/F at the archimedean places as described in the introduction.

In this paper, we only study the pull back integration of degree $q = \dim Y_{0,F}(N')$ cohomology class. Presumably the same process for degree $q' < q$ would yield another integral expression of $L_{K/F}(s, \mu, \eta)$. In this case, the Eisenstein series $E(s)$ has to be replaced by an Eisenstein differential form. When K is an imaginary quadratic field, Ghate [**Gh**] treated the case of minimal q', that is, $q' = 1$ (while $q = 2$) and obtained a rationality result for critical values of $L(s, (\mu \otimes \mu)_+)$, which implies rationality of some critical values of $L(s, \mathrm{Ad}(\lambda) \otimes \eta)$. Since our result covers the non-critical value $L(1, \mathrm{Ad}(\lambda) \otimes \alpha)$ in this case, the two results are disjoint. For a general quadratic extension, there are several values of q' between maximal q and the minimal one. It is an interesting problem to study the integral for intermediate values q'.

Here, we study extensibility of $\psi' = \alpha\psi^{-1}$ to a Hecke character φ of K up to Hecke characters of F of conductor 1. Here is a general lemma supplied by the referee of this paper:

LEMMA 2.1. *Let K/F be a Galois extension. Let J'' be the set of archimedean places of F which ramify in K. A Hecke character χ of finite order of $F^\times_{\mathbb{A}}$ extends to a Hecke character of finite order of K if and only if $\chi_\sigma = 1$ for all $\sigma \in J''$.*

PROOF. Here is a proof which is a version of the proof supplied by the referee. For a multiplicative $\mathrm{Gal}(K/F)$-module A, we write $H^r(A)$ (resp. $\widehat{H}^r(A)$) for the group cohomology group $H^r(\mathrm{Gal}(K/F), A)$ (resp. the Tate cohomology group $\widehat{H}^r(\mathrm{Gal}(K/F), A)$). Thus $\widehat{H}^r(A)$ is defined also for negative r, $\widehat{H}^r(A) = H^r(A)$ for $r > 0$ and $\widehat{H}^0(A) = H^0(A)/N_{K/F}A$. Let $C_X = X^\times_{\mathbb{A}}/X^\times$ for a number field X be the idele class group. Write D_X for the identity component of C_X. Thus by the Artin reciprocity map, C_X/D_X is isomorphic to the Galois group of the maximal abelian extension of X. Note that C_F and $C_F/C_F \cap D_K$ are closed subgroups of C_K and C_K/D_K, respectively. Thus we see

1. A character $\chi : C_F \to \mathbb{C}^\times$ is of finite order $\iff$ χ is trivial on D_F;
2. χ extends to a finite order character of C_K $\iff$ χ is trivial on $C_F \cap D_K$.

By Hilbert's theorem 90 applied to $K^\times$, we have $H^0(C_K) = C_F$, and hence $H^0(D_K) = C_F \cap D_K$. For each complex place $\sigma \in \Sigma_K(\mathbb{C})$, we write $T_\sigma = \{z \in K_\sigma = \mathbb{C} | |z|_\sigma = 1\}$ and D'_K for the image of $K^\times \prod_{\sigma \in \Sigma_K(\mathbb{C})} T_\sigma$ in C_K. By [**ArT**, Theorem 4, p. 91], $\widehat{H}^r(D'_K) = \widehat{H}^r(D_K)$ for all r. Applying this to $r = 0$, we get

$$H^0(D_K) = H^0(D'_K)N_{K/F}(D_K)$$

in D_K. Let J be the subset of Σ_K made of places above J''. Then writing $\{\pm 1\}_\sigma$ for the subgroup of order 2 in T_σ, we consider the image D''_K of $K^\times \prod_{\sigma \in J}\{\pm 1\}_\sigma$. Then by [**ArT**, Theorem 5], (or its proof) on page 92, $H^0(D_K) = D''_K D_F$, which shows the assertion. □

By this lemma, our question of the extensibility is reduced to the study of the infinity type of the characters in question. Let Ξ_X be the set of all infinity types of arithmetic Hecke characters of a number field X. We write Ξ for Ξ_F. We have a typical element $\mathbf{1}_X = \sum_{\sigma\in I_X}\sigma \in \Xi_X$. We write $\mathbf{1}$ for $\mathbf{1}_F$. If a number field X contains a CM field, we write X_{CM} for the largest CM subfield in X. Then

$$\Xi = \begin{cases} \mathrm{Inf}_{F/F_{CM}}(\Xi_{F_{CM}}) & \text{if } F \text{ contains a CM field} \\ \mathbb{Z}\mathbf{1} & \text{if } F \text{ does not contains any CM field.} \end{cases}$$

First suppose that F contains a CM field. If $K_{CM} \neq F_{CM}$, K_{CM} and F are linearly disjoint over F_{CM}, and hence,

$$\mathrm{Res}_{K/F}\Xi_K = \mathrm{Res}_{K/F}\mathrm{Inf}_{K/K_{CM}}(\Xi_{K_{CM}}) = \mathrm{Inf}_{F/F_{CM}}\mathrm{Res}_{K_{CM}/F_{CM}}(\Xi_{K_{CM}}).$$

Note that $\mathrm{Res}_{K_{CM}/F_{CM}}(\Xi_{K_{CM}}) + \mathbb{Z}\mathbf{1}_{F_{CM}} = \Xi_{F_{CM}}$, which follows from the fact that for any CM field X, $\{\tau - \tau c\}_{\tau\in I_X}$ and $1_X = \sum_{\tau\in I_X}\tau$ generate Ξ_X. Similarly, if K contains a CM field but F does not, $\mathrm{Res}^K_F\Xi_K = \Xi$. Thus we have

$$(\mathrm{Res}_{K/F}\Xi_K) + \mathbb{Z}\mathbf{1} = \Xi$$

if one of the following three conditions is satisfied: (i) $K_{CM} \neq F_{CM}$, (ii) F does not contains any CM field, and (iii) K does not contains any CM field. If K contains a CM field and $K_{CM} = F_{CM}$, we have

$$\mathrm{Res}_{K/F}\Xi_K = 2\Xi.$$

Suppose that K contains a CM field L and either $K_{CM} \neq F_{CM}$ or F does not contains a CM field. Let Φ_L be a CM type of L and $\Phi = \mathrm{Inf}^K_L\Phi_L$. Then $\Phi \in \Xi_K$. Choose any Hecke character ξ of infinity type Φ, and put $\beta = \xi_F|\ |^{-1}_{F_{\mathbb{A}}}$. Then we have $\beta_\sigma = \alpha_\sigma$ for all $\sigma \in J''$. If $\infty(\psi) \in \mathrm{Res}^K_F\Xi_K$, then we take a Hecke character φ' of K with $\mathrm{Res}^K_F\infty(\varphi') = \infty(\psi)$ and $\varphi'(x_\infty) = x_\infty^{\infty(\varphi')}$ for all $x_\infty \in K^\times_\infty$. Then $\psi'\varphi'_F = \alpha\psi^{-1}\varphi'_F$ is of finite order. Since $\psi(x_\infty) = x_\infty^{-n-2v}$, we see $(\psi^{-1}\varphi'_F)_\infty = 1$. Thus $(\alpha\psi^{-1}\varphi'_F)_\infty = \alpha_\infty$ and hence, by Lemma 2.1, we can find a finite order Hecke character φ'' of K such that $\alpha\psi^{-1}\varphi'_F\beta = \varphi''_F$. Thus $\alpha\psi^{-1} = \varphi_F\omega$ for $\varphi = (\varphi')^{-1}\varphi''\xi^{-1}$ and $\omega = |\ |_{F_{\mathbb{A}}}$. If $\infty(\psi) \notin \mathrm{Res}^K_F\Xi_K$, then $\infty(\psi^{-1}|\ |^{-1}_{F_{\mathbb{A}}}) \in \mathrm{Res}^K_F\Xi_K$. In this case, choosing an arithmetic φ' such that $\varphi'_F(x_\infty) = x_\infty^{n+2v-\mathbf{1}}$ for all $x \in F^\times_\infty$. Then $(\alpha\psi^{-1}|\ |^{-1}_{F_{\mathbb{A}}}(\varphi'_F)^{-1})_\sigma = 1$ for all $\sigma \in J''$ and hence, by Lemma 2.1, we can find a finite order character φ'' of K such that $\alpha\psi^{-1}|\ |^{-1}_{F_{\mathbb{A}}}(\varphi'_F)^{-1} = \varphi''_F$. Thus again we get $\alpha\psi^{-1} = \varphi_F\omega$ for $\varphi = \varphi'\varphi''$ and $\omega = |\ |_{F_{\mathbb{A}}}$.

Now suppose that F does not contains any CM field and $J'' \neq \emptyset$. Then $\mathrm{Res}^K_F\Xi_K = \mathbb{Z}\mathbf{1} + \Xi$ and $n + 2v = m\mathbf{1}$ for an integer m. We see that m is even if and only if $n + 2v \in \mathrm{Res}^K_F\Xi_K$. Suppose that m is odd. Then $(\alpha\psi^{-1}|\ |^{-m}_{F_{\mathbb{A}}})_\sigma = 1$ for all $\sigma \in J''$. Thus by Lemma 2.1, we can find a finite order Hecke character φ' such that $\alpha\psi^{-1}|\ |^{-m}_{F_{\mathbb{A}}} = \varphi'_F$. Thus $\alpha\psi^{-1} = \varphi_F\omega$ with $\varphi = \varphi'$ and $\omega = |\ |^m_{F_{\mathbb{A}}}$. If K contains a CM field and m is even, we can argue in the same way, replacing m by $m-1$ and requiring $\infty(\varphi') = \Phi$.

Now suppose only that $\infty(\psi) \in \mathrm{Res}^K_F\Xi_K$. Thus by the above argument, if there exists a finite order character ω of conductor 1 such that $\alpha_\infty = \omega_\infty$, we can find an arithmetic character φ of K such that $\alpha\psi^{-1} = \varphi_F\omega$. In particular, if α_∞ is trivial, then we can find φ and ω.

Note that $\infty(\psi) = -n-2v$ and $n = nc$ by the unitarity of cuspidal automorphic representations if $H^q_{\mathrm{cusp}}(Y_0(N), \mathcal{L}(\kappa, \psi; \mathbb{C})) \neq 0$. Thus we see that if $x \in K^\times_{\infty+}$, then $x^{\widehat{n}} \in F^\times_{\infty+}$ for the identity connected component $X^\times_{\infty+}$ of $X^\times_\infty$. Thus we have a unique positive square root $x^{\widehat{n}/2}$ of $x^{\widehat{n}}$, and there exists a Hecke character φ' of K such that $\varphi'_\infty(x_\infty) = x_\infty^{\widehat{n}/2+\widehat{v}}$ for all $x_\infty \in K^\times_{\infty+}$. Then $\varepsilon = \psi\varphi'_F$ is a finite order character. In particular, if $J'' = \emptyset$ (for example, if F is totally imaginary), we can find a finite order Hecke character φ'' with $\alpha\varepsilon = \varphi''_F$, and hence $\alpha\psi^{-1} = \varphi_F$ for $\varphi = \varphi'(\varphi'')^{-1}$. We record what we have proven as follows:

LEMMA 2.2. (o) *Suppose that $\infty(\psi) \in \mathrm{Res}^K_F \Xi_K$. If there exists a finite order character ω of conductor 1 such that $\alpha_\infty = \omega_\infty$, we can find an arithmetic character φ of K such that $\alpha\psi^{-1} = \varphi_F\omega$. In particular, if α_∞ is trivial, then we can find φ with $\alpha\psi^{-1} = \varphi_F$.*

(i) *Suppose that K contains a CM field. Suppose either that $K_{CM} \neq F_{CM}$ or that F does not contains a CM field. Then $\Xi = \mathbb{Z}\mathbf{1} + \mathrm{Res}^K_F \Xi_K$. Moreover we can find an arithmetic Hecke character φ of K such that $\varphi_F\psi|\,|_{F_\mathbb{A}} = \alpha$.*

(ii) *Suppose that F does not contain a CM field and there is a real place of F which extends to a complex place of K. Then $\Xi = \mathbb{Z}\mathbf{1} + \mathrm{Res}^K_F \Xi_K$. Write $n + 2v = m\mathbf{1}$ for $m \in \mathbb{Z}$. Then we can choose φ and ω, up to finite order characters of conductor 1, so that $\alpha^{-1}\psi = \varphi_F\omega$ and*

$$\omega = \begin{cases} |\,|^m_{F_\mathbb{A}} & \text{if } m \text{ is odd,} \\ |\,|^{m-1}_{F_\mathbb{A}} & \text{if } m \text{ is even and } K \text{ contains a CM field,} \end{cases}$$

$$\infty(\varphi) = \begin{cases} 0 & \text{if } m \text{ is odd,} \\ \Phi \text{ for a CM type } \Phi & \text{if } m \text{ is even and } K \text{ contains a CM field.} \end{cases}$$

(iii) *If F is totally imaginary, then we can find an algebraic Hecke character φ of K such that $\varphi_F\psi = \alpha$ and $\infty(\varphi) = 1/2(\widehat{n} + 2\widehat{v})$.*

2.5. Motivic interpretation and criticality. We already mentioned the conjecture associating λ to a compatible system $\rho(\lambda)$ of $\mathfrak{l}$-adic Galois representations. As is well known, we can state a stronger version of the conjecture in terms of pure motives of dimension 2. This is a special case of a generalization by Langlands of the Shimura-Taniyama conjecture, because H^1 of an elliptic curve is a rank 2 motive. A precise statement can be found in [**Hi94**, Conjecture 0.1]. The point of the conjecture is that there exists a finite extension $E/\mathbb{Q}(\lambda)$, which is either totally real or a CM field, and a pure simple rank 2 motive $M(\lambda)$ with coefficients in E such that $L(s, M(\lambda) \otimes \eta) = L(s, \lambda \otimes \eta)$. In particular, the system $\rho(\lambda)$ is obtained from the étale realization of $M(\lambda)$, and the Hodge type of $M(\lambda) \otimes_{F,\sigma} \mathbb{C}$ is given by

$$(n_\sigma + 1 + v_\sigma, v_{\sigma c}), \quad (v_\sigma, n_\sigma + 1 + v_{\sigma c})$$

for complex conjugation c of $\mathbb{C}$. Thus $w = n_\sigma + 1 + v_\sigma + v_{\sigma c}$ is the weight of $M(\lambda)$ and is independent of σ, which is known without supposing the conjecture. Then the Galois representation $\mathrm{Ad}(\rho(\lambda))$ corresponds to a rank 3 motive $\mathrm{Ad}(M(\lambda))$ sitting inside $M(\lambda)\otimes M(\lambda)^\vee$ for the dual $M(\lambda)^\vee$ of $M(\lambda)$. The functional equation of $L(s, \mathrm{Ad}(\lambda))$ has the Γ-factor exactly equal to that predicted by the theory of motives [**GeJ**]. To describe the Γ-factor of $L(s, \mathrm{Ad}(\lambda) \otimes \eta)$, we write $\eta_\sigma(-1) = -(-1)^{\nu_\sigma}$ with $\nu_\sigma \in \{0, 1\}$ for the restriction η_σ of η to F_σ for each $\sigma \in \Sigma_F(\mathbb{R})$. Then the Γ-factor of $L(s, \mathrm{Ad}(\lambda) \otimes \eta)$ coincides with that of $L(s, \mathrm{Ad}(M(\lambda)) \otimes \eta)$ and is given by, for a finite order character η

$$(\Gamma) \quad \Gamma(s, \mathrm{Ad}(\lambda) \otimes \eta) = \prod_{\sigma \in I} \Gamma_\mathbb{C}(s + n_\sigma + 1) \times \prod_{\sigma \in \Sigma(\mathbb{C})} \Gamma_\mathbb{C}(s)^2 \times \prod_{\sigma \in \Sigma(\mathbb{R})} \Gamma_\mathbb{R}(s + \nu_\sigma),$$

where $\Gamma_\mathbb{C}(s) = (2\pi)^{-s}\Gamma(s)$ and $\Gamma_\mathbb{R}(s) = \pi^{-s/2}\Gamma(s/2)$. Since $\mathrm{Ad}(M)$ is self dual, the functional equation gives the reflection: $s \mapsto 1 - s$, which is known to be true for $L(s, \mathrm{Ad}(\lambda) \otimes \eta)$ with finite order η. A motivic L-value $L(m, M)$ at an integer m

is called critical if the value of the Γ-factor at m and at its reflection point (of the functional equation) is finite. Therefore $L(1,\mathrm{Ad}(\lambda)\otimes\eta)$ is critical if and only if F is totally real and $\nu_\sigma = 1$ for all $\sigma\in I$ (that is, η is totally even). Moreover we have

$$\Gamma(1,\mathrm{Ad}(\lambda)\otimes\eta) = \frac{\Gamma_F(k)}{(2\pi)^{k+2\Sigma(\mathbb{C})}\pi^{\Sigma_+(\eta)}},$$
$$\mathrm{ord}_{s=0}\Gamma(s,\mathrm{Ad}(\lambda)\otimes\eta) = \#\Sigma_-(\eta)+2r_2(F),$$

where $\mathrm{ord}_{s=0}$ is the order of the pole at $s=0$; $\Sigma_-(\eta)=\{\sigma|\nu_\sigma=0\}$; $\Sigma_+(\eta)=\{\sigma|\nu_\sigma=1\}$; $k=\sum_{\sigma\in I}(n_\sigma+2)\sigma$; $\Gamma_F(k)=\prod_\sigma\Gamma(k_\sigma)$; $x^k=\prod_\sigma x^{k_\sigma}$; and for a subset X of I, $x^X=x^{\#X}$. The number $\mathrm{ord}_{s=0}\Gamma(s,\mathrm{Ad}(\lambda)\otimes\eta)$ is a good measure to know how the value $L(1,\mathrm{Ad}(\lambda)\otimes\eta)$ is far from being critical. It is interesting that the period $\Omega(\widehat{\lambda})$ defined independent of $\mathrm{ord}_{s=0}\Gamma(s,\mathrm{Ad}(\lambda)\otimes\eta)$ in purely automorphic way gives the Beilinson and Deligne period if one admits the various conjectures including the above one and the Deligne-Beilinson conjecture for motivic L-values [**RSS**]. Moreover, we can prove a very close relation of the L-value $L(1,\mathrm{Ad}(\lambda)\otimes\alpha)$ to the congruence of cohomological cusp forms and hence, presumably, to the Selmer group of $\mathrm{Ad}(\lambda)\otimes\alpha$ if K is quadratic over $\mathbb{Q}$ (see Section 5). This is striking because the definition of $\Omega(\widehat{\lambda})$ is topological (and automorphic) and has nothing to do, in an apparent way, with the algebro-geometric property of $\mathrm{Ad}(M(\lambda))\otimes\alpha$. Thus the quantity $\Omega(\widehat{\lambda})$ should be very close to the volume at infinity of the Tamagawa measure (of Bloch and Kato) of $\mathrm{Ad}(M(\lambda))\otimes\alpha$ divided by the order of $H^0(\mathbb{Q},(\mathrm{Ad}(M(\lambda))\otimes\alpha)\otimes\mathbb{Q}/\mathbb{Z})$ as in [**BK**, (5.15.1)].

3. Imaginary quadratic case

Let K be an imaginary quadratic field with discriminant $-D$ $(D>0)$. Thus the different $\mathfrak{d}$ of $K/\mathbb{Q}$ is generated by $\sqrt{-D}$. We write σ for the unique non-trivial automorphism of K. Then $H=\mathrm{GL}(2)_{/\mathbb{Q}}$ and $G=\mathrm{Res}_{K/\mathbb{Q}}\mathrm{GL}(2)_{/K}$. We write $S_{\widehat{\kappa}}(N,\chi)_{/K}$ for the space $S_{\widehat{\kappa},\emptyset}(N,\chi)$ of cohomological cusp forms on $G(\mathbb{A})$ of weight $\widehat{\kappa}$. Let $\psi:\mathbb{Q}^\times\backslash\mathbb{A}^\times\to\mathbb{C}^\times$ be a Hecke character such that $\psi_\infty(x)=x^{-n-2v}$ for all $x\in\mathbb{R}^\times$. We put $\chi=\psi\circ N_{K/\mathbb{Q}}$ and consider the isomorphism for $\widehat{\kappa}=(\widehat{n},\widehat{v})$:

$$\delta=\delta_{\emptyset,\emptyset}:S_{\widehat{\kappa}}(N,\chi)_{/K}\cong H^2_{\mathrm{cusp}}(Y_{0,K}(N),\mathcal{L}(\widehat{\kappa},\chi;\mathbb{C}))[\chi].$$

We write $I=I_{\mathbb{Q}}=\{\tau\}$ and use the same symbol τ to indicate a fixed extension of τ to K. For complex conjugation c of $\mathbb{C}$, $\sigma\tau=\tau c$. Then χ is an arithmetic Hecke character whose ∞-type is $-\widehat{n}-2\widehat{v}=-n_\tau\tau-n_\tau\tau c-2v_\tau\tau-2v_\tau\tau c$ for $0\le\widehat{n}=n_\tau\tau+n_\tau\tau c\in\mathbb{Z}[I_K]$ and $\widehat{v}=v_\tau\tau+v_\tau\tau c\in\mathbb{Z}[I_K]$. Let $f\in S_{\widehat{\kappa}}(N,\chi)$ be a cusp form, and write $f^{(a)}\in S_{\widehat{\kappa}}(\Gamma^{(a)},\chi)$ for the classical cusp form corresponding to f (see [**Hi94**, 3.5]). In other words, we consider $f^{(a)}(w)=f\left(\left(\begin{smallmatrix}a&0\\0&1\end{smallmatrix}\right)w\right)$ for $w\in\mathrm{SL}_2(\mathbb{C})\subset G(\mathbb{R})$ and $a\in R\subset T(\mathbb{A}^{(\infty)})$ (see 2.2). The Fourier expansion of $f^{(1)}$ is given as follows (cf. [**Hi94**, (6.1)]):

$$f^{(1)}(w)=|y|^{n_\tau}|y|_{K_\mathbb{A}}\left\{\sum_{\xi\in K^\times}\mathbf{a}(\xi y\mathfrak{d},f)\xi^{-\widehat{v}}W(\xi y)\mathbf{e}_K(\xi x)\right\},$$

where $w=y^{-1/2}\left(\begin{smallmatrix}y&x\\0&1\end{smallmatrix}\right)\in\mathrm{SL}_2(\mathbb{C})$ and

$$W(y)=\sum_{0\le\alpha\le n^*}\binom{n^*}{\alpha}\left(\frac{y}{\sqrt{-1}|y|}\right)^{n_\tau+1-\alpha}K_{\alpha-n_\tau-1}(4\pi|y|)S^{n^*-\alpha}T^\alpha\quad(n^*=2n_\tau+2)$$

with values in $L((n^*,0);\mathbb{C}) = \sum_\alpha \mathbb{C}S^{n^*-\alpha}T^\alpha$. We now write down $\delta(f^{(1)})$ following the process described in [**Hi94**, Section 2.5]. For that, we write $f^{(1)} = \sum_{0\le\alpha\le n^*} f_\alpha\binom{n_\tau^*}{\alpha}S^{n^*-\alpha}T^\alpha$ and

$$\begin{aligned}
&(X_\tau V - Y_\tau U)^{n_\tau}(X_{\tau c}U + Y_{\tau c}V)^{n_\tau}(AV-BU)^2\\
&\quad= \sum_{0\le j_\tau\le n}\{(-1)^{j_\tau}\binom{n_\tau}{j_\tau}X_\tau^{n_\tau-j_\tau}Y_\tau^{j_\tau}V^{n_\tau-j_\tau}U^{j_\tau}\times\sum_{0\le j_{\tau c}\le n}\binom{n_\tau}{j_{\tau c}}\\
&\quad X_{\tau c}^{n_\tau-j_{\tau c}}Y_{\tau c}^{j_{\tau c}}U^{n_\tau-j_{\tau c}}V^{j_{\tau c}}\}\times(A^2V^2-2ABUV+B^2U^2)\\
&\quad= \sum_{0\le j\le\widehat{n}}(-1)^{j_\tau}\binom{\widehat{n}}{j}X^{\widehat{n}-j}Y^j\{V^{n_\tau-j_\tau+j_{\tau c}+2}U^{n_\tau+j_\tau-j_{\tau c}}A^2\\
&\quad - 2V^{n_\tau-j_\tau+j_{\tau c}+1}U^{n_\tau+j_\tau-j_{\tau c}+1}AB + V^{n_\tau-j_\tau+j_{\tau c}}U^{n_\tau+j_\tau-j_{\tau c}+2}B^2\},
\end{aligned}$$

where $\binom{\widehat{n}}{j} = \binom{n_\tau}{j_\tau}\binom{n_\tau}{j_{\tau c}}$. Then replacing $U^\alpha V^{n^*-\alpha}$ by $(-1)^{n^*-\alpha}f_\alpha^{(1)}$, (A,B) by $(y^{-1/2}A, y^{-1/2}B)$, and (A^2,AB,B^2) by $y^{-1}(dy\wedge dx, -2dx\wedge d\overline{x}, dy\wedge d\overline{x})$, we have

$$(\delta)\qquad \begin{aligned}\delta(f^{(1)}) = w\sum_{0\le j\le\widehat{n}}&(-1)^{n_\tau-j_{\tau c}}\binom{\widehat{n}}{j}X^{\widehat{n}-j}Y^j\{f_{n_\tau+j_\tau-j_{\tau c}}y^{-2}dy\wedge dx\\
&- f_{n_\tau+j_\tau-j_{\tau c}+1}y^{-2}dx\wedge d\overline{x} + f_{n_\tau+j_\tau-j_{\tau c}+2}y^{-2}dy\wedge d\overline{x}\},\end{aligned}$$

where w acts on $(X_\tau, Y_\tau, X_{\tau c}, Y_{\tau c})$ as

$$(X_\tau, Y_\tau, X_{\tau c}, Y_{\tau c}) \mapsto (X_\tau, Y_\tau, X_{\tau c}, Y_{\tau c})\begin{pmatrix}{}^t w^\iota & 0\\ 0 & {}^t\overline{w}^\iota\end{pmatrix}.$$

Now we restrict $\delta(f^{(1)})$ to the upper half complex plane $\mathfrak{H}$ from $\mathfrak{Z} = \mathcal{H}_\tau$ and write the pull back as $\delta(f^{(1)})|_{\mathbb{Q}}$. Then we have, for w as above

$$\begin{aligned}&\delta(f^{(1)})|_{\mathbb{Q}} =\\
&\quad(-1)^{n_\tau}w\sum_{0\le j\le\widehat{n}}(-1)^{j_{\tau c}}\binom{\widehat{n}}{j}X^{\widehat{n}-j}Y^j\{f_{n_\tau+j_\tau-j_{\tau c}} + f_{n_\tau+j_\tau-j_{\tau c}+2}\}y^{-2}dy\wedge dx.\end{aligned}$$

This differential form has values in

$$L(\widehat{\kappa},\chi;\mathbb{C})|_{\mathfrak{H}} \cong L(\kappa,\chi_F;\mathbb{C})\otimes L(\kappa;\mathbb{C}).$$

Noting that $\chi_F = \psi^2$, we know that

$$L(\kappa,\chi_F;A)\otimes L(\kappa;A) \;\cong \oplus_{0\le i\le n_\tau}L((2n_\tau-2i, 2v_\tau+i),\psi^2;A)$$

for any $\mathbb{Q}(\kappa,\psi)$-algebra A [**Hi94**, (11.2a)]. In order to write down the projection to $L((0,2v_\tau+n_\tau),\chi_F;A)$, we introduce a differential operator:

$$\nabla = \frac{\partial^2}{\partial X_{\tau c}\partial Y_\tau} - \frac{\partial^2}{\partial X_\tau\partial Y_{\tau c}} = \frac{\partial^2}{\partial X_{\sigma\tau}\partial Y_\tau} - \frac{\partial^2}{\partial X_\tau\partial Y_{\sigma\tau}}.$$

Following the expression of the projection in [**Hi94**, p. 498], it is given by $(n_\tau!)^{-2}\nabla^{n_\tau}$. Thus we need to compute $(\nabla^{n_\tau}\delta(f^{(1)})|_{\mathbb{Q}})$. We see

$$\nabla^m = \sum_{0\le k\le m}(-1)^{m-k}\binom{m}{k}\left(\frac{\partial^2}{\partial X_{\sigma\tau}\partial Y_\tau}\right)^k\left(\frac{\partial^2}{\partial X_\tau\partial Y_{\sigma\tau}}\right)^{m-k}$$

and $k!^{-1}(\partial/\partial T)^k T^m = \binom{m}{k} T^{m-k}$. The differential operator ∇ decreases the degree by $(1,1)$ either in $(X_\tau, Y_{\sigma\tau})$ or in $(X_{\sigma\tau}, Y_\tau)$. Thus applying ∇ to a monomial n_τ times, we see

$$\nabla^{n_\tau} X^{n-j} Y^j = 0 \quad \text{unless} \quad n_\tau = j_\tau + j_{\sigma\tau}.$$

If $n_\tau = j_\tau + j_{\sigma\tau}$, we get

$$(n_\tau!)^{-2} \nabla^{n_\tau} X_\tau^{n_\tau - j_\tau} Y_\tau^{j_\tau} X_{\sigma\tau}^{n_{\sigma\tau} - j_{\sigma\tau}} Y_{\sigma\tau}^{j_{\sigma\tau}} = (-1)^{j_\tau} \frac{(j_\tau! j_{\sigma\tau}!)^2}{(n_\tau!)^2} \binom{n_\tau}{j_\tau} = (-1)^{j_\tau} \binom{n_\tau}{j_\tau}^{-1}.$$

Thus noting the fact that

$$\nabla^{n_\tau}(wP) = \nabla^{n_\tau} P \quad \text{if } \det(w) = 1,$$

and writing j for j_τ, we have

$$\begin{aligned}
&(n_\tau!)^{-2}(\nabla^{n_\tau} \delta(f^{(1)})|_{\mathbb{Q}}) \\
&\quad = \sum_{0 \le j \le n_\tau} \binom{n_\tau}{j}^2 (n_\tau!)^{-2} \nabla^{n_\tau} X^{n_\tau - j} Y_\tau^j X_{\tau c}^j Y_{\tau c}^{n_\tau - j} \{f_{2j} + f_{2j+2}\} y^{-2} dy \wedge dx \\
&\quad = \sum_{0 \le j \le n_\tau} \binom{n_\tau}{j} (f_{2j} + f_{2j+2}) y^{-2} dy \wedge dx,
\end{aligned}$$

which is the explicit form of the pull back differential form we have written as $\pi(\delta_{J,J'}(f))$ in Section 2.4.

Now we start computing the integral $\int_{Y_{0,\mathbb{Q}}(N')} \pi(\delta_{J,J'}(f))$. We put

$$\Phi = H(\mathbb{Q}) \cap U_0(N') H(\mathbb{R})_+ = \Gamma_0(N')$$

for $H = \mathrm{GL}(2)_{/\mathbb{Q}}$ and $N' = N \cap \mathbb{Q}$. Then $Y_{0,\mathbb{Q}}(N') = \Phi \backslash \mathfrak{H}$. We now define

$$\Phi_\infty = \left\{ \pm \begin{pmatrix} 1 & m \\ 0 & 1 \end{pmatrix} \mid m \in \mathbb{Z} \right\},$$

which is the stabilizer of the cusp ∞ in Φ. Then we look at

$$\int_{\Phi_\infty \backslash \mathfrak{H}} (n_\tau!)^{-2} \left(\nabla^{n_\tau} \delta(f^{(1)})|_{\mathbb{Q}} \right) y^s = \sum_{0 \le j \le n_\tau} \binom{n_\tau}{j} \int_0^\infty \int_0^1 (f_{2j} + f_{2j+2}) dx y^{s-2} dy.$$

Now we assume that $f|T(\mathfrak{n}) = \mu(T(\mathfrak{n}))f$ for a system of Hecke eigenvalues μ of G and f is normalized so that $\mathbf{a}(\mathfrak{n}, f) = \mu(T(\mathfrak{n}))$. We compute

$$\begin{aligned}
\int_0^\infty \int_0^1 f_{2j} dx y^{s-2} dy &= \sum_{\xi \in K^\times} \mu(T(\xi\mathfrak{d}))\xi^{-\widehat{v}} \left(\frac{\xi}{i|\xi|}\right)^{n_\tau+1-2j} \\
&\quad \cdot \int_0^\infty y^{n_\tau+s} K_{2j-n_\tau-1}(4\pi|\xi y|) \int_0^1 \mathbf{e}(\mathrm{Tr}(\xi)x) dx dy \\
&= D^{v_\tau} \sum_{0 \neq m \in \mathbb{Z}} \mu(T(m)) m^{-2v} \mathrm{sgn}(m)^{n_\tau+1-2j} \\
&\quad \cdot \int_0^\infty y^{n_\tau+s} K_{2j-n_\tau-1}(4\pi D^{-1/2}|my|) dy \\
&= D^{v_\tau}(1+(-1)^{n_\tau+1+2v}) \sum_{m=1}^\infty \mu(T(m)) m^{-2v} \\
&\quad \cdot \int_0^\infty y^{n_\tau+s} K_{2j-n_\tau-1}(4\pi D^{-1/2}|my|) dy \\
&= \frac{2^{n_\tau+s-1}(1+(-1)^{n_\tau+1+2v})D^{v_\tau}}{(4\pi D^{-1/2})^{n_\tau+s+1}} \Gamma\left(\frac{s+2j}{2}\right) \\
&\quad \cdot \Gamma\left(\frac{2n_\tau+s+2-2j}{2}\right) L_{K/\mathbb{Q}}(s, \mu, \omega),
\end{aligned}$$

where $\omega(a) = |a|_{\mathbb{A}}^{n_\tau+2v_\tau}$ and thus $\omega((\ell)) = |\ell|^{-n_\tau-2v_\tau}$ (see Lemma 2.2 (ii)). Similarly we get

$$\begin{aligned}
\int_0^\infty \int_0^1 f_{2j+2} dx y^{s-2} dy = &\frac{2^{n_\tau+s-1}(1+(-1)^{n_\tau+1+2v})D^{v_\tau}}{(4\pi D^{-1/2})^{n_\tau+s+1}} \Gamma\left(\frac{s+2j+2}{2}\right) \\
&\cdot \Gamma\left(\frac{2n_\tau+s-2j}{2}\right) L_{K/\mathbb{Q}}(s, \mu, \omega).
\end{aligned}$$

We see from $\Gamma(s+1) = s\Gamma(s)$ that

$$\begin{aligned}
\Gamma\left(\frac{s+2j}{2}\right)&\Gamma\left(\frac{2n_\tau+s+2-2j}{2}\right) + \Gamma\left(\frac{s+2j+2}{2}\right)\Gamma\left(\frac{2n_\tau+s-2j}{2}\right) \\
&= \frac{2n_\tau+s-2j}{2}\Gamma\left(\frac{s+2j}{2}\right)\Gamma\left(\frac{2n_\tau+s-2j}{2}\right) \\
&\quad + \frac{s+2j}{2}\Gamma\left(\frac{s+2j}{2}\right)\Gamma\left(\frac{2n_\tau+s-2j}{2}\right) \\
&= (n_\tau+s)\Gamma\left(\frac{s+2j}{2}\right)\Gamma\left(\frac{2n_\tau+s-2j}{2}\right).
\end{aligned}$$

Note that by [**Hi94**, p. 505]

$$\sum_{j=0}^m \binom{m}{j} \Gamma\left(\frac{s+2j}{2}\right)\Gamma\left(\frac{2m+s-2j}{2}\right) = \frac{\Gamma\left(\frac{s}{2}\right)\Gamma\left(\frac{s}{2}\right)\Gamma(s+m)}{\Gamma(s)}.$$

This shows

$$\int_{\Phi_\infty\backslash\mathfrak{H}}(n_\tau!)^{-2}(\nabla^{n_\tau}\delta(f^{(1)})|_{\mathbb{Q}})y^s$$
$$= \frac{(1+(-1)^{n_\tau+1+2v})2^{n_\tau+s-1}D^{v_\tau}}{(4\pi D^{-1/2})^{n_\tau+s+1}}\frac{\Gamma(\frac{s}{2})\Gamma(\frac{s}{2})\Gamma(s+n_\tau+1)}{\Gamma(s)}L_{K/\mathbb{Q}}(s,\mu,\omega),$$

because (cf. (Γ) in 2.5)

$$(n_\tau+s)\frac{\Gamma(\frac{s}{2})\Gamma(\frac{s}{2})\Gamma(s+n_\tau)}{\Gamma(s)} = \frac{\Gamma(\frac{s}{2})\Gamma(\frac{s}{2})\Gamma(s+n_\tau+1)}{\Gamma(s)}.$$

We now apply convolution method and rewrite the above integral as an integral of the product of $(n_\tau!)^{-2}(\nabla^{n_\tau}\delta(f^{(1)})|_{\mathbb{Q}})$ and an Eisenstein series. There are two ways of associating a Dirichlet character to a Hecke character ξ of $\mathbb{Q}$. Let C be a positive multiple of the conductor $C(\xi)$ of ξ. Restricting ξ to $\prod_{\ell|C}\mathbb{Z}_\ell^\times \subset \mathbb{A}^\times$, we get a character ξ_C of $(\mathbb{Z}/C\mathbb{Z})^\times$. On the other hand, for each prime ℓ prime to C, we consider $\ell^{(C)} \in \mathbb{A}^\times$ which is equal to 1 at all places dividing C and coincides with ℓ outside C. Then $\xi^*:\ell\mapsto\xi(\ell^{(C)})$ induces a Dirichlet character $\xi^*:(\mathbb{Z}/C\mathbb{Z})^\times\to\mathbb{C}^\times$. Since $\xi(\ell)=1$, $\xi^*=\xi_C^{-1}$. We note that $L(s-\infty(\xi),\xi)=\sum_{n=1}^\infty \xi^*(n)n^{-s}=L(s,\xi^*)$ identifying $\Xi_{\mathbb{Q}}$ with $\mathbb{Z}$ via $\tau\mapsto 1$. Let N' be a positive integer generating $N\cap\mathbb{Z}$ and $\psi_{N'}^2=(\chi_{\mathbb{Q}})_{N'}$ be the restriction of χ to $\widehat{\mathbb{Z}}^\times$. We extend $\psi_{N'}$ to Φ so that $\psi_{N'}:\left(\begin{smallmatrix}a&b\\c&d\end{smallmatrix}\right)\mapsto\psi_{N'}(d)$. We have

$$\int_{\Phi_\infty\backslash\mathfrak{H}}(n_\tau!)^{-2}(\nabla^{n_\tau}\delta(f^{(1)})|_{\mathbb{Q}})y^s = \int_{\Phi\backslash\mathfrak{H}}\sum_{\gamma\in\Phi_\infty\backslash\Phi}\{(n_\tau!)^{-2}(\nabla^{n_\tau}\delta(f^{(1)})|_{\mathbb{Q}})y^s\}\circ\gamma$$
$$= \int_{\Phi\backslash\mathfrak{H}}(n_\tau!)^{-2}(\nabla^{n_\tau}\delta(f^{(1)})|_{\mathbb{Q}})y^s\sum_{\gamma\in\Phi_\infty\backslash\Phi}\psi_{N'}^2(\gamma)|j(\gamma,z)|^{-2s}.$$

Thus writing

$$E(s)=E(s,\psi_{N'}^2)=L_{N'}(2s,(\psi^*)^2)y^s\sum_{\gamma\in\Phi_\infty\backslash\Phi}\psi_{N'}^2(\gamma)|j(\gamma,z)|^{-2s},$$

we get

$$\int_{\Phi\backslash\mathfrak{H}}(n_\tau!)^{-2}(\nabla^{n_\tau}\delta(f^{(1)})|_{\mathbb{Q}})E(s)$$
$$= \frac{(1+(-1)^{n_\tau+1+2v})2^{n_\tau+s-1}D^{v_\tau}}{(4\pi D^{-1/2})^{n_\tau+s+1}}\frac{\Gamma(\frac{s}{2})\Gamma(\frac{s}{2})\Gamma(s+n_\tau+1)}{\Gamma(s)}$$
$$\cdot L(s+1,(\mu\otimes\mu_\sigma)_+\otimes\omega).$$

This follows from $(R1)$ because $\omega=|\ |_{\mathbb{A}}^m$ for $m=n_\tau+2v_\tau$ and

$$L_{N'}(2s,(\psi^*)^2)=L_{N'}(2s,(\psi\omega)^2)=L_{N'}(2s,\chi_{\mathbb{Q}}\omega^2).$$

On the other hand, we have

$$E(z;s,\psi_{N'}^2)=\frac{1}{2}y^s\sum_{(m,n)\in\mathbb{Z}^2-(0,0)}\psi_{N'}^2(n)|mNz+n|^{-2s}.$$

As will be seen in Appendix (see also [**Hi93a**, Chapter 9, p. 293(1)]), the Fourier expansion of the Eisenstein series is given by

$$E\left(-\frac{1}{N'z}; s, \psi_{N'}^2\right) = 2^{1-2s} N'^{-s} (2\pi y)^{1-s} \frac{\Gamma(2s-1) L_{N'}(2s-1, (\psi^*)^2)}{\Gamma(s)^2}$$

$$+ \text{ Fourier expansion with coefficients in entire functions of } s.$$

This shows that

$$\mathrm{Res}_{s=1} E(z; s, \psi_{N'}^2) = 2^{-1} \pi N'^{-2} \phi(N') \delta_{\psi_{N'}^2, id},$$

where $\delta_{\varepsilon,\varepsilon'}$ is 1 or 0 according as $\varepsilon = \varepsilon'$ or not, and ϕ is the Euler function. From this we get, if $n+1+2v$ is even,

(Res3) $$\mathrm{Res}_{s=1} L(s+1, (\mu \otimes \mu_\sigma)_+ \otimes \omega) = \frac{(2\pi D^{-1/2})^{n+2} \phi(N')}{N'^2 D^{v_\tau} \Gamma(n_\tau + 2)} \delta_{\psi_{N'}^2, id} \int_{\Phi \backslash \mathfrak{H}} (n_\tau !)^{-2} (\nabla^{n_\tau} \delta(f^{(1)}))|_{\mathbb{Q}}.$$

Let $\varphi : K^\times \backslash K_{\mathbb{A}}^\times \to \mathbb{C}^\times$ be a Hecke character of infinity type $-w$ with conductor $C = C(\varphi)$. We replace f in the above formula by the twist $g = f|R(\varphi) \in S_{\widehat{\kappa}+(0,w)}(N \cap C^2, \chi\varphi^2)$ defined in [**Hi94**, p. 480] whose Fourier coefficients are given by $\mathbf{a}(\mathfrak{n}, f|R(\varphi)) = G(\varphi)\mathbf{a}(\mathfrak{n}, f)\varphi(\mathfrak{n})$. Here $G(\varphi)$ is the Gauss sum given by

$$G(\varphi) = \varphi(d)^{-1} \sum_{u \in R} \varphi_C(u) \mathbf{e}_K(d^{-1}u),$$

where R is a complete set of representatives of $C^{-1}/\mathcal{R}$ in $\prod_{\mathfrak{p}|C} K_{\mathfrak{p}}$ and $d \in K_{\mathbb{A}}^\times$ with $d^{(\mathfrak{d})} = 1$ such that $d\mathcal{R}$ is the different $\mathfrak{d} = \mathfrak{d}_K$ of $K/\mathbb{Q}$. We write N' for the positive generator of $N \cap C^2 \cap \mathbb{Z}$. Then we have from [**Hi94**, (6.8)], writing $E(s) = E(s, (\psi\varphi)_{N'}^2)$,

$$\int_{\Phi \backslash \mathfrak{H}} (n_\tau !)^{-2} (\nabla^{n_\tau} \delta(g^{(1)})|_{\mathbb{Q}}) E(s)$$
$$= \frac{(1 + (-1)^{n+1+2v+\mathrm{Res}(w)}) 2^{n+s-1} D^{v_\tau} \sqrt{-D}^w G(\varphi)}{(4\pi D^{-1/2})^{n_\tau + s + 1}} \cdot \frac{\Gamma(\frac{s}{2})\Gamma(\frac{s}{2})\Gamma(s + n_\tau + 1)}{\Gamma(s)} L_{C'}(s+1, ((\mu \otimes \varphi) \otimes (\mu \otimes \varphi)_\sigma)_+ \otimes \omega),$$

where $C' = C \cap \mathbb{Z}$, and $\omega(x) = |x|_{\mathbb{A}}^{n_\tau + 2v_\tau + \mathrm{Res}(w)}$ writing $\mathrm{Res}(w)$ for $\mathrm{Res}_{\mathbb{Q}}^K w \in \mathbb{Z}$. If $\psi\varphi_{\mathbb{Q}}\omega = \alpha$ and $\omega = |\ |_{\mathbb{A}}^{n_\tau + 2v_\tau + \mathrm{Res}(w)}$, then $1 + (-1)^{n+1+2v+\mathrm{Res}(w)} = 2$, and the above integral does not vanish trivially (it could vanish by a different reason). Now as in (Res2) in 2.4, we get

PROPOSITION 3.1. *Let $F = \mathbb{Q}$ and $K = \mathbb{Q}(\sqrt{-D})$ be an imaginary quadratic field with discriminant $-D$ and $\kappa = (n_\tau \tau, v_\tau \tau) \in \mathbb{Z}[I]$ with $I = \{\tau\}$. Let φ be an arithmetic Hecke character with $\infty(\varphi) = -w \in \mathbb{Z}[I_K]$ and the conductor C. Let $\chi = \psi \circ N_{K/\mathbb{Q}}$ for an arithmetic Hecke character ψ with $\psi_\infty(x) = x^{-n_\tau - 2v_\tau}$ for all $x \in \mathbb{A}_\infty^\times = \mathbb{R}^\times$. Suppose that $\alpha = \varphi_{\mathbb{Q}}\omega\psi$ with $\omega(x) = |x|_{\mathbb{A}}^{n_\tau + 2v_\tau + \mathrm{Res}(w)}$ and $\alpha = (\frac{-D}{})$. If a primitive system $\mu : h_{\widehat{\kappa}}(N, \chi; \mathbb{C})_{/K} \to \mathbb{C}$ is a base change lift of $\lambda : h_\kappa(N_0, \psi; \mathbb{C})_{/\mathbb{Q}} \to \mathbb{C}$, then we have*

$$\int_{Y_{0,\mathbb{Q}}(N')} (n_\tau !)^{-2} (\nabla^{n_\tau} \delta(g^{(1)})|_{\mathbb{Q}}) = C_0^{-1} L_{C'}(1, \mathrm{Ad}(\lambda) \otimes \alpha),$$

where $g = f|R(\varphi)$ *for the normalized primitive form* f *with* $f|T(\mathfrak{n}) = \mu(T(\mathfrak{n}))f$ *for all* $\mathfrak{n} \subset \mathcal{R}$, N' *is the positive generator of* $N \cap C^2 \cap \mathbb{Z}$, $C' = C \cap \mathbb{Z}$ *and*

$$\begin{aligned} C_0 &= (2\pi D^{-1/2})^{n_\tau+2} D^{-v} \sqrt{-D}^{-w} \{G(\varphi)\Gamma(n_\tau+2)\}^{-1} \phi(N') N'^{-2} \\ &= (\chi\varphi)_\infty(\sqrt{-D}) D^{-1} \{G(\varphi)\Gamma(1, \mathrm{Ad}(\lambda) \otimes \alpha)\}^{-1} \phi(N') N'^{-2}. \end{aligned}$$

If $\mu : h_{\widehat{\kappa}}(N, \chi; \mathbb{C})_{/K} \to \mathbb{C}$ *is not a twist of any base change lift, then*

$$\int_{Y_{0,\mathbb{Q}}(N')} (n_\tau!)^{-2} (\nabla^{n_\tau} \delta(g^{(1)})|_{\mathbb{Q}}) = 0.$$

PROOF. We only need to show the vanishing when μ is not a base change lift from H. We follow the argument in [**HaLR**, 3.12] (and [**Fl**, Section 5]). For that, it is sufficient to show that $L(s, ((\mu \otimes \varphi) \otimes (\mu \otimes \varphi)_\sigma)_\pm \otimes \omega)$ is holomorphic at $s = 2$. Since $\mu \otimes \varphi$ is not a base change lift, we may rewrite μ for $\mu \otimes \varphi$. Then we see, writing $\kappa' = \widehat{\kappa} + (0, w) = (n', v')$ for the weight of μ, $L(s, (\mu \otimes \mu_\sigma)_+ \otimes \omega) = L(s+m, (\mu \otimes \mu_\sigma)_+)$ for $m = n_\tau + 2v_\tau + \mathrm{Res}(w)$. Note that $m\mathbf{1}_K = n' + 2v' + n'c + 2v'c$. Looking into Euler factorization, we get

$$L(s, (\mu \otimes \mu_\sigma)_+) L(s, (\mu \otimes \mu_\sigma)_-) = L(s, \mu \otimes \mu_\sigma)$$

with the notation of 2.3. Since $\mu \neq \mu_\sigma$ for non-base change μ, $L(s, \mu \otimes \mu_\sigma)$ is an entire function of s ([**J**], [**U**] and [**GeJ**]). We show that $L(s, (\mu \otimes \mu)_-)$ is entire. For that, we recall a result in [**Gh**]. Let $\phi : h_{\kappa''}(M, \xi; \mathbb{C}) \to \mathbb{C}$ be a system of Hecke eigenvalues of weight $\kappa'' = (n'', v'')$ for G. Then writing

$$\delta' = \delta_{\emptyset, J} : S_{\kappa''}(M, \xi) \cong H^1_{\mathrm{cusp}}(Y_{0,K}(M), \mathcal{L}(\kappa'', \xi; \mathbb{C}))[\xi]$$

for $J = \{\tau\}$, we have

$$\begin{aligned} &\int_{Y_{0,\mathbb{Q}}(M')} [\delta'(h) \wedge E_2(s)] \\ &\qquad = C_1 \Gamma(s + n''_\tau + 2) \Gamma\left(\frac{s+2}{2}\right)^2 \Gamma(s+2) L(n''_\tau + \mathrm{Res}(v'') + s, (\phi \otimes \phi_\sigma)_+), \end{aligned}$$

if $n''_\tau \equiv \mathrm{Res}(v'') \bmod 2$ [**Gh**, 7.3(21) and 7.4], where h satisfies $h|T(\mathfrak{n}) = \phi(T(\mathfrak{n}))h$. Here $E_2(s)$ is a degree 1 Eisenstein differential form of weight 2 which is finite at $s = 2$ and $C_1 \neq 0$. We choose φ' such that $\varphi'_F((m)) = \alpha(m) m^{\mathrm{Res}(w')}$ for all $m \in \mathbb{Z}$. Then $\mathrm{Res}(w')$ is odd, and for $\phi = \mu \otimes \varphi'$,

$$L(n''_\tau + \mathrm{Res}(v'') + s, (\phi \otimes \phi_\sigma)_+) = L(n'_\tau + \mathrm{Res}(v') + s, (\mu \otimes \mu_\sigma)_-).$$

We can check the condition $n''_\tau \equiv \mathrm{Res}(v'') \bmod 2$ using the fact that $\mathrm{Res}(w')$ is odd and $\psi \varphi_{\mathbb{Q}} \omega = \alpha$. This shows the holomorphy of $L(s, (\mu \otimes \mu_\sigma)_-)$. On the other hand, by a result of Shahidi [**S88**, Theorem 5.1] and [**S81**, p. 564] (see also [**Fl**, Section 5] and [**FlZ**]), $L(s, (\mu \otimes \mu_\sigma)_\pm)$ does not vanish at $m + 2$. Then the result follows from the holomorphy of $L(s, \mu \otimes \mu_\sigma)$. □

Let A be a Dedekind domain in $\mathbb{C}$ containing the values $\mu(T(\mathfrak{n}))$ for ideals $\mathfrak{n} \subset \mathcal{R}$. Then it is easy to see that A contains $\chi(\mathfrak{n})$ for ideals $\mathfrak{n} \subset \mathcal{R}$. We further suppose that $H^2_{\mathrm{cusp}}(Y_{0,K}(N), \mathcal{L}(\widehat{\kappa}, \chi; A))[\mu] = A\xi(\mu)$. This condition holds for discrete valuation rings A and the integer ring A of sufficiently large finite extension of $\mathbb{Q}(\mu)$. We define $\Omega_2(\mu^\rho; A)$ by

$$\delta_2(f^\rho) = \Omega_2(\mu^\rho; A) \xi(\mu)^\rho,$$

where $\xi(\mu)^\rho$ is the Galois conjugate of $\xi(\mu)$ under the cohomological rational structure, and similarly, f^ρ is the Galois conjugate of f under the rational structure with respect to Fourier expansion. This is the definition of modular (second) periods in imaginary quadratic case. We put $\underline{\Omega}_2(\mu;\mathbb{Q}(\mu)) = (\Omega_2(\mu^\rho;\mathbb{Q}(\mu^\rho)))_\rho$ as an element of $\mathbb{Q}(\mu)\otimes_{\mathbb{Q}}\mathbb{C} = \prod_\rho \mathbb{C}$, where ρ runs over all embeddings of $\mathbb{Q}(\mu)$ into $\mathbb{C}$. Since $R(\varphi)$ sends $H^2_{\text{cusp}}(Y_{0,K}(N),\mathcal{L}(\widehat{\kappa},\chi;A))[\mu]$ into

$$H^2_{\text{cusp}}(Y_{0,K}(N\cap C^2),\mathcal{L}(\widehat{\kappa}+(0,w),\chi\varphi^2;A))[\mu\otimes\varphi]$$

as seen in the proof of Theorem 8.1 of [**Hi94**],

$$\mathbf{G}(\varphi)\underline{\Omega}_2(\mu\otimes\varphi;\mathbb{Q}(\mu\otimes\varphi)) = \underline{\Omega}_2(\mu;\mathbb{Q}(\mu))$$

in $\mathbb{Q}(\mu\otimes\varphi)\otimes_{\mathbb{Q}}\mathbb{C}$ up to factors in $\mathbb{Q}(\mu\otimes\varphi)^\times$, where $\mathbf{G}(\varphi) = (G(\varphi^\rho))_\rho \in \mathbb{Q}(\varphi)\otimes\mathbb{C}$.

For each $\rho\in\mathrm{Aut}(\mathbb{C}/\mathbb{Q})$, we take $m(\rho)\in\widehat{\mathbb{Z}}^\times$ such that the Artin symbol $[m(\rho),\mathbb{Q}]$ coincides with ρ on the maximal abelian extension of $\mathbb{Q}$. We now look into the ratio: $G(\varphi)/G(\alpha\psi^{-1})$. We conclude from $\alpha\psi^{-1} = \varphi_{\mathbb{Q}}\omega$ with ω of conductor 1 that

$$\left(\frac{G(\varphi)}{G(\alpha\psi^{-1})}\right)^\rho = \frac{\varphi(m(\rho))G(\varphi^\rho)}{\alpha\psi(m(\rho))G((\alpha\psi^{-1})^\rho)} = \frac{G(\varphi^\rho)}{G(\alpha(\psi^\rho)^{-1})}.$$

Hence $G(\varphi)/G(\alpha\psi^{-1})\in\mathbb{Q}(\varphi)$. We then put

$$L(\lambda^\rho) = \frac{(\chi\varphi)^{(\infty)}(\sqrt{-D})D\Gamma(1,\mathrm{Ad}(\lambda)\otimes\alpha)L_{C'}(1,\mathrm{Ad}(\lambda^\rho)\otimes\alpha)}{G(\alpha\psi^\rho)\Omega_2(\widehat{\lambda}^\rho;A)}.$$

Now assume that A is a valuation ring in $\mathbb{Q}(\lambda)$. In the process of proving Proposition 3.1, the only point where we might get a denominator in the L-value is through the maps: $R(\varphi)$ and $(n_\tau!)^{-2}\nabla^{n_\tau}$. Thus if $\varphi = \mathrm{id}$ and the residual characteristic of the valuation ring A is prime to $n_\tau!$, we get an A-integral value. Then from Proposition 3.1, we conclude

COROLLARY 3.2. *If $A=\mathbb{Q}(\lambda)$, we have, for all $\rho\in\mathrm{Aut}(\mathbb{C})$,*

$$L(\lambda)^\rho = L(\lambda^\rho).$$

This shows that $\{L(\lambda^\rho)\}_\rho\in\mathbb{Q}(\lambda)\otimes 1$ in $\mathbb{Q}(\lambda)\otimes_{\mathbb{Q}}\mathbb{C}$. Moreover, if A is a discrete valuation ring in $\mathbb{Q}(\lambda)$ with residual characteristic prime to $6(n!)$ and if $\varphi = \mathrm{id}$, then for the positive generator N_0 of $N\cap\mathbb{Z}$,

$$N_0{}^2\phi(N_0)^{-1}\frac{(\chi)^{(\infty)}(\sqrt{-D})D\Gamma(1,\mathrm{Ad}(\lambda)\otimes\alpha)L(1,\mathrm{Ad}(\lambda)\otimes\alpha)}{\Omega_2(\widehat{\lambda};A)}\in A.$$

4. Real quadratic case

We assume that $K=\mathbb{Q}(\sqrt{D})$ is a real quadratic field with discriminant $D>0$. Let $I_K=\{\tau,\sigma\tau\}$ be the set of real embeddings of K for the generator σ of $\mathrm{Gal}(K/\mathbb{Q})$. Let $\varphi: K^\times\backslash K^\times_{\mathbb{A}}\to\mathbb{C}^\times$ be a Hecke character of infinity type $-\widehat{w} = -w\tau - w\sigma\tau$ with conductor C. Put $\omega = |\ |_{\mathbb{A}}^{n+2v+2w}$ identifying $\mathbb{Z}[I]\cong\mathbb{Z}$ by $n\tau\leftrightarrow n$. We keep the notation for λ introduced in Section 3. Thus $\chi=\psi\circ N_{K/\mathbb{Q}}$ and $\psi_\infty(x) = x^{-n-2v}$ for $x\in\mathbb{A}^\times_\infty$. Let J be a subset of I_K with $|J|=1$. We consider a normalized primitive form $f\in S_{k,J}(N,\chi)$ with $f|T(\mathfrak{n}) = \mu(T(\mathfrak{n}))f$ for all integral ideals $\mathfrak{n}$, where $\kappa = (\widehat{n}, \widehat{v})$ with $\widehat{n} = n\sigma\tau + n\tau$ and $\widehat{v} = v\sigma\tau + v\tau$. We may assume that $J = \{\tau\}$. Then, for $y^{-1/2}\left(\begin{smallmatrix} y & x \\ 0 & 1 \end{smallmatrix}\right) \in \mathrm{SL}_2(F_\infty)$, we see

$$f^{(1)}\left(y^{-1/2}\begin{pmatrix} y & x \\ 0 & 1 \end{pmatrix}\right) = y^{\widehat{n}/2}|y|_{K_{\mathbb{A}}}\left\{\sum_{\xi\in K^\times, \xi^{\sigma\tau}<0, \xi^\tau>0} \mu(T(\xi\mathfrak{d}))|\xi|^{-\widehat{v}}\exp(-2\pi(|\xi^{\sigma\tau}y_{\sigma\tau}| + |\xi^\tau y_\tau|))\mathbf{e}(\xi x)\right\},$$

$$\delta\left(f^{(1)}\right) = \delta_J\left(f^{(1)}\right) = (y^{\widehat{n}/2}|y|_{K_{\mathbb{A}}})^{-1}f^{(1)}(-1)^n(X_\tau - z_\tau Y_\tau)^n(X_{\sigma\tau} - \overline{z}_{\sigma\tau}Y_{\sigma\tau})^n dz_\tau \wedge d\overline{z}_{\sigma\tau}.$$

This shows, as in the same manner in the previous section,

$$(n!)^{-2}(\nabla^n\delta(f^{(1)})|_{\mathbb{Q}}) = (\overline{z} - z)^n\left\{\sum_{\xi^{\sigma\tau}<0, \xi^\tau>0} \mu(T(\xi\mathfrak{d}))|\xi|^{-\widehat{v}}\exp(-2\pi(|\xi^\tau| + |\xi^{\sigma\tau}|)y)\mathbf{e}_K(\mathrm{Tr}(\xi)x)\right\}dz \wedge d\overline{z},$$

where the summation is taken over $\xi \in K^\times$ with $\xi^{\sigma\tau} < 0$ and $\xi^\tau > 0$. Replacing f by $g = f|R(\varphi)$, we get

$$(n!)^{-2}(\nabla^n\delta(g^{(1)})|_{\mathbb{Q}}) = G(\varphi)(-2iy)^n\left\{\sum_\xi \mu(T(\xi\mathfrak{d}))\frac{\varphi(\xi\mathfrak{d})}{|\xi|^{\widehat{v}+\widehat{w}}}\exp(-2\pi(|\xi^\tau| + |\xi^{\sigma\tau}|)y)\mathbf{e}_K(\mathrm{Tr}(\xi)x)\right\}dz \wedge d\overline{z}$$

Then, noting $dz \wedge d\overline{z} = 2idy \wedge dx$

$$\int_{\Phi_\infty\backslash\mathfrak{H}} (n!)^{-2}(\nabla^n\delta(g^{(1)})|_{\mathbb{Q}})y^s = -(-2i)^{n+1}G(\varphi)(4\pi D^{-1/2})^{-s-n-1}D^{v+w}\Gamma(s+n+1)L(s, \mu, \varphi_{\mathbb{Q}}\omega).$$

By Rankin convolution method, writing $E(s) = E(s, (\psi\varphi)^2_{N'})$ for $N' = N \cap C^2 \cap \mathbb{Z}$ as in the previous section, we get

$$\int_{\Phi\backslash\mathfrak{H}} (n!)^{-2}(\nabla^n\delta(g^{(1)})|_{\mathbb{Q}})E(s) = -\frac{(-2i)^{n+1}G(\varphi)D^{v+w}\Gamma(s+n+1)}{(4\pi D^{-1/2})^{s+n+1}}L_{C'}(s+1, ((\mu\otimes\varphi)\otimes(\mu\otimes\varphi)_\sigma)_+\otimes\omega),$$

where $C' = C \cap \mathbb{Z}$. In particular, if $\mu = \widehat{\lambda}$,

$$\int_{\Phi\backslash\mathfrak{H}} (n!)^{-2}(\nabla^n\delta(g^{(1)})|_{\mathbb{Q}})E(s) = -\frac{(-2i)^{n+1}G(\varphi)D^{v+w}\Gamma(s+n+1)}{(4\pi D^{-1/2})^{s+n+1}}L_{C'}(s, \mathrm{Ad}(\lambda)\otimes\psi\varphi_{\mathbb{Q}}\omega)L(s, \psi\alpha\varphi_{\mathbb{Q}}\omega).$$

Thus assuming that $\alpha = \psi\varphi_{\mathbb{Q}}\omega$ for $\alpha = (\frac{D}{})$ and comparing the residues at $s = 1$, we have

$$\int_{\Phi\backslash\mathfrak{H}} (n!)^{-2}(\nabla^n\delta((f|R(\varphi))^{(1)})|_{\mathbb{Q}})$$
$$= -(-i)^{n+1}\phi(N')^{-1}N'^2\sqrt{D}^{n+2v+2w+2}G(\varphi)\Gamma(1,\mathrm{Ad}(\lambda)\otimes\alpha)L_{C'}(1,\mathrm{Ad}(\lambda)\otimes\alpha).$$

In the real quadratic case, $H^2_{\mathrm{cusp}}(Y_{0,K}(N),\mathcal{L}(\widehat{\kappa},\chi;\mathbb{C}))[\widehat{\lambda}]$ is two dimensional. Thus we need to further decompose this space into 1 dimensional pieces. As explained in [**Hi88a**, Section 7] (see Proof of Theorem 7.2), the finite group $S = C_\infty/C_{\infty+}$ acts on $\mathfrak{Z} = G(\mathbb{R})_+/C_{\infty+}$, $S_{\widehat{\kappa},J}(N,\chi)$ and $H^2_{\mathrm{cusp}}(Y_{0,K}(N),\mathcal{L}(\kappa,\chi;A))$, where C_∞ (resp. $C_{\infty+}$) is the standard maximal compact subgroup of $G(\mathbb{R})$ (resp. $G(\mathbb{R})_+$). We can identify $S \cong \{\pm1\}^{I_K}$ by taking the determinant and S with the power set of I_K by $c \mapsto \{v|c_v = +1\}$. Thus we can think of $cJ \subset I_K$ for each $J \subset I_K$. Identifying F_∞ with $\mathbb{R}\times\mathbb{R}$ via $\xi \mapsto (\xi^\tau,\xi^{\sigma\tau})$, we identify $\mathfrak{Z}$ with $\mathcal{H}_\tau \times \mathcal{H}_{\sigma\tau}$ for copies of upper half planes $\mathcal{H}_{\sigma\tau}$ and $\mathcal{H}_\tau$. Then for $f \in S_{\widehat{\kappa},J}(N,\chi)$, $cf(x) = f(xc)$ for $c = (c_\tau, c_{\sigma\tau})$ with $c_v \in \{\left(\begin{smallmatrix}\pm1 & 0\\ 0 & 1\end{smallmatrix}\right)\}$. Then $cf \in S_{\widehat{\kappa},cJ}(N,\chi)$ with $\mathbf{a}(\mathfrak{n},cf) = \mathbf{a}(\mathfrak{n},f)$ for all $\mathfrak{n}$. Thus c takes $S_{\widehat{\kappa},J}(N,\chi)$ into $S_{\widehat{\kappa},cJ}(N,\chi)$, and this action of S commutes with Hecke operators $T(\mathfrak{n})$. When there is a unit $\varepsilon \in \mathfrak{r}^\times$ with $\varepsilon^\rho\det(c_\rho) > 0$ for both $\rho = \tau$ and $\sigma\tau$, $cf^{(a)}(z) = f^{(a)}(\varepsilon z)$, where $\varepsilon z_\rho = \varepsilon^\rho z^c_\rho$ with $z^c_\rho = \begin{cases} z_\rho & \text{if } \det(c_\rho) = 1,\\ \overline{z}_\rho & \text{if } \det(c_\rho) = -1.\end{cases}$ If there is not such a unit, we find $\varepsilon \in F^\times$ and another member a' of the complete set $\{a\}$ of representatives for Cl_F such that (i) $\varepsilon^\rho\det(c_\rho) > 0$ for both $\rho = \tau$ and $\sigma\tau$, (ii) $a\mathfrak{r} = \varepsilon a'\mathfrak{r}$, and (iii) $cf^{(a')}(z) = f^{(a)}(\varepsilon z)$.

We take a character $\varepsilon : S \to \{\pm1\}$ and consider the projection

$$\pi_\varepsilon : H^2_{\mathrm{cusp}}(Y_0(N),\mathcal{L}(\kappa,\chi;A)) \to H^2_{\mathrm{cusp}}\left(Y_0(N),\mathcal{L}\left(\kappa,\chi;A\left[\frac{1}{2}\right]\right)\right)[\varepsilon]$$

given by $\pi_\varepsilon(x) = \#(S)^{-1}\sum_{c\in S}\varepsilon(c)c(x)$. Through the Eichler-Shimura isomorphism introduced in Section 2.2

$$\delta : \bigoplus_{J\subset I_K} S_{\kappa,J}(N;\chi)_{/K} \cong H^2_{\mathrm{cusp}}(Y_0(N),\mathcal{L}(\kappa,\chi;\mathbb{C}))[\chi],$$

we can attach Fourier expansion to cohomology classes in the right-hand-side of the above formula, which we write as H^2_{cusp}. Thus $\phi \in H^2_{\mathrm{cusp}}$ has Fourier coefficients $\mathbf{a}(\mathfrak{n};\phi)$ for ideals $\mathfrak{n}$ given by $\mathbf{a}(\mathfrak{n};\delta(f)) = \mathbf{a}(\mathfrak{n};f)$. The action of S preserves the Fourier coefficients: $\mathbf{a}(\mathfrak{n};c(\phi)) = \mathbf{a}(\mathfrak{n};\phi)$. Note that

$$c\delta\left((f)^{(a)}\right)$$
$$= \{c\}(y^{\widehat{n}/2}|y|_{K_{\mathbb{A}}})^{-1}(cf)^{(a')}(-1)^n\{(X_\tau - z^c_\tau Y_\tau)^n(X_{\sigma\tau} - \overline{z}^c_{\sigma\tau}Y_{\sigma\tau})^n\}dz^c_\tau \wedge d\overline{z}^c_{\sigma\tau},$$

where identifying $S = \{\pm1\}^{I_K}$, we write $\{c\} = \prod_{v\in I_K} c_v$ and

$$z^c_v = \begin{cases} \overline{z}_v & \text{if } c_v = -1,\\ z_v & \text{if } c_v = +1.\end{cases}$$

Note that, identifying S with the power set of I_K, if either $c = J$ or $-J$, the restriction $\delta((cf)^{(a)})$ to $\mathfrak{H}$ just vanishes, where $J_\tau = +1$ and $J_{\sigma\tau} = -1$ because

$J = \{\tau\}$. Let $c = (-1_\tau, -1_{\sigma\tau})$. Then we have

$$c\delta\left((f)^{(1)}\right) = (y^{\widehat{n}/2}|y|_{K_\mathbb{A}})^{-1} f^{(1)}(-\overline{z})(-1)^n \left\{(X_\tau - z_\tau^c Y_\tau)^n (X_{\sigma\tau} - \overline{z}_{\sigma\tau}^c Y_{\sigma\tau})^n\right\} d\overline{z}_\tau \wedge dz_{\sigma\tau}.$$

We execute the computation done for $\delta(g^{(1)})$ replacing it by $c\delta(g^{(1)})$ and obtain

$$\int_{Y_{0,\mathbb{Q}}(N')} (n!)^{-2}\left(\nabla^n c\delta(g^{(1)})|_\mathbb{Q}\right) y^s = (-1)^{n+1}\int_{Y_{0,\mathbb{Q}}(N')} (n!)^{-2}\left(\nabla^n \delta(g^{(1)})|_\mathbb{Q}\right) y^s.$$

Therefore, assuming $\varepsilon(-1_\tau, -1_{\sigma\tau}) = (-1)^{n+1}$, $\mu = \widehat{\lambda}$ and $\alpha = \psi\varphi_\mathbb{Q}\omega$, we have

$$\int_{Y_{0,\mathbb{Q}}(N')} (n!)^{-2}\left(\nabla^n \pi_\varepsilon \delta((f|R(\varphi))^{(1)})|_\mathbb{Q}\right) = -(-i)^{n+1}\phi(N')^{-1}N'^2\sqrt{D}^{n+2v+2w+2} \cdot G(\varphi)\Gamma(1, \mathrm{Ad}(\lambda)\otimes\alpha)L_{C'}(1, \mathrm{Ad}(\lambda)\otimes\alpha).$$

As seen in the proof of Theorem 8.1 in [**Hi94**], the twisting operator $R(\varphi)$ takes $H^2_{\mathrm{cusp}}(Y_{0,K}(N), \mathcal{L}(\widehat{\kappa}, \chi; A))[\varepsilon\varphi_\infty]$ into $H^2_{\mathrm{cusp}}(Y_{0,K}(N\cap C^2), \mathcal{L}(\widehat{\kappa}+(0,\widehat{w}), \chi\varphi^2; A))[\varepsilon]$, where we identify S with $\{\pm1\}^{I_K} \subset K_\infty^\times$ and consider φ_∞ as a character of S. Since $\alpha = \psi\varphi_\mathbb{Q}\omega$, $\varphi_\infty((-1,-1)) = (-1)^n$. This shows that $\varepsilon_0 = \varepsilon\varphi_\infty$ satisfies $\varepsilon_0(-1,-1) = -1$. We record what we have proven:

PROPOSITION 4.1. *Let $F = \mathbb{Q}$ and $K = \mathbb{Q}(\sqrt{D})$ be a real quadratic field with discriminant $D > 0$ and $\kappa = (n\tau, v\tau) \in \mathbb{Z}[I]^2$ with $I = \{\tau\}$. Let φ be an arithmetic Hecke character with $\infty(\varphi) = -w(\tau+\sigma\tau) \in \mathbb{Z}[I_K]$ and of conductor C. Let $\chi = \psi\circ N_{K/\mathbb{Q}}$ for an arithmetic Hecke character ψ with $\psi_\infty(x) = x^{-n-2v}$ for all $x \in \mathbb{A}_\infty^\times = \mathbb{R}^\times$. Suppose that $\alpha = \varphi_\mathbb{Q}\omega\psi$ with $\omega(x) = |x|_\mathbb{A}^{n_\tau+2v_\tau+2w}$ and $\alpha = (\frac{D}{})$. If a primitive system $\mu : h_{\widehat{\kappa}}(N, \chi; \mathbb{C})_{/K} \to \mathbb{C}$ is a base change lift of $\lambda : h_\kappa(N_0, \psi; \mathbb{C})_{/\mathbb{Q}} \to \mathbb{C}$, then we have*

$$\int_{Y_{0,\mathbb{Q}}(N')} (n!)^{-2}(\nabla^n \pi_\varepsilon \delta(g^{(1)})|_\mathbb{Q}) = -(-i)^{n+1}\phi(N')^{-1}N'^2\sqrt{D}^{n+2v+2w+2} G(\varphi)\Gamma(1, \mathrm{Ad}(\lambda)\otimes\alpha)L_{C'}(1, \mathrm{Ad}(\lambda)\otimes\alpha),$$

where $g = f|R(\varphi)$ for the primitive form f with $f|T(\mathfrak{n}) = \mu(T(\mathfrak{n}))f$ for all $\mathfrak{n} \subset \mathcal{R}$, $C' = C\cap\mathbb{Z}$, and N' is the positive generator of $N\cap C^2\cap\mathbb{Z}$. If $\mu : h_{\widehat{\kappa}}(N, \chi; \mathbb{C})_{/K} \to \mathbb{C}$ is not a twist of any base change lift, then

$$\int_{Y_{0,\mathbb{Q}}(N')} (n!)^{-2}(\nabla^n \pi_\varepsilon \delta(g^{(1)})|_\mathbb{Q}) = 0.$$

PROOF. The proof of the vanishing is given in [**HaLR**, 3.12] and is basically the same as that of Proposition 3.1. □

We now define the modular periods. Let A be a Dedekind domain in $\mathbb{C}$ containing all values of $\mu(T(\mathfrak{n}))$ for ideals $\mathfrak{n} \subset \mathcal{R}$. We assume that

$$H^2_{\mathrm{cusp}}(Y_{0,K}(N), \mathcal{L}(\widehat{\kappa}, \chi; A))[\varepsilon, \mu] = A\xi_\varepsilon(\mu)$$

for a character ε of S. Then we define $\Omega_1(\varepsilon, \mu^\rho; A) \in \mathbb{C}^\times$ by

$$\pi_\varepsilon(\delta(f^\rho)) = \Omega_1(\varepsilon, \mu^\rho; A)\xi(\mu)^\rho,$$

where f^ρ is the primitive form in $S_{\widehat{\kappa},J}(N,\chi^\rho)$ with $f^\rho|T(\mathfrak{n}) = \mu(T(\mathfrak{n}))^\rho f^\rho$. This period Ω_1 is well defined up to units in A. We also define

$$\underline{\Omega}_1(\mu;\mathbb{Q}(\mu)) = (\Omega_1(\mu^\rho;\mathbb{Q}(\mu^\rho)))_\rho \in \mathbb{Q}(\mu)\otimes_{\mathbb{Q}}\mathbb{C}.$$

The period $\underline{\Omega}_1(\mu;\mathbb{Q}(\mu))$ is well defined as an element of $(\mathbb{Q}(\mu)\otimes_{\mathbb{Q}}\mathbb{C})^\times/\mathbb{Q}(\mu)^\times\otimes 1$. Again we have

$$\underline{\Omega}_1(\mu\otimes\varphi;\mathbb{Q}(\mu\otimes\varphi)) = \mathbf{G}(\varphi)^{-1}\underline{\Omega}_1(\mu;\mathbb{Q}(\mu)).$$

COROLLARY 4.2. *If $A=\mathbb{Q}(\lambda)$, we have, for all $\varepsilon : S\to\{\pm1\}$ with $\varepsilon(-1,-1)=-1$,*

$$\frac{(1\otimes i^{n+1}\sqrt{D}^n\Gamma(1,\mathrm{Ad}(\lambda)\otimes\alpha))\mathbb{L}_{C'}(1,\mathrm{Ad}(\lambda)\otimes\alpha)}{\mathbf{G}(\alpha\psi)\underline{\Omega}_1(\varepsilon,\widehat{\lambda};\mathbb{Q}(\lambda))}\in\mathbb{Q}(\lambda)$$

in $\mathbb{Q}(\lambda)\otimes_{\mathbb{Q}}\mathbb{C}$, where

$$\mathbb{L}_{C'}(s,\mathrm{Ad}(\lambda)\otimes\alpha) = (L_{C'}(s,\mathrm{Ad}(\lambda^\rho)\otimes\alpha))_\rho$$

has values in $\mathbb{Q}(\lambda)\otimes_{\mathbb{Q}}\mathbb{C}$. Moreover, if A is a discrete valuation ring in $\mathbb{Q}(\lambda)$ with residual characteristic prime to $6(n!)$ and if $\varphi=\mathrm{id}$, then for the positive generator N_0 of $N\cap\mathbb{Z}$

$$i^{n+1}{N_0}^2\phi(N_0)^{-1}\frac{\sqrt{D}^{n+2v+2}\Gamma(1,\mathrm{Ad}(\lambda)\otimes\alpha)L(1,\mathrm{Ad}(\lambda)\otimes\alpha)}{\Omega_1(\varepsilon,\widehat{\lambda};A)}\in A.$$

5. Congruence and the adjoint L-values

Here we study a simple consequence of Corollaries 3.2 and 4.2 on congruence of systems of Hecke eigenvalues. To describe such congruence among cusp forms in terms of Hecke algebras and deformation rings of Galois representations, we here introduce a general notion of congruence modules and differential modules: Let R be an algebra over a Dedekind domain A. We assume that R is an A-flat module of finite type. Let $\phi : R\to A$ be an A-algebra homomorphism. We define $C_1(\phi;A)=\Omega_{R/A}\otimes_{R,\phi}\mathrm{Im}(\phi)$, which we call the differential module of ϕ. We suppose that R is reduced (that is, the nilradical of R vanishes). Then the total quotient ring $\mathrm{Frac}(R)$ can be decomposed uniquely into $\mathrm{Frac}(R)=\mathrm{Frac}(\mathrm{Im}(\phi))\times X$ as an algebra direct product. Let $\mathfrak{a}=\mathrm{Ker}(R\to X)$. Then we put $C_0(\phi;A)=(R/\mathfrak{a})\otimes_{R,\phi}\mathrm{Im}(\phi)\cong\mathrm{Im}(\phi)/(\mathrm{Im}(\phi)\cap R)$ (cf. [**Hi88b**, Section 6]), which is called the congruence module of ϕ but is actually a ring. Here the intersection $\mathrm{Im}(\phi)\cap R$ is taken in $\mathrm{Frac}(R)$. Suppose now that A is a subring of a number field in $\overline{\mathbb{Q}}$. Since $\mathrm{Spec}(C_0(\phi;A))$ is the scheme theoretic intersection of $\mathrm{Spec}(\mathrm{Im}(\phi))$ and $\mathrm{Spec}(R/\mathfrak{a})$ in $\mathrm{Spec}(R)$, a prime $\mathfrak{p}$ is in the support of $C_0(\phi;A)$ if and only if there exists an A-algebra homomorphism $\phi' : R\to\overline{\mathbb{Q}}$ factoring through $R/\mathfrak{a}$ such that $\phi(a)\equiv\phi'(a) \bmod \mathfrak{p}$ for all $a\in R$. In other words, $\phi \bmod \mathfrak{p}$ factors through $R/\mathfrak{a}$ and can be lifted to ϕ'.

Let $K/\mathbb{Q}$ be a quadratic extension with discriminant D. Let $\alpha=\left(\frac{D}{}\right)$. We consider a system of Hecke eigenvalues $\lambda : h_\kappa(D,\psi;A)\to A$ for a discrete valuation ring A in $\mathbb{Q}(\lambda)$ with residue field $\mathbb{F}$ of characteristic $p>2$, where $\psi(x)=\alpha(x)|x|_{\mathbb{A}}^{-n}$. From our assumption that $\psi_\infty(x)=x^{-n-2v}$ for all $x\in\mathbb{R}^\times$, we conclude that $v=0$ and

$$n \text{ is } \begin{cases}\text{odd} & \text{if } K \text{ is imaginary,}\\ \text{even} & \text{if } K \text{ is real.}\end{cases}$$

The space we are looking into has the Neben character α and level D, under the classical notation, $S_{n+2}(\Gamma_0(D),\left(\frac{D}{}\right))$. Let $\mathcal{O}$ be the completion under the $\mathfrak{m}_A$-adic topology for the maximal ideal $\mathfrak{m}_A$ of A. Let $\mathfrak{m}$ be the maximal ideal of $\mathcal{O}$. We consider the Galois representation $\rho_\lambda : \mathrm{Gal}(\overline{\mathbb{Q}}/\mathbb{Q}) \to \mathrm{GL}_2(\mathcal{O})$ which is the $\mathfrak{m}_A$-adic member of the compatible system $\rho(\lambda)$ associated to λ. We define the residual representation $\overline{\rho} : \mathrm{Gal}(\overline{\mathbb{Q}}/\mathbb{Q}) \to \mathrm{GL}_2(\mathbb{F})$ by ρ_λ mod $\mathfrak{m}$. Let $\mathcal{G} = \mathrm{Gal}(K^{(p)}/\mathbb{Q})$ for the maximal extension $K^{(p)}/K$ unramified outside $\{p, \infty\}$. Then ρ_λ factors through $\mathcal{G}$. Let E be a number field in $K^{(p)}$ with integer ring $\mathfrak{r}$, and put $\mathcal{H} = \mathrm{Gal}(K^{(p)}/E)$. For any representation ρ of $\mathcal{G}$, we write ρ_E for the restriction of ρ to $\mathcal{H}$. We consider the following condition:

(AI_E) $\quad\overline{\rho}_E$ *is absolutely irreducible.*

We now consider deformations of $\overline{\rho}$, introduced by Mazur [**Ma**], over the category $CNL_\mathcal{O}$ of complete local noetherian $\mathcal{O}$-algebras with residue field $\mathbb{F}$. A Galois representation $\rho : \mathcal{H} \to \mathrm{GL}_2(B)$ for $B \in CNL_\mathcal{O}$ is a deformation of $\overline{\rho}_E$ if ρ_E mod $\mathfrak{m}_B$ coincides with $\overline{\rho}$ as matrix representations. We look into the deformation functor $\mathcal{F} : CNL_\mathcal{O} \to SETS$ given by

$$\mathcal{F}_E(B) = \{\rho : \mathcal{H} \to \mathrm{GL}_2(B) | \rho \bmod \mathfrak{m}_B = \overline{\rho}\}/\approx,$$

where "$\approx$" is the conjugation by elements in $1 + M_2(\mathfrak{m}_B)$. We impose more conditions on deformations. A deformation ρ is called p-ordinary over a number field E if $\rho|_\mathcal{D} \cong \left(\begin{smallmatrix} \delta_\mathcal{D} & * \\ 0 & \varepsilon_\mathcal{D} \end{smallmatrix}\right)$ with an unramified character $\delta_\mathcal{D}$ on every decomposition subgroup $\mathcal{D}$ of $\mathcal{H}$ at $\mathfrak{p}|p$. We like to impose some of the following two conditions depending on the situation:

(ord_E) ρ *is p-ordinary over E with two distinct characters $\delta_\mathcal{D}$ and $\varepsilon_\mathcal{D}$ for all $\mathcal{D}$;*
(fl_E) $\rho|_\mathcal{D}$ *is realized on the generic fibre of a finite flat group scheme over $\mathfrak{r}_\mathfrak{p}$ for all $\mathfrak{p}|p$.*

For primes $\mathfrak{l}|D$ outside p, we impose the following condition:

$$(\alpha_E)\rho \cong \begin{pmatrix} 1 & 0 \\ 0 & \alpha \end{pmatrix} \text{ on the inertia subgroup } I_\mathfrak{l} \subset \mathcal{H} \text{ for each prime } \mathfrak{l}|D \text{ and } \mathfrak{l} \nmid p.$$

Let ν be the p-adic cyclotomic character. By class field theory, we regard ψ as a character of $\mathcal{G}$. Then $\det(\rho_\lambda) = \nu\psi$. Hereafter we suppose that $p \nmid D$. We consider the following subfunctor of $\mathcal{F}_E$:

$$\mathcal{F}'_E(B) = \begin{cases} \{\rho \in \mathcal{F}_E(B) | \ \rho \text{ satisfies } (\mathrm{ord}_E),\ (\alpha_E) \text{ and } \det(\rho) = (\nu\psi)_E\} & \text{if } n > 0, \\ \{\rho \in \mathcal{F}_E(B) | \ \rho \text{ satisfies } (\mathrm{ord}_E),\ (fl_E),\ (\alpha_E) \text{ and } \det(\rho) = (\nu\psi)_E\} & \text{if } n = 0. \end{cases}$$

Under (AI_E) and (ord_E) for $\overline{\rho}$, this functor is representable by a universal couple (R_E, ϱ_E) (see [**Ma**] and [**Hi96**, Appendix]).

We put $h_\kappa(D, \psi; \mathcal{O}) = h_\kappa(D, \psi; A) \otimes_A \mathcal{O}$ and define $h_\kappa(Dp, \psi; \mathcal{O})$ similarly. Let

$$e_0 = \lim_{n\to\infty} T(p)^{n!} \in h_\kappa(D, \psi; \mathcal{O}) \quad \text{and} \quad e = \lim_{n\to\infty} T(p)^{n!} \in h_\kappa(Dp, \psi; \mathcal{O}).$$

Since $T(p)$ of level D and that of level Dp are different, we know $e_0 \neq e$ and we have a surjective $\mathcal{O}$-algebra homomorphism of $eh_\kappa(Dp, \psi; \mathcal{O})$ onto $e_0 h_\kappa(D, \psi; \mathcal{O})$ taking $T(n)$ to $T(n)$ for all n prime to p [**MaT**]. If $n > 0$, $e_0 h_\kappa(D, \psi; \mathcal{O}) \cong eh_\kappa(Dp, \psi; \mathcal{O})$, which is a consequence of [**Hi86**, Proposition 4.7]. We assume that λ factors through $e_0 h_\kappa(D, \psi; \mathcal{O})$, which is equivalent to $\lambda(T(p)) \in A^\times$ and implies that ρ_λ is p-ordinary. Let $h = h_\mathbb{Q}$ (resp. $h'_\mathbb{Q}$) be the unique local ring of $e_0 h_\kappa(D, \psi; \mathcal{O})$ (resp. $eh_\kappa(Dp, \psi; \mathcal{O})$) through which λ factors. Thus if $n > 0$, $h_\mathbb{Q} \cong h'_\mathbb{Q}$. As is well known, under $(AI_{\mathbb{Q}})$, there is a unique Galois representation $\rho_{\mathbb{Q}}$, up to isomorphisms, such that $\mathrm{Tr}(\rho_{\mathbb{Q}}(Frob_\ell)) = T_h(\ell)$ for all primes ℓ outside Dp (for uniqueness, see [**C**]), where $Frob_\ell$ is the Frobenius element in $\mathcal{G}$ at the prime ℓ, and $T_h(\ell)$ is the projection of $T(\ell)$ to $h_{\mathbb{Q}}$. Let $k(E) = E(\sqrt{(-1)^{(p-1)/2}p})$. It has been proven by Taylor and Wiles [**W**] and [**TW**] (see also [**Fu**] when $n = 0$) that, under $(AI_{k(\mathbb{Q})})$ and $(\mathrm{ord}_{\mathbb{Q}})$ for $\overline{\rho}$, $\mathcal{F}'_{\mathbb{Q}}$ is representable by the pair $(h_{\mathbb{Q}}, \rho_{\mathbb{Q}})$. Thus in this case, the natural morphism $\pi_{\mathbb{Q}} : R_{\mathbb{Q}} \to h_{\mathbb{Q}}$ with $\pi_{\mathbb{Q}}\varrho_{\mathbb{Q}} \approx \rho_{\mathbb{Q}}$ is a surjective isomorphism in $CNL_{\mathcal{O}}$. When E is totally real, this result is generalized by Fujiwara [**Fu**] under certain assumptions. We describe it here for real quadratic K. Since $K^{(p)}$ is unramified outside $\{p, \infty\}$, $\mathcal{F}'_K$ is (basically) defined by ramification condition (ord_K) and the determinant condition. We define in the same way the idempotent $e = \lim_{n\to\infty} T(p)^{n!} \in h_{\widehat{\kappa}}(p, \mathrm{id}; \mathcal{O})$ and $e_0 \in h_{\widehat{\kappa}}(1, \mathrm{id}; \mathcal{O})$. Let h_K (resp. h'_K) be the local ring of $e_0 h_{\widehat{\kappa}}(1, \mathrm{id}; \mathcal{O})$ (resp. $eh_{\widehat{\kappa}}(p, \mathrm{id}; \mathcal{O})$) through which the base change lift $\widehat{\lambda}$ factors through. We again have $h_K \cong h'_K$ if $n > 0$. We have a modular deformation $\rho_K : \mathcal{H} \to \mathrm{GL}_2(h_K)$. Then Fujiwara has proven that $\mathcal{F}'_K$ is representable by (h_K, ρ_K) under the following condition in addition to $(AI_{k(K)})$ and (ord_K) for $\overline{\rho}_K$:

(unr) *p is unramified in K.*

We write $\pi_K : R_K \cong h_K$ for the isomorphism inducing $\pi_K \varrho_K \approx \rho_K$.

When it is necessary to indicate the dependence on E, we write $\lambda_{\mathbb{Q}}$ for λ and λ_K for $\widehat{\lambda}$. Since the conductor of the Neben character coincides with the level, h_E and the Hecke algebras are reduced. In the course of the proof of the above result: $R_E \cong h_E$ for $E = K$ or $\mathbb{Q}$, it is shown that R_E is a local complete intersection. This fact is basically equivalent to

(C1) $$|C_1(\lambda_E; \mathcal{O})| = |C_0(\lambda_E; \mathcal{O})| \quad \text{for } E = \mathbb{Q} \text{ and real } K,$$

where $\lambda_E : h_E \to \mathcal{O}$. When $E = \mathbb{Q}$, the above fact implies that, for any given character $\varepsilon : S = \{\pm 1\} \to \{\pm 1\}$,

$(\mathrm{mlt}_{\mathbb{Q}})$ $$H^1_{\mathrm{cusp}}(Y_{0,\mathbb{Q}}(D), \mathcal{L}(\kappa, \psi; \mathcal{O}))[h_{\mathbb{Q}}, \varepsilon] \cong h_{\mathbb{Q}} \cong R_{\mathbb{Q}} \quad \text{as } h_{\mathbb{Q}}\text{-modules},$$

where the left-hand-side is the eigenspace for $h_{\mathbb{Q}}$, in other words, writing 1_h for the idempotent of $h_{\mathbb{Q}}$ in $h_\kappa(D, \psi; \mathcal{O})$,

$$H^1_{\mathrm{cusp}}(Y_{0,\mathbb{Q}}(D), \mathcal{L}(\kappa, \psi; \mathcal{O}))[h_{\mathbb{Q}}, \varepsilon] = 1_h(H^1_{\mathrm{cusp}}(Y_{0,\mathbb{Q}}(D), \mathcal{L}(\kappa, \psi; \mathcal{O}))[\varepsilon]).$$

Since $E = \mathbb{Q}$, there are only two ε; one is trivial, which we write as "+", and we write the other as "−". By virtue of $(\mathrm{mlt}_{\mathbb{Q}})$, we can compute $C_0(\lambda; \mathcal{O})$ using cohomology groups. To explain this, we write $L(B) = H^1_{\mathrm{cusp}}(Y_{0,\mathbb{Q}}(D), \mathcal{L}(\kappa, \psi; B))[\varepsilon]$. Then $L(\mathcal{O}) = L(A) \otimes_A \mathcal{O}$. Decomposing $\mathrm{Frac}(h_\kappa(D, \psi; A)) = \mathrm{Frac}(\mathrm{Im}(\lambda)) \times X$, we define $L^\lambda(A)$ to be the image of $L(A)$ in $L(A) \otimes_{h_\kappa(D,\psi;A)} \mathrm{Frac}(\mathrm{Im}(\lambda))$ and a cohomological congruence module by

$$C_0^H(\lambda; A) = L^\lambda(A)/(L(A) \cap L^\lambda(A)).$$

Then $(\mathrm{mlt}_{\mathbb{Q}})$ shows that $C_0(\lambda_E; \mathcal{O}) \cong C_0^H(\lambda; A)$. It is shown in [**Hi81**] and [**Hi88b**] that the p-adic absolute value of $\Gamma(1, \mathrm{Ad}(\lambda))L(1, \mathrm{Ad}(\lambda))/\Omega_1(+, \lambda; A)\Omega_1(-, \lambda; A)$ is the inverse of the order of the right-hand-side module of the above equation. That is, under $(AI_{\mathbb{Q}})$ and $(\mathrm{ord}_{\mathbb{Q}})$ for $\overline{\rho}$, if $p \nmid 6D$,

(C2) $$\left|\frac{\Gamma(1, \mathrm{Ad}(\lambda))L(1, \mathrm{Ad}(\lambda))}{\Omega_1(+, \lambda; A)\Omega_1(-, \lambda; A)}\right|_p^{-r} = \#(C_0^H(\lambda; A)) = \#(C_0(\lambda; A)) = \#(C_1(\lambda; A)),$$

where $r = r(\mathcal{O}) = \mathrm{rank}_{\mathbb{Z}_p} \mathcal{O}$, and $|\ |_p$ is the p-adic absolute value of A normalized so that $|p|_p = p^{-1}$. It is easy to see that for a non-zero element $\eta(\lambda) \in A$, $C_0(\lambda; A) \cong$

$A/\eta(\lambda)A$, and (C1) is equivalent to saying that the L-value is equal to $\eta(\lambda)$ up to A-units. Even if $p|D$, there is a similar formula (see [**Hi88b**] for details).

The Selmer group over E of $\mathrm{Ad}(\rho_\lambda)$ is defined as follows. Let $V(\lambda)$ be the $\mathcal{O}$-free module of rank 2 on which $\mathcal{G}$ acts via ρ_λ. For each decomposition subgroup $\mathcal{D}_{\mathfrak{p}} \subset \mathcal{H}$ at $\mathfrak{p}|p$ for primes $\mathfrak{p}$ of E, we write $V_{\mathfrak{p}}$ for the $\delta_{\mathcal{D}}$ eigenspace in $V(\lambda)$. For primes $\mathfrak{l}|D$ prime to p, we write $V_{\mathfrak{l}}$ for the subspace fixed by the inertia at $\mathfrak{l}$. Let $\mathcal{K}=\mathrm{Frac}(\mathcal{O})$. We identify $\mathrm{Ad}(\rho_\lambda)$ with trace 0 subspace W of $\mathrm{Hom}_{\mathcal{O}}(V(\lambda),V(\lambda))$. We put $W_{\mathfrak{l}} = \{\phi\in W|\phi(V_{\mathfrak{l}})=0\}$ for $\mathfrak{l}|Dp$. Define, for $\mathfrak{l}$ ramifying in $K^{(p)}/E$, writing $\mathcal{I}_{\mathfrak{l}}$ for the inertia subgroup of $\mathfrak{l}$ in $\mathcal{H}$,

$$L_{\mathfrak{l}} = \mathrm{Ker}(H^1(\mathcal{D}_{\mathfrak{l}},W^*)\to H^1(\mathcal{I}_{\mathfrak{l}},(W/W_{\mathfrak{l}})^*)),$$

where $X^* = X\otimes_{\mathcal{O}}\mathcal{K}/\mathcal{O}$. Then we put

$$\mathrm{Sel}(\mathrm{Ad}(\rho_\lambda))_{/E} = \bigcap_{\mathfrak{l}} \mathrm{Ker}(H^1(\mathcal{H},W^*)\to H^1(\mathcal{D}_{\mathfrak{l}},W^*)/L_{\mathfrak{l}}),$$

where $\mathfrak{l}$ runs over all primes ramifying in $K^{(p)}/E$. It is a general fact [**MaT**] (see also [**Hi96**, 3.2]) that

(C3) $\qquad \mathrm{Sel}(\mathrm{Ad}(\rho_\lambda))_{/E} \cong \mathrm{Hom}_{\mathbb{Z}_p}(C_1(\lambda_E\circ\pi_E : R_E\to\mathcal{O};\mathcal{O}),\mathbb{Q}_p/\mathbb{Z}_p)$ if $n>0$.

Thus combining all we said, we get the following order formula of the Selmer group under $(AI_{k(\mathbb{Q})})$, $(\mathrm{ord}_{\mathbb{Q}})$ for $\overline{\rho}$ and $p\nmid 6D$,

(CN1) $$\frac{\Gamma(1,\mathrm{Ad}(\lambda))L(1,\mathrm{Ad}(\lambda))}{\Omega_1(+,\lambda;A)\Omega_1(-,\lambda;A)} = \eta(\lambda) \quad \text{up to } A\text{-units, and}$$

(CN2) $$\left|\frac{\Gamma(1,\mathrm{Ad}(\lambda))L(1,\mathrm{Ad}(\lambda))}{\Omega_1(+,\lambda;A)\Omega_1(-,\lambda;A)}\right|_p^{-r(\mathcal{O})} = \#(\mathrm{Sel}(\mathrm{Ad}(\rho_\lambda))_{/\mathbb{Q}}) \quad \text{if } n>0.$$

This is a non-abelian generalization of a classical analytic class number formula (see [**W**, Chapter 4] and [**HiTU**]). The definition of the Selmer group can be interpreted by Fontaine's theory, and the above formula can be viewed as an example of the Tamagawa number formula of Bloch and Kato for the motive $M(\mathrm{Ad}(\lambda))$ (see [**W**, p. 466] and [**BK**, Section 5]). By using this interpretation, we can modify the definition of the Selmer group, and the formula (CN2) is valid even for $n=0$ for the modified Selmer group.

There is a partial generalization of the above facts for quadratic fields K. We define $L_K(B) = H^{r_1(K)+r_2(K)}_{\mathrm{cusp}}(Y_{0,K}(1),\mathcal{L}(\widehat{\kappa},id;B))[\varepsilon]$, where we fix for the moment a character $\varepsilon : S\to\{\pm1\}$ when K is real. If K is imaginary, we just forget about ε. Then we define the cohomological congruence module $C_0^H(\widehat{\lambda};A)$ by $L_K^{\widehat{\lambda}}(A)/(L_K^{\widehat{\lambda}}(A)\cap L_K(A))$, where decomposing $\mathrm{Frac}(h_{\widehat{\kappa}}(1,\mathrm{id};A)) = \mathrm{Frac}(\mathrm{Im}(\widehat{\lambda}))\times X$, $L_K^{\widehat{\lambda}}(A)$ is the image of $L_K(A)$ in $L_K(A)\otimes_{h_{\widehat{\kappa}}(1,\mathrm{id};A)}\mathrm{Frac}(\mathrm{Im}(\widehat{\lambda}))$. It has been proven by Urban [**U**], under $(\mathrm{ord}_{\mathbb{Q}})$ for $\overline{\rho}$, when K is imaginary and $p\nmid 6D$,

(C4) $$\left|\frac{\Gamma(1,\mathrm{Ad}(\widehat{\lambda}))L(1,\mathrm{Ad}(\widehat{\lambda}))}{\Omega_1(\widehat{\lambda};A)\Omega_2(\widehat{\lambda};A)}\right|_p^{-r(\mathcal{O})} = \#(C_0^H(\widehat{\lambda};A)).$$

Here are several remarks to be made. In [**U**], (0) the result is more general covering non-base change lift, (i) the normalization of the period is different from the one we made here by a power of 2π, (ii) there is a rational constant α_n showing up in [**U**, Theorem A], which is not explicitly computed (see [**U**, 5.5a]) as a product of factorials. The computation of α_n can be done (see $c_1(s)$ below (Δ) in Section 8 in the text), and α_n is included in the Gamma factor. Finally (iii) it is assumed in [**U**] that the level N of $\widehat{\lambda}$ is sufficiently large to ensure the smoothness of $Y_{0,K}(N)$, which guarantees the duality between (the p-ordinary parts of) $H^1_{\mathrm{cusp}}(Y_{0,K}(N), \mathcal{L}(\widehat{\lambda}, \chi; A))$ and $H^2_{\mathrm{cusp}}(Y_{0,K}(N), \mathcal{L}(\widehat{\lambda}, \chi; A))$. The duality is a key to the proof of the above formula. However, for any N, we can take a multiple N' so that $Y_{0,K}(N')$ is smooth. Then if $p \nmid [Y_{0,K}(N') : Y_{0,K}(N)]$, by using the restriction-corestriction technique, we can recover the duality. By varying N', the common divisor of the covering degree $[Y_{0,K}(N') : Y_{0,K}(N)]$ is made of primes appearing in the order of elliptic elements in $\mathrm{SL}_2(K)$, which is in turn a product of primes dividing the order of roots of unity ζ such that $[K(\zeta) : K] = 2$ (see [**Hi88a**, Lemma 7.1]). Since K is imaginary quadratic, the possibility of such primes are only 2 and 3. In this way, we get the result for $p > 3$ and for level 1.

Although only imaginary quadratic fields K are treated in [**U**], all argument can be generalized to arbitrary E at least for p-ordinary λ. In particular, we get for real K, supposing $(\mathrm{ord}_{\mathbb{Q}})$ for $\overline{\rho}$ and $p \nmid 6D$ (see Section 8 in the case of $K = F \times F$),

$$\text{(C5)} \qquad \left|\frac{\Gamma(1, \mathrm{Ad}(\widehat{\lambda}))L(1, \mathrm{Ad}(\widehat{\lambda}))}{\Omega_1(\varepsilon, \widehat{\lambda}; A)\Omega_1(-\varepsilon, \widehat{\lambda}; A)}\right|_p^{-r(\mathcal{O})} = \#(C_0^H(\widehat{\lambda}; A)),$$

where $(-\varepsilon)(-1_\tau) = -(\varepsilon(-1_\tau))$ for embeddings $\tau \in I_K$.

We consider the following condition:

$$(\mathrm{mlt}_K) \qquad H^{r_1(K)+r_2(K)}_{\mathrm{cusp}}(Y_{0,K}(1), \mathcal{L}(\widehat{\kappa}, \mathrm{id}; \mathcal{O}))[h_K, \varepsilon] \cong h_K \cong R_K \text{ as } h_K\text{-modules.}$$

When K is real, it is plausible to get (mlt_K), under (cl) and (unr), as an application of the method of Taylor, Wiles and Fujiwara [**Fu**], where a similar statement is proven for the modular cohomology group obtained from everywhere unramified definite quaternion algebra over K. When K is imaginary, the assertion (mlt_K) might follow similarly, but it is certainly more difficult. Anyway in this paper, we do not touch this point, but we would just like to remark that the generalization of (CN1-2) for quadratic K follows from (mlt_K).

As we have shown in [**DHI**], there is a natural action of $\mathrm{Gal}(K/\mathbb{Q}) = \langle\sigma\rangle$ on h_K. This action brings $T(\mathfrak{n})$ to $T(\mathfrak{n}^\sigma)$ and induces an action on $C_1(\widehat{\lambda}; \mathcal{O})$. When p is odd, we get a decomposition:

$$C_1(\widehat{\lambda}; \mathcal{O}) = C_1(\widehat{\lambda}; \mathcal{O})[\mathrm{id}] \oplus C_1(\widehat{\lambda}; \mathcal{O})[\alpha],$$

where "$[\alpha]$" indicates α-eigenspace regarding α as the unique non-trivial character of $\mathrm{Gal}(K/\mathbb{Q})$.

CONJECTURE 5.1. *Suppose* (AI_K), $\lambda(T(p)) \in A^\times$, $(\mathrm{ord}_{\mathbb{Q}})$ *for* $\overline{\rho}$ *and that* $p \nmid 6D$. *Then*

$$\text{(CN3)} \qquad \left|\frac{\Gamma(1, \mathrm{Ad}(\lambda) \otimes \alpha)L(1, \mathrm{Ad}(\lambda) \otimes \alpha)}{\Omega_d(\varepsilon, \widehat{\lambda}; A)}\right|_p^{-r(\mathcal{O})} = \#(C_1(\widehat{\lambda}; \mathcal{O})[\alpha]),$$

where $\varepsilon(-1, -1) = -1$ *when* K *is real, we disregard* ε *if* K *is imaginary, and*

$$d = \begin{cases} 2 & \textit{if } K \textit{ is imaginary,} \\ 1 & \textit{if } K \textit{ is real.} \end{cases}$$

When K is real, we have

$$\text{(P)} \qquad \frac{\Omega_1(\varepsilon,\widehat{\lambda};A)}{\Omega_1(+,\lambda;A)\Omega_1(-,\lambda;A)} \in A^\times.$$

This conjecture is made in [**DHI**] for even Dirichlet characters in place of quadratic α. Some numerical example supporting the conjecture is given there for real K. The point here is the inclusion of imaginary quadratic characters α into the scope. We prove here a consequence of the conjecture on congruence among systems of Hecke eigenvalues without referring to the conjecture.

Let M be a sufficiently large number field containing all Hecke eigenvalues on $S_{\widehat{\kappa}}(1,\mathrm{id})$ and such that $H^q_{\mathrm{cusp}}(Y_{0,K}(1),\mathcal{L}(\widehat{\kappa},\mathrm{id};O))[\varepsilon,\mu]$ is O-free of rank 1 for all Hecke eigensystems $\mu : h_{\widehat{\kappa}}(1,\mathrm{id};O) \to O$, where O is the integer ring of M. Such M always exists because the above module is an O-module projective of rank 1. For two systems $\widehat{\lambda} \neq \mu$ of Hecke eigenvalues, we write, for a prime $\mathfrak{p}$ in O, $\widehat{\lambda} \equiv \mu \bmod \mathfrak{p}$ if $\widehat{\lambda}(T(\mathfrak{n})) \equiv \mu(T(\mathfrak{n})) \bmod \mathfrak{p}$ for all integral ideals $\mathfrak{n} \subset \mathcal{R}$. If this happens, $\widehat{\lambda} \bmod \mathfrak{p}$ factors through $C_0(\widehat{\lambda};O)$. Thus we call primes in the support of $C_0(\widehat{\lambda};O)$ congruence primes of $\widehat{\lambda}$. We write $\widehat{\lambda} \equiv^H \mu \bmod \mathfrak{p}$ if for a generator $\xi(\widehat{\lambda})$ of $H^q_{\mathrm{cusp}}(Y_{0,K}(1),\mathcal{L}(\widehat{\kappa},\mathrm{id};O))[\varepsilon,\widehat{\lambda}]$, there is an element $\xi(\mu) \in H^q_{\mathrm{cusp}}(Y_{0,K}(1),\mathcal{L}(\widehat{\kappa},\mathrm{id};O))[\varepsilon,\mu]$ such that

$$\xi(\widehat{\lambda}) - \xi(\mu) \in \mathfrak{p}H^q_{\mathrm{cusp}}(Y_{0,K}(1),\mathcal{L}(\widehat{\kappa},\mathrm{id};O)),$$

where $q = r_1(K) + r_2(K)$. Of course, we have

$$\widehat{\lambda} \equiv^H \mu \bmod \mathfrak{p} \Rightarrow \widehat{\lambda} \equiv \mu \bmod \mathfrak{p}.$$

The converse follows if (mlt_K) holds. The cohomological congruence $\widehat{\lambda} \equiv^H \mu \bmod \mathfrak{p}$ is equivalent to $\mathfrak{p} \in \mathrm{Supp}(C_0^H(\widehat{\lambda};O))$. Conjecture 5.1 implies that for non-base change $\mu : h_{\widehat{\kappa}}(1,\mathrm{id};O) \to O$,

$$\widehat{\lambda} \equiv \mu \bmod \mathfrak{p} \text{ for a non-base change}\mu \iff \mathfrak{p} \Big| \frac{\Gamma(1,\mathrm{Ad}(\lambda)\otimes\alpha)L(1,\mathrm{Ad}(\lambda)\otimes\alpha)}{\Omega_d(\varepsilon,\widehat{\lambda};O)}$$

under the assumption and the notation of the conjecture.

THEOREM 5.2. *Let the notation be as in the conjecture.*

1. *Suppose that $p \nmid 6D(n!)$. If $\widehat{\lambda} \equiv^H \mu \bmod \mathfrak{p}$ for a prime $\mathfrak{p}|p$ of O and a non-base-change μ, then*
$$\mathfrak{p} \Big| \frac{\Gamma(1,\mathrm{Ad}(\lambda)\otimes\alpha)L(1,\mathrm{Ad}(\lambda)\otimes\alpha)}{\Omega_d(\varepsilon,\widehat{\lambda};O)} \in O\left[\frac{1}{6D(n!)}\right].$$
2. *Let K be a real quadratic field. Suppose the assertion (P) of the conjecture for a choice of ε in addition to $p \nmid 6D$, $\lambda(T(p)) \in A^\times$, $(AI_{k(\mathbb{Q})})$ and $(\mathrm{ord}_{\mathbb{Q}})$. Then if a prime $\mathfrak{p}$ with $\mathfrak{p}|p$ of O divides*
$$\frac{\Gamma(1,\mathrm{Ad}(\lambda)\otimes\alpha)L(1,\mathrm{Ad}(\lambda)\otimes\alpha)}{\Omega_1(\varepsilon,\widehat{\lambda};O)}$$
but is prime to $\Gamma(1,\mathrm{Ad}(\lambda))L(1,\mathrm{Ad}(\lambda))/\Omega_1(+,\lambda;O)\Omega_1(-,\lambda;O)$, then there exists a non-base-change lift μ such that $\mu \equiv \widehat{\lambda} \bmod \mathfrak{p}$.

PROOF. Let A be the valuation ring $O_{\mathfrak{p}}$. We consider the following sequence of maps:

$$i^* : H^q_{\text{cusp}}(Y_{0,K}(1), \mathcal{L}(\widehat{\kappa}, \text{id}; A)) \longrightarrow H^q_{\text{cusp}}(Y_{0,\mathbb{Q}}(1), \mathcal{L}(\widehat{\kappa}, \text{id}; A)|_{Y_{0,\mathbb{Q}}(1)});$$
$$\pi_* : H^q_{\text{cusp}}(Y_{0,\mathbb{Q}}(1), \mathcal{L}(\widehat{\kappa}, \text{id}; A)|_{Y_{0,\mathbb{Q}}(1)}) \longrightarrow H^q_c(Y_{0,\mathbb{Q}}(1), A) \cong A.$$

The map i^* is induced by the inclusion $i : Y_{0,\mathbb{Q}}(1) \hookrightarrow Y_{0,K}(1)$ and π_* is induced by the morphism of sheaves $\pi : \mathcal{L}(\widehat{\kappa}, \text{id}; A)|_{Y_{0,\mathbb{Q}}(1)} \to A$, which is induced by $P(X,Y) \mapsto (n!)^{-2}\nabla^n P(X,Y)$. Thus the composite $Ev = \pi_* \circ i^*$ is well defined integrally over A if $\mathfrak{p}$ is prime to $n!$. Suppose that $\xi(\widehat{\lambda}) - \xi(\mu) \in \mathfrak{p}H^q_{\text{cusp}}(Y_{0,K}(1), \mathcal{L}(\widehat{\kappa}, \text{id}; A))$. By Propositions 3.1 and 4.1 and their proof, $Ev(\xi(\widehat{\lambda}))$ is equal to the L-value

$$\frac{\Gamma(1, \text{Ad}(\lambda) \otimes \alpha)L(1, \text{Ad}(\lambda) \otimes \alpha)}{\Omega_d(\varepsilon, \widehat{\lambda}; O)}$$

up to A-units because $p \nmid 6D$. On the other hand, if μ is not a base change lift, twists of μ by any Hecke characters cannot be base change lift, because μ is of level 1. Thus again by Propositions 3.1 and 4.1, $Ev(\xi(\mu)) = 0$. Thus $Ev(\xi(\widehat{\lambda})) = Ev(\xi(\widehat{\lambda}) - \xi(\mu))$, which is divisible by $\mathfrak{p}$ by definition of the cohomological congruence. This proves the assertion (1).

Under the assumption (P), we see that

$$\mathfrak{p}\left|\frac{\Gamma(1, \text{Ad}(\lambda) \otimes \alpha)L(1, \text{Ad}(\lambda) \otimes \alpha)}{\Omega_1(\varepsilon, \widehat{\lambda}; O)}\right. \Rightarrow \mathfrak{p}\left|\frac{\Gamma(1, \text{Ad}(\widehat{\lambda}))L(1, \text{Ad}(\widehat{\lambda}))}{\Omega_1(\varepsilon, \widehat{\lambda}; O)\Omega_1(-\varepsilon, \widehat{\lambda}; O)}\right.,$$

because

$$\frac{\Gamma(1, \text{Ad}(\widehat{\lambda}))L(1, \text{Ad}(\widehat{\lambda}))}{\Omega_1(\varepsilon, \widehat{\lambda}; A)\Omega_1(-\varepsilon, \widehat{\lambda}; A)} = \frac{\Gamma(1, \text{Ad}(\lambda))L(1, \text{Ad}(\lambda))\Gamma(1, \text{Ad}(\lambda) \otimes \alpha)L(1, \text{Ad}(\lambda) \otimes \alpha)}{\Omega_1(+, \lambda; A)^2\Omega_1(-, \lambda; A)^2}$$

up to A-units. Thus by $(C5)$, we have a congruence $\widehat{\lambda} \equiv \mu \mod \mathfrak{p}$. As shown in [**DHI**], $C_1(\widehat{\lambda}; A)[\text{id}] \cong C_1(\lambda; A)$. This combined with $(CN1)$ shows that the factor $\mathfrak{p}$ has to be an associated prime of $C_1(\widehat{\lambda}; A)[\alpha]$, which shows the result. □

6. The case of quadratic extensions of totally real fields

Let F be a totally real field and K be a quadratic extension of F. We use the notation introduced in Section 2. Thus $H = \text{Res}_{\mathfrak{r}/\mathbb{Z}}\text{GL}(2)$ and $G = \text{Res}_{\mathcal{R}/\mathbb{Z}}\text{GL}(2)$. We consider a weight $\kappa = (n, v) \in \mathbb{Z}[I]^2$ with $n \geq 0$. Then we write $\widehat{\kappa} = (\widehat{n}, \widehat{v}) = \text{Inf}(\kappa)$ in $\mathbb{Z}[I_K]^2$. We then consider a cohomological primitive form $f_0 \in S_{\kappa,I}(N_0, \psi)_{/F}$ with $f_0|T(\mathfrak{n}) = \lambda(T(\mathfrak{n}))f_0$. We choose a subset Ψ of I_K such that $I_K = \Psi \sqcup \sigma\Psi$ for the generator σ of $\text{Gal}(K/F)$. Then we put $J = \Sigma_K(\mathbb{R}) \cap \Psi$ and choose $\Sigma_K(\mathbb{C})$ in I_K so that $\Psi = \Sigma_K(\mathbb{C}) \cup J$. Let $f \in S_{\kappa,J}(N, \chi)_{/K}$ be the base change lift of f_0 with $f|T(\mathfrak{n}) = \widehat{\lambda}(T(\mathfrak{n}))f$. Thus $\chi = \psi \circ N_{K/F}$. For each $t = \left(\begin{smallmatrix} a & 0 \\ 0 & 1 \end{smallmatrix}\right) \in H(\mathbb{A}^{(\infty)})$, we consider

$$U_a = tU_0(N)t^{-1}, \quad \Gamma^{(a)} = U_aG(\mathbb{R})_+ \cap G(\mathbb{Q}) \quad \text{and} \quad \Phi^{(a)} = H(\mathbb{Q})_+ \cap \Gamma^{(a)},$$

where $H(\mathbb{Q})_+ = H(\mathbb{Q}) \cap H(\mathbb{R})_+$. Here note that $Y_{0,F}(N') = \sqcup_a \Phi^{(a)} \backslash \mathfrak{H}$ for $\mathfrak{H} = H(\mathbb{R})_+/C_{\infty+}$, where a runs over a complete set of representatives for $F^\times_{\mathbb{A}}/F^\times\widehat{\mathfrak{r}}^\times F^\times_{\infty+}$

with the identity component $F_{\infty+}^\times$ of $F_\infty^\times$. Let $\mathfrak{d} = \mathfrak{d}_K$ be the absolute different of $K/\mathbb{Q}$. Let Cl_X be the strict class group of X, that is

$$\{fractional\ ideals\ of\ X\}/\{(\xi)|\xi \in X, \xi \gg 0\},$$

where $\xi \gg 0$ means that image of ξ is positive for every real embedding of X.

For $z = y^{-1/2}\left(\begin{smallmatrix} y & x \\ 0 & 1 \end{smallmatrix}\right) \in \mathrm{SL}_2(K_\infty) \subset G(\mathbb{R})_+$ for real positive y

$$f^{(a)}(z) = f(tz) = y^{\widehat{n}/2}|ay|_{K_\mathbb{A}}\left\{\sum_{\xi \in K^\times} \widehat{\lambda}(T(\xi a\mathfrak{d}))\xi^{-\widehat{v}}W(\xi y)\mathbf{e}(\xi x)\right\},$$

where ξ runs over all elements in $a^{-1}\mathfrak{d}^{-1}$ such that $\xi^\tau > 0$ for all $\tau \in J$ and $\xi^\tau < 0$ for all $\tau \in \Sigma_K(\mathbb{R}) - J$. Here, writing $n_\tau^* = 2n_\tau + 2$ for $\tau \in \Sigma_K(\mathbb{C})$

$$W(y) = \prod_{\tau \in \Sigma_K} W_\tau(y_\tau)$$

with

$$W_\tau(y) = \begin{cases} \sum_\alpha \binom{n_\tau^*}{\alpha}\left(\frac{y}{\sqrt{-1}|y|}\right)^{n_\tau+1-\alpha} K_{\alpha-n_\tau-1}(4\pi|y|)S_\tau^{n_\tau^*-\alpha}T_\tau^\alpha & \text{if } \tau \in \Sigma_K(\mathbb{C}), \\ \exp(-2\pi|y|) & \text{if } \tau \in \Sigma_K(\mathbb{R}), \end{cases}$$

where α runs over integers with $0 \le \alpha \le n_\tau^*$.

For each subset Y of I_X and $m \in \mathbb{Z}[I_X]$, we write $m! = \prod_\tau m_\tau!$ and $m(Y) = \sum_{\tau \in Y} m_\tau \tau$; in particular, we write $n(\mathbb{C}) = \mathrm{Res}_F^K \widehat{n}(\Sigma_K(\mathbb{C}))$ and $n(\mathbb{R}) = \mathrm{Res}_F^K \widehat{n}(J)$. For $x \in X \otimes_\mathbb{Q} \mathbb{C}$, we write $x^m = \prod_{\tau \in I_X} x^{\tau m_\tau}$ and $x^Y = \prod_{\tau \in Y} x^\tau$. For some specific number, $x = \pi$ or $x = 2i$, we write x^m and x^Y identifying x with a tuple $(x, x, \dots, x) \in \prod_{\tau \in I_X} \mathbb{C} = X \otimes_\mathbb{Q} \mathbb{C}$; thus, for example, $(2\pi i)^m = (2\pi i)^{\sum_\tau m_\tau}$. Note that $\Sigma_K(\mathbb{C})c = \sigma\Sigma_K(\mathbb{C})$. Then we consider $\nabla_\tau = \partial^2/\partial X_{\sigma\tau}\partial Y_\tau - \partial^2/\partial X_\tau \partial Y_{\sigma\tau}$ and write $\nabla^n = \prod_{\tau \in \Psi} \nabla_\tau^{n_\tau}$. We then compute the pull back $\nabla^n \delta(g^{(a)})|_F = i^*\nabla^n\delta(g^{(a)})$ for $g = f|R(\varphi)$ and $i : \mathfrak{H} \hookrightarrow \mathfrak{Z}$:

$$\begin{aligned}(n!)^{-2}\nabla^n\delta(g^{(a)})|_F &= (n!)^{-2}\nabla^n\delta_J(g^{(a)})|_F \\ &= (-1)^J(-2i)^{n(\mathbb{R})+J} \sum_{0 \le j \le n(\mathbb{C})} \binom{n(\mathbb{C})}{j}\left((g^{(a)})_{2j} + (g^{(a)})_{2j+2}\right)y^{n(\mathbb{R})}d\mu(z),\end{aligned}$$

where we write

$$g^{(a)} = \sum_{0 \le \alpha \le n^*} g_\alpha^{(a)}\binom{n^*}{\alpha}S^{n^*-\alpha}T^\alpha \text{ for } n^* = \textstyle\sum_{\tau \in \Sigma_K(\mathbb{C})}(n_\tau + n_{\tau\sigma} + 2)\tau,$$

and

$$d\mu(z) = \bigwedge_{\tau \in I_F} y_\tau^{-2}dy_\tau \wedge dx_\tau.$$

We now choose $\Delta \in K^\times$ such that $\Delta^\sigma = -\Delta$, $\Sigma_K(\mathbb{C}) = \{\tau \in I_K(\mathbb{C})|\,\mathrm{Im}(\Delta^\tau) > 0\}$ and $J = \{t \in \Sigma_K(\mathbb{R})|\Delta^\tau > 0\}$, where $I_K(\mathbb{C}) = \Sigma_K(\mathbb{C}) \sqcup \Sigma_K(\mathbb{C})c$ is the set of all complex embeddings of K into $\mathbb{C}$. Then we see

$$\xi\Delta^{-1} \in \left\{x \in a^{-1}\mathfrak{d}^{-1}|x^\sigma = -x\right\} \iff \xi \in a^{-1}\mathfrak{d}^{-1}\Delta \cap F.$$

Note that $a^{-1}\mathfrak{d}^{-1}\Delta$ is an ideal of F and $I_K = \Psi \sqcup \sigma\Psi$. For $\xi \in F$,

$$N_{F/\mathbb{Q}}(\xi\Delta^{-1}\mathfrak{d}) = |N_{F/\mathbb{Q}}(\xi)|D_F|\Delta^{-\Psi}|N_{F/\mathbb{Q}}(D_{K/F})^{1/2},$$

where $\Delta^{-\Psi} = \prod_{\tau \in \Psi}(\Delta^\tau)^{-1}$.

The stabilizer $\Phi^{(a)}_\infty$ of the cusp ∞ of $\Phi^{(a)}$ is an extension of $a\mathfrak{r}$ by $\mathfrak{r}^\times_+ = \{\varepsilon \in \mathfrak{r}^\times | \varepsilon \gg 0\}$, that is, the following sequence is exact:

$$0 \to a\mathfrak{r} \to \Phi^{(a)} \xrightarrow{\det} \mathfrak{r}^\times_+ \to 1.$$

Let φ be an arithmetic Hecke character of conductor C and with infinity type $-w$. Note that $(n!)^{-2}(\nabla^n \delta(g^{(a)})|_F)$ is a differential form on $Y_{0,F}(N')$ with values in $L((0, n+2v+\mathrm{Res}(w)), (\psi\varphi_F)^2; \mathbb{C})$ because we have (cf. [**Hi94**, 11.2a])

$$L((\widehat{n}, \widehat{v}+w), \chi\varphi^2; \mathbb{C})|_F \cong \bigoplus_{0 \le j \le 2n} L((2n-2j, 2v+\mathrm{Res}(w)+j), (\psi\varphi_F)^2; \mathbb{C}).$$

Let $\omega = |\ |^m_{F_\mathbb{A}}$ for $m\mathbf{1}_F = n + 2v + \mathrm{Res}(w)$. Thus we have

$$\int_{\Phi\backslash\mathfrak{H}} (n!)^{-2}(\nabla^n\delta(g^{(a)})|_F)|a|^s_{K_\mathbb{A}} y^{s\mathbf{1}} = (-1)^J(-2i)^{(n(J))+J}|a|^s_{K_\mathbb{A}}$$
$$\times \int_{F_\mathbb{R}/\mathfrak{r}^\times_+}\int_{F_\infty/a\mathfrak{r}} \sum_{0\le j\le n(\mathbb{C})} \binom{n(\mathbb{C})}{j}((g^{(a)})_{2j} + (g^{(a)})_{2j+2})dxy^{(s-2)\mathbf{1}+n(\mathbb{R})}dy$$
$$= \omega(a)^{-1}c_1(s)G(\varphi)L_a(s, \widehat{\lambda}, \varphi\omega),$$

where $r_2 = r_2(K)$ is the number of complex places of K and

$$L_a(s, \widehat{\lambda}, \varphi\omega) = \sum_{0 \ll \xi \in a^{-1}\mathfrak{d}^{-1}\Delta\cap F} \widehat{\lambda}(T(\xi a\Delta^{-1}\mathfrak{d}))\varphi_F\omega(\xi a\mathfrak{d}\Delta^{-1})N_{F/\mathbb{Q}}(\xi a\Delta^{-1}\mathfrak{d})^{-s-1}$$
$$c_1(s) = (-1)^{n(J)}\sqrt{-1}^{n(J)+J}2^{J(1-s)-2\mathbf{1}}|D_F|^{(3/2)+m+s}N(D_{K/F})^{-(m+s+1)/2}$$
$$\times\left\{\prod_{\tau\in\Psi(\mathbb{C})}(1+(-1)^{m+1})\right\}\left(\frac{\Gamma(\frac{s}{2})\Gamma(\frac{s}{2})}{\Gamma(s)}\right)^{r_2(K)}\left\{\prod_{\tau\in I}\Gamma_\mathbb{C}(s+n_\tau+1)\right\}.$$

Here $\xi \gg 0$ implies $\xi^\tau > 0$ for all $\tau \in I$, and we have used the fact that

$$\int_{F_\mathbb{R}/a\mathfrak{r}} dx = |D_F|^{1/2}|a|^{-1}_{F_\mathbb{A}}$$

for the discriminant D_F of $F/\mathbb{Q}$;

$$f^{(a)}(z) = y^{\widehat{n}/2}|ay|_{K_\mathbb{A}}\sum_\xi \lambda(T(\xi a\mathfrak{d}))\xi^{-\widehat{v}}W(\xi y)\mathbf{e}_K(\xi x),$$

where ξ runs over all elements in $a^{-1}\mathfrak{d}^{-1}$ such that $\xi^\tau > 0$ for all $\tau \in J$ and $\xi^\tau < 0$ for all $\tau \in \Sigma_K(\mathbb{R}) - J$, and

$$y^{\widehat{n}/2} = y^n \quad \text{and} \quad |a|_{K_\mathbb{A}} = |a|^2_{F_\mathbb{A}}.$$

We now look into

$$\sum_{a\in Cl_F}\omega(a)\int_{\Phi^{(a)}_\infty\backslash\mathfrak{H}}(n!)^{-2}(\nabla^n\delta(g^{(a)})|_F)|a|^s_{F_\mathbb{A}}y^{s\mathbf{1}}.$$

As a runs over a complete set of representatives in $F^\times_\mathbb{A}$ for Cl_F, the set of ideals $\{a\Delta\mathfrak{d}\}$ gives again a complete set of representatives for the ideal class group Cl_F

because $\Delta\mathfrak{d}$ is an ideal of F. Thus we get

$$\sum_{a\in Cl_F}\omega(a)\int_{\Phi_\infty^{(a)}\backslash\mathfrak{H}}(n!)^{-2}(\nabla^n\delta(g^{(a)})|_F)|a|_{F_\mathbb{A}}^s y^{s\mathbf{1}}=c_1(s)G(\varphi)L(s,\widehat{\lambda},\varphi_F\omega).$$

By the Rankin convolution method, we get

$$\int_{\Phi_\infty^{(a)}\backslash\mathfrak{H}}(n!)^{-2}(\nabla^n\delta(g^{(a)})|_F)y^{s\mathbf{1}}=\int_{\Phi^{(a)}\backslash\mathfrak{H}}(n!)^{-2}(\nabla^n\delta(g^{(a)})|_F)y^{s\mathbf{1}}\mathcal{E}^{(a)}(s),$$

where

$$\mathcal{E}^{(a)}(s)=\sum_{\delta\in\Phi^{(a)}/\Phi_\infty^{(a)}}\varphi_{N'}^2\psi_{N'}^2(\gamma)y^{s\mathbf{1}}\circ\gamma.$$

We now put

$$E(s)=E(x,s)=\sum_{a\in Cl_F}\omega(a)|a|_{F_\mathbb{A}}^s\mathcal{E}^{(a)}(s)$$

as a function of $x\in H(\mathbb{A})$. Writing $Y=Y_{0,F}(N')=\sqcup_{a\in Cl_F}Y^{(a)}$, if $(\omega\psi\varphi_F)^2=\mathrm{id}$, by (RES3) in Appendix, we get

$$\begin{aligned}&\mathrm{Res}_{s=1}\zeta_{F,N'}(2s)\int_Y E(s)(n!)^{-2}(\nabla^n\delta(g)|_F)\\&=N_{F/\mathbb{Q}}(N')^{-2}\phi(N')\frac{2^{[F:\mathbb{Q}]-1}\pi^{[F:\mathbb{Q}]}R_\infty}{w(F)|D_F|}\sum_a\omega(a)\int_{Y^{(a)}}(n!)^{-2}(\nabla^n\delta(g)|_F),\end{aligned}$$

where R_∞ is the regulator of F. Let $\varepsilon:S=C_\infty/C_{\infty+}=\{\pm1\}^{\Sigma_K(\mathbb{R})}\to\{\pm1\}$ be a character such that

$$\varepsilon((-1,-1)_\tau)=(-1)^{n_\tau+1}\quad\text{for all }\ \tau\in J,$$

where $(-1,-1)_\tau$ is an element of S whose component is equal to 1 outside $\{\tau,\sigma\tau\}$ and is equal to -1 at τ and $\sigma\tau$. We consider the projection π_ε to the ε-eigenspace. In the same manner as in Section 4, under the condition of the parity of ε as above, the application of π_ε on the integrand does not have any effect on the outcome of the computation. Thus if $\omega\psi\varphi_F=\alpha$,

$$\begin{aligned}\mathrm{Res}_{s=1}\zeta_{F,N'_F}(2s)&\int_Y E(s)(n!)^{-2}(\nabla^n\pi_\varepsilon\delta(g)|_F)\\&=c_1(1)\{\mathrm{Res}_{s=1}\zeta_{F,N'_F}(s)\}L(1,\mathrm{Ad}(\lambda)\otimes\alpha).\end{aligned}$$

Thus we get

$$\begin{aligned}\Gamma(1,\mathrm{Ad}(\lambda)\otimes\alpha)L_{C'}(1,\mathrm{Ad}(\lambda)\otimes\alpha)&=c_0N(D_{K/F})^{1/2}G(\varphi)^{-1}i^{n(J)+J}\\&\times h(F)^{-1}\sum_a\omega(a)\int_{Y^{(a)}}(n!)^{-2}(\nabla^n\pi_\varepsilon\delta(g)|_F),\end{aligned}$$

where

$$c_0=2^{r_2(K)-(r_1(K)/2)-1}(N_{F/\mathbb{Q}}N')^{-2}\phi(N')|D_F|^{-m-3}N_{F/\mathbb{Q}}(D_{K/F})^{(m+1)/2}\in\mathbb{Q}.$$

Let μ be a system of Hecke eigenvalues of level N and with character χ. Let $d=[F:\mathbb{Q}]$ and A be a Dedekind domain in $\overline{\mathbb{Q}}$ containing $\mu(\mathfrak{n})$ for all ideals $\mathfrak{n}\subset\mathfrak{r}$. For each character $\varepsilon:S\to\{\pm1\}$, we consider $H^d(Y_{0,K}(N),\mathcal{L}(\widehat{\kappa},\chi;A))[\varepsilon,\mu]$ which is projective of rank 1 over A. Extending A a bit, we may assume that it is free of rank 1 over A. Let ξ be the generator of this free A-module of rank 1. Let f be the normalized Hecke eigenform with eigenvalues μ. We define

$$[\pi_\varepsilon\delta_{J,\emptyset}(f)] = \Omega_{\Sigma_K(\mathbb{R}),2\Sigma_K(\mathbb{C})}(\varepsilon,\mu;A)\xi,$$

where $[\phi]$ indicates the cohomology class of a closed form ϕ. Here the subscript "$\Sigma_K(\mathbb{R}), 2\Sigma_K(\mathbb{C})$" indicates that the cohomology class is of degree 1 at archimedean places in $\Sigma_K(\mathbb{R})$ and of degree 2 at archimedean places in $\Sigma_K(\mathbb{C})$. When $A = \mathbb{Q}(\mu)$, we have the cohomological conjugate

$$\xi^\rho \in H^d(Y_{0,K}(N), \mathcal{L}(\widehat{\kappa}\rho, \chi^\rho; A))[\varepsilon, \mu^\rho]$$

and define automorphic conjugate f^ρ by $\mathbf{a}(\mathfrak{n}, f^\rho) = \mathbf{a}(\mathfrak{n}, f)^\rho$. Then we define

$$[\pi_\varepsilon\delta_{J,\emptyset}(f^\rho)] = \Omega_{\Sigma_K(\mathbb{R}),2\Sigma_K(\mathbb{C})}(\varepsilon,\mu^\rho;A)\xi^\rho,$$

and put

$$\underline{\Omega}_{\Sigma_K(\mathbb{R}),2\Sigma_K(\mathbb{C})}(\varepsilon,\mu;A) = (\Omega_{\Sigma_K(\mathbb{R}),2\Sigma_K(\mathbb{C})}(\varepsilon,\mu^\rho;A))_{\rho\in I_{\mathbb{Q}(\mu)}}$$

as an element of $(\mathbb{Q}(\mu)\otimes_{\mathbb{Q}}\mathbb{C})^\times$. Then we get similarly to Corollaries 3.2 and 4.2 the following result.

THEOREM 6.1. *Suppose that we have an arithmetic Hecke character φ of K and ω of F such that (i) the conductor of ω is 1 and (ii) $\psi\varphi_F\omega = \alpha$. Then we have*

$$\frac{(1\otimes\Gamma(1,\mathrm{Ad}(\lambda)\otimes\alpha))\mathbb{L}_{C'}(1,\mathrm{Ad}(\lambda)\otimes\alpha)}{(1\otimes N_{F/\mathbb{Q}}(D_{K/F})^{1/2}i^{n(J)+J})\mathbf{G}(\psi\alpha)\underline{\Omega}_{\Sigma_K(\mathbb{R}),2\Sigma_K(\mathbb{C})}(\varepsilon,\widehat{\lambda};\mathbb{Q}(\lambda))} \in \mathbb{Q}(\lambda)\subset\mathbb{Q}(\lambda)\otimes_{\mathbb{Q}}\mathbb{C}$$

for any ε with $\varepsilon((-1,-1)_\tau) = -1$ for all $\tau\in J$. Moreover if φ can be chosen to be the identity character, then for all valuation ring A of $\mathbb{Q}(\lambda)$ with residual characteristic prime to $2e(n!)D_K$, we have

$$\frac{(N_{F/\mathbb{Q}}N_0)^2\phi(N_0)^{-1}N_{F/\mathbb{Q}}(D_{K/F})^{1/2}i^{n(J)+J}\Gamma(1,\mathrm{Ad}(\lambda)\otimes\alpha)L(1,\mathrm{Ad}(\lambda)\otimes\alpha)}{\Omega_{\Sigma_K(\mathbb{R}),2\Sigma_K(\mathbb{C})}(\varepsilon,\widehat{\lambda};A)} \in A,$$

where $N_0 = N\cap\mathfrak{r}$ and e is the least common multiple of the order of maximal torsion subgroups of $\Gamma^{(a)}/\mathcal{R}^\times$ for all a.

If either K is totally real or a CM field, we can always find ω and φ as in the theorem (see Lemma 2.2). If K has both complex and real places, this condition really imposes a restriction. Probably one could remove this condition taking the integral over $Y(S)$ for a smaller subgroup S in $U_0(N')$ allowing ω with non-trivial conductor, but we might lose more Euler factors of the adjoint L in the process. Since this would further complicate our computation, we do not look into this point in the present paper.

We can formulate the divisibility of the L-value by congruence primes of $\widehat{\lambda}$ as is done in Section 5 for quadratic fields, which we leave to the reader.

The parity restriction on ε is explained by the following fact:

$$\varphi((-1,-1)_\tau) = \varphi_F((-1)_\tau) = \psi((-1)_\tau)\alpha((-1)_\tau) = (-1)^{n_\tau}.$$

Thus the condition that $\varepsilon((-1,-1)_\tau) = -1$ is equivalent to

$$\varepsilon\varphi_\infty((-1,-1)_\tau) = (-1)^{n_\tau+1},$$

and we have

$$\Omega_{\Sigma_K(\mathbb{R}),2\Sigma_K(\mathbb{C})}(\varepsilon\varphi_\infty,\widehat{\lambda}\otimes\varphi;\mathbb{Q}(\lambda)) = G(\varphi)^{-1}\Omega_{\Sigma_K(\mathbb{R}),2\Sigma_K(\mathbb{C})}(\varepsilon,\widehat{\lambda};\mathbb{Q}(\lambda)).$$

7. Quadratic extensions K of an imaginary quadratic F

Now we assume F to be an imaginary quadratic field and keep notation introduced in the previous sections. We write σ for the generator of $\mathrm{Gal}(K/F)$, and fix an embedding $\tau : K \to \mathbb{C}$. We then write $J' = \{\sigma\tau\}$. For each $\iota \in I_K$, we write $[\iota]$ for the archimedean place arising from ι. Thus this correspondence induces a linear map $[\ ] : \mathbb{Z}[I_K] \to \mathbb{Z}[\Sigma]$ for the set $\Sigma = \Sigma_K = I_K/\langle c\rangle$ of archimedean places. If no confusion is likely, we write Σ for Σ_K to simplify the notation. We decompose $I_K = \Psi \sqcup \sigma\Psi$ with $\Psi = \{\tau, \tau c\}$. We identify Σ with $\{\tau, \sigma\tau\}$. Let f be a cohomological modular form on G of weight $(\widehat{n}, \widehat{v})$. Then we can write for $z \in \mathfrak{Z} = G(\mathbb{R})/Z(\mathbb{R})C_\infty$ given by $y^{-1/2}\left(\begin{smallmatrix} y & x \\ 0 & 1\end{smallmatrix}\right) \in \mathrm{SL}_2(K_\infty)$

$$f^{(a)}(z) = f\left(ty^{-1/2}\begin{pmatrix} y & x \\ 0 & 1\end{pmatrix}\right) = y^{\widehat{n}/2}|ay|_{K_A}\left\{\sum_{\xi\in K^\times}\widehat{\lambda}(T(\xi a\mathfrak{d}))\xi^{-\widehat{v}}W(\xi y)\mathbf{e}(\xi x)\right\}$$

where

$$W(y) = \sum_{0\le\alpha\le\widehat{n}^*}\binom{\widehat{n}^*}{\alpha}\left(\frac{y}{\sqrt{-1}|y|}\right)^{\widehat{n}+\mathbf{1}-\alpha}K_{\alpha-\widehat{n}-\mathbf{1}}(4\pi|y|)S^{\widehat{n}^*-\alpha}T^\alpha \quad (\widehat{n}^* = [\widehat{n}] + 2\Sigma).$$

We write $f^{(a)}(z) = \sum_{0\le\alpha\le\widehat{n}^*} f_\alpha^{(a)}(z)\binom{\widehat{n}^*}{\alpha}S^{\widehat{n}^*-\alpha}T^\alpha$. Then by a computation similar to (δ) in Section 3 and [**Hi94**, Section 2], $z^{-1}\delta_{J'}(f^{(a)})$ for $z = y^{-1/2}\left(\begin{smallmatrix} y & x \\ 0 & 1\end{smallmatrix}\right)$ is given by the following formula through replacing $U_{\alpha_\rho}U_{\beta_\theta}$ by $f_{[\alpha_\rho\rho+\beta_\theta\theta]}$:

$$\begin{aligned}
&y_{\sigma\tau}^{-1}y_\tau^{-2}\sum_{0\le j\le\widehat{n}}(-1)^{n_\tau-j_{\tau c}+n_{\sigma\tau}-j_{\sigma\tau c}}\binom{\widehat{n}}{j}X^{\widehat{n}-j}Y^j\\
&\times\left\{U_{n_{\tau c}+j_\tau-j_{\tau c}}dy_\tau\wedge dx_\tau - 4U_{n_{\tau c}+j_\tau-j_{\tau c}+1}dx_\tau\wedge d\overline{x}_\tau + U_{n_{\tau c}+j_\tau-j_{\tau c}+2}dy_\tau\wedge d\overline{x}_\tau\right\}\\
&\bigwedge\left\{U_{n_{\sigma\tau c}+j_{\sigma\tau}-j_{\sigma\tau c}}dx_{\sigma\tau} - 2U_{n_{\sigma\tau c}+j_{\sigma\tau}-j_{\sigma\tau c}+1}dy_{\sigma\tau} - U_{n_{\sigma\tau c}+j_{\sigma\tau}-j_{\sigma\tau c}+2}d\overline{x}_{\sigma\tau}\right\},
\end{aligned}$$

where we have written $X^{\widehat{n}-j}Y^j$ for $\prod_{\rho\in I_K}X_\rho^{\widehat{n}_\rho-j_\rho}Y_\rho^{j_\rho}$, $\binom{\widehat{n}}{j}$ for $\prod_{\rho\in I_K}\binom{\widehat{n}_\rho}{j_\rho}$, and j runs over $j \in \mathbb{Z}[I_K]$ with $0 \le j_\rho \le \widehat{n}_\rho$ for all $\rho \in I_K$. We get a non-trivial result only for base change lift; so, we may assume that $n_\tau = n_{\tau c} = n_{\sigma\tau} = n_{\sigma\tau c}$. The action of z does not affect the outcome of the differential operator $\nabla^n = (\nabla_\tau\nabla_{\tau c})^{n_\tau}$ applied to $\delta_{J'}(f)|_F$ ($\nabla_\rho = \partial^2/\partial X_{\sigma\rho}\partial Y_\rho - \partial^2/\partial X_\rho\partial Y_{\sigma\rho}$); so, we forget about it. We then restrict the differential form to $\mathfrak{H} = H(\mathbb{R})/Z_H(\mathbb{R})C_\infty$ and hence, we may assume that $y = y_\tau = y_{\sigma\tau}$ and $x = x_\tau = x_{\sigma\tau}$. Thus we get

$$\begin{aligned}
&(n!)^{-2}\nabla^n\delta_{J'}(f)|_F\\
&\qquad = (-1)^{n_\tau}\sum_{0\le j\le n}-(-1)^j\binom{n}{j}\{f_{[n'+2J']} - 8f_{[n'+\Sigma]} + f_{[n'+2(\Sigma-J')]}\}d\mu(z),
\end{aligned}$$

where $d\mu = y^{-3}dy\wedge dx\wedge d\overline{x}$, $\binom{n}{j} = \binom{n_\tau}{j_\tau}\binom{n_{\tau c}}{j_{\tau c}}$, and $n' = \widehat{n}(\Sigma) + j^\#$ for $j^\# = (j_\tau - j_{\tau c})\tau + (j_{\tau c} - j_\tau)\sigma\tau$. We now compute for $X = J', \Sigma - J', \Sigma$, writing x for the number of elements in X and g for $f|R(\varphi)$,

$$\begin{aligned}&\int_{\Phi_\infty^{(a)}\backslash\mathfrak{Z}} g^{(a)}_{[n'+2X]}|ay|^s_{F_\mathbb{A}}d\mu(z)\\&\quad=-(-1)^{j+x}2^{2n_\tau+2s}\sqrt{-1}\mathrm{vol}(F_\mathbb{R}/a\mathfrak{r})G(\varphi)|a|^{s+2}_{F_\mathbb{A}}(|D_F|N_{F/\mathbb{Q}}(D_{K/F})^{1/2})^{1+s}\\&\quad\cdot\frac{\Gamma(n_\tau+1+s)^2\Gamma(n_\tau+s+j_\tau-j_{\tau c}+x)\Gamma(n_\tau+s+j_\tau-j_{\tau c}+2-x)}{\Gamma(2n_\tau+2+2s)}\\&\quad\cdot\sum_{\xi\in F^\times}\widehat{\lambda}(T(\xi a\Delta^{-1}\mathfrak{d}))\varphi(\xi a\Delta^{-1}\mathfrak{d})\xi^{-n-2v-\mathrm{Res}(w)}\Delta^{\widehat{v}+w+\widehat{n}/2}N_{F/\mathbb{Q}}(\xi\Delta^{-1}\mathfrak{d})^{-1-s},\end{aligned}$$

where $\Delta\in K^\times$ satisfies $\Delta^\sigma=-\Delta$ and we have used the identity:

$$N_{F/\mathbb{Q}}(\Delta^{-1}\mathfrak{d})=|D_F||\Delta^{-\tau}|^2N_{F/\mathbb{Q}}(D_{K/F})^{1/2}$$

along with the convention that $\widehat{\lambda}(T(\mathfrak{n}))=0$ for non-integral ideals $\mathfrak{n}$.

Now we suppose to have a Hecke character ω of conductor 1 of $F^\times_\mathbb{A}/F^\times$ with $\infty(\omega)=n+2v+\mathrm{Res}(w)$. For that, we may need to allow algebraic character φ not necessarily arithmetic to have such ω (see Lemma 2.2). Note that

$$\omega(\xi a\Delta^{-1}\mathfrak{d})=\omega(a\Delta^{-1}\mathfrak{d})\xi^{n+2v+\mathrm{Res}(w)}.$$

Then we have, for $g=f|R(\varphi)$,

$$\begin{aligned}&\int_{\Phi_\infty^{(a)}\backslash\mathfrak{H}}(n!)^{-2}(\nabla^n\delta_{J'}(g^{(a)})|_F)|ay|^s_{F_\mathbb{A}}(-1)^{n_\tau+1}2^{2n_\tau+2s}(4\pi)^{-2n_\tau-2-2s}\sqrt{-1}\mathrm{vol}(F_\mathbb{R}/a\mathfrak{r})\\&\qquad\cdot G(\varphi)(|D_F|N_{F/\mathbb{Q}}(D_{K/F})^{1/2})^{1+s}\times\omega(a\Delta^{-1}\mathfrak{d})^{-1}\Delta^{\widehat{v}+w+\widehat{n}/2}\Gamma(n_\tau+1+s)^2\\&\qquad\cdot\{G_0(s)+8G_1(s)+G_2(s)\}L_a(s,\widehat{\lambda},\varphi_F\omega),\end{aligned}$$

where

$$L_a(s,\widehat{\lambda},\varphi_F\omega)=\sum_{\xi\in F^\times/\mathfrak{r}^\times}\widehat{\lambda}(T(\xi a\Delta^{-1}\mathfrak{d}))\varphi\omega(\xi a\Delta^{-1}\mathfrak{d})N_{F/\mathbb{Q}}(\xi a\Delta^{-1}\mathfrak{d})^{-1-s},$$

$$G_x(s)=\sum_{0\le j\le n}\binom{n}{j}\frac{\Gamma(n_\tau+s+j_\tau-j_{\tau c}+x)\Gamma(n_\tau+s+j_\tau-j_{\tau c}+2-x)}{\Gamma(2n_\tau+2+2s)}$$

for $x=0,1,2$. Using the formula in [**Hi94**, p. 505] twice, we get

$$G_x(s)=\frac{\Gamma(s+x)\Gamma(s+2-x)}{\Gamma(2+2s)},$$

$$G_0(s)+8G_1(s)+G_2(s)=2(5s+1)\frac{\Gamma(s)\Gamma(s+1)}{\Gamma(2+2s)}.$$

Then using the fact that $\mathrm{vol}(F_\infty/\mathfrak{r})=2^{-1}|D_F|^{1/2}$, we have

$$\sum_{a\in Cl_F}\omega(a)\int_{\Phi_\infty^{(a)}\backslash\mathfrak{H}}(n!)^{-2}(\nabla^n\delta_{J'}(g^{(a)})|_F)|a|^s_{F_\mathbb{A}}y^{s\mathbf{1}}=c_1(s)L(s,\widehat{\lambda},\varphi_F\omega),$$

where

$$\begin{aligned}c_1(s)&=\sqrt{-1}|D_F|^{1/2}(|D_F|N_{F/\mathbb{Q}}(D_{K/F})^{1/2})^{1+s}G(\varphi)\omega(\Delta^{-1}\mathfrak{d})^{-1}\Delta^{\widehat{v}+w+\widehat{n}/2}\\&\quad\cdot(-1)^{n_\tau+1}2^{2n_\tau+2s}(4\pi)^{-2n_\tau-2-2s}(5s+1)\frac{\Gamma(s)\Gamma(s+1)}{\Gamma(2+2s)}\Gamma(n_\tau+1+s)^2.\end{aligned}$$

By Rankin's convolution method, we get

$$\int_{\Phi_\infty^{(a)}\backslash\mathfrak{H}}(n!)^{-2}(\nabla^n\delta_{J'}(g^{(a)})|_F)y^{s\mathbf{1}}=\int_{\Phi^{(a)}\backslash\mathfrak{H}}(n!)^{-2}(\nabla^n\delta_{J'}(g^{(a)})|_F)\mathcal{E}^{(a)}(s),$$

where

$$\mathcal{E}^{(a)}(s)=\sum_{\gamma\in\Phi^{(a)}/\Phi_\infty^{(a)}}\varphi_{N'}^2\psi_{N'}^2(\gamma)y^{s\mathbf{1}}\circ\gamma.$$

Put $E(s)=\sum_{a\in Cl_F}\omega(a)|a|_{F_\mathbb{A}}^s\mathcal{E}^{(a)}(s)$. If $(\omega\psi\varphi_F)^2=\mathrm{id}$, by (RES3) in Appendix, we get for $Y=Y_{0,F}(N')=\sqcup_{a\in Cl_F}Y^{(a)}$

$$\begin{aligned}&\mathrm{Res}_{s=1}\zeta_{F,N'}(2s)\int_Y E(s)(n!)^{-2}(\nabla^n\delta(g)|_F)\\&=N_{F/\mathbb{Q}}(N')^{-2}\phi(N')\frac{2\pi^2R_\infty}{|D_F|}\sum_{a\in Cl_F}\omega(a)\int_{Y^{(a)}}(\nabla^n\delta(g^{(a)})|_F).\end{aligned}$$

On the other hand, if $\alpha=\psi\omega\varphi_F$, we have

$$\begin{aligned}&\mathrm{Res}_{s=1}\zeta_{F,N'}(2s)\int_Y E(s)(n!)^{-2}(\nabla^n\delta(g)|_F)\\&=c_1(1)\{\mathrm{Res}_{s=1}\zeta_{F,N'}(s)\}L_{C'}(1,\mathrm{Ad}(\lambda)\otimes\alpha).\end{aligned}$$

Comparing the two expressions, we get

$$\begin{aligned}\Gamma(1,\mathrm{Ad}(\lambda)\otimes\alpha)L_{C'}(1,\mathrm{Ad}(\lambda)\otimes\alpha)&=c_0G(\varphi)^{-1}\omega(\Delta^{-1}\mathfrak{d})\Delta^{-\widehat{v}-w-\widehat{n}/2}\\&\cdot\sqrt{-1}h(F)^{-1}\sum_{a\in Cl_F}\omega(a)\int_{Y^{(a)}}(n!)^{-2}(\nabla^n\delta(g^{(a)})|_F),\end{aligned}$$

where $c_0=2(-1)^{n_\tau+1}|D_F|^{-3}N_{F/\mathbb{Q}}(D_{K/F})^{-1}\in\mathbb{Q}$.

At the beginning, we fixed an embedding τ of K into $\mathbb{C}$. We study what happens if we start with $\sigma\tau$ instead of τ. The result is the same, but J' will be replaced by $\{\tau\}$. Note that $\{\delta_{J'}(f)\}_{J'\subset\Sigma_K(\mathbb{C})}$ forms a base of $H^3_{\mathrm{cusp}}(Y_0(N),\mathcal{L}(\widehat{\kappa},\chi;\mathbb{C}))[\widehat{\lambda}]$. Thus

$$H^3(\mathbb{Q}(\lambda))[\widehat{\lambda}]=H^3_{\mathrm{cusp}}(Y_0(N),\mathcal{L}(\widehat{\kappa},\chi;\mathbb{Q}(\lambda)))[\widehat{\lambda}]\cong\mathbb{Q}(\lambda)^2.$$

Then σ acts on $Y_0(N)$ via the Galois action: $x\mapsto x^\sigma$ on $G(\mathbb{A})$, because $N^\sigma=N$. We let σ act on $L(n;\mathbb{Q}(\lambda))$ by $\sigma(X_\tau,Y_\tau)=(X_{\sigma\tau},Y_{\sigma\tau})$. Thus for a differential form ϕ on $\mathfrak{Z}$ with values in $L(\widehat{n};\mathbb{C})$ such that $\gamma^*\phi=\gamma\phi$, we see

$$\gamma^*(\sigma(\sigma^*\phi))=\sigma\sigma^*(\gamma^\sigma)^*\phi=\sigma\gamma^\sigma(\sigma^*\phi)=\gamma\sigma(\sigma^*\phi).$$

Thus via $\phi\mapsto\sigma\sigma^*\phi$, $\mathrm{Gal}(K/F)$ acts on $H^3_{\mathrm{cusp}}(Y_0(N),\mathcal{L}(\widehat{\kappa},\chi;\mathbb{C}))$. We see that $\sigma^*\delta_{\{\tau\}}(f)=\sigma\delta_{\{\sigma\tau\}}(f)$. Since $\widehat{\lambda}$ is stable under σ, $H^3[\widehat{\lambda}]$ is stable under the action of σ. Thus $H^3(\mathbb{Q}(\lambda))[\widehat{\lambda}][\sigma-1]$ is one dimensional, and we take a generator ξ such that

$$H^3(\mathbb{Q}(\lambda))[\widehat{\lambda}][\sigma-1]=\mathbb{Q}(\lambda)\xi.$$

The action of σ defined above commutes with the Galois action induced covariantly by the Galois action on $L(\widehat{\kappa},\chi;\mathbb{Q}(\lambda))$. We write ξ^ρ for the Galois conjugate of ξ under the latter action. We note that $H^3(\mathbb{C})[\widehat{\lambda}^\rho][\sigma-1]=\mathbb{C}(\delta_{\{\tau\}}(f^\rho)+\delta_{\{\sigma\tau\}}(f^\rho))$. We then define a complex number $\Omega_{(1,2)}(\widehat{\lambda}^\rho;\mathbb{Q}(\lambda))\in\mathbb{C}^\times$ by

$$(\delta_{\{\tau\}}(f^\rho)+\delta_{\{\sigma\tau\}}(f^\rho))=\Omega_{(1,2)}(\widehat{\lambda}^\rho;\mathbb{Q}(\lambda))\xi^\rho.$$

We put

$$\underline{\Omega}_{(1,2)}(\widehat{\lambda}, \mathbb{Q}(\lambda)) = (\Omega_{(1,2)}(\widehat{\lambda}^{\rho}; \mathbb{Q}(\lambda)))_{\rho \in I_{\mathbb{Q}(\lambda)}} \in (\mathbb{Q}(\lambda) \otimes_{\mathbb{Q}} \mathbb{C})^{\times}.$$

Note that $\Delta^{-1}\mathfrak{d} = \mathfrak{d}_F \xi$. Thus for a finite idele d of F generating $\mathfrak{d}_F$, we have $\omega(\Delta^{-1}\mathfrak{d}) = \omega(d)\xi^{n+2v+\mathrm{Res}(w)}$ because ω is of conductor 1. We choose φ as in Lemma 2.2. Then it is easy to see that $\xi^{n+2v+\mathrm{Res}(w)} \in \mathbb{Q}$. This shows that

$$\omega(\Delta^{-1}\mathfrak{d})^{-1} G(\varphi)/G(\psi^{-1}\alpha) \in \mathbb{Q}(\lambda).$$

Since $G(\psi^{-1}\alpha)G(\psi\alpha) \in \mathbb{Q}$, we finally get

THEOREM 7.1. *Suppose that we have an arithmetic Hecke character φ of K and ω of F such that (i) the conductor of ω is 1 and (ii) $\psi\varphi_F\omega = \alpha$. Then we have*

$$\frac{(1 \otimes \Delta^{\widehat{v}+w+\widehat{n}/2}\Gamma(1, \mathrm{Ad}(\lambda) \otimes \alpha))\mathbb{L}_{C'}(1, \mathrm{Ad}(\lambda) \otimes \alpha)}{(1 \otimes \sqrt{-1})\mathbf{G}(\psi\alpha)\underline{\Omega}_{(1,2)}(\widehat{\lambda}; \mathbb{Q}(\lambda))} \in \mathbb{Q}(\lambda) \subset \mathbb{Q}(\lambda) \otimes_{\mathbb{Q}} \mathbb{C}.$$

If φ and ω are algebraic Hecke characters satisfying (i) and (ii), then we have

$$\frac{\Gamma(1, \mathrm{Ad}(\lambda) \otimes \alpha) L_{C'}(1, \mathrm{Ad}(\lambda) \otimes \alpha)}{\Omega_{(1,2)}(\widehat{\lambda}; \mathbb{Q}(\lambda))} \in \overline{\mathbb{Q}}.$$

Since the L-value in Theorem 7.1 does not depends on the choice of φ, even if φ is not arithmetic, presumably, the $\mathbb{Q}(\lambda)$-rationality would hold in general. Further study has to be done to prove this.

The period $\Omega_{(1,2)}$ is only defined for the base change lift $\widehat{\lambda}$, while we have defined similar periods for any system μ of Hecke eigenvalues in the previous sections when F is totally real. When $\mu \neq \mu_\sigma$, the action of σ on $H^3(\mathbb{Q}(\mu))$ does not preserve $H^3(\mathbb{Q}(\mu))[\mu]$, and this causes a trouble. If F has complex places, the same problem shows up as will be seen in the following section.

8. General quadratic extensions

Let K/F be a semi-simple quadratic extension of a number field F. Thus we allow here $K = F \oplus F$. When $K = F \oplus F$, we regard $D_{K/F} = 1$ and $\alpha = \mathrm{id}$; otherwise, α denotes the quadratic character of $F_{\mathbb{A}}^{\times}/F^{\times}$ corresponding to K/F. We shall prove the rationality theorem of $L(1, \mathrm{Ad}(\lambda) \otimes \alpha)$ in this general case. Computation is the same as in the previous sections. Since the definition of $f \mapsto \delta(f)$ is given in [**Hi94**] by a procedure which is basically a tensor product of definitions over archimedean places, the computation of the pull back $\delta(f)|_F$ is again essentially the tensor product of the pull back over archimedean places of K. Thus the computation is the same as in Section 3 for a complex place of K over a real place of F, the same as in Section 4 for two real places of K over a real place of F and the same as in Section 7 for two complex places of K over a complex place of F. After computing $\delta(f)$, the computation of the Rankin product is fairly standard. Thus our exposition of the computation will be brief, but we state the result in a precise form. When $K = F \oplus F$, we regard that every archimedean place of F splits in K.

We decompose $I = I(\mathbb{C}) \sqcup I(\mathbb{R})$ and $I(\mathbb{C}) = \Sigma(\mathbb{C}) \sqcup \Sigma(\mathbb{C})c$. Similarly we decompose $I_K = I_K(\mathbb{R}) \sqcup I_K(\mathbb{C})$. We decompose $I_K = \Psi \sqcup \sigma\Psi$ for the generator $\sigma \in \mathrm{Gal}(K/F)$ and $\Psi = \Psi(\mathbb{R}) \sqcup \Psi(\mathbb{C})$, where $\Psi(\mathbb{R})$ is the subset of all real embeddings of Ψ. When $K = F \oplus F$, we have $\sigma(x \oplus y) = y \oplus x$. We write J' (resp. J'') for $\{\sigma\tau \in \sigma\Psi(\mathbb{C}) | \mathrm{Res}_{K/F}(\tau) \in \Sigma(\mathbb{C})\}$ (resp. $\{\tau \in I(\mathbb{R}) | \tau$ extends to a complex place of $K\}$) and put $J = \Psi(\mathbb{R})$. When $K = F \oplus F$, we identify I_K with $I \sqcup I$ through the right and left projections of K to F. Then we identify Ψ with the left component I in I_K. Then we shall make use of the isomorphism

$$\delta_{J,J'} : S_{\widehat{\kappa},J}(N,\chi) \hookrightarrow H^q(Y_0(N),\mathcal{L}(\widehat{\kappa},\chi;\mathbb{C})) \quad (\kappa=(n,v))$$

defined in [**Hi94**, Proposition 2.1], where $q=[F:\mathbb{Q}]+r_1(F)+r_2(F)=\dim(\mathfrak{H})$. When $K=F\oplus F$, then $Y_{0,K}(N)=Y_{0,F}(N_0)\times Y_{0,F}(N_0)$ for the ideal $N=N_0\oplus N_0\subset \mathfrak{r}\oplus\mathfrak{r}=\mathcal{R}$. In this case,

$$L((\widehat{n},\widehat{v}),\chi;\mathbb{C})\cong L((n,v),\psi;\mathbb{C})\otimes_{\mathbb{C}} L((n,v),\psi;\mathbb{C}).$$

For normalized Hecke eigenforms $f_0\in S_{\kappa,I(\mathbb{R})}(N_0,\psi;\mathbb{C})[\lambda]$ and $f_0'\in S_{\kappa,\emptyset}(N_0,\psi;\mathbb{C})[\lambda]$, the base change lift $f\in S_{\widehat{\kappa},J}(N,\chi;\mathbb{C})$ to $G=H\times H$ is just $f(z,z')=f_0(z)f_0'(z')$ for $(z,z')\in\mathfrak{H}\times\mathfrak{H}=\mathfrak{Z}$ and

$$\delta_{J,J'}(f)=\delta_{I(\mathbb{R}),I(\mathbb{C})}(f_0)(z)\wedge\delta_{\emptyset,\emptyset}(f_0')(z'),$$

which gives a cohomology class in

$$\begin{aligned} H^q(Y_{0,N}(N_0),&\mathcal{L}((n,v),\psi;\mathbb{C}))\otimes_{\mathbb{C}} H^{q'}(Y_{0,F}(N_0),\mathcal{L}((n,v),\psi;\mathbb{C}))\\ &\subset H^Q(Y_{0,K}(N),\mathcal{L}((\widehat{n},\widehat{v}),\chi;\mathbb{C}))\end{aligned}$$

for $Q=q+q'=\dim\mathfrak{H}$ as real manifolds, $q'=r_1(F)+2r_2(F)$ and $q=r_1(F)+r_2(F)$. In this case, we have a standard choice of φ: $\varphi=\mathrm{id}\times\psi^{-1}$.

We compute the integral of $\pi_\varepsilon(\nabla^n\delta(g))$ on $Y_{0,F}(N')$ for $g=f|R(\varphi)$. At each real place ρ of K, we follow the computation done in the real quadratic case, at each complex place ρ over a real place of F, we follow the computation in the imaginary quadratic case, and for each ρ over complex place of F, we follow the computation done in the case of quadratic extension of an imaginary quadratic field.

We have

$$f^{(a)}\left(y^{-1/2}\begin{pmatrix} y & x\\ 0 & 1\end{pmatrix}\right)=y^{\widehat{n}/2}|ay|_{K_{\mathbb{A}}}\left\{\sum_{\xi\in K^\times}\widehat{\lambda}(T(\xi a\mathfrak{d}))\xi^{-\widehat{v}}W(\xi y)\mathbf{e}(\xi x)\right\},$$

where, writing $n_\tau^*=2n_\tau+2$ for $\tau\in\Sigma_K(\mathbb{C})$,

$$W(y)=\prod_{\tau\in\Sigma_K}W_\tau(y_\tau),$$

$$W_\tau(y)=\begin{cases}\sum_{\alpha=0}^{n_\tau^*}\binom{n_\tau^*}{\alpha}\left(\frac{y}{i|y|}\right)^{n_\tau+1-\alpha}K_{n_\tau+1-\alpha}(4\pi|y|)S_\tau^{n_\tau^*-\alpha}T_\tau^\alpha & \text{if } \tau\in\Sigma_K(\mathbb{C}),\\ \exp(-2\pi|y|) & \text{if } \tau\in\Sigma_K(\mathbb{R}).\end{cases}$$

We consider $\nabla_\tau=\partial^2/\partial X_{\sigma\tau}\partial Y_\tau-\partial^2/\partial X_\tau\partial Y_{\sigma\tau}$ and write $\nabla^n=\prod_{\tau\in\Psi}\nabla_\tau^{n_\tau}$.

Take an algebraic Hecke character φ of K and an everywhere unramified Hecke character ω of F with $\infty(\varphi)=-w\in 1/2\Xi_K$ such that $\psi\omega\varphi_F=\alpha$. We then compute the pull back $(n!)^{-2}(\nabla^n\delta_{J,J'}(g^{(a)})|_F)=\iota^*((n!)^{-2}\nabla^n\delta_{J,J'}(g^{(a)}))$ for $g=f|R(\varphi)$ and $\iota:\mathfrak{H}\hookrightarrow\mathfrak{Z}$ induced by the inclusion $H\subset G$. Here we identify J' with a subset of I by Res_F^K and J'' with $\{\tau\in I(\mathbb{R})|\ \tau \text{ extends to a complex place of } K\}$. We put $\widehat{n}^*=\sum_{\tau\in\Sigma_K(\mathbb{C})}(2n_\tau+2)\tau$. We write I' for $I(\mathbb{C})\cup J''$. For each $j\in\mathbb{Z}[I']$, we also write n' (resp. j') for $n(I')$ (resp. $\widehat{n}(\Sigma_K(\mathbb{C}))+2j(J'')+j^\#$), where $j^\#=$

$\sum_{\tau\in J'}(j_\tau - j_{\tau c})\tau + (j_{\tau c} - j_\tau)\tau c$. Then we have

$$\begin{aligned}
&(n!)^{-2}\nabla^n\delta_{J,J'}(g^{(a)})|_F \\
&\quad = (-1)^{n(J\cup J')}(2i)^{n(J)+J} \sum_{j\in\mathbb{Z}[I'],0\le j(I')\le n(I')} (-1)^{j(I(\mathbb{C}))}\binom{n'}{j} d\mu(z) \\
&\quad\cdot \{ -g^{(a)}_{j'+2J'} + g^{(a)}_{j'+2J''+J'+\sigma J'} - g^{(a)}_{j'+2\sigma J'} - g^{(a)}_{j'+2J''+2J'} \\
&\quad + g^{(a)}_{j'+2J''+J'+\sigma J'} - g^{(a)}_{j'+2J''+2\sigma J'} \},
\end{aligned}$$

where

$$d\mu = \left\{ \bigwedge_{\tau\in I(\mathbb{R})} y_\tau^{-2} dy_\tau \wedge dx_\tau \right\} \wedge \left\{ \bigwedge_{\tau\in\Xi} y_\tau^{-3} dy_\tau \wedge dx_\tau \wedge d\overline{x}_\tau \right\},$$

$$g^{(a)} = \sum_{0\le\alpha\le\widehat{n}^*} g^{(a)}_\alpha \binom{\widehat{n}^*}{\alpha} S^{\widehat{n}^*-\alpha}T^\alpha.$$

We choose $\Delta \in K^\times$ such that

$$(\Delta) \qquad \Delta^\sigma = -\Delta \quad \text{and} \quad \Psi(\mathbb{R}) = \{\tau \in I_K(\mathbb{R}) | \Delta^\tau > 0\}.$$

If $K = F \oplus F$, as already explained, there is a standard choice of J, J'. In this case, $\Delta = (1,-1) \in K$ is an optimal choice of Δ. Anyway we see

$$\xi\Delta^{-1} \in \{x \in a^{-1}\mathfrak{d}^{-1} | x^\sigma = -x\} \iff \xi \in a^{-1}\Delta\mathfrak{d}^{-1} \bigcap F.$$

Thus we have

$$\sum_{a\in Cl_F} \omega(a) \int_{\Phi^{(a)}_\infty\backslash\mathfrak{H}} (n!)^{-2}(\nabla^n\delta_{J,J'}(g^{(a)})|_F)|a|^s_{F_\mathbb{A}} y^{s\mathbf{1}} = c_1(s)L(s,\lambda,\varphi_F\omega),$$

where

$$\begin{aligned}
c_1(s) &= \omega(\Delta^{-1}\mathfrak{d})^{-1}\Delta^{\widehat{v}+w+\widehat{n}/2}|D_F|^{1/2}G(\varphi)\left(|D_F|N_{F/\mathbb{Q}}(D_{K/F})^{1/2}\right)^{1+s} \\
&\cdot 2^{n+sI(\mathbb{C})+sJ''+J}(-1)^{n(J\cup J')+J'}\sqrt{-1}^{n(J)+J+J'}(4\pi)^{-n-(s+1)\mathbf{1}} \\
&\cdot 2^{-J''}\prod_{\tau\in J''}\left(1+(-1)^{n_\tau+1+2v_\tau+\mathrm{Res}(w)_\tau}\right) \\
&\cdot \left(\frac{(5s+1)\Gamma(s)\Gamma(s+1)}{\Gamma(2+2s)}\right)^{r_2(F)} \left(\frac{\Gamma(\frac{s}{2})^2}{\Gamma(s)}\right)^{|J''|} \prod_{\tau\in I}\Gamma(n_\tau+1+s).
\end{aligned}$$

We apply the Rankin convolution method. For that we put

$$E(s) = \sum_{a\in Cl_F} \omega(a)|a|^s_{F_\mathbb{A}}\mathcal{E}^{(a)}(s),$$

where

$$\mathcal{E}^{(a)}(s) = \sum_{\gamma\in\Phi^{(a)}/\Phi^{(a)}_\infty}{}' \varphi^2_{N'}\psi^2_{N'}(\gamma)y^{s\mathbf{1}} \circ \gamma.$$

Write $Y = Y_{0,F}(N') = \bigsqcup_{a \in Cl_F} Y^{(a)}$, if $(\omega\psi\varphi_F)^2 = \mathrm{id}$, by (RES3) in Appendix, we get in exactly the same manner as in the previous sections,

$$\begin{aligned}\mathrm{Res}_{s=1}\zeta_{F,N'}(2s)\int_Y E(s)(n!)^{-2}(\nabla^n\delta_{J,J'}(g)|_F)\\ = \frac{\phi(N')2^{[F:\mathbb{Q}]-1}\pi^{[F:\mathbb{Q}]}R_\infty}{N_{F/\mathbb{Q}}(N')^2 w(F)|D_F|}\sum_{a\in Cl_F}\omega(a)\int_{Y^{(a)}}(n!)^{-2}(\nabla^n\delta_{J,J'}(g^{(a)})|_F).\end{aligned}$$

Then by choosing a character $\varepsilon : C_\infty/C_{\infty+} \to \{\pm1\}$ with $\varepsilon((-1,-1)_\tau) = (-1)^{n_\tau+1}$ for all $\tau \in J$ as in Section 5, we compute the residue using the Euler factorization of the zeta function of the right hand side, and we get, choosing φ so that $\omega\varphi_F\psi = \alpha$

$$\begin{aligned}\Gamma(1,\mathrm{Ad}(\lambda)\otimes\alpha)L_{C'}(1,\mathrm{Ad}(\lambda)\otimes\alpha)\\ = c_0 G(\varphi)^{-1}\omega(\Delta^{-1}\mathfrak{d})^{-1}\Delta^{-\widehat{v}-w-\widehat{n}/2}\sqrt{-1}^{-n(J)+J+J'}\\ \cdot\sum_{a\in Cl_F}\omega(a)\int_{Y^{(a)}}(n!)^{-2}(\nabla^n\pi_\varepsilon\delta_{J,J'}(g^{(a)})|_F),\end{aligned}$$

where c_0 is an explicit non-zero constant in $\mathbb{Q}$.

We now define a modular transcendental factor of the above L-value. When F has more than one complex places, we cannot define the transcendental factor solely using the data of cohomology group for $Y_{0,K}(N)$ as in the previous sections. Instead we need to use cohomology groups for $Y_{0,K}(N)$ and $Y_{0,F}(N')$. Because of this, we can define the period only for $\widehat{\lambda}$ and not for general non-base-change lift. Let $\varepsilon : S = C_\infty/C_{\infty+} \to \{\pm1\}$ be a character satisfying $\varepsilon((-1)_\tau,(-1)_{\sigma\tau}) = -1$ for all $\tau \in J$. We start with a system of Hecke eigenvalues λ for H with character ψ and write its base change lift to G as $\widehat{\lambda}$, whose conductor is N. Let $q = \dim(\mathfrak{H}) = 2r_1(F) + 3r_2(F)$. We write

$$H(A)[\widehat{\lambda},\varepsilon] = H^q_{\mathrm{cusp}}(Y_{0,K}(N),\mathcal{L}((\widehat{n},\widehat{v}),\chi;A))[\widehat{\lambda},\varepsilon]$$

for the subspace on which $T(\mathfrak{n})$ acts via $\widehat{\lambda}$ and S acts via ε. Note that $\{\pi_\varepsilon\delta_{J,J'}(f)\}_{J'}$, J' running through all subsets of $\Sigma_K(\mathbb{C})$ of cardinality $\#(\Sigma(\mathbb{C}))$, forms a base of $H(\mathbb{C})[\widehat{\lambda},\varepsilon]$. Let $N_\varphi = N \cap C^2$ for the conductor C of φ. We have the following sequence of maps:

$$R(\varphi) : H(A)[\lambda,\varepsilon] \longrightarrow H^q_{\mathrm{cusp}}(Y_0(N_\varphi),\mathcal{L}((\widehat{n},\widehat{v}+w),\chi\varphi^2;A))[\lambda,\varepsilon\varphi_\infty];$$

$$i^* : H^q_{\mathrm{cusp}}(Y_0(N_\varphi),\mathcal{L}((\widehat{n},\widehat{v}+w),\chi\varphi^2;A)) \longrightarrow H^q_{\mathrm{cusp}}(Y,\mathcal{L}((\widehat{n},\widehat{v}+w),(\chi\varphi^2);A)|_Y);$$

$$\pi_* : H^q_{\mathrm{cusp}}(Y,\mathcal{L}((\widehat{n},\widehat{v}+w),(\chi\varphi^2)_F;A)) \longrightarrow H^q_c(Y,\mathcal{L}(0,\omega^{-2};A));$$

$$\omega_* : H^q_c(Y,\mathcal{L}(0,\omega^{-2};A)) \underset{\cup[\det(\omega)]}{\longrightarrow} H^q_c(Y,A) \cong A.$$

The map i^* is induced by the inclusion $i : Y \hookrightarrow Y_{0,K}(N_\varphi)$, and π_* is induced by the morphism of sheaves $\pi : \mathcal{L}((\widehat{n},\widehat{v}+w),\chi\varphi^2;A)|_Y \to \mathcal{L}(0,\omega^{-2};A)$ (here $\chi\varphi^2$ is well chosen under the condition that $\omega\varphi_F\psi = \alpha$ so that the map π exists). The last map is induced by the cup product with the global section $\det(\omega)$ of $\mathcal{L}(0,\omega^2;A)$. We consider the composition $Ev = Ev_\varphi = \omega_* \circ \pi_* \circ i^* \circ R(\varphi)$. Then by the above computation, $Ev([\pi_\varepsilon\delta_{J,J'}(f)]) \neq 0$. We write M for the image under Ev of $H(A)[\lambda,\varepsilon]$. We suppose that $M \subset A$ is free of rank 1 over A and write its generator as $\xi_\varepsilon(f)$. Since the value $Ev([\pi_\varepsilon\delta_{J,J'}(f)])$ is independent of J', we can define $\Omega'_{1,2}(\varepsilon,\widehat{\lambda};A)\in\mathbb{C}^\times$ by the following formula:

$$Ev([\pi_\varepsilon\delta_{J,J'}(f)])=\Omega'_{1,2}(\varepsilon,\widehat{\lambda};A)\xi_\varepsilon(f),$$

where f is the cohomological modular form whose Fourier coefficient at ideal $\mathfrak{n}$ is given by $\widehat{\lambda}(T(\mathfrak{n}))$. For the moment, we assume that φ and ω are arithmetic, and A is a Dedekind domain inside a finite extension of $\mathbb{Q}(\widehat{\lambda},\varphi,\omega)$. Here we note that

$$Ev([\pi_\varepsilon\delta_{J,J'}(f)])=\sum_{a\in Cl_F}\omega(a)\int_{Y(a)}(n!)^{-2}(\nabla^n\pi_\varepsilon\delta_{J,J'}(f|R(\varphi))|_F).$$

We also define for $\rho\in\mathrm{Aut}(\mathbb{C})$ a constant $\Omega'_{1,2}(\varepsilon,\lambda^\rho;A)$ by

$$Ev([\pi_\varepsilon\delta_{J,J'}(f^\rho)])=\Omega'_{1,2}(\varepsilon,\lambda^\rho;A)\xi_\varepsilon(f^\rho).$$

If we have two choices of (φ,ω), say (φ,ω) and (φ',ω'), the ratio:

$$Ev_\varphi([\pi_\varepsilon\delta_{J,J'}(f)])/Ev_{\varphi'}([\pi_\varepsilon\delta_{J,J'}(f)])$$

is just equal to the identity component of

$$(\omega(\Delta^{-1}\mathfrak{d})^{-1}\otimes 1)\mathbf{G}(\varphi)/(\omega'(\Delta^{-1}\mathfrak{d})^{-1}\otimes 1)\mathbf{G}(\varphi')$$

in $\mathbb{Q}(\lambda,\varphi,\varphi')\otimes_{\mathbb{Q}}\mathbb{C}$, and therefore, by an argument on Gauss sums close to the one given just above Theorem 7.1, $(G(\psi^\rho\alpha)^{-1}\Omega'_{1,2}(\varepsilon,\widehat{\lambda}^\rho;A))_\rho$ is a well defined element in

$$(\mathbb{Q}(\lambda)\otimes_{\mathbb{Q}}\mathbb{C})^\times/(\mathbb{Q}(\lambda)^\times\otimes 1),$$

which we write $\underline{\Omega}_{(1,2)}(\varepsilon,\widehat{\lambda};\mathbb{Q}(\lambda))=(G(\psi^\rho\alpha)^{-1}\Omega'_{1,2}(\varepsilon,\widehat{\lambda};A))_\rho$. When φ or ω is algebraic (but not arithmetic), we write simply $\Omega_{(1,2)}(\varepsilon,\widehat{\lambda};\overline{\mathbb{Q}})$ for $\Omega'_{(1,2)}(\varepsilon,\widehat{\lambda};\overline{\mathbb{Q}})$. Then we have

THEOREM 8.1. *Let α be a Hecke character of $F_{\mathbb{A}}^\times/F^\times$ with $\alpha^2=1$. We allow $\alpha=\mathrm{id}$. Suppose that we have an arithmetic Hecke character φ of K and ω of F such that (i) the conductor of ω is 1 and (ii) $\psi\varphi_F\omega=\alpha$. Then we have*

$$\frac{(1\otimes\Delta^{\widehat{v}+w+\widehat{n}/2}\Gamma(1,\mathrm{Ad}(\lambda)\otimes\alpha))\mathbb{L}_{C'}(1,\mathrm{Ad}(\lambda)\otimes\alpha)}{(1\otimes\sqrt{-1}^{n(J)+J+J'})\mathbf{G}(\psi\alpha)\underline{\Omega}_{(1,2)}(\varepsilon,\widehat{\lambda};\mathbb{Q}(\lambda))}\in\mathbb{Q}(\lambda)\subset\mathbb{Q}(\lambda)\otimes_{\mathbb{Q}}\mathbb{C}$$

for each character $\varepsilon:S\to\{\pm1\}$ with $\varepsilon((-1,-1)_\tau)=-1$ for all $\tau\in J$. If φ is an algebraic Hecke character satisfying (i) and (ii), then we have

$$\frac{\Gamma(1,\mathrm{Ad}(\lambda)\otimes\alpha)L_{C'}(1,\mathrm{Ad}(\lambda)\otimes\alpha)}{\Omega_{(1,2)}(\varepsilon,\widehat{\lambda};\overline{\mathbb{Q}})}\in\overline{\mathbb{Q}}.$$

9. Period relation

We shall list here several period relations which follow easily from the main theorems. Some of them is a partial generalization of such relations for totally real extensions studied by Shimura, Harris and Yoshida (cf. [**Y95**], [**Y94**]). For simplicity, we suppose that F is totally imaginary. Let ϕ^ρ be a unique cohomological form on H whose Mellin transform is the standard L-function of λ^ρ. We write C_0 for the conductor of ϕ. Then we choose a generator $\xi_m(\phi^\rho)$ $(m=1,2)$ of

$$H^{mq}_{\mathrm{cusp}}(Y_{0,F}(C_0),\mathcal{L}((n,v),\psi;\mathbb{Q}(\lambda^\rho))[\lambda^\rho]$$

for $q = [F : \mathbb{Q}]/2$ so that $\xi_m(\phi^\rho) = \xi_m(\phi)^\rho$, where $\rho \in \mathrm{Aut}(\mathbb{C})$ acts on the cohomology group through covariant functoriality of its action on $L((n, v), \psi; \mathbb{Q}(\lambda))$. We then define a constant $\Omega_m(\lambda^\rho; \mathbb{Q}(\lambda))$ $(m = 1, 2)$ by

$$[\delta(\phi^\rho)] = \Omega_m(\lambda^\rho; \mathbb{Q}(\lambda^\rho))\xi_m(\phi^\rho).$$

Of course, we can give more integral definition of $\Omega_m(\lambda^\rho; A)$ for, say, a discrete valuation ring A of $\mathbb{Q}(\lambda^\rho)$. However, here we just want to discuss rational and algebraic relations among modular periods Ω_j and $\Omega_{1,2}$. In other words, we like to study relations among periods of harmonic forms of different degree but belonging to the same system of Hecke eigenvalues.

First we apply Theorem 8.1 to $F \oplus F$ with the standard choice of $\varphi = \mathrm{id} \times \psi^{-1}$. Then it is easy to conclude from an argument given in Urban [**U**] for imaginary quadratic fields F that

$$\frac{\Gamma(1, \mathrm{Ad}(\lambda))L(1, \mathrm{Ad}(\lambda))}{\Omega_1(\lambda; \mathbb{Q}(\lambda))\Omega_2(\lambda; \mathbb{Q}(\lambda))} \in \overline{\mathbb{Q}}^\times.$$

Thus writing $a \sim b$ if $a/b \in \overline{\mathbb{Q}}^\times$, we see that

$$\Omega_{1,2}(\widehat{\lambda}; \mathbb{Q}(\lambda)) \sim \Omega_1(\lambda; \mathbb{Q}(\lambda))\Omega_2(\lambda; \mathbb{Q}(\lambda)).$$

This is one of the reasons why we have written the period as $\Omega_{1,2}$. This type of relation can be shown in a little more general case. These quantities Ω_j are defined depending only on F (and λ), but $\Omega_{1,2}$ is defined relative to a general quadratic extension K/F (and λ). Thus we fix λ and write $\Omega_j^F(\lambda)$ (resp. $\Omega_{1,2}^{K/F}(\lambda)$) for $\Omega_j(\lambda; \mathbb{Q}(\lambda))$ (resp. $\Omega_{1,2}(\widehat{\lambda}; \mathbb{Q}(\lambda))$ relative to K/F). When $\alpha = \mathrm{id}$, we get from the above identity

$$\Omega_{1,2}^{F\oplus F/F}(\lambda) \sim \Omega_1^F(\lambda)\Omega_2^F(\lambda). \tag{P1}$$

Applying $(P1)$ to $\widehat{\lambda}$ in place of λ, we get

$$\begin{aligned}\Omega_1^K(\widehat{\lambda})\Omega_2^K(\widehat{\lambda}) &\sim \Gamma(1, \mathrm{Ad}(\widehat{\lambda}))L(1, \mathrm{Ad}(\widehat{\lambda}))\\ &= \Gamma(1, \mathrm{Ad}(\lambda))L(1, \mathrm{Ad}(\lambda))\Gamma(1, \mathrm{Ad}(\lambda) \otimes \alpha)L(1, \mathrm{Ad}(\lambda) \otimes \alpha)\\ &\sim \Omega_1^F(\lambda)\Omega_2^F(\lambda)\Omega_{1,2}^{K/F}(\lambda),\end{aligned}$$

On the other hand, it is proven in [**Hi94**, Theorem 8.1] that

$$\begin{aligned}(2\pi)^{2j+2\mathbf{1}}\Omega_1^K(\lambda) &\sim L(0, \lambda \otimes \eta \circ N_{K/F})\\ &= L(0, \lambda \otimes \eta)L(0, \lambda \otimes \eta\alpha) \sim (2\pi)^{2j+2\mathbf{1}}\Omega_1^F(\lambda)^2\end{aligned}$$

if $0 \le j \le n$ and $\infty(\eta) = j + v + \mathbf{1}$, as long as the modular standard L-values are non-zero for some j, for example, if $n_\tau > 2$ for all τ. Thus we conclude (see (P) of Conjecture 5.1), if $n_\tau > 2$ for all τ,

$$\Omega_1^K(\widehat{\lambda}) \sim \Omega_1^F(\lambda)^2. \tag{P2}$$

This shows

$$\Omega_1^F(\lambda)^2\Omega_2^K(\widehat{\lambda}) \sim \Omega_1^K(\widehat{\lambda})\Omega_2^K(\widehat{\lambda}) \sim \Omega_1^F(\lambda)\Omega_2^F(\lambda)\Omega_{1,2}^{K/F}(\lambda),$$

and hence

$$\Omega_1^F(\lambda)\Omega_2^K(\widehat{\lambda}) \sim \Omega_2^F(\lambda)\Omega_{1,2}^{K/F}(\lambda). \tag{P3}$$

We put

$$\Omega'^F_2(\lambda) = \Omega^F_1(\lambda)^{-1}\Omega^{K/F}_{1,2}(\lambda).$$

Then we have

$$\Omega^K_2(\widehat{\lambda}) \sim \Omega^F_2(\lambda)\Omega'^F_2(\lambda). \tag{$P4$}$$

This formula seems to be a generalization of (P2) for Ω_2. It is an interesting problem to find out its motivic meaning. Suppose that $\lambda = \widehat{\lambda}_0$ for a system of Hecke eigenvalues λ_0 for $\mathrm{GL}(2)_{/L}$ for a totally real subfield L with $[F : L] = 2$. We further assume that $K = FF'$ for two totally imaginary quadratic extensions F/L and F'/L. Then one can prove easily from the above relations that, if $n_\tau > 2$ for all τ

$$\Omega'^F_2(\lambda) \sim \Omega^{F'}_2(\mu), \quad \Omega^{K/F}_{1,2}(\lambda) \sim \Omega^F_1(\lambda)\Omega^{F'}_2(\mu) \quad \text{and} \quad \Omega^K_2(\widehat{\lambda}) \sim \Omega^F_2(\lambda)\Omega^{F'}_2(\mu), \tag{P5}$$

where μ is the base change lift of λ_0 to $\mathrm{GL}(2)_{/F'}$.

Appendix A. Eisenstein series of weight 0

Here, for the reader's convenience, we shall prove the residue formula of the Eisenstein series we used in the principal text, which seems not to be found in the literature in its exact form.

Let F be a number field. We consider the algebraic group $H = \mathrm{Res}_{F/\mathbb{Q}}\mathrm{GL}(2)_{/F}$. We use the same symbol introduced in [**Hi91**, Section 4], where F is assumed to be totally real, but symbols themselves have meaning. For a finite order character χ and θ of the idele class group $F^\times_{\mathbb{A}}/F^\times$ modulo an ideal C of $\mathfrak{r}$, we consider the following Eisenstein series $\mathcal{E}(x,s) = \mathcal{E}(x,\chi,\theta;s) : H(\mathbb{A}) \to \mathbb{C}$:

$$\mathcal{E}(x,s) = \sum_{\gamma \in \mathfrak{r}^\times B(\mathbb{Q})\backslash H(\mathbb{Q})} \chi^*(\gamma x)\theta(\gamma x)\eta(\gamma x)^s,$$

where

$$B(A) = \left\{ \begin{pmatrix} a & b \\ 0 & 1 \end{pmatrix} | a \in (F \otimes_{\mathbb{Q}} A)^\times \text{ and } b \in F \otimes_{\mathbb{Q}} A \right\} \subset H(A)$$

for $\mathbb{Q}$-algebras A, and for the identity component $H(\mathbb{R})_+$ of $H(\mathbb{R})$,

$$\chi^*(x) = \begin{cases} \chi_C(d_C) \text{ for } x = \begin{pmatrix} a & b \\ c & d \end{pmatrix} & \text{if } x \in B(\mathbb{A})U_0(C)H(\mathbb{R})_+, \\ 0 & \text{otherwise,} \end{cases}$$

$$\theta(x) = \begin{cases} \theta(a) & \text{if } x \in \begin{pmatrix} a & b \\ 0 & 1 \end{pmatrix} Z_H(\mathbb{A})H(\widehat{\mathbb{Z}})C_{\infty+} \\ 0 & \text{otherwise,} \end{cases}$$

and $\eta = \theta$ for $\theta = |\ |_{F_{\mathbb{A}}}$. We normalize the Eisenstein series in the following way:

$$E^*(x,\chi,\theta;s) = N_{F/\mathbb{Q}}(C)2^{-r_2}|D_F|^{1/2}\sum_a \chi^{-1}(a)\mathcal{E}(au,\chi,\theta;s),$$

where a runs over a complete set of representatives for $F^\times_{\mathbb{A}}/F^\times\widehat{\mathfrak{r}}^\times F^\times_\infty$ and r_2 is the number of complex places of F. We further put

$$E(x,\chi,\theta,s) = L_C(2s,\chi^{-1}\theta^2)E^*(x,\chi,\theta;s).$$

We now compute the Fourier expansion of the above Eisenstein series. We put following [**Hi94**, Appendix]

$$\zeta(y,t;s)=\int_{\mathbb{C}}\exp(-2\pi i\mathrm{Tr}(t\tau))(|\tau|^2+y^2)^{-s}d\tau.$$

Then for the modified Bessel function $K_s(t)$ as in [**Hi94**, Section 6],

$$\zeta(y,t;s)=\begin{cases}(2\pi)^s\Gamma(s)^{-1}y^{1-s}|t|^{s-1}K_{s-1}(4\pi|t|y) & \text{if } t\neq 0,\\ \pi(s-1)^{-1}y^{2-2s} & \text{if } t=0.\end{cases}$$

We take $\varepsilon=\left(\begin{smallmatrix}0&-1\\1&0\end{smallmatrix}\right)$ and compute the residue at $s=1$ of $E(x,s)=\mathcal{E}(x\varepsilon^{(\infty)},s)$ via its Fourier expansion. Let w be a variable of $H(\mathbb{R})_+$. Then we write $z=w(z_0)$, where $z_{0,\sigma}=\sqrt{-1}$ if σ is real and ε if σ is complex. We write $y(z)=(y_\sigma)\in F_\infty$ when $z_\sigma=x_\sigma+\sqrt{-1}y_\sigma$ for σ real and $z_\sigma=\left(\begin{smallmatrix}x_\sigma&-y_\sigma\\y_\sigma&\overline{x}_\sigma\end{smallmatrix}\right)$ for σ complex. For $\gamma_\sigma=\left(\begin{smallmatrix}a_\sigma&b_\sigma\\c_\sigma&d_\sigma\end{smallmatrix}\right)$, we define an automorphic factor by

$$j(\gamma_\sigma,z_\sigma)=\begin{cases}c_\sigma z_\sigma+d_\sigma & \text{if } \sigma \text{ is real},\\ \det(\rho(c_\sigma)z_\sigma+\rho(d_\sigma)) & \text{if } \sigma \text{ is complex},\end{cases}$$

where $\rho(x)=\left(\begin{smallmatrix}x&0\\0&\overline{x}\end{smallmatrix}\right)$. Then by [**Hi91**, Lemma 6.4], we have

$$\chi^*((\varepsilon a\alpha(x)\begin{pmatrix}a'&0\\0&1\end{pmatrix}\varepsilon^{-1})^{(\infty)})=\chi(a_Ca'_C),$$

$$\eta((\varepsilon a\alpha(x)\begin{pmatrix}a'&0\\0&1\end{pmatrix}\varepsilon^{-1})^{(\infty)})=|(a^2a')^{(\infty)}|_{F_{\mathbb{A}}},$$

$$\eta((\varepsilon a\alpha(x)w)_\infty)=|a_\infty^2y(z)|_{F_{\mathbb{A}}}\prod_{\sigma\in I/\langle c\rangle}|j(\varepsilon a\alpha(x)_\sigma,z_\sigma)|^{-2},$$

$$\theta((\varepsilon a\alpha(x)\begin{pmatrix}a'&0\\0&1\end{pmatrix}\varepsilon^{-1})^{(\infty)})=\theta(a^2a'\mathfrak{r})\quad\text{if } a'_C=1,$$

$$\theta((\varepsilon a\alpha(x)w)_\infty)=1,$$

where $\alpha(x)=\left(\begin{smallmatrix}1&x\\0&1\end{smallmatrix}\right)$. Then the Fourier coefficient $b(\xi,u,s)$ at $\xi\in F$ of $E(x,s)$ given by

$$\int_{F_{\mathbb{A}}/F}E(\alpha(x)u,s)\mathbf{e}_F(-\xi x)dx,$$

we get from [**Hi91**, Lemma 6.6] that for $u=x\left(\begin{smallmatrix}a&0\\0&1\end{smallmatrix}\right)w$ with $a_\infty=a_C=1$ and $x\in Z_H(\mathbb{A}^{(\infty)})$

$$\xi\in\mathfrak{a}^{-1}C^{-1}\mathfrak{d}^{-1}\quad\text{if } b(\xi,u,s)\neq 0,$$

where $\mathfrak{d}$ is the absolute different of $F/\mathbb{Q}$ and $\mathfrak{a}=a\mathfrak{r}$. Then we see from [**Hi91**, (6.11)]

$$\begin{aligned}b(\xi,u,s)=&N_{F/\mathbb{Q}}(C)^{-1}|D_F|^{-1/2}\chi\theta^{-1}(a^{(\infty)})\chi(x)N_{F/\mathbb{Q}}(\mathfrak{a})^{s-1}\\&\cdot\int_{F_\infty}\eta(\varepsilon x\alpha(v_\infty)w)^s\mathbf{e}_F(-\xi v_\infty)dv_\infty\\&\cdot\sum_{\mathfrak{n}\sim x\mathfrak{a}}\theta^2\chi^{-1}(\mathfrak{n})N_{F/\mathbb{Q}}(\mathfrak{n})^{-2s}\sum_{\mathfrak{b}\supset\mathfrak{n}+\xi\mathfrak{a}C\mathfrak{d}}\mu(\mathfrak{n}/\mathfrak{b})N_{F/\mathbb{Q}}(\mathfrak{b}),\end{aligned}$$

where $\mathfrak{a}\sim\mathfrak{b}$ means that the two ideals belong to the same ideal class. As seen in [**Hi91**, 6.12b], from its definition combined with the above formula, the Fourier coefficient at ξ for $E^*(u,\chi,\theta;s)$ is given by

$$\chi\theta^{-1}(a^{(\infty)})N_{F/\mathbb{Q}}(\mathfrak{a})^{s-1}L_C(2s,\chi^{-1}\theta^2)^{-1}\int_{F_\infty}\eta(\varepsilon\alpha(v_\infty)w)^s\mathbf{e}_F(-\xi v_\infty)dv_\infty$$
$$\cdot\begin{cases}\sum_{\mathfrak{b}\supset\xi\mathfrak{a}C\mathfrak{b}}\chi^{-1}\theta^2(\mathfrak{b})N_{F/\mathbb{Q}}(\mathfrak{b})^{1-2s} & \text{if } \xi\neq 0,\\ L_C(2s-1,\chi^{-1}\theta^2) & \text{if } \xi=0.\end{cases}$$

Following [**Hi94**, Appendix], we get for a complex place σ

$$2^{-1}\int_{\mathbb{C}}\eta(\varepsilon\alpha(v_\sigma)_\sigma w_\sigma)^s\mathbf{e}_F(-\xi v_\sigma)dv_\sigma=|y_\sigma|^{2s}\zeta(y,\xi;2s,0)$$
$$=\begin{cases}(2\pi)^{2s}\Gamma(2s)^{-1}y_\sigma|\xi|^{2s-1}K_{2s-1}(4\pi|\xi y|) & \text{if } \xi\neq 0,\\ \pi(2s-1)^{-1}y_\sigma^{2-2s} & \text{if } \xi=0.\end{cases}$$

For real σ, we get from [**Hi91**, 6.9b]

$$\int_{\mathbb{R}}\eta(\varepsilon\alpha(v_\sigma)_\sigma w_\sigma)^s\mathbf{e}_F(-\xi v_\sigma)dv_\sigma$$
$$=\begin{cases}\pi^s\Gamma(s)^{-1}|\xi|^{s-1}\exp(-2\pi|\xi y|)\omega(4\pi|\xi y|;s,s) & \text{if } \xi\neq 0,\\ \pi\Gamma(s)^{-2}\Gamma(2s-1)(4y_\sigma)^{1-s} & \text{if } \xi=0,\end{cases}$$

where $\omega(t;s,s)$ is the hyper-geometric function defined in [**Hi91**, Section 6]. By this computation, we know that the Eisenstein series has meromorphic continuation to the whole complex s-plane. Moreover the non-constant term of $E(x,\chi,\mathrm{id},s)$ is an entire function of s, and hence the residue is constant. The constant term is given by

$$\chi(a^{(\infty)})N_{F/\mathbb{Q}}(\mathfrak{a})^{s-1}(\pi 4^{1-s}\Gamma(s)^{-2}\Gamma(2s-1))^{r_1}\cdot(\pi(2s-1)^{-1})^{r_2}|y|_{F_{\mathbb{A}}}^{1-s}L_C(2s-1,\chi).$$

This shows

(RES1) $\quad \mathrm{Res}_{s=1}E(x,\chi,\mathrm{id},s)$
$$=\begin{cases}0 & \text{if } \chi\neq\mathrm{id},\\ N_{F/\mathbb{Q}}(C)^{-1}\phi(C)\frac{2^{r_1+r_2-1}\pi^{[F:\mathbb{Q}]}R_\infty h(F)}{w(F)|D_F|^{1/2})} & \text{if } \chi=\mathrm{id},\end{cases}$$

where ϕ is the Euler function: $\phi(C)=\#(\mathfrak{r}/C)^\times$, R_∞ is the regulator of F and $w(F)$ is the number of roots of unity in F.

We now show that the Eisenstein series $\mathcal{E}(x,\chi;s;C)=\mathcal{E}(x,\chi,\mathrm{id};s)$ gives a section of the sheaf $L((0,0),\chi^{-1};\mathbb{C})$ over $Y_{0,F}(C)$. The identity

$$\begin{pmatrix}a&b\\c&d\end{pmatrix}=\gamma t\begin{pmatrix}y&x\\0&1\end{pmatrix}=\gamma\begin{pmatrix}y&x\\0&1\end{pmatrix}t\in B(\mathbb{A})U_0(C)H(\mathbb{R})_+$$

for $t=\left(\begin{smallmatrix}a&0\\0&1\end{smallmatrix}\right)$ implies that

$$\gamma\begin{pmatrix}y&x\\0&1\end{pmatrix}\in B(\mathbb{A})U_0(C)t^{-1}H(\mathbb{R})_+=B(\mathbb{A})tU_0(C)t^{-1}H(\mathbb{R})_+.$$

Thus (c,d) is the second row of a matrix in $\Phi^{(a)}=H(\mathbb{Q})\cap tU_0(C)t^{-1}H(\mathbb{R})_+$. This shows that we can choose the γ modulo $\mathfrak{r}^\times B(\mathbb{Q})_+$ inside $\Phi^{(a)}$. Note that $\Phi^{(a)}\cap\mathfrak{r}^\times B(\mathbb{Q})_+=\Phi^{(a)}_\infty$. Then for $w=\left(\begin{smallmatrix}y&x\\0&1\end{smallmatrix}\right)$

$$\gamma wt=\gamma_\infty wtt^{-1}\gamma^{(\infty)}t\in\gamma_\infty wtU_0(C).$$

This shows that, for $z = w(z_0)$ which is the image of $\left(\begin{smallmatrix} y & x \\ 0 & 1 \end{smallmatrix}\right)$ in $\mathfrak{H}$,

$$\eta(\gamma wt) = |a|_{F_{\mathbb{A}}} y^{\mathbf{1}} \circ \gamma \quad \left(\text{for } \mathbf{1} = \sum_{\sigma \in I} \sigma\right) \text{ and } \chi^*(\gamma \begin{pmatrix} y & x \\ 0 & 1 \end{pmatrix} t) = \chi_C(d).$$

Thus we get

$$\mathcal{E}\left(t\begin{pmatrix} y & x \\ 0 & 1 \end{pmatrix}, \chi; s, C\right) = |a|_{F_{\mathbb{A}}}^s \sum_{\gamma \in \Phi^{(a)}/\Phi^{(a)}_\infty} \chi_C(\gamma) y^{s\mathbf{1}} \circ \gamma = |a|_{F_{\mathbb{A}}}^s \mathcal{E}^{(a)}(z).$$

This shows the claim.

Out of (RES1), we compute the residue of $\mathcal{E}$. We shall make use of the following identity:

$$\begin{aligned} \sum_{\varphi} E^*(x; \chi\varphi) &= N_{F/\mathbb{Q}}(C) 2^{-r_2} |D_F|^{1/2} \sum_{\varphi} \sum_{a \in Cl_F} (\chi\varphi)^{-1}(a) \mathcal{E}(ax, \chi) \\ &= N_{F/\mathbb{Q}}(C) 2^{-r_2} |D_F|^{1/2} \sum_{a \in Cl_F} \chi(a)^{-1} \sum_{\varphi} \varphi(a^{-1}) \mathcal{E}(ax, \chi\varphi) \\ &= h(F) N_{F/\mathbb{Q}}(C) 2^{-r_2} |D_F|^{1/2} \mathcal{E}(x, \chi), \end{aligned}$$

where φ runs over all characters of the class group Cl_F of F. We know from the above residue formula that

(RES2) $$\begin{aligned} &\mathrm{Res}_{s=1} L_C(2s, \chi^{-1}) E^*(x; \chi) \\ &\qquad = \delta_{\chi,\mathrm{id}} N_{F/\mathbb{Q}}(C)^{-1} \phi(C) \frac{2^{r_1+r_2-1} \pi^{[F:\mathbb{Q}]} R_\infty h(F)}{w(F)|D_F|^{1/2}}, \end{aligned}$$

where $\delta_{\chi,\mathrm{id}} = \begin{cases} 1 & \text{if } \chi = \mathrm{id}, \\ 0 & \text{otherwise.} \end{cases}$

In the principal text, we have used the residue formula when $\chi = \omega^{-2}$ for ω chosen so that $\alpha = \psi\varphi_F\omega$ under the notation of Section 2.4. In this case, $C = N'$. Since the Euler product for $L_C(2s, \omega^2)$ converges at $s = 1$, we note that $L_C(2, \omega^2) \neq 0$. This shows that if $\chi = \omega^{-2} = (\psi\varphi_F)^2$,

(RES3) $$\begin{aligned} &\mathrm{Res}_{s=1} |a|_{F_{\mathbb{A}}}^s \mathcal{E}^{(a)}(z) \\ &\qquad = \zeta_{F,N'}(2)^{-1} N_{F/\mathbb{Q}}(N')^{-2} \phi(N') \frac{2^{[F:\mathbb{Q}]-1} \pi^{[F:\mathbb{Q}]} R_\infty}{w(F)|D_F|}. \end{aligned}$$

References

[ArT] E. Artin and J. Tate, Class Field Theory, New York: Benjamin (1967).

[As] T. Asai, *On certain Dirichlet series associated with Hilbert modular forms and Rankin method*, Math. Ann. **226** (1977), 81–94.

[BK] S. Bloch and K. Kato, *L-functions and Tamagawa numbers of motives*, Grothendieck Festschrift I, Progress in Math. **86**, Boston: Birkhäuser (1990), 333–400.

[BR] D. Blasius and J. D. Rogawski, *Motives for Hilbert modular forms*, Inventiones Math. **114** (1993), 55–87.

[C] H. Carayol, *Formes modulaires et représentations galoisiennes à valeurs dans un anneau local compact*, p-adic Monodromy and the Birch and Swinnerton-Dyer Conjecture (Boston, MA, 1991), Contemporary Math. **165**, Providence, RI: Amer. Math. Soc. (1994), 213–237.

[DHI] K. Doi, H. Hida and H. Ishii, *Discriminants of Hecke fields and the twisted adjoint L-values for* GL(2), Inventiones Math. **134** (1998), 547–577.

[Fl] Y. Z. Flicker, *Twisted tensors and Euler products*, Bull. Soc. Math. France **116** (1988), 295–313.

[FlZ] Y. Z. Flicker and D. Zinoviev, *On poles of twisted tensor L-functions*, Proc. Japan Acad. series A **71** (1995), 114–116.

[Fu] K. Fujiwara, *Deformation rings and Hecke algebras in the totally real case*, preprint (1996).

[GeJ] S. Gelbart and H. Jacquet, *A relation between automorphic representations of* GL(2) *and* GL(3), Ann. Scient. Ec. Norm. Sup. 4-th series **11** (1978), 471–542.

[Gh] E. Ghate, *Critical values of the twisted tensor L-function in the imaginary quadratic case*, to appear, Duke Math. J. **96** (1998).

[Ha] G. Harder, *Eisenstein cohomology of arithmetic groups, the case* GL_2, Inventiones Math. **89** (1987) 37–118.

[HaLR] G. Harder, R. P. Langlands and M. Rapoport, *Algebraische Zyklen auf Hilbert-Blumenthal Flächen*, J. Reine Angew. Math. **366** (1986), 53–120.

[Hi81] H. Hida, *Congruences of cusp forms and special values of their zeta functions*, Inventiones Math. **63** (1981), 225–261.

[Hi86] H. Hida, *Galois representations into* $\mathrm{GL}_2(\mathbb{Z}_p[[X]])$ *attached to ordinary cusp forms*, Inventiones Math. **85** (1986), 545–613.

[Hi88a] H. Hida, *On p-adic Hecke algebras for* GL_2 *over totally real fields*, Ann. of Math. **128** (1988), 295–384.

[Hi88b] H. Hida, *Modules of congruence of Hecke algebras and L-functions associated with cusp forms*, Amer. J. of Math. **110** (1988), 323–382.

[Hi89] H. Hida, *Theory of p-adic Hecke algebras and Galois representations*, Sugaku Expositions **2** (1989), 75–102.

[Hi91] H. Hida, *On p-adic L-functions of* $\mathrm{GL}(2) \times \mathrm{GL}(2)$ *over totally real fields*, Ann. l'institut Fourier **41** (1991), 311–391.

[Hi93a] H. Hida, Elementary Theory of L-Functions and Eisenstein Series, London Mathematical Society Student Texts **26**, Cambridge: Cambridge University Press (1993).

[Hi93b] H. Hida, *p-Ordinary cohomology groups for* $SL(2)$ *over number fields*, Duke Math. J. **69** (1993), 259–314.

[Hi94] H. Hida, *On the critical values of L-functions of* GL(2) *and* $\mathrm{GL}(2) \times \mathrm{GL}(2)$, Duke Math. J. **74** (1994), 431–529.

[Hi96] H. Hida, *On Selmer groups of adjoint modular Galois representations*, Number Theory, Sém. Théorie des Nombres de Paris, 1993-94, LMS Lecture notes series **235** (1996), 89–132.

[HiTU] H. Hida, J. Tilouine and E. Urban, *Adjoint modular Galois representations and their Selmer groups*, Proc. NAS **94** (1997), 11121–11124.

[I] J. Im, *Special values of Dirichlet series attached to Hilbert modular forms*, Amer. J. Math. **113** (1991), 975–1017.

[J] H. Jacquet, Automorphic Forms on GL(2), II, Lecture Notes in Math. **278**, Berlin-New York: Springer-Verlag (1972).

[L] R. P. Langlands, Base Change for GL(2), Annals of Math. Studies **96**, Princeton: Princeton University Press (1980).

[Ma] B. Mazur, *Deforming Galois representations*, MSRI Publ. **16** (1989), 385–437.

[MaT] B. Mazur and J. Tilouine, *Représentations Galoisiennes, différentielles de Kähler et "conjectures principales"*, Publ. IHES **71** (1990), 65–103.

[Mi] T. Miyake, Modular Forms, Berlin-New York: Springer-Verlag (1989).

[RSS] M. Rapoport, N. Schappacher and P. Schneider, Beilinson's Conjectures on Special Values of L-Functions, Perspectives in Math. **4**, Boston: Academic Press (1988).

[S81] F. Shahidi, *On certain L-functions*, Amer. J. Math. **103** (1981), 297–355.

[S88] F. Shahidi, *On the Ramanujan conjecture and finiteness of poles for certain L-functions*, Ann. of. Math. **127** (1988), 547–584.

[Sh71] G. Shimura, Introduction to the Arithmetic Theory of Automorphic Functions, Princeton: Iwanami Shoten and Princeton University Press (1971).

[Sh75] G. Shimura, *On the holomorphy of certain Dirichlet series*, Proc. London Math. Soc. **31** (1975), 79–98.

[Sh94] G. Shimura, *Euler products and Fourier coefficients of automorphic forms on symplectic groups*, Inventiones Math. **116** (1994), 531–576.

[St] J. Sturm, *Special values of zeta functions, and Eisenstein series of half integral weight*, Amer. J. Math. **102** (1980), 219–240.

[TW] R. Taylor and A. Wiles, *Ring theoretic properties of certain Hecke algebras*, Ann. of Math. **142** (1995), 553–572.

[U] E. Urban, *Formes automorphes cuspidales pour* GL_2 *sur un corps quadratiques imaginaires. Valeurs spéciales de fonctions L et congruences*, Compositio Math. **99** (1995), 283–324.

[W] A. Wiles, *Modular elliptic curves and Fermat's last theorem*, Ann. of Math. **142** (1995), 443–551.

[Y94] H. Yoshida, *On the zeta functions of Shimura varieties and periods of Hilbert modular forms*, Duke Math. J. **75** (1994), 121–191.

[Y95] H. Yoshida, *On a conjecture of Shimura concerning periods of Hilbert modular forms*, Amer. J. Math. **117** (1995), 1019–1038.

Department of Mathematics, UCLA, Los Angeles, CA 90095-1555, U.S.A.

E-mail address: hida@math.ucla.edu

Proceedings of Symposia in Pure Mathematics
Volume **66.1**, 1999

Bounds for Congruence Primes

M. Ram Murty

Dedicated to Goro Shimura

Abstract. Given a normalized newform f of weight 2 and level N and rational integral Fourier coefficients, $a_n(f)$, we will be interested in estimating primes ℓ for which there exists a cusp form (not necessarily an eigenform) g of weight 2 and level N with rational integer coefficients $a_n(g)$, such that ℓ divides the difference of all the Fourier coefficients of $a_n(f) - a_n(g)$ with n coprime to N. Such primes are called congruence primes. They are intimately related to the prime divisors of the minimal degree of the modular parametrization of the elliptic curve attached to E via the Eichler-Shimura mapping. We conjecture that this degree is $O(N^{2+\epsilon})$ and show that it is equivalent to the celebrated ABC conjecture. For CM elliptic curves, we prove that this degree is $O(N^{3/2+\epsilon})$. We also study the related question: given two non-isogenous elliptic curves E_1 and E_2, without complex multiplication and conductors N_1 and N_2 respectively, how large must ℓ be so that the Galois group of $\mathbb{Q}(E_1[\ell], E_2[\ell])$ over $\mathbb{Q}$ is as large as possible.

1. Introduction

Let f be a normalized newform of weight 2 for $\Gamma_0(N)$. Suppose that the Fourier coefficients of f are rational integers. We will be interested in those primes ℓ for which there exists a cusp form g (not necessarily an eigenform) of weight 2 and level N, also with integer coefficients such that f and g are orthogonal with respect to the Petersson inner product and

$$f \equiv g \pmod{\ell}.$$

This means that, if $a_n(f)$ and $a_n(g)$ denote the n-th Fourier coefficient of f and g respectively, then

$$a_n(f) \equiv a_n(g) \pmod{\ell}$$

for all n coprime to N. Such a prime ℓ will be called a *congruence prime* for f.

More generally, we are interested in natural numbers δ for which there is a congruence

$$f \equiv g \pmod{\delta}$$

1991 *Mathematics Subject Classification.* Primary 11F33; 11F30 Secondary 14G20.
Key words and phrases. Modular forms, congruences, Fourier coefficients.
Research partially supported by NSERC grant number OGP 0009418E and FCAR grant.

with g orthogonal to f and $a_n(g) \in \mathbb{Z}$. We will then say δ is a *congruence divisor* for f.

Let $\mathcal{S}$ denote the set of cusp forms of weight 2 on $\Gamma_0(N)$ with rational integer Fourier coefficients. Denote by L the orthogonal complement of f in $\mathcal{S}$. Then, one knows that there are only finitely many congruence divisors for f and the largest such, denote it by $D = D_f$, is the exponent of the finite group $\mathcal{S}/(\mathbb{Z}f + L)$ (see Zagier [**Z**, p. 381]). In [**Maz77**], Mazur introduces the notion of primes of fusion. That this notion essentially coincides with the concept of a congruence prime has been shown by Ribet [**Ri83**].

Our aim is to obtain bounds for congruence primes and congruence divisors of f in terms of the level of f. One conjectures that

CONJECTURE D: $D_f = \mathrm{O}(N^{2+\epsilon})$ for any $\epsilon > 0$.

A classical theorem of Eichler and Shimura associates an elliptic curve $E_f/\mathbb{Q}$ such that the L-function of E_f agrees with the L-function of f. More precisely, consider the map

$$\phi(\tau) = 2\pi i \int_{i\infty}^{\tau} f(z)dz \quad (\tau \in \mathcal{H}^*)$$

where $\mathcal{H}^* = \mathcal{H} \cup \mathbb{Q} \cup \{i\infty\}$ and $\mathcal{H}$ denotes the upper half plane. It is easily seen that ϕ is independent of the path chosen from $i\infty$ to τ. If $\gamma \in \Gamma_0(N)$, then

$$\begin{aligned}\omega(\gamma) :&= \phi(\gamma\tau) - \phi(\gamma)\\ &= 2\pi i \int_{\tau}^{\gamma\tau} f(z)dz\end{aligned}$$

is independent of τ. Moreover, $\omega : \Gamma_0(N) \to \mathbb{C}$ is a homomorphism. The image of this map is a lattice of rank 2 in $\mathbb{C}$. If we denote this lattice by Λ_f, then we have the Eichler-Shimura map ϕ_{ES} given by

$$\phi_{ES} : X_0(N) = \mathcal{H}^*/\Gamma_0(N) \to \mathbb{C}/\Lambda_f.$$

Eichler and Shimura have shown that the torus $\mathbb{C}/\Lambda_f$ is the set of complex points of an elliptic curve E_f defined over $\mathbb{Q}$. An elliptic curve $E/\mathbb{Q}$ which is isogenous to E_f for some f will be called modular.

The map $\phi = \phi_{ES}$ is a branched covering between the two Riemann surfaces $X_0(N)$ and E_f. From the work of Doi, Hida, Ohta, Mazur, Ribet and others, we know that the largest congruence divisor D for f and the topological degree of ϕ are related (see Zagier [**Z**, p. 381]). The precise relationship has not been spelled out in the literature. For a given modular elliptic curve $E/\mathbb{Q}$, of conductor N, we choose a modular parametrization

$$\phi_E : X_0(N) \to E$$

of minimal degree. In any $\mathbb{Q}$ isogeny class of modular elliptic curves, there is a curve E with $\deg \phi$ minimal. We call such an E a strong modular elliptic curve. Mazur has shown that if E and E' are isogenous over $\mathbb{Q}$, then there is some isogeny between them of degree ≤ 163. Thus, $\deg \phi_{E_f} \asymp \deg \phi_E$ for the strong modular elliptic curve in the isogeny class of E_f. We have the following conjecture of Frey [**F**]:

DEGREE CONJECTURE: For any modular elliptic curve $E/\mathbb{Q}$ of conductor N,

$$\deg \phi_E = \mathrm{O}(N^{2+\epsilon}).$$

The precise relationship between the minimal degree of a modular parametrization of E_f and the congruence divisor $D = D_f$ is given by:

PROPOSITION 0. (Mazur-Ribet) *For a strong modular elliptic curve E with corresponding newform f, we have $\deg \phi_E | D_f$. The only primes dividing $D_f / \deg \phi_E$ are primes dividing the conductor of E.*

PROOF. The proof of this theorem will appear in a forthcoming paper by Mazur and Ribet [**MazR**].

Following Cremona [**C**], the topological degree can be computed by observing that

$$\begin{aligned} 4\pi^2(f,f) &= 2\pi^2 i \int_{X_0(N)(\mathbb{C})} f(\tau)d\tau \wedge \overline{f(\tau)d\tau} \\ &= \frac{i}{2}\int_{X_0(N)(\mathbb{C})} (2\pi i f(\tau)d\tau) \wedge \overline{2\pi i f(\tau)d\tau} \\ &= \frac{i}{2}\int_{X_0(N)(\mathbb{C})} \phi^*(dz) \wedge \overline{\phi^*(dz)}, \end{aligned}$$

since $\phi^*(dz) = 2\pi i f(\tau)d\tau$, where $z = x + iy$ is the co-ordinate on E_f. Whence,

$$\begin{aligned} 4\pi^2(f,f) &= \frac{i}{2}(\deg\phi)\int_{E_f(\mathbb{C})} dz \wedge \overline{dz} \\ &= (\deg\phi)\int_{E_f(\mathbb{C})} dxdy \\ &= (\deg\phi)\mathrm{Vol}(E_f) \end{aligned}$$

where $\mathrm{Vol}(E_f)$ is the area of the fundamental parallelogram of the lattice of periods of E_f. As in Cremona [**C**], the Manin constant does not enter into this calculation.

Conjecture D is not unrelated to the celebrated ABC conjecture. In fact, as we shall show below, it is essentially equivalent to it. This relationship is similar in spirit to Szpiro's conjecture [**Sz**] and its relationship to ABC as exposed in Oesterlé [**O**] or Hindry [**H**]. In fact, as we shall see below,

$$\log D_f \sim \log N + \frac{1}{6}\log\max\left\{|c_4|^3, |c_6|^2\right\}$$

so that Conjecture D is easily seen to be equivalent to Conjecture 4' in [**O**]. The ABC conjecture predicts that if

$$A + B = C \quad \text{with} \quad A, B, C \in \mathbb{Z}$$

and mutually coprime, then

$$\max(|A|,|B|,|C|) \le \kappa(\epsilon)\left(\prod_{p|ABC} p\right)^{1+\epsilon}$$

for any $\epsilon > 0$ and $\kappa(\epsilon)$ is a constant depending only on ϵ.

To understand the ABC conjecture, Frey [**F**] had the idea of studying the associated elliptic curve

$$E_{A,B,C}: \qquad y^2 = x(x-A)(x+B).$$

We will call such curves *Frey curves*. They have conductor

$$N = 2^{\delta} \prod_{p|ABC} p$$

with δ bounded (see [**DK**]).

Rubin and Silverberg observed that the modularity of Frey curves follows from Diamond [**Di**] which is an extension of the work of Wiles [**W**]. Later, Diamond and Kramer [**DK**] showed that it follows directly from Wiles [**W**] and Taylor-Wiles [**TW**]. That is, such a curve is isogenous to a curve of the form E_f for some newform f with $a_n(f) \in \mathbb{Z}$. We have:

THEOREM 1.
(i) *The degree conjecture* $\Rightarrow$*ABC;*
(ii) *ABC*$\Rightarrow$ *the degree conjecture for all Frey curves.*

REMARK. A weaker version of (i) was first proved by Frey [**F**]. Namely, he established that some bound for the degree gives some version of ABC (that is, with $1+\epsilon$ replaced by some larger number). Mai and Murty established (ii) in [**MaM**]. As we shall see below, the precise relationship between the exponents in these two conjectures is determined by invoking a Siegel-type theorem for the lower bound of the symmetric square L-function at $s=2$ due to Hoffstein and Lockhart [**HL**].

By Proposition 0, we infer that bounding D_f would give us bounds for $\deg \phi_{E_f}$. In addition, bounding the prime divisors of D_f gives us bounds for the prime divisors of $\deg \phi_{E_f}$. It is this theme that we shall explore in this paper.

For CM elliptic curves, we establish

THEOREM 2. *Let $E/\mathbb{Q}$ be a CM elliptic curve with conductor N. Then E is modular and the minimal degree of a modular parametrization satisfies*

$$N^{3/2-\epsilon} \ll \deg \phi_E \ll N^{3/2+\epsilon}$$

for any $\epsilon > 0$. (The implied constant may depend on ϵ.)

For any Frey curve, we have:

THEOREM 3. $\log D = O(N^{1/3+\epsilon})$.

COROLLARY 4. *If ℓ is a congruence prime for f, then*

$$\ell \ll \exp(N^{1/3+\epsilon}).$$

REMARK. This result also implies that the number of congruence primes for f is

$$\ll N^{1/3+\epsilon}.$$

The degree can be estimated in several ways. We obtain the following curious properties of various determinants. Let $f_1, \dots, f_r$ be a basis of normalized eigenforms for $S_2(N)$, the space of cusp forms of weight 2 and level N. Let S be any set of r distinct natural numbers coprime to N. Consider the $r \times r$ matrix

$$A_S = (a_n(f_i))_{\substack{n \in S \\ i \le r}}.$$

THEOREM 5. $(\det A_S)^2$ *is a rational integer and*

$$D_f^2 | (\det A_S)^2.$$

Using this theorem, we will deduce:

COROLLARY 6.

$$D_f = \mathrm{O}\left((2N^2r)!^{3/2}(\log 2N^2r)^r\right)$$

where $r = \dim S_2(N)$.

Mazur [**Maz**] asked the question: given f as above, let ℓ be a prime such that there is an eigenform g with integer coefficients such that $f \equiv g \pmod{\ell}$. Is ℓ bounded? We will prove:

THEOREM 7. *Given a newform* $f \in S_2(N)$ *with* $a_n(f) \in \mathbb{Z}$, *suppose that there is an eigenform* $g \in S_2(N)$ *orthogonal to* f *and with* $a_n(g) \in \mathbb{Z}$, *and a prime* ℓ *such that* $f \equiv g \pmod{\ell}$. *Then* $\ell \leq N^{1/2+\epsilon}$.

We indicate in the last section how one can replace $1/2$ by $3/2$ using an easier argument. Recently, Frey has given examples of non-isogenous E and E' over $\mathbb{Q}$, such that $E[11] \simeq E'[11]$ so that $\ell \geq 13$ in Theorem 7.

Theorem 7 has an application to the computation of Galois groups. Recall the well-known theorem of Serre [**S72**]. Let E be an elliptic curve $/\mathbb{Q}$ without complex multiplication. Then, the field $\mathbb{Q}(E[\ell])$ obtained by adjoining the ℓ-division points of E has Galois group over $\mathbb{Q}$ isomorphic to $\mathrm{GL}_2(\mathbb{F}_\ell)$ for ℓ sufficiently large. How large ℓ should be is answered by means of the effective Chebotarev density theorem. Serre [**S81**] showed that assuming the generalized Riemann hypothesis (GRH) for the Dedekind zeta functions of $\mathbb{Q}(E[\ell])$, one can prove that $\ell \gg (\log N)^2$. It is also possible to show that unconditionally, there is a constant $c > 0$ so that the result holds for $\ell \geq N^c$. (See [**S**, p. 716].)

Given two distinct non-isogenous elliptic curves E_1 and E_2 over $\mathbb{Q}$, without complex multiplication, how large is

$$\mathrm{Gal}(\mathbb{Q}(E_1[\ell], E_2[\ell])/\mathbb{Q})?$$

The largest it can be is the subgroup

$$H_\ell = \{(a,b) \in \mathrm{GL}_2(\mathbb{F}_\ell) \times \mathrm{GL}_2(\mathbb{F}_\ell) : \det a = \det b\}$$

of

$$\mathrm{GL}_2(\mathbb{F}_\ell) \times \mathrm{GL}_2(\mathbb{F}_\ell).$$

THEOREM 8. *Suppose* E_1 *and* E_2 *are non-isogenous modular elliptic curves* $/\mathbb{Q}$ *without complex multiplication and having conductors* N_1 *and* N_2 *respectively. Let* $N = \max(N_1, N_2)$. *There is an effective constant* $c > 0$, *such that if* $\ell \geq N^c$, *then*

$$\mathrm{Gal}(\mathbb{Q}(E_1[\ell], E_2[\ell])/\mathbb{Q}) = H_\ell.$$

If we assume the GRH, we can improve this to $\ell \gg \log N$.

REMARK. Mazur's conjecture would imply the result for all $\ell \geq c_1$ for some constant c_1.

2. Proof of theorem 1

From the formula already stated:

$$4\pi^2(f,f) = (\deg\phi)\mathrm{Vol}(E_f)$$

we find by Deligne [**D**, p. 4] and as in Frey [**F**, p. 131] that

$$c^2 4\pi^2(f,f) = (\deg\phi)e^{-2h(E)}$$

where $h(E)$ is the Faltings height of E and c is Manin's constant. By recent work of Edixhoven [**E**], we know that c is bounded. Therefore

$$2h(E) + \log(f, f) = \log \deg \phi + O(1).$$

By Silverman [**Si86**, p. 258]

$$h(E) = \frac{1}{12} h(j_E) + + \frac{1}{12} \log \gamma + O(\log h(j_E)),$$

where $h(j_E)$ is the naive (Weil) height of j_E and γ is the unstable minimal discriminant (see [**Si86**, p. 257]). Writing $j_E = a/b$ with $(a, b) = 1$ and $a, b \in \mathbb{Z}$, we see that the unstable minimal discriminant is Δ/b where Δ is the discriminant of E. It is easily verified that in the case of Frey curves, the unstable minimal discriminant is bounded as the conductor tends to infinity, since

$$\Delta = (ABC)^2/2^8$$

and

$$j_E = \frac{2^8(A^2 + B^2 + AB)^3}{(ABC)^2} > 1.$$

Now $A^2 + B^2 + AB$ is coprime to ABC so that

$$h(j_E) = 3 \log(A^2 + B^2 + AB).$$

A simple calculation shows that

$$\log(A^2 + B^2 + AB) = 2 \log \max(|A|, |B|, |C|) + O(1).$$

We will now prove that for any $\epsilon > 0$ and N greater than some constant depending only on ϵ, we have the inequality

$$(1 - \epsilon) \log N < \log(f, f) < (1 + \epsilon) \log N.$$

The upper bound is an easy consequence of the Phragmén-Lindelöf theorem applied to the L-function of the symmetric square, as in Mai-Murty [**MaM**]. The L-function attached to the symmetric square of f has the special value

$$L(2, \mathrm{sym}^2(f)) = 288\pi^3 \prod_{p^2 | N} \left(1 - \frac{1}{p^2}\right) \frac{(f, f)}{N}.$$

By the Phragmén-Lindelöf theorem, we have

$$L(2, \mathrm{sym}^2(f)) = O(\log N).$$

Hence,

$$(f, f) \leq c_2 N \log N$$

for some constant c_2. By a result of Hoffstein and Lockhart [**HL**], we have

$$\log(f, f) > (1 - \epsilon) \log N.$$

Hence

$$\log \max(|A|, |B|, |C|) + O(\log \log \max(|A|, |B|, |C|)) = \log \deg \phi - \log(f, f) + O(1).$$

Therefore, if $\deg \phi \leq N^{2+\epsilon}$, we get the ABC conjecture.

For the converse,

$$\log \max(|A|, |B|, |C|) + O(\log \log \max(|A|, |B|, |C|)) \\ \geq \log \deg \phi - \log N - \log \log N + O(1).$$

Therefore, assuming the ABC conjecture gives

$$\deg \phi = \mathrm{O}(N^{2+\epsilon}).$$

This proves (ii). □

3. Proof of theorem 2

Recall that if $E/\mathbb{Q}$ has CM, and $j_E \neq 0, 1728$, then E is a quadratic twist of one of a finite set of elliptic curves E_i of conductor N_i and discriminant Δ_i (say). If $j_E = 0$, or 1728, then it has a Weierstrass equation $y^2 = x^3 + D$ or $y^2 = x^3 + Dx$ respectively (see [**Si86b**, p. 308 & p. 340]). We will discuss the case the curve is a quadratic twist by D (the other cases being handled similarly). In this case, the conductor divides $N_i D^2$ for some i. An easy calculation shows that the discriminant of E is $\mathrm{O}(\Delta_i D^6)$. We therefore obtain

$$\log \deg \phi = \log(f, f) + \mathrm{O}(1)$$

which implies by the proof of Theorem 1,

$$N_E^{3/2-\epsilon} \ll \deg \phi \ll N_E^{3/2+\epsilon}.$$

This completes the proof of Theorem 2. □

4. Proof of theorem 3

Stewart and Yu [**SY91**] have proved that for the ABC conjecture, for any $\epsilon > 0$, there is an effectively computable constant $\kappa(\epsilon)$ so that

$$\log \max(|A|, |B|, |C|) \leq \kappa(\epsilon) N^{2/3+\epsilon}.$$

The exponent has recently been improved by Stewart [**SY**] to $1/3 + \epsilon$. Hence, by the above analysis, we deduce that for some constant $\kappa_1(\epsilon)$,

$$\deg \phi < \exp\left(\kappa_1(\epsilon) N^{1/3+\epsilon}\right).$$

This completes the proof of Theorem 3. □

To deduce Corollary 4, we invoke Proposition 0 to deduce that the primes dividing the degree are precisely the congruence primes apart from primes dividing the level N. Therefore, each prime divisor ℓ of $\deg \phi$ satisfies

$$\ell \leq \exp\left(\kappa_1(\epsilon) N^{1/3+\epsilon}\right).$$

5. Proof of theorem 5

Let $f_1, \ldots, f_r$ be a normalized (i.e., $a_1(f_i) = 1$) orthogonal basis for $S_2(N)$ which are eigenforms for all the Hecke operators T_p for $(p, N) = 1$. (See Lang [**L**].) Consider $(\det A_S)^2$. Since $A_S A_S^{(t)}$ is invariant under Galois conjugation, we find that $(\det A_S)^2$ is a rational integer. Suppose it is not zero. By going to a finite algebraic extension, we can solve the system

$$\sum_{j \in S} a_j(f_i) m_j = \begin{cases} \det A_S & \text{if } i = 1 \\ 0 & \text{if } i \neq 1, \end{cases}$$

with the m_j algebraic integers. Let g be orthogonal to $f = f_1$ with $a_n(g) \in \mathbb{Z}$. Suppose that

$$f \equiv g \pmod{D}.$$

Let us write

$$g = c_1 f_1 + c_2 f_2 + \cdots c_r f_r,$$

so that $c_1 = 0$. Now consider

$$h(z) = \sum_{j \in S} m_j T_j(g)$$

where T_i denotes the i-th Hecke operator. This is easily seen to be

$$\sum_{i=1}^{r} c_i \left(\sum_{j \in S} a_j(f_i) m_j \right) f_i$$

which is equal to 0 since $c_1 = 0$. Recall that S consists of numbers coprime to N. Then the n-th Fourier coefficient of $h(z)$ is

$$\begin{aligned} 0 &= \sum_{j \in S} m_j \sum_{d|n,j} a_{nj/d^2}(g) d \\ &\equiv \sum_{j \in S} m_j \sum_{d|n,j} a_{nj/d^2}(f_1) d \pmod{D} \\ &\equiv \sum_{j \in S} m_j a_n(f_1) a_j(f_1) \pmod{D} \\ &\equiv a_n(f_1) \det A_S \pmod{D} \end{aligned}$$

so that choosing $n = 1$ gives that $D^2 | (\det A_S)^2$.

We now prove Corollary 6. By Proposition 2.16 of Shimura [**Sh**, p. 39], we deduce that the order of vanishing at $i\infty$ of any non-zero cusp form $f \in S_2(N)$ is

$$< \frac{N}{6} \prod_{p|N} \left(1 + \frac{1}{p}\right) = h \quad \text{(say)}.$$

Note that $h > 2 \dim S_2(N) = 2r$. Let χ_N be a primitive quadratic character with conductor supported at all the primes dividing N. The r vectors

$$v_i \otimes \chi_N(n) = (a_1(f_i)\chi_N(1), \dots, a_h(f_i)\chi_N(h)) \qquad 1 \le i \le r$$

corresponding to $f_i \otimes \chi_N \in S_2(M)$ for some $M | N^3$ must be linearly independent for $h = 2 \dim S_2(M)$. Otherwise, we would have a non-zero element in $S_2(M)$ with a zero of order exceeding h contrary to the above remark. Thus, the matrix whose rows are the vectors v_i has rank r. Hence, there are r independent columns. Thus, there is a subset

$$S \subseteq \{1, 2, \dots, h\}$$

consisting of elements coprime to N, so that $|S| = r$ and

$$\det A_S \neq 0.$$

This particular choice allows us to estimate D. Since $h \le 2 \dim S_2(N^3) = 2N^2 r$, we have by Deligne's estimate for the Fourier coefficients:

$$D \le (2N^2 r)!^{3/2} \prod_{i=1}^{2N^2 r} d(i)^{1/2}.$$

where $d(i)$ denotes the divisor function. By the arithmetic-geometric mean inequality, we deduce

$$\left\{ \prod_{i=1}^{2N^2 r} d(i) \right\}^{1/(2N^2 r)} \leq \frac{1}{2N^2 r} \sum_{i=1}^{2N^2 r} d(i) \leq \log 2N^2 r,$$

by well-known estimates of analytic number theory.

REMARK. If N is prime, we need not take S consisting of elements coprime to N since all eigenforms are newforms and thus eigenfunctions for all Hecke operators (including T_p for $p|N$). In this case, we can take S to be

$$\{1, 2, \ldots, r\}$$

then we would get the sharper estimate

$$\det A \leq r!^{3/2} \prod_{i=1}^{r} d(i)^{1/2} \leq r!^{3/2} (\log r)^{r/2}.$$

This is related to the question of whether $i\infty$ is a Weierstrass point for $\Gamma_0(N)$. (Recall that a point x on a compact Riemann surface X is called a Weierstrass point of X if there exists a non-zero differential form whose order of vanishing at x exceeds the genus of X.) If N is a prime, Atkin [**A**] has shown that $i\infty$ is not a Weierstrass point so that we get the sharper bound for D in this case. It is conjectured that for any squarefree N, $i\infty$ is not a Weierstrass point.

6. The Rankin-Selberg method

We collect in this section the necessary facts and estimates derived from the Rankin-Selberg method. All of these facts are contained in the original papers of Rankin and Selberg though they perhaps did not pay much attention to the dependence on the level in the estimation of the error terms. Using standard estimates of the L-function of the symmetric square obtained by invoking the Phragmén - Lindelöf theorem as in [**MaM**], we obtain:

LEMMA 9. *Let f be a newform of weight 2 and level N. Then,*

$$\frac{1}{2} \sum_{n \leq x} |a_n(f)|^2 \log^2 \frac{x}{n} = c(f) x^2 + O\left(x^{3/2} N^{1+\epsilon}\right)$$

where

$$c(f) = \frac{8\pi^3 \phi(N)(f, f)}{N^2}.$$

If g is another newform orthogonal to f, then

$$\frac{1}{2} \sum_{n \leq x} a_n(f) \overline{a_n(g)} \log^2 \frac{x}{n} = O\left(x^{3/2} N^{1+\epsilon}\right).$$

REMARK. The constant $c(f)$ is given explicitly in Stark [**St**, p. 90].

If we want estimates for

$$\sum_{\substack{n \leq x \\ (n,M)=1}} |a_n(f)|^2 \log^2 \frac{x}{n}$$

and

$$\sum_{\substack{n \leq x \\ (n,M)=1}} a_n(f)\overline{a_n(g)} \log^2 \frac{x}{n},$$

then this involves removing the Euler factors for primes $p|M$ in the Dirichlet series

$$\sum_{n=1}^{\infty} \frac{|a_n(f)|^2}{n^s}$$

and

$$\sum_{n=1}^{\infty} \frac{a_n(f)\overline{a_n(g)}}{n^s}$$

before applying the method to derive the above asymptotic formulas. This yields

LEMMA 10. *For f and g as in Lemma 9,*

$$\frac{1}{2} \sum_{\substack{n \leq x \\ (n,M)=1}} |a_n(f)|^2 \log^2 \frac{x}{n} = c(f,M)x^2 + O(x^{3/2}N^{1+\epsilon}M^{\epsilon})$$

and

$$\sum_{\substack{n \leq x \\ (n,M)=1}} a_n(f)\overline{a_n(g)} \log^2 \frac{x}{n} = O(x^{3/2}N^{1+\epsilon}M^{\epsilon})$$

where

$$c(f,M) = \frac{8\pi^3\phi(N)(f,f)}{N^2} \prod_{p|M} \left(\sum_{k=1}^{\infty} \frac{|a_{p^k}(f)|^2}{p^{2k}} \right)^{-1}.$$

We can use this to deduce:

LEMMA 11. *If f and g are as in Lemma 9, then there is an $n = O(N^{1+\epsilon}M^{\epsilon})$ such that $(n,M) = 1$ and*

$$a_n(f) \neq a_n(g).$$

PROOF. We use the estimate of Hoffstein and Lockhart for the Petersson inner product to estimate the residue from below. The result is now immediate.

We briefly indicate the proofs of these lemmas since they do not appear in the literature in the form we need it with the explicit dependence on the level N as stated above. If N is squarefree and $M = 1$, a variant of Lemma 11 is due to Goldfeld and Hoffstein [**GH**]. A simplified and generalized version of their work is found in [**Mu97**].

Let $L(s)$ be a Dirichlet series satisfying the following properties:

(i) $L(s) = \prod_p L_p(s)$ for $\sigma > 1$ where L_p is the inverse of a polynomial of degree m_p in p^{-s}. The product is over all finite primes.

(ii) There is a positive integer d so that $m_p \leq d$ for all p and $m_p = d$ for all but finitely many p.

Let us write

$$L_p(s) = \prod_{i=1}^{m_p} (1 - \pi_{i,p}p^{-s})^{-1}$$

where $\pi_{i,p} \in \mathbb{C}$, $|\pi_{i,p}| = 1$. Define

$$L_p^v(s) = \prod_{i=1}^{m_p} (1 - \overline{\pi}_{i,p} p^{-s})^{-1}$$

and set

$$L^v(s) = \prod_p L_p^v(s).$$

Let A be a positive real number and a, b non-negative integers with $a + b \leq d$. Set

$$\Lambda(s) = A^{s/2}(\pi^{-s/2}\Gamma(s/2))^a(\pi^{(s+1)/2}\Gamma((s+1)/2))^b L(s)$$
$$\Lambda^v(s) = A^{s/2}(\pi^{-s/2}\Gamma(s/2))^a(\pi^{-(1+s)/2}\Gamma((1+s)/2)^b L^v(s).$$

Suppose that

(iii) $\Lambda(s)$ and $\Lambda^v(s)$ have an analytic continuation to the entire complex plane except possibly for a pole at $s = 0$ or $s = 1$,

(iv) $\Lambda(s) = w\Lambda^v(1 - s)$ with $w \in \mathbb{C}$, $|w| = 1$. Under assumptions (i) to (iv), we have for $0 \leq \sigma \leq 1$,

$$|L(\sigma + it)| \leq (A\tau^d)^{(1-\sigma)/2}(\log A\tau^d)^d$$

where $\tau = |t| + 2$. This is an effective version of the Phragmén-Lindelöf theorem. The proof given in [**MuM**] applies to functions with no pole at $s = 0$ or 1. But it can easily be adapted to give the same result if $L(s)$ has a pole at $s = 0$ or 1. (Let us also take this opportunity to point out a misprint in [**MuM**]. On page 251 of [**MuM**] in condition (i), L_p is the *inverse* of a polynomial in p^{-s}.)

We apply this result to the series

$$L_{f,g}(s) = \zeta_N(2s) \sum_{n=1}^{\infty} \frac{a_n \overline{b_n}}{n^{s+1}}$$

where

$$\zeta_N(2s) = \prod_{p \nmid N} \left(1 - \frac{1}{p^{2s}}\right)^{-1}.$$

Rankin [**R**] established the analytic continuation of $L_{f,g}(s)$. Ogg [**Og**], derived the functional equation in the case f and g have squarefree levels. Li [**Li**] has generalized this to arbitrary level and weight.

For the sake of simplicity, we will confine our attention to squarefree level. An easy adaptation of the method to the non-squarefree level is effected by using the (slightly more complicated) functional equation of Li [**Li**].

Let f, g be normalized cusp forms of squarefree level N_1 and N_2 respectively. Let $N = lcm(N_1, N_2)$ and $M = gcd(N_1, N_2)$. For $p|M$, let

$$c_p = a_p b_p = \pm 1$$

and set

$$\Phi^+(s) = \Phi(s) \prod_{p|M} \left(1 - c_p p^{-s}\right)^{-1}$$

where

$$\Phi(s) = \left(\frac{\sqrt{N}}{2\pi}\right)^{2s} \Gamma(s)\Gamma(s+1)L_{f,g}(s).$$

Then

$$\Phi^+(1-s) = \Phi^+(s).$$

We now consider the integral

$$\frac{1}{2\pi i}\int_{c-i\infty}^{c+i\infty}\left(\sum_{n=1}^{\infty}\frac{a_n\overline{b_n}}{n^{s+1}}\right)\frac{x^s}{s^3}ds$$

which is equal to

$$\frac{1}{2}\sum_{n\le x}\frac{a_n\overline{b_n}}{n}\log^2\frac{x}{n}.$$

We write the Dirichlet series in the integrand as

$$\frac{1}{\zeta_N(2s)}\left(\zeta_N(2s)\sum_{n=1}^{\infty}\frac{a_n\overline{b_n}}{n^{s+1}}\right),$$

and apply the Phragmén-Lindelöf theorem to the paranthetic term to obtain an estimate of

$$\mathrm{O}(N^{1/2}T\log^4 NT)$$

on the line $\Re(s) = 1/2$. The term $1/\zeta_N(2s)$ can be estimated easily on $\Re(s) = 1/2$. We derive the estimate

$$\frac{1}{2}\sum_{n\le x}\frac{a_n\overline{b_n}}{n}\log^2\frac{x}{n} = \mathrm{O}(x^{1/2}N^{1/2}\log^4 N).$$

when $f = g$, we find

$$\sum_{n\le x}\frac{|a_n|^2}{n}\log^2\frac{x}{n} = 2\alpha_f x + \mathrm{O}(x^{1/2}N^{1/2}\log^4 N)$$

where

$$\alpha_f = \frac{4\pi^3\phi(N)(f,f)}{N^2}$$

Thus,

$$\frac{1}{2}\sum_{n\le x}(a_n-b_n)^2\log^2\frac{x}{n} = (\alpha_f+\alpha_g)x + \mathrm{O}(x^{1/2}N^{1/2}\log^4 N).$$

This leads to the result that

$$a_n \ne b_n$$

for some $n = \mathrm{O}(N\log^8 N)$. (In **[Mu97]**, the slightly better estimate $\mathrm{O}(N\log\log N)$ is obtained.) This completes the sketch of the proof of Lemmas 9, 10 and 11. □

7. Proof of theorem 7

If we have a congruence between newforms f and g

$$a_n(f) \equiv a_n(g) \pmod{\ell}$$

for all $(n, N) = 1$, then by Lemma 11, we deduce that for some $n = \mathrm{O}(N^{1+\epsilon}\ell^{\epsilon})$, we have

$$a_n(f) \ne a_n(g)$$

and $(n, N) = 1$. Therefore,

$$\ell \ll N^{1/2+\epsilon}.$$

This completes the proof of Theorem 7. □

8. Proof of theorem 8

By Serre [**S72**], we know that if $\ell \geq N^c$ for some constant c, then

$$\mathrm{Gal}(\mathbb{Q}(E_1[\ell])) \simeq \mathrm{GL}_2(\mathbb{F}_\ell)$$

and

$$\mathrm{Gal}(\mathbb{Q}(E_2[\ell])) \simeq \mathrm{GL}_2(\mathbb{F}_\ell)$$

We now apply results of Ribet [**Ri83**] to deduce that the only obstruction to the assertion of the theorem is the set of congruence primes linking f_1 and f_2 where f_1 and f_2 are the newforms associated to E_1 and E_2. We will need the following lemma which is a special case of Theorem 3.8 of [**Ri75**].

LEMMA 12. *For a prime $\ell \geq 5$, let e be a divisor of $\ell - 1$, let V, V' be vector spaces of dimension 2 over $\mathbb{F}_\ell$, let B, B' be the subgroups of $\mathrm{GL}(V)$ and $\mathrm{GL}(V')$ respectively consisting of all elements whose determinants are e-th powers. Let E be the subgroup of $B \times B'$ consisting of all (b, b') with $\det b = \det b'$ and suppose H is a subgroup of E in $B \times B'$ whose projections to each factor are surjective. Then, if $H \neq E$, there is an isomorphism ϕ from V to V' and a character χ of H with $\chi^2 = 1$ such that*

$$b' = \chi(h)\phi b \phi^{-1}$$

for all $(b, b') \in H$.

We apply this lemma in the case $e = 1$. We suppose ℓ is a prime such that $\ell \geq N^c$ with c sufficiently large and $G_i = \mathrm{Gal}(\mathbb{Q}(E_i[\ell])/\mathbb{Q}) \simeq \mathrm{GL}_2(\mathbb{F}_\ell)$ for $i = 1, 2$. Then, let $H = \mathrm{Gal}(\mathbb{Q}(E_1[\ell], E_2[\ell])/\mathbb{Q}) \subseteq G_1 \times G_2$ and E the subgroup of elements (b, b') of $G_1 \times G_2$ with $\det b = \det b'$. Clearly H satisfies the conditions of the lemma. If $H \neq E$, there is an isomorphism ϕ and a quadratic character χ of H such that

$$b' = \chi(h)\phi b \phi^{-1} \qquad \forall (b, b') \in H.$$

Thus, evaluating this on the powers of the Frobenius automorphism gives

$$a_{p^m}(E_1) \equiv \chi(p^m) a_{p^m}(E_2) \pmod{\ell}$$

for all

$$p \nmid \ell N_1 N_2.$$

Since E_1 and E_2 are modular, there are cusp forms f_1 and f_2 such that

$$a_{p^m}(E_i) = a_{p^m}(f_i), \qquad i = 1, 2.$$

Therefore,

$$a_n(f_1) = \chi(n) a_n(f_2) \pmod{\ell}$$

for all n coprime to $\ell N_1 N_2$. Moreover, the quadratic character χ can only be ramified at most at the prime dividing ℓ, since the kernel of χ cuts out a quadratic extension contained in $\mathbb{Q}(E_i[\ell])$ and the quadratic subfield of $\mathbb{Q}(\zeta_\ell)$ is the only such. By Lemma 11, we deduce that for some $n = O(N^{1+\epsilon}\ell^\epsilon)$, and $(n, \ell N_1 N_2) = 1$, we deduce

$$a_n(f_1) \neq \chi(n) a_n(f_2).$$

Therefore, by the Ramanujan estimate, we deduce $\ell \ll N^{1/2+\epsilon}$. The first assertion now follows from Theorem 7.

To deduce the statement on the GRH, we know from Serre that the Galois groups generated by the ℓ-division points of E_1 and E_2 are as large as possible for $\ell \geq (\log N)^2$. By the Rankin-Selberg method, we know that for $i = 1, 2$,

$$\sum_{p \leq x} |a_p(f_i)|^2 \log p = x^2 + \mathrm{O}\left(x(\log Nx)^2\right)$$

and

$$\sum_{p \leq x} a_p(f_1)\overline{a_p(f_2)} = \mathrm{O}\left(x(\log Nx)^2\right)$$

from which we deduce that

$$\sum_{p \leq x} \left(a_p(f_1) - a_p(f_2)\right)^2 = 2x^2 + \mathrm{O}\left(x(\log Nx)^2\right)$$

so that for some $p_0 \ll (\log N)^2$, we have

$$a_{p_0}(f_1) \neq a_{p_0}(f_2).$$

Hence, ℓ divides

$$|a_{p_0}(f_1) - a_{p_0}(f_2)|$$

so that

$$\ell \ll (\log N)$$

which completes the proof of Theorem 8. □

9. An alternate method

We make some remarks using an alternate method. At the outset, let us note that optimal results are obtained by the Rankin-Selberg method as in Theorem 7. Also, assuming the generalized Riemann hypothesis, one can bound the congruence prime ℓ by $\mathrm{O}(\log N)$ using the method outlined in the proof of Theorem 8. If instead of the generalized Riemann hypothesis, we assume only the generalized Lindelöf hypothesis for the Rankin-Selberg convolutions that intervene in the proof of Theorem 7, then the term $N^{1/2+\epsilon}$ appearing in the error terms in Lemmas 9, 10, and 11 can be replaced by N^{ϵ}. This would then give an estimate of $\ell \ll N^{\epsilon}$ in Theorem 7.

Without the Rankin-Selberg method, one can prove:

THEOREM 7*. *Let f and g be as in Theorem 7 and suppose that*

$$f \equiv g \pmod{\ell}.$$

Then, $\ell \ll N^{3/2+\epsilon}$.

PROOF. We may suppose $\ell > N$. We have

$$a_n(f) \equiv a_n(g) \pmod{\ell}$$

for all $(n, N) = 1$. We twist by a quadratic character χ_N whose conductor is divisible by all the primes dividing N. Then,

$$a_n(f)\chi_N(n) \equiv a_n(g)\chi_N(n) \pmod{\ell}$$

for all n. The level of the twist of f by χ_N divides N^3. Therefore, if

$$a_n(f)\chi_N(n) = a_n(g)\chi_N(n)$$

for all

$$n \leq \sum_{d|N^3} \dim S_2(d) = \mathrm{O}(N^{3+\epsilon})$$

then

$$a_n(f) = a_n(g)$$

for all $(n, N) = 1$ contradicting the assumption $(f, g) = 0$. Hence, for some $n = \mathrm{O}(N^{3+\epsilon})$, we have

$$a_n(f)\chi_N(n) \neq a_n(g)\chi_N(n).$$

Reasoning now as before, we deduce that

$$\ell \ll N^{3/2+\epsilon}.$$

This argument may be of some independent interest. It is similar to arguments used in [**Mu97**].

10. Concluding remarks

Masser and Wusthölz [**MW**] have obtained results similar to Theorem 8. Their bounds on how large ℓ should be to ensure that the Galois groups be large as possible are given in terms of the Faltings height of the elliptic curves whereas the results above are in terms of the conductor of N.

References

[A] A. O. L. Atkin, *Weierstrass points at cusps of* $\Gamma_0(N)$, Ann. of Math. (2) **85** (1967), 42–45.

[C] J. Cremona, *Computing the degree of the modular parametrization of a modular elliptic curve*, Math. Comp. **64** (1995), 1235–1250.

[D] P. Deligne, *Preuve des conjectures de Tate et de Shafarevitch (d'après G. Faltings)*, Sém. Bourbaki, **616** (1983), Astérisque No. 121-122 (1985), 25–41.

[Di] F. Diamond, *On deformation rings and Hecke rings*, Ann. of Math. (2) **144** (1996), 137–166.

[DK] F. Diamond and K. Kramer, *Modularity of a family of elliptic curves*, Math. Res. Lett. **2** (1995), 299–304.

[E] B. Edixhoven, *Modular parametrizations at primes of bad reduction*, in preparation.

[F] G. Frey, *Links between elliptic curves and solutions of* $A - B = C$, Journal of the Indian Math. Soc. **51** (1987), 117–145.

[GH] D. Goldfeld and J. Hoffstein, *On the number of Fourier coefficients that determine a modular form*, Contemp. Math. **143** (1993), 385–393.

[H] M. Hindry, "a,b,c", conducteur, discriminant, Publications Mathématiques de l'Université Pierre et Marie Curie **84** (1986-87).

[HL] J. Hoffstein and P. Lockhart, *Coefficients of Maass forms and the Siegel zero*, Ann. of Math. (2) **140** (1994), 161–181.

[L] S. Lang, Introduction to Modular Forms, Berlin: Springer-Verlag (1995).

[Li] W. Winnie Li, *L-series of Rankin type and their functional equations*, Math. Annalen **244** (1973), 135–166.

[MaM] L. Mai and R. Murty, *The Phragmén-Lindelöf theorem and modular elliptic curves*, Contemporary Math. **166** (1994), 335–340.

[MW] D. W. Masser and G. Wüstholz, *Galois properties of division fields of elliptic curves*, Bull. London Math. Soc. **25** (1993), 247–254.

[Maz77] B. Mazur, *Modular curves and the Eisenstein ideal*, Inst. Hautes Études Sci. Publ. Math. **47** (1977), 33–186.

[Maz78] B. Mazur, *Rational isogenies of prime degree*, Inv. Math. **44** (1978), 129–162.

[Maz] B. Mazur, *Questions about number*, to appear.

[MazR] B. Mazur and K. Ribet, *Degrees of modular parametrizations and congruences between modular forms*, to appear.

[Mu97] R. Murty, *Congruences between modular forms*, in Analytic Number Theory, London Mathematical Society Lecture Note Series **247**, edited by Y. Motohashi, Cambridge: Cambridge University Press (1997), 309–320.

[MuM] M. Ram Murty and V. Kumar Murty, *A variant of the Bombieri - Vinogradov theorem*, in Number Theory (Montreal, Que., 1985), edited by H. Kisilevsky and J. Labute, Providence: Amer. Math. Soc. (1987), 243–272.

[O] J. Oesterlé, *Nouvelle approches du "Théorème" du Fermat*, Sém. Bourbaki, 1987-88, exposé 694, Astérisque (1988), 161–162.

[Og] A. P. Ogg, *On a convolution of L-series*, Inv. Math. **7** (1969), 297–312.

[R] R. Rankin, *Contributions to the theory of the Ramanujan's τ-function and similar arithmetical functions, I, II*, Proc. Cambridge Phil. Soc. **35** (1939), 351–372.

[Ri75] K. Ribet, *In ℓ-adic representations attached to modular forms*, Inv. Math. **28** (1975), 245–275.

[Ri83] K. Ribet, *Mod p Hecke operators and congruences between modular forms*, Inv. Math. **71** (1983), 193–205.

[RS] K. Rubin and A. Silverberg, *On Wiles' Cambridge lectures*, Bull. Amer. Math. Soc. **31** (1994), 15–38.

[S72] J.-P. Serre, *Propriétes galoisienne des pints d'ordre fini des courbes elliptiques*, Inv. Math. **15** (1972), 259–331.

[S81] J.-P. Serre, *Quelques applications du théorème de densité de Chebotarev*, Publ. Math. IHES **54** (1981), 123–301. (See also, Volume 3 of Collected Papers, Springer-Verlag).

[S] J.-P. Serre, Collected Papers, Volume 3, Berlin-New York: Springer-Verlag (1986).

[Sh] G. Shimura, Introduction to the Arithmetic Theory of Automorphic Functions, Publ. Math. Soc. Japan **11**, Princeton: Iwanami Shoten and Princeton University Press (1971).

[Si86] J. H. Silverman, *Heights and Elliptic curves*, in Arithmetic Geometry, edited by G. Cornell and J. H. Silverman, New York-Berlin: Springer-Verlag (1986), 253–265.

[Si86b] J. H. Silverman, The Arithmetic of Elliptic Curves, New York-Berlin: Springer-Verlag (1986).

[St] H. M. Stark, *L-functions at $s = 1$, II, Artin L-functions with Rational Characters*, Advances in Math. **17** (1975), 60–92.

[SY91] C. L. Stewart and K. Yu, *On the abc conjecture*, Math. Annalen **291** (1991), 225–230.

[SY] C. L. Stewart and K. Yu, *On the abc conjecture, II*, to appear.

[Sz] L. Szpiro, *La conjecture de Mordell (d'aprés Faltings)*, Sém. Bourbaki, 1983-84, exposé 619, Astérisque 121–122 (1985), 93–103.

[TW] R. Taylor and A. Wiles, *Ring theoretic properties of certain Hecke algebras*, Ann. of Math. **141** (1995), 553–572.

[W] A. Wiles, *Modular elliptic curves and Fermat's last theorem*, Ann. of Math. **141** (1995), 443–551.

[Z] D. Zagier, *Modular parametrization of elliptic curves*, Can. Math. Bulletin **28** (1985), 372–384.

Department of Mathematics, Queen's University, Kingston, Ontario, K7L 3N6, Canada

E-mail address: murty@mast.queensu.ca

Proceedings of Symposia in Pure Mathematics
Volume **66.1**, 1999

Frobenius Distributions and Galois Representations

V. Kumar Murty

To Professor Goro Shimura

ABSTRACT. We consider a class of Galois representations and formulate a conjecture about the distribution of Frobenius elements. This is a generalization of the conjecture of Lang and Trotter about the Galois representations associated to an elliptic curve. A new conjecture which arises in the general context has to do with the frequency with which the trace of Frobenius lies in a proper subfield of the field generated by all traces. We examine these conjectures in the context of modular forms and obtain some conditional results.

TABLE OF CONTENTS

1. Introduction

Let E be an elliptic curve defined over a number field K. If $\mathfrak{p}$ is a prime of good reduction, denote by $a_\mathfrak{p}$ the integer

$$\mathbb{N}\mathfrak{p} + 1 - |E(\mathbb{F}_\mathfrak{p})|.$$

About twenty years ago, Lang and Trotter formulated some conjectures about the values taken by this integer. Let a be an integer, and let us set

$$\pi_{E,a}(x) = \#\{\mathfrak{p} : \mathbb{N}\mathfrak{p} \le x \text{ and } a_\mathfrak{p}(E) = a\}.$$

Then, Lang and Trotter conjecture [**LT**] that there is a constant $c_{E,a}$ (possibly zero) such that

$$\pi_{E,a}(x) = (c_{E,a} + \mathbf{o}(1))\sqrt{x}/\log x$$

1991 *Mathematics Subject Classification.* Primary 11R44, 11R32; Secondary 11F30, 11F80, 11N05.

Key words and phrases. Galois representations, Frobenius automorphisms, Chebotarev density theorem, modular forms, supersingular primes.

Research partially supported by NSERC grants OGP0044342 and SMF0169898.

E.W.R. Steacie Fellow.

provided $a \neq 0$ or E does not have complex multiplication. The constant $c_{E,a}$ depends on the Galois representation attached to E. It is known that if $a \neq 0$ or if E does not have complex multiplication, then $\pi_{E,a}(x) = \mathbf{o}(x/\log x)$. This is due to Serre [**S68**] and in fact, he proved [**S81**] that

$$\pi_{E,a}(x) \ll_\epsilon x/(\log x)^{5/4-\epsilon}.$$

The exponent 5/4 was improved to 2 by Daqing Wan [**W**]. A further refinement was obtained in [**Mu97**] where it is shown that

$$\pi_{E,a}(x) \ll x(\log\log x)^2/(\log x)^2.$$

Of special interest is the case that $a_\mathfrak{p} = 0$ which corresponds to E having supersingular reduction at $\mathfrak{p}$. It is well-known by results of Deuring that if E has complex multiplication by an order in the imaginary quadratic field F, then the set of supersingular primes of K has density

$$\begin{cases} \frac{1}{2} & \text{if } F \text{ is not contained in } K \\ 0 & \text{if } F \subseteq K. \end{cases}$$

On the other hand, if E does not have complex multiplication, then it is a result of Elkies, Kaneko and R. Murty [**E92**] that

$$\pi_{E,0}(x) \ll x^{3/4}.$$

The Lang-Trotter conjecture is a Chebotarev Density Theorem for certain infinite extensions. Indeed, associated to the curve E above is a Galois representation

$$\rho_\ell : \mathrm{Gal}(\overline{K}/K) \longrightarrow \mathrm{GL}_2(\mathbb{Z}_\ell)$$

for each prime ℓ. This representation gives the action of $\mathrm{Gal}(\overline{K}/K)$ on the points of ℓ-power order in $E(\bar{K})$. Denote by G_ℓ the image of this representation. It is the Galois group of L/K where L is the field obtained by adjoining to K the coordinates of all such points. Denote by C_a the conjugacy stable subset consisting of elements of G_ℓ of trace a. Then $\pi_{E,a}(x)$ is (apart from a bounded amount)

$$\pi_{C_a}(x) = \#\{\mathfrak{p} : \mathbb{N}\mathfrak{p} \leq x,\ \mathfrak{p} \text{ unramified in } \rho_\ell,\ \mathrm{Frob}_\mathfrak{p} \subset C_a\}.$$

More generally, we may consider an infinite extension L/K of number fields with Galois group G a compact ℓ-adic Lie group and a conjugacy stable subset C of G and ask for the size of $\pi_C(x)$. Artin's conjecture on primitive roots can be formulated in this way and it has been studied extensively from this point of view by R. Murty (see [**Mu**] and the references therein). In the general case, Serre [**S81**] obtains upper bounds. In all of these cases, the method is to reduce to a finite extension and use the known effective versions of the Chebotarev Density Theorem.

In this article, we will reexamine the conjectures of Lang and Trotter, and in particular, formulate them (§2) in the context of more general Galois representations. Lang and Trotter themselves suggest that this should be possible. Moreover, we will be restricting ourselves to the simplest form of a general conjecture. Namely, we will not examine the constants involved in the asymptotics. Nor will we attempt to produce a probabilistic model to support the conjecture. Undoubtedly, both of these are possible to execute. Rather, we will rest content to examine (§3) specific instances of these conjectures in the case of modular forms. We would like to mention that independently of this work, and earlier, Bayer and Rovira [**BR**] have formulated a conjecture for supersingular reductions of the Abelian varieties associated to modular forms of weight 2. They have also obtained numerical evidence for their conjecture and provided a probabilistic model.

A new conjecture which arises in the general context has to do with the frequency with which the trace of Frobenius lies in a proper subfield of the field generated by all traces. This conjecture and some conditional results are discussed in §4. In the case of a normalized eigenform f without complex multiplication, of weight 2 and having a stable trace field which is quadratic over $\mathbb{Q}$ (see §2 and §3 for the definition of this term), a special case of Conjecture 3.4 asserts that for some constant $c_{f,\mathbb{Z}}$

$$\#\{p \leq x : a_f(p) \in \mathbb{Z}\} = (c_{f,\mathbb{Z}} + \mathbf{o}(1))\sqrt{x}/\log x$$

where $a_f(p)$ denotes the p-th Fourier coefficient. Assuming the Riemann Hypothesis for Dedekind zeta functions, we prove (Theorem 3.6) that

$$\#\{p \leq x : a_f(p) \in \mathbb{Z}\} \ll x^{7/8}/(\log x)^{1/2}.$$

Finally, in §5, we briefly discuss the problem of ordinary and supersingular primes, especially in the context of Abelian varieties.

This paper is respectfully dedicated to Prof. G. Shimura. One of the ingredients in formulating the general conjecture is the notion of inner twists. It was Shimura [**Sh**] who first studied the geometric meaning of such twists (associated to quadratic characters) in the context of the endomorphisms of Abelian varieties associated to modular forms. This was later expanded and developed in the work of Momose [**M**], Ribet [**Ri80**] and [**Mu83**] Since then, inner twists have also found application in other contexts, for example in [**MuR**].

The author would like to thank Dipendra Prasad and Brooks Roberts for helpful comments on an earlier version of this paper.

2. A conjecture for primitive self-dual Galois representations

Let K and E be number fields and $\mathcal{O} = \mathcal{O}_E$ the ring of integers of E. Suppose for almost all primes ℓ we are given a *free* $\mathcal{O} \otimes \mathbb{Z}_\ell$ module M_ℓ of finite rank r, and a continuous Galois action

$$\rho_\ell : \mathrm{Gal}(\overline{K}/K) \longrightarrow \mathrm{Aut}_{\mathcal{O}\otimes\mathbb{Z}_\ell} M_\ell.$$

We have a decomposition

$$\mathcal{O} \otimes \mathbb{Z}_\ell = \prod \mathcal{O}_\lambda$$

and a corresponding decomposition

$$M_\ell = \prod M_\lambda$$

where λ ranges over the primes of E which divide ℓ. Let us also set

$$M_\ell \otimes \overline{\mathbb{Q}_\ell} = \prod_{\sigma : E \hookrightarrow \overline{\mathbb{Q}_\ell}} M_\sigma.$$

The representation ρ_ℓ is a sum of the representations

$$\rho_\lambda : \mathrm{Gal}(\overline{K}/K) \longrightarrow \mathrm{Aut}_{\mathcal{O}_\lambda} M_\lambda \simeq \mathrm{GL}_r(\mathcal{O}_\lambda).$$

We assume that this is a strictly compatible family of λ-adic representations. In particular, we assume that there is a finite set of primes S of K so that for $\mathfrak{p} \notin S_\ell = S \cup \{$ primes of K dividing $\ell\}$, we have

$$\text{tr} \ \ \rho_\lambda(\text{Frob}_\mathfrak{p}) \ \in \ \mathcal{O}$$

and is independent of λ. Let P denote the subfield of E generated by the traces

$$\{\text{tr} \ \ \rho_\lambda(\text{Frob}_\mathfrak{p}) : \mathfrak{p} \notin S_\ell\}.$$

REMARK 2.1. In general, P is a proper subfield of E. However, there are infinitely many primes λ so that $E_\lambda = P_\mu$ where μ is the prime of P below λ.

We *assume* that the following conditions hold:

(i) $M_\lambda \otimes \mathbb{Q}_\ell$ is a simple $E_\lambda[\text{Gal}(\overline{K}/K)]$-module and in particular, $M_\ell \otimes \mathbb{Q}_\ell$ is a *semisimple* $(E \otimes \mathbb{Q}_\ell)[\text{Gal}(\overline{K}/K)]$-module

(ii) $E \otimes \mathbb{Q}_\ell$ is the maximal commutative semisimple subalgebra of

$$\text{End}_{\mathbb{Q}_\ell[\text{Gal}(\overline{K}/K)]} M_\ell \otimes \mathbb{Q}_\ell.$$

(iii) $\det \rho_\lambda = \epsilon \chi_\ell^{wr/2}$ for some integer $w \geq 0$ such that wr is even, and a finite order character ϵ. (Here, χ_ℓ denotes the ℓ-adic cyclotomic character and $\chi_\ell(\text{Frob}_\mathfrak{p}) = \mathbb{N}\mathfrak{p}$.)

(iv) each M_σ is an irreducible $\overline{\mathbb{Q}_\ell}[H]$-module, for any open subgroup H of

$$\text{Gal}(\overline{K}/K).$$

REMARK 2.2. Note that ϵ is independent of ℓ. Indeed, for $\mathfrak{p} \notin S$, we have $\epsilon(\text{Frob}_\mathfrak{p}) = \det \rho_\lambda(\text{Frob}_\mathfrak{p})(\mathbb{N}\mathfrak{p})^{-wr/2}$ for any $\lambda \notin S \cup \{$ primes of E dividing $\mathbb{N}\mathfrak{p}\}$.

REMARK 2.3. It would be of interest to weaken condition (iii). In particular, we might consider one dimensional M_ℓ which arise from algebraic Hecke characters.

REMARK 2.4. The condition (iv) is equivalent to saying that M_σ is not a component of an induced representation. Indeed, suppose that for some proper open subgroup H of $\text{Gal}(\overline{K}/K)$, M_σ is *reducible*. Let N_σ be an irreducible component. Then by Frobenius reciprocity,

$$(M_\sigma, \ \text{Ind}_H^{\text{Gal}(\overline{K}/K)} N_\sigma) \ \geq \ 1.$$

As M_σ is irreducible over K, this means that M_σ is a constituent of $\text{Ind} N_\sigma$. The converse is also true: if the above inequality is true with N_σ of dimension less than r, then M_σ restricted to H is reducible. If M_σ is actually an induced representation,

$$M_\sigma = \text{Ind}_H^{\text{Gal}(\overline{K}/K)} N_\sigma$$

and if H is normal, then we have

$$\text{tr} \ \ \rho_\sigma(\text{Frob}_\mathfrak{p}) = 0 \text{ if } \text{Frob}_\mathfrak{p} \not\subset H$$

and

$$\text{tr} \ \ \rho_\sigma(\text{Frob}_\mathfrak{p}) = \sum_{\mathfrak{P}|\mathfrak{p}} \text{tr } (\text{Frob}_\mathfrak{P}|N_\sigma) \text{ if } \text{Frob}_\mathfrak{p} \subset H.$$

REMARK 2.5. Condition (i) implies that the isomorphism class of the representation ρ_λ is determined by a *finite* number of Frobenius elements. This follows by a well-known argument of Faltings using Nakayama's Lemma and the Chebotarev Density Theorem.

REMARK 2.6. If M satisfies the above hypotheses, then so does any Tate twist $M(s)$ of M with $2s \leq w$. The integer w of (iii) changes to $w - 2s$. Here, $M_\ell(s)$ denotes $M_\ell \otimes \chi_\ell^{-s}$.

DEFINITION 2.7. *We shall call M_ℓ a primitive* $\mathrm{Gal}(\overline{K}/K)$*-module if conditions (i)-(iv) are satisfied. We shall call M_ℓ self-dual if there is a finite order character*

$$\phi : \mathrm{Gal}(\overline{K}/K) \longrightarrow \bar{\mathbb{Q}}^\times$$

such that

$$\check{M}_\ell \simeq M_\ell(w) \otimes \phi.$$

Here

$$\check{M}_\ell = \mathrm{Hom}_{\mathcal{O}\otimes\mathbb{Z}_\ell}(M_\ell, \mathcal{O} \otimes \mathbb{Z}_\ell)$$

and it is a $\mathrm{Gal}(\overline{K}/K)$-module by the action $(g.f)(m) = f(g^{-1}m)$.

We have the following elementary facts.

LEMMA 2.8. *Suppose that M_ℓ is a primitive self-dual* $\mathrm{Gal}(\overline{K}/K)$*-module.*

(1) *If $M_\sigma \simeq M_\sigma \otimes \psi$ for a finite order character ψ, then $\psi = 1$.*
(2) *The characters ϵ and ϕ take values in the trace subfield P.*

PROOF. The first statement follows from the irreducibility. Indeed, the isomorphism means explicitly that there is a non-scalar matrix X such that for all $g \in \mathrm{Gal}(\overline{K}/K)$, we have

$$\rho_\sigma(g) = X\rho_\sigma(g)X^{-1}\psi(g). \tag{2.9}$$

(Here, ρ_σ is the action of $\mathrm{Gal}(\overline{K}/K)$ on M_σ.) Restricting g to lie in the kernel H of ψ shows that X gives a non-trivial H-equivariant endomorphism of M_σ and this contradicts the irreducibility as stated in (iv). This proves (1).

For (2), view P and the fields of values of ϵ and ϕ as subfields of $\bar{\mathbb{Q}}_\ell$ by means of some fixed embedding. Let σ be an automorphism of $\bar{\mathbb{Q}}_\ell$ which is the identity on P. Then for any τ,

$$M_{\sigma\tau} \simeq M_\tau \tag{2.10}$$

and in particular, the determinants agree. This proves that

$$\epsilon^\sigma = \epsilon$$

and so σ is the identity on the field of values $\mathbb{Q}(\epsilon)$ of ϵ. For ϕ, we deduce from (2.10) that

$$M_{\sigma\tau} \otimes \tau\phi \simeq M_\tau \otimes \tau\phi \simeq M_\tau(-w)\check{} \simeq M_{\sigma\tau}(-w)\check{} \simeq M_{\sigma\tau} \otimes \sigma\tau\phi.$$

It follows from (1) that for any τ,

$$\sigma\tau\phi = \tau\phi$$

and so σ fixes the field $\mathbb{Q}(\phi)$ of values of ϕ also. This proves (2).

In the remainder of this section, we shall formulate a question on the distribution of the traces of Frobenius elements for primitive self-dual modules.

Let us consider E as a subfield of $\overline{\mathbb{Q}_\ell}$ by a fixed embedding $\mathbf{1}$. In the following, we write $M = M_\ell$. Consider the set

$$\Gamma_M(H) = \{\sigma : E \hookrightarrow \overline{\mathbb{Q}_\ell} : M_\sigma \simeq M_\mathbf{1} \text{ as } \overline{\mathbb{Q}_\ell}[H] \text{ modules}\},$$

where H is an open subgroup of $\mathrm{Gal}(\overline{K}/K)$. Let us also set

$$\Gamma_M = \cup_H \Gamma_M(H).$$

Let us set

$$a_{\mathfrak{p}} = \mathrm{tr}\ \ \rho_{\mathbf{1}}(\mathrm{Frob}_{\mathfrak{p}}).$$

Then $\sigma \in \Gamma_M(H)$ if and only if

$$a_{\mathfrak{p}}^{\sigma} = a_{\mathfrak{p}}$$

for all $\mathfrak{p}$ which have a degree one factor in the fixed field of H. If H is normal, this means that for all $\mathfrak{p} \notin S$,

$$a_{\mathfrak{p}}^{\sigma} = a_{\mathfrak{p}} \chi_{\sigma}(\mathfrak{p}) \tag{2.11}$$

for some *abelian* character χ_{σ}. (Indeed, we have a map

$$\mathrm{Gal}(\overline{K}/K) \ \longrightarrow\ \mathrm{Hom}_H(M_{\sigma}, M_{\mathbf{1}}) \ \simeq \bar{\mathbb{Q}}_{\ell}^{*}$$

given by

$$\tau \ \mapsto\ (f \ \mapsto\ \tau f \tau^{-1})$$

and this character is χ_{σ}.) Thus, $\sigma \in \Gamma_M(H)$ if and only if $\sigma|_P$ satisfies (2.11).

For $\sigma \in \Gamma_M(H)$, we have from (2.11) that

$$M_{\sigma} \ \simeq\ M_{\mathbf{1}} \otimes \chi_{\sigma}.$$

By Lemma 2.8, it follows that χ_{σ} is uniquely determined by σ. Comparing determinants, it follows that

$$\epsilon^{\sigma} = \epsilon \chi_{\sigma}^{r}. \tag{2.12}$$

Though it is not clear from this, σ is in fact an automorphism of P. Let us set Γ_M^{o} to denote $\Gamma_M|_P$.

LEMMA 2.13. *We have*

$$\Gamma_M^{o} \ \subseteq\ \mathrm{Aut}(P/\mathbb{Q}).$$

PROOF. For $\sigma \in \Gamma_M$, we have from (2.11) that

$$M_{\sigma} \ \simeq\ M_{\mathbf{1}} \otimes \chi_{\sigma}.$$

It also follows that

$$\check{M}_{\sigma} \ \simeq\ \check{M}_{\mathbf{1}} \otimes \chi_{\sigma}^{-1}.$$

Using the self-dual property, it follows that

$$M_{\sigma} \otimes \phi^{\sigma} \ \simeq\ M_{\mathbf{1}} \otimes \phi \chi_{\sigma}^{-1}.$$

Hence,

$$\phi(\phi^{\sigma})^{-1} \chi_{\sigma}^{-1} = \chi_{\sigma}.$$

Thus

$$\phi(\phi^{\sigma})^{-1} = \chi_{\sigma}^{2}.$$

As ϕ is a character of finite order, its values are roots of unity and so ϕ^{σ} and ϕ take values in the same field. It follows that there is a P-valued character ϕ_0 of finite order and a quadratic character ω such that $\chi_{\sigma} = \omega \phi_0$. In particular, χ_{σ} takes values in P also. Thus, $a_{\mathfrak{p}}^{\sigma} \in P$ and $\sigma|_P$ is an automorphism of P.

We are interested in the frequency with which the trace $a_{\mathfrak{p}}$ assumes a fixed value. Some values (such as the value $a = 0$ in the case of a representation associated to an elliptic curve with complex multiplication) will be exceptional in that they may be attained more frequently than others. We will only formulate a conjecture for "general" values. We make this precise in the following definition.

Denote by G_ℓ the Zariski closure of the image of ρ_ℓ and by G_λ the projection to the λ component. Denote by G_λ^0 the connected component of the identity of G_λ. We shall consider the trace morphism

$$T : G_\lambda^0 \longrightarrow \mathbb{G}_a.$$

DEFINITION 2.14. The element $a \in \mathcal{O}_P$ is *generic* if the hypersurface in G_λ^0 defined by $T - a$ is irreducible over P and of codimension 1.

Let us define the *stable trace field* by

$$F = P^{\Gamma_M^o}.$$

Hence, the element $a_\mathfrak{p} \in E$ is determined (up to a bounded number of possibilities) by the vector

$$\{a_\mathfrak{p}^\tau\}_{\tau : F \hookrightarrow \overline{\mathbb{Q}}}.$$

Let us *assume* that the module M_ℓ is of weight w in the sense that for all $\mathfrak{p} \notin S$, the eigenvalues of $\rho_\sigma(\mathrm{Frob}_\mathfrak{p})$ all have complex absolute value $(\mathbb{N}\mathfrak{p})^{w/2}$. Then, we have

$$|a_\mathfrak{p}^\tau| \;\le\; r(\mathbb{N}\mathfrak{p})^{w/2}$$

for all $\tau : F \hookrightarrow \bar{\mathbb{Q}}$. Hence, the "probability" that $a_\mathfrak{p}$ takes a given value is proportional to

$$(\mathbb{N}\mathfrak{p})^{-w[F:\mathbb{Q}]/2}$$

Summing this over $\mathfrak{p}$ with $\mathbb{N}\mathfrak{p} \le x$, we arrive at the following conjecture.

CONJECTURE 2.15. *For $a \in \mathcal{O}_F$, let us set*

$$\pi_M(x, a) = \#\{\mathfrak{p} : \mathbb{N}_{K/\mathbb{Q}}\mathfrak{p} \le x : a_\mathfrak{p} = a\}.$$

Then, for primitive self-dual modules M_ℓ of weight w, and generic a, there is a constant $c_{M,a}$ such that

$$\pi_M(x,a) = (c_{Ma} + \mathbf{o}(1))c_{M,a}\begin{cases} \sqrt{x}/\log x & \text{if } w = 1 \text{ and } F = \mathbb{Q} \\ \log\log x & \text{if } w = 2, F = \mathbb{Q} \text{ or } w = 1, [F:\mathbb{Q}] = 2 \\ 1 & \text{otherwise.} \end{cases}$$

It may be of interest to consider explicitly the special case $a = 0$. For this, we introduce the *honest weight*.

DEFINITION 2.16. Consider the set of non-negative integers s such that the Tate twist $M(s)$ is a primitive self-dual module. This set has a maximal element, s^* say. The *honest weight* w^* of M is defined to be $w - 2s^*$.

CONJECTURE 2.17. *When 0 is generic for $M(s^*)$, we have*

$$\pi_M(x,0) = (c_{M,0} + \mathbf{o}(1))\begin{cases} \sqrt{x}/\log x & \text{if } w^* = 1 \text{ and } F = \mathbb{Q} \\ \log\log x & \text{if } w^* = 2, F = \mathbb{Q} \text{ or } w^* = 1, [F:\mathbb{Q}] = 2 \\ 1 & \text{otherwise.} \end{cases}$$

REMARK 2.18. To illustrate the meaning of genericity, consider for example, a representation which is a tensor product. There will be a corresponding factorization of the traces. In particular, let E be an elliptic curve without complex multiplication, defined over $\mathbb{Q}$, and for which the Lang-Trotter conjecture predicts that the number $\pi_{E,0}(x)$ of supersingular primes satisfies

$$\pi_{E,0}(x) = c_E + \mathbf{o}(1))\frac{\sqrt{x}}{\log x}$$

with $c_E \neq 0$. Let ρ_1 be an ℓ-adic representation associated to E and ρ_2 to another such elliptic curve which is non-isogenous to the first. Consider $\rho = \rho_1 \otimes \rho_2$. The trace morphism will factor:

$$T(g_1, g_2) = (\mathrm{tr}\ g_1)(\mathrm{tr}\ g_2).$$

The trace will vanish at any prime which is supersingular for either elliptic curve. However, the weight of the tensor product representation is 2 and under the conjecture above, there should only be $\mathbf{O}(\log\log x)$ primes $p \leq x$ for which the trace vanishes for such a representation.

Another example is obtained by considering the symmetric cube of the ℓ-adic representation associated to a modular form f of weight 2. In this case, the trace of Frobenius at a good prime p is given by

$$a_f(p)(a_f(p)^2 - 2p)$$

and in particular, this vanishes whenever $a_f(p) = 0$. (Here, $a_f(p)$ denotes the p-th Fourier coefficient of f.) The number of such primes could be $\gg \sqrt{x}/\log x$ (according to the Lang-Trotter conjectures) even though the weight of the symmetric cube representation is 3.

REMARK 2.19. It is clear that if $c_{M,a}$ exists, it is independent of ℓ and the notation is meant to emphasize this. It is also reasonable to expect that the exact determination of the constant will depend on detailed information about the image of the Galois representations ρ_λ. What is not clear is whether the image of Galois will affect the asymptotic growth (modulo constants) beyond what one deduces from the existence of the inner twists Γ_M. The above conjecture expresses the view that it does not.

REMARK 2.20. The case of one-dimensional M is also interesting though it doesn't fit into our assumptions (particularly, (iii)). In this case, it should be possible to investigate the asymptotic behavior of $\pi_M(x,a)$ using sieve methods and methods of analytic number theory. Upper bounds have been obtained by Rajan [**R**]. It would be very interesting to investigate lower bounds.

REMARK 2.21. As a general fact, effective forms of the Chebotarev Density Theorem imply that

$$\pi_M(x,a) \ll x/(\log x)^{1+\gamma}$$

for some $\gamma = \gamma(M,a) > 0$. This is due to Serre [**S81**]. Indeed, the condition of genericity implies that $T-a$ is non-zero and by [**S81**, Theorem 8], it follows that it is of Minkowski dimension less than the Minkowski dimension of G_λ^0. Hence, [**S81**, Theorem 10], applies to give the above bound.

REMARK 2.22. Though we assumed that we have a compatible family of λ-adic representations, the essential point is the algebraicity of the traces of Frobenius elements. Given the extensive study to which Galois representations have recently been subjected, it is possible to find conditions on a single λ-adic representation (semistability, for example) that would imply the algebraicity of the $a_\mathfrak{p}$.

3. Special cases

We shall consider the case of Galois representations that arise from modular forms. We recall some results of Momose [**M**] and [**Ri80**].

Let f be a normalized eigenform of weight k, level N and character ϵ. For a prime p, write $a_f(p)$ for the p-th Fourier coefficient of f. Let $\mathcal{O}_f$ denote the $\mathbb{Z}$-module generated by the $a_f(p)$ and let E_f denote the fraction field. Then, we have a strictly compatible family of representations

$$\rho_\lambda : \mathrm{Gal}(\overline{\mathbb{Q}}/\mathbb{Q}) \longrightarrow \mathrm{GL}_2(E_\lambda)$$

which is unramified outside $N\ell$. For $p \nmid N\ell$, the characteristic polynomial of $\rho_\lambda(\mathrm{Frob}_p)$ is

$$T^2 - a_f(p)T + \epsilon(p)p^{k-1}.$$

We *assume* that f is not of CM-type and $k \geq 2$. In this case, thanks to results of Ribet [**Ri80**], the underlying rank 2 module M_ℓ is primitive. It is also self-dual as

$$\check{M}_\ell = M_\ell \otimes \epsilon^{-1}.$$

By definition, the trace subfield of this module is E_f. We shall consider the problem of how often the Fourier coefficient takes a given value.

By the general construction of the previous section, we have a subfield F_f of E_f. Explicitly, it is defined as follows. We consider the embeddings

$$\sigma : E_f \hookrightarrow \bar{\mathbb{Q}}$$

for which there is a Dirichlet character χ such that

$$a_f(p)^\sigma = a_f(p)\chi(p).$$

As we are assuming that f does not have complex multiplication, such a χ is unique. Let us denote it by χ_σ. The set Γ_f of these embeddings σ forms an Abelian group of automorphisms of E_f and F_f is defined to be the fixed field. In the terminology of §2, F_f is the stable trace field. Let us set

$$J_{F_f} = \mathrm{Hom}(F_f, \mathbb{C}).$$

We note that (for f fixed), once we know the vector $\{a_f(p)^\sigma : \sigma \in J_F\}$, we know all the conjugates of $a_f(p)$ and in particular, $a_f(p)$ is completely determined. In fact, we see that to know the value of $a_f(p)$ is equivalent to determining this vector.

In the following discussion f will be fixed and so we shall drop it from the subscript to simplify the notation. In this case, the associated λ-adic representation ρ_λ has weight $w = k - 1$. Thanks to the work of Momose [**M**] and Ribet [**Ri85**], the image of ρ_λ is understood and we feel more confident in stating Conjecture 2.15 in this special case. Let us set

$$\pi_{f,a}(x) = \#\{p \leq x : a_f(p) = a\}.$$

In this case, any value of a is generic, and the conjecture of the previous section is the following.

CONJECTURE 3.1. *Let $a \in \mathcal{O}_F$, $k \geq 2$ and assume that f is not of CM-type. There is a constant $c_{f,a}$ such that*

$$\pi_{f,a}(x) \begin{cases} \sqrt{x}/\log x & \textit{if } k = 2 \textit{ and } F = \mathbb{Q} \\ \log\log x & \textit{if } k = 2 \textit{ and } [F:\mathbb{Q}] = 2 \textit{ or } k = 3 \textit{ and } F = \mathbb{Q} \\ 1 & \textit{otherwise.} \end{cases}$$

The conditions above are obtained by considering when the quantity

$$\frac{k-1}{2}[F:\mathbb{Q}]$$

is <1, $=1$ and >1 respectively. Note that if $k=2$ and $E=\mathbb{Q}$, the first part of Conjecture 3.1 is contained in the original conjecture of Lang and Trotter. In §5, we shall discuss what Conjecture 3.1 says in the case $a=0$.

REMARK 3.2. In the case that f is of weight 2 and $a=0$, this conjecture was formulated independently and earlier by P. Bayer and J. González i Rovira [**BR**]. They also give a probabilistic model to support their conjecture as well as some numerical evidence.

REMARK 3.3. In the case of elliptic curves, some results are known on average and are due to Fouvry-Murty and David-Pappalardi. For each pair of integers a, b consider the elliptic curve

$$E_{a,b} : y^2 = x^3 + ax + b.$$

A consequence of [**FM**, Theorem 4], is that if $A, B > x^{1+\epsilon}$, then

$$\sum_{|a|\le A}\sum_{|b|\le B} \pi_{E_{a,b},0}(x) \sim \frac{4\pi}{3}AB\frac{\sqrt{x}}{\log x}.$$

This has been generalized to $\pi_{E_{a,b},r}$ for $r\neq 0$ in David-Pappalardi [**DP**]. The proofs of these results rely on the fact that one is dealing with elliptic curves. It would be very interesting to prove them for modular forms.

Of course, it is possible to study Conjecture 3.1 numerically. Using a Mathematica program of David Roberts, the following numbers were computed by Vijay Patankar for a form f of weight 2 on $\Gamma_0(512)$ whose Fourier coefficients lie in the field $\mathbb{Q}(\sqrt{2})$. This form has a non-trivial inner twist (i.e., $F=\mathbb{Q}$). He computed

$$c(x,a) = \pi_{f,a}(x)\log x/\sqrt{x}$$

for $a = 0, 6, -6$ and found the following numbers.

x	$c(x,-6)$	$c(x,0)$	$c(x,6)$
97,000	2.61763	1.88027	2.47015
98,000	2.60656	1.87232	2.45971
99,000	2.63221	1.86449	2.44942
100,000	2.65772	1.89317	2.47568

To get an idea of the size of the quantities which we are comparing, we note that for $x = 100,000$, we have $\sqrt{x}/\log x \sim 27.47$.[1]

As a variant of Conjecture 3.1, one can also consider how often $a_f(p)$ lies in a proper subfield of E_f. Let M be a subfield of F. We want to study the quantity

$$\pi_{f,M}(x) = \#\{p\le x : a_f(p)\in M\}.$$

[1]With the help of Kevin James, computations have now been made up to $x = 1.6\times 10^7$. These will be presented elsewhere.

CONJECTURE 3.4. *Assume that f is not of CM-type and $k \geq 2$. There is a constant $c_{f,M}$ such that*

$$\pi_{f,M}(x) = (c_{f,M} + \mathbf{o}(1)) \begin{cases} x/\log x & \text{if } M = F \\ \sqrt{x}/\log x & \text{if } k = 2,\ M = \mathbb{Q} \text{ and } [F:\mathbb{Q}] = 2 \\ \log\log x & \text{if } k = 2 \text{ and } [M:\mathbb{Q}] = [F:M] = 2, \\ & \text{or } k = 2 \text{ and } M = \mathbb{Q},\ [F:M] = 3, \\ & \text{or } k = 3 \text{ and } M = \mathbb{Q},\ [F:\mathbb{Q}] = 2 \\ 1 & \text{otherwise.} \end{cases}$$

These conditions are obtained by considering when the quantity

$$\frac{k-1}{2}[M:\mathbb{Q}]([F:M]-1)$$

is smaller than, equal to, or larger than 1. It would be of interest to make the constant $c_{f,M}$ precise. It is possible that this conjecture may be more amenable to study than Conjecture 3.1.

A special case of the above is when $M = \mathbb{Q}$. In this case, the conjecture says that

$$\pi_{f,\mathbb{Q}}(x) = (c_{f,\mathbb{Q}} + \mathbf{o}(1)) \begin{cases} x/\log x & \text{if } F = \mathbb{Q} \\ \sqrt{x}/\log x & \text{if } k = 2 \text{ and } [F:\mathbb{Q}] = 2 \\ \log\log x & \text{if } k = 2 \text{ and } [F:\mathbb{Q}] = 3 \\ & \text{or } k = 3 \text{ and } [F:\mathbb{Q}] = 2 \\ 1 & \text{otherwise.} \end{cases}$$

In general, it is not clear whether there exists even *one* prime for which $a_f(p)$ is a rational integer.

Though it does not fit into our general setting, we can consider a pair of modular forms f, g of weight 2 and for which the stable trace field F is quadratic over $\mathbb{Q}$. We can then ask for the distribution of primes p such that *both* Fourier coefficients $a_f(p)$ *and* $a_g(p)$ lie in $\mathbb{Z}$. Heuristics similar to that used above would suggest that if g is not a quadratic twist of a Galois conjugate of f, then the number of such primes $\leq x$ should be $\mathbf{O}(\log\log x)$. This would then make plausible the following conjecture.

CONJECTURE 3.5. *Suppose that f and g are both of weight 2 and have a stable trace field which is quadratic over $\mathbb{Q}$. If the set of primes p for which $a_f(p)$ is a rational integer coincides with the similar set for g, then g is a quadratic twist of a Galois conjugate of f.*

We shall prove the following result.

THEOREM 3.6. *Suppose that f does not have complex multiplication and that F_f is not equal to $\mathbb{Q}$. If we assume the Riemann Hypothesis for Dedekind zeta functions, then*

$$\#\{p \leq x : a_f(p) \in \mathbb{Z}\} \ll x^{7/8}/(\log x)^{1/2}.$$

The main tools for proving Theorem 3.6 are the Galois representations attached to f and the effective Chebotarev Density Theorem. The proof is given in the next section.

Next, we briefly consider symmetric powers of the representations associated to f. Let $\rho_{\ell,m}$ denote the m-th symmetric power of ρ_ℓ. It is clear that if $\sigma \in \Gamma_f$

then

$$\mathrm{Sym}^m f^\sigma \;\simeq\; \mathrm{Sym}^m f \,\otimes\, \chi_\sigma^m.$$

Thus, the subfield F_m which is the subfield of E left fixed by $\Gamma_{\mathrm{Sym}^m f}$ is a subfield of F. Suppose that $\chi_\sigma^m = 1$. Then

$$\mathrm{Sym}^m f^\sigma \;\simeq\; \mathrm{Sym}^m f$$

and in particular, $\mathrm{Sym}^m M_{f,\ell}$ is not primitive (for example, it does not satisfy (i)).

The weight of $\mathrm{Sym}^m \rho_\lambda$ is $(k-1)m$ where k is the weight of the modular form f. The general conjecture predicts the following.

CONJECTURE 3.7. *Suppose that $m \geq 2$, that $\mathrm{Sym}^m M_{f,\ell}$ is a primitive module and that $a \in F_m$ is generic. Then there is a constant $c_{f,m,a}$ such that*

$$\#\{p \leq x : a(p^m) = a\} = (c_{f,m,a} + \mathbf{o}(1)) \begin{cases} \log\log x & \textit{if } k=2, m=2 \textit{ and } F_m = \mathbb{Q} \\ 1 & \textit{otherwise.} \end{cases}$$

REMARK 3.8. For $m = 2$, it follows from [**MuP**, Remark 2.10], that for forms f without complex multiplication, $\Gamma_{\mathrm{Sym}^2 f} = \Gamma_f$. The remark in question states that if ρ_1 and ρ_2 are two two-dimensional irreducible non-dihedral representations, and $\mathrm{Sym}^2\rho_1 \;\simeq\; \mathrm{Sym}^2\rho_2 \otimes \mu$ for a finite order character μ, then $\rho_1 \simeq \rho_2 \otimes \chi$ with $\mu = \chi^2$.

REMARK 3.9. Consider an f of weight 4 which has rational integral Fourier coefficients. If $a(p^2) = a$, then $(p, a(p))$ is an integral point on the elliptic curve

$$y^2 = x^3 + a.$$

Thus, the set $\{p : a(p^2) = a\}$ is finite. More generally, finiteness can be established in many cases using results from transcendence. Some results of this kind were obtained in [**MMSh**].

4. Values in a subfield

In this section, we shall prove Theorem 3.6. We begin by establishing a version of the Chebotarev Density Theorem which in a special case is sharper than the version of Lagarias and Odlyzko [**LO**]. We consider a Galois representation

$$\rho : \mathrm{Gal}(\bar{\mathbb{Q}}/\mathbb{Q}) \longrightarrow \mathrm{GL}_n(\mathbb{F})$$

where $\mathbb{F}$ is a finite field of characteristic ℓ. Let K be the fixed field of the kernel of ρ and let G denote $\mathrm{Gal}(K/\mathbb{Q})$. Clearly, G is isomorphic to the image of ρ. Let D denote a conjugacy set in G. We want to estimate

$$\pi_D(x) = \#\{p \leq x : \mathrm{Frob}_p \subset D\}.$$

For any number field F, let us set $n_F = [F : \mathbb{Q}]$ and d_F the absolute value of the discriminant of $F/\mathbb{Q}$. Denote by B the Borel subgroup of upper triangular matrices in G and by U the subgroup of B consisting of unipotent matrices. Let us call D *triangular* if every conjugacy class in D has a non-empty intersection with B. For a triangular D, define an associated conjugacy set D_0 in B as follows: For each conjugacy class C in D, let C_0 be a conjugacy class in B contained in $C \cap B$. Let D_0 be the union of the C_0. Note that there may be more than one associated conjugacy set to a given triangular D.

THEOREM 4.1. *Assume the GRH and assume that D is a triangular conjugacy set. Suppose that for some associated D_0, we have $UC_0 \subseteq C_0$ for each conjugacy class in D_0. Then,*

$$\pi_D(x) = \frac{|D|}{|G|}\text{Li } x \;+\; \mathbf{O}(|D|^{1/2}\left(\frac{[G:B]}{|U|}\right)^{1/2} x^{1/2}\log(xd_K^{1/n_K})) + \mathbf{O}\left(\frac{1}{|B|}\log d_K \;+\; \frac{|G|}{|B|}x^{1/2}\right).$$

PROOF. Let M be the subfield of K fixed by B and let $C \subseteq D$ be a conjugacy class. By a well-known principle (see [**S81**], [**MMS**, 1.41])

$$\pi_C(x) - \frac{|C|}{|G|}\text{Li } x = \lambda_C\left(\pi_{C_0}(x) - \frac{|C_0|}{|B|}\text{Li } x\right) + \mathbf{O}\left(\frac{1}{|B|}\log d_K + n_M x^{1/2}\right)$$

where

$$\lambda_C = \frac{|C|}{|G|}\frac{|B|}{|C_0|}.$$

In fact, a similar estimate holds true if C is replaced by D. Thus,

$$\begin{aligned}\pi_D(x) - \frac{|D|}{|G|}\text{Li } x &= \sum_{C\subseteq D}\left(\pi_C(x) - \frac{|C|}{|G|}\text{Li } x\right)\\ &= \sum_{C\subseteq D}\lambda_C\left(\pi_{C_0}(x) - \frac{|C_0|}{|B|}\text{Li } x\right) + \mathbf{O}\left(\frac{1}{|B|}\log d_K + \frac{|G|}{|B|}x^{1/2}\right).\end{aligned}$$

Let N denote the fixed field of U. Denote by $\bar{C}_0$ the image of C_0 in B/U. Notice that B/U is isomorphic to a group A of diagonal matrices. Since

$$UC_0 \subseteq C_0,$$

we have by [**S81**] and [**MMS**, 1.4],

$$\begin{aligned}&\sum_{C\subseteq D}\lambda_C\left(\pi_{C_0}(x) - \frac{|C_0|}{|B|}\text{Li } x\right)\\ &\qquad = \sum_{\bar{C}_0\subseteq\bar{D}_0}\lambda_C\left(\pi_{\bar{C}_0}(x) - \frac{|\bar{C}_0|}{|A|}\text{Li } x\right) + \mathbf{O}\left(\frac{1}{|B|}\log d_K + n_M x^{1/2}\right).\end{aligned}$$

For the first term, we have

$$\begin{aligned}&\sum_{\bar{C}_0\subseteq\bar{D}_0}\lambda_C\left|\pi_{\bar{C}_0}(x) - \frac{|\bar{C}_0|}{|A|}\text{Li } x\right|\\ &\qquad \le \left(\sum_{\bar{C}_0\subseteq\bar{D}_0}\lambda_C^2|\bar{C}_0|\right)^{1/2}\left(\sum_{\bar{C}_0\subseteq\bar{D}_0}\frac{1}{|\bar{C}_0|}\left|\pi_{\bar{C}_0}(x) - \frac{|\bar{C}_0|}{|A|}\text{Li } x\right|^2\right)^{1/2}.\end{aligned}$$

Since A is Abelian, we certainly know that Artin L-functions for N/M are entire. Therefore, by [**MMS**, 3.7]

$$\sum_{\bar{C}_0\subseteq\bar{D}_0}\frac{1}{|\bar{C}_0|}\left|\pi_{\bar{C}_0}(x) - \frac{|\bar{C}_0|}{|A|}\text{Li } x\right|^2 \ll xn_M^2(\log xd_K^{1/n_K})^2.$$

Moreover,

$$\lambda_C^2|\bar{C}_0| = \frac{|C|^2}{|G|^2}\frac{|B|^2}{|C_0||U|}.$$

Since

$$|C_0|[G:B] \geq |C|,$$

we have

$$\frac{|C|^2}{|G|^2}\frac{|B|^2}{|C_0||U|} \ll |C|\frac{|B|}{|G||U|}.$$

Hence, putting all of the above together, we deduce that

$$\pi_D(x) - \frac{|D|}{|G|}\mathrm{Li}\, x \ll \left(\sum |C|\right)^{1/2}\left(\frac{|G|}{|B||U|}\right)^{1/2} x^{1/2}\log x d_K^{1/n_K} + \frac{1}{|B|}\log d_K + n_M x^{1/2}$$

proving the result.

REMARK 4.2. In some cases, this result improves on the estimates of [**LO**] and [**S81**]. For example, suppose that

$$|U| \gg [G:B]$$

where the implied constant is absolute, and in particular does not depend on the residue characteristic. (This condition is satisfied, for example, if $G = \mathrm{GL}_n(\mathbb{F})$.) Then, Theorem 4.1 gives an error term of

$$\mathbf{O}\left(|D|^{1/2}x^{1/2}\log x d_K^{1/n_K}\right)$$

whereas [**LO**] gives

$$\mathbf{O}\left(|D|x^{1/2}\log x d_K^{1/n_K}\right)$$

and [**S81**, Theorem 7(ii)] gives

$$\mathbf{O}\left(\frac{|D|}{|U|}x^{1/2}\log x d_K^{1/n_K}\right).$$

Next, we need an estimate for the existence of primes of degree larger than 1.

PROPOSITION 4.3. *Let M be an imaginary quadratic extension of $\mathbb{Q}$ and F a disjoint field, not necessarily Galois over $\mathbb{Q}$ of degree n_F. Assume the GRH. The number of primes $\ell \leq x$ which split in M and which do not split completely in F is*

$$\frac{1}{2}\left(1 - \frac{1}{n_F}\right)\mathrm{Li}\, x + \mathbf{O}(n_F!x^{1/2}\log(xd_{FM})).$$

PROOF. We begin by assuming that F is Galois over $\mathbb{Q}$. Then this follows from a straightforward application of the estimate of [**LO**]. In fact the logarithmic term in the error term may be replaced by $\log(xd_{FM}^{1/2n_F})$. The general case can be reduced to the Galois case. Denote by K the normal closure of F. If ℓ does not split completely in K, then it has a factor of degree > 1 in F. Moreover, by Hensel's estimate

$$\log d_K \leq n_K \sum_{p|d_F} \log p.$$

The result follows.

We are now ready to give a proof of the main theorem.

PROOF OF THEOREM 3.6. Let λ be a prime of E_f and consider the representation

$$\rho_{f,\lambda} : \mathrm{Gal}(\bar{\mathbb{Q}}/\mathbb{Q}) \longrightarrow \mathrm{GL}_2(\mathcal{O}_{f,\lambda}).$$

Recall that it is unramified outside ℓN (where ℓ is the rational prime underlying λ) and for a prime p which does not divide ℓN, $\rho_{f,\lambda}(\mathrm{Frob}_p)$ has characteristic polynomial

$$T^2 - a_f(p)T + \epsilon(p)p^{k-1}.$$

Let

$$H = \cap_{\sigma\in\Gamma_f} \mathrm{Ker}\ \chi_\sigma$$

and let K be the fixed field of H. It is a certain Abelian extension of $\mathbb{Q}$. Then the restriction of $\rho_{f,\lambda}$ to H takes values in $\mathrm{GL}_2(\mathcal{O}_\mu)$ where μ is the restriction of λ to F_f.

Let us set

$$\mathfrak{G} = \{g \in \mathrm{GL}_2(\mathcal{O}_\mu) : \det g \in (\mathbb{Z}_\ell^\times)^{k-1}\}.$$

By a result of Momose [**M**] and Ribet [**Ri85**], $\rho_{f,\lambda}$ maps H surjectively onto $\mathfrak{G}$, provided ℓ is sufficiently large, say $\ell \geq \ell_0$. Suppose ℓ satisfies this condition. Reduce the representation modulo μ to get

$$\bar{\rho}_f : H \longrightarrow \bar{\mathfrak{G}} = \{g \in \mathrm{GL}_2(\mathbb{F}_\mu) : \det g \in (\mathbb{F}_\ell^\times)^{k-1}\}.$$

Here $\mathbb{F}_\mu$ is the residue field $\mathcal{O}_\mu/\mu$.

If p is a prime such that $a_f(p) \in \mathbb{Z}$, then for $\sigma \in \Gamma_f$,

$$a_f(p) = a_f(p)^\sigma = a_f(p)\chi_\sigma(p).$$

Thus, either $a_f(p) = 0$ or we have

$$\chi_\sigma(p) = 1 \text{ for all } \sigma \in \Gamma_f. \tag{4.4}$$

In particular

$$\epsilon(p) = 1 \tag{4.5}$$

as $\chi_\sigma = \bar{\epsilon}$ when σ is complex conjugation. The condition (4.4) means that $\mathrm{Frob}_p \in H$. We can restrict ourselves to this case as

$$\#\{p \leq x : a_f(p) = 0\} \ll x^{3/4}$$

according to a result of Serre [**S81**, p. 175].

Let y and u be parameters to be specified with $y\exp(-c(\log y)^{1/2}) \leq u \leq y$. Denote by $spl(F)$ the set of primes p which split completely in F. Let I denote the interval $[y, y+u]$ and consider the quantity

$$\sum_{\substack{\ell\in I\\ \ell\notin spl(F)}} \#\left\{p \leq x : 0 \neq a_f(p) \in \mathbb{Z}, \Delta_p \neq 0, \text{ and } \left(\frac{\Delta_p}{\ell}\right) = 1\right\}$$

where

$$\Delta_p = a_f(p)^2 - 4\epsilon(p)p^{k-1}.$$

By (4.5), and the fact that

$$|a_f(p)| \leq 2p^{(k-1)/2}$$

we see that $\mathbb{Q}(\sqrt{\Delta_p})$ is an imaginary quadratic field or $\mathbb{Q}$, whereas F is a subfield of the totally real subfield of E. Hence they are disjoint and so, the conditions that ℓ does not split in F and splits in $\mathbb{Q}(\sqrt{\Delta_p})$ are compatible.

If $\Delta_p = 0$, then (in view of (4.5)), we have

$$a_f(p) = \pm p^{(k-1)/2}.$$

By [**MMS**], the number of primes $p \leq x$ for which this equality holds is $\mathbf{O}(x^{4/5})$. We may therefore ignore these primes.

Rearranging the sum, and invoking Proposition 4.3, we see that it is equal to

$$\sum_{\substack{p\leq x\\ 0\neq a_f(p)\in\mathbb{Z},\Delta_p\neq 0}} \#\left\{\ell \in I:\ \ell \notin spl(F) \text{ and } \left(\frac{\Delta_p}{\ell}\right) = 1\right\}.$$

Thus, we have to estimate

$$\sum_{\substack{p\leq x\\ 0\neq a_f(p)\in\mathbb{Z}}}{}^{*} \left\{c(\pi(y+u)-\pi(y)) \ + \ \mathbf{O}(y^{1/2}\log \Delta_p y)\right\}$$

where the implied constant depends on F (and hence on f) but not on p and the asterisk on the summation indicates that we only consider primes p for which the discriminant Δ_p is non-zero. Moreover, $c = c_{k,F}$ is equal to

$$\frac{1}{2}\left(1-\frac{1}{n_F}\right)$$

if $k=2$ and in general is a small multiple of this depending only on k. Thus, this is

$$\gg \ (\pi(y+u)-\pi(y))\pi^*_{f,\mathbb{Q}}(x)$$

where the asterisk indicates that we are only counting primes p for which $a_f(p) \neq 0$ and $\Delta_p \neq 0$. Thus,

$$\pi^*_{f,\mathbb{Q}}(x) \ll \ \max_{\substack{\ell\in I\\ \ell\notin spl(F)}} \#\left\{p\leq x: 0\neq \ a_f(p)\in\mathbb{Z}, \Delta_p\neq 0, \text{ and } \left(\frac{\Delta_p}{\ell}\right)=1\right\}.$$

Let ℓ be a prime as above: $\ell \in I$ and $\ell \notin spl(F)$. Consider the set

$$C = \left\{\sigma \in \mathrm{GL}_2(\mathbb{F}_\mu) : \mathrm{tr}\ g \in \mathbb{F}_\ell, \det g \in \mathbb{F}_\ell^{k-1}, \text{ and } \left(\frac{\Delta_g}{\ell}\right)=1\right\}$$

where

$$\Delta_g = (\mathrm{tr}\ g)^2 - 4\det g$$

and μ is a prime of F of degree larger than 1 which divides ℓ. Then, C is stable under conjugation, and hence is a union of conjugacy classes. Hence, we can count

$$\pi_C(x) = \#\{p\leq x : \bar{\rho}_f(\sigma_p)\in C\}$$

using the Chebotarev density theorem. Moreover, by construction, every conjugacy class in C intersects the Borel subgroup and $UC_0 \subseteq C_0$ for each conjugacy class in $C\cap B$. (Indeed, by choice of ℓ, C_0 contains a diagonal matrix, γ, say, with distinct eigenvalues. One easily checks that for $u\in U$, $u\gamma$ is conjugate to γ by a matrix whose determinant is non-zero and lies in $\mathbb{F}_\ell^{k-1}$.)

Assume the GRH. Then we may apply Theorem 4.1 to deduce that

$$\pi_C(x) = \frac{|C|}{|G|}\mathrm{Li}\ x \ + \ \mathbf{O}(|C|^{1/2}x^{1/2}\log x d_K^{1/n_K}) \ + \ \mathbf{O}\left(\frac{1}{|B|}\log d_K \ + \ \frac{|G|}{|B|}x^{1/2}\right).$$

In our case,

$$|\mathfrak{G}| \sim (\mathbb{N}\mu)^3\ell, \quad |C|\sim(\mathbb{N}\mu)^2\ell^2$$

where $\mathbb{N}$ denotes norm from F to $\mathbb{Q}$. (In the above, we are writing $X \sim Y$ for

$$Y \ll X \ll Y$$

where the implied constants are absolute.) Let us say that $\mathbb{N}\mu = \ell^a$. Then, by the above and by Hensel's estimate

$$\log|d_K| \leq n_K \sum_{p|d_K} \log p$$

we deduce that

$$\pi_C(x) = \frac{1}{\ell^{a-1}}\mathrm{Li}\ x \ + \ \mathbf{O}(\ell^{a+1}x^{1/2}\log \ell x).$$

Now

$$\#\{p \leq x : \mathrm{tr}\ \rho_{f,\lambda}(\mathrm{Frob}_p) \in \mathbb{Z}\} \ \leq \ \#\{p \leq x : \bar{\rho}_f(\mathrm{Frob}_p) \in C\} \ + \ \mathbf{O}(x^{4/5})$$

for *any* μ. Choose $\ell = \ell(\mu) \geq \ell_0$ and μ of degree > 1. Then,

$$\#\{p \leq x : a_f(p) \in \mathbb{Z}\} \ \leq \ \frac{1}{\ell^{a-1}}\mathrm{Li}\ x + \mathbf{O}(\ell^{a+1}x^{1/2}\log \ell x) + \ \mathbf{O}(x^{4/5})$$

uniformly in ℓ. Now choose y so that

$$y^{2a} = x^{1/2}(\log x)^{-2}$$

and $\ell \in [y, y+u]$. Then the right hand side above is

$$\ll x^{\frac{1}{2}+\frac{a+1}{4a}}(\log x)^{1-\frac{a+1}{a}} = x^{3/4}\left(\frac{x^{1/4}}{\log x}\right)^{1/a} \ll \ x^{7/8}(\log x)^{-1/2}$$

as $a \geq 2$. This completes the proof.

5. Ordinary and supersingular primes

We make a few remarks about the set of primes of ordinary reduction of an Abelian variety A defined over a number field K. A prime p is said to be *ordinary* for A if the Frobenius endomorphism Frob_p acting on $H^1(A, \mathcal{O})$ has maximal rank (namely, the dimension of A). It is *supersingular* if Frob_p has minimal rank (namely 0).

There is a conjecture of Serre (see [**Oo**], $11A_?$) that says there is a finite extension field L of K such that the base change $A_{/L}$ of A to L has ordinary reduction at almost all primes of L. (Here 'almost all primes' means that the property holds for a set of primes of density one.) This is known for A of dimension ≤ 2 and due to Serre (in the dimension 1 case) and to Ogus [**O**, Corollary (2.9)] (in the dimension 2 case). It is also known in the CM-case. In general, it is not known whether there is even one prime of (good) ordinary reduction. Note that a prime is ordinary for an Abelian subvariety of A if it is ordinary for A. Moreover, if A is isogenous to a product $B \times C$, then p is ordinary for A if and only if it is ordinary for B and C. It follows that if A is isogenous to a product of elliptic curves and Abelian surfaces, then (possibly after a base change) it is ordinary for almost all primes.

In the case that we have an f as in §3 of weight 2, one can consider the supersingular and ordinary reductions of the associated Abelian variety A_f. Observe that $A = A_f$ has supersingular (resp. ordinary) reduction at the prime $p > 5$ if and only if $a_f(p) = 0$ (resp. $a_f(p) \neq 0$). Moreover, apart from a finite number of exceptions, if A is not supersingular at p then it is ordinary. If f is not of CM-type, Conjecture (3.1) gives

$$\pi_A^{\text{supersingular}}(x) = (c + \mathbf{o}(1)) \begin{cases} \sqrt{x}/\log x & \text{if } F = \mathbb{Q} \\ \log\log x & \text{if } [F : \mathbb{Q}] = 2 \\ 1 & \text{otherwise.} \end{cases}$$

Thus, in all cases, this conjecture predicts

$$\pi_A^{\text{ordinary}}(x) \sim x/\log x$$

and this is consistent with Serre's conjecture. In terms of the structure of A, we can interpret the above conditions using the results of Momose and Ribet on the endomorphism algebra. Their results tell us that

$$\mathrm{End}(A_f) \otimes \mathbb{Q} = \mathrm{M}_n(F_f) \text{ or } \mathrm{M}_{n/2}(D)$$

where $n = [E_f : F_f]$ and D is a quaternion division algebra over F_f. In particular, if $F_f = \mathbb{Q}$, then A_f is either a power of an elliptic curve (without complex multiplication) or is a power of a "fake elliptic curve", i.e., an Abelian surface with multiplication by a (totally definite) quaternion division algebra. And if F_f is a quadratic number field, then A_f is a power of an Abelian surface with multiplication by F_f or is a power of an Abelian fourfold with multiplication by a quaternion division algebra over F_f. In all of these cases, the conjecture above predicts that there should be infinitely many supersingular primes. It will be worth while to see if the methods of Elkies [**E87**] can be adapted to this setting.

References

[BR] P. Bayer and J. Gonzalez i Rovira, *On the Hasse-Witt invariants of modular curves*, Experimental Mathematics **6** (1997), 57–76.

[DP] C. David and F. Pappalardi, *Average Frobenius distributions of elliptic curves*, Intl. Math. Res. Notices, **4** (1999), 1–19.

[E87] N. Elkies, *The existence of infinitely many supersingular primes for every elliptic curve over* $\mathbb{Q}$, Invent. Math. **89** (1987), 561–568.

[E92] N. Elkies, *Distribution of supersingular primes*, Journées Arithmétiques (Luminy, 1989), Astérisque (1992), 127–132.

[FM] E. Fouvry and M. Ram Murty, *On the distribution of supersingular primes*, Canadian J. Math. **48** (1) (1996), 81–104.

[LO] J. Lagarias and A. Odlyzko, *Effective versions of the Chebotarev density theorem*, in Algebraic Number Fields, ed. A. Fröhlich, New York: Academic Press (1977), 409-464.

[LT] S. Lang and H. Trotter, Frobenius Distributions in GL_2 Extensions, Lecture Notes in Mathematics **504**, Heidelberg: Springer (1976).

[M] F. Momose, *On the ℓ-adic representations attached to modular forms*, J. Fac. Sci. Univ. Tokyo **28** (1981), 89–109.

[MMS] M. R. Murty, V. K. Murty and N. Saradha, *Modular forms and the Chebotarev density theorem*, Amer. J. Math. **110** (1988), 253–281.

[MMSh] M. R. Murty, V. K. Murty and T. N. Shorey, *Odd values of the Ramanujan τ-function*, Bull. Soc. Math. France **115** (1987), 391–395.

[Mu] M. R. Murty, *Artin's conjecture and elliptic analogues*, in Sieve Methods, Exponential Sums and their Applications in Number Theory, ed. G. R H. Greaves, G. Harmon, M. N. Huxley, Cambridge: Cambridge Univ. Press (1996), 325–344.

[Mu83] V. K. Murty, *Algebraic cycles on abelian varieties*, Duke Math. J. **50** (1983), 487–504.

[Mu97] V. K. Murty, *Modular forms and the Chebotarev density theorem, II*, in Analytic Number Theory, ed. Y. Motohashi, London Math. Soc. Lecture Notes **247**, Cambridge: Cambridge Univ. Press (1997), 287–308.

[MuP] V. K. Murty and D. Prasad, *Tate cycles on a product of two Hilbert modular surfaces*, preprint (1996).

[MuR] V. K. Murty and D. Ramakrishnan, *Period relations and the Tate conjecture for Hilbert modular surfaces*, Invent. Math. **89** (1987), 319–345.

[O] A. Ogus, *Hodge cycles and crystalline cohomology*, in Hodge Cycles, Motives and Shimura Varieties, Lecture Notes in Mathematics **900**, ed. P. Deligne, J. S. Milne, A. Ogus, K.-y. Shih, New York: Springer-Verlag (1982), 357–414.

[Oo] F. Oort, *Some questions in algebraic geometry*, preprint (1995).

[R] C. S. Rajan, *Distribution of values of Hecke characters of infinite order*, Acta Arith. **85** (1998), 279–291.

[Ri80] K. Ribet, *Twists of modular forms and endomorphisms of Abelian varieties*, Math. Ann. **253** (1980), 43–62.

[Ri85] K. Ribet, *On ℓ-adic representations attached to modular forms II*, Glasgow Math. J. **27** (1985), 185–194.

[S68] J.-P. Serre, Abelian ℓ-adic Representations and Elliptic Curves, New York: Benjamin (1968). Reprinted by A. K. Peters: Wellesley (1998).

[S81] J.-P. Serre, *Quelques applications du théorème de densité de Chebotarev*, Publ. Math. IHES **54** (1981), 123–201.

[Sh] G. Shimura, *On the factors of the Jacobian variety of a modular function field*, J. Math. Soc. Japan **25** (1973), 523–544.

[W] D. Wan, *On the Lang-Trotter conjecture*, J. Num. Th. **35** (1990), 247–268.

Department of Mathematics, University of Toronto 100 St. George St., Toronto, ON M5S 1A1, Canada

E-mail address: murty@math.toronto.edu

Proceedings of Symposia in Pure Mathematics
Volume **66.1**, 1999

Mod 6 Representations of Elliptic Curves

Karl Rubin and Alice Silverberg

Dedicated to Goro Shimura

ABSTRACT. We study the elliptic curves over $\mathbb{Q}$ whose mod 6 representations are symplectically isomorphic to that of a given one E. We show that if the j-invariant of E is not 0, 1728, $4\cdot 1728$, or $-8\cdot 1728$, then there are infinitely many such curves. When $j(E)$ is $4 \cdot 1728$ or $-8 \cdot 1728$, then there are only finitely many. When $j(E)$ is 0 or 1728, then for infinitely many E's the number is finite, and for infinitely many E's the number is infinite.

Introduction

Suppose E is an elliptic curve, N is a positive integer, and $E[N]$ is the kernel of multiplication by N on E. Let $X_{E,N}$ denote the moduli space parametrizing pairs (E', ψ) where E' is an elliptic curve and ψ is an isomorphism between $E[N]$ and $E'[N]$ which is compatible with the Weil pairings. If $N < 6$ then $X_{E,N}$ has genus 0 (and therefore has infinitely many rational points), and if $N > 6$ then $X_{E,N}$ has genus greater than one (and therefore has only finitely many rational points). See (1.6.4) of [**S**] for a formula for the genus.

If $N = 6$, then $X_{E,N}$ has genus one. In "A question" on p. 133 of [**M**], Mazur states that it might be interesting to consider in some detail the problem of determining all elliptic curves over $\mathbb{Q}$ whose mod 6 representation is symplectically isomorphic to that of a given one. This case is more difficult than the cases $N = 3$, 4, or 5 (see [**RS95**] and [**Si**]), since the moduli spaces are no longer of genus 0.

In Theorem 3.1 we show that if E is an elliptic curve over $\mathbb{Q}$ with discriminant D, then $X_{E,6}$ is the elliptic curve $y^2 = x^3 + D$. In the case where $E[6] \cong \mathbb{Z}/6\mathbb{Z} \times \boldsymbol{\mu}_6$, we give a model for the resulting fiber system of elliptic curves (see Theorem 2.1).

In §4 we consider the elliptic curves over $\mathbb{Q}$ whose mod 6 representation is symplectically isomorphic to that of a given one E. We show that if $j(E)$ is not 0, 1728, $4 \cdot 1728$, or $-8 \cdot 1728$, then there are infinitely many such curves. When $j(E)$ is $4 \cdot 1728$ or $-8 \cdot 1728$, then the number is 1 or 2, respectively. When $j(E)$ is 0 or

1991 *Mathematics Subject Classification.* Primary 11G05; Secondary 11G18, 11F33.
Key words and phrases. Elliptic curves, Galois representations, modular curves.
Partially supported by NSF Grant #DMS-9306287 (Rubin) and NSA Grant #MDA904-96-1-0054 (Silverberg).

1728, then for infinitely many E's the number is finite, and for infinitely many E's the number is infinite.

The authors thank the NSF and NSA. The authors also thank the IHES, where this work was carried out.

1. Notation

We review some of the notation from [**RS95**] and [**Si**]. See [**Si**] and [**KO**] for further details and additional background.

Let $\mathbb{Z}$, $\mathbb{Q}$, and $\mathbb{C}$ denote, respectively, the integers, rational numbers, and complex numbers. If $F \subseteq \bar{\mathbb{Q}}$ is a number field, let $G_F = \mathrm{Gal}(\bar{\mathbb{Q}}/F)$. If E is an elliptic curve, let $j(E)$ denote the j-invariant of E.

Denote by Y_N the (non-compact) modular curve over $\mathbb{Q}$ which parametrizes triples (E, P_N, C_N) where E is an elliptic curve, P_N is a point of exact order N on E, C_N is a cyclic subgroup of order N on E, and C_N and P_N generate $E[N]$. Let X_N denote the compactification of Y_N.

Let W_N denote the universal elliptic curve with level structure as above. Then W_N is a quasi-projective surface defined over $\mathbb{Q}$, with a projection morphism

$$\pi_N : W_N \to Y_N$$

and a zero-section $Y_N \to W_N$, both defined over $\mathbb{Q}$, such that π_N has N^2 sections defined over $\bar{\mathbb{Q}}$ of order dividing N, and such that the fibers of π_N correspond to the triples classified by Y_N.

DEFINITION 1.1. If E and E' are elliptic curves over $\mathbb{Q}$, and $\psi : E[N] \to E'[N]$ is a $G_{\mathbb{Q}}$-equivariant isomorphism which is equivariant with respect to the pairings e_N on E and E', we call ψ a *symplectic isomorphism* and say that $E[N]$ and $E'[N]$ are *symplectically isomorphic.*

If E is an elliptic curve over $\mathbb{Q}$, it was shown in §2 of [**Si**] how to obtain a twist

$$W_{E,N} \xrightarrow{\pi_{E,N}} Y_{E,N}$$

of $W_N \xrightarrow{\pi_N} Y_N$, all defined over $\mathbb{Q}$, with the following properties:

- the points of $Y_{E,N}$ correspond to isomorphism classes of pairs (E', ψ) where E' is an elliptic curve and $\psi : E[N] \to E'[N]$ is an isomorphism which respects the Weil pairings,
- the fiber over a point of $Y_{E,N}$ corresponding to (E', ψ) is E'.

If $t \in Y_{E,N}(\mathbb{C})$ and $\mathcal{E}_t$ is the fiber over t in $W_{E,N}$, then $\mathcal{E}_t[N]$ and $E[N]$ are isomorphic as $\mathrm{Gal}(\overline{\mathbb{Q}(t)}/\mathbb{Q}(t))$-modules. In particular, we can view points of $Y_{E,N}(\mathbb{Q})$ as corresponding to $\mathbb{Q}$-isomorphism classes of pairs (E', ψ) where E' is an elliptic curve over $\mathbb{Q}$ and $\psi : E[N] \to E'[N]$ is a symplectic isomorphism.

From now on we will take $N = 6$, and will write Y_E and W_E for $Y_{E,6}$ and $W_{E,6}$, respectively. Let X_E denote the compactification of Y_E.

2. Modular curve and elliptic modular surface of level 6

We give a model for the elliptic modular surface W_6. We can view W_6 as a surface over $\mathbb{Q}$ or as an elliptic curve $\mathcal{E}_t$ over the function field of X_6.

THEOREM 2.1. *An affine model for* X_6 *is*

$$s^2 = t^3 + 1.$$

We have $Y_6 = X_6 - \mathcal{S}$, *where* $\mathcal{S}$ *consists of the point at infinity and the points on* $s^2 = t^3 + 1$ *satisfying* $st(t^3 - 8) = 0$. *A model for the fiber in* W_6 *over* $(t, s) \in Y_6$ *can be given by*

$$\mathcal{E}_t : y^2 = x^3 + a(t)x + b(t)$$

where

$$a(t) = \frac{-(256 + 256t^3 + 960t^6 + 232t^9 + t^{12})}{3},$$

$$b(t) = \frac{2(4096 + 6144t^3 - 30720t^6 - 24640t^9 - 12072t^{12} - 516t^{15} + t^{18})}{27}.$$

The discriminant and j*-invariant of* $\mathcal{E}_t$ *are*

$$\Delta(\mathcal{E}_t) = 2^{12}t^6(-8 + t^3)^6(1 + t^3)^3 = (4st(-8 + t^3))^6,$$

$$j(\mathcal{E}_t) = \frac{(4 + t^3)^3(64 + 48t^3 + 228t^6 + t^9)^3}{t^6(-8 + t^3)^6(1 + t^3)^3}.$$

The points $P_1(t) = (2(8-20t^3-t^6)/3, 0)$ *and* $P_2(t, s) = ((-8-24s^3+20t^3+t^6)/3, 0)$ *are points of order* 2 *on* $\mathcal{E}_t$. *The points*

$$Q_1(t) = \left(\frac{(4 + 6t^2 + t^3)^2}{3}, 4t^2(1 - t + t^2)(4 + 2t + t^2)^2\right)$$

and

$$Q_2(t) = \left(-(4 + t^3)^2, \frac{-4(1 + t^3)(-8 + t^3)^2}{3\sqrt{-3}}\right)$$

are independent points of order 3 *on* $\mathcal{E}_t$.

PROOF. Let X denote the elliptic curve defined by $s^2 = t^3 + 1$, let $\mathcal{S}$ denote the set consisting of the origin of X and the 11 points where $st(t^3 - 8) = 0$, and let Y denote $X - \mathcal{S}$. Note that $X(\mathbb{Q})$ is the cyclic group generated by the point $(2, 3)$, and $X(\mathbb{Q}) \subset \mathcal{S}$. For $(t, s) \in Y$, it is easy to check the formulas for $\Delta(\mathcal{E}_t)$ and $j(\mathcal{E}_t)$, and to check that the points $P_1(t)$, $P_2(t, s)$, $Q_1(t)$, and $Q_2(t)$ are points on $\mathcal{E}_t$ of the given orders. The function field $\mathbb{Q}(X) = \mathbb{Q}(t, s)$ has degree 2 over $\mathbb{Q}(t)$. For $(t, s) \in Y$, the triple

$$(\mathcal{E}_t, P_1(t) + Q_1(t), \langle P_2(t, s) + Q_2(t)\rangle)$$

defines a point in Y_6. We therefore obtain a morphism $f : X \to X_6$. Let $g : X_6 \to \mathbf{P}^1$ be the (degree 72) morphism induced by $(E, P, C) \mapsto j(E)$. By the formula for $j(\mathcal{E}_t)$, we see that the function field $\mathbb{Q}(t)$ has degree 36 over $\mathbb{Q}(j(\mathcal{E}_t))$. Therefore, $\mathbb{Q}(X)$ has degree 72 over $\mathbb{Q}(\mathbf{P}^1)$, and the morphism $g \circ f$ has degree 72. Thus, f is an isomorphism.

That $\mathcal{E}_t$ is a model for W_6 now follows from the universal property of W_6. In the appendix we explain briefly how we obtained this model.

The 12 cusps of X_6 correspond to the 12 points with singular fibers, which are the points of $\mathcal{S}$. □

The model for X_6 is well-known. Fricke and Klein studied models for elliptic modular surfaces (see especially [**Fr**]).

3. Twists of X_6

THEOREM 3.1. *Suppose $E : y^2 = x^3 + ax + b$ is an elliptic curve, with $a, b \in \mathbb{Q}$. Then a model for the modular curve X_E is given by*

$$s^2 = t^3 + D,$$

where $D = -16(4a^3 + 27b^2)$ is the discriminant of E.

PROOF. Since X_E has a rational point (corresponding to E), and is isomorphic over $\mathbb{C}$ to $y^2 = x^3 + 1$, we know X_E has a model of the form $y^2 = x^3 + \delta$, with $\delta \in \mathbb{Q}$ unique up to sixth powers, where the origin $\mathcal{O}$ corresponds to the pair $(E, \text{identity})$. For $(t, s) \in X_6$, let $\mathcal{E}_t$ denote the fiber of W_6 over (t, s), as in Theorem 2.1. There exists a $\mathbb{C}$-isomorphism $X_E \to X_6$, which sends $\mathcal{O}$ to a point (t_0, s_0) such that $\mathcal{E}_{t_0}$ is $\mathbb{C}$-isomorphic to E, i.e., $j(\mathcal{E}_{t_0}) = j(E)$. Such a map must be of the form

$$(t, s) \mapsto (t', s') = (t\alpha^{-2}, s\alpha^{-3}) + (t_0, s_0),$$

where the addition on the right side is addition on the elliptic curve $X_6 : y^2 = x^3 + 1$ and where $\alpha^6 = \delta$. Let $v = t/s$ and $w = 1/s$. Then v has a simple zero at $\mathcal{O}$ and w has a triple zero at $\mathcal{O}$. Since X_E is defined over $\mathbb{Q}$, we have $j(\mathcal{E}_{t'}) \in \mathbb{Q}(v, w)$.

First, suppose both a and b are non-zero. Using that $w \in v^3 + v^9\mathbb{Q}[[v]]$, one can expand $j(\mathcal{E}_{t'})$ as a power series in v, and see that when $ab \neq 0$, the coefficient of v is

$$\frac{27j(E)\alpha b(t_0)}{s_0 t_0(-8 + t_0^3)a(t_0)},$$

where $a(t)$ and $b(t)$ are as in Theorem 2.1. (The authors used Mathematica and Pari to do this.) From the formula for $\Delta(\mathcal{E}_{t_0})$ in Theorem 2.1, and the relation

$$\frac{D}{\Delta(\mathcal{E}_{t_0})} = \left(\frac{a(t_0)b}{b(t_0)a}\right)^6,$$

we see that α^6/D is the sixth power of a rational number. We can therefore take $\delta = D$.

Now remove the restriction that a and b are non-zero. If we write

$$\mathcal{F}_{t,s} : y^2 = x^3 + a_1(t, s)x + b_1(t, s)$$

for the fiber in W_E above the point $(t, s) \in X_E$, then $\mathcal{F}_{t,s}$ is $\mathbb{C}$-isomorphic to $\mathcal{E}_{t'}$, so there exists a function $\mu(t, s) \in \mathbb{C}(t, s)$ such that

$$a_1(t, s)\mu(t, s)^2 = a(t') \text{ and } b_1(t, s)\mu(t, s)^3 = b(t').$$

Changing variables as above, we can write $a_1(t, s)$ and $b_1(t, s)$ as power series $a_1(v), b_1(v) \in \mathbb{Q}[[v]]$, and we can write $\mu(t, s)$, $a(t')$, and $b(t')$ as power series $\mu(v), A(v), B(v) \in \mathbb{C}[[v]]$.

Now suppose $a = 0$, so $j(E) = 0$. Letting $t_0 = (-4)^{1/3}$, then $j(\mathcal{E}_{t_0}) = 0$ and we can let $s_0 = \sqrt{-3}$. Using these values for t_0 and s_0, a computation shows that

$$a_1'(0) = \mu(0)^{-2}A'(0) = 36b^{2/3}\alpha/(\sqrt{-3}(-4)^{1/3}).$$

Since $a_1'(0) \in \mathbb{Q}$, we see that, up to the sixth power of a rational number, we have

$$\delta = \alpha^6 = -2^4 3^3 b^2 = D.$$

Now suppose $b = 0$, so $j(E) = 1728$. Letting $t_0 = -1 + \sqrt{3}$, then $j(\mathcal{E}_{t_0}) = 1728$ and we can let $s_0 = \sqrt{-9 + 6\sqrt{3}}$. Using these values for t_0 and s_0, a computation shows that

$$b_1'(0) = \mu(0)^{-3}B'(0) = 4a\sqrt{-a}\alpha.$$

Since $b_1'(0) \in \mathbb{Q}$, up to a sixth power of a rational number we have

$$\delta = \alpha^6 = -2^6 a^3 = D.$$

□

4. Elliptic curves with symplectically isomorphic mod 6 representations

If E is an elliptic curve over $\mathbb{Q}$, let

$$S(E) = \{E' : (E', \psi) \in Y_E(\mathbb{Q}) \text{ for some } \psi\},$$

i.e., $S(E)$ is the set of all elliptic curves E' over $\mathbb{Q}$ (up to isomorphism over $\mathbb{Q}$) such that $E[6]$ and $E'[6]$ are symplectically isomorphic. Note that $S(E)$ is infinite if and only if $X_E(\mathbb{Q})$ is infinite.

THEOREM 4.1. *Suppose E is an elliptic curve over $\mathbb{Q}$.*

(a) *If the j-invariant of E is not* 0, 1728, $4 \cdot 1728$, *or* $-8 \cdot 1728$, *then $S(E)$ is infinite.*
(b) *If $j(E) = 4 \cdot 1728$, then $S(E) = \{E\}$.*
(c) *If $j(E) = -8 \cdot 1728$, then $S(E) = \{E, E^{(-1)}\}$, where $E^{(-1)}$ is the twist of E by the quadratic character associated to the extension $\mathbb{Q}(i)/\mathbb{Q}$.*

PROOF. Let $y^2 = x^3 + ax + b$ be a model for E, with $a, b \in \mathbb{Q}$, and let $D = -16(4a^3 + 27b^2)$. Let X' denote the elliptic curve $-3s^2 = t^3 + D$. By Theorem 3.1, a model for X_E is $s^2 = t^3 + D$. The rational map

$$(t, s) \mapsto \left(\frac{t^3 + 4s^2}{t^2}, \frac{-s(3t^3 + 8s^2)}{t^3}\right)$$

defines an isogeny f from X' onto X_E. Clearly, $(4a, 12b)$ is a point on $X'(\mathbb{Q})$. Its image under f is

$$P = \left(\frac{4(a^3 + 9b^2)}{a^2}, \frac{-36b(a^3 + 6b^2)}{a^3}\right) \in X_E(\mathbb{Q}).$$

Since X_E is of the form $s^2 = t^3 + D$, the torsion subgroup of $X_E(\mathbb{Q})$ has order dividing 6 (see for example Theorem V of [**C**]). If $j(E) \neq 0$, then $a \neq 0$, so $P \neq \mathcal{O}$.

If P has order 6 then $2P$ has order 3, and it follows that the first coordinate of $2P$ vanishes. However, the first coordinate of $2P$ has numerator

$$4(a^{12} + 18b^2a^9 + 108b^4a^6 + 324b^6a^3 + 729b^8),$$

which vanishes with rational a and b only when $a^3 = -9b^2$, i.e., when P has order 3.

If D is $-2^4 3^3$ (up to a sixth power), then one can show that $a = 0$ and $j(E) = 0$. Otherwise, the point P has order 3 exactly when its first coordinate vanishes, i.e., exactly when $j(E) = 4 \cdot 1728$. In this case, E is of the form $y^2 = x^3 - 9c^2x + 9c^3$ for some $c \in \mathbb{Q}^\times$. It follows from Theorem 3.1 that a model for X_E is $y^2 = x^3 + 16$. There are exactly 3 rational points on this curve (see [**C**]). We have

$$\mathrm{Gal}(\mathbb{Q}(E[2])/\mathbb{Q}) \cong \mathbb{Z}/3\mathbb{Z},$$

since the discriminant of $x^3-9c^2x+9c^3$ is $(3c)^6$, a square. It follows that there are 3 $G_{\mathbb{Q}}$-equivariant automorphisms of $E[2]$, and therefore 3 symplectic automorphisms φ_1, φ_2, and φ_3 of $E[6]$. The (E,φ_i)'s correspond to the 3 points of $X_E(\mathbb{Q})$. We therefore have (b).

The point P has order 2 exactly when its second coordinate vanishes, i.e., exactly when $j(E) = 1728$ or $-8\cdot 1728$. Now suppose $j(E) = -8\cdot 1728$. Then E is of the form $y^2 = x^3 - 6c^2x + 6c^3$ for some $c \in \mathbb{Q}^\times$, and it follows from Theorem 3.1 that a model for X_E is $y^2 = x^3 - 27$. There are exactly 2 rational points on this curve (see [**C**]). Theorem 4.1 of [**RS95**] gives an equation for the family $\mathcal{E}_t$ of elliptic curves over $\mathbb{Q}$ whose mod 3 representation is symplectically isomorphic to that of E. When $t = -1/3$, the elliptic curve $\mathcal{E}_t$ is isomorphic over $\mathbb{Q}$ to

$$E^{(-1)} : y^2 = x^3 - 6c^2x - 6c^3,$$

the twist of E by -1 (i.e., the twist of E by the quadratic character associated to $\mathbb{Q}(i)/\mathbb{Q}$). Thus, $E[3]$ is symplectically isomorphic to $E^{(-1)}[3]$. For every elliptic curve over $\mathbb{Q}$, its mod 2 representation is symplectically isomorphic to that of any quadratic twist. Therefore, $E^{(-1)} \in S(E)$. Since E and $E^{(-1)}$ are not isomorphic over $\mathbb{Q}$, they correspond to the 2 points of $X_E(\mathbb{Q})$. We therefore have (c).

We can now conclude that if $j(E)$ is not 0, 1728, $4\cdot 1728$, or $-8\cdot 1728$, then P is a point of $X_E(\mathbb{Q})$ of infinite order, giving (a). □

Next, we consider the cases where $j(E)$ is 1728 or 0.

THEOREM 4.2. *For fourth-power-free integers a, let E_a denote the elliptic curve $y^2 = x^3 + ax$. Then for infinitely many a, $S(E_a)$ is infinite, and for infinitely many a, $S(E_a)$ is finite. If $S(E_a)$ is finite, then $S(E_a) = \{E_a\}$.*

PROOF. By Theorem 3.1, X_{E_a} is isomorphic to $y^2 = x^3 - a^3$. As shown in [**F**], if a is a prime congruent to 3 (mod 4) then $X_{E_a}(\mathbb{Q})$ has rank one, and if a is a prime congruent to 5 (mod 12) then $X_{E_a}(\mathbb{Q})$ has rank zero. For further results on $X_{E_a}(\mathbb{Q})$, see [**F**].

Suppose that $X_{E_a}(\mathbb{Q})$ is finite. If $-a$ is not a square (in $\mathbb{Q}$), then

$$\mathrm{Gal}(\mathbb{Q}(E_a[2])/\mathbb{Q}) \cong \mathbb{Z}/2\mathbb{Z},$$

and there are 2 $G_{\mathbb{Q}}$-equivariant automorphisms of $E_a[2]$, which give rise to the 2 points of $X_{E_a}(\mathbb{Q})$. If $-a$ is a square, then $E_a[2] \subset E_a(\mathbb{Q})$, and there are 6 $G_{\mathbb{Q}}$-equivariant automorphisms of $E_a[2]$, which give rise to the 6 points of $X_{E_a}(\mathbb{Q})$. □

THEOREM 4.3. *For sixth-power-free integers b, let E_b denote the elliptic curve $y^2 = x^3 + b$. Then for infinitely many b, $S(E_b)$ is infinite, and for infinitely many b, $S(E_b)$ is finite. If $S(E_b)$ is finite, then $S(E_b) = \{E_b, E_b^{(-3)}\}$, where $E_b^{(-3)}$ is the twist of E_b by the quadratic character associated to the extension $\mathbb{Q}(\sqrt{-3})/\mathbb{Q}$, if b is twice a cube, and $S(E_b) = \{E_b\}$ otherwise.*

PROOF. By Theorem 3.1, X_{E_b} is isomorphic to $y^2 = x^3 - 2^4 3^3 b^2$. The X_{E_b}'s are cubic twists of $X_0(27)$ (the Fermat cubic), and X_{E_b} is birationally isomorphic to $x^3 + y^3 = b$. Lucas and Sylvester (see Chap. XXI of [**D**]) showed that there are infinitely many cube-free integers b such that $X_{E_b}(\mathbb{Q})$ has rank zero.

That there are infinitely many b such that $X_{E_b}(\mathbb{Q})$ has rank at least 3 follows from [**ST**], where information is given on the density of such b. Since it is easy to give a short elementary proof that $X_{E_b}(\mathbb{Q})$ is infinite for infinitely many cube-free integers b, we do so here. Define positive integers x_n and b_n recursively by

$$x_1 = 2, \quad b_n = x_n^3 + 1, \quad x_{n+1} = \prod_{i=1}^{n} b_i.$$

Then the cube-free parts of the b_n's are pairwise relatively prime. For every n, the point $(x_n, 1)$ is a point of infinite order on the elliptic curve $x^3 + y^3 = b_n$. In this way, one obtains infinitely many cube-free integers b such that $X_{E_b}(\mathbb{Q})$ is infinite. Note that a similar technique works to show the analogous result in Theorem 4.2.

Suppose that $X_{E_b}(\mathbb{Q})$ is finite. If b is neither a cube nor twice a cube, then $\#X_{E_b}(\mathbb{Q}) = 1$, so $S(E_b) = \{E_b\}$. If $b = 2c^3$ with $c \in \mathbb{Q}^\times$, then $E_b^{(-3)} : y^2 = x^3 - 27b$ occurs in the family $\mathcal{E}_t$ of elliptic curves whose mod 3 representations are symplectically isomorphic to that of E_b, given in Theorem 4.6 of [**RS95**], with $t = 12/c$. It follows that E_b and $E_b^{(-3)}$ give rise to the 2 points of $X_{E_b}(\mathbb{Q})$. Suppose b is a cube. Using the isomorphism $X_{E_b} \to X_6$ given in the proof of Theorem 3.1, one checks that 1 of the 3 rational points of X_{E_b} is a cusp. Since $\mathrm{Gal}(\mathbb{Q}(E_b[2])/\mathbb{Q}) \cong \mathbb{Z}/2\mathbb{Z}$, the other 2 rational points come from the 2 $G_{\mathbb{Q}}$-equivariant automorphisms of $E_b[2]$. □

Appendix

We now explain how we obtained the model for W_6 given in Theorem 2.1. A Weierstrass model for W_3 is given by

$$A(u) : y^2 = x^3 + a_0(u)x + b_0(u)$$

where

$$a_0(u) = -27u(8 + u^3), \qquad b_0(u) = -54(8 + 20u^3 - u^6)$$

(see (1) of [**RS95**]). The points

$$q_1(u) = (3(2+u)^2, 36(1+u+u^2)), \quad q_2(u) = (-9u^2, 12\sqrt{-3}(1-u^3))$$

have order 3 on $A(u)$. Consider the modular curve that parametrizes quadruples (E, P_2, P_3, C_3) where P_n is a point of order n and C_3 is a subgroup of order 3 not containing P_3. We will denote by $X_0(18)$ the compactification of this modular curve (it is, in fact, isomorphic to the curve usually denoted $X_0(18)$). The map

$$(E, P_2, P_3, C_3) \mapsto (E, P_3, C_3)$$

induces a degree 3 covering $X_0(18) \to X_3$. Therefore, $\mathbb{Q}(X_0(18)) = \mathbb{Q}(u, x)$ where $x^3 + a_0(u)x + b_0(u) = 0$. Let

$$u_0(t) = \frac{4+t^3}{3t^2} \qquad \text{and} \qquad x_0(t) = \frac{-(16+t^3)}{t} + 3u_0(t)^2.$$

Then

$$x_0(t)^3 + a_0(u_0(t))x_0(t) + b_0(u_0(t)) = 0,$$

and t is a parameter on $X_0(18)$, i.e., $\mathbb{Q}(X_0(18)) = \mathbb{Q}(t)$. (The functions $u_0(t)$ and $x_0(t)$ were solved for using Mathematica and Pari.) In terms of t, the discriminant Δ of $A(u_0(t))$ is

$$\Delta = \frac{2^{12}(-8+t^3)^6(1+t^3)^3}{t^{18}}.$$

Since all the sections of X_6 of order 2 are defined over $\mathbb{Q}$, Δ is a square in $\mathbb{Q}(X_6)$. Therefore, t^3+1 is a square in $\mathbb{Q}(X_6)$. Since the map

$$(E, P_6, C_6) \mapsto (E, 3P_6, 2P_6, 2C_6)$$

induces a degree 2 covering $X_6 \to X_0(18)$, it follows that $\mathbb{Q}(X_6) = \mathbb{Q}(t,s)$ where $s^2 = t^3+1$. With $a(t)$ and $b(t)$ as in the statement of Theorem 2.1, we have $a(t) = a_0(u_0(t))t^8$ and $b(t) = b_0(u_0(t))t^{12}$. The point $(x_0(t)t^4, 0) = P_1(t)$ is a point on $\mathcal{E}_t$ of order 2. The other points of order 2 can be solved for using the square root of the discriminant of $\mathcal{E}_t$. The points $q_1(u)$ and $q_2(u)$ on $A(u)$ induce (after multiplying the first coordinate by t^4 and the second by t^6) the points $Q_1(t)$ and $Q_2(t)$.

References

[C] J. W. S. Cassels, *The rational solutions of the diophantine equation* $Y^2 = X^3 - D$, Acta Math. **82** (1950), 244–273.

[D] L. Dickson, History of the Theory of Numbers II: Diophantine Analysis, New York: Chelsea Publishing Co. (1966).

[F] G. Frey, *Der Rang der Lösungen von* $Y^2 = X^3 \pm p^3$ *über* $\mathbb{Q}$, Manuscripta Math. **48** (1984), 71–101.

[Fr] R. Fricke, Die Elliptischen Funktionen und ihre Anwendungen II, Leipzig-Berlin: B. G. Teubner (1922); New York: Johnson Reprint Corp. (1972).

[KO] A. Kraus, J. Oesterlé, *Sur une question de B. Mazur*, Math. Ann. **293** (1992), 259–275.

[M] B. Mazur, *Rational isogenies of prime degree*, Invent. Math. **44** (1978), 129–162.

[RS94] K. Rubin, A. Silverberg, *A report on Wiles' Cambridge lectures*, Bull. Amer. Math. Soc. (N. S.) **31** (1994), 15–38.

[RS95] K. Rubin, A. Silverberg, *Families of elliptic curves with constant mod p representations*, in Conference on Elliptic Curves and Modular Forms, Hong Kong, December 18–21, 1993, Cambridge, MA: Intl. Press (1995), 148–161.

[S] G. Shimura, Introduction to the Arithmetic Theory of Automorphic Functions, Princeton: Princeton Univ. Press (1971).

[Si] A. Silverberg, *Explicit families of elliptic curves with prescribed mod N representations*, in Modular Forms and Fermat's Last Theorem (Boston, MA, 1995), eds. Gary Cornell, Joseph H. Silverman, Glenn Stevens, New York: Springer-Verlag (1997), 447–461.

[ST] C. L. Stewart, J. Top, *On ranks of elliptic curves and power-free values of binary forms*, J. Amer. Math. Soc. **8** (1995), 943–973.

Note added in proof: Joseph Oesterlé has informed us that J-I. Papadopoulos proved Theorem 3.1 earlier, by different methods, in his thesis, *Deux questions relatives à l'arithmétique des courbes elliptiques*, thèse de doctorat de l'Université Paris 6, 16 Juillet 1992.

DEPARTMENT OF MATHEMATICS, OHIO STATE UNIVERSITY, 231 W. 18 AVENUE, COLUMBUS, OHIO 43210-1174

Current address: Department of Mathematics, Stanford University, Stanford, CA 94305-1174

E-mail address: rubin@math.stanford.edu

DEPARTMENT OF MATHEMATICS, OHIO STATE UNIVERSITY, 231 W. 18 AVENUE, COLUMBUS, OHIO 43210-1174

E-mail address: silver@math.ohio-state.edu

Proceedings of Symposia in Pure Mathematics
Volume **66.1**, 1999

On Absolute CM-Periods

Hiroyuki Yoshida

Dedicated to Professor Goro Shimura

ABSTRACT. We formulate a conjecture which expresses the derivatives of Artin L-functions at $s = 0$ using Shimura's CM-period symbol p_K. By a theorem of Shimura, this conjecture can be tested comparing with a product of critical values of L-functions attached to Grössencharacters of A_0-type. We present three convincing numerical examples. Taking the action of $\mathrm{Gal}(\overline{\mathbf{Q}}/\mathbf{Q})$ into account, we transform the conjecture into a more precise form, which raises another problem of determining the reciprocity map. We prove a fairly complete result on the reciprocity map for imaginary quadratic fields. Finally we discuss the conjectural form of the reciprocity map in the general case by the theory of motives.

Introduction

In this paper, we shall study relations among the derivatives of Artin's L-functions at $s = 0$, critical values of Hecke's L-functions with Grössencharacters of A_0-type and periods of abelian varieties with complex multiplication.

Let K be a CM-field, J_K be the set of all isomorphisms of K into $\mathbf{C}$ and I_K be the free abelian group generated by J_K. For a CM-type Φ of K, we can find an abelian variety A of type (K, Φ) defined over $\overline{\mathbf{Q}}$. For every $\sigma \in \Phi$, there is a non-zero holomorphic differential 1-form ω_σ on A such that ω_σ is multiplied by a^σ by the action of $a \in K \cap \mathrm{End}(A)$ and that ω_σ is rational over $\overline{\mathbf{Q}}$. Then there exists a constant $p_K(\sigma, \Phi) \in \mathbf{C}^\times$ such that

$$(1) \qquad \int_c \omega_\sigma \sim \pi\, p_K(\sigma, \Phi) \quad \text{for every} \quad c \in H_1(A, \mathbf{Z}).$$

Here, for a, $b \in \mathbf{C}$, we write $a \sim b$ if $b \neq 0$ and $a/b \in \overline{\mathbf{Q}}$. We know that $p_K(\sigma, \Phi)$ mod $\overline{\mathbf{Q}}^\times$ does not depend on the choice of A and ω_σ. Shimura showed [**Sh79, Sh80**] that p_K can be extended to the bilinear form from $I_K \times I_K$ to $\mathbf{C}^\times/\overline{\mathbf{Q}}^\times$, which enjoys several functorial properties (see §1). Thus we have the "CM-period" $p_K(\sigma, \tau) \in \mathbf{C}^\times$, uniquely determined mod $\overline{\mathbf{Q}}^\times$, for every σ, $\tau \in I_K$.

1991 *Mathematics Subject Classification.* Primary 11R42, 11G15; Secondary 11F67, 11Y99.
Key words and phrases. CM-periods, Artin L-functions, Grössencharacters of A_0-type, reciprocity law, motives.

Now let F be the maximal real subfield of K and χ be the Hecke character which corresponds to the quadratic extension K/F. Let $L_F(s,\chi)$ denote the Hecke L-function attached to χ and put

$$P = \exp\left(\frac{L_F'(0,\chi)}{L_F(0,\chi)}\right).$$

The first theme of this paper is

CONJECTURE A. $P \sim \pi^{[F:\mathbf{Q}]}\prod_{\sigma\in J_K} p_K(\sigma,\sigma)$.

Let L be a CM-field which is normal over $\mathbf{Q}$ and put $G = \mathrm{Gal}(L/\mathbf{Q})$. Let $\rho \in G$ be the complex conjugation. As is well known, ρ belongs to the center of G.

CONJECTURE B. *Let ψ be a representation of G and let χ_ψ be the character of ψ. We assume that $\psi(\rho) = -\mathrm{id}.$ and that χ_ψ is $\mathbf{Q}$-valued. Let $L(s,\psi)$ be the Artin L-function attached to ψ. Then*

$$\exp\left(\frac{L'(0,\psi)}{L(0,\psi)}\right) \sim \pi^{\dim\psi}\prod_{\sigma\in G} p_L(\mathrm{id},\sigma)^{\chi_\psi(\sigma^{-1})}.$$

In §2, we shall discuss formal relations of conjectures. In particular, we shall show that Conjecture B for L is equivalent to Conjecture A for all CM-subfields K of L (Proposition 2.5). We shall also formulate a stronger version of Conjecture B (Conjecture 2.2) which is essentially equivalent to assert that Conjecture B holds without the rationality of χ_ψ. (Since $p_K(\mathrm{id},\sigma)^{\chi_\psi(\sigma)}$ may not be well defined, we use the "dual" version.) Conjecture 2.2 is equivalent to a conjecture of Colmez [**C**] and is proved when K is abelian over $\mathbf{Q}$ by Anderson [**A**]. Our formulation using Shimura's period symbol p_K is simpler and more transparent.[1]

In §4, we shall submit Conjecture A to numerical test. Let us explain basic procedures. Let Φ be a CM-type of K and let $\mathfrak{f}$ be an integral ideal of K. Let λ be a Grössencharacter of K such that the conductor of λ divides $\mathfrak{f}$ and that

$$\lambda((\alpha)) = \prod_{\sigma\in\Phi}(\alpha^{\sigma\rho}/|\alpha^\sigma|)^{t_\sigma} \quad \text{if} \quad \alpha \equiv 1 \mod {}^\times\mathfrak{f}. \tag{2}$$

Here t_σ are positive integers such that $t_\sigma \mod 2$ are independent of σ. A theorem of Shimura [**Sh75, Sh80**] states that

$$L(m/2,\lambda) \sim \pi^{e/2}p_K\left(\sum_{\sigma\in\Phi} t_\sigma\cdot\sigma, \Phi\right), \quad e = m[F:\mathbf{Q}] + \sum_{\sigma\in\Phi} t_\sigma, \tag{3}$$

for $m \in \mathbf{Z}$ which satisfies $-t_\sigma < m \le t_\sigma$, $m \equiv t_\sigma \mod 2$ for every $\sigma \in \Phi$. Furthermore Shimura's method of the proof of this theorem, which uses Eisenstein series and differential operators, can be applied efficiently for numerical computation of $L(m/2,\lambda)$ (cf. §3). By virtue of (3), Conjecture A is transformed to the form

$$\frac{\prod_{i=1}^q L(m/2,\lambda_i)}{\pi^A P^e} \in \overline{\mathbf{Q}}, \tag{4}$$

[1]We obtained these conjectures independently of [**C**] and [**A**]. When K is abelian, we can express $p_K(\mathrm{id},\sigma)$ in terms of the Gamma function by solving a system of linear equations which are obtained using (1) and the properties of p_K taking sufficiently many CM-types Φ of K. When K is an imaginary quadratic field, Conjecture A is equivalent to the Chowla-Selberg formula. Shimura predicted [**Sh79**, p. 571] that his theorem would give a generalization of the Chowla-Selberg formula when K is abelian over $\mathbf{Q}$.

where A, $e \in 2^{-1}\mathbf{Z}$ and λ_i $(1 \leq i \leq q)$ are certain Grössencharacters of type (2) of K. We expect that P is a kind of "absolute period" and (4) behaves under $\sigma \in \mathrm{Aut}(\mathbf{C})$ as[2]

$$\text{(X)} \qquad \left(\frac{\prod_{i=1}^{q} L(m/2,\lambda_i)}{\pi^A P^e}\right)^{\sigma} = \frac{\prod_{i=1}^{q} L(m/2,\lambda_i^{\sigma})}{\pi^A P^e}.$$

Assuming (X), we can know all conjugates of (4) and Conjecture A can thus be tested effectively. In §4, we shall treat the case where $[K : \mathbf{Q}] = 4$ and K is not normal over $\mathbf{Q}$, i.e., the case which goes back to Hecke. We shall discuss three numerical examples in detail. These examples fit well into the above picture. However several other examples, including the one given in a footnote of §4, suggest that (X) cannot be true in general.

In §5, we shall make this picture more precise. Suppose that Conjecture A is equivalent to (4) (cf. §5 for the accurate meaning). What we should expect in general is

CONJECTURE C. $\left(\frac{\prod_{i=1}^{q} L(m/2,\lambda_i)}{\pi^A P^e}\right)^{\sigma} = \zeta \cdot \frac{\prod_{i=1}^{q} L(m/2,\lambda_i^{\sigma})}{\pi^A P^e}$ *where ζ is a root of unity.*

Now the determination of ζ arises as our second theme, which may be called a reciprocity law. In §6, we shall study this problem in detail when K is an imaginary quadratic field. In this case, taking $\Phi = \{\mathrm{id}\}$ in (2), we have

$$\lambda((\alpha)) = (\alpha^{\rho}/|\alpha|)^k, \quad \alpha \equiv 1 \mod {}^{\times}\mathfrak{f}.$$

Put $Q(\lambda) = L(m/2,\lambda)/\pi^{m/2}P^{k/2}$. Then Conjecture C states

$$Q(\lambda)^{\sigma} = \zeta(\sigma,\lambda)Q(\lambda^{\sigma}) \quad \text{for} \quad \sigma \in \mathrm{Aut}(\mathbf{C})$$

with a root of unity $\zeta(\sigma,\lambda)$. Our strategy is to relate $Q(\lambda)$ to quotients of nearly holomorphic modular forms evaluated at a CM-point and determine $Q(\lambda)^{\sigma}$ by Shimura's reciprocity law [**Sh75**, Main Theorem III]. Let h be the class number of K and set $k/2h = b/l$ with $0 < b, l \in \mathbf{Z}$. To pursue the above strategy to the end, we are naturally led to use an l-fold covering group of $\mathrm{GL}_+(2,\mathbf{R})$ (cf. §6) to study the behavior of $\eta^{2/l}(z)$, where $\eta(z)$ is the Dedekind eta function. When $l = 2$, i.e., k is divisible by h, we obtain a satisfactory expression of $\zeta(\sigma,\lambda)$ (Theorem 6.9). At present we can obtain only a partial result when $l > 2$.[3] We note that Brattström and Lichtenbaum [**BL**] obtained a result of the similar nature under the assumption that h is odd and divides k.

In §7, we shall show that Conjecture C is consistent with the theory of motives. Combined with results in §4 and §6, we can foresee what ζ should be in general up to sign under a certain divisibility condition by the class number of K; roughly speaking this condition is the divisibility of k by $2h$ when K is quadratic.

There are a few problems suggested by our conjectures. One is the connection with certain arithmetic property at CM-points of non-holomorphic automorphic forms studied by Asai [**As**]. We mention the one which is very intriguing. The most general conjecture in our present knowledge which predicts special values of motivic L-functions is Beilinson's: it gives the transcendental part of the leading

[2]Cf. §3 for precise definition of λ_i^{σ}.

[3]Cf. (6.3.16). On the other hand, if $l = 1$, i.e., k is divisible by $2h$, the problem becomes considerably easier.

coefficient of the Taylor expansion of the L-function at integral points (cf. several articles in [**JKS**] on this topic). The conjectures discussed above give the next coefficient to the leading one of the Taylor expansion, which does not seem to be immediately predictable by Beilinson's conjecture. The problem is to investigate whether such coefficients can be described in the framework of the theory of motives.

We would like to dedicate this paper most cordially to Professor Shimura who is the founder of modern arithmetic theory of automorphic forms and associated zeta functions. Though we believe that his works play a fundamental role and show their penetrating power in this paper, we still feel that we have just scratched the surface of his deep and vast works. We hope that this would be excused and our admiration would be accepted.

NOTATION AND TERMINOLOGY. Throughout the paper, we fix an algebraic closure $\overline{\mathbf{Q}}$ of $\mathbf{Q}$ in $\mathbf{C}$. By an algebraic number field, we understand an algebraic extension of $\mathbf{Q}$ of finite degree contained in $\overline{\mathbf{Q}}$. We denote by ρ the complex conjugation. For an algebraic number field K, J_K denotes the set of all isomorphisms of K into $\mathbf{C}$ and I_K denotes the free abelian group generated by J_K. The ring of integers of K is denoted by $\mathfrak{O}_K$. We denote by $I(K)$ the ideal group of K. For an integral ideal $\mathfrak{f}$ of K, $I_{\mathfrak{f}}(K)$ denotes the ideal group of K modulo $\mathfrak{f}$. We denote by K_{ab} the maximal abelian extension of K in $\overline{\mathbf{Q}}$. For a place v of K, K_v denote the completion of K at v. We denote by $K_A^\times$ the idele group of K. For $c \in K_A^\times$, $[c, K] \in \mathrm{Gal}(K_{ab}/K)$ denotes the image of c under the Artin map. For $a \in K$, $a \gg 0$ means that a is totally positive. We abbreviate $\rho|K$ to ρ if no confusion is likely. For an extension L of K of finite degree, $\mathrm{Res}_{L/K}$ denotes the restriction homomorphism from I_L to I_K; $\mathrm{Inf}_{L/K}$ denotes the homomorphism from I_K to I_L such that, for $\sigma \in J_K$, $\mathrm{Inf}_{L/K}(\sigma)$ is the sum of all elements of J_L whose restrictions to K coincide with σ. The norm (resp. trace) map from L to K is denoted by $N_{L/K}$ (resp. $\mathrm{Tr}_{L/K}$). We abbreviate $N_{K/\mathbf{Q}}$ (resp. $\mathrm{Tr}_{K/\mathbf{Q}}$) to N (resp. Tr). The relative discriminant of L over K is denoted by $D(L/K)$. For an abelian extension L of K of finite degree and for a fractional ideal $\mathfrak{X}$ of K, which is relatively prime to the conductor of L, $(\frac{L/K}{\mathfrak{X}})$ denotes the Artin symbol.

By a CM-field, we understand a totally imaginary quadratic extension of a totally real algebraic number field. For a CM-field K, $\Phi \in I_K$ is called a CM-type if $\Phi + \Phi\rho$ is the sum of all elements in J_K. If $\Phi = \sum_{i=1}^n \sigma_i$, we often identify Φ with the set of isomorphisms $\{\sigma_1, \sigma_2, \cdots, \sigma_n\}$. For a finite group G, a subgroup H of G and a representation ψ of H, the induced representation from ψ is denoted by $\mathrm{Ind}_H^G \psi$ or $\mathrm{Ind}(\psi; H \longrightarrow G)$. For an element g of a group, $\langle g \rangle$ denotes the subgroup generated by g. For a set S, $|S|$ denotes the cardinality of S. For a positive integer N, $\exp(2\pi i/N)$ is denoted by ζ_N. The principal congruence subgroup of level N of $\mathrm{SL}(2, \mathbf{Z})$ is denoted by $\Gamma(N)$. For a, $b \in \mathbf{C}$, we write $a \sim b$ if $b \neq 0$ and $a/b \in \overline{\mathbf{Q}}$. For integers $a \neq 0$, $b \in \mathbf{Z}$, we write $a\,|\,b$ (resp. $a \nmid b$) if a divides b (resp. a does not divide b).

1. CM-periods

In this section, we state a theorem of Shimura on CM-periods which are essential for succeeding sections. The following factorization theorem of the geometric period $p_K(\sigma, \Phi)$ is proved by Shimura [**Sh80**, Theorem 1.1]. (cf. also [**Sh98**, Theorem 32.5].)

THEOREM S1. *For every* CM-*field* K, *there exists a map* $p_K : I_K \times I_K \longrightarrow \mathbf{C}^\times$ *with the following properties.*

1. $p_K(\sigma, \Phi)$ *is defined by (1) in Introduction if* Φ *is a* CM-*type of* K *and* $\sigma \in \Phi$.
2. $p_K(\sigma_1 + \sigma_2, \tau) \sim p_K(\sigma_1, \tau) p_K(\sigma_2, \tau)$, $p_K(\sigma, \tau_1 + \tau_2) \sim p_K(\sigma, \tau_1) p_K(\sigma, \tau_2)$ *for every* σ, σ_1, σ_2, τ, τ_1, $\tau_2 \in I_K$.
3. $p_K(\xi\rho, \eta) \sim p_K(\xi, \eta\rho) \sim p_K(\xi, \eta)^{-1}$ *for every* ξ, $\eta \in I_K$.
4. $p_K(\xi, \mathrm{Res}_{L/K}(\zeta)) \sim p_L(\mathrm{Inf}_{L/K}(\xi), \zeta)$ if $\xi \in I_K$, $\zeta \in I_L$ and $K \subset L$, L *is a* CM-*field.*
5. $p_K(\mathrm{Res}_{L/K}(\zeta), \xi) \sim p_L(\zeta, \mathrm{Inf}_{L/K}(\xi))$ *if* $\xi \in I_K$, $\zeta \in I_L$ *and* $K \subset L$, L *is a* CM-*field.*
6. $p_{K'}(\gamma\xi, \gamma\eta) \sim p_K(\xi, \eta)$ *if* γ *is an isomorphism of* K' *onto* K.

REMARK. (1) $p_K(\sigma, \tau) \mod \overline{\mathbf{Q}}^\times$ is uniquely determined.
(2) We can take $p_K(\sigma, \tau)$ from $\mathbf{R}^\times$ for every σ, $\tau \in I_K$. This can be seen using Shimura [**Sh79**, Proposition 1.6], and following the proof of Theorem 1.1 of [**Sh80**].
(3) If we consider periods of differential 1-forms of the second kind, we can interpret 3 as a generalized Legendre's relation.

2. Conjectures (first version)

Let K be a CM-field. We assume that K is normal over $\mathbf{Q}$ and set $G = \mathrm{Gal}(K/\mathbf{Q})$. Let $\rho \in G$ be the complex conjugation. It is well known that ρ belongs to the center of G. Let ψ be a representation of G. We call ψ *odd* (resp. *even*) if $\psi(\rho) = -\mathrm{id}$ (resp. $\psi(\rho) = \mathrm{id}$). Let $\widehat{G}$ be the set of all equivalence classes of irreducible representations of G. Let $\widehat{G}_-$ (resp. $\widehat{G}_+$) be the subset of $\widehat{G}$ which consists of all equivalence classes of irreducible odd (resp. even) representations. We have $\widehat{G} = \widehat{G}_+ \cup \widehat{G}_-$. If $\eta \in \widehat{G}_-$, then $L(1, \eta) \neq 0$ and the Gamma factor to go with $L(s, \eta)$ is $\Gamma((s+1)/2)^{\dim \eta}$. Hence $L(0, \eta) \neq 0$ for $\eta \in \widehat{G}_-$.

CONJECTURE 2.1. *Let* ψ *be a virtual representation of* G *which is a* $\mathbf{Z}$*-linear combination of representations in* $\widehat{G}_-$. *We assume that the character* χ_ψ *of* ψ *is* $\mathbf{Q}$*-valued and that* $\psi \neq 0$. *Then we have*

$$\exp\left(\frac{L'(0, \psi)}{L(0, \psi)}\right) \sim \pi^{\dim \psi} \prod_{\sigma \in G} p_K(id, \sigma)^{\chi_\psi(\sigma^{-1})}. \tag{2.1}$$

We note that $\chi_\psi(\sigma) = \chi_\psi(\sigma^{-1})$ and $L(0, \psi) \in \mathbf{Q}^\times$. Conjecture 2.1 expresses $\exp(L'(0, \psi))$ in terms of p_K. It seems impossible to express p_K in terms of $\exp(L'(0, \psi))$ if G is non-abelian. This can be explained as follows. It may be conjectured that $p_K(\mathrm{id}, \sigma_i)$, $1 \leq i \leq [K : \mathbf{Q}]/2$ are algebraically independent if $\sum_{i=1}^{[K:\mathbf{Q}]/2} \sigma_i$ is a CM-type of K (cf. [**Sh80**, p. 319]). If G is not abelian, we can easily show that $|\widehat{G}_-| \leq [K : \mathbf{Q}]/2 - 3$. Hence the expression of p_K in terms of $\exp(L'(0, \psi))$, $\psi \in \widehat{G}_-$ would be impossible.

It may be the case that (2.1) remains true without assuming χ_ψ is $\mathbf{Q}$-valued, if we could define p_K more precisely so that

$$\prod_{\sigma \in G} p_K(\mathrm{id}, \sigma)^{\chi_\psi(\sigma^{-1})}$$

is well defined. A conjecture of the same strength can be formulated as follows.

CONJECTURE 2.2. *Let c be a conjugacy class in G. Then*

$$\prod_{\sigma\in c} p_K(id,\sigma) \sim \pi^{-\mu(c)/2} \prod_{\eta\in\widehat{G}_-} \exp\left(\frac{|c|\chi_\eta(c)}{[K:\mathbf{Q}]}\frac{L'(0,\eta)}{L(0,\eta)}\right), \tag{2.2}$$

where χ_η is the character of η and

$$\mu(c) = \begin{cases} 1 & \text{if} \quad c=\{1\}, \\ -1 & \text{if} \quad c=\{\rho\}, \\ 0 & \text{if} \quad c\neq\{1\},\{\rho\}. \end{cases}$$

This conjecture is stronger than Conjecture 2.1 as shown below. We note that $2|\widehat{G}_-|$ is equal to the number of conjugacy classes c of G such that $c\neq c\rho$. For a conjugacy class c of G, the order of $g\in c$ is called the order of c; for $a\in\mathbf{Z}$, the conjugacy class of g^a will be denoted by c^a.

PROPOSITION 2.3. *The following two assertions are equivalent.*

1. *For every conjugacy class c of G, we have*

$$\prod_{a=1,(a,n)=1}^{n} \prod_{\sigma\in c^a} p_K(id,\sigma) \sim \prod_{a=1,(a,n)=1}^{n} \pi^{-\mu(c^a)/2} \prod_{\eta\in\widehat{G}_-} \exp\left(\frac{|c^a|\chi_\eta(c^a)}{[K:\mathbf{Q}]}\cdot\frac{L'(0,\eta)}{L(0,\eta)}\right), \tag{2.3}$$

where n is the order of c.

2. *Conjecture 2.1 holds.*

PROOF. For $\omega\in\widehat{G}$, let M_ω be the field generated over $\mathbf{Q}$ by the values of χ_ω and put $J_\omega=J_{M_\omega}$. We divide $\widehat{G}$ into a disjoint union of orbits under the action of $\mathrm{Gal}(\overline{\mathbf{Q}}/\mathbf{Q})$: ω_1 and ω_2 belong to the same orbit if and only if $\chi_{\omega_1^\tau}=\chi_{\omega_2}$ for some $\tau\in\mathrm{Gal}(\overline{\mathbf{Q}}/\mathbf{Q})$. Each orbit is contained in $\widehat{G}_+$ or $\widehat{G}_-$. From each orbit we choose a representation ω and put $\psi=\oplus_{\tau\in J_\omega}\omega^\tau$, $J_\psi=J_\omega$. Then the character of ψ is $\mathbf{Q}$-valued. Let R be the set of equivalence classes of representations obtained in this manner. Let R_- (resp. R_+) be the subset of R which consists of equivalence classes of odd (resp. even) representations. Clearly we see that every $\mathbf{Q}$-valued virtual character of G is a $\mathbf{Z}$-linear combination of characters of representations in R.

For a conjugacy class c, let $n(c)$ denote the order of c. Let $\omega\in\widehat{G}$ and set $\psi=\oplus_{\tau\in J_\omega}\omega^\tau$. Since $\sum_{a=1,(a,n(c))=1}^{n(c)}\chi_\omega(c^a)$ is $\mathbf{Q}$-valued, we have

$$|J_\omega|\sum_{a=1,(a,n(c))=1}^{n(c)}\chi_\omega(c^a)=\sum_{\tau\in J_\omega}\sum_{a=1,(a,n(c))=1}^{n(c)}\chi_{\omega^\tau}(c^a)=\sum_{a=1,(a,n(c))=1}^{n(c)}\chi_\psi(c^a).$$

Therefore we obtain

$$\sum_{a=1,(a,n(c))=1}^{n(c)}\chi_\omega(c^a)=\frac{\varphi(n(c))}{|J_\psi|}\chi_\psi(c), \tag{2.4}$$

where φ denotes Euler's function. By this formula and noting

$$\frac{L'(0,\psi)}{L(0,\psi)}=\sum_{\omega\in J_\omega}\frac{L'(0,\omega^\tau)}{L(0,\omega^\tau)}$$

for $\omega \in \widehat{G}_-$, we see that (2.3) is equivalent to

$$\prod_{a=1,(a,n(c))=1}^{n(c)} \prod_{\sigma\in c^a} p_K(\mathrm{id},\sigma) \sim \pi^{-\mu(c)/2} \prod_{\psi\in R_-} \exp\left(\frac{|c|\varphi(n(c))}{[K:\mathbf{Q}]}\frac{\chi_\psi(c)}{|J_\psi|}\frac{L'(0,\psi)}{L(0,\psi)}\right). \tag{2.5}$$

A subset of the form $\cup_{a=1,(a,n(c))=1}^{n(c)} c^a$ of G is called an "Abteilung" in the old terminology which goes back to Frobenius [**F**]. Let $u(c)$ be the minimum positive integer such that $c^{u(c)} = c$. Now assume 1. We have

$$\begin{aligned}\prod_{\sigma\in G} p_K(\mathrm{id},\sigma)^{\chi_\psi(\sigma^{-1})} &= \prod_c \prod_{a=1,(a,n(c))=1}^{n(c)} \prod_{\sigma\in c^a} p_K(\mathrm{id},\sigma)^{\chi_\psi(\sigma^{-1})/u(c)} \\ &\sim \pi^{-(\chi_\psi(1)-\chi_\psi(\rho))/2} \prod_c \prod_{\eta\in R_-} \exp\left(\frac{|c|\varphi(n(c))}{[K:\mathbf{Q}]}\frac{\chi_\eta(c)}{|J_\eta|}\frac{\chi_\psi(c^{-1})}{u(c)}\frac{L'(0,\eta)}{L(0,\eta)}\right) \\ &= \pi^{-\dim\psi} \prod_{\eta\in R_-} \exp\left(\frac{1}{[K:\mathbf{Q}]|J_\eta|}\sum_{\sigma\in G}\chi_\eta(\sigma)\chi_\psi(\sigma^{-1})\frac{L'(0,\eta)}{L(0,\eta)}\right).\end{aligned}$$

Here the product $\prod_c$ extends over Abteilungen choosing one conjugacy class c from each Abteilung. By the orthogonality relations, we have $\sum_{\sigma\in G}\chi_\eta(\sigma)\chi_\psi(\sigma^{-1})$ is equal to $|G||J_\eta|$ if $\psi \cong \eta$ and 0 if ψ is not equivalent to η. Thus we obtain 2.

Next we assume 2. Set

$$Q = \prod_{a=1,(a,n(c))=1}^{n(c)} \prod_{\sigma\in c^a} p_K(\mathrm{id},\sigma), \quad \tilde{c} = \cup_{a=1,(a,n(c))=1}^{n(c)} c^a.$$

By the orthogonality relation and using (2.4), we have

$$\sum_{\psi\in R} \chi_\psi(\sigma)\chi_\psi(c^{-1})|c|\varphi(n(c))/|J_\psi||G| = \begin{cases} u(c) & \text{if } \sigma\in\tilde{c}, \\ 0 & \text{otherwise.}\end{cases}$$

Hence we obtain

$$Q = \prod_{\psi\in R}\prod_{\sigma\in G} p_K(\mathrm{id},\sigma)^{\chi_\psi(\sigma)\chi_\psi(c^{-1})|c|\varphi(n(c))/|J_\psi||G|}.$$

Using Theorem S1, 3 and the assumption, we have

$$Q \sim \prod_{\psi\in R_-}\left(\pi^{-\dim\psi}\exp\left(\frac{L'(0,\psi)}{L(0,\psi)}\right)\right)^{\chi_\psi(c^{-1})|c|\varphi(n(c))/|J_\psi||G|}.$$

Considering regular representations of G and $G/\langle\rho\rangle$, we have

$$\sum_{\psi\in R_-} \dim\psi\cdot\chi_\psi(\sigma)/|J_\psi| = \begin{cases} |G|/2 & \text{if } \sigma = 1, \\ -|G|/2 & \text{if } \sigma = \rho, \\ 0 & \text{if } \sigma \neq 1,\rho.\end{cases}$$

Hence we obtain (2.5). This completes the proof. □

CONJECTURE 2.4. *Let K be a CM-field (not necessarily normal over $\mathbf{Q}$) and F be its maximal real subfield. Let χ be the Hecke character of $F_A^\times$ which corresponds to the quadratic extension K/F and $L_F(s,\chi)$ be the Hecke L-function attached to χ. Then*

$$\exp\left(\frac{L'_F(0,\chi)}{L_F(0,\chi)}\right) \sim \pi^{[K:\mathbf{Q}]/2} \prod_{\sigma\in J_K} p_K(\sigma,\sigma). \tag{2.6}$$

We note that if K is normal over $\mathbf{Q}$, (2.6) can be written as

$$\exp\left(\frac{L'_F(0,\chi)}{L_F(0,\chi)}\right) \sim (\pi^{1/2}p_K(\mathrm{id},\mathrm{id}))^{[K:\mathbf{Q}]}.$$

by Theorem S1, 6.

PROPOSITION 2.5. *Let L be a CM-field normal over $\mathbf{Q}$. The following two assertions are equivalent.*

1. *Conjecture 2.1 holds for all odd representations of Gal($L/\mathbf{Q}$) with $\mathbf{Q}$-valued characters.*
2. *Conjecture 2.4 holds for all CM-subfields of L.*

PROOF. Put $G = \mathrm{Gal}(L/\mathbf{Q})$. Let K be a CM-subfield of L and set $H = \mathrm{Gal}(L/K)$, $\widetilde{H} = \mathrm{Gal}(L/F)$. Let χ be the non-trivial character of $\mathrm{Gal}(K/F) \cong \widetilde{H}/H$ and lift χ to the character $\widetilde{\chi}$ of $\widetilde{H}$. Put $\psi = \mathrm{Ind}_{\widetilde{H}}^G \widetilde{\chi}$. We shall show

$$\prod_{\sigma\in J_K} p_K(\sigma,\sigma) \sim \prod_{\sigma\in G} p_L(\mathrm{id},\sigma)^{\chi_\psi(\sigma)}. \tag{2.7}$$

For $g \in G$, set

$$n_+(g) = |\{x\in G \mid xgx^{-1}\in H\}|, \quad n_-(g) = |\{x\in G \mid xgx^{-1}\in H\rho\}|.$$

By the formula of induced characters, we have

$$\chi_\psi(g) = (n_+(g)-n_-(g))/|\widetilde{H}|.$$

We choose a subset S of G so that $G = S\cup S\rho$ is a disjoint union. By Theorem S1, 3, we have

$$\prod_{\sigma\in G} p_L(\mathrm{id},\sigma)^{\chi_\psi(\sigma)} \sim \prod_{\sigma\in S} p_L(\mathrm{id},\sigma)^{2(n_+(\sigma)-n_-(\sigma))/|\widetilde{H}|}.$$

By Theorem S1, 5, 6, we get

$$\begin{aligned}\prod_{\sigma\in J_K} p_K(\sigma,\sigma) &\sim \prod_{g\in G} p_K(g|K,g|K)^{1/|H|}\\ &\sim \prod_{g\in G}\prod_{h\in H} p_L(g,hg)^{1/|H|} \sim \prod_{g\in G}\prod_{h\in H} p_L(\mathrm{id},g^{-1}hg)^{1/|H|}.\end{aligned}$$

For $\sigma\in S$, we have

$$\begin{aligned}|\{g\in G, h\in H \mid g^{-1}hg=\sigma\}| &= n_+(\sigma),\\ |\{g\in G, h\in H \mid g^{-1}hg=\sigma\rho\}| &= n_-(\sigma).\end{aligned}$$

Hence, by Theorem S1, 3, we obtain

$$\prod_{\sigma\in J_K} p_K(\sigma,\sigma) \sim \prod_{\sigma\in S} p_L(\mathrm{id},\sigma)^{(n_+(\sigma)-n_-(\sigma))/|H|}.$$

We have proved (2.7).

Now assume 1. Identify the Hecke character χ with the non-trivial character of $\mathrm{Gal}(K/F)$ as above. Since $L_F(s,\chi) = L(s,\psi)$, we have

$$\exp\left(\frac{L'_F(0,\chi)}{L_F(0,\chi)}\right) \sim \pi^{\dim\psi} \prod_{\sigma\in G} p_L(\mathrm{id},\sigma)^{\chi_\psi(\sigma)}.$$

By (2.7), we obtain 2.

Next assume 2. Let ψ be a virtual representation of G as in Conjecture 2.1. By a theorem of Artin (cf. [**Ar**], [**T**, p. 45]), there exist a positive integer n, subgroups H_i of G and integers m_i $(1 \le i \le r)$ such that

$$n\chi_\psi = \sum_{i=1}^{r} m_i \,\mathrm{Ind}_{H_i}^{G} 1_{H_i},$$

where 1_{H_i} denotes the trivial character of H_i. If $\rho \in H_i$, then $\mathrm{Ind}_{H_i}^{G} 1_{H_i}$ is an even representation. Hence we may assume that $\rho \notin H_i$ for every i. Put $\widetilde{H}_i = H_i \cup H_i\rho$. We have

$$n\chi_\psi = \sum_{i=1}^{r} m_i \,\mathrm{Ind}_{\widetilde{H}_i}^{G} (1_{\widetilde{H}_i} \oplus \chi_{\widetilde{H}_i}),$$

where $\chi_{\widetilde{H}_i}$ denotes the non-trivial character of $\widetilde{H}_i$ which is trivial on H_i. Since $\mathrm{Ind}_{\widetilde{H}_i}^{G} 1_{\widetilde{H}_i}$ is even, we have

$$n\chi_\psi = \sum_{i=1}^{r} m_i \,\mathrm{Ind}_{\widetilde{H}_i}^{G} \chi_{\widetilde{H}_i}.$$

By (2.7) and the assumption, (2.1) holds for all representations $\mathrm{Ind}_{\widetilde{H}_i}^{G} \chi_{\widetilde{H}_i}$. Therefore (2.1) holds for χ_ψ. This completes the proof. □

For a totally real algebraic number field F, a CM-field K such that $[K:F]=2$ is called a CM-*extension* of F.

PROPOSITION 2.6. *Let F be a totally real algebraic number field. Let K_1 and K_2 be* CM*-extensions of F and K_0 be the composite field of K_1 and K_2. If Conjecture 2.4 holds for K_1 and K_2, then it holds also for K_0.*

PROOF. We may assume $K_1 \neq K_2$. Let F_0 be the maximal real subfield of K_0. Let χ_0, χ_1 and χ_2 be the Hecke characters which correspond to the quadratic extensions K_0/F_0, K_1/F and K_2/F respectively. Let α and β be the generators of $\mathrm{Gal}(K_0/K_1)$ and $\mathrm{Gal}(K_0/K_2)$ respectively. Considering an induced representation, we find

$$L_{F_0}(s,\chi_0) = L_F(s,\chi_1)L_F(s,\chi_2).$$

Hence, by the assumption and Theorem S1, 5, we obtain

$$\begin{aligned}
\exp\left(\frac{L'_{F_0}(0,\chi_0)}{L_{F_0}(0,\chi_0)}\right) &= \exp\left(\frac{L'_F(0,\chi_1)}{L_F(0,\chi_1)}\right)\exp\left(\frac{L'_F(0,\chi_2)}{L_F(0,\chi_2)}\right)\\
&\sim \pi^{[K_1:\mathbf{Q}]/2} \prod_{\sigma\in J_{K_1}} p_{K_1}(\sigma,\sigma)\ \pi^{[K_2:\mathbf{Q}]/2} \prod_{\sigma\in J_{K_2}} p_{K_2}(\sigma,\sigma)\\
&\sim \pi^{[K_0:\mathbf{Q}]/2} \left(\prod_{\sigma\in J_{K_0}} p_{K_0}(\sigma,\sigma)p_{K_0}(\sigma,\alpha\sigma)p_{K_0}(\sigma,\sigma)p_{K_0}(\sigma,\beta\sigma)\right)^{1/2}.
\end{aligned}$$

We have $\beta = \alpha\rho$ and $\beta\sigma = \alpha\rho\sigma = \alpha\sigma\rho$ for $\sigma \in J_{K_0}$. Now the assertion follows from Theorem S1, 3. □

3. Preparations on CM-fields

In the next section, we shall render Conjecture 2.1 to numerical tests. For this purpose, we collect several general facts on CM-fields in this section.

For an algebraic number field L, let D_L, h_L, E_L, W_L and R_L denote the discriminant, the class number, the group of units, the group of roots of unity in L and the regulator of L respectively. We put $w_L = |W_L|$.

Let K be a CM-field and F be the maximal real subfield of K. Put $n = [F : \mathbf{Q}]$. Let χ denote the Hecke character which corresponds to the quadratic extension K/F. By the well known formula for residues of the Dedekind zeta functions and by the functional equations of $\zeta_F(s)$ and $\zeta_K(s)$, we obtain

$$L_F(0,\chi) = \frac{2R_K}{w_K R_F} \cdot \frac{h_K}{h_F}. \tag{3.1}$$

By the definition of the regulator, we have

$$\frac{R_K}{R_F} = \frac{2^{n-1}}{[E_K : W_K E_F]}. \tag{3.2}$$

Hence we get

$$L_F(0,\chi) = \frac{2^n}{w_K[E_K : W_K E_F]} \cdot \frac{h_K}{h_F} = \frac{2^{n-1}}{[E_K : E_F]} \cdot \frac{h_K}{h_F}. \tag{3.3}$$

Let $\mathfrak{d}_{K/F}$ denote the relative different of K over F. The next Lemma is well known. We omit the proof.

LEMMA 3.1. *$[E_K : W_K E_F] = 1$ or 2. If $\mathfrak{d}_{K/F}$ does not divide (2), then $E_K = W_K E_F$. If $[E_K : W_K E_F] = 2$, then $K = F(\sqrt{-\epsilon})$ with a totally positive unit $\epsilon \in E_F$.*

Let Φ be a CM-type and $\mathfrak{f}$ be an integral ideal of K. Let λ be a Grössencharacter of $I_{\mathfrak{f}}(K)$ such that

$$\lambda((\alpha)) = \prod_{\sigma\in\Phi} (\alpha^{\sigma\rho}/|\alpha^\sigma|)^{t_\sigma} \quad \text{if} \quad \alpha \equiv 1 \mod {}^\times\mathfrak{f},$$

where t_σ, $\sigma \in \Phi$ are non-negative integers. Let $L(s,\lambda)$ denote the L-function attached to λ. We write $L(s,\lambda)$ as $L_K(s,\lambda)$ when we emphasize the dependence on K. For the later use, we explain our convention for the action of $\mathrm{Aut}(\mathbf{C})$ on λ. We put

$$t_0 = \max_{\sigma\in\Phi} t_\sigma. \tag{3.4}$$

We assume that $t_\sigma \mod 2$ is independent of σ. We define a quasicharacter λ_0 of $I_{\mathfrak{f}}(K)$ by

$$\lambda_0(\mathfrak{a}) = \lambda(\mathfrak{a})N(\mathfrak{a})^{t_0/2}, \quad \mathfrak{a} \in I_{\mathfrak{f}}(K). \tag{3.5}$$

Then we have $L(s,\lambda) = L(s + t_0/2, \lambda_0)$ and

$$\lambda_0((\alpha)) = \prod_{\sigma\in\Phi} (\alpha^{\sigma\rho})^{(t_0+t_\sigma)/2}(\alpha^\sigma)^{(t_0-t_\sigma)/2} \quad \text{if} \quad \alpha \equiv 1 \mod {}^\times\mathfrak{f}. \tag{3.6}$$

For $\sigma \in \mathrm{Aut}(\mathbf{C})$, we define λ_0^σ by $\lambda_0^\sigma(\mathfrak{a}) = \lambda_0(\mathfrak{a})^\sigma$ and put $\lambda^\sigma(\mathfrak{a}) = \lambda_0^\sigma(\mathfrak{a})N(\mathfrak{a})^{-t_0/2}$, $\mathfrak{a} \in I_\mathfrak{f}(K)$.

We quote a fundamental theorem of Shimura for the use in succeeding sections ([**Sh75**, Theorem 2], combined with [**Sh80**, Theorem 1.1]; or see [**Sh98**, Theorem 32.12]; cf. also [**Sh78b**, §5]).

THEOREM S2. *For every integer m such that $m - t_\sigma \in 2\mathbf{Z}$ and $-t_\sigma < m \le t_\sigma$ for every $\sigma \in \Phi$, we have*

$$L(m/2, \lambda) \sim \pi^{e/2} p_K\left(\sum_{\sigma\in\Phi} t_\sigma \cdot \sigma, \Phi\right),$$

where $e = m[F : \mathbf{Q}] + \sum_{\sigma\in\Phi} t_\sigma$.

To compute $L(m/2, \lambda)$, we apply Shimura's method [**Sh75**], which we are going to explain briefly. Let $k \in \mathbf{Z}$, $k > 0$, $r = (r_1, r_2, \cdots, r_n) \in \mathbf{Z}^n$, $r_i \ge 0$ for $1 \le i \le n$. Set $\{r\} = \sum_{i=1}^n r_i$, $\mathbf{1} = (1, 1, \cdots, 1) \in \mathbf{Z}^n$. For $x = (x_1, x_2, \cdots, x_n) \in \mathbf{C}^n$, $a = (a_1, a_2, \cdots, a_n) \in \mathbf{Z}^n$, put $x^a = \prod_{i=1}^n x_i^{a_i}$. ($x^a$ is defined similarly also for $x \in \mathbf{R}_+^n$, $a \in \mathbf{C}^n$.) Let $J_F = \{\tau_1, \tau_2, \cdots, \tau_n\}$. Let $\mathfrak{H}$ denote the complex upper half plane. For $z \in \mathfrak{H}^n$ and $s \in \mathbf{C}$, define an Eisenstein series $E_{k,r}(z, s)$ by

$$(3.7)\qquad E_{k,r}(z, s) = (z - \bar{z})^{-r}(2\pi i)^{-kn-\{r\}} \sum_{(c,d)/\sim} \left(\frac{c\bar{z} + d}{cz + d}\right)^r (cz + d)^{-k\cdot\mathbf{1}}|cz + d|^{-s\cdot\mathbf{1}},$$

where $cz + d = (c^{\tau_1}z_1 + d^{\tau_1}, c^{\tau_2}z_2 + d^{\tau_2}, \cdots, c^{\tau_n}z_n + d^{\tau_n}) \in \mathbf{C}^n$ and $\sum_{(c,d)/\sim}$ means that (c, d) runs over $\mathfrak{O}_F \oplus \mathfrak{O}_F \backslash \{(0, 0)\}$ under the equivalence relation

$$(c, d) \sim (c_1, d_1) \iff \exists\epsilon \in E_F, c_1 = \epsilon c, d_1 = \epsilon d.$$

This sum is well defined if $N(\epsilon)^k = 1$ for $\epsilon \in E_F$. Then $E_{k,r}(z, s)$ converges absolutely when $\Re(s) + k > 2$ and can be continued meromorphically to the whole s-plane. Furthermore $E_{k,r}(z, s)$ is holomorphic at $s = 0$. We put

$$(3.8)\qquad E_{k,r}(z) = E_{k,r}(z, 0).$$

Then we have the transformation formula

$$(3.9)\qquad E_{k,r}(\gamma z) = E_{k,r}(z)(cz + d)^{2r}(cz + d)^{k\cdot\mathbf{1}} \quad \text{for every} \quad \gamma = \begin{pmatrix} a & b \\ c & d \end{pmatrix} \in \mathrm{SL}(2, \mathfrak{O}_F).$$

Let K be a CM-extension of F. Let $\mathfrak{A}$ be a fractional ideal of K; we assume that $\mathfrak{A} = \mathfrak{O}_F\omega \oplus \mathfrak{O}_F$ with $\omega \in K$. Let Φ be a CM-type of K such that $\Im(\omega^\sigma) > 0$ for every $\sigma \in \Phi$. Regard ω as a point $(\omega_1, \omega_2, \cdots, \omega_n)$ of $\mathfrak{H}^n$ where $\omega_i = \omega^\sigma$ for $\sigma \in \Phi$ such that $\sigma|F = \tau_i$. If $\Re(s)$ is sufficiently large, then by straightforward calculation, we get

$$(3.10)\qquad E_{k,r}(\omega, s) = [E_K : E_F](2\pi i)^{-kn-\{r\}} \prod_{\sigma\in\Phi} (2\Im(\omega^\sigma)i)^{-r(\sigma)} \cdot \sum_{(\alpha)\subseteq\mathfrak{A}} \prod_{\sigma\in\Phi} \left(\frac{\alpha^{\sigma\rho}}{|\alpha^\sigma|}\right)^{2r(\sigma)+k} N(\alpha)^{-k/2-s/2},$$

where (α) extends over all principal ideals contained in $\mathfrak{A}$ and $r(\sigma) = r_j$ if $\sigma|F = \tau_j$. By analytic continuation, we can use $E_{k,r}(\omega)$ to evaluate the right hand side at $s = 0$.

For $t \in \mathbf{C}$, define a differential operator $D_{j,t}$ by $D_{j,t} = \frac{1}{2\pi i}\left(\frac{t}{2iy_j} + \frac{\partial}{\partial z_j}\right)$. Then we have

$$D_{j,k+2r_j} E_{k,r}(z) = (k + r_j) E_{k,r^*}(z), \tag{3.11}$$

where $r^* = r + (0, 0, \cdots, 0, 1, 0, \cdots, 0) \in \mathbf{Z}^n$ (added the vector whose j-th component is 1, other component 0).

If the class number of F is 1, $k \geq 2$ and $r = (0, 0, \cdots, 0)$, we have (cf. Siegel [**Si69**], (16) $\sim$ (18)), excluding the case $F = \mathbf{Q}$ and $k = 2$,

$$(3.12)\ E_{k,0}(z) = (2\pi i)^{-kn} \left\{ \zeta_F(k) + \left(\frac{(2\pi i)^k}{(k-1)!}\right)^n |D_F|^{1/2-k} \sum_{\mathfrak{d}^{-1}|\nu \gg 0} \sigma_{k-1}(\nu) e^{2\pi i S(\nu z)} \right\},$$

where $\mathfrak{d}$ denotes the different of F over $\mathbf{Q}$, $S(\nu z) = \sum_{j=1}^n \nu^{\tau_j} z_j$,

$$\sigma_{k-1}(\nu) = \sum_{\mathfrak{d}^{-1}|(\alpha)|\nu} \operatorname{sgn}(N(\alpha^k)) N((\alpha)\mathfrak{d})^{k-1}. \tag{3.13}$$

We can use (3.10) $\sim$ (3.13) to compute numerical values of $L(m/2, \lambda)$ if the conductor of λ is (1) and if the class number of F in the narrow sense is 1.[4]

4. Numerical examples

In this section, we shall examine Conjecture 2.1 numerically for some simple non-abelian CM-fields. The following example is discussed in [**ST**, p. 74].

Let $F = \mathbf{Q}(\sqrt{d})$, $0 < d \in \mathbf{Q}$ be a real quadratic field. Let $x + y\sqrt{d} \in F$, x, $y \in \mathbf{Q}$ be a totally positive element and set $\xi = \sqrt{x + y\sqrt{d}}\ i$, $\xi' = \sqrt{x - y\sqrt{d}}\ i$, $K = \mathbf{Q}(\xi)$. Then $[K : F] = 2$ and K is a CM-field. We assume that K is not normal over $\mathbf{Q}$. This assumption implies $E_K = E_F$ (cf. [**Sh77**, Proposition A.7, (iii)]). Let L be the normal closure of K over $\mathbf{Q}$. Then L is a CM-field and we have $L = \mathbf{Q}(\xi, \xi')$, $[L : \mathbf{Q}] = 8$. Define σ, $\tau \in \operatorname{Gal}(L/\mathbf{Q})$ by

$$\sigma : (\xi, \xi') \longrightarrow (\xi', -\xi), \qquad \tau : (\xi, \xi') \longrightarrow (\xi', \xi).$$

Then $\operatorname{Gal}(L/\mathbf{Q})$ is generated by σ and τ which are subject to the relations

$$\sigma^4 = \tau^2 = 1, \quad \tau\sigma = \sigma^3\tau, \quad \sigma^2 = \rho.$$

Thus $\operatorname{Gal}(L/\mathbf{Q})$ is the dihedral group of order 8. Define a CM-type Φ of K by $\Phi = \{\mathrm{id}, \sigma|K\}$. Then the reflex of (K, Φ) is (K', Φ'), where $K' = \mathbf{Q}(\xi + \xi')$, $\Phi' = \{\mathrm{id}, \sigma\tau|K'\}$. Put $d' = x^2 - y^2 d$. Since $(\xi + \xi')^2 = -2(x + \sqrt{d'})$, we have $K' = \mathbf{Q}(\sqrt{2(x + \sqrt{d'}}\ i)$ and $F' = \mathbf{Q}(\sqrt{d'})$ is the maximal real subfield of K'. Let F_0 be the maximal real subfield of L. We have $F_0 = \mathbf{Q}(\sqrt{d}, \sqrt{d'})$. We note that

$$\sqrt{d}^{\,\sigma} = -\sqrt{d}, \quad \sqrt{d'}^{\,\sigma} = -\sqrt{d'}, \quad \sqrt{d}^{\,\tau} = -\sqrt{d}, \quad \sqrt{d'}^{\,\tau} = \sqrt{d'}. \tag{4.1}$$

[4]Here we restricted ourselves to a simple case mostly sufficient for the use in §4. To deal with the general case, we must employ Eisenstein series with congruence conditions. For details, see [**Sh75**].

Put $G = \mathrm{Gal}(L/\mathbf{Q})$. We have

$$G/\langle\rho\rangle \cong \mathrm{Gal}(F_0/\mathbf{Q}) \cong \mathbf{Z}/2\mathbf{Z} \oplus \mathbf{Z}/2\mathbf{Z}.$$

Hence $\widehat{G}_+$ consists of four one dimensional representations. We see easily that $\widehat{G}_- = \{\eta\}$, where η is the unique irreducible two dimensional representation of G. Let χ be the non-trivial character of $\mathrm{Gal}(K/F)$ and regard χ as a character of $\mathrm{Gal}(L/F)$. Then we find immediately that

$$\eta \cong \mathrm{Ind}(\chi\,; \mathrm{Gal}(L/F) \longrightarrow \mathrm{Gal}(L/\mathbf{Q})).$$

Hence we have $L(s,\eta) = L_F(s,\chi)$, where $L_F(s,\chi)$ denotes the Hecke L-function attached to χ regarded as the Hecke character of $F_A^\times$ corresponding to the quadratic extension K/F. The character χ_η of η is given by $\chi_\eta(g) = 2$, -2 or 0 according as $g = 1$, $g = \rho$ or $g \neq 1, \rho$. Hence Conjecture 2.1 for $L(s,\eta)$ is equivalent to

$$\exp\left(\frac{L'(0,\eta)}{L(0,\eta)}\right) \sim \pi^2 p_L(\mathrm{id},\mathrm{id})^4 \tag{4.2}$$

in view of Theorem S1, 3. On the other hand, in view of Theorem S1, 3, Conjecture 2.4 for $L_F(s,\chi)$ is equivalent to

$$\exp\left(\frac{L_F'(0,\chi)}{L_F(0,\chi)}\right) \sim \pi^2 p_K(\mathrm{id},\mathrm{id})^2 p_K(\sigma,\sigma)^2. \tag{4.3}$$

Here we abbreviated $\sigma|K$ to σ. Similar notation will be used hereafter since no confusion is likely. Using Theorem S1, we see that (4.2) is equivalent to (4.3), i.e., Conjecture 2.1 for $L(s,\eta)$ and Conjecture 2.4 for $L_F(s,\chi)$ are equivalent. We note the following relations due to Shimura [**Sh77**, Proposition A.7].

$$\begin{gathered} h_K/h_F = h_{K'}/h_{F'}, \\ N_{F/\mathbf{Q}}(D(K/F))D(F/\mathbf{Q}) = N_{F'/\mathbf{Q}}(D(K'/F'))D(F'/\mathbf{Q}). \end{gathered} \tag{4.4}$$

For given F and χ, $\exp(L_F'(0,\chi)/L_F(0,\chi))$ can be calculated by Shintani's formulas [**Shi76, Shi77**].[5] To compute CM-periods, we apply Theorem S2. For positive integers a and b, let $\lambda_{a,b}^{(1)}$ and $\lambda_{a,b}^{(2)}$ be Grössencharacters of $I_{\mathfrak{f}}(K)$ such that

$$\lambda_{a,b}^{(1)}((\alpha)) = \left(\frac{\alpha^\rho}{|\alpha|}\right)^a \left(\frac{\alpha^{\sigma\rho}}{|\alpha^\sigma|}\right)^b, \quad \alpha \equiv 1 \mod {}^\times\mathfrak{f},$$

$$\lambda_{a,b}^{(2)}((\alpha)) = \left(\frac{\alpha^\rho}{|\alpha|}\right)^a \left(\frac{\alpha^{\sigma}}{|\alpha^{\sigma\rho}|}\right)^b, \quad \alpha \equiv 1 \mod {}^\times\mathfrak{f},$$

where $\mathfrak{f}$ is an integral ideal of K. Similarly let $\mu_{a,b}^{(1)}$ and $\mu_{a,b}^{(2)}$ be Grössencharacters of $I_{\mathfrak{f}'}(K')$ such that

$$\mu_{a,b}^{(1)}((\alpha)) = \left(\frac{\alpha^\rho}{|\alpha|}\right)^a \left(\frac{\alpha^{\sigma\tau\rho}}{|\alpha^{\sigma\tau}|}\right)^b, \quad \alpha \equiv 1 \mod {}^\times\mathfrak{f}',$$

$$\mu_{a,b}^{(2)}((\alpha)) = \left(\frac{\alpha^\rho}{|\alpha|}\right)^a \left(\frac{\alpha^{\sigma\tau}}{|\alpha^{\sigma\tau\rho}|}\right)^b, \quad \alpha \equiv 1 \mod {}^\times\mathfrak{f}',$$

[5] Shintani gave an arithmetic formula for $L_F(0,\chi)$ in [**Shi76**]. In [**Shi77**], he gave a closed formula for $L_F'(0,\chi)$ in terms of the double gamma function when $[F:\mathbf{Q}] = 2$. (See [**Shi77b**] for the general case.) Shintani's formulas for partial zeta functions (cf. (16), (17) of [**Shi78**]) have striking similarity to the case of the Hurwitz zeta function. For the evaluation of the double gamma function, we can efficiently use the asymptotic expansion given in [**Shi77**, p. 179].

where $\mathfrak{f}'$ is an integral ideal of K'. We note that $L_K(1,\lambda_{a,b}^{(i)}) \neq 0$, $L_{K'}(1,\mu_{a,b}^{(i)}) \neq 0$ for every i and a, b. By Theorem S2, we have

$$\begin{aligned} L_K\big(1,\lambda_{2,2}^{(1)}\big) &\sim \pi^4 p_K(2\cdot \mathrm{id} + 2\cdot\sigma,\ \mathrm{id}+\sigma),\\ L_K\big(1,\lambda_{2,2}^{(2)}\big) &\sim \pi^4 p_K(2\cdot \mathrm{id} + 2\cdot\sigma\rho,\ \mathrm{id}+\sigma\rho),\\ L_K\big(1,\lambda_{4,2}^{(1)}\big) &\sim \pi^5 p_K(4\cdot \mathrm{id} + 2\cdot\sigma,\ \mathrm{id}+\sigma),\\ L_K\big(1,\lambda_{2,4}^{(1)}\big) &\sim \pi^5 p_K(2\cdot \mathrm{id} + 4\cdot\sigma,\ \mathrm{id}+\sigma). \end{aligned}$$

By Theorem S1, we get

$$L_K\big(1,\lambda_{2,2}^{(1)}\big)L_K\big(1,\lambda_{2,2}^{(2)}\big) \sim \pi^8 p_K(\mathrm{id},\mathrm{id})^4 p_K(\sigma,\sigma)^4.$$

By (4.3), Conjecture 2.1 for $L(s,\eta)$ is equivalent to

$$\pi^{-4}L_K\big(1,\lambda_{2,2}^{(1)}\big)L_K\big(1,\lambda_{2,2}^{(2)}\big)\exp\left(-\frac{2L'_F(0,\chi)}{L_F(0,\chi)}\right) \quad \text{is algebraic.} \tag{4.5}$$

We see easily that (4.5) is also equivalent to

$$\pi^{-4}L_{K'}\big(1,\mu_{2,2}^{(1)}\big)L_{K'}\big(1,\mu_{2,2}^{(2)}\big)\exp\left(-\frac{2L'_F(0,\chi)}{L_F(0,\chi)}\right) \quad \text{is algebraic.} \tag{4.5$'$}$$

We shall derive another relation. By Theorems S1 and S2, we get

$$\begin{aligned} \frac{L_K\big(1,\lambda_{4,2}^{(1)}\big)}{L_K\big(1,\lambda_{2,2}^{(1)}\big)} &\sim \pi\, p_K(\mathrm{id},\ \mathrm{id}+\sigma)^2,\\ \frac{L_{K'}\big(1,\mu_{2,4}^{(2)}\big)}{L_{K'}\big(1,\mu_{2,2}^{(2)}\big)} &\sim \pi\, p_{K'}(\sigma\tau\rho,\ \mathrm{id}+\sigma\tau\rho)^2. \end{aligned} \tag{4.6}$$

Using Theorem S1, we have

$$\frac{L_K\big(1,\lambda_{4,2}^{(1)}\big)}{L_K\big(1,\lambda_{2,2}^{(1)}\big)}\cdot\frac{L_{K'}\big(1,\mu_{2,4}^{(2)}\big)}{L_{K'}\big(1,\mu_{2,2}^{(2)}\big)} \sim \pi^2 p_L(\mathrm{id},\mathrm{id})^4. \tag{4.7}$$

Therefore Conjecture 2.1 for $L(s,\eta)$ is also equivalent to

$$\frac{L_K\big(1,\lambda_{4,2}^{(1)}\big)}{L_K\big(1,\lambda_{2,2}^{(1)}\big)}\cdot\frac{L_{K'}\big(1,\mu_{2,4}^{(2)}\big)}{L_{K'}\big(1,\mu_{2,2}^{(2)}\big)}\cdot\exp\left(-\frac{L'_F(0,\chi)}{L_F(0,\chi)}\right) \quad \text{is algebraic.} \tag{4.8}$$

For a real quadratic field k, we denote the archimedean places of k by ∞_1 and ∞_2. We choose ∞_1 as the place corresponding to the identity embedding of k into $\mathbf{R}$.

EXAMPLE 1. We take $F = \mathbf{Q}(\sqrt{5})$, $K = \mathbf{Q}\Big(\sqrt{\frac{13+\sqrt{5}}{2}}\, i\Big)$. Then we have $F' = \mathbf{Q}(\sqrt{41})$, $K' = \mathbf{Q}(\sqrt{13+2\sqrt{41}}\, i)$. We have $h_F = h_{F'} = 1$. We have $(41) = \left(\frac{13+\sqrt{5}}{2}\right)\left(\frac{13-\sqrt{5}}{2}\right)$ in F. Put $\mathfrak{p} = \left(\frac{13+\sqrt{5}}{2}\right)$, $z = \Big(\sqrt{\frac{13+\sqrt{5}}{2}}\, i + \frac{1-\sqrt{5}}{2}\Big)/2$. Then we see that z is integral over $\mathfrak{O}_F$. Hence we see that $\mathfrak{p}$ ramifies in K and all the other prime ideals of F are unramified in K. Furthermore we have $\mathfrak{O}_K = \mathfrak{O}_F z \oplus \mathfrak{O}_F$. We can also see easily that K is the maximal ray class field of conductor $\mathfrak{p}\infty_1\infty_2$ of F. In F', we have $(5) = (13+2\sqrt{41})(13-2\sqrt{41})$. Put $\mathfrak{p}' = (13+2\sqrt{41})$, $z' = (\sqrt{13+2\sqrt{41}}\, i + 1)/2$. Similarly to the above, we see that $\mathfrak{p}'$ is the only prime ideal which ramifies in K', $\mathfrak{O}_{K'} = \mathfrak{O}_{F'}z' \oplus \mathfrak{O}_{F'}$ and that K' is the maximal ray class field of conductor $\mathfrak{p}'\infty_1\infty_2$ of F'.

We compute $L_F(0,\chi)$ and $L'_F(0,\chi)$ by the method of Shintani. We obtain

$$L_F(0,\chi) = 2, \quad L'_F(0,\chi) = -0.26558039348000766099171 65\cdots . \tag{4.9}$$

Since $E_K = E_F$, $E_{K'} = E_{F'}$, we have $h_K = h_{K'} = 1$ by (3.3), (4.4) and (4.9).

For every positive even integers a and b, there exist (unique) Grössencharacters $\lambda_{a,b}^{(1)}$, $\lambda_{a,b}^{(2)}$ (resp. $\mu_{a,b}^{(1)}$, $\mu_{a,b}^{(2)}$) of $I(K)$ (resp. $I(K')$) of conductor (1). By (3.12), we have [6]

$$\begin{aligned} E_{2,0}^F(z) &= \tfrac{\sqrt{5}}{2^3\cdot 3\cdot 5^3}\left\{1 + 120 \sum_{(\sqrt{5})^{-1}|\nu\gg 0} \sigma_1(\nu)e^{2\pi i S(\nu z)}\right\}, \\ F &= \mathbf{Q}(\sqrt{5}), \end{aligned} \tag{4.10}$$

$$\begin{aligned} E_{2,0}^{F'}(z) &= \tfrac{\sqrt{41}}{3\cdot 41^2}\left\{1 + 3 \sum_{(\sqrt{41})^{-1}|\nu\gg 0} \sigma_1(\nu)e^{2\pi i S(\nu z)}\right\}, \\ F' &= \mathbf{Q}(\sqrt{41}). \end{aligned} \tag{4.11}$$

By (3.10), we have

$$\begin{aligned} L_K(1,\lambda_{2,2}^{(1)}) &= (2\pi i)^4 E_{2,0}^F(\omega), \\ L_K(1,\lambda_{4,2}^{(1)}) &= (2\pi i)^5\sqrt{\frac{13+\sqrt{5}}{2}}\, i\, E_{2,\{1,0\}}^F(\omega), \\ L_K(1,\lambda_{2,4}^{(1)}) &= (2\pi i)^5\sqrt{\frac{13-\sqrt{5}}{2}}\, i\, E_{2,\{0,1\}}^F(\omega), \end{aligned} \tag{4.12}$$

where

$$\omega = \left(\frac{1}{2}\left(\sqrt{\frac{13+\sqrt{5}}{2}}\, i + \frac{1-\sqrt{5}}{2}\right), \frac{1}{2}\left(\sqrt{\frac{13-\sqrt{5}}{2}}\, i + \frac{1+\sqrt{5}}{2}\right)\right) \in \mathfrak{H}^2.$$

We get

$$\begin{aligned} L_K(1,\lambda_{2,2}^{(1)}) &= 1.13172003415883621168\cdots, \\ L_K(1,\lambda_{4,2}^{(1)}) &= 1.06744092000442833016\cdots, \\ L_K(1,\lambda_{2,4}^{(1)}) &= 1.40377042189535030964\cdots. \end{aligned}$$

Next evaluating Eisenstein series at a CM-point

$$\left(\frac{1}{2}\left(\sqrt{\frac{13+\sqrt{5}}{2}}\, i + \frac{1-\sqrt{5}}{2}\right)\cdot\frac{1+\sqrt{5}}{2}, \right.$$
$$\left. \frac{1}{2}\left(-\sqrt{\frac{13-\sqrt{5}}{2}}\, i + \frac{1+\sqrt{5}}{2}\right)\cdot\frac{1-\sqrt{5}}{2}\right) \in \mathfrak{H}^2,$$

we obtain

[6] We write $E_{k,r}(z)$ for F as $E_{k,r}^F(z)$ to indicate the dependence on F.

$$L_K\big(1,\lambda_{2,2}^{(2)}\big) = 1.05594274348607867398\cdots,$$
$$L_K\big(1,\lambda_{4,2}^{(2)}\big) = 1.48897355341553581717\cdots,$$
$$L_K\big(1,\lambda_{2,4}^{(2)}\big) = 1.36329307657845702302\cdots.$$

In this way, we obtain

$$x(1) = 0.98504294158954033508134 07\cdots,$$

where

$$x(1) = \frac{L_K\big(1,\lambda_{4,2}^{(1)}\big)}{L_K\big(1,\lambda_{2,2}^{(1)}\big)} \cdot \frac{L_{K'}(1,\mu_{2,4}^{(2)})}{L_{K'}(1,\mu_{2,2}^{(2)})} \cdot \frac{1}{\exp(L_F'(0,\chi)/2)},$$

the quantity in (4.8). It would be very hard to identify $x(1)$ with an algebraic number by just looking the numerical value. However, regarding $\exp(L_F'(0,\chi)/2)$ as "absolute period", we let $\mathrm{Gal}(L/\mathbf{Q})$ act formally on Grössencharacters and will make conjugates of $x(1)$. For $\sigma \in \mathrm{Gal}(L/\mathbf{Q})$, CM-types $\{\mathrm{id}, \sigma|K\}$ and $\{\mathrm{id}, \sigma\tau\rho|K'\}$ of K and K' change to

$$\{\sigma|K, \sigma^2|K\} = \{\sigma\rho|K, \mathrm{id}\}^\rho, \quad \{\sigma|K', \sigma\tau\rho\sigma|K'\} = \{\sigma\tau|K', \mathrm{id}\}^\rho.$$

Hence, by the action of σ, $\lambda_{2,2}^{(1)}$, $\lambda_{4,2}^{(1)}$, $\mu_{2,2}^{(2)}$, $\mu_{2,4}^{(2)}$ are transformed to $\overline{\lambda_{2,2}^{(2)}}$, $\overline{\lambda_{2,4}^{(2)}}$, $\overline{\mu_{2,2}^{(1)}}$, $\overline{\mu_{4,2}^{(1)}}$ respectively. Here we note that $\overline{L_K(1,\lambda_{a,b}^{(i)})} = L_K(1,\lambda_{a,b}^{(i)})$ and $\overline{L_K(1,\mu_{a,b}^{(i)})} = L_K(1,\mu_{a,b}^{(i)})$. Put

$$x(2) = \frac{L_K(1,\lambda_{2,4}^{(2)})}{L_K(1,\lambda_{2,2}^{(2)})} \cdot \frac{L_{K'}(1,\mu_{4,2}^{(1)})}{L_{K'}(1,\mu_{2,2}^{(1)})} \cdot \frac{1}{\exp(L_F'(0,\chi)/2)}.$$

We have

$$x(2) = 9.86656529268701848014163 28\cdots.$$

We can regard $x(2)$ as a "conjugate" of $x(1)$ under σ. Similarly for $\tau \in \mathrm{Gal}(L/\mathbf{Q})$, we get

$$x(3) = \frac{L_K\big(1,\lambda_{2,4}^{(1)}\big)}{L_K\big(1,\lambda_{2,2}^{(1)}\big)} \cdot \frac{L_{K'}\big(1,\mu_{2,4}^{(1)}\big)}{L_{K'}\big(1,\mu_{2,2}^{(1)}\big)} \cdot \frac{1}{\exp(L_F'(0,\chi)/2)},$$
$$x(3) = 2.32749724236461357673638 64\cdots.$$

For $\sigma\tau \in \mathrm{Gal}(L/\mathbf{Q})$, we get[7]

$$x(4) = \frac{L_K\big(1,\lambda_{4,2}^{(2)}\big)}{L_K\big(1,\lambda_{2,2}^{(2)}\big)} \cdot \frac{L_{K'}\big(1,\mu_{4,2}^{(2)}\big)}{L_{K'}\big(1,\mu_{2,2}^{(2)}\big)} \cdot \frac{1}{\exp(L_F'(0,\chi)/2)},$$
$$x(4) = 1.61256119002549427470730 81\cdots.$$

Now we compute the polynomial

$$f(T) = \prod_{j=1}^{4} (T - 2^3 \cdot 3^2 \cdot 41\ x(j))$$

[7]By Theorems S1 and S2, we can verify $x(1) \sim x(2) \sim x(3) \sim x(4)$.

using these numerical values of $x(j)$. We obtain

$$f(T) = T^4 - 43665.\overbrace{00\cdots0}^{20}45\cdots T^3 + 489989565.\overbrace{00\cdots0}^{15}12\cdots T^2$$
$$- 2032811110800.\overbrace{00\cdots0}^{12}89\cdots T + 2770077684038400.\overbrace{00\cdots0}^{8}16\cdots.$$

Therefore it is very plausible that

$$f(T) = T^4 - 43665T^3 + 489989565T^2 - 2032811110800T$$
$$+ 2770077684038400.$$

The roots of this polynomial are

$$\frac{43665}{4} + \frac{12669}{4}\sqrt{5} + \frac{3765}{4}\sqrt{41} + \frac{1425}{4}\sqrt{205}$$

and its conjugates. Comparing with numerical values, we obtain identifications:

$$2^3 \cdot 3^2 \cdot 41\, x(1) = \frac{43665}{4} - \frac{12669}{4}\sqrt{5} - \frac{3765}{4}\sqrt{41} + \frac{1425}{4}\sqrt{205},$$
$$2^3 \cdot 3^2 \cdot 41\, x(2) = \frac{43665}{4} + \frac{12669}{4}\sqrt{5} + \frac{3765}{4}\sqrt{41} + \frac{1425}{4}\sqrt{205},$$
$$2^3 \cdot 3^2 \cdot 41\, x(3) = \frac{43665}{4} + \frac{12669}{4}\sqrt{5} - \frac{3765}{4}\sqrt{41} - \frac{1425}{4}\sqrt{205},$$
$$2^3 \cdot 3^2 \cdot 41\, x(4) = \frac{43665}{4} - \frac{12669}{4}\sqrt{5} + \frac{3765}{4}\sqrt{41} - \frac{1425}{4}\sqrt{205}.$$

The numerical values of $x(j)$ coincide with these algebraic values up to the 23-rd decimal place. By (4.1), we also have the compatibility with

$$x(1)^\sigma = x(2), \quad x(1)^\tau = x(3), \quad x(1)^{\sigma\tau} = x(4).$$

Theorem S2 tells that

$$A = \frac{\pi^2 L_K(1, \lambda^{(1)}_{4,2}) L_K(1, \lambda^{(1)}_{2,4})}{L_K(1, \lambda^{(1)}_{2,2})^3}, \qquad B = \frac{\pi^2 L_K(1, \lambda^{(2)}_{4,2}) L_K(1, \lambda^{(2)}_{2,4})}{L_K(1, \lambda^{(2)}_{2,2})^3},$$
$$A' = \frac{\pi^2 L_{K'}(1, \mu^{(1)}_{4,2}) L_{K'}(1, \mu^{(1)}_{2,4})}{L_{K'}(1, \mu^{(1)}_{2,2})^3}, \qquad B' = \frac{\pi^2 L_{K'}(1, \mu^{(2)}_{4,2}) L_{K'}(1, \mu^{(2)}_{2,4})}{L_{K'}(1, \mu^{(2)}_{2,2})^3}$$

are algebraic. For the sake of completeness, we computed these values numerically and found that

$$A = \frac{\sqrt{5}}{2^6 \cdot 3}\left(\frac{-585 + 365\sqrt{41}}{2}\right), \quad B = \frac{\sqrt{5}}{2^6 \cdot 3}\left(\frac{585 + 365\sqrt{41}}{2}\right),$$
$$A' = \frac{\sqrt{41}}{2^4 \cdot 3}(693 + 334\sqrt{5}), \qquad B' = \frac{\sqrt{41}}{2^4 \cdot 3}(-693 + 334\sqrt{5}).$$

(The coincidence is up to the 22-nd decimal place.)

We carried out the second calculation with higher precision [8] and found that the coincidence of $x(j)$ with algebraic values was improved to the 64-th decimal place.

[8]For these calculations, we employed "UBASIC" created by Y. Kida and a desk computer PC9821-Ap2.

We presented above a roundabout way to study (4.3) numerically, since this example shows intricate relations among various quantities which appear in examples in this section. Now we shall show more direct way. Set

$$Q = \pi^{-4} L_K(1, \lambda_{2,2}^{(1)}) L_K(1, \lambda_{2,2}^{(2)}) \exp(-L'_F(0,\chi)),$$
$$R = \pi^{-4} L_{K'}(1, \mu_{2,2}^{(1)}) L_{K'}(1, \mu_{2,2}^{(2)}) \exp(-L'_F(0,\chi)),$$

the quantities in (4.5) and (4.5′). For Q and R, $\mathrm{Gal}(L/\mathbf{Q})$ "acts trivially". Hence we can guess that Q and R are rational numbers. In fact we find

$$Q = \frac{2}{5^3}, \quad R = \frac{2^4}{41^2}.$$

Numerical coincidence is to the 67-th decimal place.

EXAMPLE 2. We take $F = \mathbf{Q}(\sqrt{17})$, $K = \mathbf{Q}(\sqrt{9+2\sqrt{17}}\ i)$. Then we have $F' = \mathbf{Q}(\sqrt{13})$, $K' = \mathbf{Q}\left(\sqrt{\frac{9+\sqrt{13}}{2}}\ i\right)$. We have $h_F = h_{F'} = 1$. We have

$$\mathfrak{O}_K = \mathfrak{O}_F \cdot \frac{1}{2}\left(\sqrt{9+2\sqrt{17}}\ i + 1\right) \oplus \mathfrak{O}_F$$

and that K is the maximal ray class field of conductor $\mathfrak{p}\infty_1\infty_2$ of F. Here $\mathfrak{p} = (9+2\sqrt{17})$. We also have

$$\mathfrak{O}_{K'} = \mathfrak{O}_{F'} \cdot \frac{1}{2}\left(\sqrt{\frac{9+\sqrt{13}}{2}}\ i + \frac{-1+\sqrt{13}}{2}\right) \oplus \mathfrak{O}_{F'}$$

and that K' is the maximal ray class field of conductor $\mathfrak{p}'\infty_1\infty_2$ of F', where $\mathfrak{p}' = \left(\frac{9+\sqrt{13}}{2}\right)$. We have

$$L_F(0,\chi) = 2, \quad L'_F(0,\chi) = -0.42389018379529710565596698\cdots.$$

Hence $h_K = h_{K'} = 1$. Let Q and R be defined in the completely same manner as in Example 1. We find

$$Q = \frac{2^4}{3 \cdot 17^2}, \quad R = \frac{2}{13^2}.$$

The coincidence is up to the 35-th decimal place.

EXAMPLE 3. We take $F = \mathbf{Q}(\sqrt{5})$, $K = \mathbf{Q}(\sqrt{7+2\sqrt{5}}\ i)$. Then we have $F' = \mathbf{Q}(\sqrt{29})$, $K' = \mathbf{Q}\left(\sqrt{\frac{7+\sqrt{29}}{2}}\ i\right)$. We have $h_F = h_{F'} = 1$. In F, we have $(29) = (7+2\sqrt{5})(7-2\sqrt{5})$. Put $\mathfrak{p} = (7+2\sqrt{5})$ and let M be the maximal ray class field of conductor $(4)\mathfrak{p}\infty_1\infty_2$ of F. We have $[M : F] = 16$ and $K \subseteq M$. We find

$$L_F(0,\chi) = 4, \quad L'_F(0,\chi) = -11.95434037167457903017\cdots.$$

Hence $h_K = h_{K'} = 2$. We have

$$\mathfrak{O}_K = \mathfrak{O}_F \cdot \sqrt{7+2\sqrt{5}}\ i \oplus \mathfrak{O}_F.$$

The prime ideal (2) of F ramifies in K. Put $(2) = \mathfrak{P}_2^2$ in K. We can easily verify that $\mathfrak{P}_2$ is not a principal ideal and that

$$\mathfrak{P}_2 = \mathfrak{O}_F \cdot \left(\sqrt{7+2\sqrt{5}}\ i + 1\right) \oplus 2\mathfrak{O}_F.$$

Similarly (2) ramifies in K'. Let $\mathfrak{P}'_2$ be the prime factor of (2) in K'. Then $\mathfrak{P}'_2$ is not principal.

For every positive even integers a and b, there exist two Grössencharacters $\lambda^{(1)}_{a,b}$, $\lambda^{(2)}_{a,b}$ (resp. $\mu^{(1)}_{a,b}$, $\mu^{(2)}_{a,b}$) of $I(K)$ (resp. $I(K')$) of conductor (1). They are determined by $\lambda^{(i)}_{a,b}(\mathfrak{P}_2) = 1$ or -1 (resp. $\mu^{(i)}_{a,b}(\mathfrak{P}'_2) = 1$ or -1).

First let Q and R be the quantities of (4.5) and (4.5$'$) determined by $\lambda^{(1)}_{2,2}(\mathfrak{P}_2) = \lambda^{(2)}_{2,2}(\mathfrak{P}_2) = 1$, $\mu^{(1)}_{2,2}(\mathfrak{P}'_2) = \mu^{(2)}_{2,2}(\mathfrak{P}'_2) = 1$. Then we find

$$Q = \frac{2^4 \cdot 13}{5^2}, \quad R = \frac{2^4 \cdot 3^2 \cdot 5 \cdot 11}{29^2}.$$

Similarly define Q and R by choosing $\lambda^{(1)}_{2,2}(\mathfrak{P}_2) = \lambda^{(2)}_{2,2}(\mathfrak{P}_2) = -1$, $\mu^{(1)}_{2,2}(\mathfrak{P}'_2) = \mu^{(2)}_{2,2}(\mathfrak{P}'_2) = -1$. Then we find

$$Q = \frac{2^4}{5}, \quad R = \frac{2^4 \cdot 11}{29}.$$

The numerical coincidence is up to the 30-th decimal place.[9]

To clarify the relation with the Stark-Shintani conjecture, let us examine this example more closely. Let C denote the ideal class group of conductor $(4)\mathfrak{p}\infty_1\infty_2$. We have $C \cong \mathbf{Z}/4\mathbf{Z} \oplus \mathbf{Z}/2\mathbf{Z} \oplus \mathbf{Z}/2\mathbf{Z} \cong \mathrm{Gal}(M/F)$. Let α, β, γ be the classes of ideals (89), (59), $(106 + \sqrt{5})$ in C respectively. Then $\alpha^4 = \beta^2 = \gamma^2 = 1$ and $C = \langle\alpha\rangle \oplus \langle\beta\rangle \oplus \langle\gamma\rangle$. Let D be the character group of C. Define ω, ψ, $\eta \in D$ by

$$\omega(\alpha) = \sqrt{-1}, \quad \omega(\beta) = \omega(\gamma) = 1, \quad \psi(\alpha) = \psi(\gamma) = 1, \quad \psi(\beta) = -1,$$
$$\eta(\alpha) = \eta(\beta) = 1, \quad \eta(\gamma) = -1.$$

For $\mu \in D$, let $\mathfrak{f}(\mu)$ denote the conductor of μ. We have

$$\mathfrak{f}(\omega) = \mathfrak{p}\infty_2, \quad \mathfrak{f}(\psi) = (4)\infty_1, \quad \mathfrak{f}(\eta) = (4)\infty_1\infty_2.$$

The Hecke character χ, regarded as a character of C, is equal to $\omega^2\eta$.

For $c \in C$, let $\zeta_F(s, c) = \sum_{\mathfrak{a}\in c} N(\mathfrak{a})^{-s}$ be the partial zeta function of the class c. Let D_0, D_1 and D_2 be the subgroups of D consisting of all characters whose conductors divide $(4)\mathfrak{p}$, $(4)\mathfrak{p}\infty_1$ and $(4)\mathfrak{p}\infty_2$ respectively. Let C_i be the annihilator of D_i for $i = 0, 1, 2$. We have

$$C_0 = \{1, \beta\gamma, \alpha^2\beta, \alpha^2\gamma\}, \quad C_1 = \{1, \alpha^2\gamma\}, \quad C_2 = \{1, \beta\gamma\}.$$

From the functional equation, we see easily that

$$\sum_{c\in C} \mu(c)\zeta_F(0, c) = \sum_{c\in C} \mu(c)\zeta'_F(0, c) = 0$$

for every $\mu \in D_0$. Hence we obtain

$$\sum_{c_0\in C_0} \zeta'_F(0, cc_0) = 0 \quad \text{for every } c \in C.$$

[9]In general, Q and R are not rational numbers. For example, let $K = \mathbf{Q}(\sqrt{3+\sqrt{2}}\, i)$ and $\mathfrak{P}_2$ be the unique prime divisor of (2) in K. We have $h_K = 2$. We find $Q = \frac{3}{2^2}$, $Q = \frac{\sqrt{2}}{2^2\cdot 3}$, $Q = \frac{\sqrt{13-\sqrt{7}}}{2^2\cdot 3}$, $Q = \frac{\sqrt{13+\sqrt{7}}}{2^2\cdot 3}$ according as the cases $\epsilon_1 = \epsilon_2 = 1$, $\epsilon_1 = \epsilon_2 = -1$, $\epsilon_1 = -\epsilon_2 = 1$, $-\epsilon_1 = \epsilon_2 = 1$ where $\epsilon_1 = \lambda^{(1)}_{2,2}(\mathfrak{P}_2)$, $\epsilon_2 = \lambda^{(2)}_{2,2}(\mathfrak{P}_2)$.

If we admit the Stark-Shintani conjecture, $\exp(2(\zeta_F'(0,c)+\zeta_F'(0,c\beta\gamma)))$, $c\in C$ is a unit of the maximal ray class field of conductor $(4)\mathfrak{p}\infty_2$. In fact, following the procedure described in [**St**, III], our numerical computation suggests overwhelmingly that $\exp(2(\zeta_F'(0,c)+\zeta_F'(0,c\beta\gamma)))+\exp(-2(\zeta_F'(0,c)+\zeta_F'(0,c\beta\gamma)))$ is a root of the polynomial

$$X^4-(31+13\sqrt{5})X^3+\frac{1143+513\sqrt{5}}{2}X^2 \\ -(3711+1661\sqrt{5})X+(7649+3420\sqrt{5})$$

for every $c\in C$. Similarly it is very plausible that $\exp(2(\zeta_F'(0,c)+\zeta_F'(0,c\alpha^2\gamma)))$ $+\exp(-2(\zeta_F'(0,c)+\zeta_F'(0,c\alpha^2\gamma)))$ is a root of the polynomial

$$X^4-(476+212\sqrt{5})X^3+(18584+8312\sqrt{5})X^2 \\ -(168576+75392\sqrt{5})X+(441344+197376\sqrt{5})$$

for every $c\in C$.

Put $z(c)=\exp(\zeta_F'(0,c))$, $c\in C$. Admitting the Stark-Shintani conjecture, we have $\prod_{c_i\in C_i} z(cc_i)\sim 1$ for every $i=0,\ 1,\ 2$ and every $c\in C$. Hence we see easily that every $z(c)$ can be written as a monomial of $z(1)$, $z(\alpha)$, $z(\alpha^2)$ and $z(\alpha^3)$ up to algebraic numbers. We have $F(i)\subseteq M$ and Conjecture 2.4 holds for $F(i)$ by Proposition 2.6. Let M_0 be the maximal CM-subfield of M. We have $\mathrm{Gal}(M/M_0)=\{1,\alpha^2\beta\}$, $[M_0:F]=8$. From abelian L-functions attached to the characters η, $\omega^2\eta$, $\omega\psi$ and $\omega^3\psi$ for M_0/F, we can obtain four Artin L-functions for $\widetilde{M_0}/\mathbf{Q}$ with odd representations, where $\widetilde{M_0}$ denotes the normal closure of M_0 over $\mathbf{Q}$. If we trust Conjecture 2.2, all of $z(1)$, $z(\alpha)$, $z(\alpha^2)$ and $z(\alpha^3)$ can be expressed by CM-periods.

REMARK. This example is typical in the sense that we can always predict the transcendental nature of $L'(0,\psi)$ where $L(s,\psi)$ is Artin's L-function. In fact, let K be a finite Galois extension of $\mathbf{Q}$ and ψ be an irreducible representation of $\mathrm{Gal}(K/\mathbf{Q})$. We may assume that ψ is non-trivial and faithful. The gamma factor of $L(s,\psi)$ is of the form $\Gamma(s/2)^a\Gamma((s+1)/2)^b$, $a+b=\dim\psi$. If $a\geq 2$, then $L'(0,\psi)=0$. If $a=1$, then $L(0,\psi)=0$ and $L'(0,\psi)\neq 0$. The nature of $L'(0,\psi)$ is given by Stark's conjecture. If $a=0$, then $\psi(\rho)=-\mathrm{id}$. Since ψ is faithful, ρ belongs to the center of $\mathrm{Gal}(K/\mathbf{Q})$. Therefore K is a CM-field and essential information on the nature of $L'(0,\psi)$ is provided by Conjecture 2.2.

REMARK. Let F be a totally real algebraic number field and L be an abelian extension of F which is a CM-field. Recently the author formulated a conjecture which predicts $p_L(\mathrm{id},\alpha)$ for $\alpha\in\mathrm{Gal}(L/F)$ in terms of the multiple gamma function and obtained several convincing numerical evidence when $[F:\mathbf{Q}]=2$. Thus, for K and σ as in the beginning of this section, we can express $p_K(\mathrm{id},\mathrm{id})$ *and* $p_K(\sigma,\sigma)$ in terms of the double gamma function not only the product of them. The details will be discussed elsewhere.

5. A refined conjecture on absolute CM-periods

The numerical examples presented in §4 suggest the possibility to make conjectures in §2 in more precise forms, i.e., to formulate them in covariant forms under the action of $\mathrm{Aut}(\mathbf{C})$. In this section, we shall discuss this problem in certain simple cases.

Let K be a CM-field and F be its maximal real subfield. Let χ be the Hecke character of $F_A^\times$ which corresponds to the quadratic extension K/F. We put

$$P = \exp\left(\frac{L'_F(0,\chi)}{L_F(0,\chi)}\right).$$

Let $\mathfrak{f}$ be an integral ideal of K and let $\lambda_1, \lambda_2, \cdots, \lambda_q$ be Grössencharacters of $I_{\mathfrak{f}}(K)$. We assume that there exists a CM-type Φ_i and non-negative integers $t_\sigma^{(i)}$, $\sigma \in \Phi_i$ for every i such that

$$\lambda_i((\alpha)) = \prod_{\sigma\in\Phi_i} (\alpha^{\sigma\rho}/|\alpha^\sigma|)^{t_\sigma^{(i)}} \quad \text{if} \quad \alpha \equiv 1 \mod {}^\times\mathfrak{f}.$$

Assume that there exists an integer m which satisfies

$$m - t_\sigma^{(i)} \in 2\mathbf{Z} \quad \text{and} \quad -t_\sigma^{(i)} < m \leq t_\sigma^{(i)}$$

for every i and every $\sigma \in \Phi_i$. By Theorem S2, we have

$$L(m/2, \lambda_i) \sim \pi^{e_i/2} p_K\left(\sum_{\sigma\in\Phi_i} t_\sigma^{(i)}\cdot\sigma, \Phi_i\right), \qquad e_i = m[F:\mathbf{Q}] + \sum_{\sigma\in\Phi_i} t_\sigma^{(i)}.$$

Let L be the normal closure of K over $\mathbf{Q}$. For every i, we take $\Phi_i^0 \in I_L$ so that $\mathrm{Res}_{L/K}(\Phi_i^0) = \Phi_i$ and put $\eta_i = \mathrm{Inf}_{L/K}(\sum_{\sigma\in\Phi_i} t_\sigma^{(i)}\cdot\sigma)$. By Theorem S1, 4, 6, we have

$$p_K\left(\sum_{\sigma\in\Phi_i} t_\sigma^{(i)}\cdot\sigma, \Phi_i\right) \sim p_L\left(\eta_i, \Phi_i^0\right) \sim p_L\left(\mathrm{id}, \eta_i^{-1}\Phi_i^0\right),$$

where $\eta_i^{-1}\Phi_i^0 = \sum_{\gamma\in\eta_i,\,\delta\in\Phi_i^0}\gamma^{-1}\delta$. Take a CM-type S of L. We can write

$$\eta_i^{-1}\Phi_i^0 = \sum_{\sigma\in S} l_\sigma^{(i)}\cdot\sigma + m_\sigma^{(i)}\cdot\sigma\rho.$$

By Theorem S1, 3, we have

$$\prod_{i=1}^q p_K\left(\sum_{\sigma\in\Phi_i} t_\sigma^{(i)}\cdot\sigma, \Phi_i\right) \sim p_L\left(\mathrm{id}, \sum_{\sigma\in S} n_\sigma\cdot\sigma\right), \tag{5.1}$$

$$n_\sigma = \sum_{i=1}^q \left(l_\sigma^{(i)} - m_\sigma^{(i)}\right).$$

Regard χ as the non-trivial character of $\mathrm{Gal}(K/F)$ and lift χ to the character $\widetilde{\chi}$ of $\mathrm{Gal}(L/F)$. Put $\psi = \mathrm{Ind}(\widetilde{\chi}; \mathrm{Gal}(L/F) \longrightarrow \mathrm{Gal}(L/\mathbf{Q}))$. By Theorem S1, 3 and (2.7), Conjecture 2.4 is equivalent to

$$P \sim \pi^{[K:\mathbf{Q}]/2} \prod_{\sigma\in S} p_L(\mathrm{id}, 2\chi_\psi(\sigma)\cdot\sigma). \tag{5.2}$$

We assume that there exists $e \in 2^{-1}\mathbf{Z}$ such that

$$\sum_{\sigma\in S} n_\sigma\cdot\sigma = e\sum_{\sigma\in S} 2\chi_\psi(\sigma)\cdot\sigma. \tag{5.3}$$

Put

$$A = \sum_{i=1}^q \frac{e_i}{2} - e\cdot\frac{[K:\mathbf{Q}]}{2}. \tag{5.4}$$

By Theorem S2, Conjecture 2.4 is equivalent to

$$\prod_{i=1}^{q} L(m/2,\lambda_i) \sim \pi^A P^e. \tag{5.5}$$

Now we can state a precise version of Conjecture 2.4.

CONJECTURE 5.1. *Let $h_{\mathfrak{f}}$ be the order of the ideal class group modulo $\mathfrak{f}$ of K. For every $\sigma \in \mathrm{Aut}(\mathbf{C})$, we have*

$$\left(\frac{\prod_{i=1}^{q} L(m/2,\lambda_i)}{\pi^A P^e}\right)^{\sigma} = \zeta \cdot \frac{\prod_{i=1}^{q} L(m/2,\lambda_i^{\sigma})}{\pi^A P^e}$$

with a root of unity ζ. Furthermore we have $\zeta^{wh_{\mathfrak{f}}} = 1$, where w is a certain positive integer depending only on F.

6. The case of imaginary quadratic fields

Let K be an imaginary quadratic field with discriminant $-d$. Let h be the class number of K and w be the number of roots of unity contained in K. Let χ be the Dirichlet character which corresponds to the quadratic extension $K/\mathbf{Q}$. Using the well known formulas for the Hurwitz zeta function (cf. [**WW**, p. 271]), we have

$$P = \exp\left(\frac{L'(0,\chi)}{L(0,\chi)}\right) = \frac{1}{d}\prod_{a=1}^{d-1}\Gamma\left(\frac{a}{d}\right)^{w\chi(a)/2h}. \tag{6.0.1}$$

Let λ be a Grössencharacter of K whose conductor divides $\mathfrak{f}$ and satisfies

$$\lambda((\alpha)) = \left(\frac{\alpha^{\rho}}{|\alpha|}\right)^{k}, \quad \alpha \equiv 1 \mod {}^{\times}\mathfrak{f}.$$

For an integer m such that $m-k \in 2\mathbf{Z}$ and $-k < m \le k$, Conjecture 5.1 states

$$\left(\frac{L(m/2,\lambda)}{(2\pi)^{m/2}P^{k/2}}\right)^{\sigma} = \zeta \cdot \frac{L(m/2,\lambda^{\sigma})}{(2\pi)^{m/2}P^{k/2}} \tag{6.0.2}$$

for every $\sigma \in \mathrm{Aut}(\mathbf{C})$, where ζ is a certain root of unity. We divide the determination of ζ into five steps.

First step: An expression of $(2\pi)^{-m/2}P^{-k/2}L(m/2,\lambda)$.

We set

$$E = E_K, \quad E_{\mathfrak{f}} = \{\epsilon \in E \mid \epsilon \equiv 1 \mod \mathfrak{f}\}, \quad \mathfrak{f}\cap\mathbf{Z} = f\mathbf{Z},\ f>0.$$

Let B be a complete set of representatives of the ideal classes modulo $\mathfrak{f}$ of K. We have

$$L(s,\lambda) = \frac{1}{|E_{\mathfrak{f}}|}\sum_{\mathfrak{b}\in B}\lambda(\mathfrak{b})^{-1}N(\mathfrak{b})^{s}\sum_{\alpha\in\mathfrak{b},\alpha\equiv 1 \mod {}^{\times}\mathfrak{f}}\lambda((\alpha))N((\alpha))^{-s}. \tag{6.1.1}$$

We assume that all ideals in B are relatively prime to (f). Take $\mathfrak{b} \in B$. We can write

$$\mathfrak{b} = \mathbf{Z}\omega_{1,\mathfrak{b}} \oplus \mathbf{Z}\omega_{2,\mathfrak{b}}, \quad \Im(\omega_{1,\mathfrak{b}}/\omega_{2,\mathfrak{b}}) > 0.$$

We temporarily put $\omega_1 = \omega_{1,\mathfrak{b}}$, $\omega_2 = \omega_{2,\mathfrak{b}}$. There exists a finite subset $R_{\mathfrak{b}}$ of $\mathbf{Z}\oplus\mathbf{Z}$ such that

$$\begin{aligned}&\{\alpha \in \mathfrak{b} \mid \alpha \equiv 1 \mod {}^{\times}\mathfrak{f}\}\\ &= \cup_{(c_0,d_0)\in R_{\mathfrak{b}}}\{c\omega_1 + d\omega_2 \mid c,d\in\mathbf{Z}, c\equiv c_0, d\equiv d_0 \mod f\} \quad \text{(disjoint union)}.\end{aligned}$$

We obtain

$$(6.1.2)\qquad \sum_{\alpha\in\mathfrak{b},\alpha\equiv 1 \mod {}^{\times}\mathfrak{f}} \lambda((\alpha))N((\alpha))^{-s}$$
$$= \sum_{(c_0,d_0)\in R_{\mathfrak{b}}} \sum_{c\equiv c_0, d\equiv d_0 \mod f} (c\omega_1+d\omega_2)^{-k}|c\omega_1+d\omega_2|^{-2(s-k/2)}.$$

We assume $m \geq 1$ and let $k = m+2r$ with a non-negative integer r. We introduce an Eisenstein series

$$E_{m,r}(z,s;c_0,d_0;f) = (z-\overline{z})^{-r}(2\pi i)^{-k-r}$$
$$\sum_{c\equiv c_0,d\equiv d_0 \bmod f}{}' \left(\frac{c\overline{z}+d}{cz+d}\right)^r (cz+d)^{-k}|cz+d|^{-s}.$$

Here $\sum'$ means that the term $(c,d) = (0,0)$ is omitted. This series converges absolutely uniformly for $\Re(s)$ is sufficiently large, can be continued meromorphically to the whole s-plane and holomorphic at $s=0$. We put

$$E_{m,r}(z;c_0,d_0;f) = E_{m,r}(z,0;c_0,d_0;f).$$

By (6.1.2) and by analytic continuation, we obtain

$$(6.1.3)\qquad L(m/2,\lambda) = \frac{(2\pi i)^{m+r}}{|E_{\mathfrak{f}}|}\sum_{\mathfrak{b}\in B}\lambda(\mathfrak{b})^{-1}N(\mathfrak{b})^{m/2}\omega_{2,\mathfrak{b}}^{-k}N(\omega_{2,\mathfrak{b}})^r$$
$$\cdot\{2\Im(\omega_{1,\mathfrak{b}}/\omega_{2,\mathfrak{b}})i\}^r \sum_{(c_0,d_0)\in R_{\mathfrak{b}}} E_{m,r}(\omega_{1,\mathfrak{b}}/\omega_{2,\mathfrak{b}};c_0,d_0;f).$$

Let $\mathfrak{a}_1$, $\mathfrak{a}_2$, $\cdots$, $\mathfrak{a}_h$ be a complete set of representatives of the ideal classes of K. Let F be a holomorphic modular form of weight l with respect to $\mathrm{SL}(2,\mathbf{Z})$. For a lattice L in $\mathbf{C}$, we choose ω_1, ω_2 so that $L = \mathbf{Z}\omega_1 \oplus \mathbf{Z}\omega_2$, $\Im(\omega_1/\omega_2) > 0$ and put $F(L) = F(\omega_1/\omega_2)\omega_2^{-l}$. We see easily that this is well defined and that $F(\alpha L) = \alpha^{-l}F(L)$, $\alpha\in\mathbf{C}^\times$. Let

$$\eta(z) = e^{2\pi iz/24}\prod_{n=1}^{\infty}(1-e^{2\pi inz})$$

be the Dedekind η-function. Then the exact Chowla-Selberg formula gives (cf. Weil [**W**, p. 92])

$$(6.1.4)\qquad \prod_{i=1}^{h} N(\mathfrak{a}_i)^{12}|\eta^{24}(\mathfrak{a}_i)|^2 = (2\pi)^{-12h}P^{12h}.$$

Let $\mathfrak{O}_K = \mathbf{Z}\omega\oplus\mathbf{Z}$, $\Im(\omega)>0$. Put $\mathfrak{a}_i = \mathbf{Z}\omega_{1,i}\oplus\mathbf{Z}\omega_{2,i}$, $\Im(\omega_{1,i}/\omega_{2,i})>0$. Define $\alpha_i\in G_{\mathbf{Q}+}$ by $\binom{\omega_{1,i}}{\omega_{2,i}} = \alpha_i\binom{\omega}{1}$. Let $\lambda_0(\mathfrak{b}) = \lambda(\mathfrak{b})N(\mathfrak{b})^{k/2}$ as in §3 and put

$$Q = 2^r i^{m+r}|E_{\mathfrak{f}}|^{-1}\eta^{-2k}(\omega)\sum_{\mathfrak{b}\in B}\lambda_0(\mathfrak{b})^{-1}N(\mathfrak{b})^{m+r}\omega_{2,\mathfrak{b}}^{-k}\{\Im(\omega_{1,\mathfrak{b}}\overline{\omega_{2,\mathfrak{b}}})i\}^r$$
$$\cdot\sum_{(c_0,d_0)\in R_{\mathfrak{b}}} E_{m,r}(\omega_{1,\mathfrak{b}}/\omega_{2,\mathfrak{b}};c_0,d_0;f).$$

Then, by (6.1.3) and (6.1.4), we have

$$\text{(6.1.5)} \qquad \frac{L(m/2,\lambda)}{(2\pi)^{m/2}P^{k/2}} = Q\,\eta^{2k}(\omega)\prod_{i=1}^{h} N(\mathfrak{a}_i)^{-k/2h}|\omega_{2,i}|^{k/h}|\eta(\omega_{1,i}/\omega_{2,i})|^{-2k/h}.$$

Second step: Application of Shimura's reciprocity law

We are going to determine the action of $\mathrm{Aut}(\mathbf{C})$ on the right hand side of (6.1.5) using Shimura's reciprocity law. Let $G = \mathrm{GL}(2)$ considered as an algebraic group over $\mathbf{Q}$. Let G_A denote the adelization of G. We set

$$\begin{aligned} \mathrm{GL}_+(2,\mathbf{R}) &= \{g \in \mathrm{GL}(2,\mathbf{R}) \mid \det(g) > 0\}, \\ G_{\infty+} &= \{g \in G_\infty \mid \det(g) > 0\} \cong \mathrm{GL}_+(2,\mathbf{R}), \\ G_{A+} &= \{g \in G_A \mid g_\infty \in G_{\infty+}\}, \\ U &= \prod_p \mathrm{GL}(2,\mathbf{Z}_p) \times G_{\infty+}. \end{aligned}$$

For a positive integer N, we set

$$U_N = \left\{u \in U \mid u_p \equiv \begin{pmatrix} 1 & 0 \\ 0 & 1 \end{pmatrix} \mod NM(2,\mathbf{Z}_p) \text{ for every } p \text{ which divides } N\right\},$$

$$S_N = \left\{u \in U \mid u_p \equiv \begin{pmatrix} 1 & 0 \\ 0 & a_p \end{pmatrix} \mod NM(2,\mathbf{Z}_p) \text{ with } a_p \in \mathbf{Z}_p^\times \text{ for every } p \text{ which divides } N\right\}.$$

For $\gamma = \left(\begin{smallmatrix} a & b \\ c & d \end{smallmatrix}\right) \in \mathrm{GL}_+(2,\mathbf{R})$ and $z \in \mathfrak{H}$, we put $j(\gamma,z) = cz+d$. For a function h on $\mathfrak{H}$, $\gamma \in \mathrm{GL}_+(2,\mathbf{R})$ and an integer t, we define a function $h \mid_t \gamma$ on $\mathfrak{H}$ by $(h \mid_t \gamma)(z) = h(\gamma(z))j(\gamma,z)^{-t}$, $z \in \mathfrak{H}$. We put $h||_t\gamma = \det(\gamma)^{t/2}h|_t\gamma$. We define a subgroup $\mathfrak{G}$ of $G_{A+} \times \mathrm{Gal}(\overline{\mathbf{Q}}/\mathbf{Q})$ by

$$\mathfrak{G} = \{(x,\sigma) \in G_{A+} \times \mathrm{Gal}(\overline{\mathbf{Q}}/\mathbf{Q}) \mid [\det(x)^{-1},\mathbf{Q}] = \sigma|\mathbf{Q}_{ab}\}.$$

For a subfield Λ of $\mathbf{C}$, let $\mathcal{M}_t(\Lambda)$ denote the space of all holomorphic modular forms of weight t with respect to some $\Gamma(N)$ and whose Fourier coefficients belong to Λ. For $t \in \mathbf{Z}$, let $\mathcal{A}_t(\Lambda)$ be the space of all meromorphic functions of the form f/g with $f \in \mathcal{M}_{t+u}(\Lambda)$, $g \in \mathcal{M}_u(\Lambda)$.

In [**Sh75, Sh78b**], Shimura defined the action of $\mathfrak{G}$ on $\mathcal{A}_t(\overline{\mathbf{Q}})$ so that the action of (x,σ) on $\mathcal{A}_0(\mathbf{Q}_{ab})$ coincides with the action of x in the frame work of canonical models. We adopt the normalization of [**Sh78b**]. This action is defined as follows. For $f(z) = \sum_{n=n_0}^\infty a_n e^{2\pi i n z/N} \in \mathcal{A}_t(\overline{\mathbf{Q}})$ and $\sigma \in \mathrm{Aut}(\mathbf{C})$, we set

$$f^\sigma(z) = \sum_{n=n_0}^\infty a_n^\sigma e^{2\pi i n z/N}.$$

Take $(x,\sigma) \in \mathfrak{G}$ and $f \in \mathcal{A}_t(\overline{\mathbf{Q}})$. Then $f^\sigma \in \mathcal{A}_t(\Gamma(N))$ for some N. We can write $x = s\alpha$ with $s \in S_N$, $\alpha \in G_{\mathbf{Q}+}$. Then $f^{(x,\sigma)}$ is defined to be $f^\sigma|_t\alpha$ (cf. [**Sh78b**, p. 643]). For $t \in \mathbf{C}$, we define a differential operator D_t by $D_t = (1/2\pi i)\,(t/2iy + \partial/\partial z)$ which acts on functions of $z = x+iy$. We have

$$D_t(f|_t\,\alpha) = (\det\alpha)(D_t f)|_{t+2}\,\alpha, \quad D_t(f||_t\,\alpha) = (D_t f)||_{t+2}\,\alpha,$$

for $\alpha \in \mathrm{GL}_+(2, \mathbf{R})$, $t \in \mathbf{Z}$.

Let $\mathcal{N}_t$ be the space of nearly holomorphic forms of weight t (cf. Shimura [**Sh87**]). $g \in \mathcal{N}_t$ satisfies, for some N, $g|_t\gamma = g$ for all $\gamma \in \Gamma(N)$. Let $\mathcal{N}_t(\Gamma(N))$ denote the subspace of all such g. We have the Fourier expansion

$$g = \sum_{0 \le p \le t/2} c_p(\pi y)^{-p} \sum_{n=0}^{\infty} a_n^{(p)} e^{2\pi i n z/N}.$$

For $\sigma \in \mathrm{Aut}(\mathbf{C})$, we put

$$g^\sigma = \sum_{0 \le p \le t/2} c_p^\sigma(\pi y)^{-p} \sum_{n=0}^{\infty} a_n^{(p)\sigma} e^{2\pi i n z/N}.$$

Then we have $D_t(g^\sigma) = (D_t g)^\sigma$. Let $\mathcal{N}_t(\overline{\mathbf{Q}})$ be the space of all arithmetic nearly holomorphic forms of weight t; $g \in \mathcal{N}_t(\overline{\mathbf{Q}})$ is equivalent to that all c_p, $a_n^{(p)}$ are algebraic [**Sh87**, Proposition 3.9]. For $0 \le p \in \mathbf{Z}$, we put $D^p = D_{t+2p-2} \cdots D_{t+2} D_t$ (D^p depends on t but this will be clear from the context.) By Theorems 5.2 and 5.4 of [**Sh87**], we have the unique expression

$$g = \sum_{0 \le p \le t/2} D^p h_p + c D^{t/2-1} E_2 \tag{6.2.1}$$

where $h_p \in \mathcal{M}_{t-2p}(\overline{\mathbf{Q}})$, $c \in \overline{\mathbf{Q}}$, $E_2(z) = E_{2,0}(z; 0, 0, 1)$. The last term can appear only when t is even. We define the associative action of $(x, \sigma) \in \mathfrak{G}$ in the same way as above, i.e., $g^{(x,\sigma)} = g^\sigma|_t\alpha$. [10]

We take $w \in K$ so that $\Im(w) > 0$ and set

$$q(a)\begin{pmatrix} w \\ 1 \end{pmatrix} = \begin{pmatrix} aw \\ a \end{pmatrix}, \quad a \in K,\ q(a) \in M(2, \mathbf{Q}).$$

We have $\det(q(a)) = N_{K/\mathbf{Q}}(a)$; q defines a homomorphism of $K^\times$ into $G_{\mathbf{Q}+}$. We can extend q to a homomorphism of $K_A^\times$ into G_{A+} and denote it by the same letter q. For $s \in K_A^\times$, choose $\sigma \in \mathrm{Gal}(\overline{\mathbf{Q}}/K)$ so that $\sigma|K_{ab} = [s, K]$ and put $\mathfrak{x} = (q(s)^{-1}, \sigma) \in \mathfrak{G}$.

THEOREM S3 [**Sh75**, Main Theorem III] *Let $g_1 \in \mathcal{N}_t(\overline{\mathbf{Q}})$, $g_2 \in \mathcal{A}_{t+2r}(\overline{\mathbf{Q}})$, where $0 \le r \in \mathbf{Z}$. Assume that g_2 is finite and non-zero at w. Then we have*

$$\left((g_2^{-1} D^r g_1)(w)\right)^\sigma = \left((g_2^{\mathfrak{x}})^{-1} (D^r g_1)^{\mathfrak{x}}\right)(w).$$

Since the action $||_t\alpha$ is used in [**Sh75**] instead of $|_t\alpha$ in [**Sh78b**], some explanation is called for. If $g_1 \in \mathcal{A}_t(\overline{\mathbf{Q}})$ and finite at w, we see immediately that the statement follows from Main Theorem III of [**Sh75**]. The difference mentioned above brings $(D^r g_1)^{\mathfrak{x}}$ in place of $D^r(g_1^{\mathfrak{x}})$ in [**Sh75**]. Assume $g_1 \in \mathcal{N}_t(\overline{\mathbf{Q}})$. By (6.2.1), we can reduce to the case where $g_1 = E_2$. Then by the formula in [**Sh75**, p. 496], which expresses $D^r E_2$ in terms of the derivatives of η^{24}, we see that the statement still holds.

Let $g \in \mathcal{N}_t(\overline{\mathbf{Q}}) \cap \mathcal{N}_t(\Gamma(M))$. Let C_M denote the maximal ray class field of conductor (M) of K. Assume $s \in K_A^\times$ satisfies $q(s)_p = 1$ for every p which divides

[10] The associativity is non-trivial but it follows from (6.2.1) without difficulty.

M and let $\mathfrak{x}$ be as above. We are going to write $g^{\mathfrak{x}}$ explicitly when $w = \omega$. Put $\mathfrak{X} = s\mathfrak{O}_K$. Then we have

$$[s,K]|C_M = \left(\frac{C_M/K}{\mathfrak{X}}\right).$$

Let

$$\mathfrak{X}^{-1} = \mathbf{Z}\omega_1 \oplus \mathbf{Z}\omega_2, \quad \Im(\omega_1/\omega_2) > 0, \quad \begin{pmatrix}\omega\\1\end{pmatrix} = \xi\begin{pmatrix}\omega_1\\\omega_2\end{pmatrix}, \quad \xi \in G_{\mathbf{Q}+}.$$

Regarding $\mathbf{Z}^2$ as consisting of row vectors, we have

$$\mathbf{Z}^2 q(s)^{-1}\begin{pmatrix}\omega\\1\end{pmatrix} = \mathfrak{X}^{-1}, \quad \mathbf{Z}^2\xi^{-1}\begin{pmatrix}\omega\\1\end{pmatrix} = \mathfrak{X}^{-1}.$$

Put $t = q(s)^{-1}\xi$. Since $\mathbf{Z}^2 t = \mathbf{Z}^2$, we have $t \in U$. For $u \in \mathbf{Q}_A^\times$, put $m(u) = \left(\begin{smallmatrix}1&0\\0&u\end{smallmatrix}\right) \in G_A$. Since $m(\det t)^{-1}t \in U \cap \mathrm{SL}(2,\mathbf{Q}_A)$, we can find a $\gamma \in \mathrm{SL}(2,\mathbf{Z})$ so that $m(\det t)^{-1}t\gamma^{-1} \in U_M$. We have

$$g^{\mathfrak{x}} = g^{(m(\det t),\sigma)(m(\det t)^{-1}t\gamma^{-1},1)(\gamma\xi^{-1},1)}.$$

Note that $g^\sigma \in \mathcal{N}_t(\overline{\mathbf{Q}}) \cap \mathcal{N}_t(\Gamma(M))$. Since $m(\det t)^{-1}t\gamma^{-1} \in U_M$, we obtain, by the definition of the action of $\mathfrak{G}$, a concrete expression

$$g^{\mathfrak{x}} = (g^\sigma)^{(m(\det t)^{-1}t\gamma^{-1},1)(\gamma\xi^{-1},1)} = (g^\sigma)^{(\gamma\xi^{-1},1)} = g^\sigma|_t\gamma\xi^{-1}.$$

PROPOSITION 6.1. *Let $g_1 \in \mathcal{N}_t(\Gamma(N))$, $g_2 \in \mathcal{M}_{t+2r}(\Gamma(N))$ and we assume that the Fourier coefficients of g_1 and g_2 with respect to $e^{2\pi iz/N}$ belong to $\mathbf{Q}(\zeta_N)$. Let $\alpha \in G_{\mathbf{Q}+}$ and put $h = (D^r(g_1|_t\,\alpha))/g_2$. Choose a positive integer M so that $U_M \subseteq \alpha^{-1}U_N\alpha \cap U_N$. For $\sigma \in \mathrm{Gal}(\overline{\mathbf{Q}}/K)$, take $s \in K_A^\times$ so that $\sigma|K_{ab} = [s,K]$, $q(s)_p = 1$ for every p which divide M and define γ and ξ as above. Assume that g_2 is finite and non-zero at ω. Then we have*

$$h(\omega)^\sigma = (D^r((g_1|_t\,\alpha)^\sigma)(\gamma\xi^{-1}(\omega))/g_2^\sigma(\gamma\xi^{-1}(\omega)).$$

Furthermore we have $h(\omega) \in C_M$.

PROOF. Let $\mathfrak{x} = (q(s)^{-1},\sigma) \in \mathfrak{G}$. By Theorem S3, we have

$$h(\omega)^\sigma = ((D^r(g_1|_t\,\alpha))^{\mathfrak{x}}/g_2^{\mathfrak{x}})(\omega).$$

As shown above, we have $g_2^{\mathfrak{x}} = g_2^\sigma|_{t+2r}\,\gamma\xi^{-1}$. Since $g_1|_t\,\alpha \in \mathcal{N}_t(\overline{\mathbf{Q}}) \cap \mathcal{N}_t(\Gamma(M))$, we have

$$(D^r(g_1|_t\,\alpha))^{\mathfrak{x}} = (D^r(g_1|_t\,\alpha))^\sigma|_{t+2r}\,\gamma\xi^{-1} = (D^r(g_1|_t\,\alpha)^\sigma)|_{t+2r}\,\gamma\xi^{-1}.$$

Hence we obtain the formula for $h(\omega)^\sigma$ canceling automorphic factors.

Now assume $\sigma|C_M = \mathrm{id}$. We may assume that $q(s) \in U_M$. Since $\alpha q(s)^{-1}\alpha^{-1} \in U_N$, we have

$$(g_1|_t\,\alpha)^\sigma = g_1^{(\alpha,1)(q(s)^{-1},\sigma)} = g_1^{(\alpha q(s)^{-1}\alpha^{-1},\sigma)(\alpha,1)} = g_1^{(\alpha,1)} = g_1|_t\,\alpha.$$

Similarly we obtain $g_2^\sigma = g_2$. Put $\delta_0 = m(\det t)^{-1}t\gamma^{-1} \in U_M$. Then we have

$$\gamma\xi^{-1} = \delta_0^{-1}m(\det t)^{-1}q(s)^{-1} \in G_{\mathbf{Q}+} \cap S_M = \mathrm{SL}(2,\mathbf{Z}) \cap S_M = \Gamma(M).$$

Therefore we obtain $h(\omega)^\sigma = h(\omega)$. This completes the proof. □

We have $D_{m+2r}E_{m,r} = (m+r)E_{m,r+1}$, $0 \le r \in \mathbf{Z}$. We put

$$g(z; c_0, d_0; f) = E_{m,0}(z; c_0, d_0; f),$$
$$g_{\mathfrak{b}}(z; c_0, d_0; f) = (g(\cdot; c_0, d_0; f)|_m \alpha_{\mathfrak{b}})(z),$$
$$\varphi_{\mathfrak{b}}(z) = \sum_{(c_0,d_0)\in R_{\mathfrak{b}}} g_{\mathfrak{b}}(z; c_0, d_0; f),$$

where $\alpha_{\mathfrak{b}} \in G_{\mathbf{Q}+}$ is defined by

$$\begin{pmatrix} \omega_{1,\mathfrak{b}} \\ \omega_{2,\mathfrak{b}} \end{pmatrix} = \alpha_{\mathfrak{b}} \begin{pmatrix} \omega \\ 1 \end{pmatrix}.$$

With these notations, we have

$$Q = c\, i^{m+r}\eta^{-2k}(\omega) \sum_{\mathfrak{b}\in B} \lambda_0(\mathfrak{b})^{-1} N(\mathfrak{b})^m \{\Im(\omega_{1,\mathfrak{b}}\overline{\omega_{2,\mathfrak{b}}})i\}^r (D^r\varphi_{\mathfrak{b}})(\omega), \tag{6.2.2}$$

since $\det\alpha_{\mathfrak{b}} = N(\mathfrak{b})$, where $c = 2^r/m(m+1)\cdots(m+r-1)|E_{\mathfrak{f}}|$. We choose a positive integer M so that

$$U_M \subseteq \alpha_i^{-1}U_{24}\alpha_i \cap U_{24} \quad \text{for every } i,\ 1 \le i \le h, \tag{6.2.3}$$
$$U_M \subseteq \alpha_{\mathfrak{b}}^{-1}U_f\alpha_{\mathfrak{b}} \cap U_f \quad \text{for every } \mathfrak{b} \in B.$$

Take any $\sigma \in \mathrm{Aut}(\mathbf{C})$ so that $\sigma|K = \mathrm{id}$. Take a fractional ideal $\mathfrak{X}$ relatively prime to (M) so that $\sigma|C_M = \left(\frac{C_M/K}{\mathfrak{X}}\right)$. By Hecke's results [**H**], we see that $E_{m,0}(z; c_0, d_0; f)$ has Fourier coefficients in $\mathbf{Q}(\zeta_f)$ and that $E_{m,0}(z; c_0, d_0; f) \in \mathcal{M}_m(\Gamma(f))$ if $m \ne 2$ and $\in \mathcal{N}_2(\Gamma(f))$ if $m = 2$. Furthermore we have

$$E^\sigma_{m,0}(z; c_0, d_0; f) = E_{m,0}(z; c_0, d_1; f), \quad d_1 \equiv N(\mathfrak{X})d_0 \mod f. \tag{6.2.4}$$

We choose $\xi \in G_{\mathbf{Q}+}$ and $\gamma \in \mathrm{SL}(2,\mathbf{Z})$ as explained above. By Proposition 6.1, we have $(D^r g_{\mathfrak{b}})(\omega; c_0, d_0; f)/\eta^{2k}(\omega) \in C_M$,

$$\left(\frac{(D^r g_{\mathfrak{b}})(\omega; c_0, d_0; f)}{\eta^{2k}(\omega)}\right)^\sigma = \frac{(D^r (g_{\mathfrak{b}})^\sigma)(\gamma\xi^{-1}(\omega); c_0, d_0; f)}{\eta^{2k}(\gamma\xi^{-1}(\omega))}. \tag{6.2.5}$$

We can find β_i, $\beta_{\mathfrak{b}} \in G_{\mathbf{Q}+}$ so that

$$\alpha_i m(\det t) = s_i\beta_i, \quad s_i \in S_{24}, \quad 1 \le i \le h,$$
$$\alpha_{\mathfrak{b}} m(\det t) = s_{\mathfrak{b}}\beta_{\mathfrak{b}}, \quad s_{\mathfrak{b}} \in S_f, \quad \mathfrak{b} \in B.$$

We have

$$(g_{\mathfrak{b}})^\sigma = g^{(\alpha_{\mathfrak{b}},1)(m(\det(t),\sigma)} = g^{(\alpha_{\mathfrak{b}} m(\det t),\sigma)} = g^{(s_{\mathfrak{b}}\beta_{\mathfrak{b}},\sigma)}.$$

Since $g^\sigma \in \mathcal{N}_m(\Gamma(f))$, we get $(g_{\mathfrak{b}})^\sigma = g^\sigma|_m \beta_{\mathfrak{b}}$. We also note a similar fact $(\eta^2|_1\alpha_i)^\sigma = \eta^2|_1\beta_i$ for the use in the next step, which holds since $\eta^2 \in S_1(\Gamma(24))$. Now we obtain

$$\begin{aligned}
(D^r(g_{\mathfrak{b}})^\sigma)(\gamma\xi^{-1}(\omega); c_0, d_0, f) &= (D^r(g^\sigma(\cdot; c_0, d_0, f)|_m \beta_{\mathfrak{b}}))(\gamma\xi^{-1}(\omega)) \\
&= (\det\beta_{\mathfrak{b}})^r (D^r g^\sigma(\cdot; c_0, d_0, f)|_k\beta_{\mathfrak{b}})(\gamma\xi^{-1}(\omega)) \\
&= (\det\beta_{\mathfrak{b}})^r (D^r g^\sigma)(\beta_{\mathfrak{b}}\gamma\xi^{-1}(\omega); c_0, d_0; f) j(\beta_{\mathfrak{b}}, \gamma\xi^{-1}(\omega))^{-k}.
\end{aligned}$$

Hence we see that the quantity in (6.2.5) is equal to

$$N(\mathfrak{b})^r \frac{(D^r g^\sigma)(\cdot; c_0, d_0; f)|_k \beta_{\mathfrak{b}}\gamma\xi^{-1})(\omega)}{\eta^{2k}(\gamma\xi^{-1})(\omega))} j(\gamma\xi^{-1}, \omega)^k.$$

Therefore we obtain

(6.2.6) $$Q^\sigma = c\,(i^{m+r})^\sigma \eta^{-2k}(\gamma\xi^{-1}(\omega))j(\gamma\xi^{-1},\omega)^k \cdot \sum_{\mathfrak{b}\in B} \lambda_0^\sigma(\mathfrak{b})^{-1}N(\mathfrak{b})^{m+r}\{\Im(\omega_{1,\mathfrak{b}}\overline{\omega_{2,\mathfrak{b}}})i\}^r \sum_{(c_0,d_0)\in R_\mathfrak{b}} ((D^r g^\sigma)|_k\,\beta_\mathfrak{b}\gamma\xi^{-1})(\omega).$$

LEMMA 6.2. *For $1 \le i \le h$ and for $\mathfrak{b} \in B$,*

$$\mathbf{Z}^2\beta_i\gamma\xi^{-1}\begin{pmatrix}\omega\\1\end{pmatrix} = \mathfrak{a}_i\mathfrak{X}^{-1}, \quad \mathbf{Z}^2\beta_\mathfrak{b}\gamma\xi^{-1}\begin{pmatrix}\omega\\1\end{pmatrix} = \mathfrak{b}\mathfrak{X}^{-1}.$$

PROOF. Put $\delta_0 = m(\det t)^{-1}t\gamma^{-1} \in U_M$. By (6.2.3), we have

$$\begin{aligned}\mathbf{Z}^2\beta_i\gamma\xi^{-1} &= \mathbf{Z}^2 s_i^{-1}\alpha_i m(\det t)\gamma\xi^{-1} = \mathbf{Z}^2\alpha_i t\gamma^{-1}\delta_0^{-1}\gamma\xi^{-1}\\ &= \mathbf{Z}^2(\alpha_i t\gamma^{-1}\delta_0^{-1}\gamma t^{-1}\alpha_i^{-1})\alpha_i t\xi^{-1} = \mathbf{Z}^2\alpha_i t\xi^{-1} = \mathbf{Z}^2\alpha_i q(s)^{-1}.\end{aligned}$$

Put $\mathfrak{a}_i = a_i\mathfrak{O}_K$, $a_i \in K_A^\times$. Then we have $\mathbf{Z}^2\alpha_i = \mathbf{Z}^2q(a_i)$, $\mathbf{Z}^2\alpha_i q(s)^{-1} = \mathbf{Z}^2q(a_is^{-1})$ and $\mathbf{Z}^2q(a_is^{-1})\binom{\omega}{1} = \mathfrak{a}_i\mathfrak{X}^{-1}$. The latter assertion can be proved similarly. This completes the proof. □

We put

$$\mathfrak{b}\mathfrak{X}^{-1} = \mathbf{Z}\omega_1'' \oplus \mathbf{Z}\omega_2'',\ \Im(\omega_1''/\omega_2'') > 0, \quad \begin{pmatrix}\omega_1''\\\omega_2''\end{pmatrix} = \alpha_{\mathfrak{b}\mathfrak{X}^{-1}}\begin{pmatrix}\omega\\1\end{pmatrix}, \quad \alpha_{\mathfrak{b}\mathfrak{X}^{-1}} \in G_{\mathbf{Q}+}.$$

By Lemma 6.2, we have $\beta_\mathfrak{b}\gamma\xi^{-1}\alpha_{\mathfrak{b}\mathfrak{X}^{-1}}^{-1} \in \mathrm{SL}(2,\mathbf{Z})$. Choosing ω_1'' and ω_2'' suitably, we may assume that $\beta_\mathfrak{b}\gamma\xi^{-1}\alpha_{\mathfrak{b}\mathfrak{X}^{-1}}^{-1} \in \Gamma(f)$. Then we have

$$(D^rg^\sigma)|_k\,\beta_\mathfrak{b}\gamma\xi^{-1} = (D^rg^\sigma)|_k\,\alpha_{\mathfrak{b}\mathfrak{X}^{-1}}.$$

We have $\alpha_\mathfrak{b}$, $\alpha_{\mathfrak{b}\mathfrak{X}^{-1}} \in \prod_{p|f}U_p$. Let $\overline{\alpha_\mathfrak{b}}$ and $\overline{\alpha_{\mathfrak{b}\mathfrak{X}^{-1}}}$ denote their canonical images in $\mathrm{GL}(2,\mathbf{Z}/f\mathbf{Z})$. We put

$$R_\mathfrak{O} = \{(c,d) \in \mathbf{Z}^2 \mid c\omega + d \equiv 1 \mod \mathfrak{f}\}.$$

Let $\overline{R_\mathfrak{O}}$, $\overline{R_\mathfrak{b}}$ and $\overline{R_{\mathfrak{b}\mathfrak{X}^{-1}}}$ denote the images of $R_\mathfrak{O}$, $R_\mathfrak{b}$ and $R_{\mathfrak{b}\mathfrak{X}^{-1}}$ under the canonical map $\mathbf{Z}^2 \longrightarrow (\mathbf{Z}/f\mathbf{Z})^2$. Then we have

$$\overline{R_\mathfrak{b}} = \overline{R_\mathfrak{O}}\,\overline{\alpha_\mathfrak{b}}^{-1}, \quad \overline{R_{\mathfrak{b}\mathfrak{X}^{-1}}} = \overline{R_\mathfrak{O}}\,\overline{\alpha_{\mathfrak{b}\mathfrak{X}^{-1}}}^{-1}, \quad \overline{R_{\mathfrak{b}\mathfrak{X}^{-1}}} = \overline{R_\mathfrak{b}}\,\overline{\alpha_\mathfrak{b}}\,\overline{\alpha_{\mathfrak{b}\mathfrak{X}^{-1}}}^{-1}$$

By definition, we have

$$\alpha_\mathfrak{b}\alpha_{\mathfrak{b}\mathfrak{X}^{-1}}^{-1} = s_\mathfrak{b}\beta_\mathfrak{b}\delta_0\gamma\xi^{-1}q(s)\alpha_{\mathfrak{b}\mathfrak{X}^{-1}}^{-1}, \quad \delta_0 = m(\det t)^{-1}t\gamma^{-1}.$$

Since $\delta_0 \in U_f$ and $q(s)_p = 1$ for every p which divides M, we have

$$\beta_\mathfrak{b}\delta_0\gamma\xi^{-1}q(s)\alpha_{\mathfrak{b}\mathfrak{X}^{-1}}^{-1} \equiv \beta_\mathfrak{b}\gamma\xi^{-1}\alpha_{\mathfrak{b}\mathfrak{X}^{-1}}^{-1} \equiv 1 \mod U_f.$$

Therefore we obtain $\overline{R_{\mathfrak{b}\mathfrak{X}^{-1}}} = \overline{R_\mathfrak{b}}\,\overline{s_\mathfrak{b}}$. By definition, we see easily that we may take $s_\mathfrak{b}$ so that $\overline{s_\mathfrak{b}} = \left(\begin{smallmatrix}1 & 0\\ 0 & \overline{N(\mathfrak{X})}\end{smallmatrix}\right)$, where $\overline{N(\mathfrak{X})} = N(\mathfrak{X}) \mod f \in \mathbf{Z}/f\mathbf{Z}$. Therefore we have $\overline{R_{\mathfrak{b}\mathfrak{X}^{-1}}} = \overline{R_\mathfrak{b}}\left(\begin{smallmatrix}1 & 0\\ 0 & \overline{N(\mathfrak{X})}\end{smallmatrix}\right)$. Hence, by (6.2.4), we obtain

$$\sum_{(c_0,d_0)\in R_\mathfrak{b}} g^\sigma|_m\,\beta_\mathfrak{b}\gamma\xi^{-1} = \sum_{(c_0,d_0)\in R_{\mathfrak{b}\mathfrak{X}^{-1}}} g|_m\,\alpha_{\mathfrak{b}\mathfrak{X}^{-1}} = \varphi_{\mathfrak{b}\mathfrak{X}^{-1}}.$$

By (6.2.6), $\det(\beta_\mathfrak{b}\gamma\xi^{-1}) = N(\mathfrak{b})N(\mathfrak{X})^{-1}$ and by

$$\begin{aligned}&\lambda_0^\sigma(\mathfrak{b})^{-1}N(\mathfrak{b})^m\{\Im(\omega_{1,\mathfrak{b}}\overline{\omega_{2,\mathfrak{b}}})i\}^r(D^r\varphi_\mathfrak{b})(\omega)\\ &= \lambda_0^\sigma(\mathfrak{b}\mathfrak{X}^{-1})^{-1}N(\mathfrak{b}\mathfrak{X}^{-1})^m\{\Im(\omega_{1,\mathfrak{b}\mathfrak{X}^{-1}}\overline{\omega_{2,\mathfrak{b}\mathfrak{X}^{-1}}})i\}^r(D^r\varphi_{\mathfrak{b}\mathfrak{X}^{-1}})(\omega),\end{aligned}$$

we get

$$Q^\sigma = c\,(i^{m+r})^\sigma \eta^{-2k}(\gamma\xi^{-1}(\omega))j(\gamma\xi^{-1},\omega)^k\lambda_0^\sigma(\mathfrak{X})^{-1}N(\mathfrak{X})^k \cdot \sum_{\mathfrak{b}\in B}\lambda_0^\sigma(\mathfrak{b})^{-1}N(\mathfrak{b})^m\{\Im(\omega_{1,\mathfrak{b}}\overline{\omega_{2,\mathfrak{b}}})i\}^r(D^r\varphi_{\mathfrak{b}})(\omega). \tag{6.2.7}$$

By (6.1.5), (6.2.2) and (6.2.7), we obtain

$$\left(\frac{L(m/2,\lambda)}{(2\pi)^{m/2}P^{k/2}}\right)^\sigma = \zeta(\sigma,\lambda)\frac{L(m/2,\lambda^\sigma)}{(2\pi)^{m/2}P^{k/2}}, \tag{6.2.8}$$

$$\begin{aligned}\zeta(\sigma,\lambda) = {}& \lambda^\sigma(\mathfrak{X})^{-1}N(\mathfrak{X})^{k/2}j(\gamma\xi^{-1},\omega)^k\frac{(i^{m+r})^\sigma}{i^{m+r}} \\ &\cdot\left(\eta^{2k}(\gamma\xi^{-1}(\omega))\prod_{i=1}^h N(\mathfrak{a}_i)^{-k/2h}|\omega_{2,i}|^{k/h}|\eta(\omega_{1,i}/\omega_{2,i})|^{-2k/h}\right)^{-1} \\ &\cdot\left(\eta^{2k}(\omega)\prod_{i=1}^h N(\mathfrak{a}_i)^{-k/2h}|\omega_{2,i}|^{k/h}|\eta(\omega_{1,i}/\omega_{2,i})|^{-2k/h}\right)^\sigma.\end{aligned} \tag{6.2.9}$$

Third step: Introduction of covering groups

By (6.2.9), we have reduced the determination of ζ to a problem concerning the η-function. It is well known that

$$\eta(z+1) = \zeta_{24}\eta(z), \quad \eta(-1/z) = \eta(z)\sqrt{z/i} \tag{6.3.1}$$

where $\sqrt{z/i}$ takes positive values for $z \in i\mathbf{R}_+$. For $c \in \mathbf{Q}$, we define $\eta^c(z)$ by choosing the branch so that it takes positive values for $z \in i\mathbf{R}_+$. We are going to define covering groups of $\mathrm{GL}_+(2,\mathbf{R})$ following Shimura's idea in [**Sh73**]. Let l be a positive integer and let

$$\mathcal{G}_{\mathbf{R}}^{(l)} = \{(\alpha,\psi) \mid \alpha \in \mathrm{GL}_+(2,\mathbf{R}), \psi \text{ is a holomorphic function on } \mathfrak{H} \text{ which satisfies } \psi(z)^l = j(\alpha,z)\}.$$

Then $\mathcal{G}_{\mathbf{R}}^{(l)}$ forms a group under the composition law

$$(\alpha,\psi(z))(\alpha',\psi'(z)) = (\alpha\alpha',\psi(\alpha'(z))\psi'(z)).$$

For a function f on $\mathfrak{H}$, an integer b and $\beta = (\alpha,\psi) \in \mathcal{G}_{\mathbf{R}}^{(l)}$, we put

$$(f|_{b/l}\,\beta)(z) = f(\alpha(z))\psi(z)^{-b}.$$

Then we have

$$f|_{b/l}\,\gamma_1\gamma_2 = (f|_{b/l}\,\gamma_1)|_{b/l}\,\gamma_2, \quad \gamma_1,\gamma_2 \in \mathcal{G}_{\mathbf{R}}^{(l)}.$$

Let $\mathcal{G}_{\mathbf{Q}}^{(l)}$ be the subgroup of $\mathcal{G}_{\mathbf{R}}^{(l)}$ which consists of elements (α,ψ) such that $\alpha \in G_{\mathbf{Q}+}$.

We put $k/2h = b/l$ with $0 < b, l \in \mathbf{Z}$. We do not assume that l and b are relatively prime. For $1 \le i \le h$, take a holomorphic function ψ_i on $\mathfrak{H}$ so that $\psi_i(z)^l = j(\alpha_i,z)$ and put $\tilde{\alpha}_i = (\alpha_i,\psi_i) \in \mathcal{G}_{\mathbf{R}}^{(l)}$. Since $\omega_{1,i}/\omega_{2,i} = \alpha_i(\omega)$, $\omega_{2,i} = j(\alpha_i,\omega)$, we obtain

$$\eta^{2k}(\omega)\prod_{i=1}^h|\omega_{2,i}|^{k/h}|\eta(\omega_{1,i}/\omega_{2,i})|^{-2k/h} = \prod_{i=1}^h \frac{\eta^{k/h}(\omega)}{(\eta^{k/h}|_{b/l}\tilde{\alpha}_i)(\omega)}\overline{\frac{\eta^{k/h}(\omega)}{(\eta^{k/h}|_{b/l}\tilde{\alpha}_i)(\omega)}}.$$

We note that this formula holds independently of the choice of ψ_i. We have

$$\left(\frac{\eta^{k/h}(\omega)}{(\eta^{k/h}|_{b/l}\tilde{\alpha}_i)(\omega)}\right)^l = \frac{\eta^{2b}(\omega)}{(\eta^{2b}|_b\alpha_i)(\omega)},$$
$$\left(\frac{\eta^{2b}(\omega)}{(\eta^{2b}|_b\alpha_i)(\omega)}\right)^\sigma = \frac{\eta^{2b}(\gamma\xi^{-1}(\omega))}{(\eta^{2b}|_b\alpha_i)^\sigma(\gamma\xi^{-1}(\omega))}$$

by definition and by Proposition 6.1. Therefore we obtain

$$\left(\frac{\eta^{k/h}(\omega)}{(\eta^{k/h}|_{b/l}\tilde{\alpha}_i)(\omega)}\right)^\sigma = \mu_i(\sigma)\frac{\eta^{k/h}(\gamma\xi^{-1}(\omega))}{(\eta^{k/h}|_{b/l}\tilde{\alpha}_i)^\sigma(\gamma\xi^{-1}(\omega))}. \tag{6.3.2}$$

Here $\mu_i(\sigma)$ is a root of unity such that $\mu_i(\sigma)^l = 1$ and σ acts on $\eta^{k/h}|_{b/l}\,\tilde{\alpha}_i$ through its Fourier coefficients. [11] Hereafter we choose ω explicitly so that $\omega = \sqrt{-d_1}$ if $d \equiv 0 \mod 4$, $d_1 = d/4$, $\omega = (1+\sqrt{-d})/2$ if $d \equiv 3 \mod 4$. We define $\delta \in \mathrm{GL}(2, \mathbf{Z})$ by

$$\delta = \begin{cases} \begin{pmatrix} -1 & 0 \\ 0 & 1 \end{pmatrix} & \text{if } d \equiv 0 \mod 4, \\ \begin{pmatrix} -1 & 1 \\ 0 & 1 \end{pmatrix} & \text{if } d \equiv 3 \mod 4. \end{cases}$$

We see easily that

$$\rho\sigma\rho|C_M = \left(\frac{C_M/K}{\mathfrak{X}^\rho}\right), \quad \rho\sigma\rho|\mathbf{Q}_{ab} = \sigma|\mathbf{Q}_{ab}, \quad q(a^\rho) = \delta q(a)\delta, \quad a \in K_A^\times,$$
$$(\mathfrak{X}^\rho)^{-1} = \mathbf{Z}(-\bar{\omega}_1) \oplus \mathbf{Z}\bar{\omega}_2, \quad \Im(-\bar{\omega}_1/\bar{\omega}_2) > 0,$$

and that $\xi' = \delta\xi\delta$ satisfies

$$\begin{pmatrix} \omega \\ 1 \end{pmatrix} = \begin{cases} \xi' \begin{pmatrix} -\bar{\omega}_1 \\ \bar{\omega}_2 \end{pmatrix} & \text{if } d \equiv 0 \mod 4, \\ \xi' \begin{pmatrix} -\bar{\omega}_1 + \bar{\omega}_2 \\ \bar{\omega}_2 \end{pmatrix} & \text{if } d \equiv 3 \mod 4. \end{cases}$$

We have

$$s^\rho\mathfrak{O}_K = \mathfrak{X}^\rho, \quad q(s^\rho)_p = 1 \quad \text{for every } p \text{ which divides } M,$$
$$t' = q(s^\rho)^{-1}\xi' = \delta q(s)^{-1}\xi\delta = \delta t\delta \in U.$$

Take $v \in \mathbf{Z}$ so that $v \equiv N(\mathfrak{X}) \mod M$ and put $\gamma' = \gamma_0\delta\gamma\delta$ where

$$\gamma_0 = \begin{cases} 1_2 & \text{if } d \equiv 0 \mod 4, \\ \begin{pmatrix} 1 & -1+v \\ 0 & 1 \end{pmatrix} & \text{if } d \equiv 3 \mod 4. \end{cases}$$

Then we have $m(\det t')^{-1}t'\gamma'^{-1} \in U_M$. Hence we obtain

[11] It is easy to see that $\eta^{k/h}|_{b/l}\tilde{\alpha}_i$ has a usual Fourier expansion with respect to $e^{2\pi iz/M}$ for some M. The analycity of $(\eta^{k/h}|_{b/l}\tilde{\alpha}_i)^\sigma$ follows from (6.3.5). We note that $\mu_i(\sigma)$ does not depend on the choice of ψ_i.

$$\left(\frac{\overline{\eta^{k/h}(\omega)}}{\overline{(\eta^{k/h}|_{b/l}\tilde{\alpha}_i)(\omega)}}\right)^{\sigma} = \left(\frac{\eta^{k/h}(\omega)}{(\eta^{k/h}|_{b/l}\tilde{\alpha}_i)(\omega)}\right)^{\rho\sigma\rho\cdot\rho}$$
$$= \mu_i(\rho\sigma\rho)^{-1}\frac{\overline{\eta^{k/h}(\gamma'\xi'^{-1}(\omega))}}{\overline{(\eta^{k/h}|_{b/l}\tilde{\alpha}_i)^{\rho\sigma\rho}(\gamma'\xi'^{-1}(\omega))}}.$$

We have

$$\overline{\eta(\gamma'\xi'^{-1}(\omega))} = \overline{\eta(\gamma_0\delta\gamma\xi^{-1}\delta(\omega))}$$
$$= \overline{\eta(\gamma_0\delta\gamma\xi^{-1}(\bar{\omega}))} = \eta(-\gamma_0\delta\gamma\xi^{-1}(\omega)).$$

Hence we see easily that

$$\left(\frac{\eta(\omega)}{\overline{\eta(\omega)}}\right)^{\sigma} = \frac{\eta(\gamma\xi^{-1}(\omega))}{\overline{\eta(\gamma'\xi'^{-1}(\omega))}}. \tag{6.3.3}$$

Since $(\eta^2|_1\alpha_i)^{\rho\sigma\rho} = (\eta^2|_1\alpha_i)^{\sigma}$, we have

$$(\eta^{k/h}|_{b/l}\tilde{\alpha}_i)^{\rho\sigma\rho} = \nu_i(\sigma)(\eta^{k/h}|_{b/l}\tilde{\alpha}_i)^{\sigma} \tag{6.3.4}$$

with an l-th root of unity $\nu_i(\sigma)$. Take $\tilde{\beta}_i \in \mathcal{G}_{\mathbf{Q}}^{(l)}$ so that $\tilde{\beta}_i = (\beta_i, \varphi_i)$, $\varphi_i(z)^l = j(\beta_i, z)$. Since $(\eta^2|_1\alpha_i)^{\sigma} = \eta^2|_1\beta_i$, we have

$$(\eta^{k/h}|_{b/l}\tilde{\alpha}_i)^{\sigma} = \nu_i'(\sigma)\eta^{k/h}|_{b/l}\tilde{\beta}_i \tag{6.3.5}$$

with a root of unity $\nu_i'(\sigma)$. Therefore we obtain

$$(\eta^{k/h}|_{b/l}\tilde{\alpha}_i)^{\sigma}(\gamma\xi^{-1}(\omega))\overline{(\eta^{k/h}|_{b/l}\tilde{\alpha}_i)^{\sigma}(\gamma'\xi'^{-1}(\omega))}$$
$$= (\eta^{k/h}|_{b/l}\tilde{\beta}_i)(\gamma\xi^{-1}(\omega))\overline{(\eta^{k/h}|_{b/l}\tilde{\beta}_i)(\gamma'\xi'^{-1}(\omega))}.$$

We note that this formula holds independently of the choice of φ_i. Let $\widetilde{\gamma\xi^{-1}} = (\gamma\xi^{-1}, \varphi) \in \mathcal{G}_{\mathbf{Q}}^{(l)}$, $\widetilde{\gamma'\xi'^{-1}} = (\gamma'\xi'^{-1}, \varphi') \in \mathcal{G}_{\mathbf{Q}}^{(l)}$ where $\varphi(z)^l = j(\gamma\xi^{-1}, z)$, $\varphi'(z)^l = j(\gamma'\xi'^{-1}, z)$. Canceling automorphic factors, we get

$$(\eta^{k/h}|_{b/l}\tilde{\beta}_i)(\gamma\xi^{-1}(z))\overline{(\eta^{k/h}|_{b/l}\tilde{\beta}_i)(\gamma'\xi'^{-1}(z))}$$
$$= (\eta^{k/h}|_{b/l}\tilde{\beta}_i\widetilde{\gamma\xi^{-1}}(z))\overline{(\eta^{k/h}|_{b/l}\tilde{\beta}_i\widetilde{\gamma'\xi'^{-1}}(z))} \cdot \varphi(z)^b\overline{\varphi'(z)}^b.$$

Since $\overline{j(\gamma'\xi'^{-1}, \omega)} = j(\gamma\xi^{-1}, \omega)$, we can choose φ and φ' so that

$$\overline{\varphi'(\omega)} = \varphi(\omega). \tag{6.3.6}$$

By (6.1.4) and Lemma 6.2, we obtain

$$\prod_{i=1}^{h} N(\mathfrak{a}_i)^{1/2}|(\eta^2|_1\alpha_i)(\omega)| = \prod_{i=1}^{h} N(\mathfrak{a}_i\mathfrak{X}^{-1})^{1/2}|(\eta^2|_1\beta_i\gamma\xi^{-1})(\omega)|.$$

Hence we have

$$N(\mathfrak{X})^{k/2}\prod_{i=1}^{h}(\eta^{k/h}|_{b/l}\tilde{\alpha}_i)(\omega)\overline{(\eta^{k/h}|_{b/l}\tilde{\alpha}_i)(\omega)}$$
$$= (\eta^{k/h}|_{b/l}\tilde{\beta}_i\widetilde{\gamma\xi^{-1}})(\omega)\overline{(\eta^{k/h}|_{b/l}\tilde{\beta}_i\widetilde{\gamma\xi^{-1}})(\omega)}.$$

Therefore by (6.2.9), (6.3.3) and (6.3.6), we obtain

$$\zeta(\sigma,\lambda) = \lambda^\sigma(\mathfrak{X})^{-1}\cdot\frac{(i^{m+r})^\sigma}{i^{m+r}}\prod_{i=1}^{h}\frac{N(\mathfrak{a}_i)^{k/2h}}{(N(\mathfrak{a}_i)^{k/2h})^\sigma}\mu_i(\sigma)\mu_i(\rho\sigma\rho)^{-1}\nu_i(\sigma) \cdot\frac{\overline{(\eta^{k/h}|_{b/l}\tilde{\beta}_i\widetilde{\gamma\xi^{-1}})(\omega)}}{\overline{(\eta^{k/h}|_{b/l}\tilde{\beta}_i\widetilde{\gamma'\xi'^{-1}})(\omega)}}. \tag{6.3.7}$$

We have

$$\begin{aligned}\eta^{2/l}|_{1/l}\left(\begin{pmatrix}1&1\\0&1\end{pmatrix},1\right)&=\zeta_{12l}\eta^{2/l},\\ \eta^{2/l}|_{1/l}\left(\begin{pmatrix}0&-1\\1&0\end{pmatrix},\psi_0\right)&=\zeta_{4l}^{-1}\eta^{2/l},\end{aligned}\tag{6.3.8}$$

where ψ_0 is the holomorphic function on $\mathfrak{H}$ such that $\psi_0(z)^l = z$ and $\psi_0(i) = e^{\pi i/2l}$. We put

$$\widetilde{\mathrm{SL}(2,\mathbf{Z})}_s^{(l)} = \{(\gamma,\psi)\mid\gamma\in\mathrm{SL}(2,\mathbf{Z}),\ (\gamma,\psi)\in\mathcal{G}_{\mathbf{Q}}^{(l)}\}.$$

By (6.3.8), we have

$$\eta^{2/l}|_{1/l}\tilde{\gamma} = \tilde{\chi}_s(\tilde{\gamma})\eta^{2/l},\quad \tilde{\gamma}\in\widetilde{\mathrm{SL}(2,\mathbf{Z})}_s^{(l)}$$

with $\tilde{\chi}_s(\tilde{\gamma})\in\mathbf{C}^\times$. We see easily that $\tilde{\chi}_s$ is a character of $\widetilde{\mathrm{SL}(2,\mathbf{Z})}_s^{(l)}$ taking values in $\langle\zeta_{12l}\rangle$.

We have

$$\mathfrak{a}_i\mathfrak{X}^\rho\mathfrak{X}^{-1} = (\mu_i)\mathfrak{a}_{j(i)},\quad \mu_i\in K^\times,\quad 1\le i\le h. \tag{6.3.9}$$

Here $i\longrightarrow j(i)$ is a permutation on h letters. Similarly to the proof of Lemma 6.2, we have $\mathbf{Z}^2\beta_i\gamma'\xi'^{-1}\begin{pmatrix}\omega\\1\end{pmatrix} = \mathfrak{a}_i(\mathfrak{X}^\rho)^{-1}$. Hence

$$\mathbf{Z}^2\beta_{j(i)}\gamma'\xi'^{-1}q(\mu_i)\begin{pmatrix}\omega\\1\end{pmatrix} = \mu_i\mathbf{Z}^2\beta_{j(i)}\gamma'\xi'^{-1}\begin{pmatrix}\omega\\1\end{pmatrix} = \mathfrak{a}_i\mathfrak{X}^{-1} = \mathbf{Z}^2\beta_i\gamma\xi^{-1}\begin{pmatrix}\omega\\1\end{pmatrix}$$

and we see that there exists $\gamma_i\in\mathrm{SL}(2,\mathbf{Z})$ such that

$$\beta_i\gamma\xi^{-1} = \gamma_i\beta_{j(i)}\gamma'\xi'^{-1}q(\mu_i),\quad 1\le i\le h. \tag{6.3.10}$$

For $\mu\in K^\times$, we shall always choose $\widetilde{q(\mu)} = (q(\mu),\psi_\mu)\in\mathcal{G}_{\mathbf{Q}}^{(l)}$ in the following way. We have $\psi_\mu(\omega)^l = j(q(\mu),\omega) = \mu$. We choose $\mu^{1/l}$ so that $0\le\arg(\mu^{1/l}) < 2\pi/l$ and choose ψ_μ so that $\psi_\mu(\omega) = \mu^{1/l}$. We note that

$$(f|_{1/l}\widetilde{q(\mu)}^{-1})(\omega) = f(\omega)\mu^{1/l}$$

for a function f on $\mathfrak{H}$ and $\mu\in K^\times$. There exists $\tilde{\gamma}_i\in\widetilde{\mathrm{SL}(2,\mathbf{Z})}_s^{(l)}$ such that

$$\tilde{\beta}_i\widetilde{\gamma\xi^{-1}} = \tilde{\gamma}_i\tilde{\beta}_{j(i)}\widetilde{\gamma'\xi'^{-1}}\widetilde{q(\mu_i)},\quad 1\le i\le h. \tag{6.3.11}$$

Then we have

$$\frac{\prod_{i=1}^h\overline{(\eta^{2/l}|_{1/l}\tilde{\beta}_i\widetilde{\gamma\xi^{-1}})(\omega)}}{\prod_{i=1}^h\overline{(\eta^{2/l}|_{1/l}\tilde{\beta}_i\widetilde{\gamma'\xi'^{-1}})(\omega)}} = \prod_{i=1}^h\overline{\chi_s(\tilde{\gamma}_i)(\mu_i^{1/l})}^{-1}. \tag{6.3.12}$$

We take another model of $\mathcal{G}_{\mathbf{R}}^{(l)}$ in the following way (cf. [**Y92**]). For $g = \left(\begin{smallmatrix} a & b \\ c & d \end{smallmatrix}\right) \in \mathrm{GL}_+(2, \mathbf{R})$, put

$$x(g) = \begin{cases} c & \text{if } c \neq 0, \\ d & \text{if } c = 0, \end{cases} \qquad y(g) = \begin{cases} 1 & \text{if } x(g) > 0, \\ \zeta_l & \text{if } x(g) < 0. \end{cases}$$

and define a holomorphic function ψ_g on $\mathfrak{H}$ by

$$\psi_g(z)^l = j(g, z), \quad 0 \leq \arg \psi_g(z) < 2\pi/l \quad \text{for all } z \in \mathfrak{H}.$$

Then

$$\begin{aligned} &\xi_{\mathbf{R}}(g_1, g_2) \\ &= \psi_{g_1}(g_2(z))\psi_{g_2}(z)\psi_{g_1 g_2}(z)^{-1} y(g_1 g_2) y(g_1)^{-1} y(g_2)^{-1}, g_1, g_2 \in \mathrm{GL}_+(2, \mathbf{R}). \end{aligned}$$

is a 2-cocycle of $\mathrm{GL}_+(2, \mathbf{R})$ taking values in $\langle \zeta_l \rangle$. On the product set $\mathrm{GL}_+(2, \mathbf{R}) \times \langle \zeta_l \rangle$, we define a group law by

$$(g_1, \epsilon_1)(g_2, \epsilon_2) = (g_1 g_2, \epsilon_1 \epsilon_2 \xi_{\mathbf{R}}(g_1, g_2)).$$

We denote this group by $\mathrm{GL}_+\widetilde{(2, \mathbf{R})}^{(l)}$. We have $\mathrm{GL}_+\widetilde{(2, \mathbf{R})}^{(l)} \cong \mathcal{G}_{\mathbf{R}}^{(l)}$ the isomorphism being given by

$$(g, \epsilon) \longrightarrow (g, \epsilon y(g)^{-1} \psi_g). \tag{6.3.13}$$

Put

$$\mathrm{SL}\widetilde{(2, \mathbf{Z})}^{(l)} = \{(\gamma, \epsilon) \in \mathrm{GL}_+\widetilde{(2, \mathbf{R})}^{(l)} \mid \gamma \in \mathrm{SL}(2, \mathbf{Z})\}.$$

Let $\tilde{\chi}$ be the character of $\mathrm{SL}\widetilde{(2, \mathbf{Z})}^{(l)}$ obtained from $\tilde{\chi}_s$ using the isomorphism (6.3.13). By (6.3.8), we have

$$\tilde{\chi}\left(\begin{pmatrix} 1 & 1 \\ 0 & 1 \end{pmatrix}, 1\right) = \zeta_{12l}, \quad \tilde{\chi}\left(\begin{pmatrix} 0 & -1 \\ 1 & 0 \end{pmatrix}, 1\right) = \zeta_{4l}^{-1}. \tag{6.3.14}$$

We can verify easily that $\tilde{\chi}(\left(\begin{smallmatrix} 1 & 0 \\ 1 & 1 \end{smallmatrix}\right), 1) = \zeta_{12l}^{-1}$. Now let us consider the equality (6.3.11) in $\mathrm{GL}_+\widetilde{(2, \mathbf{R})}^{(l)}$ taking account of the relation (6.3.6). We put $\gamma\xi^{-1} = \left(\begin{smallmatrix} a & b \\ c & d \end{smallmatrix}\right)$. For simplicity, we assume that $c = 0$ and $d > 0$. Then we can choose $\varphi = \varphi' = d^{1/l}$. Hence we may assume

$$\widetilde{\gamma\xi^{-1}} = (\gamma\xi^{-1}, 1), \quad \widetilde{\gamma'\xi'^{-1}} = (\gamma'\xi'^{-1}, 1). \tag{6.3.15}$$

(The same conclusion holds when $c > 0$. But we shall not use this fact.) We note that we may assume $\tilde{\beta}_i = (\beta_i, 1)$ in (6.3.8). We obtain

$$\begin{aligned} \zeta(\sigma, \lambda) = \lambda^\sigma(\mathfrak{X})^{-1} \cdot \frac{(i^{m+r})^\sigma}{i^{m+r}} \prod_{i=1}^{h} \frac{N(\mathfrak{a}_i)^{k/2h}}{(N(\mathfrak{a}_i)^{k/2h})^\sigma} \mu_i(\sigma)\mu_i(\rho\sigma\rho)^{-1}\nu_i(\sigma) \\ \cdot \overline{[\tilde{\chi}(\tilde{\gamma}_i)(\mu_i^{1/l})^{-1}]^b}. \end{aligned} \tag{6.3.16}$$

REMARK. We can relate (6.2.9) to elliptic units studied in Siegel [**Si61**, p. 103–107].

Fourth step: Explicit computations for double covering groups

Hereafter we shall consider the case where k is divisible by h. We put $k = hn$. We may take $l = 2$. We first note that $\mu_i(\sigma) = 1$ and $\nu_i(\sigma) = 1$ for every i. To see this, let $\theta(z) = \sum_{n \in \mathbf{Z}} \exp(\pi n^2 z)$ be the theta function and put, for $t \in \mathbf{Z}$,

$$\mathcal{A}_{t/2}(\mathbf{Q}_{ab}) = \{\theta^t f \mid f \in \mathcal{A}_0(\mathbf{Q}_{ab})\}.$$

Now by Shimura [**Sh78**, p. 44], $\mathcal{A}_{t/2}(\mathbf{Q}_{ab})$ is stable under the action of $\mathcal{G}^{(2)}_{\mathbf{Q}}$. We have $\eta|_{1/2}\,\tilde{\alpha}_i \in \mathcal{A}_{1/2}(\mathbf{Q}_{ab})$ by stability. Hence $(\eta|_{1/2}\,\tilde{\alpha}_i)/\eta \in \mathcal{A}_0(\mathbf{Q}_{ab})$ by definition. Taking M sufficiently large, we may assume that

$$(6.4.1)\qquad \begin{aligned}&(\eta|_{1/2}\,\tilde{\alpha}_i)/\eta \in \mathcal{A}_0(\Gamma(M)) \quad \text{and its Fourier coefficients}\\ &\text{belong to } \mathbf{Q}(\zeta_M) \text{ for } \quad 1 \le i \le h.\end{aligned}$$

Now $\nu_i(\sigma) = 1$ is clear. By Proposition 6.1, we have $(\eta|_{1/2}\,\tilde{\alpha}_i)(\omega)/\eta(\omega) \in C_M$ and that

$$\left(\frac{(\eta|_{1/2}\,\tilde{\alpha}_i)(\omega)}{\eta(\omega)}\right)^{\sigma} = \frac{(\eta|_{1/2}\,\tilde{\alpha}_i)^{\sigma}(\gamma\xi^{-1}(\omega))}{\eta(\gamma\xi^{-1}(\omega))}, \quad 1 \le i \le h,$$

where $\sigma|C_M = \left(\frac{C_M/K}{\mathfrak{X}}\right)$ as before. Hence $\mu_i(\sigma) = 1$. Now (6.3.16) takes the simpler form

$$(6.4.2)\qquad \zeta(\sigma,\lambda) = \lambda^{\sigma}(\mathfrak{X})^{-1}\cdot\frac{(i^{m+r})^{\sigma}}{i^{m+r}}\prod_{i=1}^{h}\frac{N(\mathfrak{a}_i)^{n/2}}{(N(\mathfrak{a}_i)^{n/2})^{\sigma}}\overline{[\tilde{\chi}(\tilde{\gamma}_i)(\mu_i^{1/2})^{-1}]^{n}}.$$

We shall denote $\widetilde{\mathrm{GL}_+(2,\mathbf{R})}^{(2)}$ and $\widetilde{\mathrm{SL}(2,\mathbf{Z})}^{(2)}$ by $\widetilde{\mathrm{GL}_+(2,\mathbf{R})}$ and $\widetilde{\mathrm{SL}(2,\mathbf{Z})}$. In this case we have

$$\xi_{\mathbf{R}}(g_1,g_2) = (x(g_1),x(g_2))_{\mathbf{R}}(-x(g_1)x(g_2),x(g_1g_2))_{\mathbf{R}},$$

where $(\ ,\ )_{\mathbf{R}}$ denotes the Hilbert symbol of $\mathbf{R}$.

LEMMA 6.3. *Let $z_i^{-1} = x_i + y_i\omega \in K^\times$ for $i = 1,\,2$ and put $z^{-1} = z_1^{-1}z_2^{-1} = x + y\omega$. Here $x_i,\,y_i$ $(i = 1,2)$, $x,\,y \in \mathbf{Q}$. If y_1, y_2, and y are non-zero, then we have*

$$\mathrm{sgn}(y_1)\mathrm{sgn}(y_2)(z_1^{1/2})^{-1}(z_2^{1/2})^{-1} = -\mathrm{sgn}(y)(z^{1/2})^{-1}\xi_{\mathbf{R}}(q(z_1)^{-1},q(z_2)^{-1}).$$

Since the proof is easy, we omit it.

Let p be a prime number. For $g = \left(\begin{smallmatrix} a & b\\ c & d\end{smallmatrix}\right) \in \mathrm{SL}(2,\mathbf{Q}_p)$, we put $x(g) = c$ if $c \ne 0$, $x(g) = d$ if $c = 0$. Then

$$\xi_p(g_1,g_2) = (x(g_1),x(g_2))_p(-x(g_1)x(g_2),x(g_1g_2))_p, \quad g_1,g_2 \in \mathrm{SL}(2,\mathbf{Q}_p)$$

defines a 2-cocycle of $\mathrm{SL}(2,\mathbf{Q}_p)$ taking values in $\{\pm1\}$. Here $(\ ,\ )_p$ denotes the Hilbert symbol of $\mathbf{Q}_p$. On the product set $\mathrm{SL}(2,\mathbf{Q}_p)\times\{\pm1\}$, we define a group law

$$(g_1,\epsilon_1)(g_2,\epsilon_2) = (g_1g_2,\epsilon_1\epsilon_2\xi_p(g_1,g_2)).$$

We denote this group by $\widetilde{\mathrm{SL}(2,\mathbf{Q}_p)}$. For $g = \left(\begin{smallmatrix} a & b\\ c & d\end{smallmatrix}\right) \in \mathrm{SL}(2,\mathbf{Z}_p)$, we define

$$s_p(g) = \begin{cases} 1 & \text{if } c \in \mathbf{Z}_p^\times \text{ or } c = 0,\\ (a,c)_p & \text{if } c \in p\mathbf{Z}_p \text{ and } c \ne 0.\end{cases}$$

We can extend the 2-cocycle ξ_p of $\mathrm{SL}(2,\mathbf{Q}_p)$ to a 2-cocycle $\tilde{\xi}_p$ of $\mathrm{GL}(2,\mathbf{Q}_p)$ in the following way. For $g = \left(\begin{smallmatrix} a & b\\ c & d\end{smallmatrix}\right) \in \mathrm{GL}(2,\mathbf{Q}_p)$ and $z \in \mathbf{Q}_p^\times$, we put

$$p(g) = \begin{pmatrix}1 & 0\\ 0 & \det g\end{pmatrix}^{-1} g \in \mathrm{SL}(2,\mathbf{Q}_p), \quad g^z = \begin{pmatrix}1 & 0\\ 0 & z\end{pmatrix}^{-1} g \begin{pmatrix}1 & 0\\ 0 & z\end{pmatrix},$$

$$v_p(z,g) = \begin{cases} 1 & \text{if} \quad c \ne 0,\\ (z,d)_p & \text{if} \quad c = 0,\end{cases}$$

$$\tilde{\xi}_p(g_1, g_2) = \xi_p(p(g_1)^{\det(g_2)}, p(g_2))v_p(\det g_2,\ p(g_1)), \quad g_1, g_2 \in \mathrm{GL}(2, \mathbf{Q}_p),$$

$$\tilde{s}_p(g) = \begin{cases} 1 & \text{if } cd = 0 \text{ or } 2\,|\,\mathrm{ord}_p(c), \\ (c, d/\det g)_p & \text{if } cd \neq 0 \text{ and } 2\nmid \mathrm{ord}_p(c). \end{cases}$$

Let $\widetilde{\mathrm{GL}(2, \mathbf{Q}_p)}$ be the double covering group of $\mathrm{GL}(2, \mathbf{Q}_p)$ defined using $\tilde{\xi}_p$ in the same way as for $\mathrm{SL}(2, \mathbf{Q}_p)$. We have $\widetilde{\mathrm{SL}(2, \mathbf{Q}_p)} \subset \widetilde{\mathrm{GL}(2, \mathbf{Q}_p)}$ canonically since $\tilde{\xi}_p|\mathrm{SL}(2, \mathbf{Q}_p) = \xi_p$. We have $\tilde{s}_p|\mathrm{SL}(2, \mathbf{Z}_p) = s_p$ if $p \neq 2$ and an explicit splitting

$$\tilde{\xi}_p(g_1, g_2) = \tilde{s}_p(g_1)\tilde{s}_p(g_2)\tilde{s}_p(g_1 g_2)^{-1}, \quad g_1, g_2 \in \mathrm{GL}(2, \mathbf{Z}_p),\ p \neq 2. \tag{6.4.3}$$

(see Kubota [**K**] for these results on covering groups of p-adic groups.)

Let

$$\widetilde{\mathrm{SL}(2, \mathbf{Z}_2)} = \{(\gamma, \epsilon) \in \widetilde{\mathrm{SL}(2, \mathbf{Q}_2)} \mid \gamma \in \mathrm{SL}(2, \mathbf{Z}_2)\}.$$

Put $\tilde{K} = \widetilde{\mathrm{SL}(2, \mathbf{Z}_2)} \times \mathrm{SL}(2, \mathbf{Z}_3)$ and define a mapping φ of $\widetilde{\mathrm{SL}(2, \mathbf{Z})}$ into $\tilde{K}$ by

$$\varphi((\gamma, \epsilon)) = \left(\gamma, \epsilon \prod_{p \neq 2} s_p(\gamma)\right) \times (\gamma)_3.$$

Here $\gamma \in \mathrm{SL}(2, \mathbf{Z})$, $\epsilon = \pm 1$ and $(\gamma)_3$ denotes $\gamma \in \mathrm{SL}(2, \mathbf{Z}_3)$. Using the explicit forms of cocycles ξ_p and $\xi_{\mathbf{R}}$, (6.4.3) and the product formula for the Hilbert symbol, we can verify that φ is an injective homomorphism.

Let χ_3 be the character of $\mathrm{SL}(2, \mathbf{Z}_3)$ which factors through the canonical map $\mathrm{SL}(2, \mathbf{Z}_3) \longrightarrow \mathrm{SL}(2, \mathbf{Z}/3\mathbf{Z})$ and such that

$$\chi_3\left(\begin{pmatrix} 1 & 1 \\ 0 & 1 \end{pmatrix}\right) = \zeta_3^2, \quad \chi_3\left(\begin{pmatrix} 1 & 0 \\ 1 & 1 \end{pmatrix}\right) = \zeta_3^{-2}.$$

The existence of such a χ_3 is clear since the 2-Sylow subgroup of $\mathrm{SL}(2, \mathbf{Z}/3\mathbf{Z})$ is normal. Put $\Gamma_8 = \{\gamma \in \mathrm{SL}(2, \mathbf{Z}_2) \mid \gamma \equiv 1_2 \mod 8\}$ and $\tilde{\Gamma}_8 = \{(\gamma, 1) \mid \gamma \in \Gamma_8\}$. Then $\tilde{\Gamma}_8$ is a subgroup of $\widetilde{\mathrm{SL}(2, \mathbf{Z}_2)}$. There exists a character $\tilde{\chi}_2$ of $\widetilde{\mathrm{SL}(2, \mathbf{Z}_2)}$, which is trivial on $\tilde{\Gamma}_8$, such that

$$\tilde{\chi}_2\left(\begin{pmatrix} 1 & 1 \\ 0 & 1 \end{pmatrix}, 1\right) = \zeta_8^3,$$

$$\tilde{\chi}_2\left(\begin{pmatrix} 1 & 0 \\ 1 & 1 \end{pmatrix}, 1\right) = \zeta_8^{-3},$$

$$\tilde{\chi}_2\left(\begin{pmatrix} 1 & 0 \\ 0 & 1 \end{pmatrix}, -1\right) = -1.$$

The existence of $\tilde{\chi}_2$ is highly non-trivial but we can verify it by a standard technique. We omit the proof. We note that

$$\tilde{\chi}_2\left(\begin{pmatrix} u & 0 \\ 0 & u^{-1} \end{pmatrix}, 1\right) = \begin{cases} 1 & \text{if } u \equiv 1 \mod 8, \\ -i & \text{if } u \equiv 3 \mod 8, \\ -1 & \text{if } u \equiv 5 \mod 8, \\ i & \text{if } u \equiv 7 \mod 8. \end{cases}$$

Define a character $\tilde{\chi}_{\tilde{K}}$ of $\tilde{K}$ by $\tilde{\chi}_{\tilde{K}}((g_2, \epsilon) \times (g_3)) = \tilde{\chi}_2((g_2, \epsilon))\chi_3(g_3)$. Now, checking values for generators of $\widetilde{\mathrm{SL}(2, \mathbf{Z})}$, we obtain $\tilde{\chi}(\tilde{\gamma}) = \tilde{\chi}_{\tilde{K}}(\varphi(\tilde{\gamma}))$, $\tilde{\gamma} \in \widetilde{\mathrm{SL}(2, \mathbf{Z})}$. If

$\tilde{\gamma} = (\gamma, \epsilon)$, $\epsilon = \pm 1$, then we have

$$\tilde{\chi}(\tilde{\gamma}) = \epsilon \prod_{p \neq 2} s_p(\gamma)\tilde{\chi}_2((\gamma, 1))\chi_3(\gamma). \tag{6.4.4}$$

To simplify the computation of $\prod_{i=1}^{h} \tilde{\chi}(\tilde{\gamma}_i)(\mu_i^{1/2})^{-1}$, we choose $\mathfrak{a}_i$ in the following way. We take $\mathfrak{a}_i$ as a prime ideal of degree 1 which is unramified in $K/\mathbf{Q}$ and is relatively prime to (6). We put

$$\mathfrak{q}_i = \mathfrak{a}_i, \quad q_i = N(\mathfrak{q}_i), \quad 1 \leq i \leq h.$$

Then q_i is a prime number which does not divide $6d$. We take M so that (6.2.3) and (6.4.1) hold and that M is divisible by $24f \prod_{i=1}^{h} q_i$. We take $\mathfrak{X} = \mathfrak{l}^{-1}$ where $\mathfrak{l}$ is a prime ideal of degree 1 which is unramified in $K/\mathbf{Q}$ and is relatively prime to (M). We put $l = N(\mathfrak{l})$. When $d \equiv 3 \mod 4$, we further impose a condition:

(A) None of $q_1, q_2, \cdots, q_h$ and l divide $d+1$.

We choose $\beta \in \mathbf{Z}$ so that $\omega \equiv -\beta \mod \mathfrak{l}$. Then we have $\mathfrak{l} = \mathbf{Z}(\omega + \beta) \oplus \mathbf{Z}l$. Hence we can take ξ so that $\xi = \begin{pmatrix} 1 & -\beta/l \\ 0 & 1/l \end{pmatrix}$. We may take $s = (1, \cdots, 1, \pi_{\mathfrak{l}}^{-1}, 1, \cdots, 1) \in K_A^{\times}$ with a prime element $\pi_{\mathfrak{l}}$ of $K_{\mathfrak{l}}$. We may and shall assume that $\omega + \beta \notin \mathfrak{l}^2$. Then we may take $\pi_{\mathfrak{l}} = \omega + \beta$. Now, for $t = q(s)^{-1}\xi$, we have

$$\begin{aligned} m(\det t)^{-1} t = \Bigg(&\begin{pmatrix} 1 & -\beta/l \\ 0 & 1 \end{pmatrix}, \cdots, \begin{pmatrix} 1 & -\beta/l \\ 0 & 1 \end{pmatrix}, \\ &\begin{pmatrix} 1 & 0 \\ 0 & lN(\omega+\beta)^{-1} \end{pmatrix} q(\omega+\beta) \begin{pmatrix} 1 & -\beta/l \\ 0 & 1/l \end{pmatrix}, \begin{pmatrix} 1 & -\beta/l \\ 0 & 1 \end{pmatrix}, \cdots \Bigg). \end{aligned}$$

Choose $u \in \mathbf{Z}$ so that $-\beta/l \equiv u \mod M$ and put $\gamma = \begin{pmatrix} 1 & u \\ 0 & 1 \end{pmatrix}$. Then we have $m(\det t)^{-1} t \gamma^{-1} \in U_M$. By definition, we have

$$\gamma' = \begin{pmatrix} 1 & -u \\ 0 & 1 \end{pmatrix}, \quad \xi' = \begin{pmatrix} 1 & \beta/l \\ 0 & 1/l \end{pmatrix} \quad \text{if } d \equiv 0 \mod 4.$$

We choose $l' \in \mathbf{Z}$ so that $ll' \equiv 1 \mod M$. If $d \equiv 3 \mod 4$, we have

$$\gamma_0 = \begin{pmatrix} 1 & l'-1 \\ 0 & 1 \end{pmatrix}, \quad \gamma' = \begin{pmatrix} 1 & -u+l'-1 \\ 0 & 1 \end{pmatrix}, \quad \xi' = \begin{pmatrix} 1 & (\beta+1)/l - 1 \\ 0 & 1/l \end{pmatrix}.$$

We choose $v_i \in \mathbf{Z}$ so that $\omega \equiv -v_i \mod \mathfrak{q}_i$. Then we have $\mathfrak{q}_i = \mathbf{Z}(\omega + v_i) \oplus \mathbf{Z}q_i$ and we may take $\alpha_i = \begin{pmatrix} 1 & v_i \\ 0 & q_i \end{pmatrix}$. Put $\beta_i = \begin{pmatrix} 1 & l'v_i \\ 0 & q_i \end{pmatrix}$. Then we can easily verify that $\alpha_i m(\det t)\beta_i^{-1} \in S_{24}$.

Let C_K denote the ideal class group of K. Let h_0 be the order of the ideal class of $\mathfrak{l}(\mathfrak{l}^\rho)^{-1}$ in C_K. After a renumbering of $\mathfrak{q}_i$, we have

$$\mathfrak{q}_i \mathfrak{l}(\mathfrak{l}^\rho)^{-1} = (\mu_i)\mathfrak{q}_{i+1}, \quad 1 \leq i \leq h_0 - 1, \quad \mathfrak{q}_{h_0}\mathfrak{l}(\mathfrak{l}^\rho)^{-1} = (\mu_{h_0})\mathfrak{q}_1$$

where $\mu_i \in K^\times$. Then (6.3.10) and (6.3.11) can be written as

$$\beta_i \gamma \xi^{-1} = \gamma_i \beta_{i+1} \gamma' \xi'^{-1} q(\mu_i), \quad 1 \leq i \leq h_0 - 1, \quad \beta_{h_0} \gamma \xi^{-1} = \gamma_{h_0} \beta_1 \gamma' \xi'^{-1} q(\mu_{h_0}),$$

$$\tilde{\beta}_i \widetilde{\gamma \xi^{-1}} = \tilde{\gamma}_i \tilde{\beta}_{i+1} \widetilde{\gamma' \xi'^{-1}} \widetilde{q(\mu_i)}, \quad 1 \leq i \leq h_0 - 1, \quad \tilde{\beta}_{h_0} \widetilde{\gamma \xi^{-1}} = \tilde{\gamma}_{h_0} \tilde{\beta}_1 \widetilde{\gamma' \xi'^{-1}} \widetilde{q(\mu_{h_0})}$$

with $\gamma_i \in \mathrm{SL}(2, \mathbf{Z})$, $\tilde{\gamma}_i \in \widetilde{\mathrm{SL}(2, \mathbf{Z})}$. To compute $\prod_{i=1}^{h} \tilde{\chi}(\tilde{\gamma}_i)(\mu_i^{1/2})^{-1}$, we first compute $\prod_{i=1}^{h_0} \tilde{\chi}(\tilde{\gamma}_i)(\mu_i^{1/2})^{-1}$ and shall take the product over h/h_0 such terms.

We put

$$\mu_i^{-1} = x_i + y_i \omega, \quad x_i, y_i \in \mathbf{Q}, \quad 1 \leq i \leq h_0.$$

We have

$$(\mu_i^{-1}) = \mathfrak{q}_i^{-1}\mathfrak{q}_{i+1}\mathfrak{l}^\rho\mathfrak{l}^{-1}, \quad 1 \leq i \leq h_0 - 1, \quad (\mu_{h_0}^{-1}) = \mathfrak{q}_{h_0}^{-1}\mathfrak{q}_1\mathfrak{l}^\rho\mathfrak{l}^{-1}.$$

Hence we have

$$x_i \neq 0, \quad y_i \neq 0, \quad 2x_i + y_i \neq 0 \text{ (when } d \equiv 3 \mod 4), \quad 1 \leq i \leq h_0.$$

(Use the assumption A to see $x_i \neq 0$ when $d \equiv 3 \mod 4$.) By our choice of $\widetilde{q(\mu_i)} \in \mathcal{G}_{\mathbf{R}}^{(2)}$, we see easily that the corresponding element of $\widetilde{q(\mu_i)}$ in $\mathrm{GL}_+(2,\mathbf{R})$ is given by

$$\widetilde{q(\mu_i)}^{-1} = (q(\mu_i)^{-1}, -\mathrm{sgn}(y_i)), \quad 1 \leq i \leq h_0,$$

since $y(q(\mu_i)) = -\mathrm{sgn}(y_i)$. In view of (6.3.15), we obtain

$$\tilde{\gamma}_i = (\gamma_i, -\mathrm{sgn}(y_i)), \quad 1 \leq i \leq h_0.$$

Assume $d \equiv 0 \mod 4$. Since $q(\mu_i)^{-1} = \left(\begin{smallmatrix} x_i & -y_i d_1 \\ y_i & x_i \end{smallmatrix}\right)$, a direct computation shows $\gamma_i = \left(\begin{smallmatrix} x_i + y_i(\beta + ul + ll'v_i) & * \\ q_i l y_i & * \end{smallmatrix}\right)$, $1 \leq i \leq h_0$. We have

$$\prod_{p \neq 2} s_p(\gamma_i) = \prod_{p \neq 2, p | q_i l y_i} (x_i + y_i(\beta + ul + ll'v_i),\ q_i l y_i)_p.$$

Then after some calculations, we obtain

$$\prod_{p \neq 2} s_p(\gamma_i) = \prod_{p \neq 2, y_i \in p\mathbf{Z}_p} (x_i, y_i)_p \times \begin{cases} 1 & \text{if } h_0 \geq 2, \\ \left(\frac{\mu_1}{\mathfrak{q}_1}\right) & \text{if } h_0 = 1, \end{cases} \tag{6.4.5}$$

for $1 \leq i \leq h_0$. Here $\left(\frac{}{\mathfrak{q}_1}\right)$ denotes the quadratic residue symbol modulo $\mathfrak{q}_1$. We see that (6.4.5) holds also in the case $d \equiv 3 \mod 4$. Combining (6.4.4) and (6.4.5), we obtain

$$\tilde{\chi}(\tilde{\gamma}_i) = -\mathrm{sgn}(y_i) \prod_{p \neq 2, y_i \in p\mathbf{Z}_p} (x_i, y_i)_p \tilde{\chi}_2((\gamma_i, 1))\chi_3(\gamma_i), 1 \leq i \leq h_0 \quad \text{if } h_0 \geq 2, \tag{6.4.6}$$

$$\tilde{\chi}(\tilde{\gamma}_1) = -\left(\frac{\mu_1}{\mathfrak{q}_1}\right) \mathrm{sgn}(y_1) \prod_{p \neq 2, y_1 \in p\mathbf{Z}_p} (x_1, y_1)_p \tilde{\chi}_2((\gamma_1, 1))\chi_3(\gamma_1) \quad \text{if } h_0 = 1. \tag{6.4.7}$$

By definition, we have $\gamma \equiv m(\det t)^{-1}t \mod M$, $\gamma\xi^{-1} = \gamma t^{-1}q(s)^{-1}$. Here by a congruence of matrices, we understand congruences for all corresponding entries. Since $q(s)_p = 1$ for all $p|M$, we have $\gamma\xi^{-1} \equiv m(\det t)^{-1} \mod M$. Put $u_0 = \left(\begin{smallmatrix} 1 & 0 \\ 0 & l^{-1} \end{smallmatrix}\right)$. Then we obtain $\gamma\xi^{-1} \equiv u_0^{-1} \mod 8$. Since $\gamma'\xi'^{-1} = \gamma_0\delta\gamma\xi^{-1}\delta$ and $\gamma_0 \equiv u_0^{-1}\delta u_0 \delta \mod M$, we have $\gamma'\xi'^{-1} \equiv u_0^{-1} \mod 8$. Therefore we have

$$\gamma_i \equiv \beta_i u_0^{-1} q(\mu_i)^{-1} u_0 \beta_{i+1}^{-1} \mod 8, \quad 1 \leq i \leq h_0 - 1,$$

$$\gamma_{h_0} \equiv \beta_{h_0} u_0^{-1} q(\mu_{h_0})^{-1} u_0 \beta_1^{-1} \mod 8.$$

Call δ_i the right hand side of these congruences. We shall show

$$\tilde{\chi}_2((\gamma_i, 1)) = \tilde{\chi}_2((\delta_i, 1)), \ 1 \leq i \leq h_0.$$

Since $\tilde{\chi}_2|\tilde{\Gamma}_8$ is trivial and $(\delta_i, 1)^{-1} = (\delta_i^{-1}, 1)$ in $\widetilde{\mathrm{SL}(2,\mathbf{Z}_2)}$, it suffices to show $\xi_2(\gamma_i, \delta_i^{-1}) = 1$. This follows from the fact that $(2,1)$ entries of γ_i and δ_i^{-1} are $q_i l y_i$ and $-q_i l y_i$ respectively. Since $y_i \neq 0$, we have

$$\left(\begin{pmatrix}1&0\\0&q_i\end{pmatrix}u_0^{-1}q(\mu_i)^{-1}u_0\begin{pmatrix}1&0\\0&q_{i+1}\end{pmatrix}^{-1},1\right)$$
$$= \left(\begin{pmatrix}1&0\\0&q_i\end{pmatrix},1\right)(u_0,1)^{-1}\left(q(\mu_i)^{-1},1\right)(u_0,1)\left(\begin{pmatrix}1&0\\0&q_{i+1}\end{pmatrix},1\right)^{-1} \tag{6.4.8}$$

in $\widetilde{\mathrm{GL}(2,\mathbf{Q}_2)}$. Then we easily obtain

$$\tilde{\chi}_2((\gamma_i,1)) = \tilde{\chi}_2\left(\left(\begin{pmatrix}1&l'v_iq_i^{-1}\\0&1\end{pmatrix},1\right)\right)\tilde{\chi}_2\left(\left(\begin{pmatrix}1&-l'v_{i+1}q_{i+1}^{-1}\\0&1\end{pmatrix},1\right)\right)$$
$$\cdot\tilde{\chi}_2\left(\left(\begin{pmatrix}1&0\\0&q_i\end{pmatrix},1\right)(u_0,1)^{-1}\left(q(\mu_i)^{-1},1\right)(u_0,1)\left(\begin{pmatrix}1&0\\0&q_{i+1}\end{pmatrix},1\right)^{-1}\right) \tag{6.4.9}$$

for $1 \le i \le h_0 - 1$ and a similar formula for $i = h_0$ with obvious modifications. We shall show

$$\prod_{p\neq 2,q_i,l;\,y\in p\mathbf{Z}_p}(x_i,y_i)_p = \prod_{p\neq 2,q_i,l}\tilde{s}_p(q(\mu_i)^{-1}), \quad 1 \le i \le h_0. \tag{6.4.10}$$

We have $\det q(\mu_i)^{-1} = q_i^{-1}q_{i+1}$ for $1 \le i \le h_0-1$ and $\det q(\mu_{h_0})^{-1} = q_{h_0}^{-1}q_1$. By definition, we have

$$\tilde{s}_p(q(\mu_i)^{-1}) = \begin{cases} 1 & \text{if } 2\,|\,\mathrm{ord}_p y_i, \\ (y_i, x_iq_iq_{i+1})_p & \text{if } 2\nmid \mathrm{ord}_p(y_i) \end{cases}$$

where $(y_i, x_iq_iq_{i+1})_p$ is understood $(y_i, x_iq_{h_0}q_1)_p$ if $i = h_0$. Assume $p \neq 2$, q_i, l. Then we have x_i, $y_i \in \mathbf{Z}_p$ since $\mathfrak{q}_i(\mu_i^{-1}l)$ is an integral ideal. Assume $y_i \in p\mathbf{Z}_p$. Then we have $x_i \in \mathbf{Z}_p^\times$, since $x_i \in p\mathbf{Z}_p$ implies that $\mathfrak{q}_i(\mu_i^{-1}l)$ is divisible by (p), a contradiction. Hence $p \neq q_{i+1}$ in this case. We get $(y_i, x_iq_iq_{i+1})_p = (y_i, x_i)_p$ since $x_i^2 \equiv q_i^{-1}q_{i+1} \mod p$. Therefore we obtain (6.4.10).

Hereafter, until (6.4.13), we assume $h_0 \ge 2$. We put

$$\nu_i^{-1} = \mu_1^{-1}\mu_2^{-1}\cdots\mu_i^{-1}, \quad 1 \le i \le h_0, \quad \mu^{-1} = \mu_1^{-1}\mu_2^{-1}\cdots\mu_{h_0}^{-1} = \nu_{h_0}^{-1},$$
$$\nu_i^{-1} = X_i + Y_i\omega, \quad \mu^{-1} = x + y\omega, \quad X_i, Y_i, x, y \in \mathbf{Q}.$$

Then we have

$$(\nu_i^{-1}) = \mathfrak{q}_1^{-1}\mathfrak{q}_{i+1}(\mathfrak{l}^\rho\mathfrak{l}^{-1})^i, \quad 1 \le i \le h_0 - 1, \quad (\mu^{-1}) = (\mathfrak{l}^\rho\mathfrak{l}^{-1})^{h_0}.$$

We have

$$X_i \neq 0,\ Y_i \neq 0 \quad \text{and} \quad 2X_i + Y_i \neq 0 \text{ (when } d \equiv 3 \mod 4) \quad \text{for} \quad 1 \le i \le h_0.$$

LEMMA 6.4. *For $2 \le i \le h_0 - 1$, we have*

$$\tilde{\xi}_{q_1}\left(q(\nu_{i-1})^{-1}, q(\mu_i)^{-1}\right) = \begin{cases} 1 & \text{if } y_i \in \mathbf{Z}_{q_1}^\times, \\ (x_i, y_i)_{q_1} & \text{if } y_i \in q_1\mathbf{Z}_{q_1}, \end{cases}$$
$$\tilde{\xi}_{q_{i+1}}\left(q(\nu_{i-1})^{-1}, q(\mu_i)^{-1}\right) = \begin{cases} 1 & \text{if } Y_{i-1} \in \mathbf{Z}_{q_{i+1}}^\times, \\ (X_{i-1}, Y_{i-1})_{q_{i+1}} & \text{if } Y_{i-1} \in q_{i+1}\mathbf{Z}_{q_{i+1}}. \end{cases}$$

For $i = h_0$, we have

$$\tilde{\xi}_{q_1}\left(q(\nu_{h_0-1})^{-1}, q(\mu_{h_0})^{-1}\right) = \begin{cases} (\frac{\mu}{\mathfrak{q}_1}) & \text{if } y \in \mathbf{Z}_{q_1}^\times, \\ (\frac{\mu}{\mathfrak{q}_1})(x,y)_{q_1} & \text{if } y \in q_1\mathbf{Z}_{q_1}. \end{cases}$$

For $2 \leq i \leq h_0$, *we have*

$$\tilde{\xi}_{q_i}\left(q(\nu_{i-1})^{-1}, q(\mu_i)^{-1}\right) = \begin{cases} 1 & \text{if } Y_i \in \mathbf{Z}_{q_i}^{\times}, \\ (X_i, Y_i)_{q_i} & \text{if } Y_i \in q_i\, Z_{q_i}, \end{cases}$$

$$\tilde{\xi}_l\left(q(\nu_{i-1})^{-1}, q(\mu_i)^{-1}\right)$$
$$= \begin{cases} (l, -2X_i q_1 q_i)_l & \text{if } i \text{ is even and } d \equiv 0 \mod 4, \\ (l, -(2X_i + Y_i) q_1 q_i)_l & \text{if } i \text{ is even and } d \equiv 3 \mod 4, \\ (l, 2X_{i-1} q_1 q_i)_l & \text{if } i \text{ is odd and } d \equiv 0 \mod 4, \\ (l, (2X_{i-1} + Y_{i-1}) q_1 q_i)_l & \text{if } i \text{ is odd and } d \equiv 3 \mod 4. \end{cases}$$

PROOF. We prove only the first assertion when $d \equiv 0 \mod 4$. Remaining assertions can be shown similarly. By direct computation, we have

$$\tilde{\xi}_p(q(\nu_{i-1})^{-1}, q(\mu_i)^{-1}) = (q_1 q_{i+1} Y_{i-1}, q_i q_{i+1} y_i)_p (-q_1 q_i Y_{i-1} y_i, q_1 q_{i+1} Y_i)_p$$

for any prime number p. We have $q_1 X_{i-1}$, $q_1 Y_{i-1}$, $q_1 X_i$, $q_1 Y_i \in \mathbf{Z}_{q_1}^{\times}$, x_i, $y_i \in \mathbf{Z}_{q_1}$. Hence the above expression simplifies to

$$\tilde{\xi}_{q_1}\left(q(\nu_{i-1})^{-1}, q(\mu_i)^{-1}\right) = (Y_{i-1} Y_i, y_i)_{q_1}.$$

If $y_i \in \mathbf{Z}_{q_1}^{\times}$, this is equal to 1. Assume $y_i \in q_1 \mathbf{Z}_{q_1}$. Then we have $\mu_i^{-1} \equiv x_i \mod \mathfrak{q}_1$. We have

$$q_1 Y_{i-1} = \frac{1}{2\sqrt{-d_1}} q_1 \left(\nu_{i-1}^{-1} - \bar{\nu}_{i-1}^{-1}\right) \equiv \frac{1}{2\sqrt{-d_1}} q_1 \nu_{i-1}^{-1} \mod \mathfrak{q}_1,$$
$$q_1 Y_i \equiv \frac{1}{2\sqrt{-d_1}} q_1 \nu_i^{-1} \mod \mathfrak{q}_1.$$

Hence we have

$$q_1^2 Y_{i-1} Y_i \equiv \frac{1}{-4d_1} q_1^2 \nu_{i-1}^{-1} \nu_i^{-1} \equiv \frac{1}{-d} q_1^2 \nu_{i-1}^{-2} \mu_i^{-1} \equiv \frac{1}{-d} q_1^2 \nu_{i-1}^{-2} x_i \mod \mathfrak{q}_1.$$

Since $\left(\frac{-d}{q_1}\right) = 1$, we have $(Y_{i-1} Y_i, y_i)_{q_1} = (x_i, y_i)_{q_1}$. This completes the proof. □

For $2 \leq i \leq h_0$, we put $U_i = \tilde{\xi}_l\left(q(\nu_{i-1})^{-1}, q(\mu_i)^{-1}\right)$. Now we shall show

$$\text{(6.4.11)} \quad \prod_{i=1}^{j} \tilde{\chi}(\tilde{\gamma}_i) \left(\mu_i^{1/2}\right)^{-1} = -\operatorname{sgn}(Y_j) \left(\nu_j^{1/2}\right)^{-1} \prod_{p \neq 2, l; Y_j \in p\mathbf{Z}_p} (X_j, Y_j)_p \prod_{i=2}^{j} U_i$$
$$\cdot \tilde{\chi}_2\left(\left(\begin{pmatrix} 1 & 0 \\ 0 & q_1 \end{pmatrix}, 1\right) (u_0, 1)^{-1} \left(q(\nu_j)^{-1}, 1\right) (u_0, 1) \left(\begin{pmatrix} 1 & 0 \\ 0 & q_{j+1} \end{pmatrix}, 1\right)^{-1}\right)$$
$$\cdot \tilde{\chi}_2\left(\left(\begin{pmatrix} 1 & l' v_1 q_1^{-1} \\ 0 & 1 \end{pmatrix}, 1\right)\right) \tilde{\chi}_2\left(\left(\begin{pmatrix} 1 & -l' v_{j+1} q_{j+1}^{-1} \\ 0 & 1 \end{pmatrix}, 1\right)\right) \chi_3(\gamma_1 \gamma_2 \cdots \gamma_j)$$

for $1 \leq j \leq h_0 - 1$. Here we understand that $\prod_{i=2}^{j} U_i = 1$ if $j = 1$.

If $j = 1$, then $\nu_1 = \mu_1$ and (6.4.11) follows from (6.4.6) and (6.4.9) noting that $y_1 \notin l\mathbf{Z}_l$. Now we are going to prove (6.4.11) by induction on j. Assume that (6.4.11) holds for $j-1$. Multiplying (6.4.11) for $j-1$ by $\tilde{\chi}(\tilde{\gamma}_j)(\mu_j^{1/2})^{-1}$ using (6.4.6) and (6.4.9) and comparing with (6.4.11), we see that it suffices to show

$$
\begin{aligned}
(6.4.12)\qquad & \operatorname{sgn}(Y_{j-1})\operatorname{sgn}(y_j)\left(\nu_{j-1}^{1/2}\right)^{-1}\left(\mu_j^{1/2}\right)^{-1}\tilde{\xi}_2\left(q(\nu_{j-1})^{-1}, q(\mu_j)^{-1}\right)\\
&\cdot \prod_{p\neq 2,l;Y_{j-1}\in p\mathbf{Z}_p}(X_{j-1},Y_{j-1})_p \prod_{p\neq 2,y_j\in p\mathbf{Z}_p}(x_j,y_j)_p\\
&= -\operatorname{sgn}(Y_j)\left(\nu_j^{1/2}\right)^{-1}\prod_{p\neq 2,l;Y_j\in p\mathbf{Z}_p}(X_j,Y_j)\cdot U_j.
\end{aligned}
$$

We note that $Y_{j-1}\notin p\mathbf{Z}_p$ for $p=q_1,\ q_j,\ l$ and that $y_j\notin p\mathbf{Z}_p$ for $p=q_j,\ q_{j+1},\ l$. We have

$$
\begin{aligned}
&\prod_{p\neq 2,l;Y_{j-1}\in p\mathbf{Z}_p}(X_{j-1},Y_{j-1})_p \prod_{p\neq 2,y_j\in p\mathbf{Z}_p}(x_j,y_j)_p\\
&\quad= \prod_{p\neq 2,q_1,q_j,l}\tilde{s}_p\left(q(\nu_{j-1})^{-1}\right)\prod_{p\neq 2,q_j,q_{j+1},l}\tilde{s}_p\left(q(\mu_j)^{-1}\right)\\
&\quad= \tilde{s}_{q_{j+1}}\left(q(\nu_{j-1})^{-1}\right)\tilde{s}_{q_1}\left(q(\mu_j)^{-1}\right)\\
&\qquad \prod_{p\neq 2,q_1,q_j,q_{j+1},l}\tilde{s}_p\left(q(\nu_j)^{-1}\right)\tilde{\xi}_p\left(q(\nu_{j-1})^{-1},q(\mu_j)^{-1}\right)\\
&\quad= \tilde{s}_{q_{j+1}}\left(q(\nu_{j-1})^{-1}\right)\tilde{s}_{q_1}\left(q(\mu_j)^{-1}\right)\prod_{p\neq 2,q_1,q_j,q_{j+1},l}\tilde{s}_p\left(q(\nu_j)^{-1}\right)\\
&\cdot\tilde{\xi}_2\left(q(\nu_{j-1})^{-1},q(\mu_j)^{-1}\right)\tilde{\xi}_{q_1}\left(q(\nu_{j-1})^{-1},q(\mu_j)^{-1}\right)\tilde{\xi}_{q_j}\left(q(\nu_{j-1})^{-1},q(\mu_j)^{-1}\right)\\
&\cdot\tilde{\xi}_{q_{j+1}}\left(q(\nu_{j-1})^{-1},q(\mu_j)^{-1}\right)\tilde{\xi}_l\left(q(\nu_{j-1})^{-1},q(\mu_j)^{-1}\right)\xi_{\mathbf{R}}\left(q(\nu_{j-1})^{-1},q(\mu_j)^{-1}\right)
\end{aligned}
$$

by the product formula of the Hilbert symbol. By Lemma 6.4, this is equal to

$$
\prod_{p\neq 2,q_1,q_{j+1},l}\tilde{s}_p\left(q(\nu_j)^{-1}\right)\cdot\tilde{\xi}_2\left(q(\nu_{j-1})^{-1},q(\mu_j)^{-1}\right)U_j\xi_{\mathbf{R}}\left(q(\nu_{j-1})^{-1},q(\mu_j)^{-1}\right).
$$

Since $Y_j\notin p\mathbf{Z}_p$ for $p=q_1,\ q_{j+1},\ l$, we have

$$
\prod_{p\neq 2,q_1,q_{j+1},l}\tilde{s}_p\left(q(\nu_j)^{-1}\right)=\prod_{p\neq 2,l;Y_j\in p\mathbf{Z}_p}(X_j,Y_j)_p.
$$

Now we see that (6.4.12) holds by Lemma 6.3. We multiply (6.4.11) for the case $j=h_0-1$ by $\tilde{\chi}(\tilde{\gamma}_{h_0})(\mu_{h_0}^{1/2})^{-1}$. By (6.4.6), (6.4.9) and using Lemmas 6.3, 6.4 in the similar manner as above, we obtain

$$
\begin{aligned}
(6.4.13)\qquad & \prod_{i=1}^{h_0}\tilde{\chi}(\tilde{\gamma}_i)\left(\mu_i^{1/2}\right)^{-1} = -\operatorname{sgn}(y)\left(\mu^{1/2}\right)^{-1}\prod_{p\neq 2,y\in p\mathbf{Z}_p}(x,y)_p\left(\frac{\mu}{\mathfrak{q}_1}\right)\\
&\cdot\tilde{\chi}_2\left(\left(\begin{pmatrix}1&0\\0&q_1\end{pmatrix},1\right)\left(u_0^{-1}q(\mu)^{-1}u_0,1\right)\left(\begin{pmatrix}1&0\\0&q_1\end{pmatrix},1\right)^{-1}\right)\\
&\cdot\chi_3\left(\begin{pmatrix}1&0\\0&q_1\end{pmatrix}u_0^{-1}q(\mu)^{-1}u_0\begin{pmatrix}1&0\\0&q_1\end{pmatrix}^{-1}\right)
\end{aligned}
$$

$$\times \begin{cases} (l,-1)_l^{h_0/2}(l,2xq_1q_2\cdots q_{h_0})_l \\ \quad \text{if } h_0 \text{ is even and } d\equiv 0 \mod 4, \\ (l,-1)_l^{h_0/2}(l,(2x+y)q_1q_2\cdots q_{h_0})_l \\ \quad \text{if } h_0 \text{ is even and } d\equiv 3 \mod 4, \\ (l,-1)_l^{(h_0-1)/2}(l,q_2q_3\cdots q_{h_0})_l \\ \quad \text{if } h_0 \text{ is odd.} \end{cases}$$

To derive (6.4.13), we have assumed that $h_0 \geq 2$. Comparing with (6.4.7), we see that (6.4.13) holds also for $h_0 = 1$ if we understand the last term as $(l, q_2q_3\cdots q_{h_0})_l = 1$.

We have $(\mu^{-1}) = (\mathfrak{l}^\rho\mathfrak{l}^{-1})^{h_0}$. Hence $N(\mu) = 1$ and $\mu^{-1} = \bar{\nu}/\nu$ with $\nu \in K^\times$. We see easily that ν can be chosen so that

$$(\nu) = \mathfrak{l}^{h_0}\mathfrak{a}, \tag{6.4.14}$$

where $\mathfrak{a}$ is a square free product of prime ideals which ramify in $K/\mathbf{Q}$. We may assume that either $\mathfrak{a} = (1)$ or $\mathfrak{a}$ is a non-principal ideal, at the cost of changing μ to $-\mu$ if necessary. Let h_1 be the order of the class of $\mathfrak{l}$ in C_K. Since the class of $\mathfrak{l}(\mathfrak{l}^\rho)^{-1}$ is the square of the class of $\mathfrak{l}$, we obtain

LEMMA 6.5. *If h_1 is even, we have $h_0 = h_1/2$ and $\mathfrak{a} \neq (1)$. If h_1 is odd, we have $h_0 = h_1$ and $\mathfrak{a} = (1)$.*

We have

$$\begin{aligned}\left(\frac{\mu}{\mathfrak{q}_1}\right) &= \left(\frac{\bar{\nu}^{-2}\cdot\nu\bar{\nu}}{\mathfrak{q}_1}\right) = \left(\frac{l^{h_0}N(\mathfrak{a})}{\mathfrak{q}_1}\right) \\ &= \left(\frac{q_1}{l}\right)^{h_0}\left(\frac{N(\mathfrak{a})}{q_1}\right)(-1)^{\frac{l-1}{2}\cdot\frac{q_1-1}{2}\cdot h_0}.\end{aligned} \tag{6.4.15}$$

For $\tau \in \mathrm{Gal}(\mathbf{Q}(\zeta_{24})/\mathbf{Q})$, we can define a character $\tilde{\chi}_2^\tau$ and χ_3^τ by $\tilde{\chi}_2^\tau(x) = (\tilde{\chi}_2(x))^\tau$, $x \in \widetilde{\mathrm{SL}(2,\mathbf{Z}_2)}$, $\chi_3^\tau(x) = (\chi_3(x))^\tau$, $x \in \mathrm{SL}(2,\mathbf{Z}_3)$. Define $\tau(q_1) \in \mathrm{Gal}(\mathbf{Q}(\zeta_{24})/\mathbf{Q})$ by $\zeta_{24}^{\tau(q_1)} = \zeta_{24}^{q_1}$. We can easily verify that

$$\tilde{\chi}_2\left(\left(\begin{pmatrix}1&0\\0&q_1\end{pmatrix},1\right)x\left(\begin{pmatrix}1&0\\0&q_1\end{pmatrix},1\right)^{-1}\right) = \tilde{\chi}_2^{\tau(q_1)}(x), \quad x \in \widetilde{\mathrm{SL}(2,\mathbf{Z}_2)},$$

$$\chi_3\left(\begin{pmatrix}1&0\\0&q_1\end{pmatrix}x\begin{pmatrix}1&0\\0&q_1\end{pmatrix}^{-1}\right) = \chi_3^{\tau(q_1)}(x), \quad x \in \mathrm{SL}(2,\mathbf{Z}_3).$$

Note that

$$\sqrt{q_i}^\sigma = \left(\frac{q_i}{l}\right)\sqrt{q_i}, \qquad \frac{(i^{m+r})^\sigma}{i^{m+r}} = \left(\frac{-1}{l}\right)^{m+r} = \left(\frac{-1}{l}\right)^k\left(\frac{-1}{l}\right)^{(k-m)/2}.$$

Let C_K^0 be the subgroup of C_K generated by the class of $\mathfrak{l}^\rho\mathfrak{l}^{-1}$. To compute $\prod_{i=1}^h \tilde{\chi}(\tilde{\gamma}_i)(\mu_i^{1/2})^{-1}$, we must make the product of (6.4.13) over the cosets C_K/C_K^0. For $1 \leq i \leq h/h_0$, let $q_1^{(i)}, q_2^{(i)}, \cdots, q_{h_0}^{(i)}$ denote $q_1, q_2, \cdots, q_{h_0}$ corresponding to the i-th coset. We put

$$\text{(6.4.16)} \qquad \Xi(q_1^{(i)}) = \left(\frac{N(\mathfrak{a})}{q_1^{(i)}}\right) \tilde{\chi}_2^{\tau(q_1^{(i)})}\left((u_0^{-1}q(\mu)^{-1}u_0, 1)\right) \cdot \begin{cases} 1 & \text{if } h_0 \text{ is even,} \\ (-1)^{(l-1)/2\cdot(q_1^{(i)}-1)/2} & \text{if } h_0 \text{ is odd.} \end{cases}$$

Then by (6.4.2) and (6.4.13), we obtain

$$\text{(6.4.17)} \qquad \zeta(\sigma, \lambda) = \lambda^\sigma(\mathfrak{X})^{-1} \cdot \frac{(i^{m+r})^\sigma}{i^{m+r}} Z^n,$$

$$Z = (-\mathrm{sgn}(y)\overline{(\mu^{1/2})}^{-1} \prod_{p\neq 2, y\in p\mathbf{Z}_p} (x,y)_p)^{h/h_0}$$

$$\cdot \prod_{i=1}^{h/h_0} \overline{\Xi(q_1^{(i)})} \prod_{i=1}^{h/h_0} \overline{\chi_3^{\tau(q_1^{(i)})}\left(u_0^{-1}q(\mu)^{-1}u_0\right)}$$

$$\cdot \begin{cases} (l,-1)_l^{h/2}(l,2x)_l^{h/h_0} & \text{if } h_0 \text{ is even and } d \equiv 0 \mod 4, \\ (l,-1)_l^{h/2}(l,(2x+y))_l^{h/h_0} & \text{if } h_0 \text{ is even and } d \equiv 3 \mod 4, \\ (l,-1)_l^{(h_0-1)h/2h_0} & \text{if } h_0 \text{ is odd.} \end{cases}$$

LEMMA 6.6. *Assume $K \neq \mathbf{Q}(\sqrt{-3})$. Then we have*

$$\prod_{i=1}^{h/h_0} \chi_3^{\tau(q_1^{(i)})}\left(u_0^{-1}q(\mu)^{-1}u_0\right) = 1.$$

The proof is not difficult. We leave it to the reader.

LEMMA 6.7. *Put $\zeta = \tilde{\chi}_2((q(\mu)^{-1}, 1))$. Assume $d \equiv 3 \mod 4$ or $d \equiv 0 \mod 4$ and $d_1 \equiv 2 \mod 4$. Let q be an odd prime which does not divide $N(\mathfrak{a})$. If $N(\mathfrak{a})$ is odd, we have $\zeta \in \langle i \rangle$ and*

$$\text{(6.4.18)} \qquad \left(\frac{N(\mathfrak{a})}{q}\right)\frac{\zeta^q}{\zeta} \cdot \begin{cases} 1 & \text{if } h_0 \text{ is even} \\ (-1)^{\frac{l-1}{2}\cdot\frac{q-1}{2}} & \text{if } h_0 \text{ is odd} \end{cases} = \left(\frac{(-1)^{(N(\mathfrak{a})-1)/2}N(\mathfrak{a})}{q}\right).$$

PROOF. Since $l^{h_0}N(\mathfrak{a}) = N(\nu)$, (6.4.18) reduces to

$$\text{(6.4.19)} \qquad \zeta^q/\zeta = (-1)^{\frac{q-1}{2}\cdot\frac{N(\nu)-1}{2}}.$$

Put $\nu = s + t\omega$ with $s, t \in \mathbf{Z}$. We assume $d \equiv 0 \mod 4$, $d_1 \equiv 2 \mod 4$. Since

$$x = \frac{s^2 - t^2 d_1}{N(\nu)}, \quad y = -\frac{2st}{N(\nu)},$$

we have $y \in 2\mathbf{Z}_2$ and

$$\begin{pmatrix} x & -yd_1 \\ y & x \end{pmatrix} = \begin{pmatrix} 1 & 0 \\ x^{-1}y & 1 \end{pmatrix}\begin{pmatrix} x & 0 \\ 0 & x^{-1} \end{pmatrix}\begin{pmatrix} 1 & -x^{-1}yd_1 \\ 0 & 1 \end{pmatrix}$$

in $\mathrm{SL}(2, \mathbf{Z}_2)$. Hence we get $\zeta = \pm\tilde{\chi}_2((\begin{smallmatrix} x & 0 \\ 0 & x^{-1}\end{smallmatrix}), 1)\zeta_8^{-3xy(d_1+1)}$. This proves that $\zeta \in \langle i \rangle$. Now the assertion is clear if $q \equiv 1 \mod 4$. Assume $q \equiv 3 \mod 4$. Since $N(\nu) = s^2 + t^2 d_1 \equiv s^2 + 2t^2 \mod 4$, s must be odd. If t is even, we have $N(\nu) \equiv 1$

mod 4, $y \in 4\mathbf{Z}_2$ and $x \equiv 1 \mod 4$. Hence we have $\zeta = \pm 1$. If t is odd, we have $N(\nu) \equiv 3 \mod 4$, $x \equiv 1 \mod 4$ and $y \in 2\mathbf{Z}_2^\times$. Hence we get $\zeta = \pm i$. This settles the case $d \equiv 0 \mod 4$ and $d_1 \equiv 2 \mod 4$. The case $d \equiv 3 \mod 4$ is more complicated but can be shown similarly. This completes the proof. □

LEMMA 6.8. *Put* $X = \prod_{i=1}^{h/h_0} \Xi(q_1^{(i)})$. *Then we have:*
(1) *Assume* $d \equiv 3 \mod 4$ *or* $d \equiv 0 \mod 4$ *and* $d_1 \equiv 2 \mod 4$. *If* $N(\mathfrak{a})$ *is odd, we have*

$$X = \begin{cases} \tilde{\chi}_2((u_0^{-1}q(\mu)^{-1}u_0, 1))^{h/h_0} & \text{if} \quad \mathfrak{a} = (1), \\ (-1)^{h/2h_0}\tilde{\chi}_2((u_0^{-1}q(\mu)^{-1}u_0, 1))^{h/h_0} & \text{if} \quad \mathfrak{a} \neq (1). \end{cases}$$

(2) *Assume* $d \equiv 0 \mod 4$, $d_1 \equiv 1 \mod 4$, $K \neq \mathbf{Q}(\sqrt{-1})$ *and that* $N(\mathfrak{a})$ *is odd. Then we have* $\tilde{\chi}_2((q(\mu)^{-1}, 1)) \in \langle i \rangle$ *and*

$$X = \begin{cases} 1 & \text{if } \mathfrak{a} = (1) \\ (-1)^{h/2h_0} & \text{if } \mathfrak{a} \neq (1) \end{cases} \cdot \begin{cases} 1 & \text{if } h_0 \text{ is even,} \\ (\frac{-1}{l})^{h/2h_0} & \text{if } h_0 \text{ is odd.} \end{cases}$$

(3) *Assume that* $d \equiv 0 \mod 4$, $d_1 \equiv 1 \mod 4$, $K \neq \mathbf{Q}(\sqrt{-1})$ *and that* $N(\mathfrak{a})$ *is even. Then* $\tilde{\chi}_2((q(\mu)^{-1}, 1))$ *is a primitive* 8*-th root of unity and we have*

$$X = \tilde{\chi}_2((u_0^{-1}q(\mu)^{-1}u_0, 1))^{2h/h_0} \cdot \begin{cases} (-1)^{h/2h_0} & \text{if } N(\mathfrak{a}) = 2 \\ 1 & \text{if } N(\mathfrak{a}) \neq 2 \end{cases}$$

$$\cdot \begin{cases} 1 & \text{if } h_0 \text{ is even} \\ (\frac{-1}{l})^{h/2h_0} & \text{if } h_0 \text{ is odd.} \end{cases}$$

PROOF. We shall prove (1). The proofs of (2) and (3) are similar. The right hand side of (6.4.18) can be written as $\left(\frac{q}{N(\mathfrak{a})}\right)$ using the Jacobi symbol. For $c \in C_K$, take a prime ideal $\mathfrak{q}$ of the first degree in the class c such that $\mathfrak{q}$ is relatively prime to $\mathfrak{a}$ and put $\varphi(c) = \left(\frac{(-1)^{(N(\mathfrak{a})-1)/2}N(\mathfrak{a})}{N(\mathfrak{q})}\right)$. This is well defined and φ defines a character of C_K (a genus character). Since $C_K^2 \supseteq C_K^0$, we see that φ is trivial on C_K^0. We have

$$\prod_{i=1}^{h/h_0} \Xi(q_1^{(i)}) = \prod_{c \in C_K/C_K^0} \varphi(c) \cdot \tilde{\chi}_2((u_0^{-1}q(\mu)^{-1}u_0, 1))^{h/h_0}.$$

Since we have assumed that $\mathfrak{a} = (1)$ if $\mathfrak{a}$ is principal, φ is non-trivial if $\mathfrak{a} \neq (1)$. Hence the assertion follows. □

REMARK. If $d \equiv 3 \mod 4$, $N(\mathfrak{a})$ must be odd. If $d \equiv 0 \mod 4$ and $d_1 \equiv 2 \mod 4$, we can always reduce to the case (1) of Lemma 6.8 changing ν to $N(\mathfrak{b})\nu/\sqrt{-d_1}$ where $\mathfrak{ab} = (\sqrt{-d_1})$, at the cost of changing μ to $-\mu$.

Fifth step: Final results

THEOREM 6.9. *Let* λ *be as in the beginning of* §6 *with* $k = hn$, $0 < n \in \mathbf{Z}$ *and* m *be an integer such that* $1 \leq m \leq k$, $m \equiv k \mod 2$. *Let* $K(\lambda)$ *be the field generated by* $\lambda_0(\mathfrak{a})$, $\mathfrak{a} \in I_{\mathfrak{f}}(K)$ *over* K *and* $C_{\mathfrak{m}}$ *be the maximal ray class field of conductor* $\mathfrak{m}$ *of* K. *Then we have* $(2\pi)^{-m/2}P^{-hn/2}L(m/2, \lambda) \in K(\lambda) \vee C_{\mathfrak{m}}$, *the composite field of* $K(\lambda)$ *and* $C_{\mathfrak{m}}$. *Let* $\sigma \in \mathrm{Aut}(\mathbf{C})$ *be the automorphism which satisfies* $\sigma|K = \mathrm{id.}$,

$\sigma|C_{\mathfrak{m}} = \left(\frac{C_{\mathfrak{m}}/K}{\mathfrak{X}}\right)$. *Then we have*

$$\text{(Z)} \qquad \left(\frac{L(m/2,\lambda)}{(2\pi)^{m/2}P^{hn/2}}\right)^{\sigma} = \zeta(\sigma,\lambda)\frac{L(m/2,\lambda^{\sigma})}{(2\pi)^{m/2}P^{hn/2}}.$$

Here $\mathfrak{m}$ *and* $\zeta(\sigma,\lambda)$ *are determined according as the following cases. Let* $f(d)$ *be the number of the prime factors of* d *and* h_1 *be the order of* $\mathfrak{X}$ *in the ideal class group of* K.

Case A. We assume that $f(d) \geq 3$. *Put* $\mathfrak{X}^{h/2} = (\alpha)$. *Then we have* $\mathfrak{m} = \mathrm{LCM}((4),\mathfrak{f})$ *and*

$$\zeta(\sigma,\lambda) = \left(\frac{-1}{N(\mathfrak{X})}\right)^{(k-m)/2} \lambda^{\sigma}(\mathfrak{X})^{-1}\left(\frac{\bar{\alpha}}{\alpha}\right)^{n}.$$

Case B. We assume that $d \equiv 0 \mod 4$, $d/4 \equiv 1 \mod 4$ *and* $d/4$ *is a prime number. Put* $(2) = \mathfrak{p}_2^2$, $\mathfrak{X}^{h/2} = (\alpha)\mathfrak{p}_2^{h/h_1}$. *Then we have* $\mathfrak{m} = LCM((2),\mathfrak{f})$ *and*

$$\zeta(\sigma,\lambda) = \left(\frac{-1}{N(\mathfrak{X})}\right)^{(k-m)/2} \lambda^{\sigma}(\mathfrak{X})^{-1}\left(\frac{\bar{\alpha}}{\alpha}\right)^{n}\left(\frac{-1}{N(\mathfrak{X})}\right)^{hn/2}.$$

Case C. We assume that $d \equiv 0 \mod 4$, $d/4 \equiv 2 \mod 4$ *and* $d_2 = d/8$ *is a prime number. Put* $(d_2) = \mathfrak{p}^2$, $\mathfrak{X}^{h/2} = (\alpha)\mathfrak{p}^{h/h_1}$. *Then we have* $\mathfrak{m} = \mathrm{LCM}((2),\mathfrak{f})$ *and*

$$\zeta(\sigma,\lambda) = \left(\frac{-1}{N(\mathfrak{X})}\right)^{(k-m)/2} \lambda^{\sigma}(\mathfrak{X})^{-1}\left(\frac{\bar{\alpha}}{\alpha}\right)^{n}\left(\frac{-1}{N(\mathfrak{X})}\right)^{hn/2}\left(\frac{-1}{d_2}\right)^{hn/h_1}(-1)^{hn/h_1}.$$

Case D. We assume that $d \equiv 3 \mod 4$ *and* $f(d) = 2$. *Let* d_2 *be a prime factor of* d. *Put* $(d_2) = \mathfrak{p}^2$, $\mathfrak{X}^{h/2} = (\alpha)\mathfrak{p}^{h/h_1}$. *Then we have* $\mathfrak{m} = \mathrm{LCM}((4),\mathfrak{f})$ *and*

$$\zeta(\sigma,\lambda) = \left(\frac{-1}{N(\mathfrak{X})}\right)^{(k-m)/2} \lambda^{\sigma}(\mathfrak{X})^{-1}\left(\frac{\bar{\alpha}}{\alpha}\right)^{n}\left(\frac{-1}{N(\mathfrak{X})}\right)^{hn/2}\left(\frac{-1}{d_2}\right)^{hn/h_1}(-1)^{hn/h_1}.$$

Case E. We assume that $d \equiv 3 \mod 4$ *and* d *is a prime number or that* $d = 4$ *or that* $d = 8$. *Put* $\mathfrak{X}^{h} = (\alpha)$. *Then we have* $\mathfrak{m} = \mathrm{LCM}((8),\mathfrak{f})$. If $\mathrm{Tr}(\alpha) \neq 0$, *we have*

$$\zeta(\sigma,\lambda) = \left(\frac{-1}{N(\mathfrak{X})}\right)^{(k-m)/2} \lambda^{\sigma}(\mathfrak{X})^{-1}\left(\frac{\bar{\alpha}}{|\alpha|}\right)^{n} \cdot\left[\left(\frac{-1}{N(\mathfrak{X})}\right)^{(h+1)/2}(-1,\mathrm{Tr}(\alpha))_2\,\overline{\tilde{\chi}_2\left(\left(q\left(\frac{\alpha}{\bar{\alpha}}\right),1\right)\right)}\right]^{n}$$

if $d \neq 3,4$. *If* $d = 4$, *this formula holds by choosing* α *so that* $\alpha \equiv 1 \mod (2)$. *If* $d = 3$, *then* $\mathfrak{m} = \mathrm{LCM}((24),\mathfrak{f})$ *and*

$$\zeta(\sigma,\lambda) = \left(\frac{-1}{N(\mathfrak{X})}\right)^{(k-m)/2} \lambda^{\sigma}(\mathfrak{X})^{-1}\left(\frac{\bar{\alpha}}{|\alpha|}\right)^{n} \cdot\left[\left(\frac{-1}{N(\mathfrak{X})}\right)(-1,\mathrm{Tr}(\alpha))_2\,\overline{\tilde{\chi}_2\left(\left(q(\left(\frac{\alpha}{\bar{\alpha}}\right),1\right)\right)}\,\overline{\chi_3\left(q\left(\frac{\alpha}{\bar{\alpha}}\right)\right)}\right]^{n}.$$

In the proof given below, we shall prove (Z) with the asserted values of $\zeta(\sigma,\lambda)$ only for $\mathfrak{X} = \mathfrak{l}^{-1}$, $\sigma|C_M = ((C_M/K)/\mathfrak{l}^{-1})$, where $\mathfrak{l}$ is a prime ideal of degree 1, $\mathfrak{l}$ does not divide (M) and $l = N(\mathfrak{l})$ satisfies (A) if $d \equiv 3 \mod 4$. We shall show that the full assertion follows from this particular case. Note that $\mathfrak{m}$ divides (M). First suppose that $\sigma|C_{\mathfrak{m}} = \mathrm{id}$. We can find $\mathfrak{l}$ so that it satisfies the above conditions and that $\sigma|C_M = ((C_M/K)/\mathfrak{l}^{-1})$. We must have $\mathfrak{l}^{-1} = (\beta)$, $\beta \equiv 1 \mod {}^{\times}\mathfrak{m}$. Then we see easily that $\zeta(\sigma,\lambda) = 1$. Thus we obtain $(2\pi)^{-m/2}P^{-hn/2}L(m/2,\lambda) \in K(\lambda)\vee C_{\mathfrak{m}}$.

Next let $\sigma|C_{\mathfrak{m}} = ((C_{\mathfrak{m}}/K)/\mathfrak{X})$ as in the theorem. We can take $\mathfrak{l}$ so that it satisfies the above conditions and that $\sigma|C_M = ((C_M/K)/\mathfrak{l}^{-1})$. Then we have $\mathfrak{X} = \mathfrak{l}^{-1}(\beta)$, $\beta \equiv 1 \mod {}^{\times}\mathfrak{m}$. Now we can verify that $\zeta(\sigma, \lambda)$ calculated for $\mathfrak{l}^{-1}$ and $\mathfrak{l}^{-1}(\beta)$ gives the same value. (To check Case E, use Lemma 6.10.) Hence the assertion for $\mathfrak{X}$ holds true.

As is well known, we have $C_K/C_K^2 \cong (\mathbf{Z}/2\mathbf{Z})^{f(d)-1}$. If $f(d) \geq 2$, then h/h_0 is even and (6.4.17) simplifies to

$$\zeta(\sigma, \lambda) = \left(\frac{-1}{l}\right)^{(k-m)/2} \lambda^{\sigma}(\mathfrak{X})^{-1}\bar{\mu}^{-hn/2h_0} \left(\prod_{i=1}^{h/h_0} \overline{\Xi(q_1^{(i)})}\right)^n, \quad f(d) \geq 2. \tag{6.5.1}$$

PROOF of A. Since $f(d) \geq 3$, h/h_0 is divisible by 4 and $\mathfrak{l}^{h/2}$ is a principal ideal. By Lemma 6.8, we obtain $\prod_{i=1}^{h/h_0} \Xi(q_1^{(i)}) = 1$. Put $\mathfrak{X}^{h/2} = \mathfrak{l}^{-h/2} = (\alpha)$. Since $(\mu^{-1}) = (\mathfrak{l}^{\rho}\mathfrak{l}^{-1})^{h_0}$, we have

$$\left(\bar{\mu}^{-h/4h_0}\right) = \left(\mathfrak{l}(\mathfrak{l}^{\rho})^{-1}\right)^{h/4} = \left(l^{-1}\mathfrak{l}^2\right)^{h/4} = \left(\alpha^{-1}l^{-h/4}\right).$$

Hence $\bar{\mu}^{-h/4h_0} = \pm\alpha^{-1}l^{-h/4}$ and $\bar{\mu}^{-h/2h_0} = l^{-h/2}\alpha^{-2} = \bar{\alpha}/\alpha$. We have

$$\zeta(\sigma, \lambda) = \left(\frac{-1}{l}\right)^{(k-m)/2} \lambda^{\sigma}(\mathfrak{X})^{-1}\left(\frac{\bar{\alpha}}{\alpha}\right)^n, \quad \mathfrak{X}^{h/2} = (\alpha), \quad f(d) \geq 3. \tag{6.5.2}$$

Hence A follows. □

PROOF of B. We have $f(d) = 2$ and d_1 is a prime number. Since

$$\tilde{\chi}_2\left((u_0^{-1}q(\mu)^{-1}u_0, 1)\right)^{2h/h_0} = (-1)^{h/2h_0}$$

in Lemma 6.8, (3), we obtain

$$\prod_{i=1}^{h/h_0} \Xi(q_1^{(i)})$$
$$= \begin{cases} 1 & \text{if } N(\mathfrak{a}) = 1 \text{ or } 2 \\ (-1)^{h/2h_0} & \text{if } N(\mathfrak{a}) \neq 1 \text{ and } \neq 2 \end{cases} \times \begin{cases} 1 & \text{if } h_0 \text{ is even} \\ \left(\frac{-1}{l}\right)^{h/2h_0} & \text{if } h_0 \text{ is odd} \end{cases}$$

by Lemma 6.8, (2) and (3). By (6.5.1), we obtain

$$\begin{aligned} \zeta(\sigma, \lambda) = &\left(\frac{-1}{l}\right)^{(k-m)/2} \lambda^{\sigma}(\mathfrak{X})^{-1}\bar{\mu}^{-hn/2h_0} \\ &\cdot \left[\left(\frac{-1}{l}\right)^{h/2} \times \begin{cases} 1 & \text{if } N(\mathfrak{a}) = 1 \text{ or } 2 \\ (-1)^{h/2h_0} & \text{if } N(\mathfrak{a}) \neq 1 \text{ and } \neq 2 \end{cases}\right]^n. \end{aligned} \tag{6.5.3}$$

We have four possibilities: $N(\mathfrak{a}) = 1$, d_1, 2 or $2d_1$. Replacing ν by $\nu/\sqrt{-d_1}$ if necessary, we may assume that $N(\mathfrak{a}) = 1$ or 2. Assume h_1 is odd. Then, by Lemma 6.5, we have $h_0 = h_1$, $\mathfrak{a} = (1)$ and $\mathfrak{l}^{h_1} = (\nu)$. We also have $\mathfrak{l}^{-h/2} = (\alpha)$ with $\alpha = \nu^{-h/2h_1}$. Hence $\bar{\mu}^{-h/2h_0} = (\nu/\bar{\nu})^{h/2h_0} = \bar{\alpha}/\alpha$. Assume h_1 is even. Then we have $h_0 = h_1/2$, $\mathfrak{a} = \mathfrak{p}_2$ and $\mathfrak{l}^{h_1/2}\mathfrak{p}_2 = (\nu)$. We also have $\mathfrak{l}^{-h/2} = (\alpha)\mathfrak{p}_2^{h/h_1}$ with $\alpha = \nu^{-h/h_1}$. Hence $\bar{\mu}^{-h/2h_0} = (\nu/\bar{\nu})^{h/2h_0} = \bar{\alpha}/\alpha$. This shows that we have

$$(6.5.4)\qquad \zeta(\sigma,\lambda)=\left(\frac{-1}{l}\right)^{(k-m)/2}\lambda^{\sigma}(\mathfrak{X})^{-1}\left(\frac{\bar\alpha}{\alpha}\right)^{n}\left(\frac{-1}{l}\right)^{hn/2}$$

if we put $\mathfrak{l}^{-h/2}=(\alpha)\mathfrak{p}_2^{h/h_1}$. Note that if $\beta\equiv 1 \mod {}^{\times}(2)$, then $N(\beta)\equiv 1 \mod {}^{\times}4$ since $d\equiv 0 \mod 4$. By the principle explained above, we obtain B. □

PROOF of C and D. We have $f(d)=2$ and may assume that $\mathfrak{a}=(1)$ or $\mathfrak{p}$. By Lemma 6.8, (1) and (6.5.1), we have

$$(6.5.5)\qquad \zeta(\sigma,\lambda)=\left(\frac{-1}{l}\right)^{(k-m)/2}\lambda^{\sigma}(\mathfrak{X})^{-1}\bar\mu^{-hn/2h_0}\cdot\left[\begin{cases}\overline{\tilde\chi_2((u_0^{-1}q(\mu)^{-1}u_0,1))}^{h/h_0} & \text{if } \mathfrak{a}=(1)\\ (-1)^{h/2h_0}\overline{\tilde\chi_2((u_0^{-1}q(\mu)^{-1}u_0,1))}^{h/h_0} & \text{if } \mathfrak{a}\neq(1)\end{cases}\right]^{n}.$$

We have

$$(-1)^{h/h_1}=\begin{cases}1 & \text{if } N(\mathfrak{a})=1,\\ (-1)^{h/2h_0} & \text{if } N(\mathfrak{a})\neq 1\end{cases}$$

Recall that (cf. (6.4.19))

$$\tilde\chi_2\left((q(\mu)^{-1},1)\right)=\begin{cases}\pm 1 & \text{if } N(\nu)\equiv 1 \mod 4,\\ \pm i & \text{if } N(\nu)\equiv 3 \mod 4.\end{cases}$$

Since $N(\nu)=l^{h_0}N(\mathfrak{a})$, we get

$$\begin{aligned}\tilde\chi_2\left((q(\mu)^{-1},1)\right)^{h/h_0}&=\left(\frac{-1}{N(\nu)}\right)^{h/2h_0}=\left(\frac{-1}{l}\right)^{h/2}\left(\frac{-1}{N(\mathfrak{a})}\right)^{h/2h_0}\\&=\tilde\chi_2\left((u_0^{-1}q(\mu)^{-1}u_0,1)\right)^{h/h_0}.\end{aligned}$$

We have

$$(6.5.6)\qquad \zeta(\sigma,\lambda)=\left(\frac{-1}{l}\right)^{(k-m)/2}\lambda^{\sigma}(\mathfrak{X})^{-1}\bar\mu^{-hn/2h_0}\cdot\left[\left(\frac{-1}{N(\mathfrak{a})}\right)^{h/2h_0}\left(\frac{-1}{l}\right)^{h/2}(-1)^{h/h_1}\right]^{n}.$$

If h_1 is odd, we have $N(\mathfrak{a})=1$. If h_1 is even, we have $h_0=h_1/2$, $N(\mathfrak{a})=d_2$ and $\left(\frac{-1}{N(\mathfrak{a})}\right)^{h/2h_0}=\left(\frac{-1}{d_2}\right)^{h/h_1}$. Hence we always have $\left(\frac{-1}{N(\mathfrak{a})}\right)^{h/2h_0}=\left(\frac{-1}{d_2}\right)^{h/h_1}$. If h_1 is odd, we have $\mathfrak{l}^{-h/2}=(\alpha)$ with $\alpha=\nu^{-h/2h_1}$, $\bar\mu^{-h/2h_0}=\bar\alpha/\alpha$. If h_1 is even, we have $\mathfrak{l}^{h_1/2}\mathfrak{p}=(\nu)$, $\mathfrak{l}^{-h/2}=(\alpha)\mathfrak{p}^{h/h_1}$ with $\alpha=\nu^{-h/h_1}$, $\bar\mu^{-h/2h_0}=\bar\alpha/\alpha$. Thus we get

$$(6.5.7)\quad \zeta(\sigma,\lambda)=\left(\frac{-1}{l}\right)^{(k-m)/2}\lambda^{\sigma}(\mathfrak{X})^{-1}\left(\frac{\bar\alpha}{\alpha}\right)^{n}\left(\frac{-1}{l}\right)^{hn/2}\left(\frac{-1}{d_2}\right)^{hn/h_1}(-1)^{hn/h_1}$$

if we put $\mathfrak{l}^{-h/2}=(\alpha)\mathfrak{p}^{h/h_1}$. Hence we obtain C and D. □

PROOF of E. In this case, h is odd. First let us consider the case where $d\equiv 3 \mod 4$ and d is a prime number. We exclude the case $K=\mathbf{Q}(\sqrt{-3})$ for a moment. By (6.4.17), we obtain

(6.5.8) $$\zeta(\sigma,\lambda)=\left(\frac{-1}{l}\right)^{n}\left(\frac{-1}{l}\right)^{(k-m)/2}\lambda^{\sigma}(\mathfrak{X})^{-1}\overline{(\mu^{1/2})}^{-hn/h_0}\cdot\left[-\operatorname{sgn}(y)\prod_{p\neq 2,y\in p\mathbf{Z}_p}(x,y)_p\prod_{i=1}^{h/h_0}\overline{\Xi(q_1^{(i)})}\left(\frac{-1}{l}\right)^{(h_0-1)/2}\right]^{n}.$$

We have $h_0=h_1$ and $\mathfrak{a}=(1)$. By (6.4.19), we have

$$\tilde{\chi}_2\left((q(\mu)^{-1},1)\right)=\begin{cases}\pm 1 & \text{if } l\equiv 1 \mod 4,\\ \pm i & \text{if } l\equiv 3 \mod 4,\end{cases}$$

since $N(\nu)\equiv l \mod 4$. Hence we have

$$\tilde{\chi}_2((u_0^{-1}q(\mu)^{-1}u_0,1))=\left(\frac{-1}{l}\right)\tilde{\chi}_2\left((q(\mu)^{-1},1)\right).$$

By Lemma 6.8, (1), we obtain

(6.5.9) $$\zeta(\sigma,\lambda)=\left(\frac{-1}{l}\right)^{(k-m)/2}\lambda^{\sigma}(\mathfrak{X})^{-1}\overline{(\mu^{1/2})}^{-hn/h_1}\cdot\left[-\operatorname{sgn}(y)\prod_{p\neq 2;y\in p\mathbf{Z}_p}(x,y)_p\left(\frac{-1}{l}\right)^{(h_1-1)/2}\overline{\tilde{\chi}_2((q(\mu)^{-1},1))}^{h/h_1}\right]^{n}.$$

Recall that $(\nu)=\mathfrak{l}^{h_1}$, $\nu=s+t\omega$, $\mu^{-1}=\bar{\nu}/\nu=x+y\omega$. We have $\mu=(\nu/|\nu|)^2$. Hence $\mu^{1/2}=\operatorname{sgn} t\cdot\nu/|\nu|$. We have

(6.5.10) $$N(\nu)=s^2+st+\frac{d+1}{4}t^2=l^{h_1}.$$

(6.5.11) $$x=\frac{(s+t)^2-\frac{d+1}{4}t^2}{l^{h_1}},\quad y=-\frac{t(2s+t)}{l^{h_1}}.$$

Let p be a prime number such that $p\neq 2$, l. Then $y\in p\mathbf{Z}_p$ if and only if $p\,|\,t$ or $p\,|\,2s+t$. Assume $p\,|\,t$. Then we have $x\equiv l^{-h_1}s^2 \mod p$. By (6.5.10), we have $s^2\equiv l^{h_1}$; hence $x\equiv 1 \mod p$ and we get $(x,y)_p=1$. Assume $p\,|\,2s+t$. Then we get $x\equiv -dl^{-h_1}s^2 \mod p$ by (6.5.11) and $ds^2\equiv l^{h_1} \mod p$ by (6.5.10). Hence we get $x\equiv -1 \mod p$ and $(x,y)_p=(-1,y)_p=(-1,2s+t)_p$, since $p\nmid t$. Since $y\notin\mathbf{Z}_l$, we obtain

$$\prod_{p\neq 2,y\in p\mathbf{Z}_p}(x,y)_p=\prod_{p\neq 2,l;p|2s+t}(-1,2s+t)_p.$$

By the product formula for the Hilbert symbol, we get

$$\begin{aligned}-\operatorname{sgn}(y)\prod_{p\neq 2,y\in p\mathbf{Z}_p}(x,y)_p&=\operatorname{sgn}(t(2s+t))\prod_{p\neq 2,l;p|2s+t}(-1,2s+t)_p\\&=\operatorname{sgn}(t)(-1,2s+t)_2(-1,2s+t)_l.\end{aligned}$$

We see easily that $l\nmid 2s+t$. By (6.5.9), we obtain

$$(6.5.12)\qquad \zeta(\sigma,\lambda)=\left(\frac{-1}{l}\right)^{(k-m)/2}\lambda^{\sigma}(\mathfrak{X})^{-1}\left(\frac{\overline{\nu}}{|\nu|}\right)^{-hn/h_1}$$
$$\cdot\left[\left(\frac{-1}{l}\right)^{(h_1-1)/2}(-1,\mathrm{Tr}(\nu))_2\,\overline{\tilde{\chi}_2((q(\mu)^{-1},1))^{h/h_1}}\right]^n.$$

To simplify this formula, we need the following Lemma.

LEMMA 6.10. *Let $d\equiv 3 \mod 4$ and for $i=1,\ 2$, let $\nu_i\in K^\times\cap(\mathbf{Z}_2+\mathbf{Z}_2\omega)$, $N(\nu_i)\in\mathbf{Z}_2^\times$. We assume that $\mathrm{Tr}(\nu_1)$, $\mathrm{Tr}(\nu_2)$, $\mathrm{Tr}(\nu_1\nu_2)$, $\Im(\bar{\nu}_1/\nu_1)$, $\Im(\bar{\nu}_2/\nu_2)$, $\Im(\overline{\nu_1\nu_2}/\nu_1\nu_2)$ are non-zero. Then we have*

$$\xi_2\left(q\left(\frac{\bar{\nu}_1}{\nu_1}\right),q\left(\frac{\bar{\nu}_2}{\nu_2}\right)\right)$$
$$=(-1,\mathrm{Tr}(\nu_1))_2(-1,\mathrm{Tr}(\nu_2))_2(-1,\mathrm{Tr}(\nu_1\nu_2))_2(N(\nu_1),N(\nu_2))_2.$$

The proof is straightforward but long and tedious. We omit it.

By Lemma 6.10, we easily obtain

$$(6.5.13)\qquad \tilde{\chi}_2\left((q(\mu)^{-1},1)\right)^{2n+1}=\tilde{\chi}_2\left((q(\mu)^{-(2n+1)},1)\right)$$
$$\cdot(-1,\mathrm{Tr}(\nu))_2\left(-1,\mathrm{Tr}(\nu^{2n+1})\right)_2(N(\nu),N(\nu))_2^n$$

for a non-negative integer n, since we see easily that $\Im\left((\bar{\nu}/\nu)^i\right)\neq 0$ and $\mathrm{Tr}(\nu^i)\neq 0$ for $1\leq i\in\mathbf{Z}$.

Put $\mathfrak{X}^h=(\alpha)$. We may take $\alpha=\nu^{-h/h_1}$. We have $(\bar{\nu}/|\nu|)^{-hn/h_1}=(\bar{\alpha}/|\alpha|)^n$. Put $h/h_0=2n_0+1$. By (6.5.13), we obtain

$$\tilde{\chi}_2\left((q(\mu)^{-1},1)\right)^{h/h_1}$$
$$=\tilde{\chi}_2\left(\left(q\left(\frac{\alpha}{\bar{\alpha}}\right),1\right)\right)(-1,\mathrm{Tr}(\nu))_2\left(-1,\mathrm{Tr}(\alpha^{-1})\right)_2(N(\nu),N(\nu))_2^{n_0}.$$

We have

$$(N(\nu),N(\nu))_2=\left(l^{h_1},l^{h_1}\right)_2=(l,l)_2=(-1,l)_2=\left(\frac{-1}{l}\right),$$
$$(-1,\mathrm{Tr}(\alpha^{-1}))_2=(-1,\mathrm{Tr}(\alpha))_2\,(-1,N(\alpha))_2=(-1,\mathrm{Tr}(\alpha))_2\left(\frac{-1}{l}\right).$$

Now (6.5.12) can be written as

$$(6.5.14)\qquad \zeta(\sigma,\lambda)=\left(\frac{-1}{l}\right)^{(k-m)/2}\lambda^{\sigma}(\mathfrak{X})^{-1}\left(\frac{\bar{\alpha}}{|\alpha|}\right)^n$$
$$\cdot\left[\left(\frac{-1}{l}\right)^{(h_1+1)/2+n_0}(-1,\mathrm{Tr}(\alpha))_2\,\overline{\tilde{\chi}_2\left(\left(q\left(\frac{\alpha}{\bar{\alpha}}\right),1\right)\right)}\right]^n.$$

Since $(h_1+1)/2+n_0\equiv(h+1)/2 \mod 2$, we obtain E except for $K=\mathbf{Q}(\sqrt{-1})$, $\mathbf{Q}(\sqrt{-2})$, $\mathbf{Q}(\sqrt{-3})$. Now let us supplement these cases. If $h=1$, then we have

$$(6.5.15)\quad \zeta(\sigma,\lambda) = \lambda^{\sigma}(\mathfrak{X})^{-1}\left(\frac{-1}{l}\right)^{n}\left(\frac{-1}{l}\right)^{(k-m)/2}\overline{(\mu^{1/2})}^{-n}$$
$$\cdot\left[-\mathrm{sgn}(y)\prod_{p\neq 2, y\in p\mathbf{Z}_p}(x,y)_p\ \overline{\tilde{\chi}_2((u_0^{-1}q(\mu)^{-1}u_0,1))}\ \overline{\chi_3(u_0^{-1}q(\mu)^{-1}u_0)}\right]^{n}$$

by (6.4.2) and (6.4.13). Here $\mathfrak{l}^{\rho}\mathfrak{l}^{-1} = (\mu^{-1})$, $\mu^{-1} = x + y\omega$ as before. If $K = \mathbf{Q}(\sqrt{-3})$, we obtain the assertion by the same calculation as above.

Assume $d \equiv 0 \mod 4$. By the same argument as for the case $d \equiv 3 \mod 4$, we obtain

$$\overline{(\mu^{1/2})}^{-1}\cdot -\mathrm{sgn}(y)\prod_{p\neq 2, y\in p\mathbf{Z}_p}(x,y)_p = \frac{\nu}{|\nu|}\cdot(-1,\mathrm{Tr}(\nu))_2.$$

Here $\mathfrak{l} = (\nu)$, $\mu^{-1} = \bar{\nu}/\nu$. Hence we have

$$(6.5.16)\qquad \zeta(\sigma,\lambda) = \lambda^{\sigma}(\mathfrak{X})^{-1}\left(\frac{-1}{l}\right)^{n}\left(\frac{-1}{l}\right)^{(k-m)/2}\left(\frac{\nu}{|\nu|}\right)^{n}$$
$$\cdot\left[(-1,\mathrm{Tr}(\nu))_2\ \overline{\tilde{\chi}_2((u_0^{-1}q(\mu)^{-1}u_0,1))}\right]^{n}.$$

If $K = \mathbf{Q}(\sqrt{-1})$, we may assume $s \equiv 1 \mod 2$. Then we have $\tilde{\chi}_2((q(\mu)^{-1},1)) = \pm 1$ and

$$\zeta(\sigma,\lambda) = \lambda^{\sigma}(\mathfrak{X})^{-1}\left(\frac{\nu}{|\nu|}\right)^{n}\left[(-1,\mathrm{Tr}(\nu))_2\ \overline{\tilde{\chi}_2((q(\mu)^{-1},1))}\right]^{n}.$$

Replacing ν by α^{-1}, we see that the assertion holds for $K = \mathbf{Q}(\sqrt{-1})$.

Assume $K = \mathbf{Q}(\sqrt{-2})$. Then s is odd. If t is even, we have $l \equiv 1 \mod 4$ and $\tilde{\chi}_2((q(\mu)^{-1},1)) = \pm 1$. If t is odd, we have $l \equiv 3 \mod 4$, $x \equiv 1 \mod 4$ and $\tilde{\chi}_2((q(\mu)^{-1},1)) = \pm i$. We get

$$\zeta(\sigma,\lambda) = \lambda^{\sigma}(\mathfrak{X})^{-1}\left(\frac{-1}{l}\right)^{(k-m)/2}\left(\frac{\nu}{|\nu|}\right)^{n}\left[(-1,\mathrm{Tr}(\nu))_2\ \overline{\tilde{\chi}_2((q(\mu)^{-1},1))}\right]^{n}.$$

Replacing ν by α^{-1}, we see that the assertion holds for $K = \mathbf{Q}(\sqrt{-2})$. □

REMARK. If we put $\zeta'(\sigma,\lambda) = (-1/N(\mathfrak{X}))^{(k-m)/2}\zeta(\sigma,\lambda)$, we have

$$\zeta'(\sigma,\lambda)\zeta'(\rho\sigma\rho,\lambda) = \lambda^{\sigma}(N(\mathfrak{X}))^{-1}.$$

By this relation and the functional equation of $L(s,\lambda)$, we can see that the theorem holds for $-hn < m \leq hn$, $m \equiv k \mod 2$.

7. Consistency of the refined conjecture with the theory of motives

In this section, we shall use the notation and terminology in [**Y94**, §2], without detailed explanation. Let K be a CM-field of degree $2n$ and E be an algebraic number field. Let J_K^0 be a complete set of representatives of $J_K/\sim$, where $\sigma_1 \sim \sigma_2$ if $\sigma_1 = \sigma_2$ or $\sigma_1 = \sigma_2\rho$. Let M be a motive over K with coefficients in E. Throughout this section, we assume that M is of rank 1, i.e., the Hodge realization $H_{DR}(M)$ is a free $E\otimes_{\mathbf{Q}} K$-module of rank 1. Let

$$I^{+} : H_B^{+}(R_{K/\mathbf{Q}}(M))\otimes_{\mathbf{Q}}\mathbf{C} \cong H_{DR}^{+}(R_{K/\mathbf{Q}}(M))\otimes_{\mathbf{Q}}\mathbf{C}$$

be the canonical isomorphism as $E \otimes_{\mathbf{Q}} \mathbf{C}$-modules. We assume that $H^+_{DR}(M)$ is *not* a free $E \otimes_{\mathbf{Q}} K$-module. We know (cf. [**Y94**, §2.3]) that $H^+_{DR}(M) \oplus \rho H^+_{DR}(M)$ is a free $E \otimes_{\mathbf{Q}} K$-module. For every $\tau \in J^0_K$, we have

$$I^+_\tau : (H_{\tau,B}(M) \oplus H_{\rho\tau,B}(M))^+ \otimes_{\mathbf{Q}} \mathbf{C} \cong (H^+_{DR}(M) \oplus \rho H^+_{DR}(M)) \otimes_{K,\tau} \mathbf{C}$$

as $E \otimes_{\mathbf{Q}} \mathbf{C}$-modules. Fix a basis of $H^+_{DR}(M) \oplus \rho H^+_{DR}(M)$ as a free $E \otimes_{\mathbf{Q}} K$-module of rank 1. The covariantly defined τ-period $c^+_\tau(M)$ is defined as the determinant of I^+_τ with respect to this basis and a basis of $(H_{\tau,B}(M) \oplus H_{\rho\tau,B}(M))^+$ as an E-module. Let v_1 and v_2 be generators of $H_{DR}(M)$ and $\rho H_{DR}(M)$ as free $E \otimes_{\mathbf{Q}} K$-modules respectively. There exist orthogonal idempotents ϵ_1 and ϵ_2 of $E \otimes_{\mathbf{Q}} K$ such that

$$F^-(M) = (E \otimes_{\mathbf{Q}} K)\epsilon_2 v_1, \quad \rho F^-(M) = (E \otimes_{\mathbf{Q}} K)\epsilon_1 v_2, \quad \epsilon_1 + \epsilon_2 = 1. \tag{7.1}$$

We shall show

$$c^+(R_{K/\mathbf{Q}}(M)) = \left(1 \otimes D_F^{1/2}\right) \prod_{\tau \in J^0_K} c^+_\tau(M) \qquad \bmod E^\times, \tag{7.2}$$

where F is the maximal real subfield of K and D_F is the discriminant of F. Put $V = H^+_{DR}(M)$. We may identify V as a direct factor $(E \otimes_{\mathbf{Q}} K)\epsilon v_1$ of $(E \otimes_{\mathbf{Q}} K)v_1$. We can take $v = (\epsilon_1 v_1, \epsilon_2 v_2)$ as a generator of $V \oplus \rho V$ as a free $E \otimes_{\mathbf{Q}} K$-module. Let $\omega_1, \omega_2, \cdots, \omega_n$ be a basis of F over $\mathbf{Q}$. Then

$$(1 \otimes \omega_1)v, \quad (1 \otimes \omega_2)v, \cdots, (1 \otimes \omega_n)v$$

give a basis of $V \oplus \rho V$ as an E-module. In $(V \oplus \rho V) \otimes_{K,\tau} \mathbf{C}$, we have $(1 \otimes \omega_i)v = v \otimes \omega_i^\tau$. Since $\det(\omega_i^\tau)_{1 \le i \le n, \tau \in J^0_K} = D_F^{1/2}$, we obtain (7.2).

Now let N be a motive over K with coefficients in E. We assume that N is of rank 1. We shall determine $c^+(R_{K/\mathbf{Q}}(M \otimes N))$. We pose an assumption

$$F^-(M \otimes N) = F^-(M) \otimes N. \tag{7.3}$$

For $\tau \in J_K$, let

$$I_\tau : H_{\tau,B}(M) \otimes_{\mathbf{Q}} \mathbf{C} \cong H_{DR}(M) \otimes_{K,\tau} \mathbf{C},$$
$$J_\tau : H_{\tau,B}(N) \otimes_{\mathbf{Q}} \mathbf{C} \cong H_{DR}(N) \otimes_{K,\tau} \mathbf{C}$$

be the canonical isomorphisms as $E \otimes_{\mathbf{Q}} \mathbf{C}$-modules. Let t and u be generators of $H_{\tau,B}(M)$ and $H_{\tau,B}(N)$ as E-modules respectively. We may put

$$I_\tau(t) = P_\tau v_1, \quad I_{\rho\tau}(F_{\infty_\tau} t) = \bar{P}_\tau v_2 \tag{7.4}$$

where P_τ, $\bar{P}_\tau \in E \otimes_{\mathbf{Q}} \mathbf{C}$ and $\bar{P}_\tau$ is obtained by letting the complex conjugation act on the second argument of $E \otimes_{\mathbf{Q}} \mathbf{C}$. Hence we get

$$(P_\tau v_1, \bar{P}_\tau v_2) \equiv c^+_\tau(M)(\epsilon_1 v_1 + \epsilon_2 v_2) \quad \bmod (E \otimes_{\mathbf{Q}} \mathbf{C})\epsilon_2 v_1 + (E \otimes_{\mathbf{Q}} \mathbf{C})\epsilon_1 v_2. \tag{7.5}$$

Let w be a generator of $H_{DR}(N)$ as a free $E \otimes_{\mathbf{Q}} K$-module. We have

$$J_\tau(u) = q_\tau w, \quad J_{\rho\tau}(F_{\infty_\tau} u) = \bar{q}_\tau w, \quad q_\tau \bar{q}_\tau = \delta_\tau(N) \tag{7.6}$$

with q_τ, $\bar{q}_\tau \in E \otimes_{\mathbf{Q}} \mathbf{C}$. Now we have

$$\begin{aligned}
&((I_\tau \otimes J_\tau)(t \otimes u), ((I_{\rho\tau} \otimes J_{\rho\tau})(F_{\infty_\tau} t \otimes F_{\infty_\tau} u) \\
&\qquad \equiv c^+_\tau(M) q_\tau(\epsilon_1 v_1 \otimes w) + c^+_\tau(M)\bar{q}_\tau(\epsilon_2 v_2 \otimes w) \\
&\qquad\quad \bmod (E \otimes_{\mathbf{Q}} \mathbf{C})(\epsilon_2 v_1 \otimes w) + (E \otimes_{\mathbf{Q}} \mathbf{C})(\epsilon_1 v_2 \otimes w).
\end{aligned}$$

Let $\epsilon_{i,\tau}$ be the image of ϵ_i in $E \otimes_{\mathbf{Q}} \mathbf{C} \cong (E \otimes_{\mathbf{Q}} K \otimes_{K,\tau} \mathbf{C})$. By the above formula and (7.3), we get

$$c_\tau^+(M\otimes N)=c_\tau^+(M)(q_\tau\epsilon_{1,\tau}+\bar{q}_\tau\epsilon_{2,\tau}). \tag{7.7}$$

Using (7.2), we obtain

$$c^+(R_{K/\mathbf{Q}}(M\otimes N))=\left(\prod_{\tau\in J_K^0}(q_\tau\epsilon_{1,\tau}+\bar{q}_\tau\epsilon_{2,\tau})\right)c^+(R_{K/\mathbf{Q}}(M)). \tag{7.8}$$

Let us consider $c^+(R_{K/\mathbf{Q}}(M^{\otimes l}))$. Assume that

$$F^-\left(M^{\otimes l}\right)=\epsilon_2 H_{DR}\left(M^{\otimes l}\right). \tag{7.9}$$

Then we find easily that

$$(P_\tau v_1)^{\otimes l}\equiv c_\tau^+(M)^l\epsilon_1 v_1^{\otimes l}\qquad \bmod (E\otimes_{\mathbf{Q}}\mathbf{C})\epsilon_2 v_1^{\otimes l},$$
$$(\bar{P}_\tau v_1)^{\otimes l}\equiv c_\tau^+(M)^l\epsilon_2 v_2^{\otimes l}\qquad \bmod (E\otimes_{\mathbf{Q}}\mathbf{C})\epsilon_1 v_2^{\otimes l}.$$

Hence we obtain

$$c_\tau^+\left(M^{\otimes l}\right)=c_\tau^+(M)^l, \tag{7.10}$$

$$c^+\left(R_{K/\mathbf{Q}}\left(M^{\otimes l}\right)\right)=c^+\left(R_{K/\mathbf{Q}}(M)\right)^l\left(1\otimes D_F^{1/2}\right)^{-(l-1)}. \tag{7.11}$$

We also have

$$c_\tau^+(M)(\epsilon_1-\epsilon_2)\epsilon_1 v_1\equiv P_\tau v_1\qquad \bmod (E\otimes_{\mathbf{Q}}\mathbf{C})\epsilon_2 v_1,$$
$$c_\tau^+(M)(\epsilon_1-\epsilon_2)\epsilon_2 v_2\equiv -\bar{P}_\tau v_2\qquad \bmod (E\otimes_{\mathbf{Q}}\mathbf{C})\epsilon_1 v_2.$$

Assume that

$$F^-(M)=F^+(M). \tag{7.12}$$

Then, calculating $(I_\tau(t), I_{\rho\tau}(-F_{\infty,\tau}t))$ using (7.4) and (7.5), we obtain

$$c_\tau^-(M)=c_\tau^+(M)(\epsilon_{1,\tau}-\epsilon_{2,\tau}), \tag{7.13}$$

$$c^-(R_{K/\mathbf{Q}}(M))=\prod_{\tau\in J_K^0}(\epsilon_{1,\tau}-\epsilon_{2,\tau})c^+(R_{K/\mathbf{Q}}(M)). \tag{7.14}$$

Let Φ be a CM-type of K and $\mathfrak{f}$ be an integral ideal of K. Let λ be a primitive Grössencharacter of $I_\mathfrak{f}(K)$ of conductor $\mathfrak{f}$ which satisfies

$$\lambda((\alpha))=\prod_{\sigma\in\Phi}(\alpha^{\sigma\rho}/|\alpha^\sigma|)^{t_\sigma},\quad \alpha\equiv 1 \quad \bmod {}^\times\mathfrak{f}. \tag{7.15}$$

Here t_σ, $\sigma\in\Phi$ are positive integers such that $t_\sigma \bmod 2$ is independent of σ. We define $t_0\in\mathbf{Z}$ by (3.4) and a quasicharacter λ_0 of $I_\mathfrak{f}(K)$ by (3.5). For $\sigma\in\mathrm{Aut}(\mathbf{C})$, we define λ_0^σ and λ^σ as in §3. Let E be an algebraic number field which contains $\lambda_0(\mathfrak{a})$ for all $\mathfrak{a}\in I_\mathfrak{f}(K)$. There exists a motive $M(\lambda_0)$ over K with coefficients in E such that $L(\sigma,s,M(\lambda_0))=L(s,\lambda_0^\sigma)$, $\sigma\in J_E$; $M(\lambda_0)$ is of rank 1 and of pure weight t_0. For simplicity put $M=M(\lambda_0)$. Let $\{F^m\}$ denote the Hodge filtration on $H_{\mathrm{DR}}(M)$ by $E\otimes_{\mathbf{Q}}K$-modules. We put

$$G=\mathrm{Gal}(\overline{\mathbf{Q}}/\mathbf{Q}),\quad H=\mathrm{Gal}(\overline{\mathbf{Q}}/K),\quad H'=\mathrm{Gal}(\overline{\mathbf{Q}}/E).$$

Let ψ_m be the structure function of $F^m(H_{\mathrm{DR}}(M))$; ψ_m is a $\mathbf{Z}$-valued function on G which is left H'-invariant and right H-invariant. We have [**Y94**, §4]

$$\psi_m(g) = \begin{cases} 1 & \text{if } \ g \mod H \in \cup_{\sigma\in\Phi}(\delta_\sigma \cup \rho\delta_\sigma), \\ 0 & \text{otherwise.} \end{cases} \tag{7.16}$$

Here $\delta_\sigma = \sigma H$, $\rho\delta_\sigma = \rho\sigma H$ and δ_σ (resp. $\rho\delta_\sigma$) extends over σ such that $(t_0 - t_\sigma)/2 \geq m$ (resp. $(t_0 + t_\sigma)/2 \geq m$). If t_0 is even (resp. odd), put $t_0 = 2p$ (resp. $2p+1$). Then we have $F^-(M) = F^{p+1}(H_{\mathrm{DR}}(M))$. Let ψ be the structure function of $F^-(M)$. By (7.16), we obtain

$$\psi(g) = \begin{cases} 1 & \text{if } \ g \mod H \in \rho\Phi, \\ 0 & \text{if } \ g \mod H \notin \rho\Phi. \end{cases} \tag{7.17}$$

Hence $H^+_{\mathrm{DR}}(M)$ is not free; (7.12) is also satisfied.

Let (K', Φ') be the reflex of (K, Φ). Let S be the full inverse image of Φ under the canonical map $G \longrightarrow G/H$. We have

$$G = S \cup \rho S \quad \text{(disjoint union).} \tag{7.18}$$

Since ψ is left H'-invariant, we get $H'S = S$. Hence we have $E \supseteq K'$, which is a well known fact.

We have

$$E \otimes_{\mathbf{Q}} \mathbf{C} \cong (\mathbf{C})_{\sigma\in J_E},$$

the isomorphism being given by $x \otimes y \longrightarrow (\sigma(x)y)_{\sigma\in J_E}$. Let $\tau \in J_K$. We have

$$E \otimes_{\mathbf{Q}} K \otimes_{K,\tau} \mathbf{C} \cong (\mathbf{C})_{\sigma\in J_E}, \tag{7.19}$$

the isomorphism being given by $x \otimes y \otimes z \longrightarrow (\sigma(x)\tau(y)z)_{\sigma\in J_E}$. Let ϵ_i, $i = 1, 2$ be the idempotents defined by (7.1). Let $\epsilon_{i,\tau} = \epsilon_i \otimes 1 \in (E \otimes_{\mathbf{Q}} K) \otimes_{K,\tau} \mathbf{C}$, $i = 1, 2$. Using (7.19), we put $\epsilon_{1,\tau} = (\nu^{(1)}_{\sigma,\tau})_{\sigma\in J_E}$, $\epsilon_{2,\tau} = (\nu^{(2)}_{\sigma,\tau})_{\sigma\in J_E}$ where $\nu^{(1)}_{\sigma,\tau}$, $\nu^{(2)}_{\sigma,\tau} \in \mathbf{C}$. By (7.17) and [**Y94**, Lemma 2.1, (1)], we have

$$\nu^{(1)}_{\sigma,\tau} = \begin{cases} 1 & \text{if } \sigma^{-1}\tau \in S, \\ 0 & \text{if } \sigma^{-1}\tau \in \rho S. \end{cases} \qquad \nu^{(2)}_{\sigma,\tau} = \begin{cases} 0 & \text{if } \sigma^{-1}\tau \in S, \\ 1 & \text{if } \sigma^{-1}\tau \in \rho S. \end{cases} \tag{7.20}$$

For $\sigma \in G$, we put $N_\sigma = |\{\tau \in J^0_K \mid \tau \in \sigma\rho S\}|$. Then N_σ depends only on $\sigma H'$ and we have

$$\prod_{\tau\in J^0_K} \epsilon_{1,\tau} - \epsilon_{2,\tau} = \left((-1)^{N_\sigma}\right)_{\sigma\in J_E}. \tag{7.21}$$

Now we take Φ as J^0_K. Let

$$S = \cup_{i=1}^n \tau_i H, \quad \rho S = \cup_{i=1}^n \rho\tau_i H$$

be coset decompositions. We put

$$\sigma^{-1}\tau_i H = \rho^{\epsilon(i)}\tau_{j(i)}H, \quad \epsilon(i) = 0 \text{ or } 1, \quad 1 \leq i \leq n. \tag{7.22}$$

Then $i \longrightarrow j(i)$ gives a permutation on n letters. We put $\varphi(\sigma) = (-1)^{\sum_{i=1}^n \epsilon(i)}$. Then we have $N_\sigma = |\{i \mid \epsilon(i) = 1\}|$ and $(-1)^{N_\sigma} = \varphi(\sigma)$.

LEMMA 7.1.

1. *φ is a homomorphism of G into $\{\pm 1\}$.*
2. *Let k be the field which corresponds to* $\ker(\varphi)$. *Then k is a subfield of K'. Let $K = F(\sqrt{-\delta})$ with a totally positive element δ* of *F. Then we have $k = \mathbf{Q}(\sqrt{(-1)^n N_{F/\mathbf{Q}}(\delta)})$.*

PROOF. Let $\sigma_1, \sigma_2 \in G$ and put $\sigma_3 = \sigma_1\sigma_2$,

$$\sigma_k^{-1}\tau_i H = \rho^{\epsilon_k(i)}\tau_{j_k(i)}H, \quad 1 \le i \le n, \quad 1 \le k \le 3.$$

Since

$$(\sigma_1\sigma_2)^{-1}\tau_i H = \rho^{\epsilon_1(i)+\epsilon_2(j_1(i))}\tau_{j_2(j_1(i))},$$

we have

$$j_3 = j_2 \circ j_1, \quad \epsilon_3(i) \equiv \epsilon_1(i) + \epsilon_2(j_1(i)) \mod 2.$$

Hence we obtain

$$\sum_{i=1}^n \epsilon_3(i) \equiv \sum_{i=1}^n \epsilon_1(i) + \sum_{i=1}^n \epsilon_2(i) \mod 2$$

and 1 follows.

If $\sigma \in \mathrm{Gal}(\overline{\mathbf{Q}}/K')$, then we have $\epsilon(i) = 0$ for $1 \le i \le n$. Hence we get $k \subseteq K'$. Let $x = \prod_{\tau\in\Phi}\tau(\sqrt{-\delta})$. Then we have $x^2 = \prod_{\tau\in\Phi}\tau(-\delta) = (-1)^n N_{F/\mathbf{Q}}(\delta)$. If $\sigma \in G$ satisfies (7.22), then we have $\sigma^{-1}\tau_i(\sqrt{-\delta}) = (-1)^{\epsilon(i)}\tau_{j(i)}(\sqrt{-\delta})$. Hence we get $\sigma(x) = \varphi(\sigma)x$ and the assertion 2 follows. □

Put $d = (-1)^n N_{F/\mathbf{Q}}(\delta)$ and consider $\sqrt{d} \otimes (1/\sqrt{d}) \in E \otimes_{\mathbf{Q}} \mathbf{C}$. Then we have $\sqrt{d} \otimes (1/\sqrt{d}) = (\varphi(\sigma))_{\sigma\in J_E}$ since $\sigma(\sqrt{d}) = \varphi(\sigma)\sqrt{d}$. Therefore we obtain

$$\prod_{\tau\in\Phi} \epsilon_{1,\tau} - \epsilon_{2,\tau} = \sqrt{d} \otimes \frac{1}{\sqrt{d}} \quad \text{in} \quad E \otimes_{\mathbf{Q}} \mathbf{C}. \tag{7.23}$$

We call $m \in \mathbf{Z}$ is critical for λ_0 if

$$\frac{t_0 - t_\sigma}{2} < m \le \frac{t_0 + t_\sigma}{2} \quad \text{for every} \quad \sigma \in \Phi.$$

We call $m \in \mathbf{Z}$ is critical for λ if

$$-t_\sigma < m \le t_\sigma \quad \text{and} \quad m \equiv t_\sigma \mod 2 \quad \text{for every} \quad \sigma \in \Phi.$$

For $m \in \mathbf{Z}$, we take $N = T(m)$, the Tate motive. We have $q_\tau = 1 \otimes (2\pi\sqrt{-1})^m$ for every $\tau \in \Phi$. By (7.8) and (7.23), we obtain

$$\begin{aligned} &c^+(R_{K/\mathbf{Q}}(M \otimes T(m))) \\ &= \begin{cases} (1 \otimes (2\pi\sqrt{-1})^{mn})c^+(R_{K/\mathbf{Q}}(M)) & \text{if } m \text{ is even,} \\ (\sqrt{d} \otimes \frac{1}{\sqrt{d}}(2\pi\sqrt{-1})^{mn})c^+(R_{K/\mathbf{Q}}(M)) & \text{if } m \text{ is odd.} \end{cases} \end{aligned} \tag{7.24}$$

By Deligne's conjecture proved by Blasius [**B**] for $M = M(\lambda_0)$, we have

$$(L(m, \lambda_0^\sigma))_{\sigma\in J_E} \equiv c^+(R_{K/\mathbf{Q}}(M \otimes T(m))) \mod E \tag{7.25}$$

if m is critical for λ_0. Here, for $a, b \in E \otimes_{\mathbf{Q}} \mathbf{C}$, we write $a \equiv b \mod E$ if $b \in (E \otimes_{\mathbf{Q}} \mathbf{C})^\times$ and $a/b \in E$. Let $t^0 = \min_{\sigma\in\Phi} t_\sigma$. For $m \in \mathbf{Z}$, put

$$U(m, \lambda) = \frac{L(m/2, \lambda)}{(2\pi\sqrt{-1})^{n(m-t^0)/2}\sqrt{d}^{(m-t^0)/2}}.$$

PROPOSITION 7.2. *Let $m_1, m_2 \in \mathbf{Z}$ be critical for λ. Assume $L(m_1/2, \lambda) \neq 0$. Then we have*

$$\left(\frac{U(m_2, \lambda)}{U(m_1, \lambda)}\right)^\sigma = \frac{U(m_2, \lambda^\sigma)}{U(m_1, \lambda^\sigma)} \quad \textit{for every} \quad \sigma \in \mathrm{Aut}(\mathbf{C}).$$

PROOF. By (7.25), we have

$$\left(\frac{L(m_2/2,\lambda^\sigma)}{L(m_1/2,\lambda^\sigma)}\right)_{\sigma\in J_E} \equiv \frac{c^+(R_{K/\mathbf{Q}}(M\otimes T((t_0+m_2)/2))}{c^+(R_{K/\mathbf{Q}}(M\otimes T((t_0+m_1)/2))} \mod E.$$

Assume $m_2\equiv m_1+2 \mod 4$. By (7.24), we obtain

$$\left(\frac{L(m_2/2,\lambda^\sigma)}{L(m_1/2,\lambda^\sigma)}\right)_{\sigma\in J_E} \equiv \sqrt{d}\otimes\frac{1}{\sqrt{d}}(2\pi\sqrt{-1})^{n(m_2-m_1)/2} \mod E.$$

Hence there exists $e\in E$ such that

$$\frac{L(m_2/2,\lambda^\sigma)}{\sqrt{d}\,(2\pi\sqrt{-1})^{n(m_2-m_1)/2}L(m_1/2,\lambda^\sigma)} = e^\sigma \quad \text{for every} \quad \sigma\in J_E.$$

This shows that

$$\left(\frac{L(m_2/2,\lambda)}{\sqrt{d}\,(2\pi\sqrt{-1})^{n(m_2-m_1)/2}L(m_1/2,\lambda)}\right)^\sigma = \frac{L(m_2/2,\lambda^\sigma)}{\sqrt{d}\,(2\pi\sqrt{-1})^{n(m_2-m_1)/2}L(m_1/2,\lambda^\sigma)}$$

for every $\sigma\in\mathrm{Aut}(\mathbf{C})$. Since $\sqrt{d}^{(m_2-m_1)/2}\equiv\sqrt{d} \mod \mathbf{Q}^\times$, the assertion follows. The case $m_2\equiv m_1 \mod 4$ can be shown similarly. □

Let χ be a primitive character of the ideal class group modulo $\mathfrak{f}'$ of K where $\mathfrak{f}'$ is an integral ideal of K. We are going to investigate the transformation of $L(m/2,\lambda)$ under $\lambda\longrightarrow\lambda\chi$. We identify χ with a character of $\mathrm{Gal}(\overline{\mathbf{Q}}/K)$. We can find an abelian extension of finite degree L of K such that χ factors through the canonical map $\mathrm{Gal}(\overline{\mathbf{Q}}/K)\longrightarrow\mathrm{Gal}(L/K)$. Let E be an algebraic number field which contains $\lambda_0(\mathfrak{a})$, $\mathfrak{a}\in I_{\mathfrak{f}}(K)$ and $\chi(\mathfrak{a})$, $\mathfrak{a}\in I_{\mathfrak{f}'}(K)$. Let N be the motive over K with coefficients in E attached to χ. We have $L(s,\chi^\sigma)=L(\sigma,N,s)$ for every $\sigma\in J_E$. Then we have

$$H_{\tau,B}(N)=E \quad \text{for every } \tau\in J_K, \quad H_{\mathrm{DR}}(N)=(E\otimes_{\mathbf{Q}}L)^{\mathrm{Gal}(L/K)}.$$

Here $\mathrm{Gal}(L/K)$ acts on E by χ and acts on L as the Galois group. Put $G_L=\mathrm{Gal}(L/K)$. The E-module structure and the K-module structure of $(E\otimes_{\mathbf{Q}}L)^{G_L}$ are given by

$$x\left(\sum e_i\otimes l_i\right)=\sum xe_i\otimes l_i,$$
$$\left(\sum e_i\otimes l_i\right)y=\sum e_i\otimes l_iy, \quad e_i\in E, l_i\in L, x\in E, y\in K.$$

We note that (7.3) is satisfied. We choose $\omega\in L$ so that $\sum_{\mu\in G_L}\chi(\mu)\otimes\mu(\omega)$ gives a generator of $(E\otimes_{\mathbf{Q}}L)^{G_L}$ as a free $E\otimes_{\mathbf{Q}}K$-module of rank 1. We have

$$q_\tau=\sum_{\mu\in G_L}\chi(\mu)\otimes\tau\mu(\omega)\in E\otimes_{\mathbf{Q}}\mathbf{C}, \quad \tau\in\Phi. \tag{7.26}$$

Here we use the same letter τ for its extension to $\mathrm{Aut}(\mathbf{C})$; $q_\tau \mod E^\times$ is independent of this extension. We have

$$q_\tau=\left(\sum_{\mu\in G_L}\sigma(\chi(\mu))\tau\mu(\omega)\right)_{\sigma\in J_E}, \quad \bar{q}_\tau=\left(\sum_{\mu\in G_L}\sigma(\chi(\mu))\rho\tau\mu(\omega)\right)_{\sigma\in J_E}$$

in $(\mathbf{C})_{\sigma\in J_E}\cong E\otimes_{\mathbf{Q}}\mathbf{C}$. Hence, by (7.20), we get

$$\tag{7.27} \prod_{i=1}^{n} (q_{\tau_i}\epsilon_{1,\tau_i} + \bar{q}_{\tau_i}\epsilon_{2,\tau_i}) = \left(\prod_{\tau_i \in \sigma S} \sum_{\mu \in G_L} \sigma(\chi(\mu))\tau_i\mu(\omega) \prod_{\tau_i \in \sigma\rho S} \sum_{\mu \in G_L} \sigma(\chi(\mu))\rho\tau_i\mu(\omega) \right)_{\sigma \in J_E}.$$

Identify J_E with G/H'. Assume $\sigma \in G$ satisfies (7.22). If $\tau_i \in \sigma S$, we have $\sigma^{-1}\tau_i = \tau_{j(i)}h_i$ with $h_i \in H$. Hence we get

$$\sum_{\mu \in G_L} \sigma(\chi(\mu))\tau_i\mu(\omega) = \sigma \left(\sum_{\mu \in G_L} \chi(\mu)\sigma^{-1}\tau_i\mu(\omega) \right)$$
$$= \sigma \left(\sum_{\mu \in G_L} \chi(\mu)\tau_{j(i)}h_i\mu(\omega) \right) = \sigma(\chi(h_i)^{-1})\sigma \left(\sum_{\mu \in G_L} \chi(\mu)\tau_{j(i)}\mu(\omega) \right).$$

Assume $\tau_i \in \sigma\rho S$. Then we have $\sigma^{-1}\tau_i H = \rho\tau_{j(i)}H$; hence $\sigma^{-1}\rho\tau_i = \tau_{j(i)}h_i$ with $h_i \in H$. Then we get

$$\sum_{\mu \in G_L} \sigma(\chi(\mu))\rho\tau_i\mu(\omega) = \sigma\left(\chi(h_i)^{-1}\right)\sigma \left(\sum_{\mu \in G_L} \chi(\mu)\tau_{j(i)}\mu(\omega) \right).$$

Put $C = \prod_{i=1}^{n} \sum_{\mu \in G_L} \chi(\mu)\tau_i\mu(\omega)$. By (7.27), we obtain

$$\tag{7.28} \prod_{i=1}^{n} (q_{\tau_i}\epsilon_{1,\tau_i} + \bar{q}_{\tau_i}\epsilon_{2,\tau_i}) = \left(\prod_{i=1}^{n} \sigma(\chi(h_i)^{-1})\sigma(C) \right)_{\sigma \in J_E} \mod E.$$

PROPOSITION 7.3. *Let $m \in \mathbf{Z}$ be critical for λ it and let χ be a character of the ideal class group modulo $\mathfrak{f}'$ of K. For $\sigma \in \mathrm{Gal}(\overline{\mathbf{Q}}/\mathbf{Q})$, put*

$$\sigma^{-1}\rho^{\epsilon(i)}\tau_i = \tau_{j(i)}h_i \quad \text{with} \quad h_i \in \mathrm{Gal}(\overline{\mathbf{Q}}/K).$$

Assume $L(m/2, \lambda) \neq 0$. Then we have

$$\sigma\left(\frac{L(m/2, \lambda\chi)}{L(m/2, \lambda)} \right) = \prod_{i=1}^{n} \sigma(\chi(h_i)) \frac{L(m/2, \sigma(\lambda)\sigma(\chi))}{L(m/2, \sigma(\lambda))}$$

for every $\sigma \in \mathrm{Gal}(\overline{\mathbf{Q}}/\mathbf{Q})$.

PROOF. By (7.8), (7.24), (7.25) and (7.28), we have

$$\left(\frac{L(m/2, \sigma(\lambda)\sigma(\chi))}{L(m/2, \sigma(\lambda))} \right)_{\sigma \in J_E} \equiv \frac{c^+(R_{K/\mathbf{Q}}(M \otimes N \otimes T((t_0 + m)/2)))}{c^+(R_{K/\mathbf{Q}}(M \otimes T((t_0 + m)/2)))}$$
$$\equiv \prod_{i=1}^{n} q_{\tau_i}\epsilon_{1,\tau_i} + \bar{q}_{\tau_i}\epsilon_{2,\tau_i}$$
$$\equiv \left(\prod_{i=1}^{n} \sigma(\chi(h_i)^{-1})\sigma(C) \right)_{\sigma \in J_E} \mod E.$$

Hence there exists $e \in E$ such that

$$\frac{L(m/2,\sigma(\lambda)\sigma(\chi))}{L(m/2,\sigma(\lambda))}=\prod_{i=1}^{n}\sigma(\chi(h_i)^{-1})\sigma(C)\sigma(e).$$

Note that $\prod_{i=1}^{n}\sigma\left(\chi(h_i)^{-1}\right)=1$ if $\sigma=\mathrm{id}$. Hence we obtain

$$\sigma\left(\frac{L(m/2,\lambda\chi)}{L(m/2,\lambda)}\right)=\sigma(C)\sigma(e)=\frac{L(m/2,\sigma(\lambda)\sigma(\chi))}{L(m/2,\sigma(\lambda))}\cdot\prod_{i=1}^{n}\sigma(\chi(h_i)).$$

This completes the proof. □

We note that $h_1h_2\cdots h_n \mod [H,H]$ does not depend on the coset decomposition of S; the homomorphism $\sigma \longrightarrow h_1h_2\cdots h_n \mod [H,H]$ is called the type transfer in Lang [**L**, p. 164]. Put $\zeta(\sigma,\chi)=\prod_{i=1}^{n}\sigma(\chi(h_i))$. We can verify the "cocycle" condition

$$\zeta(\sigma_1\sigma_2,\chi)=\zeta(\sigma_1,\sigma_2\chi)\sigma_1(\zeta(\sigma_2,\chi))$$

by a direct computation, taking the CM-type $\sigma_2\Phi$ for $\sigma_2\chi$.

We can rewrite our results in terms of the right action of the Galois group. Let S be the full inverse image of Φ under the canonical map $G\longrightarrow H\backslash G$. Let $S=\cup_{i=1}^{n}H\tau_i$ be a coset decomposition and for $\sigma\in G$, put $\tau_i\sigma\rho^{\epsilon(i)}=h_i\tau_{j(i)}$, $h_i\in H$. Then we have

$$\left(\frac{L(m/2,\lambda\chi)}{L(m/2,\lambda)}\right)^{\sigma}=\prod_{i=1}^{n}\chi^{\sigma}(h_i)^{-1}\,\frac{L(m/2,\lambda^{\sigma}\chi^{\sigma})}{L(m/2,\lambda^{\sigma})}. \tag{7.29}$$

REMARK. Strictly speaking, we should have used the left action of the Galois group in the proof of Proposition 7.2. But in that case, the left action can be converted to the right action without essential modification.

Propositions 7.2 and 7.3 show that Conjecture 5.1 is consistent with the shifting the critical values and with the twisting by a character of finite order of $K_A^\times$.

In view of our results in §6, we can guess the precise form of ζ in Conjecture 5.1 though in a rather weak form. Let $C_{\mathfrak{f}}$ be the maximal ray class field of conductor $\mathfrak{f}$ of K. We can find a fractional ideal $\mathfrak{X}$ of K so that

$$\prod_{i=1}^{n}h_i \mod [H,H]|C_{\mathfrak{f}}=\left(\frac{C_{\mathfrak{f}}/K}{\mathfrak{X}}\right).$$

For every i, we assume that $t_\tau^{(i)}$ is divisible by $2h_K$ for all $\tau\in\Phi_i$ and that

$$\prod_{\tau\in\Phi_i}\left(\frac{\epsilon^{\tau\rho}}{\epsilon^{\tau}}\right)^{t_\tau^{(i)}/2h_K}=1\quad\text{for every }\epsilon\in E_K.$$

Put $\mathfrak{X}^{h_K}=(\alpha)$, $\alpha\in K$ and

$$\zeta_i=\lambda_i^{\sigma}(\mathfrak{X})^{-1}\left\{\prod_{\tau\in\Phi_i}\left(\frac{\alpha^{\tau\rho}}{\alpha^{\tau}}\right)^{t_\tau^{(i)}/2h_K}\right\}^{\sigma}$$

for every i. Then we can expect that $\zeta=\pm\prod_{i=1}^{q}\zeta_i$.

References

[A] G. W. Anderson, *Logarithmic derivatives of Dirichlet L-functions and the periods of abelian varieties*, Comp. Math. **45** (1982), 315–332.

[Ar] E. Artin, *Zur Theorie der L-Reihen mit allgemeinen Gruppencharakteren*, Hamb. Abh. **8** (1930), 292–306. (=Collected Papers, No. 7).

[As] T. Asai, *On a certain function analogous to* $\log|\eta(z)|$, Nagoya Math. J. **40** (1970), 193–211.

[B] D. Blasius, *On the critical values of Hecke L-series*, Ann. of Math. **124** (1986), 23–63.

[BL] G. Brattström and S. Lichtenbaum, *Jacobi-sum Hecke characters of imaginary quadratic fields*, Comp. Math. **53** (1984), 277–302.

[C] P. Colmez, *Périods des variétés abéliennes à multiplication complexe*, Ann. of Math. **138** (1993), 625–683.

[D] P. Deligne, *Valeurs de fonctions L et périodes d'intégrales*, Proc. Symposia Pure Math. **33** part 2, Providence, RI: Amer. Math. Soc. (1979), 313–346.

[F] F. G. Frobenius, *Über Beziehungen zwischen den Primidealen eines algebraischen Körpers und den Substitutionen seiner Gruppe*, Sitzungsber. Preußischen Akad. Berlin, (1896) 689–703. (=Gesammelte Abhandlungen II, No. 52).

[H] E. Hecke, *Theorie der Eisensteinschen Reihen höherer Stufe und ihre Anwendung auf Funktionentheorie und Arithmetik* , Hamb. Abh. **5** (1927), 199–224. (=Werke No. 24)

[JKS] U. Jannsen, S. Kleiman and J-P. Serre (eds.), Motives, Proc. Symposia Pure Math. **55** Parts 1 and 2, Providence, RI: Amer. Math. Soc. (1994).

[K] T. Kubota, On Automorphic Functions and the Reciprocity Law in a Number Field, Lectures in Mathematics **2**, Kyoto University, Tokyo: Kinokuniya Book-Store Co., Ltd. (1969).

[L] S. Lang, Complex Multiplication, Grundlehren der Mathematischen Wissenshaften **225**, New York-Berlin: Springer-Verlag (1983).

[SC] A. Selberg and S. Chowla, *On Epstein's zeta-function*, Jour. für Reine und Angew. Math. **227** (1967), 86–110. (=Collected Papers of A. Selberg, No. 34)

[Sh71] G. Shimura, Introduction to the Arithmetic Theory of Automorphic Functions, Publ. Math. Soc. Japan **11**, Tokyo-Princeton: Iwanami Shoten and Princeton University Press (1971).

[Sh73] G. Shimura, *On modular forms of half integral weight*, Ann. of Math. **97** (1973), 440–481.

[Sh75] G. Shimura, *On some arithmetic properties of modular forms of one and several variables*, Ann. of Math. **102** (1975), 491–515.

[Sh77] G. Shimura, *On abelian varieties with complex multiplication*, Proc. London Math. Soc. 3rd ser. **34** (1977), 65–86.

[Sh78] G. Shimura, *On certain reciprocity-laws for theta functions and modular forms*, Acta Math. **141** (1978), 35–71.

[Sh78b] G. Shimura, *The special values of the zeta functions associated with Hilbert modular forms*, Duke Math. J. **45** (1978), 637–679.

[Sh79] G. Shimura, *Automorphic forms and periods of abelian varieties*, J. Math. Soc. Japan **31** (1979), 561–592.

[Sh80] G. Shimura, *The arithmetic of certain zeta functions and automorphic forms on orthogonal groups*, Ann. of Math. **111** (1980), 313–375.

[Sh87] G. Shimura, *Nearly holomorphic functions on Hermitian symmetric spaces*, Math. Ann. **278** (1987), 1–28.

[Sh98] G. Shimura, Abelian Varieties with Complex Multiplication and Modular Functions, Princeton Math. Ser. **46**, Princeton: Princeton University Press (1998).

[Shi76] T. Shintani, *On evaluation of zeta functions of totally real algebraic number fields at non-positive integers*, J. Fac. Sci. Univ. Tokyo **23** (1976), 393–417.

[Shi77] T. Shintani, *On a Kronecker limit formula for real quadratic fields*, J. Fac. Sci. Univ. Tokyo **24** (1977), 167–199.

[Shi77b] T. Shintani, *On values at* $s = 1$ *of certain L functions of totally real algebraic number fields*, in Algebraic Number Theory, Proc. International Symp., Kyoto, 1976, Tokyo: Kinokuniya (1977), 201–212.

[Shi78] T. Shintani, *On certain ray class invariants of real quadratic fields*, J. Math. Soc. Japan **30** (1978), 139–167.

[Si61] C. L. Siegel, Lectures on Advanced Analytic Number Theory, Bombay: Tata Institute (1961).

[Si69] C. L. Siegel, *Berechnung von Zetafunktionen an ganzzahligen Stellen*, Göttingen Nachrichten Nr. 10 (1969), 87–102. (=Gesammelte Abhandlungen IV, No. 89)

[Si70] C. L. Siegel, *Über die Fourierschen Koefficienten von Modulformen*, Göttingen Nachrichten Nr. 3 (1970), 15–56. (=Gesammelte Abhandlungen IV, No. 90)

[ST] G. Shimura and Y. Taniyama, Complex Multiplications of Abelian Varieties and its Applications to Number Theory, Publ. Math. Soc. Japan No. **6**, Tokyo: The Mathematical Society of Japan (1961).

[St] H. M. Stark, *L-functions at* $s = 1$, *I, II, III, IV*, Advances in Math. **7** (1971), 301–343; **17** (1975), 60–92; **22** (1976), 64–84; **35** (1980), 197–235.

[T] J. Tate, Les conjectures de Stark sur les fonctions L d'Artin en $s = 0$ (Notes by D. Bernardi et N. Schappacher), Progress in Math. **47**, Boston: Birkhäuser (1984).

[W] A. Weil, Elliptic Functions According to Eisenstein and Kronecker, Ergebnisse der Mathematik und ihrer Grenzgebiete **88**, Berlin-New York: Springer Verlag (1976).

[WW] E. T. Whittaker and G. N. Watson, A Course of Modern Analysis, fourth edition, Cambridge: Cambridge University Press (1927).

[Y92] H. Yoshida, *Remarks on metaplectic representations of* SL(2), J. Math. Soc. Japan, **44** (1992), 351–373.

[Y94] H. Yoshida, *On the zeta functions of Shimura varieties and periods of Hilbert modular forms*, Duke Math. J. **75** (1994), 121–191.

DEPARTMENT OF MATHEMATICS, FACULTY OF SCIENCE, KYOTO UNIVERSITY, KYOTO 606-8502, JAPAN

E-mail address: `yoshida@kusm.kyoto-u.ac.jp`

Selected Titles in This Series

(Continued from the front of this publication)

44 **William K. Allard and Frederick J. Almgren, Jr., Editors,** Geometric measure theory and the calculus of variations (Humboldt State University, Arcata, California, July/August 1984)

43 **François Trèves, Editor,** Pseudodifferential operators and applications (University of Notre Dame, Notre Dame, Indiana, April 1984)

42 **Anil Nerode and Richard A. Shore, Editors,** Recursion theory (Cornell University, Ithaca, New York, June/July 1982)

41 **Yum-Tong Siu, Editor,** Complex analysis of several variables (Madison, Wisconsin, April 1982)

40 **Peter Orlik, Editor,** Singularities (Humboldt State University, Arcata, California, July/August 1981)

39 **Felix E. Browder, Editor,** The mathematical heritage of Henri Poincaré (Indiana University, Bloomington, April 1980)

38 **Richard V. Kadison, Editor,** Operator algebras and applications (Queens University, Kingston, Ontario, July/August 1980)

37 **Bruce Cooperstein and Geoffrey Mason, Editors,** The Santa Cruz conference on finite groups (University of California, Santa Cruz, June/July 1979)

36 **Robert Osserman and Alan Weinstein, Editors,** Geometry of the Laplace operator (University of Hawaii, Honolulu, March 1979)

35 **Guido Weiss and Stephen Wainger, Editors,** Harmonic analysis in Euclidean spaces (Williams College, Williamstown, Massachusetts, July 1978)

34 **D. K. Ray-Chaudhuri, Editor,** Relations between combinatorics and other parts of mathematics (Ohio State University, Columbus, March 1978)

33 **A. Borel and W. Casselman, Editors,** Automorphic forms, representations and L-functions (Oregon State University, Corvallis, July/August 1977)

32 **R. James Milgram, Editor,** Algebraic and geometric topology (Stanford University, Stanford, California, August 1976)

31 **Joseph L. Doob, Editor,** Probability (University of Illinois at Urbana-Champaign, Urbana, March 1976)

30 **R. O. Wells, Jr., Editor,** Several complex variables (Williams College, Williamstown, Massachusetts, July/August 1975)

29 **Robin Hartshorne, Editor,** Algebraic geometry – Arcata 1974 (Humboldt State University, Arcata, California, July/August 1974)

28 **Felix E. Browder, Editor,** Mathematical developments arising from Hilbert problems (Northern Illinois University, Dekalb, May 1974)

27 **S. S. Chern and R. Osserman, Editors,** Differential geometry (Stanford University, Stanford, California, July/August 1973)

26 **Calvin C. Moore, Editor,** Harmonic analysis on homogeneous spaces (Williams College, Williamstown, Massachusetts, July/August 1972)

25 **Leon Henkin, John Addison, C. C. Chang, William Craig, Dana Scott, and Robert Vaught, Editors,** Proceedings of the Tarski symposium (University of California, Berkeley, June 1971)

24 **Harold G. Diamond, Editor,** Analytic number theory (St. Louis University, St. Louis, Missouri, March 1972)

23 **D. C. Spencer, Editor,** Partial differential equations (University of California, Berkeley, August 1971)

22 **Arunas Liulevicius, Editor,** Algebraic topology (University of Wisconsin, Madison, June/July 1970)

21 **Irving Reiner, Editor,** Representation theory of finite groups and related topics (University of Wisconsin, Madison, April 1970)

(See the AMS catalog for earlier titles)

ISBN 0-8218-1050-2

9 780821 810507

PSPUM/66.1

Selected Titles in This Series

66 **Robert S. Doran, Ze-Li Dou, and George T. Gilbert, Editors,** Automorphic forms, automorphic representations, and arithmetic (Texas Christian University, Fort Worth, 1996)

65 **M. Giaquinta, J. Shatah, and S. R. S. Varadhan, Editors,** Differential equations: La Pietra 1996 (Villa La Pietra, Florence, Italy, 1996)

64 **G. Ferreyra, R. Gardner, H. Hermes, and H. Sussmann, Editors,** Differential geometry and control (University of Colorado, Boulder, 1997)

63 **Alejandro Adem, Jon Carlson, Stewart Priddy, and Peter Webb, Editors,** Group representations: Cohomology, group actions and topology (University of Washington, Seattle, 1996)

62 **János Kollár, Robert Lazarsfeld, and David R. Morrison, Editors,** Algebraic geometry—Santa Cruz 1995 (University of California, Santa Cruz, July 1995)

61 **T. N. Bailey and A. W. Knapp, Editors,** Representation theory and automorphic forms (International Centre for Mathematical Sciences, Edinburgh, Scotland, March 1996)

60 **David Jerison, I. M. Singer, and Daniel W. Stroock, Editors,** The legacy of Norbert Wiener: A centennial symposium (Massachusetts Institute of Technology, Cambridge, October 1994)

59 **William Arveson, Thomas Branson, and Irving Segal, Editors,** Quantization, nonlinear partial differential equations, and operator algebra (Massachusetts Institute of Technology, Cambridge, June 1994)

58 **Bill Jacob and Alex Rosenberg, Editors,** K-theory and algebraic geometry: Connections with quadratic forms and division algebras (University of California, Santa Barbara, July 1992)

57 **Michael C. Cranston and Mark A. Pinsky, Editors,** Stochastic analysis (Cornell University, Ithaca, July 1993)

56 **William J. Haboush and Brian J. Parshall, Editors,** Algebraic groups and their generalizations (Pennsylvania State University, University Park, July 1991)

55 **Uwe Jannsen, Steven L. Kleiman, and Jean-Pierre Serre, Editors,** Motives (University of Washington, Seattle, July/August 1991)

54 **Robert Greene and S. T. Yau, Editors,** Differential geometry (University of California, Los Angeles, July 1990)

53 **James A. Carlson, C. Herbert Clemens, and David R. Morrison, Editors,** Complex geometry and Lie theory (Sundance, Utah, May 1989)

52 **Eric Bedford, John P. D'Angelo, Robert E. Greene, and Steven G. Krantz, Editors,** Several complex variables and complex geometry (University of California, Santa Cruz, July 1989)

51 **William B. Arveson and Ronald G. Douglas, Editors,** Operator theory/operator algebras and applications (University of New Hampshire, July 1988)

50 **James Glimm, John Impagliazzo, and Isadore Singer, Editors,** The legacy of John von Neumann (Hofstra University, Hempstead, New York, May/June 1988)

49 **Robert C. Gunning and Leon Ehrenpreis, Editors,** Theta functions – Bowdoin 1987 (Bowdoin College, Brunswick, Maine, July 1987)

48 **R. O. Wells, Jr., Editor,** The mathematical heritage of Hermann Weyl (Duke University, Durham, May 1987)

47 **Paul Fong, Editor,** The Arcata conference on representations of finite groups (Humboldt State University, Arcata, California, July 1986)

46 **Spencer J. Bloch, Editor,** Algebraic geometry – Bowdoin 1985 (Bowdoin College, Brunswick, Maine, July 1985)

45 **Felix E. Browder, Editor,** Nonlinear functional analysis and its applications (University of California, Berkeley, July 1983)

(Continued in the back of this publication)

Automorphic Forms, Automorphic Representations, and Arithmetic

Professor Goro Shimura at the CBMS Conference
"Euler Products and Eisenstein Series,"
held at Texas Christian University, May 1996

Proceedings of Symposia in PURE MATHEMATICS

Volume 66, Part 2

Automorphic Forms, Automorphic Representations, and Arithmetic

NSF-CBMS Regional Conference in Mathematics
on Euler Products and Eisenstein Series
May 20–24, 1996
Texas Christian University

Robert S. Doran
Ze-Li Dou
George T. Gilbert
Editors

American Mathematical Society
Providence, Rhode Island

PROCEEDINGS OF AN NSF-CBMS REGIONAL CONFERENCE ON EULER PRODUCTS AND EISENSTEIN SERIES HELD AT TEXAS CHRISTIAN UNIVERSITY, FORT WORTH, TEXAS, MAY 20–24, 1996

Any opinions, findings, and conclusions or recommendations expressed in this material are those of the authors and do not necessarily reflect the view of the National Science Foundation.

1991 *Mathematics Subject Classification.* Primary 11Fxx; Secondary 14Gxx, 22Exx.

Library of Congress Cataloging-in-Publication Data

NFS-CBMS Regional Conference in Mathematics on Euler Products and Eisenstein Series (1996 : Texas Christian University)

Automorphic forms, automorphic representations, and arithmetic : NFS-CBMS Regional Conference in Mathematics on Euler Products and Eisenstein Series, May 20–24, 1996, Texas Christian University / Robert S. Doran, Ze-Li Dou, George T. Gilbert, editors.

p. cm. — (Proceedings of symposia in pure mathematics, 0082-0717 ; v. 66, pts. 1–2)

Includes bibliographical references.

ISBN 0-8218-0659-9 (set)

ISBN 0-8218-1050-2 (pt. 1)

ISBN 0-8218-1051-0 (pt. 2)

1. Automorphic functions Congresses. 2. Automorphic forms Congresses. 3. Representations of groups Congresses. I. Doran, Robert S., 1937– . II. Dou, Ze Li, 1962– . III. Gilbert, George T., 1957– . IV. Title. V. Series.

QA353.A9 N74 1996

511.3′3—dc21

99-28916

CIP

∞ The paper used in this book is acid-free and falls within the guidelines
established to ensure permanence and durability.
Visit the AMS home page at URL: http://www.ams.org/

10 9 8 7 6 5 4 3 2 1 04 03 02 01 00 99

Dedicated to Goro Shimura

Table of Contents–Part 1

Table of Contents–Part 2

Foreword

Professor Goro Shimura was the principal speaker at the CBMS regional conference held in May, 1996 at Texas Christian University, entitled "Euler Products and Eisenstein Series." His ten lectures resulted in a monograph bearing the same title, which appeared as volume 93 of the CBMS Regional Conference Series in Mathematics. In these volumes we have collected papers based on most of the other talks presented at the conference, as well as many additional contributions.

Professor Shimura's stature as a mathematical giant, based on the plethora of seminal ideas he has introduced into number theory, on the numerous and varied theories that have matured in his hands, and on the high quality of the mathematicians he has cultivated and continues to influence, is certainly secure and will long endure. It is neither our place nor our intention to comment at length on Professor Shimura's achievements, but we believe that many aspects of Professor Shimura's work as well as that of his students are reflected in these volumes. It is therefore highly fitting that they be dedicated to Professor Shimura.

We have selected a rather general title, but an informed reader will find after a glance at the tables of contents that even this title does not adequately encompass the spectrum of the topics discussed in these volumes. We have also endeavored to place the papers which consider related topics and use similar techniques in the same volume. However, the division of the papers into two volumes cannot be based on a scientific principle. All in all, we must simply allow the mathematics to speak for itself.

We thank Professor Shimura for making the conference a success, and for making these volumes possible. We also acknowledge the hard work and patience of our authors and referees. We are especially grateful to Professor Hiroyuki Yoshida for his advice and for his lending a helping hand on several occasions. We are indebted to Shirley Doran, our technical secretary, for a thoroughly professional job. Finally, our heartfelt gratitude is reserved for our families, for their cheerful long-suffering.

Robert S. Doran
Ze-Li Dou
George T. Gilbert
February 1999

Proceedings of Symposia in Pure Mathematics
Volume **66.2**, 1999

Metaplectic Generating Functions and Shimura Integrals

Daniel Bump and Solomon Friedberg

This paper is dedicated to Professor Goro Shimura.

ABSTRACT. Generalizing a now-classical construction of Shimura, Bump and Hoffstein proposed that the Rankin-Selberg convolution of an automorphic form on the n-fold metaplectic cover of $\mathrm{GL}(r)$ with a generalized theta function on the n-fold metaplectic cover of $\mathrm{GL}(n)$ is Eulerian. Though these integrals unfold to models which are not unique, Bump and Hoffstein proved their conjecture in special cases, and recently it was proved in full (and generalized) by T. Suzuki. This paper reconsiders these Rankin-Selberg integrals using generating functions for the local L-functions, a method first used for Rankin-Selberg integrals involving non-unique models by Piatetski-Shapiro and Rallis. First, the generating functions on the metaplectic group for the standard L-functions are determined, and then a (conjectural) formula expressing the Whittaker models of the exceptional representations as integrals of these generating functions is described. This conjectural formula is shown to imply the Euler products of Bump and Hoffstein, and it is proved here when $r \leq 3$.

A useful technique, having its origin in the work of Shimura, is to prove the existence of liftings of automorphic forms by a combination of the Rankin-Selberg method and the converse theorem. An early example of this technique is the paper of Shimura [**S**], where consideration of the Euler product obtained from the Rankin-Selberg convolution of a modular form of half-integral weight with a theta function leads to the existence of a lifted integral weight modular form, the celebrated Shimura correspondence. Another early example of this method is in the work of Doi and Naganuma [**DN**], who (following a suggestion of Shimura) proved quadratic base change for GL(2) by this method. Jacquet, Piatetski-Shapiro and Shalika [**JPSS**] adapted this method to obtain nonnormal cubic base change, which is an essential element in the proof of the octahedral case of Artin's conjecture by Tunnell [**T**]. This in turn is the starting point for the proof by Wiles of the semistable Shimura-Taniyama conjecture, and Fermat's Last Theorem [**W**]. In its

---

1991 *Mathematics Subject Classification.* Primary 11F66; Secondary 11F70, 22E50.

Key words and phrases. Metaplectic form, automorphic form, Euler product, Rankin-Selberg Method, generalized theta series.

We would like to thank David Ginzburg, Jeffrey Hoffstein, S. J. Patterson, I. Piatetski-Shapiro and Steve Rallis for useful discussions contributing to our understanding of this problem. This research was supported by NSA grant MDA904-95-H-1053 and by Boston College (Friedberg), and by NSF grant DMS-9622819 (Bump).

most recent development by Piatetski-Shapiro and his co-workers Cogdell, Gelbart, Ginzburg, Rallis and Soudry, this method has become a program to prove liftings from classical groups to $\mathrm{GL}(n)$.

Bump and Hoffstein [**BH87**] and [**BH86**] proposed generalizing the results of Shimura [**S**] by making use of the generalized theta series θ_n on the n-fold metaplectic cover of $\mathrm{GL}(n)$, in the sense of Kazhdan and Patterson [**KP84**]. They realized that when π is an automorphic representation of the n-fold metaplectic cover of $\mathrm{GL}(r)$, the Rankin-Selberg convolution of π with θ_n should be Eulerian, and should in fact agree with the L-function of an automorphic form on $\mathrm{GL}(r)$. This scenario, if fully realized, would be a generalization of the Shimura correspondence.

Flicker and Kazhdan [**FK**], building on earlier work of Kazhdan and Patterson [**KP86**], showed using the trace formula that such a correspondence does indeed exist, at least in some cases. The existence of the correspondence and the Eulerian nature of the integrals proposed by Bump and Hoffstein, though closely related, are two separate questions.

Toshiaki Suzuki [**Su**] has given an important extension of the generalized theta series of Kazhdan and Patterson, and of the Rankin-Selberg integrals of Bump and Hoffstein. According to Suzuki, given a cusp form on the n-fold cover of $\mathrm{GL}(k)$, a "generalized theta function" can be found as the residue of an Eisenstein series on the n-fold cover of $\mathrm{GL}(nk)$, having a unique Whittaker model, such that convolution of this form with a cusp form on the n-fold cover of $\mathrm{GL}(r)$ will result in an Euler product of degree rk.

Before Suzuki, only when $r = 2$ and n is general, or when $n = r = 3$ had the Rankin-Selberg integrals proposed by Bump and Hoffstein been shown to be Eulerian. The difficulty of this construction has its origin in the fact that these Rankin-Selberg integrals unfold to Whittaker models, and these are *not* unique in the case at hand. There are two automorphic forms involved: the cusp form on the n-fold cover of $\mathrm{GL}(r)$, and the theta function on the n-fold cover of $\mathrm{GL}(n)$. The theta function has a unique Whittaker model (as was shown by Kazhdan and Patterson [**KP84**]). However, the cusp form does not. Therefore, these examples belong to the class of Rankin-Selberg integrals which, though Eulerian, unfold to non-unique models.

Rankin-Selberg integrals involving non-unique models which are known to be Eulerian are fairly rare. When Piatetski-Shapiro and Rallis wrote their paper "A new way to get Euler products," [**PSR**], practically the only examples known were the symplectic examples of Andrianov [**A**] and Zhuravlev [**Z**] and the examples of Bump and Hoffstein. The papers of Andrianov and Zhuravlev as well as those of Bump and Hoffstein and Suzuki proved their Euler products by analyzing the complicated recursion relations satisfied by the Fourier or Whittaker coefficients in the non-unique model, coming from the Hecke algebra.

Piatetski-Shapiro and Rallis [**PSR**] analyzed the examples of Andrianov in a less *ad hoc* way using *generating functions* for the local L-functions, and proposed in the introduction to their paper that their methods could be applied to the examples of Bump and Hoffstein. Today, still only a few Eulerian Rankin-Selberg integrals based on non-unique models are known. These are surveyed in Bump, Furusawa and Ginzburg [**BFG**], which contains some of the simplest examples known, as well as some more elaborate ones. The method of proof employed by Bump, Furusawa and Ginzburg [**BFG**] is similar to that in Piatetski-Shapiro and Rallis [**PSR**].

Apart from the examples of Bump and Hoffstein (and Suzuki), Friedberg and Wong [**FW**] and Goetze [**Go**] have given interesting metaplectic examples, which may be extrapolated into the following scenario. Let π be an automorphic representation on the n-fold cover of GSp(4). There exists a Rankin-Selberg convolution $\mathrm{GSp}(4) \times \mathrm{GL}(n)$. This is essentially the $\mathrm{SO}(5) \times \mathrm{GL}(n)$ convolution, developed by Gelbart, Ginzburg, Piatetski-Shapiro and Soudry, following earlier work by Novodvorsky (see Soudry [**So**]). One may replace GSp(4) and $\mathrm{GL}(n)$ here by their n-fold metaplectic covers, and take the automorphic form on $\mathrm{GL}(n)$ to be θ_n. In this case, there is expected to be an Euler product, of degree 4 if n is odd, and of degree 5 if n is even. This dependence of the degree of the Euler products on the degree of the cover is related to a similar alternation of the structure of the Hecke algebras of the n-fold covers of GSp(4), as was found by Savin [**Sa**]. This scenario (which should generalize to odd orthogonal groups) has been verified when $n \leq 3$ by [**FW**] and [**Go**].

In this paper, we reconsider the Rankin-Selberg integrals of Bump and Hoffstein and Suzuki by a method close to that of Piatetski-Shapiro and Rallis [**PSR**] and Bump, Furusawa and Ginzburg [**BFG**]. First, we determine the generating functions on the metaplectic group for the standard L-functions. This makes essential use of the results of Kazhdan and Patterson [**KP86**]. Next, we exhibit a (conjectural) formula for the Whittaker models of the exceptional representations as integrals of these generating functions. We believe that this is an important step, because progress in cases where Rankin-Selberg integrals unfold to non-unique models will ultimately depend on this type of understanding. We show that this conjecture implies the Euler products of Bump, Hoffstein and Suzuki. Finally, we prove this conjecture when $r \leq 3$.

It has been discovered in recent years that the cocycle formulas in Kazhdan and Patterson [**KP84**] are incorrect unless it is assumed that -1 is an n-th power. Since our results in Section 3 require cocycle calculations, we will make this assumption.

1. Generating functions for local L-functions on the metaplectic group

We will rely on Kazhdan and Patterson [**KP84**] and [**KP86**] for foundational matters. Let F be a nonarchimedean local field such that the group μ_n of n-th roots of unity in F has cardinality n. We will denote by ϖ a uniformizer of F, by $\mathfrak{o}$ its ring of integers, $\mathfrak{p} = \varpi\mathfrak{o}$ the prime ideal of $\mathfrak{o}$, and by $q = |\varpi^{-1}|$ the cardinality of the residue field $\mathfrak{o}/\mathfrak{p}$.

Let $G_r = \mathrm{GL}(r)$, let B_r be the Borel subgroup of upper-triangular elements of G_r. Let T_r and U_r be the maximal torus and unipotent radical of B_r. We will abuse notation and write G_r for $G_r(F)$, etc. except during our brief discussion of the global Rankin-Selberg construction in Section 2, since no confusion can result.

Let $\widetilde{G}_r$ be the n-fold metaplectic cover of G_r, as described in [**KP84**]. We will take the parameter c in [**KP84**] to be zero. Let $\mathbf{s} : G_r \to \widetilde{G}_r$ be the standard section, as in [**KP84**]. Thus $\widetilde{G}_r$ is a central extension of G_r by μ_n. We will identify μ_n with its image in $\widetilde{G}_r$.

We will assume that n is prime to the residue characteristic of F. This implies that the metaplectic cover splits over the maximal compact subgroup $K_r = \mathrm{GL}(r, \mathfrak{o})$. Let K_r^* be the image of K_r under this splitting, and let $\widetilde{K}_r = K_r^* \mu_n$ be the full preimage of K_r. Let Z_r and $\widetilde{Z}_r$ be the centers of G_r and $\widetilde{G}_r$. Note that in general $\widetilde{Z}_r$ is *not* the full preimage in $\widetilde{G}_r$ of Z_r. Let $\widetilde{T}_r$ be the preimage of T_r in

$\widetilde{G}_r$, and let $\mathbf{t} = \mathbf{t}_r : T_r \to \widetilde{T}_r$ be the homomorphism

$$\mathbf{t}\begin{pmatrix} y_1 & & \\ & \ddots & \\ & & y_r \end{pmatrix} = \mathbf{s}\begin{pmatrix} y_1^n & & \\ & \ddots & \\ & & y_r^n \end{pmatrix}.$$

Let $U_r^* = \mathbf{s}(U_r)$ be the image of U_r under the canonical splitting.

We will fix an isomorphism of μ_n into the complex numbers, and by abuse of notation, we will identify μ_n with its image there. We will call a complex-valued function ϕ on $\widetilde{G}_r$ (or any subgroup containing μ_n) *genuine* if $\phi(\epsilon g) = \epsilon\,\phi(g)$, and *antigenuine* if $\phi(\epsilon g) = \epsilon^{-1}\,\phi(g)$. Let $\widetilde{\chi}$ be a genuine character of $\widetilde{Z}$ such that $\widetilde{\chi} \circ \mathbf{t} = \chi$.

The spherical Hecke algebra $\mathcal{H}_\chi$ is defined to be the ring of K_r-biinvariant functions ϕ on G_r, which are compactly supported modulo Z_r, and which satisfy $\phi(zg) = \chi(z)^{-1}\,\phi(g)$ when $z \in Z_r$ and $g \in G_r$. Similarly, $\widetilde{\mathcal{H}}_{\widetilde{\chi}}$ is defined to be the ring of antigenuine K_r^*-biinvariant functions $\widetilde{\phi}$ on $\widetilde{G}_r$, which are compactly supported modulo $\widetilde{Z}_r$, and which satisfy $\widetilde{\phi}(zg) = \widetilde{\chi}(z)^{-1}\,\widetilde{\phi}(g)$ when $z \in \widetilde{Z}_r$ and $g \in \widetilde{G}_r$. We will also need certain enlargements of these rings: let $\mathcal{H}_\chi^\diamond$ consist of all K_r-biinvariant functions ϕ satisfying $\phi(zg) = \chi(z)^{-1}\,\phi(g)$, such that there exists a compact set $\Omega \subset G_r$, such that the support of ϕ is contained in $\Omega\,\mathrm{Mat}_r(\mathfrak{o})$. We note that if this condition is satisfied for ϕ_1 and ϕ_2, their convolution may be defined, since in the integral

$$(\phi_1 * \phi_2)(g) = \int_{G_r/Z_r} \phi_1(gh)\,\phi_2(h^{-1})\,dh,$$

the integrand vanishes with h off a compact set. Thus $\mathcal{H}_\chi^\diamond$ becomes a ring. It may be seen that it is the completion of $\mathcal{H}_\chi$ with respect to a certain maximal ideal. Let $\widetilde{\mathcal{H}}_{\widetilde{\chi}}^\diamond$ be the enlargement of $\widetilde{\mathcal{H}}_{\widetilde{\chi}}$ obtained by relaxing the support condition in exactly the same way.

In Proposition 3.2 of Kazhdan and Patterson [**KP86**], it is shown that a K_r^*-biinvariant (anti-)genuine function is supported solely on double cosets of the form $K_r^*\,\widetilde{Z}_r\,\mathbf{t}(h)\,K_r^*$ with $h \in T_r$. It is important to bear this fact in mind throughout the following discussion. We call $h \in T_r$ *dominant* if

$$h = \begin{pmatrix} h_1 & & \\ & \ddots & \\ & & h_r \end{pmatrix}, \quad k_i = \mathrm{ord}(h_i), \quad k_1 \geq \ldots \geq k_r, \tag{1.1}$$

and we call $t \in \widetilde{T}_r$ dominant if its image in T_r is dominant.

Kazhdan and Patterson [**KP86**] described an isomorphism $\mathbf{t}_! : \mathcal{H}_\chi \sim \widetilde{\mathcal{H}}_{\widetilde{\chi}}$. If $\mathbf{t}_!\phi = \widetilde{\phi}$, then

$$\widetilde{\phi}\big(\mathbf{t}(h)\big) = \delta_r(h)^{(n-1)/2}\,\phi(h) \tag{1.2}$$

when h is dominant, where δ_r is the modular character of B_r, so with h as in (1.1), $\delta_r(h) = \prod_{i=1}^r q^{-k_i(r+1-2i)}$. The isomorphism $\mathbf{t}_!$ naturally extends to an isomorphism $\mathbf{t}_! : \mathcal{H}_\chi^\diamond \sim \widetilde{\mathcal{H}}_{\widetilde{\chi}}^\diamond$.

Let $\chi_1, \cdots, \chi_r$ be unramified quasicharacters of $F^\times$ such that

$$\chi\begin{pmatrix} z & & \\ & \ddots & \\ & & z \end{pmatrix} = (\chi_1 \cdots \chi_r)(z).$$

Let $\alpha_i = \chi_i(\varpi)$, $\alpha = \prod \alpha_i$. We consider the spherical principal series representation $\pi = \pi(\chi_1, \cdots, \chi_r)$ of G_r, obtained by normalized parabolic induction from the Borel subgroup. Similarly, we have a representation $\widetilde{\pi}$ of $\widetilde{G}_r$, which may be described as follows. There is a unique character of the group $\mathbf{t}(T_r)\,\widetilde{Z}_r$ which coincides with $\widetilde{\chi}$ on $\widetilde{Z}_r$, and having value $\chi_1(h_1) \cdots \chi_r(h_r)$ on

$$\mathbf{t}\begin{pmatrix} h_1 & & \\ & \ddots & \\ & & h_r \end{pmatrix}.$$

We extend this character to the group $\mathbf{t}(T_r)\,(\widetilde{T}_r \cap K_r^*)\widetilde{Z}_r\,U^*$ in such a way that U^* acts trivially. The representation $\widetilde{\pi}$ is obtained from this character by normalized induction. Its isomorphism class does not depend on the particular choice of the extension to $\mathbf{t}(T_r)\,(\widetilde{T}_r \cap K_r^*)\widetilde{Z}_r$. It follows from Corollary I.2.8 of Kazhdan and Patterson [**KP84**] that π and $\widetilde{\pi}$ are irreducible if $\chi_1, \cdots, \chi_r$ are in general position. We will denote

$$L(s,\pi) = (1-\alpha_1\,q^{-s})^{-1} \cdots (1-\alpha_r\,q^{-s})^{-1}.$$

Let $\sigma : G_r \to \mathbb{C}$ be the spherical function, that is, the unique K_r-biinvariant matrix coefficient of π, normalized so that $\sigma(1) = 1$, and let $\widetilde{\sigma}$ be the corresponding K_r^*-biinvariant spherical function of $\widetilde{\pi}$. By Proposition 3.6 of Kazhdan and Patterson [**KP86**], if $\widetilde{\phi} = \mathbf{t}_!\phi$, with $\phi \in \mathcal{H}_\chi$, we have

$$\int_{G_r/Z_r} \sigma(g)\,\phi(g)\,dg = \int_{\widetilde{G}_r/\widetilde{Z}_r} \widetilde{\sigma}(g)\,\widetilde{\phi}(g)\,dg. \tag{1.3}$$

Naturally this identity remains valid for $\phi \in \mathcal{H}_\chi^\circ$, provided the integrals in (1.3) are absolutely convergent.

Let s be a complex number. There exists a unique antigenuine K_r^*-biinvariant function $\widetilde{\Delta}_s(g)$ on $\widetilde{G}_r$ with the following properties. If

$$h = \mathbf{s}\begin{pmatrix} h_1 & & \\ & \ddots & \\ & & h_r \end{pmatrix}$$

then $\widetilde{\Delta}_s(h) = 0$ unless the $\mathrm{ord}(h_i)$ are nonnegative and divisible by n, and if $|h_1| \le |h_2| \le \cdots \le |h_r| \le 1$; but if this is the case, then

$$\widetilde{\Delta}_s(h) = |\det(h)|^{\frac{s}{n}+\frac{r-1}{2n}}\,\delta_r(h)^{\frac{n-1}{2n}}. \tag{1.4}$$

PROPOSITION 1.1. *We have*

$$\int_{\widetilde{G}_r} \widetilde{\Delta}_s(g)\,\sigma(g)\,dg = L(s,\pi).$$

The integral is absolutely convergent if $\mathrm{re}(s)$ *is sufficiently large.*

PROOF. Let us define

$$\Delta_s(g) = \begin{cases} |\det(g)|^{s+\frac{r-1}{2}} & \text{if } g \in \mathrm{Mat}_r(\mathfrak{o}); \\ 0 & \text{otherwise.} \end{cases}$$

Then according to Lemma 6.10 on p. 77 of Godement and Jacquet [**GJ**], we have

$$\int_{G_r} \Delta_s(g)\,\sigma(g)\,dg = L(s,\pi). \tag{1.5}$$

Let

$$\Delta_{\chi,s}(g) = \int_{Z_r} \Delta_s(zg)\,\chi(z)\,dz, \quad \widetilde{\Delta}_{\widetilde{\chi},s}(g) = \int_{\widetilde{Z}_r} \widetilde{\Delta}_s(zg)\,\widetilde{\chi}(z)\,dz.$$

Evidently $\Delta_{\chi,s} \in \mathcal{H}_\chi^\diamond$ and $\widetilde{\Delta}_{\widetilde{\chi},s} \in \widetilde{\mathcal{H}}_{\widetilde{\chi}}^\diamond$. We have

$$\int_{G_r/Z_r} \Delta_{\chi,s}(g)\,\sigma(g)\,dg = \int_{G_r} \Delta_s(g)\,\sigma(g)\,dg$$

and

$$\int_{\widetilde{G}_r/\widetilde{Z}_r} \widetilde{\Delta}_{\widetilde{\chi},s}(g)\,\widetilde{\sigma}(g)\,dg = \int_{\widetilde{G}_r} \widetilde{\Delta}_s(g)\,\widetilde{\sigma}(g)\,dg,$$

so the proposition will follow from (1.5) if we can show $\mathbf{t}_!\Delta_{\chi,s} = \widetilde{\Delta}_{\widetilde{\chi},s}$. Using the transformation property with respect to the center, it is sufficient to verify (1.2) when

$$h = \begin{pmatrix} h_1 & & & \\ & \ddots & & \\ & & h_{r-1} & \\ & & & 1 \end{pmatrix},$$

$|h_1| \leq \cdots \leq |h_{r-1}| \leq 1$. If h is of this type, it is easy to see that

$$\widetilde{\Delta}_{\widetilde{\chi},s}\big(\mathbf{t}(h)\big) = (1-\alpha\, q^{-rs-\frac{r(r-1)}{2}})^{-1}\delta_r(h)^{\frac{n-1}{2}}\,\Delta_s(h) = \delta_r(h)^{\frac{n-1}{2}}\,\Delta_{\chi,s}(h),$$

as required. □

Assume that $r < n$. We embed $G_r \to G_n$ into the upper left corner:

$$g \mapsto \begin{pmatrix} g & \\ & I_{n-r} \end{pmatrix}. \tag{1.6}$$

This embedding is covered by an embedding $\widetilde{G}_r \to \widetilde{G}_n$. Let θ_n be the exceptional representation in the sense of Kazhdan and Patterson [**KP84**] of the $\widetilde{G}_n$, which is the unique irreducible quotient of the representation obtained by genuine parabolic induction from the character $\delta_n^{\frac{1}{2}+\frac{1}{2n}}$ of the Borel subgroup B_n of G_n, where $\delta_n : B_n \to \mathbb{C}^\times$ is the modular quasicharacter. (We refer to Kazhdan and Patterson [**KP84**] for a precise description of this representation, denoted $V(\omega')$ there.)

Let ψ be a nontrivial additive character of F; we assume that the conductor of ψ is $\mathfrak{o}$. Let $\psi_U : U_n \to \mathbb{C}$ be the character $u \mapsto \prod_i \psi(u_{i,i+1})$ of U_n. According to Corollary I.3.6 of [**KP84**], θ_n has a unique Whittaker model with respect to the character ψ_U; let W_{θ_n} be the K_n^*-fixed vector in this model, normalized so that $W_{\theta_n}(1) = 1$. It is a genuine right K_n^*-invariant function on $\widetilde{G}_n$ which satisfies

$$W_{\theta_n}(ug) = \psi_U(u)\,W_{\theta_n}(g), \qquad u \in U_n^*. \tag{1.7}$$

CONJECTURE 1.2. *If $g \in G_r$, then*

$$\int_{U_r^*} \Delta_{ns-\frac{n-1}{2}}(ug)\,\psi(u)\,du = |\det(g)|^{s-\frac{n-r}{2}}\,W_{\theta_n}(g). \tag{1.8}$$

For the purpose of discussing this conjecture and its evidence, it is convenient to break it into two parts. We call a function f on $\widetilde{T}_r$ *periodic* if $f(\mathbf{t}(h)\,t) = f(t)$ for $t \in \widetilde{T}_r$, $h \in T_r$. Let $U_r^* = \mathbf{s}(U_r)$ be the image of U_r in the canonical splitting of the cover $\widetilde{G}_r$.

CONJECTURE 1.3. *For fixed s, there exists a periodic function ρ_Δ of $\widetilde{T}_r$ such that when $t \in \widetilde{T}_r$ is dominant,*

$$\int_{U_r^*} \Delta_s(ut)\,\psi_U(u)\,du = |\det(t)|^{\frac{s}{n}+\frac{r-1}{2n}}\,\delta_r(t)^{\frac{1}{2}-\frac{1}{2n}}\,\rho_\Delta(t) \tag{1.9}$$

We have proved this when $r = 2$ or 3—when $r = 3$, this is contained in Proposition 3.1. One might naively think that if this periodicity is true, then surely it could be proved by some simple argument (change of variables, etc.) but this does not seem to be the case.

It is a consequence of the Periodicity Theorem I.1.9 (e) of [**KP84**] that there exists a periodic function ρ_{θ_n} such that

$$W_{\theta_n}(t) = \delta_n(t)^{\frac{1}{2}-\frac{1}{2n}}\,\rho_{\theta_n}(t). \tag{1.10}$$

CONJECTURE 1.4. *The restriction of $\overline{\rho_{\theta_n}}$ to $\widetilde{T}_r \subset \widetilde{T}_n$ equals ρ_Δ.*

PROPOSITION 1.5. *Conjecture 1.2 is equivalent to Conjectures 1.3 and 1.4.*

PROOF. Both sides of (1.8) are right K_r^* invariant and satisfy (1.7). It is sufficient to compare them on $\widetilde{T}_r$. It is easy to see that a K_r^* invariant function satisfying (1.7) vanishes on elements of $\widetilde{T}_r$ which are not dominant. Hence Conjectures 1.3 and 1.4 are sufficient to imply Conjecture 1.2. □

2. Euler products

The Rankin-Selberg convolutions on $\mathrm{GL}(r) \times \mathrm{GL}(n)$ have a long history. The best modern reference is Cogdell and Piatetski-Shapiro [**CPS**]. There are qualitative differences between the three cases $r > n$, $r = n$ and $r < n$. We will assume that $r < n$. The adaptation to the covering groups is straightforward. We are interested in the case where the automorphic forms live on the n-fold covers $\widetilde{G}_r$ and $\widetilde{G}_n$ of G_r and G_n, respectively, the automorphic form on $\widetilde{G}_n$ being a generalized theta series in the sense of Kazhdan and Patterson [**KP84**].

Let F be a global field, and let A be its adele ring. Suppose that the group μ_n of n-th roots of unity in F has cardinality n. We will denote by $G_r(F)^* \subset G_r(A)$ the image of $G_r(F)$ under the canonical splitting. If $n-r > 1$, let P be the standard parabolic subgroup of G_n with Levi factor $\mathrm{GL}(r+1) \times \mathrm{GL}(1)^{n-r-1}$, and let U_P be its unipotent radical; if $n = r+1$, let U_P be trivial. Let U_n be the unipotent radical of the standard Borel subgroup B_n. We will denote by U_n^* and U_P^* the images of these groups under the canonical splittings. Let ψ be a nontrivial character of A/F, and let ψ_U be the character $u \mapsto \prod_i \psi(u_{i,i+1})$ of $U_n(A)^*/U_n(F)^*$. Let U_r be the unipotent radical of the Borel subgroup of G_r.

Let (π, V) and (θ_n, V_{θ_n}) be automorphic representations of the groups $\widetilde{G}_r(A)$ and $\widetilde{G}_n(A)$ respectively. Thus V and V_{θ_n} are spaces of automorphic forms. We will assume that π is cuspidal. Let $\Theta \in V_{\theta_n}$. We define a function $\Theta_{n,r+1}$ on $\widetilde{G}_{r+1}$ by

$$\Theta_{n,r+1}(g) = \int_{U_P(F)^*\backslash U_P(A)^*} \Theta\left(u\begin{pmatrix} g & \\ & I_{n-r-1}\end{pmatrix}\right)\psi_U(u)\,du.$$

This is not automorphic with respect to all of $G_{r+1}(F)^*$, but it *is* automorphic with respect to $G_r(F)^*$ embedded in the upper left corner, since this subgroup fixes the character ψ_U. Let X be the unipotent radical of the standard parabolic subgroup of G_{r+1} having Levi factor $\mathrm{GL}(r) \times \mathrm{GL}(1)$. Consider

$$(2.1)\quad \int_{G_r(F)\backslash G_r(A)} \phi(g)\,\overline{\left[\Theta_{n,r+1}\begin{pmatrix} g & \\ & 1\end{pmatrix} - \int_{X(F)^*\backslash X(A)^*}\Theta_{n,r+1}\left(x\begin{pmatrix} g & \\ & 1\end{pmatrix}\right)dx\right]} \times |\det(g)|^{s-\frac{n-r}{2}}\,dg,$$

where $\phi \in V$. This Hecke integral has a functional equation, and the center of the critical strip is at $s = \frac{1}{2}$. In writing this integral, we are making use of the fact that since ϕ and Θ are both genuine, $\phi\,\overline{\Theta}$ is μ_n-invariant, hence is a well-defined function on $G_r(A)$ (embedded in $G_n(A)$ as usual into the left corner). The second term in brackets is the "constant term" which must be subtracted in order to obtain convergence, just as one subtracts the constant term of the theta function in proving the functional equation of the zeta function after Riemann.

As in [**CPS**], this can be unfolded to

$$(2.2)\qquad \int_{U_r(A)^*\backslash G_r(A)} W_\phi(g)\,\overline{W_\Theta\begin{pmatrix} g & \\ & I_{n-r}\end{pmatrix}}\,|\det(g)|^{s-\frac{n-r}{2}}\,dg,$$

where W_ϕ and W_Θ are the global Whittaker functions attached to ϕ and Θ. The conjecture of Bump and Hoffstein [**BH87**], [**BH86**] is that this integral is Eulerian. The representation θ_n has a unique Whittaker model, though π does not.

Now let F be a nonarchimedean local field, and let π be an irreducible representation of $\widetilde{G}_r(F)$ in the spherical principal series. We assume that the additive character ψ of F has conductor $\mathfrak{o}$. Let $\mathcal{W}$ be any Whittaker model of π (with respect to ψ), and let W be a spherical vector in $\mathcal{W}$; let W_{θ_n} be the unique spherical Whittaker function for the θ_n, where the local exceptional representation of $\widetilde{G}_n(F)$ was described in Section 1. We consider

$$(2.3)\qquad \int_{U_r(F)^*\backslash G_r(F)} W(g)\,\overline{W_{\theta_n}\begin{pmatrix} g & \\ & I_{n-r}\end{pmatrix}}\,|\det(g)|^{s-\frac{n-r}{2}}\,dg.$$

PROPOSITION 2.1. *If Conjecture 1.2 is true, then the local integral (2.3) equals* $L\big(ns-(n-1)/2,\pi\big)\,W(1)$, *where* $L(s,\pi)$ *is as in Section 1.*

Note that Proposition 2.1 allows us to reduce the global integral (2.2) to an Euler product in the weak sense of Piatetski-Shapiro and Rallis [**PSR**] or Bump, Furusawa and Ginzburg [**BFG**] by pulling out an Euler factor at every nonramified place. The archimedean places may also be treated, since these are automatically complex, and the cover splits, so the archimedean theory is subsumed in the nonmetaplectic theory of [**JPSS**].

PROOF. We have

$$\int_{K_r^*} W(kg)\,dk = \sigma(g)\,W(1),$$

where as in Section 1, σ is the spherical function, since the left side is a K_r^*-biinvariant matrix coefficient of π. Thus by Proposition 1.1,

$$L\left(ns - \frac{n-1}{2}, \pi\right) W(1) = \int_{\widetilde{G}_r(F)} \Delta_{ns-\frac{n-1}{2}}(g) \int_{K_r^*} W(kg)\,dk\,dg.$$

Interchanging the order of integration and making the variable change $g \to k^{-1}g$, we see that

$$\begin{aligned} L\left(ns - \frac{n-1}{2}, \pi\right) W(1) &= \int_{\widetilde{G}_r(F)} \Delta_{ns-\frac{n-1}{2}}(g) W(g)\,dg \\ &= \int_{U_r^*\backslash\widetilde{G}_r(F)} \int_{U_r^*} \Delta_{ns-\frac{n-1}{2}}(ug)\,W(ug)\,du\,dg \\ &= \int_{U_r^*\backslash\widetilde{G}_r(F)} \int_{U_r^*} \Delta_{ns-\frac{n-1}{2}}(ug)\,\psi_U(u)\,du\,W(g)\,dg. \end{aligned}$$

Now using (1.8) we obtain (2.3). □

3. The conjecture when $r = 2$ and $r = 3$

We have verified Conjectures 1.3 and 1.4 when $r = 2$ and $r = 3$. We give complete details when $r = 3$; when $r = 2$, the calculations are similar, but much simpler. We return to the notations of Section 1. We will make use of cocycle formulas from Kazhdan and Patterson [**KP84**]. Since these are known to be unreliable without the assumption that -1 is an n-th power in F, we will make this assumption. This implies that $(-1, a) = (a, a) = 1$ for all $a \in F^\times$.

We now give the integral to be evaluated below. Let $k_1 \geq k_2 \geq k_3 \geq 0$. Then we consider the integral

$$I_s(k_1, k_2, k_3) = \int_{F^3} \Delta_s\left(\mathbf{s}\begin{pmatrix} \varpi^{k_1} & x\varpi^{k_2} & y\varpi^{k_3} \\ & \varpi^{k_2} & z\varpi^{k_3} \\ & & \varpi^{k_3} \end{pmatrix}\right) \psi(x+z)\,dx\,dy\,dz,$$

where Δ_s is as in Section 1. After the change of variables $y \mapsto xyz$, one has

$$(3.1)\quad I_s(k_1, k_2, k_3) = \int_{F^3} \Delta_s\left(\mathbf{s}\begin{pmatrix} \varpi^{k_1} & x\varpi^{k_2} & xyz\varpi^{k_3} \\ & \varpi^{k_2} & z\varpi^{k_3} \\ & & \varpi^{k_3} \end{pmatrix}\right) |xz|\,\psi(x+z)\,dx\,dy\,dz.$$

In evaluating this integral, we use g_t to denote the Gauss sum

$$g_t = \int_{|x|=q} (\varpi, x)^{-t}\,\psi(x)\,dx.$$

Observe that if $\psi(\varpi^{-1}) = \zeta$, then this integral may be rewritten as a sum

$$g_t = \sum_{a \in (\mathfrak{o}/\varpi\mathfrak{o})^\times} \chi^t(a)\zeta^a,$$

where χ is the character obtained from the Hilbert symbol. We have $|g_t| = \sqrt{q}$ if $n \nmid t$.

PROPOSITION 3.1. *Let*

$$I_s'(k_1,k_2,k_3) = q^{\frac{(k_1+k_2+k_3)s}{n}+k_1+\frac{k_2}{n}+\frac{k_3(2-n)}{n}} I_s(k_1,k_2,k_3).$$

Then

$$I_s'(k_1,k_2,k_3) = \begin{cases} 1 & \text{if } (k_1,k_2,k_3) \equiv (0,0,0) \bmod n \\ q^{1/n-1} g_1 & \text{if } (k_1,k_2,k_3) \equiv (0,-1,1) \bmod n \\ q^{1/n-1} g_1 & \text{if } (k_1,k_2,k_3) \equiv (-1,1,0) \bmod n \\ q^{3/n-2} g_1 g_2 & \text{if } (k_1,k_2,k_3) \equiv (-1,-1,2) \bmod n \\ q^{3/n-2} g_1 g_2 & \text{if } (k_1,k_2,k_3) \equiv (-2,1,1) \bmod n \\ q^{4/n-3} g_1^2 g_2 & \text{if } (k_1,k_2,k_3) \equiv (-2,0,2) \bmod n \\ 0 & \text{otherwise.} \end{cases}$$

PROOF. To evaluate the basic integral, we must compute the Cartan-Satake decomposition of the matrix

$$m = \begin{pmatrix} \varpi^{k_1} & x\varpi^{k_2} & xyz\varpi^{k_3} \\ & \varpi^{k_2} & z\varpi^{k_3} \\ & & \varpi^{k_3} \end{pmatrix}.$$

If this matrix lies in the Cartan cell $K_3 \operatorname{diag}(\varpi^{h_1},\varpi^{h_2},\varpi^{h_3}) K_3$, we will call the quantities $\varpi^{h_1},\varpi^{h_2},\varpi^{h_3}$ the *elementary divisors* of m. In evaluating the integral, there is also an n-th root of unity to be determined, since we must ultimately do this calculation on the metaplectic group. This root of unity will be expressed as a Hilbert symbol in the relevant variables.

The basic calculation which must be done is the following one on the cover of GL_2; the Cartan decompositions described in detail below will be obtained by essentially embedding this calculation into $\widetilde{G}$ in various ways. We will make use of the explicit cocycle described in Kazhdan and Patterson [**KP84**]; see also Bump and Hoffstein [**BH87**].

LEMMA 3.2. *Suppose ord(f) $= d < 0$ and $l \geq p$. Then*

$$\mathbf{s}\begin{pmatrix} \varpi^l & f\varpi^p \\ & \varpi^p \end{pmatrix} \in (\varpi^{-d}, f)\, K_3^* \,\mathbf{s}\begin{pmatrix} \varpi^{p+d} & \\ & \varpi^{l-d} \end{pmatrix} K_3^*.$$

PROOF. Let $f = \varpi^d u$. Then we have

$$\begin{pmatrix} \varpi^l & f\varpi^p \\ & \varpi^p \end{pmatrix} = \begin{pmatrix} u & 0 \\ \varpi^{-d} & u^{-1} \end{pmatrix} \begin{pmatrix} \varpi^{p+d} & \\ & \varpi^{l-d} \end{pmatrix} \begin{pmatrix} f^{-1}\varpi^{l-p} & 1 \\ -1 & 0 \end{pmatrix}.$$

Computing the cocycles which arise when one does this calculation metaplectically, one arrives at the value for ζ given above. □

We turn to the matrix m. Let us write

$$a = \operatorname{ord}(x), \quad b = \operatorname{ord}(y), \quad c = \operatorname{ord}(z). \tag{3.2}$$

The determination of the elementary divisors of m is governed by the minimum valuation of its minors. For example, computing the 1×1 minors, and recalling that $k_1 \geq k_2 \geq k_3 \geq 0$, one finds that

$$h_3 = \min(k_3, c+k_3, a+b+c+k_3, a+k_2). \tag{3.3}$$

Similarly, consideration of the 2×2 minors gives

(3.4) $$h_2+h_3 = \min\big(k_2+k_3, k_2+k_3+a, k_1+k_3+c, k_2+k_3+a+c+\operatorname{ord}(y-1)\big).$$

Also, the determinant must satisfy

(3.5) $$h_1+h_2+h_3 = k_1+k_2+k_3.$$

Then for $\Delta_s(\mathbf{s}(m))$ to be nonzero, each of the quantities (3.3)–(3.5) must be a non-negative integer multiple of n. In particular, the integral (3.1) may be restricted without loss to the domain of integration given by the conditions

(3.6) $$a+k_2, c+k_3, a+b+c+k_3 \geq 0.$$

We do so henceforth.

Now, since the basic integral (3.1) includes integration over a character in x and in z, while the dependence of Δ_s on these quantities is a function of $|x|$, $|z|$, and powers of the Hilbert symbols (ϖ, x) and (ϖ, z) (arising in some instances from the Cartan decomposition on the metaplectic group), the cancellation of characters over a group shows that $a, c \geq -1$.

We now evaluate the integral by subdividing the domain of integration into different Cartan cells, and computing the contribution of each Cartan cell to the integral. For convenience, set

$$\Delta'_s\left(\mathbf{s}\begin{pmatrix}\varpi^{k_1} & x\varpi^{k_2} & y\varpi^{k_3}\\ & \varpi^{k_2} & z\varpi^{k_3}\\ & & \varpi^{k_3}\end{pmatrix}\right) = q^{\frac{(k_1+k_2+k_3)s}{n}+k_1+\frac{k_2}{n}+\frac{k_3(2-n)}{n}}\,\Delta_s\left(\mathbf{s}\begin{pmatrix}\varpi^{k_1} & x\varpi^{k_2} & y\varpi^{k_3}\\ & \varpi^{k_2} & z\varpi^{k_3}\\ & & \varpi^{k_3}\end{pmatrix}\right).$$

Then

(3.7) $$I'_s(k_1,k_2,k_3) = \int_{F^3}\Delta'_s\left(\mathbf{s}\begin{pmatrix}\varpi^{k_1} & x\varpi^{k_2} & xyz\varpi^{k_3}\\ & \varpi^{k_2} & z\varpi^{k_3}\\ & & \varpi^{k_3}\end{pmatrix}\right)|xz|\,\psi(x+z)\,dx\,dy\,dz.$$

Let us consider first the Cartan cells such that $h_2+h_3 = k_2+k_3$. From (3.4) one concludes that $a, a+c+\operatorname{ord}(y-1) \geq 0$.

Case 1: $h_3 = k_3$, $h_2+h_3 = k_2+k_3$. Note that the integral over this cell is nonzero only when $(k_1,k_2,k_3) \equiv (0,0,0) \bmod n$, as Δ_s is nonzero only when all elementary divisors are n-th powers. Suppose that this congruence condition holds. The minimums in (3.3) and (3.4) are as specified if and only if $a, c, a+b+c \geq 0$. Thus in this cell $x, z, x(1-y)z \in \mathfrak{o}$, and one may multiply $\mathbf{s}(m)$ by a matrix in $K_3^* \cap U_3^*$ to make it diagonal. Thus $\Delta'_s(\mathbf{s}(m)) = \Delta'_s(\mathbf{s}(\operatorname{diag}(\varpi^{k_1},\varpi^{k_2},\varpi^{k_3}))) = 1$, and the contribution over this cell to I'_s is easily seen to be 1.

Case 2: $h_3 = k_3-1$, $h_2+h_3 = k_2+k_3$. On this cell the integral is nonzero only when $(k_1,k_2,k_3) \equiv (0,-1,1) \bmod n$. Suppose that this holds. To obtain $h_3 = k_3-1$, we must have either $c=-1$, $a+b \geq 0$ or $c \geq 0$, $a+b+c=-1$. (The possibility $a=-1$ and $k_2=k_3$ is ruled out by the congruence conditions.) But, as noted, the condition $h_2+h_3 = k_2+k_3$ forces $a \geq 0$, $a+c+\operatorname{ord}(y-1) \geq 0$. It is easily seen that these may hold simultaneously if and only if $a \geq 0$, $c = -1$, and $a + \operatorname{ord}(y-1) \geq 1$. Thus $x, xz(1-y) \in \mathfrak{o}$, and we have

$$\mathbf{s}\begin{pmatrix} 1 & -x & xz(1-y) \\ & 1 & \\ & & 1 \end{pmatrix} \mathbf{s}(m) = \mathbf{s}\begin{pmatrix} \varpi^{k_1} & 0 & 0 \\ & \varpi^{k_2} & z\varpi^{k_3} \\ & & \varpi^{k_3} \end{pmatrix}.$$

Combining this with a calculation similar to Lemma 3.2, one sees that on this cell, $\Delta_s'(\mathbf{s}(m)) = q^{1/n-1}(\varpi, z)^{-1}$. Thus the z integral gives q times the Gauss sum g_1, and the contribution of the entire cell to I_s' is

$$q^{1/n} g_1 \int_{|x|\leq 1, |x(y-1)|<1} |x|\, dx\, dy = q^{1/n-1} g_1.$$

There are no other Cartan cells for which $h_2 + h_3 = k_2 + k_3$. For if $h_3 < k_3 - 1$, then (3.3) would imply that $a + b + c \leq -2$, while (3.4) would imply that $a, a + c + \operatorname{ord}(y-1) \geq 0$. But since $c \geq -1$, these equations would imply that $b < 0$, hence $\operatorname{ord}(y-1) = b$, and so these are inconsistent.

Next we consider the Cartan cells such that $h_2 + h_3 = k_2 + k_3 - 1$. These cells arise from the domain of integration with $a = -1$, $c + \operatorname{ord}(y-1) \geq 0$ and from $a \geq 0$, $a + c + \operatorname{ord}(y-1) = -1$. We consider the various possibilities for h_3.

Case 3: $h_3 = k_3$, $h_2 + h_3 = k_2 + k_3 - 1$. On this cell, Δ_s is nonzero only when the congruence condition $(k_1, k_2, k_3) \equiv (-1, 1, 0) \bmod n$ holds. Suppose it does. The equality (3.3) implies that $c, a + b + c \geq 0$. The conditions $a \geq 0$, $a + c + \operatorname{ord}(y-1) = -1$, are not consistent with this. Thus Case 3 arises if and only if $a = -1$, $b + c \geq 1$, $c \geq 0$. Then, similarly to Case 2 above, multiplying by a matrix in $K_3^* \cap U_3^*$ reduces $\mathbf{s}(m)$ to

$$\mathbf{s}\begin{pmatrix} \varpi^{k_1} & x\varpi^{k_2} & 0 \\ & \varpi^{k_2} & 0 \\ & & \varpi^{k_3} \end{pmatrix},$$

and one may compute the Cartan decomposition as in Lemma 3.2. One sees that $\Delta_s'(\mathbf{s}(m)) = q^{1/n-1}(\varpi, x)^{-1}$, and this case contributes the same quantity $q^{1/n-1} g_1$ as does Case 2.

Case 4: $h_3 = k_3 - 1$, $h_2 + h_3 = k_2 + k_3 - 1$. For these conditions to hold, we must have (i) either $c = -1$ and $a + b \geq 0$; or $c \geq 0$ and $a + b + c = -1$; and (ii) either $a = -1$ and $c + \operatorname{ord}(y-1) \geq 0$; or $a \geq 0$ and $a + c + \operatorname{ord}(y-1) = -1$. Combining these, we arrive at the 3 possibilities (a) $c = -1$, $a \geq 0$, $a + \operatorname{ord}(y-1) = 0$; (b) $a = -1$, $c \geq 0$, $b + c = 0$; and (c) $a, c \geq 0$, $a + b + c = -1$. We claim that for all of these subdomains, the integral is zero. Consider first subdomain (a). Since $x \in \mathfrak{o}$, after changing by an integral upper triangular unipotent, we need to evaluate

$$\Delta_s\left(\mathbf{s}\begin{pmatrix} \varpi^{k_1} & 0 & x(y-1)z\varpi^{k_3} \\ & \varpi^{k_2} & z\varpi^{k_3} \\ & & \varpi^{k_3} \end{pmatrix}\right). \tag{3.8}$$

Computing the Cartan decomposition by a calculation similar to Lemma 3.2 (used twice), one sees that $\Delta_s'(\mathbf{s}(m))$ gives $q^{2/n-2}$ times the Hilbert symbol $\big(\varpi, x(y-1)z\big)^{-1}$. Thus, assuming $(k_1, k_2, k_3) \equiv (-1, 0, 1) \bmod n$, the contribution to I_s' from the integral over subdomain (a) is equal to

$$q^{2/n-2} \int_{|x|\leq 1, |x(y-1)|=1, |z|=q} |xz| \big(\varpi, x(y-1)z\big)^{-1} \psi(z)\, dx\, dy\, dz = 0.$$

The argument for subdomain (b) is similar. Consider then subdomain (c). Since $x, z \in \mathfrak{o}$, after changing by an upper triangular unipotent and changing y to $y+1$ we must evaluate

$$\Delta_s'\left(\mathbf{s}\begin{pmatrix} \varpi^{k_1} & 0 & xyz\varpi^{k_3} \\ & \varpi^{k_2} & 0 \\ & & \varpi^{k_3} \end{pmatrix}\right).$$

This contributes $q^{2/n-2}$ times the Hilbert symbol $(\varpi, xyz)^{-1}$. Once again the integral against this symbol is zero. More precisely, after changing $y \mapsto y/xz$, the integral over subdomain (c) is given by

$$q^{2/n-2}\int_{|x|,|z|\leq 1,|y|=q^{-1}} (\varpi, y)^{-1}\, dx\, dy\, dz = 0,$$

as claimed.

Case 5: $h_3 = k_3-2$, $h_2+h_3 = k_2+k_3-1$. The function Δ_s is zero on this Cartan cell unless one has the congruence conditions $(k_1, k_2, k_3) \equiv (-1,-1,2) \bmod n$, so let us assume this. Since $a, c \geq -1$, equality (3.3) implies that $a+b+c = -2$. If $b < 0$, then $b = \operatorname{ord}(y-1)$, and (3.3), (3.4) give incompatible conditions on $a+b+c = a+\operatorname{ord}(y-1)+c$. Hence we must have $b=0$, $a=c=-1$, and in fact $\operatorname{ord}(y-1) \geq 1$. One may once again compute the Cartan decomposition by a calculation similar to Lemma 3.2, used twice. More precisely, using it in the upper left 2×2 block reduces m to

$$\begin{pmatrix} \varpi^{k_2-1} & 0 & yz\varpi^{k_3-1} \\ & \varpi^{k_1+1} & xz(1-y)\varpi^{k_3+1} \\ & & \varpi^{k_3} \end{pmatrix}. \tag{3.9}$$

Metaplectically, this multiplication gives the symbol (ϖ, x). Then one may remove the $(2,3)$ entry by left multiplication by an element of $K_3^* \cap U_3^*$, and reduce the remaining matrix to a diagonal by a calculation similar to Lemma 3.2 in the $(1,3)\times(1,3)$ block. One finds that $\Delta_s'(\mathbf{s}(m))$ contributes a Hilbert symbol of $(\varpi, x)^{-1}(\varpi^2, yz)$ and a factor of $q^{3/n-3}$. Thus the contribution from this domain to the integral I_s' is given by

$$q^{3/n-3}\int_{|x|=|z|=q,|y-1|\leq q^{-1}} |xz|(\varpi, x)^{-1}(\varpi^2, yz)\,\psi(x+z)\, dx\, dy\, dz.$$

The integrals in x and z contribute q times the Gauss sums g_1 and g_2, respectively. Since the Hilbert symbol is independent of y, the integral is equal to $q^{3/n-2}g_1g_2$.

Note that if $h_2+h_3 = k_2+k_3-1$, then one can not have $h_3 < k_3-2$. For then (3.3) implies that $b<0$ (as $a, c \geq -1$), and we obtain the inequality $a+b+c = a+\operatorname{ord}(y-1)+c < -2$, which is incompatible with (3.4).

We turn to the Cartan cells such that $h_2+h_3 = k_2+k_3-2$. By equality (3.4), we must have $a+c+\operatorname{ord}(y-1) = -2$. If $b\geq 0$ then $a=c=-1$, $\operatorname{ord}(y-1)=0$. Thus (3.3) gives $h_3 = k_3-1$ or $h_3 = k_3-2$. If $b<0$, $b = \operatorname{ord}(y-1)$ and so $a+b+c=-2$. Then (3.3) implies that $h_3 = k_3-2$. We analyze these cases.

Case 6: $h_3 = k_3-1$, $h_2+h_3 = k_2+k_3-2$. The value of $\Delta_s(\mathbf{s}(m))$ is zero unless $(k_1,k_2,k_3) \equiv (-2,1,1) \bmod n$. Suppose this holds. As noted above, we must have $a=c=-1$. Since (3.3) forces $a+b+c\geq -1$, we must have $b \geq 1$. One obtains the Cartan decomposition by reducing to (3.9) as in Case 5 above, then using the calculation of Lemma 3.2 in the lower right block, and finally right multiplying by an element of $K_3^*\cap U_3^*$. The cocycle then gives the Hilbert symbol $\left(\varpi, x^2(1-y)z\right)$.

Since $|y| < 1$, $(\varpi, 1-y) = 1$. Thus the contribution to the integral I_s' from this Cartan cell is given by

$$q^{3/n-3} \int_{|x|=|z|=q, |y|<1} |xz| (\varpi, x)^{-2} (\varpi, z)^{-1} \psi(x+z)\, dx\, dy\, dz = q^{3/n-2} g_1 g_2.$$

Case 7: $h_3 = k_3 - 2$, $h_2 + h_3 = k_2 + k_3 - 2$. For this case to contribute a nonzero value, we must have $(k_1, k_2, k_3) \equiv (-2, 0, 2) \bmod n$. Suppose this. Here we must consider the two domains of integration (a) $a = c = -1$, $|y| = |y-1| = 1$; and (b) $b < 0$, $a + b + c = -2$.

Consider first case (a): $a = c = -1$, $|y| = |1-y| = 1$. To compute the Cartan decomposition, proceeding as in Case 5, one reduces to (3.9), with a cocycle (ϖ, x). Then using the $(1,3) \times (1,3)$ embedding and the calculation of Lemma 3.2, this matrix reduces to

$$\begin{pmatrix} \varpi^{k_3-2} & 0 & 0 \\ x(1-y)z\varpi^{k_3+1} & \varpi^{k_1+1} & xy^{-1}(1-y)\varpi^{k_2+1} \\ & & \varpi^{k_2+1} \end{pmatrix},$$

with an additional cocycle factor of (ϖ^2, yz). One may remove the $(2,1)$ entry by an integral unipotent, and finally reduce to a diagonal by another calculation similar to Lemma 3.2, in the lower right block. The calculation shows that on this cell, $\Delta_s'(\mathbf{s}(m)) = q^{4/n-4}(\varpi, x^2y(1-y)z^2)^{-1}$. Thus the contribution to I_s' is given by

$$q^{4/n-2} \int_{|x|=|z|=q, |y|=|1-y|=1} (\varpi, x^2y(1-y)z^2)^{-1} \psi(x+z)\, dx\, dy\, dz$$
$$= q^{4/n-2} g_2^2 \int_{|y|=|1-y|=1} (\varpi, y(1-y))^{-1}\, dy.$$

But the integral in y is q^{-1} times the classical Jacobi sum

$$J(\chi, \chi) = \sum_{r+s=1} \chi(r)\chi(s).$$

This Jacobi sum may be expressed as a quotient of Gauss sums: $J(\chi,\chi) = g_1^2/g_2$. Thus the contribution to I_s' from subdomain (a) equals $q^{4/n-3} g_1^2 g_2$.

There is also case (b): $b < 0$, $a + b + c = 2$. Ultimately this subdomain yields zero. To see this, it is convenient to separate this region of integration further into the subregions (b-i) $a \geq 0$, $c = -1$, $a + b = -1$; (b-ii) $c \geq 0$, $a = -1$, $b + c = -1$; and (b-iii) $a, c \geq 0$, $a + b + c = -2$. In case (b-i), $x \in \mathfrak{o}$, and $\Delta_s(\mathbf{s}(m))$ reduces to the calculation of (3.8). Changing y to $y+1$ for convenience, a calculation similar to Lemma 3.2, using the $(1,3) \times (1,3)$ block, then gives the equality

$$\Delta_s(\mathbf{s}(m)) = (\varpi^2, xyz)^{-1} \Delta_s \left(\mathbf{s} \begin{pmatrix} \varpi^{k_3-2} & 0 & 0 \\ z\varpi^{k_3} & \varpi^{k_2} & (xy)^{-1}\varpi^{k_1} \\ & & \varpi^{k_1+2} \end{pmatrix} \right).$$

Changing by integral unipotents makes the argument of the right hand side into a diagonal matrix with no further cocycles. Thus the contribution from subdomain (b-i) to the quantity I_s' is given by

$$q^{4/n-4} \int_{|x|\leq 1, |z|=|xy|=q} |xz| (\varpi^2, xyz)^{-1} \psi(z)\, dx\, dy\, dz = 0.$$

Case (b-ii) gives zero; in fact the integral is the same as that of case (b-i) after the variable change $x \leftrightarrow z$.

Finally, in case (b-iii), one changes $\mathbf{s}(m)$ by a matrix in $K_3^* \cap U_3^*$ to obtain

$$\mathbf{s}\begin{pmatrix} \varpi^{k_1} & 0 & xz(y-1)\varpi^{k_3} \\ & \varpi^{k_2} & 0 \\ & & \varpi^{k_3} \end{pmatrix}.$$

Changing y to $y+1$ for convenience, and proceeding as in Lemma 3.2 in the $(1,3)\times(1,3)$ block, one obtains a cocycle of (ϖ^2, xyz), and a contribution of

$$q^{4/n-4}\int_{|x|,|z|\leq 1, |xzy|=q^2} |xz|(\varpi^2, xyz)^{-1}\,dx\,dy\,dz = 0.$$

This completes the evaluation in Case 7.

The final possibility is $h_2+h_3 < k_2+k_3-2$. As we shall see, this case does not contribute to I_s'.

Case 8: $h_2+h_3 < k_2+k_3-2$. Thus $a+b+c < -2$, and so $b<0$. Thus $\mathrm{ord}(y-1)=b$, and so (3.3), (3.4) give $h_3 = k_3+a+b+c$, $h_2+h_3 = k_2+k_3+a+b+c$. Suppose each is congruent to 0 modulo n. In doing the Cartan decomposition, it is convenient to separate the cases (a) $a\geq 0$ and (b) $a=-1$.

The calculation of Δ_s for subdomain (a) is similar to Case 7 (b-i). Indeed, in subdomain (a), $x\in\mathfrak{o}$, and the value of Δ_s reduces to (3.8). Computing as in Lemma 3.2 (using the $(1,3)\times(1,3)$ embedding) gives

$$\Delta_s(\mathbf{s}(m)) = (\varpi^{-(a+b+c)}, xyz)^{-1}\Delta_s\left(\mathbf{s}\begin{pmatrix} \varpi^{k_3+a+b+c} & 0 & 0 \\ z\varpi^{k_3} & \varpi^{k_2} & (xy)^{-1}\varpi^{k_1} \\ & & \varpi^{k_1-(a+b+c)} \end{pmatrix}\right).$$

The matrix on the right may be diagonalized by left and right multiplication by integral unipotent matrices, and no additional Hilbert symbols arise. Now $\varpi^{-(a+b+c)} = \varpi^{-(\mathrm{ord}(x)+\mathrm{ord}(y)+\mathrm{ord}(z))}$. Recalling conditions (3.6), and changing $y\mapsto y/xz$, one sees that subdomain (a) contributes a power of q times the integral

$$\int_{|x|\leq 1, q^2<|y|\leq q^{k_3}, |z|\leq q^{k_3}} (\varpi^{-\mathrm{ord} y}, y)^{-1}\,\psi(z)\,dx\,dy\,dz.$$

But the integral in z is zero unless $k_3=0$, in which case the domain of integration is empty.

We turn to case (b). The calculation of Δ_s here is similar to Case 7 (a). To carry out the Cartan decomposition one may first reduce m to the matrix (3.9), with a Hilbert symbol (ϖ, x). Using the $(1,3)\times(1,3)$ block and reducing as in Lemma 3.2, one arrives at

$$\begin{pmatrix} \varpi^{k_3+a+b+c} & 0 & 0 \\ x(1-y)z\varpi^{k_3+1} & \varpi^{k_1+1} & xy^{-1}(1-y)\varpi^{k_2+1} \\ & & \varpi^{k_2-b-c} \end{pmatrix}.$$

This step contributes the symbol $(\varpi^{-(a+b+c)}, yz)$. Changing on the left by an integral unipotent eliminates the $(2,1)$ entry, and then carrying out one more 2×2 computation, this time on the lower right block, one obtains a diagonal matrix, with total cocycle $(\varpi,x)(\varpi^{-(a+b+c)}, yz)(\varpi^{-(b+c)}, xy^{-1}(1-y))$. But $b<0$, so $(\varpi, y^{-1}(1-y))=1$, and this product reduces to $(\varpi^{-(a+b+c)}, xyz)$. Then arguing as in case (a), the integral is zero.

This completes the proof of the Proposition 3.1. □

COROLLARY *Conjecture 1.2 is true if* $r = 2$ *or* 3.

PROOF. When $r = 3$, this follows from Proposition 3.1, together with the known values of the spherical Whittaker function, computed in Kazhdan and Patterson [**KP84**] and Bump and Hoffstein [**BH87**]. The $r = 2$ case may be checked directly by a calculation similar to the one above—details are omitted here. Alternatively, the truth of the conjecture for given r implies the truth of the conjecture for $r - 1$, which gives the $r = 2$ case from the above, except when $n = 3$. □

References

[A] A. Andrianov, *Euler expansions of theta-transforms of Siegel modular forms of degree n*, Mat. Sb. **105 (147)** (1978), 291–341.

[BFG] D. Bump, M. Furusawa and D. Ginzburg, *Non-unique models in the Rankin-Selberg method*, J. Reine Angew. Math. **468** (1995), 77–111.

[BH86] D. Bump and J. Hoffstein, *Some Euler products associated with cubic metaplectic forms on* GL(3), Duke Math. J. **53** (1986), 1047–1072.

[BH87] D. Bump and J. Hoffstein, *On Shimura's correspondence*, Duke Math. J. **55** (1987), 661–691.

[CPS] J. Cogdell and I. Piatetski-Shapiro *Converse theorems for* GL_n, Inst. Hautes Études Sci. Publ. Math. **79** (1994), 157–214.

[DN] K. Doi and H. Naganuma, *On the functional equation of certain Dirichlet series*, Invent. Math. **19** (1969), 1–14.

[FK] Y. Flicker and D. Kazhdan, *Metaplectic correspondence*, Inst. Hautes Études Sci. Publ. Math. **64** (1986), 53–110.

[FW] S. Friedberg and S.-T. Wong, *On the Shimura correspondence for* GSp(4), Math. Ann. **290** (1991), 183–207.

[GJ] R. Godement and H. Jacquet, Zeta functions of simple algebras, Springer Lecture Notes in Mathematics **260**, Berlin-New York: Springer-Verlag (1972).

[Go] T. Goetze, *Euler products associated to metaplectic automorphic forms on the* 3*-fold cover of* GSp(4), Trans. Amer. Math. Soc. **350** (1998), 975–1011.

[JPSS] H. Jacquet, I. Piatetski-Shapiro and J. Shalika, *Relèvement cubique non normal*, C. R. Paris Sér. I Math. **193** (1981), 13–18.

[JS] H. Jacquet and J. Shalika, *Rankin-Selberg convolutions: Archimedean theory*, Festschrift in honor of I. I. Piatetski-Shapiro on the occasion of his sixtieth birthday, Part I, Israel Math. Conf. Proc. **2**, Jerusalem: Weizmann (1990), 125–207.

[KP84] D. Kazhdan and S. Patterson, *Metaplectic forms*, Inst. Hautes Études Sci. Publ. Math. **59**, (1984), 35–142.

[KP86] D. Kazhdan and S. Patterson, *Towards a generalized Shimura correspondence*, Adv. in Math. **60** (1986), 161–234.

[PSR] I. Piatetski-Shapiro and S. Rallis, *A new way to get Euler products*, J. Reine Angew. Math. **392**, (1988), 110–124.

[Sa] G. Savin, *Local Shimura correspondence*, Math. Ann. **280** (1988), 185–190.

[S] G. Shimura, *On modular forms of half-integral weight*, Ann. of Math. **97** (1973), 440–481.

[So] D. Soudry, *Rankin-Selberg convolutions for* $\mathrm{SO}_{2l+1} \times \mathrm{GL}_n$*: local theory*, Mem. Amer. Math. Soc. **105** (1993).

[Su] T. Suzuki, *Distinguished representations of metaplectic groups*, preprint (1996).

[T] J. Tunnell, *Artin's conjecture for representations of octahedral type*, Bull. Amer. Math. Soc. **5** (1981), 173–175.

[W] A. Wiles, *Modular elliptic curves and Fermat's last theorem*, Ann. of Math. **141** (1995), 443–551.

[Z] V. G. Zhuravlev, *Euler expansions of theta-transforms of Siegel modular forms of half-integral weight and their analytic properties*, Mat. Sb. **123 (165)** (1984), 174–194.

Department of Mathematics, Stanford University, Stanford, CA 94305-2125
E-mail address: `bump@math.stanford.edu`

Department of Mathematics, The University of California at Santa Cruz, Santa Cruz, CA 95064 and Department of Mathematics, Boston College, Chestnut Hill, MA 02467-3806
E-mail address: `friedber@bc.edu`

Proceedings of Symposia in Pure Mathematics
Volume **66.2**, 1999

On the Plancherel Measure for the Continuous Spectrum of the Modular Group

Bill Casselman

To Goro Shimura with respect and gratitude

ABSTRACT. The Plancherel measure for Eisenstein series is derived in an apparently new way by using truncation and Fourier analysis on the real multiplicative group, combined with the notion of extended automorphic form.

Let $\Gamma = \mathrm{SL}_2(\mathbb{Z})$. In this paper I will give what is apparently a new proof of the basic Plancherel theorems on $\Gamma\backslash\mathcal{H}$. The same argument will work in much more varied circumstances.

Maass' Eisenstein series is defined by the formula

$$E_s(z) = \frac{1}{2} \sum_{(m,n)=1} \frac{y^s}{|mz+n|^{2s}}$$

for z in the upper half plane $\mathcal{H}$. This series is convergent for $\mathrm{RE}(s) > 1$, and continues meromorphically in s to all of $\mathbb{C}$. It is invariant under Γ and an eigenfunction of the Laplacian:

$$\Delta E_s = s(s-1)E_s.$$

It satisfies the functional equation

$$E_{1-s} = c(s)E_s, \quad c(s) = \frac{\xi(2s-1)}{\xi(2s)}, \quad \xi(s) = \pi^{-s/2}\Gamma(s/2)\zeta(s).$$

The only properties of $c(s)$ that we shall need are:

- The function $c(s)$ is smooth on $\mathrm{RE}(s) = 1/2$.
- For $\mathrm{RE}(s) = 1/2$ it satisfies

$$|c(s)| = 1, \quad c(s)c(1-s) = 1.$$

These or analogous properties hold for any arithmetic group of rank 1. The function E_s also satisfies the growth condition

$$E_s(z) = y^s + c(s)y^{1-s} + O(y^{-N})$$

for all positive N and large y.

1991 *Mathematics Subject Classification.* Primary 22E40, 22E30.

Key words and phrases. Eisenstein series, Plancherel formula, truncation, extended automorphic form.

For $\varphi(s)$ smooth and of compact support on $\mathrm{Re}(s) = \sigma = 1/2$ define the function

$$E_\varphi(z) = \frac{1}{2}\frac{1}{2\pi i}\int_{1/2-i\infty}^{1/2+i\infty} \varphi(s)E_s(z)\,ds.$$

For y large this is equal to

$$\frac{1}{4\pi i}\int_{\sigma=1/2} \varphi(s)\big(y^s + c(s)y^{1-s}\big)\,ds + O(y^{-N})$$

for all N. The integral is equal to

$$\frac{1}{4\pi i}\int_{\sigma=1/2} \big(\varphi(s) + c(1-s)\varphi(1-s)\big)y^s\,ds = \frac{1}{2\pi i}\int_{\sigma=1/2} \varphi_\#(s)y^s\,ds$$

where

$$\varphi_\#(s) = \frac{1}{2}\big(\varphi(s) + c(1-s)\varphi(1-s)\big)$$

is the projection of φ onto the space of functions satisfying

$$\varphi_\#(1-s) = c(s)\varphi_\#(s).$$

The usual Fourier duality of Schwartz spaces on the multiplicative group $(0,\infty)$ guarantees that

$$E_\varphi(z) = O(y^{1/2}\log^{-N} y)$$

for all $N > 0$ and large y. Therefore for any t with $\mathrm{Re}(t) = 1/2$ the Fourier-Eisenstein integral

$$\langle E_\varphi, E_{1-t}\rangle = \int_{\Gamma\backslash\mathcal{H}} E_\varphi(z)E_{1-t}(z)\,\frac{dx\,dy}{y^2}$$

is absolutely convergent.

THEOREM. (Fourier inversion) *We have*

$$\langle E_\varphi, E_{1-t}\rangle = \varphi_\#(t).$$

For φ of compact support on $\mathrm{Re}(s) = 1/2$ define

$$F_\varphi = \frac{1}{2\pi i}\int_{\sigma=1/2} \varphi(s)y^s\,ds.$$

This is a variant of the usual inverse Fourier transform for the multiplicative group $(0,\infty)$, given the measure dy/y^2.

THEOREM. (Plancherel formula) *We have*

$$\langle E_\varphi, E_\psi\rangle_{\Gamma\backslash\mathcal{H}} = \left(\frac{1}{2}\right)\frac{1}{2\pi i}\int_{\sigma=1/2} \varphi_\#(s)\psi_\#(1-s)\,ds = \left(\frac{1}{2}\right)\langle F_{\varphi_\#}, F_{\psi_\#}\rangle_{(0,\infty)}.$$

This theory (extended to include arbitrary arithmetic groups Γ) is due to Maass, Roelcke, and Selberg. The usual proof of these results (found, for example, in Langlands [**L**] or §7.4 of Kubota [**K**]) involves moving a contour integral from the domain of convergence of Eisenstein series to the critical line $\sigma = 1/2$. This will be avoided in the argument below, which concerns only Eisenstein series parametrized by s on the critical line. The usual argument, I should remark, also proves the full Plancherel theorem for the continuous spectrum and the part of the discrete spectrum arising from residues of Eisenstein series, whereas the argument below says nothing directly about completeness. Another argument, however, like that in Casselman [**C83**], can be used in some circumstances to prove an even stronger completeness result starting with the one given here. In particular, the argument given here allows one to prove the Paley-Wiener theorem of Casselman [**C83**] without moving contours twice, first one way and then the other, as was done there.

The proof below, and that of all its generalizations, proceeds in three steps: (1) *truncation*; (2) *regularization*, which yields the Maass-Selberg formula, following Casselman [**C93**]; (3) *Fourier analysis* on $(0, \infty)$. Of course the aim in all proofs of the Plancherel theorems for Eisenstein series is to reduce the calculation on $\Gamma\backslash\mathcal{H}$ to that on the multiplicative group. The only difference between various proofs is exactly how and where one does this. In particular, it is how this step is taken which is new in this paper. In the literature, when truncation is used to derive a Plancherel measure, it is followed by taking limits as the truncation parameter passes to infinity. What is apparently novel in this paper is that a well known device from harmonic analysis makes the limit process unnecessary.

The usual proof, as explained in Langlands [**L**] generalizes to other rank one groups easily. It generalizes to groups of higher rank only with a great deal of trouble, since one has to be extremely careful of how one moves contours of integration around. This part of the theory is notoriously difficult, and the difficulty is only partly alleviated by the recent exposé in Moeglin-Waldspurger [**MW**]. The proof I give below also generalizes immediately to other groups of rank one. Although the generalization to groups of higher rank is a little less obvious, it is still rather elementary. I will deal with this in another place. The proof I give below also seems to work without essential modification for real groups, p-adic groups, and symmetric spaces, but that, too, is another story.

The first section recalls well known facts, and the second contains the proofs of the Theorems.

I would like to take this opportunity to thank Goro Shimura for accepting me as a student one day many years ago in the hall of the old Fine Hall. Without his council, I might well have never taken up the subject of automorphic forms.

1. Truncation

For any function F on $\Gamma\backslash\mathcal{H}$ its **constant term** is defined to be

$$CF(z) = \int_{\mathbb{Z}\backslash\mathbb{R}} F(z + x)\, dx.$$

Since F is invariant under Γ and Γ contains all translations by integers n, this may be identified with a function of $y = \mathrm{Re}(z)$ alone, or in other words a function on $(0, \infty)$.

There is a construction formally adjoint to this. For any function f on $(0, \infty)$ we can extend it to a function $\overline{f}$ on $\mathcal{H}$ invariant with respect to real translations $z \mapsto z + x$. Then the function $\overline{f}$ is invariant under $\Gamma \cap P$, where P is the group of upper triangular elements in $\mathrm{SL}_2(\mathbb{R})$ and $\Gamma \cap P$ is thus the group of upper triangular elements

$$\pm \begin{bmatrix} 1 & x \\ 0 & 1 \end{bmatrix}$$

in Γ. Note that

$$\begin{bmatrix} 1 & x \\ 0 & 1 \end{bmatrix}$$

acts by translation on $\mathcal{H}$. We can then define, at least formally, the Eisenstein series

$$E_f(z) = \sum_{\Gamma\cap P\backslash\Gamma} \overline{f}(\gamma(z)).$$

For $f = y^s$ this construction gives Maass' series.

For $T > 0$, define the operator C^T on $(0,\infty)$ to be multiplication by the characteristic function $\chi_{(T,\infty)}$, and Λ^T to be multiplication by $\chi_{(0,T)}$. Thus we have an orthogonal decomposition

$$f = \Lambda^T f + C^T f.$$

Now we combine these constructions, to define analogous operations on $\Gamma\backslash\mathcal{H}$. Assume $T > 1$. For F on $\Gamma\backslash\mathcal{H}$, let f be its constant term CF. Then for $T > 0$, $C^T f$ has support on (T,∞). We can consider its Eisenstein series in turn, which I define to be $C^T F$:

$$C^T F = E_{\chi_{(T,\infty)}CF} = E_{C^T f}.$$

Define the **truncation** of F at T to be the difference

$$\Lambda^T F = F - C^T F.$$

Because the image of

$$\mathcal{H}_T = \{\mathrm{I\scriptsize{M}}(z) \geq T\}$$

under γ in Γ is either $\mathcal{H}_T$ itself (if γ lies in $\Gamma\cap P$) or a circle in the region $\{\mathrm{I\scriptsize{M}}(z) < 1\}$ (otherwise), the decomposition

$$F = \Lambda^T F + C^T F$$

is orthogonal whenever orthogonality makes sense. In particular, the function $C^T F$ has the property that

$$C^T C^T F = C^T F.$$

For functions F satisfying some mild growth and smoothness conditions—when F is of **uniform moderate growth** on $\Gamma\backslash\mathcal{H}$—the function $\Lambda^T F$ is rapidly decreasing at ∞, and uniformly in F.

The integral E_s has uniform moderate growth. Thus

$$E_s = \Lambda^T E_s + C^T E_s$$

is an orthogonal sum at least formally, and $\Lambda^T E_s$ is rapidly decreasing at ∞. In particular it is square-integrable on $\Gamma\backslash\mathcal{H}$.

The **Maass-Selberg formula** asserts that

$$\langle\Lambda^T E_s, \Lambda^T E_{1-t}\rangle = \frac{T^{s-t} - c(s)c(1-t)T^{-(s-t)}}{s-t} + \frac{c(s)T^{1-s-t} - c(1-t)T^{s+t-1}}{1-s-t}.$$

In Casselman [**C93**] this is derived from a curious regularization of integrals on $\Gamma\backslash\mathcal{H}$, which allows one to start with the formal identity

$$\int_{\Gamma\backslash\mathcal{H}} E_s E_{1-t}\, dx\, dy/y^2 = 0$$

which can be justified for generic values of s and t. In this scheme, the Maass-Selberg formula is expressed in the form

$$\langle\Lambda^T E_s, \Lambda^T E_{1-t}\rangle = \int_0^T \left(y^s + c(s)y^{1-s}\right)\left(y^{1-t} + c(1-t)y^t\right) dy/y^2$$

where the right hand side is evaluated by analytic continuation, writing

$$\int_0^T y^s\,dy/y = \frac{T^s}{s}.$$

Note that the right hand side of the Maass-Selberg formula cannot, and does not, have any poles, even though its individual terms do.

2. Fourier analysis

Because of orthogonality, we can write

$$\begin{aligned}
E_\varphi &= \Lambda^T E_\varphi + C^T E_\varphi \\
\langle E_\varphi, E_{1-t} \rangle &= \langle \Lambda^T E_\varphi, \Lambda^T E_{1-t} \rangle + \langle C^T E_\varphi, C^T E_{1-t} \rangle \\
&= \left\langle \frac{1}{4\pi i} \int_{\sigma=1/2} \varphi(s)\Lambda^T E_s\,ds, \Lambda^T E_{1-t} \right\rangle \\
&\quad + \left\langle \frac{1}{4\pi i} \int_{\sigma=1/2} \varphi(s) C^T E_s\,ds, C^T E_{1-t} \right\rangle.
\end{aligned}$$

The point now is that each of these two integrals has an interpretation purely in terms of harmonic analysis on $(0,\infty)$.

The second integral is simplest. The constant term of E_s is $y^s + c(s)y^{1-s}$, so we can write it as

$$\begin{aligned}
&\left\langle \frac{1}{4\pi i} \int_{\sigma=1/2} \varphi(s) C^T E_s\,ds, C^T E_{1-t} \right\rangle \\
&= \int_T^\infty (y^{1-t} + c(1-t)y^t)\,dy/y^2\, \frac{1}{4\pi i} \int_{\sigma=1/2} \varphi(s)(y^s + c(s)y^{1-s})\,ds \\
&= \int_T^\infty (y^{1-t} + c(1-t)y^t)\,dy/y^2\, \frac{1}{2\pi i} \int_{\sigma=1/2} \varphi_\#(s) y^s\,ds \\
&= \langle C^T \Phi_\#, y^{1-t} + c(1-t)y^t \rangle_{(0,\infty)}
\end{aligned}$$

where

$$\Phi_\#(y) = \frac{1}{2\pi i} \int_{\sigma=1/2} \varphi_\#(s) y^s\,ds.$$

As for the first integral, we can bring the integral outside the coupling because the integrand is square-integrable. We then have from Maass-Selberg

$$\begin{aligned}
&\frac{1}{4\pi i} \int_\sigma \varphi(s)\,\langle \Lambda^T E_s, \Lambda^T E_{1-t} \rangle\,ds \\
&= \frac{1}{4\pi i} \int_\sigma \varphi(s) \left[\frac{T^{s-t} - c(s)c(1-t)T^{-(s-t)}}{s-t} + \frac{c(s)T^{1-s-t} - c(1-t)T^{s+t-1}}{1-s-t} \right] ds.
\end{aligned}$$

Since the integrand is holomorphic in s and t we can also write it as

$$\begin{aligned}
&\lim_{\epsilon\to 0} \frac{1}{4\pi i} \int_{|s-t|\ge\epsilon, |s+t-1|\ge\epsilon} \\
&\varphi(s) \left[\frac{T^{s-t} - c(s)c(1-t)T^{-(s-t)}}{s-t} + \frac{c(s)T^{1-s-t} - c(1-t)T^{s+t-1}}{1-s-t} \right] ds.
\end{aligned}$$

But now we can separate terms in the integral to get this to be the limit as $\epsilon \to 0$ of the sum

$$\frac{1}{4\pi i} \int_{|s-t|\geq\epsilon} \varphi(s) \frac{T^{s-t}}{s-t}\, ds - \frac{1}{4\pi i} c(1-t) \int_{|s-t|\geq\epsilon} \varphi(s)c(s) \frac{T^{t-s}}{s-t}\, ds$$
$$+ \frac{1}{4\pi i} \int_{|s+t-1|\geq\epsilon} \varphi(s)c(s) \frac{T^{1-s-t}}{1-s-t}\, ds - \frac{1}{4\pi i} c(1-t) \int_{|s+t-1|\geq\epsilon} \varphi(s) \frac{T^{s+t-1}}{1-s-t}\, ds.$$

We can make a change of variables of $1-s$ for s in two of these and reassemble to make this

$$\lim_{\epsilon\to 0} -\frac{1}{2\pi i} \int_{|s-t|\geq\epsilon} \varphi_{\#}(s) \frac{T^{-(t-s)}}{t-s}\, ds - \frac{1}{2\pi i} c(1-t) \int_{|s-t|\geq\epsilon} \varphi_{\#}(s) \frac{T^{-(1-t-s)}}{1-t-s}\, ds.$$

How can this be interpreted? At this point I recall some elementary Fourier analysis on $(0,\infty)$. Recall that we are given on $(0,\infty)$ the measure dy/y^2. The Fourier transform is therefore

$$\widehat{f}(s) = \int_0^\infty f(y) y^{1-s}\, dy/y^2$$

for $\mathrm{Re}(s) = 1/2$. The inverse transform is

$$\frac{1}{2\pi i} \int_{1/2-i\infty}^{1/2+i\infty} F(s) y^s\, ds.$$

Define sgn_T to be the function

$$\mathrm{sgn}_T = \begin{cases} 1 & \text{if } y > T \\ -1 & \text{otherwise} \end{cases}$$

Its Fourier transform is twice the **principal value** of T^{-s}/s, which is the distribution

$$\langle \mathcal{PV}(T^{-s}/s), \varphi \rangle = \lim_{\epsilon\to 0} \frac{1}{2\pi i} \int_{|s|\geq\epsilon} \varphi(s) \left(\frac{T^{-s}}{s}\right) ds.$$

Recalling how convolution and multiplication relate to each other under the Fourier transform, we see that

$$\lim_{\epsilon\to 0} -\frac{1}{2\pi i} \int_{|s-t|\geq\epsilon} \varphi_{\#}(s) \frac{T^{-(t-s)}}{t-s}\, ds - \frac{1}{2\pi i} c(1-t) \int_{|s-t|\geq\epsilon} \varphi_{\#}(s) \frac{T^{-(1-t-s)}}{1-t-s}\, ds$$
$$= -(1/2) \left\langle \mathrm{sgn}_T \Phi_{\#}, y^{1-t} + c(1-t) y^t \right\rangle_{(0,\infty)}.$$

In other words, Fourier duality on $(0,\infty)$ proves this extremely useful if technical result, which is really the main result of this paper:

THEOREM. *For φ of compact support on* $\mathrm{Re}(s) = 1/2$

$$\langle \Lambda^T E_\varphi, \Lambda^T E_{1-t} \rangle = -\left(\frac{1}{2}\right) \langle \mathrm{sgn}_T \Phi_{\#}, y^{1-t} + c(1-t) y^t \rangle$$

where

$$\Phi_{\#} = \frac{1}{2\pi i} \int_{\sigma=1/2} \varphi_{\#}(s) y^s\, ds.$$

Our original expression therefore becomes

$$\begin{aligned}\big\langle -(1/2)\mathrm{sgn}_T\Phi_\#, y^{1-t} + c(1-t)y^t\big\rangle + \big\langle C^T\Phi_\#, y^{1-t} + c(1-t)y^t\big\rangle \\ = \big\langle (1/2)\Phi_\#, y^{1-t} + c(1-t)y^t\big\rangle \\ = (1/2)\big(\varphi_\#(t) + c(1-t)\varphi_\#(1-t)\big) \\ = \varphi_\#(t)\end{aligned}$$

since $-(1/2)\mathrm{sgn}_T + \chi_{(T,\infty)}$ is the constant $1/2$.

From the Theorem we can easily evaluate the inner product

$$\langle E_\varphi, E_\psi\rangle = \left(\frac{1}{2}\right)\langle \Phi_\#, \Psi_\#\rangle .$$

Note that $\Phi_\#$ is the constant term of E_φ, $\Psi_\#$ that of E_ψ.

The function sgn_T is a multiplicative translation of $\mathrm{sgn} = \mathrm{sgn}_1$. This function has exactly two important properties we require for this proof to work:

$$\begin{aligned}-(1/2)\mathrm{sgn} + \chi_{(1,\infty)} = 1/2 \\ \widehat{\mathrm{sgn}} = \mathcal{PV}\left(\frac{1}{s}\right).\end{aligned}$$

I leave it for now as interesting exercise in Euclidean truncation to figure out what function will replace it for root systems of rank > 1.

References

[C83] W. Casselman, *Automorphic forms and a Hodge theory for congruence subgroups of* $\mathrm{SL}_2(\mathbb{Z})$, in Lie Group Representations II, Lect. Notes in Math. **1041**, Berlin-New York: Springer (1984).

[C93] W. Casselman, *Extended automorphic forms in the upper half plane*, Math. Ann. **296** (1993), 755–762.

[K] T. Kubota, Elementary Theory of Eisenstein Series, New York-London-Sydney: Kodansha and Wiley (1973).

[L] R. P. Langlands, *Eisenstein series*, in Algebraic Groups and Discontinuous Subgroups (Proc. Sympos. Pure Math., Boulder, Colo., 1965), Providence, RI: Amer. Math. Soc. (1966), 235–252.

[MW] C. Moeglin and J.-L. Waldspurger, Spectral Decomposition and Eisenstein Series, Cambridge: Cambridge University Press (1995).

Department of Mathematics, University of B.C., Vancouver, Canada
E-mail address: cass@math.ubc.ca

Proceedings of Symposia in Pure Mathematics
Volume **66.2**, 1999

Average Values of Cubic L-Series

David Farmer, Jeffrey Hoffstein, and Daniel Lieman

Dedicated to Goro Shimura

ABSTRACT. We obtain average values estimates for the Fourier coefficients of a certain metaplectic form on the triple cover of GL(3). These coefficients are each a cubic L-series times a certain finite polynomial.

1. Introduction

Our main result is an estimate for the average values of Hecke cubic L-series. While there are a number of methods for estimating Hecke quadratic L-series, the problem of estimating Hecke cubic L-series at or near the center of the critical strip has remained unresolved. In general, past attempts to obtain such estimates have been blocked by the random nature of the arguments of the cubic Gauss sums associated to the functional equations of the L-series.

In order to state our main theorems, we must first set some notation, which we will use throughout this paper. We will write $\mathcal{O}$ for the ring of integers in $\mathbb{Q}(\sqrt{-3})$, λ for $\sqrt{-3}$, and $\Re(z)$ for the real part of the complex number z. Given $a \in \mathcal{O}$ with $a \equiv 1 \pmod 3$, let $\chi_n(a)$ denote the cubic residue symbol

$$\chi_n(a) = \left(\frac{a}{n}\right)_3.$$

If $n \in (1/\lambda^3)\mathcal{O}$ but $n \notin \mathcal{O}$, then by abuse of notation, we write $\chi_n(a)$ for $\chi_{\lambda^3 n}(a)$. We will write $L(\rho, \chi_n)$ for the L-series

$$\begin{aligned} L(\rho, \chi_n) &= \sum_{\substack{a \in O \\ a \equiv 1 \pmod 3}} \chi_n(a) \mathbb{N}(a)^{-\rho} \\ &= \prod_{\substack{p \in O \\ p \equiv 1 \pmod 3}} \left(1 - \chi_n(p) p^{-\rho}\right)^{-1}. \end{aligned}$$

1991 *Mathematics Subject Classification.* Primary 11M06, 11F12; Secondary 11F30.

Key words and phrases. Cubic L-series, automorphic forms, Mellin transform.

All three authors partially supported by NSF grants DMS-9206236 (Farmer), DMS-9322150 (Hoffstein), and DMS-9400882 (Lieman). In addition, all three authors would like to thank MSRI, Berkeley, for its hospitality during part of the time this research was performed.

Our main objective is to estimate the average value of these L-series by interpreting them as the Fourier coefficients of a certain metaplectic form on the cubic cover of GL(3). If we let $L^*(\rho, \overline{\chi}_n)$ denote a certain product of $L(\rho, \overline{\chi}_n)$ with a finite sum whose complexity increases with the divisibility of n by cubes, we may state our main results as follows.

MAIN THEOREM. *There exist non-zero constants $c_1(\rho)$ and $c_2(\rho)$ such that*

$$\sum_{\substack{n \in \frac{1}{\lambda^3}\mathcal{O} \\ \mathbb{N}(n) < X}} L^*(\rho, \overline{\chi}_n) = c_1(\rho)X + c_2(\rho)X^{4/3-\rho} + O\left(X^{\theta(\rho)+\epsilon}\right),$$

where $\theta(\rho)$ is the function

$$\theta(\rho) = \begin{cases} \dfrac{7 + 2\rho - 2\rho^2}{2(3+2\rho)}, & \frac{1}{2} \leq \Re(\rho) \leq 1; \\ \dfrac{(3-\rho)(5-4\rho)}{2(6-4\rho)}, & \frac{1}{3} \leq \Re(\rho) \leq \frac{1}{2}. \end{cases}$$

REMARK 0. Note that we obtain the following special cases of the Main Theorem:

$$\sum_{\substack{n \in \frac{1}{\lambda^3}\mathcal{O} \\ \mathbb{N}(n) < X}} L^*(1, \overline{\chi}_n) = c_1(1)X + O\left(X^{7/10+\epsilon}\right),$$

$$\sum_{\substack{n \in \frac{1}{\lambda^3}\mathcal{O} \\ \mathbb{N}(n) < X}} L^*\left(\frac{1}{2}, \overline{\chi}_n\right) = c_1\left(\frac{1}{2}\right)X + O\left(X^{15/16+\epsilon}\right),$$

and that

$$\sum_{\substack{n \in \frac{1}{\lambda^3}\mathcal{O} \\ \mathbb{N}(n) < X}} L^*(\rho, \overline{\chi}_n) \sim c_1(\rho)X$$

whenever $\Re(\rho) > (9 - \sqrt{33})/8 \approx 0.407$. If we used a weighted sum in the proof instead of an unweighted sum, we would obtain an estimate with an error magnitude less then X all the way back to $\Re(\rho) > 1/3$. This is somewhat unexpected. It implies that although the L-series themselves are increasing in size for $\Re(\rho) < 1/2$, cancellation among the terms prevents a change in the growth rate of the sum. This phenomenon also occurs in function fields, where the L-series are just ratios of polynomials. In that case, Rosen [**R**] has obtained estimates for the average values of cubic L-series for $\Re(\rho) > 1/3$.

REMARK 1. In order to obtain any interpretation about the values of the L-series themselves, we must verify that the contribution of terms where the index differs by a cube (i.e., terms of the form $L^*(\rho, \overline{\chi}_{nm^3})\mathbb{N}m^{w'}$ for any fixed n) is bounded. This is precisely the content of [**L94**, Lemma 4.3], for ρ with real part sufficiently large.

REMARK 2. One can also obtain an analogous theorem where the index of summation includes only rational integers. In this case, the coefficient $L^*(\rho, \overline{\chi}_n)$ will be further multiplied by a complicated power of 3 which depends on n. To complete this approach, one should modify the form studied by Bump and Hoffstein in order to enlarge the group under which it is automorphic, cf. [**L94**]. The convolution studied in [**L93**] (a generalization of work done independently by Asai [**A**] and

Patterson [**P78**]), together with the analysis of Section 2, below, then yields the desired result.

REMARK 3. Finally, it is possible to obtain a version of our main theorem where the index of summation includes only squarefree numbers. In this special case and when n is also congruent to 1 mod 3, $L^*(\rho, \overline{\chi}_n)$ is just

$$\frac{1}{2}3^{11/2}L(\rho, \overline{\chi}_n),$$

and thus it is particularly easy to interpret our main theorem as a statement about the cubic L-series themselves (cf. [**BH86**, Theorem 2.2]). To obtain such a result, one needs the functional equation for the Mellin transform of a metaplectic form on the cubic cover of GL(3) twisted by an additive character. This twisted transform is studied in [**FL**]. Using this twisted transform, however, would result in an error term with exponent less than 1, but worse than that given in the main theorem.

We should mention that this application (to cubic L-series) was the primary motivation of the analysis contained in [**FL**], and the reader should consult that paper for technical details related to the Mellin transform involved. For a general overview of metaplectic forms on cubic and higher covers (as well as the double cover, with which the reader may be more familiar), we refer the reader to the survey article by Hoffstein [**H**].

2. Proof of the main theorem

We begin by fixing some notation, which we will use throughout. Write ϕ for the metaplectic form on the cubic cover of GL(3) studied by Bump and Hoffstein [**BH86**] (note that ϕ has a parameter s, which we suppress). We may describe ϕ as follows. Let E denote the maximal parabolic Eisenstein series induced from the theta function on the cubic cover of GL(2). Then ϕ is defined by $\phi(\tau) = E(w_1\tau)$ where w_1 is the long element of the Weyl group of GL(3). We will use the standard notation $\phi_{m,n}$ (resp. ϕ_n^m) for Fourier coefficients (resp. sums of Fourier coefficients) and w_i for the elements of the Weyl group of GL(3), as in [**B84, BH86**]. Of interest to us is the remarkable property that the Fourier coefficients of ϕ contain the Hecke cubic L-series. (Recall that the Fourier coefficients of a GL(3) form are doubly indexed.) In particular, we have

THEOREM [**BH86**, Theorem 2.1]. *Let $a_{1,n}(s)$ denote the arithmetic part of the $(1,n)^{th}$ Fourier coefficient of ϕ. Then*

$$\begin{aligned}(2.1)\quad & a_{1,n}(s) \\ &= \frac{1}{2}3^{5/2}\mathbb{N}(n)^{s-1} \prod_{\substack{p \nmid n \\ p\equiv 1 (mod\ 3)}} \left(1 - \overline{\chi}_n(p)\mathbb{N}(p)^{1-3s/2}\right)^{-1} \left(1 - \mathbb{N}(p)^{3-9s/2}\right) \\ &\quad \times \sum_{\substack{C\equiv 1 (mod\ 3) \\ p|C \Rightarrow p|27n}} |C|^{1-3s} \sum_{\substack{C=cd \\ c\equiv d\equiv 1 (mod\ 3) \\ d|27nc}} \tau\left(\frac{n_1 c}{d}\right) T(c,d),\end{aligned}$$

where $\tau(r)$ is the rth Fourier coefficient of the theta function on the triple cover of GL(2)*, studied by Patterson* [**P77, P77b**]*, and where $T(c,d)$ is a finite sum of Gauss sums, given explicitly by* [**BH86**, (1.20)].

We now fix $\rho = 3s/2-1$. It is precisely the value $\mathbb{N}(n)^{1-s}L(3\rho,\chi_1)a_{1,n}(s)$ which we denote by $L^*(\rho,\overline{\chi}_n)$ throughout this paper. That the value (2.1) differs from the cubic L-series $L(\rho,\overline{\chi}_n)$ by a finite polynomial is proved in [**L94**, p. 98]. The next theorem is the primary analytic tool we will use to deduce the main theorem. It can easily be modified to apply to any form on a cover of GL(3). We state it here for our one specific form.

THEOREM 2. *Let $D(\rho,w')$ denote the Dirichlet series*

$$\sum_{n\in\frac{1}{\lambda^5}\mathcal{O}} \frac{L^*(\rho,\overline{\chi}_n)}{\mathbb{N}(n)^{w'}},$$

and let $D^\sharp(\rho,w')$ denote the nearly identical Dirichlet series

$$\sum_{n\in\frac{1}{\lambda^3}\mathcal{O}} \frac{L^\sharp(\rho,\chi_n)}{\mathbb{N}(n)^{w'}},$$

where $L^\sharp$ differs from L^* only by an Euler factor at the prime 3, and is explicitly described in Section 3, below. Finally, write $G(\rho,w')$ for the product of gamma functions

$$G(\rho,w') = (2\pi)^{-w'-\frac{2}{3}\rho-\frac{2}{3}}\,\Gamma\left(w'+\rho-\frac{1}{3}\right)\Gamma\left(w'+\rho-\frac{2}{3}\right)\Gamma\left(w'\right).$$

Then $D(\rho,w')$ (which a priori converges for w' with real part sufficiently large) has a meromorphic continuation (in w') to the entire complex plane, with simple poles at $w'=0$, $w'=1$, and $w'=4/3-\rho$, and satisfies the functional equation

$$D(\rho,w')G(\rho,w')L(3\rho,\chi_1)^{-1} = D^\sharp(1-\rho,1-w')G(1-\rho,1-w')L(3(1-\rho),\chi_1)^{-1}.$$

The residues $R_1(\rho), R_2(\rho)$, at the poles $w'=1$, $w'=4/3-\rho$, are non-zero. (Note also that the simple poles at $w'=1$ and $w'=4/3-\rho$ do not combine to yield a double pole at $w'=1$ when $\rho=1/3$.)

We defer the proof of this theorem until Section 3.

We now show how to prove the Main Theorem from Theorem 2.

PROOF OF THE MAIN THEOREM. Assume that $\Re(\rho)\ge 1/3$. Consider

$$\frac{1}{2\pi i}\int_{2-i\infty}^{2+i\infty} \frac{X^{w'}D(\rho,w')}{w'(w'+1)\cdots(w'+r)}\,dw'. \tag{2.2}$$

Shifting the line of integration to the left, we see that (2.2) is equal to

$$\frac{XR_1(\rho)}{(r+1)!} + \frac{X^{\frac{4}{3}-\rho}R_2(\rho)}{(\frac{4}{3}-\rho)(\frac{4}{3}+1-\rho)\cdots(\frac{4}{3}+r-\rho)} + \frac{1}{2\pi i}\int_{-\delta-i\infty}^{-\delta+i\infty} \frac{X^{w'}D(\rho,w')}{w'(w'+1)\cdots(w'+r)}\,dw'. \tag{2.3}$$

The series $\sum |L(\rho,\overline{\chi}_n)|^2/\mathbb{N}(n)^{w'}$ converges for $\Re(w')\ge 1$ when $\Re(\rho)\ge 1/2$ and thus

$$D^\sharp(1-\rho,1+\delta+it) \ll 1$$

when

$$\frac{1}{3} \le \Re(\rho) \le \frac{1}{2} \text{ with } \delta > \epsilon > 0,$$
$$\frac{1}{2} \le \Re(\rho) \le 1 \text{ with } \delta > \Re(\rho) - \frac{1}{2} + \epsilon.$$

(The convergence of the series $\sum |L(\rho, \overline{\chi}_n)|^2/\mathbb{N}(n)^{w'}$ and the locations of its poles follow from a certain Rankin-Selberg convolution; see [**B89**] for an introduction to these constructions.) When $1/2 \le \Re(\rho) \le 1$ we may use this estimate and the functional equation, and, fixing $\delta = \rho - 1/2 + \epsilon$, we have

$$D(\rho, -\delta - it) \ll |t|^{2+2\rho+6\epsilon}.$$

For $1/3 \le \Re(\rho) \le 1/2$, we instead fix $\delta = \epsilon$, and we see that

$$D(\rho, -\delta - it) \ll |t|^{5-4\rho+6\epsilon}.$$

Assume now that $r \ge 4$. On the one hand, (2.2) is equal to (2.3), and we may use the two estimates above to see that (2.2) is equal to

$$\frac{XR_1(\rho)}{(r+1)!} + \frac{X^{\frac{4}{3}-\rho}R_2(\rho)}{(\frac{4}{3}-\rho)(\frac{4}{3}+1-\rho)\cdots(\frac{4}{3}+r-\rho)} + \begin{cases} O_{r,\epsilon}\left(X^{+\epsilon}\right), & \frac{1}{3} \le \Re(\rho) \le \frac{1}{2}; \\ O_{r,\epsilon}\left(X^{\frac{1}{2}-\rho+\epsilon}\right), & \frac{1}{2} \le \Re(\rho) \le 1. \end{cases}$$

On the other hand, we may explicitly evaluate (2.2) to obtain

$$\frac{1}{2\pi i}\int_{2-i\infty}^{2+i\infty} \frac{X^{w'}D(\rho, w')}{w'(w'+1)\cdots(w'+r)}\,dw' = \frac{1}{r!}\sum_{\substack{n\in\frac{1}{\lambda^3}\mathcal{O}\\ \mathbb{N}(n)<X}} L^*(\rho, \overline{\chi}_n)\left(1 - \frac{\mathbb{N}(n)}{X}\right)^r.$$

Continuing this line of reasoning leads to the estimate mentioned in Remark 0.

The estimate for the unweighted sum is completed in an analogous manner. Instead of (2.2), one considers

$$\text{(2.4)} \qquad \frac{1}{2\pi i}\int_{2-iT}^{2+iT} \frac{X^{w'}D(\rho, w')}{w'}\,dw'.$$

Applying the error estimate for the integral (cf. [**D**]) along with the convexity bound for the L-series at ρ, we note that the above is equal to

$$\sum_{\substack{n\in\frac{1}{\lambda^3}\mathcal{O}\\ \mathbb{N}(n)<X}} L^*(\rho, \overline{\chi}_n) + O\left(\frac{X^{\frac{3-\rho}{2}+\epsilon}}{T}\right)$$

when $0 \le \Re(\rho) \le 1$. Shifting the line of integration, however, we see that (2.4) is equal to

$$\frac{XR_1(\rho)}{(r+1)!} + \frac{X^{\frac{4}{3}-\rho}R_2(\rho)}{(\frac{4}{3}-\rho)} + \begin{cases} O\left(X^{-\epsilon}T^{5-4\rho+6\epsilon}\right), & \frac{1}{3} \le \Re(\rho) \le \frac{1}{2}; \\ O\left(X^{\frac{1}{2}-\rho-\epsilon}T^{2+2\rho+6\epsilon}\right), & \frac{1}{2} \le \Re(\rho) \le 1. \end{cases}$$

We now obtain the main theorem by making the optimal choice of T (to minimize the exponent in the error term)

$$T = \begin{cases} X^{\frac{3-\rho}{2(6-4\rho)}}, & \frac{1}{3} \leq \Re(\rho) \leq \frac{1}{2}; \\ X^{\frac{1+\rho/2}{3+2\rho}}, & \frac{1}{2} \leq \Re(\rho) \leq 1. \end{cases}$$

3. Proof of the functional equation

The following theorem is one of two main theorems from [**FL**]. It deals with a Dirichlet series similar to that on the left hand side of Theorem 2, except that it is twisted by a character. The references to Lemmas 4A and 6A are to lemmas in [**FL**]; we include them here for completeness, although our discussion following the statement of this theorem will render them irrelevant.

In what follows, we will write $L(w, s, \mu/q)$ for the twisted series

$$L(w, s, \mu/q) = \sum_n \frac{e\left(\frac{\mu n}{q}\right) L^*(\frac{3}{2}s - 1, \overline{\chi}_n)}{\mathbb{N}(n)^{w-s+1}}.$$

THEOREM 3 [**FL**, Main Theorem A]. *Suppose q is equivalent to a root of unity mod 3, and that $\overline{q}$ is the root of unity so that $q\overline{q} \equiv 1$ mod 3. Without loss of generality, we may assume that $\mu \equiv 0$ mod 3. Then $L(w, s, \mu/q)$ converges for w with real part sufficiently large, and has a meromorphic continuation to all w with (possible) poles only at the locations $w = s - 1$, $w = s$, and $w = 4/3 - s/2$. Further, let $\psi(\tau)$ denote the automorphic form defined by $\psi(\tau) = \tilde{\phi}(w_4\tau)$, and let $\mp$ be defined as in* [**FL**, Lemma 4A] *and ρ and B defined as in the proof of* [**FL**, Lemma 6A]. *(In particular, $\mp$ depends on μ in a fixed way, and ρ and B depend on the index a, below, in a fixed way.) Let κ_a denote a cube root of unity, depending on a in a fixed way (it is the image of the Kubota homomorphism on the coset depending on a, cf. Lemma 6A). Finally, let r denote $(a, \lambda^3 q)/\lambda^3$. Then we have the functional equation*

$$\begin{aligned} &\overline{G}(w, z, s) L\left(\frac{9}{2}s - 3, \chi_1\right)^{-1} L(w, s, \mu/q) \\ &\quad = \frac{2}{\sqrt{3}\mathbb{N}(\lambda^5 q^3)} \sum_{a \in \mathcal{O}/\lambda^5 q^2} e\left(-\frac{a\overline{q} \mp -nB(a, \lambda^3 q)}{\lambda^3 q}\right) \kappa_a \\ &\quad \times \int_0^\infty \int_0^\infty \sum_{n \neq 0} \psi^\rho_{r,n} \begin{pmatrix} ry/q^3 & & \\ & t/r & \\ & & 1 \end{pmatrix} y^{-2w} t^z \frac{dt}{t} \frac{dy}{y} \end{aligned}$$

where $\overline{G}$ is a quotient involving seven gamma factors.

Note that the left-hand side of the functional equation in Theorem 3 is *almost* the left-hand side of Theorem 2. Consider Theorem 3 in the case where there is no twisting (i.e., $\mu = 0$ and $q = 1$). In this case, the functional equation simplifies considerably, as nearly all of the character sums vanish, due to the presence of the Kubota symbol (cf. [**BH86b**] for a discussion of how to compute the symbol corresponding to a given cusp ρ, which in turn is determined by the index a). In particular, we have

$$\overline{G}(w,z,s)L\left(\frac{9}{2}s-3,\chi_1\right)^{-1}L(w,s,0)$$
$$=\frac{2}{3^{5/2}}\int_0^\infty\int_0^\infty\sum_{n\neq 0}\psi_{1,n}\begin{pmatrix}y&&\\&t&\\&&1\end{pmatrix}y^{-2w}t^{z}\frac{dt}{t}\frac{dy}{y}.$$

Theorem 2 follows from the explicit evaluation of the Whittaker function of ψ given by Bump and Friedberg [**BF**]. This explicit formula is a quotient involving seven gamma functions, four of which are evaluated at points identical to four of the gamma functions in $\overline{G}$. These four functions cancel from each side of the functional equation; the three remaining gamma factors on each side are given in Theorem 2. In particular, all references to the variable z occur in the "cancelled" gamma factors. The coefficients $L^\sharp(1-\rho,\chi_n)$ appearing in Theorem 2 are precisely the Fourier coefficients of the form ψ. Indeed, the right hand side of the above equation contains a Dirichlet series built out of the $(1,n)$ Fourier coefficients of $\psi(\tau)$ evaluated at s. But by the functional equation of the Eisenstein series, these are exactly the $(1,n)$ Fourier coefficients of $\phi(w_4\tau)$ evaluated at $2-s$. These coefficients differ from $L^\sharp(1-\rho,\chi_n)$ by a finite Euler factor at 3 which is calculated explicitly in [**L94**] (where it is denoted $T(w)$). Note that we have renormalized so that $w'=w+1-s$, and the functional equation $s\to 2-s$ induces a function equation

$$\rho=\frac{3}{2}s-1\to\frac{3}{2}(2-s)-1=2-\frac{3}{2}s=1-\left(\frac{3}{2}s-1\right)=1-\rho.$$

This completes the proof of the functional equation. The residues $R_1(\rho)$ and $R_2(\rho)$ are calculated explicitly in [**FL**], and are non-zero, thus completing the proof of the theorem.

References

[A] T. Asai, *On certain Dirichlet series associated with Hilbert modular forms and Rankin's method*, Math. Ann. **226** (1977), 81–94.

[B84] D. Bump, Automorphic Forms on GL(3, **R**), Lecture Notes in Mathematics **1083**, Berlin-New York: Springer-Verlag (1984).

[B89] D. Bump, *The Rankin-Selberg method: a survey*, in Number Theory, Trace Formulas, and Discrete Groups (K. Aubert, E. Bombieri, D. Goldfeld, eds.), San Diego: Academic Press, Inc. (1989), 49–109.

[BF] D. Bump and S. Friedberg, *On Mellin transforms of unramified Whittaker functions on* GL(3, $\mathbb{C}$), J. Math. Anal. Appl. **139** (1989), 205–216.

[BH86] D. Bump and J. Hoffstein, *Cubic metaplectic forms on* GL(3), Invent. Math. **84** (1986), 481–505.

[BH86b] D. Bump and J. Hoffstein, *Some Euler products associated with cubic metaplectic forms on* GL(3), Duke Math. J. **53** (1986), 1047–1072.

[D] H. Davenport (revised by Hugh Montgomery), Multiplicative Number Theory, Graduate Texts in Mathematics **74**, New York-Berlin: Springer-Verlag (1980).

[FL] D. Farmer and D. Lieman, *The* GL(3) *Mellin transform for twisted non-cuspidal forms of higher level*, Israel J. Math., to appear.

[H] J. Hoffstein, *Eisenstein series and theta functions on the metaplectic group*, Theta functions: from the classical to the modern, CRM Proc. Lecture Notes, Amer. Math. Soc. **1** (1993), 65–104.

[L93] D. Lieman, *The* GL(3) *Rankin–Selberg convolution for functions not of rapid decay*, Duke Math. J. **69** (1993), 219–242.

[L94] D. Lieman, *Nonvanishing of L-series associated to cubic twists of elliptic curves*, Annals of Math. **140** (1994), 81–108.

[P77] S. Patterson, *A cubic analogue of the theta series I*, J. Reine Angew. Math. **296** (1977), 125–161.

[P77b] S. Patterson, *A cubic analogue of the theta series II*, J. Reine Angew. Math. **296** (1977), 217-220.

[P78] S. Patterson, *On Dirichlet series associated with cubic Gauss sums*, J. Reine Angew. Math. **303** (1978), 102–125.

[R] M. Rosen, *Average values of class numbers in cyclic extensions of the rational function field*, in Number Theory (Halifax, NS, 1994), CMS Conf. Proceedings **15**, Providence, RI: Amer. Math Soc. (1995), 307–323.

DEPARTMENT OF MATHEMATICS, BUCKNELL UNIVERSITY, LEWISBURG, PA 17837
E-mail address: `farmer@bucknell.edu`

DEPARTMENT OF MATHEMATICS, BROWN UNIVERSITY, PROVIDENCE, RI 02912
E-mail address: `jhoff@math.brown.edu`

DEPARTMENT OF MATHEMATICS, UNIVERSITY OF MISSOURI, COLUMBIA, MO 65211
E-mail address: `lieman@math.missouri.edu`

Proceedings of Symposia in Pure Mathematics
Volume **66.2**, 1999

Euler Factorization of Global Integrals

Paul Garrett

This work is dedicated to Professor Goro Shimura.

ABSTRACT. We consider integrals of cuspforms f on reductive groups G defined over number fields k against restrictions $\iota^* E$ of Eisenstein series E on "larger" reductive groups $\tilde{G}$ over k via imbeddings $\iota : G \to \tilde{G}$. We give hypotheses sufficient to assure that such global integrals have Euler products. At good primes, the local factors are shown to be rational functions in the corresponding parameters q^{-s} from the Eisenstein series and in the Satake parameters q^{-s_i} coming from the spherical representations locally generated by the cuspform. The *denominators* of the Euler factors at good primes are estimated in terms of "anomalous" intertwining operators, computable via orbit filtrations on test functions. The standard intertwining operators (attached to elements of the (spherical) Weyl group) among these unramified principal series yield symmetries among the anomalous intertwining operators, thereby both sharpening the orbit-filtration estimate on the denominator and implying corresponding symmetry in it. Finally, we note a very simple dimension-counting heuristic for fulfillment of our hypotheses, thereby giving a simple test to *exclude* configurations $\iota : G \to \tilde{G}$, E, as candidates for Euler product factorization. Some simple examples illustrate the application of these ideas.

Introduction

Application of the *analytic* properties of Eisenstein series to L-functions occurs in a variety of settings wherein L-functions are expressed as integrals of cuspforms against restrictions of Eisenstein series (and similar devices). At the very least, the analytic continuation and functional equation of an Eisenstein series translate immediately into an analytic continuation and functional equation for whatever it is that arises as an integral against the Eisenstein series.

The *arithmetic* of L-functions is subtler, and of more recent origin. The idea that special values of L-functions could inherit their *arithmetic* properties from those of Eisenstein series was made clear in **[Sh75a]** and in the many succeeding papers of Shimura and a few others. For that matter, in lectures 1975-77 at

1991 *Mathematics Subject Classification.* Primary 11F70, 11F67, 22D30.

Key words and phrases. Automorphic L-function Euler product, zeta integral, local integral, periods, spherical representation, unramified principal series, Eisenstein series, Rankin-Selberg integral, local factor.

Princeton, Shimura gave a compelling exposition of the results of [**Kl**] wherein one employs Hilbert modular Eisenstein series to prove special values results for certain zeta functions (and L-functions) of totally real number fields.

As in [**Sh75a**] and [**Sh75b**] and many other of his papers, Shimura consistently made the general arithmetic of holomorphic automorphic forms appear as a corollary of his results on canonical models [**Sh70**]. In fact, even the arithmetic of Eisenstein series can be made to be a consequence of canonical models, as in [**H81, H84**]. Although this viewpoint seems to be the broadest in scope (see [**H85**] and [**H86**]), it is also possible to prove many foundational arithmetic results directly, as done in [**G90**] and [**G92**], by using properties of Eisenstein series.

It is not a trivial matter to obtain an integral representation of an L-function, whether for analytic or arithmetic ends. In the first place, it is *not* the case that one chooses an L-function and then hunts for an integral representation of it. Rather, one "searches" among integrals of cuspforms against Eisenstein series (etc.) for those integrals which admit Euler product factorizations. Traditionally, it is only after the local integrals are computed that one discovers what L-function has been obtained. (Possible exceptions are [**L67, MS, S**], and other papers of these authors which obtain L-functions as constant terms of suitable Eisenstein series, in a less mysterious fashion than Rankin-Selberg integrals).

Certainly few among integrals of cuspforms against Eisenstein series have any sort of Euler product expansion at all, and indeed may fail to be Dirichlet series entirely (with or without tolerance for trouble with the archimedean integrals).

And even when an integral of cuspform against Eisenstein series can be shown to factor over primes in a meaningful way, there can remain an onerous computation to determine the local integrals. If a too-classical viewpoint is taken, it may be impossible to anticipate the ultimate success or failure of such a venture until the last moment. Surveys such as [**GeS**] give an idea of the variety of phenomena which appear.

Given the haphazard nature of hunting for integral representations, it is striking to this author that, whenever a global integral *does* factor over primes, the local factors are not only *rational* functions in the respective p^{-s}'s, but, quite magically, are also *automorphic*, as opposed to being meaningless Dirichlet series capriciously attached to a cuspform. In part, the present paper is this author's attempt to explain why integral representations should work as well as they do.

Now we give an overview of this paper using terminology explained only later.

The first third of the paper addresses Euler factorization of Rankin-Selberg type global integrals involving Eisenstein series and cuspforms.

In section 1.1 we do the by-now-usual unwinding of the Eisenstein series occurring in the global integral. Although it is not strictly necessary, we assume for simplicity that there is a single *cuspidal representative.*

In section 1.2 we rearrange the unwound integral to isolate all really global information in what we call the *inner integral*, which does not involve the parameter "s" from the Eisenstein series. It is observed that with a certain multiplicity-one hypothesis locally almost everywhere, the inner integral factors over primes, up to a global constant which is reasonably viewed as a *period* of the cuspform. In this case, it then follows that the *whole* integral factors over primes, with the local factors depending only upon local data, except for the single global object, the period. Of course, an important element here is that we view all the integrals involved as *values of intertwining operators.*

Section 1.3 makes a trivial but important observation that vanishing of global integrals can be due to local features of the situation.

In section 1.4 we use orbit filtrations on the relevant induced representations (from 3.2) to give an easily verifiable (when true) condition which suffices implies that the inner integral factors. To do so, we effectively replace spherical representations by unramified principal series. The natural intertwining operators among unramified principal series, together with the generic irreducibility of the unramified principal series, offer a further refinement of this multiplicity estimate.

Section 1.5 gives a heuristic criterion for success or failure of this approach in a given situation, in terms of a simple dimension-count.

The middle third of this paper addresses computation of a class local integrals including the so-called *outer integrals* which are the Euler factors of the global integrals in 1.2.

Section 2.1 sets up the notion of *parametrized family of representations*, and *parametrized family of intertwining operators.* For our purposes, the parameter spaces are affine algebraic varieties. Some simple natural examples are given, mostly referring to unramified principal series and degenerate principal series, which are all we need here.

In section 2.2 the *Rationality Lemma* is stated. This is the key device in proving the rationality of the local factors. We immediately invoke the Rationality Lemma to prove two theorems asserting the *rationality of certain types of local integrals.*

Section 2.3 uses the results on rationality of local integrals to give an analytic continuation result for the natural integral expressions for the intertwining operators arising in the *inner integral* in the Euler factors, under finite-orbit and generic multiplicity-one hypotheses.

Section 2.4 again invokes the rationality of local integrals to prove that, under finite-orbit and *generic* multiplicity-one hypotheses, the Euler factors at almost all primes are *rational functions* in the corresponding parameter q^{-s}, and in the Satake parameters q^{-s_i} of the cuspform. We also anticipate the later result which gives a limitation on the possible hypersurfaces along which these rational functions may have poles.

The newest item here is the study of the *denominators* of the local factors. In preparation for this, section 2.5 states the theorem on *strong meromorphy* of families of intertwining operators under assumptions of generic multiplicity one. As a corollary of this theorem on strong meromorphy, section 2.6 shows how to associate an *anomalous intertwining operator* to a hypersurface along which a parametrized intertwining operator has a pole by taking the residue (or, generally, leading term in a Laurent expansion). These anomalous intertwining operators can be estimated by orbit-filtration methods. Thus, as a corollary, we have a computational device to estimate the denominators of the local factors. And the intertwining operators among unramified principal series (attached to elements of the Weyl group) imply *symmetry* in this estimate.

Section 2.7 illustrates the application of these ideas to some simple situations where for other reasons we know "the answer" for comparison: Tate's thesis [**T**], and an instance of [**GoJ**].

The last third of the paper completes the proofs. In section 3.1 we prove the Rationality Lemma. This is essentially elementary, but a little complicated.

Section 3.2 settles necessary details concerning the orbit filtrations. Section 3.3 proves the exactness of Jacquet functors (and of the functors which project to

K-fixed vectors, for compact open subgroups K) in the context of parametrized families of representations. Similarly, section 3.4 proves Frobenius Reciprocity for parametrized families.

Finally, section 3.5 proves the orbit criterion for strong meromorphy, thus completing all the arguments.

The idea of viewing integrals as intertwining operators is well known, as is the idea of using multiplicity-free-ness of induced representations. [**P**] and [**JLan**] are two of the earliest examples of this.

There is the more technical theme of understanding the dependence of local integrals upon parameters. The *Rationality Lemma* here is a variation and development of a natural idea evidently originating already in [**Bru**], and systematically exploited in [**BeZ76, BeZ77**]. The fullest treatment of the idea of parametrization of families of representations seems to be [**Be84**]. In our context, we use a Rationality Lemma both to prove that various local integrals are rational expressions in parameters q^{-s} (and in Satake parameters q^{-s_i}), and (thereby) also to give *analytic continuations* of intertwining operators. The present issues are in part different from those in [**Be84**], seeking as we do to *assert* an algebraic structure on intertwining operators given by integrals. But after the algebraicization is accomplished, the present issues are an amplification of a special case of [**Be84**], aimed simply at unramified principal series, refined a bit as preparation for more delicate discussion of *the hypersurfaces along which natural intertwining operators have "poles"*.

It should be noted that throughout we systematically replace (admissible) spherical representations by the unramified principal series which map surjectively to them, invoking [**M**], [**Bo**], or [**C80**]. In this optic, the Satake parameters can be read from the data for the corresponding unramified principal series. Most importantly, modeling spherical representations as (images of) induced representations allows application of a variety of more "physical" arguments, including not only Frobenius Reciprocity, but also *orbit filtrations* on spaces of test functions, as instigated in [**Bru**] and used extensively in [**BeZ76, BeZ77, Be84**]. Thus, *generically*, the surjection of unramified principal series to a spherical representation is an *isomorphism*.

One particular sharpening of the ideas about parametrized families is necessary for detailed consideration of the *residue* (or, generally, *leading term* in a Laurent expansion) at a pole of a meromorphic family of intertwining operators. Such residues are demonstrably intertwining operators of a rather special sort. Coming from the other side, attempting to define an *anomalous* intertwining operator as one which "does not exist for *generic* parameter values", we can *estimate* the poles of the family of intertwining operators by estimating the possible anomalous intertwining operators. Orbit-filtration methods are very effective here, as well. But, of course, the phrase "does not exist for generic parameter values" must be made precise.

There is a further complication in discussion of residues of meromorphic families of intertwining operators, namely verification that the Laurent expansion along a hypersurface has *uniformly* finite order. This is a *strong meromorphy* condition, which we see follows from a natural Noetherian-ness condition. In the present context, the relevant Noetherian-ness for unramified principal series is not hard to prove. A much more general Noetherian-ness result is proven in [**Be84**].

Discussion of residues of families of intertwining operators along hypersurfaces, studied as anomalous intertwining operators, also requires that we tolerate representations over the fields of rational functions on the hypersurfaces, which typically are of positive transcendence degree over the natural base field (e.g., **C**). This complication arises because in addition to the ubiquitous q^{-s} parameter we might also want to track the Satake parameters q^{-s_i} of spherical representations, leading naturally to many-variable situations.

Incidental to the above, we prove some lemmas about filtrations of various test-function representation spaces by orbits. These are the parametrized-family version of results going back to [**Bru**]. Parametrized-family versions of facts about Jacquet functors and related matters are also necessary. It must be noted that these extensions of otherwise standard representation-theoretic facts are not entirely trivial, since many properties of representations on *vector spaces* certainly do not extend to representations on *modules*. See [**Be84**] for similar extensions of some standard results.

A small but meaningful point is the choice of category in which to do representation theory: for present purposes, we operate almost entirely in the category of *smooth* representations (of totally disconnected groups, or of Lie groups, or of adele groups). Thus, there are technical but significant differences in comparison to the somewhat more popular category of *unitary* representations. See [**Gr**] for some illustrations of effective methods in the unitary category. For effective discussion of *parametrized families*, it appears necessary to give up unitariness in order to gain the possibility of having the parameter space be an (affine) algebraic variety. This viewpoint is corroborated by [**BeZ76, BeZ77, Be84**].

1. Factorization over primes

1.1. The usual unwinding trick. In this section, we set up family of a class of global integrals, in which a cuspform is integrated against a restriction of an Eisenstein series. The only thing we accomplish is the usual *unwinding of Eisenstein series*, and explain the technical hypothesis that there is a *unique cuspidal orbit.*

Let $G \subset \tilde{G}$ be two reductive linear groups defined over a number field k. Let $\tilde{P}$ be k-rational parabolic subgroup of $\tilde{G}$. Let

$$\alpha : \tilde{P} \to \mathbf{G}_m$$

be an algebraic character on $\tilde{P}$, and for a complex parameter s let

$$\tilde{\chi} : \tilde{P}_{\mathbf{A}} \to \mathbf{C}^\times$$

be defined by

$$\tilde{\chi}(\tilde{p}) = |\alpha(\tilde{p})|^{-s}$$

where $| \, |$ is the idele norm. This character factors over primes, and we write

$$\tilde{\chi} = \Pi'_v \, \tilde{\chi}_v$$

where $\tilde{\chi}_v$ is a continuous (unramified) character of the v-adic points $\tilde{P}_v$ of $\tilde{P}$. Choose a vector ε of the special form

$$\varepsilon = \bigotimes_v \varepsilon_v$$

in the induced representation

$$\text{c-Ind}_{\tilde{P}_{\mathbf{A}}}^{\tilde{G}_{\mathbf{A}}} \tilde{\chi} \approx \bigotimes{}'_v \text{c-Ind}_{\tilde{P}_v}^{\tilde{G}_v} \tilde{\chi}_v$$

where each ε_v is in the induced representation $\text{c-Ind}_{\tilde{P}_v}^{\tilde{G}_v} \tilde{\chi}_v$ and form an Eisenstein series

$$E(\tilde{g}) = E_\varepsilon(\tilde{g}) = \sum_{\gamma \in \tilde{P}_k \backslash \tilde{G}_k} \varepsilon(\gamma \tilde{g}).$$

(The adelic compact-induced representation is defined to be the restricted product of all the local compact-induced representations).

Of course, a similar construction with several algebraic characters α_i and several complex parameters gives a more general class of Eisenstein series. One can involve a cuspform on a Levi component of $\tilde{P}$, as well.

Let G be a reductive group over k, with $G \subset \tilde{G}$. Let the center of G be Z, and let $\tilde{Z}$ be the center of $\tilde{G}$. There is a minor but unavoidable technical issue here, which is sometimes overlooked: we would wish to assume that the center Z of G is central in $\tilde{G}$, but this is far from true, generally, and is needlessly restrictive. Rather, we require that

$$(\tilde{Z}_{\mathbf{A}} \cdot Z_k) \backslash Z_{\mathbf{A}} \qquad \text{is compact.}$$

This is the case if the k-ranks of Z and $\tilde{Z}$ are the same (see [**Go63**]).

Let f be a cuspform on G. Consider **global integrals**

$$\zeta(f, \tilde{\chi}) = \zeta(f, s) = \int_{\tilde{Z}_{\mathbf{A}} G_k \backslash G_{\mathbf{A}}} E(g) \, f(g) \, dg.$$

The complex parameter s enters via the Eisenstein series.

Regarding convergence of this integral: from [**Go66**], smooth square-integrable cuspforms are of rapid decay in Siegel sets. On the other hand, Eisenstein series are of moderate growth in Siegel sets, even when analytically continued, by [**L64**] or [**MoW**]. A comparison of Siegel sets in G versus those in $\tilde{G}$ is necessary to be sure that the moderate growth of the Eisenstein series is inherited by its restriction (see [**Go63**]).

The first fundamental trick is the by-now-standard *unwinding*

$$\int_{\tilde{Z}_{\mathbf{A}}G_k\backslash G_{\mathbf{A}}} E(g)\, f(g)\, dg = \sum_{\xi\in\tilde{P}_k\backslash\tilde{G}_k/G_k} \int_{\tilde{Z}_{\mathbf{A}}G_k\backslash G_{\mathbf{A}}} \sum_{\gamma\in\Theta_k^\xi\backslash G_k} \varepsilon(\xi\gamma g)\, f(g)\, dg$$

$$= \sum_{\xi\in\tilde{P}_k\backslash\tilde{G}_k/G_k} \int_{\tilde{Z}_{\mathbf{A}}\Theta_k^\xi\backslash G_{\mathbf{A}}} \varepsilon(\xi g)\, f(g)\, dg$$

where

$$\Theta^\xi = \xi^{-1}\tilde{P}\xi\cap G$$

which is valid for s in the range so that the Eisenstein series converges.

A representative ξ or orbit $\tilde{P}\xi G$ is **non-cuspidal** (or *negligible*) if Θ^ξ has a normal subgroup which is the unipotent radical of some k-parabolic of G. So say that ξ is **cuspidal** if Θ^ξ does *not* have a normal subgroup which is a unipotent radical. If there *is* such a normal subgroup, then by the Gelfand definition of cuspform the corresponding integral is 0. Thus,

$$\int_{\tilde{Z}_{\mathbf{A}}G_k\backslash G_{\mathbf{A}}} E(g)\, f(g)\, dg = \sum_{\xi\text{ cuspidal}} \int_{\tilde{Z}_{\mathbf{A}}\Theta_k^\xi\backslash G_{\mathbf{A}}} \varepsilon(\xi g)\, f(g)\, dg.$$

REMARK: It is traditional to hypothesize that there is just one cuspidal double-coset representative $\xi\in\tilde{P}_k\backslash\tilde{G}_k/G_k$, but this may be needlessly restrictive.

REMARK: Later we will take the hypothesis that the v-adic double-coset space $\tilde{P}_v\backslash\tilde{G}_v/P_v$ is *finite* for a minimal parabolic P_v of G (over the v-adic completion k_v of k). In practice, it seems that this finiteness assumption "nearly" assures the finiteness of $\tilde{P}_k\backslash\tilde{G}_k/G_k$, but this author knows of no simple general assertion in this direction. Certainly from a dimension-counting viewpoint the global double-coset space $\tilde{P}_k\backslash\tilde{G}_k/G_k$ is "smaller"; however, *rationality* issues enter strongly in the case of $\tilde{P}_k\backslash\tilde{G}_k/G_k$. It is often necessary to make adjustments, such as replacing isometry groups by the corresponding similitude groups.

1.2. Multiplicity-one criterion for Euler factorization. In this section we rewrite the integral in a form so that the global features are isolated in what we call the **inner integral**. The (apocryphal?) theorem of this section asserts roughly that *if the inner integral factors over primes, then the whole integral factors over primes.* We clarify what should be meant by "factors over primes".

In subsequent sections this will be refined, so as to anticipate more precisely the nature of the local factors, proving *a priori* that the pth factor is a *rational function* in the usual q^{-s} *and* in the Satake parameters $\alpha_i = q^{-s_i}$ attached to the cuspform. Further, we will indicate how to estimate the set of parameter values at which the denominator of this rational function *vanishes*. In particular, we will give a criterion to assure that the denominator is a product of terms exclusively of the form

$$1 - c\,q^{-m_1 s_1}\ldots q^{-m_n s_n}\,q^{-s}$$

where the m_i are integers and c is a constant independent of the s, s_i. And we indicate computationally effective means to verify the hypotheses.

First we recall an integration identity (see [**W65**], chapter II). In temporary notation, let G be a topological group (locally compact, Hausdorff, with a countable basis). Let H be a closed subgroup. Let dg and dh denote right Haar measures on G and H. Let δ_G and δ_H be the modular functions on G and H. As usual, let $C_c^o(G)$ denote the collection of continuous compactly-supported (complex-valued) functions on G. Let $C_c^o(H\backslash G, \delta_H\delta_G^{-1})$ denote the collection of continuous functions f on G so that

$$f(hg) = \frac{\delta_H(h)}{\delta_G(h)} \cdot f(g)$$

for all $h \in H$ and $g \in G$, and whose support is compact left-modulo H. Then, given the Haar measures on G and H, there is a unique functional on $C_c^o(H\backslash G, \delta_H\delta_G^{-1})$, denoted

$$F \to \int_{H\backslash G} F(\overline{g})\, d\overline{g}$$

so that for any $f \in C_c^o(G)$ we have the identity

$$\int_G f(g)\, dg = \int_{H\backslash G} \left(\int_H f(h\overline{g})\, \frac{\delta_G(h)}{\delta_H(h)}\, dh \right) d\overline{g}.$$

It is elementary that the **averaging map**

$$\text{avg} : f \to \left(g \to \int_H f(h\overline{g})\, \frac{\delta_G(h)}{\delta_H(h)}\, dh \right)$$

is a surjection to $C_c^o(H\backslash G, \delta_H\delta_G^{-1})$. Note that unless

$$\delta_H = \delta_G|_H$$

it is not the case that the functional

$$F \to \int_{H\backslash G} F(\overline{g})\, d\overline{g}$$

is literally an *integral*. As a further abuse of notation, we will omit the overbar, thus writing the identity as

$$\int_G f(g)\, dg = \int_{H\backslash G} \left(\int_H f(hg)\, \frac{\delta_G(h)}{\delta_H(h)}\, dh \right) dg.$$

Of course, the Haar integral $f \to \int_G f(g)\, dg$ extends by continuity to a continuous functional on $L^1(G)$ (the latter with right Haar measure). The proof of Fubini's theorem (e.g., by monotone families) still works in this case, at least with base field $\tilde{k} = \mathbf{R}$ or $\mathbf{C}$, proving that for $f \in L^1(G)$ and for almost all $g \in G$

$$\int_H |f(hg)| \frac{\delta_G(h)}{\delta_H(h)}\, dh < +\infty.$$

This allows an extension of the "integral" on $H\backslash G$, and the identity above still holds.

In particular, we can define **"integrable"** functions in a generalized sense on $H\backslash G$ as being the images of elements of $L^1(G)$ under the extension of the averaging map. We can also talk about **convergence** of such generalized integrals by similar devices.

Now return to the global integral. Continue with notation and terminology from above. Consider a *cuspidal* ξ. For reductive G, the adele group $G_{\mathbf{A}}$ is unimodular, so no δ_G will appear. Further, for any linear algebraic group H defined over a number field k, the modular function $\delta_{H_{\mathbf{A}}}$ is trivial on the group of k-rational points H_k of H. (See [**W61**]: this is essentially the *product formula*). Thus, using the identity above twice, we can rewrite the global integral as

$$\int_{\tilde{Z}_{\mathbf{A}}G_k\backslash G_{\mathbf{A}}} E(g)\, f(g)\, dg = \int_{\tilde{Z}_{\mathbf{A}}\Theta^{\xi}_{\mathbf{A}}\backslash G_{\mathbf{A}}} \int_{Z'_{\mathbf{A}}\Theta^{\xi}_k\backslash\Theta^{\xi}_{\mathbf{A}}} \varepsilon(\xi\theta g)\, f(\theta g)\, \delta_{\Theta_{\mathbf{A}}}(\theta)^{-1}\, d\theta\, dg$$

where

$$Z' = \tilde{Z} \cap \Theta^{\xi}.$$

A crucial hypothesis, which is elementary to prove or disprove in any particular situation, is that

$$\xi\Theta^{\xi}_{\mathbf{A}}\xi^{-1} \subset \ker(\tilde{\chi} \text{ on } \tilde{P}_{\mathbf{A}}).$$

Granting this, we have

$$\int_{\tilde{Z}_{\mathbf{A}}G_k\backslash G_{\mathbf{A}}} E(g)\, f(g)\, dg = \int_{\tilde{Z}_{\mathbf{A}}\Theta^{\xi}_{\mathbf{A}}\backslash G_{\mathbf{A}}} \varepsilon(\xi g) \int_{Z'_{\mathbf{A}}\Theta^{\xi}_k\backslash\Theta^{\xi}_{\mathbf{A}}} f(\theta g)\, \delta^{-1}_{\Theta_{\mathbf{A}}}(\theta)\, d\theta\, dg.$$

Refer to the integral

$$\int_{Z'_{\mathbf{A}}\Theta^{\xi}_k\backslash\Theta^{\xi}_{\mathbf{A}}} f(\theta g)\, \delta^{-1}_{\Theta_{\mathbf{A}}}(\theta)\, d\theta$$

as **the inner integral**.

The hypothesis just mentioned, namely that

$$\xi\Theta^{\xi}_{\mathbf{A}}\xi^{-1} \subset \ker(\tilde{\chi} \text{ on } \tilde{P}_{\mathbf{A}})$$

assures that the function $\varepsilon(\xi g)$ acting as kernel for formation of the Eisenstein series can be moved outside the inner integral. Thus, this hypothesis is that **the Eisenstein kernel escapes from the inner integral**.

The intuitive point is that *if the inner integral factors over primes, then the whole integral factors over primes.* Indeed, by its definition, the kernel $\varepsilon(\xi g)$ for the Eisenstein series is at worst a finite sum of products over primes, so the only serious obstacle to such factorization is the *inner integral.*

And, of course, we must clarify what **factorization over primes** must mean. First, and most elementarily, a function $F(g)$ on an adele group would be said to *factor over primes* v if it can be rewritten as

$$F(g) = \Pi_v\, F_v(g_v)$$

where g_v is the vth component of g in the adele group. But, when F depends additionally upon a parameter such as a cuspform f, more must be said. Indeed, by hypothesis, (and using the strong notion of "cuspform") the representation π_f of the adele group $G_{\mathbf{A}}$ generated by f under the right regular representation is *irreducible* (and unitarizable), so factors over primes as (the completion of) a restricted product

$$\pi_f \approx \bigotimes{}'_v\, \pi_v$$

where π_v is an irreducible unitarizable representation of the v-adic points G_v of the group G. The representation π_v is the **local data** of f at v. For almost all v π_v is *spherical.*

We will say that **the inner integral factors over primes** if

$$\int_{Z'_{\mathbf{A}}\Theta^{\xi}_k\backslash\Theta^{\xi}_{\mathbf{A}}} f(\theta g)\, \delta^{-1}_{\Theta_{\mathbf{A}}}(\theta)\, d\theta = \Pi_v\, F_v(g_v)$$

where the function F_v on G_v depends only upon the local data of f at v. This notion may be modified to exclude a finite set S of 'bad' primes, saying that

$$\int_{Z'_{\mathbf{A}}\Theta_k^\xi\backslash\Theta_{\mathbf{A}}^\xi} f(\theta g)\,\delta_{\Theta_{\mathbf{A}}}^{-1}(h)\,d\theta = \Pi_{v\notin S}\,F_v(g_v)\times F_{\rm bad}(g_{\rm bad})$$

where for $v \notin S$ the function F_v depends only upon the local data of f at v, and $g_{\rm bad}$ varies on the product

$$G_{\rm bad} = \Pi_{v\in S}\,G_v.$$

In all cases known to this author, if the whole integral factors over primes, then it is an L-function or product of L-functions. Thus, the immediate issue is to *give a criterion for the inner integral to factor over primes* in the sense above. The role that representation theory plays in such factorization has been addressed many times, going back to [**P**] and [**JLan**]. By contrast, the idea of [**GoJ**] (extending [**T**] in a natural way) directly presents a product of local factors, whose *analytic continuation* is less patent.

THEOREM *(Apocryphal?): We assume that there is a single cuspidal orbit, and that the Eisenstein kernel escapes from the corresponding inner integral (in the sense above). If for all primes v outside a finite set S we have the* multiplicity-one *property*

$$\dim_{\mathbf{C}}\,\mathrm{Hom}_{G_v}(\pi_v, \mathrm{Ind}_{\Theta_v^\xi}^{G_v}\,\delta_H) = 1$$

then the inner integral factors over primes (in the sense above). And, as a consequence, in that event the whole integral factors over primes.

PROOF. For simplicity, we suppress the index ξ. Write

$$I(f,g) = \int_{Z'_{\mathbf{A}}\Theta_k\backslash\Theta_{\mathbf{A}}} f(\theta g)\,\delta_{\Theta_v}(\theta)\,d\theta.$$

And we may as well enlarge the finite set S to include all primes v so that π_v is *not* spherical, and to contain all archimedean primes.

Let $\pi_f = \bigotimes'_v\pi_v$ be the restricted tensor product of irreducible smooth representations π_v generated by f under the right regular representation of $G_{\mathbf{A}}$, with basepoint vector e_v in π_v in the restricted product. (For all but finitely-many v, the e_v must be a spherical vector in π_v, which must be a spherical representation). Fix an isomorphism

$$\iota : \bigotimes{}'_v\,\pi_v \to \pi_f.$$

At all the primes v where π_v is spherical, and where

$$\dim_{\mathbf{C}}\,\mathrm{Hom}_{G_v}(\pi_v, \mathrm{Ind}_{\Theta_v}^{G_v}\,\delta_{\Theta_v}) = 1$$

normalize an intertwining

$$\Phi_v : \pi_v \to \mathrm{Ind}_{\Theta_v}^{G_v}\,\delta_{\Theta_v}$$

by

$$\Phi_v(e_v)(1_{G_v}) = 1.$$

Let T be a finite set of primes disjoint from the set S. Let

$$\varphi = \left(\bigotimes_{v\in T}\varphi_v\right)\otimes\left(\bigotimes_{v\notin T} e_v\right)$$

Let g_T be an element of $G_{\mathbf{A}}$ with components just 1 unless $v \in T$, and g'_T an element of $G_{\mathbf{A}}$ with components 1 at $v \in T$. Then the function

$$g_T \to I(\iota\varphi, g_T)$$

is an element of $\mathrm{Ind}_{\Theta_T}^{G_T} \delta_{\Theta_T}$, so

$$\varphi \to (g_T \to I(\iota\varphi, g_T))$$

is an element of

$$\mathrm{Hom}_{G_T}\left(\bigotimes_{v\in T} \pi_v, \mathrm{Ind}_{\Theta_T}^{G_T} \delta_{\Theta_T}^{-1}\right)$$

where for an algebraic subgroup H of G, H_T denotes $\Pi_{v\in T} H_v$. Since T is *finite*, and since we are operating in the relatively algebraic category of *smooth* representations, it is elementary that there is a natural isomorphism

$$\mathrm{Hom}_{G_T}\left(\bigotimes_{v\in T} \pi_v, \mathrm{Ind}_{\Theta_T}^{G_T} \delta_{\Theta_T}^{-1}\right) \approx \bigotimes_{v\in T} \mathrm{Hom}_{G_v}\left(\pi_v, \mathrm{Ind}_{\Theta_v}^{G_v} \delta_{\Theta_T}^{-1}\right).$$

Therefore, by the multiplicity-one hypothesis,

$$I(\iota\varphi, g_T) = c_T \cdot \Pi_{v\in T} \Phi_v(\varphi_v)$$

where c_T is a constant (not depending upon g_T).

By definition (of restricted tensor product), the whole restricted direct product comprising π_f is merely a colimit (direct limit) of the finite products

$$\bigotimes_{v\in T} \pi_v \otimes \bigotimes_{v\notin T} \{e_v\}.$$

Let

$$g = g_{\mathrm{bad}} \cdot g_{\mathrm{good}}$$

where g_{bad} is the aggregate of components of g at $v \in S$, and g_{good} is the aggregate of the components of g not in S. We see that for each fixed g_{bad} the map

$$\varphi \to (g_{\mathrm{good}} \to I(\iota\varphi, g))$$

gives an intertwining operator

$$i_{g_{\mathrm{bad}}} : \bigotimes\nolimits'_{v\notin S} \pi_v \to \bigotimes\nolimits'_{v\notin S} \mathrm{Ind}_{\Theta_v}^{G_v} \delta_{\Theta_T}^{-1}.$$

Therefore, by the multiplicity-one hypothesis, with a constant $c_{\mathrm{bad}}(g_{\mathrm{bad}})$ not depending upon g_{good}, and with intertwining $\Phi_v : \pi_v \to \mathrm{Ind}_{\Theta_v}^{G_v} \mathbf{C}$ normalized by $\Phi(e_v)(1) = 1$, we have

$$i_{g_{\mathrm{bad}}} = c_{\mathrm{bad}} \cdot \bigotimes_{v\notin S} \Phi_v.$$

Thus, granting the multiplicity-one hypothesis, the global integral factors as

$$\begin{aligned}\zeta(f,s) &= \int_{\tilde{Z}_{\mathbf{A}} G_k \backslash G_{\mathbf{A}}} E(g)\, f(g)\, dg \\ &= c_{\mathrm{bad}} \cdot \Pi_{v\notin S} \int_{\tilde{Z}_v \Theta_v \backslash G_v} \varepsilon_v(\xi g) \Phi_v(e_v)(g_v)\, dg_v.\end{aligned}$$

(The elementary theory of spherical representations shows that a spherical function cannot vanish at 1_G).

Thus, the whole integral is a constant multiple (depending on bad-prime data) of a product of integrals over quotients $\tilde{Z}_v\Theta_v\backslash G_v$ of integrands respectively depending only upon the local data π_v of f at v. □

Thus, the issues are

- Criteria for dim $\mathrm{Hom}_{G_v}(\pi_v, \mathrm{Ind}_{\Theta_v}^{G_v} \delta_{\Theta_T}^{-1}) \leq 1$.
- Evaluation of the good-prime **outer (local) integrals**

$$\int_{\tilde{Z}_v\Theta_v\backslash G_v} \varepsilon_v(\xi g)\Phi_v(e_v)(g_v)\, dg_v$$

whose value is the vth Euler factor.

1.3. Trivial local criterion for period-vanishing. It can happen that the *inner integral* above vanishes identically for "most" cuspforms, for purely local reasons. The explanation is a simple application of the idea of viewing integrals as values of intertwining operators. Here we give this general but essentially trivial *local* condition for vanishing of certain *global* constants attached to cuspforms. Given the definition of these global constants (below) it is reasonable to call them **periods**.

And, again, the computational aspects of this discussion are clearest when (admissible) spherical representations are rewritten as images of unramified principal series, as in the next section.

Let f be a cuspform on a reductive group G defined over a number field k. Let $\tilde{Z}$ be a k-subgroup of the center Z of G so that

$$(\tilde{Z}_{\mathbf{A}} \cdot Z_k)\backslash Z_{\mathbf{A}} \qquad \text{is compact.}$$

Let H be a subgroup of G defined over k. The **period** Π_f of f over H is

$$\Pi_f = \int_{Z'_{\mathbf{A}}\, H_k\backslash H_{\mathbf{A}}} f(h)\, \delta_{H_{\mathbf{A}}}^{-1}(h)\, dh$$

where $Z' = \tilde{Z} \cap H$. Let π_f be the irreducible representation of the adele group $G_{\mathcal{A}}$ generated by f, and let π_v be the vth local factor of π_f (when expressed as restricted tensor product of local representations). We will not worry about convergence here.

Proposition (Apocryphal?) *If for any prime v*

$$\mathrm{Hom}_{G_v}(\pi_v,\ \mathrm{Ind}_{H_v}^{G_v}\ \delta_{H_v}) = \{0\}$$

then the period vanishes:

$$\Pi_f = 0.$$

Proof. Let $g_v \in G_v$. The function

$$g_v \to \int_{Z'_{\mathbf{A}}\, H_k\backslash G_{\mathbf{A}}} f(hg_v)\, \delta_{H_{\mathbf{A}}}^{-1}(h)\, dh$$

is a function in the induced representation $\mathrm{Ind}_{H_v}^{G_v}\ \delta_{H_v}$. Thus, the map

$$\varphi \to \left(g_v \to \int_{Z'_{\mathbf{A}}\, H_k\backslash H_{\mathbf{A}}} \varphi(hg_v)\, \delta_{H_{\mathbf{A}}}^{-1}(h)\, dh \right)$$

gives a G_v-intertwining from π_v to $\mathrm{Ind}_{H_v}^{G_v}\ \delta_{H_v}$. If there is no such intertwining operator (other than 0), then it must be that the integral is 0. □

REMARKS: Generally, after Jacquet, we would call π_f **distinguished** with respect to the subgroup H if the period of φ over H is non-zero for some vector in the representation π_f generated by f. It seems that in many cases such *distinguished* representations are in correspondence with representations on another group (and perhaps distinguished with respect to a subgroup of that other group). Such results are non-trivial. For example, in [**JR92**] and [**JR92b**] it is shown that cuspforms on $\mathrm{GL}(2n)$ have no non-zero periods for $\mathrm{Sp}(n)$, and in [**Ku**] $U(n,n)$ periods inside $O^*(4n)$ are considered. One necessary ingredient is Jacquet's relative trace formula, which appears in [**JLai, J86, J87**]. [**JLaiR**] gives one general formalism to address such problems.

1.4. Orbit-filtration estimate of multiplicity. We continue with the notations and conventions from above. At good primes, we obtain computational criteria sufficient to assure

$$\dim \mathrm{Hom}_{G_v}(\pi_v, \mathrm{Ind}_{\Theta_v}^{G_v} \delta_{H_v}) \leq 1$$

which (from section 1.2) would assure factorization of the *inner integral* over primes. (The case that this dimension is 0 is the case of *trivial vanishing* of the period for local reasons, treated in section 1.3).

First, we rewrite spherical representations as images of unramified principal series, thereby making available Frobenius Reciprocity and other standard elementary devices applicable to induced representations. In particular, we use orbit filtrations on test functions to estimate spaces of intertwining operators on filtered spaces of test functions via the intertwining operators on the graded pieces of the filtration. (See [**Bru**], [**BeZ76, BeZ77**]). It is at this point that we require the *finiteness* of a double-coset space, denoted $P_v\backslash G_v/\Theta_v$ below.

This estimate on intertwining operators can be refined somewhat by noting the symmetry implicitly required by the intertwining operators among the unramified principal series whose characters (suitably normalized) differ only by an element of the spherical Weyl group.

Thus, we make use of the non-trivial results of [**M, Bo, C80**] concerning the relations between *unramified principal series* and admissible representations with Iwahori-fixed vectors: an irreducible admissible representation with an Iwahori-fixed vector imbeds into an unramified principal series. Further, any subrepresentation or quotient of an unramified principal series is generated by its Iwahori-fixed vectors. Thus, an (admissible) spherical representation imbeds into an unramified principal series, so, by dualization, is also an *image* of an unramified principal series.

Thus, for a spherical representation π_v of a p-adic reductive group G_v, there is at least one surjection

$$\text{c-Ind}_{P_v}^{G_v} \chi_v \delta_{P_v}^{1/2} \to \pi_v$$

where P is a minimal parabolic of G (defined over the vth completion k_v of k), χ_v is an unramified character, and δ_{P_v} is the modular function on the (v-adic points of) P. (The latter is introduced as a convenient normalization for discussion of Weyl-group symmetry, among other things.) That is, χ_v is trivial on the unipotent radical of P_v, and is trivial on the unique maximal compact subgroup M_o of a Levi component M_v of P_v. Here c-Ind is compactly-supported smooth induction.

Thus, to prove that

$$\dim \mathrm{Hom}_{G_v}(\pi_v, \mathrm{Ind}_{\Theta_v}^{G_v} \delta_{\Theta_v}) \leq 1$$

it certainly suffices to prove

$$\dim \mathrm{Hom}_{G_v}\left(\text{c-Ind}_{P_v}^{G_v}\chi_v\delta_{P_v}^{1/2}, \mathrm{Ind}_{\Theta_v}^{G_v}\,\delta_{\Theta_v}\right) \le 1$$

for the appropriate χ_v. By Frobenius Reciprocity,

$$\mathrm{Hom}_{G_v}\left(\text{c-Ind}_{P_v}^{G_v}\chi_v\delta_{P_v}^{1/2}, \mathrm{Ind}_{\Theta_v}^{G_v}\,\delta_{\Theta_v}\right) \approx \mathrm{Hom}_{\Theta_v}\left(\mathrm{Res}_{\Theta_v}^{G_v}\text{c-Ind}_{P_v}^{G_v}\chi_v\delta_{P_v}^{1/2},\, \delta_{\Theta_v}\right).$$

From section 3.2 below on *orbit filtrations of test functions*, we decompose G_v by $P_v \times \Theta_v$-orbits, with P_v acting on the left and Θ_v on the right, thereby obtaining an orbit filtration of $\mathrm{Res}_{\Theta_v}^{G_v}(\text{c-Ind}_{P_v}^{G_v}\chi_v\delta_{P_v}^{1/2})$, with *graded pieces*

$$\text{c-Ind}_{H_\alpha}^{\Theta_v}\chi^\alpha$$

where $\alpha \in P_v\backslash G_v/\Theta_v$,

$$H_\alpha = \alpha^{-1}P_v\alpha_v \cap \Theta_v$$

and

$$\chi^\alpha(\theta) = \chi_v\delta_{P_v}^{1/2}(\alpha\theta\alpha^{-1})$$

for $\theta \in H_\alpha$. The orbit filtration requires that $P_v\backslash G_v/\Theta_v$ be *finite.*

Therefore, *assuming the finiteness of* $P_v\backslash G_v/\Theta_v$, to show that the dimension of the space of intertwining operators is less than or equal 1 it suffices to show that

$$\sum_\alpha \mathrm{Hom}_{\Theta_v}(\text{c-Ind}_{H_\alpha}^{\Theta_v}\chi^\alpha,\, \delta_{\Theta_v}) \le 1.$$

This condition can be simplified: dualizing and using Frobenius Reciprocity again, this is

$$\sum_\alpha \dim \mathrm{Hom}_{\Theta_v}\left(\mathbf{C},\, \delta_{\Theta_v} \otimes \mathrm{Ind}_{H_\alpha}^{\Theta_v}\check{\chi}^\alpha\frac{\delta_{H_\alpha}}{\delta_{\Theta_v}}\right) = \sum_\alpha \dim \mathrm{Hom}_{H_\alpha}\left(\mathbf{C}, \delta_{\Theta_v}\check{\chi}^\alpha\frac{\delta_{H_\alpha}}{\delta_{\Theta_v}}\right)$$

$$= \sum_\alpha \dim \mathrm{Hom}_{H_\alpha}(\chi^\alpha, \delta_{H_\alpha})$$

where $\check{\chi}^\alpha$ is the smooth dual of χ^α.

That is, we have proven

THEOREM: *Assume that for almost all places v of k the double coset space $\tilde{P}_v\backslash\tilde{G}_v/G_v$ is finite. The dimension of the space of intertwining operators from π_v to $\mathrm{Ind}_{\Theta_v}^{G_v}\,\delta_{\Theta_v}$ satisfies*

$$\dim(\pi_v, \mathrm{Ind}_{\Theta_v}^{G_v}\,\delta_{\Theta_v}) \le \sum_\alpha \dim \mathrm{Hom}_{H_\alpha}(\chi^\alpha,\, \delta_{H_\alpha}).$$

Thus, if this sum is ≤ 1, then the inner integral factors over primes, and the global integral has an Euler product.

REMARKS: In principle, for any particular choices of the characters, the simplified version of the condition on intertwining operators is effectively provable or disprovable since it involves simply questions of equality of restrictions to a common subtorus of two characters on two subgroups.

We can refine this Theorem somewhat by use of the intertwining operators among unramified principal series, as studied in [**Bru**] and [**C80**]. For an element w of the spherical Weyl group W attached to a maximal split torus A inside the minimal parabolic P_v of G_v, and for $a \in A$, define another character χ_v^w on A by

$$\chi_v^w(a) = \chi_v(w\,a\,w^{-1}).$$

Say that χ_v is *regular* if $\chi_v \neq \chi_v^w$ for $w \neq 1$.

With the unramified principal series c-$\mathrm{Ind}_{P_v}^{G_v} \chi_v \delta_{P_v}^{1/2}$ normalized by the inclusion of the square root $\delta_{P_v}^{1/2}$ of the modular function δ_{P_v} of P_v, for regular χ_v and for $w \in W$

$$\dim \mathrm{Hom}(\text{c-}\mathrm{Ind}_{P_v}^{G_v} \chi_v \delta_{P_v}^{1/2}, \text{c-}\mathrm{Ind}_{P_v}^{G_v} \chi_v^w \delta_{P_v}^{1/2}) = 1.$$

Further, **[C80]** determines finitely-many essentially elementary conditions on χ_v so that, except when these conditions are met, the unramified principal series c-$\mathrm{Ind}_{P_v}^{G_v} \chi_v \delta_{P_v}^{1/2}$ is *irreducible*. That is, for "generic" unramified principal series, the essentially unique intertwining operators among the unramified principal series with characters differing by elements of the Weyl group W are all isomorphisms.

Thus, for given regular χ_v so that c-$\mathrm{Ind}_{P_v}^{G_v} \chi_v \delta_{P_v}^{1/2}$ is *irreducible*, suppose that the αth graded piece in the orbit filtration had a non-zero intertwining operator. Above, we saw that this is equivalent to

$$\delta_{H_\alpha} = \chi^\alpha|_{H_\alpha}.$$

If the intertwining operator on the graded piece were to *extend* to an intertwining operator for the whole representation $\mathrm{Ind}_{\Theta_v}^{G_v} \delta_{\Theta_v}$, then the isomorphism of c-$\mathrm{Ind}_{P_v}^{G_v} \chi_v \delta_{P_v}^{1/2}$ with c-$\mathrm{Ind}_{P_v}^{G_v} \chi_v^w \delta_{P_v}^{1/2}$ would imply that also

$$\delta_{H_\alpha} = (\chi^w)^\alpha|_{H_\alpha}.$$

That is, we have proven

THEOREM: *Assume that for almost all places v of k the double coset space $\tilde{P}_v\backslash\tilde{G}_v/G_v$ is finite. Let χ_v be a regular character so that the unramified principal series c-$\mathrm{Ind}_{P_v}^{G_v} \chi_v \delta_{P_v}^{1/2}$ is irreducible. The dimension of the space of intertwining operators from π_v to $\mathrm{Ind}_{\Theta_v}^{G_v} \delta_{\Theta_v}$ satisfies*

$$\dim \mathrm{Hom}_{G_v}(\text{c-}\mathrm{Ind}_{P_v}^{G_v} \chi_v \delta_{P_v}^{1/2}, \mathrm{Ind}_{\Theta_v}^{G_v} \delta_{\Theta_v})$$
$$\leq \min_{w \in W} \sum_{\alpha \in P_v\backslash G_v/\Theta_v} \dim \mathrm{Hom}_{H_\alpha}((\chi^w)^\alpha, \delta_{H_\alpha}).$$

REMARKS: The chief difficulty in reaching clear conclusions in this manner is that one knows nearly nothing about *which* unramified characters χ_v can occur in such manner associated to a cuspform, apart from a few things such as square-integrability of the representation π_v. Therefore, one is apparently in the position of having to prove the above estimate on intertwinings of the graded pieces in the orbit filtration *for all unramified principal series.* But it is seldom the case that the estimate on intertwining operators of the graded pieces really holds for *all* χ_v.

Nevertheless, it does happen very often that for **generic** χ_v the estimate above holds, assuring the Euler factorization. The sense of 'generic' here is neither the colloquial one, nor is it the condition 'has a Whittaker model'. Rather, the relevant sense is with regard to the position of the complex parameters specifying the character, as in section 2.1 below.

In some examples, invocation of the *unitariness* of the representations π_v occurring in association with a square-integrable cuspform is sufficient to dispatch the non-generic cases (in the above sense) in which the sum of the dimensions of the intertwining operators on the graded pieces is greater than 1. That is, it is often possible to obtain the desired multiplicity-one conclusion simply by using the orbit filtration and the unitariness. Otherwise, one can pretend to be invoking some sort of generalized Ramanujan hypothesis to ensure that the unramified principal series are sufficiently well-behaved. On the other hand, such "pseudo-Ramanujan" hypotheses may inadvertently imply identical vanishing of the *inner integral* for trivial reasons (as in section 1.3).

1.5. Simple heuristic criterion. Now we give a simple heuristic criterion, in terms of dimension-counting, for Euler factorization of a global integral.

We do *not* assert that *failing* the dimension-count *proves* that there is no Euler product, nor that *passing* the dimension-count test proves that there *is* an Euler product. Nevertheless, it appears to be rather difficult to arrange cases in which dimension-count criterion is not met and yet there is an Euler product. And, on the other hand, in practice it seems to be that *meeting* the criterion nearly guarantees that factorization into an Euler product.

The papers [**Br**] and [**KKO**] systematically address related dimension-counting issues.

Again, let G and $\tilde{G}$ be two reductive groups defined over a number field, with G a subgroup of $\tilde{G}$. Let $\tilde{Z}$ be the center of $\tilde{G}$, and Z the center of G, and assume that

$$\tilde{Z}_{\mathbf{A}} \cdot Z_k \backslash Z_{\mathbf{A}} \qquad \text{is compact.}$$

Let $\tilde{P}$ be a parabolic subgroup of $\tilde{G}$, and P a minimal parabolic subgroup of G. Let E be an Eisenstein series attached to a one-dimensional representation of $\tilde{P}$, and let f be a cuspform on G. As above, the global integral which we would *like* to have an Euler product is

$$Z(f) = \int_{\tilde{Z}_{\mathbf{A}}\, G_k \backslash G_{\mathbf{A}}} E(g)\, f(g)\, dg.$$

OBSERVATION: If

$$\dim\, \tilde{P} + \dim\, P < \dim\, \tilde{G}$$

then our present approach will most likely fail to demonstrate that the global integral has an Euler product expansion.

Support for the observation: The unwinding trick (see section 1.1) expresses such a global integral as a sum over $\tilde{P}_k \backslash \tilde{G}_k / G_k$ of global integrals. For each $\xi \in \tilde{P}_k \backslash \tilde{G}_k / G_k$, let

$$\Theta = \Theta_\xi = G \cap \xi^{-1} \tilde{P} \xi.$$

Then the associated *inner integral* (as in section 1.2) will factor over primes if

$$\dim\, \mathrm{Hom}_{G_v}(\text{c-Ind}_{P_v}^{G_v}\, \chi_v,\ \mathrm{Ind}_{\Theta_v}^{G_v} \delta_{\Theta_v}) \leq 1$$

for almost all primes v (with suitable characters χ_v).

The orbit filtration method (3.2) to estimate this (co-)multiplicity requires that the double-coset space $P_v \backslash G_v / \Theta_v$ be *finite*. In the discussion of the orbit filtration it was shown that this implies (already at the level of the Baire category theorem) that there is an orbit $\Theta_v x P_v$ which is *open* in G_v. For this to be the case, it must be that

$$\dim\, \Theta + \dim\, P \geq \dim\, G$$

(where dimension is that of algebraic varieties).

We may reasonably call ξ *generic* if

$$\dim\, \tilde{G} - \dim\, \xi^{-1} \tilde{P} \xi = \dim\, G - \dim\, (G \cap \xi^{-1} \tilde{P} \xi).$$

Certainly

$$\dim \tilde{G} - \dim \tilde{P} = \dim \tilde{G} - \dim \xi^{-1}\tilde{P}\xi$$

for all ξ. When the double coset space $\tilde{P}_k\backslash\tilde{G}_k/G_k$ is *finite* there must exist at least one generic ξ (assuming that $\tilde{G}$ is not anisotropic over k, i.e., assuming that $\tilde{P}$ is a *proper* parabolic.)

For generic ξ, we can rewrite the dimension condition

$$\dim \Theta + \dim P \geq \dim G$$

as

$$\dim P \geq \dim G - \dim \Theta = \dim \tilde{G} - \dim \tilde{P}.$$

That is, we expect that this inequality is a *necessary* condition for success of an orbit filtration argument that the inner integral factors over primes. Thus, its negation (as in the *Observation*) would make it unlikely that we would be able to detect Euler factorization by the methods presented here. □

2. Computation of Euler factors

2.1. Parametrized families. This section sets up an essentially algebraic notion of 'family' of smooth representations of p-adic and other totally disconnected groups. For our present purposes, it suffices to consider only the case that the parametrizing spaces are affine algebraic varieties. Compare the more analytical version (due to Bernstein) sketched in [**GePR**], as well as the general purely algebraic set-up in [**BeZ76, BeZ77, Be84**].

We will discuss *meromorphic* families of intertwining operators among families of representations. The sense of "meromorphic" here is (a weak vector-valued form of) the algebraic one, designed for application to the families of representations above, parametrized by affine algebraic varieties in an algebraic manner. The notion of *locally strong meromorphy* of vector-valued function at the end of this section is important for later applications.

For a commutative ring k with a unit $1 = 1_k$, a k-**algebra** is a ring $\mathcal{O}$ with a k-module structure so that 1_k acts as the identity upon $\mathcal{O}$, and so that for $x \in k$ and $\alpha, \beta \in \mathcal{O}$

$$x(\alpha\beta) = (x\alpha)\beta = \alpha(x\beta).$$

Let k be a field. Let $\mathcal{O}$ be a commutative k-algebra with unit, and suppose that $\mathcal{O}$ is an integral domain. Let $\mathcal{M}$ be the field of fractions of $\mathcal{O}$, and let

$$X = \text{Spec } \mathcal{O}$$

be the prime spectrum of $\mathcal{O}$. For a prime ideal $x \in X$ let S_x be the multiplicative subset of $\mathcal{O}$ which is the set complement of x in $\mathcal{O}$, and define the **local ring at** x to be $\mathcal{O}_x = S_x^{-1}\mathcal{O}$. The **residue field at** x is the quotient

$$k_x = \mathcal{O}_x / x\mathcal{O}_x.$$

The residue field k_x is naturally a k-algebra, an $\mathcal{O}$-algebra, and an $\mathcal{O}_x$-algebra. Since $\mathcal{O}$ is a domain, the zero ideal is prime, and we denote its residue field $\mathcal{M}$ by k_o.

As usual, a *height one prime* in $\mathcal{O}$ is a prime ideal y so that if x is another prime ideal and $x \subset y$ then either $x = 0$ or $x = y$. The **irreducible hypersurface** $\eta = \eta_y$ attached to a height-one prime is defined to be

$$\eta = \eta_y = \{x \supset y : x \in X\}.$$

In the spirit of the Baire category theorem, a **meager** subset Y of X is a subset contained in a countable union of irreducible hypersurfaces. A **non-meager** subset of X is simply one which is not *meager*.

Let

$$q_x : \mathcal{O}_x \to k_x$$

be the natural **quotient** or **evaluation** map, and also write

$$q_x : V \to V \otimes_{\mathcal{O}} k_x$$

for any $\mathcal{O}$-module V, and further

$$q_x : V \to V \otimes_{\mathcal{O}_x} k_x$$

for any $\mathcal{O}_x$-module V.

Let G be a (locally compact, Hausdorff) totally disconnected group. Let V be an $\mathcal{O}$-module, and

$$\pi : G \to \mathrm{Hom}_{\mathcal{O}}(V, V)$$

a group homomorphism. Suppose that π is (uniformly) **smooth** in the sense that for every $v \in V$ the isotropy subgroup

$$\{g \in G : gv = v\}$$

of v is *open*. Say that π is a **$\mathcal{O}$-parametrized family** of smooth representations of G.

For $x \in X$ we have the **pointwise representation**

$$q_x\pi = \pi_x : G \to \mathrm{Hom}_{k_x}(V \otimes_{\mathcal{O}} k_x, V \otimes_{\mathcal{O}} k_x)$$

of G on the k_x-vector space $V \otimes_{\mathcal{O}} k_x$.

When $\mathcal{O}$ is Noetherian and k is algebraically closed and x is maximal, by the Nullstellensatz the residue field $q_x\mathcal{O} = k_x$ is just k itself. Thus, for k algebraically closed, the pointwise representations at maximal ideals (geometrically closed points) x form a *family* $\{q_x\pi = \pi_x : x \in X\}$ of k-linear representations in a more colloquial sense.

REMARKS: Of course, such a parametrized family is really just a single $\mathcal{O}$-linear representation of G. However, we cannot reasonably expect to develop representation theory over general commutative rings $\mathcal{O}$. Fortunately, for the most part we can skirt such issues, simply viewing the $\mathcal{O}$-linear representation as glueing together the various point-wise representations as just defined. Compare [**BeZ76, BeZ77**], and [**Be84**].

The **generic representation** in the parametrized family π is $\pi_o = \pi_{\{0\}}$, the pointwise representation corresponding to the *generic point* $\{0\}$ in the prime spectrum. The generic representation is most naturally an $\mathcal{M}$-linear representation, and typically $k_o = \mathcal{M}$ has positive transcendence degree over the base field k.

The **trivial** $\mathcal{O}$-parametrized family of smooth representations of G is the $\mathcal{O}$-module $\mathcal{O}$ itself, with trivial action of G upon it. The trivial representation of G on a residue field k_x will also be denoted simply k_x.

Let $\tilde{k}$ be an extension field of k so that $\mathcal{O} \otimes_k \tilde{k}$ is still an integral domain, with $\mathcal{M} \otimes_k \tilde{k}$ the field of fractions of $\mathcal{O} \otimes_k \tilde{k}$. Given an $\mathcal{O}$-parametrized family π of smooth representations of G with underlying $\mathcal{O}$-module V, the $\mathcal{O}\otimes_k\tilde{k}$-parametrized family $\pi \otimes_k \tilde{k}$ of representations obtained by **extending scalars** from k to $\tilde{k}$ has underlying $\mathcal{O} \otimes_k \tilde{k}$-module

$$V \otimes_k \tilde{k} \approx V \otimes_{\mathcal{O}} (\mathcal{O} \otimes_k \tilde{k})$$

and is given by the natural formula

$$(\pi \otimes_k \tilde{k})(g)(v \otimes \alpha) = \pi(g)(v) \otimes \alpha.$$

On the other hand, an $\mathcal{O} \otimes_k \tilde{k}$-parametrized family $\tilde{\pi}$ is **defined over** k or k-**rational** if there is an $\mathcal{O}$-parametrized family π so that

$$\tilde{\pi} \approx \pi \otimes_k \tilde{k}.$$

A trivial example of a parametrized family is that of a **constant** (along $\mathcal{O}$) family, by which we mean the following. Let ρ be a k-linear representation of G, and put

$$\pi = \rho \otimes_k \mathcal{O}.$$

Then π is an $\mathcal{O}$-parametrized family which is **constant** along $\mathcal{O}$, in the sense that for every $x \in X$

$$q_x \pi = \rho \otimes_k k_x.$$

The simplest meaningful example of a parametrized family of smooth representations arises from the **C**-linear representations

$$\alpha \longrightarrow |\alpha|^s$$

on the multiplicative group $F^\times$ of an ultrametric local field F, where the norm $|\,|$ is normalized (e.g.) so that

$$|\varpi| = \frac{1}{q}$$

where ϖ is a local parameter and q is the cardinality of the residue field. These are parametrized in a colloquial sense by $s \in \mathbf{C}$, or by $s \in \mathbf{C}/2\pi i \log q$. Then write

$$\pi(\alpha) = z^{\operatorname{ord} \alpha}$$

where z is an indeterminate (thought of as being q^{-s}) and *ord* is the usual ord-function. In this example, we take $k = \mathbf{C}$, let $\mathcal{O} = \mathbf{C}[z, z^{-1}]$, and let V be a copy of $\mathcal{O}$ with $\mathcal{O}$ acting by multiplication. Then we have the family

$$\pi : F^\times \longrightarrow \operatorname{Hom}_{\mathbf{C}[z,z^{-1}]}(\mathbf{C}[z, z^{-1}], \mathbf{C}[z, z^{-1}])$$

by

$$\pi(\alpha)v = z^{\operatorname{ord} \alpha} \cdot v.$$

In this example, for a maximal ideal x of $\mathcal{O} = \mathbf{C}[z, z^{-1}]$ generated by $z - q^{-s}$ for a complex number s, the pointwise representation $q_x \pi = \pi_x$ is the one-dimensional representation

$$\alpha \longrightarrow |\alpha|^s.$$

On the other hand, the *generic* representation π_o is $\mathbf{C}(z)$-linear, on the one-dimensional $\mathbf{C}(z)$-vectorspace $\mathbf{C}(z)$ itself, by

$$\pi_o(\alpha)v = z^{\operatorname{ord} \alpha} \cdot v.$$

A little more generally, let M be a product of n copies of $F^\times$ for a local field F, and define

$$\pi(m_1, \dots, m_n) = z_1^{\operatorname{ord} m_1} z_2^{\operatorname{ord} m_2} \dots z_n^{\operatorname{ord} m_n}$$

where $z_1, z_2, \dots, z_n$ are indeterminates. Here we can take

$$\mathcal{O} = \mathbf{C}[z_1, z_1^{-1}, z_2, z_2^{-1}, \dots, z_n, z_n^{-1}]$$

and take V to be a copy of $\mathcal{O}$ with module structure just given by left multiplication. Then for choice of n non-zero complex numbers $s_1, s_2, \ldots, s_n$, with $z_{1o} = q^{-s_1}, z_{2o} = q^{-s_2}, \ldots, z_{no} = q^{-s_n}$, letting x be the ideal

$$x = \mathcal{O} \cdot (z_1 - z_{1o}) + \mathcal{O} \cdot (z_2 - z_{2o}) + \ldots + \mathcal{O} \cdot (z_n - z_{no})$$

the pointwise representation π_x is

$$\pi_x(m_1, m_2, \ldots, m_n) = |m_1|^{s_1} \, |m_2|^{s_2} \ldots |m_n|^{s_n}.$$

Here the *generic* representation is simply the $\mathbf{C}(z_1, z_2, \ldots, z_n)$-linear representation

$$\pi_o : (m_1, m_2, \ldots, m_n) \to \text{ (multiplication by) } z_1^{\text{ord } m_1} z_2^{\text{ord } m_2} \ldots z_n^{\text{ord } m_n}$$

on a one-dimensional $\mathbf{C}(z_1, z_2, \ldots, z_n)$-vectorspace. Here there are non-maximal non-zero prime ideals in the parameter space for $n > 1$.

More generally, let M be a Levi component of a minimal F-parabolic in a reductive linear group G of F-rank n, defined over the local field F. Because of the minimality, there is a unique maximal compact subgroup M_o of M, and M_o is *normal*, and M/M_o is *abelian*. In fact, $M/M_o \approx \mathbf{Z}^n$ for some n. A one-dimensional representation of M is **unramified** if M_o lies inside its kernel. Let $\varpi_1, \varpi_2, \ldots, \varpi_n$ be elements in M so that their images in $M/M_o \approx \mathbf{Z}^n$ are a $\mathbf{Z}$-basis for the latter. Let $z_1, z_2, \ldots, z_n$ be indeterminates over $\mathbf{C}$. Then put

$$\pi(\varpi_1^{m_1} \varpi_2^{m_2} \ldots \varpi_n^{m_n}) = z_1^{m_1} z_2^{m_2} \ldots z_n^{m_n}.$$

Here again we take

$$\mathcal{O} = \mathbf{C}[z_1, z_1^{-1}, z_2, z_2^{-1}, \ldots, z_n, z_n^{-1}]$$

and take V to be a copy of $\mathcal{O}$ with module structure just given by left multiplication. With maximal primes x, we recover all the unramified representations of M. At the other extreme, the *generic representation* π_o in this family is the $\mathbf{C}(z_1, z_2, \ldots, z_n)$-linear representation

$$\pi_o(\varpi_1^{m_1} \varpi_2^{m_2} \ldots \varpi_n^{m_n}) = z_1^{m_1} z_2^{m_2} \ldots z_n^{m_n}$$

on any one-dimensional $\mathbf{C}(z_1, z_2, \ldots, z_n)$-vector space. And there are non-maximal non-zero primes, as well.

Generally, let P be a parabolic subgroup of a reductive p-adic group G over an ultrametric local field F, with unipotent radical N, and let $M \approx P/N$ be a copy of a Levi component. Let σ be a supercuspidal representation of M, and let π be a parametrized family of one-dimensional representations of M, as above. Then $\sigma \otimes \pi$ is certainly a parametrized family of smooth representations. We do not need such families here, but see [**BeZ76, BeZ77, Be84**] for applications.

As expected, for an $\mathcal{O}$-parametrized family σ of smooth representations of a closed subgroup H of G, the **compactly-induced** representation $\text{c-Ind}_H^G \, \sigma$ is the collection of σ-valued functions f on G which are compactly supported left modulo H, are locally constant, and so that

$$f(hg) = \sigma(h) \, f(g)$$

for $h \in H$ and $g \in G$. Likewise, the (not-necessarily compactly-) **induced** representation $\text{Ind}_H^G \, \sigma$ is the collection of σ-valued functions f on G which are *uniformly* locally constant, and so that

$$f(hg) = \sigma(h) \, f(g).$$

It is elementary but important to note that for $x \in X$ and for any $\mathcal{O}$-algebra $\mathcal{A}$ we have

$$(\text{c-Ind}_H^G\,\sigma) \otimes_{\mathcal{O}} \mathcal{A} \approx \text{c-Ind}_H^G\,(\sigma \otimes_{\mathcal{O}} \mathcal{A}).$$

(This is because compactly-supported induction itself is a tensor product, and tensor products are associative). That is, *the usual construction does yield an $\mathcal{O}$-parametrized family whose pointwise representations are the compactly-induced representations from the pointwise representations σ_x.*

By contrast,

$$(\text{Ind}_H^G\,\sigma) \otimes_{\mathcal{O}} \mathcal{A} \subset\subset \text{Ind}_H^G\,(\sigma \otimes_{\mathcal{O}} \mathcal{A})$$

with strict inclusion in general unless G/H is actually compact. Thus, in general, the usual construction of not-necessarily-compact induction does *not* yield a parametrized family whose pointwise representations are the induced representations of the pointwise representations.

The notion of **smooth dual** of an $\mathcal{O}$-parametrized family σ has similar difficulties. First, for any $\mathcal{O}$-module V upon which G acts, the collection of **smooth vectors** V^∞ in V is

$$V^\infty = \{v \in V : \ v \text{ has open isotropy group }\}.$$

For any $x \in X$ let π_x be the pointwise representation at x. Naively, one would want an $\mathcal{O}$-parametrized family $\check{\pi}$ of smooth representations of G so that for every $x \in X$

$$\text{Hom}_{k_x}(\pi_x, k_x)^\infty = (\check{\pi})_x.$$

But

$$\text{Hom}_{k_x}(\pi_x, k_x) \approx \text{Hom}_{k_x}(\pi \otimes_{\mathcal{O}} k_x, k_x) \approx \text{Hom}_{\mathcal{O}}(\pi, k_x).$$

There is the obvious natural map

$$\text{Hom}_{\mathcal{O}}(\pi, \mathcal{O}) \to \text{Hom}_{\mathcal{O}}(\pi, k_x)$$

but this is not guaranteed to be a *surjection* unless (for example) π is *projective* as an $\mathcal{O}$-module.

The simplest scenario in which this projectivity holds is the trivial *constant family* case. Less trivially, the commutative ring $\mathcal{O}$ itself (with left multiplication) is a free module over itself, hence is projective (since it has an *identity*). Thus, any $\mathcal{O}$-parametrized family χ of representations of G on $\mathcal{O}$ itself *is* projective, so has a smooth dual $\mathcal{O}$-parametrized family of representations

$$\check{\chi} = \text{Hom}_{\mathcal{O}}(\chi, \mathcal{O})^\infty$$

where G acts on the dual by the usual contragredient action

$$\check{\chi}(g)(\lambda)(v) = \lambda(\chi(g^{-1})v).$$

However, the assertion

$$(\text{c-Ind}_H^G\sigma)\check{} \approx \text{Ind}_H^G\,\check{\sigma}\,\frac{\delta_H}{\delta_G}$$

(which is true for representations on *vector spaces*) is *false* for $\mathcal{O}$-parametrized families in general, since the not-necessarily-compact induction does not commute with tensor products, as already noted above.

Let σ, τ be two $\mathcal{O}$-parametrized families of smooth representations of G. An $\mathcal{M}$-linear map

$$\Phi : \sigma \otimes_{\mathcal{O}} \mathcal{M} \to \tau \otimes_{\mathcal{O}} \mathcal{M}$$

is an $\mathcal{O}$**-parametrized (meromorphic) family of intertwining operators** from σ to τ if for all $s \in \sigma$ we have

$$\Phi(\sigma(g)s) = \tau(g)\Phi(s)$$

for all $g \in G$. That is, Φ is an $\mathcal{M}$-vectorspace homomorphism, and is a G-module homomorphism. The family Φ is **holomorphic** at $x \in X$ if

$$\Phi(\sigma) \subset \tau \otimes_{\mathcal{O}} \mathcal{O}_x.$$

The following is immediate:

LEMMA: *Let σ, τ be two $\mathcal{O}$-parametrized families of smooth representations of G. If σ is a countably-generated $\mathcal{O}$-module, then a meromorphic family of intertwining operators*

$$\Phi : \sigma \to \tau$$

fails to be holomorphic on at worst a meager *set, i.e., fails to be holomorphic on a subset of some* countable union of hypersurfaces. □

For $x \in X$ at which π is holomorphic, there is a natural **pointwise intertwining operator**

$$\Phi_x : \sigma \otimes_{\mathcal{O}} k_x \to \tau \otimes_{\mathcal{O}} k_x.$$

Taking the *generic point* $x = 0$ in X, we have the **generic intertwining operator** in the family, simply the $\mathcal{M}$-linear map

$$\Phi : \sigma \otimes_{\mathcal{O}} \mathcal{M} \to \tau \otimes_{\mathcal{O}} \mathcal{M}.$$

Let $\tilde{k}$ be an extension field of k so that $\mathcal{O} \otimes_k \tilde{k}$ is still an integral domain. Let $\tilde{\mathcal{M}}$ be the fraction field of $\mathcal{O} \otimes_k \tilde{k}$. Then a meromorphic family of intertwining operators

$$\tilde{\Phi} : \sigma \otimes_{\mathcal{O}} \tilde{\mathcal{M}} \to \tau \otimes_{\mathcal{O}} \tilde{\mathcal{M}}$$

is **rational over** k if there is a meromorphic family of intertwining operators

$$\Phi : \sigma \otimes_{\mathcal{O}} \mathcal{M} \to \tau \otimes_{\mathcal{O}} \mathcal{M}$$

so that $\tilde{\Phi}$ is the natural $\tilde{k}$-linear extension of Φ.

2.2. Rationality of local integrals. In the *Rationality Lemma* just below, the last of the conclusions is seemingly the most banal. However, this conclusion has corollaries which are considerably less obvious, such as that various local integrals are rational functions of q^{-s} or of several quantities $q^{-s}, q^{-s_1}, \ldots, q^{-s_n}$. As a very special case, in section 2.4 we obtain the rationality of the Euler factors of global integrals (under an additional convergence hypothesis for integrals involved.)

Postponing the proof of the Rationality Lemma to 3.1, we use it to prove that certain integrals on totally disconnected groups are rational functions in naturally-occurring parameters. In practice, these parameters will be the familiar q^{-s}, or several such parameters $q^{-s}, q^{-s_1}, \ldots, q^{-s_n}$. Further, we have a corresponding rationality result concerning the 'scalars' involved. We give two versions of the result.

We view the latter results as a *qualitative* computation of various classes of local integrals, under hypotheses which are very frequently met in practice. The obvious follow-up is closer determination of the local integrals, without *ad hoc* computation. The latter goal is pursued in the succeeding sections.

We recall the context for discussion of parametrized families, from the last section. Fix a field k of characteristic zero. Let G be a (locally-compact Hausdorff) totally disconnected group with a *countable dense subset.* Let $\mathcal{O}$ be a commutative k-algebra with unit, without zero divisors, with prime spectrum X. Let $\tilde{k}$ be a field extension of k, so that $\mathcal{O} \otimes_k \tilde{k}$ is still an integral domain. Let $X \otimes_k \tilde{k}$ be the prime spectrum of $\mathcal{O} \otimes_k \tilde{k}$.

Let π be an $\mathcal{O}$-parametrized family of smooth representations of G on an $\mathcal{O}$-module V, so that the underlying $\mathcal{O}$-module V is *countably-generated.* Let $\tilde{\pi} = \pi \otimes_k \tilde{k}$ be the parametrized family obtained by *extending scalars* from k to $\tilde{k}$.

Recall that a *meager* set is one contained in a countable union of hypersurfaces, while a *non-meager* set is simply not *meager.* And recall the notation that k_x is the trivial one-dimensional k_x-linear representation.

RATIONALITY LEMMA:

- *Suppose that for x on a non-meager subset Y of $X \otimes_k \tilde{k}$ (on which $\pi \otimes_k \tilde{k}$ is holomorphic)*
$$\dim_{\tilde{k}_x} \mathrm{Hom}_{G\times\tilde{k}_x}(\tilde{\pi}_x, \tilde{k}_x) \leq 1$$
- *Suppose that there is $\mu_o \in V \otimes_{\mathcal{O}} \mathcal{M}$ so that for each x in Y there is an intertwining operator $\psi_{(x)} : \tilde{\pi}_x \longrightarrow \tilde{k}_x$ with the normalization*
$$\psi_{(x)}(\mu_o) = 1 \in \tilde{k}_x.$$

Then

- *There is an $\mathcal{O}\otimes_k\tilde{k}$-parametrized meromorphic family Φ of intertwining operators from $\tilde{\pi}$ to the trivial representation $\mathcal{O}$, rational over k, so that for x in a non-meager set $Y' \subset Y$ (on which Φ is holomorphic) the pointwise intertwining operators Φ_x arising from Φ are equal to the respective intertwining operators $\psi_{(x)}$:*
$$\Phi_x = \psi_{(x)} \quad \text{for } x \in Y'.$$
- Off *a meager subset of $X \otimes_k \tilde{k}$,*
$$\dim_{\tilde{k}_x} \mathrm{Hom}_{G\times\tilde{k}_x}(\tilde{\pi}_x, \tilde{k}_x) = 1.$$
- *For any $v \in V \otimes_{\mathcal{O}} \mathcal{M}$,*
$$\Phi(v) \in \mathcal{M}.$$

(Proof given below in section 3.1, after examples of applications).

REMARKS: The hypothesis
$$\dim_{\tilde{k}_x} \mathrm{Hom}_{G\times\tilde{k}_x}(\pi_x, \tilde{k}_x) \leq 1$$
for x in a non-meager set is the **generic multiplicity one** condition, and the element $\mu_o \in V \otimes_{\mathcal{O}} \mathcal{M}$ is the **good test vector**. And note that the very last conclusion contains two kinds of **rationality assertions**: first there is the assertion that the 'function' $v \longrightarrow \Phi(v)$ lies in the *field of rational functions* $\mathcal{M}\otimes_k\tilde{k}$, but there is also the *rationality of scalars* assertion, that $\Phi(v)$ actually lies in the smaller field $\mathcal{M}$ itself, without the need to extend scalars from k to $\tilde{k}$.

And, last, it is important to realize that for k *countable* (e.g., $\mathbf{Q}$), it may well be that the whole of X is *meager.* Thus, it may well be necessary to consider the larger $X \otimes_k \tilde{k}$, by extending scalars to an *uncountable* field $\tilde{k}$ (e.g., $\mathbf{C}$ or $\mathbf{Q}_p$) in order to apply this result.

Now we will consider integrals (and somewhat more general functionals) whose integrands are $\tilde{k}_x$-valued for varying maximal ideals x in $\mathcal{O} \otimes_k \tilde{k}$, so we want to assume that $\tilde{k}$ and finite extensions $\tilde{k}_x$ of it are *complete.* Thus, taking $\tilde{k}$ to be $\mathbf{R}$

or $\mathbf{C}$ or finite extensions of $\mathbf{Q}_p$ suffices. (By the Nullstellensatz, the residue fields $\tilde{k}_x$ of $\mathcal{O} \otimes_k \tilde{k}$ for maximal ideals x are finite extensions of $\tilde{k}$).

Let A, B be closed subgroups of a totally disconnected, locally compact, Hausdorff, countably-based topological group G. For our applications, we can suppose that G is *unimodular*. We assume that there is some $\xi \in G$ so that the set

$$A\,\xi\, B \subset G$$

is an **open** subset of G. That is, considering the group action of $A \times B$ upon G by

$$(a \times b)(g) = a \cdot g \cdot b^{-1}$$

there is an **open orbit**. As catch-phrase, this is the **open-orbit condition**.

Let σ, τ be $\mathcal{O}$-parametrized families of *one-dimensional* smooth representations of A, B, respectively. Thus, we may suppose that these are simply group homomorphisms

$$\sigma : A \to \mathcal{O}^\times \qquad \tau : B \to \mathcal{O}^\times .$$

For x a *maximal* ideal in the prime spectrum of $\mathcal{O} \otimes_k \tilde{k}$ consider the ($\tilde{k}_x$-valued) **local integral**

$$I(f, \varphi)(x) = \int_{(A\cap\xi^{-1}B\xi)\backslash G} q_x f(g)\, q_x \varphi(\xi^g)\, dg$$

where f is in c-$\mathrm{Ind}_A^G \sigma$, φ is in c-$\mathrm{Ind}_B^G \tau$, q_x denotes the pointwise evaluation.

REMARKS: Without loss of generality ξ can be taken to be just 1, by conjugating B. However, if that simplification is made at the outset it becomes too easy to overlook the fact that the hypothesis is as general as it really is.

According to the notational convention of 1.2, an "integral" on a quotient $H\backslash G$ of G by a closed subgroup H is only a literal integral when for $h \in H$

$$\delta_G(h) = \delta_H(h).$$

In that case, since the groups in question are totally disconnected, we can suppose that there is a right G-invariant $\mathbf{Q}$-valued measure (denoted by dg) on the quotient $H\backslash G$. In general, we *can* always assume that there is a $\mathbf{Q}$-valued right G-invariant measure on G, and on H. We have the identity

$$\int_{H\backslash G} \int_H f(hg)\, \frac{\delta_G(h)}{\delta_H(h)}\, dh\, dg$$

and the surjectivity of the averaging map

$$\mathrm{C}_c^\infty(G) \to \mathrm{C}_c^\infty(H\backslash G, \frac{\delta_H}{\delta_G})$$

given by

$$f \to \left(g \to \int_H f(hg)\, \frac{\delta_G(h)}{\delta_H(h)}\, dh \right).$$

(See 3.2). Because of the total-disconnected-ness of H, we also know that the modular functions δ_G, δ_H are $\mathbf{Q}^\times$-valued. Therefore, we may assume quite generally that **the "integral"**

$$\int_{H\backslash G} F(g)\, dg$$

is Q-valued, in the sense that for $\mathbf{Q}$-valued $f \in C_c^\infty(G)$ and for

$$F(g) = \int_H f(hg)\,\frac{\delta_G(h)}{\delta_H(h)}\,dh$$

we have

$$\int_{H\backslash G} F(g)\,dg \in \mathbf{Q}.$$

(And, further, this function F is $\mathbf{Q}$-valued.)

REMARK: It is important to realize that even though we can assume without loss of generality that the "measure" is $\mathbf{Q}$-valued, there is no assurance that the integrands relevant to global integrals are compactly supported modulo $A \cap B$ for general f and φ. Thus, such an integral is a genuinely analytic object rather than a purely algebraic one, although there seems no obstacle to its being *p-adic* rather than real or complex.

We **assume** that for maximal ideals x in some non-meager subset of $X \otimes_k \tilde{k}$ the "integral" *converges* absolutely for *all* $f_x \in \text{c-Ind}_A^G q_x(\sigma \otimes_k \tilde{k})$ and for *all* $\varphi_x \in \text{c-Ind}_B^G\, q_x(\tau \otimes_k \tilde{k})$.

We also **assume** the **generic multiplicity-one** condition: for x in a non-meager subset of $X \otimes_k \tilde{k}$ we have the *multiplicity-one* condition

$$\dim_{\tilde{k}_x}\ \mathrm{Hom}_{G\times\tilde{k}_x}(\text{c-Ind}_A^G q_x(\sigma \otimes_k \tilde{k}) \otimes \text{c-Ind}_B^G q_x(\tau \otimes_k \tilde{k}), \tilde{k}_x) \leq 1.$$

Then we have the following corollary of the Rationality Lemma, which includes rationality assertions concerning both rationality as a function of the parameter $x \in X \otimes_k \tilde{k}$, and also rationality of scalars:

THEOREM: *With the assumptions that there is an open orbit $H\,\xi\,G$, that the "integral"*

$$I(f,\varphi)(x) = \int_{(A\cap\xi^{-1}B\xi)\backslash G} q_x f(g)\, q_x\phi(\xi^g)\,dg$$

converges on a non-meager set in $X \otimes_k \tilde{k}$, and that

$$\dim_{\tilde{k}_x}\ \mathrm{Hom}_{G\times\tilde{k}_x}(\text{c-Ind}_A^G q_x(\sigma \otimes_k \tilde{k}) \otimes \text{c-Ind}_B^G q_x(\tau \otimes_k \tilde{k}), \tilde{k}_x) \leq 1$$

for x in a non-meager set in $X \otimes_k \tilde{k}$, we conclude that the map

$$f \otimes \varphi \to \int_{(A\cap\xi^{-1}B\xi)\backslash G} q_x f(g)\, q_x\phi(g)\,dg$$

extends to an $\mathcal{O} \otimes_k \tilde{k}$-parametrized meromorphic family of intertwining operators

$$\Phi : \text{c-Ind}_A^G(\sigma \otimes_k \tilde{k}) \otimes \text{c-Ind}_B^G(\tau \otimes_k \tilde{k}) \to \mathcal{O} \otimes_k \tilde{k}$$

which is rational over k. *In particular, for each fixed $f \in \text{c-Ind}_A^G\sigma$ and $\varphi \in \text{c-Ind}_B^G\tau$, the function $\Phi(f \otimes \varphi)$ is in the field $\mathcal{M}$.*

REMARKS: Thus, in a more mundane sense, as a 'function' of the parameter x, the integral $I(f,\varphi)(x)$ is a *rational function.* Further, again in a more mundane sense, the 'coefficients' of that rational 'function' lie in k, not merely in the extension $\tilde{k}$, for f, φ in the k-rational points $\text{c-Ind}_A^G\sigma$ and $\text{c-Ind}_B^G\tau$ of the representations.

PROOF. The hypotheses above include two of the three hypotheses needed to invoke the Rationality Lemma. The crux of the present argument is proof that there is a *good test vector*, thus verifying the third of the necessary hypotheses.

For most of the proof, we may as well suppose that $k = \tilde{k}$ to simplify notation. And, throughout, there is no loss of generality in assuming that $\xi = 1$. This simplifies the notation considerably.

For maximal x so that the integral converges, the map

$$q_x f \otimes q_x \varphi \to I(f, \varphi)(x)$$

is an intertwining operator

$$\text{c-Ind}_A^G q_x \sigma \otimes \text{c-Ind}_B^G q_x \tau \to k_x.$$

Thus, we have a pointwise intertwining on a non-meager subset of the parameter space, extending scalars from k to $\tilde{k}$ if necessary, which is *one* hypothesis of the Rationality Lemma.

The third hypothesis of the Rationality Lemma, which we have *not* simply assumed away, is the presence of a *good test vector*, which here must be of the form

$$f_o \otimes \varphi_o \in \text{c-Ind}_A^G \sigma \otimes \text{c-Ind}_B^G \tau$$

which should have the property

$$I(f_o, \varphi_o)(x) = 1$$

for all x in a non-meager subset of $X \otimes_k \tilde{k}$. Before addressing this, we need:

LEMMA: *Let A, B be closed subgroups of a* separable *locally compact Hausdorff totally disconnected topological group G. Suppose that $A \cdot B$ is* open *in G. Then, for arbitrarily small compact open subgroups A_o, B_o of A, B, respectively, the product $A_o \cdot B_o$ is open in G.*

PROOF OF LEMMA: We have

$$A \cdot B = \bigcup_{\alpha, \beta} \alpha \cdot A_o B_o \cdot \beta$$

where $\alpha \in A/A_o$ and $\beta \in B_o \backslash B$. The latter collections are *countable.* Thus, by the Baire category theorem, one (hence, *all*) of the (mutually homeomorphic) sets $\alpha \cdot A_o B_o \cdot \beta$ has non-empty interior. Let $a_o \cdot b_o \in U$ with U an open set inside $A_o \cdot B_o$, with $a_o \in A_o$ and $b_o \in B_o$. Then for any $a_1 \in A$ and $b_1 \in B$, as usual we have

$$a_1 b_1 = (a_1 a_o^{-1}) a_o b_o (b_o^{-1} b_1) \in (a_1 a_o^{-1}) U (b_o^{-1} b_1) \subset A_o U B_o$$
$$\subset A_o A_o B_o B_o = A_o B_o.$$

That is, every point is in the interior, so the set is open. □

Returning to construction of the good test vector and proof of the theorem: Take a small-enough compact-open subgroup K_o of G so that σ is trivial on $A \cap K_o$ and τ is trivial on $B \cap K_o$. Shrink K_o if necessary so that

$$K_o \subset A \cdot B$$

where we now make use of the fact that $A \cdot B$ is *open* in G. Put $A_o = A \cap K_o$ and $B_o = B \cap K_o$. Finally, take a small-enough compact open subgroup K inside K_o and so that

$$K \subset A_o B_o \cap B_o A_o.$$

Consider f_o, φ_o in $\text{c-Ind}_A^G \sigma$, $\text{c-Ind}_B^G \tau$, respectively, defined by

$$f_o(a\theta) = \begin{cases} \sigma(a) & \text{for } a \in A,\ \theta \in K \\ 0 & \text{off } A \cdot K \end{cases}$$

$$\varphi_o(b\theta) = \begin{cases} \tau(b) & \text{for } b \in B,\ \theta \in K \\ 0 & \text{off } B \cdot K. \end{cases}$$

Let S be the support of $f_o\varphi_o$.

We estimate the image of

$$(A \cdot K) \cap (B \cdot K)$$

in $(A \cap B)\backslash G$. We have

$$\begin{aligned} A \cdot K \cap B \cdot K &= (A \cap B \cdot K) \cdot K \subset (A \cap B \cdot B_o A_o) \cdot K \\ &= (A \cap B \cdot A_o) \cdot K = (A \cap B) \cdot A_o K = (A \cap B) \cdot K_o \end{aligned}$$

since $K \subset A_o \subset A$, $B_o \subset B$, and A_o and B_o lie inside K_o. Thus, the image is contained in the image of K_o, which is compact.

Therefore, adjusting $f_o \otimes \varphi_o$ by a positive rational constant if necessary (depending upon choice of $\mathbf{Q}$-valued right Haar measure), by elementary integration theory (e.g., surjectivity of the *averaging maps*) we have

$$\int_{(A\cap B)\backslash G} q_x f(g)\, q_x\phi(g)\, dg = \int_{K_o} q_x f(\theta)\, q_x\phi(\theta)\, d\theta = \text{meas}(K_o)$$

since K_o was chosen to be small enough so that both σ and τ are trivial on it. This holds for all x (not only for x for which the original integral was guaranteed to converge). That is, adjusting $f_o \otimes \varphi_o$ by a non-zero rational number if necessary, we have

$$\int_{(A\cap B)\backslash G} q_x f(g)\, q_x\phi(g)\, dg = 1 = \int_{K_o} q_x f(\theta)\, q_x\phi(\theta)\, d\theta.$$

Thus, we are guaranteed that *good test vectors exist.*

Thus, we can invoke the Rationality Lemma, concluding that there is an extension of the integral to a meromorphic $\mathcal{O}$-parametrized intertwining Φ.

At this point, we return to the distinction between k and $\tilde{k}$. First, σ and τ were actually $\mathcal{O}$-parametrized, rather than merely $\mathcal{O} \otimes_k \tilde{k}$-parametrized. Also, the construction of the good test vector really yields a test vector in

$$\text{c-Ind}_A^G \sigma \otimes \text{c-Ind}_B^G \tau$$

not merely in the corresponding families of representations with scalars extended from k to $\tilde{k}$. Thus, by the Rationality Lemma the meromorphic family Φ of intertwinings just proven to exist is actually k-rational, rather than merely $\tilde{k}$-rational.

In particular, this entails that for every $f \in \text{c-Ind}_A^G\sigma$ and $\varphi \in \text{c-Ind}_B^G\tau$, the value $\Phi(f \otimes \varphi)$ is in the field $\mathcal{M}$, rather than merely in the larger field $\mathcal{M} \otimes_k \tilde{k}$. □

Now we give a different-looking rationality result for local integrals with parameters. Still, the parameters are potentially the familiar q^{-s}, or several such parameters $q^{-s}, q^{-s_1}, \ldots, q^{-s_n}$, and there is a corresponding rationality result concerning the 'scalars' involved. Again, the obvious follow-up is closer determination of the local integrals without *ad hoc* computation. (This result includes the previous one as a special case).

Let $\tilde{H}$ be a locally compact totally disconnected topological group of which H and G are closed subgroups. Let Y be a closed subset of $\tilde{H}$ stable under left multiplication by H and right multiplication by G. Let Z be a locally compact totally disconnected space on which H acts trivially and upon which G acts continuously on the right. Take

$$\Omega = Y \times Z$$

Let σ be an $\mathcal{O}$-parametrized family of *finite-dimensional* smooth representations of H. Recall the notation that, for a prime ideal x in the parameter space, σ_x is the pointwise representation at x obtained from the family σ.

Let $C_c^\infty(H\backslash\Omega, \sigma)$ be the collection of σ-valued locally constant functions f on Ω the image of whose support is *compact* in the quotient space $H\backslash\Omega$, and which have the left equivariance

$$f(hx) = \sigma(h)\, f(x)$$

for all $h \in H$ and $x \in \Omega$. These functions are **(generalized) test functions** in the sense of Bruhat. The action of G on the right on Ω makes this space of functions a smooth representation space for G.

The **open orbit** hypothesis we impose is the assumption that there is some $\xi \in \Omega$ so that

$$H\,\xi\,G \subset \Omega$$

is an *open subset* of Ω. We want a $\mathbf{Q}$-valued right-G-invariant "integral"

$$f \to \int_{H\backslash\Omega} f(\omega)\, d\omega$$

supported on $H\backslash H\xi G$. By properties of test functions on totally disconnected spaces (as in 3.2), we have a natural G-isomorphism

$$C_c^\infty(H\backslash H\xi G, \sigma) \approx C_c^\infty(\xi^{-1}H\xi \cap G\backslash G, \sigma^\xi)$$

where $\sigma^\xi(x) = \sigma(\xi x \xi^{-1})$. Thus, assuming (as in 1.2) that

$$\sigma^\xi = \frac{\delta_H}{\delta_G}$$

assures the existence of such an integral in extended sense.

For x a *maximal* ideal in the prime spectrum of $\mathcal{O}\otimes_k \tilde{k}$ consider the ($\tilde{k}_x$-valued) **local "integral"** (in the extended sense)

$$I_f(x) = \int_{H\backslash\Omega} f(\omega)\, d\omega$$

where f is in $C_c^\infty(H\backslash\Omega, \sigma)$, and q_x denotes pointwise evaluation as above. Again, this integral is an analytic object rather than an algebraic one, although it may be *p-adic* rather than real or complex. We *assume* that for maximal ideals x in some non-meager subset of $X\otimes_k \tilde{k}$ the integral *converges* absolutely for *all* $f \in C_c^\infty(H\backslash\Omega, \sigma_x)$.

Let us *assume* also the **generic multiplicity-one** condition: that for x in a non-meager subset of $X \otimes_k \tilde{k}$ we have the *multiplicity-one* condition

$$\dim_{\tilde{k}_x} \mathrm{Hom}_{G\times\tilde{k}_x}(C_c^\infty(H\backslash\Omega, \sigma_x), \tilde{k}_x) \leq 1.$$

Then we have the following corollary of the Rationality Lemma, including two rationality assertions: rationality as a function of the parameter $x \in X\otimes_k \tilde{k}$, and rationality of scalars.

THEOREM: *With the open-orbit hypothesis, with convergence on a non-meager set in $X \otimes_k \tilde{k}$, and with*

$$\dim_{\tilde{k}_x} \mathrm{Hom}_{G\times\tilde{k}_x}(C_c^\infty(H\backslash\Omega, \sigma_x), \tilde{k}_x) \leq 1$$

for x in a non-meager set in $X \otimes_k \tilde{k}$, the map

$$f \to \int_{H\backslash\Omega} f(\omega)\, d\omega$$

extends to an $\mathcal{O} \otimes_k \tilde{k}$-parametrized meromorphic family of intertwining operators

$$\Phi : C_c^\infty(H\backslash\Omega, \sigma) \otimes_k \tilde{k} \to \mathcal{M} \otimes_k \tilde{k}$$

which is rational over k. *In particular, for each fixed $f \in C_c^\infty(H\backslash\Omega, \sigma)$ the "function" $\Phi(f)$ is in the field $\mathcal{M}$.*

REMARKS: Thus, as a 'function' of the parameter x, the integral $x \to I_f(x)$ is a *rational function*, and the 'coefficients' of that rational 'function' lie in k, not merely in $\tilde{k}$, for f in the k-rational points $C_c^\infty(H\backslash\Omega, \sigma)$ of the representation.

PROOF. We can assume without loss of generality for the proof that $\tilde{k} = k$.

The explicit hypotheses give us two of the three hypotheses for the Rationality Lemma. All that remains is to find a *good test vector* $f_o \in C_c^\infty(H\backslash\Omega, \sigma)$, meaning that

$$\int_{H\backslash\Omega} f_o(\omega)\, d\omega = 1$$

(independent of x in the parameter space).

We will choose f_o in the subspace

$$C_c^\infty(H\backslash H\xi G, \sigma)$$

of the whole test space $C_c^\infty(H\backslash\Omega, \sigma)$. Let

$$\Theta_\xi = \{g \in G : \text{ there is } h_g \in H \text{ so that } h_g^{-1}\xi g = \xi\}$$

For $g \in \Theta_\xi$, define

$$\sigma^\xi(g) = \sigma(h_g).$$

From the discussion of orbit filtrations of test functions (section 3.2), we have

$$C_c^\infty(H\backslash H\xi G, \sigma) \approx \text{c-Ind}_{\Theta_\xi}^G \sigma^\xi.$$

Let K be a compact-open subgroup of G small-enough so that σ^ξ is identically 1 on $K \cap \Theta_\xi$. Then put

$$f_o(\theta\ell) = \begin{cases} \sigma^\xi(\theta) & \text{for } \theta \in \Theta_\xi,\ \ell \in K \\ 0 & \text{off } \Theta \cdot K. \end{cases}$$

The hypothesis on K, necessitating that σ be finite-dimensional, assures that this is well-defined. Then (at first only for x at which the integral is known to converge)

$$\begin{aligned} I(f_o)(x) &= \int_{H\backslash\Omega} f_o(\omega)\, d\omega = \int_{\Theta_\xi\backslash\Theta_\xi K} f_o(\omega)\, d\omega \\ &= \int_K f_o(\ell)\, d\ell = \text{meas}(K) \end{aligned}$$

where $d\ell$ is a suitably-normalized Haar measure on K. Thus, correcting f_o by dividing by meas(K), we have the desired test function. □

2.3. Analytic continuation of intertwinings. Now we return to consideration of the global integrals. So far, we have factored the *inner integral* as

$$\int_{Z'_{\mathcal{A}}\,\Theta_k\backslash\Theta_{\mathcal{A}}} f(\theta\, g)\, d\theta$$

$$= (\text{ bad-primes factor and period }) \times \prod_{v \text{ good}} \Phi_v(e_v)(g_v)$$

where e_v is the normalized spherical vector in the unramified principal series

$$\text{c-Ind}_{P_v}^{G_v}\chi_v,$$

and Φ_v is a non-zero intertwining operator from the unramified principal series to $\text{Ind}_{\Theta_v}^{G_v}\delta_{\Theta_v}$.

The goal of this section is to express the intertwining operator Φ_v as an *integral.* Ironically, the natural integral expressing the intertwining operator seldom converges for χ_v in the range of characters appearing in the unramified principal series mapping surjectively to the local representations π_v generated by a (square-integrable) cuspform.

To deal with this we enlarge the class of χ_v's to include values for which the natural integral *does* converge, and then prove that the intertwining operator has an *analytic continuation.* In fact, in the cases at hand we present the family of characters χ_v as a parametrized family (in the sense of (2.1)), and invoke the Rationality Lemma and its corollaries (2.2) to prove that the natural integral is identifiable as a meromorphic family of intertwining operators. Thus, the analytic continuation issue will be resolved trivially, by proving that the object in question is in fact *rational.*

As noted in (1.3), in general the hypothesis that the period does not vanish is not trivially fulfilled, but may require *at least* some local condition on the characters χ_v. We will make a parametrized family of characters meeting at least this local condition.

REMARK: Unfortunately, still due to the fact that we have limited information on the character χ_v occurring for a cuspform, it seems difficult to give a universal answer to the question of whether or not the intertwining operator Φ_v lies in the family of intertwining operators we construct. In any particular case somewhat more can be said, of course, and in the most fortuitous circumstances (in which there is at most a single intertwining operator for each parameter value) it certainly follows that Φ_v is given by an integral as constructed below.

We inherit several assumptions from 1.1, 1.2, 1.3, and 1.4: we are assuming that $\Theta_v\backslash G_v/P_v$ is finite, that $\xi\Theta_v\xi^{-1}$ is contained in the kernel of $\tilde{\chi}_v$ (so that ε **escapes** from the inner integral as in 1.2), and that (in the notation of 1.4)

$$\sum_{\alpha\in P_v\backslash G_v/\Theta_v} \dim \text{Hom}_{H_\alpha}(\chi^\alpha,\, \delta_{\Theta_v}\delta_{H_\alpha}) = 1.$$

The latter implies that

$$\dim \text{Hom}_{G_v}\left(\text{c-Ind}_{P_v}^{G_v}\chi_v\delta_{P_v}^{1/2}, \text{Ind}_{\Theta_v}^{G_v}\delta_{\Theta_v}\right) \leq 1.$$

To avoid trivial vanishing of the inner integral (1.3), we must assume at least that

$$\dim \text{Hom}_{G_v}\left(\text{c-Ind}_{P_v}^{G_v}\chi_v\delta_{P_v}^{1/2}, \text{Ind}_{\Theta_v}^{G_v}\delta_{\Theta_v}\right) = 1.$$

Let N_v be (the v-adic points of) the unipotent radical of P, and M_v (the v-adic points of) a Levi component of P. Let M_o be the maximal compact subgroup of P_v (unique because P_v is minimal). Let $\varpi_1, \ldots, \varpi_n$ be generators for the abelian group M_v/M_o. Let $z_1, \ldots, z_n$ be transcendental over the ground field k, and put

$$\mathcal{O}' = k[z_1, z_1^{-1}, \ldots, z_n, z_n^{-1}].$$

Define

$$\chi' : M_v \to M_v/M_o \to (\mathcal{O}')^\times$$

by

$$\chi'(\varpi_i) = z_i$$

extending by multiplicativity, and where the first map is the quotient map. Then extend χ' to P_v by taking χ' identically 1 on N_v. This χ' is the (universal) $\mathcal{O}'$-parametrized family of unramified characters on P_v. To avoid trivial vanishing, in general we must use a smaller family, as follows.

As in 3.2, partially order the orbits $P_v\alpha\Theta_v$ (or representatives α) by

$$P_v\alpha\Theta_v \leq P_v\beta\Theta_v$$

if $P_v\beta\Theta_v$ lies in the closure of $P_v\alpha\Theta_v$. We have the orbit filtration on c-$\mathrm{Ind}_{P_v}^{G_v}\chi'$ described by

$$\left(\text{c-Ind}_{P_v}^{G_v}\chi'\right)^{\leq\beta} = \left\{ f \in \text{c-Ind}_{P_v}^{G_v}\chi' : \text{ support } f \text{ is in } \bigcup_{\alpha\leq\beta} P_v\alpha\Theta_v \right\}.$$

Also put

$$\left(\text{c-Ind}_{P_v}^{G_v}\chi'\right)^{<\beta} = \left\{ f \in \text{c-Ind}_{P_v}^{G_v}\chi' : \text{ support } f \text{ is in } \bigcup_{\alpha<\beta} P_v\alpha\Theta_v \right\}.$$

From 3.2, the βth graded piece of this filtration is naturally isomorphic as a Θ_v-representation to

$$\text{c-Ind}_{H_\beta}^{\Theta_v}(\chi')^\beta$$

where

$$H_\beta = \beta^{-1}P_v\beta \cap \Theta_v$$

and for $\theta \in H_\beta$

$$(\chi')^\beta(\theta) = \chi'(\beta\,\theta\,\beta^{-1}).$$

(These are none other than the sort of graded pieces considered in using such a filtration to estimate

$$\dim_{k_x} \mathrm{Hom}_{G_v}(\text{c-Ind}_{P_v}^{G_v}\chi_v,\ \mathrm{Ind}_{\Theta_v}^{G_v}\delta_{\Theta_v})$$

as we did in 1.4.)

As in 1.4, it is an elementary computation that

$$\dim_{k_x} \mathrm{Hom}_{\Theta_v}(\text{c-Ind}_{H_\beta}^{\Theta_v}(\chi')^\beta, \delta_{\Theta_v}) \leq 1$$

and that we get equality if and only if

$$(\chi')^\beta|_{H_\beta} = \delta_{H_\beta}.$$

For representative β, for each $h \in H_\beta$ the condition

$$(\chi')^\beta(h) = \delta_{H_\beta}(h)$$

is something of the form

$$z_1^{a_1(h)} \dots z_n^{a_n(h)} = c(h)$$

where the $a_i(h)$ are integers and $c(h) \in \mathbf{Q}^\times$ (since the group is totally disconnected). Let I_β be the radical of the ideal in $\mathcal{O}'$ generated by all such relations for $h \in H_\beta$.

Let $P_v\beta\Theta_v$ be *minimal* (in the partial order above) so that $I_\beta \neq \mathcal{O}'$. (That the condition *can* ever be met is a consequence of the existence of a non-trivial intertwining operator for *some* pointwise representation obtained from χ'). Then let

$$\mathcal{O} = \mathcal{O}'/I_\beta$$

for such β. Thus, $\mathcal{O}$ parametrizes a family of characters so that the βth graded piece of the orbit filtration admits an intertwining operator to δ_{H_β}.

REMARK: In practice, it seems that usually if $P_v\beta\Theta_v < P_v\alpha\Theta_v$ and $I_\beta \neq \mathcal{O}'$ then

$$I_\beta \subset\subset I_\alpha \subset \mathcal{O}'$$

with the first containment *strict*. Further, it often happens that there is a *unique* minimal β whose associated graded piece admits an intertwining operator. In such circumstances there is of course no ambiguity in choice of $\mathcal{O}$.

Then the natural integral associated to this choice of β and $\mathcal{O}$ is

$$\Phi(f)(g) = \int_{\beta^{-1}P_v\beta\cap\Theta_v} f(\beta\,\theta\,g)\,\delta_{\Theta_v}(\theta)\,d\theta.$$

It is immediate that, *if* this converges, then it is an intertwining operator from c-$\mathrm{Ind}_{P_v}^{G_v} q_x\chi'$ to $\mathrm{Ind}_{\Theta_v}^{G_v}\delta_{\Theta_v}$. Yet, while this integral is actually a finite sum for f in $(\text{c-}\mathrm{Ind}_{P_v}^{G_v}\chi')^{\leq\beta}$, convergence for general f in c-$\mathrm{Ind}_{P_v}^{G_v}\chi'$ is an issue. Let $X = \operatorname{spec}\mathcal{O}$, and let q_x be the evaluation map(s) at x.

THEOREM: *Assume that for a non-meager collection of $x \in X \otimes_k \tilde{k}$ and for all f in* c-$\mathrm{Ind}_{P_v}^{G_v}\chi'$ *the integral for $\Phi(q_x f)$ is absolutely convergent, and that*

$$\dim_{\tilde{k}_x} \mathrm{Hom}_{G_v}(\text{c-}\mathrm{Ind}_{P_v}^{G_v} q_x\chi', \mathrm{Ind}_{\Theta_v}^{G_v}\delta_{\Theta_v}) \leq 1.$$

Then this integral extends to give a meromorphic family Φ of intertwining operators from c-$\mathrm{Ind}_{P_v}^{G_v}\chi'$ *to* $\mathrm{Ind}_{\Theta_v}^{G_v}\delta_{\Theta_v}$, *in the sense of 2.1.*

PROOF. Fix $g \in G_v$ and denote the map given by the integral whenever it converges absolutely by φ_g. This φ_g is an intertwining operator

$$\mathrm{Res}_{\Theta_v}^{G_v}\,\text{c-}\mathrm{Ind}_{P_v}^{G_v} q_x\chi' \to \delta_{\Theta_v}$$

(observing that the representation δ_{Θ_v} is a *constant* family.) The assumption on the dimension of the space of intertwining operators immediately translates by Frobenius Reciprocity to

$$\dim_{k_x} \mathrm{Hom}_{\Theta_v}(\mathrm{Res}_{\Theta_v}^{G_v}\,\text{c-}\mathrm{Ind}_{P_v}^{G_v} q_x\chi', \delta_{\Theta_v}) \leq 1.$$

This and the assumptions of convergence allow us to invoke the theorem(s) of 2.2, concluding that φ_g is a *meromorphic family* of intertwining operators from $\mathrm{Res}_{\Theta_v}^{G_v}$ c-$\mathrm{Ind}_{P_v}^{G_v} q_x\chi'$ to δ_{Θ_v}.

Since

$$\Phi(f)(g) = \varphi_g(f)$$

by the definition of meromorphy of a family of intertwining operators (from 2.1) we have the conclusion of the theorem. □

REMARK: The meromorphy asserted here might be construed as a *weak* meromorphy, since there is not necessarily any uniform estimate on poles as $g \in G_v$ varies. This flaw will be corrected in 2.5, when *strong* meromorphy is discussed.

2.4. Rationality of the Euler factors. By this point, we have managed to rewrite the global integral as a product of the *period* and an *Euler product* with v^{th} factor

$$\int_{\tilde{Z}_v \Theta_v \backslash G_v} \varepsilon_v(\xi g)\, \Phi_v(e_v)(g)\, dg$$

(using the notation of 1.2). In the last section we found an integral expression giving (when analytically continued) a meromorphic family of intertwining operators which we *assume* contains Φ_v. Namely, we put

$$\Phi(f) = \int_{\beta^{-1} P_v \beta \cap \Theta_v \backslash \Theta_v} f(\beta\, \theta\, g)\, \delta_{\Theta_v}(\theta)\, d\theta.$$

We inherit several assumptions: we are assuming that $\Theta_v \backslash G_v / P_v$ is finite, and that $\xi \Theta_v \xi^{-1}$ is contained in the kernel of $\tilde{\chi}_v$ (so that ε **escapes** from the inner integral). To use the integral representation of the last section, we must assume that the character χ_v lies among the pointwise representations from the family χ' described in the last section (as well as that the intertwining operator Φ_v lies in the family Φ).

Now we replace Φ_v in the outer integral by its integral expression from the last expression: the vth Euler factor is the outer integral, which is of the form

$$\int_{\tilde{Z}_v \Theta_v \backslash G_v} \varepsilon_v(\xi g) \int_{\beta^{-1} P_v \beta \cap \Theta_v \backslash \Theta_v} f(\beta\, \theta\, g)\, \delta_{\Theta_v}(\theta)\, d\theta\, dg$$

where instead of simply the spherical vector e_v we now use an arbitrary f in some pointwise representation arising from the family c-$\text{Ind}_{P_v}^{G_v} \chi'$ constructed in the last section. By integration theory (see 1.2), we can immediately rewrite this as a single integral

$$\int_{\tilde{Z}_v H_\beta \backslash G_v} \varepsilon_v(\xi g)\, f(\beta\, \theta\, g)\, dg$$

where

$$H_\beta = \beta^{-1} P_v \beta \cap \Theta_v.$$

Now we assume *further* that

$$\tilde{P}_v \backslash \tilde{G}_v / P_v \qquad \text{is finite}$$

for almost all primes v. We may as well note the following:

PROPOSITION: *If $\tilde{P}_v \backslash \tilde{G}_v / P_v$ is finite then $\Theta_v \backslash G_v / P_v$ is finite.*

PROOF. $\Theta = \xi^{-1} \tilde{P} \xi \cap G$. □

In addition to postulating that the local character χ_v arising from the cuspform is also a pointwise representation from the $\mathcal{O}$-parametrized family χ' constructed in the last section, we must also present the character $\tilde{\chi}_v$ (from the Eisenstein series) in the same fashion. Just as in general we must restrict the parameter space for χ' to avoid trivial vanishing of the inner integral (see 1.3), we must restrict the family in which $\tilde{\chi}_v$ lies in order to assure that the integrand is $\tilde{Z}_v$-invariant.

The corresponding construction is even simpler than that done in the last section, so without repeating it we assume now that we have the commutative Noetherian ring $\tilde{\mathcal{O}} = k[z, z^{-1}]$ whose spectrum $\tilde{X}$ parametrizes a family $\tilde{\chi}'$ of characters on $\tilde{P}_v$, so that the integrand above is $\tilde{Z}_v$-invariant.

Recall also from 2.3 that the ring $\mathcal{O}$ parametrizing the family of characters χ' is a quotient of $k[z_1, z_1^{-1}, \ldots, z_n, z_n^{-1}]$. We abusively use the symbols z_i to refer to their images in $\mathcal{O}$.

THEOREM: *Let v be a prime so that π_v is spherical. Suppose that the double-coset space*

$$\tilde{P}_v \backslash \tilde{G}_v / P_v$$

is finite. *Suppose that for x in a non-meager subset of* $\operatorname{spec} \tilde{\mathcal{O}} \otimes \mathcal{O} \otimes_k \tilde{k}$ *for all* $f \in \text{c-Ind}_{P_v}^{G_v} q_x \chi'$ *the integral*

$$\int_{\tilde{Z}_v H_\beta \backslash G_v} \varepsilon_v(\xi g)\, f(\beta\,\theta\, g)\, dg$$

is absolutely convergent, and also

$$\dim_{k_x} \operatorname{Hom}_{G_v}(\operatorname{Res}_{G_v}^{\tilde{G}_v} \left(\text{c-Ind}_{\tilde{P}_v}^{\tilde{G}_v} \tilde{\chi}_v\right) \otimes \text{c-Ind}_{P_v}^{G_v} \chi_v, \tilde{k}_x) \leq 1.$$

Then this integral extends to give a meromorphic family of intertwining operators

$$\operatorname{Res}_{G_v}^{\tilde{G}_v} \text{c-Ind}_{\tilde{P}_v}^{\tilde{G}_v} \tilde{\chi}' \otimes \text{c-Ind}_{P_v}^{G_v} \chi' \to (\tilde{\mathcal{O}} \otimes \mathcal{O} \otimes_k \tilde{k})^\times.$$

COROLLARY *Under the hypotheses of the previous theorem, the vth Euler factor is a* rational function *of $z, z_1, z_2, \ldots, z_n$.*

REMARK: We think of z as q^{-s} where this s is the "live" variable occurring in the Eisenstein series, while $z_i = q^{-s_i}$ map homomorphically to give the Satake parameters of the cuspform.

PROOFS: The hypotheses are those necessary to apply the rationality theorems of 2.2, so we obtain this theorem at once. By integration theory, the present convergence hypothesis implies convergence of the integral of 2.3 used to rewrite the inner integral. The Corollary is obtained by replacing the general f in the parametrized family of unramified principal series by the spherical vector e' in that parametrized family, uniformly normalized (as is done pointwise) by the condition

$$e'(1) = 1.$$

□

REMARK: Pursuing the quantitative issue of the computation or estimation of the rational expressions which occur in this result, we will prove later that if the double-coset space $\tilde{P}_v \backslash \tilde{G}_v / P_v$ is finite then the poles of the rational expressions so occurring have leading terms in their Laurent expansions which are interpretable as *anomalous intertwining operators* (in a sense made precise later (section 2.6)). *This will put very strong limitations on the possible denominators of the rational expressions possibly occurring in the local integrals.*

Further, much as was done in 1.4 for the inner integral, we can obtain an estimate on the dimension of the indicated space of intertwining operators using the orbit filtration method from 3.2, as follows. By dualizing and applying Frobenius

Reciprocity, the space of intertwining operators becomes (pointwise, for x in the parameter space)

$$\dim_{k_x} \mathrm{Hom}_{P_v}(\mathrm{Res}^{\tilde{G}_v}_{P_v}\text{c-Ind}^{\tilde{G}_v}_{\tilde{P}_v} q_x\tilde{\chi}', (q_x\chi')^{-1}\delta_{P_v}) = 1.$$

Using the $\tilde{P}_v \times P_v$-orbit filtration on the test functions

$$\mathrm{Res}^{\tilde{G}_v}_{P_v}\left(\text{c-Ind}^{\tilde{G}_v}_{\tilde{P}_v} q_x\tilde{\chi}'\right)$$

we obtain graded pieces

$$\text{c-Ind}^{G_v}_{H_\beta} q_x\tilde{\chi}'^{\beta}$$

where $\beta \in \tilde{P}_v\backslash\tilde{G}_v/P_v$,

$$H_\beta = \beta^{-1}\tilde{P}_v\beta \cap P_v$$

and

$$q_x\tilde{\chi}'^{\beta}(p) = q_x\tilde{\chi}'(\beta p\beta^{-1})$$

for $p \in H_\beta \subset P_v$.

Thus, we obtain an inequality

$$\dim_{k_x} \mathrm{Hom}_{G_v}(\mathrm{Res}^{\tilde{G}_v}_{G_v}(\text{c-Ind}^{\tilde{G}_v}_{\tilde{P}_v} q_x\tilde{\chi}') \otimes \text{c-Ind}^{G_v}_{P_v} q_x\chi',\ k_x)$$

$$\leq \sum_\beta \dim_{k_x} \mathrm{Hom}_{P_v}\left(\text{c-Ind}^{P_v}_{H_\beta} q_x\tilde{\chi}'^{\beta}, (q_x\chi')^{-1}\delta_{P_v}\right).$$

Dualizing and applying Frobenius Reciprocity once more, the latter sum is

$$\sum_\beta \dim_{k_x} \mathrm{Hom}_{H_\beta}(q_x\chi'\delta^{-1}_{P_v}, q_x\tilde{\chi}'^{\beta}\frac{\delta_{H_\beta}}{\delta_{P_v}}) = 1.$$

Simplifying and dualizing again, we get

$$\dim_{k_x} \mathrm{Hom}_{G_v}\left(\mathrm{Res}^{\tilde{G}_v}_{G_v}(\text{c-Ind}^{\tilde{G}_v}_{\tilde{P}_v} q_x\tilde{\chi}'\right) \otimes \text{c-Ind}^{G_v}_{P_v} q_x\chi',\ k_x)$$

$$\leq \sum_\beta \dim_{k_x} \mathrm{Hom}_{H_\beta}(q_x\chi', q_x\tilde{\chi}'^{\beta}\delta_{H_\beta}).$$

The latter sum is easy to compute, since all the representations are one-dimensional.

Further, one may notice that the the orbit-filtration estimate on multiplicity for the inner integral in 1.4 is simply a *subsum* of the present sum

$$\sum_{\beta\in\tilde{P}_v\xi\tilde{G}P_v} \dim_{k_x} \mathrm{Hom}_{H_\beta}(q_x\chi', q_x\tilde{\chi}'^{\beta}\delta_{H_\beta})$$

because $\xi^{-1}\Theta_v\xi$ was assumed to lie in the kernel of the character $q_x\tilde{\chi}'$. Thus, we have

OBSERVATION: If the graded-piece estimate *generically* fails to show multiplicity one for the inner integral, it will also fail for the outer integral.

REMARKS: Since the multiplicity-one condition need not be satisfied for *every* parameter value, but only for a sufficiently large set of such, fewer additional devices are needed to get to the desired *generic* multiplicity-one condition than a *strict* multiplicity-one condition.

If the program were successful up to this point, the discussion would begin another phase, in which the outer local integrals should be (literally) *computed.* By this point, one would already know that the integrals were rational functions of parameters such as q^{-s}, and also of less-obvious parameters such as the Satake parameters q^{-s_i}, which become visible when spherical representations are treated as quotients of unramified principal series.

Yet, rather than engage in *ad hoc* computations, in the sequel we will interpret the 'poles' of the integrals as being special kinds of intertwining operators, referred to below as *anomalous intertwining operators.*

2.5. Orbit criterion for strong meromorphy. To study local integrals further, we must attach meaning to the *zeros* and *poles* in the rational expressions proven above to occur. Not surprisingly, the study of poles is the easier, since (as in other scenarios as well) the *residue* (or more generally *leading term*) in a *Laurent expansion* of an intertwining operator at a *pole* has demonstrably special properties. Thus, to estimate the possible poles, one instead classifies the possible **anomalous intertwining operators**, in a sense described precisely below in 2.6.

One minor complication is that in general, and in many situations of interest (e.g., intertwinings involving unramified principal series) the parameter space $X = \operatorname{Spec} \mathcal{O}$ is of dimension greater than 1, so *poles*, being codimension 1, are of positive dimension.

Generally, for a $G \times \mathcal{O}$-module V and for a height-one prime x in $\mathcal{O}$, say that a (meromorphic $\mathcal{O}$-parametrized family of) intertwining(s)

$$\Phi \in \operatorname{Hom}_{G\times\mathcal{O}}(V, \mathcal{M})$$

is **locally strongly meromorphic at** x if there is some integer power ϖ^N of a local parameter $\varpi = \varpi_x$ at x so that

$$\varpi^N \Phi \in \operatorname{Hom}_{G\times\mathcal{O}}(V, \mathcal{O}_x)$$

where $\mathcal{O}_x$ is the local ring at x. If Φ is locally strongly meromorphic at every height-one prime x in $\mathcal{O}$, then Φ is **locally strongly meromorphic**.

REMARKS: The locally strong meromorphy condition at x precludes there being a sequence of vectors $v_i \in V$ so that the *rational functions* $\Phi(v_i) \in \mathcal{M}$ have deeper and deeper poles along the hypersurface given by x. On the other hand, even though our parameter space $\operatorname{Spec} \mathcal{O}$ is an affine variety, the present definition of (local) strong meromorphy makes no assertion about finiteness of the collection of hypersurfaces along which there might be a pole. Such claims will be made later, depending upon interpretation of Laurent expansions at poles as giving *anomalous intertwinings.*

Now we can describe a fairly general scenario in which locally strong meromorphy is provably assured. (The general construction of the set Ω is the same as earlier).

Let $\tilde{H}$ be a locally compact totally disconnected topological group of which H and G are closed subgroups. Let Y be a closed subset of $\tilde{H}$ stable under left multiplication by H and right multiplication by G. Let Z be a locally compact totally disconnected space on which H acts trivially and upon which G acts continuously on the right. Take

$$\Omega = Y \times Z$$

Let σ be an $\mathcal{O}$-parametrized family of *finite-dimensional* smooth representations of H.

Let $C_c^\infty(H\backslash\Omega, \sigma)$ be the collection of σ-valued locally constant functions f on Ω the image of whose support is *compact* in the quotient space $H\backslash\Omega$, and which have the left equivariance

$$f(hx) = \sigma(h)\, f(x)$$

for all $h \in H$ and $x \in \Omega$. These functions we view as **(generalized) test functions**.

Let P be a closed subgroup of G, with closed normal subgroup N, and closed subgroup M so that P is a semi-direct product of M and N. We suppose that N is an ascending union of compact open subgroups. And suppose that there is a compact open subgroup M_o of M so that with idempotent

$$e = \frac{\text{characteristic function of } M_o}{\text{measure of } M_o}$$

in the Hecke algebra $\mathcal{G}_M$ of M the subalgebra $e\mathcal{G}_M e$ is *Noetherian.*

Let χ be an $\mathcal{O}$-parametrized family of *finite-dimensional* smooth representations of $M \approx P/N$, so as to give an $\mathcal{O}$-parametrized family of representations of P, trivial on N. We denote this by the same letter χ. We suppose that the image $e \cdot \chi$ of the $\mathcal{O}$-module χ is a *Noetherian* $e\mathcal{G}_M e$-module, where the idempotent e is as just above.

The previous hypotheses are met very often in practice. In particular, the hypotheses on P, M, N and e are met for *unramified principal series*, where P is a parabolic subgroup of a p-adic reductive group G, N is the unipotent radical of P, and so on.

The critical hypothesis here is that the double coset space

$$H\backslash\Omega/P$$

be **finite**. In that case, we have

THEOREM: *Assuming finiteness of the double coset space $H\backslash\Omega/P$, any meromorphic $\mathcal{O}$-parametrized family of intertwining operators*

$$\Phi \in \mathrm{Hom}_{G\times\mathcal{O}}(C_c^\infty(H\backslash\Omega, \sigma) \otimes_{\mathcal{O}} \text{c-Ind}_P^G\, \chi, \mathcal{M})$$

is locally strongly meromorphic. *(Proof given in 3.5).*

2.6. Denominators and anomalous intertwinings. Granting that a meromorphic $\mathcal{O}$-parametrized family of intertwining operators is local strongly meromorphic allows meaning to be attached to the poles. In particular, the *residues* (or more general *leading terms*), in a sense described below, are *anomalous intertwining operators.* Thus, the representation theory of the situation gives *a priori* restrictions on the possible locations of poles of (parametrized) intertwining operators.

Thus, viewing a local integral as a certain kind of (parametrized) intertwining operator, we can give *a priori* restrictions on poles of the integral.

Now we add the hypothesis that $\mathcal{O}$ is *integrally closed,* in addition to being Noetherian (and commutative), and let x be a *height one* prime in $\mathcal{O}$. Then the local ring $\mathcal{O}_x$ is an integrally closed Noetherian integral domain in which every non-zero prime (here only $x\mathcal{O}_x$ in fact) is maximal. Thus, $\mathcal{O}_x$ is a *Dedekind domain* with one non-zero prime, so is a *valuation ring.* Thus, there exists a *local parameter* $\varpi = \varpi_x$, a generator for the unique maximal ideal in $\mathcal{O}_x$.

Suppose that a meromorphic family $\Phi : \sigma \to \tau$ of intertwinings between $\mathcal{O}$-parametrized families of representations σ and τ is (locally) **strongly meromorphic**. Let $\varpi = \varpi_x$ be a local parameter for a height-one prime x. Then for some power

ϖ^N of the local parameter at x the family $\varpi^N \Phi$ of intertwinings is *holomorphic* at x. The least integer N so that $\varpi^N \Phi$ is *holomorphic* at x is the **order of the pole** of Φ at x.

Assuming that a meromorphic family Φ of intertwining operators $\sigma \to \tau$ is locally strongly meromorphic and has a pole of order at least 1 at a height-one prime x, we can consider a new intertwining operator, the **leading term** of Φ at x. Let N be the least integer so that $\varpi^N \Phi$ is holomorphic at x. Then the *pointwise* intertwining operator

$$q_x(\varpi^N \Phi) : \sigma_x \to \tau_x$$

is the **leading term** of Φ at x. In the extreme case that the order of the pole is 1, this leading term is the **residue** of Φ at x.

REMARKS: Unless maximal primes x in $\mathcal{O}$ are height one, the corresponding residue fields k_x are of positive transcendence degree over the base field k. Thus, except in the special case that maximal primes are height one, residues or other leading term intertwining operators are intertwining operators between vectorspace representations over fields of positive transcendence degree over k.

We will look at local integrals of the sort considered in the *second* version of the *rationality of local integrals* results in 2.2 above. Fix a field extension $\tilde{k}$ of k, and suppose that $\mathcal{O} \otimes_k \tilde{k}$ is still an integral domain, with $\mathcal{M} \otimes_k \tilde{k}$ the field of fractions of $\mathcal{O} \otimes_k \tilde{k}$. Assume that $\tilde{k}$ and finite extensions $\tilde{k}_x$ of it are *complete.* (As earlier, taking $\tilde{k}$ to be $\mathbf{R}$ or $\mathbf{C}$ or finite extensions of $\mathbf{Q}_p$ suffices).

As before, let $\tilde{H}$ be a locally compact totally disconnected topological group of which H and G are closed subgroups. Let Y be a closed subset of $\tilde{H}$ stable under left multiplication by H and right multiplication by G. Let Z be a locally compact totally disconnected space on which H acts trivially and upon which G acts continuously on the right. Take

$$\Omega = Y \times Z.$$

Let σ be an $\mathcal{O}$-parametrized family of *finite-dimensional* smooth representations of H.

Again, let $\mathrm{C}_c^\infty(H\backslash\Omega, \sigma)$ be the collection of σ-valued locally constant functions f on Ω the image of whose support is *compact* in the quotient space $H\backslash\Omega$, and which have the left equivariance

$$f(hx) = \sigma(h)\, f(x)$$

for all $h \in H$ and $x \in \Omega$.

The **open orbit** hypothesis we impose is the assumption that there is some $\xi \in \Omega$ so that

$$H\,\xi\,G \subset \Omega$$

is an *open subset* of Ω.

And we require that there is a right G-invariant $\mathbf{Q}$-valued measure $d\omega$ on Ω, supported (for simplicity) on the open orbit $H\xi G$. For x a *maximal* ideal in the prime spectrum of $\mathcal{O} \otimes_k \tilde{k}$ consider the ($\tilde{k}_x$-valued) **local integral**

$$I_f(x) = \int_{H\backslash\Omega} f(\omega)\, d\omega$$

where f is in $\mathrm{C}_c^\infty(H\backslash\Omega, \sigma)$, and q_x denotes pointwise evaluation as above.

As earlier, we *assume* that for maximal ideals x in some non-meager subset of $X \otimes_k \tilde{k}$ the integral *converges* absolutely for *all* $f \in \mathrm{C}_c^\infty(H\backslash\Omega, \sigma_x)$, and that we have the *generic multiplicity one* condition

$$\dim_{\tilde{k}_x} \mathrm{Hom}_{G\times\tilde{k}_x}(\mathrm{C}_c^\infty(H\backslash\Omega, \sigma_x), \tilde{k}_x) \leq 1.$$

We have shown that

$$f \to \int_{H\backslash\Omega} f(\omega)\, d\omega$$

extends to an $\mathcal{O} \otimes_k \tilde{k}$-parametrized meromorphic family of intertwining operators

$$\Phi : \mathrm{C}_c^\infty(H\backslash\Omega, \sigma) \otimes_k \tilde{k} \to \mathcal{M} \otimes_k \tilde{k}$$

rational over k.

THEOREM (on Anomalous Intertwinings): *Now we assume that the ring $\mathcal{O}$ is integrally closed, Noetherian, commutative. Let Φ be the extension just described of the local integral*

$$\int_{H\backslash\Omega} f(\omega)\, d\omega$$

Let x be a height-one prime at which the meromorphic $\mathcal{O}$-parametrized family of intertwining operators Φ given in $\mathrm{Hom}_{G\times\mathcal{O}}(\mathrm{C}_c^\infty(H\backslash\Omega, \sigma), \mathcal{O})$ *has a pole of order $N > 0$. Let*

$$\rho_x = q_x(\varpi^N \Phi)$$

be the leading term *of Φ at x, where ϖ is a local parameter at x. Then ρ_x is a non-zero intertwining operator*

$$\rho_x \in \mathrm{Hom}_{G\times k_x}(\mathrm{C}_c^\infty(H\backslash\Omega, \sigma) \otimes_{\mathcal{O}} k_x, k_x)$$

and

$$\rho_x(\mathrm{C}_c^\infty(H\backslash H\xi G, \sigma) \otimes_{\mathcal{O}} k_x) = \{0\}.$$

In particular, there is no pole *at a hypersurface x* unless

$$\dim_{k_x} \mathrm{Hom}_{G\times k_x}(\mathrm{C}_c^\infty(H\backslash(\Omega - H\xi G), \sigma), k_x) > 0.$$

REMARKS: By definition, a k_x-linear intertwining operator from the space of test functions

$$\mathrm{C}_c^\infty(H\backslash(\Omega - H\xi G), \sigma_x)$$

on the complement of the open orbit $H\xi G$ to k_x is an **anomalous intertwining operator**. When composed with the surjective (see 3.2) map

$$\mathrm{C}_c^\infty(H\backslash\Omega, \sigma_x) \to \mathrm{C}_c^\infty(H\backslash(\Omega - H\xi G), \sigma_x)$$

we obtain (equivalently) intertwining operators which are zero on the subspace $\mathrm{C}_c^o(H\backslash H\xi G, \sigma_x)$.

REMARKS: From a slightly different viewpoint: while the associated *Mackey-Bruhat* distribution of the intertwining operators at holomorphic points of Φ have *support* with non-trivial interior, the leading-term intertwining operators such as ρ_x have support contained in much smaller sets (e.g., with empty interior).

PROOF OF THEOREM: The fact that the leading term ρ_x is an intertwining operator is straightforward. And the idea of the rest of the proof is clear: the family Φ when restricted to the smaller space of test functions does *not* have a pole at x, so when Φ is multiplied by the local parameter ϖ it *vanishes* at x.

Functions f_o in the subspace $C_c^o(H\backslash H\xi G, \sigma)$ of $C_c^o(H\backslash\Omega, \sigma)$ have compact support left-modulo H, so the integral (defining Φ pointwise)

$$I(f_o) = \int_{H\backslash\Omega} f_o(\omega)\, d\omega$$

has a compactly-supported (locally constant) integrand. Thus, for such f_o, we can integrate over $\mathcal{O}$ itself, since such integrals are really just algebraic objects.

In particular, for any $N > 0$, for such f_o,

$$\varpi^N \int_{H\backslash\Omega} f_o(\omega)\, d\omega \in \varpi\mathcal{O}_x.$$

Thus, the evaluation map q_x at q sends this to 0, as claimed. □

COROLLARY: *Suppose that we are in the situation of 2.4, with the vth Euler factor of our global integral expressed as a value of a parametrized family of intertwining operators. Let 0 denote the zero ideal in the parameter space, and k_o the associated residue field. Suppose that*

$$\dim_{k_o} \mathrm{Hom}_{G_v}(\mathrm{Res}^{\tilde{G}_v}_{G_v} \text{c-Ind}^{\tilde{G}_v}_{\tilde{P}_v} \tilde{\chi}' \otimes \text{c-Ind}^{G_v}_{P_v} \chi',\ k_o) = 1.$$

Then (up to units in the parametrizing ring $\tilde{\mathcal{O}} \otimes \mathcal{O}$) the denominator of the vth Euler factor is a product of factors of the form

$$1 - c\, z_1^{m_1} \ldots z_n^{m_n} z^m$$

for integers $m, m_1, \ldots, m_n$, and c a non-zero element of the basefield k.

REMARK: The fact that c lies in the basefield means that it is "independent" of the parameters $z, z_1, \ldots, z_n$. And we imagine that $z = q^{-s}$ and $z_i = q^{-s_i}$, so this is to assert that the denominator has factors only of the form

$$1 - c\, q^{-m_1 s_1} \ldots q^{-m_n s_n} q^{-ms}$$

with c independent of the $s, s_1, \ldots, s_n$.

PROOF. Let Ψ denote the meromorphic family of intertwining operators given by the integral

$$\int_{\tilde{Z}_v H_\beta \backslash G_v} \varepsilon_v(\xi g)\, f(\beta\, \theta\, g)\, dg$$

of 2.4. Any denominator gives rise to a pole (at least over an algebraic closure of the base field), which can only occur at a hypersurface along which there is an anomalous intertwining operator. As before, using the $\tilde{P}_v \times P_v$-orbit filtration on the test functions

$$\mathrm{Res}^{\tilde{G}_v}_{P_v}(\text{c-Ind}^{\tilde{G}_v}_{\tilde{P}_v} q_x \tilde{\chi}')$$

we obtain graded pieces

$$\text{c-Ind}^{G_v}_{H_\alpha} q_x \tilde{\chi}'^{\alpha}$$

where $\alpha \in \tilde{P}_v \backslash \tilde{G}_v / P_v$,

$$H_\alpha = \alpha^{-1} \tilde{P}_v \alpha \cap P_v$$

and

$$q_x \tilde{\chi}'^{\alpha}(p) = q_x \tilde{\chi}'(\alpha p \alpha^{-1})$$

for $p \in H_\alpha \subset P_v$. The intertwining operator Ψ is an extension from an intertwining operator on one of these graded pieces.

The anomalous intertwining operators can only occur at points x at which a *second* graded piece in the orbit filtration has a non-zero intertwining operator to k_x. The condition for this to occur is of the general form

$$(\tilde{\chi}' \otimes \chi')^{\alpha}|_{H_\alpha} = \delta_{H_\alpha}$$

as computed in both 1.4 and 2.4. As noted in 2.4, in terms of the coordinates $z, z_1, \ldots, z_n$, this condition is of the indicated form. □

REMARKS: Further, the isomorphisms among irreducible unramified principal series whose characters differ by an element of the Weyl group implies a corresponding symmetry in the poles of the intertwining operator, thus implying corresponding symmetry of the denominator. Further, the degenerate principal series (locally generated by the Eisenstein series on $\tilde{G}$) also have natural intertwining operators which give additional symmetry in the "variable" s. Thus, these intertwining operators cause symmetries which tend to assure that the denominators of such local integrals look "automorphic".

2.7. Illustrative examples. Here we apply the previous machinery to some simple examples, where one already knows that there is an Euler product, and also knows what the local integrals should be. First we look at the local zeta integrals of Tate's thesis [**T**] and [**GoJ**]. Although such integrals certainly can be understood in a direct computational manner, it is interesting that we can reach conclusions without computation.

Tate's thesis revisited:

Let q be the cardinality of the residue field of a non-archimedean local field k. Let $\mathcal{S} = C_c^\infty(k)$ be the space of $\mathbf{C}$-valued test functions on k. Following Tate's thesis, we consider the *local zeta integrals*

$$\zeta(\phi, s) = \int_{k^\times} \phi(x)\chi(x)|x|^s \, d^\times x$$

where $d^\times x$ refers to a multiplicative Haar measure, $\phi \in \mathcal{S}$, χ is a fixed character on $k^\times$, and $|\,|$ is a normalization of the norm so that $|\varpi| = q^{-1}$ for a local parameter ϖ. Write

$$\chi_s(x) = \chi(x)|x|^s.$$

We observe that this zeta integral is an intertwining operator in

$$\mathrm{Hom}_{k^\times}(\mathcal{S} \otimes \chi_s, \mathbf{C}).$$

The family of representation spaces χ_s is parametrized in the following manner: Let

$$\mathcal{O} = \mathbf{C}[z, z^{-1}].$$

Let V be a one-dimensional complex vectorspace, and for $g \in k^\times$ define

$$\pi(g)(v) = \chi(g) z^{\mathrm{ord} g} \cdot v$$

where the ord function is as usual defined by

$$g\mathbf{o} = \varpi^{\mathrm{ord} g}\mathbf{o}$$

where $\mathbf{o}$ is the valuation ring in k. Then the pointwise representations are recovered by mapping

$$z \to q^{-s}$$

From the orbit filtration on test functions, we have an exact sequence

$$0 \to C_c^\infty(k^\times) \to C_c^\infty(k) \to C_c^\infty(\{0\}) \to 0$$

And

$$\mathrm{Hom}_{k^\times}(C_c^\infty(k^\times) \otimes \chi_s, \mathbf{C}) \approx \mathrm{Hom}_{k^\times}(C_c^\infty(k^\times), \chi_{-s})$$

$$= \mathrm{Hom}_{k^\times}(\text{c-Ind}_{\{1\}}^{k^\times}\mathbf{C}, \chi_s^{-1}) \approx \mathrm{Hom}_{k^\times}(\chi_s, \mathrm{Ind}_{\{1\}}^{k^\times}\mathbf{C}).$$

By Frobenius Reciprocity this is

$$\mathrm{Hom}_{\{1\}}(\chi_s|_{\{1\}}, \mathbf{C}) \approx \mathbf{C}.$$

Thus, this space is *one-dimensional.* The Tate local zeta integral is convergent for all $s \in \mathbf{C}$ for $\phi \in C_c^\infty(k^\times)$, so gives such a $k^\times$-homomorphism.

The smaller orbit $\{0\}$ gives

$$C_c^\infty(\{0\}) \approx \text{ trivial } k^\times\text{-representation}$$

so unless $\chi_s = 1$ this smaller orbit *cannot* 'support' any intertwining operator, and so when $\chi_s = 1$ we have an *anomalous* intertwining operator.

For $\mathrm{Re}\,(s)$ sufficiently large the integral defining the zeta integral on the whole space $C_c^\infty(k)$ is absolutely convergent, yielding non-zero intertwining operators on the *non-meager* set of complex s with large real part. Thus, so far, we conclude that off at most countably many points the dimension of the space of intertwining operators is at most one-dimensional. That is, we have the *generic multiplicity-one* condition.

And taking $\phi_o \in C_c^\infty(k)$ to be a suitable scalar multiple of the characteristic function of a sufficiently small compact open subgroup, we have

$$\int_{k^\times} \phi_o(x)\chi_s(x)\, d^\times x = 1$$

for all s. This fulfills the *good test vector* condition. Note that matters are simplified here in that the representation $\mathcal{S} = C_c^\infty(k)$ is 'constant' (i.e., did not depend upon the parameters).

Thus, without any further work, we have the *qualitative* conclusion, from the results on rationality of local integrals (section 2.2), that *for any $\phi \in \mathcal{S}$ the local zeta integral is a rational function of $z = q^{-s}$.*

Quantitatively: Let $\mathcal{H}$ be the Hecke algebra of the totally disconnected group $k^\times$. Let Θ be a sufficiently small compact open subgroup of $k^\times$ so that χ_s is identically 1 on Θ, and let $e \in \mathcal{H}$ be the corresponding idempotent

$$e = \text{ characteristic function of } \Theta / \text{ measure } \Theta.$$

Then $e\mathcal{H}e$ is a commutative Noetherian ring. We suppose that χ is unramified, so that Θ is the whole unit group $\mathbf{o}^\times$, and then

$$e\mathcal{H}e \approx \mathbf{C}[x, x^{-1}].$$

In that case, we may as well take $\chi = 1$, so that the character is completely specified by the parameter s. With these observations, by the *orbit criterion for strong meromorphy* (section 2.5), as intertwining operator on parametrized families of representations *the local zeta integral is strongly meromorphic.*

Therefore, the only possible pole of the local zeta integral can be at the point where there is an *anomalous intertwining operator*, that is, at parameter values where there is an intertwining operator which is 0 on the subrepresentation $C_c^\infty(k^\times)$ of $C_c^\infty(k)$. Thus, with $z = q^{-s}$, the only possible pole is at $z = 1$.

In other words, the only possible linear factor of the denominator of the local zeta integral is

$$1 - z = 1 - q^{-s}.$$

(Of course, here z, z^{-1} are units).

And we can easily verify that the collection of all possible values of the local zeta integral is a *fractional ideal* of $\mathcal{O} = \mathbf{C}[z, z^{-1}]$ containing 1.

Thus, in summary, without computing anything directly we can see that in the unramified case there is some test function so that the local zeta integral applied to it yields

$$\frac{1}{(1-q^{-s})^m}$$

for a non-negative integer m. To prove that the 'pole' actually occurs, and with order 1, would require a little further investigation.

In the case of *ramified* character we directly conclude that there is no 'denominator', and that the greatest common divisor of all 'numerators' is just 1. Further, if we choose *any* test function, we are assured that the result will be a *polynomial* in q^{-s}, in any case.

Godement-Jacquet revisited:

Now we recall some of the local results of [**GoJ**] in the case of GL(2), and then recover several of the (non-archimedean) local results at good primes in a substantially non-computational manner. We restrict our attention to GL(2) for simplicity, although the case of GL(n) (and other simple algebras) is entirely analogous. (The general Godement-Jacquet result subsumes Tate's thesis as a special case.)

Let k be a non-archimedean local field of characteristic zero, and let $G = \mathrm{GL}(2, k)$. Given an irreducible smooth representation π of G, given $v \in \pi$ and λ in the smooth dual $\check{\pi}$ of π, given a test function ϕ on the space of 2×2 matrices over k, and given $s \in \mathbf{C}$ with $\mathrm{Re}\,(s)$ sufficiently positive, define a local integral

$$Z(v, \lambda, \phi, s) = \int_G |\det(g)|^s \, c_{v,\lambda}(g)\, \phi(g)\, dg.$$

Here $c_{v,\lambda}$ is the ('matrix'-) coefficient-function

$$c_{v,\lambda}(g) = \langle \pi(g)v, \lambda \rangle$$

and $\langle , \rangle$ is the canonical bilinear map

$$\langle , \rangle : \pi \times \check{\pi} \to \mathbf{C}.$$

This integral converges for real part of s sufficiently positive.

We assume for simplicity that the central character of π is unramified, since otherwise this particular form of zeta integral vanishes identically. The more general version with arbitrary central character is treated quite analogously.

For any list of data π, v, λ, ϕ, s, it is true that the local integral is a *rational expression in* q^s with complex coefficients (the latter depending upon π, v, λ, ϕ), where q is the cardinality of the residue field. This is not obvious.

Fixing π, the collection of all local integrals with $v \in \pi$ and $\lambda \in \check{\pi}$ is a *fractional ideal* of $\mathbf{C}[q^s, q^{-s}]$ in its fraction field $\mathbf{C}(q^s)$. This is not obvious. Further, this fractional ideal contains 1, so contains $\mathbf{C}[q^s, q^{-s}]$.

Then the L-factor attached to π is defined to be the *greatest common divisor* in $\mathbf{C}[q^s, q^{-s}]$ of all the local integrals attached to π. Since $\mathbf{C}[q^s, q^{-s}]$ is a principal ideal domain, this fractional ideal does have a single generator. Further, since this fractional ideal contains 1, the ambiguity (by units) of choice of generator is eliminated by taking the unique greatest common divisor of the form

$$\frac{1}{(\text{monic polynomial in } q^{-s})}$$

Now we reconsider these results within the present framework. Let $G \times G$ act on the space $C_c^\infty(E)$ of test functions on the space E of 2×2 matrices over k, by

$$(g_1, g_2)f(x) = L_{g_1} R_{g_2} f(x) = f(g_1^{-1} x g_2).$$

Then

$$Z(\pi(g_2)v, \check{\pi}(g_1)\lambda, L_{g_1} R_{g_2}\phi, s) = |\det g_1|^s \, |\det g_2|^{-s} \, Z(v, \lambda, \phi, s).$$

That is, the map

$$v \otimes \lambda \otimes \phi \to Z(v, \lambda, \phi, s)$$

is a $G \times G$-homomorphism

$$(\pi \otimes \check{\pi}) \otimes C_c^\infty(E) \to |\det *|^s \, |\det *|^{-s}.$$

Or, writing

$$\pi_s = \pi \otimes |\det *|^s$$

we can rearrange this all to view the map as

$$(\pi_{-s} \otimes \check{\pi}_s) \otimes C_c^\infty(E) \to \mathbf{C}.$$

Or, we can rearrange this to

$$C_c^\infty(E) \to \check{\pi}_s \otimes \pi_s$$

where the admissibility of π entails that π is *reflexive*, that is, that the natural inclusion $\pi \to \check{\check{\pi}}$ is an *isomorphism.*

We filter $C_c^\infty(E)$ by $G \times G$-orbits: there are 3 orbits E_r of $G \times G$ on E, indexed by *rank* r of the matrices. Then $C_c^\infty(E_2)$ naturally imbeds in $C_c^\infty(E)$. Let Δ be the diagonal copy of G in $G \times G$. This is the isotropy group of the point $1 \in E$. Then

$$C_c^\infty(E_2) \overset{G\times G}{\approx} \text{c-Ind}_\Delta^{G\times G}\mathbf{C}.$$

Thus, we compute

$$\begin{aligned} \text{Hom}_{G\times G}(C_c^\infty(E_2), \check{\pi}_s \otimes \pi_s) &\approx \text{Hom}_{G\times G}(\pi_{-s} \otimes \check{\pi}_s, \text{Ind}_\Delta^{G\times G}\mathbf{C}) \approx \\ &\approx \text{Hom}_G(\pi_{-s} \otimes \check{\pi}_s, \mathbf{C}) \quad \text{(by Frobenius Reciprocity)} \\ &\approx \text{Hom}_G(\pi_{-s}, \check{\check{\pi}}_s) \approx \text{Hom}_G(\pi_{-s}, \pi_{-s}) \end{aligned}$$

which is *one-dimensional* for all irreducibles π_{-s}, by Schur's Lemma for smooth representations of countable dimension (for separable groups).

We will see that, for given π, for all but finitely-many $s \in \mathbf{C}$ each intertwining operator just above has at most one extension to $C_c^\infty(E)$. This will provide the requisite *generic multiplicity-one* hypothesis toward application of our results.

Let

$$P = \left\{ \begin{pmatrix} * & * \\ 0 & * \end{pmatrix} \right\}$$

$$N = \left\{ \begin{pmatrix} 1 & * \\ 0 & 1 \end{pmatrix} \right\}$$

and let χ be a one-dimensional smooth representation of P factoring through P/N. Let δ be the modular function of P. Let

$$I_\chi = \text{Ind}_P^G \chi\delta^{1/2}$$

denote that associated *principal series* representation of G. Its smooth dual $\check{\pi}$ is

$$\check{I}_\chi \approx \mathrm{Ind}_P^G \chi^{-1}\delta^{1/2} = I_{\chi^{-1}}.$$

Again, all spherical representations occur as quotients and subrepresentations of unramified principal series. Indeed, 'generically' I_χ is irreducible, so I_χ is spherical (e.g., see [**C80**]). However, as usual, the objects which best lend themselves to examination here are *not* the spherical representations but the *unramified principal series.*

Take $\pi = I_\chi$. Let

$$\omega_s(g) = |\det(g)|^s$$

and (as before) write

$$\pi_s = \pi \otimes \omega_s.$$

We wish to show that for fixed χ and for all but finitely-many s

$$\mathrm{Hom}_{G\times G}(\mathrm{C}_c^\infty(E_1 \cup E_0), \check{\pi}_s \otimes \pi_{-s}) = 0.$$

This would imply that each intertwining operator in

$$\mathrm{Hom}_{G\times G}(\mathrm{C}_c^\infty(E_2), \check{\pi}_s \otimes \pi_{-s})$$

has at most one extension to an element of

$$\mathrm{Hom}_{G\times G}(\mathrm{C}_c^\infty(E), \check{\pi}_s \otimes \pi_{-s}).$$

Since we showed that for any π and for any s

$$\dim_{\mathbf{C}} \mathrm{Hom}_{G\times G}(\mathrm{C}_c^\infty(E_2), \check{\pi}_s \otimes \pi_{-s}) = 1$$

we will obtain the desired *generic multiplicity-one* result.

For $p \in P$ write

$$\chi_s(p) = \chi(p)\omega_s(p).$$

By Frobenius Reciprocity,

$$\mathrm{Hom}_{G\times G}(\mathrm{C}_c^\infty(E_1 \cup E_0), \check{\pi}_s \otimes \pi_{-s}) \approx$$
$$\mathrm{Hom}_{P\times P}(\mathrm{C}_c^\infty(E_1 \cup E_0), \chi_s^{-1}\delta^{1/2} \otimes \chi_{-s}).$$

Now we filter the 2×2-matrices of less than full rank by $P \times P$-orbits. Fix $G \times G$-representatives

$$\varepsilon_1 = \begin{pmatrix} 0 & 1 \\ 0 & 0 \end{pmatrix}$$
$$\varepsilon_0 = \begin{pmatrix} 0 & 0 \\ 0 & 0 \end{pmatrix}.$$

Clearly $P\varepsilon_0 P = \{\varepsilon_0\}$ is one $P \times P$-orbit. Let

$$Q = \left\{ \begin{pmatrix} a & b \\ 0 & d \end{pmatrix} \times \begin{pmatrix} a' & b' \\ 0 & a \end{pmatrix} \right\}$$

denote the isotropy group Q in $(P \times P)\backslash(G \times G)$ of $(P \times P) \cdot \varepsilon_1$. This group Q is just slightly smaller than the full product $P \times P$. To first break the $G \times G$-orbit E_1 into $P \times P$-orbits amounts to computing

$$(P \times P)\backslash(G \times G)/Q.$$

Via a Bruhat decomposition, we find four irredundant representatives

$$1 \times 1, \quad 1 \times w, \quad w \times 1, \quad w \times w$$

Indeed, $Q \supset N \times N$, and already

$$(P \times P)\backslash(G \times G)/N \times N \approx \{1 \times 1, 1 \times w, w \times 1, w \times w\}.$$

On the other hand, $Q \subset P \times P$, and we have the same representatives for

$$(P \times P)\backslash(G \times G)/(P \times P).$$

Projected to Q, the isotropy groups of these representatives are, respectively,

$$Q = \left\{ \begin{pmatrix} a & b \\ 0 & d \end{pmatrix} \times \begin{pmatrix} a' & b' \\ 0 & a \end{pmatrix} \right\} \text{ (for } 1 \times 1\text{)},$$

$$Q_1 = \left\{ \begin{pmatrix} a & b \\ 0 & d \end{pmatrix} \times \begin{pmatrix} a' & 0 \\ 0 & a \end{pmatrix} \right\} \text{ (for } 1 \times w\text{)},$$

$$Q_2 = \left\{ \begin{pmatrix} a & 0 \\ 0 & d \end{pmatrix} \times \begin{pmatrix} a' & b' \\ 0 & a \end{pmatrix} \right\} \text{ (for } w \times 1\text{)},$$

$$Q_* = \left\{ \begin{pmatrix} a & 0 \\ 0 & d \end{pmatrix} \times \begin{pmatrix} a' & 0 \\ 0 & a \end{pmatrix} \right\} \text{ (for } w \times w\text{)}.$$

Let Θ be any one of these isotropy groups. The *graded pieces* associated to this filtering are

$$\mathrm{Hom}_{P \times P}(\text{c-Ind}_\Theta^{P \times P} \mathbf{C}, \chi_s^{-1} \delta^{1/2} \otimes \chi_{-s} \delta^{1/2}).$$

By dualizing and by Frobenius Reciprocity, we have

$$\mathrm{Hom}_{P \times P}(\text{c-Ind}_\Theta^{P \times P} \mathbf{C}, \chi_s^{-1} \delta^{1/2} \otimes \chi_{-s} \delta^{1/2}) \approx$$
$$\mathrm{Hom}_{P \times P}(\chi_{-s} \delta^{-1/2} \otimes \chi_s^{-1} \delta^{-11/2}, \mathrm{Ind}_\Theta^{P \times P} \delta_\Theta \delta_{P \times P}^{-1}) \approx$$
$$\mathrm{Hom}_\Theta(\chi_{-s} \delta^{-1/2} \otimes \chi_s^{-1} \delta^{-1/2}, \delta_\Theta \delta_{P \times P}^{-1}).$$

This is $\{0\}$ unless

$$(\chi_{-s} \delta^{-1/2} \otimes \chi_s^{-1} \delta^{-1/2})|_\Theta = (\delta_\Theta \delta_{P \times P}^{-1})|_\Theta$$

in which case it is one-dimensional. The latter condition simplifies a little, to

$$(\chi_{-s} \delta^{1/2} \otimes \chi_s^{-1} \delta^{1/2})|_\Theta = \delta_\Theta.$$

We will examine this condition for each of the four orbits inside the less-than-full-rank matrices.

Write

$$\chi\left(\begin{pmatrix} a & * \\ 0 & d \end{pmatrix}\right) = \chi_1(a)\chi_2(d).$$

Then the respective conditions for the four orbits are as follows. For the first, we have

$$\chi_1(a)\chi_2(d)|ad|^{-s}|a/d|^{1/2}\chi_1^{-1}(a')\chi_2^{-1}(a)|a'a|^s|a'/a|^{1/2} = |a/d||a'/a|.$$

This is equivalent to the system of equations

$$\chi_1 = \chi_2 \qquad \text{(from the } a\text{-component)},$$
$$\chi_2 = |*|^{s-1/2} \text{(from the } d\text{-component)},$$
$$\chi_1 = |*|^{s-1/2} \text{(from the } a'\text{-component)}.$$

Unless $\chi_1 = \chi_2$ this system of equations is not satisfied for *any* value of s. And, even when $\chi_1 = \chi_2$, there is only one s which satisfies it. So *generically s does not satisfy this system.*

For the orbit with isotropy group Q_1:

$$\chi_1(a)\chi_2(d)|ad|^{-s}|ad|^{1/2}\chi_1^{-1}(a)\chi_2^{-1}(d')|ad'^{-1}|^s|a/d'|^{1/2} = |a/d|$$

which is equivalent to

$$\chi_2 = |*|^{s-1/2} \quad \text{from } d \text{ and/or from } d'$$

since the 'scalar equation' arising from a is $0 = 0$, and since the scalar equation from d is identical to that from d'. Unless χ_2 is of the form $|*|^{s'}$, i.e., is *unramified*, this system of equations cannot be satisfied by *any* s. Even when the conditions are met, there is only one s which satisfies the system. So *generically s does not satisfy this system.*

Similar (symmetrical) remarks apply to the orbit with isotropy group Q_2: the conditions simplify to

$$\chi_1 = |*|^{s-1/2}.$$

Unless χ_1 is of the form $|*|^{s'}$, i.e., is *unramified*, this system of equations cannot be satisfied by any s. Even when the conditions is met, there is only one s which satisfies the system. So *generically s does not satisfy this system.*

For the orbit with isotropy group Q_*:

$$\chi_1(a)\chi_2(d)|ad|^{-s}|a/d|^{1/2}\chi_1^{-1}(d)\chi_2^{-1}(d')|dd'|^s|d/d'|^{1/2} = 1$$

which is equivalent to

$$\begin{aligned} \chi_1 &= |*|^{s-1/2} \text{from } a, \\ \chi_1 &= \chi_2 \qquad \text{from } d, \\ \chi_2 &= |*|^{s-1/2} \text{from } d'. \end{aligned}$$

Unless $\chi_1 = \chi_2$ this system has no solutions s. Even when this condition is satisfied, there is only one s which satisfies it. So *generically s does not satisfy this system.*

(The rank-zero orbit is even simpler to treat than these four orbits, since there the representation of $G \times G$ which arises is just $\mathbf{C} \otimes \mathbf{C}$.)

We conclude that for $\chi_1 \neq \chi_2$ *generically there is at most a one-dimensional space of intertwining operators*

$$\mathrm{Hom}_{G\times G}(\mathrm{C}_c^\infty(E) \otimes \left((I_\chi)_{-s} \otimes (I_{\chi^{-1}})_s\right), \mathbf{C}).$$

(And all such intertwining operators are uniquely determined by their restrictions to the space $\mathrm{C}_c^\infty(E_2)$ of test functions supported on the open orbit E_2.)

The family of representations

$$\mathrm{C}_c^\infty(E) \otimes \left((I_\chi)_{-s} \otimes (I_{\chi^{-1}})_s\right)$$

considered above is a parametrized family of representations, over the ring

$$\mathcal{O} = \mathbf{C}[z, z^{-1}]$$

where the pointwise representations are recovered by $z \to q^{-s}$. Taking the test function ϕ_o to be a suitable scalar multiple of the characteristic function of a sufficiently small neighborhood of $1 \in E$, the local zeta integral is just

$$Z(v, \lambda, \phi_o, s) = 1.$$

Thus, we have the *generic multiplicity-one and good-test-vector properties* required to conclude that *all zeta integrals are rational functions of q^{-s}.*

Now we will estimate the possible denominators of these local zeta integrals (at good primes) in terms of *anomalous intertwining operators.* To this end we will re-use the computations just above. Specifically, we saw that for $\chi_1 \neq \chi_2$ the only possible anomalous intertwining operators occur for

$$\chi_1 = |*|^{s-1/2} \qquad \text{or} \qquad \chi_2 = |*|^{s-1/2}.$$

Indeed, at any other points the intertwining operators from $C_c^\infty(E_2)$ have unique extensions to $C_c^\infty(E)$. Let

$$\chi_1 = |*|^\alpha \qquad \chi_2 = |*|^\beta.$$

Then the general result on denominators and anomalous intertwining operators asserts that the only possible linear factors of the denominator in the unramified principal series case are

$$(1 - q^{\alpha+1/2}q^{-s}) \qquad (1 - q^{\beta+1/2}q^{-s}).$$

This matches the direct computation in [**GoJ**] of these local integrals.

Note that the Weyl group symmetry in this case would simply interchange α and β. Indeed, the expression we obtain is symmetrical in this regard.

3. Completions of proofs

3.1. Proof of the Rationality Lemma. PROOF OF THEOREM: This argument requires some preparation. Several parts of this discussion have their origins in the work of J. Bernstein mentioned without proof in [**GePR**]. This is related to, but also a bit different from, the more algebraic results from [**BeZ76, BeZ77**], and [**Be84**].

Let k be a field of characteristic zero, and V a k-vectorspace. A **linear system** (over k, with coefficients in V) is a set Ξ of ordered pairs (v_i, c_i) where $v_i \in V$ and $c_i \in k$ and $i \in I$ for some index set I. A **solution** to the linear system is $\lambda \in V^*$ so that, for all indices i,

$$\lambda(v_i) = c_i$$

where $V^* = \mathrm{Hom}_k(V, k)$ is the k-linear dual space of V. Obviously the set of all solutions is an affine subspace of V^*, so has a sense of **dimension**. A system Ξ is **homogeneous** if all the constants c_i are 0. In that case, the collection of solutions is a vector subspace of V^*. Two systems are **equivalent** if they have the same set of solutions. It is immediate that any system is equivalent to a system with at most one constant c_i non-zero.

Let $\tilde{k}$ be an extension field of k. Let W be a k-vectorspace. There is a natural inclusion $W \to W \otimes_k \tilde{k}$. For any k-subspace W' of W there is naturally associated the $\tilde{k}$-subspace $W' \otimes_k \tilde{k}$ of $W \otimes_k \tilde{k}$. Any $\tilde{k}$-subspace of $W \otimes_k \tilde{k}$ occurring as $W' \otimes_k \tilde{k}$ for a k-subspace W' of W is a **k-rational subspace** of $W \otimes_k \tilde{k}$.

Let $\Xi = \{(v_i, c_i)\}$ be a linear system over k with coefficients in a $k$vectorspace V. For an extension field $\tilde{k}$ of k we can **extend scalars** by looking for solutions λ not only in $\mathrm{Hom}_k(V, k)$ but also in the larger space

$$\mathrm{Hom}_k(V, \tilde{k}) \approx \mathrm{Hom}_{\tilde{k}}(V \otimes_k \tilde{k}, \tilde{k}).$$

A solution $\lambda \in \mathrm{Hom}_{\tilde{k}}(V \otimes_k \tilde{k}, \tilde{k})$ is **k-rational** if it is in the image of the natural inclusion

$$\mathrm{Hom}_k(V, k) \subset \mathrm{Hom}_k(V, \tilde{k}) \approx \mathrm{Hom}_{\tilde{k}}(V \otimes_k \tilde{k}, \tilde{k}).$$

We may write $\Xi \otimes_k \tilde{k}$ for the linear system obtained by extending scalars.

PROPOSITION: (Existence, Uniqueness, and Rationality)

Let $\Xi = \{(v_i, c_i) : i \in I\}$ be a k-linear system with coefficients v_i in a k-vectorspace V. Suppose that there is at most one index $i_o \in I$ so that $c_{i_o} = 0$.

- *If v_{i_o} does not lie in the k-span of $\{v_i : i \neq i_o\}$ then there is at least one solution to the linear system.*
- *If the coefficient vectors $\{v_i : i \in I\}$ span V, then there is at most one solution.*
- *If Ξ is* homogeneous *then and if the solution space N in* $\mathrm{Hom}_k(V,k)$ *is* finite-dimensional *then, for any extension field $\tilde{k}$ of k, the solution space of $\Xi \otimes_k \tilde{k}$ is the k-rational subspace $N \otimes_k \tilde{k}$ of* $\mathrm{Hom}_k(V,\tilde{k}) \approx \mathrm{Hom}_k(V,k) \otimes_k \tilde{k}$.
- *If for some field extension $\tilde{k}$ of k $\Xi \otimes_k \tilde{k}$ has a* unique *solution λ, then that solution is in fact k-rational, and the original system Ξ therefore has a unique solution.*

PROOF. If v_{i_o} does not lie in the span of the other coefficient vectors, then (via the Axiom of Choice) there is a linear functional λ in the dual space V^* so that $\lambda(v_i) = 0$ for $i \neq i_o$, but $\lambda(v_o) = 1$. Then $c_{i_o}\lambda$ is a solution of the system Ξ.

Next, for two solutions λ, λ' of Ξ the difference $\mu = \lambda - \lambda'$ is a solution of the homogeneous system $\Xi_o = \{(v_i, 0)\}$. If the v_i span V, then the condition

$$\mu(v_i) = 0 \quad \text{for all } i \in I$$

implies that $\mu = 0$. This is the uniqueness.

Now suppose that Ξ is homogeneous with finite-dimensional solution space N, and let $\tilde{k}$ be an extension field of k. Given a solution Λ of $\Xi \otimes_k \tilde{k}$, we must find finitely-many solutions λ_j of Ξ and scalars $a_j \in \tilde{k}$ so that

$$\Lambda = \sum_k a_j \lambda_j.$$

Whatever the $\tilde{k}$-span of the coefficient vectors $v_i \otimes 1$ in $V \otimes_k \tilde{k}$ may be, it is $W \otimes_k \tilde{k}$ where W is the k-span of the v_i in k. Let $e_1, \ldots, e_n$ be a k-basis for a necessarily finite-dimensional complementary subspace to W inside V. Then the $e_j \otimes 1$ form a $\tilde{k}$-basis for a $\tilde{k}$-subspace complementary to $W \otimes_k \tilde{k}$ in $V \otimes_k \tilde{k}$. Let $\lambda_1, \ldots, \lambda_n$ be in the solution space N so that $\lambda_i(e_j)$ is 1 or 0 depending on whether $i = j$ or not. Then it is easy to check that

$$\tilde{\lambda} = \sum_j \lambda(e_j)\lambda_j.$$

Finally, let $\tilde{\lambda}$ be the unique solution to $\Xi \otimes_k \tilde{k}$. By the first part of this lemma, the uniqueness of the solution implies that the $v_i \otimes 1$ span $V \otimes_k \tilde{k}$ over $\tilde{k}$, but that $v_{i_o} \otimes 1$ does not lie in the $\tilde{k}$-span of the other vectors. This immediately implies that the vectors v_i span V over k, and that v_{i_o} is not in the k-span of the other vectors. Thus, again by the first part of this lemma, there is a unique solution λ in $\mathrm{Hom}_k(V,k)$. The natural inclusion and isomorphism

$$\mathrm{Hom}_k(V,k) \subset \mathrm{Hom}_k(V,\tilde{k}) \approx \mathrm{Hom}_{\tilde{k}}(V \otimes_k \tilde{k}, \tilde{k})$$

necessarily send λ to $\tilde{\lambda}$, so $\tilde{\lambda}$ is k-rational, as asserted. □

For an extension field $\tilde{k}$ of k so that $\mathcal{O} \otimes_k \tilde{k}$ is still an integral domain, with $\mathcal{M} \otimes_k \tilde{k}$ the field of fractions of $\mathcal{O} \otimes_k \tilde{k}$, as a notational device let

$$X \otimes_k \tilde{k} = \text{ prime ideal spectrum of } \mathcal{O} \otimes_k \tilde{k}.$$

Let V be an $\mathcal{O}$-module. A **parametrized linear system** over $\mathcal{O}$ (or over X) with coefficients in $V \otimes_{\mathcal{O}} \mathcal{M}$ is a collection Ξ of ordered pairs (μ_i, f_i) with $\mu_i \in M \otimes_{\mathcal{O}} \mathcal{M}$, $f_i \in \mathcal{M}$.

A **generic solution** to such a parametrized system is

$$\lambda \in (V \otimes_{\mathcal{O}} \mathcal{M})^* = \mathrm{Hom}_{\mathcal{M}}(V \otimes_{\mathcal{O}} \mathcal{M}, \mathcal{M})$$

so that for all indices i

$$\lambda(\mu_i) = f_i.$$

That is, a generic solution is simply a *solution* (in the previous sense) to the $\mathcal{M}$-linear system on the $\mathcal{M}$-vectorspace $V \otimes_{\mathcal{O}} \mathcal{M}$.

Note that we do not require the coefficient vectors μ_i to be in the module V, but only in $V \otimes_{\mathcal{O}} \mathcal{M}$, and likewise the f_i need not be in $\mathcal{O}$, but only in $\mathcal{M}$. Of course, the same collection of generic solutions would be obtained if each (μ_i, f_i) were replaced by $(g_i\mu_i, g_i f_i)$ for non-zero $g_i \in \mathcal{O}$. Thus, one *could* assume without loss of generality that all the μ_i are in V and the f_i are in $\mathcal{O}$, but it is not necessary to do so.

For $x \in X$, $f \in \mathcal{M}$ is **holomorphic at** x if $f \in \mathcal{O}_x$. An element μ of $V \otimes_{\mathcal{O}} \mathcal{M}$ is **holomorphic at** x if $\mu \in M \otimes_{\mathcal{O}} \mathcal{O}_x$. A parametrized system $\Xi = \{(\mu_i, f_i)\}$ is **holomorphic at** x if for all indices i both f_i and μ_i are holomorphic at x. (From the definitions, *every* parametrized system is holomorphic at the generic point $0 \in X$).

For any $x \in X$ at which the parametrized system Ξ is holomorphic, we have the **associated pointwise system** Ξ_x, obtained by replacing all the elements $\mu_i \in M \otimes_{\mathcal{O}} \mathcal{O}_x$ by their images in $V \otimes_{\mathcal{O}} k_x$, and likewise by replacing the $f_i \in \mathcal{O}_x$ by their images in k_x. Thus, Ξ_x is a k_x-linear system with coefficients in $V \otimes_{\mathcal{O}} k_x$.

A **pointwise solution** λ_x **at** x to the parametrized linear system Ξ is just a solution to the linear system Ξ_x. Thus, it is

$$\lambda \in \mathrm{Hom}_{\mathcal{O}}(V, k_x) \approx \mathrm{Hom}_{k_x}(V \otimes_{\mathcal{O}} k_x, k_x)$$

so that for all indices i

$$\lambda(\mu_i) = f_i \mod x\mathcal{O}_x.$$

Let $y \subset x$ be two prime ideals in X. Then $S_y \supset S_x$, $\mathcal{O}_y \supset \mathcal{O}_x$, and there is a natural inclusion

$$\mathcal{O}_x/y\mathcal{O}_x \subset \mathcal{O}_y/y\mathcal{O}_y = k_y.$$

A solution λ to the pointwise system Ξ_y at y is **holomorphic** at x if

$$\lambda(V \otimes_{\mathcal{O}} k_y) \subset \mathcal{O}_x/y\mathcal{O}_x \subset k_y.$$

In that case, the solution λ to Ξ_y gives a solution to the pointwise system Ξ_x by taking the image of λ under the natural map

$$\mathrm{Hom}_{k_y}(V \otimes_{\mathcal{O}} k_y, \mathcal{O}_x/y\mathcal{O}_x) \approx \mathrm{Hom}_{\mathcal{O}}(V, \mathcal{O}_x/y\mathcal{O}_x) \to \mathrm{Hom}_{\mathcal{O}}(V, k_x)$$
$$\approx \mathrm{Hom}_{k_x}(V \otimes_{\mathcal{O}} k_x, k_x)$$

where the last map arises from the quotient map

$$\mathcal{O}_x/y\mathcal{O}_x \to \mathcal{O}_x/x\mathcal{O}_x = k_x.$$

LEMMA: *Suppose that $\mathcal{O}$ is Noetherian. Given $f \neq 0$ in $\mathcal{M}$, there is a finite collection $\eta_1, \ldots, \eta_n$ of irreducible hyperplanes so that $1/f \in \mathcal{O}_x$ for x not in the union $\eta_1 \cup \ldots \cup \eta_n$. In particular, the hyperplanes are those attached to the isolated primes in a primary decomposition of the ideal $f\mathcal{O}$.*

PROOF. For a prime ideal $x \in X$, if $f \notin x$ then $1/f \in \mathcal{O}_x$. Thus, we must show that if $f \in x$ then there is a height-one prime y so that $f \in y \subset x$.

Let $J = \bigcap_i Q_i$ be a primary decomposition of an ideal $J \subset x$, where Q_i is primary with associated prime x_i. Taking radicals,

$$x = \operatorname{rad} x \supset \bigcap_i \operatorname{rad} Q_i = \bigcap_i x_i$$

so, for some index i, $x \supset x_i$. Thus, x must contain some one of the minimal (i.e., isolated) primes among the primes associated to J.

Krull's Principal Ideal Theorem asserts that every prime *minimal* among those occurring in a primary decomposition of $f\mathcal{O}_x$ is height-one. This gives the result. □

LEMMA: *Let Y be a non-meager subset of X, the prime spectrum of $\mathcal{O}$, where $\mathcal{O}$ is a Noetherian integral domain. Then*

$$\bigcap_{x \in Y} x\mathcal{O}_x = \{0\}.$$

PROOF. Let $r \neq 0$ be in the indicated intersection. By the previous lemma, there would be finitely-many hypersurfaces $\eta_1, \ldots, \eta_n$ so that for $x \in X$ *off* the union of these hypersurfaces we would have both $r \in \mathcal{O}_x$ and $1/r \in \mathcal{O}_x$. Removing from Y the intersections of Y with these hypersurfaces would still leave a non-meager set Y_r. In particular, Y_r would be non-empty, and, by construction, for all $x \in Y_r$ both r and $1/r$ would lie in $\mathcal{O}_x$. But this would contradict the hypothesis $r \in x\mathcal{O}_x$. Thus, it must be that the indicated intersection is $\{0\}$, as claimed. □

LEMMA: *Let $\Xi = \{(\mu_i, f_i) : i \in I\}$ be a $\mathcal{O}$-parametrized system with coefficients in $V \otimes_{\mathcal{O}} \mathcal{M}$. Then*

- *There is a union $\bigcup_{i \in I} \eta_i$ of irreducible hyperplanes η_i off which the system Ξ is holomorphic.*
- *Let $\{m_j : j \in J\}$ be a generating set for V over $\mathcal{O}$. For a generic solution λ of Ξ, there is a union $\bigcup_{j \in J} \eta_i$ of irreducible hyperplanes η_i off which λ is holomorphic.*

PROOF. Let $\{g_i : i \in I\}$ be a collection of non-zero elements of $\mathcal{O}$ so that, for for all i, $g_i\mu_i \in M$ and $g_i f_i \in \mathcal{O}$. Then Ξ is holomorphic off the union of the hyperplanes attached to the isolated primes occurring in the primary decompositions of the ideals $g_i\mathcal{O}$.

Similarly, let $\{g_j : i \in J\}$ be a collection of non-zero elements of $\mathcal{O}$ so that, for for all $j \in J$, $g_i\lambda(m_j) \in \mathcal{O}$. Then λ is holomorphic off the union of the hyperplanes attached to the isolated primes occurring in the primary decompositions of the ideals $g_j\mathcal{O}$. □

THEOREM: *Let k be a field, and $\mathcal{O}$ a commutative Noetherian k-algebra. Let $\tilde{k}$ be an extension field of k so that $\mathcal{O} \otimes_k \tilde{k}$ is still an integral domain, with $\mathcal{M} \otimes_k \tilde{k}$ the field of fractions of $\mathcal{O} \otimes_k \tilde{k}$. Let $\Xi = \{(\mu_i, f_i) : i \in I\}$ be an $\mathcal{O}$-parametrized linear system with coefficients in $V \otimes_{\mathcal{O}} \mathcal{M}$ where V is a countably-generated $\mathcal{O}$-module, where also the index set I is countable. Let $\Xi \otimes_k \tilde{k}$ be the system obtained by extending scalars from k to $\tilde{k}$, and let $X \otimes_k \tilde{k}$ be the prime spectrum of $\mathcal{O} \otimes_k \tilde{k}$.*

- *If there is a unique generic solution* λ *to* $\Xi \otimes_k \tilde{k}$, *then for* $x \in X \otimes_k \tilde{k}$ *off a meager set there is a unique solution* λ_x *to the pointwise linear system* $(\Xi \otimes_k \tilde{k})_x$ *at* x, *and this pointwise solution is obtained from* λ *via the natural map*

$$\mathrm{Hom}_{\mathcal{O}\otimes_k \tilde{k}}(V \otimes_k \tilde{k}, (\mathcal{O} \otimes_k \tilde{k})_x) \to \mathrm{Hom}_{\mathcal{O}\otimes_k \tilde{k}}(V \otimes_k \tilde{k}, \tilde{k}_x)$$
$$\approx \mathrm{Hom}_{\tilde{k}_x}(V \otimes_{(\mathcal{O}\otimes_k \tilde{k})} \tilde{k}_x, \tilde{k}_x).$$

- *Suppose that* Y *is a non-meager subset of* $X \otimes_k \tilde{k}$ *so that for* $x \in Y$ *the system* $\Xi \otimes_k \tilde{k}$ *is holomorphic at* x, *and so that for* $x \in Y$ *the pointwise system* $(\Xi \otimes_k \tilde{k})_x$ *at* x *has a unique solution* λ_x. *Then there is a unique* generic solution *to* $\Xi \otimes_k \tilde{k}$, *and this solution is* rational *over* k.

Proof. First, we reduce to the case that at most one of the f_i is not $0 \in \mathcal{M}$. If all f_i are already 0, we are done. So suppose that some f_{i_o} is non-zero. For another index i, replace the condition

$$\lambda(v_i) = f_i$$

by the condition

$$\lambda\left(v_i - \frac{f_i}{f_{i_o}} v_{i_o}\right) = 0.$$

The collection of generic solutions is unchanged by such an adjustment, and off a meager subset of X this change gives pointwise systems equivalent to the original pointwise system Ξ_x. Thus, overlooking a meager subset of X, we can assume without loss of generality that for at most one index i_o is f_{i_o} not 0.

Also, we ignore the countably-many hypersurfaces on which the system Ξ fails to be holomorphic.

If there is a unique generic solution λ, then by the proposition the vectors μ_i span $V \otimes_{\mathcal{O}} \mathcal{M}$, and if f_{i_o} is not 0 then μ_{i_o} does not lie in the $\mathcal{M}$-span of the other vectors.

Let $v_1, v_2, \ldots$ be a countable collection of generators for the $\mathcal{O}$-module V. The images of these v_j span all the vectorspaces $V \otimes_{\mathcal{O}} k_x$ for $x \in X$. Then $v_1 \otimes 1, v_2 \otimes 1, \ldots$ is an $\mathcal{M}$-basis for $V \otimes_{\mathcal{O}} \mathcal{M}$. Express each $m_j \otimes 1$ as an $\mathcal{M}$-linear combination of the μ_i as

$$v_j \otimes 1 = \sum_i g_{ji}\, \mu_i$$

with $g_{ji} \in \mathcal{M}$. Off the meager set of points $x \in X$ where some one of the g_{ji} fails to be holomorphic, tensoring with $k_x = \mathcal{O}_x / x\mathcal{O}_x$ expresses the image of v_j as a k_x-linear combination of the images of μ_i in $V \otimes_{\mathcal{O}} k_x$. (We have already excluded the meager set on which any one of the μ_i fails to be holomorphic). That is, we conclude that off a meager set the pointwise system Ξ_x has at most one solution.

Then, given a generic solution λ, off the meager set where λ fails to be holomorphic, the image of λ under the natural map

$$\mathrm{Hom}_{\mathcal{M}}(V \otimes_{\mathcal{O}} \mathcal{M}, \mathcal{O}_x) \approx \mathrm{Hom}_{\mathcal{O}}(V, \mathcal{O}_x)$$
$$\to \mathrm{Hom}_{\mathcal{O}}(V, k_x) \approx \mathrm{Hom}_{k_x}(V \otimes_{\mathcal{O}} k_x, k_x)$$

is a solution of the pointwise system Ξ_x. Thus, *uniqueness of the generic solution implies uniqueness of solution to pointwise systems* Ξ_x *for* x *off a meager set.*

On the other hand, suppose that for x in a non-meager subset Y (on which Ξ is holomorphic) the pointwise system Ξ_x has a unique solution. To prove that a generic solution exists, we must show that μ_{i_o} does not lie in the $\mathcal{M}$-span of the other μ_i. Indeed, if

$$\mu_{i_o} = \sum_{i \neq i_o} g_i \mu_i$$

with $g_i \in \mathcal{M}$, then, off the meager set where some one of the g_i or μ_i fails to be holomorphic, the image of μ_{i_o} in $V \otimes \mathcal{O}k_x$ would be a k_x-linear combination of the images of the other elements μ_i with $i \neq i_o$. The proposition above assures that this cannot happen. Thus, there is at least one generic solution.

Suppose that there were two distinct generic solutions, and call their difference δ. The pointwise uniqueness hypothesis assures that, for x in a non-meager set Y (on which we may assume δ holomorphic), the image of δ in $\mathrm{Hom}_{k_x}(V \otimes_{\mathcal{O}} k_x, k_x)$ is 0. Thus, for all $x \in Y$, for all of the countably-many $\mathcal{O}$-generators m_j for V,

$$\delta(m_j \otimes 1) \subset x\mathcal{O}_x.$$

But we saw above that the intersection of sets $x\mathcal{O}_x$ for x ranging over any non-meager set is $\{0\}$. Thus, $\delta(m_j \otimes 1) = 0$, proving uniqueness. □

Now we can return to discussion of parametrized families of smooth representations, in the setting of linear systems and parametrized linear systems.

Let $\{t_i\}$ be a countable set of $\mathcal{O}$-generators for V. Let $\{g_j\}$ be a countable dense subset of G. The condition that an $\mathcal{O}$-linear map

$$\lambda : M \otimes_{\mathcal{O}} \mathcal{M} \to \mathcal{M}$$

be an $\mathcal{O}$-parametrized family of intertwining operators is that

$$\lambda(\pi(g_j)t_i) = \lambda(t_i)$$

for all indices i, j, since the isotropy group of each t_i is *open* (by smoothness). This is

$$\lambda(\pi(g_j)t_i - t_i) = 0.$$

The countable collection of such conditions is a homogeneous parametrized linear system Ξ with coefficients in $V \otimes_{\mathcal{O}} \mathcal{M}$.

First we prove the Rationality Lemma without concern for extending scalars. That is, we suppose that $\tilde{k} = k$. Then the hypothesis of the theorem is that the associated pointwise system Ξ_x has a solution space for dimension less than or equal 1 for x in the non-meager subset Y of X.

Consider the single further condition

$$\lambda(\mu_o) = 1$$

and let Ξ^+ be the system obtained from the homogeneous system Ξ by adjoining this condition. On a non-meager set Ξ^+ has the *unique* solution μ_o. Thus, by the previous theorem, there is a unique generic solution φ to Ξ^+. (By earlier results, necessarily $\varphi_x = \psi_{(x)}$ on a non-meager subset). And, by the previous theorem, the uniqueness of the generic solution to Ξ^+ implies that off a meager subset of X the pointwise solution is unique.

Finally, consider the issue of extension of scalars. The previous discussion applies as well to the representation $\pi \otimes_k \tilde{k}$ of G on $V \otimes_k \tilde{k}$, yielding an intertwining

$$\tilde{\varphi} : \pi \otimes_k \tilde{k} \to \mathcal{O} \otimes_k \tilde{k}$$

Further, by the theorem above, this intertwining is k-rational. Thus, in particular, rather than merely being able to assert that $\varphi(\mu_o) \in \mathcal{M} \otimes_k \tilde{k}$, we know that actually $\varphi(\mu_o) \in \mathcal{M}$. □

3.2. Orbit filtrations on test functions. The material of this section has its origin in [**Bru**], although some technical adjustments are necessary for present use. See also treatment of related matters in [**BeZ76, BeZ76, Be84**].

Let G be a locally compact, Hausdorff topological group. For a *closed* subgroup H of G, the quotient G/H is Hausdorff because H is closed. Let Ω be a Hausdorff topological space with a continuous transitive action of G upon it. Suppose that G has a *countable basis.* Let x_0 be any fixed element of Ω, and let

$$G_{x_0} := \{g \in G : gx_0 = x_0\}$$

be the *isotropy group* of x_0 in G. Recall the basic

LEMMA: *The natural map*

$$G/G_{x_0} \to \Omega$$

by $gG_{x_0} \to gx_0$ *is a homeomorphism.*

PROOF. The map $gG_{x_0} \to gx_0$ is a continuous bijection, by assumption. We need to show that it is *open.* Let U be an open subset of G, and take a compact neighborhood V of $1 \in G$ so that $V^{-1} = V$ and $gV^2 \subset U$ for fixed $g \in U$.

Since G has a countable basis, there is a countable list $g_1, g_2, \ldots$ of elements of G so that $G = \bigcup_i g_i V$. Let $W_n = g_n V x_o$. By the transitivity, $\Omega = \bigcup_i W_i$. Now W_n is compact, being a continuous image of a compact set, so is closed since it is in the Hausdorff space Ω.

Since Ω is locally compact and Hausdorff, by Urysohn's Lemma it is *regular.* In particular, if no W_n contained an open set, then there would be a sequence of non-empty open sets U_n with compact closure so that

$$U_{n-1} - W_{n-1} \supset \bar{U}_n$$

and

$$\bar{U}_1 \supset \bar{U}_2 \supset \bar{U}_3 \supset \ldots$$

Then $\bigcap \bar{U}_i \neq \emptyset$, yet this intersection fails to meet any W_n, contradiction.

Therefore, some $W_m = g_m V x_0$ contains an open set S of Ω. For $h \in V$ so that $hx_0 \in S$,

$$gx_0 = gh^{-1}hx_0 \in gh^{-1}S \subset gh^{-1}Vx_0 \subset gV^{-1} \cdot Vx_0 \subset Ux_0.$$

Therefore, gx_0 is an interior point of Ux_0, for all $g \in U$. □

Let Ω be any locally compact Hausdorff space or a complete metric space. Recall that a locally compact Hausdorff space with a countable basis *is* metrizable (and complete). Let G be a locally compact Hausdorff topological group, with a countable basis, so that G has a countable dense subset. Suppose that G acts continuously on Ω. From the Baire category theorem we can deduce several properties of the orbits of G on Ω, as follows.

LEMMA: *Suppose that the collection* $G\backslash\Omega$ *of* G*-orbits on* Ω *is* countable. *Then there is at least one open* G*-orbit in* Ω*, and every* G*-orbit is open in its own closure in* Ω*.*

PROOF. By the Baire category theorem, at least one orbit $G_o x$ (with $x_o \in \Omega$) has non-empty interior. Let U be an open subset of Gx_o, and $g_o x_o \in U$ with $g_o \in G$. Then for arbitrary $g \in G$

$$gx_o = (gg_o^{-1})(g_o x_o) \in (gg_o^{-1})U \subset Gx_o.$$

Thus, every point gx_o of Gx_o is an interior point, so Gx_o is open.

Now fix $x_1 \in \Omega$. The closure Ω_1 of Gx_1 in Ω is still locally compact, Hausdorff, and countably-based, so is complete metrizable. Also, it is still acted upon continuously by G. By the previous argument, there is at least one open orbit Gx_2 inside Ω_1. Every point of Gx_2, including interior points, is in the closure of Gx_1, so necessarily $Gx_2 = Gx_1$, as claimed. □

Let $\mathcal{O}$ be a commutative ring with identity. As earlier, for a totally disconnected space Ω let $C_c^\infty(\Omega)$ denote the collection of compactly-supported locally constant $\mathcal{O}$-valued **test functions** on Ω. The following two lemmas are elementary:

LEMMA: *Let $\Omega_1, \ldots, \Omega_n, \ldots$ be disjoint open sets in a totally disconnected space Ω. Then the direct sum of restriction maps*

$$f \to \oplus (f|_{\Omega_i})$$

$$C_c^\infty\left(\bigcup_i \Omega_i\right) = \bigoplus_i C_c^\infty(\Omega_i)$$

is an isomorphism. □

LEMMA: *For an open subset Y of a totally disconnected space Ω we have an exact sequence*

$$0 \to C_c^\infty(Y) \to C_c^\infty(\Omega) \to C_c^\infty(\Omega - Y) \to 0$$

where the map $C_c^\infty(\Omega) \to C_c^\infty(\Omega - Y)$ is by restriction. □

Analogues of the above lemmas for somewhat more general spaces of test functions are necessary. As earlier, the space $C_c^\infty(H\backslash\Omega, \sigma)$ is defined to be those σ-valued functions f on Ω which are compactly-supported left modulo H and so that

$$f(hx) = \sigma(h)(f(x)).$$

Here σ is a smooth representation of H on an $\mathcal{O}$-module (the latter also denoted σ).

The obvious analogues are not true in much greater generality than when Ω is constructed as follows, as in the theorem giving the orbit criterion for locally strong meromorphy, which will suffice for our purposes. Let $\tilde{H}$ be a locally compact totally disconnected topological group of which H is a closed subgroup. Let Y be a closed subset of $\tilde{H}$ stable under left multiplication by H. Let Z be a locally compact totally disconnected space on which H acts trivially. Then take

$$\Omega = Y \times Z.$$

PROPOSITION: *Let U be an open subset of Ω, with Ω as just above. Let σ be an $\mathcal{O}$-parametrized smooth representation of H on an $\mathcal{O}$-module V. Then we have an exact sequence*

$$0 \to C_c^\infty(H\backslash U, \sigma) \to C_c^\infty(H\backslash\Omega, \sigma) \to C_c^\infty(H\backslash(\Omega - U), \sigma) \to 0$$

where the first map is 'extend-by-zero', and the second is restriction of functions from Ω to $\Omega - U$.

PROOF. If extending elements of $\mathrm{C}_c^\infty(H\backslash U,\sigma)$ by zero off Y really were to yield elements of $\mathrm{C}_c^\infty(H\backslash\Omega,\sigma)$, then the injectivity would be clear. To justify extending by zero, we must show that a subset C of U which is compact left modulo H *in* U is also compact left modulo H *in* Ω. Here that we use the fact that all this is happening inside a larger topological group $\tilde{H}$. In particular, since H is a *closed* subgroup of $\tilde{H}$, the quotient $H\backslash\tilde{H}$ is *Hausdorff*. Then $H\backslash C$ is assumed compact in $H\backslash U$, and we are to show that it is compact in $H\backslash\Omega$. But these *quotient* topologies are the same as the *subspace* topologies from $H\backslash\tilde{H}$, from the first lemma above. Thus, compactness of $H\backslash HC$ in $H\backslash U$ implies compactness of $H\backslash HC$ in $H\backslash\tilde{H}$. Then since $H\backslash\Omega$ is closed and $H\backslash\tilde{H}$ is Hausdorff, $H\backslash HZ$ is surely compact in $H\backslash\Omega$. Thus, we have proven that the extension-by-zero map makes sense, and it is certainly an injection.

Restriction to $\Omega-U$ of a function with compact support left mod H inside U certainly gives 0; and, if the restriction to $\Omega-U$ of a function is 0, then its support must be contained in U. Again, compactness of support is assessed in the quotient $H\backslash\tilde{H}$. This proves exactness at the middle joint.

What remains to be checked is the surjectivity of the restriction map from $\mathrm{C}_c^\infty(H\backslash\Omega,\sigma)$ to $\mathrm{C}_c^\infty(H\backslash(\Omega-U),\sigma)$ for an open H-stable subset U of Ω. It is elementary that the restriction map

$$\mathrm{C}_c^\infty(\Omega,\mathcal{O})\to\mathrm{C}_c^\infty(\Omega-U,\mathcal{O})$$

is *surjective*. Thus, by the right exactness of tensor products, tensoring over $\mathcal{O}$ with the representation space V of σ still gives a surjective restriction map

$$\mathrm{C}_c^\infty(\Omega,\mathcal{O})\otimes_{\mathcal{O}}V\to\mathrm{C}_c^\infty(\Omega-U,\mathcal{O})\otimes_{\mathcal{O}}V.$$

The averaging maps

$$\alpha_\Omega:\mathrm{C}_c^\infty(\Omega,\mathcal{O})\otimes V\to\mathrm{C}_c^\infty(H\backslash\Omega,\sigma)$$
$$\alpha_{\Omega-U}:\mathrm{C}_c^\infty(\Omega-U,\mathcal{O})\otimes V\to\mathrm{C}_c^\infty(H\backslash(\Omega-U),\sigma)$$

given (in both cases) by the formula

$$f\otimes v\ \to\ (x\to\int_H f(hx)\,\sigma(h^{-1})v\,dh$$

are *surjective*.

Thus, all arrows except possibly the lower horizontal one in the commuting square

$$\begin{array}{ccc}\mathrm{C}_c^\infty(\Omega,\mathcal{O})\otimes V & \to & \mathrm{C}_c^\infty(\Omega-U,\mathcal{O})\otimes V\\ \alpha_\Omega\downarrow & & \downarrow\alpha_{\Omega-U}\\ \mathrm{C}_c^\infty(H\backslash\Omega,\sigma)\otimes V & \to & \mathrm{C}_c^\infty(H\backslash(\Omega-U),\sigma)\otimes V\end{array}$$

are *surjections*. But then it is clear that the lower horizontal one is a surjection, as well. □

Let H and Ω be as just above, and let G be another totally disconnected locally compact topological group acting (continuously) on the right on Ω. Suppose that there are *finitely*-many $G\times H$-orbits on Ω. From above, each orbit is open in its closure.

Write $z\geq w$ for $G\times H$-orbits z,w on Ω if w is in the closure $\bar{z}$ of z. Since z,w are both $G\times H$-orbits, $\bar{z}$ and $\bar{z}\cap w$ are unions of orbits, and if w meets $\bar{z}$ (non-trivially) we can conclude that

$$w\cap\bar{z}=w$$

Thus, if w merely *meets* the closure $\bar{z}$ of z, then $z \geq w$. Since the closure of the closure of a set is the closure, the relation $\leq$ is a partial ordering.

For an orbit z, let

$$[z] = \bigcup_{w \leq z} w$$

$$(z) = \bigcup_{w < z} w = [z] - z.$$

Let M be a module. Let T be a poset with order $<$, and suppose that for each $t \in T$ we have a submodule $M_t \subset M$, and that $s < t$ implies that $M_s \subset M_t$. Then the collection

$$\{M_t : t \in T\}$$

is a **filtration** of M indexed by T. For $t \in T$ the **associated graded piece** of M is the quotient object

$$\sum_{s \leq t} M_s / \sum_{s < t} M_s.$$

Thus, in the situation under consideration at present, we would consider $M = \mathrm{C}_c^\infty(H\backslash\Omega, \sigma)$, take T to be the collection of $G \times H$ orbits on Ω, and

$$M_z = \mathrm{C}_c^\infty(H\backslash[z], \sigma).$$

With the hypothesis that there are *finitely-many orbits*, the results above concerning test functions yield:

COROLLARY: *Suppose that there are finitely-many $G \times H$-orbits on Ω. Then the filtration of $\mathrm{C}_c^\infty(H\backslash\Omega, \sigma)$ indexed by the partially ordered set of $G \times H$-orbits has associated graded pieces*

$$\frac{\mathrm{C}_c^\infty(H\backslash[z], \sigma)}{\mathrm{C}_c^\infty(H\backslash(z), \sigma)} \approx \mathrm{C}_c^\infty(H\backslash z, \sigma).$$

That is, each orbit's own space of equivariant test functions appears among the graded pieces of the $G \times H$-orbit filtration on Ω, and these are all of the graded pieces. □

3.3. Exactness of some functors. Now we begin preparations for proof of the orbit criterion for locally strong meromorphy. Here we prove the exactness of some relatively elementary functors on smooth representation spaces. Some of this occurs in the unpublished notes [**C76**] in a different form.

Let $\mathcal{O}$ be a commutative ring with identity, and $\mathcal{A}$ an associative $\mathcal{O}$-algebra. Say that $\mathcal{A}$ is an *idempotented algebra* if for any finite collection $\eta_1, \ldots, \eta_n$ of elements of $\mathcal{A}$ there is an idempotent $e \in \mathcal{A}$ so that $e\eta_i = \eta_i = \eta_i e$ for all i.

A module M over an idempotented algebra $\mathcal{A}$ is *smooth* if, for every finite list $m_1, \ldots, m_n$ of elements of M there is an idempotent e in $\mathcal{A}$ so that $em_i = m_i$ for all i. For example, $\mathcal{A}$ is smooth as a (left) module over itself.

Given *any* module M over an idempotented algebra $\mathcal{A}$, we form the submodule M^∞ of *smooth vectors* in M by taking the set of $m \in M$ so that there exists some idempotent $e \in \mathcal{A}$ so that $em = m$. Then M^∞ is the unique maximal smooth submodule of M.

An $\mathcal{O}$-algebra $\mathcal{A}$ is *augmented* if it is equipped with an $\mathcal{O}$-algebra homomorphism $\varepsilon : \mathcal{A} \to \mathcal{O}$ which maps every *sufficiently small* idempotent to 1. For fixed augmentation ε, the *trivial* $\mathcal{A}$-module denoted by ε is the $\mathcal{O}$-module $\mathcal{O}$ itself on which $\eta \in \mathcal{A}$ acts by

$$\eta r = \varepsilon(\eta) \cdot r$$

for all $r \in \mathcal{O}$.

In the case of a Hecke algebra $\mathcal{G}$ of a totally disconnected group G, the *standard augmentation* is simply the *integral* of a locally-constant, compactly-supported $\mathcal{O}$-valued function: thus,

$$\varepsilon(\mathrm{ch}_{KgK}) = \mathrm{meas}(KgK)$$

for every $g \in G$, where ch_{KgK} is the characteristic function of KgK and *meas* is a $\mathbf{Q}$-valued right Haar measure. (These 'integrals' are in fact finite sums, so there is no analysis, and this makes sense over an arbitrary field of characteristic zero, at least).

PROPOSITION: *For any fixed idempotent e in an idempotented $\mathcal{O}$-algebra $\mathcal{A}$ the functor $V \to eV$ is an* exact functor *from smooth $\mathcal{A}$-modules to $e\mathcal{A}e$-modules.*

PROOF. Suppose that

$$A \xrightarrow{\phi} B \xrightarrow{\psi} C$$

is an exact sequence of $\mathcal{A}$-modules. Then for $x \in A$

$$\psi(\phi(ex)) = e(\psi(\phi(x)) = e \cdot 0 = 0$$

so the kernel of ψ on eB is contained in the image $\phi(eA) = e\phi(A)$. On the other hand, if $\psi(ey) = 0$ for $y \in B$, then invoking the exactness of the original sequence take $x \in A$ so that $\phi(x) = ey$. But then also

$$\phi(ex) = e\phi(x) = e(ey) = ey$$

since e is idempotent. This verifies the asserted exactness. □

We can abstract the notion of *Jacquet module.* Say that an augmented idempotented algebra $\mathcal{N}$ has **large idempotents** if, given a finite collection $f_1, \ldots, f_n \in \mathcal{N}$, there is an idempotent e so that

$$ef_i = f_i e = \varepsilon(f) \cdot e$$

where ε is the augmentation. (This abstracts the situation in which $\mathcal{N}$ is the Hecke algebra of a unipotent p-adic group).

The **ε-co-isotype** V_ε is the largest quotient of V so that any $\mathcal{N}$-homomorphism

$$\varphi : V \to \varepsilon$$

factors through the quotient map $V \to V_\varepsilon$. This is the natural dual notion to the more common usage of *ε-isotype*, meaning the smallest submodule V^ε so that every homomorphism $\varepsilon \to V$ factors through the inclusion $V^\varepsilon \subset V$.

We may define the **Jacquet functor** $\mathcal{J}$ on $\mathcal{N}$-modules to be that which associates to a smooth $\mathcal{N}$-module V the ε-co-isotype V_ε. Again we have an exactness result which does not depend upon the nature of the 'scalars' $\mathcal{O}$:

PROPOSITION: *For an idempotented $\mathcal{O}$-algebra $\mathcal{N}$ with 'large idempotents', and with augmentation ε, the functor $\mathcal{J}$ which takes the ε-co-isotype V_ε of a smooth $\mathcal{N}$-module V is* exact.

PROOF. First, it is elementary to check that V_ε is the quotient of V by the $\mathcal{O}$-submodule generated by all expressions of the form

$$\eta \cdot v - \varepsilon(\eta)v$$

with $\eta \in \mathcal{N}$ and $v \in V$.

Second, we claim that an element $v \in V$ is in the kernel of the quotient map $q : V \to V_\varepsilon$ of the ε co-isotype if and only if there is an idempotent e in $\mathcal{N}$ so that $ev = 0$.

If $v = \sum_i (\eta_i v - \varepsilon(\eta_i)v)$ is in the kernel of q, then take an idempotent e so that $e\eta_i = \varepsilon(\eta_i)e$ for all i. Then

$$e(\eta_i v - \varepsilon(\eta_i)v) = 0.$$

On the other hand if $ev = 0$, then

$$v = 1 \cdot v - 0 \cdot v = \varepsilon(e)v - ev$$

which lies in the kernel of q. This proves the claim.

Suppose that

$$A \xrightarrow{f} B \xrightarrow{g} C$$

is an exact sequence of $\mathcal{N}$-modules.

Suppose that $(\mathcal{J}g)(q_B b) = 0$ where $q_B b$ is the image in $\mathcal{J}B$ of $b \in B$ via the quotient map $q_B : B \to \mathcal{J}B$. Then by definition of $\mathcal{J}$ we find that $q_C(gb) = 0$, where $q_C : C \to \mathcal{J}C$ is the quotient map. Thus, by the first observation of this proof, for a sufficiently large idempotent e of $\mathcal{N}$ we have $e(gb) = 0$. Since g is $\mathcal{N}$-linear, this implies that $g(eb) = 0$. By the exactness, there is $a \in A$ so that $f(a) = eb$. Then

$$f(ea) = ef(a) = e(eb) = eb.$$

That is, the kernel of $\mathcal{J}g$ is contained in the image of $\mathcal{J}f$.

And

$$(\mathcal{J}g)(\mathcal{J}f)(q_A a) = q_C((g \circ f)(a)) = q_C(0) = 0$$

simply follows from our (correct) presumption that $\mathcal{J}$ is a functor. □

3.4. Frobenius Reciprocity for parametrized families. For a locally compact Hausdorff totally disconnected group G, let $\mathcal{G} = \mathcal{G}_{\mathbf{Q}}$ be the Hecke algebra of locally-constant compactly-supported $\mathbf{Q}$-valued functions on the totally disconnected group G. For any $\mathbf{Q}$-algebra $\mathcal{A}$, let

$$\mathcal{G}_{\mathcal{A}} = \mathcal{G} \otimes_{\mathbf{Q}} \mathcal{A}$$

which can be naturally identified with the collection of $\mathcal{A}$-valued compactly-supported locally-constant functions on G. The product is the convolution

$$(\eta * \zeta)(g) = \int_G \eta(gh^{-1})\,\zeta(h)\,dh$$

where dh denotes a $\mathbf{Q}$-valued right Haar measure on G. The latter exists since G is totally disconnected. Note that such an integral with compact support and locally-constant integrand uses only *finite* additivity, and thus we can make sense of such integrals with integrands having values in any k-vectorspace whatsoever.

For an $\mathcal{O}$-parametrized family π of smooth representations of G on an $\mathcal{O}$-module V, there is a $\mathcal{G}_{\mathcal{O}}$-module structure on V by

$$\eta \cdot m = \int_G \eta(g)\,\pi(g)m\,dg.$$

Thus, $\mathcal{O}$-linear representations of G become $\mathcal{G} \otimes \mathcal{O}$-modules.

For a compact-open subgroup K of G, let e_K be the element of $\mathcal{G}_{\mathbf{Q}} \subset \mathcal{G}_{\mathcal{O}}$ given by

$$e_K = \frac{\text{characteristic function of } K}{\text{measure of } K}$$

A $\mathcal{G}_{\mathcal{O}}$-module V is **smooth** if for every finite collection $v_1, \ldots, v_n$ of elements of V there is a compact-open subgroup K of G so that $e_K m_i = m_i$ for all indices i. (The functor of the previous paragraph attaches *smooth* $\mathcal{G}_{\mathcal{O}}$-modules to $\mathcal{O}$-parametrized families of smooth representations of G). Then from any smooth $\mathcal{G}_{\mathcal{O}}$-module V one can recover a smooth representation π of G by

$$\pi(g)m = \chi_{gK} \cdot m$$

where χ_{gK} is the characteristic function of the subset gK of G, and where K is a small-enough compact-open subgroup of G so that $e_K m = m$. This functor is the inverse of the functor of the previous paragraph.

The $\mathcal{G}$-module $\mathcal{G}$ itself has at least two natural (left) $\mathcal{G}$-module structures on it. The first is $\mathcal{G}$ itself: the multiplication is

$$\eta \cdot f = \eta * f \quad \text{(convolution)}$$

which directly reflects the ring structure. On the other hand, let M_{rt} be the left $\mathcal{G}$-module consisting of $\mathcal{G}$ itself, but with the module structure

$$\eta \cdot m = m * \check{\eta}$$

for $\eta \in \mathcal{G}$ and $m \in M_{\mathrm{rt}}$, where

$$\check{\eta}(g) = \frac{\eta(g^{-1})}{\delta_G(g)}$$

with δ_G being the modular function of G, normalized by

$$\delta_G(g_o) = \frac{d(g_o g)}{dg}.$$

Direct computation verifies the isomorphism of $\mathcal{G}$ modules

$$\Psi : M_{\mathrm{rt}} \to \mathcal{G}$$

by

$$\Psi(m) = \check{m}.$$

Let $\mathcal{H}$ be the Hecke algebra of the closed subgroup H of G. Give $\mathcal{G}$ the natural *right* $\mathcal{H}$-module structure as follows:

$$(f \cdot \zeta)(g_o) = \int_H f(g_o h^{-1})\, \zeta(h)\, dh$$

where $\zeta \in \mathcal{H}$ and dh refers to a $\mathbf{Q}$-valued right Haar measure on H. The associativity follows from the associativity of convolution on H.

PROPOSITION: *Let σ be a smooth $\mathcal{O}$-parametrized family of representations of a closed subgroup H of a totally disconnected group G. Let $\mathcal{G}$ be the Hecke algebra of G, and $\mathcal{H}$ the Hecke algebra of H. Then the map*

$$\Phi(f \otimes v) = \alpha_\sigma(\check{f} \otimes v)$$

defines a $\mathcal{G}_{\mathcal{O}}$-isomorphism

$$\Phi : \mathcal{G} \otimes_{\mathcal{H}} \left(\sigma \otimes \frac{\delta_G}{\delta_H}\right) \to c - \mathrm{Ind}_N^G \sigma$$

where $\mathcal{G}$ has the $\mathcal{G} \times \mathcal{H}$-bimodule structure as above.

PROOF. The only serious issue is that this is well-defined: we must check that

$$\Phi(f\zeta \otimes v) = \Phi(f \otimes \zeta v).$$

Computing:

$$\begin{aligned}\Phi(f\zeta \otimes v)(g) &= \alpha_\sigma(((f\zeta)\check{} \otimes v)(g) \\ &= \int_H \sigma(h^{-1}) \left[\int_H \frac{f(hg)^{-1}h_2^{-1})}{\delta_G(hg)}\, \zeta(h_2)\, dh_2\, v\right] dh \\ &= \int_H \int_H \sigma(h^{-1})\, \frac{f(g^{-1}h^{-1})}{\delta_G(hg)}\, \frac{\sigma(h_2)\delta_H(h_2^{-1})}{\delta_G(h_2^{-1})} \zeta(h_2)\end{aligned}$$

by replacing h by $h_2^{-1}h$. Then this is

$$\int_H \sigma(h^{-1})\, \check{f}(hg) \left(\sigma \frac{\delta_G}{\delta_H}\right)(\zeta)v\, dh\, dh_2 = \alpha_\sigma(\check{f} \otimes \zeta v)(g) = \Phi(f \otimes \zeta v)(g)$$

which verifies that the map is well-defined. Then the fact that the indicated map is an isomorphism follows easily. □

The group-theoretic **Frobenius Reciprocity** isomorphism

$$\mathrm{Hom}_{G\times k}(\pi, \mathrm{Ind}_H^G \sigma) \approx \mathrm{Hom}_{H\times k}(\mathrm{Res}_H^G\, \pi, \sigma)$$

does not hold in general for $\mathcal{O}$-linear representations, but only for linear representations over *fields* k, where σ is a k-linear smooth representation of H and π is a k-linear representation of G. The isomorphism is given by

$$\Phi \longrightarrow \varphi_\Phi$$

described by

$$\varphi_\Phi(v) = \Phi(v)(1_G)$$

for $v \in \pi$, with inverse

$$\varphi \longrightarrow \Phi_\varphi$$

given by

$$\Phi_\varphi(v)(g) = \varphi(\pi(g)v).$$

On the other hand, in the language of modules over Hecke algebras, we have:

PROPOSITION:

$$\begin{aligned}\mathrm{Hom}_{\mathcal{G}_\mathcal{O}}(\pi \otimes \text{c-Ind}_H^G \sigma, \mathcal{O}) &\approx \mathrm{Hom}_{\mathcal{G}_\mathcal{O}}\left(\pi \otimes \left(\mathcal{G}_\mathcal{O} \otimes_{\mathcal{H}_\mathcal{O}} \sigma \frac{\delta_G}{\delta_H}\right), \mathcal{O}\right) \\ &\approx \mathrm{Hom}_{\mathcal{H}_\mathcal{O}}\left(\pi \otimes \sigma \frac{\delta_G}{\delta_H}, \mathcal{O}\right)\end{aligned}$$

and the map is $\Phi \longrightarrow \varphi_\Phi$ described by

$$\varphi_\Phi(v \otimes x) = \Phi(v \otimes (e \otimes x))$$

where $v \in \pi$, $x \in \sigma$, and $e \in \mathcal{G}_\mathcal{O}$ is any idempotent 'small enough' so that $ev = v$. The inverse map is $\varphi \longrightarrow \Phi_\varphi$ given by

$$\Phi_\varphi(v \otimes (\eta \otimes x)) = \varphi(\check{\eta} v \otimes x)$$

for $\eta \in \mathcal{G}_\mathcal{O}$. □

3.5. Proof of the orbit criterion for strong meromorphy. Keep notation as in the statement of the orbit criterion for locally strong meromorphy. With the preparation of the previous sections, we can prove the orbit-criterion theorem.

Let Φ be a meromorphic $\mathcal{O}$-parametrized family of intertwining operators

$$\Phi \in \mathrm{Hom}_{G\times\mathcal{O}}(\mathrm{C}_c^\infty(H\backslash\Omega,\sigma)\otimes_{\mathcal{O}} \text{c-Ind}_P^G\chi, \mathcal{M}).$$

Let

$$\varphi_\Phi \in \mathrm{Hom}_{P\times\mathcal{O}}\Big(\mathrm{C}_c^\infty(H\backslash\Omega,\sigma)\otimes_{\mathcal{O}} \chi\frac{\delta_G}{\delta_H}, \mathcal{M}\Big)$$

be the element corresponding to it by Frobenius Reciprocity. That is, with $f \in \mathrm{C}_c^\infty(H\backslash\Omega,\sigma)$ and $v \in \chi\delta_G/\delta_H$,

$$\varphi_\varphi(f\otimes v) = \Phi(f\otimes(e'\otimes v))$$

for any idempotent $e' \in \mathcal{G}$ sufficiently small so that $e'f = f$. Thus, for convenience, let

$$\chi' = \chi\frac{\delta_G}{\delta_H}.$$

LEMMA: *Let x be a height one prime in $\mathcal{O}$. The meromorphic family Φ is locally strongly meromorphic at x if and only if the meromorphic family φ_Φ associated to it by Frobenius Reciprocity (as just above) is locally strongly meromorphic.*

PROOF. It is important to use the tensor product and Hecke-algebra version of Frobenius Reciprocity, rather than the more group-theoretic, in order to verify this result.

Certainly the values of φ_Φ are among the values of Φ, by the definition of φ_Φ in terms of Φ. Conversely, for $f \in \mathrm{C}_c^\infty(H\backslash\mathbf{C},\sigma)$, $\eta\in\mathcal{G}$, $v\in\delta_G/\delta_H$ we have

$$\Phi(f\otimes(\eta\otimes v)) = \phi(\check{\eta}f\otimes v)$$

Thus, the collection of values of $\psi\Phi$ is identical to the collection of values assumed by $\psi\varphi_\Phi$. Therefore, if $\varpi_x^N\Phi$ takes values in $\mathcal{O}_x$ (rather than merely $\mathcal{M}$), then $\varpi_x^N\varphi_\Phi$ has the same property, and conversely. □

Thus, we wish to prove that any meromorphic $\mathcal{O}$-parametrized family of intertwining operators

$$\varphi \in \mathrm{Hom}_{P\times\mathcal{O}}(\mathrm{C}_c^\infty(H\backslash\Omega,\sigma)\otimes_{\mathcal{O}}\chi', \mathcal{M})$$

is ineluctably *locally strongly meromorphic everywhere.*

Since we will only consider one height-one prime x at a time, we may as well suppose that $\mathcal{O}$ is *already* a discrete valuation ring with unique non-zero prime x. Thus, we may as well suppose that $\mathcal{O} = \mathcal{O}_x$ already. This simplification will allow us to write simply 'π' rather than '$\pi\otimes_{\mathcal{O}}\mathcal{O}_x$' when convenient.

LEMMA: *Let π be an $\mathcal{O}$-parametrized family of smooth representations of a totally disconnected group G. Let Φ be a meromorphic $\mathcal{O}$-parametrized family of intertwining operators Φ in* Hom)$G\times\mathcal{O}(\pi,\mathcal{M})$. *Define*

$$V_i = \{v\in\pi\otimes_{\mathcal{O}}\mathcal{O}_x : \varpi^i\Phi(v)\in\mathcal{O}_x\}$$

(with local parameter ϖ for x), obtaining a filtration

$$V_o \subset V_1 \subset V_2 \subset \ldots \subset \pi\otimes_{\mathcal{O}}\mathcal{O}_x$$

If Φ fails to be locally strongly meromorphic at x then for every index i

$$\mathrm{Hom}_{G\times\mathcal{O}_x}(V_{i+1}/V_i, k_x) \approx \mathrm{Hom}_{G\times k_x}(V_{i+1}/V_i, k_x) \neq \{0\}.$$

PROOF. (Suppress the '$\otimes_{\mathcal{O}}\mathcal{O}_x$'). Certainly $\varpi V_i \subset V_{i-1}$, so the quotients V_i/V_{i-1} are indeed $k_x$vectorspaces.

If Φ is *not* locally strongly meromorphic at x, then no one of the V_i contains π entirely. Further, since $\operatorname{ord}_x\Phi(v)$ takes arbitrarily large negative values, by simply multiplying through by powers of ϖ we can obtain *every* value of $\operatorname{ord}_x\Phi(v)$. If $\operatorname{Hom}_{G\times k_x}(V_i/V_{i-1}, k_x)$ is non-trivial then actually

$$\operatorname{Hom}_{G\times\mathcal{O}_x}(V_{i-\ell}/V_{i-1-\ell}, k_x)$$

is also non-trivial, simply by multiplying everything by ϖ. Thus, *every* quotient V_i/V_{i-1} is a non-zero k_x-vectorspace and

$$\operatorname{Hom}_{G\times\mathcal{O}_x}(V_i/V_{i-1}, k_x)$$

is non-zero for *all* indices $i \geq 1$. □

Generally, if a $G\times\mathcal{O}_x$-module V has the property that for any ascending chain

$$V_0 \subset V_1 \subset \ldots$$

of $G\times\mathcal{O}_x$-submodules only *finitely-many* of the spaces

$$\operatorname{Hom}_{G\times k_x}(V_i/V_{i-1}, k_x)$$

are non-zero, then say that V is **co-iso-Noetherian**. References to the prime x and to the group G are suppressed.

Thus, the last lemma can be paraphrased as asserting that if π is co-iso-Noetherian, then Φ is locally strongly meromorphic.

So suppose that φ is *not* locally strongly meromorphic at x, and let V_i be a filtration of

$$C_c^\infty(H\backslash\Omega,\sigma)\otimes_{\mathcal{O}}\chi'$$

as in the lemma.

Next, we use the finiteness of the double-coset space $H\backslash\Omega/P$ to obtain a *finite* $H\times P$-orbit filtration of $C_c^\infty(H\backslash\Omega,\sigma)$ as $P\times\mathcal{O}_x$-module:

LEMMA: *Let*

$$\{0\} = U_o \subset U_1 \subset \ldots \subset U_n = V$$

be a finite *filtration by $G\times\mathcal{O}_x$-modules of a $G\times\mathcal{O}_x$-module V. Let Φ be a meromorphic $\mathcal{O}$-parametrized family in*

$$\operatorname{Hom}_{G\times\mathcal{O}_x}(V,\mathcal{O}_x)$$

and let

$$V_i = \{v\in V : \varpi^i\Phi(v)\in\mathcal{O}_x\}$$

where ϖ is a local parameter for x. Suppose that

$$V_o = \{0\} \subset V_1 \subset \ldots \subset V$$

is an infinite *filtration of V. Then for some index i the filtration*

$$\frac{(U_i\cap V_o)+U_{i-1}}{U_{i-1}} \subset \frac{(U_i\cap V_1)+U_{i-1}}{U_{i-1}} \subset \ldots \subset \frac{(U_i\cap V_n)+U_{i-1}}{U_{i-1}} \subset \ldots$$

of U_i/U_{i-1} is infinite. *Let i be the smallest such index. Then, further, letting*

$$\Gamma_j = \frac{(U_i\cap V_j)+U_{i-1}}{U_{i-1}}$$

there are infinitely-many indices j so that

$$\mathrm{Hom}_{G\times k_x}(\Gamma_j/\Gamma_{j-1}, k_x) \neq \{0\}.$$

PROOF. This is an extension of a Jordan-Holder argument.

Suppose the assertion of the lemma were false, i.e., that the induced filtrations of all of the quotients U_i/U_{i-1} were finite. For each of the finitely-many indices i let $j(i)$ be the smallest index so that $j(i) \geq j(i-1)$ and so that

$$(U_i \cap V_{j(i)}) + U_{i-1} = U_i.$$

Since $U_o = \{0\}$, $U_1 \subset V_{j(1)}$. We prove $U_i \subset V_{j(i)}$ by induction:

$$V = U_i \subset (U_i \cap V_{j(i)}) + U_{i-1} \subset (U_i \cap V_{j(i)}) + (U_{i-1} \cap V_{j(i-1)})$$
$$\subset (U_i \cap V_{j(i)}) + (U_{i-1} \cap V_{j(i)}) \subset U_i \cap V_{j(i)}$$

since

$$U_{i-1} \cap V_{j(i)} \subset U_i \cap V_{j(i)}.$$

Then

$$V = U_n \subset V_{j(n)}$$

contradicting the fact that the filtration of the V_j is infinite.

Let i be the smallest index such that this filtration is infinite. The argument just given *does* prove that for $i' < i$

$$U_{i'} \subset V_{j(i')}$$

for some index j. In particular,

$$U_{i-1} \subset V_{j(i-1)}.$$

Thus, for $j \geq j(i-1)$,

$$U_i \cap V_j \supset U_i$$

and

$$(U_i \cap V_j) + U_{i-1} = U_i \cap V_j.$$

Thus, restricting $j \geq j(i-1)$, the chain

$$U_i \cap V_j \subset U_i \cap V_{j+1} \subset U_i \cap V_{j+2} \subset \ldots$$

must be infinite. Thus, for given j_o no matter how large, there is $j > j_o$ so that

$$\frac{U_i \cap V_j}{U_i \cap V_{j-1}} \neq 0.$$

In particular, some element v of V_j but not in V_{j-1} lies in U_i. Thus, $\varpi^j \Phi$ composed with reduction modulo ϖ is non-trivial on such $v \in U_i$, proving that Γ_j/Γ_{j-1} has non-trivial homomorphisms to k_x. □

We have already shown that the graded pieces of the $H \times P$-orbit filtration on $C_c^\infty(H\backslash\Omega, \sigma)$ are of the form

$$\text{c-Ind}_{\Theta^\xi}^P \sigma_\xi$$

where

$$\sigma_\xi(p) = \sigma(\xi p \xi^{-1})$$

for $p \in P \cap \xi^{-1} H \xi$. Thus, if the original Φ were *not* locally strongly meromorphic at x, then for some $\xi \in H\backslash\Omega/P$ the there would be an infinite chain of subspaces V_i of

$$\text{c-Ind}_{\Theta^\xi}^P \sigma_\xi \otimes_{k_x} \chi' \otimes k_x$$

so that

$$\mathrm{Hom}_{P\times k_x}(V_i/V_{i-1}, k_x) \neq \{0\}.$$

Now $\text{c-Ind}_{\Theta^\xi}^P \sigma_\xi$ is an image of $C_c^\infty(P, \mathcal{O}_x) \otimes \sigma_\xi$ under an averaging map, where in the latter space P acts *trivially* on σ_ξ. Thus, if

$$\text{c-Ind}_{\Theta^\xi}^P \sigma_\xi \otimes \chi'$$

fails to have the co-iso-Noetherian property so will

$$C_c^\infty(P, \mathcal{O}_x) \otimes \sigma_\xi \otimes \chi'.$$

Since σ_ξ is finite-dimensional, this can happen only if already

$$C_c^\infty(P, \mathcal{O}_x) \otimes \chi'$$

fails to have the co-iso-Noetherian property.

Next, we use the fact that χ is a finite sum of trivial representations of N (since χ is finite-dimensional and trivial on N), and the assumption that N is an ascending union of compact open subgroups. The latter hypothesis assures that the *Jacquet functor* (i.e., trivial co-isotype functor) for N is *exact*, as shown earlier. And since N is normal in P, such co-isotypes are still P-representations. The exactness assures that

$$\mathcal{J}V_i/\mathcal{J}V_{i-1} \approx \mathcal{J}(V_i/V_{i-1})$$

for submodules V_i of $C_c^\infty(P, \mathcal{O}_x) \otimes \chi'$. Since generally (by the defining property of this co-isotype functor $\mathcal{J}$)

$$\mathrm{Hom}_{P\times\mathcal{O}_x}(V, k_x) \approx \mathrm{Hom} P \times \mathcal{O}_x(\mathcal{J}V, k_x)$$

we conclude that

$$\mathcal{J}(C_c^\infty(P, \mathcal{O}_x) \otimes \chi') \approx (\mathcal{J}(C_c^\infty(P, \mathcal{O}_x)) \otimes \chi'$$

fails to have the co-iso-Noetherian property.

But now the Jacquet module (co-isotype) $\mathcal{J}C_c^\infty(P, \mathcal{O}_x)$ is readily computed to be simply $C_c^\infty(N\backslash P, \mathcal{O}_x)$. Thus, from this point we can restrict our attention to M-representations. Thus, to this point, we have concluded that the original supposed failure of locally strong meromorphy implies that

$$C_c^\infty(M, \mathcal{O}_x) \otimes \chi'$$

fails to have the co-iso-Noetherian condition as $M \times \mathcal{O}_x$-module.

By hypothesis, for a certain idempotent e in the Hecke algebra $\mathcal{G}_M$ of M, $e\chi = \chi$ (referring to the representation space of χ). Certainly δ_H and δ_G are trivial on compact subgroups, so also $e\chi' = \chi'$. Therefore,

$$\mathrm{Hom}_{M\times\mathcal{O}_x}(C_c^\infty(M, \mathcal{O}_x) \otimes \chi', k_x) \approx \mathrm{Hom}_{e\mathcal{G}_M e\times\mathcal{O}_x}(eC_c^\infty(M, \mathcal{O}_x) \otimes \chi', k_x)$$

We saw earlier that the map $E \to eE$ from $\mathcal{G}_M$-modules to $e\mathcal{G}_M e$-modules is *exact* quite generally. Thus, it must be that $eC_c^\infty(M, \mathcal{O}_x) \otimes \chi'$ fails to be co-iso-Noetherian as a $e\mathcal{G}_M e \times \mathcal{O}_x$-module.

But by hypothesis $e\mathcal{G}_M e$ and $\mathcal{O}_x$ are Noetherian. By the finite-generation hypothesis on $eC_c^\infty(M, \mathcal{O}_x) = e\mathcal{G}_M$ and on $\chi' = e\chi'$, the $e\mathcal{G}_M e \times \mathcal{O}_x$-module

$$eC_c^\infty(M, \mathcal{O}_x) \otimes \chi'$$

is Noetherian. This certainly contradicts the conclusion that it was not co-iso--Noetherian, since Noetherian-ness is a stronger condition than co-iso-Noetherian. So we conclude that, after all, Φ *was* locally strongly meromorphic.

This completes the proof of the orbit criterion for strong meromorphy. □

References

[Be84] J. Bernstein, *Le "centre" de Bernstein*, notes by P. Deligne, Representations des Groupes Reductifs sur un Corps Local, Paris: Hermann (1984).

[BeZ76] J. Bernstein and A. V. Zelevinsky, *Representations of the group* $\mathrm{GL}(n, F)$*, where* F *is a non-archimedean local field*, Russian Math. Surveys **31** (1976), 1–68.

[BeZ77] J. Bernstein and A. V. Zelevinsky, *Induced representations of reductive p-adic groups, I*, Ann. Sci. E.N.S. **4**, t. 10 (1977), 441–472.

[Bo] A. Borel, *Admissible representations of a semi-simple group over a local field with vectors fixed under an Iwahori subgroup*, Inv. Math. **35** (1976), 233–259.

[Br] M. Brion, *Classification des espaces homogenes spheriques*, Comp. Math. **63** (1987), 189-208.

[Bru] F. Bruhat, *Distributions sur un groupe localement compacte et applications a l'etude des representations des groups p-adiques*, Bull. Math. Soc. France **89** (1961), 43–75.

[C] P. Cartier, *Representations of p-adic groups: A survey*, in Automorphic Forms, Representations, and L-functions, Proc. Symp. Pure Math. **33** part I, Providence, RI: Amer. Math. Soc. (1979), 111–156.

[C76] W. Casselman, *Admissible representations of p-adic reductive groups*, duplicated notes.

[C80] *The unramified principal series of p-adic groups, I: the spherical function*, Comp. Math. **40** fasc. 2 (1980), 387–406.

[F] D. Flath, *Decomposition of representations into tensor products*, in Automorphic Forms, Representations, and L-functions, Proc. Symp. Pure Math. **33** part I, Providence, RI: Amer. Math. Soc. (1979), 179–184.

[G83] P. B. Garrett, *Pullback of Eisenstein series*; applications, in Automorphic Forms of Several Variables, ed. I. Satake and Y. Morita, Boston: Birkhauser (1984).

[G90] P. B. Garrett, Holomorphic Hilbert Modular Forms, Pacific Grove, CA: Wadsworth-Brooks-Cole (1990).

[G92] P. B. Garrett, *On the arithmetic of Siegel-Hilbert cuspforms: Petersson inner products and Fourier coefficients*, Inv. Math. **107** (1992), 453–481.

[GePR] S. Gelbart, I. Piatetski-Shapiro, and S. Rallis, Explicit Constructions of Automorphic L-functions, Lecture Notes in Mathematics **1254**, New York: Springer (1987).

[GeS] S. Gelbart and F. Shahidi, Analytic Properties of Automorphic L-functions, Boston: Academic Press (1988).

[Go63] R. Godement, Domaines Fondamentaux des Groupes Arithmetiques, Sem. Bourb. no. 257 (1962–3).

[Go66] R. Godement, *The spectral decomposition of cuspforms*, in Proc. Symp. Pure Math. **IX**, Providence, R.I: Amer. Math. Soc. (1966), 225–234.

[GoJ] R. Godement and H. Jacquet, Zeta Functions of Simple Algebras, Lecture Notes in Mathematics **260**, New York: Springer-Verlag (1972).

[Gr] B. Gross, *Some applications of Gelfand pairs to number theory*, Bull. A.M.S. 24 no. 2 (1991), 277–301.

[H81] M. Harris, *The rationality of holomorphic Eisenstein series*, Inv. Math. **63** (1981), 305–310.

[H84] M. Harris, *Eisenstein series on Shimura varieties*, Ann. of Math. **119** (1984), 59–94.

[H85] M. Harris, *Arithmetic vector bundles and automorphic forms on Shimura varieties, I*, Inv. Math. **82** (1985), 151–189.

[H86] M. Harris, *Arithmetic vector bundles and automorphic forms on Shimura varieties, II*, Comp. Math. **60** (1986), 323–378.

[J86] H. Jacquet, *Sur un resultat de Waldspurger*, Ann. Sci. E.N.S. **19** (1986), 185–289.

[J87] H. Jacquet, *On the non-vanishing of some L-functions*, Indian Acad. Sci. Proc. **97** (1987), pp. 117–155.

[JLai] H. Jacquet and K. F. Lai, *A relative trace formula*, Comp. Math. **54** (1985), 243–310.

[JLaiR] H. Jacquet, K. F. Lai, and S. Rallis, *A trace formula for symmetric spaces*, Duke Math. J. **70** (1993), 305–372.

[JLan] H. Jacquet and R. P. Langlands, Automorphic Forms on GL(2), Lecture Notes in Mathematics **114**, New York: Springer-Verlag (1970).

[JR92] H. Jacquet and S. Rallis, *Kloosterman integrals for skew-symmetric matrices*, Pac. J. Math. **154** (1992), 265–283.

[JR92b] H. Jacquet and S. Rallis, *Symplectic periods*, J. Reine und Angew. Math. **423** (1992), 175–197.

[KKO] S.-I. Kasai, T. Kimura, and S.-I. Otani, *A classification of simple weakly spherical homogeneous spaces, I*, J. of Algebra **182** (1996), 235–255.

[Kl] H. Klingen, *Über die Werte der Dedekindschen Zetafunktion*, Math. Ann. **145** (1961/1962), 265–272.

[Ku] R. Kumanduri, *Distinguished representations for unitary groups*, Pacific J. Math. **178** (1997), 293–306.

[L64] R. P. Langlands, On the Functional Equations Satisfied by Eisenstein Series, Lecture Notes in Mathematics **544**, New York: Springer-Verlag (1976).

[L67] R. P. Langlands, Euler Products, James K. Whitmore Lectures, New Haven, CT: Yale University Press (1967).

[M] H. Matsumoto, Analyse Harmonique dans les Systems de Tits Bornologiques de Type Affine, Lecture Notes in Math. **590**, Berlin-New York: Springer-Verlag (1977).

[MoW] C. Moeglin and J.-L. Waldspurger, Spectral Decomposition and Eisenstein Series, Cambridge: Cambridge University Press (1995).

[MS] C. J. Moreno and F. Shahidi, *The L-functions $L(s, Sym^r, \pi)$*. Can. Math. Bull. **28** no. 4 (1985), 405–410.

[P] I. Piatetski-Shapiro, *Euler subgroups*, in Lie Groups and their Representations, New York: Halsted (1975).

[R] R. Rankin, *Contributions to the theory of Ramanujan's function $\tau(n)$ and similar arithmetic functions, I*, Proc. Cam. Phil. Soc. **35** (1939), 351–372.

[S] F. Shahidi, *Third symmetric power L-functions for* GL(2), Comp. Math. **70** no. 3 (1989), 245–273.

[Sh70] G. Shimura, *On canonical models of arithmetic quotients of bounded symmetric domains I,II*, Ann. of Math. **91** (1970), 144–222; ibid **92** (1970), 528–549.

[Sh75a] G. Shimura, *On some arithmetic properties of modular forms in one and several variables*, Ann. of Math. **102** (1975), 491–515.

[Sh75b] G. Shimura, *On Fourier coefficients of modular forms of several variables*, Nachr. Wiss. Gott. **17** (1975), 261–268.

[T] J. Tate, Ph.D. thesis, Princeton University, 1950, reprinted in Algebraic Number Theory, ed. J. W. S. Cassels and A. Frohlich, Washington, D.C.: Thompson Book Co. (1967).

[W61] A. Weil, Adeles and Algebraic Groups, Princeton, 1961; reprinted Boston: Birkhauser (1982).

[W65] A. Weil, Integration dans les groupes topologiques et ses applications, Actualites Sci. Ind. **1145**, Hermann, Paris, 1965.

DEPARTMENT OF MATHEMATICS, UNIVERSITY OF MINNESOTA, MINNEAPOLIS, MN 55455-0100
E-mail address: garrett@math.umn.edu

Proceedings of Symposia in Pure Mathematics
Volume **66.2**, 1999

Cohomological Automorphic Forms on Unitary Groups, I: Rationality of the Theta Correspondence

Michael Harris

To Goro Shimura, with admiration

ABSTRACT. This article studies the arithmetic properties of the theta correspondence for automorphic forms on pairs of unitary groups of hermitian vector spaces over imaginary quadratic fields. We specifically consider the correspondence taking automorphic forms of general discrete series type on groups of signature $(n-1,1)$ to holomorphic automorphic forms on groups of general signature. It is shown that the correspondence preserves arithmeticity up to multiplication by an explicit scalar, which can be expressed in terms of periods of elliptic curves with complex multiplication–CM periods. The scalar depends on the normalization of the theta correspondence and on the central character of the initial representation. Arithmeticity of holomorphic automorphic forms is defined by a version of Shimura's criterion; arithmeticity of general discrete series automorphic forms is defined in terms of coherent cohomology.

The technique is recursive, and ultimately comes down to a calculation of theta lifts from $U(1)$ to $U(1)$, whose arithmetic properties are expressed in terms of special values of Hecke L-functions. A final section explains the relevance of the present results to Deligne's conjecture on special values of motivic L-functions.

This article continues the study, begun in [**H97**], of the L-functions associated to cusp forms of cohomological type on unitary groups. Let V be a hermitian space over an imaginary quadratic field $\mathcal{K}$. As in [**H97**], we adopt the point of view of Clozel [**C90, C91**] and take as our basic object a family of nearly equivalent automorphic representations of the unitary group $G = U(V)$, assumed to have signature (n-1,1) at infinity. Here by "nearly equivalent" we mean locally isomorphic at almost all finite places; in other words, L-indistinguishable; the same term can also be applied to automorphic representations of different unitary groups of the same dimension. The archimedean components of the representations in our family

---

1991 *Mathematics Subject Classification.* Primary 11F67, 11F27, 11F55, 14G10; Secondary 14G35, 11G18, 11G40.

Key words and phrases. Automorphic representations, unitary groups, periods, L-functions, theta functions, cohomology of arithmetic groups.

Partially supported by NSF Grants DMS-9203142 and DMS-9423758, Institut de Mathematiques de Jussieu-UMR CNRS 7586.

are to vary in the L-packet $\{\pi_\infty\}$ of discrete series representations with fixed infinitesimal character. Thus there is a unique finite-dimensional representation $\mathcal{V}$ of G such that the relative Lie algebra cohomology space $H^{n-1}(\mathfrak{g}, K_\infty; \pi_\infty \otimes \mathcal{V})$ does not vanish; here $\mathfrak{g} = \mathrm{Lie}(G(\mathbb{C}))$ and $K_\infty \subset G(\mathbb{R})$ is a maximal compact subgroup.

The choice of a complex structure on the symmetric space $X = G(\mathbb{R})/K_\infty$ corresponds to a Harish-Chandra decomposition of the Lie algebra

$$g \cong \mathrm{Lie}(K_\infty)_\mathbb{C} \oplus \mathfrak{p}^+ \oplus \mathfrak{p}^-,$$

where $\mathfrak{p}^+$ and $\mathfrak{p}^-$ are naturally isomorphic to the holomorphic and antiholomorphic tangent spaces, respectively, of X at the fixed point of K_∞. The elements of the L-packet $\{\pi_\infty\}$ associated to $\mathcal{V}$ as above can be numbered $\pi_{\infty,i}, i = 0, \dots, n-1$, so the $\bar\partial$-cohomology space

$$H^j_{\bar\partial}(\check\pi_{\infty,i} \otimes E) \ : \ = H^j(\mathrm{Lie}(K_\infty)_\mathbb{C} \oplus \mathfrak{p}^-, K_\infty; \check\pi_{\infty,i} \otimes E)$$

is non-trivial for some K_∞-module E if and only if $i = j$, in which case E is uniquely determined. (For $\bar\partial$-cohomology see [**H90**] and the discussion in II,§1, below.) In other words, an automorphic representation π of G with archimedean component $\check\pi_{\infty,i}$ contributes to coherent cohomology of the associated Shimura variety $\mathrm{Sh}(G, X)$ in degree i.

The distinct $\pi_{\infty,i}$ in an L-packet can also be distinguished by their behavior under the theta correspondence. Let V' be a second hermitian space over $\mathcal{K}$, with signature (r, s) at infinity, $G' = U(V')$, and suppose $r + s \geq n$. It follows from work of Li [**L90**] that, with the sign conventions used in the present article, if the representation $\check\pi_{\infty,i}$ lifts to a non-trivial holomorphic representation of $G'(\mathbb{R})$, then $i = n-s-1$. We thus obtain a theta lift from cuspidal automorphic representations π of G contributing to coherent cohomology in degree $n - s - 1$ of $\mathrm{Sh}(G, X)$ to holomorphic automorphic representations π' of G'.

The theta lift depends on the additional data of splitting characters. These are Hecke characters χ, χ' of $\mathcal{K}^\times_\mathbf{A}/\mathcal{K}^\times$ restricting on the ideles of $\mathbb{Q}$ to certain powers of the quadratic character defining the extension $\mathcal{K}/\mathbb{Q}$. Our main theorem states that, up to an explicit factor $\mathbf{p}(\pi, V, V', \chi, \chi')$, the theta lift described in the preceding paragraph preserves rationality over $\mathcal{K}$, in the sense of coherent cohomology. The reader is referred to Theorem V.1.9 for an explicit statement. As indicated, the factor depends only on the representation π on the spaces V and V', and on the splitting characters. In fact, the dependence on π is completely determined by the central character of π, the dependence on V and V' is determined by their dimensions and the signature of V', and the factor $\mathbf{p}(\pi, V, V', \chi, \chi')$ can be expressed explicitly as a product of powers of $2\pi i$ and periods of CM motives. Of course, by "rationality" of the theta lifting, we mean that the theta lifts of rational automorphic forms with respect to a specific natural family of rational theta kernels are rational, up to multiplication by the indicated factor.

Results on the rationality of theta lifts go back to the nineteenth century and the calculation of the Fourier coefficients of theta functions associated to definite quadratic forms. Results on the rationality of theta lifts of holomorphic forms from non-compact groups were proved in the 1970s by a number of people, by studying behavior of Fourier coefficients under lifting. The approach to rationality questions adopted in the present paper goes back to work of Shimura in the early 1980s on the theta lifts from Hilbert modular forms to forms on orthogonal groups [**Sh80, Sh81**]. We follow Shimura in defining rational holomorphic forms in terms of their normalized special values at CM points. The "rationality test" of a form on G' obtained by theta lift is reduced, by the method of "seesaw pairs," to a calculation of an integral on $G(\mathbf{A})$ involving the original form on G. In this we also follow Shimura, although the calculation on G is considerably more involved in the present case, since we are working with general discrete series representations.

The first results on the rationality of theta lifts of non-holomorphic forms were obtained by Shimura in [**Sh81**], working with forms of mixed type on Hilbert modular varieties. My article [**HK**] with Kudla was the first to study the rationality of forms of higher coherent cohomological type obtained by theta lift from holomorphic (or mixed holomorphic and anti-holomorphic) forms. The present article studies the lift in the opposite direction. However, a combination of the methods developed in this paper with those of [**HK**] can be used to establish rationality of theta lifts between various kinds of coherent cohomological forms on unitary groups of different signatures.

We have concentrated on the lift from coherent cohomological forms on $U(n-1,1)$ to holomorphic forms on general $U(r,s)$ because we plan to apply our results to study the special values of the L-functions of the motives that appear in the cohomology of the Shimura varieties $\mathrm{Sh}(G,X)$. The rationale for such applications was discussed at length in [**H97**] and the strategy was described in a particular case in [**H94**, §5-6]. Here it suffices to recall that, according to the general Langlands philosophy, the motives in question should include the base change to $\mathcal{K}$ of all polarized regular motives over $\mathbb{Q}$, in the sense of [**C90, H97**], of rank n over a totally real coefficient field and whose weight has parity opposite to that of n; when n is even it may also be necessary to impose a hypothesis of local ramification at one finite prime. The article [**H97**] shows that Deligne's conjecture on the critical values of such motives (up to possible factors in $\mathcal{K}^\times$) would follow if one could establish certain period relations for "nearly equivalent" automorphic forms on unitary groups of different signatures. The present article is the first step in the proof of these period relations. I expect to complete the proof in the sequel, at least for motives that are sufficiently regular (sufficiently high spacing between the Hodge types).

I now turn to a sketch of the proof. As mentioned above, we start with a cusp form F on G of coherent cohomological type and assume it has a nontrivial theta lift to a holomorphic form $\theta(F)$ on G'. We test $\theta(F)$ for rationality using Shimura's method, as extended in [**H93**] on the basis of the results of [**H86**]. Namely, we write V' as a sum of one-dimensional hermitian spaces:

$$V' = T_1 \oplus T_2 \oplus \cdots \oplus T_m,$$

where $U(T_i)$ has signature (1,0) for $1 \leq i \leq r$, and signature (0,1) for $r+1 \leq i \leq m$. We integrate $\theta(F)$ over $\Pi U(T_i)(\mathbb{Q})\backslash U(T_i)(\mathbf{A})$ against a product $\eta_1 \otimes \cdots \otimes \eta_m$, where η_i is a Hecke character of $U(T_i)(\mathbb{Q})\backslash U(T_i)(\mathbf{A})$, and compare the result to the respective CM periods corresponding to the η_i and the signatures of the $U(T_i)$. By the method of seesaw pairs, the result equals an integral (more generally, a linear combination of integrals) of the form

$$\int_{G(\mathbb{Q})\backslash G(\mathbf{A})} F \cdot \Pi\theta(\eta_i)\, dg \tag{0.1}$$

where each $\theta(\eta_i)$ is an automorphic form on $G(\mathbb{Q})\backslash G(\mathbf{A})$. The $\theta(\eta_i)$ belong to what I call "small" automorphic representations. These are representations contributing to cohomology of $\mathrm{Sh}(G,X)$ in degree one, along with certain degenerate non-cohomological representations of holomorphic type. It turns out that the $\theta(\eta_i)$ can naturally be interpreted as coherent cohomology classes in degree zero for $1 \leq i \leq r$, and in degree one for $r+1 \leq i \leq m$; of course, this dichotomy corresponds to the signatures of the $U(T_i)$. Moreover, and this is the crucial point, the integral above can be interpreted as a cup product in coherent cohomology with values in the $G(\mathbf{A}^f)$-coinvariants of the "interior cohomology" space $H_!^{n-1}(\mathrm{Sh}(G,X),\Omega^{n-1})$. This latter space is isomorphic to $\mathbb{C}$, but it has a natural $\mathcal{K}$-structure and the cup product is rational over $\mathcal{K}$.

In this way questions about rationality of $\theta(F)$ can be reduced to questions about rationality of each of the $\theta(\eta_i)$. When $1 \leq i \leq r$, $\theta(\eta_i)$ is holomorphic and its rationality can be tested by means of another seesaw diagram, based on a decomposition $V = V_1 \oplus \cdots \oplus V_n$, with each V_j of dimension one. This in turn reduces the problem to the study of the rationality of the theta lifts from $U(V_j)$ to $U(T_i)$. A formula proved by Brocco [**Br**] shows that each such lifting is rational up to the central critical value of the L-function attached to an algebraic (motivic) Hecke character of $\mathcal{K}$. Damerell's theorem and its more recent improvements express the latter in terms of CM periods.

On the other hand, when $r+1 \leq i \leq m$, $\theta(\eta_i)$ is the complex conjugate of a holomorphic theta series $\theta'(\eta_i)$ attached to a Hecke character (but for the complex conjugate splitting characters). A method developed in [**H90**] for anti-holomorphic cusp forms, and extended here to square-integrable forms in "small" automorphic representations, then shows, roughly, that $\overline{c\theta'}(\eta_i)/ < c\theta'(\eta_i), c\theta'(\eta_i) >$ defines an arithmetic degree one coherent cohomology class if $c \in \mathbb{C}$ is a constant such that $c\theta'(\eta_i)$ is an arithmetic holomorphic form. We have already seen that such a c can be expressed in terms of CM periods. On the other hand, $< \theta'(\eta_i), \theta'(\eta_i) >$ can also be expressed in terms of the critical values of the L-function attached to an algebraic Hecke character, and thus again in terms of CM periods. Thus $\theta(\eta_i)$ is arithmetic up to explicit CM periods for all i.

Tracing back the seesaw diagrams, the arithmeticity of $\theta(F)$ is thus established, up to a product of explicit CM periods. Upon multiplying these periods, we obtain the invariant $\mathbf{p}(\pi, V, V', \chi, \chi')$.

I should confess at once that the preceding sketch is inaccurate on several counts. Most significantly, we work not with the dual reductive pair $(U(V), U(V'))$ but rather with its double $(U(V) \times U(-V), U(V') \times U(V'))$; here $-V$ is the space V endowed with the negative of its original hermitian form. This doubling persists through all the seesaw pairs, and it is to the double of the pair described above that Brocco's theorem applies. The appearance of a central critical L-value in a doubling construction has been analyzed by T. Yang [**Y**], who obtains a factorization of the algebraic part of the central critical value, extending earlier work of Villegas.

The sketch is also inaccurate in that it ignores the role of similitude groups. The Shimura variety is attached, not to the unitary group $U(V)$ but rather to the group of unitary similitudes $\mathrm{GU}(V)$. The theta correspondence can be extended to similitude groups [**HK, H93, Ro**] and we work with this extension. However, the L-functions we are studying are attached to the unitary groups themselves, and in the end we show that our constructions are independent of the similitude factors. I strongly suspect that there is a systematic way to study the arithmetic of automorphic forms on unitary groups (and symplectic groups, and orthogonal groups, and quaternionic unitary groups) without making use of Shimura varieties attached to similitude groups. However, this would require the development of a complete theory of Shimura varieties, and automorphic vector bundles, attached to such groups, possibly along the lines of what Milne calls "connected Shimura varieties" [**M**]. As a practical matter, it is simpler to extend our representations to the similitude groups and use the theory already available than to develop the foundations of a new theory.

Apart from these two points, the proof sketched above represents an accurate outline of the contents of this paper. For me, the crucial point was the realization that the $\theta(\eta_i)$, for $i > r$, can be identified with coherent cohomology classes of degree one. Once that is understood, the proof is just a matter of checking the details, which boils down to keeping track of K-types and central characters under the archimedean theta correspondence. As the length of the paper indicates, there are a lot of details to check. The logic of the argument, as sketched above, is obscured by the need to coordinate the splitting characters and sign changes. However, these splitting characters are essential to making the numbers work out. The reader who actually tries to follow the proofs is likely to be struck, as I have been, by the internal rigidity of the theory. At nearly every step, the inequalities defining the set of critical values of L-functions turn out to be exactly the same as the inequalities determining the archimedean theta correspondence, and these are in turn strictly dependent on the choice of splitting characters.

Similarly, it is worth pointing out the role of an asymmetry at the level of CM periods. We assume we are given an embedding $i_0 \colon \mathcal{K} \to \mathbb{C}$. If η is a Hecke character of $\mathcal{K}_{\mathbf{A}}^{\times}/\mathcal{K}^{\times}$, and $\eta^{\iota}(a) = \eta \circ \iota(a)$, where ι denotes Galois conjugation of the ideles of $\mathcal{K}$, then the L-functions $L(s,\eta)$ and $L(s,\eta^{\iota})$ are equal. On the other hand, if η is a motivic Hecke character, then attached to η are two CM period invariants, denoted $p(\eta, i_0)$ and $p(\eta,\iota) = p(\eta^{\iota}, i_0)$, that are equal if and only if η factors through the norm $N_{\mathcal{K}/\mathbb{Q}}$. Both CM period invariants are needed in the rationality testing of holomorphic automorphic forms on unitary groups; the choice is determined by the signature of the unitary group in question. On the other hand, the critical values of $L(s,\eta)$ have to choose one or the other invariant, and this choice is determined by sign of the restriction of η to the unit circle in $\mathbb{C}^{\times}$, identified via i_0 with the norm one elements in $(\mathcal{K}_{\infty})^{\times}$. It turns out that the sign changes in the theta correspondence, at each stage, are exactly those needed to coordinate the CM periods needed for rationality testing with the critical L-values that show up in the doubling method.

Many years ago, when I began to think about the arithmetic of the theta correspondence for unitary groups, I asked Kazhdan whether he thought there might be a general pattern. His reply was to suggest a possible connection with Deligne's conjecture on critical values of L-functions. At the time I had no idea what he meant, but even the smallest details of this paper confirm his insight.

The present article represents an extension of the techniques of [**H93**], which studied the theta correspondence when $\dim(V) = \dim(V') = 2$, but when $\mathcal{K}$ is replaced by a general CM field. The algebraicity results of [**H93**] are stated up to factors in $\bar{\mathbb{Q}}$. This was mainly because the oscillator representation is only defined over the field $\mathbb{Q}^{\mathrm{ab}}$ generated over $\mathbb{Q}$ by roots of unity. The present paper makes use of a simple observation (cf. the discussion preceding IV, (4.4)) to overcome this difficulty. Briefly, the theta correspondence for adelic representation is only rational over $\mathbb{Q}^{ab}$; but an application of strong approximation allows us to show that the correspondence for *automorphic* representations is rational over each $\mathbb{Q}^{(p)}$ separately, where $\mathbb{Q}^{(p)}$ is the field generated by all p-power roots of unity. Using this remark, it should be possible to improve the results of [**H93**] and to obtain factorizations of the Petersson norms of Hilbert modular forms with the correct Galois equivariance.

The paper is organized as follows. The first three sections are preliminary in nature and treat the theta correspondence, the parameters of archimedean representations, and the coherent cohomology of Shimura varieties, respectively. In each case it is a matter of working out the consequences of a general theory in the situation of interest. The first section pays special attention to the role of the splitting characters. Fortunately, everything needed here has been worked out by Kudla in [**Ku94**] and in my joint paper with Kudla and Sweet [**HKS**]. The second section works out the details of the theta lift of discrete series of real unitary groups, following Li [**L90**], as well as the parameters for the ladder representations obtained by lifting characters of $U(1)$ to $U(n-1,1)$. Since Li works with the metaplectic double covers of unitary groups, we have to translate his results to take the splittings into account; this explains the proliferation of parameters. The second section also includes a crucial calculation of the map at the level of $\bar{\delta}$-cohomology that allows us to interpret the integral (0.1) as a cup product in coherent cohomology. Here the main ingredient is provided by a special case (Pieri's formula) of the Littlewood-Richardson rules for decomposing tensor products $V_1 \otimes V_2$ of representations of $\mathrm{GL}(n-1)$: when V_2 is a symmetric power of the standard representation, this tensor product decomposes with multiplicity one. The third section provides enough of the general theory of coherent cohomology of Shimura varieties to explain the rationality of this cup product, and to describe the Petersson inner product of "small" holomorphic automorphic forms.

The main theorems are stated and proved in IV and V. Part IV contains the global theory of small automorphic representations, and studies the rationality of the theta lifts of Hecke characters of $U(1) \times U(1)$ to holomorphic and anti-holomorphic forms on $U(V) \times U(-V)$. *Note added in proof. This includes calculation of the Petersson norms of such theta lifts. The main theorem on the rationality of lifting from G to G' is stated in V, §1 and proved in V, §2. The final section (V, §3) explains how, under a series of hypotheses, this theorem can be applied to prove the period relations which are needed in turn to prove Deligne's conjecture. The verification of these hypotheses in a large number of cases will be the substance of the sequel. The final step in the proof of the period relations is a comparison between three different expressions for special values of closely related L-functions. The first expression is based on a formula of Rallis for the norm of a theta lift, and involves the invariant $\mathbf{p}(\pi, V, V', \chi, \chi')$ introduced here; the other expressions are those derived in [**H97**]. Each of these special values formulas involves the periods of interest as well as a lot of apparently random junk — powers of $2\pi i$ and periods of auxiliary CM motives, including those attached to the splitting characters χ and χ'. None of these expressions is at all intuitive, and that the period relations drop out in the end seems to be little more than a bizarre coincidence. This is unfortunately typical of the theory of special values of L-functions in its present state.

The outline of the present paper was developed during the academic year 1992-1993, which I spent as a visitor to Paris 7 (Jussieu) and Orsay. I announced a preliminary version of the present results (for $\dim(V)$ and $\dim(V')$ even and only up to factors in $\bar{\mathbb{Q}}$) at the Isaac Newton Institute in Cambridge in April 1993. I thank these institutions for their hospitality. I also thank S. Kudla for numerous conversations over many years on the subject of the theta correspondence; our joint project with W. J. Sweet was especially important in helping to establish the general framework of the present paper. Finally, I thank G. Schwarz for explaining the Littlewood-Richardson rules, and D. Blasius, L. Clozel, S. Gelbart, J.-S. Li, S. Rallis, and J. Rogawski for helpful discussions.

The present article is an extended application of the arithmetic of the theta correspondence to compare the special values of L-functions to the periods of automorphic forms. This approach was pioneered by Goro Shimura in the 1970s and 1980s, and the informed reader will find in Shimura's work the origins of many of the techniques of the present paper. Shimura's work also anticipated the theory of coherent cohomology, the language of the present paper. Most important, perhaps, was his recognition of the central role played by Hecke characters and their associated periods in the study of automorphic forms on groups of higher dimension. Of course, Shimura also initiated the study of special L-values and that of automorphic periods, not to mention the theory of what are now called Shimura varieties, which provides the framework for practically all contemporary work on the arithmetic of automorphic forms. Finally, though I was only briefly his student in the formal sense, Shimura was one of my first teachers of mathematics, and his support and encouragement have been of inestimable importance. For all these reasons, it is with great pleasure and gratitude that I dedicate this paper to Goro Shimura.

Notation

If G is an algebraic group, we denote by Z_G the identity component of its center and by G^{der} its derived subgroup.

Let G be any algebraic group over $\mathbb{Q}$. Choose a maximal compact subgroup $K_\infty \subset G(\mathbb{R})$, and let Z_∞ be the center of $G(\mathbb{R})$. Throughout this paper, we use the term "$G(\mathbf{A})$-module" to designate a $(\mathrm{Lie}(G(\mathbb{R})), Z_\infty \cdot K_\infty) \times G(\mathbf{A}^f)$-module. Note that this definition depends on the choice of maximal compact subgroup K_∞. The categories of $G(\mathbf{A})$-modules for different choices of K_∞ are equivalent, but not canonically so.

We let $\mathcal{A}(G)$ denote the space of automorphic forms on $G(\mathbb{Q})\backslash G(\mathbf{A})$, which are assumed to be $Z_\infty \cdot K_\infty$-finite; thus $\mathcal{A}(G)$ is a $G(\mathbf{A})$-module in the sense described above. Similarly, we let $\mathcal{A}_0(G) \subset \mathcal{A}(G)$ denote the subspace of cusp forms and $\mathcal{A}_{(2)}(G) \subset \mathcal{A}(G)$ the space of forms which restrict to square-integrable forms on $G^{\mathrm{der}}(\mathbb{Q})\backslash G^{\mathrm{der}}(\mathbf{A})$.

If G is a reductive algebraic group over a local or global field F, we follow the notation of Kottwitz and denote by ${}^LG$ its L-group, and by $\hat{G}$ the connected identity component of ${}^LG$ as a complex algebraic group. We only need the finite Galois group form of the L-group, a semi-direct product $\hat{G} \rtimes \mathrm{Gal}(F'/F)$, where F' is a finite extension of F, with respect to some homomorphism $\mathrm{Gal}(F'/F) \to \mathrm{Aut}(\hat{G})$.

If χ is a Dirichlet character (mod n) and ψ is an additive character of $\mathbb{Z}/n\mathbb{Z}$, we let $g(\chi, \psi)$ denote the Gauss sum. When ψ is understood or irrelevant, we write $g(\chi)$ instead of $g(\chi, \psi)$.

Measures

We need to define normalized Haar measures on adele groups of the form $G(\mathbf{A})$, where $G = U(V)$ is the unitary group of a hermitian vector space V over an imaginary quadratic field $\mathcal{K}$. Such a measure is of the form $dg = \Pi dg_v$, where v ranges over the completions of $\mathbb{Q}$ and dg_v is a Haar measure on $G(\mathbb{Q}_v)$. At each finite place v we choose a special maximal compact subgroup K_v; at the archimedean place we choose K_∞ to be a maximal compact subgroup which is defined as algebraic group over $\mathcal{K}$. We call dg a *rational measure* if $\mathrm{vol}(G(\mathbb{Q})\backslash G(\mathbf{A}), dg) = \int_{G(Q)\backslash G(\mathbf{A})} dg \in \mathbb{Q}$. For example, the measure dg such that $\mathrm{vol}(G(\mathbb{Q})\backslash G(\mathbf{A}), dg) = 1$ is a rational measure. We use the measure of total volume 1 typically when $\dim(V) = 1$, for testing the rationality of holomorphic automorphic forms by integrating them against characters and comparing the result to standard CM periods, as in III,§3. On the other hand, for calculations involving L-functions it is more natural to choose dg_v such that, when v is finite, dg_v takes value 1 on K_v. This condition then determines dg_∞ up to a rational factor. If we let $X = G(\mathbb{R})/K_\infty$ and write $dg_\infty = dx \cdot dk_\infty$, where $\mathrm{vol}(K_\infty, dk_\infty) = 1$ and dx is a left $G(\mathbb{R})$-invariant measure on X, then dx is determined up to a rational factor. In practice, we let dx be given by the top differential form on X whose value at the fixed point of K_∞ is

$$\left(2\pi\sqrt{D_{\mathcal{K}}}\right)^{-\dim(X)} \cdot \bigwedge_{i=1}^{\dim(X)} dz_i \wedge d\bar{z}_i,$$

where $-D_{\mathcal{K}}$ is the discriminant of $\mathcal{K}$ and dz_i is a $\mathcal{K}$-rational basis of the (holomorphic) tangent space to X at the fixed point of $\mathcal{K}$. It follows from the theory of Tamagawa numbers that this gives a rational measure.

Suppose $G = \mathrm{GU}(V)$ is a unitary similitude group over $\mathbb{Q}$, let $Z_{G,+} \subset Z_G$ be the $\mathbb{Q}$-split component of the center. Thus $Z_{G,+} \cap U(V) = \{\pm Id\}$. We define a rational measure on $G(\mathbf{A})$ to be any factorizable Haar measure $dg = \Pi dg_v$ such that the restriction of dg_v to the open subgroup $Z_{G,+}(\mathbb{Q}_v)U(V)(\mathbb{Q}_v)$ is of the form $(1/2)dz_v \times du_v$. Here du_v is the v-component of a fixed rational measure du on $U(V)(\mathbf{A})$, in the sense defined above. The measure dz_v plays no special role, since it disappears in the integrals, but for definiteness we may assume dz_v is the component at v of the usual Tamagawa measure.

For finer integral properties of special values of L-functions it is essential to use Tamagawa numbers. However, our applications of the theta correspondence do not easily lend themselves to calculations over the integers.

I. Standard L-functions of U(V)

1. Notation, basic identity

We fix once and for all an imaginary quadratic field $\mathcal{K} = \mathbb{Q}(\sqrt{-D})$, where D denotes the absolute value of the discriminant, and let $\varepsilon_{\mathcal{K}} \colon \mathbf{A}^\times/\mathbb{Q}^\times \to \{\pm 1\}$ be the quadratic character attached to $\mathcal{K}$. We let ι denote the non-trivial automorphism of $\mathcal{K}$ over $\mathbb{Q}$. Let V be an n-dimensional vector space over $\mathcal{K}$, endowed with an ε-hermitian form $(\cdot,\cdot)_V$. We let $-V$ denote the ε-hermitian space $(V, -(\cdot,\cdot)_V)$, and let $2V$ be the $2n$-dimensional ε-hermitian space $V \oplus (-V)$. Let $\Delta(V)$ be the diagonal $\{(v,v)\}$ in $2V$; then $\Delta(V)$ is a totally isotropic subspace of $2V$. In what follows, we let $H = U(2V)$, $G = U(V)$, and $P \subset H$ the subgroup fixing $\Delta(V)$. Then there is a natural embedding $i_V\colon G \times G \hookrightarrow H$ as the subgroup which preserves the decomposition $2V = V \oplus (-V)$. Moreover, P is a maximal parabolic subgroup with abelian unipotent radical, denoted N, and Levi factor $L = P/N$ canonically isomorphic to $\mathrm{GL}(\Delta(V)) \cong \mathrm{GL}_{\mathcal{K}}(V)$. We will write $H(V)$, $G(V)$, $P(V)$, etc. when it is necessary to specify V.

The groups G and H are regarded as algebraic groups over $\mathbb{Q}$, by restriction of scalars. We denote by $p \mapsto A(p)$ the homomorphism $P \to \mathrm{GL}(V)$. Let v be any place of $\mathbb{Q}$, $|\cdot|_v$ the corresponding absolute value on $\mathbb{Q}_v$, and let

$$\delta_{P,v}(p) = |N_{\mathcal{K}/\mathbb{Q}} \circ \det(A(p))|_v^{n/2}, \quad p \in P(\mathbb{Q}_v).$$

Then $\delta_{P,v}\colon P(\mathbb{Q}_v) \to \mathbb{R}_+^\times$ is the modulus character. Similarly, letting $|\cdot|_{\mathbf{A}}$ denote the absolute value on the adeles of $\mathbb{Q}$, the modulus character of $P(\mathbf{A})$ is given by

$$\delta_{P,\mathbf{A}}(p) = |N_{\mathcal{K}/\mathbb{Q}} \circ \det(A(p))|_{\mathbf{A}}^{n/2}, \quad p \in P(\mathbf{A}).$$

Fix an integer $m \geq n$. Let χ (resp. χ_v) be a character of $\mathcal{K}_{\mathbf{A}}^\times$ (resp. $\mathcal{K}_v^\times$). We make the assumption that

$$\chi|_{\mathbf{A}^\times} = \varepsilon_{\mathcal{K}}^m \text{ (resp. } \chi|_{\mathbb{Q}_p^\times} = \varepsilon_{\mathcal{K},v}^m),$$

although most of this section remains true without that hypothesis. In particular, $\chi|_{\mathbf{A}^\times}$ is trivial if m is even. For any complex number s, define

$$\begin{aligned} \delta_{P,\mathbf{A}}(p,\chi,s) &= \chi(\det(A(p)) \cdot |N_{\mathcal{K}/\mathbb{Q}} \circ \det(A(p))|_{\mathbf{A}}^{n/2+s}; \\ \delta_{P,\mathbf{A}}^0(p,\chi,s) &= \chi(\det(A(p)) \cdot |N_{\mathcal{K}/\mathbb{Q}} \circ \det(A(p))|_{\mathbf{A}}^{s}. \end{aligned} \tag{1.1}$$

The local characters $\delta_{P,v}(p,\chi,s)$ and $\delta_{P,v}^0(p,\chi,s)$ are defined analogously. Let

$$\begin{aligned} I_n(s,\chi)_{\mathbf{A}} &= \mathrm{Ind}_{P(\mathbf{A})}^{H(\mathbf{A})} \delta_{P,\mathbf{A}}^0(p,\chi,s) \text{ (normalized induction)} \\ &= \{f\colon H(\mathbf{A}) \to \mathbb{C} | f(pg) = \delta_{P,\mathbf{A}}(p,\chi,s) f(g),\ p \in P(\mathbf{A}), g \in H(\mathbf{A})\}. \end{aligned} \tag{1.2}$$

The local induced representations $I_n(s,\chi_v)$ and the finite adelic representation $I_n(s,\chi)_{\mathbf{A}^f}$ are defined analogously, and we have

$$I_n(s,\chi)_{\mathbf{A}} = \bigotimes_v I_n(s,\chi_v), \tag{1.3}$$

where χ_v is the local component of χ at v, the product taken over all places of $\mathbb{Q}$. The functions in $I_n(s,\chi)_{\mathbf{A}}$ are assumed continuous and C^∞ in the archimedean variables.

Let K_∞ be a maximal compact subgroup of $H(\mathbb{R})$; then $K_\infty \cong U(n) \times U(n)$. Let $K_f = \Pi K_v$ be a maximal compact subgroup of $H(\mathbf{A}^f)$ such that K_v is a special maximal compact for all v. Then $H(\mathbf{A}^f) = P(\mathbf{A}^f) \cdot K_f$ and we have the Iwasawa decomposition $H(\mathbf{A}) = P(\mathbf{A}) \cdot K_\infty K_f$. We let

$$\delta(g) = |N_{\mathcal{K}/\mathbb{Q}}(\det A(p))|_{\mathbf{A}} \text{ whenever } g = pk,\ p \in P(\mathbf{A}),\ k \in K_\infty K_f. \tag{1.4}$$

Then $\delta(g)$ is independent of the decomposition $g = pk$. A *section* $\mathcal{F}(g,s,\chi)$ is a function which, to each $s \in \mathbb{C}$, associates a function $\mathcal{F}(g,s,\chi) \in I_n(s,\chi)_{\mathbf{A}}$ satisfying the following continuity property:

(1.5) $\delta(g)^{-s}\mathcal{F}(g,s,\chi)$ is a continuous K_∞-finite function of g, holomorphic in s.

Then for $\mathrm{Re}(s)$ sufficiently large, the Eisenstein series

$$E(g,s,\mathcal{F},\chi)=\sum_{\gamma\in P(\mathbb{Q})\backslash H(\mathbb{Q})}\mathcal{F}(\gamma g,s,\chi) \tag{1.6}$$

converges absolutely to an automorphic form on $H(\mathbf{A})$, and extends meromorphically to a function on $\mathbb{C}$, with no pole on the imaginary axis.

We assume K_∞ is chosen such that $K_\infty\cap(G\times G)(\mathbb{R})=K_{\infty,G}\times K_{\infty,G}$, for some maximal compact subgroup $K_{\infty,G}$ of G. Our automorphic forms on G will be assumed to be $K_{\infty,G}$-finite with respect to this choice. We also assume

$$K_v\cap(G\times G)(\mathbb{Q}_v)=K_{v,G}\times K_{v,G}$$

for almost all finite places, where $K_{v,G}$ is a special maximal compact of $G(\mathbb{Q}_v)$; this is certainly possible.

Let (π,H_π) be a cuspidal automorphic representation of G, $(\check{\pi},\ H_{\check{\pi}})$ its contragredient, which we assume given with compatible isomorphisms of $G(\mathbf{A})$-modules

$$\pi\xrightarrow{\sim}\bigotimes_v\pi_v,\qquad \check{\pi}\xrightarrow{\sim}\bigotimes_v\check{\pi}_v. \tag{1.7}$$

We let $(\cdot,\cdot)_{\pi_v}$ denote the canonical pairing $\pi_v\otimes\check{\pi}_v\to\mathbb{C}$.

Let p be a prime number such that π_p and the extension $\mathcal{K}_p/\mathbb{Q}_p$ are unramified. Then we can define the base change $\pi_{p,\mathcal{K}}$ as an unramified representation of $\mathrm{GL}(n,\mathcal{K}_p)$. More precisely, let Φ_n be the $n\times n$ matrix with ij-entry $(-1)^{i-1}\delta_{i,n-j+1}$. The L-group ${}^LG$ of G is the semi-direct product ${}^LG=\mathrm{GL}(n,\mathbb{C})\rtimes\mathrm{Gal}(\bar{\mathbb{Q}}/\mathbb{Q})$, where the action of $\mathrm{Gal}(\bar{\mathbb{Q}}/\mathbb{Q})$ factors through $\mathrm{Gal}(\mathcal{K}/Q)$, with the non-trivial automorphism ι of $\mathcal{K}$ acting on $\mathrm{GL}(n,\mathbb{C})$ as $g\mapsto\Phi_n{}^tg^{-1}\Phi_n^{-1}$. The L group of $R_{\mathcal{K}/\mathbb{Q}}GL(n,\mathcal{K})$ is the semi-direct product $[\mathrm{GL}(n,\mathbb{C})\times\mathrm{GL}(n,\mathbb{C})]\rtimes\mathrm{Gal}(\bar{\mathbb{Q}}/\mathbb{Q})$; again the action of $\mathrm{Gal}(\bar{\mathbb{Q}}/\mathbb{Q})$ factors through $\mathrm{Gal}(\mathcal{K}/\mathbb{Q})$, and ι acts by switching the factors. There is an obvious L-homomorphism

$$\begin{gathered}{}^LG\to{}^LR_{\mathcal{K}/\mathbb{Q}}\mathrm{GL}(n,\mathcal{K});\\ g\rtimes\gamma\mapsto(g,\Phi_n{}^tg^{-1}\Phi_n^{-1})\rtimes\gamma,\ g\in\mathrm{GL}(n,\mathbb{C}),\gamma\in\mathrm{Gal}(\bar{\mathbb{Q}}/\mathbb{Q}).\end{gathered} \tag{1.8}$$

The unramified representation π_p corresponds to a semi-simple conjugacy class $\{\sigma(\pi_p)\}$ in ${}^LG$, which defines a semi-simple conjugacy class $\{\sigma_{\mathcal{K}}(\pi_p)\}$ in

$${}^LR_{\mathcal{K}/\mathbb{Q}}\mathrm{GL}(n,\mathcal{K})$$

via (1.8); the base change $\pi_{p,\mathcal{K}}$ is the representation of $\mathrm{GL}(n,\mathcal{K}_p)$ corresponding to $\{\sigma_{\mathcal{K}}(\pi_p)\}$. Then we can define the local standard L-function of π_p, twisted by the character χ_p:

$$L_p(s,\pi_p,\chi_p,\mathrm{St})=L_p(s,\pi_{p,\mathcal{K}},\chi_p). \tag{1.9}$$

Here the right-hand side is just the standard (Godement-Jacquet) Euler factor attached to the representation $\pi_{p,\mathcal{K}}$ and the character χ_p.

Let $\mathcal{F}(g,s,\chi)$ be a section, as above, $f \in H_\pi$, $f' \in H_{\tilde{\pi}}$, and define the zeta integral, following Piatetski-Shapiro and Rallis:

$$(1.10)\quad Z(s,f,f',\mathcal{F},\chi) = \int_{(G\times G)(\mathbb{Q})\backslash(G\times G)(\mathbf{A})} E(i_V(g,g'),s,\mathcal{F},\chi)f(g)f'(g')\chi^{-1}(\det(g'))dgdg'.$$

In the applications we take $dg = \Pi dg_v$ to be a rational measure as in the Notation section, so that $\mathrm{vol}(K_{v,G}, dg_v) = 1$ for almost all v.

The local theory of this integral was worked out by Li in [**L92**]. The articles [**GPR, L92**] only considered the case of trivial χ. The presence of the Hecke character χ was first treated by Brocco in the case $n = 1$ [**Br**]. The extension to all n is routine, and is discussed in [**HKS**, §6]. In what follows, we assume $\mathcal{F} = \bigotimes_v \mathcal{F}_v$ to be factorizable relative to (1.3), with $\mathcal{F}_v \in I_n(s,\chi_v)$. We also assume $f = \bigotimes_v f_v$, $f' = \bigotimes_v f'_v$, relative to (1.7). In particular, there is a finite set of places S of $\mathbb{Q}$, including the real place, such that, for all $v \notin S$, $\mathcal{K}_v$, π_v and χ_v are unramified and f_v, f'_v are normalized spherical vectors, so that the local pairing $(f_v, f'_v)_{\pi_v} = 1$. We can then define

$$L^S(s,\pi,\chi,\mathrm{St}) = \prod_{v\notin S} L_v(s,\pi_v,\chi_v,\mathrm{St}).$$

The following basic identity relates the zeta integral to L-functions:

$$(1.11)\quad d_n(s,\chi)Z(s,f,f',\mathcal{F},\chi) = Q_V(f,f')\cdot \tilde{Z}_S(s,f,f',\mathcal{F},\chi)\cdot L^S\left(s+\frac{1}{2},\pi,\chi,\mathrm{St}\right).$$

Here the normalizing factor $d_n(s,\chi)$ depends only on the restriction $\chi|_{A^\times} = \varepsilon_{\mathcal{K}}^m$, and equals the product

$$(1.12)\quad d_n(s,\chi) = \prod_{r=0}^{n-1} L(2s+n-r, \varepsilon_{\mathcal{K}}^{m+r})$$

of classical Dirichlet L-functions. For $v \in S$, let $d_{n,v}(s,\chi)$ denote the Euler factor of $d_n(s,\chi)$ at v. Then $\tilde{Z}_S(s,f,f',\mathcal{F},\chi) = \prod_{v\in S}\tilde{Z}_v(s,f,f',\mathcal{F},\chi)$ is the product of normalized local zeta integrals $\tilde{Z}_v(s,f,f',\mathcal{F},\chi) = [d_{n,v}(s,\chi)]^{-1}Z_v(s,f,f',\mathcal{F},\chi)$, where

$$(1.13)\quad Z_v(s,f,f',\mathcal{F},\chi) = \int_{G(\mathbb{Q}_v)} \mathcal{F}_v(i_V(g_v,1),s,\chi_v)\cdot c_{f,f',v}(g_v)dg_v,$$

and $c_{f,f',v}$ is the matrix coefficient

$$(1.14)\quad c_{f,f',v}(g_v) = [(f_v,f'_v)_{\pi_v}]^{-1}\cdot(\pi_v(g_v)f_v,f'_v)_{\pi_v},$$

normalized so that $c_{f,f',v}(1) = 1$. Finally, $Q_V(f,f')$ is a global period:

$$(1.15)\quad Q_V(f,f') = \int_{G(Q)\backslash G(\mathbf{A})} f(g)f'(g)\,dg.$$

2. Conventions for theta liftings

Let E be a local field of characteristic zero, and let $\mathbb{W}$ be a $2N$-dimensional vector space over E with a non-degenerate skew-symmetric form $< \cdot, \cdot >_{\mathbb{W}}$. Let $\mathbb{W} = \mathbb{X} \oplus \mathbb{Y}$ be a complete polarization of $\mathbb{W}$ as a sum of totally isotropic subspaces. Let $\psi\colon\ E \to \mathbb{C}^\times$ be a non-trivial additive character. Let $\tilde{\mathrm{Sp}}(\mathbb{W})$ be the metaplectic covering of $\mathrm{Sp}(\mathbb{W})$, which in our normalization will be a non-trivial extension of $\mathrm{Sp}(\mathbb{W})$ by $\mathbb{C}^\times$. One obtains $\tilde{\mathrm{Sp}}(\mathbb{W})$ from the usual two-fold cover of $\mathrm{Sp}(\mathbb{W})$ by push-forward with respect to the natural inclusion of $\{\pm 1\}$ in $\mathbb{C}^\times$. Let $\mathcal{S}$ be the space $\mathcal{S}(\mathbb{X})$ of Schwartz-Bruhat functions on $\mathbb{X}$. Thus if E is archimedean, $\mathcal{S}$ is the usual space of C^∞ functions on $\mathbb{X}$, all of whose derivatives are rapidly decreasing. If E in non-archimedean, $\mathcal{S}$ is the space of locally constant compactly supported functions on $\mathbb{X}$. We consider the oscillator representation ω_ψ of $\tilde{\mathrm{Sp}}(\mathbb{W})$ on the vector space $\mathcal{S}$, associated to the additive character $\psi\colon\ E \to \mathbb{C}^\times$, in the familiar Schrödinger model associated to the complete polarization $\mathbb{W} = \mathbb{X} \oplus \mathbb{Y}$.

Let (G, G') be a dual reductive pair in $\mathrm{Sp}(\mathbb{W})$. In most cases to be discussed in this paper, G and G' will be unitary groups. Thus, let $\mathcal{K}/E$ be a quadratic extension, possibly split, and let $G = U(V)$ and $G' = U(V')$, where V is a skew-hermitian vector space over $\mathcal{K}$ of dimension n and V' is a hermitian space of dimension m (if $\mathcal{K} = E \oplus E$ a "vector space" over $\mathcal{K}$ is a free $\mathcal{K}$-module). Then $\mathbb{W} = R_{\mathcal{K}/E}(V \otimes_{\mathcal{K}} V')$ is a $2nm$-dimensional vector space over E, naturally endowed with a non-degenerate skew-symmetric form, and $G \times G'$ embeds as a dual reductive pair in $\mathrm{Sp}(\mathbb{W})$. We let $\tilde{G}$ and $\tilde{G}'$ denote the inverse images of G and G' in $\tilde{\mathrm{Sp}}(\mathbb{W})$; then $\tilde{G}$ and $\tilde{G}'$ still commute with each other in $\tilde{\mathrm{Sp}}(\mathbb{W})$.

Let π be an irreducible representation of $\tilde{G}$. If E is p-adic, we take π to be an admissible representation; if E is archimedean, π is viewed either in the category of Hilbert spaces or in the category of admissible $(\mathrm{Lie}(\tilde{G}), \tilde{K}_\infty)$-modules, where K_∞ is a maximal compact subgroup of G and $\tilde{K}_\infty$ is its inverse image in $\tilde{\mathrm{Sp}}(\mathbb{W})$. We define

$$\Theta(\pi) = \Theta(\tilde{G} \to \tilde{G}'; \pi) = (\pi \otimes \mathcal{S})_{\tilde{G}} \tag{2.1}$$

to be the maximal quotient of $(\pi \otimes \mathcal{S})$ on which $\tilde{G}$ acts trivially. Similarly, if π' is an irreducible representation of $\tilde{G}'$, we define

$$\Theta(\pi') = \Theta(\tilde{G}' \to \tilde{G}; \pi') = (\pi' \otimes \mathcal{S})_{\tilde{G}'} \tag{2.2}$$

to be the maximal quotient of $(\pi' \otimes \mathcal{S})$ on which $\tilde{G}'$ acts trivially.

We now suppose $G = U(V)$ and $G' = U(V')$, as above, with $n = \dim_{\mathcal{K}} V$ and $m = \dim_{\mathcal{K}} V'$. Let $\varepsilon_{\mathcal{K}/E}\colon E^\times \to \{\pm 1\} \subset \mathbb{C}^\times$ denote the character associated by class field theory to the quadratic extension $\mathcal{K}/E$; $\varepsilon_{\mathcal{K}/E}$ is trivial if and only if $\mathcal{K} = E \oplus E$. Let $\chi, \chi'\colon\ \mathcal{K}^\times \to \mathbb{C}^\times$ be any characters which restrict to $\varepsilon^m_{\mathcal{K}/E}$ and $\varepsilon^n_{\mathcal{K}/E}$, respectively, on $E^\times$. Then χ and χ' determine splittings $\tilde{i}_{V,V',\chi}\colon\ G \to \tilde{G}$ and $\tilde{i}_{V',V,\chi'}\colon G' \to \tilde{G}'$ of the metaplectic covers, by the procedure described in [**HKS**, §1]; however, note that, in contrast to [**HKS**], the action of G will here be written on the left.

We will be working systematically with the double of the dual reductive pair (G, G'); in other words, with the dual reductive pair $(G \times G, G' \times G')$ in $\mathrm{Sp}(\mathbb{W} \oplus \mathbb{W})$. It is more natural, as well as more useful, to define $2\mathbb{W} = \mathbb{W} \oplus (-\mathbb{W})$ in analogy with the notation of §1; thus $(-\mathbb{W})$ is the vector space $\mathbb{W}$ with skew-symmetric form $-<\cdot,\cdot>_{\mathbb{W}}$. We define the diagonal $\Delta(\mathbb{W}) \subset 2\mathbb{W}$ as in §1. Let $H = U(2V)$, $H' = U(2V')$. Then (H, G') and (G, H') are naturally dual reductive pairs in $\mathrm{Sp}(2\mathbb{W})$. We have the commutative diagram (cf. [**HKS**, 1.5])

$$(2.3)\qquad \begin{array}{ccccc} H' & \longrightarrow & Sp(2\mathbb{W}) & \longleftarrow & H \\ \uparrow & & \uparrow & & \uparrow \\ G' \times G' & \longrightarrow & Sp(\mathbb{W}) \times Sp(-\mathbb{W}) & \longleftarrow & G \times G \\ \uparrow & & \uparrow & & \uparrow \\ G' & \longrightarrow & Sp(\mathbb{W}) & \longleftarrow & G \end{array}$$

There is a corresponding diagram in which every group is replaced by its metaplectic cover. The map $\tilde{i}\colon \tilde{\mathrm{Sp}}(\mathbb{W}) \times \tilde{\mathrm{Sp}}(-\mathbb{W}) \to \tilde{\mathrm{Sp}}(2\mathbb{W})$ is determined uniquely, among liftings of the corresponding symplectic groups, by the condition that the restriction to the center $\mathbb{C}^\times \times \mathbb{C}^\times$ of $\tilde{\mathrm{Sp}}(\mathbb{W}) \times \tilde{\mathrm{Sp}}(-\mathbb{W})$ takes (z_1, z_2) to $z_1\bar{z}_2$ in the center $\mathbb{C}^\times$ of $\tilde{\mathrm{Sp}}(2\mathbb{W})$. Then the maps on the top two rows of (2.3) can be lifted as follows:

$$(2.4)\qquad \begin{array}{ccccc} H' & \xrightarrow{\tilde{i}_{2V',V,\chi'}} & \tilde{Sp}(2\mathbb{W}) & \xleftarrow{\tilde{i}_{2V,V',\chi}} & H \\ \uparrow & & \uparrow \tilde{i} & & \uparrow \\ G' \times G' & \xrightarrow{\tilde{i}_{V',V,\chi'} \times \tilde{i}_{V',V,\chi'}\cdot(\chi')^{-1}} & \tilde{Sp}(\mathbb{W}) \times \tilde{Sp}(-\mathbb{W}) & \xleftarrow{\tilde{i}_{V,V',\chi} \times \tilde{i}_{V,V',\chi}\cdot\chi^{-1}} & G \times G \end{array}$$

Diagram (2.4), where the outer vertical arrows are the natural inclusions, is a version of diagram (1.19) of [**HKS**]; the commutativity of (2.4) follows from Lemma 1.1 of [loc. cit.].

We denote by $\Delta(G) \subset G \times G$ the subgroup of elements of the form (g, g), and define $\Delta(G') \subset G' \times G'$ analogously. The pairs $(H, \Delta(G'))$ and $(H', \Delta(G))$ are dual reductive pairs in $\mathrm{Sp}(2\mathbb{W})$. Thus

$$(\tilde{i}_{2V,V',\chi}(\Delta(G)), \tilde{i}_{2V',V,\chi'}(H')) \text{ and } (\tilde{i}_{2V,V',\chi}(H), \tilde{i}_{2V',V,\chi'}(\Delta(G')))$$

are commuting pairs in $\tilde{\mathrm{Sp}}(2\mathbb{W})$. Define

$$\Delta^-(V) = \{(x, -x) \in 2V,\ x \in V\};\ \Delta^-(V') = \{(x, -x) \in 2V',\ x \in V'\}.$$

We define a complete polarization $2\mathbb{W} = \mathbb{X} \oplus \mathbb{Y}$ as follows:

$$\mathbb{X} = \mathbb{R}_{\mathcal{K}/E}(V \otimes_{\mathcal{K}} \Delta^-(V'));\ \mathbb{Y} = R_{\mathcal{K}/E}(V \otimes_{\mathcal{K}} \Delta(V')).$$

As remarked in [**HKS**, §1], we also have

$$\mathbb{X} = R_{\mathcal{K}/E}(\Delta^-(V) \otimes_{\mathcal{K}} V');\ \mathbb{Y} = R_{\mathcal{K}/E}(\Delta(V) \otimes_{\mathcal{K}} V').$$

We consider the oscillator representation ω_ψ of $\tilde{\mathrm{Sp}}(\mathbb{W})$ on the vector space $\mathcal{S}(\mathbb{X})$, associated to the additive character $\psi\colon E \to \mathbb{C}^\times$, as above. The restriction $\omega_{\psi,\chi}$ of ω_ψ to either of the pairs

$$(\tilde{i}_{2V,V',\chi}(\Delta(G)), \tilde{i}_{2V',V,\chi'}(H')) \text{ or } (\tilde{i}_{2V,V',\chi}(H), \tilde{i}_{2V',V,\chi'}(\Delta(G')))$$

is the **Weil representation** associated to ψ, χ, and χ'. We drop the notation for the splittings and speak of the Weil representations $\omega_{\psi,\chi}$ of the pairs (G, H') and (H, G'), but it should be kept in mind that the Weil representations depend on ψ

and χ in an essential way. In what follows, we identify $\mathcal{S}(\mathbb{X}) = \mathcal{S}(V \otimes_{\mathcal{K}} V')$. Then we have

LEMMA 2.5. [**HKS**, Prop. 2.2] *The representation $\omega_{\psi,\chi'} \circ \Delta$ of G' is isomorphic to the twist by χ of the linear action of G' on $\mathcal{S}(V \otimes_{\mathcal{K}} V')$, i.e., for $\varphi \in \mathcal{S}(V \otimes_{\mathcal{K}} V')$ and $v \otimes v' \in V \otimes_{\mathcal{K}} V'$,*

$$\omega_{\psi,\chi'}(i(g,g))\varphi(v \otimes v') = \chi'(\det g) \cdot \varphi(v \otimes {}^t\bar{g}^{-1}(v')).$$

Similarly, the representation $\omega_{\psi,\chi} \circ \Delta$ of G is isomorphic to the twist by χ of the linear action of G on $\mathcal{S}(V \otimes_{\mathcal{K}} V')$, i.e., for $\varphi \in \mathcal{S}(V^n)$ and $v \otimes v' \in V \otimes_{\mathcal{K}} V'$,

$$\omega_{\psi,\chi}(i(h,h))\varphi(v \otimes v') = \chi(\det h) \cdot \varphi((h^{-1}v) \otimes v').$$

The actions of H and H' via $\tilde{i}_{2V,V',\chi}(H)$ and $i_{2V',V,\chi'}(H')$, respectively, are easily written in Schrödinger models, which make the roles of the splitting characters apparent. This point is discussed in detail in §3, below.

The splittings (2.4) allow us to define the theta lifts (2.1) and (2.2) as correspondences between representations of $G' \times G'$ (resp. H', resp. $\Delta(G')$) and $G \times G$ (resp. $\Delta(G)$, resp. H). Thus if π is a representation of $G \times G$, we let

$$\Theta(\pi) = \Theta_{\psi,\chi,\chi'}(G \times G \to G' \times G'; \pi) = (\pi \otimes \mathcal{S})_{G \times G}, \tag{2.6}$$

where the actions of $G \times G$ and $G' \times G'$ are determined by the splittings. The analogous notation is used for the other two pairs.

We view χ and χ' as representations of G and G', respectively, by composing with the determinant. We may then define

$$R_n(V',\chi) = \Theta_{\psi,\chi,\chi'}(G' \to H; \check{\chi}'); \quad R_m(V,\chi') = \Theta_{\psi,\chi,\chi'}(G \to H'; \check{\chi}). \tag{2.7}$$

Then one verifies easily, using the formulas for the Schrödinger model, that

$$R_n(V',\chi) \subset I_n\left(\frac{1}{2}(m-n),\chi\right); \quad R_m(V,\chi') \subset I_m\left(\frac{1}{2}(n-m),\chi'\right). \tag{2.8}$$

The global analogue is the theory of theta functions. Our approach here is standard and we briefly review notation, restricting our attention to the imaginary quadratic extension $\mathcal{K}/\mathbb{Q}$. We now let χ and χ' be Hecke characters of $\mathcal{K}^\times$ which restrict to $\varepsilon_{\mathcal{K}}^m$ and $\varepsilon_{\mathcal{K}}^n$, respectively, on $\mathbf{A}^\times/\mathbb{Q}^\times$. If $\varphi \in \mathcal{S}(\mathbb{X}(\mathbf{A}))$ is a global Schwartz-Bruhat function we let

$$\theta_\varphi(h) = \sum_{\lambda \in X(\mathbb{Q})} r_\psi(h)\varphi(\lambda) \tag{2.9}$$

for $h \in \widetilde{\mathrm{Sp}}(\mathbb{W})(\mathbf{A})$. Via the splittings defined locally everywhere by χ and χ' this pulls back to a function $\theta_{\varphi,\chi,\chi'}(g,g')$ for $g \in (G \times G)(\mathbf{A})$, $g' \in (G' \times G')(\mathbf{A})$. If F is an automorphic form on $(G \times G)(\mathbf{A})$ we then define

$$\theta_{\varphi,\chi,\chi'}(F)(g') = \int_{(G\times G)(\mathbb{Q})\backslash(G\times G)(\mathbf{A})} \theta_{\varphi,\chi,\chi'}(g,g')F(g)dg, \tag{2.10}$$

provided the integral converges; e.g., if F is a cusp form. Here dg is a fixed Haar measure, as in the introduction. Note that we are not taking the complex conjugate of F.

By embedding G as $G \times 1 \subset G \times G$, we can use the analogous formula to define the function $\theta_{\varphi,\chi,\chi'}(F)$ on $G'(\mathbb{Q})\backslash G'(\mathbf{A})$ when F is an automorphic form on G, by replacing φ by a Schwartz-Bruhat function on a maximal isotropic subspace of $V \otimes V'$ and integrating over $G(\mathbb{Q})\backslash G(\mathbf{A})$. Of course this is how one generally defines the theta lift for a dual reductive pair; our point here is only to explain how the splittings described above allow us to work with functions on G' and G rather than on $\tilde{G}'$ and $\tilde{G}$.

Similar integrals define theta lifts from automorphic forms on $\Delta(G)$ to forms on H', and from forms on $\Delta(G')$ to forms on H. The global analogue of formulas (2.8) is the Siegel-Weil formula. In the present article we only need the right-hand formula with $m = 1$ and $n = 1$ or $n \geq 3$. This is a special case of Weil's general formula [**W**] due to Weil when $n \geq 3$, and has been verified by Brocco when $n = 1$ [**Br**]. (The case $m = 1$, $n = 2$ can be deduced from the formula for the lift from O(2) to SL(2), as noted in [**H93**].)

Since $\chi \circ \det$ is not a cusp form, the convergence of the integral defining $\theta_{\varphi,\chi',\chi}(\chi \circ \det)$ in (5.2) needs to be justified when $n \geq 3$. In fact, our hypothesis that $n \geq 3$ implies that this integral is absolutely convergent and in fact is in the range considered by Weil in [**W**]. Bearing in mind the identifications connected with our splittings, the Siegel-Weil formula specializes in this instance to

$$\theta_{\varphi,\chi',\chi}(\chi \circ \det)(h') = E\left(h', \frac{1}{2}(n-1), \mathcal{F}_\varphi, \chi'\right), \quad h' \in H'(\mathbf{A}). \tag{2.11}$$

Here $\mathcal{F}_\varphi$ is a standard Siegel section, defined as follows. The inclusion (2.8) corresponds to a homomorphism

$$\mathcal{S} \cong \mathbb{C} \cdot (\chi \circ \det) \otimes \mathcal{S} \to (\chi \otimes \mathcal{S})_G = R_1(V, \chi') \subset I_1\left(\frac{1}{2}(n-1), \chi'\right). \tag{2.12}$$

Then $\mathcal{F}_\varphi(\cdot, s, \chi') \in I_1(s, \chi')$ is a standard Siegel section whose value at $s = \frac{1}{2}(n-1)$ is given by the image of φ under the sequence of maps (2.12), and which is extended to all s as in [**KuR**].

When $n = 1$, formula (2.11) is replaced by

$$\theta_{\varphi,\chi',\chi}(\chi \circ \det)(h') = 2 \cdot E(h', \mathcal{F}_\varphi, \chi') \tag{2.13}$$

(cf. [**Br**]). The general Siegel-Weil formula for unitary groups is the subject of work in progress of Kudla, Rallis, and Sweet.

3. Mixed models and standard splittings

In this section, E is a non-archimedean local field of characteristic zero, $\mathcal{K}$ a semi-simple quadratic E-algebra, $U^1 = U^1(\mathcal{K}) = \mathrm{Ker}\ N_{\mathcal{K}/E}\colon\ \mathcal{K}^\times \to E^\times$. In §2, we have described splittings $\tilde{i}_{V,V',\chi}\colon\ G \to \tilde{G}$ and $\tilde{i}_{V',V,\chi'}\colon\ G' \to \tilde{G}'$ of the metaplectic coverings of unitary groups associated to a dual reductive pair of the form $(U(V), U(V'))$, $\dim(V) = n$, $\dim(V') = m$. Write $\tilde{i}_{V,V',\chi,\chi'}\colon\ G \times G' \to \tilde{G} \times \tilde{G}'$ for the combined splitting. The groups $SU(V)$ and $SU(V')$ being perfect, we may identify $G^{\mathrm{ab}} \cong G'^{,\mathrm{ab}} \cong U^1$. Now the splittings $i\colon\ G \times G' \to \tilde{G} \times \tilde{G}'$ form a principal homogeneous space under the group $\mathrm{Hom}(G \times G', \mathbb{C}^\times \times \mathbb{C}^\times) \cong \Xi(U^1) \times \Xi(U^1)$, where $\Xi(U^1)$ is the character group of U^1. In other words, every splitting $i\colon G \times G' \to \tilde{G} \times \tilde{G}'$ can be written in the form

$$i = (\xi \times \xi') \cdot \tilde{i}_{V,V',\chi,\chi'}\ \ (\xi \times \xi') = (\xi \circ \det) \times (\xi' \circ \det)\colon\ G \times G \to \mathbb{C}^\times \times \mathbb{C}^\times, \tag{3.1}$$

where $\xi,\ \xi' \in \Xi(U^1)$.

We will need to know how given splittings restrict to subgroups of the form $U(W) \times U(W')$ in order to apply the method of seesaw pairs. For this purpose, we introduce standard splittings in certain cases, and use them as a basis to which to compare the splittings that arise by restriction.

In what follows, $\chi|\mathcal{K}^\times \to \mathbb{C}^\times$ is a character whose restriction to $E^\times$ is the quadratic character $\varepsilon_{\mathcal{K}/E}$. In the split case $\mathcal{K} = E \oplus E$ we take χ to be the trivial character.

We first consider the case $n = m = 1$. Then $\mathbb{W} = R_{\mathcal{K}/E}(V \otimes_{\mathcal{K}} V')$ is noncanonically isomorphic to $R_{\mathcal{K}/E}(\mathcal{K})$, and $U(V)$ and $U(V')$ may be canonically identified with $U^1(\mathcal{K})$, and in particular with each other, as subgroups of $\mathrm{Sp}(\mathbb{W})$. We let U^1 denote this common subgroup of $\mathrm{Sp}(\mathbb{W})$. Similarly, we may identify $\tilde{U}(V)$ and $\tilde{U}(V')$ with the inverse image $\tilde{U}^1$ of U^1 in $\tilde{\mathrm{Sp}}(\mathbb{W})$; let $p\colon \tilde{U}(V) \to \tilde{U}^1$ and $p'\colon \tilde{U}(V') \to \tilde{U}^1$ be the natural identifications. We choose $\chi = \chi'$ and let $\hbar \in \mathcal{K}^\times$ be an element such that $\mathrm{Tr}_{\mathcal{K}/E}(\hbar) = 0$; let $\Delta = \hbar^2 \in E$. We write $\tilde{\mathrm{Sp}}(\mathbb{W})$ as the product $\mathrm{Sp}(\mathbb{W}) \times \mathbb{C}^\times$ with multiplication given by $(g,z)(g',z') = (g{\cdot}g', z{\cdot}z'{\cdot}c(g,g'))$, with c the cocycle described in §4 of [**Ku94**]. We assume $V' \cong \mathcal{K}$ with the norm form, and $V \cong \mathcal{K}$ with the skew-hermitian form $(v,w)_V = \hbar \mathbf{a} v \cdot w^\iota$, for some $\mathbf{a} \in E$. Then there is an explicit splitting $\tilde{i}_{1,\chi}\colon U(V) \times U(V') \to \tilde{U}^1$ with the property that, if $g = x + \hbar y \in \mathcal{K}^\times$, with $x, y \in E^\times$, then

$$\begin{aligned} (3.2) \qquad \tilde{i}_{1,\chi}(g,1) &= \tilde{i}_{1,\chi}(1,g) \\ &= (g, \chi(\hbar(g-1))\gamma_E(2\mathbf{a}y(1-x),\eta)(\Delta,-2y(1-x))_E \cdot \gamma_E(\eta)) \end{aligned}$$

[**Ku94**, Prop. 4.8]. Here η is the additive character $\eta(x) = \psi(\frac{1}{2}x)$, $\gamma_E(\eta)$ is shorthand for the Weil invariant $\gamma_E(\eta \circ x^2)$, $\gamma_E(\alpha,\eta) = \gamma_E(\eta\circ \alpha x^2)/\gamma_E(\eta \circ x^2)$, and $(\cdot,\cdot)_E$ is the quadratic Hilbert symbol. One can derive formulas valid when $xy = 0$, but (3.2) already suffices to determine the character χ. We write $\tilde{i}_{1,\chi}(g,1) = (g, \mu_\chi(g))$. Then obviously

$$(3.3) \qquad (\xi \times \xi') \cdot \tilde{i}_{1,\chi}(g,g') = (gg', \mu_\chi(g \cdot g') \cdot \xi(g) \cdot \xi'(g')).$$

Let ω be a character of U^1, and suppose $\Theta_{\psi,(\xi\times\xi'),\chi}(U(V) \to U(V');\omega) \neq 0$, where the subscript indicates that the theta correspondence is defined relative to the splitting $(\xi \times \xi') \cdot \tilde{i}_{1,\chi}$. Then it is clear from (3.3) that

$$(3.4) \qquad \Theta_{\psi,(\xi\times\xi')\chi}(U(V) \to U(V');\omega) = \xi' \cdot \xi^{-1} \cdot \omega^{-1}.$$

In particular, the exact form of $\mu_\chi(g)$ is irrelevant in (3.4), and only the product of the characters ξ and ξ' is important.

Although the set of splittings of the metaplectic covering of U^1 is a principal homogeneous space under $\Xi(U^1)$, the set of splitting characters χ is principal homogeneous under the set of characters $\beta\colon\ \mathcal{K}^\times \to C^\times$ such that $\beta|_{F^\times} \equiv 1$. Any such character is of the form

$$BC(\alpha)(x) = \alpha(x/x^\iota),\ \alpha \in \Xi(U^1).$$

Here BC stands for base change, since this is the standard base change map from characters of the (twisted) torus U^1 to $\mathbb{G}_{m,\mathcal{K}}$. We also write $\tilde{\alpha} = BC(\alpha)$. Say $\beta = BC(\alpha)$, $\beta' = BC(\alpha')$. It follows easily from (3.2) (cf. [**HKS**, 1.8]) that

$$i_{V,V',\beta\cdot\chi,\beta'\cdot\chi'} = (\alpha \times \alpha') \cdot i_{V,V',\chi,\chi'}.$$

This formula, and its generalization (see (3.10), below) will be important in working out the arithmetic of liftings. For example, when $\chi = \chi'$ and $\beta' = \chi^{-2}$ we can take

$\alpha' = \check{\chi}$. Then we find

$$i_{V,V',\chi,\check{\chi}} = (1 \times \check{\chi}) \cdot i_{V,V',\chi,\chi}. \tag{3.5}$$

The following lemma is a variant of [**HKS**, Lemma 5.2]:

LEMMA 3.6. *Suppose V_1 and V_2 are skew-hermitian spaces of dimension n_1 and n_2, such that $V = V_1 \oplus V_2$, $W_i = V_i \otimes V'$, $W = W_1 \oplus W_2$. Fix characters χ'_1 and χ'_2 of $\mathcal{K}^\times$ which restrict to $(\varepsilon_{\mathcal{K}/E})^{n_i}$ on $E^\times$, and suppose $\chi' = \chi'_1 \cdot \chi'_2$. The the splittings of the dual reductive pairs $(U(V_1) \times U(V'), U(V_2) \times U(V'))$ and $(U(V) \times U(V'))$ satisfy the seesaw identity:*

$$\tilde{i}_{V,V',\chi,\chi'}|_{[(U(V_1)\times U(V_2)]\times U(V')} = (\tilde{i}_{V_1,V',\chi,\chi'_1}) \times (\tilde{i}_{V_2,V',\chi,\chi'_2})|_{[(U(V_1)\times U(V_2)]\times U(V')}.$$

Here the restriction on the left-hand side corresponds to the inclusion $U(V_1) \times U(V_2) \subset U(V)$, that on the left-hand side corresponds to the diagonal inclusion of $U(V')$ in $U(V') \times U(V')$.

Next, take n even, G quasi-split, m arbitrary, $\chi' \equiv 1$. Then the splitting $\tilde{i}_{V,V',\chi,1}$ can be realized in a Schrödinger model, as described in [**Ku94**, §5]. Write $V = X \oplus Y$, where X and Y are maximal isotropic subspaces. Let $P \subset G$ be the stabilizer of Y, $M \subset P$ the stabilizer of X, and identify $M \cong \mathrm{GL}(Y)$. Let $\mathbb{X} = X \otimes V'$, $\mathbb{Y} = Y \otimes V'$. Then $\mathbb{W} = \mathbb{X} \oplus \mathbb{Y}$ is a complete polarization. The Schrödinger model $r_{\mathrm{Sch},\psi}$ of the oscillator representation of $\widetilde{\mathrm{Sp}}(\mathbb{W})$ is realized on the vector space $\mathcal{S}(\mathbb{X})$. For our purposes it suffices to write down the splitting on $M \times U(V')$, because this determines χ. The formula is as follows:

$$\begin{aligned} r_{\mathrm{Sch},\psi}(\tilde{i}_{V,V',\chi,1}(m,g')\varphi)(x \otimes v') \\ = \chi \circ \det(m) \| N_{\mathcal{K}/E} \det(m) \|^{1/2} \varphi(m^{-1}x \otimes (g')^{-1}v'), \\ m \in M, g' \in G', x \in X, v' \in V', \varphi \in \mathcal{S}(\mathbb{X}). \end{aligned} \tag{3.7}$$

The factor $\|N_{\mathcal{K}/E} \det(m)\|^{1/2}$ is trivial on the center $Z_{U(V)}$. The following formula is then obvious.

LEMMA 3.8. *Let $i = i_{V,V',\chi,1}$ as above. Let π be an irreducible representation of G with central character ξ_π, and let $\pi' = \Theta_{\psi,\chi,1}(G \to G'; \pi)$, where the theta correspondence is defined relative to the splitting i. Then the central character $\xi_{\pi'}$ of π' satisfies $\xi_{\pi'} = .\check{\xi}_\pi \cdot \check{\chi}^{n/2}$.*

PROOF. Our normalization of the theta correspondence implies that the $\check{\pi}$-isotypic quotient of $\mathcal{S}(\mathbb{X})$ has $\pi' \otimes \check{\pi}$ as unique irreducible quotient. Let $\mathcal{T}$ denote the kernel of the natural $G \times G'$-map τ: $\mathcal{S}(\mathbb{X}) \to \pi' \otimes \check{\pi}$. The conclusion of the lemma is clearly equivalent to the condition that

$$\xi_\pi(a)\check{\xi}_{\pi'}(a^{-1})\varphi - \chi(a)^{n/2} \cdot \varphi \in \mathcal{T}, \ \forall \ \varphi \in \mathcal{S}(\mathbb{X}), \ a \in Z_{U(V)} \cong Z_{U(V')}.$$

This in turn is equivalent to the condition that

$$\tau(r_{\mathrm{Sch},\psi}(\tilde{i}_{V,V',\chi,1}(a,a^{-1}))\varphi) = \chi(a)^{n/2} \cdot \varphi \ \forall \ \varphi \in \mathcal{S}(\mathbb{X}), \ a \in Z_{U(V)} \cong Z_{U(V')}.$$

But this is immediate from (3.7).

REMARK 3.9. The notation $\tilde{i}_{V,V',\chi,1}$ is ambiguous in that it also refers to the splitting obtained by doubling V. However, it is verified in the appendix of [**HKS**], especially the discussion following Corollary A.2, that these two splittings coincide; thus our notation is correct in both senses.

We now consider the relation between central characters in the general case, with $\dim(V) = n$, $\dim(V') = m$. Suppose first we choose a character χ_0 such that $\chi_0|_{E^\times} = \varepsilon_{\mathcal{K}/E}$. Then we can use $\chi = (\chi_0)^m$, $\chi' = (\chi_0)^n$ as splitting characters. Define $\pi'(\chi_0) = \Theta_{\psi,(\chi_0)^m,(\chi_0)^n}(G \to G'; \pi)$. Bearing in mind our normalizations, it follows from Corollary A.8 of [**HKS**] that $\xi_{\pi'(\chi_0)} = .\check{\xi}_\pi$. On the other hand, we have the general formula for change of splitting [**HKS**, (1.8)]:

$$i_{V,V',BC(\alpha)\cdot\chi,BC(\alpha')\cdot\chi'} = (\alpha \circ \det \times \alpha' \circ \det) \cdot i_{V,V',\chi,\chi'}. \tag{3.10}$$

PROPOSITION 3.11. *(a) Suppose n and m are even, and let*

$$\pi' = \Theta_{\psi,1,1}(G \to G'; \pi).$$

Then $\xi_{\pi'} = .\check{\xi}_\pi$.

(b) Suppose n is even, m is odd, and let

$$\pi' = \Theta_{\psi,\chi_0,1}(G \to G'; \pi).$$

Then $\xi_{\pi'} = .\check{\xi}_\pi \cdot (\check{\chi}_0)^{n/2}$.

(c) Suppose n is odd, m is even, and let

$$\pi' = \Theta_{\psi,1,\chi_0}(G \to G'; \pi).$$

Then $\xi_{\pi'} = .\check{\xi}_\pi \cdot (\chi_0)^{m/2}$.

(d) Suppose n and m are both odd. Then

If $\pi' = \Theta_{\psi,\chi_0,\chi_0}(G \to G'; \pi)$ then $\xi_{\pi'} = .\check{\xi}_\pi \cdot (\chi_0)^{(m-n)/2}$;

If $\pi' = \Theta_{\psi,\chi_0,\check{\chi}_0}(G \to G'; \pi)$ then $\xi_{\pi'} = .\check{\xi}_\pi \cdot (\check{\chi}_0)^{(m+n)/2}$.

PROOF. Arguing as in the proof of Lemma 3.8 — in particular, bearing in mind that $\mathcal{S}$ intertwines with the representation $\check{\pi} \otimes \pi'$ — it follows from (3.10) that, if $\pi' = \Theta_{\psi,BC(\alpha)\cdot\chi,BC(\alpha')\cdot\chi'}(G \to G'; \pi)$, then

$$\xi_{\pi'} = (\alpha \circ \ \det_n)^{-1} \cdot (\alpha' \circ \ \det_m).\check{\xi}_\pi = \alpha^{-n} \cdot (\alpha')^m.\check{\xi}_\pi. \tag{3.11.1}$$

Here we write $\det_n$ and $\det_m$ for the determinants in $U(V)$ and $U(V')$, respectively. In case (a), we apply (3.11.1) with $\alpha = (\check{\chi}_0)^{m/2}$, $\alpha' = (\check{\chi}_0)^{n/2}$. Then

$$i_{V,V',BC(\alpha)\cdot\chi,BC(\alpha')\cdot\chi'} = i_{V,V',1,1} \text{ and } \alpha^{-n} \cdot (\alpha')^m.\check{\xi}_\pi = .\check{\xi}_\pi,$$

as required. The remaining cases are handled similarly. In case (b) we take $\alpha = (\check{\chi}_0)^{(m-1)/2}$, $\alpha' = (\check{\chi}_0)^{n/2}$; in case (c) we take $\alpha = (\check{\chi}_0)^{m/2}$, $\alpha' = (\check{\chi}_0)^{(n-1)/2}$; finally, in case (d) we take $\alpha = (\check{\chi}_0)^{(m-1)/2}$, $\alpha' = (\check{\chi}_0)^{(n-1)/2}$ in the first line, $\alpha = (\check{\chi}_0)^{(m-1)/2}$, $\alpha = (\check{\chi}_0)^{(n+1)/2}$ in the second line. In the last instance we then have

$$\xi_{\pi'} = (\chi_0)^{\frac{1}{2}(m-1)\cdot n} \cdot (\check{\chi}_0)^{\frac{1}{2}(n+1)\cdot m}.\check{\xi}_\pi = (\check{\chi}_0)^{(m+n)/2}\check{\xi}_\pi,$$

as indicated.

For future reference, we specialize to the case $n = 1$, m arbitrary. Fix a character χ with $\chi|_{E^\times} = \varepsilon_{\mathcal{K}/E}$. Let $i = \tilde{i}_{V,V',\chi,\chi}$ (resp. $\tilde{i}_{V,V',\chi,\check{\chi}}$) if m is odd, $i = \tilde{i}_{V,V',1,\chi}$ if m is even. Let ω be a character of $G \cong U^1$, and let

$$\pi' = \begin{cases} \Theta_{\psi,\chi,\chi}(G \to G'; \omega) \text{ (resp. } \Theta_{\psi,\chi,\check{\chi}}(G \to G'; \omega)), & m \text{ odd}, \\ \Theta_{\psi,1,\chi}(G \to G'; \omega), & m \text{ even}, \end{cases}$$

where the theta correspondence is defined relative to the splitting i. Assume $\pi' \neq 0$. Then the central character $\xi_{\pi'}$ of π' satisfies

$$(3.12) \qquad \xi_{\pi'} = \begin{cases} \omega^{-1} \cdot \chi^{(m-1)/2} \text{ (resp. } \omega^{-1} \cdot \check{\chi}^{(m+1)/2}), & m \text{ odd;} \\ \xi_{\pi'} = \omega^{-1} \cdot \chi^{m/2}, & m \text{ even.} \end{cases}$$

REMARK 3.13. Suppose both V and V' are one-dimensional. We have been assuming that the hermitian space V' is $\mathcal{K}$ itself, endowed with the norm form. Changing the form on V' will certainly alter the set of characters ω of $U(V)$ for which $\Theta_{\psi,(\xi\times\xi'),\chi}(U(V) \to U(V');\omega) \neq 0$. However, the formula (3.4) remains valid whenever the left-hand side is non-trivial. The hypothesis on V' was made only to simplify the formula (3.2), and is otherwise irrelevant. Similarly, (3.12) remains true for all one-dimensional hermitian spaces V'.

(3.14). Conventions and abbreviated table of contents. The examples listed above include the three basic dual reductive pairs that will figure in the present paper. More precisely, the bulk of the paper is devoted to studying theta lifts of coherent cohomological forms on groups of the form $U(V)$, of signature $(n-1,1)$, to groups of the form $U(V')$, of signature (r,s), $r+s = m \geq n$. The technique involves auxiliary liftings from $U(W')$, $\dim W' = 1$, to $U(V)$, and then from $U(W)$ to $U(W')$, with $\dim W = 1$. In practice, we study these liftings in the reverse order.

Each lifting requires the choice of appropriate splitting characters. We make the simplifying assumption that, in the lifting from $U(W)$ to $U(W')$, the characters in the splitting $\tilde{i}_{W,W',\chi,\chi'}$ satisfy $\chi' = \chi$ or $\chi' = \chi^{-1}$. Furthermore, all splitting characters for the other pairs are also of the form χ or χ^{-1}. There is a neat way to arrange all this, but the exact formulas depend on the parity of n and take up several pages. Systematic conventions for splittings are collected in Appendix A.2, to which we refer henceforward.

(3.15). Parameters for the unramified correspondence. A conjecture of J. Adams, included as Conjecture 7.2 in [**HKS**], determines the local Arthur parameters of the representations $\Theta_{\psi,\chi,\chi'}(\pi)$ in terms of the parameters of π, χ, and χ'. As remarked in [**HKS**] (after 7.19), it is easy to verify this conjecture when π, χ, χ', and the additive character ψ are all unramified, using the techniques of [**R**] and [**Ku86**], as extended to unitary groups in [**MoVW**]. The necessary calculations can be found in [**MoVW**, pp. 66-77], although the explicit formulas for the unramified correspondence are not stated there.

For our purposes, it will be sufficient to describe the behavior of the standard L-functions under the unramified correspondence. In the following formulas we let $\mathcal{K}_p$ denote the algebra $\mathcal{K}\otimes_{\mathbb{Q}}\mathbb{Q}_p$. If $\nu\colon\ \mathcal{K}_p^\times \to \mathbb{C}^\times$ is a quasicharacter we let $L_{\mathcal{K},p}(s,\nu)$ denote the corresponding local Hecke L-function, interpreted as the product of two local factors when p splits in $\mathcal{K}$.

Let π, χ, χ', and ψ be as above. Let p be a finite rational prime which does not ramify in $\mathcal{K}$, and suppose π, χ, χ', and ψ are all unramified at p. For a positive integer k, let

$$(3.15.1)\qquad L_p(s,\chi')^{[k]} = \prod_{j=1}^{k} L_{\mathcal{K},p}\left(s,\chi'\cdot\|\cdot\|^{(k-2j+1)/2}\right) = \prod_{j=1}^{k} L_{\mathcal{K},p}\left(s+\frac{1}{2}(k+1)-j,\chi'\right).$$

Then

$$(3.15.2)\qquad L_p(s,\Theta_{\chi,\chi'}(\pi),\mathrm{St}) = L_{\mathcal{K},p}(s,\check{\pi}(\chi'/\chi),\mathrm{St})\cdot L_p(s,\chi')^{[m-n]} = L_{\mathcal{K},p}(s,\pi,(\chi/\chi'),\mathrm{St})\cdot L_p(s,\chi')^{[m-n]}.$$

The second formula is a reflection of the fact that π is unitary. Thus π and $\check{\pi}$ have the same base change to $\mathcal{K}$.

The formulas (3.15.2) make sense whether or not p splits in $\mathcal{K}$. The split case, in which $U(V)\cong \mathrm{GL}(n,\mathbb{Q}_p)$ and $U(V')\cong\mathrm{GL}(m,\mathbb{Q}_p)$, is easier and in fact is implicit in the non-split case.

4. Extensions to similitude groups

Now let $\mathcal{K}/F$ be a quadratic extension of fields of characteristic zero, and let V be an n-dimensional $\mathcal{K}$-vector space with inner product $(\cdot,\cdot)_V$. Let $\mathrm{GU}(V)$ be the group of $\mathcal{K}$-linear transformations of V that preserve inner product up to a scalar factor:

$$(4.1)\qquad \mathrm{GU}(V) = \{g\in \mathrm{GL}(V)|\ (gv,gw)_V = \nu(g)(v,w)_V,\ \forall\ v,\ w\in V\}.$$

Thus $\nu\colon\ \mathrm{GU}(V)\to\mathbb{G}_m$ is a homomorphism of algebraic groups over F. If W is a second hermitian space over $\mathcal{K}$, we let

$$(4.2)\qquad G(U(V)\times U(W)) = \{(g,g')\in \mathrm{GU}(V)\times\mathrm{GU}(W)|\ \nu(g)=\nu(g')\}.$$

Suppose F is a number field, embedded in $\mathbb{C}$ in any way. Then

$$\mathrm{GU}(V)(\mathbb{C})\cong\{(g,g')\in\mathrm{GL}(n,\mathbb{C})^2|g\cdot{}^t g' = z\in\mathbb{C}^\times\}\cong \mathrm{GL}(n,\mathbb{C})\times\mathrm{GL}(1,\mathbb{C})$$

where the second isomorphism sends (g,g') to (g,z). Correspondingly (cf. [**Rog92**]), the Langlands dual group $\hat{GU}(V)$ may be represented as $\mathrm{GL}(n,\mathbb{C})\times\mathrm{GL}(1,\mathbb{C})$. Let Φ_n be the $n\times n$ matrix with ij-entry $(-1)^{i-1}\delta_{i,n-j+i}$. Then ${}^L\mathrm{GU}(V)$ is the semi-direct product $\hat{GU}(V)\rtimes\mathrm{Gal}(\mathcal{K}/F)$, where

$$(4.3)\qquad (g,a)^\sigma = (\Phi_n{}^tg^{-1}\Phi_n^{-1}, a\cdot\det(g))$$

(cf. [**Rog92**, p. 67]). We denote by $pr\colon {}^L\mathrm{GU}(V)\to\mathrm{Gal}(\mathcal{K}/F)$ the canonical map. This description of the L-group is also valid over local fields. It extends trivially to the case of the trivial extension $\mathcal{K}=F\oplus F$, in which case $\mathrm{GU}(V)\cong\mathrm{GL}(n)\times\mathbb{G}_m$, where the projection on the second factor is given by ν.

In practice we often write $\mathrm{GU}(V)$ for the group $\mathrm{GU}(V)(F)$ of F-rational points of $\mathrm{GU}(V)$. This is an abuse of notation but it is harmless in most cases. However, in the following definition it is essential to maintain the distinction. We let

$$(4.4)\qquad \nu_V = \nu(\mathrm{GU}(V)(F))\subset F^\times.$$

To calculate ν_V we use the long exact Galois cohomology sequence attached to the short exact sequence of $\bar{F}$-valued points:

$$1 \to U(V)(F) \to \mathrm{GU}(V)(\bar{F}) \xrightarrow{\nu} \bar{F}^{\times} \to 1. \tag{4.5}$$

We write $H^1(F,G)$ for $H^1(\mathrm{Gal}(\bar{F}/F), G(\bar{F}))$. By Hilbert's theorem 90, $H^1(F,\bar{F}^{\times})$ is trivial, and we obtain

$$1 \to U(V)(F) \to \mathrm{GU}(V)(F) \xrightarrow{\nu} F^{\times} \to H^1(F,U(V)) \to H^1(F,\mathrm{GU}(V)) \to 1. \tag{4.6}$$

LEMMA 4.7. *Suppose F is a p-adic field and $\mathcal{K}/F$ is a non-trivial quadratic extension. Then the group $H^1(F,U(V))$ is of order two. If dim(V) is even, then $H^1(F,\mathrm{GU}(V))$ is of order two; if dim(V) is odd, then $H^1(F,\mathrm{GU}(V))$ is trivial.*

If $\mathcal{K} = F \oplus F$ then $H^1(F,U(V))$ is trivial.

PROOF. Of course this is well known, but we can write down a simple proof based on the formula of Kottwitz [**K**]. Write $\Gamma = \mathrm{Gal}(\bar{F}/F)$ and let $Z(H)$ denote the center of the complex algebraic group H. For $G = U(V)$ or $\mathrm{GU}(V)$, let

$$A(G) = \pi_0(Z(\hat{G})^{\Gamma})^D, \tag{4.7.1}$$

where $\hat{G}$ is the Langlands dual group of G and the superscript D denotes Pontryagin dual. Then Kottwitz constructs an isomorphism

$$H^1(F,G) \xrightarrow{\sim} A(G) \tag{4.7.2}$$

[**K**, Prop. 6.4].

It remains to compute the right-hand side of (4.7.2) for each of our groups. Note that the Γ action on $\hat{GU}(V)$ is determined by formula (4.3); the action on $\hat{U}(V)$ is obtained by restriction to the subgroup for which $a = 1$.

When $G = U(V)$, $Z(\hat{G}) = \mathbb{C}^{\times}$ and the Γ-action takes z to z^{-1}. When $G = \mathrm{GU}(V)$, $Z(\hat{G}) = \mathbb{C}^{\times} \times \mathbb{C}^{\times}$, and the Γ-action takes (z,a) to (z^{-1}, az^n). The first part of the lemma is now an easy calculation. The last assertion is an obvious consequence of Hilbert's theorem 90.

COROLLARY 4.8. *(i) Suppose F is a p-adic field. When n is even, $\nu_V = F^{\times}$, when n is odd, $\nu_V = N_{\mathcal{K}/F}(F^{\times})$.*

(ii) Suppose $F = \mathbb{R}$, $\mathcal{K} = \mathbb{C}$. Then $\nu_V = \mathbb{R}^{\times}_{+}$ unless m is even and V has an isotropic subspace of dimension $n/2$.

PROOF. It follows from the lemma and from 4.6 that ν_V is of index one in $F^{\times}$ when n is even and of index two when n is odd. By local class field theory, $N_{\mathcal{K}/F}(F^{\times})$ is of index two in $F^{\times}$. It thus remains to show that ν_V contains $N_{\mathcal{K}/F}(F^{\times})$. But the diagonal subgroup $\mathcal{K}^{\times}$ is obviously contained in $\mathrm{GU}(V)$ and the restriction of ν to $\mathcal{K}^{\times}$ is the norm map. Part (i) follows.

Part (ii) can be verified by hand. It is clear (by using the Jordan decomposition, for example) that $\nu(\mathrm{GU}(V)(\mathbb{R})) = \nu(T)$, where T is a maximally split torus of $\mathrm{GU}(V)(\mathbb{R})$. We can diagonalize T over $\mathbb{C}$, and then the assertion is obvious.

We now write $GG = \mathrm{GU}(V)$. If F is a local field, we let

$$GG_+ = \{g \in GG(F) | \nu(g) \in N_{\mathcal{K}/F}\mathcal{K}^{\times}\}.$$

If F is a number field, we let

$$GG(\mathbf{A})_+ = \{g \in GG(\mathbf{A}) | g_v \in GG(F_v)_+ \text{ for all places } v \text{ of } F\}$$

and let $GG(\mathbf{A})_0 = GG(F) \cdot GG(\mathbf{A})_+$. Then

$$GG(\mathbf{A})_0 = \{g \in GG(\mathbf{A}) | \nu(g) \in F^\times \cdot N_{\mathcal{K}/F}\mathcal{K}_{\mathbf{A}}^\times\} \tag{4.9}$$

is a normal subgroup of $GG(\mathbf{A})$ of index at most two. It follows from Corollary 4.8 that, if n is odd and F is a number field, then $GG(\mathbf{A})_0 = GG(\mathbf{A})$ and $GG(F_v)_+ = GG(F_v)$ for every finite place v. If n is even, then $GG(\mathbf{A})_0$ is of index two in $GG(\mathbf{A})$ (here we may need to invoke weak approximation to take care of the real places).

We return to the conventions of the earlier sections; thus $\mathcal{K}$ is an imaginary quadratic field, $F = \mathbb{Q}$. Suppose V and V' are two hermitian spaces over $\mathcal{K}$, as before, $GG^{(2)} = G(U(V) \times U(V))$, $GG^{\prime,(2)} = G(U(V') \times U(V'))$. We define $GG^{(2)}(\mathbf{A})_0$ by analogy with (4.9). In [**H93**, §3.8], it is explained how to extend the theta correspondence to these similitude groups. Let F be an automorphic form on $GG^{(2)}(\mathbf{A})$ and let χ, χ', and the Schwartz-Bruhat function φ be as in (2.10). The formulas given in [loc. cit] show how the integral (2.10), which we assume to be absolutely convergent, can be modified to define a function $\theta_{\varphi,\chi,\chi',0}(F)$ on $GG^{\prime,(2)}(\mathbb{Q})\backslash GG^{\prime,(2)}(\mathbf{A})_0$ if $\dim(V)$ is odd, and on $GG^{\prime,(2)}(\mathbb{Q})\backslash GG^{\prime,(2)}(\mathbf{A})$ if $\dim(V)$ is even. Note that the notation is not the same as in [loc. cit.], where the subgroup $GG^{,(2)}(\mathbf{A})_0 \subset GG^{,(2)}(\mathbf{A})$ is defined relative to V rather than to V'. The local theory has been worked out in detail by Roberts in [**Ro**], in the exactly analogous case of a symplectic-orthogonal dual reductive pair.

In particular, it follows from the previous discussion that, if either $\dim(V)$ is even or $\dim(V) \cdot \dim(V')$ is odd, then $\theta_{\varphi,\chi,\chi',0}(F)$ is an automorphic form on $GG^{\prime,(2)}$, and we denote it simply $\theta_{\varphi,\chi,\chi'}$. If $\dim(V)$ is odd and $\dim(V')$ is even then we define $\theta_{\varphi,\chi,\chi'}$ on $GG^{\prime,(2)}(\mathbb{Q})\backslash GG^{\prime,(2)}(\mathbf{A})$ by

$$\theta_{\varphi,\chi,\chi'}(F)(g) = \begin{cases} \theta_{\varphi,\chi,\chi',0}(F)(g), & \text{if } g \in GG^{\prime,(2)}(\mathbf{A})_0, \\ 0, & \text{otherwise.} \end{cases} \tag{4.10}$$

Then $\theta_{\varphi,\chi,\chi'}(F)(g)$ is again an automorphic form. In either case, if π is an automorphic representation of $GG^{(2)}$, we let $\Theta_{\psi,\chi,\chi'}(GG^{(2)} \to GG^{\prime,(2)};\pi)$ be the $GG^{\prime,(2)}(\mathbf{A})$-submodule of $\mathcal{A}(GG^{\prime,(2)})$ generated by $\{\theta_{\varphi,\chi,\chi'}(F) | F \in \pi\}$, provided the integrals defining $\theta_{\varphi,\chi,\chi',0}(F)$ converge absolutely for $F \in \pi$.

It is not presumed that $\Theta_{\psi,\chi,\chi'}(GG^{(2)} \to GG^{\prime,(2)},\pi)$ is an irreducible automorphic representation even when π is. Suppose for the moment that π is cuspidal. If we just consider the space $\Theta_{\psi,\chi,\chi'}(\pi)_0$ of forms $\theta_{\varphi,\chi,\chi',0}(F)$ as a representation of $GG^{\prime,(2)}(\mathbf{A})_0$, then the local Howe duality conjecture would imply that $\Theta_{\psi,\chi,\chi'}(\pi)_0$ is already an irreducible representation of $(G' \times G')(\mathbf{A})$, hence of $GG^{\prime,(2)}(\mathbf{A})_0$. If π is not cuspidal, then $\Theta_{\psi,\chi,\chi'}(\pi)_0$ may well be reducible but indecomposable. However, this argument is only useful under special hypotheses in residue characteristic two, where local Howe duality is not established.

Suppose $\Theta_{\psi,\chi,\chi'}(\pi)_0$ to be irreducible under $GG^{\prime,(2)}(\mathbf{A})_0$. Then the extended space $\Theta_{\psi,\chi,\chi'}(GG^{(2)};\to GG^{\prime,(2)};\pi)$ is either irreducible or the sum of two irreducible $GG^{\prime,(2)}(\mathbf{A})$-modules, which differ by twisting by the quadratic character $\varepsilon_{\mathcal{K}} \circ \nu$. In our applications the irreducibility will not be essential.

We need to explain how the Siegel-Weil formula (2.11) extends to a formula for similitude groups. In (2.11) the Eisenstein series is defined on the quasi-split unitary group H' of a two-dimensional (skew)-hermitian space. When n is even then the extended theta function on $GH'(\mathbf{A})$ can be defined as above and denoted $\theta_{\varphi,\chi',\chi}(\chi \circ \det)$, where χ is here viewed as a character of $\mathcal{K}^\times \cong \mathrm{GU}(1)$. On the other hand, $\chi' \equiv 1$ in this case and we can define

$$I_1\left(\frac{1}{2}(n-1), GH'\right) = I_1\left(\frac{1}{2}(n-1), \chi', GH'\right)$$

to be the representation of $GH'(\mathbf{A})$ unitarily induced from the character

$$\mathrm{diag}(a, ca^{\iota,-1}) \mapsto (|N_{\mathcal{K}/\mathbb{Q}}(a)|_{\mathbf{A}})^{(n-1)/2} \; a \in \mathcal{K}_{\mathbf{A}}^{\times}, \; c \in \mathbf{A}^{\times}$$

of the diagonal subgroup of $GH'(\mathbf{A})$, extended trivially to the Borel subgroup GP. Then restriction to $H'(\mathbf{A})$ naturally defines an isomorphism

$$I_1\left(\frac{1}{2}(n-1), GH'\right) \cong I_1\left(\frac{1}{2}(n-1), \chi'\right)$$

where the right-hand side is defined as in §1. Composing the series of maps (2.12) with the inverse of this isomorphism, we obtain a map $\mathcal{S} \to I_1((n-1)/2, GH')$, which we again denote $\varphi \mapsto \mathcal{F}_{\varphi}$. Then the analogue of (2.11) remains valid for similitude groups without modification:

(4.11, n even) $$\theta_{\varphi,\chi',\chi}(\chi \circ \det)(h') = E\left(h', \frac{1}{2}(n-1), \mathcal{F}_{\varphi}, \chi'\right), \quad h' \in GH'(\mathbf{A}).$$

If now n is odd, then $\theta_{\varphi,\chi',\chi}(\chi \circ \det)$ has been defined above to be identically zero off $GH'(\mathbf{A})_0$. On the other hand, we can define $I_1((n-1)/2, \chi', GH'(\mathbf{A})_0)$ to be the representation of $GH'(\mathbf{A})_0$ unitarily induced from the character

$$\mathrm{diag}(a, ca^{\iota,-1}) \mapsto \chi'(a)(|N_{\mathcal{K}/\mathbb{Q}}(a)|_{\mathbf{A}})^{(n-1)/2} \; a \in \mathcal{K}_{\mathbf{A}}^{\times}, \; c \in \mathbb{Q}^{\times} \cdot N_{\mathcal{K}/\mathbb{Q}}(\mathcal{K}_{\mathbf{A}}^{\times})$$

of $GP(\mathbf{A}) \cap GH'(\mathbf{A})_0$. As before, there is a map

$$\mathcal{S} \to I_1\left(\frac{1}{2}(n-1), \chi', GH'(\mathbf{A})_0\right); \quad \varphi \mapsto \mathcal{F}_{\varphi,0},$$

and we define $\mathcal{F}_{\varphi,+}$ to be the extension by zero of $\mathcal{F}_{\varphi,0}$ to $GH'(\mathbf{A})$. We then have

(4.11, n odd) $$\theta_{\varphi,\chi',\chi}(\chi \circ \det)(h') = E\left(h', \frac{1}{2}(n-1), \mathcal{F}_{\varphi,+}, \chi'\right), h' \in GH'(\mathbf{A}).$$

The verification of (4.11) is elementary, given the Siegel-Weil formula for H', and is left to the reader.

5. Seesaw pairs

A seesaw pair — the terminology is due to Kudla — is a pair of dual reductive pairs (G, G') and (H, H') in a fixed symplectic group $\mathrm{Sp}(\mathbb{W})$, with the property that $H \subset G$. Then necessarily H', the centralizer of H in $\mathrm{Sp}(\mathbb{W})$, contains G'. This situation is represented by the following diagram:

(5.1)

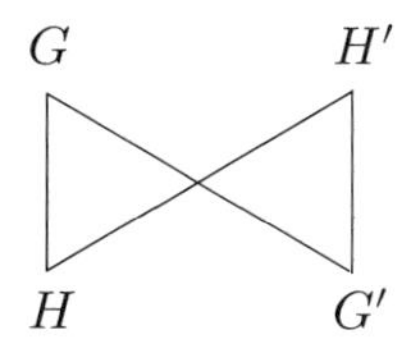

The main calculations in this article are based on formulas involving a number of seesaw pairs in which G, G', H, and H' are all unitary groups or products thereof, which we assume from now on. The pairs that appear in the article are listed in

Appendix 1. The reference [**H93**, §3] discusses these and other examples at length, and in particular explains how seesaw pairs behave with respect to splittings. Some of this is worked out in more detail in the appendix to [**HKS**], and is recalled in Lemma 3.6, above.

The principal application of seesaw pairs is the seesaw reciprocity formula, which calculates the inner product of a form lifted from G' to G against a form on H in terms of an integral on G'. For the unitary groups this is a simple application of Fubini's theorem. Proposition 3.8.4 of [**H93**] extends the seesaw reciprocity formula to compare integrals over $Z_{GH}(\mathbf{A}) \cdot GH(\mathbb{Q})\backslash GH(\mathbf{A})_0$ to integrals over $Z_{GG'}(\mathbf{A}) \cdot GG'(\mathbb{Q})\backslash GG'(\mathbf{A})_0$. Here when $H = \Pi U(V_i)$ we let

$$GH = G(\Pi U(V_i)) = (\Pi \mathrm{GU}(V_i)) \cap GG,$$

and likewise for GG'. This was sufficient for the purposes of that article. However, our Petersson inner products are expressed as integrals over

$$Z_{GH}(\mathbf{A}) \cdot GH(\mathbb{Q})\backslash GH(\mathbf{A}) \text{ and } Z_{GG'}(\mathbf{A}) \cdot GG'(\mathbb{Q})\backslash GG'(\mathbf{A}).$$

In fact, there is no difficulty extending the seesaw reciprocity formula in our situation to the full similitude groups, using the extended global theta correspondence defined in §4.

We recall the definition. Let $\mathcal{M} \in \{H, G', GH, GG'\}$ be one of the groups indicated, and let $Z_{\mathcal{M}}$ denote its center. Let F and F' be automorphic forms on $\mathcal{M}$ and suppose there is a character ξ of $Z_{\mathcal{M}}(\mathbf{A})$ such that, for all $z \in Z_{\mathcal{M}}(\mathbf{A})$ and all $m \in \mathcal{M}(\mathbf{A})$,

$$F(zm) = \xi(z)F(m), \quad F'(zm) = \xi^{-1}(z)F'(m), \tag{5.2}$$

We then let

$$< F, F' >_{\mathcal{M}} = \int_{\mathcal{M}(Q)\cdot Z_{\mathcal{M}}(A)\backslash \mathcal{M}(A)} F(m)F'(m)\, dm, \tag{5.3}$$

assuming the integral is absolutely convergent (for example, when either F or F' is a cusp form). Here dm is a rational Haar measure on $\mathcal{M}(\mathbb{Q}) \cdot Z_{\mathcal{M}}(\mathbf{A})\backslash \mathcal{M}(\mathbf{A})$, as recalled in the notation section. When $\mathcal{M} = H$ or G' we assume dm is the same measure used for the global theta lifting to H' or G, respectively. The pairing is symmetric rather than hermitian.

In the following proposition, we assume G, G', H, and H' are all unitary groups or products of unitary groups, and take F and F' to be automorphic forms on GH and GG', respectively. We choose φ in the Schwartz space $\mathcal{S}(\mathbb{X}(\mathbf{A}))$ for an appropriate complete polarization $\mathbb{W} = \mathbb{X} \oplus \mathbb{Y}$. We define $\theta_\varphi(F)$ and $\theta_\varphi(F')$ as automorphic forms on $GH'(\mathbf{A})$ and $GG(\mathbf{A})$, respectively, using the procedure introduced in §4. The splitting characters χ, χ' are omitted; we assume the splittings for the pairs (GH, GH') and (GG', GG) are chosen compatibly, as explained in [**H93**, §3].

PROPOSITION 5.4. (seesaw reciprocity). *Under the above hypotheses, we have*

$$< \theta_\varphi(F), F' >_{GG'} = < \theta_\varphi(F'), F >_{GH} . \tag{5.4.1}$$

PROOF. Proposition 3.8.4 of [**H93**] is the analogous equality where $\theta_\varphi(F)$ is replaced by the function we called $\theta_{\varphi,\chi,\chi',0}(F)$ in §4 and likewise for $\theta_\varphi(F')$, and where the integrals are taken over the appropriate subsets of $\mathcal{M}(\mathbb{Q})\cdot Z_{\mathcal{M}}(\mathbf{A})\backslash \mathcal{M}(\mathbf{A})$, $\mathcal{M} = GG'$ or GH. We denote by $< \theta_\varphi(F), F' >_{\mathcal{M},0}$ the integrals of [loc. cit.]. When G, G', H, and H' are all products of unitary groups of even-dimensional hermitian spaces, or when they are all products of unitary groups of odd-dimensional hermitian spaces, it follows from the discussion in §4 that $\theta_{\varphi,\chi,\chi',0}(F)$ equals $\theta_\varphi(F)$ (and likewise for $\theta_\varphi(F')$), and so

$$< \theta_\varphi(F), F' >_{GG',0} = < \theta_\varphi(F), F' >_{GG'}, \ < \theta_\varphi(F'), F >_{GH} = < \theta_\varphi(F'), F >_{GH,0} .$$

In this case the proposition reduces immediately to Proposition 3.8.4 of [loc. cit.].

The general case also reduces to Proposition 3.8.4 of [loc. cit.]. It suffices to remark that the process of extension by zero reduces the integrals $GG'(\mathbb{Q}) \cdot Z_{GG'}(\mathbf{A})\backslash GG'(\mathbf{A})$, (resp. $GH(\mathbb{Q}) \cdot Z_{GH}(\mathbf{A})\backslash GH(\mathbf{A})$), considered in (5.4.1), to the integrals over $GG'(\mathbb{Q}) \cdot Z_{GG'}(\mathbf{A})\backslash GG'(\mathbf{A})_0$ (resp. $GH(\mathbb{Q}) \cdot Z_{GH}(\mathbf{A})\backslash GH(\mathbf{A})_0$) considered in [loc. cit.]. This can be verified by checking the cases of different parity separately. For example, if $G = \Pi U(V_i)$, all $\dim(V_i)$ even, $G' = \Pi U(V_i')$, with at least one $\dim(V_i')$ odd, then $GG'(\mathbf{A}) = GG'(\mathbf{A})_0$, so

$$< \theta_\varphi(F), F' >_{GG',0} = < \theta_\varphi(F), F' >_{GG'} .$$

On the other hand, $\theta_\varphi(F')$ is supported on $GG(\mathbf{A})_0$, by construction, so

$$< \theta_\varphi(F'), F >_{GH} = < \theta_\varphi(F'), F >_{GH,0},$$

whatever the dimensions of the hermitian spaces associated with H and H'. We leave the remaining verifications to the reader.

II. Cohomological representations of $\mathbf{U(n-1,1)}$ and $\mathbf{U(r,s)}$

1. Parameters of cohomological representations

In this section we establish conventions for parameters of certain unitary representations of unitary groups with non-trivial $(\mathfrak{g}, K_\infty)$-cohomology. These representations are all derived functor modules, but we will not be using the parameters from the theory of cohomological induction. Instead, we will use the characterization of cohomological representations by their infinitesimal characters and lowest K-types [**VZ**, Theorem 1.4], which are better adapted to coherent cohomology. If π is a $(\mathfrak{g}, K_\infty)$-module and V is a finite-dimensional G-module such that $H^*(\mathfrak{g}, K_\infty; \pi \otimes V^*) \neq 0$, then π and V have the same infinitesimal character, which thus determines the highest weight of V by a familiar formula. Cohomological representations also have *$\bar{\partial}$-cohomology* with coefficients in a certain K_∞-module τ, which is uniquely determined when π is in the discrete series. We determine the highest weight of τ in the discrete series cases of interest to us and indicate a preferred τ in the case of ladder representations. The reader is invited to skip this section and to refer to it as needed.

We let $G = U(r,s)$, with $n = r+s$, $K_\infty = U(r) \times U(s)$, H the diagonal maximal torus of K_∞. We write $K_{r,s}$ for K_∞ when attention needs to be paid to the signature. The corresponding Lie algebras are denoted by the lower case Gothic letters $\mathfrak{g}, \mathfrak{k}, \mathfrak{h}$. The characters of H are parametrized by n-tuples $(a_1, \dots, a_r; a_{r+1}, \dots, a_n)$ with $a_i \in \mathbb{Z}$, $i = 1, \dots, n$. These n-tuples are to be viewed as linear forms on $\mathfrak{h}$; in other words, we are given an embedding of the group $X(H)$ of characters as a lattice in $\mathfrak{h}^*$, and an isomorphism $X(H) \cong \mathbb{Z}^n$. The parametrization of the characters determines the standard ordering of the set of complex roots of $(\mathfrak{g}_{\mathbb{C}}, \mathfrak{h}_{\mathbb{C}})$, in which the positive roots correspond to the upper triangular matrices. We let R^+ (resp. R_c^+) denote the set of positive roots of $(\mathfrak{g}, \mathfrak{h})$ (resp. $(\mathfrak{k}, \mathfrak{h})$). We write R^+ as the disjoint union $R_c^+ \coprod R_n^+$, where R_n^+ is the set of positive non-compact roots. Then the highest weights of finite-dimensional irreducible representations of K_∞ (resp. of G) are the n-tuples with $a_1 \geq \cdots \geq a_r; a_{r+1} \geq \cdots \geq a_n$ (resp. $a_1 \geq \cdots \geq a_r \geq a_{r+1} \geq \cdots \geq a_n$). An element $(a_1, \ldots, a_r; a_{r+1}, \ldots, a_n) \in \mathbb{R}^n \cong \mathfrak{h}^*$ is *regular* if $a_i \neq a_j$ whenever $i \neq j$.

We write $\mathfrak{k}_{\mathbb{C}} = \mathfrak{f}_{\mathbb{C}} \oplus \mathfrak{p}^+ \oplus \mathfrak{p}^-$, where $\mathfrak{p}^+$ (resp. $\mathfrak{p}^-$) is the (abelian) subalgebra spanned by the positive (resp. negative) non-compact root vectors. This defines a unique G-invariant complex structure on the quotient $X_+^{r,s} := G/K_\infty$ such that, under the isomorphism of $\mathfrak{g}_{\mathbb{C}}/\mathfrak{f}_{\mathbb{C}}$ with the complexified tangent space to $X_+^{r,s}$ at the fixed point of $K_\infty, \mathfrak{p}^+$ (resp. $\mathfrak{p}^-$) corresponds to the holomorphic (resp. anti-holomorphic) tangent space.

Let $\mathfrak{P}$ denote the subalgebra $\mathfrak{p}^- \oplus \mathfrak{k}_{\mathbb{C}}$ of $\mathfrak{g}_{\mathbb{C}}$. If π is a (g, K_∞)-module and W is a finite-dimensional K_∞-module, then the $\bar{\partial}$-cohomology of π with coefficients in W is the space $H^{\cdot}(\mathfrak{P}, K_\infty; \pi \otimes W)$. We denote this $H_{\bar{\partial}}^{\cdot}(\pi \otimes W)$ when K_∞ is understood.

Let W denote the Weyl group $W(\mathfrak{g}_{\mathbb{C}}, \mathfrak{h}_{\mathbb{C}})$, $W_c = W(\mathfrak{k}_{\mathbb{C}}, \mathfrak{h}_{\mathbb{C}})$. Let $\mathfrak{z} = \mathfrak{z}(\mathfrak{g}_{\mathbb{C}})$ denote the center of the universal enveloping algebra $U(\mathfrak{g}_{\mathbb{C}})$. If π is an irreducible $(\mathfrak{g}, K_\infty)$-module, then $\mathfrak{z}(\mathfrak{g}_{\mathbb{C}})$ acts as scalars on the space of π, and the corresponding homomorphism $\mathfrak{z}(\mathfrak{g}_{\mathbb{C}}) \to \mathbb{C}$ is called the *infinitesimal character* of π. The infinitesimal characters correspond to the linear forms on $\mathfrak{h}_{\mathbb{C}}$, modulo the action of W; they can thus be indexed by n-tuples of complex numbers.

Define $\rho = \sum_{\alpha \in R^+} \alpha$; then

$$\rho = \left(\frac{1}{2}(n-1), \frac{1}{2}(n-3), \ldots, \frac{1}{2}(n-2r+1); \frac{1}{2}(n-2r-1), \ldots, \frac{1}{2}(1-n)\right).$$

If $\Lambda \in \mathbb{Z}^n$ is the highest weight of a finite dimensional representation V_Λ of $\mathfrak{g}$, then the infinitesimal character of V_Λ is $\Lambda + \rho$, which is necessarily regular. The discrete series representations of G, or the discrete series $(\mathfrak{g}, K_\infty)$-modules, are parametrized by regular elements of the subset $X(H) + \rho \subset (\mathfrak{h}_{\mathbb{C}})^*$, modulo the action of W_c. For $\lambda \in X(H) + \rho$ we let π_λ denote the corresponding discrete series representation; λ is the *Harish-Chandra parameter* of π_λ. Then the infinitesimal character of π_λ is λ (mod W). In particular, to each finite-dimensional representation V_Λ there is a collection of $|W/W_c| = n!/[r!(n-r)!]$ discrete series $(\mathfrak{g}, K_\infty)$-modules which share the infinitesimal character $\Lambda + \rho$ with V_Λ. The collection of all discrete series $(\mathfrak{g}, K_\infty)$-modules with given infinitesimal character $\Lambda + \rho$ is called the *L-packet* associated to the dominant weight Λ.

We identify a regular infinitesimal character with the unique R^+-dominant element in its W-orbit and we identify a Harish-Chandra parameter with the unique R_c^+-dominant element in its W_c-orbit. A Harish-Chandra parameter λ is dominant with respect to a unique positive root system $R^+(\lambda) = R_c^+ \cup R_n^+(\lambda)$, where $R_n^+(\lambda)$ is the set of non-compact roots positive for λ.

The discrete series representation π_λ is uniquely characterized, among unitary representations of G, by its infinitesimal character and its lowest (or Blattner) K_∞-type τ_λ [**S**]. The K_∞-type τ_λ is distinguished by the fact that it occurs in π_λ with multiplicity one, and that every other K_∞-type has longer highest weight. We identify τ_λ with the n-tuple defining its highest weight; then $\tau_\lambda = \lambda - \rho_c + \rho_n(\lambda)$, where ρ_c is the half-sum of positive compact roots and $\rho_n(\lambda) = (1/2)\sum_{\alpha \in R_n^+} \alpha$.

We need to specify τ_λ in the following two situations:

(1.1) Relative to the complex structure defined above on $X_+^{r,s}$, π_λ is a holomorphic discrete series representation of $U(r,s)$. In that case, λ is in the *negative* Weyl chamber and $-\rho_c + \rho_n(\lambda) = -\rho$; $\tau_\lambda = \lambda - \rho$.

(1.2) The rank one cases: $(r,s) = (n-1,1)$. Then $|W/W_c| = n$ is the cardinality of a discrete series L-packet.

We follow the conventions of [**H90, HL**]; thus π_λ is a holomorphic discrete series representation of $U(r,s)$ precisely when the occurrences of its *contragredient* $\check{\pi}_\lambda$ in the space of automorphic forms correspond to holomorphic modular forms on the symmetric space $X_+^{r,s}$. Equivalently, $H^0(\mathfrak{P}, K_{r,s}; \check{\pi}_\lambda \otimes W_\Lambda) \neq 0$ for a certain irreducible $K_{r,s}$-module W_Λ with the highest weight Λ. When $(r,s) = (n-1,1)$, we say π_λ *belongs to the* j-th *chamber* if $H^{j-1}(\mathfrak{P}, K_{r,s}; \check{\pi}_\lambda \otimes W_\Lambda) \neq 0$ for some W_Λ. In either case, the degree $j-1$ and the weight Λ are uniquely determined by λ, and the unique non-vanishing space $H^{j-1}(\mathfrak{P}, K_{r,s}; \check{\pi}_\lambda \otimes W_\Lambda)$ is of dimension 1. Indeed, we have $\Lambda = \lambda - \rho$ and $j-1$ is the cardinality of $R_n^+ \cap R_n^+(\lambda)$ (cf. [**H90**, §4]).

In the holomorphic case we have an isomorphism of 1-dimensional spaces

$$H^0(\mathfrak{P}, K_{r,s}; \check{\pi}_\lambda \otimes W_\Lambda) \cong (\check{\pi}_\lambda \otimes W_\Lambda)^{K_{r,s}}. \tag{1.3}$$

Recall that the lowest K-type τ_λ of π_λ has highest weight $\lambda - \rho$ in this case; thus $\tau_\lambda \cong W_\Lambda$ and the right-hand side of (1.3) equals the image of $(\check{\tau}_\lambda \otimes W_\Lambda)^{K_{r,s}}$ in $(\check{\pi}_\lambda \otimes W_\Lambda)^{K_{r,s}}$.

In the rank one case, suppose π_λ belongs to the j-th chamber. Then

$$\begin{aligned} H^{j-1}(\mathfrak{P}, K_{n-1,1}; \check{\pi}_\lambda \otimes W_\Lambda) &\cong (\wedge^{j-1}\mathfrak{p}^{-,*} \otimes \check{\pi}_\lambda \otimes W_\Lambda)^{K_{n-1,1}} \\ &= (\wedge^{j-1}\mathfrak{p}^{-,*} \otimes \check{\tau}_\lambda \otimes W_\Lambda)^{K_{n-1,1}}, \end{aligned} \tag{1.4}$$

with $\Lambda = \lambda - \rho$ and (cf. [**HL**, (22)])

$$\begin{aligned} \tau_\lambda &= \lambda - \rho_c + \rho_n(\lambda) = \Lambda + \rho_n + \rho_n(\lambda) \\ &= \Lambda + (\underbrace{1,\dots,1}_{j-1},\underbrace{0,\dots,0}_{n+1-j}; 1-j). \end{aligned} \tag{1.5}$$

In applications of the theta correspondence to the rank one case, we will let (r', s) be auxiliary variables, with $r' + s = n$, and take π_λ in the r'-th chamber. We use τ_λ or, equivalently, Λ, to parametrize π_λ. We assume we are given a collection of non-negative integers $\{a_1, a_2, \dots, a_{r'}, b_1, \dots, b_s\}$ and an integer $r \geq r'$ such that

$$a_1 \geq a_2 \geq \cdots \geq a_{r'-1} \geq a_{r'}; \ b_1 \geq \cdots \geq b_s > a_{r'} + 1 + (r - r'). \tag{1.6}$$

We let $q = (r-s)/2$ and take τ_λ in the form

$$\tau_\lambda = (a_1 + q, \dots, a_{r'-1} + q, -b_s + q, \dots, -b_1 + q; -a_{r'} - q) + \frac{1}{2}(\alpha, \dots, \alpha) \tag{1.7}$$

where $\alpha = \pm 1$ if $r+s$ is odd, $\alpha = 0$ if $r+s$ is even, in order to make the parameters integral. Then (1.5) yields

$$\begin{aligned} \Lambda = (a_1 + q - 1, \dots, a_{r'-1} + q - 1, -b_s + q, \dots, -b_1 + q; -a_{r'} - q + r' - 1) \\ + \frac{1}{2}(\alpha, \dots, \alpha). \end{aligned} \tag{1.8}$$

To verify that these parameters do in fact correspond to a representation in the r'-th chamber, we need to determine λ and to check that $r'-1$ is the cardinality of $R_n^+ \cap R_n^+(\lambda)$. Write $\mathbf{v} = \frac{1}{2}(\alpha, \dots, \alpha)$. Then

$$
\begin{aligned}
(1.9)\quad \lambda &= \Lambda + \rho \\
&= (a_1 + *, \dots, a_{r'-1} + q - 1 + \frac{1}{2}(n - 2r' + 3), -b_s + q + \frac{1}{2}(n - 2r' + 1), \\
&\quad \dots, -b_1 + *; -a_{r'} - q + \frac{1}{2}(r' - s - 1)) + \mathbf{v},
\end{aligned}
$$

where the *'s don't matter to the determination of $R_n^+ \cap R_n^+(\lambda)$. Then

$$
\begin{aligned}
(1.10)\qquad & r' - 1 = |R_n^+ \cap R_n^+(\lambda)| \\
& \Leftrightarrow -a_{r'} - q + \frac{1}{2}(r' - s - 1) > -b_s + q + \frac{1}{2}(n - 2r' + 1).
\end{aligned}
$$

Indeed, our conventions are that $R_n^+ = \{e_i - e_n, i = 1, \dots, n-1\}$, where e_i are the standard basis vectors of $\mathbb{Z}^n$, and it is easy to see that the only inequality to check is the one involving $a_{r'}$ and b_s. But (1.10) can be rewritten

$$
(1.11)\qquad r' - 1 = |R_n^+ \cap R_n^+(\lambda)| \Leftrightarrow b_s > a_{r'} + 1 + (r - r'),
$$

which is just (1.6). Note that the spacing between $a_{r'}$ and b_s must be greater than the difference between r and r'. This corresponds on the level of the theta correspondence to the condition that the theta lift of $\check{\pi}_\lambda$ to $U(r,s)$ be of *cohomological type*, as we verify in §3, below.

We next consider the ladder representations of $U(n-1,1)$, the unique $(\mathfrak{g}, K_\infty)$-modules which have cohomology in degree 1. Say π is a $(\mathfrak{g}, K_\infty)$-module and V_Λ is a finite-dimensional representation of G such that $H^1(\mathfrak{g}, K_{n-1,1}; \pi \otimes V_\Lambda) \neq 0$. Then Λ is necessarily of one of the following forms:

$$
\begin{aligned}
\Lambda &= \Lambda(\alpha, \beta) = (\alpha + \beta, \beta, \dots, \beta); \\
\Lambda &= \Lambda(-\alpha, \beta) = (\beta, \dots, \beta, \beta - \alpha)
\end{aligned}
$$

for a pair of integers $\alpha \geq 0$ and β [**VZ**]; the sign distinction is irrelevant when $\alpha = 0$. If $\alpha > 0$ and $\Lambda = \Lambda(\alpha, \beta)$ or $\Lambda = \Lambda(-\alpha, \beta)$, there is then exactly one $(\mathfrak{g}, K_{n-1,1})$-module π such that $H^1(\mathfrak{g}, K_{n-1,1}; \check{\pi} \otimes V_\Lambda) \neq 0$; if $\alpha = 0$ there are two modules π such that $H^1(\mathfrak{g}, K_{n-1,1}; \check{\pi} \otimes V_\Lambda) \neq 0$. We denote them $\mathbb{L}^+(-\alpha, \beta)$ and $\mathbb{L}^-(\alpha, \beta)$, respectively. They are distinguished by their $\bar{\partial}$-cohomology, as follows. We let $\Lambda^-(\alpha, \beta)$ (resp. $\Lambda^+(-\alpha, \beta)$) be the highest weight $(\alpha + \beta, \beta, \dots \beta; \beta)$ (resp. $(\beta, \dots, \beta, \beta - \alpha - 1; \beta + 1)$) for $K_{n-1,1}$. Let $W^-(\alpha, \beta)$ and $W^+(-\alpha, \beta)$ be the corresponding $K_{n-1,1}$ modules. Then

$$
(1.12)\quad \dim H^0_{\bar{\partial}}(\mathbb{L}^+(-\alpha, \beta)\check{} \otimes W^+(-\alpha, \beta)) = \dim H^1_{\bar{\partial}}(\mathbb{L}^-(\alpha, \beta)\check{} \otimes W^-(\alpha, \beta)) = 1.
$$

The $\bar{\partial}$-cohomology spaces above are supported on specific minimal $K_{n-1,1}$-types of the corresponding $(\mathfrak{g}, K_{n-1,1})$-modules. We let $\tau^+(-\alpha, \beta)$ and $\tau^-(\alpha, \beta)$ be the minimal $K_{n-1,1}$-types of $\mathbb{L}^+(-\alpha, \beta)$ and $\mathbb{L}^-(\alpha, \beta)$, respectively. Then (identifying the representation with its highest weight)

$$
\begin{aligned}
(1.13)\qquad \tau^+(-\alpha, \beta) &= \Lambda^+(-\alpha, \beta) = (\beta, \dots, \beta, \beta - \alpha - 1; \beta + 1); \\
\tau^-(\alpha, \beta) &= (\alpha + \beta + 1, \beta, \dots, \beta; \beta - 1).
\end{aligned}
$$

The representation $\mathbb{L}^+(-\alpha,\beta)$ (resp. $\mathbb{L}^-(\alpha,\beta)$) is characterized uniquely among unitary representations by its infinitesimal character $\Lambda(-\alpha,\beta)+\rho$ (resp. $\Lambda(\alpha,\beta)+\rho$) and its minimal $K_{n-1,1}$-type $\tau^+(-\alpha,\beta)$ (resp. $\tau^-(\alpha,\beta)$) [**VZ**].

(1.14). We also need notation for certain non-cohomological holomorphic $(\mathfrak{g},K_{n-1,1})$-modules. First, when $\alpha=-1$ in the previous discussion, the $K_{n-1,1}$-type $\tau^+(1,\beta)=(\beta,\dots,\beta;\beta+1)$ is not the minimal type of a representation with cohomology, but there is an irreducible $(\mathfrak{g},K_{n-1,1})$-module $\mathbb{L}^+(1,\beta)$ with minimal $K_{n-1,1}$-type $\tau^+(1,\beta)$ such that

$$\dim H^0_{\bar\partial}(\mathbb{L}^+(1,\beta)^\vee\otimes W^+(1,\beta))\neq 0,$$

where $W^+(1,\beta)=\tau^+(1,\beta)$ is defined by the same formula as before. Namely, $\mathbb{L}^+(1,\beta)^\vee$ is the irreducible highest weight module with the indicated type. When $n=2$ the module $\mathbb{L}^+(1,\beta)$ corresponds to elliptic modular forms of weight 1.

Similarly, for any positive integer a, there is an irreducible holomorphic highest weight module $\mathbb{M}^+(a,\beta)^\vee$ such that

$$\dim H^0_{\bar\partial}(\mathbb{M}^+(a,\beta)^\vee\otimes T^+(a,\beta))\neq 0,$$

where $T^+(a,\beta)$ is the one-dimensional $K_{n-1,1}$-type $(\beta,\dots,\beta;a+\beta)$. Obviously $\mathbb{L}^+(1,\beta)=\mathbb{M}^+(1,\beta)$, so the former notation is unnecessary, but the two families arise by different constructions, and it seems useful to maintain this separation. In fact, one easily verifies that $\mathbb{M}^+(a,\beta)$ is a holomorphic discrete series module when $a>n-1$.

(1.15). In applications to Shimura varieties it is necessary to work with the similitude groups rather than the symmetry groups of hermitian spaces. We write $GG=\mathrm{GU}(r,s)$, with notation as in I, §4, let $Z\subset GG$ denote the subgroup of real diagonal matrices, identified with $\mathbb{R}^\times$ by the natural identification, and let $GK_{r,s}=Z\cdot K_{r,s}$. The irreducible finite-dimensional algebraic representations of $GK_{r,s}$ are parametrized by $(r;s;1)$-tuples of integers $(a_1,\dots,a_r;a_{r+1},\dots,a_{r+s};c)$ where c stands for the character $t\mapsto t^c$ of Z. The $(r;s;1)$ tuple $(a_1,\dots,a_r;a_{r+1},\dots,a_{r+s};c)$ must satisfy the following consistency condition on $Z\cap K_{r,s}=\{\pm1\}$:

$$c\equiv a_1+a_2+\dots+a_{r+s}\pmod 2. \tag{1.15.1}$$

Apart from this, the additional parameter c causes no trouble: it just tags along unvarying through all the parametrizations. For example, the parameters for discrete series of $U(n-1,1)$ in (1.7)-(1.9) are replaced by

$$\begin{aligned}\tau_\lambda&=(a_1+q,\dots,a_{r'-1}+q,-b_s+q,\dots,-b_1+q;-a_{r'}-q;c),\\ \Lambda&=(a_1+q-1,\dots,a_{r'-1}+q-1,-b_s+q,\dots,-b_1+q;\\ &\quad-a_{r'}-q+r'-1;c),\\ \lambda&=(a_1+*,\dots,a_{r'-1}+q-1+\frac{1}{2}(n-2r'+3),\\ &\quad-b_s+q+\frac{1}{2}(n-2r'+1),\dots,\\ &\quad-b_1-*;-a_{r'}-q+\frac{1}{2}(r'-s-1);c).\end{aligned} \tag{1.15.2}$$

Similarly, we write $\Lambda^\pm(\alpha,\beta;c),\mathbb{L}^\pm(\alpha,\beta;c)$, etc. for the ladder representation of similitude groups.

2. Cup products in $\bar{\partial}$-cohomology

We first treat the cup-product of $\bar{\partial}$-cohomology of ladder representations of $U(n-1,1)$, which is straightforward. We then describe a multiple cup-product which will be essential in our study of the rationality of the theta-correspondence. We begin by establishing some conventions. Let St denote the standard representation of $U(n-1)$ (or $\mathrm{GL}(n-1)$), and let St^a denote the a-th symmetric power of St for any non-negative integer a. Thus St^a has highest weight $(a,0,\dots,0)$. Similarly, let St^* denote the dual of St, and let $\mathrm{St}^{*,a}$ denote the a-th symmetric power of St^* for any non-negative integer a. Thus $\mathrm{St}^{*,a}$ has highest weight $(0,\dots,0,-a)$ and is dual to St^a.

Finally, for any $b \in \mathbb{Z}$, let $\det^b$ denote the b-th power of the determinant character of $U(n-1)$ (or $\mathrm{GL}(n-1)$), and let $[b]$ denote the b-th power of the identity character of $U(1)$ (or $\mathrm{GL}(1)$). In these terms, we have

$$\begin{aligned}
\tau^+(-\alpha,\beta) &\cong \left(\mathrm{St}^{*,\alpha+1}\otimes \det{}^\beta\right)\boxtimes[\beta+1],\\
\tau^-(\alpha,\beta) &\cong \left(\mathrm{St}^{\alpha+1}\otimes \det{}^\beta\right)\boxtimes[\beta-1],\\
W^+(-\alpha,\beta) &\cong \left(\mathrm{St}^{*,\alpha+1}\otimes \det{}^\beta\right)\boxtimes[\beta+1],\\
W^-(\alpha,\beta) &\cong \left(\mathrm{St}^{\alpha}\otimes \det{}^\beta\right)\boxtimes[\beta],
\end{aligned}\tag{2.1}$$

where the symbol $\boxtimes$ is used to separate representations of $U(n-1)$ and $U(1)$.

It then follows that

$$\tau^+(-\alpha,\beta)\cong\tau^-(\alpha,-\beta)^\vee.$$

This isomorphism extends to a duality of $(\mathfrak{g},K_{n-1,1})$-modules

$$\mathbb{L}^+(-\alpha,\beta)\cong\mathbb{L}^-(\alpha,-\beta)^\vee\tag{2.2}$$

and to a perfect duality of $\bar{\partial}$-cohomology spaces

$$H^0_{\bar{\partial}}(\mathbb{L}^+(-\alpha,\beta)^\vee\otimes W^+(-\alpha,\beta))\otimes H^1_{\bar{\partial}}(\mathbb{L}^-(\alpha,-\beta)^\vee\otimes W^-(\alpha,-\beta))\to\mathbb{C}.\tag{2.3}$$

On the other hand, (2.1) also yields a natural homomorphism

$$W^+(-\alpha,\beta)\otimes W^-(\alpha,-\beta)\to\mathrm{St}^*\boxtimes[1]\cong\mathfrak{p}^{+,*},\tag{2.4}$$

whence a canonical pairing

$$\begin{aligned}
&H^0_{\bar{\partial}}(\mathbb{L}^+(-\alpha,\beta)^\vee\otimes W^+(-\alpha,\beta))\otimes H^1_{\bar{\partial}}(\mathbb{L}^-(\alpha,-\beta)^\vee\otimes W^-(\alpha,-\beta))\to\\
&\quad H^1_{\bar{\partial}}(W^+(-\alpha,\beta))\otimes W^-(\alpha,-\beta))\to H^1_{\bar{\partial}}(\mathfrak{p}^{+,*})\\
&\quad\cong(\mathfrak{p}^{-,*}\otimes\mathfrak{p}^{+,*})^{K_{n-1,1}}\cong\mathbb{C}.
\end{aligned}\tag{2.5}$$

The composition of the last two isomorphisms in (2.5) determines the natural $U(n-1,1)$-invariant Kähler class on the symmetric space $X^{n-1,1}$. We return to this pairing in III, §4.

Next, let $V=V_\sigma$ be the irreducible representation of $U(n-1)$ with highest weight $\sigma=(a_1,\dots,a_{n-1})$, with $a_1\geq\cdots\geq a_{n-1}$, and consider the tensor product $V_\sigma\otimes\ \mathrm{St}^b$ as a representation of $U(n-1)$. As a special case of the Littlewood-Richardson rules (Pieri's formula, [**HF**, p. 455], we have

Proposition 2.6. *Let $\mathcal{L}(\sigma,b)$ denote the set of $(n-1)$-tuples*

$$\{(a_1+b_1,\dots,a_{n-1}+b_{n-1})|a_i+b_i\leq a_{i-1}\ \forall\ i, b_1+\cdots+b_{n-1}=b, \textit{all } b_i\geq 0\}$$

The tensor product $V_\sigma\otimes \mathrm{St}^b$ *decomposes as the direct sum*

$$V_\sigma \otimes \mathrm{St}^b \cong \bigoplus_{\mu\in\mathcal{L}(\sigma,b)} V_\mu$$

with multiplicity one.

By dualizing, it follows formally that

COROLLARY 2.7. *Let* $\mathcal{L}^*(\sigma,b)$ *denote the set of dominant* $(n-1)$*-tuples*

$\{(a_1-b_1\geq\cdots\geq a_{n-1}-b_{n-1})|a_i\leq a_{i-1}-b_{i-1}\ \forall\ i, b_1+\cdots+b_{n-1}=b, \text{all } b_i\geq 0\}$

The tensor product $V_\sigma\otimes \mathrm{St}^{*,b}$ *decomposes as the direct sum*

$$V_\sigma \otimes \mathrm{St}^{*,b} \cong \bigoplus_{\mu\in\mathcal{L}(\sigma,b)} V_\mu.$$

with multiplicity one.

(2.8). Let $q=(r-s)/2$. If $q\in\mathbb{Z}$, let $\alpha=0$; if not, let $\alpha=\pm1/2$ (we change notation from §1 to save space). As in (1.8), let

$$\begin{aligned}\Lambda = (&a_1+q-1+\alpha,\dots,a_{r'-1}+q-1+\alpha,-b_s+q+\alpha,\dots,\\ &-b_1+q+\alpha;-a_{r'}-q+r'-1+\alpha),\end{aligned} \tag{2.8.1}$$

corresponding to a discrete series representation π_λ in the r'th chamber, with $\lambda=\Lambda+\rho$. Thus $H^{r'-1}_{\bar\partial}(\breve\pi\otimes W_\Lambda)\neq 0$. We write

$$\begin{aligned}W_\Lambda = &\det{}^{q+\alpha}\otimes W(a_1-1,\dots,a_{r'-1}-1,-b_s,\dots,-b_1)\\ &\boxtimes[-a_{r'}-q+\alpha+r'-1],\end{aligned} \tag{2.8.2}$$

where $W(a_1-1,\dots,a_{r-1}-1,-b_s,\dots,-b_1)$ is the representation of $U(n-1)$ with the indicated weight. Let e be a non-negative integer, $r=r'+e$. For each $i\in\{1,\dots,r\}$ choose $\varepsilon_i\in\{-1,0\}$, and let

$$W_i=\begin{cases}\det^{\varepsilon_i}\otimes\mathrm{St}^{*,a_i}\boxtimes[\varepsilon_i+1], & i=1,\dots,r'-1;\\ \det^{\varepsilon_i}\boxtimes[\varepsilon_i+1], & i=r',\dots,r-1;\\ \det^{\varepsilon_i}\boxtimes[a_{r'}+\varepsilon_i+1], & i=r.\end{cases} \tag{2.8.3}$$

Similarly, for each $j\in\{1,\dots,s\}$, let $\delta_j\in\{0,1\}$, and let

$$V_j=\det{}^{\delta_j}\otimes\mathrm{St}^{b_j-1}\boxtimes[\delta_j]. \tag{2.8.4}$$

We assume $\Sigma\varepsilon_i+\Sigma\delta_j=-q-\alpha$. Then

$$\begin{aligned}&W_1\otimes\cdots\otimes W_r\otimes V_1\otimes\cdots\otimes V_s\\ &\quad\cong\det{}^{-q-\alpha}\otimes\left(\mathrm{St}^{*,a_1}\otimes\cdots\otimes\mathrm{St}^{*,a_{r'}-1}\right)\otimes\left(\mathrm{St}^{b_1-1}\otimes\cdots\otimes\mathrm{St}^{b_s-1}\right)\\ &\quad\quad\boxtimes[a_{r'}-q-\alpha+r]\\ &\quad\cong\det{}^{-q-\alpha}\otimes\left(\mathrm{St}^{*,a_1}\otimes\cdots\otimes\mathrm{St}^{*,a_{r'}-1}\right)\otimes\left(\mathrm{St}^{b_1-1}\otimes\cdots\otimes\mathrm{St}^{b_s-1}\right)\\ &\quad\quad\boxtimes[a_{r'}+q-\alpha+s]\end{aligned}$$

COROLLARY 2.9. *Up to scalars, there is a unique homomorphism of* $U(n-1)\times U(1)$*-modules:*

$$\begin{aligned}&W_1\otimes\cdots\otimes W_r\otimes V_1\otimes\cdots\otimes V_s\\ &\quad\to\det{}^{-q-\alpha}\otimes W(b_1-1,\dots,b_s-1,-a_{r'-1},\dots,-a_1)\boxtimes[a_{r'}+q-\alpha+s]\\ &\quad\cong\det{}^{-1}\otimes W_{\breve\Lambda}\boxtimes[n-1].\end{aligned}$$

Moreover, restricted to $U(n-1)$-representations, the homomorphism is non-trivial on the weight vector $w_1 \otimes \cdots \otimes w_r \otimes v_1 \otimes \cdots \otimes v_s$, where

$$w_i = (0, \dots, 0, -a_i, 0, \dots, 0) \quad \text{(non-zero in } (n-i)\text{th place);}$$
$$v_j = (0, \dots, 0, b_j - 1, 0, \dots, 0) \quad \text{(non-zero in } j\text{th place).}$$

PROOF. The first assertion follows by induction from Proposition 2.6 and Corollary 2.7. Write $(c_1, \dots, c_{n-1}) = (b_1 - 1, \dots, b_s - 1, -a_{r'-1}, \dots, -a_1)$ and let $W(j) = W(c_1, \dots, c_j, 0, \dots, 0)$, for $j = 1, \dots, n-1$, with highest weight vector $w(j)$. If $j \leq s$, we claim that the map

$$\varphi_j \colon W(j-1) \otimes \mathrm{St}^{b_j - 1} \to W(j)$$

is non-trivial on $w(j-1) \otimes v_j$. But it is obvious that φ_j is non-trivial on the subspace $w(j-1) \otimes \mathrm{St}^{b_j-1}$, and the claim then follows by comparing weights. The second assertion then follows by induction from the claim (and the analogous claim for the St^{*,a_i}).

In what follows, we let

$$\mathbb{L}_i^+ = \mathbb{L}^+(1 - a_i, \varepsilon_i), i = 1, \dots, r'-1; \; \mathbb{L}_i^+ = \mathbb{L}^+(1, \varepsilon_i), i = r', \dots, r-1;$$
$$\mathbb{L}_r^+ = \mathbb{M}^+(a_{r'} + 1, \varepsilon_i);$$
$$\mathbb{L}_j^- = \mathbb{L}_j^-(b_j - 1, \delta_j), j = 1, \dots, s.$$

The proof of Corollary 2.9 also applies, with a change of notation, to the minimal K_∞-types (τ_i^+, A_i^+) and (τ_j^-, A_j^-) of the representations $\mathbb{L}_i^+$ and $\mathbb{L}_j^-$, respectively; here A_i^+ and A_j^- denote the spaces of the corresponding representations. More precisely, let $(\tau_\lambda, V_\lambda)$ be the minimal K_∞-type of π_λ, given by (1.7) but with $\alpha/2$ now denoted α. As in (2.8.2), we write

(2.9.1) $$V_\lambda \cong \det{}^{q+\alpha} \otimes W(a_1, \dots, a_{r'-1}, -b_S, \dots, -b_1) \boxtimes [-a_{r'} - q + \alpha].$$

On the other hand, by (1.13) and 1.14 we have

(2.9.2) $$A_i^+ \cong \begin{cases} \det^{\varepsilon_i} \otimes \mathrm{St}^{*,a_i} \boxtimes [\varepsilon_i + 1], & i = 1, \dots, r'-1, \\ \det^{\varepsilon_i} \boxtimes [\varepsilon_i + 1], & i = r', \dots, r-1. \\ \det^{\varepsilon_i} \boxtimes [a_{r'} + \varepsilon_i + 1], & i = r. \end{cases}$$

$$A_j^- \cong \det{}^{\delta_j} \otimes \mathrm{St}^{b_j} \boxtimes [\delta_j - 1], j = 1, \dots, s.$$

Then we find

COROLLARY 2.9.3. *Up to scalars, there is a unique homomorphism of $U(n-1) \times U(1)$-modules:*

$$A_1^+ \otimes \cdots \otimes A_r^+ \otimes A_1^- \otimes \cdots \otimes A_s^-$$
$$\to \det{}^{-q-\alpha} \otimes W(b_1, \dots, b_s, -a_{r'-1}, \dots, -a_1) \boxtimes [a_{r'} + q - \alpha] \cong V_\lambda^\vee.$$

PROPOSITION 2.10. *The contraction of Corollary 2.9 induces a non-trivial map of $\bar\partial$-cohomology groups:*

$$C \colon H_{\bar\partial}^{r'-1}(\check\pi \otimes W_\Lambda) \otimes \bigotimes_{i=1}^{r} H_{\bar\partial}^0(\check{\mathbb{L}}_i^+ \otimes W_i) \otimes \bigotimes_{j=1}^{s} H_{\bar\partial}^1(\check{\mathbb{L}}_j^- \otimes V_j)$$
$$\to H_{\bar\partial}^{n-1}(\det{}^{-1} \boxtimes [n-1]) \cong H_{\bar\partial}^{n-1}(\wedge^{n-1}\mathfrak{p}^{+,*}).$$

PROOF. Corollary 2.9 shows the existence of the map C. We replace the representation $\check{\pi}$, $\mathbb{L}_i^+$, and $\check{\mathbb{L}}_j^-$ in the left-hand side by their minimal K_∞-types, which carry the cohomology. Then the map C can be rewritten

$$(\wedge^{r'-1}\mathfrak{p}^{-,*} \otimes V_{\check{\lambda}} \otimes W_\Lambda)^{K_\infty} \otimes \Big(\bigotimes_{i=1}^{r} \check{A}_i^+ \otimes W_i\Big)^{K_\infty} \otimes \Big(\bigotimes_{j=1}^{s} \mathfrak{p}^{-,*} \otimes \check{A}_j^- \otimes V_j\Big)^{K_\infty}$$

$$\downarrow$$

$$(2.10.1) \qquad (\wedge^{n-1}\mathfrak{p}^{-,*} \otimes \Big(V_{\check{\lambda}} \otimes \bigotimes_{i=1}^{r} \check{A}_i^+ \otimes \bigotimes_{j=1}^{s} \check{A}_j^-\Big) \otimes (W_\Lambda \otimes \bigotimes_{i=1}^{r} W_i \otimes \bigotimes_{j=1}^{s} V_j))^{K_\infty}$$

$$\downarrow$$

$$\Big(\wedge^{n-1}\mathfrak{p}^{-,*} \otimes \det{}^{-1} \boxtimes [n-1]\Big)^{K_\infty}$$

The first map is the contraction of the differentials, the second is given by Corollaries 2.9 and 2.9.3. Both the domain and the range of C are one-dimensional; we have to show that the map is non-trivial. This is a matter of showing compatibility of the maps of Corollaries 2.9 and 2.9.3 with the contraction of the differentials, and for this it suffices to look at the maps on representation of $U(n-1)$, since the $U(1)$ representation takes care of itself. We thus let $\mathcal{W} = W(b_1, -1, \ldots, b_s - 1, -a_{r'-1}, \ldots, -a_1), \mathcal{V} = W(b_1, \ldots, b_s, -a_{r'-1}, \ldots, -a_1)$. In what follows, we write "unique" for "non-trivial and unique up to scalar multiples."

Corollary 2.9, restricted to the $U(n-1)$-representations, yields a unique map

$$(2.10.2) \quad \det \otimes \wedge^{r'-1} \mathrm{St}^* \otimes \bigotimes_{i=1}^{r'-1} \mathrm{St}^{*,a_i} \otimes \bigotimes_{j=1}^{s} \mathrm{St}^{b_j - 1} \to \det \otimes \wedge^{r'-1} \mathrm{St}^* \otimes \mathcal{W} \to \mathcal{V}.$$

The uniqueness of the second map is equivalent to the one-dimensionality of the space $H_{\bar{\partial}}^{r'-1}(\check{\pi} \otimes W_\Lambda)$.

On the other hand, there is a natural isomorphism

$$(2.10.3) \qquad \det \otimes \wedge^{r'-1} \mathrm{St}^* \cong \wedge^s \mathrm{St}.$$

Moreover, Proposition 2.6, applied to $V_\sigma = \wedge^s \mathrm{St}$, yields a unique equivariant map

$$(2.10.4) \qquad \wedge^s \mathrm{St} \otimes \bigotimes_{j=1}^{s} \mathrm{St}^{b_j - 1} \to \bigotimes_{j=1}^{s} \mathrm{St}^{b_j}$$

whose image is the multiplicity one summand of the right-hand side with highest weight $(b_1, \ldots, b_s, 0, \ldots 0)$.

Via (2.10.4), we thus obtain a canonical non-trivial map

$$(2.10.5) \qquad \wedge^s \mathrm{St} \otimes \bigotimes_{i=1}^{r'-1} \mathrm{St}^{*,a_i} \otimes \bigotimes_{j=1}^{s} \mathrm{St}^{b_j-1} \to \bigotimes_{i=1}^{r'-1} \mathrm{St}^{*,a_i} \otimes \bigotimes_{j=1}^{s} \mathrm{St}^{b_j} \to \mathcal{V},$$

where the second arrow is given by Corollary 2.9.3. But (2.10.3) identifies the left-hand sides of (2.10.2) and (2.10.5). Since (2.10.2) is unique, it follows that (2.10.2) and (2.10.5) are the same, up to scalar multiples.

In view of the factorization (2.10.1) and the second part of Corollary 2.9, the assertion of the proposition is now an easy consequence of the commutativity (up to scalar multiples) of the following diagram:

$$\begin{array}{ccc} \mathrm{St}^{\otimes s} \otimes \bigotimes_{j=1}^{s} \mathrm{St}^{b_j-1} & \cong & \bigotimes_{j=1}^{s} \mathrm{St} \otimes \mathrm{St}^{b_j-1} \\ \downarrow & & \downarrow \\ \wedge^s \mathrm{St} \otimes \bigotimes_{j=1}^{s} \mathrm{St}^{b_j-1} & \longrightarrow & \bigotimes_{j=1}^{s} \mathrm{St}^{b_j} \end{array} \tag{2.10.6}$$

where the vertical maps are the obvious ones and the bottom map is (2.10.4). But this obvious, since, following either way around the diagram, the vector:

$$(1,0,\dots,0) \otimes \cdots \otimes (0,\dots,\underset{s\text{th place}}{0,1,0},\dots,0)$$
$$\otimes (b_1-1,0,\dots,0) \otimes \cdots \otimes (0,\dots,0,\underset{s\text{th place}}{b_s-1},0,\dots,0)$$

in $\mathrm{St}^{\otimes s} \bigotimes_{j=1}^{s} \mathrm{St}^{b_j-1}$ has non-trivial image in the highest weight space

$$(b_1,\dots,b_s,0,\dots 0)$$

of the image of the bottom map.

3. The archimedean theta correspondence (following J.-S. Li)

J.-S. Li has worked out most of the archimedean theta correspondence for the dual reductive pair $(G,G') = (U(a,b),U(r,s))$ when the representation of G lies in the discrete series and that of G' has non-zero cohomology. He provides expressions for the correspondence in terms of Vogan-Zuckerman parameters and in terms of minimal K-types. We recall his results in the latter expression when $(a,b) = (n-1,1)$ and when the representation of $U(r,s)$ is holomorphic or anti-holomorphic, where now $r+s \geq n$. We also consider the ladder representations of $U(n-1,1)$, associated to characters of $U(r,s)$ with $r+s=1$. The correspondence in fact pairs representations of metaplectic double covers of G and G', and when either $a+b$ or $c+d$ is odd the double covers are non-trivial. However, the push-forward to covers of G and G' by $\mathbb{C}^\times$ can always be split, and we reformulate Li's results taking splittings into account.

Let V be a finite dimensional hermitian (or skew-hermitian) space over $\mathbb{C}$, and let $U(V)$ be its unitary group. We begin by working with the algebraic double cover $\tilde{U}(V)$ of $U(V)$, defined to make the following diagram cartesian:

$$\begin{array}{ccc} \tilde{U}(V) & \longrightarrow & U(1) \\ \downarrow & & \downarrow \; u \mapsto u^2 \\ U(V) & \xrightarrow{\det} & U(1). \end{array}$$

If $H \subset U(V)$ is a maximal compact torus, as in §1, we let $\tilde{H}$ denote its inverse image in $\tilde{U}(V)$. Then the character group $X(\tilde{H})$ of $\tilde{H}$ is generated by $X(H)$, identified as above with $\mathbb{Z}^n$, and a character denoted $\det^{1/2}$, represented by the n-tuple $(1/2,\dots,1/2)$. If $U(V) \cong U(r,s)$, we denote by $\tilde{K}_{r,s}$ the inverse image of $K_{r,s}$ in $\tilde{U}(V)$, and index finite-dimensional representations of $\tilde{K}_{r,s}$ by their highest weights in $X(\tilde{H}) = \mathbb{Z}^n + \mathbb{Z} \cdot (1/2,\dots,1/2)$.

Now let V be a skew-hermitian space of signature $(n-1,1)$, and let V' be a hermitian space of signature (r,s). We consider the oscillator representation ω_ψ of $\tilde{G}\times\tilde{G}'$ on the vector space $\mathcal{S}$, associated to the additive character $\psi(x)=e^{2\pi ix}$. For $\mathcal{S}$ we may take the Schwartz space $\mathcal{S}(\mathbb{X})$ of $\mathbb{X}$ for some polarization of the symplectic vector space $R_{\mathbb{C}/\mathbb{R}}(V\otimes V')=\mathbb{X}\oplus\mathbb{Y}$. Let π be an irreducible representation of $\tilde{G}$, considered either in the category of Hilbert spaces or in the category of admissible $(\mathrm{Lie}(\tilde{G}),\tilde{K}_{n-1,1})$-modules, and define $\Theta(\pi)=(\pi\otimes\mathcal{S})_{\tilde{G}}$ as in I.2.1. Then $\Theta(\pi)$ is a representation of $\tilde{G}'$, and when π is in the discrete series $\Theta(\pi)$ is known to be irreducible (cf. [**L90**]). Similarly, if π' is an irreducible representation of $\tilde{G}'$, we define $\Theta(\pi')$.

Suppose $\pi=\check{\pi}_\lambda$ is in the discrete series. Let $m=r+s$, $r'=n-s$, and $\varepsilon=r-r'=m-n$. Let $\tau=\tau_\lambda$, $\check{\tau}=\check{\tau}_\lambda$, and write

$$\tau=(a_1+q,\dots,a_{r'-1}+q,-b_s+q,\dots,-b_1+q;-a_{r'}-q)\ \text{ with } q=\frac{1}{2}(r-s).$$

Suppose π_λ is in the r'-th chamber; thus $H^{r'-1}(\mathfrak{P},\tilde{K}_{n-1,1};\check{\pi}_\lambda\otimes W_\Lambda)\neq 0$ with $\Lambda=\lambda-\rho$. Recall (1.11) that this implies that $-a_{r'}-\varepsilon-1>-b_s$. Then $\pi'=\Theta(\pi)$ contains the $\tilde{K}_{r,s}$-type τ' with highest weight

$$\tau'=\Big(a_1+\frac{n-2}{2},\dots,a_{r'-1}+\frac{n-2}{2},\underbrace{\frac{n-2}{2},\dots,\frac{n-2}{2}}_{r-r'},-a_{r'}+\frac{n-2}{2};\quad -b_s-\frac{n-2}{2},\dots,-b_1-\frac{n-2}{2}\Big).\tag{3.1}$$

It is easy to see that π' is always a holomorphic representation, though not necessarily in the holomorphic discrete series. Indeed, the Harish-Chandra parameter $\lambda=\Lambda+\rho$ of π_λ is given by (1.9). The infinitesimal character of π_λ is then the permutation $\tilde{\lambda}$ of the entries of λ which is dominant for $(\mathfrak{g},\mathfrak{h})$ (cf. (1.10)):

$$\tilde{\lambda}=\Big(a_1+*>\dots>a_{r'-1}+\frac{\varepsilon+1}{2}>0>-a_{r'}-\frac{\varepsilon+1}{2}>-b_s+\frac{\varepsilon+1}{2}>\dots>-b_1+*\Big).\tag{3.2}$$

Define $\tilde{c}=(c_1,\dots,c_\varepsilon)$ as follows. If ε is even, then

$$\tilde{c}=\Big(\frac{\varepsilon}{2},\frac{\varepsilon-1}{2},\dots,\frac{1-\varepsilon}{2},-\frac{\varepsilon}{2}\Big).$$

If ε is odd, then

$$\tilde{c}=\Big(\frac{\varepsilon}{2}-1,\dots,-\frac{\varepsilon}{2}\Big).$$

As in [**L90**], the infinitesimal character of π' is

$$\tilde{\lambda}'=\Big(a_1+*,\dots,a_{r'-1}+\frac{\varepsilon+1}{2},c_1,\dots,c_\varepsilon,-a_{r'}-\frac{\varepsilon+1}{2},\quad -b_s+\frac{\varepsilon+1}{2},\dots,-b_1+*\Big).\tag{3.3}$$

Then π' is cohomological if and only if $\tilde{\lambda}'$ is regular, which is true if and only if it is the infinitesimal character associated to a dominant weight of $\mathrm{GL}(r+s)$. This is true if and only if

$$a_{r'} + \frac{\varepsilon+1}{2} > \frac{\varepsilon}{2}; \qquad a_{r'-1} + \frac{\varepsilon+1}{2} > \begin{cases} \frac{\varepsilon}{2} & \varepsilon \text{ even.} \\ \frac{\varepsilon}{2} - 1 & \varepsilon \text{ odd.} \end{cases}$$

But these conditions are automatic; the crucial condition is (1.11), which guarantees that π_λ is in the r'-th chamber. Moreover, Li determines π' explicitly as a derived functor module and verifies that π' is in the discrete series if and only if $0 \leq \varepsilon \leq 1$ (cf. [**L90**, 51]).

We next consider the theta lift in the reverse direction when $\dim(V') = 1$. The representations of $\tilde{U}(V')$ are then characters, parametrized by expressions of the form $(k+(n-2)/2;)$ if $(r,s) = (1,0)$ (resp. $(;k-(n-2)/2)$ if $(r,s) = (0,1)$), where $k \in (1/2)\mathbb{Z}$. Thus, let $\pi' = \omega$ be a character of $\tilde{U}(V')$, with parameter as above, and write $\Theta_{(r,s)}(\omega)$ to emphasize the signature.

There are four cases to consider. Note the inversion of the parameter in the following table!

(3.4) Suppose $(r,s) = (1,0)$, $\omega = (k+(n-2)/2;), k > 0$. Then $\Theta_{(1,0)}(\omega^{-1})$ is a ladder representation with minimal $\tilde{K}_{n-1,1}$-type $(k+1/2, 1/2, \dots, 1/2; -1/2) = \tau^-(k-1, 1/2)$, so $\Theta_{(1,0)}(\omega^{-1}) = \mathbb{L}^-(k-1,1/2) \cong \mathbb{L}^+(1-k,-1/2)\check{}$;

(3.5) Suppose $(r,s) = (1,0)$, $\omega = (k+(n-2)/2;), k \leq 0$. Then $\Theta_{(1,0)}(\omega^{-1})$ is a ladder representation with minimal $\tilde{K}_{n-1,1}$-type $(1/2, 1/2, \dots, 1/2; k-1/2)$.

(3.6) Suppose $(r,s) = (0,1)$, $\omega = (;k-(n-2)/2), k > 0$. Then $\Theta_{(0,1)}(\omega^{-1})$ is a ladder representation with minimal $\tilde{K}_{n-1,1}$-type $(-1/2, -1/2, \dots, -1/2; k+1/2)$.

(3.7) Suppose $(r,s) = (0,1)$, $\omega = (;k-(n-2)/2), k < 0$. Then $\Theta_{(0,1)}(\omega^{-1})$ is a ladder representation with minimal $\tilde{K}_{n-1,1}$-type $(-1/2, -1/2, \dots, k-1/2; 1/2) = \tau^+(k+1, -1/2)$; $\Theta_{(0,1)}(\omega^{-1}) = \mathbb{L}^+(k+1,-1/2)$.

For all this, see [**L90**, pp. 930ff.]. The case $k = 0$ in (3.5) is not discussed in [**L90**] but it can be demonstrated in the same way. Combining (3.4) and (3.7) with (2.2) and (1.12) we find that, provided $k > 0$,

$$\text{(3.8)} \qquad \begin{aligned} \dim H^0_{\bar\partial}\left(\Theta_{(1,0)}(\omega_{-k-(n-2)/2}) \otimes W^+(1-k; -1/2)\right) &= 1 \\ \dim H^1_{\bar\partial}\left(\Theta_{(0,1)}(\omega_{k+(n-2)/2}) \otimes W^-(k-1; 1/2)\right) &= 1. \end{aligned}$$

These formulas become more useful when $\tilde{U}(V)$ and $\tilde{U}(V')$ are replaced by the metaplectic coverings of $U(V)$, $U(V')$, respectively, with kernel $\mathbb{C}^\times$, and when the theta lifts are pulled back to unitary groups by means of the archimedean versions of the splittings described in I, §3. These are again parametrized by pairs of characters (χ, χ') of $\mathbb{C}^\times$ such that $\chi|_{\mathbb{R}^\times} = (\varepsilon_{\mathbb{C}/\mathbb{R}})^m, \chi'|_{\mathbb{R}^\times} = (\varepsilon_{\mathbb{C}/\mathbb{R}})^n$, where $\varepsilon_{\mathbb{C}/\mathbb{R}}$ is the quadratic character associated to the extension $\mathbb{C}/\mathbb{R}$. This means that χ and χ' are trivial on the group $\mathbb{R}^\times_+$ of positive real numbers and that $\chi(-1) = (-1)^m$, $\chi'(-1) = (-1)^n$. In particular, χ and χ' are determined by their restrictions to $S^1 \cong U(1) \subset \mathbb{C}^\times$. We identify the characters of S^1 with $\mathbb{Z}$ so that (1) corresponds to the standard inclusion in $\mathbb{C}^\times$, and write $\chi = (\alpha(\chi))$, $\chi' = (\alpha(\chi'))$, so that $\alpha(\chi) \equiv m \pmod 2$, $\alpha(\chi') \equiv n \pmod 2$.

The restrictions of the splittings to $SU(V)$ and $SU(V')$ are unique. It follows that the theta lift relative to the splitting is determined by the formulas above and by its central character. But the relation of central characters is given by

Proposition I.3.11. Indeed, though the reference [**HKS**] is only concerned with p-adic fields, Proposition I.3.11 is based on the appendix of [**HKS**], which in turn is based on Kudla's article [**Ku94**], whose formulas are valid in the real case as well.

The following formulas are based on the conventions of Appendix 1 relative to splittings. In particular, we always assume $|\alpha(\chi)|, |\alpha(\chi')| \leq 1$.

First assume $\dim V' = 1$, $\dim V = n$ even; thus $\alpha(\chi') = 0$. Lemma I.3.8 and formula (I.3.12) are still valid. Then

(3.9, n even). For $\omega = (k + (n-2)/2;), k > 0$, $\Theta_{(1,0),1,\chi}(\omega^{-1})$ has minimal $K_{n-1,1}$-type $(k+1, 1, \ldots, 1; 0)$ if $\alpha(\chi) = 1$, $(k, 0, \ldots, 0; -1)$ if $\alpha(\chi) = -1$. Hence

$$\Theta_{(1,0),1,\chi}(\omega^{-1}) = \begin{cases} \mathbb{L}^-(k-1, 1) \cong \mathbb{L}^+(1-k, -1)\check{} & \text{if } \alpha(\chi) = 1, \\ \mathbb{L}^-(k-1, 0) \cong \mathbb{L}^+(1-k, 0)\check{} & \text{if } \alpha(\chi) = -1. \end{cases}$$

Indeed, write $\Theta = \Theta_{(1,0),1,\chi}(\omega^{-1})$. We are in case (c) of Proposition I.3.1, with n playing the role of m there. Thus $\xi_\Theta = \omega \cdot \chi^{n/2} = k + (n-2)/2 + (n/2)\alpha(\chi)$. But it follows from (3.4) that Θ has minimal $K_{n-1,1}$-type $(k+1/2, 1/2, \ldots, 1/2; -1/2) + (1/2)(a, \ldots, a)$ for some integer a. Comparing the two formulas, we find that

$$k + \frac{1}{2}(n-2) + \frac{1}{2}n\alpha(\chi) = k + \frac{1}{2} + (n-2) \cdot \frac{1}{2} - \frac{1}{2} + n \cdot \frac{1}{2}a.$$

Thus $a = \alpha(\chi)$ and the minimal $K_{n-1,1}$-types are as indicated.

In the same way, we find

(3.10, n even). For $\omega = (k + (n-2)/2;), k \leq 0$, $\Theta_{(1,0),1,\chi}(\omega^{-1})$ has minimal $K_{n-1,1}$-type $(1, 1, \ldots, 1; k)$ if $\alpha(\chi) = 1$, $(0, 0, \ldots, 0; -1+k)$ if $\alpha(\chi) = -1$. Hence

$$\Theta_{(1,0),1,\chi}(\omega^{-1}) = \begin{cases} \mathbb{L}^+(1, -1)\check{} & (k = 0, \alpha(\chi) = 1), \\ \mathbb{L}^+(1, 0)\check{} & (k = 0, \alpha(\chi) = -1), \\ \mathbb{M}^+(1-k, -1)\check{} & (k < 0, \alpha(\chi) = 1), \\ \mathbb{M}^+(1-k, 0)\check{} & (k < 0, \alpha(\chi) = -1). \end{cases}$$

(3.11, n even). For $\omega = (; k - (n-2)/2), k < 0$, $\Theta_{(0,1),\chi,1}(\omega^{-1})$ has minimal $K_{n-1,1}$-type $(0, \ldots, 0, k; 1)$ if $\alpha(\chi) = 1$, $(-1, \ldots, -1, -1+k; 0)$ if $\alpha(\chi) = -1$. Hence

$$\Theta_{(0,1),1,\chi}(\omega^{-1}) = \begin{cases} \mathbb{L}^+(k+1, 0) \cong \mathbb{L}^-(-k-1, 0)\check{} & \text{if } \alpha(\chi) = 1, \\ \mathbb{L}^+(k+1, -1) \cong \mathbb{L}^-(-k-1, 1)\check{} & \text{if } \alpha(\chi) = -1. \end{cases}$$

When n is odd, the answer depends on the two constants $\alpha(\chi)$ and $\alpha(\chi')$. We fix $\alpha(\chi') = 1$ and obtain the following formulas, which differ from the even case only in the parametrization of ω:

(3.9, n odd). For $\omega = (k + (n-1)/2;)$ $k > 0$, $\Theta_{(1,0),\chi',\chi}(\omega^{-1})$has minimal $K_{n-1,1}$-type $(k+1, 1, \ldots, 1; 0)$ if $\alpha(\chi) = 1$, $(k, 0, \ldots, 0; -1)$ if $\alpha(\chi) = -1$. Hence

$$\Theta_{(1,0),\chi',\chi}(\omega^{-1}) = \begin{cases} \mathbb{L}^-(k-1, 1) \cong \mathbb{L}^+(1-k, -1)\check{} & \text{if } \alpha(\chi) = 1, \\ \mathbb{L}^-(k-1, 0) \cong \mathbb{L}^+(1-k, 0)\check{} & \text{if } \alpha(\chi) = -1. \end{cases}$$

(3.10, n odd). For $\omega = (k+(n-1)/2;\,)$, $k \leq 0$, $\Theta_{(1,0),\chi',\chi}(\omega^{-1})$ has minimal $K_{n-1,1}$-type $(1,1,\dots,1;k)$ if $\alpha(\chi)=1$, $(0,0,\dots,0;-1+k)$ if $\alpha(\chi)=-1$. Hence

$$\Theta_{(1,0),\chi',\chi}(\omega^{-1}) = \begin{cases} \mathbb{L}^+(1,-1)\check{} & (k=0,\alpha(\chi)=1),\\ \mathbb{L}^+(1,0)\check{} & (k=0,\alpha(\chi)=-1),\\ \mathbb{M}^+(1-k,-1)\check{} & (k<0,\alpha(\chi)=1),\\ \mathbb{M}^+(1-k,0)\check{} & (k<0,\alpha(\chi)=-1).\end{cases}$$

(3.11, n odd). For $\omega = (;k-(n-3)/2)$, $k<0$, $\Theta_{(0,1),\chi',\chi}(\omega^{-1})$ has minimal $K_{n-1,1}$-type $(0,\dots,0,k;1)$ if $\alpha(\chi)=1$, $(-1,\dots,-1,-1+k;0)$ if $\alpha(\chi)=-1$. Hence

$$\Theta_{(0,1),\chi',\chi}(\omega^{-1}) = \begin{cases} \mathbb{L}^+(k+1,0) \cong \mathbb{L}^-(-k-1,0)\check{} & \text{if } \alpha(\chi)=1,\\ \mathbb{L}^+(k+1,-1) \cong \mathbb{L}^-(-k-1,1)\check{} & \text{if } \alpha(\chi)=-1.\end{cases}$$

Here we are in case (d) of Proposition I.3.11. Letting $\Theta = \Theta_{(\cdot,\cdot),\chi',\chi}(\omega^{-1})$, we thus have $\xi_\Theta = \omega\cdot\chi'^{(n-1)/2}$ if $\alpha(\chi)=1$, $\xi_\Theta = \omega\cdot\chi'^{(-n-1)/2}$ if $\alpha(\chi)=-1$. i.e.,

$$\xi_\Theta = \begin{cases} \omega + \frac{1}{2}(n-1) & \text{if } \alpha(\chi)=1,\\ \omega - \frac{1}{2}(n+1) & \text{if } \alpha(\chi)=-1.\end{cases}$$

In case (3.10) we have $\xi_\Theta = k+n-1$ if $\alpha(\chi)=1$, $\xi_\Theta = k-1$ if $\alpha(\chi)=-1$, which corresponds to the indicated minimal $K_{n-1,1}$-types. The other two cases are analogous.

Finally, we return to the theta lift $\pi' = \Theta(\check{\pi}_\lambda)$ discussed at the beginning of this section. We abbreviate $q=(r-s)/2$, $q'=(n-2)/2$. For π_λ we take the representation of $U(V)$ (rather than $\tilde{U}(V)$) with $\tau=\tau_\lambda$ given by

$$\tau = (a_1+q,\dots,a_{r'-1}+q,-b_s+q,\dots,-b_1+q;-a_{r'}-q) + \frac{1}{2}\alpha(\chi)(1,\dots,1;1). \tag{3.12}$$

Then 3.1 and Proposition I.3.11 combine to imply that $\pi' = \Theta(\check{\pi}_\lambda)$ contains the (minimal) $K_{r,s}$-type with highest weight

$$\tau' = (a_1+q',\dots,a_{r'-1}+q',\underbrace{q',\dots,q'}_{r-r'},-a_{r'}+q'; -b_s-q',\dots,-b_1-q') + \frac{1}{2}\alpha(\chi')(1,\dots,1). \tag{3.13}$$

Both $q+(1/2)\alpha(\chi)$ and $q'+(1/2)\alpha(\chi')$ are necessarily integral.

In applications to coherent cohomology, we also attach to π_λ a bundle parameter Λ, as in the previous sections. The formula for Λ is given by (2.8.1), above, where for α we naturally take $(1/2)\alpha(\chi)$.

4. Change of complex structure

In §1 we have defined the G-invariant complex structure on $X^{r,s}$ through our choice of a set R_n^+ of positive non-compact roots. Let e_i be the character of H with parameter $(a_1, \dots, a_r; a_{r+1}, \dots a_n)$, as above, such that $a_j = \delta_{ij}$. Then R_n^+ is the set of characters

$$R_n^+ = \{e_i - e_j | 1 \leq i \leq r, \ r+1 \leq j \leq s\}.$$

The conjugate complex structure is obtained by interchanging R_n^+ by $-R_n^+$. By permuting the first r with the last r parameters, keeping the order, it is clear that this corresponds to moving the semicolon in our parametrization from the r-th to the s-th place. In other words there is a canonical equivariant isomorphism $\bar{X}^{r,s} \cong X^{s,r}$ preserving the complex structures. Composing this with complex conjugation, we obtain an anti-holomorphic isomorphism (cf. [**H97**, (2.5.1)].

$$\varphi_{r,s} \colon X^{r,s} \xrightarrow{\sim} X^{s,r}. \tag{4.1}$$

The groups $G = U(r,s)$ and $U(s,r)$ are canonically isomorphic, and (4.1) respects the action of G. But the change in complex structure corresponds to interchanging $\mathfrak{p}^+$ and $\mathfrak{p}^-$. Thus a $(\mathfrak{g}, K_{r,s})$-module with non-vanishing $\bar{\partial}$-cohomology in degree i is identified, via this anti-holomorphic isomorphism, with a $(\mathfrak{g}, K_{s,r})$-module with $\bar{\partial}$-cohomology in degree $rs - i$, rs being the dimension of $X^{r,s}$. To make this more precise, we define three operations on characters of H, as in [loc. cit., §2.5]. We work with the similitude parameters of 1.15, with the understanding that the parameter c is to be ignored in applications to unitary groups.

$$(a_1, \dots, a_r; a_{r+1}, \dots, a_n; c)^{\iota} = (-a_n, \dots, -a_{r+1}; -a_r, \dots, -a_1; c) \tag{4.2}$$

$$(a_1, \dots, a_r; a_{r+1}, \dots, a_n; c)^{op} = (-a_n, \dots, -a_{r+1}; -a_r, \dots, -a_1; -c) \tag{4.3}$$

$$(a_1, \dots, a_r; a_{r+1}, \dots, a_n; c)^{\#} = (-a_r, \dots, -a_1; -a_n, \dots, -a_{r+1}; c) \tag{4.4}$$

The semicolon indicates that (4.2) and (4.3) interchange parameters for $GK_{r,s}$ with those for $GK_{s,r}$. The operation (4.3) corresponds to replacing a representation of $GK_{r,s}$ by its dual and identifying $GK_{r,s}$ with $GK_{s,r}$. The other two operations correspond to duality at the level of $K_{r,s}$.

Let π_λ and W_Λ be as in (1.3), so that

$$H^0(\mathfrak{P}(r,s), K_{r,s}; \check{\pi}_\lambda \otimes W_\Lambda) = [(\check{\pi}_\lambda)^{\mathfrak{p}^-(r,s)} \otimes W_\Lambda]^{K_{r,s}} \cong (\check{\pi}_\lambda \otimes W_\Lambda)^{K_{r,s}} \neq 0.$$

Here for clarity we are writing $\mathfrak{p}^-(r,s)$ for the anti-holomorphic tangent space associated to $X^{r,s}$ and $\mathfrak{P}(r,s)$ for the corresponding maximal parabolic subalgebra. Then (ignoring the parameter c) $\check{\pi}_\lambda \cong \bar{\pi}_\lambda, \bar{W}_\Lambda \cong W_{\Lambda^\iota}$, and by complex conjugation we obtain

$$\begin{aligned} \bar{H}^0(\mathfrak{P}(r,s), K_{r,s}; \check{\pi}_\lambda \otimes W_\Lambda) &\cong [(\pi_\lambda)^{\mathfrak{p}^+(r,s)} \otimes W_{\Lambda^\iota}]^{K_{r,s}} \\ &\cong [(\pi_\lambda)^{\mathfrak{p}^-(s,r)} \otimes W_{\Lambda^\iota}]^{K_{s,r}} \cong H^0(\mathfrak{P}(s,r), K_{s,r}; \pi_\lambda \otimes W_{\Lambda^\iota}) \neq 0. \end{aligned} \tag{4.5}$$

In other words, if W_Λ corresponds to a representation with $\bar{\partial}$-cohomology in degree zero for $X^{r,s}$, then W_{Λ^ι} corresponds to one with $\bar{\partial}$-cohomology in degree zero for $X^{r,s}$. This corresponds to the fact that $H^{rs}(\mathfrak{P}(r,s), K_{r,s}; \pi_\lambda \times (W_{\Lambda^\#} \otimes \wedge^{rs}\mathfrak{p}^-)) \neq 0$, an expression of Serre duality at the level of $\bar{\partial}$-cohomology, and indeed by contracting $\wedge^{rs}\mathfrak{p}^-$ with $\wedge^{rs}(\mathfrak{p}^-)^*$ we can rewrite the conclusion of (4.5) as follows:

$$H^0(\mathfrak{P}(s,r), K_{s,r}; \pi_\lambda \otimes W_{\Lambda^\iota}) \cong H^{rs}(\mathfrak{P}(r,s), K_{r,s}; \pi_\lambda \otimes (W_{\Lambda^\#} \otimes \wedge^{rs}\mathfrak{p}^-)). \tag{4.6}$$

This reasoning applies as well to the ladder representations of $G = U(n-1,1)$, although $\wedge^{rs}\mathfrak{p}^-$ has to be replaced by $\mathfrak{p}^-$ in the above discussion. They are still ladder representations when viewed as $(\mathfrak{g}, K_{1,n-1})$-modules, but the cohomological degrees have been interchanged. Indeed, letting $^-$ denote complex conjugation, it is clear that

$$(4.7)\qquad \bar{\mathbb{L}}^+(-\alpha,\beta,c) \cong \mathbb{L}^-(\alpha,-\beta;c),\ \bar{\tau}^+(-\alpha,\beta;c) \cong \tau^-(\alpha,-\beta;c)\ (cf.\ (1.13)).$$

Thus, recalling that $W^-(\alpha,\beta;c) = W_{\Lambda^-(\alpha,\beta;c)}$, we have

$$(4.8)\qquad \begin{aligned} &H^0(\mathfrak{P}(1,n-1),K_{1,n-1};\mathbb{L}^-(\alpha,-\beta;c)^\check{} \otimes W_{\Lambda^+(-\alpha,\beta;c)^\iota})\\ &\cong \bar{H}^0(\mathfrak{P}(n-1,1),K_{n-1,1};\mathbb{L}^+(-\alpha,\beta;c)^\check{} \otimes W_{\Lambda^+(-\alpha,\beta;c)}).\end{aligned}$$

The analogue of (4.6) is the isomorphism

$$(4.9)\qquad \begin{aligned} &H^0(\mathfrak{P}(1,n-1),K_{1,n-1};\mathbb{L}^-(\alpha,-\beta;c)^\check{}\otimes W_{\Lambda^+(-\alpha,\beta;c)^\iota})\\ &\cong H^1(\mathfrak{P}(n-1,1),K_{n-1,1};\mathbb{L}^-(\alpha,-\beta;c)^\check{}\otimes W^-(\alpha,-\beta;c)).\end{aligned}$$

III. Shimura varieties for unitary groups

1. Generalities
2. Review of basic CM periods
3. Coherent cohomology and rationality testing
4. Cup products in coherent cohomology
5. Complex conjugation and small anti-holomorphic forms

1. Generalities

Let $\mathcal{K}$ be an imaginary quadratic field, $\mathrm{Gal}(\mathcal{K}/\mathbb{Q}) = \{1,\iota\}$. Let $\Sigma_\mathcal{K}$ denote the set of complex embeddings of $\mathcal{K}$; then $\Sigma_\mathcal{K} = \{i_0,\iota\}$, where i_0 is the given embedding and ι is i_0 followed by complex conjugation; of course ι is also i_0 composed with Galois conjugation (by the non-trivial element ι) on $\mathcal{K}$.

Let V be an n-dimensional hermitian $\mathcal{K}$-vector space. We define $U(V)$ and $\mathrm{GU}(V)$ as in Part I, and write $G = U(V)$, $GG = \mathrm{GU}(V)$, and $\nu\colon GG \to \mathbb{G}_m$ the $\mathbb{Q}$-rational similitude homomorphism, as in the preceding chapters.

The Weil group $W_\mathbb{R}$ of $\mathbb{C}/\mathbb{R}$ is given by

$$W_\mathbb{R} = \{(z,\tau), z\in\mathbb{C}^\times, \tau\in\mathrm{Gal}(\mathcal{K}/\mathbb{Q})\}$$

with relations $\sigma z\sigma^{-1} = \bar{z}, \sigma^2 = -1 \in \mathbb{C}^\times$, with $\sigma = (1,\iota)$. The map $(z,\tau)\to\tau$ defines a homomorphism $pr_W\colon W_\mathbb{R} \to \mathrm{Gal}(\mathcal{K}/\mathbb{Q})$ with kernel $\mathbb{C}^\times$. Any homomorphism $\varphi\colon W_\mathbb{R} \to {}^LGG$ such that $pr\circ\varphi = pr_W$, with respect to the obvious identification, defines an L-packet Π_φ of admissible representations of $GG_\mathbb{R}$ (cf. [**B, Rog92**]), and Π_φ depends on φ only up to conjugation by ${}^LGG$.

We associate to GG a Shimura variety as follows. The inclusion $GG \to \mathrm{GL}(V)$ defines a representation

$$GG_\mathbb{R} \to \mathrm{GL}((R_{\mathcal{K}/\mathbb{Q}}V)\otimes_\mathbb{Q}\mathbb{R}) \cong \mathrm{GL}(V\otimes_{\mathcal{K},i_0}\mathbb{C})\times\mathrm{GL}(V\otimes_{\mathcal{K},\iota}\mathbb{C}).$$

We write $V_{i_0} = V\otimes_{\mathcal{K},i_0}\mathbb{C}, V_\iota = V\otimes_{\mathcal{K},\iota}\mathbb{C}$. Both V_{i_0} and V_ι are endowed with hermitian forms $(\cdot,\cdot)_{i_0}$ and $(\cdot,\cdot)_\iota$, respectively, of signatures (r,s) and (s,r) for some pair of integers such that $r+s=n$. Thus we are in the situation of II, §1.

Choose a maximal connected compact subgroup $K_\infty \cong U(r)\times U(s)$ of $GG(\mathbb{R})$; then K_∞ stabilizes a pair of conjugate decompositions.

$$V_{i_0} = V_{i_0}^+ \oplus V_{i_0}^-, V_\iota = V_\iota^+ \oplus V_\iota^-,$$

$\dim(V_{i_0}^+) = \dim(V_\iota^-) = r$; $\dim(V_{i_0}^-) = \dim(V_\iota^+) = s$.

Endow $\mathcal{V} = V_{i_0} \oplus V_\iota$ with a real Hodge structure of weight -1 for which

$$\mathcal{V}^{-1,0} = V_{i_0}^+ \oplus V_\iota^+, \mathcal{V}^{0,-1} = V_{i_0}^- \oplus V_\iota^-. \tag{1.1}$$

This Hodge structure corresponds to a homomorphism

$$h = h_{K_\infty} : \underline{S} = R_{\mathbb{C}/\mathbb{R}}\mathbb{G}_{m,\mathbb{R}} \to \mathrm{GL}(\mathcal{V}).$$

Let $Z \cong \mathbb{R}^\times \subset GG(\mathbb{R})$ be the subgroup of real diagonal matrices, as in II, §1. It is clear that the image of h is contained in $Z\cdot K_\infty$; in particular h is a homomorphism $\underline{S} \to GG_{\mathbb{R}}$. Let $X^{r,s}$ denote the $GG(\mathbb{R})$-conjugacy class of this homomorphism; then the pair $(GG, X^{r,s})$ satisfies the axioms defining a Shimura variety (cf. [**D72, M**]), denoted $\mathrm{Sh}(V) = \mathrm{Sh}(GG, X^{r,s})$:

$$\mathrm{Sh}(V) = \varprojlim{}_K\, {}_K\mathrm{Sh}(V), \text{where } {}_K\mathrm{Sh}(V) = GG(\mathbb{Q})\backslash X^{r,s} \times GG(\mathbf{A}^f)/K, \tag{1.2}$$

where K runs through the open compact subgroups of $GG(\mathbf{A}^f)$. Here we are identifying the varieties $\mathrm{Sh}(V)$ with the associated complex-analytic varieties. The notation $\mathrm{Sh}(V)$ is, however, more properly reserved for the canonical model, defined over the *reflex field* $E(GG, X^{r,s})$, whose definition is given in [loc. cit]. In the special case of unitary groups for imaginary quadratic fields, we know that $E(GG, X^{r,s}) = \mathcal{K}$ unless $r = s$, in which case $E(GG, X^{r,r}) = \mathbb{Q}$ [**Sh66**]. The canonical models are preserved under right translation by $GG(\mathbf{A}^f)$.

We note that the quotient $G(\mathbb{R})/U(r)\times U(s)$, denoted $X_+^{r,s}$ in Chapter II, can be identified with the $G(\mathbb{R})$-orbit of h in $X^{r,s}$. In fact, $X^{r,s} = X_+^{r,s}$ is connected, unless $r = s$, in which case $X^{r,s}$ has two connected components.

The complex manifold $X^{r,s}$ may be identified with an open subset of the Grassmanian $\mathrm{Grass}_r(\mathbb{C}^n)$ of r-planes in complex n-space. More precisely, we consider the n-dimensional complex vector space V_{i_0}; then $X^{r,s}$ is the subset of r-planes in V_{i_0} on which the form $(\cdot,\cdot)_{i_0}$ is positive-definite. The corresponding embedding

$$\beta\colon X^{r,s} \to \mathrm{Grass}_r(\mathbb{C}^n) = \mathrm{Grass}_r(V_{i_0}) \tag{1.3}$$

is the *Borel embedding.*

In particular, when $(r,s) = (n-1,1), X^{n-1,1}$ is open subset of the projective space $\mathbb{P}^{n-1}$, and in fact in its bounded realization is equal to the unit ball in $\mathbb{C}^{n-1}$ (or to two copies of the latter when $n = 2$).

We take $V_{i_0} = \mathbb{C}^n, (\cdot,\cdot)_{i_0}$ in the standard form $\left(\begin{smallmatrix} I_r & 0 \\ 0 & -I_s \end{smallmatrix}\right)$, so that $GG(\mathbb{R}) \cong \mathrm{GU}(r,s)$, and $GK_\infty := K_\infty(r,s) := Z\cdot K_\infty$ is the subgroup of $GG(\mathbb{R})$ of matrices of the form $\left(\begin{smallmatrix} a\cdot u_r & 0 \\ 0 & a\cdot u_s \end{smallmatrix}\right)$, $a \in \mathbb{R}^\times, u_r \in U(r), u_s \in U(s)$. Let GH denote the subgroup of diagonal matrices in $GK_\infty, H = H \cap U(V,\mathbb{R})$, so that $GH = Z\cdot H$. Then GH is a maximal torus of $G(\mathbb{R})$, and we order the roots of $GH_{\mathbb{C}}$ in $GG(\mathbb{C})$ by decreeing the positive root spaces to be those corresponding to upper triangular matrices. Then we recover the positive root system R^+ of II, §1, and we retain the notation of that section. In particular, the algebraic characters of GH are given by $n+1$ tuples $(a_1,\dots,a_r;a_{r+1},\dots,a_n;c)$ satisfying the parity condition II (1.15.1).

(1.4). For future reference, we assume GK_∞ to be the stabilizer of a specific CM point $h \in X^{r,s}$. Namely, let $H_{\mathcal{K}} = R_{\mathcal{K}/\mathbb{Q}}\mathbb{G}_{m,\mathcal{K}}$, and let $X(H_{\mathcal{K}})$ denote its character group (characters over an algebraic closure of $\mathbb{Q}$). We may identify $X(H_{\mathcal{K}})$ with the free abelian group generated by $\Sigma_{\mathcal{K}}$. Then the base change $H_{\mathcal{K},\mathbb{R}}$ of $H_{\mathcal{K}}$ to $\mathbb{R}$ and the Hodge torus $\underline{S}$ are isomorphic, and this isomorphism can be normalized so that the character $i_0 \in \sum_{\mathcal{K}} \subset X(H_{\mathcal{K}})$ (resp. $\iota \in X(H_{\mathcal{K}})$) is taken to the character $z \mapsto z$ of $\underline{S}$ (resp. the character $z \mapsto \bar{z}$). Then we assume $h\colon \underline{S} \to GG_{\mathbb{R}}$ comes from a $\mathbb{Q}$-rational homomorphism $H_{\mathcal{K}} \to GG$. Moreover, we may assume GH is a $\mathbb{Q}$-rational maximal torus of GG that splits over $\mathcal{K}$; concretely, we may assume GH is the intersection with GG of the diagonal maximal torus of $R_{\mathcal{K}/\mathbb{Q}}\mathrm{GL}(n)_{\mathcal{K}}$. Then the pair (GH, h) satisfies the axioms defining a Shimura variety. To emphasize the dimension and signature we write $GH = GH^n_{\mathcal{K}}, h = h^{r,s}$. Then the inclusion of $GH^n_{\mathcal{K}}$ in GG makes $\mathrm{Sh}(GH^n_{\mathcal{K}}, h^{r,s})$ a CM Shimura subvariety of $\mathrm{Sh}(V)$.

When $n = 1$ the inclusions $GH^n_{\mathcal{K}} \subset H_{\mathcal{K}} \subset GG$ are all equalities. Thus we have attached the two Shimura varieties $\mathrm{Sh}(H_{\mathcal{K}}, h^{1,0})$ and $\mathrm{Sh}(H_{\mathcal{K}}, h^{0,1})$ to $H_{\mathcal{K}}$. More generally, for any integers u, v, we may define $h^{u,v} = (h^{1,0})^u \cdot (h^{0,1})^v$; then the Shimura variety $\mathrm{Sh}(H_{\mathcal{K}}, h^{u,v})$ is defined and has a canonical model over the reflex field $E(H_{\mathcal{K}}, h^{u,v})$, which equals $\mathbb{Q}$ when $u = v$ and $\mathcal{K}$ otherwise. These canonical models are described in [**H97**].

2. Review of basic CM periods

Matters will be relatively simple in this paper because we only have to deal with Hecke characters of a single imaginary quadratic field $\mathcal{K}$. Let $\omega\colon \mathcal{K}^\times_{\mathbf{A}}/\mathcal{K}^\times \to \mathbb{C}^\times$ be an algebraic Hecke character, and let $\mathbb{Q}(\omega)$ be the field generated by the values of ω on finite ideles; it is a finite extension of $\mathbb{Q}$ and is either a totally real field or a CM field. Letting $* = i_0$ or ι, we associate to ω the invariants

$$\mathbf{p}(\omega, *) = \big(p(\sigma(\omega), *)_{\sigma \in \mathrm{Hom}_{\mathcal{K}}(\mathbb{Q}(\omega),\mathbb{C})}\big) \in (\mathbb{Q}(\omega) \otimes_{\mathcal{K}\cap\mathbb{Q}(\omega)} \mathbb{C})^\times/\mathbb{Q}(\omega)^\times,$$

as in [**H97**], where they were denoted $p_{\mathcal{K}}(\omega, *)$. Here $\sigma(\omega)$ is the Hecke character with the same infinity type as σ (since $\mathcal{K}$ is imaginary quadratic) such that $\sigma(\omega)(a) = \sigma(\omega(a))$ for any finite idele a. The definitions in terms of coherent cohomology are recalled in §3, below. In what follows, we let $\omega^\iota(z) = \omega(\iota(z))$ for $z \in \mathcal{K}^\times_{\mathbf{A}}/\mathcal{K}^\times$; evidently $\mathbb{Q}(\omega^\iota) = \mathbb{Q}(\omega)$. Proofs of the following properties of these invariants can be found in [**H93**, §1] and the references indicated there:

$$p(\omega_1\omega_2, *) \sim_{\mathbb{Q}(\omega_1)\cdot\mathbb{Q}(\omega_2)} p(\omega_1, *)p(\omega_2, *), \tag{2.1}$$

where $(\mathbb{Q}(\omega_1\omega_2) \otimes_{\mathcal{K}} \mathbb{C})^\times/\mathbb{Q}(\omega_2\omega_2)^\times$ is viewed as a subgroup of $(\mathbb{Q}(\omega_1) \cdot \mathbb{Q}(\omega_2) \otimes_{\mathcal{K}} \mathbb{C})^\times/(\mathbb{Q}(\omega_1) \cdot \mathbb{Q}(\omega_2))^\times$;

$$p(\omega^\iota, i_0) = p(\omega, \iota) \in (\mathbb{Q}(\omega) \otimes_{\mathcal{K}\cap\mathbb{Q}(\omega)} \mathbb{C})^\times/\mathbb{Q}(\omega)^\times, \tag{2.2}$$

$$p(\|\cdot\|_{\mathbf{A}}, *) = (2\pi i)^{-1} \in \mathbb{C}^\times/\mathbb{Q}^\times. \tag{2.3}$$

The period invariants will mostly appear in the guise of special values of the Hecke L-functions $L(s, \omega)$, as in the following formula, due to [**GS**]. Define $H_{\mathcal{K}}$ and $X(H_{\mathcal{K}})$ as in §1. Then the infinity type ω_∞ of ω belongs to $X(H_{\mathcal{K}})$. We write $\omega_\infty = \omega_\lambda$ where $\lambda\colon \Sigma_{\mathcal{K}} \to \mathbb{Z}$ is the parameter of the algebraic character $z \mapsto z^{\lambda(i_0)}\iota(z)^{\lambda(\iota)}$ of $H_{\mathcal{K}}$. The character ω_λ is *critical*, in the sense defined in [**H93**], if $\lambda(i_0) \neq \lambda(\iota)$. Then we say ω_λ is *of type* i_0 if $\lambda(i_0) < \lambda(\iota)$, *of type* ι if $\lambda(i_0) > \lambda(\iota)$. If the infinity type of ω is ω_λ then we say ω is critical and of type such-and-such if ω_λ has the corresponding property. If ω_λ is critical then $\mathcal{K} \subset \mathbb{Q}(\omega)$, and we will assume this to be the case in the remainder of this section.

If ω is a critical Hecke character, we let

$$\mathbf{L}(s,\omega) = (L(s,\sigma(\omega))_{\sigma\in \mathrm{Hom}_{\mathcal{K}}(\mathbb{Q}(\omega),\mathbb{C})}) \in (\mathbb{Q}(\omega)\otimes_{\mathcal{K}}\mathbb{C})^{\times}.$$

(2.4). Suppose ω is critical and of type Φ, where $\Phi \in \Sigma_{\mathcal{K}}$. If $s = m$ is a critical value of $L(s,\omega)$, then $\mathbf{L}(m,\omega) \sim (2\pi i)^m \mathbf{p}(\omega^{-1}, \iota\Phi)$ in $(\mathbb{Q}(\omega)\otimes_{\mathcal{K}}\mathbb{C})^{\times}/\mathbb{Q}(\omega)^{\times}$.

This is a weaker version of the theorem of Goldstein and Schappacher. See the introduction to [**H97**] for comments on (2.4).

(2.5). The type of a critical ω depends only on its restriction to the subgroup $U(1)$ of norm 1 elements of $\mathbb{C}^{\times}$, determined by the integer k such that $\omega(e^{i\theta}) = e^{ik\theta}$. If $k < 0$ then ω is of type i_0, if $k > 0$ then ω is of type ι, and if $k = 0$ then ω is not critical.

We recall for future reference that the critical values of $L(s,\omega)$ are determined by the infinity type ω_λ of ω as follows (cf. [**D79**]).

(2.6). If ω is critical of type i_0 (resp. of type ι) then $m \in \mathbb{Z}$ is critical for $L(s,\omega)$ if and only if $-\lambda(\iota) < m < 1 - \lambda(i_0)$ (resp. $-\lambda(I_0) < m < 1 - \lambda(\iota)$).

3. Coherent cohomology and rationality testing

In this section we recall Shimura's method for characterizing rational holomorphic automorphic forms in terms of their values at CM points, in the version developed in [**H86, H93**]. We retain the notation of §1, with $GG = \mathrm{GU}(V)$ of signature (r,s) at infinity, and $\mathrm{Sh}(V) = \mathrm{Sh}(GG, X^{r,s})$. To any finite dimensional representation (ρ, W) of GK_∞ we can associate an automorphic vector bundle $[W]$ on $\mathrm{Sh}(V)$. As a complex analytic vector bundle over $\mathrm{Sh}(V)(\mathbb{C})$, $[W]$ is given by the following formula, to be compared with (1.2):

$$[W] = \varprojlim K \quad (GG(\mathbb{Q})\backslash GG(\mathbf{A})) \times W/K_\infty K \tag{3.1}$$

where K runs over compact subgroups of $GG(\mathbf{A}^f)$ and K_∞ acts on the right on $GG(\mathbf{A})$ and via ρ on W. If $\Lambda = (a_1, \dots, a_r; a_{r+1}, \dots, a_s; c)$ is the highest weight of (ρ, W), in the notation of II, 1.15, we will write $W = W_\Lambda$ and $E_\Lambda = [W]$. In either case, the automorphic vector bundles can be viewed just as well over the Shimura varieties ${}_K\mathrm{Sh}(V)$ of finite level.

The Shimura varieties $\mathrm{Sh}(V)$ are moduli spaces for abelian varieties with PEL structures, in the sense of Shimura's work of the 1960s, and the automorphic vector bundles $[W]$ can all be defined in terms of the algebraic de Rham cohomology of these families of abelian varieties, by a familiar procedure. This construction provides canonical models of [W] over the reflex field $E(W)$, defined in [**H86**]; the general construction of canonical models of automorphic vector bundles is described in [**M**]. Define Λ^ι by II(4.2). The following lemma is easily verified.

LEMMA 3.2. *Let* $[W] = E_\Lambda$ *be the automorphic vector bundle over* $\mathrm{Sh}(V)$, *defined as above.*

(i) *Suppose* $r = s$ *and* W *is self-conjugate (i.e., the highest weight* $\Lambda = \Lambda^\iota$*). Then* $[W]$ *has a canonical model over the reflex field* $E(W) = \mathbb{Q}$.
(ii) *If* $r \neq s$ *or* $\Lambda \neq \Lambda^\iota$, $[W]$ *has a canonical model over the reflex field* $E(W) = \mathcal{K}$.

(3.3). For each level subgroup $K \subset GG(\mathbf{A}^f)$ we consider the family of toroidal compactifications, as in [**H89, P**].

$$j_\Sigma\colon {}_K\mathrm{Sh}(V) \hookrightarrow {}_K\mathrm{Sh}(V)_\Sigma. \tag{3.3.1}$$

For simplicity we assume K to be *neat*, in the sense of [loc.cit.], and we consider only toroidal compactifications which are smooth and projective and such that the complement Z_Σ of ${}_K\mathrm{Sh}(V)$ in ${}_K\mathrm{Sh}(V)_\Sigma$ is a divisor with normal crossings, each of whose irreducible components is smooth. We may further assume that ${}_K\mathrm{Sh}(V)_\Sigma$ has a canonical model over the reflex field $E(GG, X^{r,s})$ (cf. [loc. cit.]).

When $\min(r, s) = 1$ there is a unique such toroidal compactification. It has the property that Z_Σ is smooth and isomorphic to the disjoint union of a finite number of abelian varieties isogenous to $\mathcal{E}^{n-2}$, where $\mathcal{E}$ is an elliptic curve with complex multiplication by $\mathcal{K}$.

Each automorphic vector bundle $[W]$ at finite level K has a canonical extension $[W]^{\mathrm{can}} = [W]^{\mathrm{can}}_\Sigma$ to a vector bundle on ${}_K\mathrm{Sh}(V)_\Sigma$ and the functor $[W] \mapsto [W]^{\mathrm{can}}$ commutes with all tensor operations. The construction of the canonical extension is due to Mumford [**Mu**], and is discussed further in [**H89**]. Let $\mathfrak{I}_\Sigma$ be the sheaf of ideals defining Z_Σ. We let $[W]^{\mathrm{sub}} = [W]^{\mathrm{can}} \otimes_{\mathcal{O}} \mathfrak{I}_\Sigma$, where $\mathcal{O}$ is the structure sheaf of ${}_K\mathrm{Sh}(V)_\Sigma$. As in [**H90**], we define

$$\begin{aligned} \tilde{H}^{\cdot}([W]^{\mathrm{can}}) &= \varinjlim_{K,\Sigma} H^{\cdot}({}_K\mathrm{Sh}(V)_\Sigma, [W]^{\mathrm{can}}), \\ \tilde{H}^{\cdot}([W]^{\mathrm{sub}}) &= \varinjlim_{K,\Sigma} H^{\cdot}({}_K\mathrm{Sh}(V)_\Sigma, [W]^{\mathrm{sub}}); \\ H^{\cdot}_!([W]) &= \mathrm{Im}(\tilde{H}^{\cdot}([W]^{\mathrm{sub}}) \to \tilde{H}^{\cdot}([W]^{\mathrm{can}})). \end{aligned} \tag{3.3.2}$$

The following lemma summarizes some of the results of [**H90**]:

LEMMA 3.4. *The spaces $\tilde{H}^{\cdot}([W]^{\mathrm{can}})$, $H^{\cdot}([W]^{\mathrm{sub}})$, and $H^{\cdot}_!([W])$ are admissible $GG(\mathbf{A}^f)$-modules, and the natural morphisms $H^{\cdot}([W]^{\mathrm{sub}}) \to \tilde{H}^{\cdot}([W]^{\mathrm{can}})$ are morphisms of $GG(\mathbf{A}^f)$-modules. Moreover, these spaces all possess canonical $E(W)$-rational structures, which are invariant under the action of $GG(\mathbf{A}^f)$.*

The following is a deep theorem of Franke [**F**], using the earlier and more elementary results of [**H90**]. We recall the notation of II, §1. In what follows, we write $G\mathfrak{g} = \mathrm{Lie}(GG(\mathbb{R}))$.

THEOREM 3.5. *There is a canonical isomorphism*

$$H^{\cdot}(\mathfrak{P}, GK_\infty; \mathcal{A}(GG) \otimes W) \xrightarrow{\sim} \tilde{H}^{\cdot}([W]^{\mathrm{can}})$$

of $GG(\mathbf{A}^f)$-modules.

The following theorem had already been proved in [**H90**].

THEOREM 3.6. *Let q be a non-negative integer. Suppose $W = W_\Lambda$ and suppose either (i) $\Lambda = (a_1, \ldots, a_r; a_{r+1}, \ldots, a_s; c)$ is a* regular *weight of* $\mathrm{Lie}(GG(\mathbb{R}))$: $a_i \neq a_j$ *unless $i = j$, or (ii) $q = 0$ or $\dim(X^{r,s}) = rs$. Then the canonical isomorphism of 3.5 restricts to a canonical isomorphism of $GG(\mathbf{A}^f)$-modules*

$$H^q(\mathfrak{P}, GK_\infty; \mathcal{A}_0(GG) \otimes W_\Lambda) \xrightarrow{\sim} H^q_!(E_\Lambda).$$

In any case, the map

$$H^q(\mathfrak{P}, GK_\infty; \mathcal{A}_0(GG) \otimes W_\Lambda) \to H^q_!(E_\Lambda)$$

is injective.

REMARK 3.7. Case (ii) is Prop. 5.4.2 of **[H90]**. Corollary 5.3.3 of **[H90]** asserts the existence of a constant $b > 0$ such that isomorphism of Theorem 3.6 holds for all q, provided $|a_i - a_j| \geq b$. Franke pointed out to me that the hypotheses of Corollary 5.3.3 (and Lemma 4.7, on which it is based) are satisfied with $b = 1$.

(3.8). In order to interpret the rational structures on the left-hand sides of the isomorphisms of Theorems 3.5 and 3.6 as properties of the automorphic forms on the right-hand sides, we need to be more precise about the space W_Λ which appears there. In fact, the symmetric space $X^{r,s}$ has a natural Borel embedding $\beta\colon X^{r,s} \to \mathrm{Grass}_r(\mathbb{C}^n) = \mathrm{Grass}_r(V_{i_0})$ (cf. 1.3). The space $\mathrm{Grass}_r(V_{i_0})$, together with its structure as homogeneous space under GG, has a natural rational structure over $\mathcal{K}$, which can be naturally descended to $\mathbb{Q}$ if $r = s$. Any representation (ρ, W) of GK_∞ defines a GG-homogeneous vector bundle $\mathcal{W}$ on $\mathrm{Grass}_r(V_{i_0})$, defined over the reflex field $E(W)$. The construction of rational models of automorphic vector bundles in **[H90, M]** is in fact the construction of a functor

$$\left\{ \begin{array}{c} GG - \text{homogeneous vector} \\ \text{bundles on } \mathrm{Grass}_r(V_{i_0}) \end{array} \right\} \to \left\{ \begin{array}{c} GG(\mathbf{A}^f) - \text{homogeneous} \\ \text{vector bundles on } \mathrm{Sh}(V) \end{array} \right\}$$

Let $\mathcal{W}_\Lambda$ be the GG-homogeneous vector bundle on $\mathrm{Grass}_r(V_{i_0})$ attached to (ρ, W_Λ). The space W_Λ in Theorems 3.5 and 3.6 is naturally the fiber over the fixed point of GK_∞ in $X^{r,s}$ of $\mathcal{W}_\Lambda$, unless $r = s$, in which case there are two fixed points and the isomorphism depends on the choice of one of them.

(3.9). We work out the theory for $n = 1$. Then $GG = H_\mathcal{K}$, $GK_\infty = \mathcal{K}_\infty^\times \cong \mathbb{C}^\times$, $\dim \mathrm{Sh}(V) = 0$ and coherent cohomology is concentrated in degree 0. We may write

$$\mathcal{A}(H_\mathcal{K}) = \mathcal{A}_0(H_\mathcal{K}) = \oplus \mathcal{A}_\omega,$$

where $\mathcal{A}_\omega$ is the (one-dimensional) space generated by the Hecke character

$$\omega\colon\ \mathcal{K}_\mathbf{A}^\times / \mathcal{K}^\times \to \mathbb{C}$$

and the sum runs over all Hecke characters. Thus, for any $\Lambda \in X(H_\mathcal{K})$, Theorem 3.5 yields an isomorphism

(3.9.1)
$$\begin{aligned} \oplus H^0(\mathfrak{P}, GK_\infty; \mathcal{A}_\omega \otimes W_\Lambda) &= \oplus(\mathcal{A}_\omega \otimes W_\Lambda)^{\mathcal{K}_\infty^\times} \\ &\tilde{\to} H^0(\mathrm{Sh}(H_\mathcal{K}, h^{r,s}), E_\Lambda), \quad r + s = 1. \end{aligned}$$

We write $\omega_\Lambda\colon\ \mathcal{K}_\infty^\times \to \mathbb{C}^\times$ for the complex character associated to Λ. The only non-trivial summands are evidently those for which the infinity type ω_∞ of ω satisfies $\omega_\infty = \omega_\Lambda^{-1}$; i.e., for ω of type $-\Lambda$. More generally, we say $f \in \mathcal{A}(H_\mathcal{K})$ is of type $-\Lambda$ if it is a linear combination of Hecke characters ω of type $-\Lambda$, and let $\mathcal{A}(H_\mathcal{K})_{-\Lambda} \subset \mathcal{A}(H_\mathcal{K})$ denote the space of automorphic forms of type $-\Lambda$. Then the inverse of (3.9.1) restricts to an isomorphism

(3.9.2)
$$\begin{aligned} \underline{\mathrm{Lift}}\colon\ H^0(\mathrm{Sh}(H_\mathcal{K}, h^{r,s}), E_\Lambda) &\tilde{\to} \mathcal{A}(H_\mathcal{K})_{-\Lambda} \\ &(= \mathcal{A}(H_\mathcal{K})_\Lambda \otimes \omega_\Lambda \subset \mathcal{A}(H_\mathcal{K}) \otimes W_\Lambda). \end{aligned}$$

For any Hecke character ω of type Λ (i.e., $\omega_\infty = \omega_\Lambda$) and any $f \in \mathcal{A}(H_\mathcal{K})_{-\Lambda}$, we set

$$L_\omega(f) = \int_{\mathcal{K}_\mathbf{A}^\times/\mathcal{K}_\infty^\times\mathcal{K}^\times} \omega(g)f(g)dg \tag{3.9.3}$$

where the measure is normalized as in the notation. Next, let

$$\mathbf{L}_\omega(f) = \left(L_\omega(f)_{\sigma\in\mathrm{Hom}(\mathbb{Q}(\omega),\mathbb{C})}\right) \in \mathbb{Q}(\omega)\otimes\mathbb{C}. \tag{3.9.4}$$

The defining property of the CM periods introduced in §2 is

DEFINITION-PROPOSITION 3.9.5. Let $L \subset \mathbb{C}$ be an extension of $E(W_\Lambda)$. Let

$$\mathbf{p}(\omega, h^{r,s}) = \mathbf{p}(\omega, i_0)^r \cdot \mathbf{p}(\omega, s)^s \in (\mathbb{Q}(\omega)\otimes\mathbb{C})^\times/\mathbb{Q}(\omega)^\times.$$

The section $\varphi \in H^0(\mathrm{Sh}(H_\mathcal{K}, h^{r,s}), E_\Lambda)(\mathbb{C})$ is rational over L if and only if

$$\mathbf{p}(\omega, h^{r,s})^{-1} \cdot \mathbf{L}_\omega(\underline{\mathrm{Lift}}(\varphi)) \in (\mathbb{Q}(\omega)\otimes L)^\times/\mathbb{Q}(\omega)^\times$$

for every Hecke character ω of type Λ.

This is a proposition insofar as it fits into a consistent theory of rational structures on automorphic vector bundles (cf. [**H93**, §1]). Note that the definition makes sense for arbitrary pairs of integers (r, s), and this definition is indeed the correct one for the Shimura varieties $\mathrm{Sh}(H_\mathcal{K}, h^{r,s})$.

(3.10). We next extend the rationality criterion to the Shimura varieties

$$\mathrm{Sh}(GH_\mathcal{K}^n, h^{r,s}).$$

The group $GH_\mathcal{K}^n$ is again a torus, the Shimura varieties are again zero-dimensional, and we again have $\mathcal{A}(GH_\mathcal{K}^n) = \mathcal{A}_0(GH_\mathcal{K}^n) = \oplus\mathcal{A}_\omega$, where $\mathcal{A}_\omega$ is the space generated by the Hecke character $\omega\colon\ GH_\mathcal{K}^n(\mathbf{A})/GH_\mathcal{K}^n(\mathbb{Q}) \to \mathbb{C}^\times$ and the sum runs over all Hecke characters. The character group $X(GH_\mathcal{K}^n)$ is given by the $(n+1)$-tuples $\Lambda = (a_1, \ldots, a_r; a_{r+1}, \ldots, a_s; c)$. We define the character $\omega_\Lambda\colon\ GH_\mathcal{K}^n(\mathbb{R}) \to \mathbb{C}^\times$ as before; then the Hecke character ω is of type Λ if $\omega_\infty = \omega_\Lambda$.

The field $\mathbb{Q}(\omega)$ denotes the extension of $\mathbb{Q}$ generated by the values of ω on finite adeles, as before. We define $\mathcal{A}(GH_\mathcal{K}^n)_{-\Lambda} \subset \mathcal{A}(GH_\mathcal{K}^n)$, the isomorphism

$$\underline{\mathrm{Lift}} : H^0(\mathrm{Sh}(GH_\mathcal{K}^n, h^{r,s}), E_\Lambda) \xrightarrow{\sim} \mathcal{A}(GH_\mathcal{K}^n)_{-\Lambda},$$

and the maps

$$L_\omega\colon\ \mathcal{A}(GH_\mathcal{K}^n)_{-\Lambda} \to \mathbb{C};\quad \mathbf{L}_\omega\colon\ \mathcal{A}(GH_\mathcal{K}^n)_{-\Lambda} \to \mathbb{Q}(\omega)\otimes\mathbb{C}$$

by analogy with (3.9.2–3.9.4). Then the analogue 3.9.5 remains valid, when the periods $\mathbf{p}(\omega, *)$ are replaced by periods adapted to $\mathrm{Sh}(GH_\mathcal{K}^n, h^{r,s})$.

However, we would like to compute these periods in terms of the $p(\omega, *)$ with which we are already familiar. This is easy and is done as follows. Let $N_{\mathcal{K}/\mathbb{Q}}\colon\ H_\mathcal{K} \to \mathbb{G}_{m,\mathbb{Q}}$ be the norm map, $N_{\mathcal{K}/\mathbb{Q}}^n H_\mathcal{K}^n \to \mathbb{G}_{m,\mathbb{Q}}^n$ its n-fold product with itself. We may naturally identify

$$GH_\mathcal{K}^n = \{(z_1, \ldots, z_n) \in H_\mathcal{K}^n | N_{\mathcal{K}/\mathbb{Q}}(z_1) = \cdots = N_{\mathcal{K}/\mathbb{Q}}(z_n)\} = (N_{\mathcal{K}/\mathbb{Q}}^n)^{-1}(\Delta),$$

where Δ is the diagonal in $\mathbb{G}^n_{m,\mathbb{Q}}$. This identification underlies a map of data defining Shimura varieties:

$$(3.10.1)\qquad i^{r,s}\colon (GH^n_{\mathcal{K}}, h^{r,s}) \to \underbrace{(H_{\mathcal{K}}, h^{1,0}) \times \cdots \times (H_{\mathcal{K}}, h^{1,0})}_{r\text{ times}} \times \underbrace{(H_{\mathcal{K}}, h^{0,1}) \times \cdots \times (H_{\mathcal{K}}, h^{0,1})}_{s\text{ times}}$$

Denote the product on the right-hand side of (3.10.1) by $(H^n_{\mathcal{K}}, (h^{1,0})^r \times (h^{0,1})^s)$. The canonical models of automorphic vector bundles satisfy certain functorial properties, which imply that, for any extension of $\Lambda \in X(GH^n_{\mathcal{K}})$ to $\Lambda^+ \in X(H^n_{\mathcal{K}})$, the natural pullback

$$H^0(\mathrm{Sh}(H^n_{\mathcal{K}}, (h^{1,0})^r \times (h^{0,1})^s), E_{\Lambda^+}) \to H^0(\mathrm{Sh}(GH^n_{\mathcal{K}}, h^{r,s}), E_\Lambda)$$

is rational over $E(W_{\Lambda^+})$, which is easily checked to be the same as $E(W_\Lambda)$. The following Lemma follows formally from this fact:

LEMMA 3.10.2. *(a) Let* $\omega\colon\ GH^n_{\mathcal{K}}(\mathbf{A})/GH^n_{\mathcal{K}}(\mathbb{Q}) \to \mathbb{C}^\times$ *be a Hecke character of type* Λ, *for some* $\Lambda \in X(GH^n_{\mathcal{K}})$, *and let* $\omega^+\colon\ H^n_{\mathcal{K}}(\mathbf{A})/H^n_{\mathcal{K}}(\mathbb{Q}) \to \mathbb{C}^\times$ *be any Hecke character extending* ω. *Write* $\omega^+ = (\omega_1, \ldots, \omega_n)$, *where* ω_i *is a Hecke character of* $H_{\mathcal{K}}$, *and let* $\mathbf{p}(\omega^+, r, s) = \prod_{i=1}^{r} \mathbf{p}(\omega_i, i_0) \cdot \prod_{i=r+1}^{n} \mathbf{p}(\omega_i, \iota) \in (\mathbb{Q}(\omega^+) \otimes \mathbb{C})^\times/\mathbb{Q}(\omega^+)^\times$. *Then there is a unique invariant* $\mathbf{p}(\omega, h^{r,s}) \in (\mathbb{Q}(\omega) \otimes \mathbb{C})^\times/\mathbb{Q}(\omega)^\times$ *such that, for all extensions* ω^+ *as above,* $\mathbf{p}(\omega, h^{r,s}) \sim_{\mathbb{Q}(\omega^+)} \mathbf{p}(\omega^+, r, s) \in (\mathbb{Q}(\omega^+) \otimes \mathbb{C})^\times/\mathbb{Q}(\omega^+)^\times$.

(b) Let $L \subset \mathbb{C}$ *be an extension of* $E(W_\Lambda)$. *Then the section*

$$\varphi \in H^0(\mathrm{Sh}(GH^n_{\mathcal{K}}, h^{r,s}), E_\Lambda)(\mathbb{C})$$

is rational over L *if and only if*

$$\mathbf{p}(\omega, h^{r,s})^{-1} \cdot \mathbf{L}_\omega(\underline{\mathrm{Lift}}(\varphi)) \in (\mathbb{Q}(\omega) \otimes L)^\times/\mathbb{Q}(\omega)^\times$$

for every Hecke character ω *of type* Λ.

(3.11). Now let $GG = \mathrm{GU}(V)$, of signature (r, s) as in the beginning of this section. We take GH to be a CM torus isomorphic to $GH^n_{\mathcal{K}}$, as in 1.4, and let GK_∞ be the stabilizer of the corresponding point in $X^{r,s}$. We consider L-rational elements of $H^0(\mathfrak{P}, GK_\infty; \mathcal{A}(GG) \otimes W_\Lambda) \xrightarrow{\sim} \tilde{H}^0(E^{\mathrm{can}}_\Lambda)$, when L is an extension of $E(W_\Lambda)$. The space

$$\begin{aligned} H^0(\mathfrak{P}, GK_\infty; \mathcal{A}(GG) \otimes W_\Lambda) &= (\mathcal{A}(GG) \otimes W_\Lambda)^{\mathfrak{P}} \\ &= \{\mathfrak{p}^-\text{-torsion in}(\mathcal{A}(GG) \otimes W_\Lambda)^{GK_\infty}\} \end{aligned}$$

is just the adelization of the classical theory of holomorphic automorphic forms with values in W_Λ. In particular, let $\mathbb{D}_\Lambda = U(G\mathfrak{g}) \otimes_{U\mathfrak{P}} W_\Lambda^{\check{}}$ be the generalized Verma module (highest weight module) associated to the dual $W_\Lambda^{\check{}}$ of W_Λ, and let $\mathcal{A}(GG)^{\mathrm{hol}}_\Lambda$ denote the subspace of $\mathcal{A}(GG)$ spanned by $\{\pi | \pi_\infty \text{ is a quotient of } \mathbb{D}_\Lambda\}$. Then we can replace $\mathcal{A}(GG)$ by $\mathcal{A}(GG)^{\mathrm{hol}}_\Lambda$ in the above formulas. We obtain a canonical isomorphism

$$(3.11.1)\qquad \underline{\mathrm{Lift}}\colon\ \tilde{H}^0(E^{\mathrm{can}}_\Lambda) \xrightarrow{\sim} \bigoplus_{\pi \subset \mathcal{A}(GG)^{\mathrm{hol}}_\Lambda} \pi_f \otimes (\pi_\infty \otimes W_\Lambda)^{GK_\infty}.$$

For future reference, we let

$$\mathcal{A}_0(GG)^{\mathrm{hol}}_\Lambda = \mathcal{A}(GG)^{\mathrm{hol}}_\Lambda \cap \mathcal{A}_0(GG), \mathcal{A}_{(2)}(GG)^{\mathrm{hol}}_\Lambda = \mathcal{A}(GG)^{\mathrm{hol}}_\Lambda \cap \mathcal{A}_{(2)}(GG).$$

In particular, any $\pi \subset \mathcal{A}(GG)^{\text{hol}}_{\Lambda}$ is generated as admissible $GG(\mathbf{A})$-module by the $GH(\mathbb{R})$-type -Λ, which occurs with multiplicity one in the archimedean component of π. Let $pr_{\Lambda}\colon \ \pi \to \pi$ be orthogonal projection on the $(-\Lambda)$-isotypic component for $GH(\mathbb{R})$, and let $R_{GG/GH}$ denote restriction of functions from $GG(\mathbf{A})$ to $GH(\mathbf{A})$. Let $\omega\colon \ GH^{n}_{\mathcal{K}}(\mathbf{A})/GH^{n}_{\mathcal{K}}(\mathbb{Q}) \to \mathbb{C}^{\times}$ be a Hecke character of type Λ. For any $f \in \pi$, let

(3.11.2) $L_{\omega,GG}(f) = L_{\omega}(R_{GG/GH}(pr_{\Lambda}(f))), \ \mathbf{L}_{\omega,GG}(f) = \mathbf{L}_{\omega}(R_{GG/GH}(pr_{\Lambda}(f))),$

where in each case the right-hand side is defined as in 3.10.

For any automorphic form f on GG and any $\gamma \in GG(\mathbf{A}^{f})$, we let $r_{\gamma}(f)(g) = f(g\gamma)$ be the right regular action. In view of the discussion in §1 of [**H93**], the following proposition is a special case of the theory of [**H86**, Part 1].

PROPOSITION 3.11.3. *Let $L \subset \mathbb{C}$ be an extension of $\mathcal{K}$. Then the section $\varphi \in \tilde{H}^{0}(E^{\text{can}}_{\Lambda})(\mathbb{C})$ is rational over L if and only if*

$$\mathbf{p}(\omega, h^{r,s})^{-1} \cdot \mathbf{L}_{\omega,GG}(\underline{\text{Lift}}(r_{\gamma}(\varphi))) \in (\mathbb{Q}(\omega) \otimes L)^{\times}/\mathbb{Q}(\omega)^{\times}$$

for every Hecke character ω of type Λ and every $\gamma \in GG(\mathbf{A}^{f})$.

REMARK 3.11.4. The rationality criterion in [**H86**] is expressed in terms of vector-valued forms, and is thus not obviously a consequence of the criterion in 3.11.3. However, the automorphic vector bundle E_{Λ} is associated to an irreducible representation of GK_{∞}. We may therefore appeal to the argument used to prove the last assertion of [**H93**, Cor. 2.3.3] to verify that it suffices to look at the highest-weight component of $\underline{\text{Lift}}(r_{\gamma}(\varphi))$ for all γ.

REMARK 3.11.5. If $r = s$ and $\Lambda = \Lambda^{\iota}$, then $E(W_{\Lambda}) = \mathbb{Q}$. In this case the proposition can be strengthened to a criterion for rationality over extensions of $\mathbb{Q}$, by "testing" the coefficients of $r_{\gamma}(\varphi)$ in the lowest weight space as well as the highest weight space, and by extending the $\text{Gal}(\bar{\mathbb{Q}}/\mathcal{K})$ action on the quantities $\mathbf{p}(\omega, h^{r,s})^{-1} \cdot \mathbf{L}_{\omega,GG}(\underline{\text{Lift}}(r_{\gamma}(\varphi))$ to a $\text{Gal}(\bar{\mathbb{Q}}/\mathbb{Q})$-action by allowing interchange of ω and ω^{ι}. We will not be making use of this criterion in the present paper and we therefore omit the details.

(3.12). Finally, we will primarily be working with holomorphic forms on the doubled group $GG^{(2)} = G(U(V) \times U(-V))$, in the notation of I, §4 with V of signature (r,s). Consider the homomorphism $h^{r,s} \times h^{s,r}\colon \underline{S} \to [\text{GU}(V) \times \text{GU}(-V)]_{\mathbb{R}}$. The image is evidently contained in $GG^{(2)}_{\mathbb{R}}$, and we let $X^{r,s,(2)}$ denote the $GG^{(2)}(\mathbb{R})$-conjugacy class of maps $h\colon \underline{S} \to [\text{GU}(V) \times \text{GU}(-V)]_{\mathbb{R}}$ containing $h^{r,s} \times h^{s,r}$. Then $(GG^{(2)}, X^{r,s,(2)})$ is a Shimura datum defining a Shimura variety

$$\text{Sh}(GG^{(2)}, X^{r,s,(2)}) = \text{Sh}(V^{(2)}).$$

As a complex analytic space, $X^{r,s,(2)}$ is nothing more than $X^{r,s} \times X^{s,r}$ unless $r = s$, in which case $X^{r,r,(2)}$ is a union of two of the four connected components of $X^{r,r} \times X^{r,r}$. We let $GK_{\infty}(2) = GG^{(2)}_{\mathbb{R}} \cap GK_{\infty}(r,s) \times GK_{\infty}(s,r)$.

We may view $\text{Sh}(V^{(2)})$ as a Shimura subvariety of $\text{Sh}(V)\times$ $\text{Sh}(-V)$. Define the operation $\Lambda \mapsto \Lambda^{\text{op}}$ by II (4.3). Let $E_{\Lambda(2)}$ denote the pullback to $\text{Sh}(V^{(2)})$ of the automorphic vector bundle $E_{\Lambda} \otimes E_{\Lambda^{\text{op}}}$ (external tensor product) on $\text{Sh}(V)\times$ $\text{Sh}(-V)$. Let $GH^{n,(2)}_{\mathcal{K}} = GG^{(2)} \cap GH^{n}_{\mathcal{K}} \times GH^{n}_{\mathcal{K}}$, the intersection being taken in $\text{GU}(V) \times \text{GU}(-V)$. If ω is a Hecke character of $GH^{n}_{\mathcal{K}}$ of type Λ, then $\omega \times \omega^{-1}$ is a

Hecke character of $GH^n_{\mathcal{K}} \times GH^n_{\mathcal{K}}$ of type $\Lambda \times \Lambda^{\rm op}$. We let $\omega(2)$ denote the restriction of $\omega \times \omega^{-1}$ to $GH^{n,(2)}_{\mathcal{K}}$.

We let $R_{GG(2)/GH(2)}$ denote restriction of functions from $GG^{(2)}(\mathbf{A})$ to

$$GH^{n,(2)}_{\mathcal{K}}(\mathbf{A}).$$

If λ is a Hecke character of $GH^n_{\mathcal{K}} \times GH^n_{\mathcal{K}}$ of type $\Lambda \times \Lambda^{\rm op}$ and f is an automorphic form on $GG^{(2)}(\mathbf{A})$, we define $L_{\lambda,GG^{(2)}}(f)$ and $\mathbf{L}_{\lambda,GG^{(2)}}(f)$ by analogy with (3.11.2), using projection on the $\Lambda\times\Lambda^{\rm op}$-isotypic subspace of the automorphic representation generated by f. If $\gamma \in GG^{(2)}(\mathbf{A}^f)$ we let $r_\gamma(f)(g) = f(g\gamma)$ as before.

Let $\pi \subset \mathcal{A}_{(2)}(\mathrm{GU}(V))^{\rm hol}_{\Lambda}$ and fix an irreducible constituent π_0 of the restriction of π to $U(V)(\mathbb{Q})\backslash U(V)(\mathbf{A})$. Assume for the moment $r \neq s$. We may of course identify $\mathrm{GU}(V) \cong \mathrm{GU}(-V)$ as subgroups of $\mathrm{GL}(V)$; however, π is contained naturally in the space generated by anti-holomorphic forms on $\mathrm{GU}(-V)$, with regard to the holomorphic structure defined by $X^{s,r}$. By the same token, the contragredient $\check{\pi}$ of π is contained in the space generated by holomorphic forms on $\mathrm{GU}(-V)$. Since we are not assuming π to have multiplicity one in $\mathcal{A}(GG)$, this needs to be made more precise. Let ξ_π denote the central character of π, $|\xi_\pi|$ its absolute value. Since π consists of square-integrable forms, the representation π_0 is unitary, and we define $\check{\pi}_0$ as the twisted conjugate subspace $\bar{\pi} \subset \mathcal{A}(\mathrm{GU}(V))$. Thus $\pi_0 \otimes \check{\pi}_0 \subset \mathcal{A}_{(2)}(\mathrm{GU}(V) \times \mathrm{GU}(-V))^{\rm hol}_{\Lambda\times\Lambda^{\rm op}}$, in the obvious notation, and we define $[\pi \otimes \check{\pi}]_0$ to be the irreducible subspace of $\mathcal{A}_{(2)}(GG^{(2)})^{\rm hol}_{\Lambda\times\Lambda^{\rm op}}$ generated by $\pi_0 \otimes \check{\pi}_0$. When $r = s$ the same automorphic representation π contains holomorphic and anti-holomorphic vectors. However, π decomposes as a direct sum $\pi^+ \oplus \pi^-$ of two $(\mathrm{Lie}(GG(\mathbb{R})), K_\infty) \times GG(\mathbf{A}^f)$-modules (recall that K_∞ is a <u>connected</u> subgroup of $GG(\mathbb{R})$), with π^+ generated by holomorphic vectors and π^- generated by anti-holomorphic vectors. The natural pairing between π and its contragredient places π^+ in duality with $\check{\pi}^-$ and vice versa. Then

$$\pi^{(2)} = \pi^+ \otimes \check{\pi}^- \oplus \pi^- \otimes \check{\pi}^+$$

is a (not necessarily irreducible) automorphic representation of $GG^{(2)}$ and $\pi^+ \otimes \check{\pi}^-$ is the subspace generated by holomorphic vectors for the chosen holomorphic structure. We choose an irreducible $GG^{(2)}(\mathbf{A})$ constituent $[\pi\otimes\check{\pi}]_0 \subset \pi^{(2)}$ as before.

PROPOSITION 3.12.1. *Let $\tilde{\omega}$ be the Hecke character ω/ω^ι of $GH^n_{\mathcal{K}}$. Let $\pi \subset \mathcal{A}_{(2)}(\mathrm{GU}(V))^{\rm hol}_{\Lambda}$ be an irreducible automorphic representation and let*

$$\begin{aligned}\tilde{H}^0(\mathrm{Sh}(V^{(2)}),&E_{\Lambda(2)})([\pi\otimes\check{\pi}]_0)\\ &= \underline{\mathrm{Lift}}^{-1}([\pi\otimes\check{\pi}]_{0,f} \otimes ([\pi\otimes\check{\pi}]_{0,\infty} \otimes [W_\Lambda \otimes W_{\Lambda^{op}}])^{GK_\infty(2)}\end{aligned}$$

be the subspace of sections φ such that $\underline{\mathrm{Lift}}(\varphi) \in [\pi\otimes\check{\pi}]_0$. Let $\mathcal{G} \subset GG^{(2)}(\mathbf{A}^f)$ be a subset such that $GH^n_{\mathcal{K}}(\mathbf{A})\cdot\mathcal{G}$ has dense image under the natural map to $\mathrm{Sh}(V^{(2)})$.

Let $L \subset \mathbb{C}$ be an extension of $E(W_\Lambda)$. Then the section

$$\varphi \in \tilde{H}^0(\mathrm{Sh}(V^{(2)}), E_{\Lambda(2)})([\pi\otimes\check{\pi}]_0)$$

is rational over L if and only if

$$\mathbf{p}(\tilde{\omega}, h^{r,s})^{-1}\cdot \mathbf{L}_{\omega(2)}(\underline{\mathrm{Lift}}(r_\gamma(\varphi))) \in (\mathbb{Q}(\omega)\otimes L)^\times/\mathbb{Q}(\omega)^\times$$

for every Hecke character ω of type Λ and every $\gamma \in \mathcal{G}$.

REMARK. Since $GG^{(2)}(\mathbb{Q}) \cdot GG^{(2)}(\mathbf{A}^f)$ is dense in $GG^{(2)}(\mathbf{A})$, we can take $\mathcal{G} = GG^{(2)}(\mathbf{A}^f)$ in the above; in particular, the criterion is not vacuous.

PROOF. As in the proof of Lemma 3.10.2, we can verify that rationality of a form φ_0 of type $\Lambda \times \Lambda^{\text{op}}$ on $\text{Sh}(GH_{\mathcal{K}}^{n,(2)}, h^{r,s} \times h^{s,r})$ is determined by comparing $\mathbf{L}_{\omega\times\omega'}(\underline{\text{Lift}}(\varphi_0))$ to $\mathbf{p}(\omega, h^{r,s}) \cdot \mathbf{p}(\omega', h^{s,r})$. In particular, $\varphi \in \tilde{H}^0(\text{Sh}(V^{(2)}), E_{\Lambda(2)})(\mathbb{C})$ is rational over L if and only if

$$(3.12.2) \qquad [\mathbf{p}(\omega, h^{r,s}) \cdot \mathbf{p}(\omega', h^{s,r})]^{-1} \cdot \mathbf{L}_{\omega\times\omega'}(\underline{\text{Lift}}(r_\gamma(\varphi))) \in (\mathbb{Q}(\omega,\omega') \otimes L)^\times / \mathbb{Q}(\omega,\omega')^\times$$

for all ω of type Λ and ω' of type Λ^{op} and all γ.

When $\omega' = \omega^{-1}$ it follows from (2.1) and (2.2) that

$$\mathbf{p}(\omega, h^{r,s}) \cdot \mathbf{p}(\omega^{-1}, h^{s,r}) \sim_{\mathbb{Q}(\omega);\mathcal{K}} \mathbf{p}(\tilde{\omega}, h^{r,s}).$$

It thus remains to show that, if $\varphi \in \tilde{H}^0(\text{Sh}(V^{(2)}), E_{\Lambda(2)})([\pi \otimes \check{\pi}]_0)$, it suffices to verify (3.12.2) for $\omega' = \omega^{-1}$. In fact, we show that, if $\mathbf{L}_{\omega(2)}(\underline{\text{Lift}}(r_\gamma(\varphi))) = 0$ for all ω and all γ then $\varphi = 0$. Let $\mathcal{V}(\pi) \subset \tilde{H}^0(\text{Sh}(V^{(2)}), E_{\Lambda(2)})([\pi \otimes \check{\pi}]_0)$ be the subspace of φ such that $\mathbf{L}_{\omega(2)}(r_\gamma(\underline{\text{Lift}}(\varphi))) = 0$ for all ω and all $\gamma \in \mathcal{G}$. It then follows by continuity and our hypothesis on $\mathcal{G}$ that, for such $\varphi, \mathbf{L}_{\omega(2)}(r_\gamma(\underline{\text{Lift}}(\varphi))) = 0$ for all ω and all $\gamma \in GG^{(2)}(\mathbf{A}^f)$. Thus $\mathcal{V}(\pi)$ is a $GG^{(2)}(\mathbf{A}^f)$-invariant subspace. Since

$$\tilde{H}^0\big(\text{Sh}(V^{(2)}), E_{\Lambda(2)}\big)\big([\pi \otimes \check{\pi}]_0\big) \cong [\pi \otimes \check{\pi}]_{0,f}$$

is an irreducible $GG^{(2)}(\mathbf{A}^f)$-module it follows that it suffices to construct a vector $\varphi \in \tilde{H}^0(\text{Sh}(V^{(2)}), E_{\Lambda(2)})([\pi \otimes \check{\pi}]_0)$ with $\varphi \notin \mathcal{V}(\pi)$.

Let f be a holomorphic vector in the chosen $U(V)(\mathbf{A})$-constituent of π_0. We view f as a function on $GG(\mathbf{A})$ and let $f^{(2)} \in [\pi \otimes \check{\pi}]_0$ be the restriction of $f \otimes \bar{f}$ to $GG^{(2)}(\mathbf{A})$; let $\varphi = \underline{\text{Lift}}^{-1}(f^{(2)})$. We may assume $pr_\Lambda(f) \neq 0$. Since $GG(\mathbb{Q}) \cdot GG(\mathbf{A}^f)$ is dense in $GG(\mathbf{A})$, the restriction of $r_\gamma(f)$ to $GH_{\mathcal{K}}^n(\mathbf{A})$ must be non-zero for some $\gamma \in GG(\mathbf{A}^f)$. Since $pr_\Lambda(f) \neq 0$, it then follows that $\mathbf{L}_\omega(r_\gamma(f)) \neq 0$ for some ω. On the other hand, since $GG^{(2)}(\mathbf{A}) \subset \text{GU}(V)(\mathbf{A}) \times \text{GU}(-V)(\mathbf{A})$, it is easy to see that if $\mathbf{L}_{\omega(2)}(r_\gamma(f^{(2)})) = 0$ then $\mathbf{L}_\omega(r_\gamma(f)) \cdot \mathbf{L}_{\omega^{-1}}(r_\gamma(\bar{f})) = 0$. But

$$\mathbf{L}_\omega(r_\gamma(F)) \cdot \mathbf{L}_{\omega^{-1}}(r_\gamma(\bar{f})) = |\mathbf{L}_\omega(r_\gamma(f))|^2,$$

so we are done.

REMARK 3.13. In practice we will only be working with representations of $GG^{(2)}$ of the form $[\pi \otimes \check{\pi}]_0$. Thus the character of the split component of the center, isomorphic to $\mathbf{A}^\times$, is always equal to $\xi_\pi \cdot \check{\xi}_\pi \equiv 1$. On the other hand, the rationality criterion provided by Proposition 3.12.1 does not notice when π is twisted by a Hecke character of $\text{GU}(V)/U(V)$. This is apparent from the formula, since the period invariant $p(\tilde{\omega}, h^{r,s})$ is unchanged when $\tilde{\omega}$ is multiplied by a character factoring through $N_{\mathcal{K}/\mathbb{Q}}$. But it is also obvious because the period invariants used to characterize rational forms in π and in $\check{\pi}$ are multiplied by complementary factors when π is twisted as above; here the essential point is that $N_{\mathbb{C}/\mathbb{R}} \circ h_{r,s} = N_{\mathbb{C}/\mathbb{R}} \circ h_{s,r}$. In this sense, we can say that rationality of the form φ in $[\pi \otimes \check{\pi}]_0$ depends only on the restriction of φ to $U(V)(\mathbf{A}) \times U(-V)(\mathbf{A})$.

4. Cup products and trace maps in coherent cohomology

In this section we fix $GG = \mathrm{GU}(V)$, of signature $(n-1,1)$. The cup products in $\bar\partial$-cohomology of $U(n-1,1)$, introduced in II, §2, have their global counterparts in coherent cohomology of the Shimura varieties $\mathrm{Sh}(V)$. We begin with the analogue of II.2.5. The parameters $\Lambda = \Lambda^+(-\alpha,\beta)$, etc. in Part II are adapted to unitary groups and do not involve a central character. In what follows, we assume an integer c has been chosen as the parameter of the central character, as in §3, above. Thus c and the remaining parameters must satisfy the parity condition II, (1.15.1). However, to simplify notation we omit c from the bundle parameters. Thus we write $E^+_{(-\alpha,\beta)}$ (resp. $E_{(\alpha,-\beta)}$) for E_Λ when $\Lambda = \Lambda^+(-\alpha,\beta;c)$ (resp. $\Lambda = \Lambda^-(\alpha,\beta;-c)$; note the sign change!). The pairing II.2.4 gives rise to a canonical $\mathcal{K}$-rational pairing

$$(4.1)\quad \tilde H^0(\mathrm{Sh}(V),(E^+_{(-\alpha,\beta)})^{\mathrm{can}})\otimes \tilde H^1(\mathrm{Sh}(V),(E^-_{(\alpha,-\beta)})^{\mathrm{can}}) \to \tilde H^1(\mathrm{Sh}(V),\Omega^{1,\mathrm{can}}).$$

Here $\Omega^{1,\mathrm{can}}$ is the bundle of holomorphic 1-forms on $\mathrm{Sh}(V)$, canonically extended to the toroidal compactification of $\mathrm{Sh}(V)$; this is just the bundle of 1-forms on the toroidal compactification of $\mathrm{Sh}(V)$ with logarithmic poles among the divisor at infinity. It so happens that the toroidal compactification of $\mathrm{Sh}(V)$ is unique and canonical, because $GG(\mathbb{R})^{\mathrm{der}} \cong SU(n-1,1)$; however, we do not use this fact.

We want to define a $\mathcal{K}$-rational "trace map" from the right-hand side of (4.1) to $\mathbb{C}$, to complete the analogy with II.2.5. We return to the notation $H^\cdot_{\bar\partial}$ of Part II. In principle, it is obvious what to do. Consider the isomorphism of Theorem 3.5:

$$(4.2)\quad H^1_{\bar\partial}(\mathcal{A}(GG)\otimes \mathfrak{p}^{+,*}) \cong (\mathfrak{p}^{-,*}\otimes \mathcal{A}(GG)\otimes \mathfrak{p}^{+,*})^{GK_\infty} \xrightarrow{\sim} H^1(\mathrm{Sh}(V),\Omega^{1,\mathrm{can}}).$$

Now $\mathfrak{p}^{\pm}$ are $\mathcal{K}$-rational subspaces of $\mathfrak{g}$. We identify $\mathfrak{p}^+ \cong \mathfrak{p}^{-,*}$ via the Killing form; this duality is $\mathcal{K}$-rational. Thus the natural trace map

$$(4.3)\qquad \mathbb{C} \cong (\mathfrak{p}^{-,*}\otimes \mathfrak{p}^{+,*})^{GK_\infty}$$

is $\mathcal{K}$-rational. The constant function $1 \in \mathcal{A}(GG)$ defines an element

$$\mathcal{L} \in \tilde H^1(\mathrm{Sh}(V),\Omega^{1,\mathrm{can}})$$

by composing (4.2) with (4.3). More generally, for $j = 1,2,\dots,n-1$, $\mathcal{L}^j = \Lambda^j\mathcal{L}$ is canonically a (j,j) form on $\mathrm{Sh}(V)$. Now we have

LEMMA 4.4. *Suppose $n > 3$. Then $\mathcal{L}^{n-2}$ defines a non-trivial $\mathcal{K}$-rational element of $\tilde H^{n-2}(\mathrm{Sh}(V),\Omega^{n-2,\mathrm{sub}})$.*

PROOF. Let $\mathrm{Sh}(V)^*$ denote the Baily-Borel Satake compactification of $\mathrm{Sh}(V)$. We first note that $\mathcal{L}^{n-2}$ defines a non-trivial element of the intersection cohomology group $IH^{2n-4}(\mathrm{Sh}(V)^*,\mathbb{C})$. Indeed, $\mathcal{L}^{n-1}$ is obviously harmonic and square integrable because $\mathrm{Sh}(V)$ has finite volume, and $IH^{2n-4}(\mathrm{Sh}(V)^*,\mathbb{C})$ coincides with the space of square-integrable harmonic forms by [**Z**]. On the other hand, $\mathrm{Sh}(V)^*$ has isolated singularities, hence $IH^j(\mathrm{Sh}(V)^*,\mathbb{C}) \cong H^j_c(\mathrm{Sh}(V),\mathbb{C})$ for $j > \dim(\mathrm{Sh}(V)) = n-1$ (dimension as an algebraic variety). The hypothesis $n > 3$ thus implies that $IH^{2n-4}(\mathrm{Sh}(V)^*,\mathbb{C}) \cong H^{2n-4}_c(\mathrm{Sh}(V),\mathbb{C})$. Thus $\mathcal{L}^{n-2} \in H^{2n-4}_c(\mathrm{Sh}(V),\mathbb{C})$.

On the other hand, it is obvious that

$$(4.4.1)\qquad \mathbb{C}\mathcal{L}^{n-2} = IH^{2n-4}(\mathrm{Sh}(V)^*,\mathbb{C})^{GG(\mathbf{A}^f)} = H^{2n-4}_c(\mathrm{Sh}(V),\mathbb{C})^{GG(\mathbf{A}^f)},$$

because the constants are the only $GG(\mathbf{A}^f)$-invariant harmonic functions on $GG(\mathbf{A})$. Now since $H^{2n-4}_c(\mathrm{Sh}(V),\mathbb{C})$ is pure and represented by square integrable harmonic forms, it follows from [**H94**, Theorem 2.2.7] that it has a Hodge decomposition as

$$H_c^{2n-4}(\mathrm{Sh}(V),\mathbb{C}) \cong \bigoplus_{p+q=2n-4} \tilde{H}^q(\mathrm{Sh}(V),\Omega^{p,\mathrm{sub}}), \tag{4.4.2}$$

which is compatible (under Poincaré duality and Serre duality) with the analogous Hodge decomposition

$$H^2(\mathrm{Sh}(V),\mathbb{C}) \cong \bigoplus_{p+q=2} \tilde{H}^q(\mathrm{Sh}(V),\Omega^{p,\mathrm{can}}), \tag{4.4.3}$$

It is obvious that the $GG(\mathbf{A}^f)$-invariants in (4.4.3) are concentrated in degree (1,1), and in fact coincide with the subspace $\mathbb{C}\mathfrak{L}$. (For example, we can identify $H^2(\mathrm{Sh}(V),\mathbb{C}) \cong IH^2(\mathrm{Sh}(V)^*,\mathbb{C})$, and reason as before to show that $\mathfrak{L}$, a priori in $\tilde{H}^1(\mathrm{Sh}(V),\Omega^{1,\mathrm{can}})$, defines a non-trivial class in $H^2(\mathrm{Sh}(V),\mathbb{C})$.) It follows that

$$\mathbb{C}\mathfrak{L}^{n-2} = H^{n-2}(\mathrm{Sh}(V),\Omega^{n-2,\mathrm{sub}})^{GG(\mathbf{A}^f)}.$$

We write $\mathcal{H}^{n-2,n-2}$ in place of $H^{n-2}(\mathrm{Sh}(V),\Omega^{n-2,\mathrm{sub}})$. Now the subspace

$$(\mathcal{H}^{n-2,n-2})^{GG(\mathbf{A}^f)} \subset \mathcal{H}^{n-2,n-2}$$

is clearly $\mathcal{K}$-rational. Thus, in order to verify the $\mathcal{K}$-rationality of $\mathfrak{L}^{n-2} \in \mathcal{H}^{n-2,n-2}$, it suffices to construct a $\mathcal{K}$-rational linear form $I\colon \mathcal{H}^{n-2,n-2} \to \mathbb{C}$ such that

$$I(\mathfrak{L}^{n-2}) \in \mathcal{K}.$$

Recall that our subgroup $GK_\infty \subset GG(\mathbb{R})$ is the centralizer of a $\mathbb{Q}$-rational homomorphism $h\colon H_{\mathcal{K}} \to GG$. Let $V' \subset V$ be a one-dimensional $H_{\mathcal{K}}$-invariant $\mathcal{K}$-subspace with signature (1,0), and let V'' denote the orthogonal complement of V' in V. Let

$$G' = G(U(V') \times U(V'')) = (\mathrm{GU}(V') \times \mathrm{GU}(V'')) \cap \mathrm{GU}(V).$$

Then $h(H_{\mathcal{K}}) \subset G'$. We let X' be the $G'(\mathbb{R})$-conjugacy class of homomorphisms from the Hodge torus $\underline{S}$ to $G'(\mathbb{R})$ containing h. Then (G',X') is a sub-Shimura datum of (G,X) such that $h \in X', GK'_\infty = GK_\infty \cap G'(\mathbb{R})$ is the stabilizer of h, and $\mathrm{Sh}(G',X')$ is a sub-Shimura variety of $\mathrm{Sh}(V)$ of codimension one. Modulo $U(V')$,$\mathrm{Sh}(G',X')$ is just $\mathrm{Sh}(V'')$.

Let $\mathrm{Tr}\colon H^{n-2}(\mathrm{Sh}(G',X'),\Omega^{n-2,\mathrm{sub}})^{G'(\mathbf{A}^f)} \xrightarrow{\sim} \mathbb{C}$ be the usual trace map:

$$\mathrm{Tr}(\omega) = (2\pi i)^{-(n-2)} \int_{\mathrm{Sh}(G',X')} \omega. \tag{4.4.4}$$

We let $I\colon \mathcal{H}^{n-2,n-2} \to \mathbb{C}$ be Tr composed with the restriction (cf. [**H90**, 2.5.3])

$$R\colon \mathcal{H}^{n-2,n-2} \to H^{n-2}(\mathrm{Sh}(G',X'),\Omega^{n-2,\mathrm{sub}})^{G'(\mathbf{A}^f)}.$$

In other words, $I(\omega)$ is just the integral of ω over the subvariety $\mathrm{Sh}(G',X')$; this roundabout definition guarantees its $\mathcal{K}$-rationality. It remains to show that $I(\mathfrak{L}^{n-2}) \in \mathcal{K}^\times$. But (4.4.4) is $\mathcal{K}$-rational (cf. [**H90**, 3.8]), so it suffices to show that $R(\mathfrak{L}^{n-2})$ is a non-trivial $\mathcal{K}$-rational element of $H^{n-2}(\mathrm{Sh}(G',X'),\Omega^{n-2,\mathrm{sub}})^{G'(\mathbf{A}^f)}$.

Let $\mathfrak{p}'^{,-}$ and $\mathfrak{p}'^{,+}$ $\subset\mathrm{Lie}(G'(\mathbb{R}))$ be the GK'_∞-invariant subalgebras of anti-holomorphic and holomorphic tangent vectors, respectively. Then the trace element $T_{n-2} \in (\mathfrak{p}'^{,-})^* \otimes (\mathfrak{p}'^{,+})^*$ is $\mathcal{K}$-rational (cf. (4.3)) and $\Lambda^{n-2}T_{n-2}$ is a $\mathcal{K}$-rational generator of $\Lambda^{n-2}(\mathfrak{p}'^{,-})^* \otimes \Lambda^{n-2}(\mathfrak{p}'^{,+})^*$. Let $T_{n-1} \in (\mathfrak{p}^-)^* \otimes (\mathfrak{p}^+)^*$ be the image of

$1 \in \mathbb{C}$ under (4.3). Now $\mathfrak{p}^{\prime,\pm} \subset \mathfrak{p}^{\pm}$ are subspaces of codimension one. We need to show that the natural map

$$S\colon \Lambda^{n-2}(\mathfrak{p}^-)^* \otimes \Lambda^{n-2}(\mathfrak{p}^+)^* \to \Lambda^{n-2}(\mathfrak{p}^{\prime,-})^* \otimes \Lambda^{n-2}(\mathfrak{p}^{\prime,+})^*$$

is $\mathcal{K}$-rational and does not annihilate T_{n-1}. But the $\mathcal{K}$-rationality is obvious, and the non-vanishing of $S(T_{n-1})$ is a simple exercise in the invariant theory of exterior powers.

We now define a map

$$\smile \mathfrak{L}^{n-2}\colon \tilde{H}^1(\mathrm{Sh}(V), \Omega^{1,\mathrm{can}}) \to \tilde{H}^{n-1}(\mathrm{Sh}(V), \Omega^{n-1,\mathrm{sub}})$$

by cup-product with $\mathfrak{L}^{n-2}$. It is $\mathcal{K}$-rational by Lemma 4.2. Composing with (4.1), we obtain a $\mathcal{K}$-rational pairing

$$\tilde{H}^0\big(\mathrm{Sh}(V), \big(E^+_{(-\alpha,\beta)}\big)^{\mathrm{can}}\big) \otimes \tilde{H}^1(\mathrm{Sh}(V), (E^-_{(\alpha,-\beta)})^{\mathrm{can}}) \to \tilde{H}^{n-1}\big(\mathrm{Sh}(V), \Omega^{n-1,\mathrm{sub}}\big). \tag{4.5}$$

Composing (4.5) with the trace map (4.4.4) (with G' replaced by GG) we finally obtain a $\mathcal{K}$-rational pairing

$$\mathrm{Tr}(\cdot \smile \cdot)\colon \tilde{H}^0\big(\mathrm{Sh}(V), \big(E^+_{(-\alpha,\beta)}\big)^{\mathrm{can}}\big) \otimes \tilde{H}^1\big(\mathrm{Sh}(V), (E^-_{(\alpha,-\beta)})^{\mathrm{can}}\big) \to \mathbb{C}. \tag{4.6}$$

In terms of automorphic forms, this is just the bilinear version of the L_2 pairing. Namely, let $\omega_1 \in \tilde{H}^0(\mathrm{Sh}(V), (E^+_{(-\alpha,\beta)})^{\mathrm{can}}), \omega_2 \in \tilde{H}^1(\mathrm{Sh}(V), (E_{(\alpha,-\beta)})^{\mathrm{can}})$, and suppose

$$\underline{\mathrm{Lift}}(\omega_1) = f_1 \in H^0_{\bar\partial}(\mathcal{A}(GG) \otimes W^+(-\alpha,\beta)) \subset (\mathcal{A}(GG) \otimes W^+(-\alpha,\beta))^{GK_\infty},$$

$$\underline{\mathrm{Lift}}(\omega_2) = f_2 \in H^1_{\bar\partial}(\mathcal{A}(GG) \otimes W^-(\alpha,-\beta)) \subset (\mathfrak{p}^{-,*} \otimes \mathcal{A}(GG) \otimes W^-(\alpha,-\beta))^{GK_\infty}.$$

We will see in IV, Proposition 1.1 and Corollary 5.6, that the $\mathcal{A}(GG)$ components of f_1 and f_2 are actually square-integrable; assume this for the time being. We write $\int_{GG} dg$ for the integral over $Z_{GG}(\mathbf{A})GG(\mathbb{Q})\backslash GG(\mathbf{A})$, with respect to a rational measure, as in the notation section. We integrate the $\mathcal{A}(GG)$-components of $f_1 \otimes f_2$ over $Z_{GG}(\mathbf{A})GG(\mathbb{Q})\backslash GG(\mathbf{A})$ and obtain

$$\int_{GG} f_1 \otimes f_2 \in [W^+(-\alpha,\beta) \otimes \mathfrak{p}^{-,*} \otimes W^-(\alpha,-\beta)]^{GK_\infty}.$$

The left-hand side maps naturally to $[\mathfrak{p}^{-,*} \otimes \mathfrak{p}^{+,*}]^{GK_\infty}$, via II.2.4, and thence to $\mathbb{C}$, via (4.3). Let $[\cdot]\colon [W^+(-\alpha,\beta) \otimes \mathfrak{p}^{-,*} \otimes W^-(\alpha,-\beta)]^{GK_\infty} \to \mathbb{C}$ be the map thus obtained. Then,

LEMMA 4.7. *For some $D \in \mathcal{K}^\times$, we have*

$$\mathrm{Tr}(\omega_1 \smile \omega_2) = D\left[\int_{GG} f_1 \otimes f_2 dg\right].$$

This is obvious. The number D depends on the choice of trivialization for the liftings to the group and on the choice of the rational measure dg.

We now turn to the global version of Proposition II.2.10. Define the representations W_Λ, W_i, and V_j as in II, §2, and let $[W_\Lambda]$, $[W_i]$, and $[V_j]$ be the corresponding automorphic vector bundles, where $i = 1, \dots, r$ and $j = 1, \dots, s$. We write Ω^{n-1} for the line bundle of top differentials on $\mathrm{Sh}(V)$, which is naturally an automorphic vector bundle. It then follows from Corollary II.2.9 that

LEMMA 4.8. *There is a natural $\mathcal{K}$-rational homomorphism of automorphic vector bundles*

$$I_{\Lambda,r,s}\colon [W_\Lambda] \otimes \bigotimes_{i=1}^{r} [W_i] \otimes \bigotimes_{j=1}^{s} [V_j] \to \Omega^{n-1}.$$

This induces a $\mathcal{K}$-rational homomorphism on cohomology, also denoted $I_{\Lambda,r,s}$:

$$(4.9)\quad I_{\Lambda,r,s}\colon H_!^{r'-1}([W_\Lambda]) \otimes \bigotimes_{i=1}^{r} \tilde{H}^0([W_i]^{\mathrm{can}}) \otimes \bigotimes_{j=1}^{s} \tilde{H}^1([V_j]^{\mathrm{can}}) \to H_!^{n-1}(\Omega^{n-1}),$$

the global analogue of the map in Proposition 2.10.

Now suppose $\Phi \in H_!^{r'-1}([W_\Lambda]), \omega_i \in H^0([W_i]^{\mathrm{can}})$, *and* $\nu_j \in \tilde{H}^1([V_j]^{\mathrm{can}})$, *for* $i = 1, \dots, r, j = 1, \dots, s$. *Let*

$$F = \underline{\mathrm{Lift}}(\Phi) \in H_{\bar\partial}^{r'-1}(\mathcal{A}(GG) \otimes W_\Lambda),$$

$f_i = \underline{\mathrm{Lift}}(\omega_i) \in H_{\bar\partial}^0(\mathcal{A}(GG) \otimes W_i), h_j = \underline{\mathrm{Lift}}(\nu_j) \in H_{\bar\partial}^1(\mathcal{A}(GG) \otimes V_j), i, j$ as above.

Then the analogue of Lemma 4.7 is the following statement, whose proof is immediate.

LEMMA 4.10. *For some $D = D_{\Lambda,r,s} \in \mathcal{K}^\times$, we have*

$$\mathrm{Tr} \circ I_{\Lambda,r,s}\Big(\Phi \otimes \bigotimes_{i=1}^{r} \omega_i \otimes \bigotimes_{j=1}^{s} \nu_j\Big) = D\left[\int_{GG} F \otimes \bigotimes_{i=1}^{r} f_i \otimes \bigotimes_{j=1}^{s} h_j dg\right].$$

Here $[\cdot]\colon \bigwedge^{r'-1} \mathfrak{p}^{-,*} \otimes W_\Lambda \otimes \bigotimes_{i=1}^{r} V_i \otimes \bigotimes_{j=1}^{s} (\mathfrak{p}^{-,*} \otimes W_j)] \to \bigwedge^{n-1} \mathfrak{p}^{-,*} \otimes \bigwedge^{n-1} \mathfrak{p}^{+,*} \to \mathbb{C}$ *is the composition of contraction of coefficients implicit in the statement of Proposition II,2.10 with the obvious isomorphism* $\bigwedge^{n-1} \mathfrak{p}^{-,*} \otimes \bigwedge^{n-1} \mathfrak{p}^{+,*} \xrightarrow{\sim} \mathbb{C}$.

(4.11). In applications involving the theta correspondence, it is more convenient to work with scalar-valued automorphic forms than with vector-valued functions on some $G(\mathbb{Q})\backslash G(\mathbf{A})$. The dictionary between the two points of view is discussed in [**H97**, 26.11] to which we refer for notation. In what follows, G will be one of the unitary or unitary similitude groups we consider, $\mathrm{Sh}(G) = \mathrm{Sh}(G, X)$ the Shimura variety associated to the Shimura datum (G, X), GK_∞ the stabilizer of a point in X, and $H \subset GK_\infty$ a maximal torus, given with a system R_c^+ of positive roots as in II, §1.

Let $\mathcal{E}$ be the homogeneous bundle associated to an irreducible representation (τ, V) of GK_∞, and identify

$$(4.11.1)\qquad \Gamma^\infty(\mathrm{Sh}(G), \mathcal{E}) \cong \mathrm{Hom}_{GK_\infty}(\check{V}, C^\infty(G(\mathbb{Q})\backslash G(\mathbf{A}))),$$

where $\Gamma^\infty(\cdot,\cdot)$ denotes global C^∞ sections. Say Λ is the highest weight of τ for our chosen maximal torus of GK_∞, and let $\mathbb{C}_\Lambda \subset V$ and $\mathbb{C}_{-\Lambda} \subset \check{V}$ be the corresponding highest and lowest weight spaces, respectively. Restriction to $\mathbb{C}_{-\Lambda}$ yields an injective map [**H97**, 2.6.11.2].

$$(4.11.2)\qquad \mathrm{Hom}_{GK_\infty}(\check{V}, C^\infty(G(\mathbb{Q})\backslash G(\mathbf{A}))) \to \mathrm{Hom}_H(\mathbb{C}_{-\Lambda}, C^\infty(G(\mathbb{Q})\backslash G(\mathbf{A}))).$$

with image

$$(4.11.3)\qquad \mathrm{Hom}_H(\mathbb{C}_{-\Lambda}, C^\infty(G(\mathbb{Q})\backslash G(\mathbf{A}))_{\check{V}}),$$

where the subscript $_{\check{V}}$ designates the $\check{V}$-isotypic subspace for GK_∞.

The identification $\Gamma^\infty(\mathrm{Sh}(G), \mathcal{E}) \cong \mathrm{Hom}_H(\mathbb{C}_{-\Lambda}, C^\infty(G(\mathbb{Q})\backslash G(\mathbf{A}))_{\check{V}})$ takes holomorphic sections to (highest-weight components of) holomorphic vector-valued automorphic forms. More generally, if $\mathcal{E}$ is a direct factor of a Dolbeault-type bundle

$\Omega^{0,q}\otimes\mathcal{F}$, then the identification takes coherent cohomology classes to automorphic forms whose archimedean components lie in the (highest-weight component of) the lowest GK_∞-type of certain representations; in the applications these will be ladder representations or discrete series representations of $U(n-1,1)$.

Our scalar products generally pair $\Gamma(\mathrm{Sh}(G),\mathcal{E})$, with $\Gamma(\mathrm{Sh}(G),\mathcal{E}^{\check{}})$, although they have hermitian versions (as in §5, below). In terms of (4.11.1), this is a pairing (cf. [**H97**, 2.6.11.4]):

(4.11.4) $\mathrm{Hom}_{GK_\infty}(\check{V},C^\infty(G(\mathbb{Q})\backslash G(\mathbf{A})))\otimes\mathrm{Hom}_{GK_\infty}(V,C^\infty(G(\mathbb{Q})\backslash G(\mathbf{A})))\to\mathbb{C}$,

given by multiplication, contraction of $\check{V}\otimes V$, followed by integration. On the other hand, there is a pairing

(4.11.5) $\mathrm{Hom}_H(\mathbb{C}_{-\Lambda},C^\infty(G(\mathbb{Q})\backslash G(\mathbf{A}))_{\check{V}})\otimes\mathrm{Hom}_H(\mathbb{C}_\Lambda,C^\infty(G(\mathbb{Q})\backslash G(\mathbf{A}))_V)\to\mathbb{C}$

constructed simply as integration of the product over the diagonal. We recall Proposition 2.6.12 of [**H97**]:

PROPOSITION 4.12. *Under the identifications (4.11.2) and (4.11.3), the pairings (4.11.4) and (4.11.5) are equal up to a rational factor.*

There is a variant of Proposition 4.12 in the situation of Lemma 4.10. Let τ_λ be the minimal K_∞-type of π_λ, as in II, §1, and let $-\tau_\lambda$ be the lowest weight vector of the dual $\check{\tau}_\lambda$. Similarly, let τ_i^+ and τ_j^-, $i=1,\dots,r, j=1,\dots s$, be the minimal K_∞-types of the spaces $\check{\mathbb{L}}_i^+$ and $\check{\mathbb{L}}_j^-$, respectively, that appear in Proposition II.2.10, and let ν_i^+ and ν_j^- be the corresponding highest weight vectors. Then, instead of taking the full vector-valued forms $F,\{f_i\}$, and $\{h_j\}$, it's enough to integrate the lowest weight coefficient $F_{-\tau_\lambda}$ of F against the product of the highest weight coefficients $\{f_i^+\}$ and $\{h_j^-\}$ of $\{f_i\}$ and $\{h_j\}$, relative to ν_i^+ and ν_j^-, respectively. Indeed, this reduces to Proposition 4.12, given that

$$\Sigma_i\nu_i^+ + \Sigma_j\nu_j^- \text{ is the highest weight } \tau_\lambda,$$

which is clear by construction. We record a precise version of this formula:

PROPOSITION 4.13. *For some $\tilde{D}=\tilde{D}_{\Lambda,r,s}\in\mathcal{K}^\times$, we have*

$$\mathrm{Tr}\circ I_{\Lambda,r,s}\left(\Phi\otimes\bigotimes_{i=1}^r\omega_i\otimes\bigotimes_{j=1}^s\nu_j\right)=D\int_{GG}F_{-\tau_\lambda}\cdot\prod_{i=1}^r f_i^+\cdot\prod_{j=1}^s h_j^-\,dg.$$

REMARK 4.14. The inverse of (4.11.2), restricted to the subspace

$$\mathrm{Hom}_H(\mathbb{C}_{-\Lambda},C^\infty(G(\mathbb{Q})\backslash G(\mathbf{A}))_{\check{V}})$$

of the left-hand side, yields an isomorphism

(4.14.1) $\underline{\mathrm{Lift}}^{-1}\colon\mathrm{Hom}_H(\mathbb{C}_{-\Lambda},C^\infty(G(\mathbb{Q})\backslash G(\mathbf{A}))_{\check{V}}\xrightarrow{\sim}\Gamma^\infty(\mathrm{Sh}(G),\mathcal{E})$.

In applications of the rationality criteria of §3, we will use this map $\underline{\mathrm{Lift}}^{-1}$ rather than the one defined by (3.11.1).

5. Complex conjugation and small anti-holomorphic forms

We retain the conventions of the previous section; in particular, a central parameter c is chosen to go along with the parameters (α,β). Lemma 4.7 contains a criterion for rationality of automorphic forms that represent classes in $\tilde{H}^1(\mathrm{Sh}(V),(E^-_{(\alpha,-\beta)})^{\mathrm{can}})$. First, we observe

LEMMA 5.1. *The pairing (4.6) is non-degenerate and identifies*

$$\tilde{H}^1\Big(\mathrm{Sh}(V), (E^-_{(\alpha,-\beta)})^{\mathrm{can}}\Big)$$

$\mathcal{K}$-rationally with the dual of $\tilde{H}^0(\mathrm{Sh}(V), (E^+_{(-\alpha,\beta)})^{\mathrm{can}}$.

PROOF. This follows from Lemma 4.7, once it is observed that both spaces are isomorphic to the corresponding spaces of L_2 forms; see IV, Proposition 1.1 (and Corollary 5.6) for the latter identification.

COROLLARY 5.2. *Let* $f \in H^1_{\bar\partial}(\mathcal{A}(GG) \otimes W^-(\alpha, -\beta))$ *and let L be any extension of $\mathcal{K}$ contained in $\mathbb{C}$. Then* $\underline{\mathrm{Lift}}^{-1}(f)$ *is an L-rational class in*

$$\tilde{H}^1(\mathrm{Sh}(V), \Big(E^-_{(\alpha,-\beta)})^{\mathrm{can}}\Big)$$

if and only if, for all $f' \in H^0_{\bar\partial}(\mathcal{A}(GG) \otimes W^+(-\alpha, \beta))$ *such that* $\underline{\mathrm{Lift}}^{-1}(f')$ *is an L-rational class in* $\tilde{H}^0(\mathrm{Sh}(V), (E^+_{(-\alpha,\beta)})^{\mathrm{can}})$,

$$\left[\int_{GG} f \otimes f'\, dg\right] \in L.$$

Moreover, suppose f^+ (resp. f^-) is the highest (resp. lowest) weight coefficient of f, and let μ^+ (resp. μ^-) be the corresponding weight. Write $W = W^+(-\alpha, \beta)$. Then, $\underline{\mathit{Lift}}^{-1}(f)$*, in the sense of (4.14.1), is an L-rational class in*

$$\tilde{H}^1\Big(\mathrm{Sh}(V), (E^-_{(\alpha,-\beta)})^{\mathrm{can}}\Big),$$

if and only if, for all

$$f'^{,+} \in \mathrm{Hom}_H(\mathbb{C}_{-\mu^+}, C^\infty(G(\mathbb{Q})\backslash G(\mathbf{A}))_{\check{W}})$$
$$(\textit{resp. } f'^{,-} \in \mathrm{Hom}_H(\mathbb{C}_{-\mu^-}, C^\infty(G(\mathbb{Q})\backslash G(\mathbf{A}))_{\check{W}},$$

we have

$$\int_{GG} f^+ \otimes f'^{,+} dg \in L\Big(\textit{resp. } \int_{GG} f^- \otimes f'^{,-} dg \in L\Big).$$

PROOF. The first assertion is an immediate consequence of Lemma 5.1, and is equivalent to the second by Proposition 4.12; it suffices to note that the roles of the highest and lowest weight spaces in the proof of Proposition 4.12 are completely symmetric.

In the applications in Part IV, $\underline{\mathrm{Lift}}(f)$ will be a theta lift of a Hecke character and it will suffice to restrict our attention to only f' that are themselves theta lifts. More precisely, both f and f' will actually be forms on the doubled groups; in particular, f will define a class in an appropriate $\tilde{H}^2$. The modification needed to treat this case is clear and will be made explicit when it arises.

In the remainder of this section, we recall from [**H97**] how to obtain a Petersson norm, defined as an integral of a holomorphic form against an anti-holomorphic form, from an integral of a holomorphic form on the doubled group $G(G(V) \times G(-V))$ along its diagonal. As explained in §2.5 and §2.8 of [loc. cit.], this passage involves multiplication by a certain power of i.

Here, for clarity, we have to make the central parameter explicit. We begin, as in [**H97**, 2.5], with the observation that complex conjugation, operating on $\mathcal{A}(GG)$, defines a $\mathbb{C}$-antilinear isomorphism

$$c_{\mathbb{R}}\colon H^0_{\bar\partial}(\mathcal{A}(GG)\otimes W^+(-\alpha,\beta;c)) \xrightarrow{\sim} H^1_{\bar\partial}(\mathcal{A}(GG)\otimes W^-(\alpha,-\beta;c)). \tag{5.3}$$

This identity is correct at the level of representations of GK_∞, but in order to transfer (5.3) to an isomorphism

$$c_B\colon \tilde{H}^0\Big(\mathrm{Sh}(V),(E^+_{(-\alpha,\beta)})^{\mathrm{can}}\Big) \xrightarrow{\sim} \tilde{H}^1\Big(\mathrm{Sh}(V),(E^-_{(\alpha,-\beta)})^{\mathrm{can}}\Big) \tag{5.4}$$

we need to multiply $c_{\mathbb{R}}$ by an appropriate power of i, namely by $\pm i^c$ (see also [loc. cit.], Lemma 2.8.8). This is because the naive identification of $E^-_{(\alpha,-\beta)}$ with a quotient of $\Omega^1\otimes \bar{E}^+_{(-\alpha,\beta)}$ ($^-$= complex conjugate) arising from a $\mathcal{K}$-antilinear isomorphism of the corresponding fibers at the fixed point of GK_∞, does not respect the canonical trivializations. This is justified in detail in §2.5 of [loc. cit.]. To summarize, there is a commutative diagram

$$\begin{array}{ccc} \pm i^c\cdot c_{\mathbb{R}}\colon H^0_{\bar\partial}(\mathcal{A}(GG)\otimes W^+(-\alpha,\beta;c)) & \xrightarrow{\sim} & H^1_{\bar\partial}(\mathcal{A}(GG)\otimes W^-(\alpha,-\beta;c)) \\ \downarrow \wr\| & & \downarrow \wr\| \\ c_B\colon \tilde{H}^0\Big(\mathrm{Sh}(V),(E^+_{(-\alpha,\beta)})^{\mathrm{can}}\Big) & \xrightarrow{\sim} & \tilde{H}^1\Big(\mathrm{Sh}(V),(E^+_{(\alpha,-\beta)})^{\mathrm{can}}\Big), \end{array} \tag{5.5}$$

where the vertical isomorphisms are given by $\underline{\mathrm{Lift}}^{-1}$.

Since $\underline{\mathrm{Lift}}$ is only given up to a $\mathcal{K}$-rational factor, it suffices for our purposes to have the correct isomorphism up to $\mathcal{K}$-rational factors; in particular, the sign of the top line of (5.5) doesn't matter. The power of i is included only for purposes of comparison with conjectures involving interpretations of spaces of automorphic forms as realizations of motives.

On the other hand, we can use the map $\varphi_{n-1,1}$ of II (4.1) to transfer degree one coherent cohomology classes on $\mathrm{Sh}(V)$ to degree zero classes on $\mathrm{Sh}(-V)$. The archimedean theory is discussed in II, §4. In terms of the parameters defined there, we have a $\mathcal{K}$-linear isomorphism

$$\varphi_{n-1,1}\colon \tilde{H}^1\Big(\mathrm{Sh}(V),(E^-_{(\alpha,-\beta)})^{\mathrm{can}})\Big) \xrightarrow{\sim} \tilde{H}^0\Big(\mathrm{Sh}(-V),(E_{\Lambda^+(-\alpha,\beta)^\iota})^{\mathrm{can}}\Big). \tag{5.6}$$

In the motivic theory, this map plays the role of "real Frobenius." The analogue of (5.5) is the commutative diagram (cf. [loc. cit.], Corollary 2.5.11.2)

(5.7)

$$\begin{array}{ccc} \pm i^c\cdot\phi_{n-1,1}\colon H^1_{\bar\partial}(\mathcal{A}(GG)\otimes W^-(\alpha,-\beta;c)) & \xrightarrow{\sim} & H^0_{\bar\partial}(\mathcal{A}(GG)\otimes W^+(-\alpha,\beta;c)^\iota) \\ \downarrow \wr\| & & \downarrow \wr\| \\ (\phi_{n-1,1})^*\colon \tilde{H}^1\Big(\mathrm{Sh}(V),(E^-_{(\alpha,-\beta)})^{\mathrm{can}})\Big) & \xrightarrow{\sim} & \tilde{H}^0\left(\mathrm{Sh}(-V),(E_{\Lambda^+(-\alpha,\beta)^\iota})^{\mathrm{can}})\right), \end{array}$$

Here the map $(\varphi_{n-1,1})^*$ can be defined by this diagram, and the substantive assertion is that $(\varphi_{n-1,1})^*$ so defined comes from a bundle map with the rationality property described in Proposition 2.5.11 of [loc. cit.].

Define $(E^+_{(-\alpha,\beta)})^{(2)}$ to be the pullback from $\mathrm{Sh}(V)\times$ $\mathrm{Sh}(-V)$ to $\mathrm{Sh}(V^{(2)})$ of the external tensor product $E^+_{(-\alpha,\beta)}\otimes E_{\Lambda^+(-\alpha,\beta)^\iota}$. Let $\mathrm{Sh}(V^2)$ be the Shimura subvariety of $\mathrm{Sh}(V)\times$ $\mathrm{Sh}(V)$ associated to $G(U(V)\times U(V))$ and the homomorphism $h^{r,s}\times h^{r,s}$, by analogy with the discussion in 3.12. Thus $\mathrm{Sh}(V^2)$ is obtained from $\mathrm{Sh}(V^{(2)})$ by complex conjugation in the second factor. The external tensor product $E^+_{(-\alpha,\beta)}\otimes E^-_{(\alpha,-\beta)}$ on $\mathrm{Sh}(V)\times$ $\mathrm{Sh}(V)$ defines by pullback an automorphic vector bundle, also denoted $E^+_{(-\alpha,\beta)} \otimes E^-_{(\alpha,-\beta)}$ on $\mathrm{Sh}(V^2)$. The map $(\varphi_{n-1,1})^*$ of (5.7) then provides an isomorphism (upon pullback from $\mathrm{Sh}(V)\times \mathrm{Sh}(V)$)

$$(5.8) \qquad 1 \times (\varphi_{n-1,1})^* \colon \tilde{H}^1\Big(\mathrm{Sh}(V^2), \big(E^+_{(-\alpha,\beta)} \otimes E^-_{(\alpha,-\beta)}\big)^{\mathrm{can}}\Big) \xrightarrow{\sim} \tilde{H}^0\Big(\mathrm{Sh}(V^{(2)}), \big(E^+_{(-\alpha,\beta)}\big)^{(2),\mathrm{can}}\Big).$$

The map (5.8) is $\mathbb{C}$-linear but definitely does not preserve the $\mathcal{K}$-rational structures of the two sides.

But now the pairing (4.1) defines a canonical $\mathcal{K}$-valued linear form on the left-hand side of (5.8). The corresponding linear form on the right-hand side is denoted

$$(5.9) \qquad I_\Delta \colon \tilde{H}^0\Big(\mathrm{Sh}(V^{(2)}), \big(E^+_{(-\alpha,\beta)}\big)^{(2),\mathrm{can}}\Big) \to \mathbb{C}.$$

IV. Small automorphic representations of $\mathbf{U(n-1,1)}$

1. Definitions

In this section, $\mathcal{K}$ is an imaginary quadratic field and V is a skew-hermitian space over $\mathcal{K}$ of signature $(n-1,1)$ at the unique archimedean place. We let $G = U(V)$ and $G' = U(W)$ where W is a one-dimensional hermitian space over $\mathcal{K}$. The automorphic representations of $G = U(V)$ to be described here are, in one sense, the natural generalization to higher $U(n-1,1)$ of the extensions to $U(1,1)$ of the classical binary theta functions on $SL(2)$ attached to algebraic Hecke characters η of $\mathcal{K}^\times$. In particular, they contribute to cohomology of the associated Shimura variety $\mathrm{Sh}(V)$ (with twisted coefficients) in degree 1, unless η is of finite order. So do the "small automorphic representations" considered here, and their local components at infinity are the ladder representations described in II. However, the binary theta functions on $U(1,1)$ correspond to automorphic representations all of whose local components are tempered. When $n > 1$ the small automorphic representations are square-integrable but everywhere non-tempered, and are in fact the most degenerate of all infinite-dimensional representations. When $n = 3$, these representations have been extensively studied by Rogawski in [**Rog90, Rog92b**]; they are the representations which occur in A-packets, and they can be either cuspidal or residual. For $n \geq 4$, the small representations are generally residual, rather than cuspidal; however, we will not be making use of this fact.

For our purposes, a small automorphic representation of G is any representation of the form $\Theta_{\psi,\chi,\chi'}(G' \to G; \eta)$, where the notation $\Theta_{\psi,\chi,\chi'}$ is as in I, §2. Here G' runs over the set of all unitary groups of dimension 1 (equivalently, ψ runs over the set of all additive characters of $\mathbf{A}/\mathbb{Q}$), and η runs over the set of all characters of $G'(\mathbb{Q})\backslash G'(\mathbf{A})$.

When the weight η_∞ satisfies a certain inequality, recalled at the beginning of §5, the local component at ∞ of $\Theta_{\psi,\chi,\chi'}(\eta)$ is cohomological, as we have already seen. When $n > 2$ the Baily-Borel compactification $\mathrm{Sh}(V)^*$ of $\mathrm{Sh}(V)$ has isolated singularities. Letting $j\colon \mathrm{Sh}(V) \to \mathrm{Sh}(V)^*$ denote the natural embedding, a standard computation (cf. [**B**]) shows that, for any local system $\mathcal{L}$ on $\mathrm{Sh}(V)$, the natural map

$$H^1(\mathrm{Sh}(V)^*, j_{*!}\mathcal{L}) \to H^1(\mathrm{Sh}(V), \mathcal{L})$$

is an isomorphism, where $j_{*!}\mathcal{L}$ is the middle perverse extension of $\mathcal{L}$ (the intersection complex of $\mathcal{L}$). It follows from the Zucker conjecture (due in this case to Zucker [**Z**]) that $H^1(\mathrm{Sh}(V), \mathcal{L})$ consists of square-integrable forms. It follows that

PROPOSITION 1.1. *Suppose the local component $\Theta_{\psi,\chi,\chi'}(\eta)_\infty$ is of cohomological type. Then the automorphic representation $\Theta_{\psi,\chi,\chi'}(\eta)$ is contained in the space of square-integrable automorphic forms on $G(\mathbb{Q})\backslash G(\mathbf{A})$.*

We will see this directly in §5. The local hypothesis translates into a hypothesis on the parameter on η_∞ and the local components of χ and χ', which we make explicit in §5. When $n \geq 3$ the hypothesis on η_∞ is not even necessary, although for small $|k|$ the representation $\Theta_{\varphi,\chi,\chi'}(\eta)$ is no longer cohomological.

(1.2). Formulas for the unramified correspondence. We specialize the formulas of I,3.15 to the present situation, in which π is a character. The formulas look slightly different for n even and n odd. When n is even we take $\chi' \equiv 1$, as before. When n is odd we take $\chi' = \chi^{\pm 1}$. The answer is most conveniently expressed in terms of the standard L-function at a prime p at which all data are unramified.

PROPOSITION 1.3. *Let p be a prime at which χ, η, and ψ are all unramified. Then the standard L-function of $\Theta_{\chi,\chi'}(\eta)\colon = \Theta_{\psi,\chi,\chi'}(\eta)$ is given by the following formula*

(i) If n is even, $\chi' \equiv 1$ and

$$L_p(s, \Theta_{\chi,\chi'}(\eta), St) = L_{\mathcal{K},p}(s, \tilde{\eta}^{-1}\chi)\cdot L_{\mathcal{K},p}(s,\chi)\cdot \prod_{j=1}^{\frac{n}{2}-1} [L_{\mathcal{K},p}(s,\chi)\|\cdot\|^j) L_{\mathcal{K},p}(s,\chi\|\cdot\|^{-j})]$$

(ii) If n is odd then

$$L_p(s, \Theta_{\chi,\chi'}(\eta), \mathrm{St}) = L_{\mathcal{K},p}(s, \tilde{\eta}^{-1}\cdot\chi/\chi')\cdot \prod_{j=1}^{\frac{n-1}{2}} [L_{\mathcal{K},p}(s,\chi\|\cdot\|^{j/2}) L_{\mathcal{K},p}(s,\chi\|\cdot\|^{-j/2})]$$

We leave it to the reader as an exercise in base change to verify that the central characters in Proposition 1.3 are compatible with the formulas in I,§3. Bearing in mind that $\tilde{\chi} = \chi^2$, we also obtain the following relation between $\Theta_{\chi,\chi'}(\check{\eta})$ and $\check{\Theta}_{\chi,\chi'}(\eta)$, which is consistent with Lemma I.2.5:

COROLLARY 1.4. *Under the hypotheses of Proposition 1.3, we have*

$$\Theta_{\chi,\chi'}(\check{\chi}'\check{\eta}) = \chi\circ\det\otimes\check{\Theta}_{\chi,\chi'}(\eta).$$

2. The $(\mathbf{U(1)}, \mathbf{U(1)})$ lift

The local and global theory of the theta lift from $U(1)$ to $U(1)$ has been worked out in several places, including [**Mo**], [**Rog92b**], [**GR**], [**Ku94**], [**Br**], and [**HKS**]. The article [**HKS**] analyzes the local lifting from the point of view of the doubling method, as in I, §2. The most detailed treatment of the global theory is contained in [**Br**], who also uses the doubling method. We review these results and add a few details on the archimedean pairing, which will be necessary in the arithmetic applications. In what follows, V and V' are one-dimensional vector spaces over $\mathcal{K}$, with V skew-hermitian and V' hermitian, as usual. The doubled spaces $2V$ and $2V'$ are defined as in I, §1. The notation G, G', H, and H' are as in I, §1.

First fix a place v of $\mathbb{Q}$. We work with the local seesaw pair I.2.3 and with the splitting I.2.4 associated to a pair of characters (χ, χ') of $\mathcal{K}_v^\times$, both of which restrict to $\varepsilon_{\mathcal{K}_v/\mathbb{Q}_v}$ on $\mathbb{Q}_v^\times$. We further assume that our choice of splittings conforms to our convention (cf. A.2) that $\chi' = \chi$ or $\chi' = \check{\chi} = \chi^{-1}$. We will only be considering the v-component of the global theta lift $\Theta(G \times G \to G' \times G'; *) = \Theta_{\psi,\chi,\chi'}(G \times G \to G' \times G'; *)$, where for $*$ we take a character adapted to pairing with the character of $\Delta(G')$ that lifts to the Eisenstein series on H. More precisely, if $G = U(V)$, we let $G^- = U(-V)$. Comparing with I.2.8, we see that $*$ has to be of the form $\beta \otimes \check{\chi} \cdot \check{\beta}$, where β is a character of G_v. The following lemma is then an immediate consequence of I.3.4 and I.3.5.

LEMMA 2.1. Let β be a character of G_v such that

$$\Theta(G_v \times G_v^- \to G'_v \times G'_v; \beta \otimes \check{\chi} \cdot \check{\beta}) \neq 0.$$

(i) If $\chi' = \chi$, then $\Theta(G_v \times G_v^- \to G'_v \times G'_v; \beta \otimes \check{\chi} \cdot \check{\beta}) = \check{\beta} \otimes \chi \cdot \beta$.

(ii) If $\chi' = \check{\chi}$, then $\Theta(G_v \times G_v^- \to G'_v \times G'_v; \beta \otimes \check{\chi} \cdot \check{\beta}) = \check{\chi} \cdot \check{\beta} \otimes \beta$.

Here it is understood that the right-hand sides of (i) and (ii) denote the representation spaces generated by the respective characters.

(2.2). We now look more closely at the archimedean pairing. The starting point, as always, is the doubled pair $(U(2V_\infty), U(V'_\infty)) \cong (U(1,1), U(1))$. Here we write $U(1)$ for $U(0,1)$ or $U(1,0)$, although these two groups have very different behavior. The relevant maximal compact subgroup of $U(1,1)$ is $U(1,0) \times U(0,1) = U(V_\infty) \times U(-V_\infty)$, which we denote K_∞. The theta correspondence for $(U(2V_\infty), U(V'_\infty))$ is realized on the space of K_∞-finite vectors in the Schwartz space $\mathcal{S}(\mathbb{C}) = \mathcal{S}(\mathcal{K}_\infty)$; the theta correspondence for $(U(V_\infty) \times U(-V_\infty), U(V'_\infty) \times U(V'_\infty))$ is realized on the same space. Denote by $\mathcal{S}_0(\mathbb{C})$ the space of K_∞-finite vectors. Howe duality for the latter pair, which is easy in the present case, implies that the action of $U(1,0) \times U(0,1)$ on $\mathcal{S}_0(\mathbb{C})$ is multiplicity-free. In other words, $\mathcal{S}_0(\mathbb{C})$ is canonically isomorphic to a sum of one-dimensional eigenspaces for $(U(V_\infty) \times U(-V_\infty), U(V'_\infty) \times U(V'_\infty))$ with distinct eigenvalues.

In fact, the decomposition in this case is classical. We work out the numbers on the basis of our chosen splittings. Let $\chi_n \colon U(1) \to U(1)$ be the nth power of the identity character, $n \in \mathbb{Z}$. We assume the archimedean component χ_∞ of the splittings character χ is normalized so that its restriction to $U(1)$ equals χ_1 or χ_{-1}; this certainly has the correct restriction to $\mathbb{R}^\times \cap U(1)$. We write $\Theta_{i,j}(\cdot) = \Theta_{\chi_i,\chi_j}(\cdot)$ for $i, j = \pm 1$. The formula above yields:

(2.2.1) $$\Theta_{i,i}(\chi_n \otimes \chi_{-i-n}) = \chi_{-n} \otimes \chi_{i+n}; \quad \Theta_{i,-i}(\chi_n \otimes \chi_{-i-n}) = \chi_{-i-n} \otimes \chi_n.$$

These equations are to be read conditionally: if $\Theta_{i,j}(\chi_n \otimes \chi_{i-n}) \neq 0$, then it is given by the formula above. The vanishing of $\Theta_{i,j}(\cdot)$ for given depends on n and on the signatures of V_∞ and V'_∞, where the signature of the skew-hermitian space V_∞ is determined by the choice of $i = \sqrt{-1}$ used to define the additive character $\psi_\infty(x) = e^{2\pi i x}$.

Thus, for each n such that $\Theta_{i,j}(\chi_n \otimes \chi_{-i-n}) \neq 0$, there is a unique K_∞-finite Schwartz function $\varphi_{i,j}^n(z) \in \mathcal{S}_0(\mathbb{C})$ which is an eigenfunction for $(U(V_\infty) \times U(-V_\infty), U(V'_\infty) \times U(V'_\infty))$ with eigenvalue $(\chi_{-n} \otimes \chi_{i+n}, \chi_{-n} \otimes \chi_{i+n})$ (if $i = j$) or with eigenvalue $(\chi_{-n} \otimes \chi_{i+n}, \chi_{-i-n} \otimes \chi_n)$ (if $i = -j$). This only corresponds to a part of $\mathcal{S}_0(\mathbb{C})$, namely, to those functions on which the diagonal subgroup $\Delta(U(V'_\infty))$ acts via the character χ_j. Thus

$$\bigoplus_n \mathbb{C}\varphi_{i,j}^n = \Theta(U(V'_\infty)) \to U(2V_\infty), \chi_{-j}). \tag{2.2.2}$$

The Schwartz functions $\varphi_{i,j}^n$ will be determined in the sequel to this paper as eigenfunctions of an explicit second-order differential operator. In the following section we will determine the set of n that occur. This is not difficult but the explicit formulas will be needed in calculating the archimedean zeta integrals.

(2.3). Rationality. The global theta lift $\Theta(G \times G^- \to G' \times G'; \beta \otimes \chi \cdot \check{\beta})$ is one-dimensional and generated by a character. Brocco [**Br**] has obtained an exact formula for the value at the identity of the theta lift $\theta_\varphi(G \times G^- \to G^- \times G'; \beta \otimes \chi \cdot \check{\beta})$, depending on the choices of splittings and, of course, on the Schwartz-Bruhat function φ. The value at the identity measures the rationality of the lifting which can be viewed either as an automorphic form on a zero-dimensional Shimura variety associated to $G' \times G'$ or simply as a function. The former point of view is more natural in some respects, because the value turns out to be a period invariant for zero-dimensional Shimura varieties, but it is easier to work with functions at this stage.

Here is Brocco's theorem, a special case of the Rallis inner product formula for unitary groups (see also [**HKS**, §8], for the local calculation):

THEOREM 2.3.1. *Let $\varphi = \bigotimes_v \varphi_v \in \mathcal{S}(V \otimes_\mathcal{K} V'(\mathbf{A}))$. Let β be a Hecke character of G. Let S be a finite set of places of $\mathbb{Q}$, including ∞, every place at which $\beta, \chi, \mathcal{K}$, or ψ ramifies, and every place at which φ_v is not the characteristic function of the ring of integers $\mathcal{O}_{\mathcal{K},v}$. Then*

$$\begin{aligned}\mathcal{O}_{\varphi,\chi,\chi^{\pm1}}&(G \times G^- \to G' \times G'; \beta \otimes \check{\chi} \cdot \check{\beta})(1) \\ &= \left[\prod_{v \in S} \tilde{Z}_v\left(\frac{1}{2}, \varphi, \chi, \beta\right)\right] L\left(\frac{1}{2}, \chi \cdot \tilde{\beta}\right) \cdot L(1, \varepsilon_\mathcal{K})^{-1}.\end{aligned}$$

Here

$$\tilde{Z}_v\left(\frac{1}{2}, \varphi, \chi, \beta\right) = Z_v\left(\frac{1}{2}, \beta, \check{\beta}, \mathcal{F}_\varphi, \chi\right) \cdot L_v(1, \varepsilon_\mathcal{K}) \cdot L_v\left(\frac{1}{2}, \chi \cdot \tilde{\beta}\right)^{-1} \quad \text{if } v \text{ is finite,}$$

$$\tilde{Z}_\infty\left(\frac{1}{2}, \varphi, \chi, \beta\right) = Z_\infty\left(\frac{1}{2}, \beta, \check{\beta}, \mathcal{F}_\varphi, \chi\right),$$

where Z_v is a standard zeta integral as in I.1.13 and $\mathcal{F}_\varphi$ is a Siegel-Weil section defined as in I.2.12.

Brocco expresses his theorem as a formula for the inner product of two theta lifts from G to G', but the present version is equivalent to his formulation. As indicated, the result is the same whether $\chi' = \chi$ or χ^{-1}.

The next lemma clarifies the rationality properties of Brocco's formula. We write $\varphi = \varphi_\infty \otimes \varphi_f$, where $\varphi_f \in \mathcal{S}(V \otimes_\mathcal{K} V'(\mathbf{A}^f))$. We assume φ_∞ is a scalar multiple of one of the $\varphi_{i,j}$'s described above; specifically, we assume φ_∞ is the multiple with the property that $\varphi_{i,j}(0) = 1$. (It will be recalled in the proof of the following lemma that $\varphi_{i,j}(0) \neq 0$, so this is consistent.) The indices i, j are uniquely determined by β_∞ and χ_∞ and the hypothesis, which we make here, that the archimedean theta lift is non-trivial. Say φ is *algebraic* if φ_f takes values in $\bar{\mathbb{Q}}$. Then $\mathrm{Gal}(\bar{\mathbb{Q}}/\mathbb{Q})$ acts on the space of algebraic φ by acting on the values of φ_f. Likewise, $\mathrm{Gal}(\bar{\mathbb{Q}}/\mathbb{Q})$ acts on the set of algebraic Hecke characters as in III, §2. We recall that the splitting character χ is not algebraic, since its restriction to $\mathbb{R}^\times$ is the sign character, but $\chi^+ := \chi \cdot \|\cdot\|^{1/2}$ is algebraic. We thus let $\mathrm{Gal}(\bar{\mathbb{Q}}/\mathcal{K})$ act on χ by setting $\chi^\sigma = \|\cdot\|^{-1/2} \cdot \chi^{+,\sigma}$.

LEMMA 2.3.2. *Assume φ is algebraic. Then $\tilde{Z}(\varphi, \chi, \beta) = \prod_{v \in S} \tilde{Z}_v(1/2, \varphi, \chi, \beta) \in \bar{\mathbb{Q}}$, and, for any $\sigma \in \mathrm{Gal}(\bar{\mathbb{Q}}/\mathbb{Q})$, we have*

$$\tilde{Z}(\varphi, \chi, \beta)^\sigma = \tilde{Z}(\varphi^\sigma, \chi^\sigma, \beta^\sigma).$$

In particular, if φ_f takes values in $\mathbb{Q}(\beta, \chi) := \mathbb{Q}(\beta) \cdot \mathbb{Q}(\chi)$, then $\tilde{Z}(\varphi, \chi, \beta)$ can be taken to be an element of $\mathbb{Q}(\beta, \chi)$ in the sense indicated at the beginning of III, §2.

PROOF. The rationality of the non-archimedean factors has already been treated in [**H93**, 6.2.7]. The assertion for Z_∞ is easy, since φ_∞ is taken to be an eigenvector; it can be found in [**B**, §2.7], taking into account the different conventions regarding measures. Brocco's proof also implies that $\varphi_{i,j}(0) \neq 0$, as mentioned above.

Recall that $\chi^+(-1) = -1$ whereas $\tilde{\beta}(-1) = 1$. It follows that $\chi^+ \cdot \tilde{\beta}$ is a critical Hecke character, in the sense of III, §2. Moreover, $\chi^+ \cdot \tilde{\beta}|_{\mathbb{R}^\times}$ is the character $t \mapsto t$. It follows from the standard normalizations (cf. [**H97**, §1.6]) that $L(s, \chi^+ \cdot \tilde{\beta})$ is the L- function of a CM motive of weight -1, and thus 0 is a critical value of $L(s, \chi^+ \cdot \tilde{\beta})$. Since $L(1/2, \chi \cdot \tilde{\beta}) = L(0, \chi^+ \cdot \tilde{\beta})$, the corollary below is thus a consequence of III, (2.1) and (2.4):

COROLLARY 2.3.3. *Let $\Phi \in \sum_{\mathcal{K}}$ be the embedding with respect to which $\chi^+ \cdot \tilde{\beta}$ is critical. Suppose φ is algebraic and φ_f takes values in $\mathbb{Q}(\beta, \chi)$. Let $\chi' = \chi^{\pm 1}$. Set*

$$\theta_{\varphi,\chi,\chi'}(\beta \otimes \check{\chi} \cdot \check{\beta}) = \theta_{\varphi,\chi,\chi'}(G \times G^- \to G' \times G'; \beta \otimes \check{\chi} \cdot \check{\beta}).$$

Then

$$(2\pi i \cdot g(\varepsilon_{\mathcal{K}})) \cdot p((\chi^+ \cdot \tilde{\beta}), \iota\Phi) \cdot \theta_{\varphi,\chi,\chi'}(\beta \otimes \check{\chi} \cdot \check{\beta})(1) \in \mathbb{Q}(\beta, \chi)$$

and its $\mathrm{Gal}(\bar{\mathbb{Q}}/\mathcal{K})$-conjugates are obtained by the natural action on φ, β, and χ:

$$\begin{aligned}&[(2\pi i \cdot g(\varepsilon_{\mathcal{K}})) \cdot p((\chi^+ \cdot \tilde{\beta}), \iota\Phi) \cdot \theta_{\varphi,\chi,\chi'}(\beta \otimes \check{\chi} \cdot \check{\beta})(1)]^\sigma \\ &\qquad = (2\pi i \cdot g(\varepsilon_{\mathcal{K}})) \cdot p((\chi^+ \cdot \tilde{\beta})^\sigma, \iota\Phi) \cdot \theta_{\varphi^\sigma,\chi^\sigma,\chi'^{,\sigma}}(\beta^\sigma \otimes \check{\chi}^\sigma \cdot \check{\beta}^\sigma)(1).\end{aligned}$$

Here $g(\varepsilon_{\mathcal{K}})$ is a standard Gauss sum attached to the quadratic character $\varepsilon_{\mathcal{K}}$.

REMARK 2.3.4. It has already been noted in the discussion preceding Lemma I.2.5 that the Schwartz-Bruhat space $\mathcal{S}(V \otimes_{\mathcal{K}} V'(\mathbf{A}))$ is the same whether the doubling is done with respect to V or V', i.e., whether the lifting is from $G \times G^-$ to $G' \times G'$ or from $G' \times G'^{,-}$ to $G \times G$. In particular, the definition of algebraic Schwartz-Bruhat functions is *independent of the direction of lifting*. We will be using this fact constantly when we invoke the seesaw reciprocity formula to determine rationality of holomorphic automorphic forms.

(2.4). We have written our results in terms of theta lifts from unitary groups, but in the applications we will need theta lifts on unitary similitude groups. In the present instance, both hermitian spaces are odd-dimensional, so the discussion in I, §4 shows that the lift from characters of $(G\times G^{-})(\mathbb{Q})\backslash(G\times G^{-})(\mathbf{A})$ to characters of $(G'\times G')(\mathbb{Q})\backslash(G'\times G')(\mathbf{A})$ extends to a lift from characters of $G(G\times G^{-})(\mathbb{Q})\backslash G(G\times G^{-})(\mathbf{A})$ to characters of $G(G'\times G')(\mathbb{Q})\backslash G(G'\times G')(\mathbf{A})$. The actual integral defining the similitude lifting is the same as the integral defining the usual theta lifting [**H93**, §3.8]. In particular, the value of the theta lift at $g'=1$ is the same either way. It follows that, if $\tilde{\beta}$ is an extension of $\beta\otimes\check{\chi}\cdot\check{\beta}$ to a character of $(G\times G^{-})(\mathbb{Q})\backslash(G\times G^{-})(\mathbf{A})$, then

$$(2\pi i\cdot g(\varepsilon_K))\cdot p((\chi^{+}\cdot\tilde{\beta}),\iota\Phi)\cdot\theta_{\varphi,\chi,\chi'}(\tilde{\beta})(1)\in\mathbb{Q}(\beta,\chi)$$

and the transformation law is the same as in Corollary 2.3.3. We use this fact without further comment in what follows.

3. The archimedean theta lift for $\mathbf{U(1)}\times\mathbf{U(1)}$

Since $U(1)$ is compact, the theory of the archimedean theta correspondence is elementary in this case. In fact, it comes down to decomposing the oscillator representation of $\tilde{S}p(2,\mathbb{R})$ over the subgroup $U(1,1)$, embedded by means of the splitting associated to the character χ_∞. The table that follows can be worked out by the reader as an exercise in holomorphic representations, or it can be found as a special case of [**L90**, (32)].

Thus we consider the theta lift from $G=U(p,q)$ to $G'=U(r,s)$, where $p+q=r+s=1$. We first consider the lift on the double covers $\tilde{G}$ and $\tilde{G}'$, with half-integral parameters as in II, §3. The various combinations are worked out in turn. We write $\Theta(\omega)=\Theta(G\to G';\omega)$, when ω is a character of G.

(3.1) $(p,q)=(r,s)=(1,0)$ (resp. $(0,1)$) $\omega=(k+1/2;)$ (resp. $(;k+1/2)$). Then $\Theta(\omega^{-1})=0$ if $k<0, \Theta(\omega^{-1})=\omega$ if $k\geq 0$.

(3.2) $(p,q)=(1,0),(r,s)=(0,1),\omega=(k-1/2;)$. Then $\Theta(\omega^{-1})=0$ if $k>0, \Theta(\omega^{-1})=(;k-1/2)$ if $k\leq 0$.

The remaining case is symmetric for (3.2) and is omitted. Note that, for G fixed, every character with weight in $\mathbb{Z}+1/2$ lifts to either $U(1,0)$ or $U(0,1)$, but not both; this is an archimedean example of the *theta dichotomy* worked out for p-adic unitary groups in [**HKS**].

Given the choice of splitting characters χ,χ', these formulas translate into pairings between integral characters, as follows:

(3.3a). As in (3.1), with $\chi=\chi'$. Then $\Theta(\omega^{-1})$ is either 0 or ω. If $a(\chi)=a(\chi')=1$ then $\Theta(\omega^{-1})\neq 0$ if and only if $\omega=(k+1;)$ with $k\geq 0$. If $a(\chi)=a(\chi')=-1$ then $\Theta(\omega^{-1})\neq 0$ if and only if $\omega=(k+1;)$ with $k\geq -1$.

(3.3.b). As in (3.1), with $\chi=\check{\chi}'$. Then $\Theta(\omega^{-1})$ is either 0 or $\check{\chi}\cdot\omega$. The non-triviality of $\Theta(\omega^{-1})$ is determined by χ as in (3.3.a).

(3.4.a). As in (3.2), with $\chi=\chi'$. Then $\Theta(\omega^{-1})$ is either 0 or ω. If $a(\chi)=a(\chi')=1$ then $\Theta(\omega^{-1})\neq 0$ if and only if $\omega=(k-1;)$ with $k\leq 1$. If $a(\chi)=a(\chi')=-1$ then $\Theta(\omega^{-1})\neq 0$ if and only if $\omega=(k-1;)$ with $k\leq 0$.

(3.4b). As in (3.2), with $\chi = \check{\chi}'$. Then $\Theta(\omega^{-1})$ is either 0 or $\check{\chi}\cdot\omega$. The non-triviality of $\Theta(\omega^{-1})$ is determined by χ as in (3.4.a).

4. Rationality of the lift from $\mathbf{U(1)\times U(1)}$ to $\mathbf{U(n-1,1)\times U(1,n-1)}$

Let G' be as above, and define $G = U(V)$, $G^- = U(-V)$. We now give ourselves a Hecke character η of G' and splitting characters χ, χ' as in 1.2; here $\chi' \equiv 1$ if n is even, $\chi' = \chi^{\pm 1}$ if χ is odd. In this section G' is assumed to be a unitary group of signature (1,0). We suppose the parameter of η^{-1}, in the convention of II, §3, to be equal to $k+(n-2)/2$ if n is even, resp. $k+(n-1)/2$ if n is odd, with $k \in \mathbb{Z}$. It then follows from the formulas in II, §3 that the theta lift $\Theta_{\chi,\chi'}(\eta) := \Theta_{\psi,\chi,\chi'}(G' \to G;\eta)$ is a space of holomorphic modular forms, as is

$$\begin{aligned}(4.1)\qquad \Theta_{\chi,\chi'}(\eta\otimes\check{\chi}'\cdot\check{\eta}) &:= \Theta_{\psi,\chi,\chi'}(G'\times G' \to G\times G^-;\eta\otimes\check{\chi}'\cdot\check{\eta})\\ &\cong \Theta_{\chi,\chi'}(\eta)\otimes\chi\circ\det\otimes\check{\Theta}_{\chi,\chi'}(\eta).\end{aligned}$$

Our purpose here is to show that, for φ an appropriately normalized Schwartz-Bruhat function, $\theta_{\varphi,\chi,\chi'}(\eta\otimes\check{\chi}'\cdot\check{\eta}) = \theta_{\varphi,\chi,\chi'}(G'\times G' \to G\times G^-;\eta\otimes\check{\chi}'\cdot\check{\eta})$ is an arithmetic modular form over the field of coefficients of η and χ, multiplied by an explicit complex factor.

Note that twisting by $1\otimes\chi\circ\det$ yields an isomorphism of vector spaces

$$(4.2)\qquad \Theta_{\chi,\chi'}(\eta)\otimes\check{\Theta}_{\chi,\chi'}(\eta) \cong \Theta_{\chi,\chi'}(\eta)\otimes\chi\circ\det\otimes\check{\Theta}_{\chi,\chi'}(\eta)$$

preserving the subspaces of holomorphic forms. Since χ is not a motivic Hecke character — the restriction of χ_∞ to $\mathbb{R}^\times$ is $\varepsilon_{\mathbb{C}/\mathbb{R}}$ = the sign character — it is simplest to ignore this twist and to calculate the rationality of forms on the right-hand side of (4.2) in terms of the left-hand side, via this isomorphism. In particular, we will assume without further comment that the forms $\theta_{\varphi,\chi,\chi'}(\eta\otimes\check{\chi}\cdot\check{\eta})$ have been twisted back and belong to $\Theta_{\chi,\chi'}(\eta)\otimes\check{\Theta}_{\chi,\chi'}(\eta)$.

For this we appeal to the theory recalled in III, §3, especially Proposition 3.12.1. For the purposes of notation, we let $\pi_{\chi,\chi'}(\eta)$ be some extension of $\Theta_{\chi,\chi'}(\eta)$ to a space of automorphic forms on $\mathrm{GU}(V)$ of the type considered in III: namely, the space generated by the lifts of holomorphic sections of some automorphic vector bundle, say E_Λ. This last condition, which depends only on the choice of central character at ∞, can always be satisfied, since $\Theta_{\chi,\chi'}(\eta)$ is a space of holomorphic modular forms. Alternatively, we can start with a Hecke character η' of the similitude group $G(G'\times G')$ whose restriction to $G'\times G'$ equals $\eta\otimes\check{\chi}'\cdot\check{\eta}$, and let

$$\pi_{\chi,\chi'}(\eta)^{(2)} = \Theta_{\psi,\chi,\chi'}(G(G'\times G') \to G(G\times G);\eta') \text{ (notation as in I, §4),}$$

twisted back as in (4.2) by $1\otimes(\chi\circ\det)^{-1}$. We can take $[\pi_{\chi,\chi'}(\eta)\otimes\check{\pi}_{\chi,\chi'}(\eta)]_0$, in the notation of III, §3.12, to be an irreducible constituent of $\pi_{\chi,\chi'}(\eta)^{(2)}$. Remark III.3.13 shows that the rationality of forms in $[\pi_{\chi,\chi'}(\eta)\otimes\check{\pi}_{\chi,\chi'}(\eta)]_0$ depends only on $\Theta_{\chi,\chi'}(\eta)\otimes\check{\Theta}_{\chi,\chi'}(\eta)$.

The bundle parameter Λ introduced in the previous paragraph is only a place-keeper. Indeed, in the calculations below, it suffices to know that the holomorphic sections of E_Λ whose rationality we test are obtained by lifting from $U(1)$. This presupposes some sign conditions, resumed in (4.3), below. Of course, Λ can be determined explicitly in terms of the parameter k of η and the integer $\alpha(\chi)$, as in

II, §3. Specifically, if $k \geq 0$, then II (3.9) implies that, in the notation of II, (1.12),

$$\begin{aligned}\Lambda &= \Lambda^+\Big(1-k, -\frac{1}{2}(\alpha(\chi)+1), c\Big)\\ &= \Big(-\frac{1}{2}(\alpha(\chi)+1), \dots, -\frac{1}{2}(\alpha(\chi)+1)-k; -\frac{1}{2}(1-\alpha(\chi)); c\Big)\end{aligned}$$

for some c (determined by the restriction of η_∞ to $\mathbb{R}^\times$, but of no importance here). If $k < 0$ then formula II (3.10) implies that, in the notation of II, (1.14)

$$\begin{aligned}\Lambda &= T^+\Big(1-k, -\frac{1}{2}(\alpha(\chi)+1), c\Big)\\ &= \Big(-\frac{1}{2}(\alpha(\chi)+1), \dots, -\frac{1}{2}(\alpha(\chi)+1); 1-k-\frac{1}{2}(\alpha(\chi)+1); c\Big).\end{aligned}$$

For obvious reasons, we prefer to write Λ in what follows without specifying whether it is a Λ^+ or a T^+. For applications to rationality it suffices to note that, writing $\Lambda = (\lambda(1), \dots, \lambda(n-1); \lambda(n); c)$,

$$\begin{aligned}2\lambda(i) + \alpha(\chi) &< 0 \quad \text{if } 1 \leq i \leq n-1,\\ 2\lambda(n) + \alpha(\chi) &> 0\end{aligned} \tag{4.3}$$

in every instance.

In the applications to the theta lift from $U(n-1,1)$ to $U(r,s)$, E_Λ will be one of the bundles $[W_i]$ introduced just before Lemma III.4.8. Once Λ has been determined, the forms $\theta_{\varphi,\chi,\chi'}(\eta \otimes \check{\chi}' \cdot \check{\eta})$ define holomorphic sections of the bundle denoted $E_{\Lambda(2)}$ in III, 3.12.

Note first that, in order to show that a form θ is rational over a given number field L, it suffices to show that, if p is a prime number and $\mathbb{Q}^{(p)}$ is the field of all p-th power roots of unity, then θ is rational over $L \cdot \mathbb{Q}^{(p)}$ for two distinct primes p. We next observe that, if p is any prime number, then strong approximation for $SU(V)$ implies that we can take $\mathfrak{G} = GG^{(2)}(\mathbb{Q}_p)$ in the statement of Prop. III.3.12.1. Summarizing, it suffices to show that, for there are two primes p such that, for any $\gamma \in GG^{(2)}(\mathbb{Q}_p)$, and any Hecke character ω of type Λ, we have

$$\mathbf{p}(\tilde{\omega}, h^{n-1,1}) \cdot \mathbf{L}_{\omega(2)}(r_\gamma(\theta_{\varphi,\chi,\chi'}(\eta \otimes \check{\chi}' \cdot \check{\eta}))) \in (\mathbb{Q}(\omega) \otimes L \cdot \mathbb{Q}^{(p)})^\times / \mathbb{Q}(\omega)^\times. \tag{4.4}$$

As in [**H93**, §5], the next step is to observe that the action r_γ on $\theta_{\varphi,\chi,\chi'}(\eta \otimes \check{\chi}' \cdot \check{\eta})$ can be shifted to an action on the Schwartz-Bruhat function φ in a given Schrödinger model, which we have yet to specify. In other words, at the cost of varying φ, we can assume $\gamma = 1$ in (4.4). This works exactly as in [loc. cit.] and should require no additional explanation. However, here our theta correspondences are defined in terms of unitary groups alone, whereas in the application of Prop. III.3.12.1 we have to allow γ to run through $GG^{(2)}(\mathbb{Q}_p)$. We thus suppose the oscillator representation has been extended to the similitude groups, as in I, §4. For our purposes, it is simplest to consider only primes p that split in $\mathcal{K}$; otherwise the natural extension of the oscillator representation may only be to a subgroup of $GG^{(2)}(\mathbb{Q}_p)$ of index two, as explained in I, §4. Since any two primes will do for application of our criterion, this is no problem.

We now carry out the computation of (4.4) with $\gamma = 1$. As in [**H93**], we use the principle of seesaw pairs to translate (4.4) into an integral involving the theta lift of $\omega(2)$ to

$$\begin{aligned}G(G' \times G')^n &= \{(g_1, \dots, g_n) \in GG'^{(2)} \times \cdots \times GG'^{(2)} \ (n \text{ times})\\ &\quad \mid \nu(g_1) = \cdots = \nu(g_n)\}.\end{aligned}$$

The seesaw diagram, ignoring the similitude factors, is as follows:

(4.5)

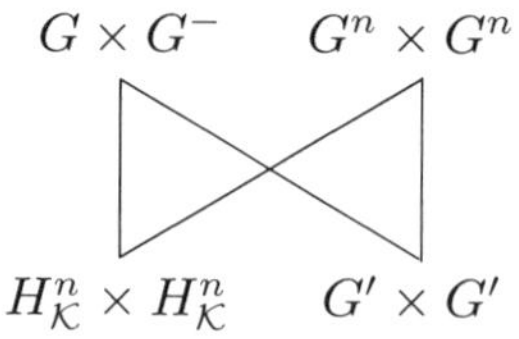

where the first and second $H^n_{\mathcal{K}}$ are associated with complex-conjugate Shimura varieties. In terms of the diagrams in Appendix 1, (4.5) is a part of the diagram involving lines (3) and (4).

As in I, §2, we write $\mathbb{W} = R_{\mathcal{K}/E}(V \otimes_{\mathcal{K}} V')$ and define the complete polarization $2\mathbb{W} = \mathbb{X} \oplus \mathbb{Y}$, with $\mathcal{S}(\mathbb{X}) \cong \mathcal{S}(V \otimes_{\mathcal{K}} V')$. The subgroup $H^n_{\mathcal{K}} \subset G$ corresponds to a decomposition $V \cong V_1 \oplus \cdots \oplus V_n$ with $\dim V_\beta = 1$, and thus to a decomposition

$$\mathbb{X} \cong (V_1 \otimes_{\mathcal{K}} V') \oplus \cdots \oplus (V_n \otimes_{\mathcal{K}} V'),\ \mathcal{S}(\mathbb{X}) \cong \mathcal{S}(V_1 \otimes_{\mathcal{K}} V') \otimes \cdots \otimes \mathcal{S}(V_n \otimes_{\mathcal{K}} V')$$

(completed tensor product). The seesaw diagram (4.5) allows us to assume our Schwartz-Bruhat function φ is factorizable as a function of the form $\varphi_1 \otimes \cdots \otimes \varphi_n$, where $\varphi_\beta \in \mathcal{S}(V_\beta \otimes_{\mathcal{K}} V')$. Moreover, we can assume

(4.6) (i) Each φ_β is algebraic and $\varphi_{\beta,f}$ takes values in $\mathbb{Q}(\eta, \chi^+)$.
(ii) $\varphi_{\beta,\infty} = \varphi_{i,j}^{\lambda(\beta)}, \beta = 1, \ldots, n$ (notation §2.2).

Here the splitting characters $\chi_i = \chi_i(\beta), \chi_j = \chi_j(\beta)$ for the pair $(U(V_\beta), U(V'))$ are chosen compatibly with the pair (χ, χ'), as explained in Appendix 2. Any $\mathbb{Q}(\eta, \chi^+)$-linear combination of factorizable functions $\varphi \in \mathcal{S}(\mathbb{X})$ satisfying (i) and (ii) of (4.6) will be called *algebraic* over $\mathbb{Q}(\eta, \chi^+)$ of type Λ.

We let $\omega(2)(\chi) = \omega(2) \cdot (1 \otimes \check{\chi} \circ \det|_{GH^n_{\mathcal{K}}})$ where we view $GH^n_{\mathcal{K}}$ as a subgroup of GG and $GH^{n,(2)}_{\mathcal{K}}$ as a subgroup of $GH^n_{\mathcal{K}} \times GH^n_{\mathcal{K}}$. Write $Z = Z_{GG'^{(2)}}$. Proposition I.5.4 then yields

$$\begin{aligned}
(4.7) \qquad & \mathbf{L}_{\omega(2)}(\theta_{\varphi,\chi,\chi'}(\eta \otimes \check{\chi}' \cdot \check{\eta})) \\
&= \int_{GG'^{(2)}(\mathbb{Q}) \cdot Z(A) \backslash GG'^{(2)}(A)} \theta_{\varphi,\chi',\chi}(\omega(2)(\chi)) \cdot (\eta \otimes \check{\chi}' \cdot \check{\eta}) \\
&= \begin{cases} \theta_{\varphi,\chi',\chi}(\omega(2)(\chi))(1) & \text{if } \theta_{\varphi,\chi',\chi}(\omega(2)(\chi)) \in \mathbb{C} \cdot (\check{\eta} \otimes \chi' \cdot \eta), \\ 0 & \text{otherwise.} \end{cases}
\end{aligned}$$

The last condition translates via Lemma 2.1 into a condition on ω. At this point it is convenient to regard ω as the restriction to $GH^n_{\mathcal{K}}$ of an n-tuple $(\omega_1, \ldots, \omega_n)$ of Hecke characters of $\mathcal{K}^\times$. Then the condition is that

$$(4.8) \qquad \omega_1 \cdot \ldots \cdot \omega_n = \eta \cdot \chi^e,$$

for some integer e that depends on whether or not $\chi = \chi'$.

We may view $H^n_{\mathcal{K}}$ as the stabilizer in G of the decomposition $V = V_1 \oplus \cdots \oplus V_n$ of one-dimensional hermitian spaces. Thus $H^n_{\mathcal{K}} \cong U(V_1) \times \cdots \times U(V_n)$, where V_i is positive definite for $i = 1, \ldots, n-1$ and V_n is negative-definite. Then $\theta_{\varphi,\chi',\chi}(\omega(2)(\chi))$, restricted to $GG'(\mathbf{A})$, is obtained as the naive product of the theta lifts of the $\omega_i \otimes \check{\chi}\check{\omega}_i$ for the dual reductive pairs $(GG^{(2)}(V_i), GG')$.

Comparing III, 2.5 with (4.3) above, we see that, if ω is of type Λ, then the following condition is satisfies:

$$\begin{aligned}(4.9)\qquad & \chi^+\tilde{\omega}_i \text{ is critical for the CM type } \{i_0\}i = 1,\dots,n-1;\\ & \chi^+\tilde{\omega}_n \text{ is critical for the CM type } \{\iota\}.\end{aligned}$$

Applying Corollary 2.3.3, we thus obtain

$$(4.10)\qquad (2\pi i\cdot g(\varepsilon_{\mathcal{K}}))^n\cdot\prod_{i=1}^{n-1}p((\chi^+\cdot\tilde{\omega}_i),\iota)p((\chi^+\cdot\tilde{\omega}_n),i_0)$$
$$\cdot\,\mathbf{L}_{\omega(2)}(\theta_{\varphi,\chi,\chi'}(\eta\otimes\check{\chi}'\cdot\check{\eta}))\in\mathbb{Q}(\omega,\chi^+).$$

Here it is essential that we have chosen our Schwartz-Bruhat functions as indicated in (4.6). Using the formulas in II, §3, and recalling that $\tilde{\beta}^\iota = \tilde{\beta}^{-1}$ for any β, we may rewrite (4.10) as

$$(4.11)\qquad [(2\pi i\cdot g(\varepsilon_{\mathcal{K}}))^n\cdot p(\chi^+,\iota)^{n-1}p(\chi^+,i_0)]\mathbf{p}(\tilde{\omega},h^{n-1,1})^{-1}$$
$$\cdot\,\mathbf{L}_{\omega(2)}(\theta_{\varphi,\chi,\chi'}(\eta\otimes\check{\chi}'\cdot\check{\eta}))\in\mathbb{Q}(\omega,\chi^+).$$

Moreover, these values transform in the obvious way under the action of $\mathrm{Gal}(\bar{\mathbb{Q}}/\mathcal{K})$ (a precise statement is given below).

Let $p(\chi^+,h^{1,n-1}) = p(\chi^+,\iota)^{n-1}p(\chi^+,i_0)$. Using Proposition 3.12.1, we can summarize this discussion as follows:

PROPOSITION 4.12. *Under the above hypotheses, the cohomology class*

$$[(2\pi i\cdot g(\varepsilon_{\mathcal{K}}))^n\cdot p(\chi^+,h^{1,n-1})]\cdot\underline{\mathrm{Lift}}^{-1}\theta_{\varphi,\chi,\chi'}(\eta\otimes\check{\chi}'\cdot\check{\eta})\in\tilde{H}^0\Big(\mathrm{Sh}\big(V^{(2)}\big),E_{\Lambda(2)}\Big)$$

is rational over $\mathbb{Q}(\eta,\chi^+)$. Here Λ is the parameter described above and $\underline{\mathrm{Lift}}^{-1}$ is the isomorphism of III.4.14 relative to the highest weight Λ. Moreover, if $\sigma\in\mathrm{Gal}(\bar{\mathbb{Q}}/\mathcal{K})$, then

$$\begin{aligned}&([(2\pi i\cdot g(\varepsilon_{\mathcal{K}}))^n\cdot p(\chi^+,h^{1,n-1})]\cdot\underline{\mathrm{Lift}}^{-1}\theta_{\varphi,\chi,\chi'}(\eta\otimes\check{\chi}'\cdot\check{\eta}))^\sigma\\ &\quad=[(2\pi i\cdot g(\varepsilon_{\mathcal{K}}))^n\cdot p(\chi^{+,\sigma},h^{1,n-1})]\cdot\underline{\mathrm{Lift}}^{-1}\theta_{\varphi^\sigma,\chi^\sigma,\chi'^{,\sigma}}(\eta^\sigma\otimes\check{\chi}'^{,\sigma}\cdot\check{\eta}^\sigma).\end{aligned}$$

PROOF. By Proposition 3.12.1 it suffices to verify that the intersection of the $\mathbb{Q}(\omega,\chi^+)$, as ω runs through the set of Hecke characters of type Λ, is contained in $\mathbb{Q}(\eta,\chi^+)$. But the ω_i's range over all n-tuples of type Λ, subject only to condition (4.8). The first statement is now clear, and the second follows in the same way from the description of the Galois action in Corollary 2.3.3.

REMARK 4.13. Our rationality criterion in Proposition 3.12.1 is stated in terms of irreducible $GG'(\mathbf{A}^f)$-modules. However, as explained in IV, §4, the representation $\Theta_{\chi,\chi'}(\eta\otimes\check{\chi}'\cdot\check{\eta})$ may not be irreducible, for the (spurious) reason that local Howe duality has not been established in residue characteristic two, and (more seriously) because of the unnatural procedure of extension by zero when n is even. This procedure is not necessarily stable under the right action of $GG'(\mathbf{A}^f)$, but only under a subgroup of index two.

However, suppose $F\in\tilde{H}^0(\mathrm{Sh}(V^{(2)}),E_{\Lambda(2)})$ belongs to the $\mathbb{Q}(\eta,\chi^+)$-linear span of the $GG'(\mathbf{A}^f)$-translates of $\underline{\mathrm{Lift}}^{-1}(\theta_{\varphi,\chi,\chi'}(\eta\otimes\check{\chi}'\cdot\check{\eta}))$, where φ is as (4.6). It is obvious that the $\mathbf{L}_{\omega(2)}(r_\gamma(F))$, for $\gamma\in GG'(\mathbf{A}^f)$, are $\mathbb{Q}(\eta,\chi^+)$-linear combinations of expressions of the form $\mathbf{L}_{\omega(2)}(\theta_{\varphi,\chi,\chi'}(\eta\otimes\check{\chi}'\cdot\check{\eta}))$, as φ varies among the algebraic Schwartz-Bruhat functions (as in 4.6). It then follows that

$$[(2\pi i\cdot g(\varepsilon_{\mathcal{K}}))^n\cdot p(\chi^+,h^{1,n-1})]\cdot F$$

is again rational over $\mathbb{Q}(\eta,\chi^+)$.

The $\mathbb{Q}(\eta,\chi^+)$-linear span of the $GG'(\mathbf{A}^f)$-translates of $\Theta_{\chi,\chi'}(\eta\otimes\check{\chi}'\cdot\check{\eta})$ breaks up as a finite sum of irreducible $GG'(\mathbf{A}^f)$-modules (finite because the theta lift in residue characteristic two is of finite type [**MoVW**, p. 48] and because $GG'(\mathbf{A}^f)\cap GG'(\mathbf{A})_0$ is of finite index in $GG'(\mathbf{A}^f)$). It follows from the above discussion that Proposition 3.12.1 applies to the "rational" elements of each of these irreducible $GG'(\mathbf{A}^f)$-modules.

(4.14). It is easy to see that (4.9) remains valid for any ω such that

$$\mathbf{L}_{\omega(2)}(\theta_{\varphi,\chi,\chi'}(\eta\otimes\check{\chi}'\cdot\check{\eta}))\neq 0;$$

i.e., if Λ is replaced by any of the weights of the representation defining E_Λ. As we have noted, it suffices to test rationality using characters of type Λ. However, for the final calculations, we need to know that the algebraic Schwartz-Bruhat functions of the type defined by (4.6) work for the lowest weight space as well as the highest weight space. But this is especially easy. The lowest weight vector of the representation with highest weight Λ is

$$\tilde{\Lambda}=\Big(-\frac{1}{2}(\alpha(\chi)+1)-k,-\frac{1}{2}(\alpha(\chi)+1),\dots,\\ -\frac{1}{2}(\alpha(\chi)+1);-\frac{1}{2}(1-\alpha(\chi));c\Big) \tag{4.14.1}$$

when $\Lambda=\Lambda^+(1-k,-(1/2)(\alpha(\chi)+1),c)$. When $\Lambda=T^+(1-k,-(1/2)(\alpha(\chi)+1),c)$, E_Λ is a line bundle and the lowest weight $\tilde{\Lambda}$ equals the highest weight Λ. In both cases, if $\Lambda=(\lambda(1),\dots,\lambda(n-1);\lambda(n);c)$ as before, then

$$\tilde{\Lambda}=(\lambda(n-1),\lambda(n-2),\dots,\lambda(1);\lambda(n);c).$$

In particular, (4.9) is still valid.

It follows that (4.10) remains true if ω is taken to be a character of type $\tilde{\Lambda}$, provided the Schwartz-Bruhat function φ satisfies

$$\varphi_{\beta,\infty}=\varphi_{i,j}^{\lambda(n-1-\beta)},\beta=1,\dots,n-1;\varphi_{n,\infty}=\varphi_{i,j}^{\lambda(n)}, \tag{4.14.2}$$

instead of (4.6)(ii). Therefore, Proposition 4.12 remains valid if φ is taken to be algebraic in this modified sense, and if $\underline{\mathrm{Lift}}^{-1}$ is taken to be the isomorphism of III.4.14 relative to the lowest weight $\tilde{\Lambda}$.

5. Petersson norms of small holomorphic rational forms

We have implicitly assumed that the theta lift $\Theta_{\chi,\chi'}(\eta\otimes\check{\chi}'\cdot\check{\eta})$ is a non-trivial space of automorphic forms. We now assume explicitly that $\Theta_{\chi,\chi'}(\eta\otimes\check{\chi}'\cdot\check{\eta})$ is a space of automorphic forms of cohomological type, not just a space with $\bar{\partial}$-cohomology. In other words, assuming $\alpha(\chi')=1$, as at the end of II, §3, we take the parameter of η^{-1}, in the convention of II, §3, to be equal to $\pm(k+(n-2)/2)$ if n is even, resp. either $k+(n-1)/2$ or $-k-(n-3)/2$ if n is odd, with k a *positive* integer.

We assume φ is algebraic over $\mathbb{Q}(\eta,\chi^+)$ and of type Λ (or $\tilde{\Lambda}$, cf, 4.14), as above. The next step is to compute the integral $\theta_{\varphi,\chi,\chi'}(\eta\otimes\check{\chi}'\cdot\check{\eta})$ along the diagonal $\Delta(GG)(\mathbb{Q})Z_{GG}(\mathbf{A})\backslash\Delta(GG)(\mathbf{A})$ of $GG^{(2)}(\mathbf{A})$; we use the shorthand notation

$\int_{GG} dg$ as in III, §4 for this integral. Just as in §2, the calculation is based on the doubling L-function, this time via the seesaw diagram

(5.1)

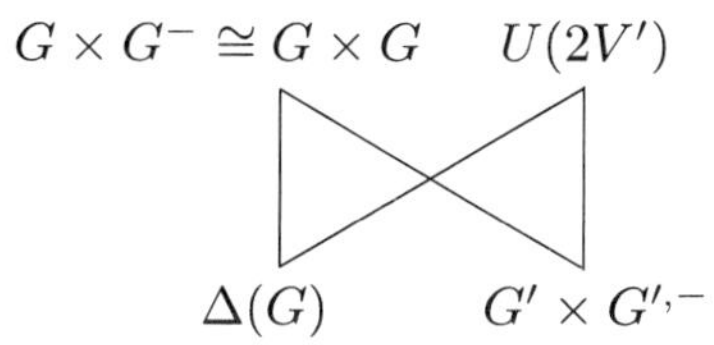

(lines 4 and 5 of Appendix 1). Here we are doubling V' rather than V, as in §4 (and, implicitly, in §3); as already observed in I, §2, this does not change the polarization. So our theta functions naturally live on $G \times G$ rather than on the isomorphic group $G \times G^-$. The isomorphism on the upper left corner of (5.1) corresponds at the level of coherent cohomology to the map $1 \times (\varphi_{n-1,1})^*$ of III (5.8). In other words, for the appropriate choice of η, $\theta_{\varphi,\chi,\chi'}(\eta \otimes \check{\chi}' \cdot \check{\eta})$ is naturally (holomorphic) $\times$ (anti-holomorphic), resp. (anti-holomorphic) $\times$ (holomorphic), on $G \times G$; via the transfer to $G \times G^-$ the form becomes pure holomorphic, resp. pure anti-holomorphic. In the discussion of rationality in §4 we were viewing $\theta_{\varphi,\chi,\chi'}(\eta \otimes \check{\chi}' \cdot \check{\eta})$ as a holomorphic form on $G \times G^-$; now we view it as a form of mixed type on $G \times G$. The effect of this change of complex structure on rationality has been discussed in III, §5 and will be reviewed below.

Seesaw reciprocity (I, Proposition 5.4) yields the formula

$$\begin{aligned}(5.2)\qquad &\int_{GG} \theta_{\varphi,\chi,\chi'}(\eta \otimes \check{\chi}' \cdot \check{\eta}) \cdot \chi \circ \det\, dg \\ &\qquad = \int_{G(G' \times G',^-)} \theta_{\varphi,\chi',\chi}(\chi \circ \det) \cdot (\eta \otimes \check{\chi}' \cdot \check{\eta})\, dg.\end{aligned}$$

Here, of course, $\int_{G(G' \times G',^-)}$ is also used to denote an adelic integral. We let $I_{\Delta,\chi}(\theta_{\varphi,\chi,\chi'}(\eta \otimes \check{\chi}' \cdot \check{\eta}))$ denote the left-hand side of (5.2).

Using the Siegel-Weil formula I (2.11), the right-hand side of (5.2) can be rewritten

$$(5.3)\qquad \int_{G(G' \times G',^-)} E\left(\cdot, \frac{1}{2}(n-1), \mathcal{F}_\varphi, \chi'\right) (\eta \otimes \check{\chi}' \cdot \check{\eta})\, dg$$

where $\mathcal{F}_\varphi$ is the standard Siegel section defined in I, §2. Applying the basic identity $I(1.11)$, we thus find that

$$\begin{aligned}(5.4)\qquad &I_{\Delta,\chi}(\theta_{\varphi,\chi,\chi'}(\eta \otimes \check{\chi}' \cdot \check{\eta})) \\ &= \tilde{Z}_S(s, f, f', \mathcal{F}, \chi) \cdot L\left(\frac{1}{2}n, \chi' \cdot \tilde{\eta}\right) / d_1\left(\frac{1}{2}(n-1), \chi'\right) \\ &= \tilde{Z}_S(s, \eta, \check{\eta}, \mathcal{F}_\varphi, \chi') \cdot L\left(\frac{1}{2}n, \chi' \cdot \tilde{\eta}\right) / L(n, \varepsilon_{\mathcal{K}}^n).\end{aligned}$$

The same reasoning as in §2 then shows that the local factor $\tilde{Z}_S$ is rational over $\mathbb{Q}(\eta, \chi)$, and we obtain

$$(5.5)\qquad I_{\Delta,\chi}(\theta_{\varphi,\chi,\chi'}(\eta \otimes \check{\chi}' \cdot \check{\eta})) \sim_{\mathbb{Q}(\eta,\chi);\mathcal{K}} L\left(\frac{1}{2}n, \chi' \cdot \tilde{\eta}\right) / (2\pi i)^n.$$

LEMMA 5.6. *Suppose the automorphic representation* $\Theta_{\chi,\chi'}(\eta \otimes \check{\chi}' \cdot \check{\eta})$ *is of cohomological type at infinity. Then* $n/2$ *is a critical value of the L-function* $L(s, \tilde{\eta})$ *if* n *is even, and* $(n-1)/2$ *is a critical value of the L-function* $L(s, \chi'^{,+}\tilde{\eta})$ *if* n *is odd.*

PROOF. At the beginning of this section we recalled that the hypothesis that $\Theta_{\chi,\chi'}(\eta \otimes \check{\chi}' \cdot \check{\eta})$ is of cohomological type corresponds to an inequality that must be satisfied by the parameter of η, and thus of $\chi'^{,+} \cdot \tilde{\eta}$. The lemma then follows from a simple calculation and from the inequalities II (2.6).

REMARK 5.7. We note that, if the parameters of η is taken to be equal to $(n-2)/2$, then $n/2$ is no longer a critical value for $L(s, \tilde{\eta})$ in the even case, and that the analogous statement holds true in the odd case. In other words, the boundary of the critical strip precisely corresponds to the boundary of the range of cohomological lifts. Thus, when $s = m$ is an integer to the right of the center of symmetry, the Siegel-Weil formula interprets $L(m, \tilde{\eta})$ as the Petersson norm of a lifted form, via Rallis' inner product formula, as we have seen in §2 and as we will see again presently. However, these Petersson norms are of forms of *non-cohomological* type, and thus have no obvious motivic interpretation.

This phenomenon was pointed out to me by J.-S. Li, who observed that it is true more generally for all theta lifts of discrete series representations of unitary groups.

Since the hypothesis Lemma 5.6 is assumed to be in force, we can continue (5.5). We obtain

$$(5.8) \quad I_{\Delta,\chi}(\theta_{\varphi,\chi,\chi'}(\eta \otimes \check{\chi}' \cdot \check{\eta})) \sim_{\mathbb{Q}(\eta,\chi);\mathcal{K}} \begin{cases} (2\pi i)^{n/2} p(\tilde{\eta}, \Phi_\eta)/(2\pi i)^n = p(\tilde{\eta}, \Phi_\eta)/(2\pi i)^{n/2}, & n \text{ even} \\ p(\chi'^{,+} \cdot \tilde{\eta}, \iota\Phi_\eta)^{-1}/(2\pi i)^{(n+1)/2} g(\varepsilon_{\mathcal{K}})^n), & n \text{ odd,} \end{cases}$$

where $\chi'^{,+} = \chi' \cdot \| \cdot \|^{1/2}$, as before. Here $\Phi \in \{i_0, \iota\}$ is the embedding with respect to which $\tilde{\eta}$, (resp. $\chi'^{,+} \cdot \tilde{\eta}$) is critical.

We note the following consequence of the above calculation:

COROLLARY 5.9. *Under the hypothesis* $n \geq 3$, *the theta-lifts* $\theta_{\varphi,\chi,\chi'}(\eta \otimes \check{\chi}' \cdot \check{\eta})$ *are square-integrable for any triple of characters* η, χ, and χ'.

In particular, this provides an alternative justification of Proposition 1.1, without the hypothesis on weights, for the small automorphic representations that interest us.

We next double the preceding doubled diagram and consider

$$\theta^{(2)}(\eta \otimes \check{\chi}' \cdot \check{\eta}) := \theta_{\varphi(2),\chi,\chi'}(G' \times G'^{,-} \times G'^{,-} \times G' \to G \times G^- \times G \times G^-; \\ \eta \otimes \check{\chi}' \cdot \check{\eta} \otimes \check{\chi}' \cdot \check{\eta} \otimes \eta).$$

Here $\varphi(2)$ is $\varphi_1 \otimes \varphi_2$ where φ_1 is a function used to lift $\eta \otimes \check{\chi}' \cdot \check{\eta}$ and φ_2 is the function used to lift $\check{\chi}' \cdot \check{\eta} \otimes \eta$ in the preceding discussion. Here we have twice used the isomorphism $G \cong G^-$ to define the theta lift in this non-standard form. It follows from III,5.7 that, under the identification of forms with coherent cohomology classes, this change of complex structure corresponds to multiplying by $(\pm i^c)^2 = \pm 1$. Up to $\mathcal{K}$-rational (but not necessarily $\mathbb{Q}$-rational!) factors, this sign can and will be ignored.

Before we continue, we need to specify the splittings. Write

$$G(G \times G^- \times G \times G^-)) = G(G(1) \times G(1)^- \times G(2) \times G(2)^-))$$

and let $G(1) = U(V(1))$, $G(2) = U(V(2))$, where $V(1)$ and $V(2)$ are two copies of the same hermitian space. We will be calculating the integral of a lifted form over the subgroup $G(G\times G^-)$, viewed as the diagonal in $G(G(1)\times G(1)^-\times G(2)\times G(2)^-))$, via the diagonal embeddings

$$GG \to G(G(1) \times G(2)),\ GG^- \to G(G^-(1) \times G^-(2));$$

i.e., the signs are assumed to match up. In order to relate this integral to the calculation (5.8), we therefore need to assume that the splittings of $G(1)$ and $G(2)$ (resp. $G^-(1)$ and $G^-(2)$) are related as in doubling diagram I.2.4. In other words, if $G(1) = U(V(1))$, $G(2) = U(V(2))$, we must have

$$i_{V(2),V',\chi,\chi'} = (\chi')^{-1} \cdot i_{V(1),V',\chi,\chi'}. \tag{5.10}$$

It is then easy to see from I.3.5 that we have

$$\begin{aligned}\theta^{(2)}(\eta \otimes \check{\chi}' \cdot \check{\eta}) = {} & \theta_{\varphi_1,\chi,\chi'}(G \times G'^{,-} \to G \times G^-; \eta \otimes \check{\chi}' \cdot \check{\eta}) \\ & \otimes \theta_{\varphi_2,\check{\chi},\check{\chi}'}(G' \times G'^{,-} \to G \times G^-; \check{\chi}' \cdot \check{\eta} \otimes \eta).\end{aligned} \tag{5.11}$$

Note that the splitting character has been inverted in the second factor.

With our conventions for complex structures, it follows from the discussion in II, §3 that the function $\theta^{(2)}$ is either of the type (holo)⊗(holo)⊗(anti-holo)⊗(anti-holo) or of type (anti-holo)⊗(anti-holo)⊗(holo)⊗(holo), depending on the sign of η. We define $\int_{G(G\times G^-)} dgdg^-$ by analogy with $\int_{GG} dg$ and let

$$I_{\Delta(2),\chi}(\theta^{(2)}(\eta \otimes \check{\chi}' \cdot \check{\eta})) = \int_{G(G\times G^-)} \theta^{(2)}(\eta \otimes \check{\chi}' \cdot \check{\eta}) \cdot \chi \circ \det(g \cdot g^-)\, dgdg^-.$$

Then it follows from (5.8) that if n is even then (writing η^{-1} instead of $\check{\eta}$)

$$\begin{aligned}I_{\Delta(2),\chi}(\theta^{(2)}(\eta \otimes \check{\chi}' \cdot \check{\eta})) & \sim_{\mathbb{Q}(\eta,\chi);\mathcal{K}} [p(\tilde{\eta}, \Phi_\eta)/(2\pi i)^{n/2}] \\ & \cdot [p(\tilde{\eta}^{-1}, \Phi_{\eta^{-1}})/(2\pi i)^{n/2}] \sim_{\mathbb{Q}(\eta,\chi);\mathcal{K}} p(\tilde{\eta}, \Phi_\eta)^2/(2\pi i)^n,\end{aligned} \tag{5.12}$$

since $\Phi_{\eta^{-1}} = \iota\Phi_\eta$. Similarly, if n is odd then

$$\begin{aligned}I_{\Delta(2)\chi}(\theta^{(2)}(\eta \otimes \check{\chi}' \cdot \check{\eta})) & \sim_{\mathbb{Q}(\eta,\chi);\mathcal{K}} [p(\chi'^{,+} \cdot \tilde{\eta}, \iota\Phi_\eta)^{-1}] \\ & \cdot [p(\chi'^{,+} \cdot (\tilde{\chi}' \cdot \tilde{\eta})^{-1}, \iota\Phi_{(\chi'\cdot\eta)^{-1}})^{-1}]/((2\pi i)^{(n+1)} g(\varepsilon_{\mathcal{K}})^{2n}) \\ & \sim_{\mathbb{Q}(\eta,\chi);\mathcal{K}} [p(\chi'^{,+} \cdot \tilde{\eta}, \iota\Phi_\eta)^{-1}] \cdot [p((\chi'^{,\iota})^+ \cdot \tilde{\eta}^\iota, \iota\Phi_{(\chi'\cdot\eta)^{-1}})^{-1}]/(2\pi i)^{(n+1)} \\ & \sim_{\mathbb{Q}(\eta,\chi),\mathcal{K}} [p(\chi'^{,+} \cdot \tilde{\eta}, \iota\Phi_\eta)^{-1}] \cdot [p((\chi'^{,+} \cdot \tilde{\eta}, \Phi_{(\chi'\cdot\eta)^{-1}})^{-1}]/(2\pi i)^{(n+1)} \\ & \sim_{\mathbb{Q}(\eta,\chi);\mathcal{K}} p(\chi'^{,+} \cdot \tilde{\eta}, \iota\Phi_\eta)^{-2}/(2\pi i)^{(n+1)}\end{aligned} \tag{5.13}$$

by III, (2.2), since it is easy to see that $\Phi_{(\chi'\cdot\eta)^{-1}} = \iota\Phi_\eta$. Applying (2.1) and (2.3) several times, and recalling that $(\chi'^{,+})^2 = (\chi')^2 \cdot \|\cdot\|$, where $(\chi')^2$ is a motivic character trivial on the ideles of $\mathbb{Q}$, we can simplify further to obtain

$$I_{\Delta(2),\chi}\big(\theta^{(2)}(\eta \otimes \check{\chi}' \cdot \check{\eta})\big) \sim_{\mathbb{Q}(\eta,\chi);\mathcal{K}} p\big((\chi')^2, \Phi_\eta\big) \cdot p(\tilde{\eta}, \Phi_\eta)^2/(2\pi i)^n, n \text{ odd}. \tag{5.14}$$

In §4 we determined that, if the theta lift $\theta_{\varphi,\chi,\chi'}(\eta \otimes \check{\chi}' \cdot \check{\eta})$ is holomorphic on $G \times G^-$, then, up to an explicit product of CM periods, it is arithmetic over the appropriate field, provided φ is algebraic over the same field. The remainder of this section is devoted to the corresponding assertion for anti-holomorphic lifts.

We thus return to the situation of a theta lift from $G' \times G'^{,-}$ to $G \times G^-$, as at the beginning of this section. We suppose $\Theta_{\chi,\chi'}(\eta \otimes \check{\chi}' \cdot \check{\eta})$ to be an anti-holomorphic representation of $G \times G^-$ of cohomological type. It follows from the discussion in II, §3 that this requires V' to be of signature (0,1), and then the parameters of $\Theta_{\chi,\chi'}(\eta \otimes \check{\chi}' \cdot \check{\eta})_\infty$ are given by II (3.11). In particular, the parameter of η^{-1} is given by $-k-(n-2)/2$ (n even), or $-k-(n-3)/2$ (n odd), with $k > 0$. As in §4, we twist by $(1 \otimes \chi \circ \det)^{-1}$ to identify $\Theta_{\chi,\chi'}(\eta \otimes \check{\chi}' \cdot \check{\eta}) \cong \Theta_{\chi,\chi'}(\eta) \otimes \check{\Theta}_{\chi,\chi'}(\eta)$, for purposes of rationality testing. Via this identification, $\Theta_{\chi,\chi'}(\eta \otimes \check{\chi}' \cdot \check{\eta})$ defines a space of coherent cohomology in degree 2 of the bundle $E_{\Lambda(2)}$, where

$$\Lambda = \Lambda^-(k-1, \frac{1}{2}(1-\alpha(\chi)); c) \quad \text{for some } c$$

in the notation of II, §1; note that what is here denoted $-k$ corresponds to what we called k in II, (1.12) and (3.11). Here we should remark that the construction of the bundle $E_{\Lambda(2)}$ on the Shimura variety associated to a group with real points isomorphic to $G(U(n-1,1) \times U(1,n-1))$ works as well for degree 2 coherent cohomology as for holomorphic global sections; this is because the change of complex structure entails complex conjugation of the holomorphic tangent space as well as the fiber of the bundle E_Λ.

We recall from III, §5 that, for any extension L of $\mathbb{Q}(\eta, \chi^+)$, the L-rational elements of $\tilde{H}^2(\mathrm{Sh}(V^{(2)}), E_{\Lambda(2)})$ are those that yield L-rational values upon pairing with L-rational elements of $\tilde{H}^0(\mathrm{Sh}(V^{(2)}), E_{\Lambda^+(2)})$, where

$$\begin{aligned}\Lambda^+ &= \Lambda^+\left(1-k, -\frac{1}{2}(1-\alpha(\chi)); -c\right) \\ &= \Lambda^+\left(1-k, -\frac{1}{2}(1+\alpha(\chi)); -c-n\alpha(\chi)\right) + (\alpha(\chi), \dots, \alpha(\chi); n\alpha(\chi)).\end{aligned}$$

Recall that the split central parameter $-c$ has to be congruent to the sum of the remaining parameters; this is why we have introduced the term $-n\alpha(\chi)$ on the right-hand side of the above equality. Recall also that the central character of the split part of the center of $\mathrm{GU}(V)$ makes no difference in the doubling process. We let $\Lambda^+(\chi) = \Lambda^+(1-k, -1/2(1+\alpha(\chi)); -c-n\alpha(\chi)), I(\chi) = (\alpha(\chi), \dots, \alpha(\chi); n\alpha(\chi))$. Thus $\Lambda^+ = \Lambda^+(\chi) + I(\chi)$.

As in Corollary III.5.2, it suffices to check this criterion by pairing coefficients of the highest weight subspaces; more precisely, to pair a highest weight coefficient of $\tilde{H}^2(\mathrm{Sh}(V^{(2)}), E_{\Lambda(2)})$ with a lowest weight coefficient of $\tilde{H}^0(\mathrm{Sh}(V^{(2)}), E_{\Lambda^+(2)})$. In what follows, $\underline{\mathrm{Lift}}^{-1}$ designates the isomorphism III.4.14.1 that identifies scalar-valued automorphic forms with coherent cohomology classes, and of course $\underline{\mathrm{Lift}}$ denotes its inverse.

LEMMA 5.15. *Suppose $A \in \tilde{H}^0(\mathrm{Sh}(V^{(2)}), E_{\Lambda^+(2)})$ pairs non-trivially with a class in $\tilde{H}^2(\mathrm{Sh}(V^{(2)}), E_{\Lambda(2)})$ of the form $\underline{\mathrm{Lift}}^{-1}((1 \otimes \chi \circ \det)^{-1}\theta_{\varphi,\chi,\chi'}(\eta \otimes \check{\chi}' \cdot \check{\eta}))$. Then A is of the form*

$$A = \underline{\mathrm{Lift}}^{-1}((1 \otimes \check{\chi} \circ \det)^{-1} \otimes \theta_{\varphi,\check{\chi},\check{\chi}'}(\check{\eta} \otimes \chi' \cdot \eta)).$$

PROOF. This is a consequence of adjunction. Given a dual pair (G, G') and an automorphic form F on G, any automorphic form F' on G' that pairs non-trivially with a theta lift $\theta(G' \to G; F)$ has a non-trivial theta lift $\theta(G' \to G; F)$ to G. In this case, $\theta(\eta \otimes \check{\chi}' \cdot \check{\eta})$ is a theta lift from an abelian group, hence $\underline{\mathrm{Lift}}(A)$ has a theta lift to a multiple of a character ν of $G(G' \times G')$ that pairs non-trivially with $\eta \otimes \check{\chi}' \cdot \check{\eta}$. Of course we must have $\nu = \check{\eta} \otimes \chi' \cdot \eta$. Another application of adjunction implies that A is itself a theta lift of some character ξ of $G(G' \times G')$.

On the other hand, suppose A is contained in a finite sum of automorphic representations unramified outside a finite set S; we assume S contains all the places at which η, χ, and χ' are ramified. Let τ be one of these representations, and let τ_v be its local constituent at a place v not in S. Then

$$\tau_v \cong \check{\Theta}_{\chi,\chi'}(\eta \otimes \check{\chi}' \cdot \check{\eta}) \cong \Theta_{\check{\chi},\check{\chi}'}(\check{\eta}) \otimes [\chi \circ \det \otimes \Theta_{\check{\chi},\check{\chi}'}(\chi' \cdot \eta)]$$

by Proposition 1.3. This is enough to determine the local component ξ_v of ξ for v not in S, and thus determine ξ by weak approximation.

Combining this with Proposition 4.12 and the remarks on lowest weight liftings in 4.14, we find that

LEMMA 5.16. *Suppose the function φ is algebraic as in (4.6), or as in the modified sense (4.14.2). Then*

$$(2\pi i \cdot g(\varepsilon_{\mathcal{K}}))^n p(\check{\chi}^+, h^{1,n-1})\underline{\mathrm{Lift}}((1 \otimes \check{\chi} \circ \det)^{-1} \otimes \theta_{\varphi,\check{\chi},\check{\chi}'}(\check{\eta} \otimes \chi' \cdot \eta))$$

is a $\mathbb{Q}(\chi^+, \eta)$-rational element of $\tilde{H}^0(\mathrm{Sh}(V^{(2)}), E_{\Lambda^+(2)})$.

PROPOSITION 5.17. *Let η be a Hecke character of G'. We assume the parameter of η^{-1} is given by $-k-(n-2)/2$ (n even), or $-k-(n-1)/2$ (n odd), with $k > 0$, where $n = \dim(V)$. Let $\Lambda = \Lambda^-(k-1, (1/2)(1-\alpha(\chi)); c)$, and let $\tau = \tau^-(k-1, (1/2)(1-\alpha(\chi)); c)$ be the highest weight of the corresponding ladder representation, in the notation of II (1.13) and II.1.15. Suppose φ is an algebraic Schwartz-Bruhat function relative to τ, as in (4.6), and write $\theta(\eta \otimes \check{\chi}' \cdot \check{\eta})$* for *$\theta_{\varphi,\chi,\chi'}(\eta \otimes \check{\chi}' \cdot \check{\eta})$.*

Define

$$\mathbf{P}(\chi,\eta,\chi') = \begin{cases} (p(\check{\chi}^+, h^{1,n-1})p(\check{\eta}, \Phi_\eta)^2)^{-1} & \text{if } n \text{ is even,} \\ (p(\check{\chi}^+, h^{1,n-1})p(\check{\eta}, \Phi_\eta)^2 p((\chi')^2, \Phi_\eta) \cdot g(\varepsilon_{\mathcal{K}}))^{-1} & \text{if } n \text{ is odd.} \end{cases}$$

Then

(i) $\mathbf{P}(\chi,\eta,\chi')\underline{\mathrm{Lift}}^{-1}(\theta(\eta \otimes \check{\chi}' \cdot \check{\eta})) \in \tilde{H}^2(\mathrm{Sh}(V^{(2)}), E_{\Lambda(2)})$ *is rational over* $\mathbb{Q}(\eta, \chi^+)$.

(ii) *For $\sigma \in \mathrm{Gal}(\bar{\mathbb{Q}}/\mathcal{K})$, we have*

$$[\mathbf{P}(\chi,\eta,\chi')\underline{\mathrm{Lift}}^{-1}(\theta(\eta \otimes \check{\chi}' \cdot \check{\eta}))]^\sigma = \mathbf{P}(\chi^\sigma, \eta^\sigma, \chi'^{,\sigma})\underline{\mathrm{Lift}}^{-1}(\theta(\eta^\sigma \otimes \check{\chi}'^{,\sigma} \cdot \check{\eta}^\sigma)).$$

PROOF. We have to test the $\mathbb{Q}(\eta, \chi^+)$-rationality of $\theta(\eta \otimes \check{\chi}' \cdot \check{\eta})$ by pairing with $\mathbb{Q}(\eta, \chi^+)$-rational elements $A \in \tilde{H}^0(\mathrm{Sh}(V^{(2)}), E_{\Lambda^+(2)})$. It follows from Lemma 5.15 that A is itself a theta lift of the kind described in Lemma 5.16, in this instance a lift with respect to the lowest weight component of $E_{\Lambda^+(2)}$. Thus we have to calculate

$$(2\pi i \cdot g(\varepsilon_{\mathcal{K}}))^n p(\chi^+, h^{n-1,1}) \cdot I_{\Delta(2),\chi}(\theta(\eta \otimes \check{\chi}' \cdot \check{\eta}) \otimes (\chi \circ \det \otimes \chi \circ \det)^{-1} \otimes \theta_{\varphi',\chi,\chi'}(\check{\chi}' \cdot \check{\eta} \otimes \eta)),$$

where φ' is an algebraic Schwartz-Bruhat function of the appropriate type. But we have already computed the term $I_{\Delta(2),\chi}(*)$ in (5.12) and (5.14) up to $\mathbb{Q}(\eta, \chi^+)$-rational factors. Assertion (i) follows immediately, and (ii) follows as in the proof of Proposition 4.12.

V. Rationality of the lifts $(\mathbf{U(n-1,1)}, \mathbf{U(r,s)})$

1. Statement of the main theorem
2. Proof (seesaw diagram and cup products)
3. Application to conjectured period relations

1. Statement of the main theorem

We now place ourselves in the global analogue of the situation discussed at the beginning of II, §3. Thus V and V' are hermitian spaces over $\mathcal{K}$ of signature $(n-1,1)$ and (r,s), respectively, at the real place, with $\dim(V') = m = r+s \geq n = \dim(V)$. Define $r' = n-s$ as above. We start with a cuspidal automorphic representation π of $G = U(V)$ with $\pi_\infty \cong \check{\pi}_\lambda$, where λ has the parameter II.1.9 in the r'-th chamber. We fix a Hecke character χ_0 of $\mathcal{K}^\times$ such that $\chi_0|_{\mathbf{A}^\times} = \varepsilon_\mathcal{K}$ and $a(\chi_0) = 1$. Eventually χ_0 will be allowed to vary in order to obtain the correct rationality statements. We then choose splitting characters χ, χ' such that

$$\chi|_{\mathbf{A}^\times} = \varepsilon_\mathcal{K}^n, \chi'|_{\mathbf{A}^\times} = \varepsilon_\mathcal{K}^n,$$

as usual, making the assumption that χ (resp. χ') is trivial if m (resp. n) is even, and that $\chi = \chi_0$ (resp. $\chi' = \chi_0$) if m (resp. n) is odd. Then the theta correspondence attached to the splitting characters χ, χ' lifts π to an automorphic representation $\pi' = \Theta_{\psi,\chi,\chi'}(\pi)$ of $G' = U(V')$. We assume further that the parameters $\{a_i, b_j\}$, $i = 1,\dots,r'$, $j = 1,\dots,s$ associated in II, §1 to λ, or equivalently to the bundle parameter Λ, satisfy the inequality II.1.6. As explained in II, §3, π'_∞ is then of cohomological type and corresponds to a space of holomorphic automorphic forms on $G'(\mathbf{A})$.

As in the previous sections, we actually work with the theta lift of forms on the doubled similitude groups $G(G\times G^-)$ and $G(G'\times G^{\prime,-})$, where we identify $G \cong G^-$ and $G' \cong G^{\prime,-}$ as groups, as necessary, but retain the notation to distinguish the associated Shimura varieties. We thus define the automorphic representation $\pi(2)$ of $G(G\times G^-)$ as some irreducible extension to the similitude group of the representation $\pi \otimes [\check{\chi} \circ \det \otimes \check{\pi}]$. Then $\Theta(\pi(2))$ is defined as in I, §4. We recall that the theta lift to the similitude group is defined by extending to the largest possible subgroup of $G(G'\times G^{\prime,-})(\mathbf{A})$ containing $G(G'\times G^{\prime,-})(\mathbb{Q})$, then extending the lifted functions by zero, if necessary, according to the recipe (4.10). For the sake of simplicity, we make the assumption that $\Theta(\pi(2))$ is a space of cusp forms, although most of our arguments work just as well if we take $\Theta(\pi(2))$ to be a space of square-integrable forms.

For the sake of rationality computations, we regard π to begin with as an irreducible cuspidal automorphic representation of $\mathrm{GU}(V)$ and choose an irreducible $G(G\times G^-)(\mathbf{A})$-constituent $[\pi\otimes\check{\pi}]_0 \subset \pi\otimes\check{\pi}$, as in III (3.12). As in IV, §5, we fix $[\pi\otimes\check{\pi}]_0$ so that the map

$$(1.1) \qquad (1\otimes\check{\chi}\circ\det{}^{-1})\colon \pi(2) \xrightarrow{\sim} [\pi\otimes\check{\pi}]_0$$

identifies $\pi(2)$ with $[\pi\otimes\check{\pi}_0]$.

Define Λ and W_Λ in terms of λ as in II,2.8. Via the map of Theorem III.3.6, the cuspidal representation $[\pi\otimes\check{\pi}]_0$ contributes to the coherent cohomology space $H_!^{2(r'-1)}(\mathrm{Sh}(V^{(2)}), [W_{\Lambda(2)}])$ where $[W_{\Lambda(2)}]$ is the restriction to $\mathrm{Sh}(V^{(2)}) \subset \mathrm{Sh}(V)\times$

$\mathrm{Sh}(-V)$ of $[W_\Lambda]\otimes[W_{\Lambda^{\mathrm{op}}}]$, just as in III, 3.12. Via (1.1), we thus obtain a map

$$\text{(1.2)}\qquad \underline{\mathrm{Lift}}^{-1}\colon [H^{2(r'-1)}(\mathfrak{P}(2),GK_\infty(2);\pi(2)\otimes(W_\Lambda\otimes W_{\Lambda^{\mathrm{op}}})] \to H_!^{2(r'-1)}\Big(\mathrm{Sh}\big(V^{(2)}\big),[W_{\Lambda(2)}]\Big).$$

Here, evidently, the action of $GK_\infty(2)$ on $\pi(2)$ is transported from the right-hand side of (1.1). The parabolic subalgebra $\mathfrak{P}(2)$ is the one that defines the complex structure on $\mathrm{Sh}(V^{(2)})$. As in III, 4.11 — more specifically, as in the discussion preceding Proposition III.4.13 — we may identify the left-hand side of (1.2) with a space of scalar-valued automorphic forms of infinity type $\pi(2)$, which lie in the lowest weight space of the $GK_\infty(2)$-type carrying the cohomology.

Similarly, we define τ' by II.3.13, let Λ' be the highest weight of the contragredient $\check{\tau}'$ of τ', and define the automorphic vector bundles $E_{\Lambda'}$ and $E_{\Lambda'(2)}=E_{\Lambda'}\otimes E_{\Lambda',\mathrm{op}}$ on $\mathrm{Sh}(V')$ and $\mathrm{Sh}(V'^{(2)})$, respectively, as in II, §3. Thus

$$\text{(1.3)}\qquad \Lambda'=\check{\tau}'=(a_{r'}-q',\underbrace{-q',\dots,-q'}_{r-r'},-a_{r'-1}-q', \dots,-a_1-q';b_1+q',\dots,b_s+q')-(1/2)\alpha(\chi')(1,\dots,1).$$

where $q'=(n-2)/2$ and where we have neglected to specify the additional parameter for the split component of the center.

Because the Howe duality conjecture is unknown in residue characteristic two, and because of the issues discussed in I, §4, we don't know *a priori* that $\Theta(\pi(2))$ is irreducible. Thus we choose an irreducible constituent $\Theta(\pi(2))_0$ containing a $U(V'^{(2)})(\mathbf{A})$-subspace of the form $\pi'\otimes(\chi'\circ\det\otimes\check{\pi}')$. Then the analogue of (1.1) is simply a map

$$\text{(1.4)}\qquad (1\otimes\chi'\circ\det{}^{-1})\colon\Theta(\pi(2))_0\xrightarrow{\sim}[\pi'\otimes\check{\pi}']_0,$$

where the right-hand side is defined as in III 3.12. The analogue of (1.2) is a map

$$\text{(1.5)}\qquad \underline{\mathrm{Lift}}^{-1}\colon[\Theta(\pi(2))_0\otimes(W_{\Lambda'}\otimes W_{\Lambda',\mathrm{op}})]^{GK_\infty(2)'}\to H^0\Big(\mathrm{Sh}\big(V'^{,(2)}\big),[E_{\Lambda(2)}]\Big).$$

Here (1.5) is the composition of (1.4) with the map of Theorem III.3.6. As before, we can identify the left-hand side with a space of highest weight components of vector-valued holomorphic automorphic forms, as in III, 4.11.

In order to study the rationality of theta lifts, we need to introduce a class of rational theta kernels, as in Part IV. We let $\mathrm{Sp}(\mathbb{W}\oplus\mathbb{W})$ be the big symplectic group containing $U(V^{(2)})\times U(V'^{(2)})$, as in I, §2, and consider a polarization of the form $\mathbb{W}\oplus\mathbb{W}=\mathbb{X}\oplus\mathbb{Y}$ where $\mathbb{X}=R_{\mathcal{K}/\mathbb{Q}}(V\otimes_{\mathcal{K}}\Delta^-(V'))$. Then, via the seesaw diagrams of Appendix 1, we can identify $\mathcal{S}(\mathbb{X})$ as a tensor product of $n(r+s)$ Schwartz-Bruhat spaces $\mathcal{S}(\mathbb{X}_{ij})$, where

$$\text{(1.6)}\qquad \mathbb{X}_{ij}=R_{\mathcal{K}/\mathbb{Q}}(V_i\otimes_{\mathcal{K}}\Delta^-(V_j')),\ \dim_{\mathcal{K}}V_i=\dim_{\mathcal{K}}V_j'=1.$$

The algebraic elements of $\mathcal{S}(\mathbb{X}_{ij})$ have already been defined in IV, §2. We then define the algebraic elements of $\mathcal{S}(\mathbb{X})$ to be the $\bar{\mathbb{Q}}$-linear combinations of tensor products of algebraic elements of the $\mathcal{S}(\mathbb{X}_{ij})$. We can similarly define elements rational over specific number fields, as in IV, Corollary 2.3.3 and IV (4.6).

Before we can state the main theorem, we need to attach certain period invariants to the data. Let $\alpha(m)=0$ if m is even, $\alpha(m)=1$ if m is odd. If n is even, we let

$$\mathbf{p}(\pi, V, V', \chi_0) = (2\pi i)^{([m/2]-m)n} \cdot p(\chi_0^+, h^{1,n-1})^{-\alpha(m)} \cdot p(\tilde{\xi}_\pi \cdot (\tilde{\chi}_0)^{(n/2)\alpha(m)}, \iota). \tag{1.7}$$

If n is odd and m is even, we let

$$\mathbf{p}(\pi, V, V', \chi_0) = (2\pi i)^{-mn/2} \cdot p(\tilde{\xi}_\pi \cdot (\tilde{\chi}_0)^{-m/2+s}, \iota). \tag{1.8}$$

Finally, if n and m are both odd, we let

$$\mathbf{p}(\pi, V, V', \chi_0) = (2\pi i)^{-(m+1)n/2} \cdot (g(\varepsilon_{\mathcal{K}})p(\chi_0^+, h^{1,n-1}))^{-1} \cdot p(\tilde{\xi}_\pi \cdot (\tilde{\chi}_0)^{(n-m)/2+s}, \iota). \tag{1.9}$$

(In 1.8 and 1.9, it may be supposed that the signature (r, s) is determined by the archimedean component of π.)

Here is the main theorem.

THEOREM 1.10. *Let $\alpha \in H_!^{2(r'-1)}(\mathrm{Sh}(V^{(2)}), [W_{\Lambda(2)}])$ be a coherent cohomology class rational over the number field L. Suppose $\alpha = \underline{\mathrm{Lift}}^{-1}(\mathcal{F})$, and let $F \in \pi(2)$ be a coefficient of $\mathcal{F}$, relative to a canonical trivialization, in the lowest weight space of the $GK_\infty(2)$-type carrying the cohomology. Let $\varphi \in \mathcal{S}(\mathbb{X})$ be an algebraic Schwartz-Bruhat function, rational over $\mathbb{Q}(\pi, \chi^+, \chi'^{,+})$. Then*

$$\mathbf{p}(\pi, V, V', \chi_0)^{-1} \cdot \underline{\mathrm{Lift}}^{-1}(\theta_{\varphi,\chi,\chi'}(F))$$

is a $\mathbb{Q}(\pi, \chi^+, \chi'^{,+})$-rational class in $H^0(\mathrm{Sh}(V'^{,(2)}), [E_{\Lambda(2)}])$. Moreover, for any $\sigma \in \mathrm{Gal}(\bar{\mathbb{Q}}/\mathcal{K})$, we have

$$[\mathbf{p}(\pi^\sigma, V, V', \chi_0^\sigma)^{-1}] \cdot \underline{\mathrm{Lift}}^{-1}(\theta_{\varphi^\sigma,\chi^\sigma,\chi'^{,\sigma}}(F^\sigma)) = [\mathbf{p}(\pi, V, V', \chi_0)^{-1} \cdot \underline{\mathrm{Lift}}^{-1}(\theta_{\varphi,\chi,\chi'}(F))]^\sigma$$

REMARK 1.11. If n and m are both even, then $\mathbb{Q}(\pi, \chi^+, \chi'^{,+}) = \mathbb{Q}(\pi)$ and $\mathbf{p}(\pi, V, V', \chi_0)$ is independent of χ_0. Otherwise, $\mathbb{Q}(\pi, \chi^+, \chi'^{,+}) = \mathbb{Q}(\pi, \chi_0^+)$.

2. Proof (seesaw diagrams and cup products)

We write $\theta(F) = \theta_{\varphi,\chi,\chi'}(F)$. Let $m = r + s = \dim(V')$. To test the rationality of $\underline{\mathrm{Lift}}^{-1}(\theta(F))$, we proceed as for the holomorphic small representations of

$$U(n-1, 1) \times U(1, n-1)$$

in IV, §4. Namely, we let η be a character of $GH_{\mathcal{K}}^m$ of type Λ' and compute $L_{\eta(2)}(\theta(F))$ be the seesaw reciprocity formula. This amounts to integrating F against the theta lift of $\eta(2)$, which is a product of small automorphic forms whose rationality was determined in IV.

The subgroup $GH_{\mathcal{K}}^m \subset \mathrm{GU}(V')$ may be chosen to be the stabilizer of a decomposition

$$V' = T_1 \oplus T_2 \oplus \cdots \oplus T_m \tag{2.1}$$

of hermitian spaces of dimension one; then we have

$$GH_{\mathcal{K}}^m \cong G(U(T_1) \times \cdots \times U(T_m)).$$

At the archimedean places, $U(T_i)$ has signature (1,0) for $1 \leq i \leq r$, and signature (0,1) for $r + 1 \leq i \leq m$.

We regard η as in IV, §4 as the restriction to $GH_{\mathcal{K}}^m$ of an m-tuple $(\eta_1, \ldots, \eta_m)$ of Hecke characters of $\mathcal{K}^\times \cong \mathrm{GU}(T_i), i = 1, \ldots, m$. The archimedean parameter of η_j, or rather the parameter of its restriction to $U(T_j)_\infty$, is given by the j-th entry in the highest weight Λ' written in (1.3). In other words, writing $\Lambda' = (\mu_1, \ldots, \mu_m)$ (ignoring the semicolon), the archimedean parameter of η_j is μ_j where

(2.1a)
$$\mu_1 = a_{r'} - \frac{1}{2}(n - 2 + \alpha(\chi')),$$
$$\mu_2 = \mu_3 = \cdots = \mu_{1+r-r'} = -\frac{1}{2}(n - 2 + \alpha(\chi'));$$
$$\mu_{r-i+1} = -a_i - \frac{1}{2}(n - 2 + \alpha(\chi')), i = 1, \ldots, r' - 1;$$
$$\mu_{r+i} = b_i + \frac{1}{2}(n - 2 - \alpha(\chi')), i = 1, \ldots, s.$$

Recall (II.1.6) that

(2.2)
$$b_1 \geq \cdots \geq b_s > a_{r'} + 1 + (r - r') \geq 1.$$

It follows that, if n is even (resp. if n is odd) then, for $j \geq r+1, \mu_j = k_j + (n-2)/2$ (resp. $= k_j + (n-3)/2$) for $k_j \in \mathbb{Z}, k_j > 0$. Thus

(2.3). Observation. Under our assumptions on λ, the characters η_j satisfy the hypotheses of IV, §5 for $j = r + 1, \ldots, m$. In particular, if $(\chi(j), \chi')$ is a pair of splitting characters for the dual reductive pair $(U(T_j), U(V'))$, then the theta lift $\Theta_{\chi(j),\chi'}(\eta_j \otimes \check{\chi}' \cdot \check{\eta}_j)$ is a space of anti-holomorphic automorphic forms of cohomological type for $r + 1 \leq j \leq m$.

That the forms in $\Theta_{\chi(j),\chi'}(\eta_j \otimes \check{\chi}' \cdot \check{\eta}_j)$ are anti-holomorphic follows from II.3.11 and the hypothesis that $U(T_j)$ has signature (0,1) for $j > r$. We may thus make use of the results of IV, §5, especially Proposition IV.5.16, to express the rationality of forms in $\Theta_{\chi(j),\chi'}(\eta_j \otimes \check{\chi}' \cdot \check{\eta}_j)$ in terms of CM periods.

In order to continue the calculation, we need to choose splitting characters $\chi(j)$, as above, for the theta lifts from $G(U(T_i) \times U(-T_i))$ to $G(G \times G^-)$, compatible with the splittings already chosen for the lifts from $G(G \times G^-)$ to $G(G' \times G'^{,-})$. We do this as in Appendix 2. Thus we let

(2.4)
$$\chi(j) = \chi_0 \text{ for } j \text{ odd}, \chi(j) = \check{\chi}_0 \text{ for } j \text{ even}.$$

We are now ready to apply the seesaw reciprocity formula (Proposition I.5.4). We choose an algebraic Schwartz-Bruhat function $\varphi \in \mathcal{S}(\mathbb{X})$ and assume it factors as

(2.5)
$$\varphi = \varphi(1) \otimes \cdots \otimes \varphi(m), \ \varphi(j) \in \mathcal{S}(\mathbb{X}_j),$$

where $\mathbb{X}_j = R_{\mathcal{K}/\mathbb{Q}}(V \otimes_{\mathcal{K}} \Delta^-(T_j))$ and φ_j is the kind of algebraic Schwartz-Bruhat function considered in IV, §4-5. Letting Z denote the center of $GG^{(2)}$ and reasoning as in IV, 4.7, we find

(2.6)
$$L_{\eta(2)}(\theta(F)) = \int_{GG^{(2)}(\mathbb{Q})\cdot Z(\mathbf{A})\backslash GG^{(2)}(\mathbf{A})} F \cdot \prod \theta_{\varphi(j),\chi',\chi(j)}(\eta_j \otimes \check{\chi}' \cdot \check{\eta}_j)\, dg$$

Write

$$f_i^+ = \theta_{\varphi(i),\chi',\chi(i)}(\eta_i \otimes \check{\chi}' \cdot \check{\eta}_i), \ i = 1, \ldots r;$$
$$h_j^- = \theta_{\varphi(r+j),\chi',\chi(r+j)}(\eta_{r+j} \otimes \check{\chi}' \cdot \check{\eta}_{r+j}), \ j = 1, \ldots s.$$

We may identify the f_i^+ and h_j^-, as in IV, §§4-5, as lifts of coherent cohomology classes ω_i and ν_j, respectively, on $\mathrm{Sh}(V^{(2)})$. Namely, we let

$$\ell_i = -\mu_i + \frac{1}{2}(n-2+\alpha(\chi')),\ i=1,\dots,r;\ k_j = b_j,\ j=1,\dots,s.$$

In the notation of II.1.12 and IV, §4, we set

$$\Lambda_i = \begin{cases} \Lambda^+\left(1-\ell_i, -\frac{1}{2}(\alpha(\chi)+1), c\right) & \text{if } \ell_i \geq 0, \text{ (i.e., for } 1+r-r' < i < r); \\ T^+\left(1-\ell_i, -\frac{1}{2}(\alpha(\chi)+1), c\right) & \text{if } \ell_i < 0, \text{ (i.e., } i=1,\dots,1+r-r'); \end{cases}$$

$$\Psi_j = \Lambda^-\left(k_j - 1, \frac{1}{2}(1-\alpha(\chi)); c\right), \qquad j=1,\dots,s.$$

We let $[W_i]$ and $[V_j]$ be the automorphic vector bundles corresponding to Λ_i and Ψ_j, respectively, $i=1,\dots,r$, $j=1,\dots,s$. Then we may write $f_i^+ = \underline{\mathrm{Lift}}(\omega_i), h_j^- = \underline{\mathrm{Lift}}(\nu_j)$, where

$$\omega_i \in \tilde{H}^0(\mathrm{Sh}(V^{(2)}), [W_{i(2)}]);$$
$$\nu_j \in \tilde{H}^2(\mathrm{Sh}(V^{(2)}), [V_{j(2)}]).$$

Here the doubled vector bundles $[W_{i(2)}]$ and $[V_{j(2)}]$ are defined by the same procedure we have already used, and $\underline{\mathrm{Lift}}$ is intended in the sense of highest weight spaces, as in III.4.11.

Moreover, we know that ω_i and ν_j are multiples by specific CM periods of rational elements of their respective coherent cohomology spaces. More precisely,

(2.7) $[(2\pi i \cdot g(\varepsilon_{\mathcal{K}}))^n \cdot p(\chi(i)^+, h^{1,n-1})]\omega_i \in \tilde{H}^0(\mathrm{Sh}(V^{(2)}), [W_{i(2)}])$ is rational over $\mathbb{Q}(\eta_i, \chi(i)^+)$;

(2.8) (n even) $(p(\check{\chi}(r+j)^+, h^{1,n-1})p(\tilde{\eta}_{r+j}, \Phi_j)^2)^{-1}\nu_j \in \tilde{H}^2(\mathrm{Sh}(V^{(2)}), [V_{j(2)}])$ is rational over $\mathbb{Q}(\eta_j, \check{\chi}(r+j)^+)$;

(2.8) (n odd) $(p(\check{\chi}^+, h^{1,n-1})p(\tilde{\eta}_j, \Phi_j)^2 p((\chi')^2, \Phi_j) g(\varepsilon_{\mathcal{K}}))^{-1}\nu_j \in \tilde{H}^2(\mathrm{Sh}(V^{(2)}), [V_{j(2)}])$ is rational over $\mathbb{Q}(\eta_j, \check{\chi}(r+j)^+)$;

Here we have written Φ_j for $\Phi_{\eta_{r+j}}$. Moreover, in (2.7) and (2.8), the $\mathrm{Gal}(\bar{\mathbb{Q}}/\mathcal{K})$-action on the indicated classes is given by the formulas in Proposition IV.4.12 and IV.5.17.

Now the right-hand side of (2.6) is exactly the double of the right-hand side of Proposition III.4.13. In other words, writing $F = \underline{\mathrm{Lift}}(\Phi)$ as in III, §4, the integral is equal to

$$\mathrm{Tr} \circ I_{\Lambda,r,s(2)}\left(\Phi \otimes \bigotimes_{i=1}^{r} \omega_i \otimes \bigotimes_{j=1}^{s} \nu_j\right), \tag{2.9}$$

up to multiplication by a scalar in $\mathcal{K}^\times$. Here Tr is the trace map from

$$H_!^{2n-2}\left(\mathrm{Sh}(V^{(2)}), \Omega^{2n-2}\right)$$

to $\mathbb{C}$ and $I_{\Lambda,r,s(2)}$ is the double of the map defined in Lemma IV.4.8.

In particular, both Tr and $I_{\Lambda,r,s(2)}$ are rational with respect to the structures defined by coherent cohomology. Note that the superscript $^+$ modifies the character $\check{\chi}(r+j)$, so that for any j we have

$$\check{\chi}(r+j)^+ = \|\cdot\|(\chi(r+j)^+)\check{}\ \left(= \|\cdot\|(\chi(r+j)^+)^{-1}\right). \tag{2.10}$$

Write $\mathbf{h} = h^{1,n-1}$ for short. Combining (2.6-2.10), we obtain

$$L_{\eta(2)}(\theta(F)) \sim_{\mathbb{Q}(\eta,\chi_0^+,\pi);\mathcal{K}} \prod_{i=1}^{r}[(2\pi i \cdot g(\varepsilon_{\mathcal{K}}))^n \cdot p(\chi(i)^+, \mathbf{h})]^{-1} \cdot \prod_{j=1}^{s}(p(\|\cdot\|(\chi(r+j)^+)\check{}, \mathbf{h})p(\tilde{\eta}_{r+j}, \Phi_j)^2) \tag{2.11}$$

if n is even;

$$L_{\eta(2)}(\theta(F)) \sim_{\mathbb{Q}(\eta,\chi_0^+,\pi);\mathcal{K}} \prod_{i=1}^{r}[(2\pi i \cdot g(\varepsilon_{\mathcal{K}}))^n \cdot p(\chi(i)^+, \mathbf{h})]^{-1} \cdot \prod_{j=1}^{s}(p(\|\cdot\|(\chi(r+j)^+)\check{}, \mathbf{h})p((\tilde{\eta}_{r+j}\cdot\chi')^2, \Phi_j)^2) \cdot g(\varepsilon_{\mathcal{K}}) \tag{2.12}$$

if n is odd.

We can simplify (2.11), up to $\mathcal{K}$-rational factors, as follows. First, $g(\varepsilon_{\mathcal{K}})^n \in \mathbb{Q}$, so it can be dropped from the formula. Moreover, it follows immediately from Observation 2.3 and III.2.5 that $\Phi_j = \iota$ for $1 \leq j \leq s$. Then the right-hand side of (2.11) is

$$\sim_{\mathbb{Q}(\eta,\chi_0^+,\pi);\mathcal{K}} (2\pi i)^{-rn} \cdot \prod_{i=1}^{m} p(\chi(i)^+, \mathbf{h})^{-1} \cdot p(\|\cdot\|, \mathbf{h})^s \prod_{j=1}^{s} p(\tilde{\eta}_{r+j}, \iota)^2.$$

But the $\chi(i)$ alternate between χ_0 and $\check{\chi}_0$. Define $\alpha(m)$ as above. It thus follows from (2.10) and III.2.3 that

$$\begin{aligned} L_{\eta(2)}(\theta(F)) &\sim_{\mathbb{Q}(\eta,\chi_0^+,\pi);\mathcal{K}} (2\pi i)^{-(r+s)n} p(\|\cdot\|, \mathbf{h})^{[-m/2]} p(\chi_0^+, h)^{-\alpha(m)} \prod_{j=1}^{s} p(\tilde{\eta}_{r+j}, \iota)^2, \\ &\sim_{\mathbb{Q}(\eta,\chi_0^+,\pi);\mathcal{K}} (2\pi i)^{([m/2]-m)n} \cdot p(\chi_0^+, \mathbf{h})^{-\alpha(m)} \prod_{j=1}^{s} p(\tilde{\eta}_{r+j}, \iota)^2, \end{aligned} \tag{2.13}$$

Similarly, if n is odd, the number of occurrences of the factor $g(\varepsilon_{\mathcal{K}})$ is $rn + s \equiv m \pmod 2$. Up to rational factors, we can replace $g(\varepsilon_{\mathcal{K}})^{rn+s}$ by $g(\varepsilon_{\mathcal{K}})^{-\alpha(m)}$, with $\alpha(m)$ as above. We may and do assume that $\chi' = \chi_0$ in this case. The above arguments then show that, when n is odd, we have

$$L_{\eta(2)}(\theta(F)) \sim_{\mathbb{Q}(\eta,\chi_0^+,\pi);\mathcal{K}} (2\pi i)^{([m/2]-m)n} \cdot (g(\varepsilon_{\mathcal{K}})p(\chi_0^+, \mathbf{h}))^{-\alpha(m)} \cdot \prod_{j=1}^{s} p((\tilde{\eta}_{r+j}\cdot\chi_0)^2, \iota). \tag{2.14}$$

Now what we really need to compute, according to Proposition III.3.12.1, is

$$\mathbf{p}(\tilde{\eta}, h^{r,s})^{-1} \cdot L_{\eta(2)}(\theta(F)).$$

When n is even, (2.13) and the relation $\tilde{\eta}^{\iota} = \tilde{\eta}^{-1}$ shows that this is

$$\begin{aligned}
(2.15)\qquad & \mathbf{p}(\tilde{\eta}, h^{r,s})^{-1} \cdot L_{\eta(2)}(\theta(F)) \\
&\sim_{\mathbb{Q}(\eta,\chi_0^+,\pi);\mathcal{K}} (2\pi i)^{([m/2]-m)n} \cdot p(\chi_0^+, \mathbf{h})^{-\alpha(m)} \prod_{i=1}^{m} p(\tilde{\eta}_i, \iota), \\
&\sim_{\mathbb{Q}(\eta,\chi_0^+,\pi);\mathcal{K}} (2\pi i)^{([m/2]-m)n} \cdot p(\chi_0^+, \mathbf{h})^{-\alpha(m)} p\Big(\prod_{i=1}^{m} \tilde{\eta}, \iota\Big), \\
&\sim_{\mathbb{Q}(\eta,\chi_0^+,\pi);\mathcal{K}} (2\pi i)^{([m/2]-m)n} \cdot p(\chi_0^+, \mathbf{h})^{-\alpha(m)} p(\tilde{\xi}_{\pi'}, \iota)^{-1},
\end{aligned}$$

since $\prod_{i=1}^{m} \eta_i = \check{\xi}_{\pi'}$.

Applying Proposition I.3.11, we can express $\xi_{\pi'}$ in terms of the initial data. This yields

$$\begin{aligned}
(2.16)\qquad \mathbf{p}(\tilde{\eta}, h^{r,s}) \cdot L_{\eta(2)}(\theta(F)) &\sim_{\mathbb{Q}(\eta,\chi_0^+,\pi);\mathcal{K}} (2\pi i)^{([m/2]-m)n} \\
&\cdot p(\chi_0^+, \mathbf{h})^{-\alpha(m)} p(\tilde{\xi}_{\pi} \cdot (\tilde{\chi}_0)^{(n/2)\alpha(m)}, \iota).
\end{aligned}$$

Similarly, when n is odd, (2.14) gives us

$$\begin{aligned}
(2.17)\quad & \mathbf{p}(\tilde{\eta}, h^{r,s}) \cdot L_{\eta(2)}(\theta(F)) \\
&\sim_{\mathbb{Q}(\eta,\chi_0^+,\pi);\mathcal{K}} (2\pi i)^{([m/2]-m)n} (g(\varepsilon_{\mathcal{K}}) p(\chi_0^+, \mathbf{h}))^{-\alpha(m)} p(\tilde{\xi}_{\pi'}, \iota)^{-1} p((\chi_0)^2, \iota)^s \\
&\sim_{\mathbb{Q}(\eta,\chi_0^+,\pi);\mathcal{K}} \begin{cases} (2\pi i)^{-mn/2} p(\tilde{\xi}_{\pi}(\tilde{\chi}_0)^{-m/2+s}, \iota), & m \text{ even}, \\ (2\pi i)^{-(m+1)n/2} (g(\varepsilon_{\mathcal{K}}) p(\chi_0^+, \mathbf{h}))^{-1} p(\tilde{\xi}_{\pi}(\tilde{\chi}_0)^{(n-m)/2+s}, \iota), & m \text{ odd}. \end{cases}
\end{aligned}$$

Here we have used the obvious identity $(\chi_0)^2 = \tilde{\chi}_0$.

This completes the proof of the first assertion of Theorem 1.10. The $\mathrm{Gal}(\bar{\mathbb{Q}}/\mathcal{K})$-equivariance is then a consequence of the Galois equivariance of formulas (2.7) and (2.8).

3. Application to conjectural period relations

Since Theorem 1.10 lacks a punch line, I include this section as a preview of what I expect to prove in the second part of this paper. As indicated in the introduction, the intention is to carry out the program sketched in [**H97**] and in §6 of [**H94**]; namely, to prove certain period relations that, in turn, will imply the Deligne conjecture for a large class of motives. In what follows, we assume the expression for the critical values of the motives considered in [**H97**] is valid at all integer points to the right of (and possibly including) the center of symmetry of the functional equation. We make additional hypotheses along the way.

Consider the first seesaw pair in Appendix 1, involving lines 1 and 2:

(3.1)

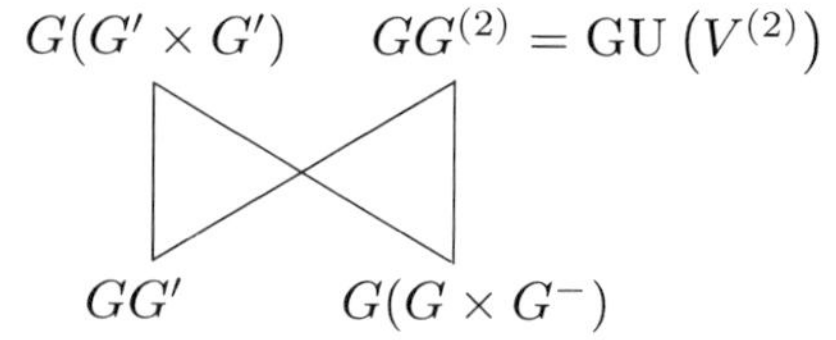

We start with an automorphic representation π of GG that satisfies the hypotheses of Theorem 1.9. We assume π to be of the form $\pi_0 \otimes \beta \circ \det$, where π_0 is self-dual and β is a Hecke character of $\mathcal{K}^\times$. We let $-k$ denote the parameter of the restriction of the archimedean component β_∞ to $U(1)$; in the terminology of [**H97**, §2.9], this means β is of type η_k^{-1}. Denote by $M(\pi_0)$ the absolute Hodge motive (over $\mathcal{K}$) associated to π_0 by the construction of [**H97**, 2.7]. It is of weight $n-1$ and rank n over its coefficient field, and has Hodge types (p_i, q_i), $i = 1, \dots, n$, where p_i can be expressed in terms of the integers (a_i, b_j) as follows: We define

$$\begin{aligned}(3.2)\quad &(a_{0,1}, \dots, a_{0,r'-1}, a_{0,r'+1} - 1, \dots, a_{0,n} - 1; a_{0,r'} + n - r') + (k, \dots, k; k)\\ &= (a_1 + q - 1, \dots, a_{r'-1} + q - 1, -b_s + q, \dots, -b_1 + q; -a_{r'} - q + r' - 1)\\ &+ \frac{1}{2}\alpha(\chi)(1, \dots, 1; 1)\end{aligned}$$

The right-hand side is the bundle parameter Λ, as in II (2.8.1), with $\alpha = (1/2)\alpha(\chi)$, and $q = (r-s)/2$, as before. The left-hand side is taken from [**H97**, 2.2.5.1], where we have replaced the notation a_j of [**H97**] by $a_{0,j} + k$, to avoid confusion. Then the first term

$$(a_{0,1}, \dots, a_{0,r'-1}, a_{0,r'+1} - 1, \dots, a_{0,n} - 1; a_{0,r'} + n - r')$$

is the bundle parameter attached to the self-dual representation $\pi_{0,\infty}$ and the shift by $(k, \dots, k; k)$ corresponds on the bundle side to twisting by β_∞. The Hodge types of $M(\pi_0)$ are then given by

$$(3.3)\qquad p_i = a_{0,i} + n - i,\ q_i = p_{n-i+1} = n - 1 - p_i.$$

(cf. [**H97**, 2.2.5.2])

Let $F \in \pi(2)$ and $\theta_{\varphi,\chi,\chi'}(F)$ be as in the statement of Theorem 1.9. Our first goal is to compute

$$(3.4)\qquad I_{\Delta,\chi'}(\theta_{\varphi,\chi,\chi'}(F)) := \int_{GG'} \theta_{\varphi,\chi,\chi'}(F) \cdot \chi' \circ \det\, dg'.$$

Here, as in III, §4 and IV, §5, $\int_{GG'}$ and $\int_{GG}$ are shorthand for the adelic integrals

$$\int_{Z_{GG'}(\mathbf{A})\cdot GG'(\mathbb{Q})\backslash GG'(\mathbf{A})} \quad \text{and} \quad \int_{Z_{GG}(\mathbf{A})\cdot GG(\mathbb{Q})\backslash GG(\mathbf{A})},$$

respectively. We also define

$$(3.5)\qquad I_{\Delta,\chi}(F) = \int_{GG} F \cdot \chi \circ \det\, dg.$$

To calculate (3.4) we apply the Rallis inner product formula. For this to be valid, we need to know that the Siegel-Weil formula is valid for the lift of the constant function 1 from GG' to $GG^{(2)}$. For $m-n$ sufficiently large this is already due to Weil, but in general this is the subject of work in progress of Kudla, Rallis, and Sweet. Indeed, the theta kernel is not generally integrable over

$$Z_{GG'}(\mathbf{A}) \cdot GG'(\mathbb{Q})\backslash GG'(\mathbf{A})$$

for $|m-n|$ small. In order to state the Siegel-Weil formula in general, we need to introduce certain differential operators, as in [**KuR**], and to keep track of possible poles of Eisenstein series. We skip the statement of the Siegel-Weil formula and assume simply that the Rallis inner product formula holds. Assume π, χ, χ', and

ψ to be unramified outside a finite set $S \cup \{\infty\}$ of primes of $\mathbb{Q}$. Then the Rallis inner product formula yields (cf. [**H94**, §6])

$$(3.6)\quad I_{\Delta,\chi'}(\theta_{\varphi,\chi,\chi'}(F)) = \delta I_{\Delta,\chi}(F) \cdot Z_\infty \cdot Z_S \cdot L^{\mathrm{mot},S}\left(\frac{1}{2}m, \pi_{0,\mathcal{K}}, \tilde{\beta}\cdot\chi\right)/d_n(*m).$$

Here $\delta = 1$ if $m > n$, $\delta = 2$ if $m = n$. By $L^{\mathrm{mot},S}(m/2, \pi_{0,\mathcal{K}}, \tilde{\beta}\cdot\chi)$ we mean the standard partial L-function (with the factors at S removed) of the base change $\pi_{0,\mathcal{K}}$ to $\mathrm{GL}(n,\mathcal{K})$, twisted by the Hecke character $\tilde{\beta}\cdot\chi$, and normalized to have center of symmetry $s = n/2$. If the base change $\pi_{0,\mathcal{K}}$ is not known to exist as an automorphic representation, we can still define

$$L^{\mathrm{mot},S}\left(\frac{1}{2}m, \pi_{0,\mathcal{K}}, \tilde{\beta}\cdot\chi\right) = L^{\mathrm{mot},S}\left(\frac{1}{2}m, \pi, \chi, \mathrm{St}\right),$$

the standard L-function of the unitary group, as described in I, §1, with the functional equation normalized as above. The factor $d_n(*m)$ is a product of special values of Dirichlet L-functions that will be recalled below.

The factor Z_∞ is an archimedean zeta integral which depends only on the local component π_∞ of π and the archimedean component φ_∞ of the Schwartz-Bruhat function used to define the theta lift. The factor Z_S is the product of normalized zeta integrals at finite primes in S. We are implicitly assuming that $L^{\mathrm{mot},S}(s, \pi_{0,\mathcal{K}}, \tilde{\beta}\cdot\chi)$ and the local factors have no pole at $s = m/2$. This can certainly be arranged by varying β and/or χ.

On the other hand, we can define

$$(3.7)\qquad \theta_{\varphi,\chi,\chi'}(F)^{\mathrm{arith}} = \mathbf{p}(\pi, V, V', \chi_0)^{-1}\theta_{\varphi,\chi,\chi'}(F).$$

Then

$$(3.8)\qquad I_{\Delta,\chi'}(\theta_{\varphi,\chi,\chi'}(F)) = \mathbf{p}(\pi, V, V', \chi_0) I_{\Delta,\chi'}(\theta_{\varphi,\chi,\chi'}(F)^{\mathrm{arith}}),$$

where $I_{\Delta,\chi'}(\theta_{\varphi,\chi,\chi'}(F)^{\mathrm{arith}})$ is defined by the integral analogous to (3.4). We write

$$P^{(s),m}(\pi,\chi,\chi') = I_{\Delta,\chi'}(\theta_{\varphi,\chi,\chi'}(F)^{\mathrm{arith}}).$$

LEMMA 3.9. *The point $s = m/2$ is a critical value, in Deligne's sense, of the L-function $L^{\mathrm{mot},S}(s, \pi_{0,\mathcal{K}}, \tilde{\beta}\cdot\chi)$. Moreover, the weight of the character $\tilde{\beta}\cdot\chi$ is in the $(r'-1)$st critical interval for the Hodge type of the motive $M(\pi_0)$.*

PROOF. The meaning of the lemma is clear when m is even and $\chi \equiv 1$. When m is odd, we mean simply that $s = (m-1)/2$ is a critical value of the L-function $L^{\mathrm{mot},S}(s, \pi_{0,\mathcal{K}}, \tilde{\beta}\cdot\chi^+)$, since $L^{\mathrm{mot},S}(s, \pi_{0,\mathcal{K}}, \tilde{\beta}\cdot\chi) = L^{\mathrm{mot},S}(s-1/2, \pi_{0,\mathcal{K}}, \tilde{\beta}\cdot\chi^+)$; by the weight of $\tilde{\beta}\cdot\chi$ we then mean the weight of the motivic Hecke character $\tilde{\beta}\cdot\chi^+$, which is easily seen to equal $2k - \alpha(\chi)$.

According to [**H97**, Corollary 3.3.8], both assertions follow from the pair of inequalities

$$(3.9.1)\qquad \frac{1}{2}m \leq p_{s+1} - k \ (m \text{ even}), \quad \frac{1}{2}(m-1) \leq p_{s+1} - k \ (m \text{ odd}),$$

$$(3.9.2)\qquad \frac{1}{2}m \leq p_{r'-1} + k \ (m \text{ even}), \quad \frac{1}{2}(m-1) \leq p_{r'-1} + k - \alpha(\chi) \ (m \text{ odd}).$$

Now (3.2) and (3.3) yield the following formula for $p_{r'}$:

$$-a_{r'} - \frac{1}{2}(r-s) + r' - 1 = a_{0,r'} + n - r' + k - \frac{1}{2}\alpha(\chi) \tag{3.9.3}$$
$$= p_{r'} + k - \frac{1}{2}\alpha(\chi).$$

Since $a_{r'} \geq 0$, and $r' - r = n - m$, this implies

$$2k - \alpha(\chi) \leq -(r-s) + 2r' - 2 - 2p_{r'} = n - m + w - 2p_{r'} - 1.$$

Applying the equality (3.3)

$$p_j + p_{n-j+1} = w = n - 1,$$

we obtain

$$m - n \leq 2p_{s+1} - n - (2k - \alpha(\chi)).$$

This is equivalent to (3.9.1), since $\alpha(\chi) = 1$ when m is odd.

Next, we apply (3.2) and (3.3) to compute $p_{r'-1}$ and obtain

$$a_{r'-1} + \frac{1}{2}(r-s) - 1 = a_{0,r'-1} + k - \frac{1}{2}\alpha(\chi) \tag{3.9.4}$$
$$= p_{r'-1} - (n - r' + 1) + k - \frac{1}{2}\alpha(\chi).$$

Since $a_{r'-1} \geq 0$, this implies

$$2k - \alpha(\chi) \geq (r-s) - 2 + 2n - 2r' + 2 - 2p_{r'-1} = m - 2p_{r'-1},$$

and this is equivalent to (3.9.2).

We now make three hypotheses:

(3.10) Hypothesis. (a) $L^{\mathrm{mot},S}(m/2, \pi_{0,\mathcal{K}}, \tilde{\beta}\cdot\chi) \neq 0$.

(b) Theorem 3.5.13 of [**H97**] remains valid at the point $s = m/2$.

(c) There exists a hermitian space V'' over $\mathcal{K}$, of signature $(r'-1, s+1)$, and a holomorphic automorphic representation π'' of $GU(V'')$, such that, for almost all p, $\pi''_p \cong \pi_{0,p}$ as representations of $GU(V'')_p \cong GU(V)_p$.

Hypothesis (a) follows from the theorem of Jacquet-Shalika [**JS**] if $m > n$, provided $\pi_{0,\mathcal{K}}$ exists as a cuspidal automorphic representation of $GL(n, \mathcal{K})$. Even if $m = n$, one can hope that (a) can be arranged by varying β (cf. the Non-Vanishing Conjecture 1.5.2 of [**H94**]); however, I know of no way to guarantee this for $n > 2$.

Theorem 3.5.13 of [**H97**], which expresses the critical values of standard L-functions in terms of certain periods, is proved in [**H97**] for critical values s such that $s > n$. This corresponds to the range of absolute convergence of the Eisenstein series used in the integral representation. The extension to $s \geq n/2$ ($= w/2$, where w is the motivic weight of the L-function, cf. [**H97**]) should follow fairly easily from the general Siegel-Weil formula, using some old ideas of Shimura; I intend to carry this out in the sequel to this paper.

Finally, if V'' is any hermitian space of dimension n, $GU(V'')_p \cong GU(V)_p$ for almost all p; the content of hypothesis (c) is thus the existence of π''. In a forthcoming paper with Labesse [**HLa**], we prove the existence of such a π'', assuming the local component of π is supercuspidal at some prime p that splits in $\mathcal{K}$. Under the same hypothesis, [**HLa**] also proves the existence of the base change $\pi_{0,\mathcal{K}}$ as an automorphic representation, and therefore verifies condition (a) provided $m > n$. In any case, when n is even, it follows from considerations of local invariants that some local hypothesis is necessary for (c) to be true for general signatures $(\tilde{r}, \tilde{s})$.

Assuming these hypotheses, Lemma 3.9 implies that we can express the right-hand side of (3.6) in terms of period invariants attached to π''. The shape of the formula depends on the parity of m and n. First, we recall the definition of $d_n(m)$. When m is even, $\chi \equiv 1$ and in the notation of I.1.12 we have

$$d_n(*m) = d_n\left(\frac{1}{2}m - \frac{1}{2}n, 1\right) = \prod_{r=0}^{n-1} L\left(2\left(\frac{1}{2}m - \frac{1}{2}n\right) + n - r, \varepsilon_{\mathcal{K}}^r\right)$$
$$= \prod_{r=0}^{n-1} L(m - r, \varepsilon_{\mathcal{K}}^r).$$

When m is odd, $\chi = \chi_0$ and we have

$$d_n(*m) = d_n\left(\frac{1}{2}m - \frac{1}{2}n, \chi_0\right) = \prod_{r=0}^{n-1} L\left(2\left(\frac{1}{2}m - \frac{1}{2}n\right) + n - r, \varepsilon_{\mathcal{K}}^{1+r}\right)$$
$$= \prod_{r=0}^{n-1} L(m - r, \varepsilon_{\mathcal{K}}^{1+r}).$$

Thus, setting $w = n - 1$,

$$d_n(*m) \sim_{\mathbb{Q};\mathcal{K}} \begin{cases} (2\pi i)^{mn-nw/2} g(\varepsilon_{\mathcal{K}})^{[n/2]} & (m \text{ even}), \\ (2\pi i)^{mn-nw/2} g(\varepsilon_{\mathcal{K}})^{[(n+1)/2]} & (m \text{ odd}). \end{cases}$$

Henceforward we ignore the powers of $g(\varepsilon_{\mathcal{K}})$ in formulas involving $\sim_{*;\mathcal{K}}$; this is legitimate because $g(\varepsilon_{\mathcal{K}}) \in \mathcal{K}^\times$. We thus have

$$d_n(*m) \sim_{\mathbb{Q};\mathcal{K}} (2\pi i)^{mn-nw/2} \tag{3.11}$$

in all cases.

We write $\sim$ for $\sim_{\mathbb{Q}(\pi_0, \beta^{(2)}, \chi_0^+);\mathcal{K}}$. Theorem 3.5.13 of [**H97**] gives

$$\begin{aligned} L^{\mathrm{mot},S}\left(\frac{1}{2}m, \pi_{0,\mathcal{K}}, \tilde{\beta}\cdot\chi\right) &\sim (2\pi i)^{mn/2-nw/2+k(r'-1-(s+1))} \\ &\cdot P^{(s+1)}(\pi_0, V'', \cdot) p(\check{\beta}^{(2)}, \iota)^{r'-1-(s+1)} \qquad (m \text{ even}), \\ L^{\mathrm{mot},S}\left(\frac{1}{2}m, \pi_{0,\mathcal{K}}, \tilde{\beta}\cdot\chi\right) &\sim (2\pi i)^{(m-1)n/2-nw/2+k(r'-s-2))+s+1} \\ &\cdot P^{(s+1)}(\pi_0, V'', \cdot) p(\beta^{(2)}\cdot\chi^+, \iota)^{(s+1)-r'+1} \qquad (m \text{ odd}). \end{aligned} \tag{3.12}$$

The notation is explained in [loc. cit]. Here we simply note that the notation $\check{\chi}$ used for Hecke characters in [loc. cit] corresponds to the notation $\check{\chi}^\iota$ used here. The term $P^{(s+1)}(\pi_0, V'', \cdot)$ (the $\cdot$ was denoted β in [loc. cit.], but was simply a place marker) is a quadratic period of an arithmetic automorphic form in the holomorphic representation π''. For our purposes here, this is unfortunate notation, and we write

$$P^{(r'-1)}(\pi_0) = P^{(s+1)}(\pi_0, V'', \cdot).$$

Moreover, it follows from [**H97**, Lemma 3.5.6] that

$$Z_S \sim_{\mathbb{Q}(\pi_0, \beta^{(2)}\cdot, \chi_0^+);\mathcal{K}} 1. \tag{3.13}$$

Here we need to be careful, because Z_S is defined in terms of the local components of the Schwartz-Bruhat function φ. The necessary verifications follow the pattern of [**H93**] and will be included in the sequel.

Combining (3.11), (3.12), and (3.13), the right-hand side of (3.6) is

$$\sim I_{\Delta,\chi}(F)\cdot Z_\infty\cdot(2\pi i)^{k(r'-s-2)-mn/2}\cdot P^{(r'-1)}(\pi_0)p(\check{\beta}^{(2)},\iota)^{r'-s-2}\ (m\text{ even});$$

$$\sim I_{\Delta,\chi}(F)\cdot Z_\infty\cdot(2\pi i)^{k(r'-s-2)-(m+1)n/2+s+1}\cdot P^{(r'-1)}(\pi_0)p(\beta^{(2)}\cdot\chi^+,\iota)^{s+2-r'}\ (m\text{ odd}).$$

Define

$$Q_{r'}(\pi_0)=p(\tilde{\beta},i_0)^{n-2}\cdot I_{\Delta,\chi}(F); \tag{3.14}$$

$$P^{(r'),*}(\pi_0,\beta)=Q_{r'}(\pi_0)\cdot Z_\infty\cdot P^{(r'-1)}(\pi_0).$$

We view $Q_{r'}(\pi_0)$ as a quadratic period, in the sense of [**H97**], attached to π_0. I recall that if ω_j is a $\mathcal{K}$-rational generator, over the coefficient field $\mathbb{Q}(\pi_0)$, of the rank-one Hodge subspace

$$F^{p_j}M(\pi_0)_{DR}/F^{p_j+1}M(\pi_0)_{DR}\cong M(\pi_0)^{p_j,q_j}\subset M(\pi_0)_{DR}\otimes\mathbb{C},$$

then the j-th quadratic period of $M(\pi_0)$, defined up to $(\mathbb{Q}(\pi_0)\otimes\mathcal{K})^\times$, is

$$Q_{r'}=\langle\omega_j,F_\infty\omega_j\rangle$$

[**H97**, (1.4.4)]. Here F_∞ is infinite Frobenius and $\langle\cdot,\cdot\rangle$ is the de Rham polarization of $M(\pi_0)$. A more explicit definition of Q_j, in the setting of motives attached to automorphic forms on unitary groups, is given in [**H97**, §2.8]; $Q_{r'}$ is the normalized Petersson inner product of an automorphic form lifting ω_j with itself.

It is thus clear that, in the notation of [**H97**, §2.9]

$$I_{\Delta,\chi}(F)\sim\langle\omega_{r'}\otimes\beta^{\mathrm{arith}},F_\infty(\omega_{r'}\otimes\beta^{\mathrm{arith}})\rangle=\langle\omega_{r'}(\beta),F_\infty(\omega_{r'}(\beta))\rangle.$$

If $Q_{r'}(\pi_0)$ is defined as in [**H97**], then (3.14) is a special case of [**H97**, (2.9.12), (2.9.13)]. (Though [**H97**] only treats character twists of holomorphic and anti-holomorphic forms, the same results hold for general cohomological degree.)

Note that

$$(2\pi i)^{k(r'-s-2)}p(\check{\beta}^{(2)},\iota)^{r'-s-2}\sim p(\check{\beta}^{(2)}\cdot\|\cdot\|^{-k},\iota)^{r'-s-2}\sim p(\tilde{\beta},i_0)^{r'-s-2}. \tag{3.15}$$

Indeed, this follows easily from the formulas in III, §2 and the definitions, as in the discussion preceding [**H97**, (2.9.15)] (but remember the change of notation).

Then the preceding formula simplifies and combines with (3.6) to give:

$$I_{\Delta,\chi'}(\theta_{\varphi,\chi,\chi'}(F))\sim(2\pi i)^{-mn/2}\cdot p(\tilde{\beta},i_0)^{-2s}p^{(r'),*}(\pi_0)\ (m\text{ even}) \tag{3.16}$$

$$I_{\Delta,\chi'}(\theta_{\varphi,\chi,\chi'}(F))\sim(2\pi i)^{-(m+1)/2n+s+1}$$

$$\cdot\, p(\tilde{\beta},i_0)^{-2s}p(\chi^+,\iota)^{s+2-r'}p^{(r'),*}(\pi_0)\ (m\text{ odd}).$$

What we really want is a formula for $I_{\Delta,\chi'}(\theta_{\varphi,\chi,\chi'}(F)^{\mathrm{arith}})$, since we will have another expression for the latter in terms of special values of L-functions. Since the definition of $I_{\Delta,\chi'}(\theta_{\varphi,\chi,\chi'}(F)^{\mathrm{arith}})$ depends on $\mathbf{p}(\pi,V,V',\chi_0)$, which depends in turn on the parity of n as well as m, we break this up into four cases.

First we note the following obvious formula:

$$p(\tilde{\xi}_\pi,\iota)^{-1}p(\tilde{\beta},i_0)^{-2s}\sim p(\tilde{\xi}_\pi,i_0)p(\tilde{\beta},i_0)^{-2s} \tag{3.17}$$

$$\sim p(\tilde{\beta},i_0)^{n-2s}\cdot p(\tilde{\xi}_{\pi_0},i_0)\sim p(\tilde{\beta},i_0)^{r'-s}\cdot p(\tilde{\xi}_{\pi_0},i_0).$$

LEMMA 3.18. *Define $d = d(\pi_0) = p(\tilde{\xi}_{\pi_0}, i_0)$. We have the following expressions for $I_{\Delta,\chi'}(\theta_{\varphi,\chi,\chi'}(F)^{\mathrm{arith}})$:*

(a) If n and m are both even,

$$I_{\Delta,\chi'}(\theta_{\varphi,\chi,\chi'}(F)^{\mathrm{arith}}) \sim_{\mathbb{Q}(\pi_0,\beta^{(2)}\cdot\chi_0^+);\mathcal{K}} d\cdot p(\tilde{\beta}, i_0)^{r'-s} P^{(r'),*}(\pi_0).$$

(b) If n is even, m is odd,

$$I_{\Delta,\chi'}(\theta_{\varphi,\chi,\chi'}(F)^{\mathrm{arith}}) \sim_{\mathbb{Q}(\pi_0,\beta^{(2)},\chi_0^+);\mathcal{K}} d\cdot(2\pi i)^{s+1}\cdot p(\chi_0^+,\mathbf{h})p(\tilde{\beta}, i_0)^{r'-s} \cdot p(\tilde{\chi}_0,\iota)^{-n/2}\cdot p(\chi_0^+,\iota)^{s+2-r'}P^{(r'),*}(\pi_0).$$

(c) If n is odd, m is even,

$$I_{\Delta,\chi'}(\theta_{\varphi,\chi,\chi'}(F)^{\mathrm{arith}}) \sim_{\mathbb{Q}(\pi_0,\beta^{(2)},\chi_0^+);\mathcal{K}} d\cdot p((\tilde{\chi}_0)^{s-m/2}, i_0)\cdot p(\tilde{\beta}, i_0)^{s-r'}P^{(r'),*}(\pi_0).$$

(d) If n and m are both odd,

$$I_{\Delta,\chi'}(\theta_{\varphi,\chi,\chi'}(F)^{\mathrm{arith}}) \sim_{\mathbb{Q}(\pi_0,\beta^{(2)},\chi_0^+);\mathcal{K}} d\cdot(2\pi i)^{s+1}\cdot p(\chi_0^+,\mathbf{h}) \cdot p((\tilde{\chi}_0)^{(n-m)/2+s}, i_0)p(\tilde{\beta}, i_0)^{r'-s}p(\chi_0^+,\iota)^{s+2-r'}P^{(r'),*}(\pi_0).$$

PROOF. In each case, one multiplies the right-hand side of (3.16) by the invariant $\mathbf{p}(\pi, V, V', \chi_0)^{-1}$ and simplifies the result using (3.17).

(3.19). The representation π_0 is self dual, so its central character ξ_{π_0} is quadratic, and the invariant $d = p(\tilde{\xi}_{\pi_0}, i_0)$ therefore belongs to a quadratic field. In [**H97**, 3.7.9] we showed that d could be identified with $d(M(\pi_0))^{1/2}$, the square root of the Hodge-de Rham discriminant of the motive $M(\pi_0)$, assuming the Tate conjecture. The invariant $d(M(\pi_0))^{1/2}$ is trivial if $M(\pi_0)$ as odd weight (i.e., if n is even and $M(\pi_0)$ is symplectic) and quadratic if $M(\pi_0)$ has even weight.

When n is even, we will assume ξ_{π_0} factors through the norm $N_{\mathcal{K}/\mathbb{Q}}$. Thus

(3.19.1) $$\tilde{\xi}_{\pi_0} \equiv 1 \text{ and } d = 1.$$

In the cases considered in [**HLa**], the Galois representation attached to π_0 has a symplectic polarization, so this hypothesis is automatic. We now make a second series of hypotheses.

(3.20) Hypothesis. (a) Theorem 3.5.13 of [**H97**] remains valid for the holomorphic automorphic representation π' of $\mathrm{GU}(V')$, for every Hecke character ν of $\mathcal{K}$ and every critical value s_0 strictly to the right of the center of symmetry of the L-function $L(s, \pi', \nu, \mathrm{St})$, assuming s_0 is not a pole of $L(s, \pi', \nu, \mathrm{St})$.

(b) For any Hecke character α of $\mathcal{K}$ and any finite set S of finite places, let $L^{\mathrm{mot},S}(s, \pi_0, \alpha, St)$ be the partial L-function with the Euler factors at S removed, normalized so that $\mathrm{Re}(s) = n/2$ is the axis of symmetry. The hypothesis is that $L^{\mathrm{mot},S}(s, \pi_0, \alpha, \mathrm{St})$ is holomorphic for $\mathrm{Re}(s) > n/2$ and that $L^{\mathrm{mot},S}(s_0, \pi_0, \alpha, \mathrm{St}) \neq 0$ for any integer $s_0 > n/2$.

(c) $p_{r'} - p_{r'+1} \geq 2$

(d) There exists a hermitian space $V^{(r',s)}$ over $\mathcal{K}$, of signature (r', s), and a holomorphic automorphic representation $\pi^{(r',s)}$ of $\mathrm{GU}(V^{(r',s)})$, such that, for almost all p, $\pi_p^{(r',s)} \cong \pi_{0,p}$ as representations of $\mathrm{GU}(V^{(r',s)})_p \cong \mathrm{GU}(V)_p$.

These hypotheses are certainly stronger than necessary, but they can be verified in a large class of cases. The verification of (a) under additional hypotheses will be included in the sequel. Both hypotheses in (b) are automatic if $\pi_{0,\mathcal{K}}$ exists as a cuspidal automorphic representation of $\mathrm{GL}(n,\mathcal{K})$. Hypothesis 3.20 (d) is completely analogous to Hypothesis 3.10 (c), and the two can be simultaneously verified under appropriate same local assumptions (cf. [**HLa**]).

We define $P^{(r')}(\pi_0)$ to be the quadratic period denoted $P^{(s)}(\pi^{(r',s)}, V^{(r',s)}, \cdot)$ in [**H97**, §2.8]; it is the Petersson inner product, normalized as in [loc. cit.], of a $\mathcal{K}$-rational holomorphic form in $\pi^{(r',s)}$ with itself. Here by a $\mathcal{K}$-rational form we actually mean a family $\{f_\sigma | \sigma\colon \mathbb{Q}(\pi_0) \to \mathbb{C}\}$ where each f_σ is rational over $\sigma(\mathbb{Q}(\pi_0)) \cdot \mathcal{K}$ and where $\tau(f_\sigma) = f_{\tau\cdot\sigma}$ for all $\sigma \in \mathrm{Aut}(\mathbb{C}/\mathcal{K})$. The quadratic period thus naturally takes values in $(\mathbb{Q}(\pi_0)\otimes\mathbb{C})^\times/(\mathbb{Q}(\pi_0)\otimes\mathcal{K})^\times$. Hypothesis 3.20 (c) is needed in order to provide a critical value s_0 strictly to the right of the center of symmetry, as in (b), when α is a character in the r'-th critical interval for $M(\pi_0)$. Hypothesis (d) then allows us to use the techniques of [**H97**] to express the L-value $L^{\mathrm{mot},S}(s_0,\pi_0,\alpha,\mathrm{St})$ in terms of $P^{(r')}(\pi_0)$.

LEMMA 3.21. *We write $\sim$ for $\sim_{\mathbb{Q}(\pi_0,\beta^{(2)},\chi_0^+);\mathcal{K}}$ and $d = d(\pi_0)$ as in Lemma 3.18. Under Hypothesis 3.20, we have the following expressions for*

$$I_{\Delta,\chi'}(\theta_{\varphi,\chi,\chi'}(F)^{\mathrm{arith}}):$$

(a) If n and m are both even,

$$I_{\Delta,\chi'}(\theta_{\varphi,\chi,\chi'}(F)^{\mathrm{arith}}) \sim p(\tilde{\beta}, i_0)^{r'-s}\cdot P^{(r')}(\pi_0).$$

(b) If n is even, m is odd,

$$I_{\Delta,\chi'}(\theta_{\varphi,\chi,\chi'}(F)^{\mathrm{arith}}) \sim (2\pi i)^{-n/2}p(\tilde{\beta}, i_0)^{r'-s}p(\chi_0^+,\iota)^{-r'}p(\chi_0^+,i_0)^{-s}\cdot P^{(r')}(\pi_0).$$

(c) If n is odd, m is even,

$$I_{\Delta,\chi'}(\theta_{\varphi,\chi,\chi'}(F)^{\mathrm{arith}}) \sim d\cdot(2\pi i)^{-m/2}p(\tilde{\beta}, i_0)^{r'-s}p(\chi_0^+,\iota)^{-s}p(\chi_0^+,i_0)^{s-m}\cdot P^{(r')}(\pi_0).$$

(d) If n and m are both odd,

$$I_{\Delta,\chi'}(\theta_{\varphi,\chi,\chi'}(F)^{\mathrm{arith}}) \sim d\cdot(2\pi i)^{(n-m)/2}p(\chi_0^+,i_0)^{n-m}\cdot p(\tilde{\beta}, i_0)^{r'-s}\cdot P^{(r')}(\pi_0).$$

We postpone the proof of Lemma 3.21. to 3.26-27, below, and turn immediately to the proof of our second main theorem.

THEOREM 3.22. *Under Hypotheses 3.10 and 3.20 and assuming (3.19.1), we have in every case*

$$P^{(r')}(\pi_0)\sim_{\mathbb{Q}(\pi_0,\beta^{(2)},\chi_0^+);\mathcal{K}} P^{(r'),*}(\pi_0).$$

PROOF. When n and m are both even, the theorem follows immediately from the two lemmas and (3.19.1). We treat the remaining cases separately. For readability, we let $Q = P^{(r')}(\pi_0)/P^{(r'),*}(\pi_0)$ and show that $Q\sim 1$. We also write $\sim$ for $\sim_{\mathbb{Q}(\pi_0,\beta^{(2)},\chi_0^+);\mathcal{K}}$.

<u>*Case(b)*</u> (n even, m odd): After cancelling the terms $p(\tilde{\beta}, i_0)^{r'-s}$ and $p(\chi_0^+,\iota)^{-r'}$ the two lemmas and (3.19.1) yield the relation

$$(2\pi i)^{s+1+n/2}\cdot p(\chi_0^+,\mathbf{h})\cdot p(\tilde{\chi}_0,\iota)^{-n/2}\cdot p(\chi^+,i_0)^s p(\chi_0^+,\iota)^{s+2}\sim Q.$$

Since $\tilde{\chi} = \tilde{\chi}_0^+$, the left-hand side can be rewritten

$$\begin{aligned}(2\pi i)^{n/2+s+1}&\,p(\chi_0^+,i_0)\,p(\chi_0^+,\iota)^{n-1}\,p(\chi_0^+,i_0)^{n/2}p(\chi_0^+,\iota)^{-n/2}\,p(\chi^+,i_0)^s p(\chi_0^+,\iota)^{s+2}\\ &\sim [(2\pi i)\,p(\chi_0^+,i_0)\,p(\chi_0^+,\iota)]^{n/2+s+1}\sim[(2\pi i)\,p(\chi_0^+\circ N_{\mathcal{K}/\mathbb{Q}},i_0)]^{n/2+s+1}\\ &\sim[(2\pi i)\,p(\|\cdot\|,i_0)]^{n/2+s+1}\sim 1\end{aligned}$$

by III.2.3

Case(c) (n odd, m even): The two lemmas provide the relation

$$d \cdot p((\tilde{\chi}_0)^{s-m/2}, i_0) \cdot p(\tilde{\beta}, i_0)^{s-r'} P^{(r'),*}(\pi_0)$$
$$\sim d \cdot (2\pi i)^{-m/2} P^{(r')}(\pi_0) p(\tilde{\beta}, i_0)^{r'-s} p(\chi_0^+, \iota)^{-s} p(\chi_0^+, i_0)^{s-m}.$$

Cancelling the common terms d and $p(\tilde{\beta}, i_0)^{s-r'}$, and expanding the $\tilde{\chi}_0$ term as in case (b), this yields

$$p(\chi_0^+, i_0)^{s-m/2} p(\chi_0^+, \iota)^{m/2-s} \sim (2\pi i)^{-m/2} p(\chi_0^+, \iota)^{-s} p(\chi_0^+, i_0)^{s-m} \cdot Q.$$

We can rewrite this

$$Q \sim [(2\pi i)\, p(\chi_0^+, i_0)\, p(\chi_0^+, \iota)]^{m/2}$$

and then conclude as in case (b).

Case(d) (n odd, m odd): Cancelling the terms common to the two sides, we find

$$(2\pi i)^{s+1} \cdot p(\chi_0^+, \mathbf{h}) \cdot p((\tilde{\chi}_0)^{(n-m)/2+s}, i_0) p(\chi_0^+, \iota)^{s+2-r'}$$
$$\sim (2\pi i)^{(n-m)/2} \cdot p(\chi_0^+, i_0)^{n-m} \cdot Q.$$

Proceeding as before, the left-hand side can be rewritten

$$(2\pi i)^{s+1} \cdot p(\chi_0^+, i_0)^{1+(n-m)/2+s} \cdot p(\chi_0^+, \iota)^{n-1-(n-m)/2-s+s+2-r'}$$
$$\sim (2\pi i)^{s+1} \cdot p(\chi_0^+, i_0)^{1+(n-m)/2+s} p(\chi_0^+, \iota)^{1-(n-m)/2+s}.$$

Simplifying, we thus find

$$Q \sim [(2\pi i) \cdot p(\chi_0^+, i_0) \cdot p(\chi_0^+, \iota)]^{1-(n-m)/2+s}$$

and we conclude as in the other two cases.

(3.23). Choice of an auxiliary Hecke character. The proof of Lemma 3.21 consists in using the results of [**H97**] to rewrite the left-hand side in terms of a special value of an L-function of the form $L(s, \pi', \alpha, \mathrm{St})$, for an appropriately chosen Hecke character α. The formula I.3.15.2 for $L(s, \pi', \mathrm{St})$ expresses its special values in terms of special values of $L(s, \pi, \mathrm{St})$ and these in turn can be rewritten in terms of periods of forms in π'', again by [**H97**].

The calculation is rather long. Define $\Lambda'^{,\mathrm{op}}$ by II.4.3, where Λ' is the bundle parameter Λ' associated to π'. We begin by letting v and μ be two integers such that $\Lambda'^{,op}$ can be written in the form

$$\Lambda'^{,\mathrm{op}} = (-b_s - q', \dots, -b_1 - q'; a_1 + q', \dots, a_{r'-1} + q', q', \dots, q', q' - a_{r'})$$
$$+ \frac{1}{2}\alpha(\chi')(1, \dots, 1) \tag{3.23.1}$$
$$= (-\mu - v - \beta_r, \dots, -\mu - v - \beta_1; \mu + \gamma_1, \dots, \mu + \gamma_s),$$

where $\beta_1 \geq \beta_2 \cdots \geq \beta_r \geq 0, \gamma_1 \geq \gamma_2 \geq \cdots \gamma_s \geq 0, \beta_i, \gamma_j \in \mathbb{Z}$ for all i, j. We suppose ν is a Hecke character of $\mathcal{K}^\times$ such that $\nu_\infty = z^{-v}$. The condition (3.23.1) is taken from [**H97**], Lemma 3.3.7 and implies, under Hypothesis 3.20 (a), that the special value $L^{\mathrm{mot}}(\mu, \pi', \check{\nu}, \mathrm{St}) = L(\mu - (m-1)/2, \pi', \check{\nu}, \mathrm{St})$ can be expressed in terms of $I_{\Delta,\chi'}(\theta_{\varphi,\chi,\chi'}(F)^{\mathrm{arith}})$ and the periods of ν, provided $\mu - (m-1)/2 > 1/2$ and is not a pole. The condition is applied to $\Lambda'^{,\mathrm{op}}$, rather than to Λ' itself, because π' is related to the dual $\check{\pi}$ rather than to π. The techniques of [**H97**] therefore provide L-values for a group of signature (s, r) rather than (r, s). We recall the exact relation below in Proposition 3.24. Note that $(m-v)/2$ is the center of symmetry (unitary point) of $L^{\mathrm{mot}}(\mu, \pi', \check{\nu}, \mathrm{St})$.

Substituting $q' = (n-2)/2$, the existence of an expression (3.23.1) is equivalent to the pair of inequalities

$$(3.23.2) \qquad -a_{r'} + \frac{1}{2}(n-2+\alpha(\chi')) \geq \mu, \qquad -b_s - \frac{1}{2}(n-2-\alpha(\chi')) \leq \mu + v.$$

We choose v to be an integer, almost as large as possible, such that the character $\tilde{\beta} \cdot (\chi/\chi') \cdot \nu$ is in the r'-th critical interval for the motive $M(\pi_0)$, in the sense of [**H97**]. Concretely, we assume $2k - \alpha(\chi) + \alpha(\chi') + v = n - 2p_{r'+1} - 3$, where $p_{r'+1}$ is given in terms of the integers $\{a_i, b_j\}$ by (3.3) and (3.2). Equivalently,

$$(3.23.3) \qquad \alpha(\chi') + v = n - 2p_{r'+1} - 3 - 2k + \alpha(\chi).$$

Note that $v - \alpha(\chi) \equiv n - \alpha(\chi') \pmod 2$. But $\alpha(\chi)$ and m have the same parity. It follows that $v \equiv m+1 \pmod 2$. Under hypothesis (3.20)(c), we have

$$2k - \alpha(\chi) + \alpha(\chi') + v = n - 2p_{r'+1} - 3 \geq n - 2(p_{r'} - 2) - 3 = w - 2p_{r'} + 2,$$

so $2k - \alpha(\chi) + \alpha(\chi') + v$ is indeed in the r'-th critical interval for $M(\pi_0)$.

We take $\mu = (m - v + 1)/2$; thus $\mu \in \mathbb{Z}$, and we have

LEMMA 3.23.4. *The integer $\mu = (m-v+1)/2$ satisfies the inequalities (3.23.2). Moreover, μ is strictly to the right of the center of symmetry of the L-function $L^{\mathrm{mot}}(\mu, \pi', \nu^{\iota}, \mathrm{St})$.*

PROOF. This is a simple calculation. On the one hand, (3.2) and (3.3) yield

$$b_s = -p_{r'+1} - k + \frac{1}{2}\alpha(\chi) + \frac{1}{2}m.$$

Thus

$$\begin{aligned} -b_s - \frac{1}{2}(n-2-\alpha(\chi')) &= p_{r'+1} + k - \frac{1}{2}\alpha(\chi) - \frac{1}{2}m - \frac{1}{2}(n-2-\alpha(\chi')) \\ &= -\frac{1}{2}(n - 2p_{r'+1} - 2 + 2k - \alpha(\chi) + \alpha(\chi')) - \frac{1}{2}m \\ &= -\frac{1}{2}(m+v+1) = -\mu - v. \end{aligned}$$

So our choice of μ and v yields a strict equality in the second expression in (3.23.2). On the other hand, $n = r' + s$, so

$$\begin{aligned} &-a_{r'} + \frac{1}{2}(n-2+\alpha(\chi')) \\ &= p_{r'} + k - \frac{1}{2}\alpha(\chi) + \left[\frac{1}{2}(r-s) - r' + 1\right] + \frac{1}{2}(n-2+\alpha(\chi')) \\ &= \frac{1}{2}[2p_{r'} + 2k - \alpha(\chi) + r - r' + \alpha(\chi')]. \end{aligned}$$

We need to show

$$\frac{1}{2}(m-v+1) = \mu \leq -a_{r'} + \frac{1}{2}(n-2+\alpha(\chi'))),$$

$$\text{i.e.,} \qquad m+1 \leq 2p_{r'} + 2k - \alpha(\chi) + r - r' + \alpha(\chi') + v.$$

Since $r - r' = m - n$, this is equivalent to

$$v + \alpha(\chi') + 2p_{r'} + 2k - \alpha(\chi) - n - 1 \geq 0.$$

Substituting our expression for $v + \alpha(\chi')$, this becomes

$$2(p_{r'} - p_{r'+1}) - 4 \geq 0,$$

which is Hypothesis 3.20(c).

The final assertion is obvious.

Now recall from the proof of Lemma 3.9 that

$$2k - \alpha(\chi) \leq n - m + w - 2p_{r'} - 1,$$

i.e., that

$$n - 2p_{r'} - 2 - (2k - \alpha(\chi)) \geq m - n.$$

Then Hypothesis 3.20 (c) implies

$$\begin{aligned} \alpha(\chi') + v = n - 2p_{r'+1} - 3 - (2k - \alpha(\chi)) &> n - 2p_{r'} - (2k - \alpha(\chi)) \\ &\geq m - n + 1. \end{aligned} \tag{3.23.5}$$

(3.24). We let $\theta_{\varphi,\chi,\chi'}(F)^{\mathrm{sw}}$ denote $\theta_{\varphi,\chi,\chi'}(F)$, viewed as an automorphic form in $\check{\pi}' \otimes \pi' \subset \mathcal{A}(G(U(-V') \times U(V'))$ by switching the variables. Then we can identify

$$\theta_{\varphi,\chi,\chi'}(F)^{\mathrm{sw,arith}} = \theta_{\varphi,\chi,\chi'}(F)^{\mathrm{arith}}. \tag{3.24.1}$$

in the obvious notation. Now in [**H97**], the quadratic period $I_{\Delta,\chi'}(\theta_{\varphi,\chi,\chi'}(F)^{\mathrm{arith}})$ was denoted $\pm i^c \cdot P^{(s)}(\pi', V', \cdot)$, where c is the split central parameter of the central character of π' and where the $\cdot$ is a placekeeper designating an arithmetic vector in π'. Via (3.24.1), we can write

$$\pm i^c \cdot P^{(s)}(\pi', V', \cdot) = \pm i^c \cdot P^{(r)}(\check{\pi}', -V', \cdot). \tag{3.24.2}$$

PROPOSITION 3.25. *Let c denote the split central parameter of the central character of π'. Write $\sim$ for $\sim_{\mathbb{Q}(\pi_0,\beta^{(2)},\chi_0^+,\nu);\mathcal{K}}$. Under Hypothesis 3.20,*

$$L^{\mathrm{mot}}(\mu, \pi', \check{\nu}, \mathrm{St}) \sim (2\pi i)^{(2-v)m/2-c} I_{\Delta,\chi'}(\theta_{\varphi,\chi,\chi'}(F)^{\mathrm{arith}}) \cdot p(\nu, \iota_0)^r p(\nu, \iota)^s,$$

provided μ is not a pole of $L^{\mathrm{mot}}(s, \pi', \check{\nu}, \mathrm{St})$.

PROOF. This is essentially Corollary 3.5.9 of [**H97**], which calculates the special value of such a function in terms of a quadratic period of an arithmetic holomorphic form in $\check{\pi}'$ on the group $\mathrm{GU}(-V')$. As mentioned before, we are working with $\Lambda'^{,\mathrm{op}}$. Therefore Corollary 3.5.9 of [**H97**] applies to the group $\mathrm{GU}(-V')$ of signature (s, r).

Note first that $\mathbb{Q}(\pi') = \mathbb{Q}(\pi_0, \beta^{(2)}, \chi_0^+)$. Next, (3.23.2) is equivalent to [**H97**], Lemma 3.3.7, which is the necessary and sufficient condition for the reasoning leading to Corollary 3.5.9 in the range of absolute convergence of the Eisenstein series. Hypothesis 3.20 (a) asserts that this reasoning remains valid down to the center of symmetry. Finally, ignoring powers of the Gauss sum $g(\varepsilon_{\mathcal{K}})$, which belongs to $\mathcal{K}^\times$, Corollary 3.5.9 writes the L-value as a product of the form

$$\begin{aligned} &(2\pi i)^{\mu m - m(m-1)/2} \pi^{-c} \cdot P^{(r)}(\check{\pi}', -V; \cdot, \nu^\iota) \\ &= (2\pi i)^{(2-v)m/2} \pi^{-c} \cdot P^{(r)}(\check{\pi}', -V; \cdot, \nu^\iota), \end{aligned} \tag{3.25.1}$$

with our choice of μ. But the discussion leading up to (2.9.15) of [**H97**] is derived from the relation

$$P^{(r)}(\check{\pi}', -V; \cdot, \check{\nu}) \sim_{\mathbb{Q}(\pi',\nu);\mathcal{K}} P^{(r)}(\check{\pi}', -V, \cdot) \cdot p(\nu, \iota_0)^r p(\nu, \iota)^s. \tag{3.25.2}$$

Combining this with (3.24.2), we find the assertion of the Proposition.

(3.26). Now I.3.15.2 provides another expression for the left-hand expression in Proposition 3.24. Recalling that $\mu - (m-1)/2 = (2-v)/2$, we have

$$(3.26.1) \qquad L^S\left(\frac{1}{2}(2-v), \pi', \check{\nu}, \mathrm{St}\right) = L^S\left(\frac{1}{2}(2-v), \pi, (\chi/\chi')\cdot \check{\nu}^{\iota}, \mathrm{St}\right) \cdot L^S\left(\frac{1}{2}(2-v), \chi', \cdot\check{\nu}\right)^{[m-n]}.$$

In particular, under Hypothesis 3.20 (b), the right-hand side has no pole at $s = (2-v)/2$, which is to the right of the center of symmetry $(1-v)/2$ of the L-function $L(s, \pi, (\chi/\chi')\cdot\check{\nu}^{\iota}, \mathrm{St})$. We therefore can use the formula in Proposition 3.25.

We first consider the second factor in the right-hand side. This equals

$$(3.26.2) \qquad \prod_{j=1}^{m-n} L^S\left(\frac{1}{2}(v+2+m-n+1) - j, \chi'\cdot\check{\nu}\right).$$

When n is even, this is a product of special values of the partial Hecke L-series $L^S(s, \check{\nu})$. By our hypothesis on the type of ν, $\check{\nu}$ is critical of type ι and the integers i with $1-v \leq i \leq 0$ are all critical for $L(s, \check{\nu})$. But

$$\left\{\frac{1}{2}(2-v+m-n+1) - j \mid j = 1, \dots, m-n\right\} = \left\{\frac{1}{2}(3-v-(m-n)), \dots, \frac{1}{2}(1-v+m-n)\right\}.$$

It thus follows from (3.23.5) that (3.26.2) is a product of critical values. Thus when n is even, it follows from III.2.4 that (3.26.2) is

$$(3.26.3,\ n \text{ even}) \qquad \sim_{\mathbb{Q}(\nu);\mathcal{K}} (2\pi i)^{(3-v-(m-n))(m-n)/2+(m-n-1)(m-n)/2} p(\nu, i_0)^{m-n} \sim_{\mathbb{Q}(\nu);\mathcal{K}} (2\pi i)^{(2-v)(m-n)/2} p(\nu, i_0)^{r-r'}.$$

Similarly, if n is odd, then (3.26.2) can be rewritten

$$\prod_{j=1}^{m-n} L^S\left(\frac{1}{2}(2-v+m-n) - j, \chi'^{,+}\cdot\check{\nu}\right).$$

It again follows easily from (3.23.5) that this is a product of critical values of type ι, and we obtain that, in this case (3.26.2) is

$$(3.26.3,\ n \text{ odd}) \qquad \sim_{\mathbb{Q}(\nu);\mathcal{K}} (2\pi i)^{(1-v)(m-n)/2} p((\chi'^{,+})^{\vee}\cdot\nu, i_0)^{r-r'}.$$

Combining (3.26.3) with Proposition 3.24 and (3.26.1), and writing $\sim$ for

$$\sim_{\mathbb{Q}(\pi_0, \beta^{(2)}, \chi_0^+, \nu);\mathcal{K}},$$

we obtain

$$(3.26.4,\ n \text{ even}) \qquad (2\pi i)^{(2-v)n/2}\pi^{-c}\cdot I_{\Delta,\chi'}(\theta_{\varphi,\chi,\chi'}(F)^{\mathrm{arith}})\cdot p(\nu, i_0)^{r'} p(\nu,\iota)^s \sim L^S\left(\frac{1}{2}(2-v), \pi, (\chi/\chi')\cdot\check{\nu}^{\iota}, \mathrm{St}\right);$$

$$(3.26.4,\ n \text{ odd}) \qquad (2\pi i)^{((1-v)n+m)/2}\pi^{-c} I_{\Delta,\chi'}(\theta_{\varphi,\chi,\chi'}(F)^{\mathrm{arith}})\cdot p(\nu, i_0)^{r'} p(\nu,\iota)^s \sim L^S\left(\frac{1}{2}(2-v), \pi, (\chi/\chi')\cdot\check{\nu}^{\iota}, \mathrm{St}\right) p(\chi'^{,+}, i_0)^{n-m},$$

provided the product (3.26.2) does not vanish. However, the complete L-function $L(s,\chi'\cdot\check{\nu})$ can only vanish at the central point $s=(1-v)/2$. It follows from Rohrlich's theorem [**Roh**] that, up to twisting by a Dirichlet character, we can assume $L((1-v)/2,\chi'\cdot\check{\nu})\neq 0$. Moreover, Rohrlich's theorem allows us to choose the local data at the places S. Thus we can arrange that the missing Euler factors have no poles at the values in question. We henceforth assume ν to have been chosen in this way. Then relation (3.26.4) is valid.

It remains to evaluate

$$L^S\left(\frac{1}{2}(2-v),\pi,(\chi/\chi')\cdot\check{\nu}^\iota,\mathrm{St}\right)=L^{\mathrm{mot},S}\left(\frac{1}{2}(n-v+1),\pi_0,\tilde{\beta}\cdot(\chi/\chi')\cdot\check{\nu}^\iota,\mathrm{St}\right).$$

We again apply the results of [**H97**], specifically Theorem 3.5.13. We have seen that the weight $2k-\alpha(\chi)+\alpha(\chi')+v$ of the character $\tilde{\beta}\cdot(\chi/\chi')\cdot\nu^\iota$ is indeed in the r'th critical interval for $M(\pi_0)$. Under Hypothesis 3.20 (c), we can thus calculate the indicated special value in terms of the quadratic period $P^{(r')}(\pi_0)$ of an arithmetic form in $\pi^{(r',s)}$.

We write $\sim$ for $\sim_{\mathbb{Q}(\pi_0,\beta^{(2)},\chi_0^+,\nu);\mathcal{K}}$. In calculating the period, we use the analogues for the group $\mathrm{GU}(V^{(r',s)})$ of the formulas (3.25.1) and (3.25.2), rather than the expression [**H97**, 2.9.15] that intervenes in Theorem 3.5.13 of loc. cit. In place of the factor $p(\nu,i_0)^r p(\nu,\iota)^s$ that appears in (3.25.2) with the character $\check{\nu}$ and a group of signature (s,r), we obtain $p(\nu^\iota,i_0)^s p(\nu^\iota,\iota)^{r'}$ for the character $\check{\nu}^\iota$ and a group of signature (r',s) when $n\equiv m \pmod 2$, with the obvious modifications in the other cases. With this in mind, and under Hypothesis 3.20 (b), (d), the result is as follows:

$$L^{\mathrm{mot},S}\left(\frac{1}{2}(n-v+1),\pi,(\chi/\chi')\cdot\check{\nu}^\iota,\mathrm{St}\right)\sim(2\pi i)^{(n-v+1-w)n/2}\cdot\pi^{-c}\cdot P^{(r')}(\pi_0)p(\tilde{\beta},i_0)^{r'-s}p(\nu,i_0)^{r'}p(\nu,\iota)^s,\ n\text{ even},\ m\text{ even};\tag{3.26.5}$$

$$L^{\mathrm{mot},S}\left(\frac{1}{2}(n-v+1),\pi,(\chi/\chi')\cdot\check{\nu}^\iota,\mathrm{St}\right)\sim d\cdot(2\pi i)^{(n-v+1-w)n/2}\cdot\pi^{-c}\cdot P^{(r')}(\pi_0)p(\tilde{\beta},i_0)^{r'-s}p(\nu,i_0)^{r'}p(\nu,\iota)^s,\ n\text{ odd},m\text{ odd};\tag{3.26.6}$$

$$L^{\mathrm{mot},S}\left(\frac{1}{2}(n-v+1),\pi,\chi^+\cdot\nu^\iota,\mathrm{St}\right)\sim(2\pi i)^{(n-v-w)n/2}\cdot\pi^{-c}\cdot P^{(r')}(\pi_0)p(\tilde{\beta},i_0)^{r'-s}p((\chi^{+,\iota})^\vee\cdot\nu,i_0)^{r'}P((\chi^{+,\iota})^\vee\cdot\nu,\iota)^s,\ n\text{ even},m\text{ odd};\tag{3.26.7}$$

(3.26.8)

$$L^{\mathrm{mot},S}\left(\frac{1}{2}(v+n),\pi,(\check{\chi}')^+\cdot\nu^\iota,\mathrm{St}\right)\sim d\cdot(2\pi i)^{(n-v-w)n/2}\cdot\pi^{-c}\cdot P^{(r')}(\pi_0)p(\tilde{\beta},i_0)^{r'-s}p((\chi^{\prime,+})^\vee\cdot\nu,i_0)^{r'}p((\chi^{\prime,+})^\vee\cdot\nu,\iota)^s,\ n\text{ odd},m\text{ even}.$$

As always, $w=n-1$ in the above formulas and d is as in Lemma 3.18. In the last formula, we used the identity $\check{\chi}'=\chi^{\prime,\iota}$.

(3.27). End of the proof of Lemma 3.21. Hypothesis 3.20 (b) guarantees that the left-hand side in each of the relations (3.26.5-8) is non-zero. The proof of Lemma 3.21 now follows immediately by comparing (3.26.4) with (3.26.5-8) and using the period relations of III, §2. For example, in Case (b) we use the period relations to rewrite

$$\begin{aligned} p((\chi^{+,\iota})^{\vee}\cdot\nu,i_0)^{r'}p((\chi^{+,\iota})^{\vee}\cdot\nu,\iota)^{s} &\sim p((\chi^{+,\iota})^{\vee},i_0)^{r'}p((\chi^{+,\iota})^{\vee},\iota)^{s}p(\nu,i_0)^{r'}\cdot p(\nu,\iota)^{s} \\ &\sim p(\chi^{+,\iota},i_0)^{-r'}p(\chi^{+,\iota},\iota)^{-s}p(\nu,i_0)^{r'}\cdot p(\nu,\iota)^{s} \\ &\sim p(\chi^{+},\iota)^{-r'}p(\chi^{+},i_0)^{-s}p(\nu,i_0)^{r'}\cdot p(\nu,\iota)^{s}. \end{aligned}$$

The remaining simplifications are analogous.

(3.28). Theorem 3.22 can be rewritten

$$P^{(r')}(\pi_0)\sim_{\mathbb{Q}(\pi_0,\beta^{(2)},\chi_0^+);\mathcal{K}} Z_\infty\cdot Q_{r'}(\pi_0)\cdot P^{(r'-1)}(\pi_0). \tag{3.28.1}$$

This can be repeated for all signatures (r',s). For definiteness we write $Z_{\infty,r'}=Z_\infty$ in (3.28.1). Then by induction, we obtain

$$P^{(r')}(\pi_0)\sim_{\mathbb{Q}(\pi_0,\beta^{(2)},\chi_0^+);\mathcal{K}}\prod_{j=1}^{r'} Z_{\infty,j}\cdot Q_j(\pi_0). \tag{3.28.2}$$

Corollary 2.8.5 of [**H97**] states that, when $U(V),U(V^{(r',s)})$, etc., are replaced by the unitary groups of division algebras over $\mathcal{K}$ with involutions of the second kind, then the Tate conjecture, combined with Kottwitz' formula for the zeta functions of the associated Shimura varieties, implies

$$P^{(r')}(\pi_0)\sim_{\mathbb{Q}(\pi_0,\beta^{(2)},\chi_0^+);\mathcal{K}}\prod_{j=1}^{r'} Q_j(\pi_0). \tag{3.28.3}$$

Relation (3.28.3), in turn, for all $r', 0\le r'\le n$, implies the Deligne conjecture for the motives $R_{\mathcal{K}/\mathbb{Q}}(M(\pi_0)\otimes M(\alpha))$, up to scalars in $\mathcal{K}^\times$, for any algebraic Hecke character α of $\mathcal{K}$. Here $M(\alpha)$ is the absolute Hodge motive of CM type attached to α.

One of the purposes of [**HLa**] is to show that Kottwitz' formula holds for the motives $M(\pi_0)$ attached to untwisted unitary groups, under appropriate local hypotheses. Then the arguments of [**H97**] derive the Deligne conjecture, up to scalars in $\mathcal{K}^\times$, from the Tate conjecture for motives of this kind.

On the other hand, we can try to prove (3.28.3) directly. By (3.28.2), it suffices to show that

$$Z_{\infty,j}\in\mathbb{Q}, j=1,\dots,n. \tag{3.28.4}$$

The local zeta integral $Z_{\infty,j}$ can be made completely explicit. I hope to prove (3.28.4) in the sequel to this paper.

Appendix 1: Seesaw diagrams

The left-hand side consists of unitary groups of skew-hermitian spaces, the right-hand side of hermitian spaces. The groups are labelled (a.L) or (a.R), where a is the index and L or R refers to the left or right side of the diagram. Groups are labeled by their real points, for ease of references.

(1.L) $= U(n-1,1) \times U(1,n-1)$
(2.L) $= U(n,n)$
(2.L) $= [U(n-1,1) \times U(1,n-1)]^r \times [U(n-1,1) \times U(1,n-1)]^s$
(4.L) $= [U(1,0)^{n-1} \times U(0,1) \times U(1,0) \times U(0,1)^{n-1}]^{r+s}$
(5.L) $= U(n,n)^r \times U(n,n)^s$
(6.L) $= [U(1,1)^{n-1} \times U(1,1)]^r \times [U(1,1)^{n-1} \times U(1,1)]^s$

(1.R) $= U(r,s) \times U(r,s)$
(2.R) $= U(r,s)$
(3.R) $= U(1,0)^r \times U(1,0)^r \times U(0,1)^s \times U(0,1)^s$
(4.R) $= U(1,0)^{nr} \times U(1,0)^{nr} \times U(0,1)^{ns} \times U(0,1)^{ns}$
(5.R) $= U(1,0)^r \times U(0,1)^s$
(6.R) $= [U(1,0)^{n-1} \times U(1,0)]^r \times [U(0,1)^{n-1} \times U(0,1)]^s$

The diagram looks as follows:

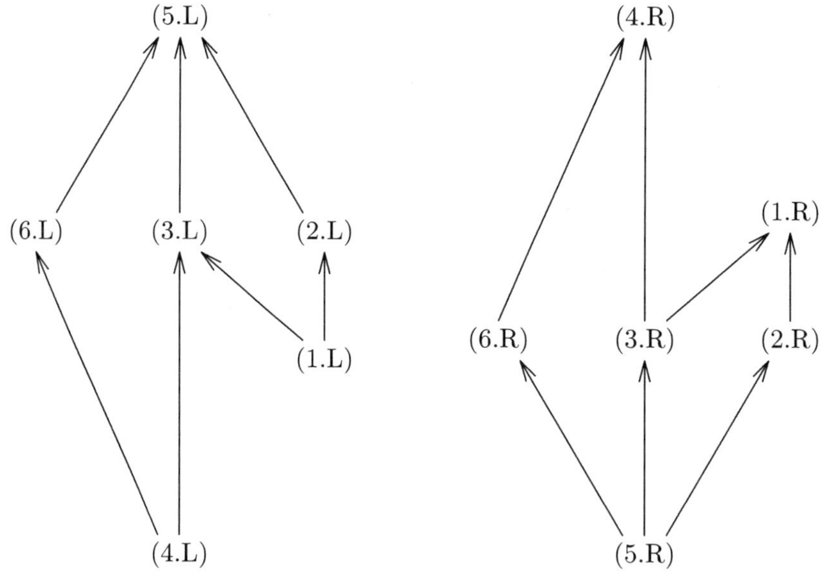

The following subdiagrams are significant:

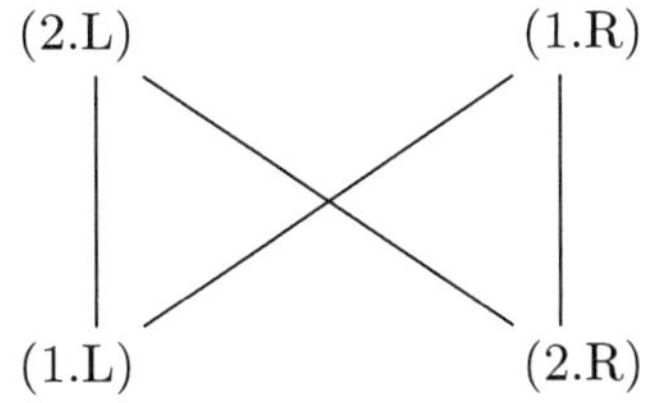

This computes the norm of theta lifts from cohomological forms on $U(n-1,1)$ to holomorphic on $U(r,s)$.

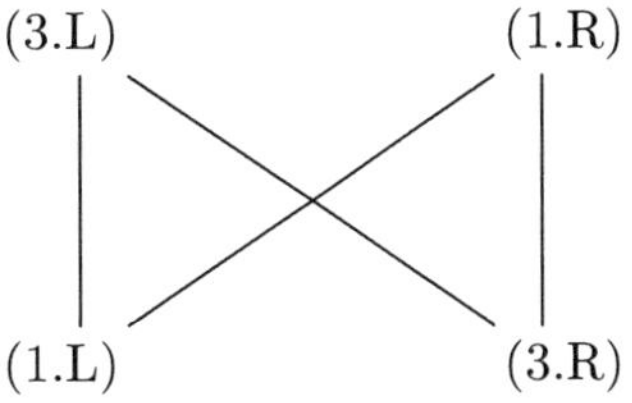

This determines the rationality of theta lifts from cohomological on $U(n-1,1)$ to holomorphic on $U(r,s)$.

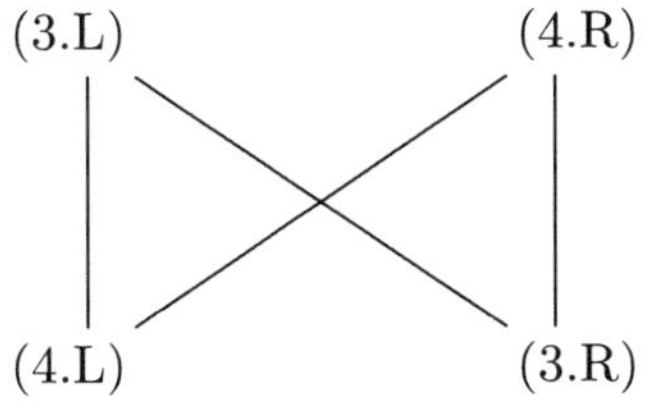

This determines the rationality of theta lifts from characters of $U(1)$ to small representations of $U(n-1,1)$.

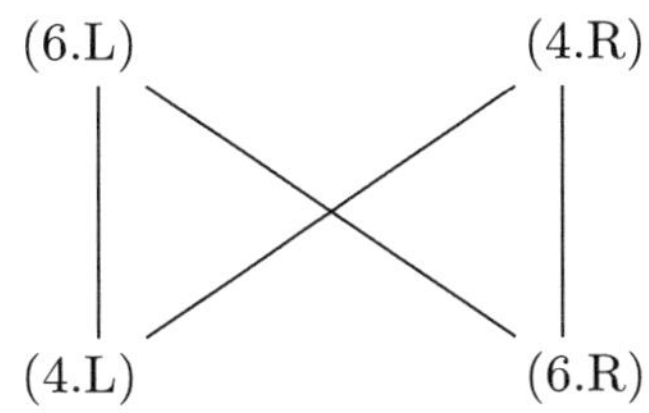

This determines the rationality of theta lifts from characters of $U(1)$ to characters of $U(1)$.

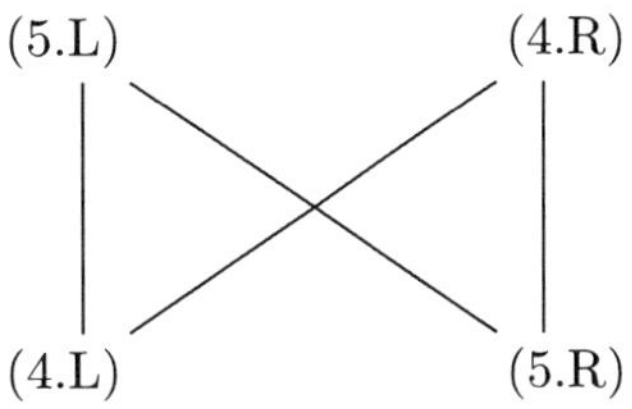

This determines (by adjunction) the norms of the lifts from $U(1)$ to small representations of $U(n-1,1)$.

Appendix 2. Splittings and seesaw pairs

In what follows, we have dual reductive pairs

$$(U(V)\times U(-V), U(V')\times U(V')),$$

with $\dim(V)=n$, $\dim(V')=m$. We choose a Hecke character χ_0 of $\mathcal{K}^\times$ such that $\chi_0|_{A^\times}=\varepsilon_{\mathcal{K}}$ and splitting characters $\chi,\chi'\in\{1,\chi_0\}$ such that $\chi=1$ (resp. $\chi'=1$) if and only if m (resp. n) is even. Write

$$V=V_1\oplus V_2\oplus\cdots\oplus V_n,\ \ V'=T_1\oplus T_2\oplus\cdots\oplus T_m,$$

where $\dim(V_i) = \dim(T_j) = 1$ for all i, j. We need to choose pairs of splitting characters $(\chi(j), \chi'(j)), j = 1, \dots, m$, for the dual reductive pairs

$$(U(V) \times U(-V), U(T_j) \times U(T_j)),$$

and pairs of splitting characters $(\chi(i,j), \chi'(i,j)), i = 1, \dots, n, j = 1, \dots, m$, for the dual reductive pairs

$$(U(V_i) \times U(-V_i), U(T_j) \times U(T_j)).$$

These characters have to satisfy various compatibility conditions, summarized by Lemma 3.6 of I. For $(\chi(j), \chi'(j))$, these take the form

$$\prod_{j=1}^{m} \chi(j) = \chi, \chi'(j) = \chi' \text{for all } j. \tag{A.2.1}$$

To ensure (A.2.1), we take

$$\chi(j) = \chi_0 \quad \text{for } j \text{ odd}, \ \chi(j) = \check{\chi}_0 \text{ for } j \text{ even}. \tag{A.2.2}$$

Then the indicated product equals 1 or χ_0, according as m is even or odd, and this is as it should be.

Next, the $(\chi(i,j), \chi'(i,j))$ have to satisfy the relation

$$\chi(i,j) = \chi(j) \quad \text{and } \prod_{i=1}^{n} \chi'(i,j) = \chi'(j) = \chi' \text{ for all } i, j. \tag{A.2.3}$$

Then we set

$$\chi(j) = \chi_0 \quad \text{for } j \text{ odd}, \ \chi(j) = \check{\chi}_0 \text{ for } j \text{ even}. \tag{A.2.4}$$

References

[B] A. Borel, *Automorphic L-functions*, in Automorphic Forms, Representations, and L-functions, Proc. Symp. Pure Math. **33** part I, Providence, RI: Amer. Math. Soc. (1979), 27–61.

[Br] S. Brocco, *On the non-vanishing of θ-liftings from $U(1)$ to $U(1)$*, Thesis, Brandeis University (1992).

[C90] L. Clozel, *Motifs et formes automorphes: applications du principe de fonctorialité*, in Automorphic Forms, Shimura Varieties, and L-functions, Perspectives in Mathematics, **10** vol. I, Boston: Academic Press (1990), 77–160.

[C91] L. Clozel, *Représentations Galoisiennes associées aux représentations automorphes autoduales de* GL(n), Publ. Math. I.H.E.S. **73** (1991), 97–145.

[D72] P. Deligne, *Travaux de Shimura*, Sém. Bourbaki (1970/71), exp. 389, Lecture Notes in Math. **244** (1972), 123–165.

[D79] P. Deligne, *Valeurs de fonctions L et périodes d'intégrales*, in Proc. Symp. Pure Math. **XXXIII** part 2 (1979), 313–346.

[F] J. Franke, *Harmonic analysis in weighted L_2 spaces*, manuscript, 1991.

[GPR] S. Gelbart, I. Piatetski-Shapiro and S. Rallis, Explicit Constructions of Automorphic L-Functions, Lecture Notes in Math. **1254**, New York: Springer (1987).

[GR] S. Gelbart and J. Rogawski, *L-functions and Fourier-Jacobi coefficients for the unitary group $U(3)$*, Inv. Math. **105** (1991), 445–472.

[GS] C. Goldstein and N. Schappacher, *Conjecture de Deligne et Γ-hypothèse de Lichtenbaum sur les corps quadratiques imaginaires*, C. R. Acad. Sci. Paris **296** (1983), 615–618.

[H86] M. Harris, *Arithmetic vector bundles and automorphic forms on Shimura varieties*, I. Inventiones Math. **82** (1985), 151–189; II. Compositio Math. **60** (1986), 323–378.

[H89] M. Harris, *Functorial properties of toroidal compactifications of locally symmetric varieties*, Proc. Lon. Math. Soc. **59** (1989), 1–22.

[H90] M. Harris, *Automorphic forms of cohomology type as coherent cohomology classes*, J. Diff. Geom. **32** (1990), 1–63.

[H93] M. Harris, *L-functions of* 2 *by* 2 *unitary groups and factorization of periods of Hilbert modular forms*, J. Am. Math. Soc. **6** (1993), 637–719.

[H94] M. Harris, *Hodge-de-Rham structures and periods of automorphic forms*, in Motives, Proc. Symp. Pure Math. AMS, **55**, Part 2, Providence, RI: Amer. Math. Soc. (1994), 573–624.

[H97] M. Harris, *L-functions and periods of polarized regular motives*, J. Reine Angew. Math. **483** (1997), 75–161.

[HF] J. Harris and W. Fulton, Representation Theory, Berlin: Springer-Verlag (1991).

[HK] M. Harris and S. S. Kudla, *Arithmetic automorphic forms for the nonholomorphic discrete series of* GSp(2), Duke Math. J. **66** (1992), 59–121.

[HKS] M. Harris, S. S. Kudla, and W. J. Sweet, Jr., *Theta dichotomy for unitary groups*, J. Am. Math. Soc. **9** (1996), 941–1004.

[HLa] M. Harris, J.-P. Labesse, *Conditional base change for unitary groups*, in preparation.

[HL] M. Harris, J.-S. Li, *A Lefschetz property for subvarieties of Shimura varieties*, J. Alg. Geom. **7** (1998), 77–122.

[JS] H. Jacquet and J. Shalika, *A non-vanishing theorem for zeta-functions of* GL_n, Invent. Math. **38** (1976), 1–16.

[K] R. Kottwitz, *Stable trace formula: cuspidal tempered terms*, Duke Math. J. **51** (1984), 611–650.

[Ku86] S. S. Kudla, *On the local theta-correspondence*, Invent. Math, **83** (1986), 229–255.

[Ku94] S. S. Kudla, *Splitting metaplectic covers of dual reductive pairs*, Israel J. Math. **87** (1994), 361–401.

[KuR] S. S. Kudla and S. Rallis, *A regularized Siegel-Weil formula: the first term identity*, Annals of Math. **140** (1994), 1–80.

[L90] J.-S. Li, *Theta liftings for unitary representations with non-zero cohomology*, Duke Math. **61** (1990), 913–937.

[L92] J.-S. Li, *Non-vanishing theorems for the cohomology of certain arithmetic quotients*, J. Reine Angew. Math. **428** (1992), 177–217.

[M] J. S. Milne, *Canonical models of (mixed) Shimura varieties and automorphic vector bundles*, Proceedings of the Ann Arbor Conference on Shimura Varieties and Automorphic Forms, Progress in Mathematics, New York: Academic Press, vol. I (1989), 283–414.

[Mo] C. Moen, *The dual pair* $(U(1), U(1))$ *over a p-adic field*, Pacific J. Math. **127** (1987), 141–154.

[MoVW] C. Moeglin, M.-F. Vigneras and J.-L. Waldspurger, Correspondence de Howe sur un corps p-adique, Lecture Notes in Math. **1291**, Berlin: Springer-Verlag (1987).

[Mu] D. Mumford, *Hirzebruch's proportionality theorem in the non-compact case*, Invent. Math. **42** (1977), 239–272.

[P] R. Pink, *Arithmetical compactification of mixed Shimura varieties*, Bonn. Math. Schrift., Nr. **209** (1990).

[R] S. Rallis, *Langlands functoriality and the Weil representation*, Am. J. Math. **104** (1982), 469–515.

[Ro] B. Roberts, *The theta correspondence for similitudes*, Israel J. Math., in press.

[Rog90] J. Rogawski, Automorphic representations of unitary groups in three variables, Ann. of Math. Studies **123**, Princeton, NJ: Princeton University Press (1990).

[Rog92] J. Rogawski, *Analytic expression for the number of points mod p*, in R. P. Langlands and D. Ramakrishnan, Eds., The Zeta Functions of Picard Modular Surfaces, Montreal: Les Publications CRM (1992), 65–110.

[Rog92b] J. Rogawski, *The multiplicity formula for A-packets*, in R. P. Langlands and D. Ramakrishnan, Eds., The Zeta Functions of Picard Modular Surfaces, Montreal: Les Publications CRM (1992), 395–419.

[Roh] D. Rohrlich, *Non-vanishing of L-functions for* GL(2), Inv. Math. **97** (1989), 381–404.

[S] W. Schmid, L^2 *cohomology and the discrete series*, Ann. of Math. **103** (1976), 375–394.

[Sh66] G. Shimura, *Moduli of abelian varieties and number theory*, in Algebraic Groups and Discontinuous Subgroups, Proc. Symp. Pure Math. **IX**, Providence, RI: Amer. Math. Soc. (1966), 312–332.

[Sh80] G. Shimura, *The arithmetic of certain zeta functions and automorphic forms on orthogonal groups*, Ann. of Math. **111** (1980), 313–375.

[Sh81] G. Shimura, *On certain zeta functions attached to two Hilbert modular forms*, I, Ann. of Math. **114** (1981), 569–607; II, Ann. of Math. **114** (1981), 127–164.

[V] F. R. Villegas, *On the square root of special values of certain L-series*, Invent. Math. **106** (1991), 549–573.

[VZ] D. Vogan and G. Zuckerman, *Unitary Representations with Non-Zero Cohomology*, Compositio Math. **53** (1984), 51–573.

[W] A. Weil, *Sur la Formule de Siegel dans la théorie des groupes classiques*, Acta Math. **113** (1965), 1–87.

[Y] T. Yang, *Theta liftings and L-functions of elliptic curves*, Ph.D. Thesis, University of Maryland (1995).

[Z] S. Zucker, *L^2 cohomology of warped products and arithmetic groups*, Invent. Math. **70** (1982), 169–218.

Institute de Mathématiques de Jussieu, U.F.R. de Mathematiques, Universite Paris 7, 2, Pl. Jussieu, 75251 Paris Cedex 05

E-mail address: harris@math.jussieu.fr

Proceedings of Symposia in Pure Mathematics
Volume **66.2**, 1999

Pullbacks of Eisenstein Series, Hecke-Jacobi Theory and Automorphic L-Functions

Bernhard E. Heim

Dedicated to Professor Goro Shimura

Abstract. In this paper a new Rankin-Selberg integral is presented, which is a Petersson scalar product of a diagonal-restricted Eisenstein series against an elliptic modular form and two Siegel modular forms of degree n. This generalizes Paul Garrett's work on the Rankin triple product L-function. Surprisingly the case $n = 2$ leads, as $n = 1$, to an Euler product. We prove, by choosing one of the Siegel forms as a Saito-Kurokawa lift, that our approach gives a new integral representation of a degree eight L-function for $\mathrm{GSp}_2 \times \mathrm{GL}_2$, the twisted spinor L-function.

Introduction

The purpose of this paper is to present new insights into the connection between pullbacks of Eisenstein series and automorphic L-functions. Surprisingly, special properties of a certain Hecke-Jacobi algebra will play a decisive role. The paper has its origin in a question asked by S. Böcherer to the author.

Let us consider the following situation. Let $(n_i)_{i=1}^l$ be a partition of a natural number n and let $F_i \in S_{n_i}^k$ be Siegel cusp forms of degree n_i and weight k. Let further $E_n^k(z;s)$ be the analytic Eisenstein series of Siegel type, here z is in the Siegel half space $\mathcal{H}_n$ of degree n and $s \in \mathbb{C}$ with $2\mathrm{Re}(s)+k > n+1$. Let us further denote by $\langle\,;\,\rangle$ the Petersson scalar product and let us assume that all the F_i's are Hecke eigenforms. Then P. Garrett [**G84**], S. Böcherer [**Bo85b**] and I. Piatetski-Shapiro and S. Rallis [**PR84**], [**GPR**] have shown, that the iterated scalar product

$$(1)\qquad \Phi^n(F_1,\dots,F_l;s) := \Big\langle \dots \Big\langle E_n^k\left(\begin{pmatrix} z_1 & & 0 \\ & \ddots & \\ 0 & & z_l \end{pmatrix}; s\right); F_1(z_1)\Big\rangle \dots F_l(z_l)\Big\rangle$$

for $l = 2$ is essentially the standard L-function $D_{F_1}^{n_1}$ of F_1 multiplied with $\langle F_1, F_2\rangle$, if $n_1 = n_2$ and zero otherwise. By considering the case $l = 3$ with $n_1 = n_2 = n_3 = 1$,

1991 *Mathematics Subject Classification.* Primary 11F66, 11F67; Secondary 11F46, 11F60.

Key words and phrases. Pullbacks of Eisenstein series, Hecke-Jacobi theory, automorphic L-functions, Rankin triple-product L-function, special values of L-functions.

P. Garrett [**G87**] succeeded in refining his previous work to get an integral representation of the triple product L-function for the involved elliptic modular forms. This was generalized by I. Piatetski-Shapiro and S. Rallis [**PR87**] to an integral representation of the "internal tensor product L-function" of an automorphic form on a semi-simple cubic abelian algebra over a number field (see also [**G85**]). (Of course, to get these integral representations a profound study of the pullbacks of the Eisenstein series via the embeddings $\mathcal{H}_{n_1} \times \ldots \times \mathcal{H}_{n_l} \hookrightarrow \mathcal{H}_n$ is needed.) These results were the starting point and a basis for a lot of important mathematical work, e.g. [**O**], [**S**], [**HK**], [**GH**], [**GroK**], [**BoS**], [**Sh95**]. A good overview of the results concerning the triple product L-function can be found in the introduction to the paper [**GH**] by P. Garrett and M. Harris. Since the approach in (1) has been so successful it is certainly natural to ask:

> *Are there any other partitions of a natural number n, which lead to integral representations of an automorphic L-functions in the just described way?*

Already in 1985, at the conference of modular functions in Oberwolfach, Garrett mentioned at the end of his talk, that it seems to be unlikely. Nevertheless, this paper gives a positive answer. We produce a family of new pullbacks of Eisenstein series attached to partitions $1+n+n$ of $2n+1$, calculate $\Phi(F_1, F_2, F_3; s)$ explicitly as a Dirichlet series and show that in a special case we have established a new integral representation of an automorphic L-series.

To state our main results we have to introduce the following notation. Let Γ_n be the Siegel modular group. Writing an element of Γ_n in the form $\gamma = \left(\begin{smallmatrix} a & b \\ c & d \end{smallmatrix}\right)$, with matrices a, b, c, d of size n, we can define the action of Γ_n on $\mathcal{H}_n$ by $\gamma < z >= (az+b)(cz+d)^{-1}$. Let $P_{n,0}$ the subgroup of Γ_n consisting of elements γ for which $c = 0$. Then we can introduce the Eisenstein series

$$E_n^k(z;s) = \sum_{\gamma \in P_{n,0} \backslash \Gamma_n} \det(cz+d)^{-k} \det(\mathrm{Im}\gamma < z >)^s.$$

Here $z \in \mathcal{H}_n, s \in \mathbb{C}$ and $k \in 2\mathbb{Z}$. Let $\mathcal{H}_1 \times \mathcal{H}_n \times \mathcal{H}_n$ be diagonally embedded in $\mathcal{H}_{2n+1}$. Let $\langle\, ; \,\rangle, \langle\, ; \,\rangle_J$ be the Petersson scalar products of Siegel modular and Jacobi forms. In the following, we assume for convenience the condition *totally real* for the Fourier coefficients of Siegel cusp forms, which can be easily dropped [**Sh76**], [**Bo85a**].

THEOREM 1. *Let $F, G \in S_n^k$ and $h \in S_1^k$ be Hecke eigenforms with totally real Fourier coefficients and k even. Then for $s \in \mathbb{C}$ with $2Re(s) + k > 2n+2$*

$$\Phi^{2n+1}(h, G, F; s) = a(n, s, k)\mathcal{L}_{h,G,F}^n(s + 2k - n - 1).$$

Here $a(n, s, k)$ is essentially a product of gamma functions and

$$\mathcal{L}_{h,G,F}^n(s+2k-n-1) = \left\{ \zeta(2s+k) \prod_{j=1}^{n} \zeta(4s+2k-2j) \right\}^{-1} \cdot D_F^n(2s+k-n) D_{h,G;F}(s),$$

where D_F^n is the standard zeta function of F and $D_{h,G;F}(s)$ denotes a certain Dirichlet series (see section 2.4) involving the Fourier coefficients of h and the Fourier-Jacobi coefficients of F and G amalgamated together by the scalar product $\langle\, ; \,\rangle_J$.

Has this anything to do with Euler products? If $n = 2$ (the case $n = 1$ corresponds to the triple product L-function [**G87**]) we can prove the following

MAIN THEOREM. *Let $F, G \in S_2^k$ and $h \in S_1^k$ be Hecke eigenforms with totally real Fourier coefficients. Let k be even and h normalized. Further we assume that F is in the Maass Spezialschar with corresponding $f \in S_1^{2k-2}$. Then for $s \in \mathbb{C}$ with $2Re(s) + k > 6$*

$$\Phi^5(h, G, F; s) = a(2, s, k)\langle \Phi_1^F; \Phi_1^G \rangle_J \, \zeta(2s+k)^{-1}\zeta(4s+2k-2)^{-1}\zeta(4s+2k-4)^{-1} \cdot L_f(2s+2k-4) Z_{G\otimes h}(s+2k-3).$$

Here L_f denotes the Mellin transform of f and $Z_{G\otimes h}$ the convolution of the spinor zeta functions attached to G and h, in other words the L-functions to $\mathrm{GSp}_2 \times \mathrm{GSp}_1$. *And with Φ_N^H we denote the N-th Fourier-Jacobi coefficient of a cusp form H.*

If we assume that the first Fourier-Jacobi coefficient of G is not identically zero, we can conclude that $Z_{G\otimes h}(s)$ has a meromorphic continuation to the whole s-plane and get a functional equation (see Corollary 5.2). Taking G with totally real algebraic Fourier coefficients and of weight $k > 6$, and $\sigma \in \mathrm{Gal}(\overline{\mathbb{Q}}/\mathbb{Q})$ we get

$$\left(\pi^{8-5k}\frac{Z_{G\otimes h}(2k-3)}{\langle G; G\rangle\langle h; h\rangle}\right)^{\sigma} = \pi^{8-5k}\frac{Z_{G^\sigma\otimes h^\sigma}(2k-3)}{\langle G^\sigma; G^\sigma\rangle\langle h^\sigma; h^\sigma\rangle} \in \overline{\mathbb{Q}}.$$

RELATED WORK: In [**KS**], W. Kohnen and N.-P Skoruppa studied analytic and arithmetic properties of $D_{F,G}(s) = \sum_{N=1}^{\infty}\langle \Phi_N^F; \Phi_N^G\rangle_J N^{-s}$ with $F, G \in S_2^k$. They proved that $D_{F,G}(s)$ has a meromorphic continuation to the whole s-plane. This was generalized in [**Y**] with respect to $F, G \in S_n^k$. Let F be in the Maass Spezialschar and G a Hecke eigenform of degree 2, then

$$D_{F,G}(s) = \langle \Phi_1^F; \Phi_1^G\rangle_J Z_G^2(s). \tag{2}$$

Here, $Z_G^2(s)$ denotes the spinor zeta function of G. If Φ_1^G is identically zero, then one replaces F by a certain paramodular cusp form such that one can still use the approach of Kohnen and Skoruppa to study $Z_G^2(s)$ for an arbitrary eigenform G. This refinement is due to V. Gritsenko [**Gr93**].

A different approach to study the L-function $Z_{G\otimes h}(s)$ was given in 1993 by M. Furusawa in [**Fu**]. He showed that the L-function is essentially a scalar product of a Hermitian Eisenstein series attached to h and G, if G satisfies certain conditions. His results are compatible with our results concerning the meromorphic continuation, functional equation and the special value property of $Z_{G\otimes h}(s)$. For certain reasons it was impossible for him with his approach to state the Galois behavior of the special value.

BRIEF OUTLINE OF THE ARTICLE: After reviewing some known results on Siegel modular forms, analytic Eisenstein series and a Hecke theory for the symplectic group, we state the Theorem of Garrett and Böcherer, which gives a first insight in the correspondence of pullbacks of Eisenstein series of Siegel and Klingen type. This leads us to the problem of understanding $P_{n+1,n}(\mathbb{Z})\backslash \mathrm{Sp}_{n+1}(\mathbb{Z})$ with respect to the action of $\mathrm{Sp}_1(\mathbb{Z}) \times \mathrm{Sp}_n(\mathbb{Z}) \hookrightarrow \mathrm{Sp}_{n+1}(\mathbb{Z})$ from the right, here $P_{n+1,n}(\mathbb{Q}) = \{ \left(\begin{smallmatrix} a & * \\ 0 & * \end{smallmatrix}\right) \in \mathrm{Sp}_{n+1}(\mathbb{Q}) \mid a \in \mathbb{Q}^\times \}$.

In section 2, we present briefly a viewpoint, which reduces this problem to the calculation of the double coset space

$$P_{n+1,n}(\mathbb{Q})\backslash \mathrm{Sp}_{n+1}(\mathbb{Q})/\mathrm{Sp}_1(\mathbb{Q}) \times \mathrm{Sp}_n(\mathbb{Q})$$

and indicate how one can do that in a systematic manner. Behind all that stands a new method, which allows us to understand orbit structures of much more general shape. This will be the main subject of a forthcoming paper. After stating a pullback formula of the relevant Eisenstein series of Klingen type, we prove Theorem 1 and give some first applications.

In the next two sections we show, that $\mathcal{L}^n_{h,G,F}(s)$ has an Euler product for $n = 2$ (the case $n = 1$ is due to Garrett [**G87**]) and that Fourier-Jacobi coefficients of eigenfunctions of Siegel modular forms have certain multiplicative properties. For that, we have chosen the following strategy:

In section 3, we construct a modified Hecke-Jacobi algebra $\tilde{\mathcal{H}}^J$ in which one can embed the usual symplectic Hecke algebra $\mathcal{H}^2$ of degree 2. Highlights of our considerations are factorizations of the Hecke and Rankin polynomials of degree 2 (cf. [**Gr82**], [**Gr93**] and [**A87**]). We interpret Jacobi forms as automorphic forms (P-forms) on a higher dimensional space to maximal parabolic subgroups of the symplectic group (Klingen [**K89**]), which explains many properties of Jacobi forms in a transparent way. With the results developed in section 2 and 3 we prove in section 4, that $\mathcal{L}^2_{h,G,F}(s)$ has an Euler product and calculate the local factors explicitly.

In section 5, we summarize our investigations and deduce important properties of the L-function $Z_{G\otimes h}(s)$, assuming that the first Fourier-Jacobi coefficient of G is not identically zero. The reader should not be surprised about that condition, because already in the case $n = 1$ [**G87**], the Fourier(-Jacobi) coefficients play a normalizing role. This paper is rounded out by a unifying presentation of some properties of the triple product L-function and the L-function $Z_{G\otimes h}(s)$. This makes it possible to translate and apply techniques, that have been developed to analyze the triple product L-function, to $Z_{G\otimes h}(s)$ and get a deeper understanding of it in the future.

This paper contains the main parts of my thesis (Mannheim 97) which was written under the supervision of Prof. Dr. S. Böcherer. I would like to thank him for introducing me to the subject and for his advice. I am grateful to the Max-Planck-Institut für Mathematik in Bonn for its hospitality and support. Finally I would like to take the opportunity to thank the organizers of the conference, Robert S. Doran, Ze-Li Dou and George T. Gilbert for their excellent job and their warm hospitality.

NOTATION: For an associative ring R with identity element, we denote by $R^\times$ the group of all invertible elements of R and by $M(m, n; R)$ the module of all $m \times n$ matrices with entries in R. If M is a matrix, M^t, $\det(M)$, and $\mathrm{tr}(M)$ stand for its transpose, determinant, and trace. We put $M_n(R) = M(n, n; R)$, $\mathrm{GL}_n(R) = M_n(R)^\times$, $\mathrm{SL}_n(R) = \{M \in \mathrm{GL}_n(R) | \det(M) = 1\}$ (when R is commutative). The identity and zero elements of $M_n(R)$ are denoted by 1_n and 0_n respectively (when

n needs to be stressed). Let $J_n = \left(\begin{smallmatrix} 0 & 1 \\ -1 & 0 \end{smallmatrix}\right)$; then

$$\mathrm{Sp}_n(R) = \{M \in \mathrm{GL}_{2n}(R) \mid M^t J_n M = J_n\},$$

$$P_{n,r}(R) = \left\{ \begin{pmatrix} * & * \\ 0_{n+r,n-r} & * \end{pmatrix} \in \mathrm{Sp}_n(R) \right\},$$

$$C_{n,r}(R) = \left\{ \begin{pmatrix} * & * \\ 0_{n-r,n+r} & * \end{pmatrix} \in \mathrm{Sp}_n(R) \right\},$$

$$C^+_{n,r}(R) = \left\{ \begin{pmatrix} * & * \\ 0_{n-r,n+r} & 1_{n-r} \end{pmatrix} \in \mathrm{Sp}_n(R) \right\}.$$

For convenience let $\Gamma_n = \mathrm{Sp}_n(\mathbb{Z})$, $P_{n,r} = P_{n,r}(\mathbb{Z})$, $\Gamma = \Gamma_1$ and $\Gamma_\infty = P_{1,0}$. For a subring R of $\mathbb{R}$ we put $R^+ = \{r \in R \mid r > 0\}$ and define

$$G^+\mathrm{Sp}_n(R) = \{M \in \mathrm{GL}_{2n}(R) \mid M^t J_n M = \mu(M) J_n \text{ with } \mu(M) \in R^+\}.$$

For real symmetric matrices A and B, we write $A > B$ ($A \geq B$) to indicate that $A - B$ is positive definite (positive semi-definite) and $A[B] = B^t A B$ if A, B are suitable. If $A_1, A_2, \ldots, A_n$ are square matrices, $[A_1, A_2, \ldots, A_n]$ denotes the matrix with $A_1, A_2, \ldots, A_n$ in the diagonal blocks and 0 in all other blocks. Further we write $[A_1, \ldots, A_m; A_{m+1} \ldots, A_{2m}]$ for numbers A_j, if $n = 2m$. The disjoint union of sets is denoted by $\bigsqcup$. Let $T, Z \in M_n(\mathbb{C})$, then we put $e\{TZ\} = e^{2\pi i (\mathrm{tr} TZ)}$ and $\mathrm{Re}(Z), \mathrm{Im}(Z)$ for the real and imaginary part of Z and for $M \in M(m, n; \mathbb{C})$ we put occasionally $x_M = \mathrm{Re}(M)$ and $y_M = \mathrm{Im}(M)$. Further $\delta(Z) = \det(\mathrm{Im}(Z))$. By $\overline{\mathbb{Q}}$ we denote the algebraic closure of $\mathbb{Q}$, which we assume to be embedded into $\mathbb{C}$ and $\bar{a}$ denotes the complex conjugation of an element $a \in \mathbb{C}$. Let us always assume that p is a prime number.

1. Preliminaries

1.1. Siegel modular forms. The group of positive symplectic similitudes $G^+\mathrm{Sp}_n(\mathbb{R})$ acts on Siegel's half space $\mathcal{H}_n = \{Z = Z^t \in M_n(\mathbb{C}) \mid \mathrm{Im}(Z) > 0\}$ of degree n as a group of biholomorphic automorphisms by

$$(M, Z) \mapsto M < Z >= (AZ + B)(CZ + D)^{-1},$$

where $M = \left(\begin{smallmatrix} A & B \\ C & D \end{smallmatrix}\right)$ with $A, B, C, D \in M_n(\mathbb{R})$. We denote the factor of automorphy by $j(M, Z) = \det(CZ + D)$. For $F : \mathcal{H}_n \to \mathbb{C}$ and $k \in \mathbb{Z}$ we define the Petersson operator

$$(F|_k M)(Z) := \mu(M)^{nk - n(n+1)/2} j(M, Z)^{-k} F(M < Z >). \tag{3}$$

The real positive number $\mu(M)$ is called factor of similitude. Let M_n^k be the space of Siegel modular forms and S_n^k the subspace of cusp forms of degree n and weight k for the Siegel modular group Γ_n. Every $F \in M_n^k$ satisfies the transformation law

$$F|_k M = F, \text{for all } M \in \Gamma_n = \mathrm{Sp}_n(\mathbb{Z}) \tag{4}$$

and has a Fourier expansion of the form

$$F(Z) = \sum_{T \in A_n} a(T) e\{TZ\},$$

with $a(T) \in \mathbb{C}$ and A_n denotes the set of semi-integral positive semidefinite matrices of size n. For cusp forms we have only to sum over $A_n^+ = \{T \in A_n | T > 0\}$. It is well known [**Sh75**] that, if σ is an automorphism of $\mathbb{C}$ and $F \in S_n^k$, then we have

$$F^\sigma(Z) = \sum_{T \in A_n^+} a(T)^\sigma e\{TZ\} \in S_n^k. \tag{5}$$

For arbitrary complex valued functions F and G on $\mathcal{H}_n$, which satisfy (4), one can define the Petersson integral, convergence assumed,

$$\langle F, G \rangle = \int_{\Gamma_n \backslash \mathcal{H}_n} F(Z)\overline{G(Z)} \det(\mathrm{Im} Z)^k \, d^*Z. \tag{6}$$

Here $d^*Z = \det(Y)^{-(n+1)} dXdY$ denotes the symplectic volume element. For more details the reader is referred to Andrianov [**A87**] or Klingen [**K90**].

1.2. Classical Hecke theory for the symplectic group. We start by recalling the basic facts on abstract Hecke rings [**A87**, §3.1], [**Sh71**, Chapter 3]. Let (R, S) be a Hecke pair, this means R is a subgroup of the group S and for $s \in S$ the coset space $R\backslash RsR$ is finite. Let A be a principal ideal domain, then R acts from the right on the $A-$module

$$L_A(R, S) = \left\{ X = \sum_{i=1}^n a_i R s_i \mid a_i \in A, s_i \in S \text{ and } n \in \mathbb{N} \right\}.$$

The subset of elements $H_A(R, S)$ invariant under this action forms a ring with

$$\left(\sum_i a_i R g_i \right) \circ \left(\sum_j b_j R h_j \right) = \sum_{i,j} a_i b_j R g_i h_j.$$

It is called Hecke ring or Hecke algebra to stress either the ring or algebra properties. Identifying double cosets $RsR = \bigsqcup_{i=1}^n Rs_i$ with $\sum_{i=1}^n Rs_i \in H_A(R, S)$ makes it possible to see them as a basis of the A-module $H_A(R, S)$.

LEMMA 1.1. *Let (Γ_1, S_1) and (Γ_2, S_2) be two Hecke pairs included in a bigger group. If we assume that*

$$\Gamma_2 \subset \Gamma_1, S_1 \subset \Gamma_1 S_2, \text{ and } \Gamma_1 \cap S_2 S_2^{-1} \subset \Gamma_2,$$

then we one can define an embedding i of $H_A(\Gamma_1, S_1)$ into $H_A(\Gamma_1, S_1)$ in the following way: write $X \in L_A(\Gamma_1, S_1)$ in the form $\sum_i a_i \Gamma_1 s_{i,2}$ with $s_{i,2} \in S_2$ and $a_i \in A$, then $i(X) := \sum_i a_i \Gamma_2 s_{i,2}$.

LEMMA 1.2. *If (R, S) is a Hecke pair, then (R, S) is also a Hecke pair and*

$$j : \begin{cases} H_A(R, S) & \longrightarrow & H_A(R, S^{-1}) \\ RsR & \mapsto & Rs^{-1}R \end{cases} \tag{7}$$

is an anti-isomorphism which commutes with the embedding i defined in the last lemma.

1.2.1. *Symplectic Hecke algebra $\mathcal{H}^n$.* By applying the elementary divisor theorem, it is not difficult to see that the symplectic Hecke algebra

$$\mathcal{H}^n := H_{\mathbb{Q}}(\Gamma_n, G^+\mathrm{Sp}_n(\mathbb{Q}))$$

decomposes into an infinite tensor product

$$\mathcal{H}^n = \otimes_p \mathcal{H}_p^n.$$

The local Hecke algebra $\mathcal{H}_p^n = H_{\mathbb{Q}}(\Gamma_n, \mathrm{GSp}_n^+(\mathbb{Z}[p^{-1}]))$ is isomorphic by the Satake-isomorphism Φ_p^n to $\mathbb{Q}[x_0^{\pm 1}, \dots, x_n^{\pm 1}]^{W_n}$. Here W_n denotes the Weyl group of Sp_n (for more details see Freitag [**F**, Kapitel 4]). The Hecke algebra $\mathcal{H}_p^n$ is generated by

$$T^{(n)}(p) = \Gamma_n[1, \dots, 1; p, \dots, p]\Gamma_n, T_n^{(n)}(p)^{-1}$$
$$T_i^{(n)}(p^2) = \Gamma_n[\underbrace{1, \dots, 1}_{n-i}, \underbrace{p, \dots, p}_{i}; p^2, \dots, p^2, p, \dots, p]\Gamma_n$$

for $i = 0, \dots, n$. Furthermore we define $T^{(n)}(a) = \sum_{M \in SD_n(a)} \Gamma_n M \Gamma_n$ with

$$SD_n(a) = \Big\{[e_1, \dots, e_n; f_1, \dots, f_n]\Big| e_i, f_j \in \mathbb{N}; e_i | e_{i+1}; e_n | f_n \text{ and } e_i f_i = a\Big\}.$$

DEFINITION 1.3. The Hecke polynomial $Q_p^n(X)$ and Rankin polynomial $D_p^n(X)$ of degree n are defined by

$$Q_p^n(X) := (1 - x_{0,p}X) \prod_{r=1}^{n} \prod_{1 \le i_1 < \dots < i_r \le n} (1 - x_{0,p} x_{i_1,p} \cdot \dots \cdot x_{i_r,p} X)$$
$$D_p^n(X) := (1 - X) \prod_{i=1}^{n} (1 - x_{i,p} X)(1 - x_{i,p}^{-1} X).$$

Taking the pre-images of the coefficients of these polynomials under Φ_p^n, we get polynomials in $\mathcal{H}_p^n[X]$ which we denote by $q_p^n(X)$ and $d_p^n(X)$. We denote the formal spinor- and standard zeta function of degree n by $Z^n(s) = \prod_p \left(Q_p^n(p^{-s})\right)^{-1}$ and $D^n(s) = \prod_p \left(D_p^n(p^{-s})\right)^{-1}$ for $s \in \mathbb{C}$.

By modifying the map j slightly we get an (anti-)automorphism

$$(8) \qquad * : \mathcal{H}^n \longrightarrow \mathcal{H}^n, \Gamma_n M \Gamma_n \mapsto (\Gamma_n M \Gamma_n)^* = \Gamma_n \mu(M) M^{-1} \Gamma_n.$$

To simplify our notation in the case $n = 1$, we put $T(a, d) = \Gamma \left(\begin{smallmatrix} a & 0 \\ 0 & d \end{smallmatrix}\right) \Gamma$ and $T(n) = T^{(1)}(n)$. Then

$$T(p) = \sum_{a \bmod p} \Gamma \begin{pmatrix} 1 & a \\ 0 & p \end{pmatrix} + \Gamma \begin{pmatrix} p & 0 \\ 0 & 1 \end{pmatrix} = \Gamma \begin{pmatrix} 1 & 0 \\ 0 & p \end{pmatrix} U_{[1,p]},$$

where $U_{[a,b]}$ denotes a representative system of $\left(\left(\begin{smallmatrix} a & 0 \\ 0 & b \end{smallmatrix}\right)^{-1} \Gamma \left(\begin{smallmatrix} a & 0 \\ 0 & b \end{smallmatrix}\right) \cap \Gamma\right) \backslash \Gamma$. In $\mathcal{H}_p^n[X]$, we have the following formal identity

$$(9) \qquad \sum_{n=0}^{\infty} T(p^n) X^n = (1 - T(p)X + pT(p,p)X^2)^{-1} = q_p^1(X)^{-1}.$$

It is not difficult to see that

$$(10) \qquad pT_1^{(2)}(p^2) + p(p^2+1)\Delta_p^{(2)} = T^{(2)}(p)^2 - T^{(2)}(p^2) - p^2 \Delta_p^{(2)}$$

holds in $\mathcal{H}_p^2$. Important for our purpose is the explicit form of $q_p^2(X)$ and $d_p^2(X)$.

LEMMA 1.4. *In $\mathcal{H}_p^2[X]$, we have*

$$(11) \qquad q_p^2(X) = 1 - A_1 X + A_2 X^2 - p^3 \Delta_p A_1 + p^6 (\Delta_p)^2 X^4$$
$$(12) \qquad d_p^2(X) = (1 - X)(1 - B_1 X + B_2 X^2 - B_1 X^3 + X^4),$$

with $A_1 = T^{(2)}(p), A_2 = p(T_1^{(2)}(p^2) + (p^2+1)\Delta_p)$ *and*

$$B_1 = \Delta_p^{-1}p^{-3}\Big(T^{(2)}(p)^2 - T^{(2)}(p^2) - p^2\Delta_p - 2\Delta_p p^3\Big),$$

$$B_2 = \Delta_p^{-1}p^{-3}\Big(2T^{(2)}(p^2) - T^{(2)}(p)^2 + 2p^3\Delta_p + 2p^2\Delta_p\Big).$$

The space S_n^k can be seen as an $\mathcal{H}^n$-module. Let $X = \sum a_i\Gamma_n g_i \in \mathcal{H}^n$, then

$$F \mapsto F|_k\mathcal{D}(X) := \sum_i a_i F|_k g_i. \tag{13}$$

It is well known that there exists a basis of S_n^k consisting of eigenforms. An eigenform F induces for every local algebra $\mathcal{H}_p^n$ an algebra homomorphism to $\mathbb{C}$, which can be characterized by $(\alpha_{0,p}, \alpha_{1,p}, \dots, \alpha_{n,p}) \in (\mathbb{C}^\times)^{n+1}$, the Satake-parameters of F. The eigenvalue $p^{nk-n(n+1)}$ of Δ_p^n gives $\alpha_{0,p}^2\alpha_{1,p}\dots\alpha_{n,p} = p^{nk-(n(n+1))/2}$. (The parameters are unique up to the action of the Weyl group.) To such a F we can define $Q_{p,F}^n(X), D_{p,F}^n(X), Z_F^n(s), D_F^n(s)$. It is known that the last two functions are absolutely convergent for Re(s) large.

In the case $n = 1$, it is convenient to use a different notation. One defines $\alpha_p = \alpha_{0,p}$ and $\beta_p = \alpha_{0,p}\alpha_{1,p}$ for Satake-parameters $\alpha_{0,p}, \alpha_{1,p}$ of a normalized $(a_1^f = 1)$ Hecke eigenform $f \in S_1^k$ where a_m^f denotes the mth Fourier coefficient of f. We usually want to assume that an eigenform in S_1^k is normalized. Then one has

$$\alpha_p\beta_p = p^{k-1} \text{ and } \alpha_p + \beta_p = a_p^f.$$

We write $L_f(s)$ for $Z_f^1(s)$ and $L_{p,f}(X)$ for $Q_{p,f}^1(X)$. With this notation we introduce the L-function $Z_{G\otimes h}$ corresponding to two eigenforms $G \in S_2^k$ and $h \in S_1^k$ with parameters $\alpha_{0,p}, \alpha_{1,p}, \alpha_{2,p}$ and α_p, β_p for every prime number p:

$$Z_{G\otimes h}(s) = \prod_p \left\{Q_{p,G}^2(\alpha_p p^{-s})Q_{p,G}^2(\beta_p p^{-s})\right\}^{-1}.$$

The triple product L-function of eigenforms $f_1, f_2, f_3 \in S_1^k$ is

$$L_{f_1\otimes f_2\otimes f_3}(s) = \prod_p \det\left(1_8 - \begin{pmatrix}\alpha_p^{f_1} & 0\\ 0 & \beta_p^{f_1}\end{pmatrix}\otimes\begin{pmatrix}\alpha_p^{f_2} & 0\\ 0 & \beta_p^{f_2}\end{pmatrix}\otimes\begin{pmatrix}\alpha_p^{f_3} & 0\\ 0 & \beta_p^{f_3}\end{pmatrix}p^{-s}\right)^{-1}.$$

REMARK 1.5. For each subring $A \subset \mathbb{C}$, let $S_n^k(A)$ be the $A-$module of all $F \in S_n^k$ for which all the Fourier coefficients are in A. Let F be an eigenform, especially we have $F|_kT^{(n)}(m) = \lambda_m^F F$ for all $m \in \mathbb{N}$ and define a subfield $\mathbb{Q}(F)$ of $\overline{\mathbb{Q}}$ generated by $\mathbb{Q}$ and the set of all λ_m^F. Then $\mathbb{Q}(F)$ is a totally real finite extension of $\mathbb{Q}$ and for $k \geq 2n+2$ there exists an orthogonal eigenbasis $\{F_j\}_{j=1}^d$ of S_n^k such that $F_j \in S_n^k(\mathbb{Q}(F_j))$ and the F_j are permuted by the action of Gal$(\overline{\mathbb{Q}}/\mathbb{Q})$, the absolute Galois group of $\mathbb{Q}$. Such a basis is called a distinguished basis. Every eigenform $F \in S_n^k(\mathbb{Q}(F))$ can be extended to such a basis [**M**, Appendix A].

1.3. Eisenstein series. In this section, we recall some basic facts on Eisenstein series of Klingen type and study some abstract properties of their pullbacks.

DEFINITION 1.6. Let r, k be rational integers and n a natural number. Further let k be even and $0 \leq r \leq n$. If $s \in \mathbb{C}$ with $k + 2Re(s) > n + r + 1$ and $F \in S_r^k$, then we denote by

$$E^k_{n,r}(Z, F; s) = \sum_{\gamma \in P_{n,r}\backslash \Gamma_n} F(\gamma < Z >_*) j(\gamma, Z)^{-k} \left(\frac{\det \operatorname{Im}\gamma < Z >}{\det \operatorname{Im}\gamma < Z >_*} \right)^s ,$$

the Eisenstein series of Klingen type on $\mathcal{H}_n$.

The series is absolutely convergent and admits a meromorphic continuation to the whole s-plane [**L**]. With regard to k and Γ_n, it satisfies the transformation law (4) and it is orthogonal to S^k_n for $r < n$. When $s = 0$ and $k > n + r + 1$, we get the holomorphic Klingen Eisenstein series [**K90**, chapter 5]. Siegel [**Si39**] has already proved that the Fourier coefficients of $E^k_{n,0}(Z, 1; 0)$ are rational numbers. For convenience we put $E^k_n(Z; s) = E^k_{n,0}(Z, 1; s)$, $\Gamma_m(s) = \prod_{j=1}^m \Gamma(s - (j-1)/2)$ and $\xi(s) = \pi^{-s/2}\Gamma(s/2)\zeta(s)$.

LEMMA 1.7. *If we put* $\mathbb{E}^k_n(Z; s) := \epsilon(n, s) E^k_n(Z; s - k/2)$ *with*

$$\epsilon(n, s) = \xi(2s) \frac{\Gamma_n(s + k/2)}{\Gamma_n(s)} \prod_{j=1}^{[n/2]} \xi(4s - 2j).$$

Then we have

$$\mathbb{E}^k_n(Z; s) = \mathbb{E}^k_n \left(Z; \frac{n+1}{2} - s \right).$$

Now we associate to a given partition $(n_i)_{i=1}^l$ of n (i.e., $\sum_{i=1}^l n_i = n$ with $n_i \in \mathbb{N}$) the embeddings $\mathcal{H}_{n_1} \times \ldots \times \mathcal{H}_{n_l} \hookrightarrow \mathcal{H}_n$ and $\Gamma_{n_1} \times \ldots \times \Gamma_{n_l} \hookrightarrow \Gamma_n$ given by

$$(z_1, \ldots, z_l) \mapsto [z_1, \ldots, z_l]$$

$$\begin{pmatrix} a_1 & b_1 \\ c_1 & d_1 \end{pmatrix} \times \ldots \times \begin{pmatrix} a_l & b_l \\ c_l & d_l \end{pmatrix} \mapsto \begin{pmatrix} a_1 & & 0 & b_1 & & 0 \\ & \ddots & & & \ddots & \\ 0 & & a_l & 0 & & b_l \\ c_1 & & 0 & d_1 & & 0 \\ & \ddots & & & \ddots & \\ 0 & & c_l & 0 & & d_l \end{pmatrix}.$$

It is convenient to use the same notation for the images and pre-images.

DEFINITION 1.8. Let n, r, k be as before and $(n_i)_{i=1}^l$ a partition of n. Assume further that $F \in S^k_r$ and $s \in \mathbb{C}$ with $Re(s)$ big enough. Then we call

$$\phi_F(z_1, \ldots, z_l; s) = E^k_{n,r} \left(\begin{pmatrix} z_1 & & 0 \\ & \ddots & \\ 0 & & z_l \end{pmatrix}, F; s \right)$$

a diagonal-restricted Eisenstein series attached to F of type $((n_i)_{i=1}^l; r)$.

It is easy to see that such functions satisfy (4) in each variable. If we can describe $\phi_F(z_1, \ldots, z_l; s)$ explicitly as a function in the variables $z_1, \ldots, z_l$, then we will call such an expression a **pullback formula** [**G84**, Main formula]. Let us put $(n_1, \ldots, n_l) = (n_1, \ldots, n_l; 0)$ and $\phi := \phi_1$. The following theorem has served as the main motivation for us to study pullback formulas.

THEOREM 1.9. *Let k be an even rational integer and $s \in \mathbb{C}$ with $k + 2Re(s) > n + 1$. Further let $(n_i)_{i=1}^l$ be a partition of n and $F_i \in S_{n_i}^k$. Then*

$$\Phi(F_1, \dots, F_l; s) := \langle \dots \langle \phi(z_1, \dots, z_l; s); F_1(z_1) \rangle \dots F(z_l) \rangle$$

satisfies

(a) *$\Phi(F_1, \dots, F_l; s)$ has an analytic continuation to a meromorphic function on $\mathbb{C}$ and inherits a functional equation from Lemma 1.7.*

(b) *If we assume that the F_i's are Hecke eigenforms with $F_i \in S_{n_i}^k(\mathbb{Q}(F_i))$ and $k > max\{n+1, max\{n_i\}+1\}$, then*

$$\frac{\Phi(F_1^\sigma, \dots, F_l^\sigma; 0)}{\prod_{i=1}^l \langle F_i^\sigma; F_i^\sigma \rangle} = \left(\frac{\Phi(F_1, \dots, F_l; 0)}{\prod_{i=1}^l \langle F_i; F_i \rangle} \right)^\sigma \in \overline{\mathbb{Q}} \textit{ with } \sigma \in \mathrm{Gal}(\bar{\mathbb{Q}}/\mathbb{Q}).$$

PROOF. Part (a) is clear. To prove (b), we assume for convenience that $l = 2$, for arbitrary l one proceeds in the same way. We can complete $f_1 = F_1$ and $g_1 = F_2$ to a *distinguished* basis of $S_{n_1}^k$ and $S_{n_2}^k$ respectively and then to an orthogonal basis of $M_{n_1}^k$ and $M_{n_2}^k$. This can be done in such a way that the orthogonal complements of $S_{n_1}^k$ and $S_{n_2}^k$ are invariant under all automorphisms of $\mathbb{C}$ (cf. [**Bo85a**, section 6]). From $\phi(z_1, z_2; 0) \in M_{n_1}^k \otimes M_{n_2}^k$ we can deduce

$$\phi(z_1, z_2; 0) = \sum_{i,j} c_{i,j} f_i(z_1) g_j(z_2) \qquad (c_{ij} \in \mathbb{C}).$$

The rationality of the Fourier coefficients of $\phi(z_1, z_2; 0)$ leads to

$$\left(\frac{\Phi(F_1, F_2; 0)}{\langle F_1; F_1 \rangle \langle F_2; F_2 \rangle} \right)^\sigma = c_{11}^\sigma = \frac{\langle \langle \phi(z_1, z_2; 0); F_1^\sigma F_2^\sigma \rangle \rangle}{\langle F_1^\sigma; F_1^\sigma \rangle \langle F_2^\sigma; F_2^\sigma \rangle},$$

for all automorphisms σ of $\mathbb{C}$. The assumptions in (b) imply that c_{11} is algebraic. □

REMARK 1.10. If we denote by $\mathbb{Q}(F_1, \dots, F_l)$ the totally real number field which is generated by $\mathbb{Q}(F_1), \dots, \mathbb{Q}(F_l)$, then

$$\frac{\Phi(F_1, \dots, F_l; 0)}{\prod \langle F_i; F_i \rangle} \in \mathbb{Q}(F_1, \dots, F_l).$$

It is of course possible to state the theorem under weaker assumptions. For example, the purely technical condition totally real *can easily be dropped (e.g.* [**Bo85a**, Satz 14]*).*

A first insight into the connection between $\Phi(F_1, \dots, F_l; s)$ and automorphic L-functions is given by the following

THEOREM 1.11. *(Garrett, Böcherer) Let $m, n \in \mathbb{N}, k \in 2\mathbb{Z}$ and $s \in \mathbb{C}$ with $s \in \mathbb{C}$ with $k + 2Re(s) > m + n + 1$. If $F \in S_n^k$ is an eigenform then the following assertion holds*

$$\langle E_{m+n}^k \left(\begin{pmatrix} z & 0 \\ 0 & * \end{pmatrix}; s \right); F \rangle$$

$$= \begin{cases} \mu(n,k,s) \{ \zeta(2s+k) \prod_{j=1}^n \zeta(4s+2k-2j) \}^{-1} & \\ \cdot D_F^n(2s+k-n) E_{m,n}^k(z, F^\#; s) & \textit{if } m \geq n, \\ & \\ 0 & \textit{otherwise.} \end{cases}$$

Here # denotes complex-conjugation and

$$\mu(n,k,s) = 2^{(n^2+3n)/2-2ns-nk+1}(-1)^{nk/2}\pi^{n(n+1)/2}\frac{\Gamma_n(k+s-(n+1)/2)}{\Gamma_n(k+s)}. \tag{14}$$

See [**G84**] and [**Bo85b**] for more details. In the special case $n = m$, $s = 0$ we have

COROLLARY 1.12. *Let $F \in S_n^k(\mathbb{Q}(F))$ be a Hecke eigenform and $k > 2n$ even. Then*

$$\left(\frac{\pi^{3n(n+1)/2-2nk-k}D_F^n(k-n)}{\langle F;F\rangle}\right)^\sigma = \frac{\pi^{3n(n+1)/2-2nk-k}D_{F^\sigma}^n(k-n)}{\langle F^\sigma;F^\sigma\rangle} \in \mathbb{Q}(F) \tag{15}$$

for all $\sigma \in \mathrm{Gal}(\overline{\mathbb{Q}}/\mathbb{Q})$.

2. A family of Dirichlet series

This section centers around the explicit calculation of the integral of an Eisenstein series of type $(1,n,n)$ against three cusp forms. To do this, we start with some LINEARE ZAHLENTHEORIE (cf. Siegel [**Si68**, 3. Abschnitt]).

2.1. Representatives for $P_{n+1,n}\backslash\Gamma_{n+1}$. The fact, that for every $g \in \mathrm{Sp}_m(\mathbb{Q})$ there exists $p_g \in \bigcap_{r=0}^{m-1} P_{m,r}(\mathbb{Q})$ with

$$p_g \cdot g \in \Gamma_m \tag{16}$$

allows us to find a suitable system of representatives for $RS^{\mathbb{Z}}$ of $P_{n+1,n}\backslash\Gamma_{n+1}$ in a rather simple way, that suits our purposes.

PROPOSITION 2.1. *The left cosets space $P_{n+1,n}(\mathbb{Q})\backslash\mathrm{Sp}_{n+1}(\mathbb{Q})$ has representatives*

$$RS^{\mathbb{Q}} = R_1^{\mathbb{Q}} \sqcup R_2^{\mathbb{Q}} \sqcup R_3^{\mathbb{Q}}.$$

$$R_1^{\mathbb{Q}} = P_{1,0}(\mathbb{Q})\backslash\mathrm{Sp}_1(\mathbb{Q}) \times 1_{2n}, \tag{17}$$

$$R_2^{\mathbb{Q}} = \pi_n \cdot (1_2 \times P_{n,n-1}(\mathbb{Q})\backslash\mathrm{Sp}_n(\mathbb{Q})), \tag{18}$$

$$R_3^{\mathbb{Q}} = \xi_n \cdot \left(P_{1,0}(\mathbb{Q})\backslash\mathrm{Sp}_1(\mathbb{Q}) \times ((T \times 1_{2(n-1)}) \cdot P_{n,n-1}(\mathbb{Q})\backslash\mathrm{Sp}_n(\mathbb{Q}))\right) \tag{19}$$

$$\textit{with } \pi_n = \begin{pmatrix} \begin{matrix} 0 & 1 \\ 1 & 0 \end{matrix} & 0 & & \\ 0 & 1_{n-1} & & 0 \\ & & \begin{matrix} 0 & 1 \\ 1 & 0 \end{matrix} & 0 \\ & 0 & 0 & 1_{n-1} \end{pmatrix}$$

$$\textit{and } \xi_n = \begin{pmatrix} \begin{matrix} 1 & 0 \\ 1 & 1 \end{matrix} & 0 & & \\ 0 & 1_{n-1} & & 0 \\ & & \begin{matrix} 1 & -1 \\ 0 & 1 \end{matrix} & 0 \\ & 0 & 0 & 1_{n-1} \end{pmatrix}.$$

Furthermore $T = T(\mathbb{Q}) = \left\{ \left(\begin{smallmatrix} a & 0 \\ 0 & a^{-1} \end{smallmatrix}\right) \mid a \in \mathbb{Q}^\times \right\}$.

PROOF. By Witt's Theorem [**Bou**, §4, $n^0 3$; Corollaire 2], we can identify $P_{n+1,n}(\mathbb{Q})\backslash\mathrm{Sp}_{n+1}(\mathbb{Q})$ with the set X of one-dimensional (isotropic) subspaces of $\mathbb{Q}^{2(n+1)}$. To prove the assertion of the proposition, we have to determine the orbits of the action of $\mathrm{Sp}_1(\mathbb{Q}) \times \mathrm{Sp}_n(\mathbb{Q}) \hookrightarrow \mathrm{GL}_{2(n+1)}(\mathbb{Q})$ on X. We decompose $\mathbb{Q}^{2(n+1)} = \mathbb{Q}^2 \oplus \mathbb{Q}^{2n}$ and denote by $p_1(U)$, $p_2(U)$ the projection of $U \in X$ on $\mathbb{Q}^2$ and $\mathbb{Q}^{2n}$ respectively. Then it is clear, that the orbits are characterized by the ranks $\big(\mathrm{rank}(p_1(U)), \mathrm{rank}(p_2(U))\big)$ of the projection: $(1,0)$, $(0,1)$ $(1,1)$. So we get three orbits. Then we fix an element in each orbit and determine its isotropy group. Translating these observations into the language of matrices proves the proposition. □

To get the desired representatives over $\mathbb{Z}$, we have to alter $RS^{\mathbb{Q}}$ by suitable elements of $P_{n+1,n}(\mathbb{Q})$ from the left.

COROLLARY 2.2. *The left cosets space $P_{n+1,n}\backslash\Gamma_{n+1}$ has representatives*

$$RS^{\mathbb{Z}} = R_1^{\mathbb{Z}} \sqcup R_2^{\mathbb{Z}} \sqcup R_3^{\mathbb{Z}}.$$

$$R_1^{\mathbb{Z}} = P_{1,0}\backslash\Gamma_1 \times 1_{2n}, \tag{20}$$

$$R_2^{\mathbb{Z}} = \pi_n \cdot (1_2 \times P_{n,n-1}\backslash\Gamma_n), \tag{21}$$

$$R_3^{\mathbb{Z}} = \bigsqcup_{\substack{\varepsilon\in\mathbb{N},\nu\in\mathbb{Z}\backslash\{0\}\\(\varepsilon,\nu)=1}} \xi_n^{\varepsilon,\nu} \cdot (P_{1,0}\backslash\Gamma \times P_{n,n-1}\backslash\Gamma_n) \tag{22}$$

$$\textit{with } \xi_n^{\varepsilon,\nu} = \begin{pmatrix} * & * & 0 & 0 \\ \varepsilon & \nu & 0 & 0 \\ 0 & 0 & \nu & -\varepsilon \\ 0 & 0 & * & * \end{pmatrix} \times 1_{2(n-1)} \in \Gamma_{n+1}.$$

PROOF. The property (16) implies that there is an 1-1 correspondence between $P_{n,r}(\mathbb{Q})\backslash\mathrm{Sp}_{n,r}(\mathbb{Q})$ and $P_{n,r}\backslash\Gamma_n$. So it remains only to translate $R_3^{\mathbb{Q}}$ back to $R_3^{\mathbb{Z}}$. That means we are looking for a $p_a \in P_{n,r}(\mathbb{Q})$ with

$$p_a \cdot \left(1_2 \times \begin{pmatrix} a & 0 \\ 0 & a^{-1} \end{pmatrix} \times 1_{2(n-1)}\right) \in \Gamma_{n+1}.$$

If we parametrize $a \in \mathbb{Q}^\times$ with $\frac{\nu}{\varepsilon}$ $(\nu \in \mathbb{Z}\backslash\{0\}, \varepsilon \in \mathbb{N})$, then

$$p_{\varepsilon,\nu} = \begin{pmatrix} \begin{matrix} \nu^{-1} & \mu' \\ 0 & \varepsilon \end{matrix} & 0 \\ 0 & \begin{matrix} \nu & 0 \\ \mu & \varepsilon^{-1} \end{matrix} \end{pmatrix} \times 1_{2(n-1)}$$

does the job, if we choose $\mu, \mu' \in \mathbb{Z}$ in such a way that $\mu\varepsilon \equiv 1 \mod \nu$. □

The assertion of the Proposition 2.1 still holds for arbitrary fields. The corollary is of course not true for every integral domain.

2.2. Pullback formulas. This subsection is devoted to explain how one can use Corollary 2.2 to produce pullback formulas. We achieve this by giving two examples.

THEOREM 2.3. *Let $k \in 2\mathbb{Z}$ and $s \in \mathbb{C}$ with $2Re(s) + k > 2n+2$. If $F \in S_n^k$ and $\tau \in \mathcal{H}$, $\tau' \in \mathcal{H}_n$, then we get*

$$E_{n+1,n}^k\left(\begin{pmatrix} \tau & 0 \\ 0 & \tau' \end{pmatrix}, F; s\right) = E_1^k(\tau, s)F(\tau') + \begin{cases} E_1^k(\tau'; s)F(\tau) & n = 1 \\ E_{n,n-1}^k(\tau', F_\tau; s) & n > 1 \end{cases}$$

$$+2\delta(\tau)^s\delta(\tau')^s \sum_{\substack{\epsilon,\nu=1 \\ (\epsilon,\nu)=1}}^{\infty} \sum_{\substack{\gamma_1 \in \Gamma_\infty\backslash\Gamma \\ \gamma_2 \in P_{n,n-1}\backslash\Gamma_n}} \chi_j^{k,s}(\gamma_1,\tau)\chi_j^{k,s}(\gamma_2,\tau')$$

$$\cdot F\left(\begin{pmatrix} \varepsilon^2\gamma_1 < \tau > & 0 \\ 0 & 0 \end{pmatrix} + (\gamma_2 < \tau' >)\left[\begin{pmatrix} \nu & 0 \\ 0 & 1_{n-1} \end{pmatrix}\right]\right)$$

$$\cdot\delta\left(\begin{pmatrix} \varepsilon^2\gamma_1 < \tau > & 0 \\ 0 & 0 \end{pmatrix} + (\gamma_2 < \tau' >)\left[\begin{pmatrix} \nu & 0 \\ 0 & 1_{n-1} \end{pmatrix}\right]\right)^{-s}.$$

Here $F_\tau(z) = F\left(\begin{smallmatrix} \tau & 0 \\ 0 & z \end{smallmatrix}\right) \in S_{n-1}^k$ and $\chi_j^{k,s}(\gamma_j,\tau) = j(\gamma_j,\tau)^{-k}|j(\gamma_j,\tau)|^{-2s}$.

PROOF. It is convenient to rewrite the Eisenstein series in the following manner

$$E_{n+1,n}^k\left(\begin{pmatrix} \tau & 0 \\ 0 & \tau' \end{pmatrix}, F; s\right) = \delta(\tau)^s\delta(\tau')^s \sum_{g \in P_{n+1,n}\backslash\Gamma_{n+1}} \chi_j^{k,s}\left(g, \begin{pmatrix} \tau & 0 \\ 0 & \tau' \end{pmatrix}\right)$$

$$\cdot F\left(g < \begin{pmatrix} \tau & 0 \\ 0 & \tau' \end{pmatrix} >_*\right)\delta\left(g < \begin{pmatrix} \tau & 0 \\ 0 & \tau' \end{pmatrix} >_*\right)^{-s},$$

because $\chi_j^{k,s}$ satisfies the same cocyle relation as the factor of automorphy j. Then it is easy to see that the subseries $E_1^k(\tau, s)F(\tau')$ corresponds to $R_1^{\mathbb{Z}}$. If $n = 1$, then $R_2^{\mathbb{Z}}$ leads to $E_1^k(\tau')F(\tau)$ and to $E_{n,n-1}^k(\tau', F_\tau; s)$ otherwise, because the $\left(\begin{smallmatrix} 0 & 1 \\ 1 & 0 \end{smallmatrix}\right)$ block of π_n interchanges τ and τ'. To interpret $R_3^{\mathbb{Z}}$ it suffices to look at

$$\left(\begin{pmatrix} * & * & 0 \\ \varepsilon & \nu & \\ 0 & & 1_{n-1} \end{pmatrix}\begin{pmatrix} \tilde{\tau} & 0 & \\ 0 & \tilde{z}_1 & \tilde{z}_2 \\ & \tilde{z}_2^t & \tilde{z}_4 \end{pmatrix}\begin{pmatrix} * & \varepsilon & 0 \\ * & \nu & \\ 0 & & 1_{n-1} \end{pmatrix}\right)_*$$

$$= \begin{pmatrix} \varepsilon^2\tau + \nu^2\tilde{z}_1 & \nu\tilde{z}_2 \\ \nu\tilde{z}_2^t & \tilde{z}_4 \end{pmatrix}.$$

□

REMARK 2.4.

(1) The case $n = 1$ is essentially due to Garrett [**G87**].
(2) For $n > 1$ a new phenomenon shows up, namely the occurrence of diagonal-restricted cusp forms. This will play no role in this paper.
(3) If we put $U_+ = \{\left(\begin{smallmatrix} * & * \\ c & d \end{smallmatrix}\right) \in \Gamma_\infty\backslash\Gamma | c, d > 0\}$, then we can essentially describe $R_3^{\mathbb{Z}}$ by

$$\bigsqcup_{u \in U_+}\left(\begin{pmatrix} u & 0 \\ 0 & u^\tau \end{pmatrix} \times 1_{2(n-1)}\right)(P_{1,0}\backslash\Gamma \times P_{n,n-1}\backslash\Gamma_n).$$

We are now in a good position. The pullback formula in Theorem 2.3 can easily be generalized to get pullback formulas of type $(n_1, \dots, n_l, 1; n)$. It is further possible to calculate $\Phi_F(F_1, \dots, F_{l+1}; s)$ in an explicit way (This vein will be developed in a forthcoming paper). To convince the reader, we present here the case $(1, 1, 1; 2)$.

COROLLARY 2.5. *Let $k \in 2\mathbb{Z}$ and $F \in S_2^k$. Then for $s \in \mathbb{C}$ with $Re(s) > 6$ we have*

$$E_{3,2}^k([z_1, z_2, z_3], F; s) = \sum_{(a,b,c)\in P} E_1^k(z_a; s) F\begin{pmatrix} z_b & 0 \\ 0 & z_c \end{pmatrix}$$

$$+ 2\delta(z_a)^s \delta(z_b)^s \sum_{u\in U_+} \sum_{\gamma_1,\gamma_2\in\Gamma_\infty\backslash\Gamma} \chi_j^{k,s}(\gamma_1, z_a)\chi_j^{k,s}(\gamma_1, z_b)$$

$$\cdot F_{z_c}\left(\begin{pmatrix} \gamma_1 < z_a > & 0 \\ 0 & \gamma_2 < z_b > \end{pmatrix}[u^t]_*\right) \delta\left(\begin{pmatrix} \gamma_1 < z_a > & 0 \\ 0 & \gamma_2 < z_b > \end{pmatrix}[u^t]_*\right)^{-s}$$

$$+ 2\delta(z_1)^s\delta(z_2)^s\delta(z_3)^s \sum_{u_1,u_2\in U_+} \sum_{\gamma_1,\gamma_2,\gamma_3\in\Gamma_\infty\backslash\Gamma} \prod_{i=1}^{3} \chi_j^{k,s}(\gamma_i, z_i)$$

$$\cdot F\left(\begin{pmatrix} \gamma_1 < z_1 > & 0 & 0 \\ 0 & \gamma_2 < z_2 > & 0 \\ 0 & 0 & \gamma_3 < z_3 > \end{pmatrix}\left[\begin{pmatrix} 1 & 0 \\ 0 & u_1^t \end{pmatrix}\begin{pmatrix} u_2^t & 0 \\ 0 & 1 \end{pmatrix}\right]_*\right)$$

$$\cdot \delta\left(\begin{pmatrix} \gamma_1 < z_1 > & 0 & 0 \\ 0 & \gamma_2 < z_2 > & 0 \\ 0 & 0 & \gamma_3 < z_3 > \end{pmatrix}\left[\begin{pmatrix} 1 & 0 \\ 0 & u_1^t \end{pmatrix}\begin{pmatrix} u_2^t & 0 \\ 0 & 1 \end{pmatrix}\right]_*\right)^{-s},$$

with $P = \{(1, 2, 3), (3, 1, 2), (2, 3, 1)\}$.

2.3. The integral representation. We calculate the integral of a diagonal-restricted Eisenstein series of type $(1, n, n)$ against three cusp forms. To sharpen our result, we recall some fundamental principles on Jacobi forms (see [**EZ**], [**K89**] or section 3.4). Let $F \in S_n^k$, then one denotes by

$$F\begin{pmatrix} \tau & z \\ z^t & \tau' \end{pmatrix} = \sum_{N=1}^{\infty} \Phi_N^F(\tau, z) e\{N\tau'\} (\tau' \in \mathcal{H}) \tag{23}$$

the Fourier-Jacobi expansion of F with respect to τ'. The Fourier-Jacobi coefficients $\Phi_N^F(\tau, z)$ are Jacobi forms $J_{n-1,1}^k(N)$ of degree $(n-1, 1)$, weight k and index N. By $\langle\, ; \,\rangle_J = \langle\, ; \,\rangle_{J,N}$ we denote the Petersson scalar product for such spaces. Further we put

$$U_l : \begin{cases} J_{m,1}^k(N) & \longrightarrow & J_{m,1}(Nl^2) \\ \Phi_N^F(\tau, z) & \mapsto & (\Phi_N^F | U_l)(\tau, z) = \Phi_N^F(\tau, lz). \end{cases} \tag{24}$$

In the case $n = 1$, we interpret the Fourier-Jacobi coefficients as the Fourier coefficients, the map U_l as the identity map and $\langle\, ; \,\rangle_J$ as the standard scalar product in $\mathbb{C}^2$.

THEOREM 2.6. *Let k be an even number, $F, G \in S_n^k$ and $h \in S_1^k$ with totally real Fourier coefficients. Then for $s \in \mathbb{C}$ with $k + 2Re(s) > 2n + 2$*

$$\langle\langle E_{n+1,n}^k\left(\begin{pmatrix} W & 0 \\ 0 & Z \end{pmatrix}, F; s\right); G(Z)\rangle; h(W)\rangle \tag{25}$$

$$= 2(4\pi)^{-(2k+s-n-1)} \frac{\Gamma(s + 2k - n - 1)\Gamma(s + k - n)\Gamma(s + k - 1)}{\Gamma(2k - n - 1 + 2s)}$$

$$\cdot \sum_{\substack{N,\varepsilon,l=1\\(\varepsilon,l)=1}}^{\infty} \langle \Phi_N^F | U_l ; \Phi_{Nl^2}^G \rangle_J a_{N\varepsilon^2}^h \varepsilon^{-2(k+s-1)} l^{-2(k+s-n)} N^{-(2k+s-n-1)}.$$

PROOF. To calculate (25), we insert the pullback formula developed in Theorem 2.3. After interchanging the integration and summation, we get by "unwinding"

$$\langle\langle E_{n+1,n}^k \left(\left(\begin{smallmatrix} W & 0 \\ 0 & Z \end{smallmatrix} \right), F; s \right); G(Z) \rangle; h(W) \rangle \tag{26}$$

$$= 2 \int_{\Gamma_\infty \backslash \mathcal{H}} d^*W \int_{P_{n,n-1}\backslash \mathcal{H}_n} d^*Z \sum_{\substack{\varepsilon,l=1\\(\varepsilon,l)=1}}^{\infty} F\left(\left(\begin{smallmatrix} \varepsilon^2 W & 0 \\ 0 & 0 \end{smallmatrix} \right) + Z \left[\left(\begin{smallmatrix} l & 0 \\ 0 & 1_{n-1} \end{smallmatrix} \right) \right] \right)$$

$$\cdot \delta \left(\left(\begin{smallmatrix} \varepsilon^2 W & 0 \\ 0 & 0 \end{smallmatrix} \right) + Z \left[\left(\begin{smallmatrix} l & 0 \\ 0 & 1_{n-1} \end{smallmatrix} \right) \right] \right)^{-s} \delta(W)^{k+s} \delta(Z)^{k+s} \overline{G(Z) h(W)}.$$

To use the notation in [**K89**] we rewrite this as

$$2 \int_{\Gamma_\infty \backslash \mathcal{H}} d^*W \int_{C_{n,n-1}\backslash \mathcal{H}_n} d^*Z \sum_{\substack{\varepsilon,l=1\\(\varepsilon,l)=1}}^{\infty} F\left(\left(\begin{smallmatrix} 0 & 0 \\ 0 & \varepsilon^2 W \end{smallmatrix} \right) + Z \left[\left(\begin{smallmatrix} 1_{n-1} & 0 \\ 0 & l \end{smallmatrix} \right) \right] \right)$$

$$\cdot \delta \left(\left(\begin{smallmatrix} 0 & 0 \\ 0 & \varepsilon^2 W \end{smallmatrix} \right) + Z \left[\left(\begin{smallmatrix} 1_{n-1} & 0 \\ 0 & l \end{smallmatrix} \right) \right] \right)^{-s} \delta(W)^{k+s} \delta(Z)^{k+s} \overline{G(Z) h(W)}.$$

If we denote by $\mathcal{B}_{n-1,1}$ a fundamental domain of the action of the group

$$C_{n,n-1}^+ / Z^{n,n-1}$$

on $\mathcal{H}_{n-1} \times \mathbb{C}^{n-1}$, here we denote by $Z^{n,n-1}$ the center of $C_{n,n-1}^+$, then

$$\mathcal{Q}_{n,1} = \left\{ Z = \left(\begin{smallmatrix} \tau & z \\ z^t & \tau' \end{smallmatrix} \right) \in \mathcal{H}_n \mid (\tau, z) \in \mathcal{B}_{n-1,1} \text{ and } 0 \le x_{\tau'} < 1 \right\} \tag{27}$$

is a fundamental domain of $C_{n,n-1}^+ \backslash \mathcal{H}_n$ [**K89**]. This leads to

$$\langle\langle E_{n+1,n}^k \left(\left(\begin{smallmatrix} W & 0 \\ 0 & Z \end{smallmatrix} \right), F; s \right); G(Z) \rangle; h(W) \rangle$$

$$= 2 \sum_{\substack{\varepsilon,l=1\\(\varepsilon,l)=1}}^{\infty} \int_{\mathcal{B}_{n-1,1}} d\tau\, dz \int_{y_{\tau'} > y_\tau^{-1}[y_z]} dy_{\tau'} \int_0^1 dx_{\tau'} \int_0^1 dx_W \int_0^\infty dy_W\, \delta(Z)^{k+s-(n+1)} y_W^{k+s-2}$$

$$\cdot \sum_{N=1}^{\infty} \Phi_N^F(\tau, lz) e\{N(\varepsilon^2 W + l^2 \tau')\} \overline{\sum_{M=1}^{\infty} a_M^h e\{MW\} \sum_{L=1}^{\infty} \Phi_L^G(\tau, z) e\{L\tau'\}}$$

$$\cdot \delta \left(\left(\begin{smallmatrix} 0 & 0 \\ 0 & \varepsilon^2 W \end{smallmatrix} \right) + Z \left[\left(\begin{smallmatrix} 1 & 0 \\ 0 & l \end{smallmatrix} \right) \right] \right)^{-s}$$

$$= 2 \int_{\mathcal{B}_{n-1,1}} d\tau dz \sum_{\substack{N,\varepsilon,l=1\\(\varepsilon,l)=1}}^{\infty} \Phi_N^F(\tau, lz)\overline{\Phi_{Nl^2}^G(\tau, z) a_{N\varepsilon^2}^h} \int_{y_{\tau'} > y_\tau^{-1}[y_z]} dy_{\tau'} \int_0^\infty dy_W$$

$$\delta\left(\begin{pmatrix} 0 & 0 \\ 0 & \varepsilon^2 W \end{pmatrix} + Z\left[\begin{pmatrix} 1 & 0 \\ 0 & l \end{pmatrix}\right]\right)^{-s} \delta(Z)^{k+s-(n+1)} y_W^{k+s-2} e^{-4\pi N(\varepsilon^2 y_W + l^2 y_{\tau'})}.$$

In the last step, we have integrated over the real parts of τ' and W. It remains to calculate the integration over dy_W and $dy_{\tau'}$. We have

$$\int_{y_{\tau'} > y_\tau^{-1}[y_z]} dy_{\tau'} \int_0^\infty dy_W \delta\begin{pmatrix} y_\tau & ly_z \\ ly_z^t & \varepsilon^2 W + l^2 y_{\tau'} \end{pmatrix}^{-s} \delta\begin{pmatrix} y_\tau & y_z \\ y_z^t & y_{\tau'} \end{pmatrix}^{k+s-(n+1)}$$

$$y_W^{k+s-2} e^{-4\pi N(\varepsilon^2 y_W + l^2 y_{\tau'})}$$

$$\overset{(1)}{=} \int_0^\infty dt \int_0^\infty dy_W \delta\begin{pmatrix} y_\tau & ly_z \\ ly_z^t & \varepsilon^2 y_W + l^2(t + y_\tau^{-1}[y_z]) \end{pmatrix}^{-s} \delta$$

$$\begin{pmatrix} y_\tau & y_z \\ y_z^t & t + y_\tau^{-1}[y_z] \end{pmatrix}^{k+s-n-1} y_W^{k+s-2} e^{-4\pi N\varepsilon^2 y_W} e^{-4\pi N l^2 t} e^{-4\pi N l^2 y_\tau^{-1}[y_z]}$$

$$= \int_0^\infty dt \int_0^\infty dy_W \delta\begin{pmatrix} y_\tau & ly_z \\ ly_z^t & \varepsilon^2 y_W + l^2(t + y_\tau^{-1}[y_z]) \end{pmatrix}^{-s} \delta$$

$$\begin{pmatrix} y_\tau & y_z \\ y_z^t & l^2 t + l^2 y_\tau^{-1}[y_z] \end{pmatrix}^{k+s-n-1} l^{-2(k+s-(n+1))} y_W^{k+s-2}$$

$$e^{-4\pi N\varepsilon^2 y_W} e^{-4\pi N l^2 t} e^{-4\pi N l^2 y_\tau^{-1}[y_z]}$$

$$\overset{(2)}{=} \int_0^\infty dt \int_0^\infty dy_W t^{k+s-(n+1)} (\varepsilon^2 y_W + l^2 t)^{-s} \delta(y_\tau)^{k-(n+1)} y_W^{k+s-2}$$

$$e^{-4\pi N\varepsilon^2 y_W} e^{-4\pi N l^2 t} e^{-4\pi N l^2 y_\tau^{-1}[y_z]}.$$

In (1) we have substituted $y_{\tau'}$ by $t + y_\tau^{-1}[y_z]$ and in (2) we have used

$$\delta\begin{pmatrix} y_\tau & ly_z \\ ly_z^t & \varepsilon^2 y_W + l^2(t + y_\tau^{-1}[y_z]) \end{pmatrix}^{-s} \delta\begin{pmatrix} y_\tau & ly_z \\ ly_z^t & l^2 t + l^2 y_\tau^{-1}[y_z] \end{pmatrix}^{k+s-n-1}$$
$$= (l^2 t)^{k+s-(n+1)} (\varepsilon^2 y_W + l^2 t)^{-s} \delta(y_\tau)^{k-(n+1)}$$

Recall that the invariant measure for the Petersson scalar product in $\mathcal{J}_{n-1,1}^k(m)$ is

$$\delta(y_\tau)^{k-((n-1)+2)} e^{-4\pi m y_\tau^{-1}[y_z]} d\tau dz. \tag{28}$$

So it remains only to calculate

$$\int_0^\infty \frac{dt}{t} \int_0^\infty \frac{dy_W}{y_W} t^{k+s-n} y_W^{k+s-1} (\varepsilon^2 y_W + l^2 t)^{-s} e^{-4\pi N\varepsilon^2 y_W} e^{-4\pi N l^2 t},$$

which is an Eulerian integral of first kind. Applying the formula

$$\int_0^\infty \int_0^\infty \frac{dx}{x} \frac{dy}{y} x^\alpha y^\beta \left(\frac{xy}{x+y}\right)^\gamma e^{-(x+y)} = \frac{\Gamma(\alpha+\beta+\gamma)\Gamma(\alpha+\gamma)\Gamma(\beta+\gamma)}{\Gamma(\alpha+\beta+2\gamma)}$$

gives the desired result. □

2.4. First results. The Dirichlet series we obtained in the last subsection we now denote by

$$D_{h,G;F}(s) = \sum_{\substack{N,\varepsilon,l=1\\(\varepsilon,l)=1}}^{\infty} \langle \Phi_N^F | U_l; \Phi_{Nl^2}^G \rangle_J a_{N\varepsilon^2}^h \varepsilon^{-2(k+s-1)} l^{-2(k+s-n)} N^{-(2k+s-n-1)}. \tag{29}$$

Combining results of the Theorem 2.6 with the Theorem of Garrett and Böcherer and Theorem 1.9 gives the following first résumé.

THEOREM 2.7. *Let $F, G \in S_n^k$, $h \in S_1^k$ Hecke eigenforms with totally real Fourier coefficients. Let $s \in \mathbb{C}$ with $2Re(s) + k > 2n+2$ and k even. Then we have*

$$\Phi(h, G, F; s) = a(n, s, k)\mathcal{L}_{h,G,F}^n(s + 2k - n - 1) \text{ with} \tag{30}$$

$$\begin{aligned}\mathcal{L}_{h,G,F}^n(s + 2k - n - 1) &= \left\{\zeta(2s+k)\prod_{j=1}^n \zeta(4s + 2k - 2j)\right\}^{-1} \\ &\quad \cdot D_F^n(2s + k - n) D_{h,G;F}(s)\end{aligned}$$

and $a(n,s,k) = a^*(n,s,k)\Gamma^*(n,s,k)$ with

$$\begin{aligned} a^*(n,s,k) &= 2^{-2s(n+1)-nk-4k+3+n^2+7n/2}(-1)^{nk/2} 2\pi^{n^2+3n/2-s-2k+1}, \\ \Gamma^*(n,s,k) &= \frac{\Gamma(s+2k-1-n)\Gamma(s+k-n)\Gamma(s+k-1)}{\Gamma(2k-n-1+2s)} \frac{\Gamma_n(k+s-\frac{n+1}{2})}{\Gamma_n(k+s)}. \end{aligned}$$

COROLLARY 2.8. *With the same assumptions as above, we have:*

(i) *$\mathcal{L}_{h,G,F}^n$ has meromorphic continuation to $\mathbb{C}$ and satisfies the functional equation*

$$\mathcal{L}_{h,G,F}^{n,*}(s) = \mathcal{L}_{h,G,F}^{n,*}(3k - n - 1 - s), \tag{31}$$

where

$$\mathcal{L}_{h,G,F}^{n,*}(s) = \epsilon(2n+1, s+n+1-3/2k)a(n, n+1+s-2k, k)\mathcal{L}_{h,G,F}^n(s).$$

(ii) *Let $k > 2n$, $h \in S_1^k(\mathbb{Q}(h))$, $G \in S_2^k(\mathbb{Q}(G))$ and $F \in S_2^k(\mathbb{Q}(F))$. Then we have for all $\sigma \in \mathrm{Gal}(\overline{\mathbb{Q}}/\mathbb{Q})$*

$$\left(\frac{a(n,0,k)\mathcal{L}_{h,G,F}^n(2k-n-1)}{\langle h;h\rangle\langle G;G\rangle\langle F;F\rangle}\right)^\sigma = \left(\frac{a(n,0,k)\mathcal{L}_{h^\sigma,G^\sigma,F^\sigma}^{(n)}(2k-n-1)}{\langle h^\sigma;h^\sigma\rangle\langle G^\sigma;G^\sigma\rangle\langle F^\sigma;F^\sigma\rangle}\right) \in \overline{\mathbb{Q}}.$$

The restriction $k > 2n$ can be weakened. The assumption *totally real* is only of technical nature. It can easily be removed (see [**Sh76**], [**Bo85a**]). All of the results stated so far have only been known for $n = 1$. In that case, P. Garrett [**G87**] has shown that $\mathcal{L}_{h,G,F}^1$ is essentially the triple product L-function for three elliptic eigenforms $h, G, F \in S_1^k$. So there is some hope that $\mathcal{L}_{h,G,F}^n$ has something to do with Euler products in general. The rest of this paper is devoted to show that one really gets an Euler product in the case of $n = 2$, if one chooses the cusp forms in an appropriate way.

3. Hecke-Jacobi theory

In this section, a certain Hecke-Jacobi theory [**A77b**], [**A87**], [**Gr82**], [**Gr93**] is developed, which serves mainly as an aid for studying the arithmetic properties of the Dirichlet series $D_{h,G;F}(s)$ introduced in (29). We proceed in a self-contained way to make the underlying structure as transparent as possible and to fix a good notation. This enables us to prove new results which are essential in getting an Euler product for the Dirichlet series.

3.1. The Jacobi group $G^J_{\mathcal{R}}$. For $\mathcal{R} = \mathbb{Z}$, $\mathbb{Z}[1/p]$ or $\mathbb{Q}$, we define the Heisenberg group $H_{\mathcal{R}}$, the group $\Gamma_{\mathcal{R}}$ and the Jacobi group $G^J_{\mathcal{R}}$. Let

$$H_{\mathcal{R}} = \{(X;\kappa) \,|\, X \in \mathcal{R}\times\mathcal{R}, \kappa \in \mathcal{R}\}$$

with group law $(X_1;\kappa_1)\circ(X_2;\kappa_2) = (X_1+X_2;\kappa_1+\kappa_2+\det\left(\begin{smallmatrix}X_1\\X_2\end{smallmatrix}\right))$, further let $\Gamma_{\mathcal{R}}$ be the direct product of the groups $G^+\mathrm{Sp}_1(\mathcal{R})$ and $\mathcal{R}^+$, and

$$G^J_{\mathcal{R}} = \left\{ M = \begin{pmatrix} a & 0 & b & \mu' \\ \lambda & \alpha & \mu & \kappa \\ c & 0 & d & -\lambda' \\ 0 & 0 & 0 & \beta \end{pmatrix} \;\middle|\; \begin{array}{l} \left(\begin{smallmatrix}a&b\\c&d\end{smallmatrix}\right) \in G^+\mathrm{Sp}_1(\mathcal{R}), \beta > 0, \\ \alpha\beta = \mu(M) = \mu\left(\begin{smallmatrix}a&b\\c&d\end{smallmatrix}\right), \\ \text{and } (\lambda,\mu;\kappa)\in\mathcal{R}^3 \\ \text{with } \beta^{-1}(\lambda',\mu')\left(\begin{smallmatrix}a&b\\c&d\end{smallmatrix}\right) = (\lambda,\mu) \end{array} \right\}$$

a subgroup of $G^+\mathrm{Sp}_2(\mathcal{R})$. The exact sequence

$$1 \longrightarrow H_{\mathcal{R}} \xrightarrow{\;i\;} G^J_{\mathcal{R}} \xrightarrow{\;p\;} \Gamma_{\mathcal{R}} \longrightarrow 1 \tag{32}$$

with $i(\lambda,\mu;\kappa) = \begin{pmatrix} 1 & 0 & 0 & \mu \\ \lambda & 1 & \mu & \kappa \\ 0 & 0 & 1 & -\lambda \\ 0 & 0 & 0 & 1 \end{pmatrix}$ and $p\begin{pmatrix} a & 0 & b & \mu' \\ \lambda & \alpha & \mu & \kappa \\ c & 0 & d & -\lambda' \\ 0 & 0 & 0 & \beta \end{pmatrix} = \left(\begin{pmatrix} a & b \\ c & d\end{pmatrix}, \beta\right)$

splits by

$$q : \left\{ \begin{array}{ccc} \Gamma_{\mathcal{R}} & \longrightarrow & G^J_{\mathcal{R}} \\ \left(\begin{pmatrix} a & b \\ c & d\end{pmatrix}, \beta\right) & \mapsto & \begin{pmatrix} a & b \\ c & d\end{pmatrix} \times \begin{pmatrix} * & 0 \\ 0 & \beta\end{pmatrix} \end{array}\right. .$$

That is the same as to say that $G^J_{\mathcal{R}}$ can be seen as a semi-direct product of $H_{\mathcal{R}}$ and $\Gamma_{\mathcal{R}}$. If $\gamma = (M,\beta)$, then ρ_γ given by $\rho_\gamma(X;\kappa) = (\alpha X M^{-1}; (\alpha/\beta)\kappa)$ is an automorphism of $H_{\mathcal{R}}$. The semi-direct product $H_{\mathcal{R}}\ \Gamma_{\mathcal{R}}$ is the set $H_{\mathcal{R}} \times \Gamma_{\mathcal{R}}$ with the group law

$$(h,\gamma)(\tilde{h},\tilde{\gamma}) := (h\rho_\gamma(\tilde{h}), \gamma\tilde{\gamma}). \tag{33}$$

PROPOSITION 3.1. *The groups $H_{\mathcal{R}}\ \Gamma_{\mathcal{R}}$ and $G^J_{\mathcal{R}}$ are isomorphic by*

$$\phi : \left\{ \begin{array}{ccc} H_{\mathcal{R}}\ \Gamma_{\mathcal{R}} & \longrightarrow & G^J_{\mathcal{R}} \\ \left((x_1,x_2;\kappa); \left(\begin{pmatrix} a & b \\ c & d\end{pmatrix}, \beta\right)\right) & \mapsto & \begin{pmatrix} a & 0 & b & \beta x_2 \\ y_1 & \alpha & y_2 & \beta\kappa \\ c & 0 & d & -\beta x_1 \\ 0 & 0 & 0 & \beta \end{pmatrix} \end{array}\right. . \tag{34}$$

It is convenient not to distinguish between image and pre-image of the embeddings i, q and ϕ. Then the group law in $G^J_{\mathcal{R}}$ becomes

$$\Big((X;\kappa);(M;\beta)\Big)\circ\Big((\tilde{X};\tilde{\kappa});(\tilde{M},\tilde{\beta})\Big)$$
$$= \left(\left(X+\alpha\tilde{X}M^{-1};\kappa+\frac{\alpha}{\beta}\tilde{\kappa}+\det\begin{pmatrix} X \\ \alpha\tilde{X}M^{-1}\end{pmatrix}\right);(M\tilde{M},\beta\tilde{\beta})\right)$$

and for $h=(X;\kappa)\in H_{\mathcal{R}}$ and $\gamma=(M,\beta)\in\Gamma_{\mathcal{R}}$ we have

$$(h;\gamma)=h\gamma=\gamma(\alpha^{-1}XM;\frac{\beta}{\alpha}\kappa).$$

3.2. Hecke-Jacobi algebra. In the seventies, Andrianov (e.g., [**A77b**]) recognized that it is possible to factorize the Hecke and Rankin polynomials in a convenient way, if one embeds the Hecke algebra $\mathcal{H}^n$ in a Hecke algebra belonging to the Siegel parabolic $C_{n,0}$ of Sp_n. This embedding has the advantage that one can see the action of $\mathcal{H}^n$ on M^k_n directly on the Fourier coefficients of the underlying modular form. So it is rather obvious that one should exchange, in the proceeding of Andrianov, the Siegel parabolic by another maximal parabolic, if one is interested in analyzing the arithmetic properties of the Fourier-Jacobi coefficients of a Siegel eigenform. This has been done by Gritsenko [**Gr82**], [**Gr88**], [**Gr95**] in some special cases. We want to recall some of his results by stressing the underlying general structures already developed by Andrianov and modify and generalize them for our purposes.

Let us now introduce a Hecke theory to the Hecke pair

$$\left(\Gamma^J := C^+_{2,1}(\mathbb{Z}), G^J := G^J_{\mathbb{Q}}\right).$$

That this is really a Hecke pair can be proven by a slight modification of the proof of Prop 1.1 in [**SZ**]. We call the corresponding Hecke algebra $\mathcal{H}^J$ the Hecke-Jacobi algebra and the local Hecke algebra $\mathcal{H}^J_p$ associated to the pair $(\Gamma^J, G^J_p = G^J_{\mathbb{Z}[p^{-1}]})$ the local Hecke-Jacobi algebra. It is an important question how a double coset decomposes into left cosets. The following result gives a first answer.

LEMMA 3.2. a) *Let r be a positive rational number and $g\in G^J$. Then*

$$\Gamma^J([r,r;r,r])\Gamma^J = \Gamma^J([r,r;r,r]) \quad \textit{and}$$
$$\Gamma^J g\Gamma^J\circ\Gamma^J([r,r;r,r])\Gamma^J = \Gamma^J g([r,r;r,r])\Gamma^J.$$

b) *Let a,d,α,β be natural numbers with $a\mid d$ and $ad=\alpha\beta$. Then*

$$\bigsqcup_j\left(\begin{pmatrix} a & 0\\ 0 & d\end{pmatrix},\beta\right)\bigsqcup_{\substack{\lambda \ \bmod a/(a,\alpha)\\ \mu \ \bmod d/(d,\alpha)\\ \kappa \ \bmod \beta/(\beta,\alpha)}}(\lambda,\mu;\kappa)\,(U_j\times 1_2) \tag{35}$$

is a representative system of $\Gamma^J\backslash\Gamma^J\left(\left(\begin{smallmatrix} a & 0\\ 0 & d\end{smallmatrix}\right),\beta\right)\Gamma^J$. Here U_j runs through a representative system of $\left(\left(\begin{smallmatrix} a & 0\\ 0 & d\end{smallmatrix}\right)^{-1}\Gamma\left(\begin{smallmatrix} a & 0\\ 0 & d\end{smallmatrix}\right)\bigcap\Gamma\right)\backslash\Gamma$.

The first condition is clear and the second can easily be proven (cf. [**Gr82**, Lemma 3.1]).

By Lemma 1.1, it is easy to see that

$$i:\mathcal{H}^2\hookrightarrow\mathcal{H}^J;\sum_j a_j\Gamma_2 g_j\mapsto\sum_j a_j\Gamma^J g_j$$

gives an embedding if one chooses $g_j \in \Gamma^J$. This map commutes with the anti-automorphism $*$ of Γ^J obtained by

$$\Gamma^J\,(h;(M,\beta))\,\Gamma^J \mapsto \left(\Gamma^J\,(h;(M,\beta))\,\Gamma^J\right)^* = \Gamma^J\mu(M)\,(h;(M,\beta))^{-1}\,\Gamma^J. \tag{36}$$

With (35) in Lemma 3.2, one sees that

$$j_-: \begin{cases} \mathcal{H}^1 & \longrightarrow & \mathcal{H}^J \\ \Gamma\begin{pmatrix} a & 0 \\ 0 & d \end{pmatrix}\Gamma & \mapsto & \Gamma^J\left(\begin{pmatrix} a & 0 \\ 0 & d \end{pmatrix};1\right)\Gamma^J \end{cases} \tag{37}$$

$$j_+: \begin{cases} \mathcal{H}^1 & \longrightarrow & \mathcal{H}^J \\ \Gamma\begin{pmatrix} a & 0 \\ 0 & d \end{pmatrix}\Gamma & \mapsto & \Gamma^J\left(\begin{pmatrix} a & 0 \\ 0 & d \end{pmatrix};ad\right)\Gamma^J \end{cases} \tag{38}$$

lead to two algebra monomorphisms and that $j_+(X) = (j_-(X))^*$, if $X \in \mathcal{H}^1$. We denote the images of these maps by $\mathcal{H}_-^J$ and $\mathcal{H}_+^J$. One has $(\mathcal{H}_-^J)^* = \mathcal{H}_+^J$ and all elements of $\mathcal{H}^2$ are fixed by $*$. By these constructions we have distinguished certain subalgebras $\mathcal{H}_-^J, \mathcal{H}_+^J$ and $i(\mathcal{H}^2)$ of $\mathcal{H}^J$, which are commutative, have no zero divisors and decompose in their local components.

Next we want to introduce some important elements of $\mathcal{H}^J$. Let $r \in \mathbb{Q}$ and $n \in \mathbb{N}$:

$$T_-^g(n) := j_-(T(n)), \qquad T_+^g(n) := j_+(T(n)),$$

$$\Lambda_-^g(n) := j_-(T(n,n)), \qquad \Lambda_+^g(n) := j_+(T(n,n)),$$

$$\nabla(r) := \Gamma^J\,((0;r);(1_2,1))\,\Gamma^J, \quad \Delta_n := \Gamma^J[n,n;n,n]\Gamma^J,$$

$$\nabla_n^g := \Delta_n \sum_{b \bmod n} \nabla(\tfrac{b}{n}), \qquad \Xi_n^g := \Delta_n \sum_{\lambda,\mu,\kappa \bmod n} \Gamma^J(\tfrac{\lambda}{n},\tfrac{\mu}{n};\tfrac{\kappa}{n}),$$

$$T^{J,g}(n) := \Gamma^J[1,n;n^2,n]\Gamma^J.$$

For $T(a,b) \in \mathcal{H}^1$ with $a,b \in \mathbb{N}$, we put $T_-^g(a,b) = \Gamma^J[a,ab;b,1]\Gamma^J$ and $T_+^g(a,b) = \Gamma^J[a,1;b,ab]\Gamma^J$. If $ab = n^2$ with $n \in \mathbb{N}$, we put $T_0^g(a,b) = \Gamma^J[a,n;b,n]\Gamma^J$. The double cosets $\nabla(a)$ and Δ_n decompose into only one left coset and $\Xi_n^g, \nabla_n^g, \Delta_n$ and $T^{J,g}(n)$ are fixed by $*$. Furthermore, we have $(T_-^g(n))^* = T_+^g(n)$ and $(\Lambda_-^g(n))^* = \Lambda_+^g(n)$. Further we put

$$\nabla_n^r = (\Delta_n)^{-1}\nabla_n^g, \Xi_n^r = (\Delta_n)^{-1}\Xi_n^g \text{ and } T^{J,r}(n) = (\Delta_n)^{-1}T^{J,g}(n)$$

(as a rule, we do this only when it is very convenient). By abuse of notation we put Δ_n for $\Gamma_2[n,n;n,n]\Gamma_2$ and $\Gamma^J[n,n;n,n]\Gamma^J$.

REMARK 3.3. (1) The Hecke-Jacobi algebra $\mathcal{H}^J$ is not commutative and has zero divisors, because $\Lambda_-^g(p)\cdot(\nabla_p^r - p) = 0$ and $\Lambda_-^g(p)T_+^g(p) \neq T_+^g(p)\Lambda_-^g(p)$.

(2) It is not clear how one can decompose $\mathcal{H}^J$ into its local components. Let for example p,q be prime numbers, then

$$\nabla\left(\frac{1}{q}\right)\nabla\left(\frac{1}{p}\right) = \nabla\left(\frac{p+q}{pq}\right),$$

and if we assume that $p^2 \not\equiv 1(q)$ we have $\Lambda_-^g(p)\nabla(1/q) \neq \nabla(1/q)\Lambda_-^g(p)$.

The following proposition indicates how one can get a local-global result.

PROPOSITION 3.4. *The elements T_-^g, T_+^g, $T^{J,g}$, ∇^g, Ξ^g, Λ_-^g, Λ_+^g and Δ in $\mathcal{H}^J$ commute with each other when we only allow coprime arguments. For Λ_-^g, Λ_+^g and Δ, the assumption on the coprimeness can be dropped.*

PROOF. Case by case. To indicate how one can do that, we prove that

$$T_-^g(n)T_+^g(m) = T_+^g(m)T_-^g(n) \text{for } (n,m)=1.$$

Since $\mathcal{H}^1 = \otimes_p \mathcal{H}_p^1$ and $\mathcal{H}_p^1$ is generated by $T(p), T(p,p)$ and $T(p,p)^{-1}$, it is enough to show that $T_-^g(p)T_+^g(q) = T_+^g(q)T_-^g(p)$ for two different prime numbers p and q. But this can be deduced directly from Lemma 3.2. □

COROLLARY 3.5. *If we denote by $\tilde{\mathcal{H}}^J$ the subalgebra of $\mathcal{H}^J$ generated by*

$$T_-^g, T_+^g, T^{Jg}, \nabla^g \Xi^g, \Lambda_-^g, \Lambda_+^g, \Delta$$

and if we put $\tilde{\mathcal{H}}_p^J = \mathcal{H}_p^J \bigcap \tilde{\mathcal{H}}^J$, then we have

$$\tilde{\mathcal{H}}^J = \otimes_p \tilde{\mathcal{H}}_p^J. \tag{39}$$

3.3. The Hecke algebra $\tilde{\mathcal{H}}_p^J$. First we want to show that $\mathcal{H}^2$ can be seen as a subalgebra of $\tilde{\mathcal{H}}^J$. To do that, it is enough to determine the image of the generators of $\mathcal{H}_p^2$ under the embedding i. A slight modification of Lemma 3.3.32 in [**A87**] gives us

$$\begin{aligned} i(T^{(2)}(p)) &= T_-^g(p) + T_+^g(p) \\ i(T_1^{(2)}(p^2)) &= \Lambda_-^g(p) + \Lambda_+^g(p) + T^{J,g}(p) + \nabla_p^g - \Delta_p \\ i(T_0^{(2)}(p^2)) &= T_-^g(1,p^2) + T_+^g(1,p^2) + T^{J,g}(p)(\nabla_p^r - 1) + \Xi_p^g - \nabla_p^g \\ i(\Delta_p^{(2)}) &= \Delta_p, \end{aligned}$$

which proves our claim. To simplify our notation we put $T_-^g, T_+^g, \Lambda_-^g, \Lambda_+^g, T^{J,g}$ for $T_-^g(p), T_+^g(p), \Lambda_-^g(p), \Lambda_+^g(p), T^{J,g}(p)$, if we are in a fixed local situation.

Gritsenko [**Gr95**, Abschnitt 5.1] introduced the following very convenient invariants. Let $X = \Gamma^J(h;(M,\beta))\Gamma^J \in \tilde{\mathcal{H}}_p^J$, then we denote by $\text{Sig}(X) = (\beta^2)/\mu(M)$ the signature of X andfor $r \in \mathbb{Q}^\times$ and $Y = \sum_i a_i X_i \in \tilde{\mathcal{H}}_p^J$ we put $\text{Sig}_r(Y) = \sum_{i,\text{Sig}_(X_i)=r} a_i X_i$. For $M_1, M_2 \in G^J$ we have

$$\text{Sig}\left(\Gamma^J M_1 \Gamma^J \circ \Gamma^J M_2 \Gamma^J\right) = \text{Sig}\left(\Gamma^J M_1 \Gamma^J\right) \cdot \text{Sig}\left(\Gamma^J M_2 \Gamma^J\right) \tag{40}$$

and for $X = \sum_u \text{Sig}_u(X)$ and $X = \sum_v \text{Sig}_v(X)$ we have

$$\begin{aligned} \text{Sig}_r(XY) &= \sum_{uv=r} \text{Sig}_u(X)\text{Sig}_v(Y) \\ &= \sum_{uv=r} \text{Sig}_v(Y)\text{Sig}_u(X). \end{aligned} \tag{41}$$

Applying this for example to the case $X = T^{(2)}(p)$ and $Y = T_1^{(2)}(p^2)$, leads us to

$$\begin{aligned} T_+^g \Lambda_+^g &= \Lambda_+^g T_+^g \\ T_-^g \Lambda_+^g + T_+^g T^{J,g} + T_+^g(\nabla_p^g + \Delta_p) &= \Lambda_+^g T_-^g + T^{J,g} T_+^g + (\nabla_p^g + \Delta_p) T_+^g \\ T_+^g \Lambda_-^g + T_-^g T^{J,g} + T_-^g(\nabla_p^g + \Delta_p) &= \Lambda_-^g T_+^g + T^{J,g} T_-^g + (\nabla_p^g + \Delta_p) T_-^g \\ T_-^g \Lambda_-^g &= \Lambda_-^g T_-^g. \end{aligned}$$

Such identities can be used to determine commutators in $\tilde{\mathcal{H}}_p^J$. Next, we want to simplify terms of the form $T_-^g \Lambda_+^g$, $\Lambda_-^g \Lambda_+^g$ and so on.

LEMMA 3.6. *The following relations in $\tilde{\mathcal{H}}_p^J$ hold:*

$$T_-^g \Lambda_+^g = p^2 \Delta_p T_+^g \tag{42}$$

$$\Lambda_-^g T_+^g = p^2 \Delta_p T_-^g \tag{43}$$

$$\Lambda_-^g \Lambda_+^g = p^4 (\Delta_p)^2 \tag{44}$$

$$T_-^g T_+^g = p T^{J,g} + (p^2 + p^3)\Delta_p. \tag{45}$$

PROOF. By Lemma 3.2, we have

$$\begin{aligned} T_-^g &= \Gamma^J[1,p;p,1]\left(U_{[1,p]} \times 1_2\right) \\ \Lambda_-^g &= \Gamma^J[p,p^2;p,1] \\ T_+^g &= \Gamma^J[1,1;p,p] \sum_{\mu,\kappa \bmod p} (0,\mu;\kappa)\left(U_{[1,p]} \times 1_2\right) \\ \Lambda_+^g &= \Gamma^J[p,1;p,p^2] \sum_{\substack{\lambda,\mu \bmod p \\ \kappa \bmod p^2}} (\lambda,\mu;\kappa). \end{aligned}$$

This leads to

$$\begin{aligned} T_-^g \Lambda_+^g &= p^{-2} \Gamma^J[p,p;p^2,p^2]\left(U_{[1,p]} \times 1_2\right) \sum_{\lambda,\mu,\kappa \bmod p^2} (\lambda,\mu;\kappa) \\ &= p^{-2} \Delta_p \Gamma^J[1,1;p,p] \sum_{\lambda,\mu,\kappa \bmod p^2} (\lambda,\mu;\kappa)\left(U_{[1,p]} \times 1_2\right) \end{aligned} \tag{46}$$

which proves (42). Here we used that $\mathrm{GL}_2(\mathbb{Z})$ normalizes the Heisenberg group $H_{\mathbb{Z}/p^2\mathbb{Z}}$. If we apply the anti-automorphism $*$ to equation (42) we get (43).

It is easy to see that

$$\begin{aligned} \Lambda_-^g \Lambda_+^g &= \Gamma^J[p,p^2;p,1][p,1;p,p^2] \sum_{\substack{\lambda,\mu \bmod p \\ \kappa \bmod p^2}} (\lambda,\mu;\kappa) \\ &= \Gamma^J[p^2,p^2;p^2,p^2] \sum_{\substack{\lambda,\mu \bmod p \\ \kappa \bmod p^2}} (\lambda,\mu;\kappa) = p^4(\Delta_p)^2. \end{aligned}$$

One can prove the last assumption of the lemma as follows:

$$\begin{aligned}
T^g_- T^g_+ &= \left(\Gamma^J[1,p;p,1](U_{[1,p]} \times 1_2)\right)\left(\Gamma^J[1,1;p,p] \sum_{\mu,\kappa \mod p} (0,\mu;\kappa)(U_{[1,p]} \times 1_2)\right) \\
&= p^{-1}\Gamma^J\left((T^{(1)}(p) \cdot T^{(1)}(p)) \times [p;p]\right) \sum_{\lambda,\mu,\kappa \mod p} (0,\mu;\kappa) \\
&= p^{-1}\Gamma^J\left((T^{(1)}(1,p^2) + T^{(1)}(p,p) + pT^{(1)}(p,p)) \times [p;p]\right) \sum_{\mu,\kappa \mod p} (0,\mu;\kappa) \\
&= p^{-1}\Gamma^J\left(([1,p^2]U_{[1,p^2]}) \times [p;p]\right)\left(\sum_{\mu,\kappa \mod p} (0,\mu;\kappa)\right) + (p^2+p^3)\Delta_p
\end{aligned}$$

which gives the desired result. □

We hope to have given the reader an impression of how one can prove such statements. The following list contains relations between the generators of $\tilde{\mathcal{H}}^J_p$. At the (i,j) position we have put the product of the corresponding row and column elements. With the subscript we want to indicate, which kind of relation is listed in the second chart. In the left column the position, in the middle one the relation and in the right eventually a reference is given. The number in the reference indicates another equation, which is the same as the given one. If the number comes with $*$, one has to apply $*$ to the corresponding equation. Finally, 15^m indicates a connection with equation 15.

(47)

$\circ$	T^g_-	Λ^g_-	T^g_+	Λ^g_+	$T^{J,g}$	∇^g_p	Ξ^g_p
T^g_-	11_F	12_F	$13_{\surd}$	$14_{\surd}$	$15_{[]}$	$16_{\surd}$	17_U
Λ^g_-	21_F	22_F	$23_{\surd}$	$24_{\surd}$	$25_{\surd}$	$26_{\surd}$	$27_{\surd}$
T^g_+	31_U	32_U	33_F	34_F	35_U	$36_?$	$37_?$
Λ^g_+	41_U	$42_{\surd}$	43_F	44_F	$45_?$	$46_?$	$47_?$
$T^{J,g}$	$51_{[]}$	$52_?$	$53_{[]}$	$54_{\surd}$	55_F	$56_?$	$57_?$
∇^g_p	$61_?$	$62_?$	$63_{\surd}$	$64_{\surd}$	$65_?$	66_F	$67_{\surd}$
Ξ^g_p	$71_?$	$72_?$	73_U	$74_{\surd}$	$75_?$	$76_{\surd}$	77_U

- $\surd$ simplification
- F formula
- U easy transformation
- [] essentially the commutator
- ? no real simplification

11	$T^g_- T^g_-$	$=$	$T_-(p^2) + p\Lambda^g_-$	
12	$T^g_- \Lambda^g_-$	$=$	$T^g_-(p,p^2)$	
13	$T^g_- T^g_+$	$=$	$pT^{J,g} + (p^2+p^3)\Delta_p$	
14	$T^g_- \Lambda^g_+$	$=$	$p^2\Delta_p T^g_+$	
15	$T^g_- T^{J,g}$	$=$	$p(p-1)\Delta_p T^g_- + \nabla^g_p T^g_- - T^g_+ \Lambda^g_- + T^{J,g} T^g_-$	
16	$T^g_- \nabla^g_p$	$=$	$pT^g_- \Delta^g_p$	
17	$T^g_- \Xi^g_p$	$=$	$p^2(\Delta_p - p^{-1}\nabla^g_p)T^g_- + pT^g_+ \Lambda^g_-$	

21	$\Lambda^g_- T^g_-$	$=$	$T^g_-(p,p^2)$	12
22	$\Lambda^g_- \Lambda^g_-$	$=$	$\Lambda^g_-(p^2)$	
23	$\Lambda^g_- T^g_+$	$=$	$p^2\Delta_p T^g_-$	14*
24	$\Lambda^g_- \Lambda^g_+$	$=$	$p^4(\Delta_p)^2$	
25	$\Lambda^g_- T^{J,g}$	$=$	$p\Delta_p T^g_-(1,p^2)$	
26	$\Lambda^g_- \nabla^g_p$	$=$	$p\Delta_p \Lambda^g_-$	
27	$\Lambda^g_- \Xi^g_p$	$=$	$p^3\Delta_p \Lambda^g_-$	
31	$T^g_+ T^g_-$	$=$	$\Delta_p^{-1} T^{J,g}\nabla^g_p + p\nabla^g + \Xi^g_p$	
32	$T^g_+ \Lambda^g_-$	$=$	$p^{-1}T^g_-\Xi^g_p - (p\Delta_p - \nabla^g_p)T^g_-$	
33	$T^g_+ T^g_+$	$=$	$T^g_+(p^2) + p\Lambda^g_+$	11*
34	$T^g_+ \Lambda^g_+$	$=$	$T^g_+(p,p^2)$	12*
35	$T^g_+ T^{J,g}$	$=$	$p(1-p)\Delta_p T^g_+ - T^g_+\nabla^g_p + \Lambda^g_+ T^g_- + T^{J,g}T^g_+$	15^m
36	$T^g_+ \nabla^g_p$	$=$	$---$	
37	$T^g_+ \Xi^g_p$	$=$	$---$	
41	$\Lambda^g_+ T^g_-$	$=$	$p^{-1}\Xi^g_p T^g_+ - T^g_+(p\Delta_p - \nabla^g_p)$	32*
42	$\Lambda^g_+ \Lambda^g_-$	$=$	$p^{-1}\Delta_p^{-1}\nabla^g_{p^2}\Xi^g_p$	
43	$\Lambda^g_+ T^g_+$	$=$	$T^g_+(p,p^2)$	34
44	$\Lambda^g_+ \Lambda^g_+$	$=$	$\Lambda^g_+(p^2)$	22*
45	$\Lambda^g_+ T^{J,g}$	$=$	$---$	
46	$\Lambda^g_+ \nabla^g_p$	$=$	$---$	
47	$\Lambda^g_+ \Xi^g_p$	$=$	$---$	
51	$T^{J,g} T^g_-$	$=$	$T^g_+\Lambda^g_- - p(p-1)\Delta_p T^g_- - \nabla^g_p T^g_- + T^g_- T^{J,g}$	35*
52	$T^{J,g} \Lambda^g_-$	$=$	$---$	45*
53	$T^{J,g} T^g_+$	$=$	$p(p-1)\Delta_p T^g_+\nabla^g_p - \Lambda^g_+ T^G_- + T^g_+ T^{J,g}$	15*
54	$T^{J,g} \Lambda^g_+$	$=$	$p\Delta_p T^g_+(1,p^2)$	25*
55	$T^{J,g} T^{J,g}$	$=$	$T^{J,g}(p^2) + (p-1)\nabla^g_p T^{J,g}(p) + p\Delta_p(\Xi^g_p - \nabla^g_p + p^2 + p)$	
56	$T^{J,g} \nabla^g_p$	$=$	$---$	
57	$T^{J,g} \Xi^g_p$	$=$	$---$	
61	$\nabla^g_p T^g_-$	$=$	$---$	36*
62	$\nabla^g_p \Lambda^g_-$	$=$	$---$	46*
63	$\nabla^g_p T^g_+$	$=$	$p\Delta_p T^g_+$	16*
64	$\nabla^g_p \Lambda^g_+$	$=$	$p\Delta_p \Lambda^g_+$	26*
65	$\nabla^g_p T^{J,g}$	$=$	$---$	56
66	$\nabla^g_p \nabla^g_p$	$=$	$\nabla^g_{p^2}$	
67	$\nabla^g_p \Xi^g_p$	$=$	$p\Delta_p \Xi^g_p$	
71	$\Xi^g_p T^g_-$	$=$	$---$	37*
72	$\Xi^g_p \Lambda^g_-$	$=$	$---$	47*
73	$\Xi^g_p T^g_+$	$=$	$p^2 T^g_+(\Delta_p - p^{-1}\nabla^g_p) + p\Lambda^g_+ T^g_-$	17*
74	$\Xi^g_p \Lambda^g_+$	$=$	$p^3\Delta_p \Lambda^g_+$	27*
75	$\Xi^g_p T^{J,g}$	$=$	$---$	57*
76	$\Xi^g_p \nabla^g_p$	$=$	$p\Delta_p \Xi^g_p$	67*
77	$\Xi^g_p \Xi^g_p$	$=$	$p\Delta_p^{-1}\Xi^g_p\nabla^g_{p^2} + p^3\Delta_p\nabla^g_p - p^2\nabla^g_{p^2}$	

Now we are able to state and prove the main results of this section, the announced factorization of the Hecke and Rankin polynomials $q_p^2(X)$ and $d_p^2(X)$. It is tricky to construct such decompositions, in contrast it is easy to prove the correctness of them. A good overview over this topic is given in [**A87**, §3.5]. For a proof of the next theorem the reader is referred to [**Gr93**].

THEOREM 3.7. *(Gritsenko) The Hecke polynomial $q_p^2(X)$ has the following factorization in $\tilde{\mathcal{H}}_p^J[X]$:*

$$q_p^2(X) = j_-\left(q_p^1(X)\right)\left(1 + p\Delta_p(\nabla_p^r - p)X^2\right) j_+\left(q_p^1(X)\right)$$

with $q_p^1(X) = 1 - T^{(1)}(p)X + p\Delta_p^{(1)}X^2$.

Next, we want to present a new factorization of the Rankin polynomial. It is an unpleasant fact, that the suggested formula in [**Gr88**, Theorem 3], is not correct.

THEOREM 3.8. *The Rankin polynomial $d_p^2(X)$ has the following factorization in $\tilde{\mathcal{H}}_p^J[X]$:*

$$d_p^2(X) = (1 - X)(1 - p^{-2}\Lambda_-^r X)S^{(2)}(X)(1 - p^{-2}\Lambda_+^r X) \tag{48}$$

with $S^{(2)}(X) = \sum_{j=0}^{3}(-1)^j S_j^{(2)} X^j$. Here

$$\begin{aligned} S_0^{(2)} &= 1 \\ S_1^{(2)} &= p^{-2}\left(T^{J,r} + \nabla_p^r - p^2\right) \\ S_2^{(2)} &= p^{-3}\left(T^{J,r}(\nabla_p^r - p) + \Xi^r - p\nabla_p^r + p^2\right) \\ S_3^{(2)} &= p^{-2}\left(\nabla_p^r - p\right). \end{aligned}$$

PROOF. Most of the work has already been done. We multiply the right side of the equation (48) out and get a polynomial of degree 5. With the aid of our charts, the stated formulas given in section 1.2 and at the beginning of section 3.3, we see that this polynomial is nothing else but the Rankin polynomial of degree 4. □

3.4. Jacobi forms as P-forms. We begin with a very convenient definition.

DEFINITION 3.9. Let n, s and k be natural numbers. A P-Jacobi form of degree (n, s) and weight k is a holomorphic function F on $\mathcal{H}_{n+s}$, which satisfies

(i) $F|_k\gamma = F$ for all $\gamma \in C_{n+s,n}^+(\mathbb{Z})$

(ii) There exists $t \in A_s$ such that F has the Fourier expansion

$$F(Z) = \sum_{T=\left(\begin{smallmatrix} * & * \\ * & t\end{smallmatrix}\right)\in A_{n+s}} a(T)e\{TZ\}. \tag{49}$$

We call such t the index of F. A P-Jacobi form is called P-Jacobi cusp form, (if the index is positive definite and) if the summation in (49) may be restricted to positive definite matrices. The space which is generated by all P-Jacobi forms (P-Jacobi cusp forms) of degree (n, s) and weight k is denoted by the space of P-forms $\mathcal{PJ}_{n,s}^k$ (space of P-cusp forms $\mathcal{PJ}_{n,s}^{k,\text{cusp}}$). Their elements are called P-forms and P-cusp forms respectively. Finally we denote by $\mathcal{PJ}_{n,s}^k(t)$ the subspace of P-Jacobi forms and by $\mathcal{PJ}_{n,s}^{k,\text{cusp}}(t)$ of P-Jacobi cusp forms for fixed index t. We use the convention $\mathcal{PJ}_{n,s}^{k,\text{cusp}}(t) = \{0\}$, if t is not half-integral.

To introduce the Petersson scalar product for P-forms, we put $\mathbb{C}^{n,s} = \text{Mat}(n, s; \mathbb{C}), \mathcal{H}_{n,s} = \mathcal{H}_n \times \mathbb{C}^{n,s}, \Gamma_{n,s}^J = C_{n+s,n}^+(\mathbb{Z})$ and $\Gamma_{n,s}^{J,red} = \Gamma_{n,s}^J / Z^{n,s}$, where $Z^{n,s}$ denotes the center of $\Gamma^J_{n,s}$. Let $\gamma \in \Gamma^J_{n,s}$, then

$$\gamma\left\langle\begin{pmatrix} z_1 & z_2 \\ * & * \end{pmatrix}\right\rangle = \begin{pmatrix} z_1' & z_2' \\ * & * \end{pmatrix}$$

induces the operation $(z_1, z_2) \longmapsto \gamma < z_1, z_2 >= (z_1', z_2')$ of $\Gamma^J_{n,s}$ on $\mathcal{H}_{n,s}$, which is in fact an action of $\Gamma^{J,red}_{n,s}$. Then

$$\mathcal{Q}_{n,s} = \{Z \in \mathcal{H}_{n+s} \,|\, (z_1, z_2) \in \mathcal{B}_{n,s} \text{ and } |\mathrm{Re}(z_4)_{ij}| \leq \tfrac{1}{2}\}$$

is a fundamental domain of the action of $\Gamma^J_{n,s}$ on $\mathcal{H}_{n+s}$, if $\mathcal{B}_{n,s}$ is a fundamental domain of the action of $\Gamma^{J,red}_{n,s}$ on $\mathcal{H}_{n,s}$. Every $F \in \mathcal{PJ}^k_{n,s}(t)$ has a decomposition of the form $F(Z) = \phi^F(z_1, z_2)e\{tz_4\}$.

DEFINITION 3.10. Let $F, G \in \mathcal{PJ}^k_{n,s}$. Then we denote by the value of

$$\langle F, G\rangle_{\mathcal{A}} = \int_{\mathcal{Q}_{n,s}} F(Z)\overline{G(Z)}(\det Y)^k \, d^*Z \tag{50}$$

the Petersson scalar product of F and G for $d^*Z = dXdY/(\det y)^{n+s+1}$, if it exists.

REMARK 3.11. (i) The space of Jacobi forms of degree (n, s), weight k and index t is given by

$$\mathcal{J}^k_{n,s}(t) = \{\phi : \mathcal{H}_n \times \mathbb{C}^{n,s} \longrightarrow \mathbb{C} \,|\, \tilde{\phi}_t(Z) := \phi(z_1, z_2)e\{tz_4\} \in \mathcal{PJ}^k_{n,s}\}. \tag{51}$$

In the same way one defines the space of Jacobi cusp forms $\mathcal{J}^{k,\mathrm{cusp}}_{n,s}(t)$ of index t, and more generally the spaces $\mathcal{J}^k_{n,s}$ and $\mathcal{J}^{k,\mathrm{cusp}}_{n,s}$.

(ii) Let $F, G \in \mathcal{PJ}^{k,\mathrm{cusp}}_{n,s}(t)$ with $k > n + s$ and $y_i = \mathrm{Im}(z_i)$. Then

$$\begin{aligned}\langle F, G\rangle_{\mathcal{A}} &= \beta_{s,k-n}(\det t)^{(s+1)/2+n-k} \\ &\quad \int_{\mathcal{B}_{n,s}} \phi^F(z_1, z_2)\overline{\phi^G(z_1, z_2)}e^{-4\pi \mathrm{tr}(ty_1^{-1}[y_2])}(\det y_1)^k \, d^*Z_{n,s}.\end{aligned}$$

Here $\beta_{s,k-n} = \pi^{s(s-1)/4}(4\pi)^{s(s+1)/2-s(k-n)} \prod_{\nu=1}^s \Gamma(k - n - (s + \nu)/2)$ and $d^*Z_{n,s} = dz_1dz_2/(\det y_1)^{n+s+1}$ (cf. Klingen [**K89**, Satz 3], Dulinski [**Du**]).

Let $\phi, \psi \in \mathcal{J}^{k,\mathrm{cusp}}_{n,s}(t)$ and $k > n + s$, then we get the Petersson scalar product $\langle \phi, \psi\rangle_{J,t}$ for Jacobi forms by

$$\langle \tilde{\phi}; \tilde{\psi}\rangle_{\mathcal{A}} = \beta_{s,k-n}(\det\, t)^{(s+1)/2+n-k}\langle \phi, \psi\rangle_{J,t}.$$

Let us fix in the case $s = n = 1$ the following notation. For $Z = \left(\begin{smallmatrix} \tau & z \\ z & \tau' \end{smallmatrix}\right) \in \mathcal{H}_2$, we put

$$\begin{pmatrix} \tau_x & z_x \\ z_x & \tau_x' \end{pmatrix} = \mathrm{Re}(Z) \text{ and } \begin{pmatrix} \tau_y & z_y \\ z_y & \tau_y' \end{pmatrix} = \mathrm{Im}(Z).$$

Let $F, G \in \mathcal{PJ}^{k,\mathrm{cusp}}_{1,1}(t)$, $k > 2$ and $\beta_k = (4\pi)^{k-2}\Gamma(k-2)^{-1}$, then

$$\langle \phi^F, \phi^G\rangle_{J,t} = \beta_k t^{k-2}\langle F, G\rangle_{\mathcal{A}}. \tag{52}$$

If $F : \mathcal{H}_2 \longrightarrow \mathbb{C}$ is invariant under the $|_k$-action of Γ^J, then for $X = \sum a_j\Gamma^J g_j \in \tilde{\mathcal{H}}^J$

$$F|_k\mathcal{D}(X) := \sum_j a_j F|_k g_j \tag{53}$$

has the same property. Let $F(Z) = \sum_{n=1}^\infty f_n(\tau, z)e\{n\tau'\} \in S^k_2$ and $X \in \mathcal{H}^2$, then (53) is compatible with the action introduced in (13) and the embedding i. More generally, let $X \in \tilde{\mathcal{H}}^J$, then we put $f_m\|_kX$ for the m-th Fourier-Jacobi coefficient of $F|_k\mathcal{D}(X)$ resp., τ'. Especially the action in (53) can be seen as an action of $\tilde{\mathcal{H}}^J$ on $\mathcal{P}\mathcal{J}_{1,1}^{k,\text{cusp}}$ $(\mathcal{J}_{1,1}^{k,\text{cusp}})$. To be more precise, if $X \in \tilde{\mathcal{H}}^J$ is a double coset with signature $r \in \mathbb{Q}^+$ and $F \in \mathcal{P}\mathcal{J}_{1,1}^{k,\text{cusp}}(t)$, then $F|_k\mathcal{D}(X) \in \mathcal{P}\mathcal{J}_{1,1}^{k,\text{cusp}}(tr)$. By Proposition 4.1.7, page 218 [**A87**], it is clear that

$$\langle F|_k\mathcal{D}(X); G\rangle_{\mathcal{A}} = \langle F; G|_k\mathcal{D}(X^*)\rangle_{\mathcal{A}}, \tag{54}$$

with $X \in \tilde{\mathcal{H}}^J$ and $F, G \in \mathcal{P}\mathcal{J}_{1,1}^{k,\text{cusp}}$. Further, if $\tilde{\phi} \in \mathcal{P}\mathcal{J}_{1,1}^{k,\text{cusp}}(t)$ and $(p,t) = 1$, then $\tilde{\phi}|_k\mathcal{D}(X) = 0$, if $X = \nabla_n^g, \Xi_n^g, T_+^g(n)$ or $\Lambda_+^g(n)$.

Finally, we are able to compute the adjoint operators of the classical Hecke-Jacobi operators (e.g. U_l, V_l, T_l) introduced in [**EZ**, §4], in a rather simple way. For T_p, we put T_p^{EZ}. Let A be a linear operator on $\mathcal{J}_{1,1}^{k,\text{cusp}}$, then we denote by $\tilde{A}$ the corresponding operator on $\mathcal{P}\mathcal{J}_{1,1}^{k,\text{cusp}}$, further by A^{ad} resp., $(\tilde{A})^{\text{ad}}$ the adjoint operators. Assuming that $\tilde{A}$ has signature r, we have $\widetilde{A^{\text{ad}}} = r^{k-2}(\tilde{A})^{\text{ad}}$. Then the dictionary

$$\begin{array}{rcl} \mathcal{D}(\Lambda_-^g(l)) &=& l^{3(k-2)}\tilde{U}_l \\ \mathcal{D}(T_-^g(l)) &=& l^{k-2}\tilde{V}_l \\ \mathcal{D}(T^{J,g}(p)) &=& p^{3-k}\widetilde{T_p^{EZ}} \end{array} \quad \text{leads to} \quad \begin{array}{rcl} \widetilde{U_l^{\text{ad}}} &=& l^{2-k}\mathcal{D}(\Lambda_+^g(l)) \\ \widetilde{V_l^{\text{ad}}} &=& \mathcal{D}(T_+^g(l)) \\ \widetilde{(T_p^{EZ})^{\text{ad}}} &=& \widetilde{T_p^{EZ}} \end{array}. \tag{55}$$

The following notation and observation will turn out to be convenient later on. If we put

$$S^{(2)}(X)^{\text{factor}} = 1 - p^{-2}T^{J,r}(p)X + p^{-1}X^2 \tag{56}$$

$$\begin{aligned} S^{(2)}(X)^{\text{prim}} &= 1 - (p^{-2}T^{J,r}(p) - 1)X + (p^{-1} - p^{-2})T^{J,r}(p)X^2 + p^{-1}X^3 \\ &= S^{(2)}(X)^{\text{factor}}(1+X) \end{aligned} \tag{57}$$

$$S^{(2)}(X)^{n.\text{prim}} = 1 - (p^{-2}T^{J,r}(p) + p^{-1} - 1)X + p^{-3}\Xi_p^rX^2, \tag{58}$$

then $\mathcal{D}(S^{(2)}(X)) = \mathcal{D}(S^{(2)}(X)^{\text{prim}})$ on $\mathcal{P}\mathcal{J}_{1,1}^{k,\text{cusp}}(t)$, if $(p,t) = 1$ and $\mathcal{D}(S^{(2)}(X)) = \mathcal{D}(S^{(2)}(X)^{n.\text{prim}})$ otherwise. Further we have for all $F \in \mathcal{P}\mathcal{J}_{1,1}^{k,\text{cusp}}(p)$:

$$F|_k\mathcal{D}(S^{(2)}(X)T_+^g(p)) = F|_k\mathcal{D}(T_+^g(p)(1 - p^{-2}T^{J,r}(p)X + p^{-1}X^2)). \tag{59}$$

3.5. Weak rationality theorems. Let $F \in S_2^k$ be a Hecke eigenform and $(f_n)_{n\in\mathbb{N}}$ the Fourier-Jacobi coefficients of F. Further let $Q_{p,F}^2$ and $D_{p,F}^2$ the corresponding Hecke and Rankin polynomials. Then the following identities hold in the ring of formal power series.

PROPOSITION 3.12. *(Gritsenko) Let $F \in S_2^k$ be a Hecke eigenform and $m \in \mathbb{N}$, then*

$$\begin{aligned} &Q_{p,F}^2(X)\sum_{\delta=0}^{\infty} f_{mp^\delta}|_k\mathcal{D}\left(T_+^g(p^\delta)X^\delta\right) \\ &\quad= \Big(f_m - f_{\frac{m}{p}}|_k\mathcal{D}\big(T_-^g(p)X\big) + pf_{\frac{m}{p^2}}|_k\mathcal{D}\big(\Lambda_-^g(p)X^2\big)\Big)|_k\mathcal{D}\big(1 + p(\nabla_p^r - p)\Delta_pX^2\big) \end{aligned}$$

with $f_m|_k\mathcal{D}\big(1 + p(\nabla_p^r - 1)\Delta_pX^2\big) = \begin{cases} f_m & \textit{if } m \equiv 0(p) \\ (1 - p^{2k-4}X^2)f_m & \textit{otherwise.} \end{cases}$

PROPOSITION 3.13. *Let $F \in S_2^k$ be a Hecke eigenform and $m \in \mathbb{N}$, then*

$$D_{p,F}^2(X)\sum_{\delta=0}^{\infty} f_{mp^{2\delta}}|_k\mathcal{D}\left(\Lambda_+^r(p^\delta)(p^{-2}X)^\delta\right) = (1-X)\left(f_m|_k\mathcal{D}(S^{(2)}(X)) - f_{m/p^2}|_k\mathcal{D}\big(\Lambda_-^r(p)S^{(2)}(X)p^{-2}X\big)\right).$$

Theorem 3.7, Theorem 3.8 and results from our charts give the desired results. Also a proof of the Proposition 3.12 is given in [**Gr93**], the proof of the second one proceeds in the same way. It may seem that the two propositions are of the same type, but that is not the case. Let for example $(m,p)=1$: then one can describe $f_{mp^\delta}|_k\mathcal{D}(T_+^g(p^\delta))$ in terms of eigenvalues and the function f_m, but this is not the case for $f_{mp^{2\delta}}|_k\mathcal{D}(\Lambda_+^g(p^\delta))$, where one needs to know more about the nature of $f_m|_k\mathcal{D}(T^{J,g}(p))$.

3.6. The Maass Spezialschar. There is a correspondence between certain elliptic modular forms, Jacobi forms and Siegel modular forms of degree 2, which can be described in the language of L-functions. To do that, we recall the definition of a certain subspace of S_2^k.

DEFINITION 3.14. The Maass Spezialschar $S_2^{k,M}$ of weight k is the space

$$\left\{F(Z)=\sum_{l=1}^{\infty} l^{2-k}\left(\tilde{\phi}_1|_k\mathcal{D}(T_-^g(l)\right)(Z)\,\middle|\,\tilde{\phi}_1\in\mathcal{PJ}_{1,1}^{k,\mathrm{cusp}}(1)\right\}. \tag{60}$$

It is well known that $S_2^{k,M}$ is a subspace of S_2^k and that it is stable under the action of $\mathcal{H}^2$. For simplicity we have omitted reference to Eisenstein series. But much more is known [**EZ**, §6]:

THEOREM 3.15. *(Saito-Kurokawa Conjecture) The space $S_2^{k,M}$ is spanned by Hecke eigenforms. These are in 1-1 correspondence with normalized Hecke eigenforms $f \in S_1^{2k-2}$, the correspondence being such that*

$$Z_F^2(s)=\zeta(s-k+1)\zeta(s-k+2)L_f(s). \tag{61}$$

The degeneration of the spinor zeta function leads to the degeneration of the standard zeta function.

THEOREM 3.16. *Let us choose F and f as in Theorem 3.15. Then*

$$D_F^2(s)=\zeta(s)L_f(s+k-1)L_f(s+k-2). \tag{62}$$

PROOF. To prove the latter theorem one has to determine the Satake-parameters $\alpha_{0,p},\alpha_{1,p},\alpha_{2,p}$, that is to say, one has to determine the tuple

$$(\alpha_{1,p},\alpha_{1,p}^{-1},\alpha_{2,p},\alpha_{2,p}^{-1}),$$

which is unique up to permutations of the entries. If α_p and β_p denote the p-parameters of f, then it turns out, that one can choose $\alpha_{1,p}=\alpha_p p^{1-k}$ and $\alpha_{2,p}=\beta_p p^{1-k}$, which leads to the desired result. □

REMARK 3.17. There exists a distinguished basis $(F_j)_j$ of $S_2^{k,M}$ with corresponding $f_j \in S_1^{2k-2}(\mathbb{Q}(F_j))$. Let $\sigma \in \mathrm{Gal}(\overline{\mathbb{Q}}/\mathbb{Q})$, then f_j^σ corresponds to F_j^σ.

This remark, the explicit form of D_F^2 and Corollary 1.12 lead to

PROPOSITION 3.18. *Let $F \in S_2^{k,M}(\mathbb{Q}(F))$ be a Hecke eigenform with corresponding $f \in S_1^{2k-2}$ and $k > 4$ even, then for all $\sigma \in \mathrm{Gal}(\overline{\mathbb{Q}}/\mathbb{Q})$ we have*

$$(63)\quad \left(\frac{\pi^{7-4k}L_f(2k-3)L_f(2k-4)}{\langle F;F\rangle}\right)^{\sigma} = \frac{\pi^{7-4k}L_{f^\sigma}(2k-3)L_{f^\sigma}(2k-4)}{\langle F^\sigma;F^\sigma\rangle} \in \mathbb{Q}(F).$$

4. Euler product of a Dirichlet series of type (1,2,2)

From now on let $F, G \in S_2^k, h \in S_1^k$ Hecke eigenforms with h normalized and F in the Maass Spezialschar with corresponding $f \in S_1^{2k-2}$. We show that

$$(64)\quad D_{h,G;F}(s) = \sum_{\substack{n,\varepsilon,l=1\\(\varepsilon,l)=1}}^{\infty} \langle \Phi_n^F|_k U_l; \Phi_{nl^2}^G\rangle_J a_{n\varepsilon^2}^h \varepsilon^{-2(k+s-1)} l^{-2(k+s-2)} n^{-(2k+s-3)}$$

has an Euler product by using the language of P-forms and the Hecke-Jacobi theory developed in the last section.

4.1. Preparation of the Dirichlet series. First interpreting the Fourier-Jacobi coefficients of F and G as P-forms, then using our definition of the Maass Spezialschar and the translation rules in (55), leads to the formula

$$(65)\quad D_{h,G;F}(s) = \beta_k \sum_{\substack{n,\varepsilon,l=1\\(\varepsilon,l)=1}}^{\infty} \langle \tilde{\Phi}_1^F|_k \mathcal{D}(T_-^g(n)\Lambda_-^g(l)); \tilde{\Phi}_{nl^2}^G\rangle_{\mathcal{A}} a_{n\varepsilon^2}^h \cdot \varepsilon^{-2(k+s-1)} l^{-3k-2s+6} n^{-(2k+s-3)}.$$

Next we put $X = p^{-(s+k-1)}, N = p^{k-1}$ and define

$$(66)\quad D_{h,G;F}^{(p)}(s) = \sum_{\substack{n,\varepsilon,l=0\\min(\varepsilon,l)=0}}^{\infty} \langle \tilde{\Phi}_1^F|_k \mathcal{D}\left(T_-^g(p^n)\Lambda_-^g(p^l)\right); \tilde{\Phi}_{p^{n+2l}}^G\rangle_{\mathcal{A}} a_{p^{n+2\varepsilon}}^h \cdot X^{2\varepsilon}(X^2N^{-1}p^3)^l(pXN^{-1})^n,$$

which will later lead to $D_{h,G;F}(s) = \beta_k \langle \tilde{\Phi}_1^F; \tilde{\Phi}_1^G\rangle_{\mathcal{A}} \prod_p (D_{h,G;F}^{(p)}(s)/\langle \tilde{\Phi}_1^F; \tilde{\Phi}_1^G\rangle_{\mathcal{A}})$. Our first goal is to calculate $D^{(p)}$ as explicitly as possible. To do that, we decompose $D^{(p)}$ by the condition $\min(\varepsilon, l) = 0$ into three sums:

$$(67)\quad D_{h,G;F}^{(p)}(s) = D_{(\varepsilon)}(s) + D_{(l)}(s) - D_{(\varepsilon,l)}(s).$$

Let α and α' be the p-parameters of h determined by $\alpha + \alpha' = a_p^h$ and $\alpha\alpha' = N$. Then

$$D_{(\varepsilon)}(s)(\alpha-\alpha') = \alpha \sum_{n,l=0}^{\infty} \langle \tilde{\Phi}_1^F; \tilde{\Phi}_{p^{n+2l}}^G|_k \mathcal{D}(T_+^g(p^n)\Lambda_+^g(p^l))\rangle_{\mathcal{A}} (X^2N^{-1}p^3)^l(\alpha pXN^{-1})^n$$
$$-\alpha' \sum_{n,l=0}^{\infty} \langle \tilde{\Phi}_1^F; \tilde{\Phi}_{p^{n+2l}}^G|_k \mathcal{D}(T_+^g(p^n)\Lambda_+^g(p^l))\rangle_{\mathcal{A}} (X^2N^{-1}p^3)^l(\alpha' pXN^{-1})^n$$

$$D_{(l)}(s)(\alpha-\alpha') = \alpha \sum_{n,\varepsilon=0}^{\infty} \langle \tilde{\Phi}_1^F; \tilde{\Phi}_{p^n}^G|_k \mathcal{D}(T_+^g(p^n))\rangle_{\mathcal{A}} (\alpha X)^{2\varepsilon}(\alpha pXN^{-1})^n$$
$$-\alpha' \sum_{n,\varepsilon=0}^{\infty} \langle \tilde{\Phi}_1^F; \tilde{\Phi}_{p^n}^G|_k \mathcal{D}(T_+^g(p^n))\rangle_{\mathcal{A}} (\alpha' X)^{2\varepsilon}(\alpha' pXN^{-1})^n$$

$$D_{(\varepsilon,l)}(s)(\alpha-\alpha')=\alpha\sum_{n=0}^{\infty}\langle\tilde{\Phi}_1^F;\tilde{\Phi}_{p^n}^G|_k\mathcal{D}(T_+^g(p^n))\rangle_{\mathcal{A}}(\alpha pXN^{-1})^n$$
$$-\alpha'\sum_{n=0}^{\infty}\langle\tilde{\Phi}_1^F;\tilde{\Phi}_{p^n}^G|_k\mathcal{D}(T_+^g(p^n))\rangle_{\mathcal{A}}(\alpha' pXN^{-1})^n.$$

4.2. Local calculations. Using Proposition 3.12, we get the following two lemmas.

LEMMA 4.1. *Let $X_1=\alpha pXN^{-1}$ and $X_1'=\alpha' pXN^{-1}$, then*

$$D_{(l)}-D_{(\varepsilon,l)}=\frac{\langle\tilde{\Phi}_1^F;\tilde{\Phi}_1^G\rangle_{\mathcal{A}}}{\alpha-\alpha'}\left(\frac{\alpha^3X^2}{Q_{p,G}^2(X_1)}-\frac{\alpha'^3X^2}{Q_{p,G}^2(X_1')}\right).\tag{68}$$

LEMMA 4.2. *Let $Y=pNX^2$ and l a non-negative rational integer, then*

$$\sum_{n=0}^{\infty}\tilde{\Phi}_{p^{n+2l}}^G|_k\mathcal{D}\Big(T_+^g(p^n)\Lambda_+^g(p^l)\Big)X_1^n(X^2N^{-1}p^3)^l\tag{69}$$
$$=Q_{p,G}^2(X_1)^{-1}\Big(\Big\{\tilde{\Phi}_{p^{2l}}^G-\tilde{\Phi}_{p^{2l-1}}^G|_k\mathcal{D}(T_-^g(p)X_1)+p\tilde{\Phi}_{p^{2l-2}}^G|_k\mathcal{D}(\Lambda_-^g(p)X_1^2)\Big\}\Big|_k$$
$$\mathcal{D}\left((1+p(\nabla_p^r-p)\Delta_pX_1^2)\,\Lambda_+^r(p^l)(p^{-2}Y)^l\right)\Big).$$

Notice that the eigenvalue of $\Delta_pp^5N^{-1}X^2$ is Y and that X,Y are independent of h, in contrast to X_1 and X_1'. For simplicity, let us use T_-,T_+,Λ_- and Λ_+ for $T_-(p),T_+(p)$, $\Lambda_-(p)$ and $\Lambda_+(p)$. To calculate $D_{(\varepsilon)}$, we have to analyze the infinite sums over l corresponding to $\tilde{\Phi}_{p^{2l}}^G,\tilde{\Phi}_{p^{2l-1}}^G,\tilde{\Phi}_{p^{2l-2}}^G$ as suggested in the last lemma.

PROPOSITION 4.3. *We have*

$$\alpha Q_{p,G}^2(X_1)^{-1}\sum_{l=0}^{\infty}\langle\tilde{\Phi}_1^F;\tilde{\Phi}_{p^{2l}}^G|_k\mathcal{D}\left((1+p(\nabla_p^r-p)\Delta_pX_1^2)\Lambda_+^r(p^l)\right)\rangle_{\mathcal{A}}(p^{-2}Y)^l\tag{70}$$
$$-\alpha' Q_{p,G}^2(X_1')^{-1}\sum_{l=0}^{\infty}\langle\tilde{\Phi}_1^F;\tilde{\Phi}_{p^{2l}}^G|_k\mathcal{D}\left((1+p(\nabla_p^r-p)\Delta_p(X_1')^2)\Lambda_+^r(p^l)\right)\rangle_{\mathcal{A}}(p^{-2}Y)^l$$
$$=\left(\alpha Q_{p,G}^2(X_1)^{-1}-\alpha' Q_{p,G}^2(X_1')^{-1}\right)\frac{S_F^{(2)}(Y)^{\mathrm{prim}}\langle\tilde{\Phi}_1^F;\tilde{\Phi}_1^G\rangle_{\mathcal{A}}}{D_{p,G}^{2,*}(Y)}$$
$$+\left(\frac{\alpha'^3X^2}{Q_{p,G}^2(X_1')}-\frac{\alpha^3X^2}{Q_{p,G}^2(X_1)}\right)\langle\tilde{\Phi}_1^F;\tilde{\Phi}_1^G\rangle_{\mathcal{A}}$$

with $S_F^{(2)}(Y)^{\mathrm{prim}}\tilde{\Phi}_1^F=\tilde{\Phi}_1^F|_k\mathcal{D}(S^{(2)}(Y)^{\mathrm{prim}})$ and $D_{p,G}^{2,*}(Y)=D_{p,G}^2(Y)/(1-Y)$.

PROOF. Let X_* be X_1 or X_1', then

$$\sum\nolimits_1^G(X_*):=\sum_{l=0}^{\infty}\tilde{\Phi}_{p^{2l}}^G|_k\mathcal{D}\left((1+p(\nabla_p^r-p)\Delta_pX_*^2)\Lambda_+^r(p^l)\right)(p^{-2}Y)^l$$
$$=\sum_{l=0}^{\infty}\tilde{\Phi}_{p^{2l}}^G|_k\mathcal{D}(\Lambda_+^r(p^l))(p^{-2}Y)^l-p^2\tilde{\Phi}_1^G|_k\mathcal{D}(\Delta_pX_*^2).$$

Here we used the peculiarity of $l=0$.

Applying Proposition 3.13 leads to

$$\sum\nolimits_1^G(X_*) = \left(\tilde{\Phi}_1^G|_k\mathcal{D}(S^{(2)}(Y)\right) D_{p,G}^{2,*}(Y)^{-1} - (\alpha_*)^2X^2\tilde{\Phi}_1^G,$$

where α_* is adapted to the meaning of X_*, that means $\alpha_* = \alpha$, if $X_* = X_1$ and α' otherwise. The property that $\tilde{\Phi}_1^F$ is an eigenform of $\tilde{\mathcal{H}}_0^J$ (the subalgebra of all elements of $\tilde{\mathcal{H}}^J$ with signature one) and that $S^{(2)}(Y)^{\text{prim}}$ is hermitian proves the proposition. □

We want to point out the cancellations caused by Lemma 4.1 and Proposition 4.3. By using again Proposition 3.13 and the already calculated relations between the generators of $\tilde{\mathcal{H}}_p^J$, we get

PROPOSITION 4.4. *We have*

$$\alpha Q_{p,G}^2(X_1)^{-1}\sum_{l=0}^{\infty} p\langle\tilde{\Phi}_1^F;\tilde{\Phi}_{p^{2l-2}}^G|_k\mathcal{D}\left((\Lambda_-^g X_1^2)(1+p(\nabla_p^r-p)\Delta_p X_1^2)(\Lambda_+^r p^{-2}Y)^l)\right)\rangle_{\mathcal{A}}$$

$$-\alpha' Q_{p,G}^2(X_1')^{-1}\sum_{l=0}^{\infty} p\langle\tilde{\Phi}_1^F;\tilde{\Phi}_{p^{2l-2}}^G|_k\mathcal{D}((\Lambda_-^g X_1'^2)(1+p(\nabla_p^r-p)\Delta_p X_1'^2)(\Lambda_+^r p^{-2}Y)^l))\rangle_{\mathcal{A}}$$

$$= \left(\frac{\alpha^3Np^2X^4}{Q_{p,G}^2(X_1)} - \frac{\alpha'^3Np^2X^4}{Q_{p,G}^2(X_1')}\right)\left(\frac{S_F^{(2)}(Y)^{\text{prim}}\langle\tilde{\Phi}_1^F;\tilde{\Phi}_1^G\rangle_{\mathcal{A}}}{D_G^{2,*}(Y)}\right).$$

To calculate the sum corresponding to $\tilde{\Phi}_{p^{2l-1}}^G$ is the main point here. Of course the specialness of F is essential.

PROPOSITION 4.5. *Let $\lambda_p^{F,J}$ be defined by $\tilde{\Phi}_1^F\ |_k\ \mathcal{D}(T^{J,r}(p)) = \lambda_p^{F,J}\tilde{\Phi}_1^F$. Then*

$$\alpha Q_{p,G}^2(X_1)^{-1}\sum_{l=0}^{\infty}\langle\tilde{\Phi}_1^F;\tilde{\Phi}_{p^{2l-1}}^G|_k\mathcal{D}\left((T_-^g X_1)(1+p(\nabla_p^r-p)\Delta_p X_1^2)\Lambda_+^r(p^l)(p^{-2}Y)^l)\right)\rangle_{\mathcal{A}}$$

$$-\alpha' Q_{p,G}^2(X_1')^{-1}\sum_{l=0}^{\infty}\langle\tilde{\Phi}_1^F;\tilde{\Phi}_{p^{2l-1}}^G|_k\mathcal{D}\left((T_-^g X_1')(1+p(\nabla_p^r-p)\Delta_p (X_1')^2)(\Lambda_+^r p^{-2}Y)^l)\right)\rangle_{\mathcal{A}}$$

$$= \left(\alpha^2p^2X^3Q_{p,G}^2(X_1)^{-1} - (\alpha')^2p^2X^3Q_{p,G}^2(X_1')^{-1}\right)$$

$$\cdot\,\lambda_p^G(1-p^{-2}\lambda_p^{F,J}Y+p^{-1}Y^2)\langle\tilde{\Phi}_1^F,\tilde{\Phi}_1^G\rangle_{\mathcal{A}}\left(D_{p,G}^{2,*}(Y)\right)^{-1}.$$

PROOF. First step
We start with the calculation of

$$(71)\quad \sum\nolimits_3^G(X_*) := \sum_{l=0}^{\infty}\tilde{\Phi}_{p^{2l-1}}^G|_k\mathcal{D}\left((T_-^g X_*)(1+p(\nabla_p^r-p)\Delta_p X_*^2)\Lambda_+^r(p^l)(p^{-2}Y)^l\right).$$

By $T_-^g\Lambda_+^r = p^2T_+^g$, we have

$$\begin{aligned}\sum\nolimits_3^G(X_*) &= \sum_{l=1}^{\infty}\tilde{\Phi}_{p^{2l-1}}^G|_k\mathcal{D}\left(\Lambda_+^r(p^{l-1})(p^{-2}Y)^{l-1}YT_+X_*\right)\\ &= \sum_{l=0}^{\infty}\tilde{\Phi}_{p^{2l+1}}^G|_k\mathcal{D}\left(\Lambda_+^r(p^l)(p^{-2}Y)^lYX_*T_+^g\right)\\ &= D_{p,G}^{2,*}(Y)^{-1}(\tilde{\Phi}_p^G|_k\mathcal{D}(S^{(2)}(Y)T_+^g)YX_*.\end{aligned}$$

Here we used Proposition 3.13 and that the terms corresponding to Λ_-^r provide no contribution. With the identity given in (59), we have

$$\sum\nolimits_3^G(X_*) = \alpha_* p^2 X^3 D_{p,G}^{2,*}(Y)^{-1}\tilde{\Phi}_p^G|_k\mathcal{D}\Big(T_+^g(p)S^{(2)}(Y)^{\mathrm{prim}}(1+Y)^{-1}\Big). \tag{72}$$

Second step

The P-form $\tilde{\Phi}_1^F$ is an eigenfunction of $\tilde{\mathcal{H}}_0^J$ and $S^{(2)}(Y)^{\mathrm{prim}}$ is self adjoint. This gives us

$$\begin{aligned}
&\langle\tilde{\Phi}_1^F;\tilde{\Phi}_p^G|_k\mathcal{D}\left(T_+^g S^{(2)}(Y)^{\mathrm{prim}}\right)\rangle_{\mathcal{A}}/(1+Y)\\
&= \langle\tilde{\Phi}_1^F|_k\mathcal{D}(1-p^{-2}T^{J,r}(p)Y+p^{-1}Y^2);\tilde{\Phi}_p^G|_k\mathcal{D}(T_+^g)\rangle_{\mathcal{A}}(1+Y)^{-1}\\
&= (1-p^{-2}\lambda_p^{F,J}Y+p^{-1}Y^2)\langle\tilde{\Phi}_1^F;\tilde{\Phi}_p^G|_k\mathcal{D}(T_+^g)\rangle_{\mathcal{A}}(1+Y)^{-1}\\
&= \lambda_p^G(1-p^{-2}\lambda_p^{F,J}Y+p^{-1}Y^2)\langle\tilde{\Phi}_1^F;\tilde{\Phi}_1^G\rangle_{\mathcal{A}}.
\end{aligned}$$

Here $\tilde{\Phi}_1^F|_k\mathcal{D}(T^{J,r}(p)) = \lambda_p^{F,J}\tilde{\Phi}_1^F$ and $\tilde{\Phi}_p^G|_k\mathcal{D}(T_+^g) = \lambda_p^G\tilde{\Phi}_1^G$. Thus the proposition is proven. □

We have proven the last lemmas and propositions in such a way that one can easily extract the following crucial fact.

COROLLARY 4.6. *Let n', l', ε' be coprime to p. Then we have*

$$\begin{aligned}
&\sum_{\substack{n,\varepsilon,l=0\\ min(\varepsilon,l)=0}}^{\infty} \langle\tilde{\Phi}_1^F|_k\mathcal{D}\left(T_-^g(n'p^n)\Lambda_-^g(l'p^l)\right);\tilde{\Phi}_{n'l'^2p^{n+2l}}^G\rangle_{\mathcal{A}}\\
&\qquad a^h_{n'(\varepsilon')^2p^{n+2\varepsilon}}X^{2\varepsilon}(X^2N^{-1}p^3)^l(pXN^{-1})^n\\
&\qquad = \langle\tilde{\Phi}_1^F|_k\mathcal{D}\left(T_-^g(n')\Lambda_-^g(l')\right);\tilde{\Phi}_{n'l'^2}^G\rangle_{\mathcal{A}}a^h_{n'(\varepsilon')^2}\left(D^{(p)}_{h,G;F}(s)\langle\tilde{\Phi}_1^F;\tilde{\Phi}_1^G\rangle_{\mathcal{A}}^{-1}\right).
\end{aligned}$$

Thus we have proven the following

THEOREM 4.7. *Let $F, G \in S_2^k$ and $h \in S_1^k$ Hecke eigenforms, with h normalized. Let further F be in the Maass Spezialschar. Then we have for k even and s big enough:*

$$D_{h,G;F}(s) = \beta_k\langle\tilde{\Phi}_1^F,\tilde{\Phi}_1^G\rangle_{\mathcal{A}}\prod_p\left(D^{(p)}_{h,G;F}(s)/\langle\tilde{\Phi}_1^F,\tilde{\Phi}_1^G\rangle_{\mathcal{A}}\right) \tag{73}$$

$$\begin{aligned}
&\textit{with } D^{(p)}_{h,G;F}(s) = \frac{\langle\tilde{\Phi}_1^F;\tilde{\Phi}_1^G\rangle_{\mathcal{A}}S_F^{(2)}(Y)^{\mathrm{prim}}}{(\alpha-\alpha')D_{p,G}^{2,*}(Y)}\\
&\cdot\left(\frac{\alpha}{Q^2_{p,G}(X_1)}-\frac{\alpha'}{Q^2_{p,G}(X_1')}+\frac{\alpha^3Np^2X^4}{Q^2_{p,G}(X_1)}\right.\\
&\left.-\frac{\alpha'^3Np^2X^4}{Q^2_{p,G}(X_1')}+\frac{\alpha'^2p^2X^3\lambda_p^G}{Q^2_{p,G}(X_1')(1+Y)}-\frac{\alpha^2p^2X^3\lambda_p^G}{Q^2_{p,G}(X_1)(1+Y)}\right).
\end{aligned} \tag{74}$$

It is easy to see, that the expression in the big brackets in (74), is equal to

$$\begin{aligned}
&\{\left((\alpha+\alpha^3Np^2X^4)(1+Y)-\alpha^2p^2X^3\lambda_p^G\right)Q^2_{p,G}(X_1')\\
&\quad-\left((\alpha'+\alpha'^3Np^2X^4)(1+Y)-\alpha'^2p^2X^3\lambda_p^G\right)Q^2_{p,G}(X_1)\}\\
&Q^2_{p,G}(X_1)^{-1}Q^2_{p,G}(X_1')^{-1}(1+Y)^{-1}.
\end{aligned}$$

Inserting $Y = pNX^2, N = p^{k-1}$ and $X_* = \alpha_* pN^{-1}X$ gives us

$$\Big\{ \big(\alpha + \alpha pNX^2 - \alpha^2 p^2 X^3 \lambda_p^G + \alpha^3 Np^2 X^4 + \alpha^3 N^2 p^3 X^6\big) Q_{p,G}^2(\alpha' pN^{-1}X) \\ - \left(\alpha' + \alpha' pNX^2 - \alpha'^2 p^2 X^3 \lambda_p^G + \alpha'^3 Np^2 X^4 + \alpha'^3 N^2 p^3 X^6\right) Q_{p,G}^2(\alpha pN^{-1}X) \Big\} \\ \left\{ Q_{p,G}^2(X_1) Q_{p,G}^2(X_1')(1+Y) \right\}^{-1}.$$

If we multiply out the numerator and rearrange the terms according to the powers of y, we get

$$D_{h,G;F}^{(p)}(s) = \frac{\langle \tilde{\Phi}_1^F ; \tilde{\Phi}_1^G \rangle_{\mathcal{A}} S_F^{(2)}(Y)^{\text{prim}}}{D_{p,G}^{2,*}(Y)} \left\{ Q_{p,G}^2(X_1) Q_{p,G}^2(X_1')(1+Y) \right\}^{-1}$$

$$\cdot (1-y)\Big(1 + \big(2pN - p^2N^{-1}A_2^G\big)Y + \big(2 - 2pN^{-2}A_2^G + pN^{-2}(\lambda_P^G)^2\big)Y^2 \\ + \big(2pN - p^2N^{-1}A_2^G\big)Y^3 + Y^4\Big). \tag{75}$$

Here $A_2^G = p\left(\lambda_{T_1(p^2)}^G + (p^2+1)p^{2k-6}\right)$, with $G|_k T_1^{(2)}(p^2) = \lambda_{T_1(p^2)}^G G$. The expression (75) in the last big bracket turns out to be just $D_{p,G}^2(pNX^2)$.

LEMMA 4.8. *Let $F \in S_2^{k,M}$ be a Hecke eigenform with corresponding $f \in S_1^{2k-2}$. Then*

$$S_F^{(2)}(p^{-2s-k+2})^{\text{factor}} = L_{p,f}(p^{-(2s+2k-3)}).$$

PROOF. We have $f|_k T(p) = \lambda_f(p) f \iff \tilde{\Phi}_1^F|_k \mathcal{D}(\widetilde{T_p}^{EZ}) = \lambda_f(p)\tilde{\Phi}_1^F$ (see for example [**KS**, §2]) and $T_p^{J,r} = p^{3-k}\tilde{T}_p^{EZ}$, which gives the desired result. □

Thus we have

THEOREM 4.9. *Let $F, G \in S_2^k, h \in S_1^k$ be Hecke eigenforms with totally real Fourier coefficients, h normalized and k even. Let F be in the Maass space with corresponding $f \in S_1^{2k-2}$. Then we have for $s \in \mathbb{C}$ with $k + 2Re(s) > 6$ and $X = p^{-(s+k-1)}$:*

$$D_{h,G;F}(s) = \langle \Phi_1^F ; \Phi_1^G \rangle_J \times \prod_p \frac{S_F^{(2)}(pNX^2)^{\text{factor}}(1 - pNX^2)}{Q_{p,G}^2(\alpha pN^{-1}X) Q_{p,G}^2(\alpha' pN^{-1}X)} \tag{76}$$

with $S_F^{(2)}(pNX^2)^{\text{factor}} = L_{p,f}(pX^2)$.

5. The L-function $Z_{G\otimes h}(s)$

In this last section we present our final result, a new integral representation of $Z_{G\otimes h}(s)$ for $G \in S_2^k$ and $h \in S_1^k$. Assuming that the first Fourier-Jacobi coefficient of G is not identically zero enables us to read off properties of the L-function. For example, we show that there is a close connection between the triple product L-function and $Z_{G\otimes h}(s)$.

THEOREM 5.1. *Let $F, G \in S_2^k$ and $h \in S_1^k$ be Hecke eigenforms with totally real algebraic Fourier coefficients. Let k be even and h normalized. Further we assume* *that F is in the Maass Spezialschar with corresponding $f \in S_1^{2k-2}$. Then for $s \in \mathbb{C}$ with $2Re(s) + k > 6$:*

$$\begin{aligned}
&\langle\langle\langle E_5^k([z_1, z_2, z_3]; s); h(z_1)G(z_2)F(z_3)\rangle\rangle\rangle \\
&\quad = a(2, s, k)\langle\Phi_1^F; \Phi_1^G\rangle_J \\
&\qquad \cdot\zeta(2s+k)^{-1}\zeta(4s+2k-2)^{-1}\zeta(4s+2k-4)^{-1} \\
&\qquad \cdot L_f(2s+2k-4)Z_{G\otimes h}(s+2k-3).
\end{aligned}$$

PROOF. By Theorem 2.7, we know that

$$\Phi(h, G, F; s) = a(2, s, k)\mathcal{L}_{h,G,F}^2(2k+s-3).$$

Here $\mathcal{L}_{h,G,F}^2(2k+s-3)$ is equal to

$$\left\{\zeta(2s+k)\prod_{j=1}^{2}\zeta(4s+2k-2j)\right\}^{-1} D_F^2(2s+k-2)D_{h,G;F}(s).$$

Since F is a Hecke eigenform in the Maass Spezialschar, we have

$$D_F^2(2s+k-2) = \zeta(2s+k-2)L_f(2s+\tilde{k}-1)L_f(2s+\tilde{k}-2),$$

with $\tilde{k} = 2k-2$ and $f \in S_1^{\tilde{k}}$. Finally, with the explicit form of $D_{h,G;F}(s)$ obtained in the last section the theorem becomes obvious. □

Supplementarily we want to note that $a(2, s, k)$ is equal to

$$2 \cdot 2^{-6s-6k+12}\pi^{-s-2k+6}$$

$$\frac{\Gamma_2(k+s-\frac{3}{2})\Gamma(s+2k-3)\Gamma(s+k-2)\Gamma(s+k-1)}{\Gamma_2(k+s)\Gamma(2k-3+2s)}. \tag{77}$$

From now on, we assume that $\tilde{\Phi}_1^G$ is not identically zero. It is obvious that one can choose a Hecke eigenform $F \in S_2^{k,M}$ in such a way (use Definition 3.14 and Theorem 3.15), that

$$\langle\tilde{\Phi}_1^F; \tilde{\Phi}_1^G\rangle \neq 0 \text{ and } F \in S_2^{k,M}(\mathbb{Q}(F)).$$

COROLLARY 5.2. *The L-function $Z_{G\otimes h}(s)$ has an analytic continuation to a meromorphic function on $\mathbb{C}$. If we put*

$$Z_{G\otimes h}^*(s) := (2\pi)^{-4s}\Gamma(s-k+1)^2\Gamma(s-k+2)\Gamma(s)Z_{G\otimes h}(s), \tag{78}$$

then we have the functional equation

$$Z_{G\otimes h}^*(s) = Z_{G\otimes h}^*(3k-3-s). \tag{79}$$

This follows from elementary calculations of the already proven general functional equation stated in (31) and the well known functional equation of $L_f(s)$.

REMARK 5.3. It is pleasant that in our formula the help function F has no longer any influence on the functional equation of $Z_{G\otimes h}(s)$. Further, in contrast to the situation in the case of the triple product L-function, the value of $Z_{G\otimes h}(s)$ at the center of symmetry is not forced to be zero (the minus sign of the functional equation of $L_f(s)$ plays a decisive role).

COROLLARY 5.4. *Let $k > 6$ be even, $G \in S_2^k(\mathbb{Q}(G))$ and $h \in S_1^k$ be Hecke eigenforms with h normalized. If the first Fourier-Jacobi coefficient of G is not zero, then*

$$Z'_{G\otimes h}(2k-3) := \pi^{8-5k} \frac{Z_{G\otimes h}(2k-3)}{\langle G; G\rangle\langle h; h\rangle} \in \overline{\mathbb{Q}}. \tag{80}$$

COROLLARY 5.5. *Let h and G be as in Corollary 5.4. Then*

$$\left(Z'_{G\otimes h}(2k-3)\right)^{\sigma} = Z'_{G^\sigma\otimes h^\sigma}(2k-3); \forall \sigma \in \mathrm{Gal}(\overline{\mathbb{Q}}/\mathbb{Q}). \tag{81}$$

PROOF. By Theorem 1.9 and Theorem 5.1, it is enough to show that

$$\left(\frac{a(2,0,k)\langle \phi_1^F; \phi_1^G\rangle_J L_f(2k-4)}{\zeta(2s+k)\zeta(4s+2k-2)\zeta(4s+2k-4)\langle F; F\rangle}\right)^{\sigma}$$
$$= \frac{a(2,0,k)\langle \Phi_1^{F^\sigma}; \Phi_1^{G^\sigma}\rangle_J L_{f^\sigma}(2k-4)}{\zeta(2s+k)\zeta(4s+2k-2)\zeta(4s+2k-4)\langle F^\sigma; F^\sigma\rangle} \in \overline{\mathbb{Q}}.$$

We put $a \sim b$ for $a, b \in \mathbb{C}, b \neq 0$, if $ab^{-1} \in \overline{\mathbb{Q}}$. Then $a(2,0,k) \sim \pi^{6-2k}$ and $\left(\zeta(2s+k)\zeta(4s+2k-2)\zeta(4s+2k-4)\right)^{-1} \sim \pi^{-5k+6}$, thus

$$\pi^{5k-8} a(2,0,k) \left(\zeta(2s+k)\zeta(4s+2k-2)\zeta(4s+2k-4)\right)^{-1} \sim \pi^{4-2k}.$$

By Proposition 3.18 and $L_f(2k-3)L_f(2k-4) \neq 0$, it is enough to prove:

$$\left(\frac{\pi^{-3+2k}\langle \Phi_1^F; \Phi_1^G\rangle_J}{L_f(2k-3)}\right)^{\sigma} = \frac{\pi^{-3+2k}\langle \Phi_1^{F^\sigma}; \Phi_1^{G^\sigma}\rangle_J}{L_{f^\sigma}(2k-3)} \in \overline{\mathbb{Q}}.$$

But this is an easy consequence of the properties 1.12a and 1.12b and Theorem 1 stated in [**Sh81**]. Of course, we have used the well known translation of Jacobi forms of index one into modular forms of half-index weight, where the corresponding scalar products are equal up to a non-zero rational number [**EZ**], [**Z80**]. □

The L-function $Z_{G\otimes h}(s)$ has already been investigated by M. Furusawa [**Fu**]. He constructed an integral representation by integrating a Hermitian Klingen-Eisenstein series for h against a Siegel cusp form G. Under the assumption that certain non-vanishing properties of the Fourier coefficients of G hold, he was able to deduce the meromorphic continuation, the functional equation and the algebraicity of $Z'_{G\otimes h}(2k-3)$. Because of the not sufficiently known Galois behavior of the Fourier coefficients of a Hermitian Klingen-Eisenstein series [**H**], he was not able to study the Galois action at the special value.

We want to conclude our work with the following observation. For simplification let $F \in S_n^k(\mathbb{Q}(F)), G \in S_n^k(\mathbb{Q}(G))$ and $h \in S_1^k(\mathbb{Q}(h))$ be Hecke eigenforms (and every elliptic cusp form is assumed to be normalized). Let $\sigma \in \mathrm{Gal}(\overline{\mathbb{Q}}/\mathbb{Q})$, then

$$\mathcal{L}_{h,G,F}^{n,*}(s) = \mathcal{L}_{h,G,F}^{n,*}(3k-n-1-s),$$
$$\left(\frac{\pi^{(n^2+3n-4k+2)/2}\mathcal{L}_{h,G,F}^{n}(2k-n-1)}{\langle h; h\rangle\langle G; G\rangle\langle F; F\rangle}\right)^{\sigma}$$
$$= \frac{\pi^{(n^2+3n-4k+2)/2}\mathcal{L}_{h^\sigma,G^\sigma,F^\sigma}^{n}(2k-n-1)}{\langle h^\sigma; h^\sigma\rangle\langle G^\sigma; G^\sigma\rangle\langle F^\sigma; F^\sigma\rangle} \in \bar{\mathbb{Q}}.$$

This enables us to give an unique approach to and description of the triple product L-function $L_{h,G,F}(s)$ for $h, G, F \in S_1^k$ (P. Garrett [**G87**]) and the L-function $Z_{G\otimes h}(s)$.

Let us assume that n is equal to 1 or 2 and that the first Fourier-Jacobi coefficient of G is not identically zero. Then we have

$$\mathcal{L}^n_{h,G,F}(s+2k-n-1)$$
$$= \begin{cases} \zeta(2s+k)^{-1}\zeta(4s+2k-2)^{-1}L_{h,G,F}(s+2k-2) & \text{for } n=1 \\ \zeta(2s+k)^{-1}\zeta(4s+2k-2)^{-1}\zeta(4s+2k-4)^{-1} & \\ \cdot\langle \Phi_1^F;\Phi_1^G\rangle_J L_f(2s+2k-4)Z_{G\otimes h}(s+2k-3) & \text{for } n=2, \end{cases}$$

and

$$\begin{array}{rcll} L^*_{h,G,F}(s) & = & (-1)L^*_{h,G,F}(3k-2-s) & \text{for } n=1, \\ Z^*_{G\otimes h}(s) & = & Z^*_{G\otimes h}(3k-3-s) & \text{for } n=2 \end{array}$$

with $L^*_{h,G,F}(s) = (2\pi)^{-4s}\Gamma(s-k+1)^3\Gamma(s)L_{h,G,F}(s)$ and

$$Z^*_{G\otimes h}(s) = (2\pi)^{-4s}\Gamma(s-k+1)^2\Gamma(s-k+2)\Gamma(s)Z^*_{G\otimes h}(s).$$

The central value of $\mathcal{L}^n_{h,G,F}(s)$ is equal to $(3k-n-1)/2$, which coincides for $k=n+1$ with our described special value $s=2k-n-1$. The Galois behavior shows up as

$$\left(\frac{\pi^{-5k+5}L_{h,G,F}(2k-2)}{\langle h;h\rangle\langle G;G\rangle\langle F;F\rangle}\right)^\sigma = \frac{\pi^{-5k+5}L_{h^\sigma,G^\sigma,F^\sigma}(2k-2)}{\langle h^\sigma;h^\sigma\rangle\langle G^\sigma;G^\sigma\rangle\langle F^\sigma;F^\sigma\rangle} \in \bar{\mathbb{Q}} \text{ and}$$
$$\left(\frac{\pi^{-5k+8}Z_{G\otimes h}(2k-3)}{\langle h;h\rangle\langle G;G\rangle}\right)^\sigma = \frac{\pi^{-5k+8}Z_{G^\sigma\otimes h^\sigma}(2k-3)}{\langle h^\sigma;h^\sigma\rangle\langle G^\sigma;G^\sigma\rangle} \in \bar{\mathbb{Q}}.$$

References

[A74] A. N. Andrianov, *Euler products corresponding to Siegel modular forms of genus 2*, Russian Math. Survey **29** (1974), 45–116.

[A76] A. N. Andrianov, *Symmetric squares of zeta functions of Siegel modular forms of degree 2*, Trudy Mat. Inst. Steklov **142** (1976), 22–45.

[A77a] A. N. Andrianov, *On zeta-functions of Rankin type associated with Siegel modular forms*, in Modular Functions of One Variable VI, Springer Lecture Notes **627**, Berlin-Heidelberg-New York: Springer-Verlag (1977), 325–338.

[A77b] A. N. Andrianov, *On factorization of Hecke polynomial for the symplectic group of genus n*, Math. USSR Sbornik **33** (1977), 343–373.

[A87] A. N. Andrianov, Quadratic Forms and Hecke Operators, Grundlehren der Mathematischen Wissenschaften **286**, Berlin-Heidelberg-New York: Springer (1987).

[Ar] T. Arakawa, *Jacobi Eisenstein series and a basis problem for Jacobi forms*, Commentarii Mathematici Universitatis Sancti Pauli **43** no. 2 (1994), 181–216.

[BBo] R. Berndt and S. Böcherer, *Jacobi forms and discrete series representations of the Jacobi group*, Math. Z. **204** (1990), 13–44.

[Bo85a] S. Böcherer, *Über die Fourier-Jacobi-Entwicklung der Siegelschen Eisensteinreihen II*, Math. Z. **189** (1985), 81–100.

[Bo85b] S. Böcherer, *Über die Funktionalgleichung automorpher L-Funktionen zur Siegelschen Modulgruppe*, J. Reine Angew. Math. **362** (1985), 146–168.

[Bo86] S. Böcherer, *Ein Rationalitätssatz für formale Heckereihen zur Siegelschen Modulgruppe*, Abh. Math. Sem. Univ. Hamburg **56**, (1986), 35–47.

[BoS] S. Böcherer and R. Schulze-Pillot, *On the central critical value of the triple product L-function*, preprint (1995).

[Bou] N. Bourbaki, *Formes sesquilinéaires et formes quadratiques*, Élements de Mathematique II, Chapt. 9, Hermann: Paris (1959).

[Du] J. Dulinski, *A decomposition theorem for Jacobi forms*, Math. Ann. **303** (1995), 473–498.

[EZ] M. Eichler and D. Zagier, The Theory of Jacobi Forms, Progress in Mathematics **55**, Boston: Birkhäuser Verlag (1985).

[F] E. Freitag, Siegelsche Modulfunktionen, Grundlehren der Mathematischen Wissenschaften **254**, Berlin-Heidelberg-New York: Springer Verlag (1983).

[Fu] M. Furusawa, *On L-functions for* $\mathrm{GSp}(4) \times \mathrm{GL}(2)$ *and their special values*, J. Reine Angew. Math. **438** (1993), 187–218.

[G84] P. Garrett, *Pullbacks of Eisenstein series; Applications*, in Automorphic Forms of Several Variables, Taniguchi symposium 1983, Boston: Birkhäuser (1984).

[G85] P. Garrett, *Integral representations of certain L-functions attached to one, two and three modular forms*, preprint (1985).

[G87] P. Garrett, *Decomposition of Eisenstein series: triple product L-functions*, Ann. Math. **125** (1987), 209–235.

[GH] P. Garrett and M. Harris, *Special values of triple product L-functions*, Am. Journal of Math. **115** (1993), 159–238.

[GPR] S. Gelbart, I. Piatetski-Shapiro, S. Rallis, Explicit Constructions of Automorphic L-functions, Springer Lecture Notes in Math. **1254**, Berlin-Heidelberg-New York: Springer (1987).

[Gr82] V. A. Gritsenko, *The action of modular operators on the Fourier-Jacobi coefficients of modular forms*, Matem. Sbornik **119** (1982), 248-277 (Russian); English transl. in Math. USSR Sbornik **47** (1984), 237–268.

[Gr88] V. A. Gritsenko, *Expansion of Hecke polynomials of classical groups*, Matem. Sbornik **137** (1988), 328–351; English transl. in Math. USSR Sbornik **65** (1990), 333–356.

[Gr93] V. A. Gritsenko, Arithmetical Lifting and its Applications, London Mathematical Society Lecture Note Series **215**, Number Theory, Paris (1992-3), Cambridge: Cambridge Univ. Press (1995).

[Gr95] V. A. Gritsenko, Modulformen zur Paramodulgruppe und Modulräume der Abelschen Varietäten, Mathematica Gottingensis, Schriftreihe des Sonderforschungsbereichs Geometrie und Analysis Heft **12** (1995).

[GroK] B. Gross and S. Kudla, *Heights and the central values of triple product L-functions*, Compositio Math. **81** (1992), 143–209.

[H] M. Harris, *Eisenstein series on Shimura varieties*, Ann. Math. **119** (1984), 59–94.

[HK] M. Harris and S. Kudla, *The central critical value of a triple product L-function*, Annals of Math. **133** (1991), 605–672.

[K87] H. Klingen, *Metrisierungstheorie und Jacobiformen*, Abh. Math. Sem. Univ. Hamburg **57** (1987), 165–178.

[K89] H. Klingen, *Über Kernfunktionen für Jacobiformen und Siegelsche Modulformen*, Math. Ann. **285** (1989), 405–416.

[K90] H. Klingen, Introductory Lectures on Siegel Modular Forms, Cambridge: Cambridge University Press (1990).

[KS] W. Kohnen and N.-P. Skoruppa, *A certain Dirichlet series attached to Siegel modular forms of degree two*, Invent. Math. **95** (1989), 449–476.

[L] R. Langlands, On the Functional Equations Satisfied by Eisenstein Series, Lect. Notes Math. **544**, Berlin-Heidelberg-New York: Springer (1976).

[M] S. Mizumoto, *Poles and residues of standard L-functions attached to Siegel modular forms*, Math. Ann **289** (1991), 589–612.

[Mu] A. Murase, *L-functions attached to Jacobi forms of degree n, Part I. The basic identity*, J. Reine Angew. Math. **401** (1989), 122–156.

[O] T. Orloff, *Special values and mixed weight triple products (with an appendix by Don Blasius)*, Invent. Math. **90** (1987), 169–180.

[PR84] I. I. Piatetski-Shapiro and S. Rallis, *L-functions of automorphic forms on simple classical groups*, Modular Forms (Durham, 1983), ed. R. Rankin, Chichester, England: Ellis Horwood (1984), 251–262.

[PR87] I. I. Piatetski-Shapiro and S. Rallis, *Rankin triple L-functions*, Compositio Math. **64** (1987), 31–115.

[S] T. Satoh, *Some remarks on triple L-functions*, Math. Ann. **276** (1987), 687–698.

[Sh71] G. Shimura, Introduction to the Arithmetical Theory of Automorphic Functions, Princeton: Iwanami Shoten and Princeton Univ. Press (1971).

[Sh75] G. Shimura, *On the Fourier coefficients of modular forms of several variables*, Nachr. Akad. Wiss. Göttingen Math.- Phys. Kl II. **17** (1975), 261–268.

[Sh76] G. Shimura, *The special values of the zeta functions associated with cusp forms*, Comm. Pure Appl. Math. **29** (1976), 783–804.

[Sh81] G. Shimura, *The critical values of certain zeta functions associated with modular forms of half-integral weight*, J. Math. Soc. Japan **33** (1981), 649–672.

[Sh94] G. Shimura, *Euler products and Fourier coefficients of automorphic forms on symplectic groups*, Inv. Math. **116** (1994), 531–576.

[Sh95] G. Shimura, *Eisenstein series and zeta functions on symplectic groups*, Inv. Math. **119** (1995), 539–584.

[Si39] C. L. Siegel, *Einführung in die Theorie der Modulformen n-ten Grades*, Math. Ann. **116** (1939), 617–657.

[Si68] C. L. Siegel, *Über die Fourierschen Koeffizienten von Eisensteinschen Reihen der Stufe T*, Math. Z. **105** (1968), 257–266.

[SZ] N-P. Skoruppa and D. Zagier, *A trace formula for Jacobi forms*, J. Reine Angew. Math. **393** (1989),168–198.

[Y] T. Yamazaki, *Rankin-Selberg method for Siegel cusp forms*, Nagoya Math. J. **120** (1990), 35–49.

[Z77] D. Zagier, *Modular forms whose Fourier coefficients involve zeta functions of quadratic fields*, in Modular Functions of One Variable VI, Springer Lecture Notes **627**, Berlin-Heidelberg-New York: Springer Verlag (1977), 105–169.

[Z80] D. Zagier, *Sur la conjecture de Saito-Kurokawa (d'après H. Maass)*, Sém. Delange-Pisot-Poitou 1979/1980, Progress in Math. **12**, Boston-Basel-Stuttgart (1980), 371–394.

[Zi] C. Ziegler, *Jacobi forms of higher degree*, Abh. Math. Sem. Univ. Hamburg **59** (1989), 191–224.

MAX-PLANCK-INSTITUT FÜR MATHEMATIK, GOTTFRIED-CLAREN-STRASSE 26, 53225 BONN, GERMANY

E-mail address: `heim@mpim-bonn.mpg.de, heim@euklid.math.uni-mannheim.de`

Proceedings of Symposia in Pure Mathematics
Volume **66.2**, 1999

Omega Results for Automorphic L-Functions

Jeffrey Hoffstein and Paul Lockhart

Dedicated to Goro Shimura

ABSTRACT. We prove an omega result for the values of quadratic twists of automorphic L-series at the center of the critical strip. This result is an analog of classical results concerning values of the Riemann zeta function at $1/2+it$ for large values of t. We then show that as a consequence of this, certain Rankin-Selberg convolutions of half-integral weight forms with themselves have Siegel zeros.

Introduction

The Lindelöf Hypothesis for the Riemann zeta-function states that for any $\varepsilon > 0$, $\zeta(1/2+it) \ll |t|^{\varepsilon}$ as $|t| \to \infty$, where the implied constant depends on ε. It has been known for some time that nothing much better than this can be true, as omega results of the form

$$\zeta(1/2+it) = \Omega(\exp(c\sqrt{\log t}/\log\log t))$$

have been proved with c a fixed positive constant. (See [**T**] for this, and some improvements.) In particular, this implies that for $|t|$ sufficiently large $\zeta(1/2+it)$ will be larger than any fixed power of $\log t$ infinitely often.

Properties of $\zeta(\sigma+it)$ in the t-direction often have analogs in the d-direction, where one fixes a special point s, and considers twists of $\zeta(s)$ by quadratic characters of conductor d. Thus, for example, at the point $s = 1/2$, the d-analog of the Lindelöf Hypothesis states that as $|d| \to \infty$,

$$L\left(\frac{1}{2}, \chi_d\right) \ll |d|^{\varepsilon},$$

where on the left hand side we have the Dirichlet L-series formed with χ_d, the quadratic character of conductor d. It is natural to ask whether an analogous omega result holds:

$$L\left(\frac{1}{2}, \chi_d\right) = \Omega(\exp(c\sqrt{\log d}/\log\log d)).$$

1991 *Mathematics Subject Classification.* Primary 11F67: Secondary 11F37.
Key words and phrases. Omega results, Siegel zeros, automorphic L-series.
Research by the first author was partially supported by NSF Grant #DMS-9023202.

Curiously enough, this result does not appear to be in the literature. However, Heath-Brown, in unpublished work, has proved precisely this statement. We are indebted to him for describing his proof to us [**H**].

Heath-Brown's result may be thought of as an omega result in the d-aspect for automorphic forms on GL(1). For reasons which will be explained momentarily, we became interested in extending this to certain automorphic forms on GL(2). Specifically, let f be a holomorphic newform for $\Gamma_0(N)$ of weight $2k$. Let $L(s,f)$ denote the L-series of f, normalized to have a functional equation as $s \mapsto 1-s$, and let $L(s,f,\chi_d)$ denote the twist of $L(s,f)$ by χ_d. Then one can ask if an analogous omega result holds for $L(1/2,f,\chi_d)$. This indeed proved to be the case. By adapting the ideas in Heath-Brown's proof to cusp forms, and using a technique of Iwaniec [**I**] for averaging quadratic twists of automorphic L-functions, we succeeded in proving the following:

THEOREM. *Let f be as above. Then there exists an absolute effective constant $c>0$ such that for squarefree $d \to \infty$,*

$$L\left(\frac{1}{2},f,\chi_d\right) = \Omega(\exp(c\sqrt{\log d}/\log\log d)).$$

In addition to their intrinsic interest, we began studying these omega results because of certain connections with Rankin-Selberg convolutions. In [**HL**] we showed that if f is a newform for $\Gamma_0(N)$, normalized so that the Petersson norm $\|f\|=1$, then in most cases the (properly normalized) first Fourier coefficient of f cannot be very large. To take a particular instance, let f be holomorphic of weight $2k$ with Fourier expansion

$$f(z) = \sum_{n=1}^{\infty} \rho(n) n^{k-1/2} e^{2\pi i n z}$$

and let

$$L(s, f\times f) = \sum_{n=1}^{\infty} \frac{\rho(n)^2}{n^s}$$

be the Rankin-Selberg convolution of f with itself. We showed that if $L(s,f\times f)$ is nonzero for real s in the range $1-c/\log N < s < 1$ for some constant $c>0$, then $|\rho(1)|^2 \le c' \log N$ with an effective constant c' depending on c. The contrapositive of this, of course, states that if $\rho(1)$ is large then $L(s,f\times f)$ must have a real zero in the given range.

The result just described was proved originally for non-metaplectic cusp forms on GL(2), in which case the L-series $L(s,f\times f)$ has an Euler product. The existence of this Euler product was of key importance in controlling the possible real zero. However, the relation given above between the size of $\rho(1)$ and the existence of a real zero can still be proved when the form involved is metaplectic, for example a holomorphic cusp form of half-integral weight for GL(2). The main difference is that in this case the convolution L-series will *not* have an Euler product. By applying our argument to the half-integral weight Shimura correspondent of a holomorphic cusp form, we obtain, thanks to results of Waldspurger and Kohnen, an interesting relation between omega results for $L(1/2,f,\chi_d)$ and the existence of certain real zeros of convolution L-series.

Now assume that N is odd, and that f is an element of Kohnen's "plus space" [**K82**]. We will also assume for convenience that k is even. Then there exists a form g of weight $k+1/2$ which is mapped to f by the Shimura correspondence. The assumption that f is in the plus space is made to insure that g is unique.

The Fourier coefficients of g are given explicitly by Kohnen in [**K82**], [**K85**], [**K88**]. Essentially, up to normalization, the square of the dth coefficient of g is equal to $L(1/2, f, \chi_d)$. Unfortunately, while this is precisely the case when d is relatively prime to $2N$, the situation when $(d, 2N) \neq 1$ is considerably murkier. For this reason we will restrict ourselves to the consideration of f for which the corresponding g have the property that the dth Fourier coefficients vanish when $(d, 2N) \neq 1$. This is easily guaranteed by choosing N so that $p|N$ implies $p^2|N$, or by taking any fixed f of odd level and twisting by an appropriate quadratic character. Let us stress that these restrictions on f are made simply to make it easier to apply existing results on the nature of the coefficients of forms of half-integral weight.

Now for each d, $(d, 2N) = 1$, there exists a form g_d (which is also a Shimura correspondent of f) whose first coefficient is the dth coefficient of g. This g_d has central character χ_d and level $4Nd$. By the argument described above, a zero-free region for $L(s, g_d \times g_d)$ would imply an upper bound of the form

$$L(1/2, f, \chi_d) \ll \log d,$$

with the implied constant depending only on N, k, and f. However, an omega result for the left-hand side, with d varying over a suitable class of discriminants permitting the Shimura correspondent to exist, would contradict this, forcing the existence of a real zero of $L(s, g_d \times g_d)$. Thus we obtain:

THEOREM. *Let f be as above. Then there are absolute, effective constants $c_1, c_2 > 0$ such that there exist infinitely many squarefree d for which $L(s, g_d \times g_d)$ has a real zero β, with $1 - \beta < c_1 \exp(-c_2\sqrt{\log d}/\log\log d)$.*

This is of interest for two main reasons. First, it provides yet another philosophical underpinning for the belief that any proof of the Riemann Hypothesis must rely heavily on the existence of an Euler product. Presumably the Riemann Hypothesis is true for the convolution L-series $L(s, f \times f)$, when f is non-metaplectic and an eigenfunction of all the Hecke operators. One believes this because $L(s, f \times f)$ has an Euler product under these circumstances. However, the cusp forms g_d are also eigenfunctions of the Hecke operators and their convolution L-series $L(s, g_d \times g_d)$ is indistinguishible from $L(s, f \times f)$ except for the fact that there is no Euler product.

Secondly, it is interesting to note that the relation between the first coefficient and the location of a real zero works in reverse. If the zero of $L(s, g_d \times g_d)$ could be identified and bounded away from 1 in some way, it would lead to corresponding upper bounds for $L(1/2, f, \chi_d)$.

We should remark that the two GL(1) omega results mentioned earlier also have relations with real zeros. The Rankin-Selberg convolutions in these cases involve non-holomorphic half-integral weight Eisenstein series. For example, considering the Eisenstein series of weight $1/2$ given in [**GJ**] at the point $s = 1/4 + it/2$, the first Fourier coefficient will be essentially $\zeta(1/2 + it)$. Alternatively, one can take $s = 1/4$ and introduce a quadratic character, making the first coefficient $L(1/2, \chi_d)$. The omega results of either [**T**] or Heath-Brown [**H**] then lead to analogous real-zero results. They are, however, a bit more difficult to describe easily, owing to the non-cuspidal nature of the Eisenstein series, and to the fact that they are non-holomorphic.

We would like to thank R. Heath-Brown for his extremely interesting letter, and B. Conrey, N. Jochnowitz, and J. Pipher for some valuable discussions and suggestions.

1. A mean-value estimate and an omega result

In this section we prove Theorem 1. Suppose we are given a holomorphic newform f of weight $2k$ and level N with Fourier expansion

$$f(z) = \sum_{n=1}^{\infty} a_n n^{k-1/2} e^{2\pi i n z}.$$

Let χ_d denote the Kronecker symbol $\left(\frac{d}{\cdot}\right)$ and let

$$L(s, f, \chi_d) = \sum_{n=1}^{\infty} \frac{a_n \chi_d(n)}{n^s}$$

be the L-series for f twisted by the quadratic character χ_d. We then have an Euler product expansion

$$L(s, f, \chi_d) = \prod_p \left(1 - \frac{\alpha_p \chi_d(p)}{p^s}\right)^{-1} \left(1 - \frac{\beta_p \chi_d(p)}{p^s}\right)^{-1}$$

where α_p and β_p are the usual parameters associated to f. The functional equation of $L(s, f, \chi_d)$ is given by

$$\begin{aligned} \Lambda(s) &= \left(\frac{|d|\sqrt{N}}{2\pi}\right)^{s+k-1/2} \Gamma\left(s + k - \frac{1}{2}\right) L(s, f, \chi_d) \\ &= w_d \Lambda(1-s) \end{aligned}$$

where $w_d = \pm\chi_d(-N)$, the sign depending only on f. We will in fact consider only those d for which $w_d = 1$, and hence $\Lambda(1-s) = \Lambda(s)$.

The idea of the proof is to average $L(1/2, f, \chi_d)$ over a special set of d's. Let P be a positive integer to be chosen later and d_0 a fixed integer relatively prime to $M = 4NP$ satisfying $d_0 \equiv 1 \pmod 4$. Let

$$\mathcal{D} = \{\, d > 0 \mid d \equiv d_0 v^2 \pmod M \text{ for some } v \text{ prime to } M \,\},$$

and let $\mathcal{D}^*$ denote the set of squarefree elements of $\mathcal{D}$. Thus χ_d is a primitive Dirichlet character mod d for all $d \in \mathcal{D}^*$. Note that to insure $w_d = 1$ for all $d \in \mathcal{D}$ it suffices to choose d_0 in an appropriate residue class mod $4N$.

Now let $F(Y)$ be any smooth compactly supported test function on the positive reals, and consider the sum

$$S(Y) = \sum_{d \in \mathcal{D}^*} L\left(\frac{1}{2}, f, \chi_d\right) F(d/Y). \tag{1}$$

Theorem 1 will follow from lower bounds on $S(Y)$ together with a judicious choice of P and d_0.

REMARK. In [**I**], Iwaniec considers the average value of the derivative

$$L'(1/2, f, \chi_d)$$

over squarefree d. We simply imitate his proof, keeping careful track of the new parameters M and d_0. In fact, at many points in the argument (especially where sums over D are replaced by sums over all integers) we obtain identical estimates.

The first step is to obtain a rapidly convergent series for $L(1/2, f, \chi_d)$. Consider the (normalized) incomplete gamma function

$$V(x) = \frac{1}{\Gamma(k)} \int_x^\infty e^{-t} t^k \frac{dt}{t}.$$

This function is positive, has monotonic exponential decay as x tends to infinity, and satisfies $V(x) = 1 + O(x^k)$ for x near zero. Mellin transforms yield the formula

$$V(x) = \frac{1}{2\pi i} \int_{2-i\infty}^{2+i\infty} \frac{\Gamma(s+k)}{\Gamma(k)} x^{-s} \frac{ds}{s}.$$

Now define for any $x > 0$, and any character χ,

$$\mathcal{A}(x, \chi) = \sum_{n=1}^{\infty} \frac{a_n \chi(n)}{\sqrt{n}} V\left(\frac{2\pi n}{x}\right).$$

The approximate functional equation for $L(s, f, \chi_d)$ may then be expressed as

$$L\left(\frac{1}{2}, f, \chi_d\right) = \mathcal{A}(x, \chi_d) + \mathcal{A}(d^2N/x, \chi_d), \quad d \in \mathcal{D}^*.$$

In particular, for $d \in \mathcal{D}^*$ we have the formula

$$L\left(\frac{1}{2}, f, \chi_d\right) = 2\mathcal{A}(d\sqrt{N}, \chi_d).$$

We now proceed as in [**I**] to obtain an approximate formula for $S(Y)$. We have

$$\begin{aligned} S(Y) &= 2 \sum_{d \in \mathcal{D}^*} \mathcal{A}(d\sqrt{N}, \chi_d) F(d/Y) \\ &= 2 \sum_{(a,M)=1} \mu(a) \sum_{d \in \mathcal{D}} \mathcal{A}(a^2 d\sqrt{N}, \chi_{a^2 d}) F(a^2 d/Y). \end{aligned}$$

At this point the sum breaks up into two pieces depending on the size of a. Let A be a parameter, equal to a fixed power of Y, to be determined later. For $a > A$ we are led to the sum

$$R = 2 \sum_{(b,M)=1} \sum_{\substack{a|b \\ a>A}} \mu(a) \sum_{d \in \mathcal{D}^*} \mathcal{A}(b^2 d\sqrt{N}, \chi_{b^2 d}) F(b^2 d/Y).$$

This is bounded above using a fourth-power moment result for $L(1/2, f, \chi_d)$ analogous to the one employed in Iwaniec's original argument, and we arrive at the same estimate, namely

$$R \ll (A^{-3/2} Y^{5/4} + A^{-3} Y^{3/2}) Y^{\varepsilon}.$$

The other half of the sum is

$$S = 2 \sum_{\substack{a \le A \\ (a,M)=1}} \mu(a) \sum_{d \in \mathcal{D}} F(a^2 d/Y) \sum_{(n,a)=1} \frac{a_n \chi_d(n)}{\sqrt{n}} V\left(\frac{2\pi n}{a^2 d\sqrt{N}}\right).$$

Write $n = hl^2m$, where h has prime factors dividing M, lm is prime to M, and m is squarefree. Note that $\chi_d(n) = \chi_d(h)\chi_d(m)$ for $(l,d) = 1$. Thus

$$S = 2 \sum_{\substack{a \le A \\ (a,M)=1}} \sum_{\substack{n=hl^2m \\ (n,a)=1}} \mu(a)\chi_{d_0}(h)\frac{a_n}{\sqrt{n}} \cdot \sum_{q|l} \mu(q) \sum_{dq \in \mathcal{D}} F(a^2dq/Y)\chi_{dq}(m)\left(\frac{2\pi n}{a^2dq\sqrt{N}}\right). \tag{2}$$

Here we have used the fact that if $d \in \mathcal{D}$ and h is a product of primes dividing M then $\chi_d(h) = \chi_{d_0}(h)$.

Let χ_0 denote the function which is 1 at 1 and 0 otherwise. Then for all odd squarefree $m > 0$ we have

$$\chi_d(m) = \frac{\bar{\varepsilon}_m}{\sqrt{m}} \sum_{2|r|<m} \chi_{Mr}(m)e(\bar{M}rd/m),$$

where $e(x) = e^{2\pi i x}$, ε_m equals 1 or i depending on whether m is $\pm 1 \pmod 4$, and $M\bar{M} + m\bar{m} = 1$. Substituting this into (2) gives

$$S = 2 \sum_{\substack{a \le A, n=hl^2m \\ (a,Mn)=1}} \mu(a)\chi_{d_0}(h)\frac{a_n}{\sqrt{n}}\frac{\bar{\varepsilon}_m}{\sqrt{m}} \sum_{q|l} \mu(q) \sum_{2|r|<m} \chi_{Mrq}(m) \sum_d \tag{3}$$

where

$$\sum_d = \sum_{dq \in \mathcal{D}} F(a^2dq/Y)\left(\frac{2\pi n}{a^2dq\sqrt{N}}\right) e(\bar{M}rd/m).$$

Let $\Delta = \min(1/2, Ma^2qY^{-1+\varepsilon})$. We now split up the sum as $S = S_0 + S_1 + S_2$, according to whether $r = 0$, $0 < |r| < \Delta m$, $\Delta m \le |r| < m/2$. The main term is S_0. The error terms S_1 and S_2 are treated as in [**I**].

For S_2, we use the fact that $a^2y \in \mathcal{D}$ if and only if $y \in \mathcal{D}$ to write

$$\sum_d = \sum_{\substack{y \in \mathcal{D} \\ y \equiv 0 \pmod{a^2q}}} F(y/Y)V\left(\frac{2\pi n}{y\sqrt{N}}\right) e(\bar{M}ry/a^2qm). \tag{4}$$

Now Lemma 2 of [**I**] says that if $g(x)$ satisfies $g^{(j)}(x) \ll (|x| + X)^{-j}$ for all $j \ge 1$, and α is real, then

$$\sum_{y \equiv \nu \pmod{Q}} g(y)e(\alpha y) \ll \frac{X}{Q}\left(\frac{Q}{X\|\alpha Q\|}\right)^j \tag{5}$$

for all $j \ge 2$, provided that αQ is not an integer. We apply this to (4), breaking the sum up over congruence classes mod $Q = Ma^2q$. The number of classes we denote by $\gamma(M) \le M$, and each sum has the form (5) with $g(y) = F(y/Y)V(2\pi n/(y\sqrt{N}))$ and $\alpha = \bar{M}r/a^2qm$. In this case we may take $X = Y + n$, and hence

$$\sum_d \ll \gamma(M) \cdot \frac{Y+n}{Ma^2q}\left(\frac{Ma^2q}{(Y+n)\Delta}\right)^j.$$

Choosing j sufficiently large, and assuming we have chosen A so that $MA^2 \ll Y^{1/2-\varepsilon}$, we get $\sum_d \ll (Y+n)^{-99}$. Putting this in (3) yields $S_2 \ll 1$.

To estimate S_1, we sum first over m to get

$$S_1 = 2 \sum_{\substack{a \le A \\ (a,\bar{M})=1}} \mu(a) \sum_{(hl^2,a)=1} a_h \chi_{d_0}(h) h^{-1/2} l^{-1} \sum_{q|l} \mu(q) \sum_{dq \in \mathcal{D}} F(a^2dq/Y) \sum_{r=1}^{\infty} \sum_m$$

where

$$\sum_m = \sum_{\substack{m > |r|/\Delta \\ (m,aM)=1 \\ m \text{ squarefree}}} a_{l^2m} m^{-1} \chi_{Mrq}(m) \bar{\varepsilon}_m \left(\frac{2\pi n}{a^2dq\sqrt{N}} \right) e(\bar{M}rd/m).$$

Now write $e(\bar{M}rd/m) = e(rd/Mm)e(-\bar{m}rd/M)$ and let

$$\psi(m) = \bar{\varepsilon}_m \chi_{Mrq}(m) e(-\bar{m}rd/M)$$
$$g(x) = x^{-1} V \left(\frac{2\pi hl^2 x}{a^2qd\sqrt{N}} \right) e(rd/Mx).$$

Then

$$\sum_m = \sum_{\substack{m > |r|/\Delta \\ (m,aM)=1 \\ m \text{ squarefree}}} a_{l^2m} \psi(m) g(m).$$

Now $\psi(m)$ has absolute value 1 and period Mrq, so using Lemma 1 from [**I**] we get the estimate

$$\sum_{\substack{m \le x \\ (m,a\bar{M})=1 \\ m \text{ squarefree}}} a_{l^2m} \psi(m) \ll \tau(l) \tau(aM) (Mrq)^{1/2} x^{1/2+\varepsilon}.$$

Partial summation now gives

$$\sum_m \ll \tau(l)\tau(aM)(Mq)^{1/2} \Big\{ r^{1/2} |g(r/\Delta)| (r/\Delta)^{1/2+\varepsilon} + r^{1/2} \int_{r/\Delta}^{\infty} |g'(x)| x^{1/2+\varepsilon}\, dx \Big\}.$$

Substituting the expression for $g(x)$ and summing trivially over d and r yields

$$\sum_{d,r,m} \ll \tau(l) l^{-2} aqh^{-1} M^2 Y^{1/2+\varepsilon}$$

and therefore $S_1 \ll M^2 A^2 Y^{1/2+\varepsilon}$.

We now compute the main term. When $r = 0$, only the terms with $m = 1$ contribute to the sum, and

$$S_0 = 2 \sum_{\substack{a \le A, n = hl^2 \\ (a,Mn)=1}} \mu(a) \chi_{d_0}(h) \frac{a_n}{\sqrt{n}} \sum_{q|l} \mu(q) \sum_d$$

where

$$\sum_d = \sum_{dq \in \mathcal{D}} F(a^2dq/Y) \left(\frac{2\pi n}{a^2dq\sqrt{N}} \right).$$

As with S_2, this breaks up into classes mod Ma^2q, and we get

$$\sum_d = \frac{\gamma(M)Y}{Ma^2q} \int F(t) V \left(\frac{2\pi n}{tY\sqrt{N}} \right) dt + O\left(\gamma(M) \left(1 + \tfrac{n}{Y}\right)^{-99} \right).$$

The error term contributes only $O(MAY^{1/2+\varepsilon})$ to S_0. The main term then becomes

$$(6)\qquad 2Y\frac{\gamma(M)}{M}\sum_{n=hl^2}\chi_{d_0}(h)\frac{a_n}{\sqrt{n}}\frac{\varphi(l)}{l}\sum_{\substack{a\leq A\\(a,Ml)=1}}\frac{\mu(a)}{a^2}\int F(t)V\left(\frac{2\pi n}{tY\sqrt{N}}\right)dt.$$

Now

$$\sum_{\substack{a\leq A\\(a,Ml)=1}}\frac{\mu(a)}{a^2}=\frac{6}{\pi^2}\prod_{p|Ml}(1-p^{-2})^{-1}+O(A^{-1}).$$

Substituting this is in (6) gives

$$(7)\qquad S_0=2C_MY\int F(t)B(tY\sqrt{N})\,dt+O((MAY^{1/2}+A^{-1}Y)Y^{\varepsilon})$$

where

$$C_M=\frac{6\gamma(M)}{\pi^2M}\prod_{p|M}(1-p^{-2})^{-1}$$

$$B(x)=\sum_{n=hl^2}\frac{b_n\chi_{d_0}(n)}{\sqrt{n}}V\left(\frac{2\pi n}{x}\right)$$

with

$$b_n=a_n\prod_{p|n,p\nmid M}\left(1+\frac{1}{p}\right)^{-1}.$$

Returning to the definition of $V(x)$ as an inverse Mellin transform, we get

$$B(x)=\frac{1}{2\pi i}\int_{2-i\infty}^{2+i\infty}\frac{\Gamma(s+k)}{\Gamma(k)}L\left(s+\frac{1}{2}\right)\left(\frac{x}{2\pi}\right)^s\frac{ds}{s}$$

where

$$L(s)=\sum_{n=hl^2}\frac{b_n\chi_{d_0}(n)}{n^s}=\sum_{h|M^\infty}\frac{a_h\chi_{d_0}(h)}{h^s}\sum_{(l,M)=1}\frac{b_{l^2}}{l^{2s}}.$$

Now $L(s)$ differs from $L(2s,\mathrm{sym}^2 f)/\zeta(4s)$ by a product of Euler factors which converges absolutely for $\mathrm{Re}(s)\geq 1/4$, and it follows that $L(s)$ is analytic in this range and has polynomial growth in $\mathrm{Im}(s)$. Moving the line of integration to $\mathrm{Re}(s)=-1/4$, we get

$$B(x)=L(1/2)+O(M^{\varepsilon}x^{-1/4}).$$

Substituting this into (7), we obtain

$$S_0=2C_ML(1/2)Y\int F(t)\,dt+O(MAY^{1/2}+Y^{3/4}+A^{-1}Y)Y^{\varepsilon}).$$

Putting all of our estimates together, we find (assuming $A^2M\ll Y^{1/2-\varepsilon}$)

$$\begin{aligned}S(Y)&=R+S_0+S_1+S_2\\&=\alpha Y+O((M^2A^2Y^{1/2}+Y^{3/4}+A^{-1}Y+A^{-3}Y^{3/2}+A^{-3/2}Y^{5/4})Y^{\varepsilon})\end{aligned}$$

where $\alpha=2C_ML(1/2)\int F(t)\,dt$. Let us take $A=Y^{5/28}$ and $M\approx Y^{1/16}$. Then the error terms above are $O(Y^{55/56+\varepsilon})$, and we get finally

$$\sum_{d\in\mathcal{D}^*}L\left(\frac{1}{2},f,\chi_d\right)F(d/Y)=2C_ML(1/2)Y\int F(t)\,dt+O(Y^{55/56+\varepsilon}).$$

On the other hand, the sum $\sum_{d\in\mathcal{D}^*} F(d/Y)$ is asymptotic to $C_M Y \int F(t)\,dt$, so that the quantity $2L(1/2)$ is in fact the average value of $L(1/2, f, \chi_d)$ over squarefree $d \in \mathcal{D}$. To complete the proof of Theorem 1, it suffices to choose M and d_0 in such a way that for some constant $c > 0$,

$$L(1/2) \gg \exp(c\sqrt{\log Y}/\log\log Y). \tag{8}$$

From the Euler product for $L(s)$ we find

$$L(1/2) = C\prod_{p|M}\left(1 + \frac{a_p\chi_{d_0}(p)}{\sqrt{p}} + O\left(\frac{1}{p}\right)\right)$$

where C is a constant depending only on f. To make this product as large as possible, we set $M = 4NP$ with P equal to the product of all primes less than y which do not divide $4N$. Taking $y = \frac{1}{16}\log Y$ insures that M has the correct size. Now choose d_0 so that $\chi_{d_0}(p)$ has the same sign as a_p for all $p|P$ and so that the sign of the functional equation is positive. Then

$$\log L(1/2) = \sum_{p<y} |a_p|/\sqrt{p} + O(\log\log y)$$

Since $|a_p| \le 2$, we have $|a_p| \ge a_p^2/2$. Applying Ikehara's Theorem to the logarithmic derivative of $\sum a_n^2 n^{-s}$, we see that $\sum_{p<y} a_p^2 \log p$ is asymptotic to y. Partial summation gives $\log L(1/2) \gg y^{1/2}/\log y$, which yields (8) as desired.

2. Siegel zeros and half-integral weight L-functions

We now combine the omega result of the previous section with a theorem of Kohnen to prove Theorem 2. Let f be a normalized newform of weight $2k$ and level N. Let w_p denote the eigenvalue of f under the Atkin-Lehner involution W_{p^ν} where $p^\nu \| N$. Suppose that $w_p = 1$ whenever ν is even. Fix an integer n_0 so that $\left(\frac{(-1)^k n_0}{p^\nu}\right) = w_p$ for all $p|N$. Kohnen's result says that there exists a cusp form g of weight $k + 1/2$ and level $4N$ with Fourier expansion

$$g(z) = \sum c(n) n^{k/2-1/4} e^{2\pi i n z}$$

such that if d is a fundamental discriminant with $(-1)^k d > 0$, $(d, N) = 1$, and dn_0 congruent to a square modulo N, then

$$\frac{c(|d|)^2}{\langle g, g\rangle} = \pi^{-k} 2^{\nu(N)}(k-1)!\,\frac{L(1/2, f, \chi_d)}{\langle f, f\rangle}. \tag{9}$$

Here if $h(z)$ is a form of weight $l \in 1/2\mathbb{Z}$ for a congruence subgroup Γ the inner product is defined as

$$\langle h, h\rangle = \frac{1}{\mathrm{Vol}(\Gamma\backslash\mathfrak{h})}\int_{\Gamma\backslash\mathfrak{h}} |h(z)|^2 y^l\,\frac{dx\,dy}{y^2}.$$

To apply Kohnen's result, we make the following choices: Let N be odd, and let f have the property that there exists a Shimura correspondent g whose coefficients $c(n)$ satisfy $c(n) = 0$ whenever $(n, N) \ne 1$. As mentioned in the introduction, this is easily guaranteed by choosing N so that $p|N$ implies $p^2|N$, or by taking any fixed f of odd level and twisting by an appropriate quadratic character with conductor dividing N. Finally, we'll suppose for convenience that k is even.

Now let g_d be the form obtained from g by applying the dth "shifting" operator:

$$g_d(z) = d^{-k/2-3/4} \sum_{a=0}^{d-1} g\left(\frac{z+a}{d}\right).$$

Thus g_d is a form of weight $k+1/2$ and level $4Nd$, with nebentypus χ_d, whose first Fourier coefficient is $c(d)$. In fact, g_d is also a Shimura correspondent of f. Our object is to show that for an infinite family of d, the L-series $L(s, g_d \times g_d)$ must have a Siegel zero.

First, suppose that $L(s, g_d \times g_d)$ is nonzero for real s in the range $\beta < s < 1$, where $1-\beta \ll 1/\log d$, the implied constant depending only on N. Then it follows immediately from the proof of Theorem 0.1 of [**HL**] that

$$\operatorname{Res}_{s=1} L(s, g_d \times g_d) \gg c(d)^2(1-\beta), \tag{10}$$

the implied constant depending only on N.

On the other hand, the residue at $s=1$ of $L(s, g_d \times g_d)$ can be controlled by recourse to a result of [**FH**]. We have the following

LEMMA. $\operatorname{Res}_{s=1} L(s, g_d \times g_d) \ll 1$, *where the implied constant depends only on f.*

PROOF. By definition,

$$L(s, g_d \times g_d) = \sum_{1}^{\infty} \frac{c(nd)^2}{n^s},$$

where the $c(nd)$ are the coefficients of the original g with indices divisible by d. When nd is a fundamental discriminant, the value of $c(nd)^2$ is given by (9) (with d replaced by nd), and for square multiples of such nd the values are related to $c(nd)^2$ by means of the Hecke relations.

There is a related series that arises in [**FH**]. This is, for any N_1 odd, $(N_1, N)=1$:

$$Z_1(s, f, N_1) = \sum_{\substack{M,n>0 \\ (Mn, 2NN_1)=1}} \frac{b(n)L(1/2, f, \psi_M)}{(Mn^2)^s},$$

where

$$b(n) = \prod_{\substack{p|n \\ p^\gamma || n}} \left(a(p^{2\gamma}) - \psi_M(p)p^{-1/2}a(p^{2\gamma-1})\right)$$

and $\psi_M = \left(\frac{\cdot}{M}\right)$.

Summing over $m = Mn^2$, the above transforms to

$$Z_1(s, f, N_1) = \sum_{(m, 2NN_1)=1} \frac{C(m)}{m^s},$$

where for $m = m_0 m_1^2$, m_0 square free,

$$C(m) = L(1/2, f, \psi_{m_0}) \prod_{p^i || m_1} C(p^i)$$

and $C(p^i) \ll a(p^{2i})$.

It follows from [**FH**] that for any N_1, $Z_1(s,f,N_1)$ has a simple pole at $s=1$ and

$$\operatorname{Res}_{s=1} Z_1(s,f,N_1) = c_1(f) \prod_{p|2NN_1} \left(1-\frac{1}{p}\right),$$

where $c_1(f)$ is a constant depending only on f. Thus if we set

$$Z_2(s,f,N_1) = Z_1(s,f,1) - Z_1(s,f,N_1)$$

we then have

$$Z_2(s,f,N_1) = \sum_{\substack{(m,2N)=1\\(m,N_1)\neq 1}} \frac{C(m)}{m^s}$$

and

$$\operatorname{Res}_{s=1} Z_2(s,f,N_1) = c_1(f) \prod_{p|2N} \left(1-\frac{1}{p}\right)\left(1-\prod_{p|N_1}\left(1-\frac{1}{p}\right)\right).$$

Now let us suppose N_1 is squarefree, and define

$$Z(s,f,N_1) = -N_1^s \sum_{n|N_1} \mu(n) Z_2(s,f,n).$$

Then

$$Z(s,f,N_1) = \sum_{(m,2N)=1} \frac{C(mN_1)}{m^s}$$

and

$$\begin{aligned}\operatorname{Res}_{s=1} Z(s,f,N_1) &= N_1 c_1(f) \prod_{p|2N}\left(1-\frac{1}{p}\right)\sum_{n|N_1}\mu(n)\left(-1+\prod_{p|n}\left(1-\frac{1}{p}\right)\right)\\ &= N_1 c_1(f) \prod_{p|2N}\left(1-\frac{1}{p}\right)\prod_{p|N_1}\frac{1}{p}\\ &= c_1(f)\prod_{p|2N}\left(1-\frac{1}{p}\right).\end{aligned}$$

Now it is a consequence of the Hecke relations that essentially $c(n)^2 = C(n)$. More precisely, there is a constant $c_2(f)$ such that if $n \equiv 1(4)$ and $c(n) \neq 0$ then $c(n)^2 = c_2(f)C(n)$. If $n = 4n'$, with $n' \equiv 3(4)$ then $c(4n')^2 = c_3(f')C(n')$, where we have applied the construction of [**FH**] to f twisted by the non-trivial quadratic character mod 4. In general, if $n = 4^t n'$ with n' odd, then there is a constant $c_4(f)$ such that $c(n)^2 \leq c_4(f)C(n')$. Here $C(n)$ comes from the construction associated to either f or the mod 4 quadratic twist of f, according to whether n' is congruent to 1 or 3 mod 4. A similar result holds if $n = 2(4^t)n'$, after one twists f by the appropriate non-trivial primitive character mod 8. Referring to the original f as f_1, and the succeeding two twists as f_2, f_3 we have, for all $s > 1$,

$$L(s, g_d \times g_d) \ll \sum_{i=1}^{3} Z(s, f_i, d).$$

Thus after taking the residue the lemma follows. □

Combining the Lemma with (10) we obtain

$$c(d)^2 \ll (1-\beta)^{-1}, \tag{11}$$

where the implied constant depends only on f. Now use Theorem 1 to obtain an infinite sequence of d such that

$$L\left(\frac{1}{2}, f, \chi_c\right) \gg \exp(c\sqrt{\log d}/\log\log d).$$

By (9), this then contradicts (11) whenever

$$1-\beta \gg \exp(-c/2\sqrt{\log d}/\log\log d)$$

and Theorem 2 follows.

References

[FH] S. Friedberg and J. Hoffstein, *Nonvanishing Theorems for Automorphic L-Functions on* GL(2), Ann. Math. (to appear).

[GJ] D. Goldfeld and J. Hoffstein, *Eisenstein series of* 1/2*-integral weight and the mean value of real Dirichlet L-series*, Invent. Math. **80** (1985), 185–208.

[H] R. Heath-Brown, *Letter to J. Hoffstein of October 12, 1993.*

[HL] J. Hoffstein and P. Lockhart, *Coefficients of Maass forms and the Siegel zero*, Ann. Math. **140** (1994), 161–181.

[I] H. Iwaniec, *On the order of vanishing of modular L-functions at the critical point*, Séminaire de Théorie des Nombres, Bordeaux **2** (1990), 365–376.

[K82] W. Kohnen, *Newforms of Half-Integral Weight*, Math. **333** (1982), 32–72.

[K85] W. Kohnen, *Fourier Coefficients of Modular Forms of Half-Integral Weight*, Math. Ann. **271** (1985), 237–268.

[K88] W. Kohnen, *A remark on the Shimura correspondence*, Glasgow Math. J. **30** (1988), 285–291.

[T] E. C. Titchmarsh, The Theory of the Riemann Zeta-Function, Oxford Univ. Press: Oxford (1951).

[W] J.-L. Waldspurger, *Sur les coefficients de Fourier des formes modulaires de poids demi-entier*, J. Math. Pures Appl. **60** (1981), 375–484.

Mathematics Department, Brown University, Providence, RI 02912
E-mail address: jhoff@math.brown.edu

Mathematics Department, University of California, Santa Cruz, CA 95064
E-mail address: lockhart@cats.ucsc.edu

Proceedings of Symposia in Pure Mathematics
Volume **66.2**, 1999

Representations of SL(2) $\times G$

Kamal Khuri-Makdisi

This paper is dedicated to my teacher, Goro Shimura.

ABSTRACT. Let $\mathfrak{g}$ be a complex reductive Lie algebra. We investigate a formalism for producing a natural model for representations of $\mathfrak{sl}(2) \times \mathfrak{g}$ which decompose in such a way so as to produce interesting correspondences between representations of $\mathfrak{sl}(2)$ and of $\mathfrak{g}$. Our formalism generalizes the standard dual pair SL(2) $\times$ O(2n). We use it to obtain a formally unitary representation of $\mathfrak{sl}(2, \mathbb{R}) \times \mathfrak{u}(3)$ that essentially realizes the symmetric square lift of Gelbart and Jacquet. We lift this representation of Lie algebras to a unitary representation of SL(2, $\mathbb{R}$) $\times$ U(3) on a certain L^2 space, with quite explicit formulas.

1. Introduction

We are motivated by the problem of seeking a natural realization of Gelbart and Jacquet's symmetric square lift of automorphic forms (or automorphic representations) on GL(2) to automorphic forms on GL(3) [**GJ**] in terms of something akin to, but not quite identical with, a theta-correspondence. This would lead to an integral kernel for the lifting, and perhaps even some insights into whether one could realize some other cases of Langlands functoriality similarly. Actually, it is more convenient to consider a variant of the symmetric square lift, namely the adjoint lift, corresponding to the map on the dual groups PGL(2) $\to$ SL(3); so the goal is to relate forms on SL(2) to forms on PGL(3), or rather forms on SL(2) to forms on GL(3) with trivial central character. A first step in this direction is to look for reasonable unitary representations V of SL(2, F) $\times$ GL(3, F), F a local field, with properties similar to local Howe duality (see section 6 of [**Ho**]): i.e., the irreducible quotients of V should have the form $V_2 \otimes V_3$, where V_2 is an irreducible representation of SL(2, F) and V_3 is the corresponding representation of GL(3, F) under the adjoint lift. The usual construction of V by a theta-correspondence would involve taking V to be the minimal representation of a big reductive group G containing

1991 *Mathematics Subject Classification.* Primary 11F, 22E.

Key words and phrases. Theta-correspondence, Howe correspondence, Gelbart-Jacquet lift, geometric quantization.

Part of this work was done at the Institute for Advanced Study, where the author was supported by NSF grant number DMS 9304580 to the Institute. The rest of this work was done at Harvard University.

$\mathrm{SL}(2) \times \mathrm{GL}(3)$ as a dual pair. Unfortunately those few cases in which groups of type $A_1 \times A_2$ arise as dual pairs seem to yield less interesting correspondences. We must therefore look for a representation V of $\mathrm{SL}(2) \times \mathrm{GL}(3)$ in the absence of such a big group.

Our results in this paper are all for $F = \mathbb{R}$, but we begin by describing a technique that should work for arbitrary F. Our aim in this introduction is to first explain some general ideas which motivated the construction in this paper, without worrying about certain technical details. We then summarize the contents of Sections 2–4, which contain the actual results of this paper. Those results are motivated by the considerations discussed in this introduction, but end up taking a somewhat different form.

We first describe the effect of the adjoint lift on "most" principal series representations. Let $\chi : F^\times \to \mathbb{C}^\times$ be a multiplicative character. Let PS_2^χ be the unitarily induced principal series representation of $\mathrm{SL}(2, F)$, obtained by unitarily inducing the character $\left(\begin{smallmatrix} a & b \\ 0 & a^{-1}\end{smallmatrix}\right) \mapsto \chi(a)$ from the Borel subgroup B of $\mathrm{SL}(2, F)$ up to the whole group. (In fact, even if χ is not unitary, $\mathrm{PS}_2^\chi \cong \mathrm{PS}_2^{\chi^{-1}}$ for "generic" χ.) According to the local adjoint lift, PS_2^χ corresponds to the representation PS_3^χ of $\mathrm{GL}(3, F)$ obtained by unitarily inducing the character $\chi \circ \lambda$ of its Borel subgroup up to the whole group, where λ is the following rational character of the Borel subgroup:

$$\lambda\left(\begin{pmatrix} a & * & * \\ & b & * \\ & & c \end{pmatrix}\right) = a/c. \tag{1.1}$$

One can also view λ as an integral dominant weight of $\mathrm{GL}(3)$. We note that $\mathrm{PS}_3^\chi \cong \mathrm{PS}_3^{\chi^{-1}}$ for generic χ, just as for PS_2^χ.

Write $N = \{\left(\begin{smallmatrix} 1 & b \\ 0 & 1\end{smallmatrix}\right) \mid b \in F\}$ for the unipotent radical of the Borel subgroup B. Given a representation W of $\mathrm{SL}(2)$, write ${}^N W$ for its Jacquet module of N-coinvariants; ${}^N W$ carries a representation of the multiplicative group $F^\times$, identified with the maximal torus in B. PS_2^χ is characterized by the fact that ${}^N\mathrm{PS}_2^\chi$ is two-dimensional, and that $\mathbf{G}_m$ essentially acts on it by $\chi \oplus \chi^{-1}$. To be precise, this is slightly inaccurate, due to the presence of normalizing factors from the unitary induction up to $\mathrm{SL}(2)$. These factors are thankfully absorbed into the normalizing factors of the induction up to $\mathrm{SL}(3)$ and shall henceforth be ignored.

Our desired representation V of $\mathrm{SL}(2) \times \mathrm{GL}(3)$ must therefore "mostly" satisfy

$$ {}^N V \cong \int_{\chi : F^\times \to \mathbb{C}^\times} \chi^{-1} \otimes \mathrm{PS}_3^\chi, \qquad \text{as an } F^\times \times \mathrm{GL}(3)\text{-representation.} \tag{1.2}$$

(By "mostly" we mean that the subspace of ${}^N V$ transforming under $F^\times$ by χ need only be isomorphic to PS_3^χ for generic χ.) On the other hand, there is a rather natural space on which $F^\times \times \mathrm{GL}(3)$ acts by $\int_\chi \chi^{-1} \otimes \mathrm{PS}_3^\chi$. It is the space $\mathcal{S}(\mathcal{O}_\lambda)$ of Schwartz class functions on the orbit $\mathcal{O}_\lambda$ of a highest weight vector in the rational representation V_λ of $\mathrm{GL}(3)$ with highest weight λ. We describe V_λ and $\mathcal{O}_\lambda$ explicitly in Section 3. The action of $a \times g = (a, g) \in F^\times \times \mathrm{GL}(3)$ on a function $\varphi \in \mathcal{S}(\mathcal{O}_\lambda)$ is

$$(a \times g)\varphi(x) = \varphi(a g^{-1} x), \qquad x \in V_\lambda; \tag{1.3}$$

note that $\mathcal{O}_\lambda \subset V_\lambda$ is stable under scaling by an element $a \in F^\times$. So we are led to look for candidates for V such that ${}^N V$ is isomorphic to $\mathcal{S}(\mathcal{O}_\lambda)$.

Taking a hint from the well-known theta-correspondence of $\mathrm{SL}(2) \times \mathrm{O}(2n)$, we try to realize V as $\mathcal{S}(X)$, where X has the following properties:

1. X carries an action of $F^\times \times \mathrm{GL}(3)$.
2. There exists a $\mathrm{GL}(3)$-invariant regular function f on X of "degree 2" with respect to $F^\times$, i.e.,

$$f(agx) = a^2 f(x), \qquad \text{for } a \in F^\times,\ g \in \mathrm{GL}(3), \text{ and } x \in X. \tag{1.4}$$

3. The zero set of f, $X_0 = \{x \in X \mid f(x) = 0\}$, is isomorphic to $\mathcal{O}_\lambda$, as a variety with an action of $F^\times \times \mathrm{GL}(3)$.

Let $e_F : F \to \mathbb{C}^\times$ be a nontrivial additive character of F. We define a representation of $B \times \mathrm{GL}(3)$ on $\mathcal{S}(X)$ by

$$\left(\begin{pmatrix} a & b \\ 0 & a^{-1} \end{pmatrix} \times g\right) \varphi(x) = e_F(abf(x))\varphi(ag^{-1}x). \tag{1.5}$$

In keeping with our wish to suppress normalization factors and other technical details in this motivational section, the above equation is not entirely accurate. But note that the reason for requiring f to have degree 2 is so that 1.5 gives a genuine representation of $B \times \mathrm{GL}(3)$. Then, heuristically, the space of N-coinvariants ${}^N\mathcal{S}(X)$ should be isomorphic to $\mathcal{S}(X_0)$. Indeed, ${}^N\mathcal{S}(X)$ is dual to the space of distributions that are invariant under N; by 1.5, these distributions should have support on those x such that for all $b \in F$, $e_F(bf(x)) = 1$; in other words, on the zero set of f. So if we can extend the above action of $B \times \mathrm{GL}(3)$ to all of $\mathrm{SL}(2) \times \mathrm{GL}(3)$, we obtain a representation that "generically" realizes the local adjoint lift. It should be noted that the above ideas are not limited to $\mathrm{GL}(3)$ but can be generalized to other reductive groups G; one just needs to replace the rational character λ of the Borel subgroup of $\mathrm{GL}(3)$ with a rational character of a parabolic subgroup $P \subset G$, such that P is the stabilizer of the line through a highest weight vector in V_λ. Assuming that one can find a suitable X and extend the resulting representation of $B \times G$ to one of $\mathrm{SL}(2) \times G$, one then obtains a correspondence between PS_2^χ and the induced representation $\mathrm{Ind}_P^G \chi \circ \lambda$.

We now describe our actual results, which, as mentioned above, are only for the case $F = \mathbb{R}$. In that case, we can obtain results by using the Lie algebras $\mathfrak{sl}(2)$ and $\mathfrak{gl}(3)$, or more generally $\mathfrak{sl}(2)$ and $\mathfrak{g}$, and working with infinitesimal representations of these Lie algebras, instead of with representations of the actual groups. We begin in Section 2 with a method to extend a representation of the Lie algebra $\mathfrak{b}$ of B to a representation of $\mathfrak{sl}(2)$, if one knows the action of the Casimir element $C_{\mathfrak{sl}(2)}$ of the universal enveloping algebra $U(\mathfrak{sl}(2))$. If one can find a suitable X as suggested by 1–3 above, this method then allows us to construct a candidate representation of $\mathfrak{sl}(2) \times \mathfrak{g}$ on a certain space of functions on X. The idea is to start with the representation of $\mathfrak{b} \times \mathfrak{g}$ resulting from 1.5, suitably match up the action of the Casimir elements $C_{\mathfrak{sl}(2)} \in U(\mathfrak{sl}(2))$ and $C_{\mathfrak{g}} \in U(\mathfrak{g})$, and use the aforementioned method to obtain the action of all of $\mathfrak{sl}(2)$. This is done in Section 3, where as an interesting (and perhaps unexpected) technical simplification we do not concentrate on the principal series representations PS_2^χ of $\mathfrak{sl}(2)$, but instead concentrate on making the discrete series representations DS_{k+2} of $\mathfrak{sl}(2)$ correspond to *finite-dimensional* representations of $\mathfrak{g}$. This essentially amounts to working with the compact torus $\{\left(\begin{smallmatrix} \cos\theta & -\sin\theta \\ \sin\theta & \cos\theta \end{smallmatrix}\right)\} \subset \mathrm{SL}(2,\mathbb{R})$, instead of the split torus $\{\left(\begin{smallmatrix} a & \\ & a^{-1} \end{smallmatrix}\right)\}$. The role played by the Jacquet module in the preceding motivational discussion is now played by the set of vectors annihilated by a lowering operator ε for the action of the compact torus. We specialize to the case $\mathfrak{g} = \mathfrak{gl}(3)$ and then describe a suitable approximation X' to the space X that we really want, and exhibit the desired representation of $\mathfrak{sl}(2) \times \mathfrak{g}$. This representation yields a correspondence that is still a form of the adjoint lift. Namely, it pairs off discrete series representations of $\mathfrak{sl}(2)$ coming from holomorphic modular forms of weight $k+2$ to finite-dimensional representations of $\mathfrak{g}$ with highest weight $k\rho$, where ρ is half the sum of the positive roots. On the level of groups, this pairs off discrete series representations of $\mathrm{SL}(2,\mathbb{R})$ with representations of the *compact* group $U(3)$. The resulting correspondence is given by the map of L-groups described by the following slight variation on the adjoint map from PGL(2) to GL(3):

$$(1.6) \qquad {}^{L}(\mathrm{SL}(2)) = \mathrm{PGL}(2,\mathbb{C}) \hookrightarrow \mathrm{SO}(3,\mathbb{C}) \to \mathrm{GL}(3,\mathbb{C}) \rtimes \mathbb{Z}/2\mathbb{Z} = {}^{L}(\mathrm{U}(3)).$$

In Section 4 we finally produce the "correct" form of $X_{\mathbb{R}}$, which is no longer quite an algebraic variety, but which is still a manifold. We then realize our representation by explicit unitary transformations on $L^2(X_{\mathbb{R}})$. As in our motivational section, it is easy to write down the action of $B \times \mathrm{U}(3))$, or more precisely of its identity component $B^+ \times \mathrm{U}(3))$, on $L^2(X_{\mathbb{R}})$. The formulas are more or less the ones in 1.5, and we again extend to an infinitesimal representation of $\mathfrak{sl}(2) \times \mathfrak{g}$ using the results of Section 2. We spend most of Section 4 deriving an explicit integral kernel for the operator giving the action of $\left(\begin{smallmatrix} 0 & -1 \\ 1 & 0 \end{smallmatrix}\right)$. Since $\left(\begin{smallmatrix} 0 & -1 \\ 1 & 0 \end{smallmatrix}\right)$ and B^+ generate SL(2), we have thus obtained explicit formulas for our representation. The action of $\left(\begin{smallmatrix} 0 & -1 \\ 1 & 0 \end{smallmatrix}\right)$ is similar to that of a Fourier transform; we essentially assemble the integral kernel piece by piece, by first calculating the effect of this analog of the Fourier transform on those functions analogous to the product of a Gaussian $e^{-\pi\sum x_j^2}$ with a polynomial in the $\{x_j\}$. We conclude Section 4 with some discussion of possible directions for further investigation, which can be viewed as a continuation of this introduction. As our motivation for conducting this investigation comes from automorphic forms, we touch on what might be needed in order to be able to extend our construction p-adically, and (although this seems distinctly more difficult) to be able to prove that the resulting local constructions, if they exist, fit together in a globally coherent way.

It is a pleasure to acknowledge several helpful conversations and much advice from a number of people while this work was in progress. Often the influence was indirect: a chance comment from one of these people about a related topic suddenly clarified matters at a point when I was stuck on this calculation. Of course, any errors in this paper are my own. I found a preprint [**Bl**] by and discussions with A. Bluher helpful in understanding the Fock-Bargmann model. R. Brylinski explained to me some points about geometric quantization and gave me reprints of her work (e.g., [**BrK**], and the references in it) with B. Kostant; although this paper only makes minimal reference to their work, the overall picture of their work was quite inspiring and set me off on the right track. B. Gross and I had several enlightening and encouraging discussions, and the idea of taking G to be compact grew out of them; he also suggested the candidate for the p-adic form of $X_{\mathbb{R}}$ mentioned at the end of Section 4. D. Kazhdan also gave encouraging advice, and explained a lot in his lectures on representation theory and algebraic geometry at Harvard in the spring of 1996. R. Langlands made helpful suggestions about extending a representation from B to SL(2), as did L. Mantini, who also explained some aspects of representation theory of Lie groups to me. K. Oden and I had several helpful conversations on the topic of spherical harmonics. D. Prasad made the crucial suggestion to look at the discrete series representations of SL(2) instead of the principal series. S. Rallis pointed out a mistake in my early investigations and clarified the motivating construction described in the introduction. S. Sahi explained several points about representations of Lie groups, and pointed me in the direction of [**BaSS**], which describes a different representation of GL(2) × GL(3) on the space of rank 1 matrices, which it was helpful to study. G. Savin and I discussed the material in this paper on more than one occasion, including the matching up of the actions of the centers of the universal enveloping algebras $Z(\mathfrak{sl}(2))$ and $Z(\mathfrak{g})$; his preprint [**S**] ultimately inspired a lot of the philosophy behind this paper. J. Stalker explained to me a fair amount of complex analysis. D. Vogan gave me some advice on "exponentiating" a representation of a Lie algebra to a Lie group, and also made some remarks that helped me understand the decomposition 4.13.

Finally, it is a pleasure and an honor to dedicate this paper to my Ph.D. advisor, Goro Shimura, for leading me into the subject of automorphic forms. In particular, he introduced me to the theta-correspondence between the double cover of SL(2) and O(2, 1), which, as is well-known, gives rise to Shimura's correspondence [**Sh**] between modular forms of half-integral weight and of even weight.

2. Representations of $\mathfrak{sl}(2)$

In this section we describe a way to extend representations of "part" of the Lie algebra $\mathfrak{sl}(2,\mathbb{C})$ to actual representations. Most importantly, we prove Lemma 2.2, which is crucial to this paper; here the simplicity of the expression for the Casimir element of $\mathfrak{sl}(2)$ plays an important role. To begin, we fix two bases $\{e, f, h\}$ and $\{\delta, \varepsilon, K\}$ for $\mathfrak{sl}(2)$:

$$e = \begin{pmatrix} 0 & 1 \\ 0 & 0 \end{pmatrix}, \quad f = \begin{pmatrix} 0 & 0 \\ 1 & 0 \end{pmatrix}, \quad h = \begin{pmatrix} 1 & 0 \\ 0 & -1 \end{pmatrix}, \tag{2.1}$$

$$\delta = \begin{pmatrix} 1 & i \\ i & -1 \end{pmatrix}, \quad \varepsilon = \begin{pmatrix} 1 & -i \\ -i & -1 \end{pmatrix}, \quad K = \begin{pmatrix} 0 & -1 \\ 1 & 0 \end{pmatrix}. \tag{2.2}$$

$\mathbb{R}h$ and $\mathbb{R}K$ are the split and nonsplit toral subalgebras of $\mathfrak{sl}(2,\mathbb{R})$; $\mathbb{C}h + \mathbb{C}e$ and $\mathbb{C}K + \mathbb{C}\delta$ are corresponding Borel subalgebras of $\mathfrak{sl}(2,\mathbb{C})$. The Casimir element in the universal enveloping algebra $U(\mathfrak{sl}(2))$ of $\mathfrak{sl}(2)$ is

$$C_{\mathfrak{sl}(2)} = h^2 - 2h + 4ef = -K^2 - 2iK + \delta\varepsilon. \tag{2.3}$$

DEFINITION 2.1. Let X be a manifold. We define a *formal differential operator* on X to be a $\mathcal{C}^\infty$ differential operator defined on $X - X_0$, where X_0 is a finite union of lower-dimensional submanifolds of X. (X will usually be an algebraic variety, and a formal differential operator is then one defined at the generic point of X.)

Write $\mathcal{FD}(X)$ for the ring of formal differential operators on X. A *formal representation* of a Lie algebra $\mathfrak{g}$ on X is a homomorphism of Lie algebras $r : \mathfrak{g} \to \mathcal{FD}(X)$, or equivalently a ring homomorphism $r : U(\mathfrak{g}) \to \mathcal{FD}(X)$. Typically, X will be an algebraic variety, in which case $\mathfrak{g}$ acts on the field of rational functions on X.

LEMMA 2.2. *Let $\mathfrak{b} = \mathbb{C}h + \mathbb{C}e$ be a Borel subalgebra of $\mathfrak{sl}(2)$. Let r be a formal representation of $\mathfrak{b}$ on a manifold X, and let $A \in \mathcal{FD}(X)$ be a formal differential operator. Suppose that*

1. *$r(e)$ is an invertible formal differential operator (typically, multiplication by some nonzero function on X);*

2. *A commutes with the image of* r.

Then r *can be uniquely extended to a formal representation of* $\mathfrak{sl}(2)$ *on* X *in such a way that* $r(C_{\mathfrak{sl}(2)}) = A$.

PROOF. Essentially, we must define $r(f)$. One just solves for $r(f)$ in the equation

$$r(h)^2 - 2r(h) + 4r(e)r(f) = A = r(C_{\mathfrak{sl}(2)}). \tag{2.4}$$

This involves inverting $r(e)$. One immediately checks that this works: e.g., to verify that $[r(h), r(f)] = -2r(f)$, one evaluates

$$\begin{aligned} 0 = [r(h), A] &= 4\,[r(h), r(e)]\,r(f) + 4r(e)\,[r(h), r(f)] \\ &= 8r(e)r(f) + 4r(e)\,[r(h), r(f)], \end{aligned} \tag{2.5}$$

because r is a formal representation of $\mathfrak{b}$. One can then cancel the invertible operator $r(e)$. □

REMARK 2.3. An analogous result holds if one uses K and δ instead of h and e above. Here one solves for $r(\varepsilon)$ in

$$-r(K)^2 - 2ir(K) + r(\delta)r(\varepsilon) = A = r(C_{\mathfrak{sl}(2)}). \tag{2.6}$$

EXAMPLE 2.4 (The discrete series of $\mathfrak{sl}(2)$). Given an integer $k \geq 1$, there exists a unique irreducible Harish-Chandra module DS_k with an action of $\mathfrak{sl}(2)$, with the property that DS_k contains a vector v such that $\varepsilon v = 0$ and $Kv = -ikv$. Namely,

$$\mathrm{DS}_k = \bigoplus_{l \geq 0} \mathbb{C} \cdot \delta^l v, \tag{2.7}$$

$$K(\delta^l v) = -i(k + 2l)\delta^l v, \tag{2.8}$$

$$\varepsilon(\delta^l v) = -4l(l + k - 1)\delta^{l-1} v. \tag{2.9}$$

DS_k is the type at ∞ of an automorphic representation of $\mathrm{GL}(2, \mathbb{Q})$ corresponding to a classical holomorphic modular form of weight k. (To be precise, DS_1 is really a limit of discrete series; this distinction plays no role in this paper, and will henceforth be ignored.) We shall use Lemma 2.2 to obtain a "natural" realization of DS_k.

Let X be the open interval $(0, +\infty)$, fix $\alpha > 0$ (usually $\alpha = \pi$), and consider the formal representation r_k of $\mathfrak{sl}(2)$ acting on functions $\varphi(u)$ for $u \in X$, characterized by:

$$r_k(e) = i\alpha u \qquad (\text{i.e., } r_k(e)\varphi(u) = i\alpha u \varphi(u)); \tag{2.10}$$

$$r_k(h) = 2u\frac{\partial}{\partial u} + k; \tag{2.11}$$

$$r_k(C_{\mathfrak{sl}(2)}) = k(k-2); \tag{2.12}$$

$$r_k(f) = \frac{i}{\alpha}\left(u\frac{\partial^2}{\partial u^2} + k\frac{\partial}{\partial u}\right). \tag{2.13}$$

(The last equation follows from Lemma 2.2.)

PROPOSITION 2.5. r_k *leaves stable the space of functions on* $(0, +\infty)$ *of the form* $\varphi(u) = p(u)e^{-\alpha u}$, *where* p *is a polynomial. The representation of* r_k *on this space is isomorphic to* DS_k. *In particular, the function* $e^{-\alpha u}$ *corresponds to the vector* v *above.*

PROOF. This boils down to checking that:

1. $r_k(\varepsilon)(e^{-\alpha u}) = 0$;
2. $r_k(K)(e^{-\alpha u}) = -ike^{-\alpha u}$;
3. $r_k(\delta)^l(e^{-\alpha u})$ has the form $p_l(u)e^{-\alpha u}$, where $p_l(u)$ is a polynomial with leading term $(-4\alpha u)^l$.

□

So far we have described DS_k only as a representation of the Lie algebra $\mathfrak{sl}(2)$. As is well known, DS_k comes from a unitary representation of $\mathrm{SL}(2,\mathbb{R})$. We now describe the action of $\mathrm{SL}(2,\mathbb{R})$ on the above model; this will be useful in Section 4. To do this, we recall the familiar theta-correspondence between $\mathrm{SL}(2,\mathbb{R})$ and $\mathrm{O}(2)$.

THEOREM 2.6. *Fix a parameter $\alpha > 0$. Then there exists a unique unitary representation R of $\mathrm{SL}(2,\mathbb{R}) \times \mathrm{O}(2)$ on $L^2(\mathbb{R}^2)$, given by the formulas below. To simplify notation, we view $\mathrm{O}(2)$ and $\mathrm{SL}(2,\mathbb{R})$ as subgroups of the product; e.g., $R(g)$ below really refers to $R(1 \times g)$. We also view $v \in \mathbb{R}^2$ as a row vector, and write $\langle v, w\rangle$ for the inner product of v and w.*

$$R(g)f(v) = f(vg), \qquad \text{for } g \in O(2); \tag{2.14}$$

$$R\left(\begin{pmatrix} a & b \\ 0 & a^{-1}\end{pmatrix}\right) f(v) = ae^{i\alpha ab\langle v,v\rangle} f(av); \tag{2.15}$$

$$R\left(\begin{pmatrix} 0 & -1 \\ 1 & 0\end{pmatrix}\right) f(v) = \frac{-i\alpha}{\pi}\int_{\mathbb{R}^2} f(w)e^{-2i\alpha\langle v,w\rangle}\,dw. \tag{2.16}$$

Here $\left(\begin{smallmatrix} a & b \\ 0 & a^{-1}\end{smallmatrix}\right)$ and $\left(\begin{smallmatrix} 0 & -1 \\ 1 & 0\end{smallmatrix}\right)$ are elements of $\mathrm{SL}(2,\mathbb{R})$. Under the above action, $L^2(\mathbb{R}^2)$ decomposes into the Hilbert direct sum $\widehat{\bigoplus}_{k\geq 0} \mathrm{DS}_{k+1} \otimes V_k$, where V_0 is the trivial representation of $\mathrm{O}(2)$, and V_k, for $k \geq 1$, is the irreducible two-dimensional representation of $\mathrm{O}(2)$ whose character χ_k, evaluated at $g = \left(\begin{smallmatrix} \cos\theta & -\sin\theta \\ \sin\theta & \cos\theta\end{smallmatrix}\right) \in \mathrm{O}(2)$, is $\chi_k(g) = e^{ik\theta} + e^{-ik\theta}$.

PROOF. This arises from restricting the metaplectic representation of Segal-Shale-Weil to the dual pair $\mathrm{SL}(2,\mathbb{R}) \times \mathrm{O}(2) \subset \mathrm{Sp}(4)$. See for example [**JL**], Proposition 1.3 and Lemma 5.12. Note that matrices of the form $\left(\begin{smallmatrix} a & b \\ 0 & a^{-1}\end{smallmatrix}\right)$ and $\left(\begin{smallmatrix} 0 & -1 \\ 1 & 0\end{smallmatrix}\right)$ generate $\mathrm{SL}(2,\mathbb{R})$. □

It follows from the above theorem that for $k \geq 1$, DS_k is isomorphic to the representation obtained from the action of SL(2) on the subspace

$$L^2(\mathbb{R}^2)_{k-1} = \{f \in L^2(\mathbb{R}^2) \mid f(v_1, v_2) = (v_1 + iv_2)^{k-1}\varphi(v_1^2 + v_2^2)\}. \tag{2.17}$$

Requiring f to be in L^2 is equivalent to requiring $\varphi(u) \in L^2((0,+\infty), u^{k-1}\,du)$. As a matter of fact, the representation r_k of Proposition 2.5 is exactly the infinitesimal version of the restriction of R to $L^2(\mathbb{R}^2)_{k-1}$. The theta-correspondence thus gives a way of constructing r_k without using Lemma 2.2. We nonetheless wished to illustrate the use of Lemma 2.2 on a familiar example, before embarking on the main construction of this paper.

Let us call R_k the resulting representation of SL(2) on $L^2((0,+\infty), u^{k-1}\,du)$. Here are explicit formulas for R_k, that we shall use in Section 4.

$$(2.18)\quad R_k\left(\begin{pmatrix} a & b \\ 0 & a^{-1}\end{pmatrix}\right)\varphi(u) = a^k\varphi(a^2u)e^{i\alpha abu};$$

$$(2.19)\quad R_k\left(\begin{pmatrix} 0 & -1 \\ 1 & 0\end{pmatrix}\right)\varphi(u) = (-i)^k\alpha\int_0^\infty (ul)^{-(k-1)/2}J_{k-1}(2\alpha\sqrt{lu})\varphi(l)l^{k-1}\,dl.$$

Here J_n is the usual Bessel function

$$(2.20)\qquad J_n(z) = \frac{1}{2\pi}\int_0^{2\pi} e^{iz\sin\theta}e^{-in\theta}\,d\theta = \frac{i^n}{2\pi}\int_0^{2\pi} e^{-iz\cos\theta}e^{in\theta}\,d\theta.$$

(See (4.5–5) of [**H**].) This follows from expressing the action of $\left(\begin{smallmatrix} 0 & -1 \\ 1 & 0\end{smallmatrix}\right)$, essentially a Fourier transform, in polar coordinates.

REMARK 2.7. Here is a heuristic way to understand 2.19, which may be of interest. Make the *Ansatz*

$$(2.21)\qquad R_k\left(\begin{pmatrix} 0 & -1 \\ 1 & 0\end{pmatrix}\right)\varphi(u) = \int_0^\infty K(l,u)\varphi(l)l^{k-1}\,dl,$$

for a kernel function K. Now

$$(2.22)\quad R_k\left(\begin{pmatrix} a & 0 \\ 0 & a^{-1}\end{pmatrix}\right)R_k\left(\begin{pmatrix} 0 & -1 \\ 1 & 0\end{pmatrix}\right) = R_k\left(\begin{pmatrix} 0 & -1 \\ 1 & 0\end{pmatrix}\right)R_k\left(\begin{pmatrix} a^{-1} & 0 \\ 0 & a\end{pmatrix}\right),$$

so $K(l, a^2u) = K(a^2l, u)$; thus K can depend only on the product lu: $K(l,u) = K(lu)$. Furthermore,

$$(2.23)\qquad R_k\left(\begin{pmatrix} 0 & -1 \\ 1 & 0\end{pmatrix}\right)r_k(e)R_k\left(\begin{pmatrix} 0 & -1 \\ 1 & 0\end{pmatrix}\right)^{-1} = r_k(-f);$$

hence $i\alpha lK(lu) = -i/\alpha\cdot(u\partial^2/\partial u^2 + k\partial/\partial u)K(lu)$, i.e., $zK''(z)+kK'(z)+\alpha^2K(z) = 0$. Now $z^{-(k-1)/2}J_{k-1}(2\alpha\sqrt{z})$ is one solution of this differential equation (see [**H**], (4.5–4)), and the "other" solution behaves near $z=0$ like z^{1-k} (or even $\log z$, if $k=1$); presumably the first solution is the only one small enough to be the kernel of a meaningful integral operator on $L^2((0,+\infty), u^{k-1}\,du)$.

3. Generalized Fock-Bargmann space

We recall that the Fock-Bargmann space for the dual pair SL(2) × O(2n) is the set of holomorphic functions on $\mathbb{C}^{2n}$ which are L^2 with respect to a certain measure ([**F89**], sections 1.6 and 4.2; we restrict to even-dimensional orthogonal groups in order to avoid the double cover of SL(2)). Let R be the representation of SL(2) × O(2n) on this space. We shall restrict ourselves to the subspace of SO(2) × O($2n$)-finite vectors, which is simply the space $\mathbb{C}[X] = \mathbb{C}[z_1,\dots,z_{2n}]$ of polynomials on $X = \mathbb{C}^{2n}$. Then the element $\left(\begin{smallmatrix} \cos\theta & -\sin\theta \\ \sin\theta & \cos\theta\end{smallmatrix}\right)\times g \in \mathrm{SO}(2)\times\mathrm{O}(2n)$ acts on $f\in\mathbb{C}[X]$ by

$$(3.1)\qquad R\left(\begin{pmatrix} \cos\theta & -\sin\theta \\ \sin\theta & \cos\theta\end{pmatrix}\times g\right)\varphi(v) = e^{-in\theta}\varphi(e^{-i\theta}vg),$$

where $v = (z_1,\dots,z_{2n})\in X$ is viewed as a row vector. The resulting infinitesimal representation r of $\mathfrak{so}(2)\times\mathfrak{o}(2n)$ extends to one of $\mathfrak{sl}(2)\times\mathfrak{o}(2n)$. We give the action of $\mathfrak{sl}(2)$, using the notation from 2.2:

$$r(K)\varphi = -i\left(n + \sum_{j=1}^{2n} z_j \frac{\partial}{\partial z_j}\right)\varphi, \tag{3.2}$$

$$r(\delta)\varphi = \left(\sum_{j=1}^{2n} z_j^2\right)\varphi, \tag{3.3}$$

$$r(\varepsilon)\varphi = -\left(\sum_{j=1}^{2n} \frac{\partial^2}{\partial z_j^2}\right)\varphi. \tag{3.4}$$

The following theorem is very well known.

THEOREM 3.1. *Under the action of* $\mathfrak{sl}(2) \times \mathfrak{o}(2n)$,

$$\mathbb{C}[X] \cong \bigoplus_{k\geq 0} \mathrm{DS}_{k+n} \otimes \mathcal{H}_k, \tag{3.5}$$

where $\mathcal{H}_k$ *is the representation of* $\mathrm{O}(2n)$ *on degree* k *harmonic polynomials in* $\{z_1, \dots z_{2n}\}$. *Specifically,*

$$\mathrm{DS}_{k+n} \otimes \mathcal{H}_k = \bigoplus_{l\geq 0}\left(\sum_{j=1}^{2n} z_j^2\right)^l \mathcal{H}_k. \tag{3.6}$$

Furthermore, write $Z(\mathfrak{sl}(2))$ *and* $Z(\mathfrak{o}(2n))$ *for the centers of the universal enveloping algebras* $U(\mathfrak{sl}(2))$ *and* $U(\mathfrak{o}(2n))$. *Then the images of* $r(Z(\mathfrak{sl}(2)))$ *and* $r(Z(\mathfrak{o}(2n)))$ *in* $\mathrm{End}\,\mathbb{C}[X]$ *are the same.*

In this section we present a generalization of the Fock-Bargmann model of Theorem 3.1, in which $\mathrm{O}(2n)$ is replaced by a general compact Lie group G. To this end we observe two features of the above representation r. The first is that $r(Z(\mathfrak{o}(2n)))$ is the set of polynomials in $r(C_{\mathfrak{o}(2n)})$, where $C_{\mathfrak{o}(2n)}$ is the Casimir element in $U(\mathfrak{o}(2n))$. We shall expand on this point in Lemma 3.2. The second feature to observe is that the isomorphism 3.5 boils down to asserting that every element of $\mathbb{C}[X]$ is annihilated by a sufficiently high power of $r(e)$, and that $\mathbb{C}[X]/r(\delta)\mathbb{C}[X] \cong \bigoplus_{k\geq 0} W_{k+n} \otimes \mathcal{H}_k$ as an $\mathrm{SO}(2) \times \mathrm{O}(n)$ representation. Here $W_{k+n} = \mathrm{DS}_{k+n}/\delta(\mathrm{DS}_{k+n})$ is the one-dimensional $\mathrm{SO}(2)$-representation on which $\left(\begin{smallmatrix}\cos\theta & -\sin\theta\\ \sin\theta & \cos\theta\end{smallmatrix}\right)$ acts as multiplication by $e^{-i(k+n)\theta}$. The point to note here is that $\mathbb{C}[X]/r(\delta)\mathbb{C}[X] = \mathbb{C}[X]/(\sum_{j=1}^{2n} z_j^2)\mathbb{C}[X]$ is the same as the coordinate ring $\mathbb{C}[X_0]$ of $X_0 = \{(z_1, \dots, z_{2n}) \in X \mid \sum_{j=1}^{2n} z_j^2 = 0\}$. Moreover, X_0 is the Zariski closure of the orbit of any nonzero isotropic vector in $\mathbb{C}^{2n}$ under the complex group $\mathrm{O}(2n, \mathbb{C})$. We shall come back to this point in Lemmas 3.4 and 3.5. For now, let us deal with the first observation above. Recall that $C_{\mathfrak{sl}(2)}$ is the Casimir element in $Z(\mathfrak{sl}(2))$; more generally, if $\mathfrak{g}$ is a complex reductive Lie algebra, write $C_{\mathfrak{g}}$ for the Casimir element in $Z(\mathfrak{g})$.

LEMMA 3.2. *Let* $\mathfrak{g}$ *be a complex reductive Lie algebra, with maximal torus* $\mathfrak{t}$. *Fix a set of positive roots and let* ρ *be half of their sum. Let* $0 \neq \lambda \in \mathfrak{t}^*$ *be a dominant weight with respect to the choice of positive roots, and assume that there exists an element* w *of the Weyl group* W *of* $\mathfrak{g}$, *such that*

1. *$w^2 = 1$;*
2. *$w\lambda = -\lambda$;*
3. *one of the following conditions holds (they are all equivalent, given conditions 1 and 2):*
   (a) *$\rho - w\rho$ is a multiple of λ,*
   (b) *w stabilizes the line $\{\rho + k\lambda \mid k \in \mathbb{R}\} \subset \mathfrak{t}^*$,*
   (c) *$w(\rho) = \rho - 2(\langle\rho,\lambda\rangle/\langle\lambda,\lambda\rangle)\lambda$ (since w preserves the Killing form $\langle\ ,\ \rangle$).*

Assume that r is a representation of $\mathfrak{g}$ on a vector space V that decomposes into a direct sum of finite-dimensional irreducible representations, each of which has highest weight a multiple of λ. Then $Z(\mathfrak{g})$ acts on V by polynomials in $C_{\mathfrak{g}}$, i.e., the image $r(Z(\mathfrak{g})) \subset \operatorname{End} V$ consists of the polynomials in $r(C_{\mathfrak{g}})$.

PROOF. Given $A \in Z(\mathfrak{g})$, we exhibit a polynomial $q(z)$ depending only on A and λ, but not on V, such that A and $q(C_{\mathfrak{g}})$ have the same action on V. This is an easy consequence of Harish-Chandra's theorem that $Z(\mathfrak{g})$ is isomorphic to the ring of W-invariant polynomial functions on $\mathfrak{t}^*$ (see, e.g., [**Hu**], section 23.3). Namely, A corresponds to the W-invariant polynomial function a on $\mathfrak{t}^*$ such that

$$v \in V_\mu \implies r_\mu(A)v = a(\mu + \rho)v, \tag{3.7}$$

where r_μ is a representation of $\mathfrak{g}$ with highest weight μ on the vector space V_μ (for instance, the Verma module or its irreducible quotient). Hence for all k, A acts on $V_{k\lambda}$ by the scalar $a(k\lambda + \rho)$, a polynomial function of k. Now a is W-invariant, so $a(k\lambda+\rho) = a(w(k\lambda+\rho))$, for the element $w \in W$ described above. Thus $a(k\lambda+\rho)$ is invariant under replacing k by $-k - 2\langle\rho,\lambda\rangle/\langle\lambda,\lambda\rangle$, and we obtain that

$$a(k\lambda + \rho) = q(\langle\lambda,\lambda\rangle k^2 + 2\langle\rho,\lambda\rangle k), \tag{3.8}$$

for some polynomial q. (This is analogous to saying that polynomials in z invariant under replacing z by $-z$ are themselves polynomial functions of z^2.) But $\langle\lambda,\lambda\rangle k^2 + 2\langle\rho,\lambda\rangle k$ is precisely the constant by which $C_{\mathfrak{g}}$ acts on $V_{k\lambda}$. Hence $A - q(C_{\mathfrak{g}})$ annihilates all the $V_{k\lambda}$, and hence annihilates V, as desired. □

REMARK 3.3. We shall apply Lemma 3.2 to the case $\lambda = \rho$. Then one can take w to be the longest element $w_0 \in W$. (w_0 sends the set of positive roots to the set of negative roots, so $w_0\rho = -\rho$.) Another fundamental example is $\mathfrak{g} = \mathfrak{o}(n)$ and λ such that V_λ is the standard (n-dimensional) representation of $\mathfrak{o}(n)$. We leave it to the reader to find the corresponding w.

LEMMA 3.4. *Let G be a complex reductive algebraic group, $\mathfrak{g}$ its Lie algebra, and $\lambda \in \mathfrak{t}^*$ a dominant integral weight. Let V_λ be the irreducible representation of G with highest weight λ. Let $\mathcal{O}_\lambda \subset V_\lambda$ be the orbit under G of a highest weight vector of V_λ, and call its Zariski closure $\overline{\mathcal{O}_\lambda}$. Then G acts on the coordinate ring $\mathbb{C}[\overline{\mathcal{O}_\lambda}]$ via $g\varphi(x) = \varphi(g^{-1}x)$. As a G-representation,*

$$\mathbb{C}[\overline{\mathcal{O}_\lambda}] \cong \bigoplus_{k\geq 0} V_{k\lambda}^*, \tag{3.9}$$

where $V_{k\lambda}^*$ is the representation of G dual to $V_{k\lambda}$. More precisely, $\overline{\mathcal{O}_\lambda}$ is a cone, and hence $\mathbb{C}[\overline{\mathcal{O}_\lambda}]$ inherits the structure of a graded algebra from the affine algebra $\mathbb{C}[V_\lambda]$ (which surjects onto $\mathbb{C}[\overline{\mathcal{O}_\lambda}]$). $V_{k\lambda}^*$ is then precisely the degree k part of $\mathbb{C}[\overline{\mathcal{O}_\lambda}]$.

PROOF. See Proposition 2.5 of [**BrK**] and the references cited there. Another way to view 3.9 is to consider the action of $\mathbf{G}_m \times G$ on $\overline{\mathcal{O}_\lambda}$. ($\mathbf{G}_m$ is the multiplicative group $\mathbb{C}^\times$, which acts on V_λ by scaling the vectors; this action preserves the cone $\overline{\mathcal{O}_\lambda}$.) Then under the resulting action of $\mathbf{G}_m \times G$, $\mathbb{C}[\overline{\mathcal{O}_\lambda}]$ decomposes into a direct sum of $W_k \otimes V^*_{k\lambda}$, where W_k is the one-dimensional representation of $\mathbf{G}_m$ on which $\zeta \in \mathbf{G}_m$ acts by multiplication by ζ^k. Also note that the grading on $\mathbb{C}[\overline{\mathcal{O}_\lambda}]$ arises from the action of $\mathbf{G}_m$ on $\overline{\mathcal{O}_\lambda}$. □

LEMMA 3.5. *In the situation of Lemma 3.4, suppose given an irreducible affine algebraic variety X on which $\mathbf{G}_m \times G$ acts, such that X is a cone on which $\mathbf{G}_m$ acts by dilations (so given $\zeta \in \mathbf{G}_m$ and $x \in X$, one obtains $\zeta x \in X$). Further assume that*

1. *The action of $\mathbf{G}_m$ turns the coordinate ring $\mathbb{C}[X]$ into a graded algebra in degrees ≥ 0: i.e., $\mathbb{C}[X] = \bigoplus_{k\geq 0} \mathbb{C}[X]_k$, where*

$$\text{for } \varphi_k \in \mathbb{C}[X]_k,\ x \in X, \text{ and } \zeta \in \mathbf{G}_m, \qquad \varphi_k(\zeta x) = \zeta^k \varphi(x). \tag{3.10}$$

2. *There exists $d > 0$ and a degree d G-invariant function $f \in \mathbb{C}[X]_d$ such that the zero set $X_0 = \{x \in X \mid f(x) = 0\}$ of f is isomorphic to $\overline{\mathcal{O}_\lambda}$, as a variety with an action of $\mathbf{G}_m \times G$.*

Then, as a representation of G,

$$\mathbb{C}[X]_k \cong \bigoplus_{0\leq l\leq k/d} V^*_{(k-ld)\lambda}. \tag{3.11}$$

Equivalently, let $W_k \subset \mathbb{C}[X]_k$ be the unique subspace isomorphic to $V^*_{k\lambda}$. Then

$$\mathbb{C}[X] = \bigoplus_{k,l\geq 0} f^l W_k, \qquad f^l W_k \cong V^*_{k\lambda}, \qquad f^l W_k \subset \mathbb{C}[X]_{k+ld}. \tag{3.12}$$

PROOF. Observe that $\mathbb{C}[X_0]$ is $\mathbb{C}[X]/f\mathbb{C}[X]$. By Lemma 3.4, its degree k part, $\mathbb{C}[X_0]_k$, is isomorphic to $V^*_{k\lambda}$. On the other hand, we have the short exact sequence of G-modules

$$0 \longrightarrow \mathbb{C}[X]_{k-d} \xrightarrow{\cdot f} \mathbb{C}[X]_k \longrightarrow \mathbb{C}[X_0]_k \longrightarrow 0, \qquad \mathbb{C}[X_0]_k \cong V^*_{k\lambda} \tag{3.13}$$

where $\mathbb{C}[X]_{k-d}$ is taken to be 0 if $k - d < 0$. The above sequence splits, as G is reductive, and we obtain 3.11 by induction. (We used the assumption that X was irreducible to conclude that multiplication by f was injective.) □

COROLLARY 3.6. *In the situation of Lemma 3.5, assume that λ satisfies the conditions of Lemma 3.2. Consider the representation of $\mathfrak{g}$ on $\mathbb{C}[X]$. Then $Z(\mathfrak{g})$ acts on $\mathbb{C}[X]$ by polynomials in $C_\mathfrak{g}$.*

PROOF. This is essentially immediate from Lemmas 3.2 and 3.5, but one must pay attention to the distinction between $V_{k\lambda}$ and $V^*_{k\lambda}$. Namely, let w_0 be the longest element in W. Then $V^*_{k\lambda} \cong V_{k\lambda'}$, where $\lambda' = w_0(-\lambda)$. λ' then also satisfies the conditions of Lemma 3.2, using $w_0 w w_0^{-1}$. Incidentally, note that $\langle \lambda', \lambda' \rangle = \langle \lambda, \lambda \rangle$, and hence $C_\mathfrak{g}$ acts by the same scalar on $V_{k\lambda}$ and $V^*_{k\lambda}$. □

THEOREM 3.7 (Generalized Fock-Bargmann model). *Let X be a variety as in Lemma 3.5, and λ be as in Lemma 3.2. Assume furthermore that the G-invariant function $f \in \mathbb{C}[X]$ has degree $d = 2$, and that $k_0 = 1 + \langle \rho, \lambda \rangle / \langle \lambda, \lambda \rangle$ is an integer. Then there exists a representation r of $\mathfrak{sl}(2) \times \mathfrak{g}$ on $\mathbb{C}[X]$, such that*

1. *$r(\delta)\varphi = f\varphi$, for $\varphi \in \mathbb{C}[X]$.*

2. $r(K)\varphi = -i(k_0 + E)\varphi$, *where* $E\varphi(x) = \frac{\partial}{\partial\zeta}\Big|_{\zeta=1} \varphi(\zeta x)$. *(E is usually called the Euler operator on X.)*
3. $\mathfrak{g}$ *acts by the infinitesimal action arising from the action of G.*
4. $\mathbb{C}[X] \cong \bigoplus_{k\geq 0} \mathrm{DS}_{k+k_0} \otimes V^*_{k\lambda}$.

Moreover, the images $r(Z(\mathfrak{sl}(2)))$ *and* $r(Z(\mathfrak{g}))$ *of the centers of the universal enveloping algebras in* $\operatorname{End}\mathbb{C}[X]$ *coincide. In other words,* $Z(\mathfrak{g})$ *acts by polynomials in* $C_{\mathfrak{sl}(2)} \in Z(\mathfrak{sl}(2))$.

PROOF. $C_{\mathfrak{g}}$ acts on $V^*_{k\lambda}$ by the scalar $\langle\lambda,\lambda\rangle k^2+2\langle\rho,\lambda\rangle k = \langle\lambda,\lambda\rangle(k+k_0)(k+k_0-2)-\langle\lambda,\lambda\rangle k_0(k_0-2)$. On the other hand, $C_{\mathfrak{sl}(2)}$ acts on DS_{k+k_0} by $(k+k_0)(k+k_0-2)$. Thus the desired representation r must satisfy

5. $r(C_{\mathfrak{sl}(2)}) = \frac{1}{\langle\lambda,\lambda\rangle} r(C_{\mathfrak{g}}) + k_0(k_0-2)$.

We now build up the representation r in stages. First note that requirements 1 and 2 define a representation of the Borel subalgebra $\mathfrak{b} = \mathbb{C}K + \mathbb{C}\delta$ of $\mathfrak{sl}(2)$ on $\mathbb{C}[X]$; this is where we need f to have degree 2. Next define the action of $\mathfrak{g}$ by requirement 3, and note that the actions of $\mathfrak{b}$ and of $\mathfrak{g}$ commute (since the actions of $\mathfrak{b}$ and G commute). The action of $\mathfrak{g}$ tells us how the operator $r(C_{\mathfrak{g}})$ acts, and hence by requirement 5 we obtain the action of $C_{\mathfrak{sl}(2)}$. Note that $r(C_{\mathfrak{sl}(2)})$ therefore commutes with the action of $\mathfrak{b}$, since $r(C_{\mathfrak{sl}(2)})$ can be expressed in terms of $\mathfrak{g}$. Of course, $r(C_{\mathfrak{sl}(2)})$ also commutes with the action of $\mathfrak{g}$, because $r(C_{\mathfrak{g}})$ does.

So far we have obtained a representation of $\mathfrak{b}\times\mathfrak{g}$ on $\mathbb{C}[X]$, and have an operator $r(C_{\mathfrak{sl}(2)})$ that commutes with the action of $\mathfrak{b}$ and $\mathfrak{g}$. We now use Lemma 2.2 to extend the action of $\mathfrak{b}$ to a formal representation of all of $\mathfrak{sl}(2)$. By the proof of that lemma, the resulting formal action of $\mathfrak{sl}(2)$ will commute with the existing action of $\mathfrak{g}$. By Corollary 3.6, $Z(\mathfrak{g})$ acts on $\mathbb{C}[X]$ by polynomials in $C_{\mathfrak{g}}$, and this action can be reexpressed in terms of $C_{\mathfrak{sl}(2)}$, using requirement 5. We now show that requirement 4 holds, and with it, the fact that r is a genuine (and not just formal) representation of $\mathfrak{sl}(2)$; specifically, we must check that $r(\varepsilon)$ sends $\mathbb{C}[X]$ to itself, instead of introducing a denominator of f. Let $W_k \subset \mathbb{C}[X]_k$ be the subspace isomorphic to $V^*_{k\lambda}$, as guaranteed by Lemma 3.5. Then $\bigoplus_{l\geq 0} f^l W_k$ is the $V^*_{k\lambda}$-isotypic component of $\mathbb{C}[X]$ under the action of G. Let us check that $r(\varepsilon)$ preserves this isotypic component. We have concocted r so that $C_{\mathfrak{sl}(2)}$ acts on this isotypic component by the scalar $(k+k_0)(k+k_0-2)$. We also see explicitly that if $\varphi \in W_k$, then the action of K on a typical element $f^l\varphi$ in this isotypic component is

$$r(K)(f^l\varphi) = -i(k_0+k+2l)f^l\varphi. \tag{3.14}$$

Lastly, $r(\delta)$ is multiplication by f. Putting this together and using 2.6 from the proof of Lemma 2.2, we obtain that

$$r(\varepsilon)(f^l\varphi) = -4l(l+k+k_0-1)f^{l-1}\varphi \begin{cases} \in f^{l-1}W_k & \text{if } l \geq 1, \\ = 0 & \text{if } l = 0. \end{cases} \tag{3.15}$$

It follows immediately that the isotypic component, $\bigoplus_{l\geq 0} f^l W_k$, is stable under $\mathfrak{sl}(2)\times\mathfrak{g}$. By comparison with 2.7–2.9, this isotypic component is isomorphic to $\mathrm{DS}_{k+k_0}\otimes W_k$, as desired. □

REMARK 3.8. The hypothesis that k_0 be an integer merely serves to ensure that the discrete series representations DS_{k+k_0} come from genuine representations of SL(2) and not, say, of its double cover.

REMARK 3.9. One might say that the above representation was one of $\mathfrak{sl}(2)\times G$, since the action of $\mathfrak{g}$ comes from one of the group G. The notation $\mathfrak{sl}(2)\times G$ is perhaps illogical, since it refers neither to a Lie algebra nor to a Lie group. Also note that $\mathbb{C}[X]$ is an explicit realization of the Harish-Chandra module of the unitary representation of $\mathrm{SL}(2,\mathbb{R})\times G_c$ on $\widehat{\bigoplus}_{k\geq 0}\mathrm{DS}_{k+k_0}\otimes V_{k\lambda}^*$, when G_c is the compact real form of G.

We give three examples of the results of this section.

EXAMPLE 3.10. This is essentially Theorem 3.1. Let $G=\mathrm{O}(2n)$, let V_λ be its standard $2n$-dimensional representation, and let $X=V_\lambda$ with the obvious action of $\mathbf{G}_m\times G$. Not surprisingly, we take $f\in\mathbb{C}[X]_2=\mathbb{C}[V_\lambda]_2$ to be the invariant quadratic form. Then the zero set of f is the set of isotropic vectors, and this is precisely $\overline{\mathcal{O}_\lambda}$. Then, as we have seen, $\mathbb{C}[X]$ decomposes under G into $\bigoplus_{k,l\geq 0}f^l\mathcal{H}_k$, where $\mathcal{H}_k\cong V_{k\lambda}^*\cong V_{k\lambda}$. The action of $\mathbf{G}_m\times G$ on $\mathbb{C}[X]$ extends to the aforementioned action of $\mathfrak{sl}(2)\times G$ and yields 3.5. We have therefore reconstructed a proof of Theorem 3.1.

EXAMPLE 3.11. This example works for arbitrary G, and seems rather naive: one just takes the space X to be the product of $\overline{\mathcal{O}_\lambda}$ with an affine line, so $\mathbb{C}[X]=\mathbb{C}[\overline{\mathcal{O}_\lambda}][t]$. Here G acts on the first factor alone, and $\mathbf{G}_m$ acts on both factors by acting on $\overline{\mathcal{O}_\lambda}$ in the usual way, and on the affine line by dilations (i.e., in the resulting grading on $\mathbb{C}[X]$, t has degree 1.) Then one can take the G-invariant function f to be t. Then so long as λ is suitable (e.g., if $\lambda=\rho$), as mentioned in Remark 3.3), then one can apply Lemmas 3.4 and 3.5. One can modify this slightly to have $\zeta\in\mathbf{G}_m$ act on the affine line by dilating it by ζ^2; then t has degree 2 and one can conclude the result of Theorem 3.7. In the case of $G=\mathrm{O}(2n)$ and V_λ the standard representation, it would be interesting to compare the explicit form of the representation of $\mathfrak{sl}(2)\times G$ from Theorem 3.1 with the one obtained in this setting.

EXAMPLE 3.12. This example concerns us most in this paper. Let $G=\mathrm{GL}(3)$, and let $\lambda=\rho$. Then $V_\lambda=V_\rho$ happens to be isomorphic to the 8-dimensional adjoint representation of G on $\mathfrak{sl}(3)$, and is self-dual via the Killing form. Explicitly, $\mathcal{O}_\lambda$ is the set of 3×3 complex matrices of trace 0 and rank 1, with G acting by conjugating the matrices, and $\mathbf{G}_m$ acting by scaling them. $\overline{\mathcal{O}_\lambda}$ is just $\mathcal{O}_\lambda\cup\{0\}$, and so consists of matrices of trace 0 and rank at most 1. Instead of picking X as in Theorem 3.7, we take a substitute X' to be the space of 3×3 complex matrices of rank at most 1, but with arbitrary trace. Then $\mathbb{C}[X']=\mathbb{C}[\{x_{ij}\mid 1\leq i,j\leq 3\}]/I$, where I is the ideal generated by the 2×2 minors of the matrix $(x_{ij})_{1\leq i,j\leq 3}$. With this choice of X' we can take $t=x_{11}+x_{22}+x_{33}$, the trace function, to be our G-invariant function; it then has degree 1. We then obtain the following theorem.

THEOREM 3.13. *Let $G=\mathrm{GL}(3)$, let X' be the above space, and let $t\in\mathbb{C}[X']$ be the trace function. Then $Z(\mathfrak{g})$ acts on $\mathbb{C}[X']$ by polynomials in $C_\mathfrak{g}$. Furthermore, $\mathbb{C}[X']=\bigoplus_{k,l\geq 0}t^lW_k$, where $W_k\subset\mathbb{C}[X']_k$ is the subspace isomorphic to $V_{k\rho}$.*

REMARK 3.14. There is no need to distinguish between $V_{k\rho}$ and its dual, as they are isomorphic by the observation in the proof of Corollary 3.6. Note that $\dim V_{k\rho}=(k+1)^3$ by the Weyl character formula, and that these representations are all orthogonal. For other G (e.g., $\mathrm{SL}(6)$), V_ρ may well be a symplectic representation of G.

REMARK 3.15. $Z(\mathfrak{g}) = Z(\mathfrak{gl}(3))$ is generated by elements in degrees 1, 2, and 3. The degree 1 element is the infinitesimal action of the center of G; it acts by 0 on $\mathbb{C}[X']$, since $V_{k\rho}$ is actually a representation of PGL(3). The degree 2 element is $C_{\mathfrak{g}}$. The degree 3 element can be written as the symmetrization of the determinant polynomial in the basis elements of $\mathfrak{gl}(3)$; it acts on $V_{k\rho}$ by 0.

The result of Theorem 3.13 does not immediately lead to a representation of $\mathfrak{sl}(2)\times\mathfrak{g}$. Indeed, our generalization of Fock-Bargmann space requires a G-invariant function of degree 2; so we try $f = t^2$, i.e., $f(x) = \operatorname{tr} x^2 = (\operatorname{tr} x)^2$ (this last equality follows from the fact that x has rank at most 1). Unfortunately, the zero set of $f = t^2$ is then no longer reduced. We nonetheless proceed to salvage something.

THEOREM 3.16. *Let $G = \mathrm{GL}(3)$ and let X', t, and $f = t^2$ be as above. Let $\lambda = \rho$ and $k_0 = 2$. Then there exists a formal representation r of $\mathfrak{sl}(2) \times \mathfrak{g}$ (or even of $\mathfrak{sl}(2) \times G$, in the notation of Remark 3.9) on $\mathbb{C}[X']$, satisfying requirements 1–3 and 5 of Theorem 3.7, such that furthermore*

1. *$Z(\mathfrak{sl}(2))$ and $Z(\mathfrak{g})$ have the same image in the ring of formal differential operators $\mathcal{FD}(X')$.*
2. *$\mathfrak{sl}(2) \times \mathfrak{g}$ genuinely operates on the ring $\mathbb{C}[X'][t^{-1}]$ (the coordinate ring on $X' - \overline{\mathcal{O}_\lambda}$).*
3. *Let W_k be as in Lemma 3.5. Define the subspace (not subring) $V \subset \mathbb{C}[X']$ by*

$$V = \bigoplus_{k,l\geq 0} f^l W_k = \bigoplus_{k,l\geq 0} t^{2l} W_k. \tag{3.16}$$

Then V is stable under $\mathfrak{sl}(2) \times \mathfrak{g}$ and decomposes into $V \cong \bigoplus_{k\geq 0} \mathrm{DS}_{k+2} \otimes V_{k\rho}$. Furthermore, we can view V as the Harish-Chandra module of a representation of $\mathrm{SL}(2) \times \mathrm{GL}(3)$, as in Remark 3.9.

PROOF. This is analogous to the proof of Theorem 3.7. We simply find $r(\varepsilon)$ by matching up Casimir operators as usual; this involves dividing by f at some point. □

REMARK 3.17. The reader may be interested in an explicit description of r. To this end, let us introduce coordinates on X'. There is a surjective map $c : \mathbb{C}^3 \times \mathbb{C}^3 \to X'$, namely $c(\vec{x}, \vec{\xi}) = \vec{x}\,{}^t\vec{\xi}$:

$$c(x_1, x_2, x_3, \xi_1, \xi_2, \xi_3) = \begin{pmatrix} x_1 \\ x_2 \\ x_3 \end{pmatrix} \begin{pmatrix} \xi_1 & \xi_2 & \xi_3 \end{pmatrix}. \tag{3.17}$$

This allows us to view $\varphi \in \mathbb{C}[X']$ as a function on $\mathbb{C}^3 \times \mathbb{C}^3$ such that for all $\lambda \in \mathbb{C}^\times$, $\varphi(\vec{x}, \lambda\vec{\xi}) = \varphi(\lambda\vec{x}, \vec{\xi})$. We shall describe operators on X' by giving operators in the variables x_i and ξ_i that preserve functions of the above type. We then have that

$$r(A)\varphi(\vec{x}, \vec{\xi}) = \frac{\partial}{\partial\zeta}\Big|_{\zeta=0} \varphi(\exp(-\zeta A)\vec{x}, \exp(\zeta\,{}^t\!A)\vec{\xi}), \qquad (A \in \mathfrak{g}); \tag{3.18}$$

$$r(K)\varphi = -i\left(2 + \sum_{j=1}^{3} x_j \frac{\partial}{\partial x_j}\right)\varphi = -i\left(2 + \sum_{j=1}^{3} \xi_j \frac{\partial}{\partial \xi_j}\right)\varphi; \tag{3.19}$$

$$r(\delta)\varphi = (x_1\xi_1 + x_2\xi_2 + x_3\xi_3)^2\varphi = t^2\varphi; \tag{3.20}$$

$$(3.21) \qquad r(\varepsilon)\varphi = -(x_1\xi_1 + x_2\xi_2 + x_3\xi_3)^{-1}\left(\frac{\partial^2}{\partial x_1 \partial \xi_1} + \frac{\partial^2}{\partial x_2 \partial \xi_2} + \frac{\partial^2}{\partial x_3 \partial \xi_3}\right)\varphi.$$

Note that the expression for $r(\varepsilon)$ involves a denominator of only t, not of t^2, even though we had to invert the operator $r(\delta)$ to obtain $r(\varepsilon)$ in the proof of Lemma 2.2. Equation 3.18 is the result of differentiating the action of G on functions on X', given by $R(g)\varphi(\vec{x},\vec{\xi}) = \varphi(g^{-1}\vec{x}, {}^tg\vec{\xi})$; viewing elements of X' as matrices, this amounts to the same as 4.5 below.

As noted above, one must select the subspace V, which is "half" of $\mathbb{C}[X']$, to serve as the space on which r is a genuine representation. Heuristically one may want to find a space X which is "half" of X', such that V is somehow the space of functions on X. This will come out more cleanly in the following section.

4. The Schrödinger model for SL(2) × U(3)

We recall that the Fock-Bargmann space for SL(2) × O(2n) does not conveniently generalize to the p-adics. However, the Schrödinger model of that representation does so quite satisfactorily, as originally shown by Weil [**W**]. Over $\mathbb{R}$, this model is explicitly realized as an action of $\mathrm{SL}(2,\mathbb{R})\times \mathrm{O(2n)}$ on $L^2(\mathbb{R}^{2n})$. For general n, the formulas are essentially identical to those given in Theorem 2.6, and make sense p-adically to give a representation of $\mathrm{SL}(2,\mathbb{Q}_p)\times \mathrm{O}(2n,\mathbb{Q}_p)$ on $L^2(\mathbb{Q}_p^{2n})$. (The formulas involve the function $e(x) = e^{2i\alpha x}$, usually with $\alpha = \pi$; this gets replaced by an additive character $e_p : \mathbb{Q}_p \to \mathbb{C}^\times$.) Our aim in this section is to construct an analog of the Schrödinger model for the representation of $\mathrm{SL}(2,\mathbb{R}) \times \mathrm{U}(3)$ described in Theorem 3.16, such that this Schrödinger-like model has some chance of eventually being generalized to the p-adics. At the moment, though, we can only produce the representation over $\mathbb{R}$ and speculate on possible further progress. The shift from $G = \mathrm{GL}(3)$ to the *compact* form $G_{\mathbb{R}} = \mathrm{U}(3)$ allows us to write down many explicit formulas that are analogous to classical calculations with spherical harmonics and representations of $\mathrm{O}(2n)$. We note that our representation is unitary, as in the case of the classical Schrödinger model. Our results are contained in Theorems 4.2 and 4.9.

Our starting point is an infinitesimal realization of a Schrödinger model as a unitary representation of $\mathfrak{sl}(2)\times\mathfrak{u}(3)$ on the space of smooth vectors in a certain L^2 space. We shall then explain how to (indirectly) exponentiate this representation up to the groups SL(2) × U(3). This essentially boils down to computing one operator analogous to the Fourier transform in the case of SL(2) × O(2n). We first describe the space of functions on which our representation acts. By analogy with the Fock-Bargmann model, we try a space of functions on a variety $X'_{\mathbb{R}}$ defined over $\mathbb{R}$, with an action of $G_{\mathbb{R}} = \mathrm{U}(3)$, such that the complexification X' of $X'_{\mathbb{R}}$ is the space in Example 3.12. Rather, we only want "half" of $X'_{\mathbb{R}}$.

DEFINITION 4.1. From now on we shall drop the subscript $\mathbb{R}$ from the group G, which now denotes the group $\mathrm{U}(3) = \{g \in \mathrm{GL}(3,\mathbb{C}) \mid {}^t\overline{g} = g^{-1}\}$. We shall also define $X'_{\mathbb{R}}$ to be the space of Hermitian matrices of rank at most 1:

$$(4.1) \qquad X'_{\mathbb{R}} = \{3\times 3 \text{ complex matrices } M \text{ with } \overline{{}^tM} = M \text{ and } \operatorname{rank} M \le 1\}.$$

(We could have scaled by i and had $X'_{\mathbb{R}}$ be the rank one antihermitian matrices instead, if we had wished to emphasize the parallelism between the set of rank 1 elements of $\mathfrak{gl}(3)$ from the previous section, and the set of rank 1 elements of $\mathfrak{u}(3)$ in this section. Our present choice of $X'_{\mathbb{R}}$ simplifies the notation somewhat.) We also define $X_{\mathbb{R}}$ to be the following half of $X'_{\mathbb{R}}$ (note that the trace of a Hermitian matrix is a real number):

$$(4.2)\qquad \begin{aligned} X_{\mathbb{R}} &= \{M \in X'_{\mathbb{R}} \mid \operatorname{tr} M \geq 0\} \\ &= \{M \text{ of the form } \begin{pmatrix} \overline{z}_1 \\ \overline{z}_2 \\ \overline{z}_3 \end{pmatrix} \begin{pmatrix} z_1 & z_2 & z_3 \end{pmatrix}\} \cong \mathbb{C}^3/\mathbf{T}. \end{aligned}$$

Here $\mathbf{T} = \{\zeta \in \mathbb{C}^\times \mid |\zeta| = 1\}$ is the group of complex numbers of norm one.

We define the action of G on $X'_{\mathbb{R}}$ and $X_{\mathbb{R}}$ by conjugation as before; again, it commutes with the action of $\mathbf{G}_m = \mathbb{R}^\times$ by scaling. We again take the G-invariant quadratic function $f(M) = \operatorname{tr} M^2 = (\operatorname{tr} M)^2$. Note that f is now positive definite (being $(|z_1|^2 + |z_2|^2 + |z_3|^2)^2$) and "anisotropic," in analogy to the case of the compact group $\mathrm{O}(2n)$ and its invariant quadratic form $\sum x_j^2$ on $\mathbb{R}^{2n}$. We define a G-invariant measure dM on $X_{\mathbb{R}}$ by writing $z_j = x_j + y_j$ and defining

$$(4.3)\qquad \begin{aligned} &\int_{X_{\mathbb{R}}} F(M)\, dM \\ &= \int_{\mathbb{C}^3} F\left(\begin{pmatrix} \overline{z}_1 \\ \overline{z}_2 \\ \overline{z}_3 \end{pmatrix} \begin{pmatrix} z_1 & z_2 & z_3 \end{pmatrix}\right) (|z_1|^2 + |z_2|^2 + |z_3|^2)\, dx_1 dy_1 dx_2 dy_2 dx_3 dy_3. \end{aligned}$$

We then obtain a unitary representation R of $B^+ \times G$ on $L^2(X_{\mathbb{R}}, dM)$, where $B^+ = \{\left(\begin{smallmatrix} a & b \\ 0 & a^{-1} \end{smallmatrix}\right) \mid a > 0\}$ is the connected component of the identity in the Borel subgroup of $\mathrm{SL}(2, \mathbb{R})$:

$$(4.4)\qquad R\left(\begin{pmatrix} a & b \\ 0 & a^{-1} \end{pmatrix}\right) F(M) = e^{\pi i a b \operatorname{tr} M^2} a^2 F(aM), \qquad \text{for } \begin{pmatrix} a & b \\ 0 & a^{-1} \end{pmatrix} \in B^+;$$

$$(4.5)\qquad R(g)F(M) = F(g^{-1}Mg), \qquad \text{for } g \in G.$$

We aim to extend R to all of $\mathrm{SL}(2, \mathbb{R}) \times G$. To this end, we begin by finding an "infinitesimal extension" to a representation r of the Lie algebras $\mathfrak{sl}(2) \times \mathfrak{g}$.

As a first step, R gives rise to a representation of the Lie algebras $\mathfrak{b} \times \mathfrak{g}$, which we extend to a formal representation r of $\mathfrak{sl}(2) \times \mathfrak{g}$ by matching up Casimir operators as in 5 of Theorem 3.7. This time, though, we work with respect to the split torus of $\mathrm{SL}(2)$ (which is infinitesimally generated by h), as opposed to the compact torus (infinitesimally generated by K). When convenient, we use the "coordinates" (z_1, z_2, z_3) from 4.2. This allows us to view a function $F(M)$ on $X_{\mathbb{R}}$ as a function $F(z_1, z_2, z_3)$ on $\mathbb{C}^3/\mathbf{T}$; i.e., for $\zeta \in \mathbf{T}$, $F(\zeta z_1, \zeta z_2, \zeta z_3) = F(z_1, z_2, z_3)$. Let us give the action of $\mathfrak{sl}(2)$. The formulas are similar to those of the Fock-Bargmann model in the previous section.

$$(4.6)\qquad r(e)F(M) = \pi i (\operatorname{tr} M^2) F(M);$$

$$(4.7)\qquad r(h)\,F = \left(2 + \sum_{j=1}^{3} z_j \frac{\partial}{\partial z_j}\right) F = \left(2 + \sum_{j=1}^{3} \overline{z}_j \frac{\partial}{\partial \overline{z}_j}\right) F;$$

$$(4.8)\qquad r(f)F = \frac{i}{4\pi \operatorname{tr} M} \sum_{j=1}^{3} \frac{\partial^2 F}{\partial z_j \partial \overline{z}_j} = \frac{i}{4\pi(|z_1|^2 + |z_2|^2 + |z_3|^2)} \sum_{j=1}^{3} \frac{\partial^2 F}{\partial z_j \partial \overline{z}_j}.$$

THEOREM 4.2. *The above formal representation on $L^2(X_{\mathbb{R}})$ arises from a genuine unitary representation of* $\mathrm{SL}(2,\mathbb{R}) \times G$, *which decomposes into*

$$L^2(X_{\mathbb{R}}) \cong \widehat{\bigoplus_{k\geq 0}} \mathrm{DS}_{k+2} \otimes V_{k\rho}. \tag{4.9}$$

Here $V_{k\rho}$ is, as before, the $(k+1)^3$-dimensional representation of $G = \mathrm{U}(3)$, with highest weight $k\rho$. By a slight abuse of notation we have written DS_{k+2} to mean the unitary discrete series representation of $\mathrm{SL}(2,\mathbb{R})$, *instead of its Harish-Chandra module. The decomposition 4.9 implies that the images $r(Z(\mathfrak{sl}(2)))$ and $r(Z(\mathfrak{g}))$ coincide on the Harish-Chandra module of $L^2(X_{\mathbb{R}})$, as in Theorem 3.7.*

PROOF. Ignoring the zero matrix which does not affect $L^2(X_{\mathbb{R}})$, we have the "polar decomposition"

$$X_{\mathbb{R}} - \{0\} \cong S \times (0, +\infty), \tag{4.10}$$

where

$$S = \{M \in X_{\mathbb{R}} \mid \operatorname{tr} M = 1\}, \tag{4.11}$$

and we identify $(t, M) \in (0, +\infty) \times S$ with the matrix $tM \in X_{\mathbb{R}}$. Note that S is isomorphic to the complex projective plane, because in (z_1, z_2, z_3) coordinates,

$$S = \left\{ (z_1, z_2, z_3) \in \mathbb{C}^3 \,\middle|\, \sum_{j=1}^{3} |z_j|^2 = 1 \right\} / \mathbf{T}. \tag{4.12}$$

Moreover, $S \cong G/G_0$, where $G_0 \cong \mathrm{U}(2) \times \mathrm{U}(1)$ is the subgroup of $\mathrm{U}(3)$ stabilizing the line through $(0,0,1)$. (In the classical case, if G were $\mathrm{O}(2n)$, this would be analogous to the polar decomposition $\mathbb{R}^{2n} - \{0\} \cong S^{2n-1} \times (0, +\infty)$, where $S^{2n-1} \cong \mathrm{O}(2n)/\mathrm{O}(2n-1)$ is the unit sphere in $\mathbb{R}^{2n}$.)

The action of G on $X_{\mathbb{R}}$ is entirely on the factor S, which carries a unique G-invariant measure up to scaling. Moreover, $L^2(X_{\mathbb{R}}) \cong L^2((0, +\infty), t^3\, dt) \otimes L^2(S)$. Now $L^2(S)$ decomposes orthogonally under the action of G into

$$L^2(S) \cong \widehat{\bigoplus_{k\geq 0}} V_{k\rho}. \tag{4.13}$$

(This is analogous to the classical decomposition $L^2(S^{2n-1}) \cong \widehat{\bigoplus}_{k\geq 0} \mathcal{H}_k$ under the action of $\mathrm{O}(2n)$.) 4.13 holds because $L^2(G/G_0)$ decomposes by the Peter-Weyl theorem into the sum of those representations of G which have a G_0-fixed covector, each V_λ appearing with multiplicity the dimension of $(V_\lambda^*)^{G_0}$. But G_0 is a spherical subgroup of G (equivalently, (G, G_0) is a Gelfand pair), and the V_λ's involved are precisely the $V_{k\rho}$; see for instance [**K**], Tabelle I, line 2 with $n = 2$ and $m = 1$. In particular, $V_{k\rho} \cong V_{k\rho}^*$ has up to scaling precisely one G_0-fixed vector, as can also be seen from the character of $V_{k\rho}$. Thus we obtain that the $V_{k\rho}$-isotypic subspace of $L^2(X_{\mathbb{R}})$ is $L^2((0, +\infty), t^3\, dt) \otimes V_{k\rho}$. Call this subspace E_k; it is stable under $\mathrm{SL}(2) \times G$. We claim that $E_k \cong \mathrm{DS}_{k+2} \otimes V_{k\rho}$. This will imply both the decomposition 4.9 and the fact that our representation is unitary.

Let us first investigate the formal action r of $\mathfrak{sl}(2)$ on this subspace. By virtue of our having matched up the action of the Casimir operators $C_{\mathfrak{sl}(2)}$ and $C_{\mathfrak{g}}$, we know that $C_{\mathfrak{sl}(2)}$ acts on E_k by the scalar $(k+2)k$. Now in "polar coordinates" (t, σ) on $X_{\mathbb{R}}$ (where $t \in (0, +\infty)$ and $\sigma \in S$), $r(e)$ acts by multiplication by $\pi i t^2$, and $r(h)$ by the operator $2 + t\,\partial/\partial t$. Before computing $r(f)$ explicitly on E_k, let us change coordinates to make this look more like the model for DS_{k+2} described in 2.10 through 2.13 and Proposition 2.5. We can explicitly view $V_{k\rho}$ as the space spanned by the G-translates of the function $(z_1\bar{z}_3)^k$, viewed as a function on S. Let $p_S(\sigma)$ be a linear combination of such translates, where $\sigma \in S$. p_S extends to a polynomial function $p(M)$ on $X_{\mathbb{R}}$ (namely to a linear combination of translates of the function $(z_1\bar{z}_3)^k$, viewed as a function on $X_{\mathbb{R}}$; this is a polynomial in the entries of M). Specifically, $p(t\sigma) = t^k p_S(\sigma)$, where $t \in (0, +\infty)$ and $\sigma \in S$. Hence we can view E_k as the space of functions of the form $F(M) = p(M)\varphi(\operatorname{tr} M^2)$, or in "polar coordinates" $(t, \sigma) \in (0, +\infty) \times S$, $F(t\sigma) = p_S(\sigma) \cdot t^k \varphi(t^2)$. So our change of coordinates amounts to taking functions φ such that $t^k\varphi(t^2) \in L^2((0, +\infty), t^3\,dt)$. In other words, writing $u = t^2 \in (0, +\infty)$, we are using functions $\varphi(u) \in L^2((0, +\infty), u^{k+1}\,du)$. The upshot of all this is that we obtain

$$E_k = L^2((0, +\infty), u^{k+1}\,du) \otimes V_{k\rho} \tag{4.14}$$

We are now considering $V_{k\rho}$ explicitly as a space of polynomial functions $p(M)$ on all of $X_{\mathbb{R}}$ (not just on S), of degree k in the entries of M. We immediately obtain the effect of $\mathfrak{sl}(2)$ on a function of the form $p(M)\varphi(u)$, where $p \in V_{k\rho} \subset L^2(X_{\mathbb{R}})$, and $u = \operatorname{tr} M^2$. We obtain the following formulas (the action of $\mathfrak{sl}(2)$ only affects the function φ):

$$r(e)p(M)\varphi(u) = p(M) \cdot \pi i u\varphi(u); \tag{4.15}$$

$$r(h)p(M)\varphi(u) = p(M) \cdot \left(k + 2 + 2u\frac{\partial}{\partial u}\right)\varphi(u); \tag{4.16}$$

$$r(f)p(M)\varphi(u) = p(M) \cdot \frac{i}{\pi}\left(u\frac{\partial^2}{\partial u^2} + (k+2)\frac{\partial}{\partial u}\right)\varphi(u). \tag{4.17}$$

Here we can deduce 4.17 from 4.8 directly, or from knowing that $C_{\mathfrak{sl}(2)}$ acts by $(k+2)k$, and using Lemma 2.2.

Comparing 4.15–4.17 to 2.10–2.13, we immediately obtain that $E_k \cong \mathrm{DS}_{k+2} \otimes V_{k\rho}$. Here the action of $\mathfrak{sl}(2)$ is entirely on the first factor, and is expressed in terms of the above formulas (which depend on k, and are exactly the same as those for r_k from Section 2). We note that the Harish-Chandra module of our realization of DS_{k+2} was on functions of the form $\varphi(u) = q(u)e^{-\pi u}$, where q is a polynomial. Thus functions of the form $p(M)\varphi(u)$ are polynomials in the entries of M, multiplied by the "Gaussian" $e^{-\pi \operatorname{tr} M^2}$. We have essentially identified r by observing its action on such functions. $\square$

We have thus identified the representation r of $\mathfrak{sl}(2) \times G$. Now we know already that the action of $\mathfrak{sl}(2)$ on DS_{k+2} "exponentiates" into a genuine representation of $\mathrm{SL}(2)$; so we can use this action to extend our representation R from $B^+ \times G$ to all of $\mathrm{SL}(2) \times G$. More explicitly, we can define the action of all of $\mathrm{SL}(2)$ on functions in E_k, i.e., functions of the form $F(M) = p(M)\varphi(\operatorname{tr} M^2)$, (again, $p(M) \in V_{k\rho}$ is a polynomial):

$$R(\gamma)F(M) = p(M)R_k(\gamma)\varphi(\operatorname{tr} M^2), \qquad \text{for } \gamma \in \mathrm{SL}(2). \tag{4.18}$$

Here $R_k(\gamma)$ is given by equations 2.18 and 2.19, and is the explicit "exponentiation" of r_k. Now $L^2(X_{\mathbb{R}}) \cong \widehat{\bigoplus}_{k\geq 0} E_k$, by the analogous decomposition for $L^2(S)$. Hence we have obtained our desired representation R of all of $\mathrm{SL}(2) \times G$ on $L^2(X_{\mathbb{R}})$.

We now turn to the question of giving explicit formulas for this representation. Equations 4.4 and 4.5 describe the action of $B^+ \times G$. Now B^+ and $\left(\begin{smallmatrix} 0 & -1 \\ 1 & 0 \end{smallmatrix}\right)$ generate SL(2), so it is enough for us to compute the operator $R(\left(\begin{smallmatrix} 0 & -1 \\ 1 & 0 \end{smallmatrix}\right))$. We shall spend the rest of this paper doing so.

We begin by defining $K_k : S \times S \to \mathbb{C}$ to be the integral kernel for the projection operator from $L^2(S)$ to its subspace isomorphic to $V_{k\rho}$. In other words, if $p_S \in L^2(S) \cong \widehat{\bigoplus}_{k\geq 0} V_{k\rho}$, and q_S is the projection of p_S onto the subspace isomorphic to $V_{k\rho}$, then

$$q_S(\sigma) = \int_{\sigma' \in S} p_S(\sigma') K_k(\sigma, \sigma')\, d\sigma'. \tag{4.19}$$

We combine this with 2.19, which describes $R_k(\left(\begin{smallmatrix} 0 & -1 \\ 1 & 0 \end{smallmatrix}\right))$ on our explicit model for DS_{k+2}, and formally obtain that $R(\left(\begin{smallmatrix} 0 & -1 \\ 1 & 0 \end{smallmatrix}\right))$ is given by the following integral operator, expressed in polar coordinates $(t, \sigma) \in (0, +\infty) \times S$:

$$\begin{aligned} &R\left(\begin{pmatrix} 0 & -1 \\ 1 & 0 \end{pmatrix}\right) F(t,\sigma) \\ &\quad = \sum_{k\geq 0} \int_{t'>0} \int_{\sigma' \in S} F(t', \sigma')(-i)^{k+2} \cdot 2\pi \frac{J_{k+1}(2\pi t t')}{tt'} K_k(\sigma, \sigma')\, d\sigma'\, t'^3\, dt. \end{aligned} \tag{4.20}$$

Ignoring matters of convergence, one checks this formally on functions of the form $p_S(\sigma) t^k \varphi(t^2) = p(M)\varphi(u)$, where $p_S \in V_{k\rho} \subset L^2(S)$, and $\varphi \in L^2((0, +\infty), u^{k+1}\, du)$. Hence $R(\left(\begin{smallmatrix} 0 & -1 \\ 1 & 0 \end{smallmatrix}\right))$ should be given by integrating against a kernel whose expression in polar coordinates is

$$\frac{-2\pi}{tt'} \sum_{k\geq 0} (-i)^k K_k(\sigma, \sigma') J_{k+1}(2\pi t t'). \tag{4.21}$$

This will be justified once we show in Lemma 4.5 that the above sum converges sufficiently well. We must therefore determine K_k explicitly, and sum the above series.

PROPOSITION 4.3. *Let us view elements $\sigma \in S$ as (Hermitian rank 1) matrices of trace 1. Then*

$$K_k(\sigma, \sigma') = \frac{(k+1)^3}{\operatorname{vol} S} P_k(\operatorname{tr} \sigma\sigma'), \tag{4.22}$$

where

$$P_0(x) = 1, \qquad P_1(x) = \frac{1}{2}(3x - 1), \qquad P_2(x) = \frac{1}{3}(10x^2 - 8x + 1), \quad \cdots \tag{4.23}$$

are the orthogonal polynomials on $[0,1]$ with respect to the measure $(1-x)\,dx$, normalized so that $P_k(1) = 1$.

REMARK 4.4. The P_k are essentially Jacobi polynomials; see (12.12–3) of [**H**]. Their general term is

$$\begin{aligned} P_k(x) &= \frac{1}{k+1} \sum_{j=0}^{k} (-1)^{k-j} \binom{k}{j} \binom{k+1+j}{j} x^j \\ &= \frac{(-1)^k}{k+1}\, {}_2F_1(-k, k+2; 1; x). \end{aligned} \tag{4.24}$$

The fact that $P_k(1) = 1$ is the Vandermonde theorem ([**H**], (1.4–11)).

Note also that we can view elements of S as belonging to $\mathbb{C}^3/\mathbf{T}$, and represent them by vectors $\vec{z} = (z_1, z_2, z_3) \in \mathbb{C}^3$ with $\sum_{j=1}^3 |z_j|^2 = 1$. Here $\vec{z}$ is determined up to scaling by $\zeta \in \mathbf{T}$; the corresponding matrix is $\sigma = \begin{pmatrix} \overline{z_1} \\ \overline{z_2} \\ \overline{z_3} \end{pmatrix} \begin{pmatrix} z_1 & z_2 & z_3 \end{pmatrix}$. Similarly, let σ' be the matrix corresponding to $\vec{w} = (w_1, w_2, w_3)$. Then in these coordinates, the "inner product" $\operatorname{tr} \sigma\sigma'$ is just $|\langle \vec{z}, \vec{w} \rangle|^2$, where $\langle \vec{z}, \vec{w} \rangle = \sum_{j=1}^3 z_j \overline{w_j}$. We shall feel free to use these coordinates when convenient.

PROOF OF PROPOSITION 4.3. This is entirely analogous to the theory of zonal spherical harmonics. $(1-x)\,dx$ is the measure on $[0,1] \cong G_0 \backslash S \cong G_0 \backslash G / G_0$ induced from the measure on S; the identification map $G_0 \backslash S \to [0,1]$ is $(z_1, z_2, z_3) \mapsto |z_3|^2$. To be thorough, we recall the proof (see [**F76**], Chapter 2, Section G, especially Theorem (2.55) to Proposition (2.57)).

Let $f_1, \dots f_N$ be an orthonormal basis for $V_{k\rho} \subset L^2(S)$, where $N = \dim V_{k\rho} = (k+1)^3$. Then $K_k(\sigma, \sigma') = \sum_{j=1}^N f_j(\sigma)\overline{f_j(\sigma')}$. Now $K_k(g^{-1}\sigma g, g^{-1}\sigma' g) = K_k(\sigma, \sigma')$ for $g \in G$, because the projection respects the action of G. Viewing $\vec{z}$ and $\vec{w}$ as row vectors, this means that $K_k(\vec{z}g, \vec{w}g) = K_k(\vec{z}, \vec{w})$ for $g \in G$; hence $K_k(\vec{z}, \vec{w})$ can depend only on $|\langle \vec{z}, \vec{w} \rangle|^2$. (We could have made K_k depend only on $\langle \vec{z}, \vec{w} \rangle$, but this last quantity is only defined up to a factor of $\zeta \in \mathbf{T}$, so we only care about $|\langle \vec{z}, \vec{w} \rangle|$. Using the square of the absolute value is even better, since this is now a polynomial in the entries of the matrices σ and σ'.)

Let us arrange our orthonormal basis so that f_1 spans the one-dimensional space of G_0-fixed vectors in $V_{k\rho}$. (f_1 is analogous to a zonal spherical harmonic.) Also observe that G_0 is the stabilizer of the "north pole" $\vec{n} = (0,0,1) \in \mathbb{C}^3/\mathbf{T}$. Consider the nonzero G_0-invariant linear functional on $V_{k\rho}$, sending p_S to $p_S(\vec{n})$. $V_{k\rho}$ is self-dual, so this functional must be obtained by taking an inner product with a G_0-fixed vector; so for some nonzero constant C, we obtain that

$$p_S(\vec{n}) = C \int_{\sigma \in S} p_S(\sigma)\overline{f_1(\sigma)}\,d\sigma, \qquad \text{for all } p_S \in V_{k\rho}. \tag{4.25}$$

Testing on f_1, we find that $C = f_1(\vec{n})$. Similarly, $f_2(\vec{n}) = \cdots = f_N(\vec{n}) = 0$. Hence $K_k(\sigma, \vec{n}) = f_1(\sigma)\overline{f_1(\vec{n})}$. Also, for all σ, $K_k(\sigma,\sigma) = K_k(\vec{n},\vec{n}) = |f_1(\vec{n})|^2$, by the G-invariance of K_k. This yields

$$(k+1)^3 = \int_{\sigma \in S} K_k(\sigma,\sigma)\,d\sigma = (\operatorname{vol} S)\,|f_1(\vec{n})|^2. \tag{4.26}$$

Putting this together we obtain that K_k has the form

$$K_k(\vec{z}, \vec{w}) = \frac{(k+1)^3}{\operatorname{vol} S} P_k(|\langle \vec{z}, \vec{w} \rangle|^2), \tag{4.27}$$

for some function P_k; furthermore, by evaluating $K_k(\vec{n}, \vec{n})$, we see that $P_k(1) = 1$.

We now show that P_k is a polynomial of degree at most k. For $x \in [0,1]$ and arbitrary $\zeta \in \mathbf{T}$, $P_k(x)$ equals a constant multiple of $K_k\big((\sqrt{1-x}, 0, \zeta\sqrt{x}), \vec{n}\big)$, which in turn is a constant multiple of $f_1\big((\sqrt{1-x}, 0, \zeta\sqrt{x})\big)$. But $f_1(\sigma) \in V_{k\rho}$ is a degree k polynomial in the entries of the matrix σ, i.e., $P_k(x)$ is a polynomial of degree k in the expressions x, $1-x$, $\zeta\sqrt{x(1-x)}$, and $\zeta^{-1}\sqrt{x(1-x)}$. But $P_k(x)$ is independent of ζ, and so the latter two expressions can only appear if multiplied by each other. Looking at the degrees of the resulting expressions, we see that $P_k(x)$ is a polynomial in x of degree at most k.

Finally, the P_k's are orthogonal with respect to the measure $(1-x)\,dx$ on $[0,1]$. To see this, note first that for a given k, the G_0-invariant polynomial $f_1(\vec{z}) \in V_{k\rho}$ is a constant multiple of $P_k(|z_3|^2)$. But now, if $k \neq l$, then the two polynomials $P_k(|z_3|^2) \in V_{k\rho}$ and $P_l(|z_3|^2) \in V_{l\rho}$ are orthogonal functions in $L^2(S)$, because 4.13 is an orthogonal decomposition. In coordinates, this is equivalent to

$$\int_{\substack{(z_1,z_2,z_3)\in\mathbb{C}^3\\ |z_1|^2+|z_2|^2+|z_3|^2=1}} P_k(|z_3|^2)P_l(|z_3|^2)\,d\mu = 0, \qquad \text{if } k \neq l. \tag{4.28}$$

Here $d\mu$ corresponds to the G-invariant measure on S. Up to a constant, it must be the standard measure on the unit sphere in $\mathbb{C}^3$. We evaluate the above integral along subsets where $|z_3|$ is constant to obtain the desired orthogonality property

$$\int_{x=0}^{1} P_k(x)P_l(x)(1-x)\,dx = 0, \qquad \text{if } k \neq l. \tag{4.29}$$

Specifically, change coordinates to

$$\begin{aligned} z_1 &= x_1 + iy_1 = e^{i\alpha}\sin A \sin B, \\ z_2 &= x_2 + iy_2 = e^{i\beta}\sin A \cos B, \\ z_3 &= x_3 + iy_3 = e^{i\gamma}\cos A, \end{aligned} \tag{4.30}$$

and use $d\mu = (1/x_1)dy_1 \wedge dx_2 \wedge dy_2 \wedge dx_3 \wedge dy_3$. □

We now work on the sum 4.21. This amounts to evaluating the function $\Sigma(x,y)$, where

$$\Sigma(x,y) = \sum_{k\geq 0}(-i)^k(k+1)^3P_k(x)J_{k+1}(y), \tag{4.31}$$

for $x \in [0,1]$ and $y \in (0,+\infty)$. We begin with the generating function for the P_k's, obtained from the well-known generating function for the Jacobi polynomials (see for instance [**H**], section 2.5, exercise 7). It is valid for $x \in [0,1]$ and $|q| < 1$, or more generally for x in a compact set C and $|q|$ smaller than a constant depending on C.

$$\sum_{k\geq 0}(k+1)P_k(x)q^k = \frac{2}{A(1-q+A)} = q^{-1}\left(\frac{q-1}{2(1-x)A} + \frac{1}{2(1-x)}\right), \tag{4.32}$$

$$A = \sqrt{1+2q(1-2x)+q^2}. \tag{4.33}$$

The branch of the square root is the one that equals 1 when $q=0$.

LEMMA 4.5. *Σ is an entire function of x and y.*

PROOF. By (4.5–9) of [**H**], $|J_n(y)| \leq |y|^n e^n (2n)^{-n} e^{|y|}$. So for y in a compact set, and for every $M_0 > 0$, we can estimate $|J_{k+1}(y)| \leq M_0^{-k}$ once k is sufficiently large (depending on the compact set and on M_0). We can also bound $P_k(x)$ in terms of k: given a compact $C \subset \mathbb{C}$, the range of validity of 4.32 implies the existence of $M_1 > 0$ such that $x \in C \implies |P_k(x)| \leq M_1^k$. Thus the sum in Σ converges absolutely and uniformly on compact sets. (A specific case, which will be useful later, is that $P_k(x)$ is $O((1+\epsilon)^k)$ uniformly on $[0,1]$ — this is because for $x \in [0,1]$, the poles of the generating function are on the unit circle $\{q \mid |q| = 1\}$.) □

REMARK 4.6. As mentioned before, this lemma justifies the manipulations that we perform in constructing the integral operator for $R((\begin{smallmatrix}0&-1\\1&0\end{smallmatrix}))$. Namely, the integral expression that we shall write down for $R((\begin{smallmatrix}0&-1\\1&0\end{smallmatrix}))F$ will be correct if F is a smooth function on $X_{\mathbb{R}}$ with compact support. More precisely, the estimates in Lemma 4.5 imply that the integral of the sum in 4.21 against a smooth F of compact support can be evaluated term by term. (Only the values of $\Sigma(x,y)$ for bounded y come into play.) Let F_k be the projection of F onto E_k. One sees that the integral of F against the kth term in 4.21 is indeed $R((\begin{smallmatrix}0&-1\\1&0\end{smallmatrix}))F_k$, because integrating out σ' first gives the projection of F to $F_k \in E_k$. Then the remaining integral in t' is correct because we explicitly know the action of $R((\begin{smallmatrix}0&-1\\1&0\end{smallmatrix}))$ on F_k. Thus our term-by-term integration yields that $\sum_k R((\begin{smallmatrix}0&-1\\1&0\end{smallmatrix}))F_k$ converges pointwise to the integral of F against the proposed kernel. On the other hand, in terms of the L^2 norm, $\|F\|^2 = \sum_k \|F_k\|^2 < \infty$, and moreover $\|R((\begin{smallmatrix}0&-1\\1&0\end{smallmatrix}))F_k\| = \|F_k\|$; so $\sum_k R((\begin{smallmatrix}0&-1\\1&0\end{smallmatrix}))F_k$ converges in L^2 to $R((\begin{smallmatrix}0&-1\\1&0\end{smallmatrix}))F$. But if a sequence of functions converges both in L^2 and pointwise, then the L^2 limit equals the pointwise limit, because L^2 convergence of a sequence implies that some subsequence converges pointwise to the L^2 limit. Hence the result of integrating F against the proposed kernel is indeed $R((\begin{smallmatrix}0&-1\\1&0\end{smallmatrix}))F$, if F has compact support. Since $R((\begin{smallmatrix}0&-1\\1&0\end{smallmatrix}))$ is unitary, its extension to all of L^2 is now straightforward.

We are now in a position to evaluate $\Sigma(x,y)$.

PROPOSITION 4.7.

$$\Sigma(x,y) = \frac{y}{2} e^{iy}\, {}_1F_1(3/2; 1; -2ixy), \tag{4.34}$$

where ${}_1F_1$ is the confluent hypergeometric function

$${}_1F_1(a;b;t) = 1 + \frac{a}{b}\frac{t}{1} + \frac{a(a+1)}{b(b+1)}\frac{t^2}{2!} + \frac{a(a+1)(a+2)}{b(b+1)(b+2)}\frac{t^3}{3!} + \cdots. \tag{4.35}$$

PROOF. Without loss of generality, we can assume that $0 < x \leq 1$ and that y lies in a compact set. The general result will then follow by analytic continuation. Our estimates on $\Sigma(x,y)$ show that

$$\Sigma(x,y) = \lim_{r\to 1^-} r \sum_{k\geq 0} (k+1)^3 (-ir)^k P_k(x) J_{k+1}(y/r). \tag{4.36}$$

(For $1/2 < r < 1$, some tail $\sum_{k>N}$ of the above sum is small uniformly in r. Then use the continuity of the first N terms.) Using the integral expression for J_{k+1} from 2.20, we obtain

$$\Sigma(x,y) = \lim_{r\to 1^-} \frac{i}{2\pi} \int_{\theta=0}^{2\pi} \sum_{k\geq 0} (k+1)^3 (re^{i\theta})^{k+1} P_k(x) e^{-i(y/r)\cos\theta}\, d\theta. \tag{4.37}$$

Interchanging the summation with the integration is justified by our estimates on the size of P_k and J_{k+1} in terms of k. For convenience, write $q = re^{i\theta}$. To evaluate the sum $\sum_{k\geq 0}(k+1)^3 P_k(x) q^{k+1}$, we apply the differential operator

$$q\frac{\partial}{\partial q} q \frac{\partial}{\partial q} q$$

to the identity 4.32. We obtain

$$\sum_{k\geq 0}(k+1)^3 P_k(x)q^{k+1} = \frac{q(1-q)((1+q)^2+2qx)}{((1+q)^2-4qx)^{5/2}}, \tag{4.38}$$

where the denominator is just A^5. Since $q = re^{i\theta}$, $(\cos\theta)/r = (1/2)(1/q + q/r^2)$. Hence $\Sigma(x,y)$ equals the limit, as $r \to 1^-$, of

$$\frac{i}{2\pi}\int_{\theta=0}^{2\pi} \frac{q(1-q)((1+q)^2+2qx)}{((1+q)^2-4qx)^{5/2}} e^{-i(y/2)(1/q+q+q(1/r^2-1))}\, d\theta. \tag{4.39}$$

Now make the change of variable $b = (1+q)^2/q = 1/q+q+2$, $db = i((q^2-1)/q)\, d\theta$. As θ traverses $[0, 2\pi]$, b moves *clockwise* around a thin and flat ellipse Ω_r in the complex plane, enclosing the real line segment $[0,4]$. With some care about the branches of the square roots (or by checking the limit as x approaches 0) we obtain $\Sigma(x,y)$ as the limit as $r \to 1^-$ of

$$\frac{-1}{2\pi}\int_{b \text{ on } \Omega_r} \frac{b+2x}{(b-4x)^2\sqrt{b(b-4x)}} e^{-i(y/2)(b-2+q(1/r^2-1))}\, db, \tag{4.40}$$

where $q = (b-2-\sqrt{b(b-4)})/2$. We have chosen the branches of $\sqrt{b(b-4x)}$ and of $\sqrt{b(b-4)}$ for $b \in \mathbb{C} - [0,4]$, by requiring that if b is real and greater than 4, then the above square roots will be positive. At this point we note that the integral is independent of the path Ω_r, and so we can replace Ω_r by, say, $\Omega_{1/2}$, and *then* let r tend to 1. Hence

$$\Sigma(x,y) = \frac{-e^{iy}}{2\pi}\int_{b \text{ on } \Omega_{1/2}} \frac{b+2x}{(b-4x)^2\sqrt{b(b-4x)}} e^{-iyb/2}\, db. \tag{4.41}$$

This last integral now depends on x in a simple way, and so we can replace b with xb. The new contour goes clockwise around the segment $[0, 4/x]$, but can again be replaced with $\Omega_{1/2}$. We obtain

$$\Sigma(x,y) = \frac{-e^{iy}}{2\pi x}\int_{b \text{ on } \Omega_{1/2}} \frac{b+2}{(b-4)^2\sqrt{b(b-4)}} e^{-ixyb/2}\, db \tag{4.42}$$

$$= \frac{-e^{iy}}{2\pi x}\sum_{n\geq -1}\frac{(-ixy/2)^{n+1}}{(n+1)!}\int_{b \text{ on } \Omega_{1/2}} \frac{(b+2)b^{n+1}}{(b-4)^2\sqrt{b(b-4)}}\, db, \tag{4.43}$$

where we recall that b traverses $\Omega_{1/2}$ *clockwise.* (We have begun the summation with $n=-1$ because the first term turns out to vanish.) One easily evaluates this last integral in terms of the "residue at ∞" (i.e., change variables so that $1/b$ is the variable of integration). The value of the integral turns out to be $-2\pi i(n+1)(3/2)_n 4^n/n!$, which vanishes for $n = -1$. (For $n \geq 1$, we have used the Pochhammer symbol $(3/2)_n = (3/2)(3/2+1)\cdots(3/2+n-1)$. For $n = 0$, we take $(3/2)_0 = 1$.) Proposition 4.7 finally follows. □

REMARK 4.8. Once we obtain 4.42, we know that the sum $\Sigma(x,y)$ has the form $(e^{iy}/x)f(xy)$ for some function f. Then another way to find that f is the stated confluent hypergeometric function is to observe that $R(\left(\begin{smallmatrix}0 & -1\\ 1 & 0\end{smallmatrix}\right))r(e) = -r(f)R(\left(\begin{smallmatrix}0 & -1\\ 1 & 0\end{smallmatrix}\right))$, which leads to an easily solvable differential equation for f. This is analogous to Remark 2.7.

We put together all these ingredients to obtain

THEOREM 4.9. *In polar coordinates, $R((\begin{smallmatrix} 0 & -1 \\ 1 & 0 \end{smallmatrix}))$ acts on $F \in \mathcal{C}_c^\infty(X_\mathbb{R})$ by*

$$(4.44)\quad R\left(\begin{pmatrix} 0 & -1 \\ 1 & 0 \end{pmatrix}\right) F(t,\sigma) = \frac{-2\pi^2}{\operatorname{vol} S} \iint_{\substack{t' \in (0,+\infty) \\ \sigma' \in S}} F(t',\sigma') e^{2\pi i t t'} {}_1F_1(3/2; 1; -2\pi i t t' \operatorname{tr} \sigma\sigma')\, d\sigma'\, t'^3\, dt'.$$

In terms of the matrices $M = t\sigma$ and $N = t\sigma'$,

$$(4.45)\quad R\left(\begin{pmatrix} 0 & -1 \\ 1 & 0 \end{pmatrix}\right) F(M) = \frac{-4}{\pi} \int_{N \in X_\mathbb{R}} F(N) e^{2\pi i \operatorname{tr} M \operatorname{tr} N} {}_1F_1(3/2; 1; -2\pi i \operatorname{tr} MN)\, dN,$$

with the measure dN defined in 4.3.

By Theorems 4.2 and 4.9, our formulas 4.4, 4.5, and 4.45 define a representation of $\mathrm{SL}(2) \times \mathrm{U}(3)$ on $L^2(X_\mathbb{R})$, which decomposes according to

$$(4.46)\quad L^2(X_\mathbb{R}) \cong \widehat{\bigoplus_{k \geq 0}} \mathrm{DS}_{k+2} \otimes V_{k\rho}.$$

As mentioned in the introduction, this decomposition realizes the adjoint lift from $\mathrm{SL}(2,\mathbb{R})$ to $\mathrm{U}(3)$, and the formulas are sufficiently explicit that they might possibly generalize to the p-adic case. This would involve addressing several points. The first point is that there are no anisotropic forms of $\mathrm{U}(3)$ over a p-adic field. Moreover, for global applications one would need to deal with the split case anyhow. It seems reasonable to first try out the split case over $\mathbb{R}$, using the techniques of this paper. These techniques easily give rise to a formal representation of $\mathfrak{sl}(2,\mathbb{R}) \times \mathfrak{gl}(3,\mathbb{R})$ on the space of functions on a "split" form X'_{split} of X', namely the 3×3 real matrices M of rank at most 1. (It is nonetheless unclear whether e.g., restricting $\operatorname{tr} M$ to be nonnegative provides a suitable "half" of this space to obtain an analog of X.) In the terminology of Remark 3.9, we then obtain a formal representation of $\mathfrak{sl}(2) \times \mathrm{GL}(3)$ on the space of functions on X'_{split}. We namely begin with the representation of $B^+ \times \mathrm{GL}(3)$ defined by the same formulas as 4.4 and 4.5. We can then match up the Casimir operators $C_{\mathfrak{sl}(2)}$ and $C_{\mathfrak{gl}(3)}$, and use Lemma 2.2, to obtain a formal action of all of $\mathfrak{sl}(2)$. It seems more difficult to "exponentiate" this resulting representation to all of $\mathrm{SL}(2)$; the main complication is that now the structure of $L^2(X'_{\mathrm{split}})$ as a representation of $\mathrm{GL}(3)$ seems less transparent now that the group is no longer compact. Even assuming some analog of Theorem 4.2, telling us that the formal representation comes from a genuine representation of $\mathrm{SL}(2)$, we expect to run into some hairy calculations in trying to generalize our computation of $\Sigma(x,y)$. Essentially, the sum in 4.31 will get replaced by an integral, because the analog of $L^2(S)$ will now decompose continuously under $\mathrm{GL}(3)$. Instead of trying to evaluate such an integral, it may be worth first investigating if the formula 4.45 still works. This would be analogous to the fact that for the dual pair $\mathrm{SL}(2) \times \mathrm{O}(2n)$, the action of $(\begin{smallmatrix} 0 & -1 \\ 1 & 0 \end{smallmatrix})$ is essentially via a Fourier transform no matter what the signature of the quadratic form fixed by $\mathrm{O}(2n)$.

The second complication that arises in extending our results to p-adic fields is how to translate our formulas for $\mathbb{R}$ to a p-adic setting. It is straightforward to generalize the definition of $X'_\mathbb{R}$, whether we are dealing with $\mathrm{GL}(3)$ or with $\mathrm{U}(3)$

over a p-adic field. Taking "half" of the resulting space seems more mysterious. Perhaps in the case of a unitary group over a p-adic field F, one could require the trace of M to belong to the index 2 subgroup of $F^\times$, consisting of norms from the quadratic extension of F used to construct the unitary group. In that case, the analog of the action of $B^+ \times G$ is easy to write down, using an additive character $e_F : F \to \mathbb{C}^\times$ instead of the function $e(x) = e^{2\pi i x}$. The next step, which is slightly more difficult, is to write down (or guess) a formula giving the action of $R(\left(\begin{smallmatrix} 0 & -1 \\ 1 & 0 \end{smallmatrix}\right))$ in terms of an integral kernel. (X'_{split} still carries more or less the same measure as before.) In this respect the expression $e^{2\pi i \operatorname{tr} M \operatorname{tr} N}$ in the integral should correspond to $e_F(\operatorname{tr} M \operatorname{tr} N)$, but the confluent hypergeometric function seems less tractable. Of course, it would be a difficult calculation to verify that the resulting formulas, even if correct, do indeed fit together to yield a genuine representation of $\mathrm{SL}(2) \times G$.

Even if these complications are overcome, and we obtain representations of (say) $\mathrm{SL}(2) \times \mathrm{GL}(3)$ over $\mathbb{R}$ and every $\mathbb{Q}_p$, that yield the local adjoint lift, it will probably still be a difficult problem to show that these representations are globally coherent, in the sense that they match up *automorphic* representations of $\mathrm{SL}(2)$ with *automorphic* representations of $\mathrm{GL}(3)$. Again bearing in mind the case of $\mathrm{SL}(2) \times \mathrm{O}(2n)$, one could hope that there might exist some analog of the Poisson summation formula. Such an analog might well be rather difficult to find. Another direction for further investigation is to search for explicit constructions in the style of Section 4 for other groups G. Even if one sticks to compact G, it may be nontrivial to find a suitable X and $X_{\mathbb{R}}$, since Example 3.11 seems too simplistic a construction, and may not yield suitable formulas. An even more serious problem in generalizing Section 4 to other G is the fact that our "polar decomposition," 4.10, is rather special in that the representations of G occurring in $L^2(S)$ are all ones with highest weight a multiple of a single weight, ρ. This is one of the reasons why our calculations in this paper remained tractable.

References

[BaSS] D. Barbasch, S. Sahi and B. Speh, *Degenerate series representations for* $\mathrm{GL}(2n, \mathbb{R})$ *and Fourier analysis*, Symposia Math. **31** (Rome, 1988), London: Academic Press (1990), 45–69.

[Bl] A. Bluher, *The theta-correspondence over* $\mathbb{R}$, preprint.

[BrK] R. Brylinski and B. Kostant, *Differential operators on conical Lagrangian manifolds*, in Lie Theory and Geometry in Honor of Bertram Kostant, Progress in Math. **123**, Boston: Birkhäuser (1994).

[F76] G. Folland, Introduction to Partial Differential Equations, Math. Notes **17**, Princeton: Princeton University Press (1976).

[F89] G. Folland, Harmonic Analysis in Phase Space, Annals of Math. Studies **122**, Princeton: Princeton University Press (1989).

[GJ] S. Gelbart and H. Jacquet, *A relation between automorphic representations of* $\mathrm{GL}(2)$ *and* $\mathrm{GL}(3)$, Annales Scientifiques ENS 4^e serie, **11** (1978), 471–542.

[H] P. Henrici, Applied and Computational Complex Analysis, Volumes 1 and 2, New York: Wiley-Interscience (1974).

[Ho] R. Howe, *θ-series and invariant theory*, Proceedings of Symposia in Pure Math. part 1 **33** (1979), 275–285.

[Hu] J. Humphreys, Introduction to Lie Algebras and Representation Theory, Graduate Texts in Math. **9**, New York-Berlin: Springer-Verlag (1972).

[JL] H. Jacquet and R. Langlands, Automorphic Forms on GL(2), Lecture Notes in Math. **114**, Berlin-New York: Springer-Verlag (1970).

[K] M. Krämer, *Sphärische Untergruppen in kompakten zusammenhängenden Liegruppen*, Compositio Math. **38** (1979), 129–153.

[S] G. Savin, *Some new algebras and correspondences of representations of reductive Lie groups*, preprint.

[Sh] G. Shimura, *On modular forms of half-integral weight*, Annals of Math. **97** (1973), 440–481.

[W] A. Weil, *Sur certains groupes d'opérateurs unitaires*, Acta Math. **111** (1964), 143–211.

Department of Mathematics and Statistics, McGill University, 805 Sherbrooke St. West, Montréal, Québec H3A 2K6, Canada

E-mail address: makdisi@scylla.math.mcgill.ca

Proceedings of Symposia in Pure Mathematics
Volume **66.2**, 1999

Truncated Integrals and the Shintani Zeta Function for the Space of Binary Quartic Forms

Jason Levy

Dedicated to Professor Shimura

ABSTRACT. We study the Shintani zeta function for the space of binary quartic forms under the action of SL(2), and compute some of its poles. We do this via truncated integrals, using methods similar to those of Jacquet-Langlands [**JL**] and Labesse-Langlands [**LaL**].

0. Introduction

The Poisson summation formula is a simple formula that has had vast applications to number theory. It says that given a number field k with adele ring $\mathbb{A}$ and a Schwartz-Bruhat function f on the adelic points $V(\mathbb{A})$ of a finite dimensional vector space V defined over k, one has the equality

$$\sum_{\gamma \in V(k)} f(\gamma) = \sum_{\hat{\gamma} \in \widehat{V}(k)} \hat{f}(\hat{\gamma}), \tag{0.0}$$

where $\widehat{V}$ is the vector space dual to V and $\hat{f}$ is the Fourier transform of f. (The classical, non-adelic, equation that equation (0.0) generalizes is $\sum_{n \in \mathbb{Z}} f(n) = \sum_{n \in \mathbb{Z}} \hat{f}(n)$ for a Schwartz function f on $\mathbb{R}$.) Many recent applications in number theory of the Poisson summation formula have been obtained by a more general "averaging" of equation (0.0), given as follows: Let π be a rational representation of a rational reductive algebraic group G on V, and let $\hat{\pi}$ be its contragredient. Equation (0.0) implies that

$$\sum_{\gamma \in V(k)} f(\pi(g^{-1})\gamma) = |\det \pi(g)| \sum_{\hat{\gamma} \in \widehat{V}(k)} \hat{f}(\hat{\pi}(g^{-1})\hat{\gamma}).$$

The "averaged" equation is the equality

1991 *Mathematics Subject Classification.* Primary 11M41, 11E45, 11F72.
Key words and phrases. Shintani zeta function, quartic forms.
Supported by NSERC grant #2560149.

$$\int_{G_V(\mathbb{A})G(k)\backslash G(\mathbb{A})} \varphi(g) \sum_{\gamma \in V(k)} f(\pi(g^{-1})\gamma)dg \tag{0.1}$$
$$= \int_{G_V(\mathbb{A})G(k)\backslash G(\mathbb{A})} \varphi(g)|\det \pi(g)| \sum_{\hat{\gamma} \in \widehat{V}(k)} \hat{f}(\hat{\pi}(g^{-1})\hat{\gamma})dg,$$

valid when both integrals converge, with G_V the kernel of the representation π, and φ a slowly increasing function on the left quotient $G_V(\mathbb{A})G(k)\backslash G(\mathbb{A})$.

The formula (0.1), for specific groups G, representations π, and functions φ such that both integrals converge, plays an essential role in many important results. Tate's thesis proves the analytic continuation and functional equation of abelian Hecke L-functions by expressing the L-function in terms of the left hand side of equation (0.1) with $G = \mathrm{GL}(1)$ acting by multiplication on a one-dimensional vector space, with $\varphi = 1$. Godement and Jacquet [**GJ**] proved the analytic continuation and functional equation of the standard L-functions of cuspidal automorphic representations on $\mathrm{GL}(n)$ by rewriting the L-function similarly to Tate, and then applying equation (0.1) with $G = \mathrm{GL}(n)$ acting by multiplication on the vector space of n-dimensional column vectors, with φ a matrix coefficient of π. The Siegel-Weil formula in the adelic form given by Weil [**W85**], uses equation (0.1) where G is an orthogonal group; there the action of π on the space of Schwartz-Bruhat functions on $V(\mathbb{A})$ is given by the Weil representation and φ is the constant function 1.

For general G, π, and φ, the integrals in formula (0.1) do not converge, and some effort is required to produce a suitable regularization of the integrals. Where such a regularization has been obtained, it has proven to be useful.

Regularization of the integrals in (0.1) was first done in [**S**] by Shintani, who used a classical (rather than adelic) formulation of the problem. He took φ to be a smoothed Eisenstein series depending on a complex variable s to produce a convergent integral that varied meromorphically with s, and then examined the residue of this integral at $s = 1$ to obtain a regularized identity based on (0.1) with φ the constant function 1. This method of Eisenstein series regularization has been put into adelic form and extended by Datskovsky, Wright, and Yukie, and used to count extensions of number fields. See [**D**], [**W85**], [**DW86**], [**DW88**], [**Yu92**], [**Yu98**] and especially [**Yu93**] for details.

In [**Le95**], an alternative form of regularization, first introduced by Arthur [**A78**], [**A85**] in the context of the Selberg trace formula, was shown to produce a regularized identity in the case that the group G is semisimple of rank at most two and φ is the constant function 1. In this paper, we will explicitly produce the regularized identity in the case that $G = \mathrm{SL}(2)$ and V is the space of binary quartic forms. For $G = \mathrm{SL}(2)$, the regularization in [**Le95**] is quite similar to that used in [**JL**], [**Lan**], and [**LaL**] and we will adapt other ideas from these papers to product our explicit regularized identity. In the last section, we will produce a zeta function from this identity and discuss its analytic continuation and poles.

The main results of this paper, in section 3, are contained in the author's thesis [**Le93**, ch. 5]. The thesis dealt with arbitrary representations of SL(2) but contains errors, related to convergence, for larger representations. Yukie [**Yu98**] subsequently published similar work that had been done much earlier, using Eisenstein series regularization, and the methods of this paper can trivially be extended to reproduce all the results of that paper. This paper proves only marginally more than Yukie's, but it serves as an illustration of Arthur's truncation.

The author gratefully thanks D. Wright and A. Yukie for many informative discussions during his stay at Oklahoma State University.

1. General definitions

For a finite set A, the cardinality of A will be denoted by $|A|$.

We denote the fields of rational, real, complex numbers by $\mathbb{Q}, \mathbb{R}, \mathbb{C}$ respectively. We denote the ring of rational integers by $\mathbb{Z}$. The set of positive real numbers is denoted by $\mathbb{R}_+$. For any ring R, we write $R^\times$ for the set of invertible elements of R. Let k be a number field, and write $\mathbb{A}$ for its ring of adeles. If V is a vector space over k, we define $V(\mathbb{A})$ to be the restricted direct product of the vector spaces $V(k_v)$, v ranging over all places of k, and for R a local completion k_v of k or the ring of adeles $\mathbb{A}$ let $\mathcal{S}(V(R))$ be the space of Schwartz-Bruhat functions on $V(R)$.

For any place v, write k_v for the completion of k at v. If v is a finite place, write $o_v \subset k_v$ for the ring of integers of k_v. We normalize the norm $|\ |_v$ on k_v so that if v is a finite place and π_v is a uniformizer, then $|\pi_v|_v = q_v^{-1}$, if v is real then $|\ |_v$ is the usual absolute value, and if v is imaginary and $|x|$ is the usual absolute value, $|x|_v = |x|^2$.

We normalize Haar measure and Fourier transform as in [**W74**]. The discrete groups k, $k^\times$ are given the counting measure. We normalize the Haar measure dx on $\mathbb{A}$ so that $\int_{\mathbb{A}/k} dx = 1$. For any finite place v, we choose a Haar measure dx_v on k_v so that $\int_{o_v} dx_v = 1$. We use the ordinary Lebesgue measure dx_v for v real and $dx_v \wedge d\bar{x}_v$ for v imaginary. For $\lambda \in \mathbb{R}_+$, let $\underline{\lambda}$ be the idele whose component at all real places is $\lambda^{1/[k:\mathbb{Q}]}$ and at all finite places is 1, so that $|\underline{\lambda}| = \lambda$. We identify $\mathbb{R}_+$ with a subgroup of $\mathbb{A}^\times$ by the map $\lambda \to \underline{\lambda}$. Let $\mathbb{A}^1 = \{x \in \mathbb{A}^\times \mid |x| = 1\}$. Then $\mathbb{A}^\times \cong \mathbb{A}^1 \times \mathbb{R}_+$, and $\mathbb{A}^1/k^\times$ is compact. We choose a Haar measure $d^\times t^1$ on $\mathbb{A}^1$ so that $\int_{\mathbb{A}^1/k^\times} d^\times t^1 = 1$. Using this measure, we choose a Haar measure $d^\times t$ on $\mathbb{A}^\times$ so that

$$\int_{\mathbb{A}^\times} f(t) d^\times t = \int_0^\infty \int_{\mathbb{A}^1} f(\underline{\lambda} t^1) d^\times \lambda d^\times t^1,$$

where $d^\times \lambda = \lambda^{-1} d\lambda$. For any finite place v, we normalize the Haar measure $d^\times t_v$ on $k_v^\times$ so that $\int_{o_v^\times} d^\times t_v = 1$. At the infinite places, we let $d^\times t_v(x) = |x|_v^{-1} dx$ if v is real and $d^\times t_v(x) = |x|_v^{-1} dx \wedge d\bar{x}$ if v is imaginary.

The relation between the local and global measure so chosen is given as follows [**W74**, p. 91,95]: Let r_1, r_2 be the numbers of real and imaginary places respectively. Let h, R, e be the class number, regulator and the number of units of k respectively. Let Δ_k be the discriminant of k. Let $\mathfrak{C}_k = 2^{r_1}(2\pi)^{r_2} hRe^{-1}$. Then

$$dx = |\Delta_k|^{-1/2} \prod_v dx_v \quad \text{and} \quad d^\times t = \mathfrak{C}_k^{-1} \prod_v d^\times t_v.$$

Let $\zeta_k(s)$ be the Dedekind zeta function. As in [**W74**], we define

$$Z_k(s) = |\Delta_k|^{s/2} \left(\pi^{-s/2}\Gamma\left(\frac{s}{2}\right)\right)^{r_1} \left((2\pi)^{-s}\Gamma(s)\right)^{r_2} \zeta_k(s).$$

We define $\mathfrak{R}_k = \mathrm{Res}_{s=1} Z_k(s) = \mathfrak{C}_k / |\Delta_k|^{1/2}$, and $\mathfrak{B}_2 = Z_k(2)/\mathfrak{R}_k$. For $s \in \mathbb{C}$ we write ω_s for the character of $\mathbb{A}^\times/k^\times$ or of $k_v^\times$ given by $\omega_s(x) = |x|^s$.

Throughout this paper, we will write G for the group $\mathrm{SL}(2)$. In addition, we will write B for the Borel subgroup $\left\{\left(\begin{smallmatrix} t & x \\ 0 & t^{-1} \end{smallmatrix}\right)\right\}$, A for the Levi subgroup and maximal torus $\{\left(\begin{smallmatrix} t & 0 \\ 0 & t^{-1}\end{smallmatrix}\right)\}$, and N for the unipotent subgroup $\{\left(\begin{smallmatrix} 1 & * \\ 0 & 1\end{smallmatrix}\right)\}$. All these are algebraic groups defined over k. We define also a maximal compact subgroup K of $G(\mathbb{A})$ as follows: for v an infinite place, let $K_{k_v} = K_v$ be the subgroup of all unitary matrices in $G(k_v)$, and for v a finite place let K_v be $G(O_v)$. Then $K = K_{\mathbb{A}}$ is defined to be the product $\prod_v K_v$. Given $t \in \mathbb{A}^\times$, write $a(t) = \left(\begin{smallmatrix} t & 0 \\ 0 & t^{-1}\end{smallmatrix}\right)$, and given $x \in \mathbb{A}$, write $n(x) = \left(\begin{smallmatrix} 1 & n \\ 0 & 1\end{smallmatrix}\right)$. We identify $\mathfrak{a}^* = X^*(A) \otimes_{\mathbb{Z}} \mathbb{R}$ with $\mathbb{R}$ by sending the character $\chi \in X^*(A)$ defined by $\chi(a(t)) = t$ to the number 1.

Given an element $g = na(t)k, n \in N(\mathbb{A}), a \in A(\mathbb{A}), k \in K$, define $H(g) = \log|t| \in \mathbb{R}$. This is a well-defined map from $G(\mathbb{A})$ to $\mathbb{R}$. Write τ for the characteristic function of the interval $(0, \infty)$ and define

$$F(g,T) = 1 - \sum_{\delta \in B(k)\backslash G(k)} \tau(H(\delta g) - T), \quad g \in G(k)\backslash G(\mathbb{A}), T \in \mathbb{R}^*. \tag{1.1}$$

It is well-known (see [**A78**]) that for T large enough, $F^G(g,T)$ is the characteristic function of a relatively compact set in $G(k)\backslash G(\mathbb{A})$.

We normalize Haar measures in the various groups that will appear as follows: all discrete groups, such as $G(k)$ and $B(k)$, have counting measure, the compact groups K and K_v have unit measure. For R a local field or the ring of adeles, the maps $a : R^\times \to A(R)$ and $n : R \to N(R)$ given above preserve measure. The (left-invariant) measure on $B(R)$ is given by

$$\begin{aligned} \int_{B(R)} f(b)db &= \int_{A(R)} da \int_{N(R)} dnf(an) \\ &= \int_{N(R)} dn \int_{A(R)} da\; e^{2H(a)} f(na), \end{aligned} \tag{1.2}$$

and the measure on $G(R)$ is given by

$$\int_{G(R)} f(g)dg = \int_{B(R)} db \int_{K_R} dk\; f(bk), \tag{1.3}$$

where K_R is K_v if $R = k_v$ and K if $R = \mathbb{A}$. The measure on the quotient of a group by a unimodular subgroup will be the quotient measure and the measure on the direct product of two groups will be the product measure, if the measures on the two groups are given.

The measures we have chosen are not Tamagawa measures, in fact with our normalization, $\mathrm{vol}(\mathrm{SL}_2(k)\backslash \mathrm{SL}_2(\mathbb{A})) = \mathfrak{B}_2$; they were chosen for easier comparison with other works on Shintani zeta functions [**W85**], [**Yu98**], [**Yu92**], [**Yu93**].

2. Definitions on V

Although we will only produce a formula in section 3 for the vector space of binary quartic forms, the framework in which our calculations are made is almost equally easily stated in the generality of arbitrary finite-dimensional representations of SL(2). This is what we do here.

Let $\pi : \mathrm{SL}(2) \to \mathrm{GL}(V)$ be a finite-dimensional rational representation of $G = \mathrm{SL}(2)$. Write $\widehat{V}$ for the vector space dual to V and $\hat{\pi} : G \to \mathrm{GL}(\widehat{V})$ for the representation contragredient to π. (The representations π and $\hat{\pi}$ are isomorphic when $G = \mathrm{SL}(2)$.) Normalize Haar measures on $V(\mathbb{A}), \hat{V}(\mathbb{A})$ so that any linear maps

$$V(\mathbb{A}) \to \mathbb{A}^{\dim V}$$

$$\hat{V}(\mathbb{A}) \to \mathbb{A}^{\dim V}$$

given by isomorphisms defined over k preserve measure. Write $\mathcal{S}(V(\mathbb{A})), \mathcal{S}(\widehat{V}(\mathbb{A}))$ for the spaces of Schwartz-Bruhat functions on $V(\mathbb{A}), \widehat{V}(\mathbb{A})$, respectively. Fix throughout this paper a character $x \mapsto \langle x \rangle = \prod_v \langle x_v \rangle_v$ of $\mathbb{A}/k$. Define the Fourier transform $\widehat{\Phi} \in \mathcal{S}(\widehat{V}(\mathbb{A}))$ of $\Phi \in \mathcal{S}(V(R))$ by

$$\widehat{\Phi}(x) = \int_{V(\mathbb{A})} \Phi(v) \langle x(v) \rangle_R \, dv;$$

the Fourier transform on $\mathcal{S}(\hat{V}(\mathbb{A}))$ is defined similarly and yields a function in $\mathcal{S}(V(\mathbb{A}))$, using the canonical identification $V \cong \widehat{\widehat{V}}$. One has the following identities for $\Phi \in \mathcal{S}(V\mathbb{A}))$:

$$\widehat{\widehat{\Phi}}(x) = \Phi(-x), \qquad \text{(Fourier inversion)}$$

$$\sum_{\gamma \in V(k)} \Phi(\gamma) = \sum_{\hat{\gamma} \in \widehat{V}(k)} \widehat{\Phi}(\hat{\gamma}). \qquad \text{(Poisson summation formula)}$$

Furthermore, since the Jacobian of the change of variables $v \mapsto g \cdot v, v \in V(\mathbb{A})$ is 1 for $g \in G$, given $\Phi \in \mathcal{S}(V(\mathbb{A}))$ and $g \in G(\mathbb{A})$, if we define ${}^g\Phi \in \mathcal{S}(V(\mathbb{A}))$ by ${}^g\Phi(x) = \Phi(\pi(g^{-1})x)$, then $({}^g\Phi)^\wedge = {}^g(\widehat{\Phi})$. This gives us an identity,

$$\sum_{\gamma \in V(k)} \Phi(\pi(g^{-1})\gamma) = \sum_{\hat{\gamma} \in \widehat{V}(k)} \widehat{\Phi}(\hat{\pi}(g^{-1}\hat{\gamma}), \tag{2.1}$$

for each $g \in G(k)\backslash G(\mathbb{A})$.

The primary goal of this paper is to evaluate a certain regularization (from [**Le95**]) of the integral over $G(k)\backslash G(\mathbb{A})$ of these identities, at least for the space of binary quartic forms.

If V is irreducible, the space $\mathrm{Hom}_G(V, \widehat{V})$ is 1-dimensional. An element of $\mathrm{Hom}_G(V, \widehat{V})$ is equivalent to a bilinear form $[\ ,\ ]$ on V that is G-invariant, that is, $[\pi(g)v, \pi(g)v'] = [v, v']$ for all $g \in G, v, v' \in V'$. A non-degenerate form of this type can be constructed explicitly as follows.

Recall that for $i \geq 0$ an integer, the unique irreducible rational representation of SL(2) of dimension $i+1$ is $\mathrm{Sym}^i k^2$, the vector space of degree i forms $f\binom{u}{v}$ in two variables, with $G = \mathrm{SL}(2)$-action given by $(g \cdot f)\binom{u}{v} = f(g^{-1}\binom{u}{v})$; as a vector space it can be identified with k^{i+1} by the following map:

$$f\begin{pmatrix} u \\ v \end{pmatrix} = \sum_{m=0}^{i} x_m u^{i-m} v^m \qquad \mapsto (x_0, \ldots, x_i).$$

Define a non-degenerate bilinear form $[\ ,\ ]$ on $\mathrm{Sym}^i R$ by the formula

$$[x, y] = \sum_{m=0}^{i} \binom{i}{m}^{-1} (-1)^m x_m y_{i-m},$$

for $x = (x_0, \ldots, x_i), y = (y_0, \ldots, y_i)$. A short computation yields that for $n \in N$ and $v, v' \in \mathrm{Sym}^i k^2, [n \cdot v, n \cdot v'] = [v, v'] = [{}^t n \cdot v, {}^t n \cdot v']$. Since ${}^t N \cdot N$ is dense in SL(2), one concludes that $[g \cdot v, g \cdot v']_i = [v, v']_i$ for all $g \in G, v, v' \in \mathrm{Sym}^i k^2$.

We will use this form $[\ ,\ ]$ to identify V and $\widehat{V}$ when V is the space of binary quartic forms.

The regularization used in [**Le95**] works as follows. It is proven in [**Le95**] that there exist functions $J^T(\Phi, \pi)$ and $J^T(\widehat{\Phi}, \hat{\pi})$ of $T \in \mathbb{R}$ that are expressible of finite sums of products $P_\lambda^{\Phi,\pi}(T)e^{\lambda T}$ with $\lambda \in \mathbb{R}$ positive and $P_\lambda^{\Phi,f}$ a polynomial (we call these t-finite functions), such that

$$\left|J^T(\Phi,\pi) - \int_{G(k)\backslash G(\mathbb{A})} F(g,T) \sum_{\gamma\in V(k)} \Phi(\pi(g^{-1})\gamma)\, dg\right| = O(e^{-T})$$

and

$$\left|J^T(\widehat{\Phi},\hat{\pi}) - \int_{G(k)\backslash G(\mathbb{A})} F(g,T) \sum_{\hat{\gamma}\in \widehat{V}(k)} \widehat{\Phi}(\hat{\pi}(g^{-1})\hat{\gamma})\, dg\right| = O(e^{-T}).$$

The function F was defined in (1.1).

By the identity (2.1) we conclude that $J^T(\Phi,\pi) = J^T(\widehat{\Phi},\hat{\pi})$, and we define the regularized integral of the identity (2.1) to be the equality $P_0^{\Phi,\pi}(0) = P_0^{\widehat{\Phi},\hat{\pi}}(0)$ of constant terms.

We write $J(\Phi,\pi)$ for $P_0(0)$ so that the identity is $J(\Phi,\pi) = J(\widehat{\Phi},\hat{\pi})$.

In fact, one can prove a stronger result. It is proven in [**Le96**], subject to an assumption that was subsequently proven in [**Le97**], that if σ is an equivalence class of $G(k)$-orbits in $V(k)$ (a notion to be defined momentarily,) then there exists a function

$$J_\sigma^T(\Phi,\pi) = \sum_{\lambda\in\mathbb{R}} P_{\lambda,\sigma}^{\Phi,\pi}(T)e^{\lambda T}$$

that is t-finite in T such that

$$\sum_\sigma \left|J_\sigma^T(\Phi,\pi) - \int_{G(k)\backslash G(\mathbb{A})} F(g,T) \sum_{\gamma\in\sigma} \Phi(\pi(g^{-1})\gamma)\, dg\right| = O(e^{-T}), \tag{2.2}$$

and similarly with equivalence classes $\hat{\sigma} \subset \hat{V}(R)$. The identity

$$\sum_\sigma P_{0,\sigma}^{\Phi,\pi}(0) = \sum_{\hat{\sigma}} P_{0,\hat{\sigma}}^{\widehat{\Phi},\hat{\pi}}(0) \tag{2.3}$$

of constant terms corresponds to the "coarse σ-expansion" (see [**M**]) in Arthur's development of the trace formula.

Equivalence classes in $V(k)$ are defined as follows: every equivalence class contains exactly one $G(k)$ orbit $\pi(G(k))v$ of a point $v \in V(k)$ such that the variety $\pi(G)v$ is Zariski closed. Given such a point v, it is well-known [**BHC**, Theorem 3.8] that the stabilizer G_v of v is a reductive group. If the connected component G_v^0 of the identity in G_v is anisotropic over k, then the equivalence class of v is just the orbit $\pi(G(k))v$. We call such orbits and their elements, *regular elliptic* in analogy to the trace formula. Yukie [**Yu93**] calls them k-stable. The group G_v^0 is not anisotropic if and only if some element of $\pi(G(k))v$ lies in $\mathcal{P}(k) = \oplus_{\lambda\geq 0}V^\lambda(k)$, the direct sum of the non-negative weight spaces of V under the action of A. We may suppose that v is such an element. Then the equivalence class of v is defined to be $\pi(G(k))(v + \oplus_{\lambda>0}V^\lambda(k))$.

The analogue of Arthur's "fine σ-expansion" would be an expression like (2.3), following from an expression like (2.2), where σ now denotes any stratum in the Morse stratification (or in the general case, some as yet undetermined refinement of the Morse "stratification" that is a Whitney stratification) of any equivalence class. We will produce explicitly in section 3 such a "fine σ-expression" when $V = \mathrm{Sym}^4 k^2$. The Morse stratification is discussed in [**Yu93**], [**K84**], [**K85**] (see [**K85**] for discussion of stratifications not just of the null-cone), and is particularly simple if $V = \mathrm{Sym}^i k^2$. In this case, it is described as follows: every regular elliptic orbit comprises a single stratum. Given a vector v in the zero weight space $V^0(k)$, the equivalence class $\pi(G(k))(v+\oplus_{\lambda>0}V^\lambda(k))$ decomposes into $i/2$, if v is nonzero, and $[i+1/2]+1$, if v is zero, strata (notice that non-zero v exist only if i is even), given as follows. The 0th stratum is given by the orbit $\pi(G(k))v$. For

$$j' = i, i-2, i-4, \ldots, \left\{ \begin{array}{c} 1 \\ 2 \\ 4 \end{array} \right\},$$

where 2 is included only if $v=0$, write $j = i - j'/2+1$. For $j \geq 1$, the jth stratum is given by

$$\pi(G(k))\big(v + (V^{j'}(k) \setminus \{0\}) + \oplus_{\lambda>j'} V^\lambda(k)\big).$$

We have three last general remarks about the regularized integrals.

First, the proof of [**Le95**, Theorem 6.1] implies the bound

$$\int_{B(k)\backslash G(\mathbb{A})} \tau(H(g)-T) \sum_{\gamma \in V(k)\backslash \mathcal{P}(k)} |\Phi(\pi(g^{-1})\gamma)|\, dg = O(e^{-T}),$$

so for each stratum $\mathcal{S}$ of an equivalence class it is sufficient to find a t-finite approximation of the integral over $G(k)\backslash G(\mathbb{A})$ of

$$\sum_{\gamma\in\mathcal{S}} \Phi(\pi(g^{-1})\gamma) - \sum_{\delta \in B(k)\backslash G(k)} \tau(H(\delta g)-T) \sum_{\gamma \in \mathcal{S}\cap\mathcal{P}(k)} \Phi(\pi(\delta g)^{-1}\gamma) \tag{2.4}$$

instead of the integral of

$$\begin{aligned} F(g,T) &\sum_{\gamma\in\mathcal{S}} \Phi(\pi(g^{-1})\gamma) \\ &= \sum_{\gamma\in\mathcal{S}} \Phi(\pi(g^{-1})\gamma) - \sum_{\delta \in B(k)\backslash G(k)} \tau(H(\delta g)-T) \sum_{\gamma\in\mathcal{S}} \Phi(\pi(\delta g)^{-1}\gamma). \end{aligned}$$

An expression like (2.4) can be used to evaluate regularized integrals for representations (H,π,V) whenever the algebraic group H is semisimple with k-rank 1.

Second, the first remark implies that for $\mathcal{S} = \sigma = \pi(G(k))v$ a regular elliptic orbit, the integral

$$\int_{G(k)\backslash G(\mathbb{A})} \sum_{\gamma\in\sigma} \Phi(\pi(g^{-1})\gamma)\, dg$$

converges absolutely and equals $J_\sigma^T(\Phi,\pi)$ for any T, since the second term of (2.4) has a sum over the empty set $\mathcal{S}\cap\mathcal{P}(k)$. The bound (2.2) then implies that the sum

$$\sum_{\sigma \text{ regular elliptic}} \int_{G(k)\backslash G(\mathbb{A})} \sum_{\gamma\in\sigma} \Phi(\pi(g^{-1})\gamma)\, dg$$

converges absolutely and hence can be rewritten in the form (see [**Le95**])

$$\sum_{\{\gamma\}_{r.e.}} \operatorname{vol}(G_\gamma(k)\backslash G_\gamma(\mathbb{A})) \int_{G_\gamma(\mathbb{A})\backslash G(\mathbb{A})} \Phi(\pi(g^{-1})\gamma)\, dg, \tag{2.5}$$

where the sum is over a set $\{\gamma\}_{r.e.}$ of representatives of the regular elliptic $G(k)$-orbits and the same Haar measure on $G_\gamma(\mathbb{A})$ is used for the two factors.

Third, if the stratum $\mathcal{S} = \sigma$ consists of a single G-invariant vector γ (for example if $\gamma = 0$), then

$$J_\sigma^T(\Phi,\pi) = \operatorname{vol}(G(k)\backslash G(\mathbb{A}))\Phi(\gamma) = \mathfrak{B}_2\Phi(\gamma). \tag{2.6}$$

Define, for any representation (H, π, V) of a reductive algebraic group H, an *elliptic* element to be any $\gamma \in V(k)$ such that the center $Z(H_\gamma^0)\backslash Z(H)$ is anisotropic over k. Then one can show that for representations of $G = \mathrm{SL}(2)$ every elliptic element is either regular elliptic or G-invariant and we have proven that both sides of the equality

$$\int_{G(k)\backslash G(\mathbb{A})} \sum_{\substack{\gamma\in V(k)\\ \text{elliptic}}} \Phi(\pi(g^{-1})\gamma)\, dg \tag{2.7}$$
$$= \sum_{\{\gamma\}_e} \mathrm{vol}(G_\gamma(k)\backslash G_\gamma(\mathbb{A})) \int_{G_\gamma(\mathbb{A})\backslash G(\mathbb{A})} \Phi(\pi(g^{-1})\gamma)\, dg,$$

converge absolutely, where the sum is over a set $\{\gamma\}_e$ of representatives of all the elliptic $G(k)$-orbits. The validity of this statement in general has been studied by many people, see for example [**Y**], [**Yu95**].

3. Computations

At this point, we restrict to the case where $V = \mathrm{Sym}^4 k^2$, which as above, we identify with k^5 by

$$f\begin{pmatrix} u \\ v \end{pmatrix} = \sum_{i=0}^{4} x_i u^{4-i} v^i \mapsto (x_0, \dots, x_4).$$

We identify $\hat{V}$ with V via $[,]$ as in section 2. We will write $g \cdot v$ for $\pi(g)v$, for $g \in G, v \in V$. (There will be no confusion with the variable v in the line above because throughout this section we will write elements $v \in V$ as row vectors in k^5.) Notice that

$$a(t) \cdot (x_0, x_1, x_2, x_3, x_4) = (t^{-4}x_0, t^{-2}x_1, x_2, t^2 x_3, t^4 x_4).$$

The equivalence classes that are not regular elliptic orbits can be parametrized by k by setting $\sigma_x,\ x \in k$ to be the equivalence class

$$\sigma_x = G(k) \cdot \{(0, 0, x, y, z) \mid y, z \in k\}.$$

If x is 0, σ_0 is the null-cone $\mathcal{N}$ and is the union of three strata:

$$\begin{aligned} \mathcal{S}(0)_0 &= \{0\}, \\ \mathcal{S}(0)_1 &= G(k) \cdot \{(0, 0, 0, 0, y) | y \in k, y \neq 0\}, \\ \mathcal{S}(0)_2 &= G(k) \cdot \{(0, 0, 0, y, z) | y, z \in k, y \neq 0\}. \end{aligned}$$

If x is non-zero, σ_x is the union of two strata:

$$\begin{aligned} \mathcal{S}(x)_0 &= G(k) \cdot (0, 0, x, 0, 0), \\ \mathcal{S}(x)_1 &= G(k) \cdot \{(0, 0, x, 0, y) | y \in k, y \neq 0\}. \end{aligned}$$

Notice that every orbit in every equivalence class σ_x that is not regular elliptic contains a point in $\{(0, 0, x, *, *)\} \subset \mathcal{P}(k)$. We will find it useful to give names to the following points in V.

DEFINITION 1. Let $\gamma_0 = 0, \gamma_1 = (0, 0, 0, 0, 1), \gamma_2 = (0, 0, 0, 1, 0)$, and for $x \in k^\times$, let $\gamma^x = (0, 0, x, 0, 0)$. For a stratum $\mathcal{S}$ of the form $\mathcal{S}(0)_i$, $i = 1, 2, 3$ or $\mathcal{S}(x)_0$, $x \in k^\times$, let $\gamma_{\mathcal{S}}$ be the element of the above list lying in $\mathcal{S}$.

We will prove that for each stratum $\mathcal{S}$, there exists a t-finite function $J_{\mathcal{S}}^T(\Phi, \mathrm{Sym}^4k^2)$ of T such that the limit as T approaches ∞ of the difference of $J_{\mathcal{S}}^T(\Phi, \mathrm{Sym}^4k^2)$ and (2.4) is 0. This property uniquely specifies each $J_{\mathcal{S}}^T(\Phi, \mathrm{Sym}^4k^2)$ and if we write $J_{\mathcal{S}}(\Phi, \mathrm{Sym}^4k^2)$ for the constant term of the t-finite function $J_{\mathcal{S}}^T(\Phi, \mathrm{Sym}^4k^2)$, the relation (2.2) and the finiteness of the stratification of each equivalence class imply that

$$\sum_{\mathcal{S}} J_{\mathcal{S}}(\Phi, \mathrm{Sym}^4k^2) = \sum_{\mathcal{S}} J_{\mathcal{S}}(\widehat{\Phi}, \mathrm{Sym}^4k^2),$$

where $\mathcal{S}$ ranges over all strata of all equivalence classes in $V(k)$. We have already found $J_{\mathcal{S}}^T(\Phi, \mathrm{Sym}^4k^2) = J_{\mathcal{S}}(\Phi, \mathrm{Sym}^4k^2)$ for $\mathcal{S}$ an elliptic orbit, that is, if $\mathcal{S}$ is a regular elliptic orbit or $\mathcal{S}(0)_0$. In this section we will examine each of the remaining strata in turn.

We will need the following Lemma, whose proof follows from simple calculations.

LEMMA 1. *(a) Suppose that $\gamma \in V(k)$ is elliptic. If the form γ has a double root (in $\bar{k}$), then γ is of the form cq^2 with $c \in k$ and q a quadratic form with coefficients in k, corresponding to a quadratic extension F/k, and $G_\gamma \cong \mathbf{U}_{F/k}(1)$. If the form γ has no double roots, then $G_\gamma = \{1\}$.*

(b) Suppose that $\gamma \in \mathcal{P}(k)$ is a non-elliptic vector. Then G_γ depends only on the stratum $\mathcal{S}$ in which γ lies and is given by the following table, where we write $G_{\mathcal{S}}$ for the group G_γ, $\gamma \in \mathcal{S} \cap \mathcal{P}(k)$:

$\mathcal{S}$	$G_{\mathcal{S}}$
$\mathcal{S}(0)_0$	G
$\mathcal{S}(0)_1$	$a(\mu_4)N$
$\mathcal{S}(0)_2$	$a(\mu_2)$
$\mathcal{S}(x)_0\ x \neq 0$	$N_G(A)$
$\mathcal{S}(x)_1\ x \neq 0$	$a(\mu_4)$

Here μ_n denotes the group of nth roots of unity in k and $N_G(A)$ is the normalizer of the torus A.
(c) For any function f on $V(k)$ and non-elliptic stratum $\mathcal{S}$,

$$\sum_{\gamma \in \mathcal{S} \cap \mathcal{P}(k)} f(\gamma) = \frac{1}{n_{\mathcal{S}}} \sum_{q \in k^\times / k^{\times n_{\mathcal{S}}}} \sum_{\delta \in (B \cap G_{\mathcal{S}})^0(k) \backslash B(k)} f(\delta^{-1} \cdot \gamma_{\mathcal{S}}(q)),$$

where $\gamma_{\mathcal{S}}(q)$ is the point $q\gamma_{\mathcal{S}}$ unless $\mathcal{S} = \mathcal{S}(x)_1, x \neq 0$, when it is $\gamma^x + q\gamma_1$, $(B \cap G_{\mathcal{S}})^0$ is the connected component of $B \cap G_{\mathcal{S}}$, and $n_{\mathcal{S}}$ is the number of connected components of $G_{\mathcal{S}} \cap B$.

(d) For any function f on $V(k)$ and non-elliptic stratum $\mathcal{S}$,

$$\sum_{\gamma \in \mathcal{S}} f(\gamma) = \frac{1}{m_{\mathcal{S}}} \sum_{\delta \in B(k) \backslash G(k)} \sum_{\gamma \in \mathcal{S} \cap \mathcal{P}(k)} f(\delta^{-1} \cdot \gamma),$$

where $m_{\mathcal{S}} = |G_{\mathcal{S}}/G_{\mathcal{S}} \cap B|$.

DEFINITION 2. Let $\Phi \in \mathcal{S}(V(R))$, R a local field or the ring of adeles. Define $\Phi^{K_R} \in \mathcal{S}(V(R))$ by

$$\Phi^{K_R}(V) = \int_{K_R} \Phi(k \cdot v)\, dk, v \in V(R),$$

$\Phi_1 \in \mathcal{S}(R)$ *by*

$$\Phi_1(x) = \Phi^{K_R}(x\gamma_1) = \Phi^{K_R}(0,0,0,0,x), x \in R,$$

$\Phi_2 \in \mathcal{S}(R)$ *by*

$$\Phi_2(x) = |x| \int_{N(R)} \Phi^{K_R}(n^{-1} \cdot x\gamma_2) dn = \int_R \Phi^{K_R}(0,0,0,x,y) dy, x \in R,$$

$\Phi^x \in \mathcal{S}(R)$ *by*

$$\begin{aligned} \Phi^x(y) &= \int_{N(R)} \Phi^{K_R}(n^{-1} \cdot (0,0,x,0,y)) dn \\ &= \int_R \Phi^{K_R}(0,0,x,2nx,n^2x+y) dn, y \in R, x \in R^\times. \end{aligned}$$

REMARK 1. Verifying that the functions we have defined are Schwartz-Bruhat is necessary but elementary.

We will assume, for simplicity, that Φ can be written in the form

$$\Phi(\gamma) = \prod_v \Phi_v(\gamma_v),$$

with each $\Phi_v \in \mathcal{S}(V(k_v))$. Since every function in $\mathcal{S}(V(\mathbb{A}))$ is a finite sum of such functions, this will result in no loss of generality.

Consider the stratum $\mathcal{S} = \mathcal{S}(0)_1$. By Lemma 1(c),(d), the expression (2.4) equals

$$\begin{aligned} &\sum_{\delta \in B(k)\backslash G(k)} \big(1 - \tau(H(\delta g) - T)\big) \sum_{\gamma \in \mathcal{S} \cap \mathcal{P}(k)} \Phi((\delta g)^{-1} \cdot \gamma) \\ &\qquad = \frac{1}{|\mu_4|} \sum_{q \in k^\times/k^{\times 4}} \sum_{\delta \in N(k)\backslash G(k)} (1 - \tau(H(\delta g) - T)\ \Phi((\delta g)^{-1} \cdot \gamma_1) \end{aligned}$$

When integrating this over $G(k)\backslash G(\mathbb{A})$, we can absorb the sum over δ into the integral, then use the Iwasawa decomposition $N(k)\backslash G(\mathbb{A}) = (N(k)\backslash N(\mathbb{A}))A(\mathbb{A})K$ to obtain the integral

$$\begin{aligned} (3.1) \qquad &\frac{1}{|\mu_4|} \sum_{q \in k^\times/k^{\times 4}} \int_{\mathbb{A}^\times} |a|^{-2}(1 - \tau(\log|a| - T)) \\ &\quad \cdot \int_{N(k)\backslash N(\mathbb{A})} \int_K \Phi((na(a)k)^{-1} \cdot q\gamma_1) dk\ dn\ d^\times a \\ &= \frac{1}{|\mu_4|} \sum_{q \in k^\times/k^{\times 4}} \int_{\mathbb{A}^\times} |a|^{-2}\big(1 - \tau(\log|qa^4| - 4T)\big)\Phi_1(qa^{-4})\ d^\times a, \end{aligned}$$

where we recall that $N \subset G_{\gamma_1}$ and $\mathrm{vol}(N(k)\backslash N(\mathbb{A})) = 1$. (Notice that the integrals all have compact support, so converge absolutely and can be manipulated without worry. We will not mention this again.)

By Fourier inversion on $\mathbb{A}^\times/k^\times\mathbb{A}^{\times 4}$ (see [**W89**]), (3.1) equals (recall that $\Phi_1(a^{-1}) = \Phi^K(a^{-1}\gamma_1)$)

$$(3.2) \qquad \frac{1}{4} \sum_{\chi \in \Omega(\mathbb{A}^\times/k^\times)_4} \int_{\mathbb{A}} |a|^{-1/2}\chi(a)\big(1 - \tau(\log|a| - 4T)\big)\Phi_1(a^{-1})\ d^\times a,$$

where $\Omega(\mathbb{A}^\times/k^\times)_4$ denotes the (infinite) set of continuous quasi-characters of $\mathbb{A}^\times/k^\times$ of order 4; for only finitely many χ is the integral non-zero. Fix one character χ.

Introduce a factor of $|a|^{-s}$ to the integrand, with s a complex number, to obtain the entire function

$$\frac{1}{4}\int_{\mathbb{A}^\times} |a|^{-s-1/2}\chi(a)(1-\tau(\log|a|-4T))\Phi_1(a^{-1})\, d^\times a. \tag{3.3}$$

(It is entire because the integral converges absolutely, because the last factor deals with $|a|$ small and the previous factor with $|a|$ large.) We can formally separate (3.3) as 1/4 the difference of two integrals:

$$\int_{\mathbb{A}^\times} |a|^{-s-1/2}\chi(a)\Phi_1(a^{-1})\, d^\times a, \tag{3.4}$$

and

$$\int_{\mathbb{A}^\times} |a|^{-s-1/2}\chi(a)\tau(\log|a|-4T)\Phi_1(a^{-1})\, d^\times a; \tag{3.5}$$

both converge absolutely for $\Re(s) > 1/2$, and we will analytically continue them both to $s=0$.

For $\Re(s) > 1/2$, the integral (3.4) is just the Tate zeta function $\zeta(\omega_{s+1/2}\chi^{-1}, \Phi_1)$ of Φ_1, which is well-known to meromorphically continue to $s \in \mathbb{C}$ and does not have a pole at $s=0$.

The integral (3.5) equals

$$\begin{aligned}
&\int_{\mathbb{A}^\times/k^\times} |a|^{-s-1/2}\chi(a)\tau(\log|a|-4T)\sum_{q\in k^\times}\Phi_1(qa^{-1})\, d^\times a \qquad (3.6)\\
&= \int_{\mathbb{A}^\times} |a|^{-s+1/2}\chi(a)\tau(\log|a|-4T)\widehat{\Phi_1}(a)\, d^\times a\\
&\quad - \Phi_1(0)\int_{\mathbb{A}^\times/k^\times} |a|^{-s-1/2}\chi(a)\tau(\log|a|-4T)\, d^\times a\\
&\quad + \widehat{\Phi_1}(0)\int_{\mathbb{A}^\times/k^\times} |a|^{-s+1/2}\chi(a)\tau(\log|a|-4T)\, d^\times a\\
&= \int_{\mathbb{A}^\times} |a|^{-s+1/2}\chi(a)\tau(\log|a|-4T)\widehat{\Phi_1}(a)\, d^\times a\\
&\quad - \Phi_1(0)\left(\int_{4t}^{\infty} e^{(-s-1/2)x}\, dx\right)\left(\int_{\mathbb{A}^1/k^\times}\chi(a)\, d^\times a\right)\\
&\quad + \widehat{\Phi_1}(0)\left(\int_{4t}^{\infty} e^{(-s+1/2)x}\, dx\right)\left(\int_{\mathbb{A}^1/k^\times}\chi(a)\, d^\times a\right),
\end{aligned}$$

where we have used the Poisson summation formula, $\widehat{\Phi_1}$ means the Fourier transformation. in $\mathcal{S}(\mathbb{A})$ of Φ_1. The last two terms of (3.6) are zero unless χ is trivial, when they equal

$$\Phi_1(0)\frac{1}{s+1/2}\, e^{(-s-1/2)4T}, \tag{3.7}$$

and

$$-\widehat{\Phi_1}(0)\frac{1}{s-1/2}\, e^{(-s+1/2)4T}, \tag{3.8}$$

for $\Re(s) > -1/2$, and $\Re(s) > 1/2$, respectively. Both (3.7) and (3.8) are meromorphic functions of $s \in \mathbb{C}$. At $s=0$, (3.7) tends to 0 as $T \to \infty$ and (3.8) is a t-finite function in T with zero constant term. Additionally, a routine computation shows that since Φ_1 is a Schwartz-Bruhat function on $\mathbb{A}$, the first term of (3.6) converges for all $s \in \mathbb{C}$ and when $s = 0$, its value as T tends to infinity is zero. We have therefore shown that (3.5) has meromorphic continuation to the complex plane and its value at $s = 0$ approaches t-finite function of T as T tends to infinity and that the constant term of this t-finite function is zero. Therefore,

$$J_{\mathcal{S}(0)_1}(\Phi,\ \mathrm{Sym}^4 k^2) = \frac{1}{4} \sum_{\chi \in \Omega(\mathbb{A}^\times/k^\times)_4} \zeta(\omega_{1/2}\chi, \Phi_1).$$

Next consider the stratum $\mathcal{S} = \mathcal{S}(0)_2$. By Lemma 1(c),(d), the expression (2.4) equals

$$\begin{aligned} &\sum_{\delta \in B(k)\backslash G(k)} (1 - \tau(H(\delta g) - T)) \sum_{\gamma \in \mathcal{S}\cap\mathcal{P}(k)} \Phi((\delta g)^{-1} \cdot \gamma) \\ &= \frac{1}{2} \sum_{\delta \in G(k)} (1 - \tau(H(\delta g) - T)) \sum_{q \in k^\times/k^{\times 2}} \Phi((\delta g)^{-1} \cdot q\gamma_2). \end{aligned}$$

When integrating this over $G(k)\backslash G(\mathbb{A})$, we absorb the sum over δ into the integral, then use the Iwasawa decomposition to obtain the expression

$$\begin{aligned} &\frac{1}{2} \int_{\mathbb{A}^\times} (1 - \tau(\log|a| - T)) \int_{\mathbb{A}} \int_K \sum_{q \in k^\times/k^{\times 2}} \Phi((a(a)n(x)k)^{-1} \cdot q\gamma_2)\, dk\, dx\, d^\times a \\ &= \frac{1}{2} \int_{\mathbb{A}^\times} (1 - \tau(\log|a| - T))\, |a|^2 \Phi_2(qa^{-2})\, d^\times a. \end{aligned}$$

Now, we apply Fourier inversion on $\mathbb{A}^\times/k^\times\mathbb{A}^{\times 2}$, introduce a factor of $|a|^{-s}$, and separate the resulting integral as 1/2 of the difference of

$$\sum_{\chi \in \Omega(\mathbb{A}^\times/k^\times)_2} \int_{\mathbb{A}^\times} |a|^{1-s} \chi(a) \Phi_2(a^{-1})\, d^\times a \tag{3.9}$$

and

$$\sum_{\chi \in \Omega(\mathbb{A}^\times/k^\times)_2} \int_{\mathbb{A}^\times} |a|^{1-s} \chi(a) \tau(\log|a| - 2T) \Phi_2(a^{-1})\, d^\times a; \tag{3.10}$$

both expressions converge absolutely for $\Re(s) > 1$. The expression (3.9) is

$$\sum_{\chi \in \Omega(\mathbb{A}^\times/k^\times)_2} \zeta(\omega_{s-1}\chi, \Phi_2),$$

which meromorphically continues to the whole plane, taking the value

$$\sum_{\chi \in \Omega(\mathbb{A}^\times/k^\times)_2} \zeta(\omega_{-1}\chi, \Phi_2),$$

at $s = 0$. We deal with (3.10) exactly as with (3.5) and find that it can be meromorphically continued to the whole complex plane and that when $s = 0$, as T approaches ∞, (3.10) approaches

$$\Phi_2(0)\, e^{2T} + \frac{1}{2}\widehat{\Phi_2}(0)\, e^{4T}. \tag{3.11}$$

Notice that the e^{2T} term in (3.11), being multiplied by $1/2$, cancels with the e^{2T} term from (3.8), being multiplied by $1/4$, since $\widehat{\Phi_1}(0) = \Phi_2(0)$. The expression (3.8) is t-finite in T with constant term 0, so we conclude that

$$J_{\mathcal{S}(0)_2}(\Phi,\ \mathrm{Sym}^4 k^2) = \frac{1}{2} \sum_{\chi \in \Omega(\mathbb{A}^\times/k^\times)_2} \zeta(\omega_{-1}\chi, \Phi_2).$$

Next consider a stratum $\mathcal{S} = \mathcal{S}(x)_0, x \neq 0$. By Lemma 1(c),(d), the expression (2.4) equals

$$\frac{1}{2} \sum_{\delta \in A(k)\backslash G(k)} \Phi((\delta g)^{-1} \cdot \gamma^x) - \sum_{\delta \in A(k)\backslash G(k)} \tau(H(\delta g) - T)\Phi((\delta g)^{-1} \cdot \gamma^x)$$
$$= \frac{1}{2} \sum_{\delta \in A(k)\backslash G(k)} (1 - \tau(H(\delta g) - T) - \tau(H(w\delta g) - T))\Phi((\delta g)^{-1} \cdot \gamma^x),$$

where w is the matrix $\left(\begin{smallmatrix} 0 & -1 \\ 1 & 0 \end{smallmatrix}\right) \in G(k)$, and its integral over $G(k)\backslash G(\mathbb{A})$ is seen to be, after absorbing the sum over δ and using the Iwasawa decomposition,

$$\begin{aligned}(3.12)\qquad \frac{1}{2} \int_{N(A)} \Big[\int_{A(k)\backslash A(\mathbb{A})} &(1 - \tau(H(an) - T) - \\ &\tau(H(wan) - T))\, da \Big] \Phi^K(n^{-1} \cdot \gamma^x)\, dn.\end{aligned}$$

If we write $\lambda(n)$ for the number $H(wn), n \in N(\mathbb{A})$, the inner integral equals

$$\int_{A(k)\backslash A(\mathbb{A})} (1 - \tau(H(a) - T) - \tau(\lambda(n) - H(a) - T))\, da$$
$$= \int_{e^{\lambda(n)-T}}^{e^T} \frac{dx}{\lambda} = 2T - \lambda(n),$$

so (3.12) equals

$$(3.13)\qquad T \int_{N(\mathbb{A})} \Phi^K(n^{-1} \cdot \gamma^x)\, dn - \frac{1}{2} \int_{N(\mathbb{A})} \lambda(n)\Phi^K(n^{-1} \cdot \gamma^x)\, dn.$$

This is obviously a t-finite function in T with constant term

$$J_{\mathcal{S}(x)_0}(\Phi,\ \mathrm{Sym}^4 k^2) = -\frac{1}{2} \int_{N(\mathbb{A})} \lambda(n)\Phi^K(n^{-1} \cdot \gamma^x)\, dn.$$

REMARK 2. $\mathcal{S}(x)_0$ is the first stratum we have encountered where

$$J_{\mathcal{S}}(\Phi,\ \mathrm{Sym}^4 k^2)$$

is not a G-invariant distribution in Φ. Its non-invariance will cancel with that of $J_{\mathcal{S}(x)_1}(\Phi,\ \mathrm{Sym}^4 k^2)$, the contribution from the other stratum in the same equivalence class. Details will appear elsewhere.

REMARK 3. Notice that $J_{\mathcal{S}(x)_0}(\Phi,\ \mathrm{Sym}^4 k^2)$ can also be written as

$$-\frac{1}{2} \frac{d}{ds} \left(\int_{N(\mathbb{A})} \alpha(n)^s \Phi^K(n^{-1} \cdot \gamma^x)\, dn \right) \Big|_{s=0^+},$$

where $\alpha(n)$ is defined as in [**Yu92**] by

$$\alpha\left(\begin{pmatrix} 1 & n \\ 0 & 1 \end{pmatrix}\right) = \prod_v \alpha_v(n_v), \quad \textit{where} \quad \alpha_v(n_v) = \begin{cases} \sup(1, |n_v|_v)^{-1} & \text{for } v \text{ finite} \\ (1 + |n_v|_v^2)^{-1/2} & \text{for } v \text{ real} \\ (1 + |n_v|_v)^{-1} & \text{for } v \text{ complex} \end{cases}$$

and $f(s)|_{s=0^+}$ means the limit of f as s in the right half-plane $\Re(s) > 0$ tends to 0.

Consider lastly a stratum $\mathcal{S} = \mathcal{S}(x)_1, x \neq 0$. By Lemma 1(c),(d), the expression (2.4) equals

$$\frac{1}{|\mu_4|} \sum_{q \in k^\times / k^{\times 4}} \sum_{\delta \in G(k)} (1 - \tau(H(\delta g) - T)) \; \Phi((\delta g)^{-1} \cdot (0,0,x,0,q))$$

and its integral over $G(k)\backslash G(\mathbb{A})$ can be seen, after absorbing the sum over δ into the integral and using the Iwasawa decomposition, to equal

$$\frac{1}{|\mu_4|} \sum_{q \in k^\times / k^{\times 4}} \int_{\mathbb{A}^\times} (1 - \tau(\log|a| - T)) \int_{\mathbb{A}} \Phi^K((a(a)n(n))^{-1} \cdot (0,0,x,0,q)) \; dn \; d^\times a$$

$$= \frac{1}{|\mu_4|} \sum_{q \in k^\times / k^{\times 4}} \int_{\mathbb{A}^\times} (1 - \tau(\log|a| - T)) \; \Phi^x(qa^{-4}) \; d^\times a \tag{3.14}$$

$$= \frac{1}{4} \sum_{\chi \in \Omega(\mathbb{A}^\times / k^\times)_4} \int_{\mathbb{A}^\times} \chi(a)(1 - \tau(\log|a| - 4T)) \; \Phi^x(a^{-1}) \; d^\times a.$$

The sum over χ has only many finitely many non-zero terms. Fix one character $\chi \in \Omega(A^\times / k^\times)_4$, introduce a factor of $|a|^{-s}$ to the integral and separate it as 1/4 of the difference of

$$\int_{\mathbb{A}^\times} |a|^{-s} \chi(a) \Phi^x(a^{-1}) \; d^\times a \tag{3.15}$$

and

$$\int_{\mathbb{A}^\times} |a|^{-s} \tau(\log|a| - 4T) \; \Phi^x(a^{-1}) \; d^\times a, \tag{3.16}$$

where both integrals converge absolutely for $\Re(s) > 0$. This time the meromorphic continuation of both integrals will have a simple pole at $s = 0$ if χ is trivial, and (3.14) is the difference of the constant terms of their Laurent expansions at $s = 0$.

The integral (3.15) equals $\zeta(\omega_s \chi^{-1}, \Phi^x)$. If the character χ is non-trivial, this zeta function is defined at $s = 0$. Furthermore, the arguments above imply that (3.16) has holomorphic continuation to the whole complex plane, and that when $s = 0$, its limit as T approaches ∞ is zero. We have thus dealt with all non-trivial characters.

Suppose χ is the trivial character. Then (3.15) equals

$$\zeta(\omega_s, \Phi^x) = \zeta(\omega_s) \mathfrak{C}_k^{-1} \prod_v \frac{\zeta_v(\omega_s, \Phi_v^x)}{\zeta_v(\omega_s)} = \zeta(\omega_{1-s}) \mathfrak{C}_k^{-1} \prod_v \frac{\zeta_v(\omega_s, \Phi_v^x)}{\zeta_v(\omega_s)}$$

where the product is over all places of a quotient of local Tate zeta functions; each of these quotients is originally defined for $\Re(s) > 0$, but extends to an entire function of s, and is the constant function 1 for almost all v. The constant $\mathfrak{C}_k^{-1}$ appears because of the normalizations of our local and global measures. Using the functional equation of the local Tate zeta functions we find that the constant term of the Laurent expansion of (3.15) about $s = 0$ is the limit as s approaches 0 of

$$\mathfrak{C}_k^{-1} \lambda_0 \prod_v \frac{\zeta_v(\omega_s, \Phi_v^x)}{\zeta_v(\omega_s)} - |\Delta_k|^{-1/2} \frac{d}{ds} \prod_v \frac{\zeta_v(\omega_s, \Phi_v^x)}{\zeta_v(\omega_s)} \tag{3.17}$$

where the first two terms of the Laurent expansion of $\zeta(\omega_{1+s}) = Z_k(1+s)$ are $(\mathfrak{C}_k/|\Delta_k|^{1/2})s^{-1} + \lambda_0$. If χ is trivial, (3.16) can be shown, as with (3.5), to have a meromorphic continuation to all $s \in \mathbb{C}$, and its difference from

$$\Phi^x(0)\frac{1}{s}\, e^{-4sT} - \widehat{\Phi^x}(0)\frac{1}{s-1}\, e^{(-s+1)4T} \tag{3.18}$$

is an entire function $f(s,T)$ of s satisfying

$$\lim_{T\to\infty} f(0,T) = 0$$

The expression (3.18) has a pole at $s=0$ that cancels the pole of (3.15) and the constant term in its Laurent expansion about 0 is

$$-4T\Phi^x(0) + \widehat{\Phi^x}(0)\, e^{4T},$$

a t-finite function on T with constant term zero. Notice that the term $-4T\Phi^x(0)$, being multiplied by $1/4$, cancels the first term of (3.13). Therefore,

$$J_{\mathcal{S}(x)_1}(\Phi, \mathrm{Sym}^4 k^2)$$

is $1/4$ of the sum of (3.17) and

$$\sum_{\chi\in\Omega(\mathbb{A}^\times/k^\times)_4 \chi\neq 1} \zeta(\chi,\Phi^x) = \sum_{\substack{\chi\in\Omega(\mathbb{A}^\times/k^\times)_4\\ \chi\neq 1}} \mathfrak{C}_k^{-1}\zeta(\chi)\left(\prod_v \frac{\zeta_v(\chi_v\omega_s,\Phi_v^x)}{\zeta_v(\chi_v\omega_s)}\right)\Big|_{s=0^+}.$$

Notice that the summand is zero for all but finitely many terms, the number depending on Φ. Given a character $\chi\in\Omega(\mathbb{A}^\times/k^\times)_4$, write C_χ for $\mathfrak{C}_k^{-1}$ times the constant term of the Laurent expansion about 0 of the zeta function $\zeta(\chi\omega_s)$, that is, the number $\mathfrak{C}_k^{-1}\zeta(\chi)$ if χ is nontrivial and $\mathfrak{C}_k^{-1}\lambda_0$ if χ is trivial. Therefore

$$J_{\mathcal{S}(x)_1}(\Phi,\mathrm{Sym}^4k^2) = \frac{1}{4}\left[\sum_{\chi\in\Omega(\mathbb{A}^\times/k^\times)_4} C_\chi\prod_v\frac{\zeta_v(\chi_v\omega_s,\Phi_v^x)}{\zeta_v(\chi_v\omega_s)} - |\Delta_k|^{-1/2}\frac{d}{ds}\prod_v\frac{\zeta_v(\omega_s,\Phi_v^x)}{\zeta_v(\omega_s)}\right]_{s=0^+}.$$

Let us summarize our results to date.

THEOREM 1. *Given $\Phi\in\mathcal{S}(V(\mathbb{A}))$, the identity*

$$\sum_{\mathcal{S}} J_{\mathcal{S}}(\Phi,\ \mathrm{Sym}^4k^2) = \sum_{\mathcal{S}} J_{\mathcal{S}}(\widehat{\Phi},\ \mathrm{Sym}^4k^2)$$

holds, where $\mathcal{S}$ ranges over all strata in all equivalence classes in $V(k)$ and

$$J_{\mathcal{S}}(\Phi,\ \mathrm{Sym}^4k^2)$$

is given by

$$\mathfrak{B}_2\Phi(0)$$

if $\mathcal{S}=\mathcal{S}(0)_0=\{0\}$, by

$$\int_{G(\mathbb{A})}\Phi(g^{-1}\cdot\gamma)\,dg$$

if $\mathcal{S}=G(k)\cdot\gamma$ is a regular elliptic orbit of a form γ without double roots, by

$$\mathrm{vol}(G_\gamma(k)\backslash G_\gamma(\mathbb{A}))\int_{G_\gamma(\mathbb{A})\backslash G(\mathbb{A})}\Phi(g^{-1}\cdot\gamma)\,dg$$

if $\mathcal{S} = G(k)\cdot\gamma$ is a regular elliptic orbit of a form γ whose (double) roots generate the quadratic extension F/k, by

$$\frac{1}{4}\sum_{\chi\in\Omega(\mathbb{A}^\times/k^\times)_4}\zeta(\omega_{1/2}\chi,\Phi_1)$$

if $\mathcal{S} = \mathcal{S}(0)_1$, by

$$\frac{1}{2}\sum_{\chi\in\Omega(\mathbb{A}^\times/k^\times)_2}\zeta(\omega_{-1}\chi,\Phi_2)$$

if $\mathcal{S} = \mathcal{S}(0)_2$, by

$$-\frac{1}{2}\int_{N(\mathbb{A})}\lambda(n)\Phi^K(n^{-1}\cdot\gamma^x)\,dn$$

if $\mathcal{S} = \mathcal{S}(x)_0$, and by

$$\frac{1}{4}\left[\sum_{\chi\in\Omega(\mathbb{A}^\times/k^\times)_4}C_\chi\prod_v\frac{\zeta_v(\chi_v\omega_s,\Phi_v^x)}{\zeta_v(\chi_v\omega_s)}-|\Delta_k|^{-1/2}\frac{d}{ds}\prod_v\frac{\zeta_v(\omega_s,\Phi_v^x)}{\zeta_v(\omega_s)}\right]_{s=0^+}$$

if $\mathcal{S} = \mathcal{S}(x)_1$, where the functions Φ_1, Φ_2, Φ^x are defined in Definition 2, λ is defined after (3.12), and C_χ is defined just above (see section 1 for other notation).

4. The zeta function

Write $V^s(k)$ for the set of regular elliptic elements of $V(k)$; this is standard notation, as the superscript s stands for k-stable in Yukie's notation [**Yu93**]. The Shintani zeta function for the space of binary quartic forms is defined to be the integral

$$Z(\Phi,\omega)=\int_{\mathbb{A}^\times/k^\times}\int_{G(k)\backslash G(\mathbb{A})}\omega(x)\sum_{\gamma\in V^s(k)}\Phi(g^{-1}\cdot x\gamma)dg\,d^\times x,$$

with $\Phi\in\mathcal{S}(V(\mathbb{A}))$ and ω in $\Omega(\mathbb{A}^\times/k^\times)$, the group of quasi-characters on $\mathbb{A}^\times/k^\times$. This integral corresponds to the representation $(\mathrm{GL}(1)\times\mathrm{SL}(2),V)$ where $\mathrm{GL}(1)$ acts by scalar multiplication. Since the Jacobian of the action of $x\in\mathrm{GL}(1,\mathbb{A})$ on $V(\mathbb{A})$ is $|x|^5$, this integral converges absolutely if the unramified part of ω has real part $c(\omega)$ greater than 5. Theorem 3.1 gives analytic information about $Z(\Phi,\omega)$, just as the Poisson summation formula gives analytic information about the Riemann zeta function.

Define

$$Z_+(\Phi,\omega)=\int_{\substack{\mathbb{A}^\times/k^\times\\|x|\geq 1}}\int_{G(k)\backslash G(\mathbb{A})}\omega(x)\sum_{\gamma\in V^s(k)}\Phi(g^{-1}\cdot x\gamma)dg\,d^\times x;$$

this integral converges absolutely to an entire function of ω. By Theorem 3.1, if $c(\omega)>5$, we have the following:

$$Z(\Phi,\omega)=Z_+(\Phi,\omega)+\int_{\substack{\mathbb{A}^\times/k^\times\\|x|\leq 1}}\int_{G(k)\backslash G(\mathbb{A})}\omega(x)\sum_{\gamma\in V^s(k)}\Phi(g^{-1}\cdot x\gamma)dg\,d^\times x$$

$$=Z_+(\Phi,\omega)-\int_{\substack{\mathbb{A}^\times/k^\times\\|x|\leq 1}}\omega(x)\left(\sum_{\mathcal{S}\text{ not r.e.}}J_{\mathcal{S}}(x\cdot\Phi,\mathrm{Sym}^4k^2)\right)d^\times x \tag{4.1}$$

$$+ Z_+(\widehat{\Phi}, \omega_5\omega^{-1}) + \int_{\substack{\mathbb{A}^\times/k^\times \\ |x|\le 1}} \omega(x) \left(\sum_{\mathcal{S} \text{ not r.e.}} J_{\mathcal{S}}(\widehat{x\cdot\Phi}, \mathrm{Sym}^4 k^2) \right) d^\times x,$$

where $x \cdot \Phi$ is the Schwartz-Bruhat function given by $x \cdot \Phi(v) = \Phi(xv)$ and "r.e." stands for regular elliptic. Notice that for each stratum $\mathcal{S}$, $J_{\mathcal{S}}(\widehat{x\cdot\Phi}, \mathrm{Sym}^4 k^2) = |x|^{-5} J_{\mathcal{S}}(x^{-1}\cdot\widehat{\Phi}, \mathrm{Sym}^4 k^2)$.

We will pull that part of the sum corresponding to the strata $\mathcal{S}(0)_i$, $i = 0, 1, 2$ of the null-cone out of the integral; this is justified because these three integrals can easily be shown to converge absolutely for $c(\omega) > 5$. The integrals are very easy to calculate because each of the strata of the null-cone in V are preserved by the GL(1)-action.

It is easily checked that for $x \in \mathbb{A}^\times$ we have

$$(x\cdot\Phi)(0) = \Phi(0), \quad (x\cdot\Phi)_1(y) = \Phi_1(xy), \quad (x\cdot\Phi)_2(y) = |x|^{-1}\Phi_2(xy).$$

Define $\delta(\omega)$, for a character ω of $\mathbb{A}^\times/k^\times$, to be 1 if ω is trivial and 0 otherwise. Then

$$\int_{\substack{\mathbb{A}^\times/k^\times \\ |x|\le 1}} \omega(x) J_{\mathcal{S}(0)_0}(x\cdot\Phi, \mathrm{Sym}^4 k^2) d^\times x = \mathrm{vol}(G(k)\backslash G(\mathbb{A}))\Phi(0) \int_{\substack{\mathbb{A}^\times/k^\times \\ |x|\le 1}} \omega(x) d^\times x$$

$$= \mathfrak{B}_2\Phi(0)\delta(\omega\omega_s^{-1}) \int_0^1 |x|^s \frac{dx}{x} = \delta(\omega\omega_s^{-1}) \frac{\mathfrak{B}_2\Phi(0)}{s},$$

$$\int_{\substack{\mathbb{A}^\times/k^\times \\ |x|\le 1}} |x|^{-5}\omega(x) J_{\mathcal{S}(0)_0}(x^{-1}\cdot\widehat{\Phi}, \mathrm{Sym}^4 k^2) d^\times x = \delta(\omega\omega_s^{-1}) \frac{\mathfrak{B}_2\widehat{\Phi}(0)}{s-5}, \tag{4.2}$$

$$\int_{\substack{\mathbb{A}^\times/k^\times \\ |x|\le 1}} \omega(x) J_{\mathcal{S}(0)_1}(x\cdot\Phi, \mathrm{Sym}^4 k^2) d^\times x = \frac{1}{4}\delta\left((\omega\omega_s^{-1})^4\right) \frac{\zeta(\omega\omega_{s-1/2}^{-1}, \Phi_1)}{s-1/2},$$

$$\int_{\substack{\mathbb{A}^\times/k^\times \\ |x|\le 1}} |x|^{-5}\omega(x) J_{\mathcal{S}(0)_1}(x^{-1}\cdot\widehat{\Phi}, \mathrm{Sym}^4 k^2) d^\times x = \frac{1}{4}\delta\left((\omega\omega_s^{-1})^4\right) \frac{\zeta(\omega\omega_{s-1/2}^{-1}, (\widehat{\Phi})_1)}{s-9/2}, \tag{4.3}$$

$$\int_{\substack{\mathbb{A}^\times/k^\times \\ |x|\le 1}} \omega(x) J_{\mathcal{S}(0)_2}(x\cdot\Phi, \mathrm{Sym}^4 k^2) d^\times x = \frac{1}{2}\delta\left((\omega\omega_s^{-1})^2\right) \frac{\zeta(\omega\omega_{s+1}^{-1}, \Phi_2)}{s},$$

and

$$\int_{\substack{\mathbb{A}^\times/k^\times \\ |x|\le 1}} |x|^{-5}\omega(x) J_{\mathcal{S}(0)_2}(x^{-1}\cdot\widehat{\Phi}, \mathrm{Sym}^4 k^2) d^\times x = \frac{1}{2}\delta\left((\omega\omega_s^{-1})^2\right) \frac{\zeta(\omega\omega_{s+1}^{-1}, (\widehat{\Phi})_2)}{s-5}. \tag{4.4}$$

All of these integrals converge absolutely, and the resulting functions of s can be continued to meromorphic functions of the whole complex plane, with obvious poles.

What remains is the difference of

$$(4.5)\quad \int_{\substack{\mathbb{A}^\times/k^\times\\ |x|\le 1}} \omega_{-5}(x)\omega(x) \sum_{y\in k^\times}\sum_{i=0,1} J_{\mathcal{S}(y)_i}(x^{-1}\cdot \widehat{\Phi}, \mathrm{Sym}^4 k^2) d^\times x = D_+(\widehat{\Phi}, \omega_5\omega^{-1}),$$

where this equality defines the distribution D_+, and

$$(4.6)\quad \int_{\substack{\mathbb{A}^\times/k^\times\\ |x|\le 1}} \omega(x) \sum_{y\in k^\times}\sum_{i=0,1} J_{\mathcal{S}(y)_i}(x\cdot \Phi, \mathrm{Sym}^4 k^2) d^\times x.$$

It will be straightforward, after the coming calculations, to see that the integral (4.5) converges absolutely to an entire function of ω.

Break up the integral (4.6) as the difference of the integral over all $\mathbb{A}^\times/k^\times$ and the integral over those $x \in \mathbb{A}^\times/k^\times$ with $|x| \ge 1$. The latter integral, after a change of variables $x \to x^{-1}$ is seen to be $D_+(\Phi, \omega)$ in the notation of (4.5) and is therefore entire. The former integral,

$$\int_{\mathbb{A}^\times/k^\times} \omega(x) \sum_{y\in k^\times}\sum_{i=0,1} J_{\mathcal{S}(y)_i}(x\cdot \Phi, \mathrm{Sym}^4 k^2) d^\times x,$$

is the limiting value as s with $\Re(s) > 0$ tends to 0 of the expression

$$\int_{\mathbb{A}^\times/k^\times} \omega(x) \sum_{y\in k^\times} \Bigg[-\frac{1}{2}\frac{d}{ds}\int_{N(\mathbb{A})} \alpha(n)^s (x\cdot\Phi)^K (n^{-1}\cdot \gamma^y)\, dn$$
$$+\frac{1}{4}\sum_{\chi\in\Omega(\mathbb{A}^\times/k^\times)_4} C_\chi \prod_v \frac{\zeta_v(\omega_s, (x_v\cdot\Phi^y)_v)}{\zeta_v(\omega_s)} - \frac{1}{4}|\Delta_k|^{-1/2}\frac{d}{ds}\prod_v \frac{\zeta_v(\omega_s, (x\cdot\Phi^y)_v)}{\zeta_v(\omega_s)}\Bigg] d^\times x,$$

defined for $\Re(s) > 0$, where C_χ is 1 if χ is not trivial and $\mathfrak{C}_k^{-1}\lambda_0$ if χ is trivial. Call this expression $D(\Phi, \omega, s)$. Notice that $x\gamma^y = \gamma^{xy}$ and that $(x\cdot\Phi)^y(z) = \Phi^{xy}(xz)$, so we can absorb the sum over y into the integral over x to obtain that

$$(4.7)\quad D(\Phi,\omega,s) = \int_{\mathbb{A}^\times} \omega(x) \Bigg[-\frac{1}{2}\frac{d}{ds}\int_{N(\mathbb{A})} \alpha(n)^s \Phi^K (n^{-1}\cdot\gamma^x)\, dn$$
$$+\frac{1}{4}\sum_{\chi\in\Omega(\mathbb{A}^\times/k^\times)_4} C_\chi \chi^{-1}(x)|x|^{-s}\prod_v \frac{\zeta_v(\omega_s, \Phi_v^x)}{\zeta_v(\omega_s)}$$
$$-\frac{1}{4}|\Delta_k|^{-1/2}\frac{d}{ds}|x|^{-s}\prod_v \frac{\zeta_v(\omega_s, \Phi_v^x)}{\zeta_v(\omega_s)}\Bigg] d^\times x.$$

We know this converges absolutely for $c(\omega) > 5$ and $\Re(s) > 0$. We next bring all the sums and all the differentiation in (4.7) outside the integral over x—right now this can only be done formally, but we will prove that each resulting integral converges absolutely and has a limit as s tends to 0, whenever $c(\omega) > 2$. This will prove that the zeta function $Z(\Phi, \omega)$ has a meromorphic continuation to the half-plane $c(\omega) > 2$, with poles at 9/2 and 5. By comparison, Yukie [**Yu98**] proved meromorphic continuation only on the half-plane $c(\omega) > 4$.

The first term to consider is $-1/2$ of the derivative with respect to s of the integral

$$\text{(4.8)}\qquad \int_{\mathbb{A}^\times} \omega(x) \int_{N(\mathbb{A})} \alpha(n)^s \Phi^K(n^{-1}\cdot\gamma^x)\, dn\, d^\times x = \prod_v \int_{k_v^\times} \omega_v(x) \int_{N(k_v)} \alpha_v(n)^s \Phi_v^K(n^{-1}\cdot\gamma^x)\, dn\, d^\times x.$$

LEMMA 2. *(a) Suppose that v is a finite place with residue field $\mathbb{F}_q$, q odd, such that $\omega_v(x) = |x|_v^{s_1}$ and Φ_v is the characteristic function of $V(o_v)$. Then*

$$\int_{k_v^\times} \omega_v(x) \int_{N(k_v)} \alpha_v(n)^s \Phi_v^K(n^{-1}\cdot\gamma^x)\, dn\, d^\times x = \frac{1-q^{-(2s_1+s)}}{(1-q^{-s_1})(1-q^{-(2s_1+s-1)})}.$$

(b) For any place v, the integral

$$\int_{k_v^\times} \omega_v(x) \int_{N(k_v)} \alpha_v(n)^s \Phi_v^K(n^{-1}\cdot\gamma^x)\, dn\, d^\times x$$

converges absolutely for $\Re(s) > -\epsilon$, $c(\omega) > 1/2+\epsilon$.

REMARK 4. Compare this to Propositions (2.8)–(2.10) in [**Yu92**]. Notice that s and s_1 are switched with respect to that paper.

PROOF. (a) Write $k = v(x)$, $\ell = v(n)$. Then

$$\int_{k_v^\times} \omega_v(x) \int_{N(k_v)} \alpha_v(n)^s \Phi_v^K(n^{-1}\cdot\gamma^x)\, dn\, d^\times x = \sum_{k=0}^{\infty} q^{-ks_1} \sum_{\ell=-[k/2]}^{\infty} (1-q^{-1})q^{-\ell}\sup(1,q^{-\ell})^{-s},$$

where $[k/2]$ means the greatest integer that is at most $k/2$. Break up the sum over ℓ into negative and nonnegative ℓ and in the former sum replace ℓ with $-\ell$. This gives

$$\begin{aligned}
&\sum_{k=0}^{\infty} q^{-ks_1}(1-q^{-1})\left[\sum_{\ell=1}^{[k/2]} q^{\ell(1-s)} + \sum_{\ell=0}^{\infty} q^{\ell}\right] \\
&= (1-q^{-1})\sum_{k=0}^{\infty} q^{-ks_1}\left[q^{1-s}\frac{q^{[k/2](1-s)}-1}{q^{1-s}-1} + \frac{1}{1-q^{-1}}\right] \\
&= (1-q^{-1})\left[\frac{\sum_{k=0}^{\infty} q^{-ks_1}q^{[k/2](1-s)}}{1-q^{-(1-s)}} + \left(\sum_{k=0}^{\infty} q^{-ks_1}\right)\left(-\frac{1}{1-q^{-(1-s)}}+\frac{1}{1-q^{-1}}\right)\right] \\
&= \frac{1-q^{-1}}{1-q^{-(1-s)}}\left(\sum_{k-0}^{\infty} q^{-2ks_1}q^{k(1-s)} + \sum_{k=0}^{\infty} q^{-(2k+1)s_1}q^{k(1-s)}\right) \\
&\quad + \frac{1}{1-q^{-s_1}}\left(1-\frac{1-q^{-1}}{1-q^{-(1-s)}}\right) \\
&= \frac{1-q^{-1}}{1-q^{-(1-s)}}\left(\frac{1+q^{-s_1}}{1-q^{-(2s_1+s-1)}}\right) + \frac{q^{-1}(1-q^s)}{(1-q^{-s_1})(1-q^{-(1-s)})} \\
&= \frac{(1-q^{-1})(1-q^{-2s_1}) + q^{-1}(1-q^s)(1-q^{-(2s_1+s-1)})}{(1-q^{-s_1})(1-q^{-(1-s)})(1-q^{-(2s_1+s-1)})}
\end{aligned}$$

$$= \frac{(1-q^{-(1-s)})(1-q^{-(2s_1+s)})}{(1-q^{-s_1})(1-q^{-(1-s)})(1-q^{-(2s_1+s-1)})}.$$

Cancelling the term $(1-q^{-(1-s)})$ gives the desired result.

Part (b) follows from similar calculations for the finite places and a simple change of variables in the infinite ones, see for example the proofs of (2.9) and (2.10) of [**Yu92**] for a template for such proofs. □

Given this lemma, it is clear that the expression (4.8) converges absolutely to a function holomorphic on the region $\Re(s) > -\epsilon$, $c(\omega) > 2+\epsilon$, so that its derivative at $s=0$ is holomorphic on the half-plane $c(\omega) > 2$.

The remaining terms to calculate all involve the Euler product

$$\prod_v \frac{\zeta_v(\chi_v\omega_s, \Phi_v^x)}{\zeta_v(\chi_v\omega_s)}.$$

Again, we will explicitly calculate the unramified term of this, and determine the domain where the Euler product converges absolutely.

LEMMA 3. *(a) Suppose that v is a finite place with residue field $\mathbb{F}_q$, q odd, such that $\omega_v(x) = |x|_v^{s_1}$ and Φ_v is the characteristic function of $V(o_v)$. Then*

$$\int_{k_v^\times} \omega_v(x)|x|_v^{-s} \frac{\zeta_v(\chi_v\omega_s, \Phi_v^x)}{\zeta_v(\chi_v\omega_s)}\, d^\times x = \frac{f(s_1,s,q)}{(1-q^{-(s_1-2s)})(1-q^{-(2s_1-2s-1)})},$$

where

$$f(s_1,s,q) = (1-q^{-s})(1-q^{-1})^2 q^{-(s_1-2s+2)} + (1-q^{-(s_1-2s)})(1+q^{-(s_1-s)}).$$

(b) The product

$$\prod_v \int_{k_v^\times} \omega_v(x_v)|x_v|_v^{-s} \prod_v \frac{\zeta_v(\chi_v\omega_s, \Phi_v^{x_v})}{\zeta_v(\chi_v\omega_s)}\, d^\times x_v$$

converges absolutely for $|\Re(s)| < \epsilon$, $c(\omega) > 2+2\epsilon$.

PROOF. (a) Write 1_{o_v} for the characteristic function of o_v. Then

$$(4.9)\quad \int_{k_v^\times} \omega_v(x)|x|_v^{-s} \frac{\zeta_v(\chi_v\omega_s, \Phi_v^x)}{\zeta_v(\chi_v\omega_s)}\, d^\times x$$

$$= \int_{k_v^\times}\int_{k_v^\times}\int_{k_v} |x|_v^{s_1-s}(1-q^{-s})|y|^s 1_{o_v}(x) 1_{o_v}(2nx) 1_{o_v}(n^2x+y)\, dn\, d^\times y\, d^\times x$$

Switch integrals so that the innermost integral is that over y. Let $k = v(x)$ and $\ell = v(n)+v(x)$, so that both k and ℓ must be non-negative for the integrand to be non-zero. If $2\ell - k$ is non-negative, then n^2x lies in o_v, so the integrand is non-zero exactly for $y \in o_v$, and the integral over y is $\sum_{m=0}^\infty q^{-ms} = (1-q^{-s})^{-1}$. If $2\ell-k$ is negative, then the only values of y that contribute to the integral satisfy $v(y) = 2\ell - k$, and the integral over y is $q^{-(2\ell-k)s}(q-1)^{-1}q^{2\ell-k-1}$. Therefore (4.9) equals

$$(1-q^{-s})\sum_{\ell=0}^\infty q^{-(\ell-k)}(1-q^{-1})\left(\sum_{k=2\ell+1}^\infty q^{-k(s_1-s)}q^{-(k-2\ell)(1-s)}(q-1)^{-1}q^{-1}\right.$$

$$\left. + \sum_{k=0}^{2\ell} q^{-k(s_1-s)}\frac{1}{1-q^{-s}}\right)$$

$$= (1-q^{-s})(1-q^{-1})^2 q^{-2}\left(\sum_{\ell=0}^{\infty} q^{-\ell(2s-1)} \sum_{k=2\ell+1}^{\infty} q^{-k(s_1-2s)}\right)$$
$$+ (1-q^{-1})\left(\sum_{\ell=0}^{\infty} q^{-\ell} \sum_{k=0}^{2\ell} q^{-k(s_1-s-1)}\right)$$
$$= (1-q^{-s})(1-q^{-1})^2 q^{-2}\left(\sum_{\ell=0}^{\infty} q^{-\ell(2s-1)} q^{-(2\ell+1)(s_1-2s)} \frac{1}{1-q^{-(s_1-2s)}}\right)$$
$$+ (1-q^{-1})\left(\sum_{\ell=0}^{\infty} q^{-\ell} \frac{1-q^{-(2\ell+1)(s_1-s-1)}}{1-q^{-(s_1-s-1)}}\right)$$
$$= \frac{(1-q^{-s})(1-q^{-1})^2 q^{-(s_1-2s+2)}}{1-q^{-(s_1-2s)}} \times \frac{1}{1-q^{-(2s_1-2s-1)}}$$
$$+ \frac{1-q^{-1}}{1-q^{-(s_1-s-1)}}\left(\frac{1}{1-q^{-1}} - \frac{q^{-(s_1-s-1)}}{1-q^{-(2s_1-2s-1)}}\right)$$
$$= \frac{1}{(1-q^{-(s_1-2s)})(1-q^{-(2s_1-2s-1)})}$$
$$\times \Bigg[(1-q^{-s})(1-q^{-1})^2 q^{-(s_1-2s+2)}(1-q^{-(s_1-s-1)})$$
$$+ \frac{(1-q^{-(s_1-2s)})(1-q^{-(2s_1-2s-1)}) + (1-q^{-1})q^{-(s_1-s-1)}(1-q^{-(s_1-2s)})}{(1-q^{-(s_1-s-1)})}\Bigg]$$
$$= \frac{f(s_1,s,q)}{(1-q^{-(s_1-2s)})(1-q^{-(2s_1-2s-1)})}.$$

(b) Convergence of the ramified terms works just as in the previous Lemma. To see convergence of the Euler product of the unramified terms, notice that for $|\Re(s)| < \epsilon$, $\Re(s_1) > 2+2\epsilon$, both the numerator and denominator of the Euler factor are the values of certain fixed polynomials $p(x) = 1 + \sum_i c_i x^{e_i}$ (with complex exponents e_i) at q^{-1}, where the smallest value of $\Re(e_i)$ is greater than 1. □

Write $D(\Phi,\omega) = D(\Phi,\omega,s)|_{s=0^+}$. We have proven the following.

THEOREM 2. *Given $\Phi \in \mathcal{S}(V(\mathbb{A}))$, the identity*

$$Z(\Phi,\omega) = Z_+(\Phi,\omega) + Z_+(\widehat{\Phi},\omega_5\omega^{-1}) + \delta(\omega\omega_s^{-1})\mathfrak{B}_2\left(\frac{\widehat{\Phi}(0)}{s-5} - \frac{\Phi(0)}{s}\right)$$
$$+ \frac{1}{4}\delta\left((\omega\omega_s^{-1})^4\right)\left(\frac{\zeta(\omega\omega_{s-1/2}^{-1},(\widehat{\Phi})_1)}{s-9/2} - \frac{\zeta(\omega\omega_{s-1/2}^{-1},\Phi_1)}{s-1/2}\right)$$
$$+ \frac{1}{2}\delta\left((\omega\omega_s^{-1})^2\right)\left(\frac{\zeta(\omega\omega_{s+1}^{-1},(\widehat{\Phi})_2)}{s-5} - \frac{\zeta(\omega\omega_{s+1}^{-1},\Phi_2)}{s}\right)$$
$$+ D_+(\Phi,\omega) + D_+(\widehat{\Phi},\omega_5\omega^{-1}) - D(\Phi,\omega),$$

with D_+ entire and $D(\Phi,\omega)$ holomorphic for $c(\omega) > 2$, holds for quasi-characters ω of $\mathbb{A}^\times/k^\times$ with $c(\omega) > 5$ and defines a meromorphic continuation of $Z(\Phi,\omega)$ to the half-plane $c(\omega) > 2$. Its poles in that region are at $\omega = \omega_5\chi$, $\chi \in \Omega(\mathbb{A}^\times/k^\times)_2$

and $\omega = \omega_{9/2}\chi$, $\chi \in \Omega(\mathbb{A}^\times/k^\times)_4$, *with residues*

$$\operatorname*{Res}_{s=5} Z(\Phi, \omega_s\chi) = \delta(\chi)\mathfrak{B}_2\widehat{\Phi}(0) + \frac{1}{2}\zeta(\chi\omega_{-1}, (\widehat{\Phi})_2), \quad \chi \in \Omega(\mathbb{A}^\times/k^\times)_2$$

and

$$\operatorname*{Res}_{s=9/2} Z(\Phi, \omega_s\chi) = \frac{1}{4}\zeta(\chi\omega_{-1/2}, (\widehat{\Phi})_1), \quad \chi \in \Omega(\mathbb{A}^\times/k^\times)_4.$$

REMARK 5. Another zeta function one could analyze is formed by taking the sum, not over all regular elliptic elements, but only over those without double roots. This zeta function is perhaps more interesting from a number theoretical perspective. The difference between this zeta function and $Z(\Phi, \omega)$ is

$$\int_{\mathbb{A}^\times/k^\times} \int_{G(k)\backslash G(\mathbb{A})} \omega(x) \sum_{y \in k^\times/k^{\times 2}} \sum_{\gamma \in W^s(k)} \Phi\big(y(g^{-1} \cdot x\gamma)^2\big) dg\, d^\times x$$

where W is the space of binary quadratic forms, with similarly defined G-action; one expects that this integral is related to the known [**Yu92**] zeta function for the space of binary quadratic forms.

References

[A78] J. Arthur, *A trace formula for reductive groups. I. terms associated to classes in* $G(\mathbb{Q})$, Duke Math. J. **45** (1978), 911–952.

[A85] J. Arthur, *A measure on the unipotent variety*, Canad. J. Math. **37** (1985), 1237–1274.

[BHC] A. Borel and Harish-Chandra, *Arithmetic subgroups of algebraic groups*, Ann. of Math. (2) **75** (1962), 485–535.

[D] B. A. Datskovsky, *A mean-value theorem for class numbers of quadratic extensions*, in A tribute to Emil Grosswald: Number Theory and Related Analysis, Contemp. Math. **143**, Providence, RI: Amer. Math. Soc. (1993), 179–242.

[DW86] B. Datskovsky and D. J. Wright, *The adelic zeta function associated to the space of binary cubic forms. II. Local theory*, J. Reine Angew. Math. **367** (1986), 27–75.

[DW88] B. Datskovsky and D. J. Wright, *Density of discriminants of cubic extensions*, J. Reine Angew. Math. **386** (1988), 116–138.

[GJ] R. Godement and H. Jacquet, Zeta Functions of Simple Algebras, Lecture Notes in Mathematics **260**, Berlin: Springer-Verlag (1972).

[JL] H. Jacquet and R. P. Langlands, Automorphic Forms on GL(2), Lecture Notes in Mathematics **114**, New York: Springer-Verlag (1970).

[K84] F. C. Kirwan, Cohomology of Quotients in Symplectic and Algebraic Geometry, Mathematical Notes **31**, Princeton: Princeton University Press (1984).

[K85] F. C. Kirwan, *Partial desingularisations of quotients of nonsingular varieties and their Betti numbers*, Ann. of Math. (2) **122** (1985), 41–85.

[LaL] J.-P. Labesse and R. P. Langlands, *ℓ-indistinguishability for* sl(2), Canad. J. Math. **31** (1979), 726–785.

[Lan] R. P. Langlands, Base Change for GL(2), Annals of Mathematics Studies **96**, Princeton: Princeton University Press, (1980).

[Le93] J. Levy, *A modified trace formula*, Ph.D. thesis, University of Toronto (1993).

[Le95] J. Levy, *A truncated Poisson formula for groups of rank at most two*, Amer. J. Math. **117** (1995), 1371–1408.

[Le96] J. Levy, *A truncated poisson summation formula*, preprint (1998).

[Le97] J. Levy, *Hilbert-Mumford closures of orbits*, in preparation.

[M] Morning seminar on the trace formula, IAS notes (1993).

[S] T. Shintani, *On Dirichlet series whose coefficients are class numbers of integral binary cubic forms*, J. Math. Soc. Japan **24** (1972), 132–188.

[W65] A. Weil, *Sur la formule de Siegel dans la théorie des groupes classiques*, Acta Math. **113** (1965), 1–87.

[W74] A. Weil, Basic Number Theory, third edition, Die Grundlehren der Mathematischen Wissenschaften, Band 144, New York: Springer-Verlag (1974).

[W85] D. J. Wright, *The adelic zeta function associated to the space of binary cubic forms. I. Global theory*, Math. Ann. **270** (1985), 503–534.

[W89] D. J. Wright, *Twists of the Iwasawa-Tate zeta function*, Math. Z. **200** (1989), 209–231.

[Y] K. F. Ying, *On the convergence of the adelic zeta functions associated to irreducible regular prehomogeneous vector spaces*, Amer. J. Math. **117** (1995), 457–490.

[Yu92] A. Yukie, *On the Shintani zeta function for the space of binary quadratic forms*, Math. Ann. **292** (1992), 355–374.

[Yu93] A. Yukie, Shintani Zeta Functions, London Mathematical Society Lecture Note Series **183**, Cambridge University Press: Cambridge (1993).

[Yu95] A. Yukie, *On the convergence of the zeta function for certain prehomogeneous vector spaces*, Nagoya Math. J., **140** (1995), 1–31.

[Yu98] A. Yukie, *On Shintani zeta functions for* GL(2), Trans. AMS **350** (1998), 5067–5094.

DEPARTMENT OF MATHEMATICS, OKLAHOMA STATE UNIVERSITY, STILLWATER, OK 74078.

Current address: Department of Mathematics and Statistics, University of Ottawa, 585 King Edward, Ottawa, ON, K1N 6N5, Canada.

E-mail address: jlevy@science.uottawa.ca

Proceedings of Symposia in Pure Mathematics
Volume **66.2**, 1999

On the Generalized Ramanujan Conjecture for GL(n)

Wenzhi Luo, Zeév Rudnick, and Peter Sarnak

To G. Shimura with admiration

ABSTRACT. We establish new uniform bound towards the generalized Ramanujan conjecture for GL(n) over an arbitrary number field.

1. Introduction

In this paper we describe a modification to the method that we introduced in [**LRS**] for giving bounds towards the generalized Ramanujan conjectures. The modification allows us to treat the case of a general number field. The main new difficulty in dealing with a number field is the presence of an infinite group of units. The method in [**LRS**] and the earlier related method of [**DI**] made use of Dirichlet characters of varying conductors; in general the unit group limits the construction of the analogous ray class characters. The modification mentioned above allows us to exploit in a crucial way the positivity of the coefficients of the Rankin-Selberg L-function. This, together with Rohrlich's construction [**R**] of special moduli which have many ray class characters, enable us to overcome this difficulty.

Let F be a number field of degree $[F:\mathbf{Q}] = d$ with ring of integers $O = O_F$, unit group $U = U_F$, and $D = D_F$ is the discriminant of F. For $n \geq 2$ let π be an cuspidal automorphic form on $\mathrm{GL}_n(\mathbf{A}_F)$ with unitary central character and contragredient representation $\tilde{\pi}$. The standard L-function $L(s,\ \pi)$ associated with π is of the form (for $\mathrm{Re}(s) > 1$)

$$L(s,\ \pi) = \prod_v L(s,\ \pi_v).$$

The product is over all the places of F and for each such v, $L(s,\ \pi_v)$ is the local Euler factor. At an unramified finite place of π, this local factor is given by

$$L(s,\ \pi_v) = \prod_{i=1}^{n} (1 - \alpha_{i,\pi}(v)(Nv)^{-s})^{-1},$$

1991 *Mathematics Subject Classification.* Primary 11F70; Secondary 11F60.
Key words and phrases. Automorphic form, L-function, Ramanujan conjecture.
Research partially supported by NSF grant DMS-9401571 and the Bi-National Science Foundation.

for suitable complex numbers $\alpha_{i,\pi}(v)$, $i = 1, \cdots, n$, and where $N(v)$ is the absolute norm of v viewed as a prime ideal of F. At an unramified archimedean place v for π, the local factor is given in the form

$$L(s, \pi_v) = \prod_{i=1}^{n} \Gamma_v(s - \mu_{j,\pi}(v)),$$

where

$$\Gamma_v(s) = \pi^{-s/2}\Gamma(s/2), \quad \text{if } v \text{ is real,}$$

and

$$\Gamma_v(s) = (2\pi)^{-s}\Gamma(s), \quad \text{if } v \text{ is complex.}$$

In this notation the generalized Ramanujan conjecture for π asserts that for such unramified v,

$$|\alpha_{j,\pi}(v)| = 1, \quad \text{if } v \text{ is finite ;}$$
$$\mathrm{Re}(\mu_{j,\pi}(v)) = 0, \quad \text{if } v \text{ is archimedean.}$$

Jacquet and Shalika [**JS**], using the genericity of π and local methods, showed that

$$|\alpha_{j,\pi}(v)| < N(v)^{1/2}, \quad \text{if } v \text{ is finite;} \tag{1}$$

$$|\mathrm{Re}(\mu_{j,\pi}(v))| < 1/2, \quad \text{if } v \text{ is archimedean.} \tag{2}$$

For v finite, as was first observed by Serre [**S**], one can do better than the above bound by using the Rankin-Selberg L-function (assuming its analytic properties) in conjunction with Landau's lemma [**L**]. This leads to

$$|\alpha_{j,\pi}(v)| \le N(v)^{1/2-1/(dn^2+1)}$$

To describe our main results let S be any finite set of places of F. For B an integral ideal of F let $X_{\mathcal{B}}$ denote the set of all wide ray class characters χ (see [**Si**] for definitions) of conductor $\mathcal{B}$ such that $\chi(v) = 1$, for any $v \in S$. Let $X^*_{\mathcal{B}}$ be the subset of $X_{\mathcal{B}}$ consisting of primitive characters. Set $X = \cup_{\mathcal{B}} X_{\mathcal{B}}$ and $X^* = \cup_{\mathcal{B}} X^*_{\mathcal{B}}$. For $\chi \in X^*_{\mathcal{B}}$, the partial Rankin-Selberg L-function $L_S(s, (\pi \otimes \chi) \times \tilde{\pi})$ is defined to be

$$L_S(s, (\pi \otimes \chi) \times \tilde{\pi}) = \prod_{v \notin S} L(s, (\pi_v \otimes \chi_v) \times \tilde{\pi}_v).$$

At an unramified place v of π the local factor is given by

$$L(s, (\pi_v \otimes \chi_v) \times \tilde{\pi}_v) = \prod_{i,j=1}^{n} (1 - \chi(v)\alpha_{i,\pi}(v)\bar{\alpha}_{j,\pi}(v)(Nv)^{-s})^{-1}, \text{ if } v \text{ is finite;} \tag{3}$$

$$L(s, (\pi_v \otimes \chi_v) \times \tilde{\pi}_v) = \prod_{i,j=1}^{n} \Gamma_v(s - \mu_{i,\pi}(v) - \bar{\mu}_{j,\pi}(v)), \text{ if } v \text{ is archimedean.} \tag{4}$$

With these notations our main result is:

THEOREM 1. *Given n, π, S as the above and any $\beta > 1 - 2/(n^2+1)$, there are infinitely many $\chi \in X^*$ such that*

$$L_S(\beta, (\pi \otimes \chi) \times \tilde{\pi}) \neq 0.$$

Combining Theorem 1 with the known analytic properties of the Rankin-Selberg L-functions, we establish

THEOREM 2. *Given n, π as the above and let v be a place of F at which π is unramified. Then for any j, $1 \leq j \leq n$,*

(a) $|\log_{N(v)} |\alpha_{j,\pi}(v)|| \leq \frac{1}{2} - \frac{1}{n^2+1}$, if v is finite;

(b) $|\mathrm{Re}(\mu_{j,\pi}(v)| \leq \frac{1}{2} - \frac{1}{n^2+1}$, if v is archimedean.

For $n \geq 3$, part (b) is the first general result going beyond the bound (2). Part (a) has the feature of being uniform (i.e. independent of the degree d).

In the case $n = 2$, we can do better than the bound in Theorem 2 by combining the $n = 3$ case of Theorem 2 together with the Gelbart-Jacquet lift [**GJ**] from $\mathrm{GL}_2(\mathbf{A}_F)$ to $\mathrm{GL}_3(\mathbf{A}_F)$ (which is in turn based on Shimura's symmetric square theory [**Shi**]) to obtain

THEOREM 3. *For π an unitary cuspidal automorphic representation of $\mathrm{GL}_2(\mathbf{A}_F)$, unramified at v and $j = 1, 2$, we have*

(a) $|\log_{N(v)} |\alpha_{j,\pi}(v)|| \leq 1/5$, if v is finite;

(b) $|\mathrm{Re}(\mu_{j,\pi}(v))| \leq 1/5$, if v is archimedean.

REMARK 1. The bound in part (a) of Theorem 3 is not new. It coincides essentially with the best known bound at the finite places obtained by Shahidi [**Sh2**] (note that he also showed that the equality cannot occur in (a)) by a very different method. Part (b) gives a uniform generalization of our result [**LRS**] over $\mathbf{Q}$ to an arbitrary number field F.

REMARK 2. In the special case that $F = \mathbf{Q}$ or F is an imaginary quadratic field (so that the group of units U is finite), one can apply the method of [**LRS**] to the symmetric square L-function on $\mathrm{GL}_n(\mathbf{A}_F)$ (note that the method in [**LRS**] does not use the positivity of the coefficients of the Rankin-Selberg L-functions). This leads to a slight improvement of Theorem 2 and hence in Theorem 3 for such F's. We will not pursue this here.

It is expected that the Ramanujan conjecture on $\mathrm{GL}_n(\mathbf{A}_F)$ will play a fundamental role in understanding more general groups. In a few cases where this connection has been established we can apply Theorems 2 and 3. We exhibit a couple of examples:

(A). The Jacquet-Langlands correspondence together with Theorem 3 implies the following. Let $X = \Gamma\backslash\mathbf{H}^2$ be a Shimura curve, by which we mean that Γ is a lattice in $\mathrm{SL}_2(\mathbf{R})$ coming from a congruence subgroup of a quaternion algebra D defined over a totally real number field K (if $v_1, \cdots, v_l$ are the archimedean places of K, then we require that $v_1 \otimes D \cong \mathrm{SL}_2(\mathbf{R})$ while $v_j \otimes D \cong$ Hamilton quaternion algebra for $j = 2, \cdots, l$). Then

$$\lambda_1(X) \geq \frac{21}{100}, \tag{5}$$

where λ_1 is the first positive eigenvalue of the Laplacian on the hyperbolic surface X. Knowing the full Ramanujan conjectures would allow us to replace 0.21 by 0.25.

(B). Using the method of Ramanujan duals (see [**BLS, BS**]) together with Theorem 3 we obtain the following result. Let Γ be a congruence subgroup of an orthogonal group G defined by a quadratic form over totally real number field. Assume that the form is of signature $(n,\ 1)$ at one such real place and is definite at all the other places. Then Γ acts discontinuously on hyperbolic n-space $\mathbf{H}^n$ and $X = \Gamma\backslash\mathbf{H}^n$ is a hyperbolic n-manifold (i.e. it has constant sectional curvature $\equiv -1$). For $n \geq 3$

$$\lambda_1(X) \geq \frac{3n}{5} - \frac{24}{25}, \tag{6}$$

where again λ_1 is the first positive eigenvalue of the Laplacian. Previously it was known that (see [**BS**])

$$\lambda_1(X) \geq \frac{n}{2} - \frac{3}{4}, \tag{7}$$

while the Ramanujan conjecture for these X's asserts that

$$\lambda_1(X) \geq n - 2. \tag{8}$$

Again (7) would follows from the Ramanujan conjecture for $\mathrm{GL}_2(\mathbf{A}_F)$.

2. Proofs

We begin by showing that Theorem 3 follows from Theorem 2 (applied with $n = 3$), using the Gelbart-Jacquet lift [**GJ**]. Indeed if π is a unitary cuspidal automorphic representation on $\mathrm{GL}_2(\mathbf{A}_F)$, let Π be its lift to $\mathrm{GL}_3(\mathbf{A}_F)$. Those π's for which Π is not cuspidal satisfy the full Ramanujan conjectures, and so we can assume that Π is a unitary cuspidal automorphic representation on $\mathrm{GL}_3(\mathbf{A}_F)$. Its local Euler factor $L(s, \Pi_w)$ at an unramified w of π is given in terms of the local parameters of π at w:

$$\begin{aligned} L(s,\Pi_w) &= (1-\alpha_{1,\pi}^2(w)(Nw)^{-s})^{-1}(1-\alpha_{1,\pi}(w)\alpha_{2,\pi}(w)(Nw)^{-s})^{-1} \\ &\quad \cdot (1-\alpha_{2,\pi}^2(w)(Nw)^{-s})^{-1} \quad \text{if } w \text{ is finite}, \\ L(s,\Pi_w) &= \Gamma_w(s-2\mu_{1,\pi}(w))\Gamma_w(s-\mu_{1,\pi}(w)+\mu_{2,\pi}(w))\Gamma_w(s-2\mu_{2,\pi}(w)), \\ &\quad \text{if } w \text{ is archimedean}. \end{aligned}$$

Inspecting the local parameters and applying Theorem 2 to Π yields Theorem 3.

REMARK 3. Theorem 3 can also be made to apply to other π, such as odd Maass forms for the modular group, which are ramified at infinity. The point is that their lifts to GL(3) are still unramified at infinity and so we may apply Theorem 2.

Next we show that Theorem 2 follows from Theorem 1. The global Rankin-Selberg L-function $L(s,\ (\pi\otimes\chi)\times\tilde{\pi})$ is known to be holomorphic on the whole complex plane except for simple poles at $s = 1$ and $s = 0$ if $\chi \equiv 1$. This is a consequence of the works of Jacquet, Shalika, Piatetski-Shapiro, Shahidi, Mœglin and Waldspurger [**J, JS, JSR-S, MW, Sh1**]. Now

$$L(s,\ (\pi\otimes\chi)\times\tilde{\pi}) = L^S(s,\ (\pi\otimes\chi)\times\tilde{\pi})\cdot L_S(s,\ (\pi\otimes\chi)\times\tilde{\pi}), \tag{9}$$

where

$$L^S(s,\ (\pi\otimes\chi)\times\tilde{\pi}) = \prod_{v\in S} L(s,\ (\pi_v\otimes\chi_v)\times\tilde{\pi}_v). \tag{10}$$

Let v be a place of F at which π is unramified. Choose $S = \{v\}$. For $\chi \in X^*$, $\chi(v) = 1$ so that (see (3),(4))

$$L(s,\ (\pi_v\otimes\chi_v)\times\tilde{\pi}_v) = L(s,\ \pi_v\times\tilde{\pi}_v).$$

It follows that any pole of $L(s,\ \pi_v\times\tilde{\pi}_v)$ must be a zero of $L_S(s,\ (\pi\otimes\chi)\times\tilde{\pi})$. Again from (3) and (4) we see that if β is chosen so that for some j, $|\alpha_{j,\pi}(v)|^2 =$

$N(v)^\beta$ if v is finite, and so that $2\mathrm{Re}(\mu_{j,\pi}(v)) = \beta$ if v is archimedean, then for any $\chi \in X^*$, we have

$$L_S(\beta,\ (\pi \otimes \chi) \times \tilde{\pi}) = 0. \tag{11}$$

According Theorem 1, we must therefore have

$$\beta \leq 1 - \frac{2}{n^2+1}.$$

That is, for any $j = 1, \cdots, n$,

$$|\alpha_{j,\pi}(v)| \leq N(v)^{1/2-1/(n^2+1)}, \text{ if } v \text{ is finite;}$$

$$\mathrm{Re}(\mu_{j,\pi}(v)) \leq \frac{1}{2} - \frac{1}{n^2+1}, \text{ if } v \text{ is archimedean.}$$

The unitarity of π_v ensures that

$$\left\{\alpha_{j,\pi}^{-1}(v)\right\}_{j=1}^n = \left\{\bar{\alpha}_{j,\pi}(v)\right\}_{j=1}^n;$$

and

$$\left\{-\mu_{j,\pi}(v)\right\}_{j=1}^n = \left\{\bar{\mu}_{j,\pi}(v)\right\}_{j=1}^n.$$

Hence Theorem 2 is established.

It remains to prove Theorem 1. For any place v and $s \in \mathbf{C}$, $L(s,\ (\pi_v \otimes \chi_v) \times \tilde{\pi}_v) \neq 0$ [**JS**]. Hence if we can establish Theorem 1 for S, then it holds for any $S' \subset S$. We may therefore assume that S contains all the archimedean places of F as well as the places at which π is ramified and those at which F ramifies. This assumption about S will be made from now on.

We need to know that there are many $\mathcal{Q}$'s for which $|X_\mathcal{Q}|$ is large (that is, it should be almost of the order of $N(\mathcal{Q})$). This is accomplished using Rohrlich's construction [**R**]. For $\epsilon > 0$ (fixed but arbitrarily small) he shows using a variant of the Bombieri-Vinogradov theorem due to Murty-Murty [**MM**], that there are infinitely many integral ideals $\mathcal{Q}$ of F such that

(i) The integer $q = N(\mathcal{Q})$ is a product of distinct primes, and $(\mathcal{Q},\ S) = 1$.

(ii) Any element in $(O/\mathcal{Q})^*$ has order at most q^ϵ.

As Rohrlich shows, these imply that the group $C(\mathcal{Q})$ of ray classes (mod $\mathcal{Q}$) has order $\gg q^{1-\epsilon}$. In fact from (ii) and the fact that for any ideal $\mathcal{B}, \mathcal{B}^h$ is principal (here $h = |C(1)|$ is the class number of F), we see that the order of any subgroup of $C(\mathcal{Q})$ generated by a fixed number r of elements, is $\ll q^{r\epsilon}$. Choosing these r elements to be the finite primes v of S, we conclude that for the above sequence of $\mathcal{Q}$'s

$$|X_\mathcal{Q}| \gg q^{1-r\epsilon}. \tag{12}$$

As mentioned above, the L-function $L(s,\ \pi \times \tilde{\pi})$ is holomorphic except for simple poles at $s = 1$ and $s = 0$. Also it has non-negative coefficients in its series expansion (see [**RS**]):

$$\sum_{\mathcal{A}} a(\mathcal{A})(N\mathcal{A})^{-s}, \tag{13}$$

with $a(\mathcal{A}) \geq 0$. From this one has

$$\sum_{N(\mathcal{A}) \leq X} a(\mathcal{A}) \sim C_\pi X, \quad C_\pi \neq 0. \tag{14}$$

For $\chi \in X_{\mathcal{B}}^*$, the L-function $L(s,\ (\pi \otimes \chi) \times \tilde{\pi})$ satisfies the following functional equation [**LRS**]:

$$L(s,\ (\pi \otimes \chi) \times \tilde{\pi}) = \chi(\mathrm{f}(\pi \times \tilde{\pi}))W(\chi)^{n^2}W(\pi \times \tilde{\pi})N^{1/2-s}L(1-s,\ (\tilde{\pi} \otimes \bar{\chi}) \times \pi),$$

where $N = D^{n^2}N(\mathrm{f}(\pi \times \tilde{\pi}))N(\mathcal{B})^{n^2}$, $W(\pi \times \tilde{\pi})$ is the root number of $\pi \times \tilde{\pi}$ (it is non-zero), and $W(\chi)$ is the normalized Gauss sum corresponding to χ, and $\mathrm{f}(\pi \times \tilde{\pi})$ is the conductor of $\pi \times \tilde{\pi}$.

By our choice of S, $\chi \equiv 1$ at the places of ramification of $\pi \times \tilde{\pi}$, so that $\chi(\mathrm{f}(\pi \times \tilde{\pi})) = 1$. Denote by $L^S(s,\ \pi \times \tilde{\pi})$ the product

$$L^S(s,\ \pi \times \tilde{\pi}) = \prod_{v \in S} L(s,\ \pi_v \times \tilde{\pi}_v).$$

Then since $\chi(v) = 1$ for $v \in S$, we have

$$L(s,\ (\pi \otimes \chi) \times \tilde{\pi}) = L^S(s,\ \pi \times \tilde{\pi}) \cdot L_S(s,\ (\pi \otimes \chi) \times \tilde{\pi}). \tag{15}$$

Denote by $L_{S,\chi}(s)$ the partial L-function $L_S(s,\ (\pi \otimes \chi) \times \tilde{\pi})$ and by $G(s)$ the quotient

$$G(s) = \frac{L^S(1-s,\ \pi \times \tilde{\pi})}{L^S(s,\ \pi \times \tilde{\pi})}.$$

The local bounds of [**JS**] which are applicable to any place v, assert that $L(s,\ \pi_v \times \tilde{\pi}_v)$ is analytic in $\mathrm{Re}(s) > \beta_v$, where $\beta_v < 1$ (and depends on π). S being finite, it follows that $L^S(s,\ \pi \times \tilde{\pi})$ is analytic in $\mathrm{Re}(s) > \beta_0$ for some $\beta_0 = \beta_0(\pi) < 1$. Since $L^S(s,\ \pi \times \tilde{\pi})$ is nowhere zero, we have that

$$G(1-s) \quad \text{is analytic in} \quad \mathrm{Re}(s) > \beta_0, \tag{16}$$

for some $\beta_0 < 1$.

The functional equation for $L_{S,\chi}(s)$ when $\chi \in X_{\mathcal{B}}^*$ becomes

$$L_{S,\chi}(s) = W(\chi)^{n^2}W(\pi \times \tilde{\pi})N^{1/2-s}G(s)L_{S,\bar{\chi}}(1-s). \tag{17}$$

The series representation for $L_{S,\chi}(s)$ is

$$\sum_{(\mathcal{A},S)=1} a(\mathcal{A})\chi(\mathcal{A})(N\mathcal{A})^{-s}. \tag{18}$$

We are ready to prove Theorem 1. For $\beta \geq 1$, $L(\beta,\ (\pi \otimes \chi) \times \tilde{\pi}) \neq 0$ (see [**Sh1**]), so we can assume $1 - 2/(n^2+1) < \beta < 1$. The idea is to show that for $N(\mathcal{Q}) = q$ large enough (as in (i), (ii) above)

$$\left| \sum_{\mathcal{B}|\mathcal{Q},\mathcal{B}\neq O} \sum_{\chi \in X_{\mathcal{B}}^*} L_{S,\chi}(\beta) \right| \gg q^{1-\epsilon}. \tag{19}$$

This immediately implies Theorem 1.

Let $f \in C_c^\infty(0,\ \infty)$ with $\mathrm{supp}(f) = [A,\ B]$, $0 < A < B$, such that $f(y) \geq 0$ and $\int_0^\infty f(y)\frac{dy}{y} = 1$. Define

$$k(s) = \int_0^\infty f(y)y^{s-1}dy, \tag{20}$$

and

$$f_1(y) = \frac{1}{2\pi i}\int_{\mathrm{Re}(s)=2} k(s)y^{-s}\frac{ds}{s}, \tag{21}$$

for $y > 0$. Then, $k(s)$ is entire and for $s = \sigma + it$, $|\sigma| \leq a$, we infer, by partial integration several times, that

(22) $$k(s) \ll_{a,m} \frac{1}{(|t|+1)^m}, \quad \text{for any positive integer } m.$$

It is easy to see, by means of Mellin inversion formula, that

(23) $$f_1(y) = \int_y^\infty \frac{f(x)}{x} dx.$$

Thus, we have: $0 \leq f_1(y) \leq 1$; $f_1(y) = 1$, if $0 < y \leq A$; $f_1(y) = 0$, if $y \geq B$. Next we set $f_2(y)$ to be

(24) $$f_2(y) = \frac{1}{2\pi i} \int_{\mathrm{Re}(s)=2} k(-s)G(-s+\beta)y^{-s}\frac{ds}{s},$$

for $y > 0$. Then, as in our previous paper [**LRS**] and using (16), we have

(i) $f_2(y) \ll_m y^{-m}$ for $y \geq 1$ and any positive integer m.

(ii) $f_2(y) \ll_\epsilon 1 + y^{1-\beta_0-\beta-\epsilon}$ for $0 < y \leq 1$ and any $\epsilon > 0$.

The proofs are the same except in (ii) the integrand might have infinitely many poles on the line $\mathrm{Re}(s) = \max(0, \beta + \beta_0 - 1)$, coming from $\prod_{v \in S} L(1+s-\beta, \pi_v \times \tilde{\pi}_v)$. However it causes no problem in view of the estimate of $k(s)$ in (22).

The functional equation (17) leads to the following approximate functional equation: For $\chi \in X^*_{\mathcal{B}}$, $B \neq O$, $0 < \beta < 1$, and any $Y > 1$,

$$\begin{aligned} L_{S,\chi}(\beta) = &\sum_{(\mathcal{A},S)=1} \frac{a(\mathcal{A})\chi(\mathcal{A})}{(N\mathcal{A})^\beta} f_1\left(\frac{N(\mathcal{A})}{Y}\right) + W(\pi\times\tilde{\pi})(D^{n^2}N(\mathrm{f})N(\mathcal{B})^{n^2})^{1/2-\beta} \\ &\cdot \sum_{(\mathcal{A},S)=1} \frac{a(\mathcal{A})\bar{\chi}(\mathcal{A})}{(N\mathcal{A})^{1-\beta}} W(\chi)^{n^2} f_2\left(\frac{N(\mathcal{A})Y}{D^{n^2}N(\mathrm{f})N(\mathcal{B})^{n^2}}\right) \\ = &\sum\nolimits_{(1)} + \sum\nolimits_{(2)}, \quad \text{say.} \end{aligned}$$

Since the derivation of the above formula is almost identical to the Lemma 3.2 in [**LRS**], we omit the proof.

Now for any ideal $\mathcal{A}$,

$$\sum_{\mathcal{B}|\mathcal{Q}} \sum_{\chi \in X^*_{\mathcal{B}}} \chi(\mathcal{A}) \geq 0.$$

Indeed, write $\mathcal{Q} = \mathcal{Q}_1\mathcal{Q}_2$, where

$$\mathcal{Q}_1 = \prod_{\mathcal{P}^a \| \mathcal{Q}, \mathcal{P}|(\mathcal{A},\mathcal{Q})} \mathcal{P}^a.$$

Then $(\mathcal{Q}_2, \mathcal{Q}_1\mathcal{A}) = 1$ and we have

(25) $$\sum_{\mathcal{B}|\mathcal{Q}} \sum_{\chi \in X^*_{\mathcal{B}}} \chi(\mathcal{A}) = \sum_{\mathcal{B}|\mathcal{Q}_2} \sum_{\chi \in X^*_{\mathcal{B}}} \chi(\mathcal{A}) = \sum_{\chi \in X_{\mathcal{Q}_2}} \chi(\mathcal{A}) \geq 0,$$

since $X_{\mathcal{Q}_2}$ is a subgroup of $(C(\mathcal{Q}_2))^\wedge$.

We see that the contribution from $\mathcal{A} = O$ and the positivity imply that

(26) $$\sum_{\mathcal{B}|\mathcal{Q}} \sum_{\chi \in X^*_{\mathcal{B}}} \sum\nolimits_{(1)} \geq \sum_{\mathcal{B}|\mathcal{Q}} \sum_{\chi \in X^*_{\mathcal{B}}} 1 = |X_\mathcal{Q}| > q^{1-r\epsilon},$$

by (12). Moreover the contribution from $\mathcal{B} = O$ to (26) is estimated by

$$\begin{aligned}\sum_{\chi\in X_O}\sum\nolimits_{(1)} &= \sum_{\chi\in X_O}\sum_{(\mathcal{A},S)=1}\frac{a(\mathcal{A})\chi(\mathcal{A})}{(N\mathcal{A})^\beta}f_1\left(\frac{N\mathcal{A}}{Y}\right)\\ &\ll \sum_{i\geq 0,\ 2^i\ll Y}\ \sum_{2^i\leq N(\mathcal{A})<2^{i+1}}\frac{a(\mathcal{A})}{(N\mathcal{A})^\beta}f_1\left(\frac{N\mathcal{A}}{Y}\right)\\ &\ll \sum_{i\geq 0,\ 2^i\ll Y}(2^i)^{1-\beta}\ll Y^{1-\beta+\epsilon},\end{aligned}$$

using (14).

Hence we deduce that for q large,

$$\sum_{\mathcal{B}|\mathcal{Q},\mathcal{B}\neq O}\ \sum_{\chi\in X_{\mathcal{B}}^*}\sum\nolimits_{(1)} > q^{1-r\epsilon}+O(Y^{1-\beta+\epsilon}). \tag{27}$$

Next we treat $\sum_{(2)}$. For a fractional ideal $\mathcal{M} = \mathcal{A}^{-1}$, and $\mathcal{B}|\mathcal{Q}$, $\mathcal{B}\neq O$, we have, as in (66), (67) of [**R**] and also (32) and (33) in [**BR**] (which appeals to Deligne's bound [**D**] on Kloosterman sums), that

$$\sum_{\chi\in X_{\mathcal{B}}^*}\chi(\mathcal{M})W(\chi)^{n^2}\ll q^{-1/2+\epsilon}\sum_{k \pmod q}\left|\sum_{\chi\in X_{\mathcal{B}}^*}\chi(k\mathcal{T})\right|,$$

where $\mathcal{T}$ is some fractional ideal relatively prime to $\mathcal{B}$. But we have

$$\begin{aligned}\sum_{k \pmod q}\left|\sum_{\chi\in X_{\mathcal{B}}^*}\chi(k\mathcal{T})\right| &\leq |<\{\mathcal{P},\ v_{\mathcal{P}}\in S,\ v_{\mathcal{P}}<\infty\}>|\\ \cdot|\text{image of } U \text{ in } (\mathcal{O}/\mathcal{Q})^*|\cdot\sum_{c\in C_1(\mathcal{B})}\left|\sum_{\chi\in X_{\mathcal{B}}^*}\chi(c)\right| &\ll q^\epsilon\sum_{c\in C_1(\mathcal{B})}\left|\sum_{\chi\in X_{\mathcal{B}}^*}\chi(c)\right|,\end{aligned}$$

where $C_1(\mathcal{B}) = C(\mathcal{B})/<\{\mathcal{P},\ v_{\mathcal{P}}\in S,\ v_{\mathcal{P}}<\infty\}>$. Now from the relation (χ^0 stands for the trivial character)

$$\sum_{\mathcal{A}|\mathcal{B}}\left(\sum_{\chi\in X_{\mathcal{A}}^*}\chi\chi^0\right) = \sum_{\chi\in X_{\mathcal{B}}}\chi,$$

and Möbius inversion, we have

$$\sum_{\chi\in X_{\mathcal{B}}^*}\chi\chi^0 = \sum_{\mathcal{A}|\mathcal{B}}\left(\sum_{\chi\in X_{\mathcal{B}/\mathcal{A}}}\chi\right)\mu(\mathcal{A}),$$

where $\mu(\cdot)$ is the Möbius function defined on ideals. Thus,

$$\begin{aligned}\sum_{c\in C_1(\mathcal{B})}\left|\sum_{\chi\in X_{\mathcal{B}}^*}\chi(c)\right| &\leq \sum_{\mathcal{A}|\mathcal{B}}\sum_{c\in C_1(\mathcal{B})}\left|\sum_{\chi\in X_{\mathcal{B}/\mathcal{A}}}\chi(c)\right|\\ &\ll \sum_{\mathcal{A}|\mathcal{B}}|X_{\mathcal{B}/\mathcal{A}}|\frac{|X_{\mathcal{B}}|}{|X_{\mathcal{B}/\mathcal{A}}|}\ll d(\mathcal{B})|X_{\mathcal{B}}|.\end{aligned}$$

Hence, we have

$$\sum_{\chi\in X^*_{\mathcal{B}}} \chi(\mathcal{M})W(\chi)^{n^2} \ll q^{-1/2+\epsilon}N(\mathcal{B}). \tag{28}$$

We infer from this, using the estimates for $f_2(y)$, (14) and (28), that

$$\begin{aligned}\sum_{\mathcal{B}|\mathcal{Q}}\sum_{\chi\in X^*_{\mathcal{B}}}\sum\nolimits^{(2)} \\ &\ll (q^{n^2})^{\frac{1}{2}-\beta}q^{\frac{1}{2}+\epsilon}\max_{\mathcal{B}|\mathcal{Q},\mathcal{B}\neq O}\left(\sum_{(\mathcal{A},S)=1}\frac{a(\mathcal{A})}{(N\mathcal{A})^{1-\beta}}\left|f_2\left(\frac{N(\mathcal{A})Y}{D^{n^2}N(\mathfrak{f})N(\mathcal{B})^{n^2}}\right)\right|\right)\\ &\ll (q^{n^2})^{\frac{1}{2}-\beta}q^{\frac{1}{2}+\epsilon}q^{n^2\beta}Y^{-\beta} \quad \ll q^{(n^2+1)/2+\epsilon}Y^{-\beta}.\end{aligned}$$

Choosing $Y = q^{(n^2+1)/2}$, we conclude that

$$\left|\sum_{\mathcal{B}|\mathcal{Q},\mathcal{B}\neq O}\sum_{\chi\in X^*_{\mathcal{B}}} L_{S,\chi}(\beta)\right| > q^{1-r\epsilon} + O\left(Y^{1-\beta+\epsilon} + q^{(n^2+1)/2+\epsilon}Y^{-\beta}\right) \gg q^{1-r\epsilon},$$

provided ϵ is sufficiently small, q is sufficiently large and $\beta > 1 - 2/(n^2+1)$. This completes our proof of (19) and hence of Theorem 1.

References

[BR] L. Barthel and D. Ramakrishnan, *A nonvanishing result for twists of L-functions of* GL(n), Duke Math. J. **74** (1994), 681–700.

[BLS] M. Burger, J. S. Li and P. Sarnak, *Ramanujan duals and automorphic spectrum*, Bull. of the Amer. Math. Soc. **26** (1992), 253–257.

[BS] M. Burger and P. Sarnak, *Ramanujan duals II*, Invent. Math. **106** (1991), 1–11.

[D] P. Deligne, SGA $4\frac{1}{2}$ Cohomologie Etale, Lecture Notes in Math. **569** Berlin-New York: Springer-Verlag (1977).

[DI] W. Duke and H. Iwaniec, *Estimates for coefficients of L-functions*, Proc. C.R.M. Conf. on Automorphic Forms and Analytic Number Theory, Montréal (1989), 43–48.

[GJ] S. Gelbart and H. Jacquet, *A relation between automorphic representations of* GL(2) *and* GL(3), Ann. Sci. École Norm. Sup. 4ᵉ série, **11** (1978), 471–552.

[J] H. Jacquet, I. I. Piatetski-Shapiro and J. A. Shalika, *Rankin-Selberg convolutions*, Amer. J. Math. **105** (1983), 367–464.

[JS] H. Jacquet and J. A. Shalika, *On Euler products and the classification of automorphic representations I,II*, Amer. J. Math. **103** (1981), 499–557, 777–815.

[JSR-S] H. Jacquet and J. A. Shalika, *Rankin-Selberg convolutions: archimedean theory*, Festschrift in honor of I. I. Piatetski-Shapiro on the occasion of his sixtieth birthday, Part I (Ramat Aviv, 1989), Israel Math. Conf. Proceedings, **2** (1990), 125–208.

[L] E. Landau, *Über die Anzahl der Gitterpunkte in gewissen Bereichen*, Gött. Nachr. (1915), 209–243.

[LRS] W. Luo, Z. Rudnick and P. Sarnak, *On Selberg's eigenvalue conjecture*, Geom. Funct. Anal. **5** (1995), 387–401.

[MW] C. Mœglin and J.-L. Waldspurger, *Le spectre résiduel de* GL(n), Ann. Sci. École Norm. Sup. 4ᵉ série, 22, (1989), 605–674.

[MM] M. R. Murty and V. K. Murty, *A variant of the Bombieri-Vinogradov theorem*, Number theory (Montreal, Que., 1985), CMS Conf. Proc., Vol. 7, Providence: Amer. Math. Soc. (1987).

[R] D. Rohrlich, *Nonvanishing of L-functions for* GL(2), Invent. Math. **97** (1989), 383–401.

[RS] Z. Rudnick and P. Sarnak, *Zeros of principal L-functions and random matrix theory*, Duke Math. J. **81** (1996), 269–322.

[S] J.-P. Serre, Letter to J.-M. Deshouillers (1981).

[Sh1] F. Shahidi, *On certain L-functions*, Amer. J. Math. **103**, (1981), 297–355.

[Sh2] F. Shahidi, *On the Ramanujan conjecture and finiteness of poles for certain L-functions*, Ann. of Math. **127** (1988), 547–584.

[Shi] G. Shimura, *On the holomorphy of certain Dirichlet series*, Proc. London Math. Soc. (3) **31** (1975), 79–98.

[Si] C. L. Siegel, Lectures on Advanced Analytic Number Theory, Bombay: Tata Inst. (1961).

DEPARTMENT OF MATHEMATICS, PRINCETON UNIVERSITY, PRINCETON, NEW JERSEY 08544, USA.

E-mail address: luo@math.Princeton.EDU

RAYMOND AND BEVERLY SACKLER FACULTY OF EXACT SCIENCES, SCHOOL OF MATHEMATICAL SCIENCES, TEL AVIV UNIVERSITY, RAMAT AVIV 69978 TEL AVIV, ISRAEL.

E-mail address: rudnick@math.tau.ac.il

DEPARTMENT OF MATHEMATICS, PRINCETON UNIVERSITY, PRINCETON, NEW JERSEY 08544, USA.

E-mail address: sarnak@math.Princeton.EDU

Proceedings of Symposia in Pure Mathematics
Volume **66.2**, 1999

On the Global Root Numbers of GL(n) $\times$ GL(m)

Dipendra Prasad and Dinakar Ramakrishnan

To Professor G. Shimura

ABSTRACT. Let F be a number field, and let π, π' be self-dual Hecke eigencusp forms on $\mathrm{GL}(n)/F$, $\mathrm{GL}(m)/F$ respectively. In this paper we study the *sign of the functional equation* $W(\pi \times \pi')$ of the L-function of the pair (π, π'). By analogy with the situation for orthogonal Galois representations, we conjecture that this sign should be $+1$ when π, π' are of symplectic type. We prove this for $m, n \leq 4$ under a local condition at 2.

1. Introduction

Let F be a number field, $n, m \geq 1$, and $\pi = \otimes_v \pi_v$, $\pi' = \otimes_v \pi'_v$ cuspidal, unitary automorphic representations of $\mathrm{GL}(n, \mathbb{A}_F)$, $\mathrm{GL}(m, \mathbb{A}_F)$ respectively. To this data is associated an Euler product

$$L(s, \pi \times \pi') = \prod_v L(s, \pi_v \times \pi'_v),$$

which converges absolutely in $\Re(s) > 1$. One knows by Jacquet, Piatetski-Shapiro and Shalika [**JPSS**] and Shahidi [**Sh88**], that this L-function extends to a meromorphic function on all of $\mathbb{C}$ and admits a functional equation

$$L(s, \pi \times \pi') = W(\pi \times \pi')(N(\pi \times \pi')d_F^n)^{1/2-s} L(1-s, \pi^\vee \times \pi'^\vee).$$

Here $\pi^\vee$ (resp. $\pi'^\vee$) denotes the contragredient of π (resp. π'), d_F the discriminant of F, and $N(\pi \times \pi')$ a positive rational number ("the conductor"). The arithmetically important **global root number** $W(\pi \times \pi')$ is non-zero and satisfies $W(\pi \times \pi')W(\pi^\vee \times \pi'^\vee) = 1$. We will be particularly interested in the **self-dual** situation, when

$$W(\pi \times \pi') = \pm 1.$$

It is of importance to determine the sign. Our object here is to make some modest progress on this question.

1991 *Mathematics Subject Classification.* Primary 11F70, 11F85, 11R39 and 11S40.

Key words and phrases. Root numbers, GL(n), cusp forms, automorphic representations, symplectic representations, orthogonal representations, L-functions, and lifting.

The second author would like to thank the NSF for support through the grant DMS-9501151.

We will say that a cuspidal automorphic representation η of $\mathrm{GL}(r, \mathbb{A}_F)$ is of **symplectic type** if, for a finite set S of places, the (incomplete) exterior square L-function $L^S(s, \eta, \Lambda^2)$ (see section 3 below for a definition) admits a pole at $s = 1$. One knows that this cannot happen if r is odd [**JS90**].

CONJECTURE I. *Suppose n, m are even, and π, π' are both cusp forms of symplectic type. Then $W(\pi \times \pi') = 1$.*

As positive evidence, we will prove (in section 5) the following

THEOREM A. *Let π, π' be as in the conjecture with $n, m \leq 4$. If n or m is 4, assume that the conductor of the representation is prime to 2. Then*

$$W(\pi \times \pi') = 1.$$

In fact, our proof will show a bit more. If n or m is 4, it suffices to assume that, at every place v dividing 2, the base change of the local component at v to some quadratic extension of F_v is not supercuspidal.

If π (resp. π') corresponds to a Galois representation σ (resp. σ') of dimension $2n$ (resp. $2m$) of Artin type, then π (resp. π') being symplectic is equivalent to the image of σ (resp. σ') being contained in $\mathrm{Sp}(2n, \mathbb{C})$ (resp. $\mathrm{Sp}(2m, \mathbb{C})$). So $\sigma \otimes \sigma'$ would then be orthogonal, and one can deduce the fact that $W(\sigma \otimes \sigma') = 1$ by appealing to a well known result due to Fröhlich-Queyrut [**FQ**] and Deligne [**D76**]. The problem in attempting such an argument in the automorphic setup is that there is **no global group** $\mathfrak{L}_F$ available (as of yet) whose n-dimensional representations parametrize automorphic representations of $\mathrm{GL}(n, \mathbb{A}_F)$. In (the motivational) section 2, we indicate how to get the result on $W(\sigma \otimes \sigma')$ by using only a local result of Deligne relating root numbers to Stiefel-Whitney classes. (Recently, we have come to know that this local argument has already been found by D. Rohrlich in [**R**, section 1, Prop. 2]. Putting this in a general framework leads to Conjecture II.

After some preliminaries in section 3, we discuss the status of the local Langlands conjecture for GL(4) in section 4. In particular we show that the local correspondence works (see Prop. 4.1) for the class of representations of GL(4) satisfying the ramification condition at primes above 2 which was alluded to above. (In the odd residual characteristic case, the correspondence for GL(4) is a consequence of the recent works of M. Harris [**Ha**] and J. Chen [**Ch**].

The heart of this paper is in section 5, where we study $W(\pi \times \pi')$ by some global arguments for automorphic forms on $\mathrm{GL}(4)/F$ and lifting to $\mathrm{GSp}(4)/F$. The key problem (in proving Theorem A) becomes one of showing that, locally at the ramified places, the representations σ_v of the Weil-Deligne group defined by a global π of symplectic type have images in the symplectic group (see Theorem 5.1 and Propositions 5.1 - 5.3). Another problem is to know that at such a place, $W(\pi_v \times \pi'_v) = W(\sigma_v \otimes \sigma'_v)$, which is addressed earlier in section 4 (cf. Proposition 4.2).

When $n = 2$, it is easy to see that π is of symplectic type iff its central character is trivial. Thus Theorem A proves in particular the triviality of the root number $W(\pi \times \pi')$ associated to a pair of cuspidal automorphic representations of π, π' $\mathrm{PGL}(2, \mathbb{A}_F)$. However, the difficult part of the Theorem deals with $\mathrm{GL}(4) \times \mathrm{GL}(2)$ and $\mathrm{GL}(4) \times \mathrm{GL}(4)$.

Given any cuspidal automorphic representation π of $\mathrm{GL}(2, \mathbb{A}_F)$, it is natural to look at the symmetric power L-functions $\{L(s, \pi, \mathrm{Sym}^r) | r \geq 1\}$ (see section 4). These were first defined by Langlands [**L70**] at almost all places. Thanks to the work of Shahidi [**Sh88, Sh94**], one knows, for each $r \leq 5$, how to extend $L(s, \pi, \mathrm{Sym}^{\mathrm{r}})$ to a meromorphic function satisfying a functional equation, with a good definition of bad factors as well. For $r = 2$, one knows much more by the work of Gelbart and Jacquet [**GJ**], namely that $L(s, \pi, \mathrm{Sym}^2)$ is the standard L-function of an automorphic form $\mathrm{Sym}^2(\pi)$ on $\mathrm{GL}(3)/F$. For $r = 3$, one also knows, by the work of Bump, Ginzburg and Hoffstein, that the (symmetric cube) L-function is holomorphic in $\{\Re(s) \geq 3/4\}$; but we will not have occasion to use this.

Using [**Sh88, Sh94**], we can therefore define the global root number

$$W(\pi, \mathrm{Sym}^r),$$

for each $r \leq 5$. In section 6, we will prove

THEOREM B. *Let π be a cuspidal automorphic representation of* $\mathrm{GL}(2, \mathbb{A}_F)$ *of trivial central character. Then*

$$W(\pi, \mathrm{Sym}^2) = W(\pi, \mathrm{Sym}^4) = 1.$$

It should be noted that if π corresponds to a compatible system $\{\sigma_\ell\}$ of ℓ-adic representations of $\mathrm{Gal}(\overline{\mathbb{Q}}/F)$, then $\mathrm{Sym}^{2k}(\sigma_\ell)$ will be an orthogonal similitude representation of even (motivic) weight, and the assertion that the global root number is 1 already follows from the work of T. Saito [**Sa**], who generalized the results of Fröhlich and Deligne [**D76**] on Galois representations of Artin type, and also the result of Coates and Schmidt [**CoS**] on the symmetric square of a modular elliptic curve. But our result appears to be new for Maass forms over $\mathbb{Q}$ and for forms of arithmetic type over number fields with a complex place.

Professor Shimura established a number of important special value results for L-series of pairs (π, π') of arithmetic type. He also was the first one to prove the holomorphy of the symmetric square L-function associated to elliptic modular newforms. Here we study problems of a different sort, but for the same types of L-series, and we are honoured to dedicate this paper to him.

We would like to thank F. Shahidi for helpful discussions concerning his results on L-functions, and for his help with Lemma 5 of section 5. We would also like to thank M. Harris, H. Jacquet, D. Rohrlich and J.-P. Serre for useful conversations/correspondence. The first author would like to thank Caltech for an invitation to visit during February 97, when this work was begun.

2. Motivation and generalization

Suppose σ (resp. σ') is a continuous, irreducible $\mathbb{C}$-representation of the absolute Galois group $\mathrm{Gal}(\overline{\mathbb{Q}}/F)$ of dimension n (resp. m). Then the analogue of $W(\pi \times \pi')$ is the **global Artin root number** $W(\sigma \otimes \sigma')$. One says that σ is symplectic if the image is in fact contained in $\mathrm{Sp}(n, \mathbb{C})$, or in other words, if the exterior square of σ contains the trivial representation. This is known to be equivalent to the existence of a pole at $s = 1$ of the Artin L-function $L(s, \Lambda^2(\sigma))$. Indeed, this is so because $L(s, \tau)$ is invertible at $s = 1$ for any non-trivial irreducible τ. (This can be seen, for example, by Prop. 3.4 of [**T**] together with the functional equation.) Clearly, every symplectic σ is self-dual and of even dimension. Similarly for σ'.

Though we will stick to Galois representations of Artin type below, much of what we say will also be valid for compatible systems. However, it should be noted that, for an irreducible ℓ-adic representation σ_ℓ, the equivalence between being symplectic and having the exterior square L-function admit a pole is not known, though predicted by the Tate conjectures.

One knows by a theorem of Langlands (see [**D73**] for an elegant global proof) that there is a factorization

$$W(\sigma \otimes \sigma') = \prod_v W_v(\sigma \otimes \sigma'),$$

where each local constant $W_v(\sigma \otimes \sigma')$ depends only on the restriction $\sigma_v \otimes \sigma'_v$ to the decomposition group D_v at v.

Now fix a place v, and suppose τ is a virtual **orthogonal** representation of $\mathrm{Gal}(\overline{F}_v/F_v)$ of determinant 1 and dimension 0. Then, by a fundamental theorem of Deligne [**D76**], one has the equality

$$W_v(\tau) = e^{\pi i w_2(\tau)}, \tag{2.1}$$

where $w_2(\tau) \in \frac{1}{2}\mathbb{Z}/\mathbb{Z}$ is the image of the second Stiefel-Whitney class of τ under the isomorphism

$$H^2(F_v, \{\pm 1\}) = {}_2\mathrm{Br}(F_v) \simeq \frac{1}{2}\mathbb{Z}/\mathbb{Z}.$$

The motivation for the conjecture made in the introduction arises from the following

PROPOSITION 2.1. *Let σ, σ' be both symplectic of respective dimensions $2n, 2m$. Then*

$$W(\sigma \otimes \sigma') = 1.$$

When σ, σ' are symplectic, their tensor product is clearly orthogonal, and so this Proposition is an immediate consequence of a theorem of Fröhlich-Queyrut [**FQ**] and Deligne [**D76**]. But in the automorphic context, there is (as of yet) no analog of the global Galois group, and a starting point for this paper is supplied by the following

ALTERNATE PROOF. Let v be any place. Since σ_v and σ'_v are both symplectic, their tensor product must be orthogonal and of determinant 1. Put

$$\tau = (\sigma_v \otimes \sigma'_v) \ominus 4nm[1],$$

where [1] denotes the trivial representation. Then τ is of dimension 0, determinant 1, and of orthogonal type, and so by Deligne, $W_v(\tau) = e^{\pi i w_2(\tau)}$. Note that $W_v([1])^4 = 1 = e^{\pi i w_2(4[1])}$. For any $N \geq 1$, let $\mathrm{Spin}(N, \mathbb{C})$ denote the two-fold covering group of $\mathrm{SO}(N, \mathbb{C})$. Clearly, $\sigma_v \otimes \sigma'_v$ lies in $\mathrm{SO}(4nm, \mathbb{C})$. It follows then, by the definition and properties of Stiefel-Whitney classes (cf. [**D76**], that $W_v(\sigma \otimes \sigma')$ can be -1 iff $\sigma_v \otimes \sigma'_v$ cannot be lifted to a representation into $\mathrm{Spin}(4nm, \mathbb{C})$. But this representation factors through $\mathrm{Sp}(2n, \mathbb{C}) \times \mathrm{Sp}(2m, \mathbb{C})$, which is *simply connected* as an algebraic group. Now we appeal to the following basic Lemma, whose proof is left to the reader.

LEMMA 1. *Let $\phi : H \to H'$ be a morphism of semisimple algebraic groups over $\mathbb{C}$, and let H be connected and simply connected as an algebraic group. Let $\tilde{H}'$ denote the universal cover of H' (in the sense of algebraic groups), with covering map $p : \tilde{H}' \to H'$. Then there exists a lifting $\tilde{\phi} : H \to \tilde{H}'$ such that $\phi = p \circ \tilde{\phi}$.*

We apply this with $H = \mathrm{Sp}(2n, \mathbb{C}) \times \mathrm{Sp}(2m, \mathbb{C})$ and $H' = \mathrm{SO}(4nm, \mathbb{C})$, and note that $\tilde{H}'$ is then none other than $\mathrm{Spin}(4nm, \mathbb{C})$. Consequently, $w_2(\sigma_v \otimes \sigma'_v)$ is 1. So is $W_v(\sigma_v \otimes \sigma'_v)$ by (2.1). Now we are done as *every* local factor of $W(\sigma \otimes \sigma')$ is 1. □

REMARK: We have recently learned from D. Rohrlich that this argument above has already been found by him, see [**R**, section 1, Prop. 2] and the following remark.

More generally, let σ be a Galois representation with values in the $\mathbb{C}$-points of a simply connected semisimple group $\hat{G}$, and let

$$r : \hat{G} \to \mathrm{GL}(N, \mathbb{C}),$$

be an algebraic representation whose image is contained in $\mathrm{SO}(N, \mathbb{C})$. Then the argument proves that $W_v(r(\sigma)) = 1$, for every v.

This suggests a *general conjecture* about the root numbers of automorphic L-functions associated to a split semisimple group G over F when the (Langlands) dual group $\hat{G}$ is simply connected. The reason is this. The automorphic representations of $G(\mathbb{A})$ are expected, modulo some anomalous ones, to be parametrized by the conjugacy classes of homomorphisms ϕ into $\hat{G}$ from a conjectural, pro-reductive group $\mathfrak{L}_F$, which should be an extension of $\mathrm{Gal}(\overline{F}/F)$ by its connected component. (See [**L79**] for a discussion of anomaly.) The cuspidal ones among them should correspond to those (classes of) ϕ with $\mathrm{im}(\phi)$ not contained in any Levi subgroup.

CONJECTURE II. *Let G be a split semisimple group over F with a simply connected (Langlands) dual group $\hat{G}$, and let π be a non-anomalous, cuspidal automorphic representation of $G(\mathbb{A}_F)$. Suppose r is an algebraic representation of $\hat{G}$ which is orthogonal. Then one should have $W_v(\pi, r) = 1$ for every v.*

Whether or not $\hat{G}$ is simply connected, the global root number of π relative to an orthogonal representation r of $\hat{G}$ is expected to be 1; but it will be very subtle to establish, requiring a product formula, as the local root numbers need not be 1.

When $G = \mathrm{PGL}(n, \mathbb{A}_F)$, $\hat{G}$ is $\mathrm{SL}(n, \mathbb{C})$, and every cuspidal automorphic representation of $G(\mathbb{A}_F)$ is non-anomalous, called *isobaric* in this context. When $G = \mathrm{SO}(2n+1) \times \mathrm{SO}(2m+1)$, $\hat{G}$ is $\mathrm{Sp}(2n, \mathbb{C}) \times \mathrm{Sp}(2m, \mathbb{C})$, and if r is the tensor product representation of $\hat{G}$ into $\mathrm{GL}(4nm, \mathbb{C})$, then $\mathrm{im}(r)$ lies in $\mathrm{SO}(4nm, \mathbb{C})$, and the Conjecture II applies, giving a strengthening of Conjecture I stated in the introduction. The connection is understood via the functoriality principle, which predicts that isobaric automorphic forms of symplectic type on $\mathrm{GL}(2n)/F$ correspond to packets of non-anomalous ones on $\mathrm{SO}(2n+1)/F$.

3. Preliminaries

Given any unitary, cuspidal automorphic representation π of $\mathrm{GL}(n, \mathbb{A}_F)$, one has, at every finite place v where π_v is unramified, a (Langlands) conjugacy class $A_v(\pi)$ in $\mathrm{GL}(n, \mathbb{C})$, represented by a diagonal matrix $[\alpha_{1,v}, \dots, \alpha_{n,v}]$. Let S be any finite set of places containing the archimedean and ramified places (for π). If r is any algebraic representation of $\mathrm{GL}(n, \mathbb{C})$, we put

$$L^S(s, \pi, r) = \prod_{v \notin S} L(s, \pi_v, r),$$

where

$$L(s, \pi_v, r) = \det(I - (Nv)^{-s} r(A_v(\pi)))^{-1}.$$

Of particular interest is when r is the exterior square Λ^2, or the symmetric kth power Sym^k, for some $k \geq 1$, of the standard representation.

One knows (cf. [**JS90, BF, Sh88**] that $L^S(s,\pi,\Lambda^2)$ converges absolutely in $\Re(s) > 1$ and admits a meromorphic continuation with a functional equation of the usual type relating s and $1-s$.

DEFINITION. π is of symplectic type iff $L^S(s,\pi,\Lambda^2)$ has a pole at $s=1$.

Here is a simple Lemma which will be needed later.

LEMMA 2. *If π is of symplectic type, then it is self-dual.*

PROOF. We have the factorization

$$L^S(s,\pi\times\pi) = L^S(s,\pi,\Lambda^2)L^S(s,\pi,\mathrm{Sym}^2).$$

One knows by Shahidi [**Sh88**] that $L^S(s,\pi,\mathrm{Sym}^2)$ has no zero at $s=1$. Thus, if π is of symplectic type, the pole at $s=1$ of $L^S(s,\pi,\Lambda^2)$ will introduce a pole of $L^S(s,\pi\times\pi)$ at $s=1$. This implies, by a theorem of Jacquet and Shalika [**JS81**], that π is equivalent to $\pi^\vee$. □

Thus, for every place v, the local component π_v of a cusp form π on $\mathrm{GL}(n)/F$ of symplectic type is self-dual. It is not clear, however, that the conjecturally associated representation (by the local Langlands conjecture)

$$\sigma_v : W_{F_v'} \to \mathrm{GL}(n,\mathbb{C})$$

should be symplectic. (One could equally well use the modified Weil group $\tilde{W}_{F_v} = W_{F_v}\times \mathrm{SL}(2,\mathbb{C})$; in this setting, the indecomposable module $\mathrm{sp}(n)$ corresponds to $1\otimes \mathrm{Sym}^{n-1}(st)$, where st denotes the standard representation of $\mathrm{SL}(2,\mathbb{C})$.) This leads to the following

CONJECTURE III. *Let $n\geq 1$, and let π be a cuspidal automorphic representation of $\mathrm{GL}(2n,\mathbb{A}_F)$ of symplectic type. Let v be a place such that π_v is functorially associated to a $2n$-dimensional representation σ_v of W_{F_v}'. Then σ_v is symplectic, i.e., the image of σ_v lies in $\mathrm{Sp}(2n,\mathbb{C})$.*

If $n=2$, it is easy to see that this conjecture is true. Indeed, as remarked earlier, a cusp form on $\mathrm{GL}(2)/F$ is of symplectic type iff its central character ω is trivial. Now if v is a place, then the determinant of σ_v is equal to ω_v, which is trivial. Since σ_v is two-dimensional, it must then be symplectic. In the next section, on the way to proving Theorem A, we will prove this conjecture for cusp forms (of symplectic type) on $\mathrm{GL}(4)/F$ satisfying a condition above 2.

We end this section discussing **base change and automorphic induction** for $\mathrm{GL}(n)/F$. We will rely on the results of Arthur and Clozel in [**AC**]. We will use the modern terminology of *isobaric representations* together with the "sum operation" $\boxplus$ [**L79, JS81**]. These representations are suitable subquotients of parabolically induced representations from essentially unitary cuspidal (resp. essentially square-integrable) representations when F is global (resp. local). For every $n\geq 1$, denote by $\mathrm{Isob}(n,F)$ (resp. $\mathrm{Isob}(n,F_v)$) the set of irreducible, isobaric automorphic (resp. admissible) representations of $\mathrm{GL}(n,\mathbb{A}_F)$ (resp. $\mathrm{GL}(n,F_v)$). Then in [**AC**, chapter 3, sec. 3–6,], one finds a construction, for any cyclic extension K/F with $[K:F]$ a prime ℓ, of maps

$$b_{K/F} : \mathrm{Isob}(n,F)\to \mathrm{Isob}(n,K), \pi\to\pi_K \quad \text{(base change)}$$

and

$$I_{K/F} : \mathrm{Isob}(n,K)\to \mathrm{Isob}(n\ell,F), \pi\to I(\pi) \quad \text{(automorphic induction)},$$

such that at every place v of F which is finite and unramified for the representations and K/F (or archimedean) and a place w of K above v, we have (respectively)

$$\mathrm{res}^{F_v}_{K_w}(\sigma(\pi_v)) \simeq \sigma((\pi_K)_w), \tag{3.1}$$

and

$$\sigma(I(\pi)_v) \simeq \mathrm{ind}^{F_v}_{K_w}(\pi_w).$$

Given any local field E, if β an unramified representation of $\mathrm{GL}(r, E)$ (or if E is archimedean), we write $\sigma(\beta)$ to signify the associated r-dimensional representation of W'_E (or W_E).

There are also local analogs of these maps, customarily called "local base change" and "local automorphic induction". The former is constructed and discussed in great detail in chapter 1 of [**AC**]. The later is discussed briefly in [**Cl**] and the relevant assertions are consequences of the results of [**AC**]. An alternate construction of the local automorphic induction for essentially tempered representations, which works also in characteristic p, is given in the paper [**HH**] of Henniart and Herb. The properties we will need are summarized below:

PROPOSITION 3.1. *Let K/F be a cyclic extension of number fields or local fields of degree ℓ, a prime. Let θ be a generator of Gal(K/F), and let χ denote the character of the idele class group (resp. multiplicative group) of F in the global (resp. local) case associated to K. Then*

1. *The image of $b_{K/F}$ consists precisely of those $\beta \in$Isob(n, K) such that $\beta \simeq \beta \circ \theta$.*
2. *The image of $I_{K/F}$ consists precisely of those $\pi \in$Isob$(n\ell, F)$ such that $\pi \simeq \pi \otimes \chi$.*
3. *For every $\pi \in$Isob(m, F) and $\beta \in$Isob(n, K), we have the* **adjointness** *property:*
$$L(s, \pi \times I(\beta)) = L(s, \pi_K \times \beta),$$
and
$$\varepsilon(s, \pi \times I(\beta))\varepsilon(s, 1_K)^{nm} = \varepsilon(s, \pi_K \times \beta) \prod_{j=0}^{\ell-1} \varepsilon(s, \chi^j)^{nm}.$$
4. *Suppose β is cuspidal (resp. supercuspidal) in Isob(n, K), for K global (resp. local). Then*
$$I(\beta)_K \simeq \boxplus_{j=0}^{\ell-1} \beta \circ \theta^j.$$
Moreover, $I(\beta)$ is cuspidal (resp. supercuspidal) iff β is **not** *isomorphic to $\beta \circ \theta$.*
5. *Suppose π is cuspidal (resp. supercuspidal) in Isob(n, F), for F global (resp. local). Then*
$$I(\pi_K) \simeq \boxplus_{j=0}^{\ell-1} \pi \otimes \chi^j.$$
Moreover, π_K is cuspidal (resp. supercuspidal) iff π is not isomorphic to $\pi \otimes \chi$.

For a proof of 1, 2 and the second half of 5, see [**AC, Cl, HH**]. There one also finds formulae for the L- and ε-factors of pairs (π_K, π'_K) and $(I(\beta'), I(\beta))$, which do not quite imply the identities of 3. In any case, one gets the adjointness formulae, as it is well known to experts, by a similar global argument, which we will briefly indicate here for completeness. First let us consider the global case. By using the identities (3.1), one sees easily that for almost all v (including the unramified and archimedean ones), the local factors at v of the functions $L(s, \pi \times I(\beta))$ and $L(s, \pi_K \times \beta)$, both viewed as Euler products over F, coincide. Such a relationship holds with π replaced by $\pi \otimes \mu$, for any character μ. Now fix a ramified place v_0. We may choose a finite order character μ which is 1 at v_0 and is sufficiently ramified at all the other ramified places u to ensure that the L-factors on both sides (and their contragredient factors) are 1 at each u. (This is easily done by using the formula in [**JS81**] for the L-factors for $\mathrm{GL}(m)\times\mathrm{GL}(n\ell)$.) Consequently, one gets, by comparing the functional equations of both global L-functions, the desired identity of the local factors at v_0. The L-function identity of 3. follows in the global case. To get this locally, one reduces the problem, by inductivity, to the case when the representations are supercuspidal, which can then be realized as the local components of cuspidal automorphic representations by using the trace formula. The local identity then follows from the global one. (This argument is very similar to the proof of Proposition 6.9 in chapter 1 of [**AC**].) For epsilon factors, which are non-trivial only at a finite set S of places, we exploit the following trick used in [**He86**]. (We may assume that S consists only of finite places as the archimedean comparison poses no problem.) By a theorem of Jacquet and Shalika [**JS85**], if μ is sufficiently ramified at a place v, then $\epsilon(s, \pi_v \times \pi_v' \otimes \mu_v)$ has a simple expression, depending only on the central characters of π, π' and the twisting character. Now pick any place v_0 in S, and pick μ to be 1 at v_0, but sufficiently ramified at every other place u in S. Then, by comparing the functional equations of the μ-twisted L-functions in question, we can isolate the epsilon factors at v_0 and deduce the assertion. (It should be noted that the extraneous looking factors appear for the epsilon identity as these factors are inductive only in degree zero.)

Now we prove part 4. In the global case, once again, the L-functions of both sides of the identity agree at all the unramified (and archimedean) places. Twisting by a suitable character and comparing functional equations, we get the assertion at every place. (Locally, we employ the same trick as above.) Suppose first that β is not isomorphic to $\beta \circ \theta$. Let π be a cuspidal element of $\mathrm{Isob}(m, F)$ which occurs in the isobaric sum decomposition [**L80**] of $I(\beta)$. Then $L(s, \pi^\vee \times I(\beta))$, and hence $L(s, \pi_K^\vee \times \beta)$ by the adjointness formula, admits a pole at $s = 1$ by [**JS81**]. (In the local case, one exploits the pole at $s = 0$.) This means that π_K contains β in its isobaric sum decomposition. But π_K is invariant under θ by part 1. So it must contain $\beta \circ \theta^j$ for all j, and so $I(\beta)_K$. Since π occurs in $I(\beta)$, we must have $\pi \simeq I(\beta)$, and so $I(\beta)$ is cuspidal. Conversely, suppose $I(\beta)$ is cuspidal. If we had $\beta \simeq \beta \circ \theta$, then by part 1, we can find some $\eta \in \mathrm{Isob}(n, F)$ such that $\beta \simeq \eta_K$. Now we interject and observe that the identity of part 5. can be established by the same idea used for the identity of part 4. Consequently,

$$I(\beta) \simeq I(\eta_K) \simeq \boxplus_{j=1}^{\ell-1} \eta \otimes \chi^j,$$

which contradicts the cuspidality of $I(\beta)$. The local argument is the same.

Thus the assertions of the Proposition hold.

4. On the local correspondence for GL(4)

We begin with a technical definition restricting the amount of ramification, to be used at places abcve 2. This is necessitated by the lack of knowledge of the local Langlands conjecture for GL(4).

DEFINITION. Let E be a non-archimedean local field. An irreducible, admissible representation η of $\mathrm{GL}(4, E)$ is **allowable** iff the following condition is satisfied:

$(R(E))$ *There exists a quadratic extension K of E such that the base change η_K of η to K is not supercuspidal.*

PROPOSITION 4.1. *Let E be a non-archimedean local field of characteristic zero. Then the local Langlands conjecture holds for* GL(4) *if the residual characteristic is odd. In the even residual characteristic case, it holds for the class of irreducible admissible representations β of* $\mathrm{GL}(4, E)$ *satisfying the condition $(R(E))$. To be explicit, this class is in bijection, preserving local factors, with the class of representations τ of W'_E such that the restriction of τ to W'_K, for some quadratic extension K of E, is not irreducible.*

PROOF. Let p denote the residual characteristic of E. By using inductivity and the proof of the local Langlands conjecture for GL(2) and GL(3) due respectively to Kutzko [**K**] and Henniart [**He84**], we may reduce to proving the bijection between supercuspidals of $\mathrm{GL}(4, E)$ and irreducibles of W_E of dimension 4, satisfying the hypothesis for $p = 2$. A recent theorem of Michael Harris [**Ha**] furnishes, for any $n \geq 1$, a family of bijections $\phi_m : \beta \to \tau$ between supercuspidals β of $\mathrm{GL}(m, E)$ and irreducibles τ of W_E of dimension m, for *every* $m \leq n$, compatible with taking contragredients, such that, *if m, n are prime to p*, the following identity of epsilon factors holds, for all pairs (β, β') of supercuspidals of $\mathrm{GL}(n, E)$ and $\mathrm{GL}(m, E)$ respectively:

$$\varepsilon(s, \beta \times \beta') = \varepsilon(s, \tau \otimes \tau'). \tag{4.1}$$

From this, using a criterion of Henniart [**He93**], he concludes the local Langlands conjecture for $n < p$, which of course gives the Proposition for GL(4) if $p \geq 5$. Another recent result, due to Jeff Chen [**Ch**], gives a finer criterion than that of Henniart for $n = 4$, and says that it suffices, in order to verify that a given bijection is the right one, to check the equality of epsilon factors of pairs for $(n, m) = (4, 1)$ and $(4, 2)$. So for any odd p, we have the needed identity thanks to Harris (namely (4.1)), and so the local Langlands conjecture follows for GL(4) for *any* odd p.

Now we begin the proof for $p = 2$. Since β satisfies $(R(E))$, there exists a quadratic extension K/E such that the base change β_K is not supercuspidal. Appealing to properties of base change [**AC**], more precisely Proposition 3.1 of this article, we then see that β must be a local automorphic induction $I_{K/E}(\lambda)$, for a supercuspidal λ of $\mathrm{GL}(2, K)$. Let μ be the associated irreducible, two-dimensional representation of W_K, given by [**K**]. Since β is cuspidal, λ must be inequivalent to $\lambda \circ \rho$, where ρ denotes the non-trivial automorphism of K over E (see part 4. of the same Prop. 3.1). This implies that μ is not isomorphic to the representation $\mu^{[\rho]}$, defined by $w \to \mu(\rho w \rho^{-1})$ $(\forall w \in W_K)$. Let τ be the representation of W_E induced by μ. Then it must be irreducible by Mackey theory. Having defined $\beta \to \tau$, we must check that it is well defined, i.e., independent of K, and has the right functorial properties. For this, it suffices, by Chen [**Ch**], to show that

$$\varepsilon(s, \beta \times \beta') = \varepsilon(s, \tau \otimes \tau'), \tag{$*$}$$

for all supercuspidal representations β' of $\mathrm{GL}(k, E)$, $k \leq 2$, with corresponding irreducibles τ' of W_E of dimension k. But $\tau \otimes \tau'$ is isomorphic to $\mathrm{Ind}_K^E(\mu \otimes \tau'_K)$, where Ind_K^E denotes the induction from W_K to W_E and τ'_K the restriction of τ' to

W_K. The epsilon factor is additive, but inductive only in dimension zero; so we get

$$\varepsilon(s,\tau\otimes\tau')\varepsilon(s,\mathrm{Ind}_K^E(1_K))^{-2k}=\varepsilon(s,\mu\otimes\tau'_K)\varepsilon(s,1_K)^{-2k},$$

where 1_K denotes the trivial representation of W_K. By the base change theory (see Prop. 3.1), we know that a similar formula holds with $\tau\otimes\tau'$ (resp. $\mu\otimes\tau'_K$) replaced by $\pi\times\pi'$ (resp. $\lambda\times\beta'_K$).

Then the desired identity (*), and hence Proposition 4.1, is a consequence of knowing that the factor $\varepsilon(s,\lambda\times\beta'_K)$ is the same as $\varepsilon(s,\mu\otimes\beta'_K)$. If $k=1$, or if $k=2$ with β'_K reducible, this is clear. The case when β_K is irreducible is settled by appealing to the $n=m=2$ case of the following

PROPOSITION 4.2. *Let E be a non-archimedean local field, $n,m\in\{2,4\}$, and λ (resp. λ') a supercuspidal representation of* $\mathrm{GL}(n,E)$ *(resp.* $\mathrm{GL}(m,E)$*), functorially associated to an irreducible μ (resp. μ') of W_E. If n or m is 4 and E has residual characteristic 2, assume that the supercuspidal representation in question is allowable. Then we have*

$$\varepsilon(s,\lambda\times\lambda')=\varepsilon(s,\mu\otimes\mu').$$

PROOF. If the residual characteristic p is odd, this is the content of Harris's identity (4.1). So assume that $p=2$. If μ and μ' are local automorphic inductions of characters of cyclic extensions, the assertion is a special case of a general result of Henniart (see part (iii) of Theorem on page 146 of [**He86**]. In the general case, the proof is still similar, the key point being the existence of global (automorphic) representations Π, Π' with local components λ, λ' respectively. We give the complete argument.

To begin, we may assume that $n\geq m$, and after twisting by suitable unramified characters, that μ and μ' are continuous (irreducible) representations of $\mathrm{Gal}(\overline{E}/E)$, where $\overline{E}$ denotes an algebraic closure of E. Let L (resp. L') denote the Galois extension of E cut out by the kernel of τ (resp. τ'), and let $\tilde{L}$ denote the compositum of L and L'.

The following simple, but useful, lemma was shown to the second author some time ago by J.-P. Serre.

LEMMA 3. *Let E be a non-archimedean local field of characteristic zero and of residual characteristic p. Fix a finite set S of rational primes other than p, but possibly including ∞. Let E'/E be a finite Galois extension. Then there exists a finite Galois extension k'/k of number fields, and a place v of k extending to a place v' of k', such that (i) $k_v=E$, (ii) $k'_{v'}=E'$, and (iii) the decomposition group of v' in* $\mathrm{Gal}(k'/k)$ *is the whole group. Moreover, all the primes in S split completely in k'.*

This lemma is a consequence of Krasner's lemma and its proof will be left to the reader. A slightly weaker version can be found in [**D73**, page 544], as Lemme 4.13.

Applying this lemma to our setup, we get a Galois extension $\tilde{M}/k$ of number fields with local extension $\tilde{L}/E$, such that $\mathrm{Gal}(\tilde{M}/k)=\mathrm{Gal}(\tilde{L}/E)$. Then $\mathrm{Gal}(\tilde{E}/L)$ and $\mathrm{Gal}(\tilde{E}/L')$ are subgroups of $\mathrm{Gal}(\tilde{M}/k)$, and let us denote their fixed fields in $\tilde{M}$ by M' and M respectively. It is easy to see that L (resp. L') is a local completion of M (resp. M') at a place u (resp. u') with $\mathrm{Gal}(L/E)=\mathrm{Gal}(M/k)$ (resp. $\mathrm{Gal}(L'/E)=\mathrm{Gal}(M'/k)$. Thus we get continuous, irreducible representations

β and β' of $\mathrm{Gal}(\overline{\mathbb{Q}}/k)$, acting via the respective quotients $\mathrm{Gal}(L/k)$ and $\mathrm{Gal}(L'/k)$, such that $\mu = \beta_u$ and $\mu' = \beta'_{u'}$.

Note that the image of β (resp. β') in $\mathrm{GL}_n(\mathbb{C})$ (resp. $\mathrm{GL}_m(\mathbb{C})$) is solvable.

Now suppose $n = m = 2$. Then, by the theorems of Langlands [**L80**] and Tunnell [**Tu**] on tetrahedral and octahedral Galois representations, there exist cuspidal automorphic representations Π and Π' of $\mathrm{GL}(n, \mathbb{A}_k)$ and $\mathrm{GL}(m, \mathbb{A}_k)$ respectively, such that we have the global L-function identities $L(s, \Pi \otimes \chi) = L(s, \beta \otimes \chi)$ and $L(s, \Pi' \otimes \chi) = L(s, \beta' \otimes \chi)$, for all idele class characters χ of k.

It then follows [**L80**] that, for all quasi-characters ν of E^*,

$$\varepsilon(s, \Pi_u \otimes \nu) = \varepsilon(s, \mu \otimes \nu),$$

and

$$\varepsilon(s, \Pi'_u \otimes \nu) = \varepsilon(s, \mu' \otimes \nu).$$

Then, by uniqueness, Π_u must be isomorphic to λ, and $\Pi'_{u'}$ to λ'. Now applying Theorem 4.1 of [**He93**], we get

$$\varepsilon(s, \lambda \times \lambda') \frac{L(1 - s, \lambda^\vee \times \lambda'^\vee)}{L(s, \lambda \times \lambda')} = \varepsilon(s, \mu \otimes \mu') \frac{L(1 - s, \mu^\vee \otimes \mu'^\vee)}{L(s, \mu \otimes \mu')}.$$

Suppose μ' is not of the form $\mu \otimes \nu$, for a quasi-character ν. Then $L(s, \mu \otimes \mu')$ and $L(1 - s, \mu^\vee \otimes \mu'^\vee)$ are both 1. In this case, λ' cannot be a character twist of λ either, as μ (resp. μ') is functorially associated to λ (resp. λ'). Then one sees by Jacquet-Shalika [**JS81**] that the corresponding $L-$factors are also 1. So, to prove Proposition 4.2, we may assume that $\mu' \simeq \mu \otimes \nu$ and $\lambda' \simeq \lambda \otimes \nu$, for some ν. Then we have the factorizations

$$L(s, \mu \otimes \mu') = L(s, \mathrm{Sym}^2(\mu) \otimes \nu) L(s, \omega\nu)$$

and

$$L(s, \lambda \times \lambda') = L(s, \mathrm{Sym}^2(\lambda) \otimes \nu) L(s, \omega\nu),$$

where $\mathrm{Sym}^2(\lambda)$ is the representation of $\mathrm{GL}(3, K)$ associated to λ by the symmetric square lifting (cf. [**GJ**], and ω the central character of λ. Since μ is associated to λ, it follows that $\mathrm{Sym}^2(\mu)$ is associated to $\mathrm{Sym}^2(\lambda)$, and so we get the equality of $L(s, \mu \otimes \mu') = L(s, \lambda \times \lambda')$. Similarly for their contragredients. This proves Proposition 4.2 in this ($n = m = 2$) case, and hence also Proposition 4.1.

Next suppose $(n, m) = (4, 2)$. Then, since λ satisfies $(R(E))$, $\mu = \beta_u$ must be, by an earlier argument, induced by an irreducible θ of $\mathrm{Gal}(\overline{E}/K)$, for some quadratic extension K of E. Clearly, K must be subfield of L, and so must correspond, by construction, to a quadratic extension N of k contained in M; in other words, there is a place w of N extending u, and lying below u', such that $N_w = K$. Then β must be induced by an irreducible δ of $\mathrm{Gal}(\overline{k}/N)$, with $\delta_w = \theta$. Since δ is two-dimensional and has solvable image, we can apply Langlands and Tunnell once again to get a cuspidal automorphic representation α of $\mathrm{GL}(2, \mathbb{A}_N)$ functorially associated to δ. Then we can conclude that α_w is a supercuspidal representation of $\mathrm{GL}(2, K)$ whose induction to E is isomorphic to λ. Then λ is the component of Π at u. Since $m = 2$, we already knew that there exists cuspidal Π' with $\Pi'_u = \lambda'$. Applying what was proved above for the $(2, 2)$ case, we get

$$\varepsilon(s, \alpha_w \times \lambda'_K) = \varepsilon(s, \theta \otimes \mu'_K).$$

The Proposition then follows (in this case) by using the way the epsilon factors change under induction, since $\mathrm{Ind}_K^E(\theta) = \mu$ and $\mathrm{Ind}_K^E(\alpha_w) = \lambda$.

Finally, when $n = m = 4$, we get the same identity of epsilon factors over K, but with μ'_K four-dimensional. But then we can apply the $(4,2)$ case (just proved) and deduce what we want. □

5. Proof of Theorem A

In this section we will prove the following strengthening of Theorem A:

THEOREM A'. *Let $m, n \leq 4$, and π, π' unitary, cuspidal automorphic representations of* $\mathrm{GL}(n, \mathbb{A}_F)$, $\mathrm{GL}(m, \mathbb{A}_F)$ *respectively, which are of symplectic type. If n or m is 4, assume that, for every place v above 2, the local component at v of the representation in question is allowable. Then*

$$W(\pi \times \pi') = 1.$$

Our proof of theorem A' depends crucially on the following

THEOREM 5.1. *Let π be a cuspidal automorphic representation of* $\mathrm{GL}(4, \mathbb{A}_F)$ *of symplectic type, and at each v, let σ_v be the W'_{F_v}-module associated to π_v by Prop. 4.1. Assume that for every v above 2, π_v is allowable. Then σ_v is* **symplectic** *at* **every** *v.*

THEOREM 5.1 $\Longrightarrow$ THEOREM A'. If π and π' are as in the Theorem A', we can, by Proposition 4.1, functorially associate at every v, representations σ_v, σ'_v of W'_{F_v} to the local components π_v, π'_v respectively. These are symplectic by Theorem 5.1. Setting $s = 1/2$ in the epsilon factor identity of Prop. 4.2, we get $(\forall v)$

$$W(\pi_v \times \pi'_v) = W(\sigma_v \otimes \sigma'_v). \tag{5.1}$$

Since we have the product formula $W(\pi \times \pi') = \prod_v W(\pi_v \times \pi'_v)$, it suffices to show that $W(\pi_v \times \pi'_v) = 1$ at each v. So we get what we want by appealing to the reasoning in section 2 (see the proof of Prop. 2.1). Note that, though we considered only the representations of W_{F_v} in section 2, the crucial result (2.1) of Deligne applies to those of W'_{F_v} as well, as seen by the discussion in section 5.4 of [**D76**]. □

PROOF OF THEOREM 5.1. First we need the following

LEMMA 4. *Let E be a non-archimedean local field of odd residual characteristic, β a self-dual supercuspidal representation of* $\mathrm{GL}(4, E)$ *of trivial central character, and τ the irreducible 4-dimensional representation of W_E associated to β by the local Langlands correspondence (cf. Prop. 4.1). Suppose τ is not symplectic. Then β is allowable.*

PROOF. Since the local correspondence is compatible with taking contragredients, τ is self-dual. Moreover, since the central character of β is trivial, the determinant of τ must be trivial. Thus the image of τ lands in either $\mathrm{Sp}(4, \mathbb{C})$ or $\mathrm{SO}(4, \mathbb{C})$. By hypothesis, we are not in the former case. We may view τ as a representation of $\Gamma_E = \mathrm{Gal}(\overline{E}/E)$.

Consider $M_2(\mathbb{C})$ as a four dimensional quadratic space under the non-degenerate symmetric bilinear form $B(X, Y) = \mathrm{tr}({}^tXY)$. Then $\mathrm{GO}(4, \mathbb{C})$ identifies with the group of similitudes of B. Let $\mu(g)$ denote the similitude factor of g in $\mathrm{GO}(4, \mathbb{C})$. Define $\mathrm{GSO}(4, \mathbb{C})$ to be the kernel of the map $g \to \mu(g)^{-2}\det(g)$. (Some authors write SGO(4) instead of GSO(4).) Then there is a well known short exact sequence of $\mathbb{C}$-algebraic groups

$$1 \to \mathbb{C}^* \to \mathrm{GL}(2,\mathbb{C}) \times \mathrm{GL}(2,\mathbb{C}) \to \mathrm{GSO}(4,\mathbb{C}) \to 1,$$

where the map on $\mathbb{C}^*$ sends t to $(tI, t^{-1}I)$, and the one on $\mathrm{GL}(2,\mathbb{C})\times\mathrm{GL}(2,\mathbb{C})$ sends (g_1, g_2) to the similitude $(X \to {}^t g_1 X g_2)$. Viewing this as an exact sequence of trivial Γ_E-modules, we get the following (part of the) associated long exact sequence in cohomology:

$$\mathrm{Hom}(\Gamma_E, \mathrm{GL}(2,\mathbb{C}) \times \mathrm{GL}(2,\mathbb{C})) \to \mathrm{Hom}(\Gamma_E, \mathrm{GSO}(4,\mathbb{C})) \to H^2(\Gamma_E, \mathbb{C}^*),$$

where the left arrow identifies with the tensor product map. But $H^2(\Gamma_E, \mathbb{C}^*)$ is trivial by a theorem of Tate (see [**Se**] for a proof). Note that $\mathrm{SO}(4,\mathbb{C})$ is a subgroup of $\mathrm{GSO}(4,\mathbb{C})$; it is the image of the subgroup of $\mathrm{GL}(2,\mathbb{C})\times\mathrm{GL}(2,\mathbb{C})$ consisting of (g_1, g_2) such that $\det(g_1)^2\det(g_2)^2 = 1$. So we may view τ as an element of $\mathrm{Hom}(\Gamma_E, \mathrm{GSO}(4,\mathbb{C}))$. Then, by Tate's theorem, we can find 2-dimensional representations τ_j, $j = 1, 2$, of Γ_E such that $\tau \simeq \tau_1 \otimes \tau_2$. (The choice of (τ_1, τ_2) is not unique, as we can replace it by $(\tau_1 \otimes \nu, \tau_2 \otimes \nu^{-1})$, for a character ν, but this will not matter to us.) Since τ is irreducible, τ_1 and τ_2 must be irreducible.

Since the residual characteristic is odd, we know that τ_1 must be induced by a (linear) character χ of Γ_K, for a quadratic extension K of E. Then the restriction of τ_1, and hence τ, to Γ_K is reducible. This implies that β_K is not supercuspidal. □

REMARK: We may realize $\mathrm{SO}(4,\mathbb{C})$ as a quotient of $\mathrm{SL}(2,\mathbb{C})\times\mathrm{SL}(2,\mathbb{C})$ by $\{\pm 1\}$. But $H^2(\Gamma_E, \{\pm 1\}) = {}_2\mathrm{Br}(E)$ is not trivial, and so we cannot hope to write τ as $\tau_1 \otimes \tau_2$ with τ_1, τ_2 of determinant 1, unless $W(\tau) = 1$.

Now let π be as in the theorem, and fix a place v of F. By Lemma 2 of section 3, it is self-dual. We will see later (cf. Remark following Theorem 5.2) that it has trivial central character as well. By Lemma 4 we may then assume that π_v is *allowable* at *every* finite v. The assertion of Theorem 5.3 will be a consequence, for v finite, of putting together the following three propositions. The archimedean case will use a different, but simpler argument.

PROPOSITION 5.1. *Suppose v is finite. Then the following are equivalent:*

1. *σ_v is symplectic.*
2. *$L(s, \Lambda^2(\sigma_v))$ has a pole at $s = 0$.*

PROOF. This is clear when σ_v is a true representation of the Weil group W_{F_v}, as $\Lambda^2(\sigma_v)$ admits a trivial summand iff its L-function has a pole at $s = 0$. Indeed, if β is an irreducible non-trivial summand, then $L(s, \beta) = 1$ if β has dimension > 1 or is ramified of dimension 1; otherwise, there is a non-trivial, unramified quasi-character ν of F_v^* such that $L(s, \beta) = (1 - \nu(\varpi_v) N v^{-s})^{-1}$, which is holomorphic at $s = 0$. The converse direction is also clear. (Here ϖ_v denotes a uniformizer at v, and Nv the norm.) By linearity, we are then reduced to treating representations of W'_{F_v} of the following form: $\sigma_v = \beta \otimes \mathrm{sp}(m)$, where β is an irreducible, unitary representation of W_{F_v} pulled back to W'_{F_v}. (See [**D73**] for the definition and properties of the indecomposable module $\mathrm{sp}(m)$, which corresponds to the symmetric $(m-1)$th power representation of the standard representation of $\mathrm{SL}(2,\mathbb{C})$ and, in the local Langlands correspondence, to the Steinberg module of $\mathrm{GL}(m, F_v)$.) In such a case, we have

$$\Lambda^2(\sigma_v) \simeq \mathrm{Sym}^2(\beta) \otimes \Lambda^2(\mathrm{sp}(m)) \oplus \Lambda^2(\beta) \otimes \mathrm{Sym}^2(\mathrm{sp}(m)).$$

Since $\mathrm{sp}(m)$ is symplectic (resp. orthogonal) iff m is even (resp. odd), we see that σ_v is symplectic iff β is orthogonal (resp. symplectic) when m is even (resp. odd). On the other hand, if δ is an irreducible summand of $\beta \otimes \beta = \mathrm{Sym}^2(\beta) \oplus \Lambda^2(\beta)$, we have

$$L(s, \delta \otimes \mathrm{sp}(k)) = L(s, \delta \otimes |.|^{k-1/2}),$$

for any k. Since δ is unitary, this L-function has a pole at $s = 0$ iff $k = 1$ and $\delta = 1$. The Proposition now follows as any irreducible summand of the tensor square of $\mathrm{sp}(m)$ is of the form $\mathrm{sp}(k)$ for some k. □

Let $L_1(s, \pi_v, \Lambda^2)$ denote the local factor associated to π_v and the exterior square representation of ${}^LG = \mathrm{GL}(4, \mathbb{C})$ by the theory of Eisenstein series relative to the realization of GL(4) as a Levi subgroup of SO(8) [**Sh88**]. To be precise, he associates a factor $\gamma(s, \pi_v, \Lambda^2)$ to this situation, which occurs naturally in the functional equation, and one has

$$\gamma(s, \pi_v, \Lambda^2) = \epsilon_1(s, \pi_v, \Lambda^2) \frac{L_1(1-s, \pi_v^\vee, \Lambda^2)}{L_1(s, \pi_v, \Lambda^2)}. \tag{5.2}$$

The ε_1-factor is invertible, and the L_1-factors have the usual shape, in particular having no zeros. It should also be noted that by definition there are no common factors between the numerator and the denominator. They need not be the root number and the L-factor appearing in the definition of $\gamma(s, \Lambda^2(\sigma_v)$ as dictated by the parametrization problem, if π_v is not tempered. If v is archimedean, one knows [**Sh85**] that $\gamma(s, \pi_v, \Lambda^2)$ equals $\gamma(s, \Lambda^2(\sigma_v))$.

PROPOSITION 5.2. *Let v be finite. Then the following are equivalent:*

1. *$L_1(s, \pi_v, \Lambda^2)$ has a pole at $s = 0$.*
2. *$L(s, \Lambda^2(\sigma_v))$ has a pole at $s = 0$.*

PROOF. First consider the case when π_v is supercuspidal. Then, since π_v is allowable by hypothesis, we can find, by Prop. 4.1, a quadratic extension K/F_v and a supercuspidal λ of $\mathrm{GL}(2, K)$ such that π_v is $I(\lambda)$ (local automorphic induction). Let τ be the associated irreducible two dimensional W_K-module. Then σ_v is the induction of τ to W_{F_v}. Arguing as in the proof of Proposition 4.2, we can find an irreducible two dimensional representation α of the Weil group of a quadratic extension N/F with local extension K/F_v such that τ is its restriction to W_K. Let β be the corresponding automorphically induced cuspidal of $\mathrm{GL}(4, \mathbb{A}_F)$ with local component π_v, corresponding globally to the induction $I(\alpha)$ of α to W_F. Denote by $L(s, \Lambda^2(I(\alpha)))$ the (completed) Artin L-function of the exterior square of $I(\alpha)$, and let $L_1(s, \beta, \Lambda^2)$ be the global exterior square L-function of β considered by Shahidi [**Sh88**]. It is known that both these functions have meromorphic continuations and admit functional equations of the standard type. If $f(s), g(s)$ are two functions whose quotient is invertible, we will write $f(s) \equiv g(s)$; this is clearly an equivalence relation. At the places v where the representations are unramified, one knows that Shahidi's local factors coincide with the Langlands factors and hence with those of $\Lambda^2(I(\alpha))$. Moreover, the factors also agree at the archimedean places [**Sh85**]. This leads to the following equivalence:

$$\prod_{u \in S} L_u(s, \Lambda^2(I(\alpha))) L_1(1-s, \beta_u^\vee, \Lambda^2) \equiv \prod_{u \in S} L_u(1-s, \Lambda^2(I(\alpha)^\vee)) L_1(s, \beta_u, \Lambda^2).$$

Suppose that v lies in S, for otherwise there is nothing to prove. By Brauer's theorem, $L(s, \Lambda^2(I(\alpha)))$ is a ratio of abelian L-functions, and this allows us to find a linear character μ of W_N such that (i) $\mu_v = 1$, and (ii) $L_u(s, \Lambda^2(I(\alpha) \otimes \mu_u)) = 1$, at each $u \in S - \{v\}$; for this, we just have to make μ_u sufficiently ramified. (We will, by abuse of notation, use μ to denote also the idele class character defined by class field theory.) One also has the following automorphic analog:

LEMMA 5. *(Shahidi) Let u be any finite place. Then there exists a positive number $C = C(u, \beta)$ such that, for every idele class character μ whose local component μ_u is non-quadratic and has conductor larger than C, $L_1(s, \beta_u \otimes \mu_u, \Lambda^2) = 1$.*

Since this is not in print, we indicate how to prove it. Let μ be any idele class character with μ_u non-quadratic. Viewing $M =$GL(4) as a Levi subgroup of $G =$ SO(8), Shahidi realizes the exterior square L-function of $\beta \otimes \mu$ via the Eisenstein series on $G(\mathbb{A}_F)$ defined by inducing $\beta \otimes \mu$. We need to analyze the poles of $L_1(s, \beta_u \otimes \mu_u, \Lambda^2)$. These are contained in the set of poles of the corresponding (local) intertwining operator, and this set is empty if the local induced representation is irreducible, for any unramified twist of π_u. By inductivity and the factorization formula [**Sh90a, Sh90b**] of Shahidi, we may assume that the inducing representations are supercuspidal. If w_0 denotes the longest element in the Weyl group of G modulo that of M, reducibility can happen only if $w_0(\beta_u \otimes \mu_u) \simeq \beta_u \otimes \mu_u \nu$, for some unramified character ν. (This is a folklore assertion, discussed in [**Sh88**], whose main theorem says that the commuting algebra of the induced representation has dimension 1 otherwise, leading to irreducibility.) Suppose such an identity holds for $\beta_u \otimes \mu_u$ and for $\beta_u \otimes \mu'_u$, for a second character μ'. Then the criterion above implies that there exists an unramified character λ such that the following holds:

$$\beta_u \simeq \beta_u \otimes \mu_u {\mu'_u}^{-1} w_0(\mu_u^{-1} \mu'_u)\lambda.$$

In other words, β_u admits a self-twist. Note that, since w_0 is an involution, it cannot fix any non-quadratic character. Once we fix μ', we see then that, even with variable χ, there are only a finite number of *ramified* non-quadratic μ_u for which this can happen. We are done by choosing C to be larger than the conductors of all such exceptional μ_u. □

Now applying this lemma and the earlier remark dealing with the Galois side, we can easily find a unitary idele class character μ such that

$$\begin{aligned} L_1(s, \beta_u \otimes \mu_u, \Lambda^2) &= L_1(s, \beta_u^\vee \otimes \overline{\mu}_u, \Lambda^2) = L(s, \Lambda^2(I(\alpha) \otimes \mu)_u) \\ &= L(s, \Lambda^2(I(\alpha) \otimes \mu)_u^\vee) = 1, \end{aligned}$$

for every u in $S - \{v\}$.

Implicit at this point, and later on in similar situations, is the assertion that a relevant identity of L-factors at an unramified place continues to hold after twisting by a possibly ramified character. This is known to be true for the functions we consider.

Consequently, remembering that $\beta_v = \pi_v$ and $I(\alpha)_v = \sigma_v$ by construction, we deduce the equivalence

$$\begin{aligned} &L_1(s, \pi_v \otimes \mu_v, \Lambda^2) L(1-s, \Lambda^2(\sigma_v^\vee \otimes \overline{\mu}_v)) \\ &\qquad \equiv L(s, \Lambda^2(\sigma_v \otimes \mu_v)) L_1(1-s, \pi_v^\vee \otimes \overline{\mu}_v, \Lambda^2). \end{aligned} \tag{5.3}$$

None of these L-factors can have zeros. Moreover, the poles of $L(1-s,\Lambda^2(\sigma_v^\vee \otimes \overline{\mu}_v))$ all occur on the line $\Re(s) = 1$, since the corresponding inverse roots have absolute value 1. We do not have such a strong result in general on the automorphic side, but for supercuspidals we do by Shahidi [**Sh88**]. (In fact, all we need is that $L_1(1-s,\pi_v^\vee \otimes \overline{\mu}_v,\Lambda^2)$ has no pole at $s=0$, which works in general, see *loc. cit.*) The assertion of Proposition 5.2 thus follows for π_v supercuspidal.

Next suppose that π_v is not supercuspidal. Then one knows that there is a maximal parabolic subgroup P' of M =GL(4) with Levi M' =GL$(a)\times$GL(b), $a \geq b \geq 1$, $a+b=4$, and a generic representation $\eta = \eta_1 \otimes \eta_2$ of $M(F_v)$ such that π_v is a subrepresentation of the representation of $M(F_v)$ parabolically induced by η. (π_v is the unique generic constituent of this induced module.) An important factorization result due to Shahidi (see [**Sh90b**], last formula on p. 284) gives the following identity:

$$\gamma(s,\pi_v,\Lambda^2) = \gamma(s,\eta_1,\Lambda^2)\gamma(s,\eta_2,\Lambda^2)\gamma(s,\eta_1\times\eta_2), \tag{5.4}$$

where the first two factors on the right are the ones associated to the exterior square representations of GL$(a,\mathbb{C})$ and GL$(b,\mathbb{C})$ respectively, and the last one is the Rankin-Selberg factor on GL$(a)\times$GL(b). (If $b=1$, the middle factor is taken to be 1. It may also be useful to note that there may be cancellations, up to invertible factors, in this deceptively uniform formula when π_v is not the full induced representation, for example when it is the Steinberg module.)

Let τ_1 (resp. τ_2) be the representation of W'_{F_v} of dimension a (resp. b) associated to η_1 (resp. η_2) by the local Langlands correspondence, which we can apply as $a,b\leq 3$. Then it is easy to see that there is a similar factorization on the Galois side, with π_v replaced by σ_v and η_j by τ_j, $j=1,2$. Note that we have $\gamma(s,\eta_1\times\eta_2) = \gamma(s,\tau_1\otimes\tau_2)$ because the epsilon and L-factors of these pairs agree. (For this we use [**He84**] and Prop. 4.2, which we can apply as $a+b=4$.) On the other hand, the exterior square representation of GL(3) identifies with the map $g \to g(\mathrm{det}g)^{-1}$. Consequently, if $a=3$, then we have

$$\gamma(s,\eta_1,\Lambda^2) = \gamma(s,\eta_1\otimes\omega_1^{-1}) = \gamma(s,\tau_1\otimes\omega_1^{-1}) = \gamma(s,\Lambda^2(\tau_1)),$$

where ω_j denotes the central character of η_j, identified with the determinant of τ_j. When $a=b=2$, $L(s,\eta_j,\Lambda^2)$ is none other than $L(s,\omega_j)$, agreeing with $L(s,\Lambda^2(\tau_j))$. One also knows by the Rankin-Selberg theory that $L(s,\eta_1\times\eta_2) = L(s,\tau_1\otimes\tau_2)$. So in either case we get the equality of $\gamma(s,\pi_v,\Lambda^2)$ and $\gamma(s,\Lambda^2(\sigma_v))$. This once again leads to the equivalence (5.3), and the remainder of the argument is as in the supercuspidal case. □

The next step is to make use of the following

THEOREM 5.2. *(Jacquet, Piatetski-Shapiro and Shalika) Let π be a cuspidal automorphic representation of* GL$(4,\mathbb{A}_F)$ *such that, for some unitary idele class character ν and a finite set S of places, $L^S(s,\pi,\Lambda^2\otimes\nu^{-1})$ has a pole at $s=1$. Then there exists a globally generic, cuspidal automorphic representation Π of* GSp$(4,\mathbb{A}_F)$ *of central character ν. Moreover, one has the following properties:*

1. *At every unramified place v, the Langlands classes $A_v(\Pi)$ and $A_v(\pi)$ agree under the natural embedding $r_4 : {}^L\mathrm{GSp}(4) = \mathrm{GSp}(4,\mathbb{C}) \hookrightarrow \mathrm{GL}(4,\mathbb{C})$.*
2. *At every archimedean place v, the representation of W_{F_v} with values in* GSp$(4,\mathbb{C})$ *associated to Π_v by Langlands agrees with σ_v under r_4.*

This beautiful result is unfortunately unpublished still, though a key step is achieved in [**JS90**] and the strategy is sketched in [**So**]. Efforts are under way to remedy this situation, and we ask the reader's indulgence on this part. (It should perhaps be mentioned that some people, including one of the authors, have gone through the details of proof.)

We will apply this theorem to our π and its twists. One immediate consequence of this result is that σ_v is symplectic if v is archimedean. Indeed, σ_v lands in $\mathrm{GSp}(4,\mathbb{C})$ by the above, and moreover, its polarization identifies, under the archimedean Langlands correspondence, with the central character of Π_v, which is trivial as $\nu = 1$ for $\beta = \pi$. Thus we may, and we will, assume henceforth that v is finite.

REMARK: Let v be a finite place where π is unramified. Since π is of symplectic type, the central character of Π is trivial. Thus we see that the Langlands class of π_v is, thanks to the Theorem 5.2, represented by a diagonal matrix of the form $[a_v, b_v, a_v^{-1}, b_v^{-1}]$, which has determinant 1; so π_v has trivial central character. Now, if ω is the central character of π, it is an idele class character which is trivial at almost all places, hence must be trivial by the "strong multiplicity one" for GL(1) due to Hecke. (In fact, Hecke's theorem says that ω is trivial if $\omega_v = 1$ for all v in a set of primes of density $> 1/2$.)

Now let r_5 denote the natural five dimensional representation of $\mathrm{GSp}(4,\mathbb{C})$. Then, as is well known, we have the identity

$$\Lambda^2 \circ r_4 = r_5 \oplus \lambda, \tag{5.5}$$

where λ is the polarization (symplectic similitude) character. Denote by $L_1(s, \Pi, r_5)$ the global L-function associated to (Π, r_5) by Shahidi's theory, by using the embedding of GSp(4) in GSp(6). Again, it has meromorphic continuation and functional equation, with the archimedean and unramified factors agreeing with the recipe of Langlands. (This L-function, some times called the standard L-function of GSp(4), can also be approached via the integral representations of Rallis and Piatetski-Shapiro [**PSR**], but we do not seem to have good information at the ramified places.)

Let μ be a unitary idele class character. Apply Theorem 5.2 to the representation $\pi \otimes \mu$, and write the corresponding representation of $\mathrm{GSp}(4)/F$ as $\Pi(\mu)$, which has central character μ^2. Appealing to (5.5), we then get the following identity at every finite place u (with norm Nu) where $\pi \otimes \mu$ is *unramified*:

$$L_1(s, \pi_u \otimes \mu_u, \Lambda^2) = L_1(s, \Pi(\mu)_u, r_5) L(s, \mu_u^2). \tag{5.6}$$

Such an identity also holds at archimedean places. Thus, in view of Propositions 5.1 and 5.2, Theorem 5.1 (and hence Theorem A') will follow once we prove the following

PROPOSITION 5.3. *Write $\Pi = \Pi(1)$. Then $L_1(s, \pi_v, \Lambda^2)$ has a pole at $s = 0$.*

PROOF. For any μ, associate the following global L-function:

$$L_6(s, \Pi(\mu)) = L_1(s, \Pi(\mu), r_5) L(s, \mu^2).$$

Then it has a functional equation, with almost all the factors agreeing with those of $L(s, \pi \otimes \mu, \Lambda^2)$, which has its own functional equation. Noting that π and Π are self-dual, this leads to the equivalence

$$\prod_{u\in S} L_6(s,\Pi(\mu)_u)L_1(1-s,\pi_u\otimes\overline{\mu}_u,\Lambda^2) \equiv \prod_{u\in S} L_6(1-s,\Pi(\overline{\mu}))L_1(s,\pi_u\otimes\mu_u,\Lambda^2), \tag{5.7}$$

where S is the set of finite places where $\pi\otimes\mu$ is ramified.

We may suppose that π is ramified at v, as otherwise the assertion is clear. Arguing as in Lemma 5, we can find a unitary idele class character μ, trivial at v, such that, at each u in $S-\{v\}$, μ_u is sufficiently ramified that the factors $L_1(s,\pi_u\otimes\mu_u,\Lambda^2)$, $L_1(s,\pi_u\otimes\overline{\mu}_u,\Lambda^2)$, $L_1(s,\Pi(\mu)_u,r_5)$, $L_1(s,\Pi(\overline{\mu}),r_5)$, $L(s,\mu_u^2)$, and $L(s,\overline{\mu}_u^2)$ are all 1. Since $\mu_v=1$, this leads to the equivalence

$$L_1(s,\Pi_v,r_5)L(s,1_v)L_1(1-s,\pi_v,\Lambda^2) \equiv L_1(1-s,\Pi,r_5)L(1-s,1_v)L_1(s,\pi_v,\Lambda^2). \tag{5.8}$$

Since $L(s,1_v) = (1-(Nv)^{-s})^{-1}$, it has a unique pole at $s=0$. So the right hand side (of (5.8)) also has a pole at $s=0$, which could not be accounted for by $L(1-s,1_v)$. Furthermore, it is known that at any place u, $L_1(s,\Pi_u,r_5)$ has no pole at $s=1$. (This is because its poles must come from those of the corresponding intertwining operator, which by [**Sh88**, Theorem 5.2], is holomorphic in $\Re(s)\geq 1$.) So the pole at $s=0$ of the expression on the right of (5.8) must come from that of $L_1(s,\pi_v,\Lambda^2)$. □

6. Proof of Theorem B

Let π be a cuspidal automorphic representation of $\mathrm{GL}(2,\mathbb{A}_F)$ of trivial central character. Then π is self-dual and of symplectic type. Let $L(s,\pi,\mathrm{Sym}^2)$ be the L-function associated [**GJ**] to the symmetric square lift $\mathrm{Sym}^2(\pi)$, which is an automorphic representation of $\mathrm{GL}(3,\mathbb{A}_F)$. Then one knows that there is a factorization

$$L(s,\pi\times\pi) = L(s,\pi,\mathrm{Sym}^2)L(s,1), \tag{6.1}$$

where $L(s,1)$ denotes the Dedekind Zeta function of F with the archimedean factors added. It should be noted that from the published proofs, one gets this factorization at almost all places. But then we can use the fact that both sides have functional equations, and that (by using [**JS85**], given any place u, we can twist by a highly ramified character to trivialize the L and ϵ-factors at u. The exact identity (6.1) then follows by checking it at the ramified places, one by one, as in sections 4, 5.

Thus we get

$$W(\pi,\mathrm{Sym}^2) = W(\pi\times\pi)/W(1).$$

It is well known that $W(1)=1$, and moreover, $W(\pi\times\pi)=1$ by Theorem A. This proves the triviality of $W(\pi,\mathrm{Sym}^2)$.

Now consider the symmetric fourth power root number of π. One defines the symmetric fourth power L-function as follows:

$$L(s,\pi,\mathrm{Sym}^4) = L(s,\mathrm{Sym}^2(\pi)\times\mathrm{Sym}^2(\pi))/L(s,\mathrm{Sym}^2(\pi))L(s,1).$$

We have just seen that $W(\mathrm{Sym}^2(\pi)) = W(1) = 1$. So the desired triviality of $W(\pi,\mathrm{Sym}^4)$ will result from the following

PROPOSITION 6.1. *$W(\mathrm{Sym}^2(\pi)\times\mathrm{Sym}^2(\pi))=1$.*

Note that this cannot be proved by appealing to Theorem A as $\mathrm{Sym}^2(\pi)$ is not of symplectic type.

PROOF. At any place v, let σ_v denote the representation of W'_{F_v} associated to π_v by the local Langlands correspondence. It has determinant 1 as π has trivial central character. Under the local correspondence for GL(3), $\mathrm{Sym}^2(\sigma_v)$ is associated to $\mathrm{Sym}^2(\pi_v)$. We claim that, at any place v, we have

$$L(s, \mathrm{Sym}^2(\pi_v) \times \mathrm{Sym}^2(\pi_v)) = L(s, \mathrm{Sym}^2(\sigma_v) \otimes \mathrm{Sym}^2(\sigma_v)) \tag{6.2}$$

and

$$\epsilon(s, \mathrm{Sym}^2(\pi_v) \times \mathrm{Sym}^2(\pi_v)) = \epsilon(s, \mathrm{Sym}^2(\sigma_v) \otimes \mathrm{Sym}^2(\sigma_v)). \tag{6.2}$$

Indeed, applying Theorem 4.1 of [**He86**] we get the equality of the associated gamma factors. Then, twisting by a highly ramified character, and arguing as in section 4.2 of *loc. cit.*, we get the assertion of the claim.

Consequently, it suffices to prove that for each v,

$$W(\mathrm{Sym}^2(\sigma_v) \otimes \mathrm{Sym}^2(\sigma_v)) = 1. \tag{6.3}$$

Write τ_v for $\mathrm{Sym}^2(\sigma_v) \otimes \mathrm{Sym}^2(\sigma_v)$. Then τ_v is the composite

$$W'_{F_v} \to \mathrm{SL}(2, \mathbb{C}) \times \mathrm{SL}(2, \mathbb{C}) \to \mathrm{GL}(9, \mathbb{C}),$$

where the first arrow is (σ_v, σ_v), and the second arrow, r say, is the composition of taking the symmetric square of each factor, landing in $\mathrm{GL}(3, \mathbb{C})\times \mathrm{GL}(3, \mathbb{C})$, and then taking the tensor product. Since $\mathrm{Sym}^2(\sigma_v)$ is orthogonal, so is its tensor square. Thus r takes values in $\mathrm{O}(9, \mathbb{C})$. Since $\mathrm{SL}(2, \mathbb{C})\times \mathrm{SL}(2, \mathbb{C})$ is simply connected, we may apply the discussion following Lemma 1 of section 2 and obtain (6.3). □

References

[AC] J. Arthur and L. Clozel, Simple Algebras, Base Change and the Advanced Theory of the Trace Formula, Ann. Math. Studies **120**, Princeton: Princeton University Press (1989).

[BF] D. Bump, S. Friedberg, *The exterior square automorphic L-functions on* GL(n), in Festschrift in Honor of I. I. Piatetski-Shapiro, part II, Israel Math. Conf. Proc. **3**, Jerusalem: Weizmann (1990), 47–65.

[Ch] J.-P. Jeff Chen, *Local Factors, Central Characters, and Representations of the General Linear Group over Non-archimedean Fields*, doctoral dissertation, Yale University (1996).

[Cl] L. Clozel, *Base change for* GL(n), Proc. ICM, Berkeley (1986), Providence, RI: Amer. Math Soc. (1987), 791–797.

[CoS] J. Coates and C.-G. Schmidt, *Iwasawa theory for symmetric square of an elliptic curve*, J. Reine Angew. Math. **375/ 376** (1987), 104–156.

[D73] P. Deligne, *Les constantes des équations fonctionnelles des fonctions L*, in Modular Functions of One Variable II, Springer Lecture Notes **349**, Berlin-New York: Springer-Verlag (1973), 501–597.

[D76] P. Deligne, *Les constantes locales de l'équation fonctionnelle de la fonction L d'Artin d'une représentation orthogonale*, Inventiones Math. **35** (1976), 299–316.

[FQ] A. Fröhlich and J. Queyrut, *On the functional equation of the Artin L-function for characters of real representations*, Inventiones Math. **20** (1973), 125–138.

[GJ] S. Gelbart and H. Jacquet, *A relation between automorphic representations of* GL(2) *and* GL(3), Ann, Scient. ENS (4) **11** (1979), 471–542.

[Ha] M. Harris, *The local Langlands conjecture for* GL(n) *over a p-adic field, $n < p$*, Invent. Math. **134** (1998), 177–210.

[He84] G. Henniart, La Conjecture Locale pour GL(3), Mémoires S.M.F. (nouvelle série), no. 11/12 (1984).

[He86] G. Henniart, *On the local Langlands conjecture for* GL(n)*, the cyclic case*, Ann. Math. **123** (1986), 145–203.

[He93] G. Henniart, *Caractérisation de la correspondence de Langlands locale pour les facteurs* ε *de paires*, Inventiones Math. **113** (1993), 339–350.

[HH] G. Henniart and R. Herb, *Automorphic induction for* GL(n) *(over local non-Archimedean fields)*, Duke Math. Journal **78** (1995), 131–192.

[JPSS] H. Jacquet, I. Piatetski-Shapiro and J. Shalika, *Rankin-Selberg convolutions*, Amer. J. of Math. **105** (1983), 367–464.

[JS81] H. Jacquet and J. A. Shalika, *Euler products and the classification of automorphic forms I & II*, Amer. J. of Math. **103** (1981), 499–558 & 777–815.

[JS85] H. Jacquet and J. A. Shalika, *A lemma on highly ramified epsilon factors*, Math. Annalen **271** (1985), 319–332.

[JS90] H. Jacquet and J. A. Shalika, *Exterior square L-functions*, in Automorphic Forms, Shimura Varieties, and L-functions, vol. 2, ed. L. Clozel and J. Milne, Boston: Academic Press (1990), 143–226.

[K] P. Kutzko, *The Langlands conjecture for* GL(2) *over a local field*, Ann. Math. **112** (1980), 381–412.

[L70] R. P. Langlands, *Problems in the theory of automorphic forms*, in Lectures in Modern Analysis and Applications, Springer Lecture Notes **170**, Berlin: Springer (1970), 18–61.

[L79] R. P. Langlands, *Automorphic representations, Shimura varieties and motives. Ein Märchen*, Proc. Symp. Pure Math. **33**, ed. A. Borel and W. Casselman, Part 2, Providence, RI: Amer. Math. Soc. (1979), 205–246.

[L80] R. P. Langlands, Base Change for GL(2), Ann. Math. Studies **96**, Princeton: Princeton University Press (1980).

[PSR] I. Piatetski-Shapiro and S. Rallis, *A new way to get Euler products*, J. Reine Angew. Math. **392** (1988), 110–124

[R] D. Rohrlich, *Galois theory, elliptic curves, and root numbers*, Compositio Mathematica **100** (1996), 311–349.

[Sa] T. Saito, *The sign of the functional equation of the L-function of an orthogonal motive*, Inventiones Math. **120** (1995), 119–142.

[Se] J.-P. Serre, *Modular forms of weight one and Galois representations*, in Algebraic Number Fields, ed. A. Fröhlich, London-New York: Acad. Press (1977), 193–268.

[Sh85] F. Shahidi, *Local coefficients as Artin factors for real groups*, Duke Math. Journal **52** (1985), 973–1007.

[Sh88] F. Shahidi, *On the Ramanujan conjecture and the finiteness of poles for certain L-functions*, Ann. of Math. (2) **127** (1988), 547–584.

[Sh90a] F. Shahidi, *A proof of the Langlands conjecture on Plancherel measures; complementary series for p-adic groups*, Ann. of Math. **132** (1990), 273–330.

[Sh90b] F. Shahidi, *On multiplicativity of local factors*, in Festschrift in honor of I. I. Piatetski-Shapiro, Israel Math. Conf. Proc. **3**, Jerusalem: Weizmann (1990), 279–289.

[Sh94] F. Shahidi, Symmetric power L-functions for GL(2), CRM Proceedings & Lecture Notes **4**, ed. M. R. Murty, Providence, RI: Amer. Math Soc. (1994).

[So] D. Soudry, *Automorphic forms on* GSp(4), in Festschrift in Honor of I. I. Piatetski-Shapiro, Israel Math. Conf. Proc. **3**, Jerusalem: Weizmann (1990), 291–303.

[T] J. Tate, Les conjectures de Stark sur les Fonctions L d'Artin en $s = 0$, Progress in Math. **47**, Boston: Birkhäuser (1984).

[Tu] J. Tunnell, *Artin's conjecture for representations of octahedral type*, Bulletin AMS **5** (1981), 173–175.

Mehta Research Institute, 10 Kasturba Marg, Allahabad 21102, India.
E-mail address: dprasad@mri.ernet.in

Department of Mathematics, California Institute of Technology, Pasadena, CA 91125.
E-mail address: dinakar@cco.caltech.edu

Selected Titles in This Series

(Continued from the front of this publication)

44 **William K. Allard and Frederick J. Almgren, Jr., Editors,** Geometric measure theory and the calculus of variations (Humboldt State University, Arcata, California, July/August 1984)

43 **François Trèves, Editor,** Pseudodifferential operators and applications (University of Notre Dame, Notre Dame, Indiana, April 1984)

42 **Anil Nerode and Richard A. Shore, Editors,** Recursion theory (Cornell University, Ithaca, New York, June/July 1982)

41 **Yum-Tong Siu, Editor,** Complex analysis of several variables (Madison, Wisconsin, April 1982)

40 **Peter Orlik, Editor,** Singularities (Humboldt State University, Arcata, California, July/August 1981)

39 **Felix E. Browder, Editor,** The mathematical heritage of Henri Poincaré (Indiana University, Bloomington, April 1980)

38 **Richard V. Kadison, Editor,** Operator algebras and applications (Queens University, Kingston, Ontario, July/August 1980)

37 **Bruce Cooperstein and Geoffrey Mason, Editors,** The Santa Cruz conference on finite groups (University of California, Santa Cruz, June/July 1979)

36 **Robert Osserman and Alan Weinstein, Editors,** Geometry of the Laplace operator (University of Hawaii, Honolulu, March 1979)

35 **Guido Weiss and Stephen Wainger, Editors,** Harmonic analysis in Euclidean spaces (Williams College, Williamstown, Massachusetts, July 1978)

34 **D. K. Ray-Chaudhuri, Editor,** Relations between combinatorics and other parts of mathematics (Ohio State University, Columbus, March 1978)

33 **A. Borel and W. Casselman, Editors,** Automorphic forms, representations and L-functions (Oregon State University, Corvallis, July/August 1977)

32 **R. James Milgram, Editor,** Algebraic and geometric topology (Stanford University, Stanford, California, August 1976)

31 **Joseph L. Doob, Editor,** Probability (University of Illinois at Urbana-Champaign, Urbana, March 1976)

30 **R. O. Wells, Jr., Editor,** Several complex variables (Williams College, Williamstown, Massachusetts, July/August 1975)

29 **Robin Hartshorne, Editor,** Algebraic geometry – Arcata 1974 (Humboldt State University, Arcata, California, July/August 1974)

28 **Felix E. Browder, Editor,** Mathematical developments arising from Hilbert problems (Northern Illinois University, Dekalb, May 1974)

27 **S. S. Chern and R. Osserman, Editors,** Differential geometry (Stanford University, Stanford, California, July/August 1973)

26 **Calvin C. Moore, Editor,** Harmonic analysis on homogeneous spaces (Williams College, Williamstown, Massachusetts, July/August 1972)

25 **Leon Henkin, John Addison, C. C. Chang, William Craig, Dana Scott, and Robert Vaught, Editors,** Proceedings of the Tarski symposium (University of California, Berkeley, June 1971)

24 **Harold G. Diamond, Editor,** Analytic number theory (St. Louis University, St. Louis, Missouri, March 1972)

23 **D. C. Spencer, Editor,** Partial differential equations (University of California, Berkeley, August 1971)

22 **Arunas Liulevicius, Editor,** Algebraic topology (University of Wisconsin, Madison, June/July 1970)

21 **Irving Reiner, Editor,** Representation theory of finite groups and related topics (University of Wisconsin, Madison, April 1970)

(See the AMS catalog for earlier titles)

ISBN 0-8218-1051-0

9 780821 810514

PSPUM/66.2